Live Tutoring

Students use NetTutor™ to access an online tutoring service to get help when they need it. This is especially great for commuter schools and distance education schools.

NetTutor™ is a revolutionary new online learning environment for the live distribution of mathematical content. NetTutor™ includes web-based graphical chat, threaded and platform independent, allowing students to use their own computers to access learning materials in a nonlinear fashion or in real-time with live corresponding tutors.

Some features of NetTutor™ include

- **Whiteboard.** The whiteboard enables students and tutors to "speak" math. They are able to use mathematical symbols as well as circle, highlight, and pull out important steps. **Real-time discussion on the whiteboard** is the key to this service by allowing tutors to explain in mathematical symbols as well as to "converse" in real time with the student to answer questions.

- **Students register online** and there are no passwords for instructors or McGraw-Hill to maintain.

- **Free!** Students simply need to tell the tutor which McGraw-Hill text they are using, the page number, and the exercise.

- Tutors will work on **only odd-numbered problems.** The even-numbered problems remain available for faculty for assigned homework.

- **Seventy-two hours weekly** of live tutors online. Tutors are placed at peak hours, and usage is monitored frequently to see if hours need to be changed.

- **Students may access archives** of previous sessions saved by other students to see if there is a similar problem/exercise.

- If students are not able to enter a tutorial session (not enough time or the tutor is not available), they can **post a question and it will be answered within 24 hours.**

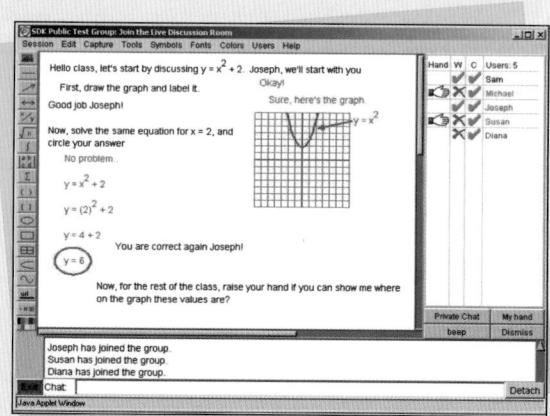

To learn more about NetTutor™, or any other McGraw-Hill product, please contact your sales representative. To locate your sales representative's contact information, visit **www.mhhe.com** and click on "find my sales rep" in the help center.

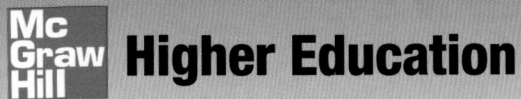 **Higher Education**

ALEKS® Makes the Grade!

Would your students appreciate their own personal math tutor available 24 hours per day, 7 days a week? A tutor who teaches what they're most ready to learn? And a tutor who analyzes their answers to problems and responds with specific advice when they make a mistake?

ALEKS individualizes assessment and learning, is customizable for each course topic, and provides a course management system telling you exactly what your students know and don't know.

Be Our Guest

To request your FREE 24-hour trial of ALEKS, visit our website at www.highed.aleks.com/guest.html. Click on "login" under the "as a student" heading.

To learn more about ALEKS, visit the website at www.highed.aleks.com or contact your local McGraw-Hill representative. To locate your sales representative's contact information, visit www.mhhe.com and click on "find my sales rep" in the help center.

Here's what professors are saying about ALEKS . . .

"The system appears to offer many advantages over anything else currently offered."

"Good product. It's practical and logical—based on a model that makes sense."

"This is the first time I have gotten excited about using a computer-based approach in our classes. ALEKS eliminates the equipment and cost barriers we face with other systems."

Here's what students are saying about ALEKS . . .

"This program helped me dramatically with math. from getting D's on tests to receiving an A+ on my I actually scored a 200/200 on a final."

"My math abilities really have progressed. ALE makes math a much easier course."

Models of Classroom Integration

- Supervised Math Lab
- Math Lab in Structured Course
- Small-Group Instruction
- Self-Paced Learning
- Distance Learning

In some ways, planning a course in which ALEKS is to be used is simpler than planning other kinds of courses. The instructor may assume complete freedom in planning lectures, lessons, and assignments, while ALEKS ensures that students can progress toward mastery, regardless of their level of preparation.

ALEKS® is a registered trademark of ALEKS Corporation.

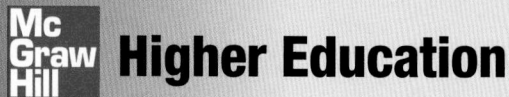 **Higher Education**

 **MathZone**

MathZone™   Free   Easy   Has It All

MathZone's™ powerful feature set includes book-specific, assignable algorithmic content; ADA-compliant videos; and e-Professor animated solutions. MathZone provides students virtually unlimited practice through algorithmic quizzing and testing, and free live tutoring via NetTutor's™ whiteboard technology, and instructors can track their students' progress in the online gradebook.

Free to you and your students

MathZone is provided for free with the purchase of a new McGraw-Hill math textbook. Each text is packaged with a registration code for the website at www.mathzone.com and comes with a MathZone CD-ROM.

Easy to use

All functions for students and instructors are located within one login. The online course management system offers simple functionality. The gradebook provides detailed feedback on student responses and the problem-solving process that students performed to select those responses. In addition to tracking student performance on problems, the gradebook also tracks students' usage of nonassessed elements, such as video lecture or e-Professor. The gradebook can be easily exported to Excel, allowing the instructor the flexibility to use MathZone in a variety of ways.

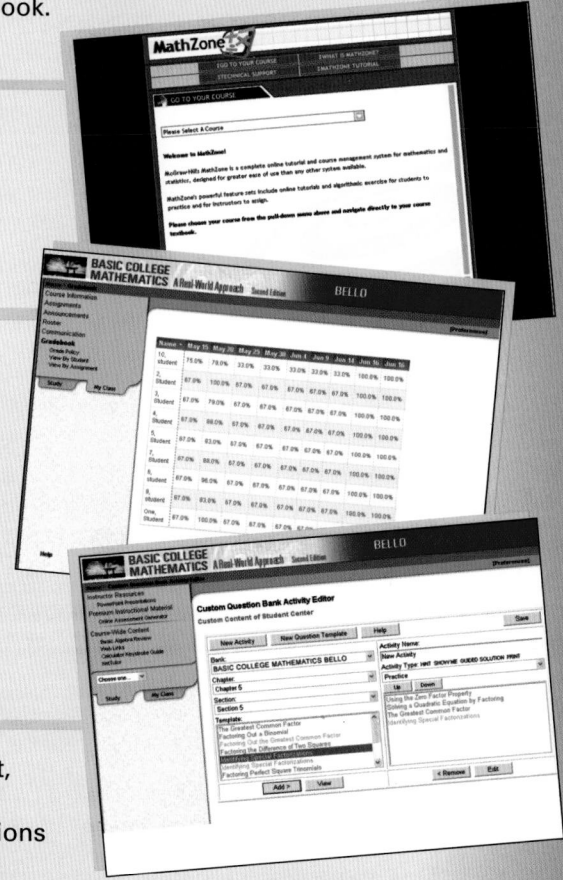

Has it all

MathZone is tailored to a McGraw-Hill textbook, so every assignment, question, e-Professor tutorial, and video-lecture piece is derived directly from text-specific materials. Instructors can modify questions and assignments and create their own from scratch.

   With just a few clicks of the mouse, you can share your MathZone course with an unlimited number of colleagues. Instructors can share assignments, algorithmically generated questions, or static questions with simplicity and speed.

Instructor benefits

- Offers complete course management.
- Track students' progress with unprecedented depth.
- Includes 100% algorithmic homework, quizzing, and testing.
- Edit, change, and create algorithmic questions.
- Assign all course resources, including homework, video lectures and e-Professor.
- Live, book-specific tutoring saves instructors time.
- Provides unmatched ease-of-use in one system.

Student benefits

- Includes video-lectures that are ADA-compliant.
- Provides e-Professor-guided solutions.
- Provides prompted, step-by-step problem solving.
- Includes online algorithmic practice and assessment.
- Students have access to all resources, whether assigned or not.
- The end result is a better understanding and an improved grade.

Intermediate Algebra

Intermediate Algebra

A Real-World Approach

Second Edition

Ignacio Bello

Fran Hopf

Hillsborough Community College
Tampa, Florida

Boston   Burr Ridge, IL   Dubuque, IA   Madison, WI   New York   San Francisco   St. Louis
Bangkok   Bogotá   Caracas   Kuala Lumpur   Lisbon   London   Madrid   Mexico City
Milan   Montreal   New Delhi   Santiago   Seoul   Singapore   Sydney   Taipei   Toronto

Higher Education

INTERMEDIATE ALGEBRA: A REAL-WORLD APPROACH, SECOND EDITION

 This book is printed on recycled, acid-free paper containing 10% postconsumer waste.

1 2 3 4 5 6 7 8 9 0 QPD/QPD 0 9 8 7 6 5
1 2 3 4 5 6 7 8 9 0 QPD/QPD 0 9 8 7 6 5

ISBN 0–07–283106–5
ISBN 0–07–298486–4 (Annotated Instructor's Edition)

Publisher, Mathematics and Statistics: *William K. Barter*
Publisher, Developmental Mathematics: *Elizabeth J. Haefele*
Director of Development: *David Dietz*
Senior Developmental Editor: *Randy Welch*
Marketing Manager: *Steven R. Stembridge*
Senior Project Manager: *Vicki Krug*
Senior Production Supervisor: *Sherry L. Kane*
Senior Media Project Manager: *Sandra M. Schnee*
Lead Media Technology Producer: *Jeff Huettman*
Senior Designer: *David W. Hash*
Cover/Interior Designer: *Rokusek Design*
(USE) Cover Image: *©Reuters/CORBIS, Fireworks burst from the Eiffel Tower in Paris to mark the new millennium*
Senior Photo Research Coordinator: *Lori Hancock*
Photo Research: *David Tietz*
Supplement Producer: *Brenda A. Ernzen*
Compositor: *Interactive Composition Corporation*
Typeface: *10/12 New Times Roman*
Printer: *Quebecor World Dubuque, IA*

The credits section for this book begins on page C-1 and is considered an extension of the copyright page.

About the Authors

Ignacio Bello attended the University of South Florida (USF), where he earned a B.A. and M.A. in Mathematics. He began teaching at USF in 1967, and in 1971 he became a member of the faculty and Coordinator of the Math and Sciences Department at Hillsborough Community College (HCC). Professor Bello instituted the USF/HCC remedial program, which started with 17 students taking Intermediate Algebra and grew to more than 800 students with courses covering Developmental English, Reading, and Mathematics.

Aside from the present series of books (*Basic College Mathematics, Introductory Algebra,* and *Intermediate Algebra*), Professor Bello is the author of more than 40 textbooks, including *Topics in Contemporary Mathematics, College Algebra, Algebra and Trigonometry,* and *Business Mathematics.* Many of these textbooks have been translated into Spanish. With Professor Fran Hopf, Bello started the Algebra Hotline, the only live, college-level television help program in Florida.

Professor Bello is featured in three television programs on the award-winning Education Channel. He has helped create and develop the USF Mathematics Department website (http://mathcenter.usf.edu), which serves as support for the Finite Math, College Algebra, Intermediate Algebra, Introductory Algebra, and CLAST classes at USF. You can see Professor Bello's presentations and streaming videos at that website, as well as at http://www.ibello.com. Professor Bello is a member of the MAA and AMATYC. He has given many presentations regarding the teaching of mathematics at the local, state, and national levels.

Fran Hopf has been teaching mathematics for Hillsborough Community College for 27 years. She teaches algebra and finite mathematics both on campus and as distance learning courses for the University of South Florida where she is presently a PhD student in higher education with a cognate in mathematics. Fran has a bachelor's degree from Florida State University and a master's in education from the University of South Florida.

In addition to this textbook she has authored course supplements for intermediate algebra and has reviewed texts for developmental and liberal arts mathematics. Fran and Ignacio were co-hosts for the live, call-in Algebra Hotline program that aired weekly for three years on the Education Channel in Tampa, Florida, that supported a college algebra telecourse. Fran has developed curriculum and videos for a math review for the College Level Academic Skills Test (CLAST), a mandatory test given to college students attending state universities in Florida. The CLAST material can be found at http://mathcenter.usf.edu/. She is a member of AMATYC and has co-presented with Bello at the AMATYC national convention. Whether in the classroom, on the Internet, or writing curriculum, Fran enjoys sharing her enthusiasm for mathematics.

The Team

Bello and Hopf, Hopf and Bello. They have been working together for more than 20 years. As founders of the Algebra Hotline they helped students on live television every Wednesday at 8 P.M. for three years. To reach more students, they created and taught telecourses at both HCC and USF. The idea was expanded to cover college algebra and now it is used in intermediate algebra by other instructors. The team has also worked in creating instructor's manuals, solutions manuals, and video tapes as well as spreading the joy of mathematics at local, regional, and national mathematics meetings. Finally, through the auspices of McGraw-Hill, you can see their work, ideas, and techniques right here: *Intermediate Algebra.* The word to students from the team? **MATH IS FUN, AND YOU CAN DO IT!**

Contents

1 The Real Numbers

2 Linear Equations and Inequalities

3 Graphs and Functions

4 Solving Systems of Linear Equations and Inequalities

5 Polynomials

6 Rational Expressions

7 Rational Exponents and Radicals

8 Quadratic Equations and Inequalities

9 Quadratic Functions and the Conic Sections

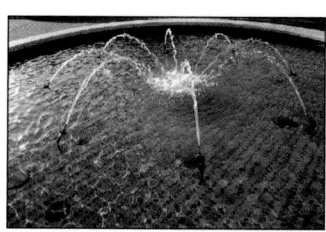

10 Functions—Inverse, Exponential, and Logarithmic

Appendix A   Sequences and Series

Preface

FROM THE AUTHORS

The Inspiration for Our Teaching

Bello: I was born in Havana, Cuba and encountered the same challenges of mathematics that many other students do: I failed freshman math. However, perseverance was one of my traits: I made 100% on the final exam the second time around.

I might still be in Cuba except that a police officer kindly informed my family that the members of a club to which I belonged were in jeopardy. Figuring out the obvious was another one of my traits: I left for the United States.

I came to the U.S. and, yes, I did know some English. After working in various jobs (roofer, sheetrock installer, dock worker), I went back to school, finishing high school in one year and receiving a college academic scholarship. I enrolled in Calculus and made a C. Never one to be discouraged, my perseverance again took over, so I became a math major and learned to excel in the courses that had previously frustrated me. While a graduate student at the University of South Florida (USF), I taught at a technical school, a decision that contributed to my resolve to teach math and make it come alive for my students the way brilliant instructors such as Jack Britton, Donald Rose, and Frank Cleaver had done for me.

A Lively Approach to Reach Today's Students

Teaching math at the University of South Florida and Hillsborough Community College was a great new career for both of us, but we were disappointed by the materials we had to use. A rather imposing, mathematically correct but boring book was in vogue. Students hated it, professors hated it, and administrators hated it. Bello took the challenge to write a better book, a book that was not only mathematically correct, but student-oriented with **interesting applications**—many suggested by the students themselves—and even, dare we say, entertaining! That book's approach and philosophy proved an instant success and was a precursor to our current series.

Students fondly called Bello's class "The Bello Comedy Hour," but they worked hard, and they performed well. Because our students always ranked among the highest on the common final exam at USF/HCC, we knew we had found a way to motivate them through **common-sense language** and humorous, **realistic math applications.** We also wanted to show students they could overcome the same obstacles we had in math and become successful too.

If math has just never been a subject that some of your students have felt comfortable with, then they're not alone! We wrote this book with the **math-anxious** student in mind, so they'll find our tone is jovial, our explanations are patient, and instead of making math seem mysterious, we make it down-to-earth and easily digestible. For example, after we've explained the different methods for simplifying fractions, we speak directly to readers: "Which way should you simplify fractions? The way you understand!" Once students realize that math is within their grasp and not a foreign language, they'll be surprised at how much more confident they feel.

A Real-World Approach: Applications, Student Motivation, and Problem Solving

What is a "real-world approach"? We found that most textbooks put forth "real-world" applications that meant nothing to the "real world" of our students. How many of our students would really need to calculate the speed of a bullet (unless they're in its way) or cared to know when two trains traveling in different directions would pass by each other (disaster will certainly occur if they are on the same track)? For our students, both traditional and non-traditional, the real world consists of questions such as "How do I find the best cell phone plan?" and "How will I pay my tuition and fees if they increase by x%?" That is why we introduce mathematical concepts through everyday applications with **real data** and give homework using similar, well-grounded situations (see the "Getting Started" application that introduces every section's topic and the word problems in every exercise section).

Putting math in a real-world context has helped us to overcome one of the problems we all face as math educators: **student motivation.** Seeing math in the real world makes students perk up in a math class in a way we have never seen before, and realism has proven to be the best motivator we've ever used. In addition, the real-world approach has enabled us to enhance students' **problem-solving skills** since they are far more likely to tackle a real-world problem that matters to them than they would attempt a problem that seems contrived.

Diverse Students and Multiple Learning Styles

We know we live in a pluralistic society, so how do you write one textbook for everyone? The answer is to build a flexible set of teaching tools that instructors and students can adapt to their own situation. Are any of your students members of a **cultural minority?** So is Bello! Did they learn **English as a second language?** So did Bello! You'll find our book speaks directly to them in a way that no other book ever has, and fuzzy explanations in other books will be clear and comprehensible in ours.

Do your students all have the same **learning style?** Of course not. That's why we wrote a book that will help students learn mathematics no matter what their personal learning style is. **Visual learners** will benefit from the text's clean page layout, careful use of color highlighting, "Web Its," and the video lectures on the text's website. **Auditory learners** will profit from the audio "e-Professor" lectures on the text's website, and both **auditory** and **social learners** will be aided by the Collaborative Learning projects. **Applied** and **pragmatic learners** will find a bonanza of features geared to help them: Pretests, practice problems by every example, and Mastery Tests, to name just a few. **Spatial learners** will find the chapter Summary is designed especially for them, while **creative learners** will find the Research Questions to be a natural fit. Finally, **conceptual learners** will feel at home with features like "The Human Side of Algebra" and the "Write On" exercises. Every student who is accustomed to opening a math book and feeling like they've run into a brick wall will find in our book that a number of doors are standing open and inviting them inside.

Listening to Student and Instructor Concerns

McGraw-Hill has given us a wonderful resource for making our textbook more responsive to the immediate concerns of students and faculty. In addition to sending our manuscript out for review by instructors at many different colleges, several times a year McGraw-Hill holds symposia and focus groups with math instructors where the emphasis is *not* on selling products but instead on the **publisher listening** to the needs of faculty and their students. These encounters have provided us with a wealth of ideas on how to improve my chapter organization, make the page layout of my books more readable, and fine tune exercises in every chapter so that students and faculty will feel comfortable using my book because it incorporates their specific suggestions and anticipates their needs.

IMPROVEMENTS IN THE SECOND EDITION

Based on the valuable feedback of numerous reviewers and users over the years, the following improvements were made to the Second Edition of *Intermediate Algebra.*

Organizational Changes:

- Chapter 7 has been thoroughly re-written and reorganized to cover Graphs and Functions in Chapter 3.

- Solving Systems of Linear Equations now includes Systems of Linear Inequalities and has been moved from Chapter 8 to Chapter 4 and precedes the chapter on Polynomials
- Variation is now included at the end of Chapter 6 on Rational Expressions.
- Sequences and Series has been moved from a chapter to an appendix since we found that many instructors do not cover this material as part of their regular course.

Pedagogical Changes:

- Many examples, applications, and real-data problems have been added or updated to keep the book's content current.
- The book is now produced as a paperback workbook to encourage students to write in their books as they do their homework.
- *Practice problems* with answers at the bottom of the page now appear adjacent to each example to give students immediate reinforcement of their own skills after they have read through the step-by-step solutions of the example.
- *Web Its* have been added to encourage students to visit math sites while they're web surfing and discover the many informative and creative sites that are dedicated to stimulating better education in math.
- *Pretests* with answer grids immediately following them have been added to the beginning of each chapter to serve as a diagnostic tool.
- *Calculate Its* and *Calculator Corners* have been updated with recent information and keystrokes relevant to currently popular calculators.
- *Applications* have been titled where appropriate to help orient students to the kind of word problem they are about to solve.
- The RSTUV approach to problem-solving has been expanded in this edition due to positive user response from the previous edition.
- Two *Collaborative Learning* exercises have been added to every chapter to encourage students to work in teams to solve fun and thought-provoking projects.
- A *Cumulative Review* has been added to the end of each chapter to continually reinforce material students have previously learned.

I.B. and F.H.

ACKNOWLEDGMENTS

We would like to thank the following people at McGraw-Hill (in order of appearance):

David Dietz, our sponsoring editor, who provided the necessary incentives and encouragement for creating this series with the cooperation of Bill Barter; Christien Shangraw, our first developmental editor who worked many hours getting reviewers and gathering responses into concise and usable reports; Randy Welch who continued and expanded the Christien tradition into a well honed editing engine with many features, including humor, organization, and very hard work; Liz Haefele, our new editor and publisher, who was encouraging and always on the lookout for new markets; Lori Hancock and her many helpers (LouAnn, Emily, David), who always gets the picture; Dr. Tom Porter, of Photos at Your Place, who improved on the pictures we provided; Vicki Krug, one of the most exacting persons at McGraw-Hill, who will always give you the time of day, and then solve the problem; to Cindy Trimble, for the accuracy of the text; Jeff Huettman, one of the best 100 producers in the United States, who learned Spanish in anticipation of this project; Marie Bova, for her detective work in tracking down permission rights; Steve Stembridge and Barbara Owca, for their help and enthusiasm in marketing the Bello series; and to Professor Nancy Mills, for her expert advice on how our book addresses multiple learning styles. Finally, thanks to our attack secretary, Beverly DeVine, who managed to send all materials back to the publisher on time. To all of them, our many thanks.

We would also like to extend our gratitude to the following reviewers of the Bello series for their many helpful suggestions and insights, which helped us to write better textbooks:

Tony Akhlaghi, *Bellevue Community College*

Theresa Allen, *University of Idaho*

John Anderson, *San Jacinto College–South Campus*

Keith A. Austin, *Devry University–Arlington*

Sohrab Bakhtyari, *St. Petersburg College–Clearwater*

Fatemah Bicksler, *Delgado Community College*

Ann Brackebusch, *Olympic College*

Gail G. Burkett, *Palm Beach Community College*

Linda Burton, *Miami Dade College*

Judy Carlson, *Indiana University–Purdue University Indianapolis*

Randall Crist, *Creighton University*

Mark Czerniak, *Moraine Valley Community College*

Parsla Dineen, *University of Nebraska–Omaha*

Sue Duff, *Guilford Technical Community College*

Lynda Fish, *St. Louis Community College–Forest Park*

Donna Foster, *Piedmont Technical College*

Jeanne H. Gagliano, *Delgado Community College*

Debbie Garrison, *Valencia Community College*

Donald K. Gooden, *Northern Virginia Community College–Woodbridge*

Ken Harrelson, *Oklahoma City Community College*

Joseph Lloyd Harris, *Gulf Coast Community College*

Tony Hartman, *Texarkana College*

Susan Hitchcock, *Palm Beach Community College*

Patricia Carey Horacek, *Pensacola Junior College*

Peter Intarapanich, *Southern Connecticut State University*

Judy Ann Jones, *Madison Area Technical College*

Linda Kass, *Bergen Community College*

Joe Kemble, *Lamar University*

Joanne Kendall, *Blinn College–Brenham*

Bernadette Kocyba, *J S Reynolds Community College*

Marie Agnes Langston, *Palm Beach Community College*

Kathryn Lavelle, *Westchester Community College*

Angela Lawrenz, *Blinn College–Bryan*

Richard Leedy, *Polk Community College*

Judith L. Maggiore, *Holyoke Community College*

Timothy Magnavita, *Bucks Community College*

Tsun-Zee Mai, *University of Alabama*

Harold Mardones, *Community College of Denver*

Lois Martin, *Massasoit Community College*

Gary McCracken, *Shelton State Community College*

Tania McNutt, *Community College of Aurora*

Barbara Miller, *Lexington Community College*

Danielle Morgan, *San Jacinto College–South Campus*

Joanne Peeples, *El Paso Community College*

Faith Peters, *Miami Dade College–Wolfson*

Jane Pinnow, *University of Wisconsin–Parkside*

Janice F. Rech, *University of Nebraska–Omaha*

Libbie Reeves, *Mitchell Community College*

Karen Roothaan, *Harold Washington College*

Don Rose, *College of the Sequoias*

Pascal Roubides, *Miami Dade College–Wolfson*

Juan Saavedra, *Albuquerque Technical Vocational Institute*

Judith Salmon, *Fitchburg State College*

Mansour Samimi, *Winston-Salem State University*

Susan Santolucito, *Delgado Community College*

Ellen Sawyer, *College of DuPage*

Laura Schaben, *University of Nebraska—Omaha*

Sandra Siegrist, *Central Ohio Technical College*

Judith Smalling, *St. Petersburg College—Gibbs*

Carol E. Smith, *Bakersfield College*

Ray Stanton, *Fresno City College*

Bryan Stewart, *Tarrant County College—Southwest*

Ann Thrower, *Kilgore College*

Nguyen Vu, *Rio Hondo College*

Betty Weinberger, *Delgado Community College*

Jadwiga Weyant, *Edmonds Community College*

Denise Widup, *University of Wisconsin—Parkside*

Cheryll Wingard, *Community College of Aurora*

Jeff Young, *Delaware Valley College*

Marilyn A. Zopp, *McHenry County College*

A COMMITMENT TO ACCURACY

You have a right to expect an accurate textbook, and McGraw-Hill invests considerable time and effort to make sure that we deliver one. Listed below are the many steps we take to make sure this happens.

OUR ACCURACY VERIFICATION PROCESS

First Round

Step 1: Numerous **college math instructors** review the manuscript and report on any errors that they may find, and the authors make these corrections in their final manuscript.

Second Round

Step 2: Once the manuscript has been typeset, the **authors** check their manuscript against the first page proofs to ensure that all illustrations, graphs, examples, exercises, solutions, and answers have been correctly laid out on the pages, and that all notation is correctly used.

Step 3: An outside, **professional mathematician** works through every example and exercise in the page proofs to verify the accuracy of the answers.

Step 4: A **proofreader** adds a triple layer of accuracy assurance in the first pages by hunting for errors, then a second, corrected round of page proofs is produced.

Third Round

Step 5: The **author team** reviews the second round of page proofs for two reasons: 1) to make certain that any previous corrections were properly made, and 2) to look for any errors they might have missed on the first round.

Step 6: A **second proofreader** is added to the project to examine the new round of page proofs to double check the author team's work and to lend a fresh, critical eye to the book before the third round of paging.

Fourth Round

Step 7: A **third proofreader** inspects the third round of page proofs to verify that all previous corrections have been properly made and that there are no new or remaining errors.

Step 8: Meanwhile, in partnership with **independent mathematicians,** the text accuracy is verified from a variety of fresh perspectives:

- The **test bank author** checks for consistency and accuracy as they prepare the computerized test item file.
- The **solutions manual author** works every single exercise and verifies their answers, reporting any errors to the publisher.
- A **consulting group of mathematicians,** who write material for the text's MathZone site, notifies the publisher of any errors they encounter in the page proofs.
- A video production company employing **expert math instructors** for the text's videos will alert the publisher of any errors they might find in the page proofs.

Final Round

Step 9: The **project manager,** who has overseen the book from the beginning, performs a **fourth proofread** of the textbook during the printing process, providing a final accuracy review.

⇒ What results is a mathematics textbook that is as accurate and error-free as is humanly possible, and our authors and publishing staff are confident that our many layers of quality assurance have produced textbooks that are the leaders of the industry for their integrity and correctness.

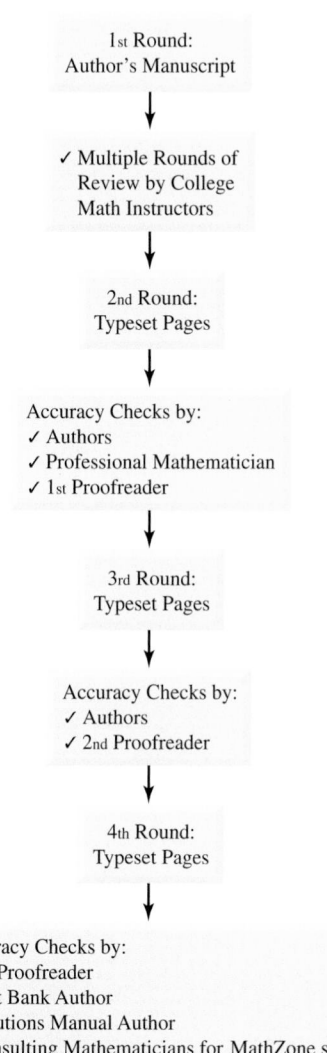

1st Round:
Author's Manuscript

↓

✓ Multiple Rounds of Review by College Math Instructors

↓

2nd Round:
Typeset Pages

↓

Accuracy Checks by:
✓ Authors
✓ Professional Mathematician
✓ 1st Proofreader

↓

3rd Round:
Typeset Pages

↓

Accuracy Checks by:
✓ Authors
✓ 2nd Proofreader

↓

4th Round:
Typeset Pages

↓

Accuracy Checks by:
✓ 3rd Proofreader
✓ Test Bank Author
✓ Solutions Manual Author
✓ Consulting Mathematicians for MathZone site
✓ Math Instructors for text's video series

↓

Final Round:
Printing

↓

✓ Accuracy Check by 4th Proofreader

Features and Supplements

• *Motivation for a Diverse Student Audience*

A number of features exist in every chapter to motivate students' interest in the topic and thereby increase their performance in the course:

The Human Side of Algebra

To personalize the subject of algebra, the origins of algebraic notation, concepts, and methods are introduced through the lives of real people solving ordinary problems.

The Human Side of Algebra

The inventor of the Cartesian coordinate system, Rene Descartes, was born March 31, 1596, near Tours, France. Trained as a gentleman and educated in Latin, Greek, and rhetoric, he maintained a healthy skepticism toward all he was taught. Disenchanted with these studies, he took an interlude of pleasure in Paris, later moving to the suburb of St. Germain, where he worked on mathematics for 2 years. In 1637, at age 41, his friends persuaded him to print

---

GETTING STARTED   **Sloping Salaries**

Beginning Teacher Salaries and Job Offers for New Graduates

A line segment joining two points in a plane has a certain "steepness" or inclination. For example, suppose you want to find the average yearly raise in beginning teacher's average salaries. According to the graph, the salary in 1990 was about $22 (thousand); in 2000, it was about $28 (thousand). The change is

$$28 - 22 = 6 \text{ (thousand)}$$

during the 10-year period, and the average yearly raise for beginning

Getting Started

Each topic is introduced in a setting familiar to students' daily lives, making the subject personally relevant and more easily understood.

Web It

Appearing in margins where relevant, these boxes refer students to the abundance of resources available on the Web that can show them fun, alternative explanations and demonstrations of important topics.

Web It

For another explanation of finding slopes along with an interactive practice, go to link 3-3-1 at mhhe.com/bello.

Write On

Writing exercises give students the opportunity to express mathematical concepts and procedures in their own words, thereby internalizing what they have learned.

WRITE ON

What is the difference between:

79. A function and a relation when they are written as a set of ordered pairs?

80. The graph of a function and the graph of a relation?

Write your own definition of:

81. The domain of a function.

82. The range of a function.

Collaborative Learning

Concluding the chapter are exercises for collaborative learning that promote teamwork by students on interesting and enjoyable exploration projects.

COLLABORATIVE LEARNING 3A

Suppose you interviewed for a job with the two companies shown here, and received a job offer from each of the companies. Should you be greedy and take the job with the higher base salary or is there more to it? Let's do some collaboration to help with your decision.

Divide into two groups and have each group select one of the two companies to investigate. To conduct the investigations do the following:

1. Make a table of values for potential sales (x-values) and their respective salary (y-values) based on that company's compensation plan. Let the potential sales begin with $0 and increase by increments of $50,000 up to $500,000.

2. Make a graph plotting the pairs of values from the table.

After each group has completed their company's graph, have someone from each group put their group's graph on the same large poster board graph for all to view. Have a group discussion about the comparison of these two graphs including the following questions:

1. Does the salary from the $80,000/year job always yield more income?

2. Is there a point where the amount of sales will yield the same income for both jobs? If so, when?

3. Does the income from the $60,000/year job ever yield more income? If so, when?

4. Which job should you take? Explain.

Company:	Sales Consultants of Tacoma	Job Type:	Banking
Location:	US-Washington		Business Development
Base Pay:	$80,000.00/Year		Finance
Other Pay:	10% of all sales		Marketing
Employee Type:	Full-Time Employee		Sales
Industry:	Banking–Financial Services	Req'd Education:	BS degree
	Consulting	Req'd Experience:	At Least 3 Years
	Sales–Marketing	Req'd Travel:	Negligible
		Relocation Covered:	No

Research Questions

Research questions provide students with additional opportunities to explore interesting areas of math where they may find that the questions can lead to surprising answers.

Research Questions

1. Some historians claim that the official birthday of analytic geometry is November 10, 1619. Investigate and write a report on why this is so and the events that led Descartes to the discovery of analytic geometry.

2. Find out what led Descartes to make his famous pronouncement, "*Je pense, donc je suis*" (I think, therefore I am), and write a report about the contents of one of his works, *La Geometrie*.

3. She was 19, a capable ruler, a good classicist, a remarkable athlete, and an expert hunter and horsewoman. Find out who this queen was and what connections she had with Descartes.

• *Abundant Practice and Problem Solving*

Bello/Hopf offer students numerous opportunities and different paths for developing their problem-solving skills.

Pretest

An optional pretest that begins each chapter is especially helpful for students taking the course as a review who may remember some concepts but not others. **The answer grid** that follows immediately afterward gives students the page number, section, and example to study in case they missed a question.

Pretest for Chapter 3

(Answers on pages 161–162)

1. Graph the solutions to $2x + y = 2$.

2. Find the x- and y-intercepts of $y = -x + 3$.

3. Graph the solutions to $-5x = -10$.

4. Find the domain and range of the relation
$\{(2, -1), (4, 0), (1, 1), (-2, 1)\}$

5. a. Find the domain and range of the relation.
b. Is the relation a function?

6. Given the function, $f(x) = \frac{3x}{x-5}$
a. Find the domain of the function.
b. Find $f(2) - f(6)$.

Answers to Pretest

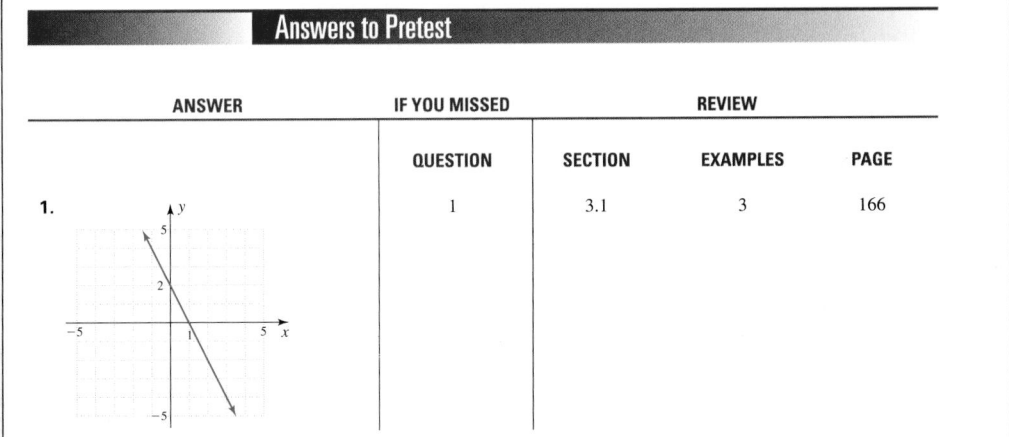

ANSWER	IF YOU MISSED		REVIEW	
	QUESTION	SECTION	EXAMPLES	PAGE
1.	1	3.1	3	166

Paired Examples/Problems

Examples are placed adjacent to similar problems intended for students to obtain immediate reinforcement of the skill they have just observed. These are especially effective for students who learn by doing and who benefit from frequent practice of important methods. Answers to the problems appear at the bottom of the page.

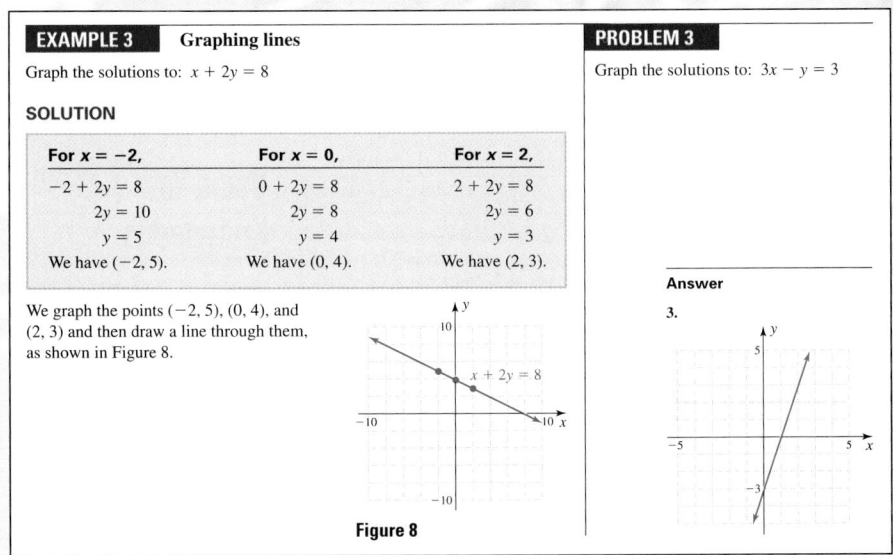

EXAMPLE 3   Graphing lines

Graph the solutions to: $x + 2y = 8$

SOLUTION

For $x = -2$,	For $x = 0$,	For $x = 2$,
$-2 + 2y = 8$	$0 + 2y = 8$	$2 + 2y = 8$
$2y = 10$	$2y = 8$	$2y = 6$
$y = 5$	$y = 4$	$y = 3$
We have $(-2, 5)$.	We have $(0, 4)$.	We have $(2, 3)$.

We graph the points $(-2, 5)$, $(0, 4)$, and $(2, 3)$ and then draw a line through them, as shown in Figure 8.

$x + 2y = 8$

Figure 8

PROBLEM 3

Graph the solutions to: $3x - y = 3$

Answer

3.

GUIDED TOUR

PROCEDURE

RSTUV Method for Solving Word Problems

1. **R**ead the problem carefully and decide what is asked for (the unknown).
2. **S**elect a variable to represent this unknown.
3. **T**hink of a plan to help you write an equation.
4. **U**se algebra to solve the resulting equation.
5. **V**erify the answer.

RSTUV Method

The easy-to-remember **"RSTUV" method** gives students a reliable and helpful tool in demystifying word problems so that they can more readily translate them into equations they can recognize and solve.

- **R**ead the problem carefully and decide what is being asked for (the unknown).
- **S**elect a variable to represent this unknown.
- **T**hink of a plan to help you write an equation.
- **U**se algebra to solve the resulting equation.
- **V**erify the answer.

Exercises

A wealth of exercises for each section is organized according to the learning objectives for that section, giving students a reference to study if they need extra help.

Exercises 3.2

A In Problems 1–10, find the domain and range, and determine whether the relation is a function.

1. $\{(-3, 0), (-2, 1), (-1, 2)\}$

2. $\{(-1, -2), (0, -1), (1, 0)\}$

3. $\{(3, 0), (4, 0), (5, 0)\}$

4. $\{(0, 1), (0, 2), (0, 3)\}$

Applications

Students will enjoy the exceptionally creative applications in most sections that bring math alive and demonstrate that it can even be performed with a sense of humor.

APPLICATIONS

Practicing Math with Real Word Problems

65. The revenue obtained from selling x textbooks is given by $R(x) = 30x - 0.0005x^2$. The cost of producing the books is $C(x) = 100,000 + 6x$.
   a. Find the profit function $P(x) = R(x) - C(x)$.
   b. Find the profit when 10,000 books are sold.

66. The Fahrenheit temperature reading F is a function of the Celsius temperature reading C. This function is given by

$$F(C) = \frac{9}{5}C + 32$$
   a. If the temperature is 15°C, what is the Fahrenheit temperature?
   b. Water boils at 100°C. What is the corresponding Fahrenheit temperature?
   c. The freezing point of water is 0°C or 32°F. How many Fahrenheit degrees below freezing is a temperature of −10°C?

67. When you exercise, your pulse rate should be within a certain *target zone*. The *upper limit U* of your target zone when exercising is a function of age a (in years) and is given by

USING YOUR KNOWLEDGE

Everything You Always Wanted to Know about Relations (in Mathematics)

A special kind of relation that's important in mathematics is called an **equivalence relation.** A relation R is an equivalence relation if it has the following three properties:

a. Reflexive: If a is an element of the domain of R, then (a, a) is an element of R.

b. Symmetric: If (a, b) is an element of R, then (b, a) is an element of R.

c. Transitive: If (a, b) and (b, c) are both elements of R, then (a, c) is an element of R.

A simple example of an equivalence relation is

We check the three properties as before:

a. Reflexive: Given a person A, is (A, A) an element of R? Yes, A is obviously a member of the same immediate family as A.

b. Symmetric: Suppose that (A, B) is an element of R. Then B is a member of the same immediate family as A. But then A is a member of the same immediate family as B, so (B, A) is an element of R.

c. Transitive: Suppose that (A, B) and (B, C) both belong to R. Then A, B, and C are all members of the same immediate family. Thus (A, C) is an element of R.

Using Your Knowledge

Optional, extended applications give students an opportunity to practice what they've learned in a multistep problem requiring reasoning skills in addition to algebraic operations.

• *Study Aids to Make Math Accessible*

Since some students confront math anxiety as soon as they sign up for the course, the Bello/Hopf system provides numerous study aids to make their learning easier.

3.3 USING SLOPES TO GRAPH LINES

To Succeed, Review How To . . .

1. Add, subtract, multiply, and divide integers (pp. 17–23).

2. Find the reciprocal of a number (p. 24).

Objectives

A Find the slope of a line passing through two given points.

B Use the definition of slope to decide whether two lines are perpendicular, parallel, or neither.

C Graph a line given its slope and a point on the line.

Reviews

Every section begins with "To succeed, review how to . . . ," which directs students to specific pages to study key topics they need to understand to successfully begin that section.

Objectives

The objectives for each section not only identify the specific tasks students should be able to perform, but they organize the section itself with letters corresponding to each section heading, making it easy to follow.

Calculate It

Adding Integers

To enter $-3 + (-2)$ using a scientific calculator, enter

| 3 | +/− | + | 2 | +/− | ENTER |

If your calculator has a set of parentheses, then you can enter parentheses around the -3 and -2.

Calculate It

Appearing in margins where relevant, these boxes give students optional advice on how to use calculators to reinforce their understanding of algebra and check their work.

Skill Checkers

These brief exercises help students keep their math skills well honed in preparation for the next section.

SKILL CHECKER

Try the "Skill Checker" exercises so you'll be ready for the next section.

Solve for y:

63. $6x + 3y = 12$

64. $2x + 3y = 6$

65. $3y - 2x = 12$

66. $5y - 2x = 10$

MASTERY TEST

If you know how to do these problems, you have learned your lesson!

91. If $f = \{(4, 3), (5, -1), (6, 0)\}$, find:

  a. $f(4)$ **b.** $f(5)$ **c.** $f(4) - f(5)$

92. If $f(x) = 2x - 3$, find:

  a. $f(4)$ **b.** $f(2)$ **c.** $f(x + 1)$

Find the domain of:

93. $f(x) = \dfrac{1}{x - 2}$

94. $y = \sqrt{x - 3}$

Mastery Tests

Brief tests in every section give students a quick checkup to make sure they're ready to go on to the next topic.

Summary

SECTION	ITEM	MEANING	EXAMPLE
5.1A	Monomial	A constant times a product of variables with whole-number exponents	$3x^2y$, $-7x$, $0.5x^3$, $\frac{2}{3}x^2yz^4$
	Polynomial	A sum or difference of monomials	$3x^2 - 7x + 8$, $x^2y + y^3$
	Terms	The individual monomials in a polynomial	The terms of $3x^2 - 7x + 8$ are $3x^2$, $-7x$, and 8.
	Coefficient	The numerical factor of a term	The coefficient of $3x^2$ is 3.
	Binomial	A polynomial with two terms	$5x^2 - 7$ is a binomial.
	Trinomial	A polynomial with three terms	$-3 + x^2 + x$ is a trinomial.
5.1B	Degree of a polynomial	Largest sum of the exponents in any term	The degree of $x^3 + 7x$ is 3. The degree of $-2x^3yz^2$ is 6.
	Descending order	Polynomial in one variable ordered from highest exponent to lowest	$7x^3 - 5x^2 + x - 3$ is in descending order.

Summary

An easy-to-read grid summarizes the essential chapter information by section, providing an item, its meaning, and an example to help students connect concepts with their concrete occurrences.

Review Exercises

Chapter review exercises are coded by section number and give students extra reinforcement and practice to boost their confidence.

Review Exercises

(If you need help with these exercises, look in the section indicated in brackets.)

1. [5.1A, B] Classify as a monomial, binomial, or trinomial and give the degree.

   a. $x^3 + x^2y^3z$

   b. $x^3y^2z^3$

2. [5.1B] Write the polynomials in descending order.

   a. $-x^2 + 3x^4 - 5x + 2$

   b. $3x - x^2 + 4x^3$

3. [5.1C, E] If a $50,000 computer depreciates 20% each year and its value $v(t)$ after t years is given by $v(t) = 50,000(1 - 0.20t)$.

   a. Find the value of the computer after 3 yr.

Practice Test 5

(Answers on page 417)

1. Classify as a monomial, binomial, or trinomial and give the degree of $xy^3z^4 - x^7$.

2. Write in descending order: $-4 + 3x^2 - x^4 + 2x^3$

3. The total dollar cost $C(x)$ of manufacturing x units of a product each week is given by $C(x) = 15x + 300$.

   a. Find the cost of manufacturing 400 units.

   b. Find the cost of manufacturing 1500 units.

4. Let $P(x) = x^2 - 3x + 2$. Find $P(-2)$.

Practice Test with Answers

The chapter Practice Test offers students a nonthreatening environment to review the material and determine whether they are ready to take a test given by their instructor. The answers to the Practice Test give students immediate feedback on their performance, and the answer grid gives them specific guidance on which section, example, and pages to review for any answers they may have missed.

Answers to Practice Test

ANSWER		IF YOU MISSED	REVIEW		
		QUESTION	SECTION	EXAMPLES	PAGE
1. Binomial; 8		1	5.1A, B	1, 2	345 & 346
2. $-x^4 + 2x^3 + 3x^2 - 4$		2	5.1B	2	346
3. a. $6300 **b.** $22,800		3	5.1C, E	3, 8	347 & 350

Cumulative Review

The Cumulative Review covers material from the present chapter and any prior chapters. It can be used for extra homework or for student review to improve retention of important skills and concepts.

Cumulative Review Chapters 1–10

1. Simplify: $[(3x^2 - 2) + (8x + 3)] - [(x - 2) + (2x^2 - 6)]$

2. Simplify: $(4x^4y^{-2})^2$

3. Solve: $0.02P + 0.04(1700 - P) = 65$

4. Solve: $|x - 4| = |x - 8|$

5. Graph on a number line: $\{\,x \mid x < -4 \text{ or } x \geq 4\,\}$

6. Graph on a number line: $|6x - 9| \leq 3$

• *Supplements for Instructors*

Annotated Instructor's Edition

This version of the student text contains **answers** to all odd- and even-numbered exercises in addition to helpful **teaching tips.** The answers are printed on the same page as the exercises themselves so that there is no need to consult a separate appendix or answer key.

Instructor's Testing and Resource CD

This cross-platform CD-ROM provides a wealth of resources for the instructor. Supplements featured on this CD-ROM include a **computerized test bank** utilizing Brownstone Diploma® **algorithm-based** testing software to quickly create customized exams. This user-friendly program allows instructors to search for questions by topic, format, or difficulty level; edit existing questions or add new ones; and scramble questions and answer keys for multiple versions of the same test.

Instructor's Solutions Manual

This supplement contains detailed solutions to **all** exercises in the text. The methods used to solve the problems in the manual are the same as those used to solve the examples in the textbook.

 www.mathzone.com*

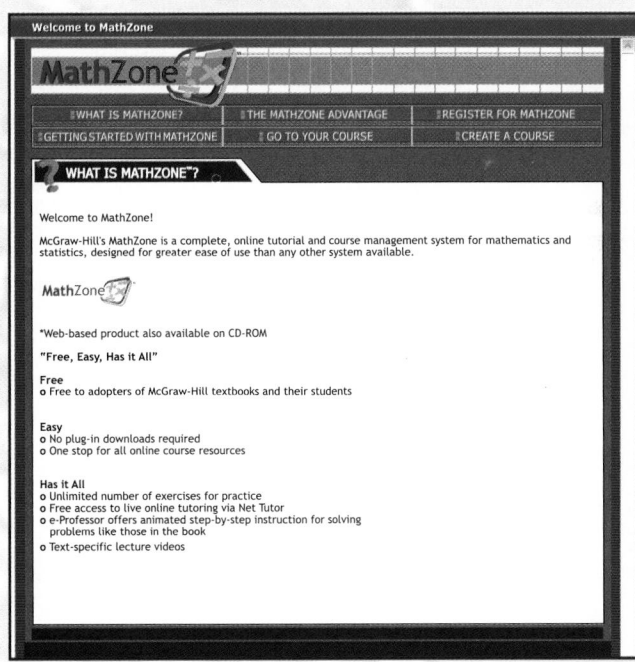

McGraw-Hill's **MathZone** is a complete, online tutorial and course management system for mathematics and statistics, designed for greater ease of use than any other system available. **Free** upon adoption of a McGraw-Hill title, the system allows instructors to create and share courses and assignments with colleagues and adjuncts with only a few

---

*Web-based product also available on CD-ROM

clicks of the mouse. All assignments, questions, e-Professors, online tutoring, and video lectures are directly tied to text-specific materials in Bello, *Intermediate Algebra, 2nd Edition*. MathZone courses are customized to your textbook, but you can **edit** questions and algorithms, **import** your own content, and **create** announcements and due dates for assignments. MathZone has **automatic grading** and reporting of easy-to-assign algorithmically generated homework, quizzing, and testing. All student activity within MathZone is automatically recorded and available to you through a **fully integrated grade book** that can be downloaded to Excel.

ALEKS® v2.0

ALEKS® (**A**ssessment and **LE**arning in **K**nowledge **S**paces) is an artificial intelligence-based system for individualized math learning, available over the Web. ALEKS delivers precise, qualitative diagnostic assessments of students' math knowledge, guides them in the selection of appropriate new study material, and records their progress toward mastery of curricular goals in a robust classroom management system. See page xxix for more details regarding ALEKS.

PageOut

PageOut is McGraw-Hill's unique, intuitive tool enabling instructors to create a full-featured, professional quality course website *without* being a technical expert. With PageOut you can post your syllabus online, assign content from the Bello MathZone site, add links to important off-site resources, and maintain student results in the online grade book. PageOut is free for every McGraw-Hill Higher Education user and, if you're short on time, we even have a team ready to help you create your site. Contact your McGraw-Hill representative for further information.

• *Supplements for Students*

Student's Solutions Manual

This supplement contains complete worked-out solutions to all odd-numbered exercises and all odd- and even-numbered problems in the Review Exercises and Cumulative Reviews in the textbook. The methods used to solve the problems in the manual are the same as those used to solve the examples in the textbook. This tool can be an invaluable aid to students who want to check their work and improve their grades by comparing their own solutions to those found in the manual and finding specific areas where they can do better.

 www.mathzone.com*

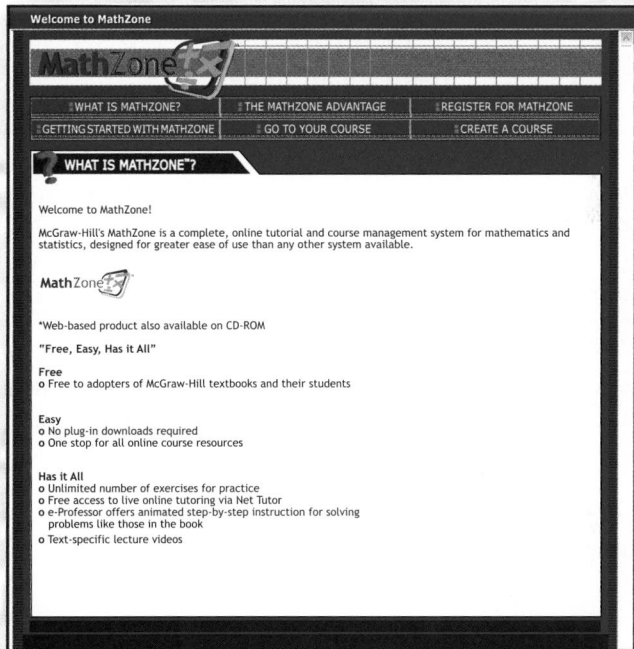

McGraw-Hill's MathZone is a powerful new online tutorial for homework, quizzing, testing, and interactive applications. MathZone offers:

- **Practice exercises** based on the text and generated in an unlimited number for as much practice as needed to master any topic you study.

- **Videos** of classroom instructors giving lectures and showing you how to solve exercises from the text.

- **e-Professors** to take you through animated, step-by-step instructions (delivered via on-screen text and synchronized audio) for solving problems in the book, allowing you to digest each step at your own pace.

- **NetTutor,** which offers live, personalized tutoring via the Internet.

- Every assignment, question, e-Professor, and video lecture is derived directly from Bello, *Intermediate Algebra, 2ⁿᵈ Edition.*

---

*Web-based product also available on CD-ROM

NetTutor

Also available separately from MathZone, NetTutor is a revolutionary system that enables students to interact with a live tutor over the Web by using NetTutor's Web-based, graphical chat capabilities. Students can also submit questions and receive answers, browse previously answered questions, and view previous live chat sessions. NetTutor can be accessed on the text's MathZone site through the Student Edition.

ALEKS®

ALEKS® (**A**ssessment and **LE**arning in **K**nowledge **S**paces) is an artificial intelligence-based system for individualized math learning, available over the Web. ALEKS delivers precise, qualitative diagnostic assessments of students' math knowledge, guides them in the selection of appropriate new study material, and records their progress toward mastery of curricular goals in a robust classroom management system. See page xxix for more details regarding ALEKS.

Bello Video Series

The video series is available on DVD and VHS tape and features an instructor introducing topics and working through selected exercises from the text, explaining how to complete them step-by-step. The DVDs are **closed-captioned** for the hearing-impaired and also **subtitled in Spanish.**

Math for the Anxious: Building Basic Skills, by Rosanne Proga

Math for the Anxious: Building Basic Skills is written to provide a practical approach to the problem of math anxiety. By combining strategies for success with a pain-free introduction to basic math content, students will overcome their anxiety and find greater success in their math courses.

ALEKS is an artificial intelligence-based system for individualized math learning, available for Higher Education from McGraw-Hill over the World Wide Web.

ALEKS delivers precise assessments of math knowledge, guides the student in the selection of appropriate new study material, and records student progress toward mastery of goals.

ALEKS interacts with a student much as a skilled human tutor would, moving between explanation and practice as needed, correcting and analyzing errors, defining terms and changing topics on request. By accurately assessing a student's knowledge, ALEKS can focus clearly on what the student is ready to learn next, helping to master the course content more quickly and easily.

ALEKS is:

- **A comprehensive course management system.** It tells the instructor exactly what students know and don't know.

- **Artificial intelligence.** It totally individualizes assessment and learning.

- **Customizable.** ALEKS can be set to cover the material in your course.

- **Web-based.** It uses a standard browser for easy Internet access.

- **Inexpensive.** There are no setup fees or site license fees.

ALEKS 2.0 adds the following new features:

- **Automatic Textbook Integration**
- **New Instructor Module**
- **Instructor-Created Quizzes**
- **New Message Center**

ALEKS maintains the features that have made it so popular including:

- **Web-Based Delivery** No complicated network or lab setup

- **Immediate Feedback** for students in learning mode

- **Integrated Tracking of Student Progress and Activity**

- **Individualized Instruction** which gives students problems they are *Ready to Learn*

For more information please contact your McGraw-Hill Sales Representative or visit ALEKS at http://www.highedmath.aleks.com.

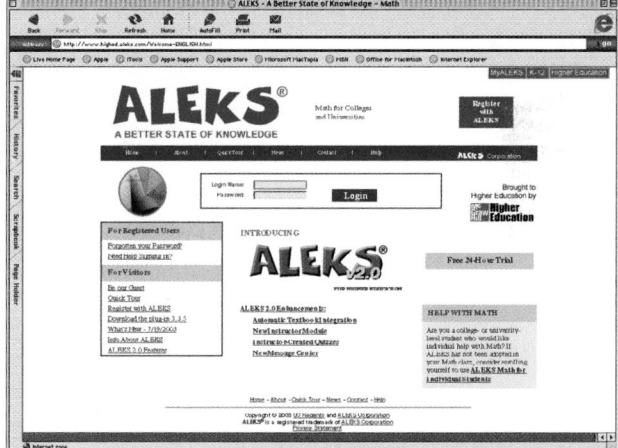

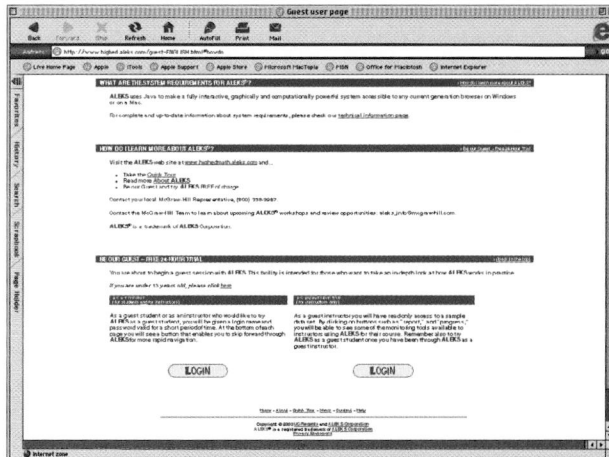

Applications Index

doubling every year, A31
principal *vs.* annual interest, 500
saving by putting aside money,
     A16–A17, A19, A44, A45
stocks and bonds, 114

Food and Drink

apples, complex word problem
     involving, 709
barbecue grill made from cylindrical
     drum, 681
beef, per capita consumption of, 602–603
calories and fat in daily diet, 148–149
Coca-cola
     consumption of, in U.S. *vs.* Mexico, 270
     percentage of water in, 15
coffee, price of Jamaican Blue, 119
eggs, complex word problem
     involving, 709
french fries, fat content of, 320
hamburgers, fat content of, 15, 320
juices, mixing for healthy
     results, 278–279
milk, per capita consumption of, 604
plastic plates, shape of, 683
poultry, per capita consumption
     of, 602–603
shortcake, calculating dimensions
     of, 704
shrimp, curried, multiplication of
     recipe for, 513
tea, price of Oolong, 119
tomatoes, stewed, in expanded
     recipe, 483

Geography

Calaveras Jumping Jubilee, 665
Canada
     Calgary, Alberta, 29
     University of Guelph, Ontario, 507
     value of Canadian dollar, 750–752
Capitol building, Washington, D.C., 718
China, speakers of Mandarin Chinese, 135
Delos, oracle at, 643
Denmark
     taxation of cigarettes in, 110
     Tycho Brahe Planetarium,
          Copenhagen, 669
Mt. Everest, height of, 135, 291
Hawaii, fish catch in, 746
Japan, hyperboloid tower in Kobe, 686
Longs Peak, 291
maps, using scale to calculate distance, 483
mountains, highest, 135, 291, 773
New England
     clam catch in, 746
     crab catch in, 746
Persia, Shah of, A10
Pikes Peak, 291
St. Ives, A6
St. Louis Science Center, 686

Geometry

cone, radius of, calculated from volume and
     height, 558
crosses on graph paper, of increasing
     size, A18
cube, doubling the size of, 643
pyramid, square, surface area of, 374
rectangle
     area of, 443
     calculating dimensions of, 104–105,
          333, 708, 827
     calculating perimeter of, 90,
          104–105, 153, 157, 577
     golden, approximating dimensions
          of, 488–489
     maximizing area of
          for a given perimeter, 659
          for a given perimeter on three
               sides, 665
     Washington Monument, perimeter
          of base of, 289
right triangle(s)
     calculating sizes of, 408, 416
     finding one whose sides are consecutive
          even integers, 408
     finding one whose sides are consecutive
          integers, 405
sphere
     radius of
          calculated from surface
               area, 558
          calculated from volume, 538
     radius of curvature of, formula
          for, 496
spiral, constructed from right
     triangles, 570
staircases on graph paper, of increasing
     size, A18
trapezoid, area of, 487, 513
trigonometric identities
     a formula for cosines, 497
     a formula for summing tangents, 497
volume
     of a rectangular container, 503
     of a Rubik's Cube, 31

Government

Fairfax County, Virginia
     government and school system salary
          trends, 202
     health and welfare spending,
          195–196
     public safety spending, 195–196
Medicare costs, 1990–1995
     average, per over-65 citizen, 746
     total, 746
patents, annual number of, 217
*Statistical Abstract of the
     United States,* 773, 813
taxation
     by assessed value and millage
          rate, 739–740

of cigarettes in Denmark, 110
Homestead Exemption, in Florida, 739
IRS form 1040, 119
IRS form 1040A, 490, 497
United States
     Department of Transportation, 813
     federal income *vs.* federal spending, 119
     salaries of president, vice president, and
          chief justice, 281

History

Adams, John Quincy, eavesdropping habit
     of, 718
Babylonian rings, broken and
     weighed, 270
Carroll, Lewis, 271
SS *Constitution,* 504
Delos, oracle at, 643
Diophantus, 110
Goddard, Robert, 594
Malthus, Thomas Robert, 769
Manhattan, purchase of, 781
Naismith, Dr. James, 270
Persia, Shah of, A10, A20
Pythagoras, death of, 517

Home and Garden

fertilizer, ingredients of, 304
firewood, splitting logs for, 614
garden, grassy border of, 407
ladder, leaning against a house, 405
landscaping, 301
lawn mowers, speed of, 513
nails, 302–303
painting a room
     how long it will take with one painter or
          two, 512, 516, 731
     how much paint will be needed, 411
picture frame, calculating dimensions
     of, 704
snow, shoveling, 620

Law Enforcement

blood alcohol level
     vs. number of drinks, weight, sex, and time
          elapsed, 189
     vs. probability of having an
          accident, 802, 805
drug, concentration of in bloodstream
     over time, 799
height, estimating from size of leg
     bone, 88, 96
highway curves, radius and maximum safe
     speed on, 520, 552, 559
robberies, annual number of, 188
skid marks
     calculating speed from, 527,
          530, 571, 622
     longest recorded, 633

The Real Numbers

1

1.1 Numbers and Their Properties

1.2 Operations and Properties of Real Numbers

1.3 Properties of Exponents

1.4 Algebraic Expressions and the Order of Operations

The Human Side of Algebra

The development of the number system used in algebra has been a multicultural undertaking. More than 20,000 years ago, our ancestors needed to count their possessions, their livestock, and the passage of days. Australian aborigines counted to two, South American Indians near the Amazon counted to six, and the Bushmen of South Africa were able to count to ten ($10 = 2 + 2 + 2 + 2 + 2$).

The earliest technique for visibly expressing a number is tallying (from the French verb *tailler* "to cut"). Tallying, a practice that reached its highest level of development in the British Exchequer tallies, used flat pieces of hazelwood about 6–9 inches long and about an inch thick, with notches of varying sizes and types. When a loan was made, the appropriate notches were cut and the stick split into two pieces, one for the debtor, one for the Exchequer. In this manner, transactions could easily be verified by fitting the two halves together and noticing whether the notches coincided, hence the expression "our accounts tallied."

The development of written numbers is due mainly to the Egyptians (about 3000 B.C.), the Babylonians (about 2000 B.C.), the early Greeks (about 400 B.C.), the Hindus (about 250 B.C.), and the Arabs (about 200 B.C.). Here are the numbers three of these civilizations used:

Egyptian, about 3000 B.C.

1	10	100	1000	10,000	100,000	1,000,000

Babylonian, about 2000 B.C.

0	1	10	12	20	60	600

Early Greek, about 400 B.C.

1	5	10	50	100	500	5000

Pretest for Chapter 1

(Answers on page 3)

1. Use roster notation to list the whole numbers between 7 and 12.

2. Write $\frac{1}{6}$ as a decimal.

3. Classify the given number using one or more of the classifications—natural number, whole number, integer, rational number, irrational number, and real number.

    a. $1\frac{1}{2}$ **b.** -7 **c.** $\sqrt{11}$

4. Find the additive inverse of -1.2

5. Find $\left|\dfrac{-2}{7}\right|$

6. Fill in the blank with $<$, $>$, or $=$ to make the resulting statement true.

$$0.7 \underline{\hspace{1cm}} \frac{7}{100}$$

7. Find $-0.5 - (-1.2)$

8. Find $-18 - 9$

9. Find $\dfrac{1}{4} - \dfrac{3}{8}$

10. Find $(-6)(-2.9)$

11. Find $\dfrac{-2}{5} \div \dfrac{9}{5}$

12. Name the property illustrated in this statement.

$$8 + (2 + 4) = (2 + 4) + 8$$

13. Evaluate $(-3)^3 + \dfrac{5 - 13}{-2} + 18 \div (-6)$

14. Write x^{-5} without a negative exponent.

15. Perform the indicated operation and simplify.

    a. $(-7x^3y^5)(-6x^{-4}y)$ **b.** $\dfrac{-36x^6}{18x^{-3}}$

16. Simplify.

    a. $(-3x^{-5}y)^4$ **b.** $\left(\dfrac{x^2}{y^{-4}}\right)^{-3}$

17. Perform the calculation and write your answer in scientific notation.

$$(8.4 \times 10^{-5}) \times (3 \times 10^2)$$

18. $\pi r^2 + \pi rs$ gives the surface area of a circular cone. Find the value of the surface area if

$$\pi = \frac{22}{7}, \quad r = 7, \quad \text{and} \quad s = 3.$$

19. Simplify.

$$9x - (4x + 5) + (3x + 3)$$

20. Simplify.

$$[(7x^2 - 2) + (5x + 6)] - [(1 - 3x) + (x^2 - 9)]$$

Answers to Pretest

ANSWER	IF YOU MISSED	REVIEW		
	QUESTION	SECTION	EXAMPLES	PAGE
1. $\{8, 9, 10, 11\}$	1	1.1	1	6
2. $0.1666\ldots$	2	1.1	2	7
3. a. Real, rational	3	1.1	3	8
b. Real, rational, integer				
c. Real, irrational				
4. 1.2	4	1.1	4	10
5. $\dfrac{2}{7}$	5	1.1	5	11
6. $>$	6	1.1	6	12–13
7. 0.7	7	1.2	2	19
8. -27	8	1.2	2	19
9. $-\dfrac{1}{8}$	9	1.2	2	19
10. 17.4	10	1.2	3, 4	20–21
11. $-\dfrac{2}{9}$	11	1.2	6, 7	24–25
12. Commutative property of addition	12	1.2	8	25
13. -26	13	1.4	4	49
14. $\dfrac{1}{x^5}$	14	1.3	2	34
15. a. $\dfrac{42y^6}{x}$ **b.** $-2x^9$	15	1.3	3, 4	36–37
16. a. $\dfrac{81y^4}{x^{20}}$ **b.** $\dfrac{1}{x^6 y^{12}}$	16	1.3	5, 6	39–40
17. 2.52×10^{-2}	17	1.3	7, 8, 9	41–42
18. 220	18	1.4	5	50
19. $8x - 2$	19	1.4	9	55
20. $6x^2 + 8x + 12$	20	1.4	10	55

1.1 NUMBERS AND THEIR PROPERTIES

To Succeed, Review How To ...

1. Write the fraction $\frac{a}{b}$ as a decimal by dividing a by b.

2. Distinguish between a positive and a negative number.

Note: The *To Succeed* section tells you what you need to know or review *before* you go on.

Objectives

A Write a set of numbers using roster or set-builder notation.

B Write a rational number as a decimal.

C Classify a number as natural, whole, integer, rational, irrational, or real.

D Find the additive inverse of a number.

E Find the absolute value of a number.

F Given two numbers, use the correct notation to indicate equality or which is larger.

GETTING STARTED   Algebra and Wages by Educational Degree

Higher Education Means Higher Income
How the average earnings of those with High School Diplomas, Bachelor's Degrees, and Advanced Degrees compare over a 20-year period.

Average Earnings of Workers 18 Years and Older, By Educational Attainment: Selected Years

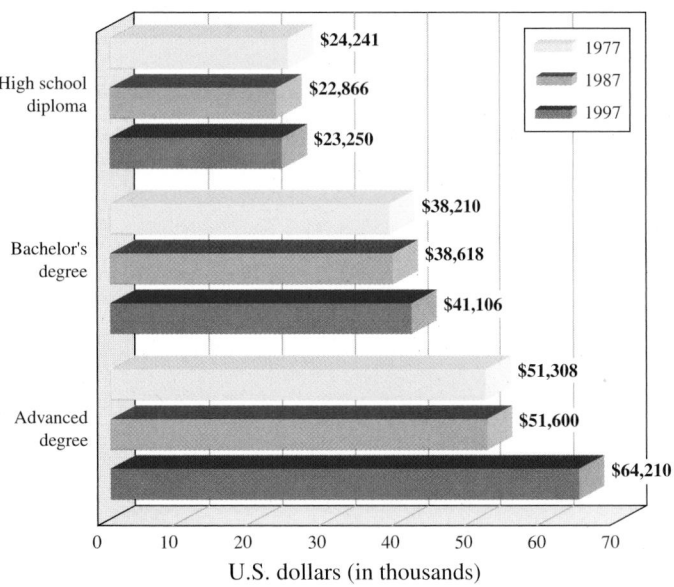

Source: ACE Fact Sheet of Higher Education, March 1999. Information contained in the fact sheet is from the U.S. Census Bureau Current Population Survey.

Look at the graph. What was the difference in earnings of a person with an advanced degree in 1997 as compared to one with a bachelor's degree in 1997? The answer is \$64,210 − \$41,106; or \$23,104.

In arithmetic we are accustomed to expressions such as

$$26 + 18, \quad 8 \times 32, \quad 26 - 18, \quad \text{and} \quad \frac{40}{5}$$

In algebra we use expressions such as

$$26 + x, \quad 2\pi r, \quad 26 - w, \quad \text{and} \quad \frac{d}{t}$$

The letters x, r, w, d, and t are **variables** that stand for different numbers. When a letter stands for *one* number, it's called a **constant.** For example, if in the graph b were to represent the average earnings for a person with a bachelor's degree in 1997, then b would be a constant. However, earnings change from year to year, so in some other algebra problem we could say the variable s will represent the salary of a person with a bachelor's degree for a given year.

What kind of numbers can we use in place of x, r, w, d, and t? In algebra we start with the real numbers and then study complex numbers. We begin this section by discussing different sets of numbers.

R-I-S-E TO SUCCESS IN MATH

Why are some students more successful in math than others? Often it is because they know how to manage their time and have a plan for action. Use models similar to the tables on the next page to make a weekly schedule of your time (classes, study, work, personal, etc.)

Weekly Time Schedule							
Time	S	M	T	W	R	F	S
8:00							
9:00							
10:00							
11:00							
12:00							
1:00							
2:00							
3:00							
4:00							
5:00							
6:00							
7:00							
8:00							
9:00							
10:00							
11:00							

Semester Calendar					
Wk	M	T	W	R	F
1					
2					
3					
4					
5					
6					
7					
8					
9					
10					
11					
12					
13					
14					
15					
16					

and a semester calendar indicating major course events like tests, papers, and so on. Then, try to do as many of the suggestions on the "**R-I-S-E**" list as possible.

R—Read and/or view the material before and after each class. This includes the textbook, the videos that come with the book, and any special material given to you by your instructor.

I—Interact and/or practice using the CD that comes with the book, Web exercises suggested in the sections, or seeking tutoring from your school.

S—Study and/or discuss your homework and class notes with a study partner/group, with your instructor, or on a discussion board if available.

E—Evaluate your progress by checking the odd homework questions with the answer key in the back of the book, using the mastery questions in each section of the book as a self-test, and using the Chapter Reviews and Chapter Practice Tests as practice before taking the actual test.

As the items on this list become part of your regular study habits, you will be ready to "R–I–S–E" to success in math.

Sets of Numbers

The idea of a **set** is familiar in everyday life. You may own a set of dishes, a tool set, or a set of books. In algebra we use sets of numbers.

We use capital letters to denote sets and lowercase letters to denote elements (or members) of these sets. If possible, we "list" the elements of a set in braces { } and separate them by commas. Thus $A = \{1, 2, 3\}$ is the set A that has 1, 2, and 3 for its elements. This

Web It

To learn different ways to describe a set, including set builder notation, try link 1-1-1 at mhhe.com/bello.

For a lesson about sets, including writing sets using roster notation, see link 1-1-2 at mhhe.com/bello.

notation is known as **roster notation.** If a set has no elements, it is called the **empty,** or **null,** set and is denoted by { } or by the symbol $\varnothing$ (used without braces and read "the empty or null set").

> **NOTE**
>
> { } is empty but the set $\{\varnothing\}$ is not the empty set because $\{\varnothing\}$ has one element, $\varnothing$.

Here are three sets of numbers frequently used in algebra.

Natural Numbers	The set of numbers used for counting.
	$$N = \{1, 2, 3, \ldots\}$$

The three dots (called an ellipsis) mean that the pattern continues in the indicated direction.

Whole Numbers	The set of natural numbers and zero.
	$$W = \{0, 1, 2, 3, \ldots\}$$

Integers	The set of whole numbers and their opposites (negatives).
	$$I = \{\ldots, -2, -1, 0, 1, 2, \ldots\}$$

The Greek letter $\in$ (epsilon) is used to indicate that an element *belongs* to a set. Thus $-5 \in I$ indicates that -5 is in the set I; that is, -5 *is an integer*. On the other hand, $0 \notin N$ indicates that 0 *is not a natural number*.

EXAMPLE 1   Roster notation

Use roster notation to write the following sets:

a. The natural numbers between 2 and 7.

b. The first three whole numbers.

c. The first two negative integers.

d. The only number that is neither positive nor negative.

SOLUTION

a. The set of natural numbers between 2 and 7 is $\{3, 4, 5, 6\}$. (2 and 7 are not included.)

b. The set of the first three whole numbers is $\{0, 1, 2\}$.

c. The set of the first two negative integers is $\{-1, -2\}$.

d. The set containing the only number that is neither positive nor negative is $\{0\}$.

PROBLEM 1

Use roster notation to write the following sets:

a. The natural numbers between 3 and 8.

b. The first four whole numbers.

c. The first four negative integers.

d. The first whole number.

If a number can be written in the form $\frac{a}{b}$, where a and b are integers and b is not 0 ($b \neq 0$), the number is called a *rational number* (because it is a *ratio* of two integers). Thus

$$\frac{1}{5}, \qquad \frac{-4}{3}, \qquad \frac{4}{1} = 4, \qquad \text{and} \qquad \frac{-7}{1} = -7$$

are rational numbers. There is no obvious pattern with which we can list all the rational numbers, so we use a new notation, called **set-builder notation,** to define this set.

Answers

1. a. $\{4, 5, 6, 7\}$ **b.** $\{0, 1, 2, 3\}$
c. $\{-1, -2, -3, -4\}$ **d.** $\{0\}$

Web It

For an excellent lesson on converting fractions to decimals, try link 1-1-3 at mhhe.com/bello.

Web It

If you prefer an interactive lesson on the conversion of a fraction to a decimal, try link 1-1-4 at mhhe.com/bello.

RATIONAL NUMBERS

The set Q of **rational numbers** consists of all the numbers that can be written as the ratio of two integers. Thus,

$$Q = \left\{ r \,\middle|\, r = \frac{a}{b},\ a \text{ and } b \text{ integers},\ b \neq 0 \right\}$$

This is read "Q equals the set of all r such that r equals a divided by b, a and b integers and b not equal to 0." The symbol $\mid$ is read as "such that."

Using set-builder notation, the sets N, W, and I can be written as

$$N = \{x \mid x \text{ is a counting number}\}$$
$$W = \{x \mid x \text{ is a whole number}\}$$
$$I = \{x \mid x \text{ is an integer}\}$$

B Writing Rational Numbers as Decimals

The rational number $\frac{a}{b}$ can also be written as a decimal by dividing a by b. The result is either a **terminating decimal** (as in $\frac{1}{2} = 0.5$) or a **nonterminating, repeating decimal** (as in $\frac{1}{3} = 0.333\ldots$). We often place a bar over the repeating digits in a nonterminating, repeating decimal. Thus $\frac{1}{3} = 0.\overline{3}$, and $\frac{2}{11} = 0.181818\ldots = 0.\overline{18}$.

EXAMPLE 2 Writing fractions as decimals

Write as a decimal:

a. $\dfrac{4}{5}$ **b.** $\dfrac{3}{11}$ **c.** $\dfrac{95}{30}$

PROBLEM 2

Write as a decimal:

a. $\dfrac{3}{8}$ **b.** $\dfrac{5}{11}$ **c.** $\dfrac{95}{60}$

SOLUTION

a. Dividing 4 by 5, we obtain $\frac{4}{5} = 0.8$, a terminating decimal.

b. Dividing 3 by 11, we have

$$\frac{3}{11} = 0.272727\ldots = 0.\overline{27}$$

a nonterminating, repeating decimal.

c. Dividing 95 by 30, we have

$$\frac{95}{30} = 3.1666\ldots = 3.1\overline{6}$$

a nonterminating, repeating decimal.

Calculate It Numerical Calculations

The numerical calculations in this section can be done with a calculator, but be aware that calculator procedures vary. (When in doubt, read the manual.) Start at the "Home Screen" and do Example 2(a), by dividing 4 by 5: that is, find $4 \div 5$. On a TI-83 Plus, press 4 [÷] 5 [ENTER]. The answer is shown as .8. Do parts b and c of Example 2.

```
4/5
                    .8
```

Answers

2. a. 0.375
b. 0.454545 . . . or $0.\overline{45}$
c. 1.58333 . . . or $1.58\overline{3}$

Since any rational number of the form $\frac{a}{b}$ ($b \neq 0$) is either a terminating or a repeating decimal, the set Q of rational numbers can also be defined as follows.

Alternative Definition for the Set Q	$Q = \{x \mid x \text{ is a terminating or a repeating decimal}\}$

There are some numbers such as $\sqrt{2}$ (the square root of 2), π, and $\sqrt{10}$ that are *not* rational numbers. These are called *irrational* numbers because they cannot be written as the ratio of two integers. When written as decimals, irrational numbers are nonterminating and nonrepeating. For example, $0.101001000\ldots$ and $3.1234567\ldots$ are irrational. Here is the definition of irrational numbers.

IRRATIONAL NUMBERS

Irrational numbers are numbers that *cannot* be written as ratios of two integers. The set of irrational numbers is

$$H = \{x \mid x \text{ is a number that is not rational}\}$$

The **real numbers** include both the rational numbers and the irrational numbers.

Web It

For practice classifying numbers, try link 1-1-5 at mhhe.com/bello.

REAL NUMBERS

Numbers that are either rational or irrational are called **real numbers.** The set of real numbers R is defined by

$$R = \{x \mid x \text{ is a number that is rational or irrational}\}$$

 **C   Classifying Numbers**

Here are some real numbers:

$$5,\ 17,\ -4,\ -9,\ 0,\ \frac{3}{5},\ 0.6,\ \frac{1}{-10},\ -0.\overline{1},\ \frac{4}{-3},\ \sqrt{3},\ \pi,\ 0.345\ldots$$

EXAMPLE 3   **Classifying numbers**

Classify the given number by making a check mark (✓) in the appropriate row(s).

Set	0	$-\frac{4}{5}$	-4	$\sqrt{2}$	7	$-\pi$	$0.\overline{8}$	$0.01001000\ldots$
Natural numbers					✓			
Whole numbers	✓				✓			
Integers	✓		✓		✓			
Rational numbers	✓	✓	✓		✓		✓	
Irrational numbers				✓		✓		✓
Real numbers	✓	✓	✓	✓	✓	✓	✓	✓

PROBLEM 3

Classify the given numbers by making a check mark (✓) in the appropriate row(s).

Set	$\frac{8}{5}$	π	6	$\sqrt{3}$	-11
Natural numbers					
Whole numbers					
Integers					
Rational numbers					
Irrational numbers					
Real numbers					

Do you have a good idea of the relationship between the sets of numbers we have discussed? We can clarify the situation by using the idea of a *subset*. We say that A is a **subset** of B, denoted by $A \subseteq B$, when all the elements in A are also in B. Thus, because all natural numbers N are whole numbers, $N \subseteq W$ (read "N is a subset of W"). Also, because all whole numbers are integers, $W \subseteq I$. Here is the complete picture:

$$N \subseteq W \subseteq I \subseteq Q \subseteq R$$

The diagram in Figure 1 shows the sets involved.

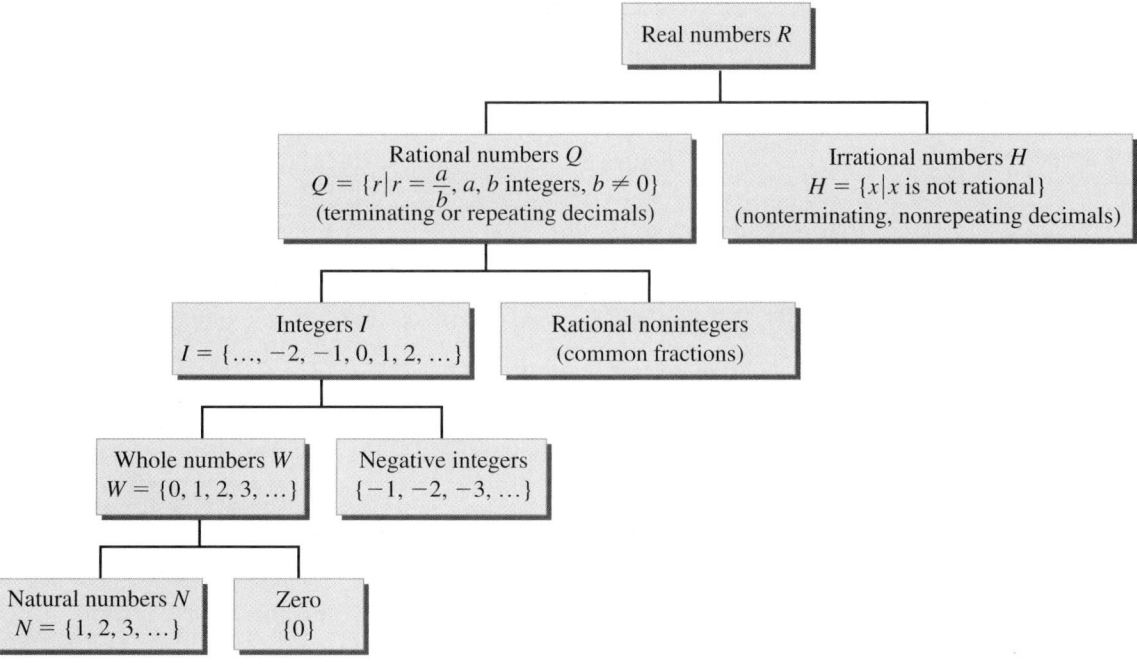

Figure 1

To make a picture or graph of some real numbers, we can think of a number line completely filled with all the real numbers. In this line the *positive* real numbers are to the *right* of zero; the *negative* real numbers are to the *left* of zero; and zero, in the center of the number line, is called the *origin*, as shown in Figure 2.

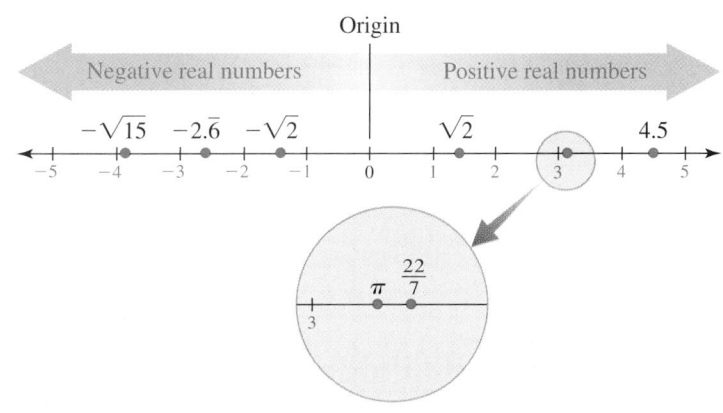

Figure 2

The number corresponding to a point on the number line is the **coordinate** of the point, and the point is called the **graph** of the number. In Figure 2, some irrational numbers and some rational numbers are shown. There is exactly one real number for each point graphed on the number line and one point for each real number.

Answer

3.

Set	$\frac{8}{5}$	π	6	$\sqrt{3}$	−11
Natural numbers			✓		
Whole numbers			✓		
Integers			✓		✓
Rational numbers	✓		✓		✓
Irrational numbers		✓		✓	
Real numbers	✓	✓	✓	✓	✓

D   Additive Inverses (Opposites)

Each point on the number line has another point *opposite* it with respect to zero. The numbers corresponding to these two points are called the *additive inverses (opposites)* of each other. Thus 3 and -3 (read "the additive inverse of 3," or "negative 3") are additive inverses, as are -4 and 4. (See Figure 3.) A number and its additive inverse are always the same distance from zero.

Web It

For a well illustrated site discussing additive inverses, try link 1-1-6 at mhhe.com/bello.

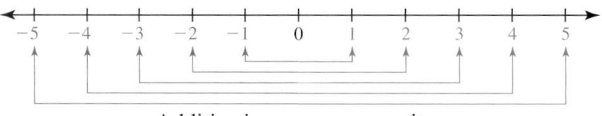

Additive inverses or opposites

Figure 3

Here is the definition.

ADDITIVE INVERSE

The **additive inverse (opposite)** of a is $-a$.

The sum of a number and its additive inverse is always zero—that is,
$a + (-a) = (-a) + a = 0$.

EXAMPLE 4 **Finding additive inverses**	**PROBLEM 4**

Find the additive inverse:

a. 5   **b.** -3.5 **c.** $\dfrac{2}{3}$

Find the additive inverse:

a. 3   **b.** -2.5 **c.** $\dfrac{3}{5}$

SOLUTION

a. The additive inverse of 5 is -5. (See Figure 4.)

b. The additive inverse of -3.5 is 3.5. (See Figure 4.) In symbols, $-(-3.5) = 3.5$ (read "the additive inverse of negative 3.5 is 3.5," or "the additive inverse of the inverse of 3.5 is 3.5").

c. The additive inverse of $\dfrac{2}{3}$ is $-\dfrac{2}{3}$. (See Figure 4.)

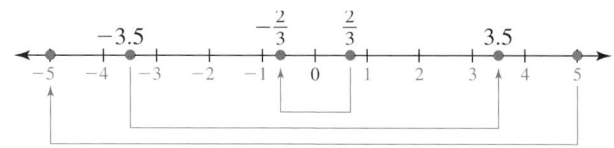

Additive inverses

Figure 4

Calculate It   Negative Sign versus Subtraction Key

Texas Instrument (TI) calculators use a gray key **(−)** to find the additive inverse of a number and a subtraction key **—** to indicate subtraction. To find the *additive inverse* of -3.5, you have to find the additive inverse of a negative number. With a TI calculator, enter **(−)** **(−)** 3.5 and then press **ENTER** to find the answer shown. What does the calculator indicate for **—** **—** 3.5?

```
- - 3.5
                3.5
```

Answers

4. a. -3 **b.** 2.5   **c.** $-\dfrac{3}{5}$

 **E**   **Absolute Values**

Now, let's go back to the number line. What is the distance between 3 and 0? The answer is 3 units. What about the distance between -3 and 0? The answer is *still* 3 units. The distance between any number a and 0 is called the *absolute value* of a and is denoted by $|a|$. Thus $|-3| = 3$ and $|3| = 3$. In general, we have the following definition.

ABSOLUTE VALUE OF A REAL NUMBER

The **absolute value** of a real number a, denoted by $|a|$, is defined as the distance between a and 0 on the real-number line. In general,

$$|a| = \begin{cases} a & \text{if } a \text{ is positive} \\ 0 & \text{if } a \text{ is zero} \\ -a & \text{if } a \text{ is negative} \end{cases}$$

$|11| = 11 \quad$ and $\quad \left|\frac{1}{2}\right| = \frac{1}{2}$

$|0| = 0$

$|-5| = -(-5) = 5 \quad$ and $\quad \left|-\frac{1}{3}\right| = -\left(-\frac{1}{3}\right) = \frac{1}{3}$

CAUTION

The absolute value of a number represents a distance and a distance is *never* negative, therefore the absolute value of a number is *never* negative. It is always positive or zero. However, if a is not 0, $-|a|$ is *always* negative. Thus

$$-|-3| = -3, \quad -|4.2| = -4.2, \quad \text{and} \quad -|0.\overline{3}| = -0.\overline{3}$$

EXAMPLE 5   **Finding absolute values**

Find the absolute values:

a. $|-8|$ **b.** $\left|\frac{1}{7}\right|$ **c.** $|0|$ **d.** $|4.2|$ **e.** $|0.\overline{3}|$ **f.** $-\left|-\frac{5}{9}\right|$

SOLUTION

a. $|-8| = 8$ -8 is 8 units from 0. $|-8| = -(-8) = 8.$

b. $\left|\frac{1}{7}\right| = \frac{1}{7}$ $\frac{1}{7}$ is $\frac{1}{7}$ units from 0.

c. $|0| = 0$ 0 is 0 units from 0.

d. $|4.2| = 4.2$ 4.2 is 4.2 units from 0.

e. $|0.\overline{3}| = 0.\overline{3}$ $0.\overline{3}$ is $0.\overline{3}$ units from 0.

f. $-\left|-\frac{5}{9}\right| = -\frac{5}{9}$ $-\frac{5}{9}$ is $\frac{5}{9}$ units from 0 and the opposite of $\frac{5}{9}$ is $-\frac{5}{9}$.

PROBLEM 5

Find the absolute value:

a. $|-19|$ **b.** $\left|\frac{1}{6}\right|$ **c.** $|-0|$

d. $|3.1|$ **e.** $|0.\overline{6}|$ **f.** $-\left|-\frac{5}{7}\right|$

Calculate It   Finding the Absolute Value

To find the absolute value of a number with a TI-83 Plus, tell the calculator you are doing math involving a number by pressing (MATH) (▶). Press 1 to select absolute value and enter the -8 (remember the $-$ sign is entered with the key marked (−)). Press) and (ENTER). You will get the absolute value of -8, which is 8.

abs(-8)
               8

Answers

5. a. 19   **b.** $\frac{1}{6}$ **c.** 0   **d.** 3.1

e. $0.\overline{6}$ **f.** $-\frac{5}{7}$

F Equality and Inequality

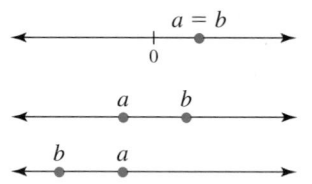

Web It

For a review of sets and numbers including a practice on ordering numbers, try link 1-1-8 at mhhe.com/bello.

TRICHOTOMY LAW

If you are given any two real numbers a and b, only one of three things can be true:

1. a is **equal to b,** denoted by $a = b$, or

2. a is **less than b,** denoted by $a < b$, or

3. a is **greater than b,** denoted by $a > b$.

On a number line, numbers are shown in order; they *increase* as you move right and *decrease* as you move left. Thus

1. $a = b$ means that the graphs of a and b coincide.

2. $a < b$ means that a is to the left of b on the number line.

3. $a > b$ means that a is to the right of b on the number line.

The symbols $<$ and $>$ are called **inequality** signs and statements such as $a > b$ or $b < a$ are called **inequalities.** For example, $3 < 4$ (and $4 > 3$) because 3 is to the left of 4 on the number line, and $-3 < -2$ (and $-2 > -3$) because -3 is to the left of -2. Similarly, $3.14 > 3.13$ (and $3.13 < 3.14$) because 3.14 is to the right of 3.13 on the number line as shown in Figure 5.

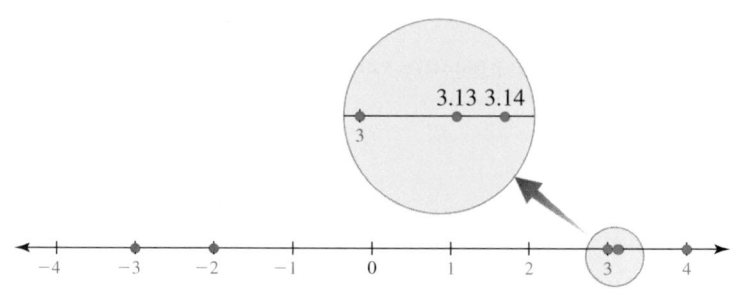

Figure 5

EXAMPLE 6 Determining relationships between numbers

Fill in the blank with $<$, $>$, or $=$ to make the resulting statement true:

a. -3 _____ -2 **b.** -5 _____ -6 **c.** $\dfrac{1}{2}$ _____ 0

d. $\dfrac{1}{5}$ _____ $\dfrac{1}{3}$ **e.** -2.3 _____ -2.2 **f.** $\dfrac{1}{3}$ _____ $0.\overline{3}$

SOLUTION

a. $-3 < -2$

b. $-5 > -6$

c. $\dfrac{1}{2} > 0$

PROBLEM 6

Fill in the blank with $<$ or $>$ to make the resulting statement true. (*Hint:* The arrow always points to the smaller number.)

a. -8 _____ -4

b. -7 _____ -6

c. 0 _____ $-\dfrac{1}{3}$

d. $\dfrac{1}{3}$ _____ $\dfrac{1}{2}$

e. -2.1 _____ -2.3

Answers

6. a. $<$ **b.** $<$ **c.** $>$ **d.** $<$
e. $>$

d.

$$\frac{1}{5} < \frac{1}{3}$$

$\frac{1}{5} = 0.2$ and $\frac{1}{3} = 0.333\ldots$
Thus $\frac{1}{5} < \frac{1}{3}$.

e.

$$-2.3 < -2.2$$

f.

$$\frac{1}{3} = 0.\overline{3}$$

Dividing 1 by 3, we obtain $\frac{1}{3} = 0.333\ldots = 0.\overline{3}$. Thus, $\frac{1}{3}$ and $0.\overline{3}$ correspond to the same point.

Teaching Tip

$\frac{1}{5}$ and $\frac{1}{3}$ can also be compared by comparing the cross products in the following way:

$$1(3) = ③ \qquad ⑤ = 1(5)$$

$$\frac{1}{5} \qquad \frac{1}{3}$$

Since $3 < 5$, then $\frac{1}{5} < \frac{1}{3}$.

Calculate It π: **Rational?**

Let us consider this question: Can you see the *exact* decimal representation of an irrational number like π or $\sqrt{2}$ on your calculator screen? Try entering π as shown. You get 3.141592654, terminating and thus rational. But π is not rational! Can you explain what the problem is? The π symbol is above the [∧] key. To engage it, press [2nd] and then [∧]. Your screen should look like this.

Note: Make sure that your floating decimal (under the [MODE] menu) is set at "Float."

```
π
                    3.141592654
```

Exercises 1.1

A In Problems 1–10, use roster notation to write the indicated set.

1. The first two natural numbers $\{1, 2\}$

2. The first six natural numbers $\{1, 2, 3, 4, 5, 6\}$

3. The natural numbers between 4 and 8 $\{5, 6, 7\}$

4. The natural numbers between 7 and 10 $\{8, 9\}$

5. The first three negative integers $\{-1, -2, -3\}$

6. The negative integers between -4 and 7 $\{-3, -2, -1\}$

7. The whole numbers between -3 and 4 $\{0, 1, 2, 3\}$

8. The integers less than 0 $\{-1, -2, -3, \ldots\}$

9. The integers greater than 0 $\{1, 2, 3, \ldots\}$

10. The nonnegative integers $\{0, 1, 2, 3, \ldots\}$

In Problems 11–16, write the set using set-builder notation.

11. $\{1, 2, 3\}$ $\{x \mid x$ is an integer between 0 and 4$\}$

12. $\{6, 7, 8\}$ $\{x \mid x$ is an integer between 5 and 9$\}$

13. $\{-2, -1, 0, 1, 2\}$ $\{x \mid x$ is an integer between -3 and 3$\}$

14. $\{-8, -7, -6, -5\}$ $\{x \mid x$ is an integer between -9 and $-4\}$

15. The set of even numbers between 19 and 78 $\{x \mid x$ is an even number between 19 and 78$\}$

16. The set of all multiples of 3 between 8 and 78 $\{x \mid x$ is a multiple of 3 between 8 and 78$\}$

In Problems 17–26, classify each statement as true or false. (Recall that N, W, I, Q, H, and R are the sets of natural, whole, integer, rational, irrational, and real numbers, respectively.)

17. $0 \in N$ False

18. $0 \in W$ True

19. $-0.3 \notin I$ True

20. $-0.\overline{8} \in Q$ True

21. $\sqrt{3} \in H$ True

22. $8.101001000\ldots \in Q$ False

23. $8.112233 \notin H$ True

24. $\sqrt{7} \in H$ True

25. $\dfrac{3}{8} \in R$ True

26. $-0 \in W$ True

B In Problems 27–34, write the given number as a decimal.

27. $\dfrac{2}{3}$ $0.\overline{6}$

28. $\dfrac{1}{6}$ $0.1\overline{6}$

29. $\dfrac{7}{8}$ 0.875

30. $\dfrac{5}{6}$ $0.8\overline{3}$

31. $\dfrac{5}{2}$ 2.5

32. $\dfrac{4}{3}$ $1.\overline{3}$

33. $\dfrac{7}{6}$ $1.1\overline{6}$

34. $\dfrac{9}{8}$ 1.125

C In Problems 35–44, classify the given numbers by placing a check mark in the appropriate row(s).

Set	35. $\frac{-3}{8}$	36. 0	37. $\sqrt{8}$	38. $\frac{3}{7}$	39. $0.\overline{3}$	40. -9	41. 0.9	42. -3.4	43. 3.1416	44. $3.141618\ldots$
Natural numbers										
Whole numbers		✓								
Integers		✓				✓				
Rational numbers	✓	✓		✓	✓	✓	✓	✓	✓	
Irrational numbers			✓							✓
Real numbers	✓	✓	✓	✓	✓	✓	✓	✓	✓	✓

In Problems 45–48, classify each statement as true or false.

45. $N \subseteq Q$ True

46. $I \subseteq W$ False

47. $R \subseteq W$ False

48. $Q \subseteq R$ True

D In Problems 49–64, find the additive inverse of the given number.

49. 8 -8

50. -9 9

51. -7 7

52. 6 -6

53. $\dfrac{3}{4}$ $-\dfrac{3}{4}$

54. $-\dfrac{1}{4}$ $\dfrac{1}{4}$

55. $-\dfrac{1}{5}$ $\dfrac{1}{5}$

56. $\dfrac{2}{5}$ $-\dfrac{2}{5}$

57. 0.5 -0.5

58. -0.6 0.6

59. $0.\overline{2}$ $-0.\overline{2}$

60. $-0.\overline{3}$ $0.\overline{3}$

61. $-1.\overline{36}$ $1.\overline{36}$

62. $2.\overline{38}$ $-2.\overline{38}$

63. π $-\pi$

64. $-\pi$ π

E In Problems 65–78, find each value.

65. $|10|$ 10

66. $|-11|$ 11

67. $|-17|$ 17

68. $|18|$ 18

69. $\left|\dfrac{3}{5}\right|$ $\dfrac{3}{5}$

70. $\left|-\dfrac{5}{7}\right|$ $\dfrac{5}{7}$

71. $|0.\overline{5}|$ $0.\overline{5}$

72. $|-0.\overline{7}|$ $0.\overline{7}$

73. $|-3.\overline{61}|$ $3.\overline{61}$ **74.** $|2.\overline{48}|$ $2.\overline{48}$ **75.** $-|\sqrt{2}|$ $-\sqrt{2}$ **76.** $-|-\sqrt{3}|$ $-\sqrt{3}$

77. $|-\pi|$ π **78.** $|\pi|$ π

F   In Problems 79–88, fill in the blanks with $<$ or $>$ to make the resulting statement true.

79. $-5 \underline{<} 2$ **80.** $3 \underline{>} -4$ **81.** $-6 \underline{>} -8$ **82.** $-7 \underline{<} -5$

83. $\dfrac{1}{2} \underline{>} \dfrac{1}{4}$ **84.** $\dfrac{1}{3} \underline{<} \dfrac{1}{2}$ **85.** $-\dfrac{3}{5} \underline{<} -\dfrac{1}{4}$ **86.** $-\dfrac{1}{3} \underline{<} -\dfrac{1}{4}$

87. $-3.5 \underline{<} -3.4$ **88.** $-3.2 \underline{<} -3.1$

USING YOUR KNOWLEDGE

Hamburgers, Lawyers, and Wages

In this section we learned how to compare integers and decimals; that is, we learned that $0.33 > 0.32$ and that $\frac{1}{3} < \frac{1}{2}$. To compare 0.33 and $\frac{1}{3}$, we write $\frac{1}{3}$ as a decimal by dividing the numerator by the denominator, obtaining $0.333\ldots$ We then write 0.33 as $0.33\mathbf{0}$ (note the extra zero) and write both numbers in a column with the decimal points aligned.

$$0.333\ldots$$

$$0.330$$

         $\hookrightarrow$ $3 > 0$, so $0.333\ldots > 0.330$

Thus $0.333\ldots > 0.330$.

We can use this knowledge in solving problems.

89. A McDonald's® hamburger weighs 100 grams (g) and contains 11 g of fat; that is, $\frac{11}{100}$ is fat. A Burger King® hamburger is 0.11009 fat. Write $\frac{11}{100}$ as a decimal and determine which hamburger has the larger percentage of fat. 0.11; Burger King

90. Lawyers have to know about fractions and decimals, too. In a court case called the *U.S. v. Forty Barrels and Twenty Kegs of Coca-Cola*,® a chemical analysis indicated that $\frac{3}{7}$ of the Coca-Cola was water. A second analysis showed that 0.41 was water. Which of the two analyses indicated more water in the Coke? The first ($\frac{3}{7} = 0.428\ldots$)

91. Using the graph in the *Getting Started* it can be determined that the average earnings of a worker with an advanced degree increased by approximately 25% or 0.25 from 1977 to 1997. Writing 2896/38,210 as a decimal will indicate the increase from 1977 to 1997 in the average earnings of a worker with a bachelor's degree. Determine which of the two degrees had the better increase in average earnings from 1977 to 1997 by comparing the two decimals. The advanced degree, since $0.25 > 0.0757\ldots$

Here are the heights of three of the tallest people in the world:

Name	Feet	Inches
(a) Sulaiman Ali Nashnush	8	$\frac{1}{25}$
(b) Gabriel Estavao Monjane	8	$\frac{3}{4}$
(c) Constantine	8	0.8

92. Which one is the tallest of the three?
(c) Constantine 8′0.8″

93. Which one is the second tallest?
(b) Monjane 8′0.75″

94. Which one is the shortest?
(a) Nashnush 8′0.04″

WRITE ON

95. Consider the statement, "$5 \subseteq N$, where N is the set of natural numbers." Is this statement true or false? Explain. False. 5 is not a set. To be correct, $\{5\} \subseteq N$ or $5 \in N$.

96. Consider the statement, "$N \in W$, where W is the set of whole numbers." Is this statement true or false? Explain. False. N is a set that is a subset of W. To be correct, $N \subseteq W$.

97. Write in your words a definition for the set of rational numbers and the set of irrational numbers using the idea of a fraction. Answers may vary.

98. Write in your words a definition of the set of rational numbers and the set of irrational numbers using the idea of a decimal. Answers may vary.

99. Explain why every integer is a rational number but not every rational number is an integer.
Answers may vary.

MASTERY TEST

If you know how to do these problems, you have learned your lesson!

Write as a decimal:

100. $\dfrac{5}{7}$ $0.\overline{714285}$

101. $\dfrac{1}{8}$ 0.125

Find the additive inverse of:

102. 0 0

103. 8 -8

Find the value of each:

104. $-\left|-\sqrt{17}\right|$ $-\sqrt{17}$

105. $|-x|$ if x is -2 2

Write the following using set-builder notation:

106. $\{3, 6, 9, \ldots\}$ $\{x \mid x$ is a positive multiple of 3$\}$

107. $\{3, 6, 10, 15, \ldots\}$ $\{x \mid x$ is 3 or an integer obtained when 3, 4, 5 . . . is added in succession$\}$

108. Classify the numbers in the following set as rational, Q, or irrational, H.

$$\left\{\frac{22}{7}, 3.1415, \pi, 3, -5, 0.\overline{77}, 34.010010001 \ldots\right\}$$
$\quad Q \quad\quad Q \quad\; H \; Q \; Q \quad Q \quad\quad\quad\quad H$

Fill in the blank with $<$ or $>$ so that the result is a true statement:

109. $\dfrac{1}{3}$ __$>$__ $0.331332333334 \ldots$

110. 3.1416 __$<$__ $\dfrac{22}{7}$

| **1.2** | **OPERATIONS AND PROPERTIES OF REAL NUMBERS** |

To Succeed, Review How To ...

1. Find the additive inverse of a number (p. 10).

2. Find the absolute value of a number (p. 11).

3. Recognize positive and negative numbers (p. 9).

Objectives

A Add, subtract, multiply, and divide signed numbers.

B Identify uses of the properties of the real numbers.

GETTING STARTED   **Signed Numbers and Growth Patterns**

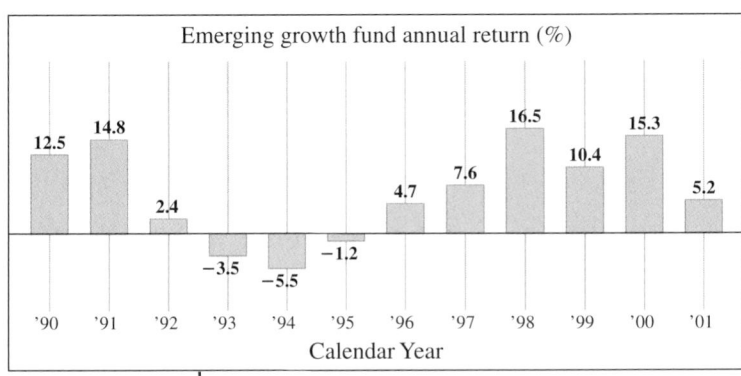

Money Grows on Trees Company

Look at the graph. What is the net percentage of change in this company's growth fund annual return for the years 1995 though 1997? To find the answer, we have to add the percentages for 1995 (-1.2), 1996 ($+4.7$), and 1997 ($+7.6$). The result is

$$-1.2 + 4.7 + 7.6 =$$
$$-1.2 + 12.3 = \quad \text{Adding 4.7 and 7.6}$$
$$+ 11.1 \quad \text{Subtracting 1.2 from 12.3}$$

This answer means that the company had a positive growth fund annual return of 11.1 percent for the years 1995 through 1997. You could find out whether the company has an overall positive or negative growth pattern from 1990 to 2001 by adding all the signed numbers together.

The following are two concepts used when adding and subtracting signed numbers:

1. When adding $-1.2 + 4.7$ it can also be written as $4.7 + (-1.2)$ because of the commutative law for addition. Notice the use of parentheses around the negative number in this case. It makes it less confusing than $4.7 + -1.2$.

2. To add -1.2 and 12.3, we can subtract 1.2 from 12.3; subtracting 1.2 from 12.3 is the same as adding -1.2 and 12.3.

In this section we shall learn to perform additions, subtractions, multiplications, and divisions using real numbers, and we shall study the rules for these four fundamental operations.

 **Operations with Signed Numbers**

In algebra we use the idea of absolute value to define the addition and subtraction of "signed" numbers.

PROCEDURE

To Add Two Numbers with the Same Sign

Add their absolute values and give the sum the common sign ($+$ if both numbers are positive, $-$ if both numbers are negative).

Teaching Tip

Positive numbers can be thought of as "wins" and negatives as "losses."

Example: If you play one round of a game and lose 2 points (-2), then play the second round and lose 5 points (-5), you are losing by 7 points (-7). Thus, $(-2) + (-5) = (-7)$. Have students discuss some other examples.

PROCEDURE

To Add Two Numbers with Different Signs

1. Find the absolute value of the numbers.

2. Subtract the number with the smaller absolute value from the one with the greater absolute value.

3. Use the sign of the number with the greater absolute value for the result obtained in step 2.

If zero is added to any number a, the result is the number a. Thus $3 + 0 = 3$ and $0 + \frac{1}{3} = \frac{1}{3}$. The number 0 is called the **identity** for addition or the **additive identity.**

Additive Identity	For any real number a, $$a + 0 = a = 0 + a$$ (Zero is the identity for addition.)

EXAMPLE 1	**Adding signed numbers**

Find:

a. $-5 + (-11)$ **b.** $0.8 + (-0.5)$ **c.** $-0.7 + 0.4$

d. $\dfrac{4}{5} + \left(-\dfrac{2}{5}\right)$ **e.** $-\dfrac{3}{8} + \dfrac{1}{4}$ **f.** $-0.2 + 0.6$

SOLUTION

a. $-5 + (-11) = -11$ Add and keep the negative sign

b. $0.8 + (-0.5) = 0.3$ Subtract and use the positive sign

c. $-0.7 + 0.4 = -0.3$ Subtract and use the negative sign

d. $\dfrac{4}{5} + \left(\dfrac{-2}{5}\right) = \dfrac{4 + (-2)}{5} = \dfrac{2}{5}$

The fractions have the same denominator, so keep it and add the numerators. The numerators have different signs, so subtract and use the positive sign.

e. $\dfrac{-3}{8} + \left(\dfrac{1}{4}\right) \cdot \dfrac{2}{2} = \dfrac{-3 + 2}{8} = \dfrac{-1}{8}$

The lowest common denominator (LCD) for the fractions is 8, so we rename $\frac{1}{4}$ by multiplying the numerator and denominator by 2. The numerators have different signs, so subtract and use the negative sign.

f. $-0.2 + 0.6 = 0.4$ Subtract and use the positive sign. Remember to line up decimal points.

PROBLEM 1

Find:

a. $-3 + (-16)$ **b.** $0.7 + (-0.2)$

c. $-0.5 + 0.3$ **d.** $\dfrac{5}{7} + \left(-\dfrac{3}{7}\right)$

e. $-\dfrac{5}{6} + \dfrac{2}{3}$ **f.** $-0.4 + 0.8$

Calculate It   Fractions on the Calculator

Some calculators have special keys to enter fractions. If yours does not, you can always enter them as indicated divisions. To do Example 1(e), enter $-\frac{3}{8} + \frac{1}{4}$ as shown on the screen. The answer is given as $-.125$.

Is this the same as the $-\frac{1}{8}$ given in Example 1(e)? If you divide 1 by 8, you will see that the answers are identical.

With a TI-83 Plus, the decimal $-.125$ can be converted to a fraction by pressing MATH 1 ENTER to obtain $-\frac{1}{8}$ as shown.

```
-3/8+1/4
                -.125
Ans ▸ Frac
                 -1/8
```

Answers

1. **a.** -19 **b.** 0.5 **c.** -0.2
d. $\frac{2}{7}$ **e.** $-\frac{1}{6}$ **f.** 0.4

We have developed a procedure for adding real numbers, so it will be convenient to describe subtraction in terms of addition.

Subtraction of Signed Numbers	If a and b are real numbers, $a - b = a + (-b)$.

This means that we can *subtract* by adding the additive inverse or "opposite." Thus,

Web It

For interactive practice with subtraction of signed numbers, try link 1-2-1 at mhhe.com/bello.

$$6 - (-3) = 6 + 3 = 9$$

$$-0.7 - 0.2 = -0.7 + (-0.2) = -0.9 \qquad \text{Subtract by adding the opposite.}$$

$$-\frac{1}{5} - \left(-\frac{4}{5}\right) = -\frac{1}{5} + \frac{4}{5} = \frac{3}{5}$$

EXAMPLE 2 Subtracting signed numbers	**PROBLEM 2**

Find:

a. $-20 - 5$　　　　**b.** $-0.6 - (-0.2)$　　　　**c.** $-\frac{1}{7} - \left(-\frac{2}{7}\right)$

SOLUTION

a. $-20 - 5 = -20 + (-5) = -25$ 　　Add the opposite of 5

b. $-0.6 - (-0.2) = -0.6 + 0.2 = -0.4$ 　　Add the opposite of -0.2

c. $-\frac{1}{7} - \left(-\frac{2}{7}\right) = -\frac{1}{7} + \frac{2}{7} = \frac{1}{7}$ 　　Add the opposite of $-\frac{2}{7}$

In actual practice most of these operations are carried out mentally.

Problem 2: Find:
a. $-15 - 3$　　**b.** $-0.7 - (-0.3)$
c. $-\frac{1}{5} - \left(-\frac{1}{5}\right)$

Web It

For a good mental practice and drill in subtraction of signed numbers, try link 1-2-2 at mhhe.com/bello.

Note that $-3 + 3 = 0$, $\frac{1}{2} + (-\frac{1}{2}) = 0$, and $-2.1 + 2.1 = 0$. Thus when **opposites (additive inverses)** are added, their sum is zero.

Additive Inverses (Opposites)	
For any real number a, $a + (-a) = (-a) + a = 0$.	

Teaching Tip

For a visual example to illustrate how subtraction can become addition, use a temperature example, such as the difference between 10° above 0 and 3° below 0 or $10 - (-3)$.

In arithmetic the product of a and b is written as $a \times b$. In algebra, however, the multiplication sign ($\times$) can be mistaken for the letter x; thus the product of a and b is written in one of the following ways.

PROCEDURE

How to Signify Multiplication

Using a raised dot, $\cdot$	$a \cdot b$
Writing a and b next to each other	ab
Using parentheses	$(a)(b), \quad a(b), \quad \text{or} \quad (a)b$

Answers

2. **a.** -18　**b.** -0.4　**c.** 0

The numbers represented by a and b are called **factors,** and the result of the multiplication is called the **product** of a and b.

Now suppose you own four shares of stock and the price is *down* \$3, written as -3. Your loss that day would be the product $4 \cdot (-3)$, with factors 4 and -3. What is this product? This multiplication is the repeated addition of -3.

$$4 \cdot (-3) = \underbrace{(-3) + (-3) + (-3) + (-3)}_{4 \text{ negative threes}} = -12$$

Next, look at $(-3) \cdot 4$. In Section B, you will see that multiplication of real numbers has the commutative property. Thus

$$(-3) \cdot 4 = 4 \cdot (-3) = -12$$

We can generalize this idea to show that the product of any two numbers, one *positive* and the other *negative,* is a *negative number.* As in addition, we can state this result in terms of absolute values.

Teaching Tip

You can also demonstrate this rule by the following pattern obtained by multiplying 3, 2, 1, 0, -1, -2, -3, by 2:

$$\left.\begin{array}{l} +3 \cdot (2) = +6 \\ +2 \cdot (2) = +4 \\ +1 \cdot (2) = +2 \end{array}\right\} \begin{array}{l}\text{Answers decrease} \\ \text{by 2}\end{array}$$
$$\begin{array}{l} 0 \cdot (2) = 0 \\ -1 \cdot (2) = -2 \\ -2 \cdot (2) = -4 \\ -3 \cdot (2) = -6 \end{array}$$

Give the students the first 3 products of this list and let them discover the pattern and continue the chart to $-3 \cdot (2)$.

PROCEDURE

Multiplying Numbers with Different Signs

To multiply a *positive* number by a *negative* number, multiply their absolute values. Make the product *negative.*

Thus, $3 \cdot (-2) = -6$, $-4 \cdot 8 = -32$, and $2 \cdot (-3.5) = -7$. The factors in each case, 3 and -2, -4 and 8, and 2 and -3.5, have *different* (*unlike*) signs.

EXAMPLE 3	**Multiplying signed numbers**	**PROBLEM 3**

Find:

a. $7 \cdot (-8)$ **b.** $(-2.5) \cdot 4$

Find:

a. $9 \cdot (-7)$ **b.** $(-4.5) \cdot 2$

SOLUTION

a. $7 \cdot (-8) = -56$ **b.** $-2.5 \cdot 4 = -10.0$ or -10

Web It

What about the product of two negative integers such as $-2 \cdot (-3)$? First, look for a pattern.

This number decreases by 1.   This number increases by 3.

$$2 \cdot (-3) = -6$$
$$1 \cdot (-3) = -3$$
$$0 \cdot (-3) = 0$$
$$-1 \cdot (-3) = 3$$
$$-2 \cdot (-3) = 6$$

Thus $-2 \cdot (-3) = 6$. In this case -2 and -3 have the *same* sign ($-$). When we multiply $2 \cdot 3$, 2 and 3 are both positive, so they also have the *same* sign ($+$). We can summarize this discussion by the following rules.

Answers

3. a. -63 **b.** -9.0 or -9

Signs of Multiplication Products	When Multiplying Two Numbers with	The Product Is
	Same (like) signs	Positive $(+)$
	Different (unlike) signs	Negative $(-)$

Web It

For an interesting lesson dealing with reciprocals and the multiplicative identity through an application, try link 1-2-6 at mhhe.com/bello.

Thus,

$$(-3) \cdot (-4.2) = 12.6 \qquad 4 \cdot (-2.1) = -8.4$$

Same signs → Positive answer   Different signs → Negative answer

Multiplying by 1 leaves the number unchanged ($3 \cdot 1 = 3$ and $-7 \cdot 1 = -7$). Thus, 1 is the identity for multiplication.

Identity for Multiplication	
	For any real number a, $$a \cdot 1 = 1 \cdot a = a.$$ (1 is the identity element for multiplication.)

What about fractions? To multiply fractions, we need the following definition.

Multiplication of Fractions	
	$$\frac{a}{b} \cdot \frac{c}{d} = \frac{a \cdot c}{b \cdot d} \qquad (b, d \neq 0)$$

The same laws of signs apply. Thus,

$$\left(-\frac{9}{5}\right) \cdot \frac{3}{4} = -\frac{9 \cdot 3}{5 \cdot 4} = -\frac{27}{20}$$

Different signs   Negative answer

When multiplying fractions, it saves time if common factors are divided out before you multiply. Thus to multiply $\frac{5}{7} \cdot \left(-\frac{2}{5}\right)$, we write

$$\frac{\overset{1}{5}}{7} \cdot \left(-\frac{2}{\underset{1}{5}}\right) = -\frac{1 \cdot 2}{7 \cdot 1} = -\frac{2}{7} \qquad \frac{5}{5} = 1$$

EXAMPLE 4   **Multiplying signed numbers**

Find:

a. $\left(-\dfrac{3}{7}\right) \cdot \dfrac{7}{8}$ **b.** $\left(-\dfrac{5}{8}\right) \cdot \left(-\dfrac{4}{15}\right)$

SOLUTION

a. $\left(-\dfrac{3}{\underset{1}{7}}\right) \cdot \dfrac{\overset{1}{7}}{8} = -\dfrac{3 \cdot 1}{1 \cdot 8} = -\dfrac{3}{8}$ Different (unlike) signs; the answer is negative.

b. $\left(-\dfrac{5}{\underset{2}{8}}\right) \cdot \left(-\dfrac{\overset{1}{4}}{\underset{3}{15}}\right) = \dfrac{1 \cdot 1}{2 \cdot 3} = \dfrac{1}{6}$ Same (like) signs; the answer is positive.

PROBLEM 4

Find:

a. $\left(-\dfrac{2}{5}\right) \cdot \dfrac{5}{7}$ **b.** $\left(\dfrac{3}{14}\right) \cdot \left(\dfrac{7}{6}\right)$

Answers

4. a. $-\frac{2}{7}$ **b.** $\frac{1}{4}$

Just as we were able to define subtraction in terms of addition, we can also define division in terms of multiplication.

Division of Real Numbers	If a and b are real numbers and b is not zero, $$\frac{a}{b} = q \quad \text{means that} \quad a = b \cdot q$$ where a is called the **dividend,** b is the **divisor,** and q is the **quotient.**

Calculate It

Verify Fraction Sign

Use your TI-83 Plus to verify that even though the numerator and denominator of the fraction $\frac{-3}{-5}$ do not have a common factor other than 1, you must rename it as positive $\frac{3}{5}$. Press (−) 3 ÷ (−) 5 ENTER

$-3/-5$

.6

Then convert to a fraction by pressing

MATH 1 ENTER

Ans ▸ Frac   3/5

Teaching Tip

Have a class discussion on why $\frac{-3}{-5}$ is $\frac{3}{5}$.

Thus,

$$\frac{48}{-6} = -8 \quad \text{means that} \quad 48 = -6 \cdot (-8)$$

and

$$\frac{-28}{-7} = 4 \quad \text{means that} \quad -28 = -7 \cdot (4)$$

When Dividing Two Numbers with	The Quotient Is
Same (like) signs	Positive (+)
Different (unlike) signs	Negative (−)

Because of this, the same rules of sign that apply to the multiplication of real numbers also apply to the division of real numbers; that is, the quotient of two numbers with the *same* sign is *positive,* and the quotient of two numbers with *different* signs is *negative,* as shown in the table. Here are some examples:

$$\frac{24}{6} = 4 \qquad \frac{3.2}{1.6} = 2$$

$$\frac{-18}{-9} = 2 \qquad \frac{-3.3}{-1.1} = 3$$

\right\} Same signs, positive answers

$$\frac{-32}{4} = -8 \qquad \frac{-6.3}{0.9} = -7$$

$$\frac{35}{-7} = -5 \qquad \frac{4.5}{-0.5} = -9$$

\right\} Different signs, negative answers

Because

$$\frac{-32}{4} = \frac{32}{-4} = -\frac{32}{4} = -8$$

the following holds true.

Signs of a Fraction	For any real number a and nonzero real number b, there are two cases of signs of a fraction: **1.** $\dfrac{-a}{b} = \dfrac{a}{-b} = -\dfrac{a}{b}$ **2.** $\dfrac{-a}{-b} = \dfrac{a}{b}$

There are *three* signs associated with every fraction: the sign of the numerator, the sign of the denominator, and the sign of the fraction.

| **EXAMPLE 5**   Dividing signed numbers | | | **PROBLEM 5** |

Find:

a. $48 \div 6$ **b.** $\dfrac{54}{-9}$ **c.** $\dfrac{-63}{-7}$

d. $-28 \div 4$ **e.** $5 \div 0$ **f.** $3.4 \div 1.7$

g. $\dfrac{4.8}{-1.2}$ **h.** $\dfrac{-5.6}{-0.8}$ **i.** $0 \div 3.5$

SOLUTION

a. $48 \div 6 = 8$ 48 and 6 have the same sign; the answer is positive.

b. $\dfrac{54}{-9} = -6$ 54 and −9 have different signs; the answer is negative.

c. $\dfrac{-63}{-7} = 9$ −63 and −7 have the same sign; the answer is positive.

d. $-28 \div 4 = -7$ −28 and 4 have different signs; the answer is negative.

e. $5 \div 0$ is not defined. If you make $5 \div 0$ equal any number, say a, you will have

$$\dfrac{5}{0} = a, \quad \text{which means} \quad 5 = a \cdot 0 = 0$$

This says that 5 = 0, which is impossible.

Thus, $\frac{5}{0}$ is not defined.

f. $3.4 \div 1.7 = 2$ 3.4 and 1.7 have the same sign; the answer is positive.

g. $\dfrac{4.8}{-1.2} = -4$ 4.8 and −1.2 have different signs; the answer is negative.

h. $\dfrac{-5.6}{-0.8} = 7$ −5.6 and −0.8 have the same sign; the answer is positive.

i. In this case, $0 \div 3.5 = 0$. We can check this using the definition of division: $0 \div 3.5 = \frac{0}{3.5} = 0$ means $0 = 3.5 \cdot 0$, which is true.

PROBLEM 5

Find:

a. $60 \div 10$ **b.** $\dfrac{48}{-3}$

c. $\dfrac{-18}{-2}$ **d.** $-14 \div 2$

e. $-4 \div 0$ **f.** $4.8 \div 1.6$

g. $\dfrac{4.2}{-2.1}$ **h.** $\dfrac{-3.8}{-1.9}$

i. $0 \div 9.2$

Web It

For a lesson on how to divide signed numbers, visit link 1-2-7 at mhhe.com/bello.

Here are three rules to help you out.

Zero in Division

For $a \neq 0$, $\dfrac{0}{a} = 0$ and $\dfrac{a}{0}$ is *not* defined. Moreover, $\dfrac{0}{0}$ is indeterminate.

CAUTION

$\frac{0}{k}$ is okay but $\frac{n}{0}$ is a no-no!

Answers

5. a. 6   **b.** −16   **c.** 9   **d.** −7
e. Undefined   **f.** 3   **g.** −2
h. 2   **i.** 0

Let's look at the division problem $2 \div 5$. We can write $2 \div 5 = \frac{2}{5} = 2 \cdot \frac{1}{5}$. Thus, to *divide* 2 *by* 5, we *multiply* 2 *by* $\frac{1}{5}$. The numbers 5 and $\frac{1}{5}$ are *reciprocals* or *multiplicative inverses*. Here is the definition.

Multiplicative Inverse (Reciprocal)	Every nonzero real number a has a **reciprocal (multiplicative inverse)** $\frac{1}{a}$ such that $$a \cdot \frac{1}{a} = 1$$

Web It

To study reciprocals and dividing fractions, go to link 1-2-8 at mhhe.com/bello.

The reciprocal of 3 is $\frac{1}{3}$, the reciprocal of -6 is $\frac{1}{-6} = -\frac{1}{6}$, and the reciprocal of $\frac{2}{3}$ is $\frac{3}{2}$. The reciprocal of a *positive* number is *positive* and the reciprocal of a *negative* number is *negative*.

EXAMPLE 6 Finding reciprocals	**PROBLEM 6**
Find the reciprocal:	Find the reciprocal of:
a. $\frac{2}{7}$ **b.** $-\frac{4}{5}$ **c.** 0.2	**a.** $\frac{5}{9}$ **b.** $-\frac{7}{8}$ **c.** 0.5

SOLUTION

a. The reciprocal of $\frac{2}{7}$ is $\frac{7}{2}$.

b. The reciprocal of $-\frac{4}{5}$ is $-\frac{5}{4}$. (Remember that the reciprocal of a negative number is negative.)

c. The reciprocal of 0.2 is

$$\frac{1}{0.2} = \frac{1}{\frac{2}{10}} = 1 \cdot \frac{10}{2} = 5$$

Or you can rename 0.2 as $\frac{2}{10}$, reduce it to $\frac{1}{5}$, and then the reciprocal is 5.

Division of fractions is done in terms of reciprocals. Since

$$\frac{a}{b} = a \cdot \frac{1}{b}$$

to divide by a number (such as b) we multiply by its reciprocal $\frac{1}{b}$. Here is the general definition.

Division of Fractions	$$\frac{a}{b} \div \frac{c}{d} = \frac{a}{b} \cdot \frac{d}{c}$$ (b, c, and d are not zero.)

Thus to divide by $\frac{c}{d}$, we multiply by the reciprocal $\frac{d}{c}$. For example, to divide $\frac{4}{5}$ by $\frac{2}{3}$, we multiply $\frac{4}{5}$ by the reciprocal of $\frac{2}{3}$, that is, by $\frac{3}{2}$. Thus,

$$\frac{4}{5} \div \frac{2}{3} = \frac{4}{5} \cdot \frac{3}{2}$$

Answers

6. a. $\frac{9}{5}$ **b.** $-\frac{8}{7}$ **c.** 2

EXAMPLE 7   **Dividing fractions**

Find the following using reciprocals:

a. $\dfrac{2}{5} \div \left(-\dfrac{3}{4}\right)$ **b.** $\left(-\dfrac{5}{6}\right) \div \left(-\dfrac{7}{2}\right)$ **c.** $\left(-\dfrac{3}{7}\right) \div \dfrac{6}{7}$

SOLUTION

a. $\dfrac{2}{5} \div \left(-\dfrac{3}{4}\right) = \dfrac{2}{5} \cdot \left(-\dfrac{4}{3}\right) = -\dfrac{8}{15}$

b. $\left(-\dfrac{5}{6}\right) \div \left(-\dfrac{7}{2}\right) = \left(-\dfrac{5}{6}\right) \cdot \left(-\dfrac{2}{7}\right) = \dfrac{10}{42} = \dfrac{5}{21}$

It is easier to "reduce," or divide common factors, before multiplying like this

$$\left(-\dfrac{5}{\overset{}{\underset{3}{6}}}\right) \cdot \left(-\dfrac{\overset{1}{2}}{7}\right) = \dfrac{5}{21}$$

c. $\left(-\dfrac{3}{7}\right) \div \dfrac{6}{7} = \left(-\dfrac{3}{7}\right) \cdot \dfrac{7}{6} = -\dfrac{21}{42} = -\dfrac{1}{2}$

You can also "reduce," or divide common factors, before multiplying by writing

$$\left(-\dfrac{\overset{1}{3}}{7}\right) \cdot \dfrac{\overset{1}{7}}{\underset{2}{6}} = -\dfrac{1}{2}$$

PROBLEM 7

Find:

a. $\dfrac{3}{5} \div \left(-\dfrac{4}{7}\right)$ **b.** $\left(-\dfrac{6}{7}\right) \div \left(-\dfrac{3}{5}\right)$

c. $\left(-\dfrac{4}{5}\right) \div \dfrac{8}{5}$

Teaching Tip

Use this example to emphasize placement of a negative sign in a fraction:

$$\dfrac{-2}{3} = \dfrac{2}{-3} = -\dfrac{2}{3}$$

See if students can acknowledge the sign rule for division as verification of this.

Calculate It   Divide Fractions

Your calculator does division of fractions. To do Example 7(c), enter

$$(-3 \div 7) \div (6 \div 7)$$

The result is as shown.

    Would you get the same results if the parentheses were not used around one or both of the fractions being divided? Try the problem without the parentheses, and discuss your findings.

```
(-3/7)/(6/7)
                    -.5
Ans▶Frac
                   -1/2
```

B   Real-Number Properties

Web It

For a colorful summary of the properties of the real numbers, visit link 1-2-9 at mhhe.com/bello.

We have already mentioned two properties of the real numbers: zero (0) is the **identity for addition,** and one (1) is the **identity for multiplication.** We end this section by summarizing some additional real-number properties in Table 1 and then examining how they can help us in our work with algebra.

    The **commutative properties** tell us that the *order* in which we add or multiply two numbers does not affect the result; thus, for the sum

$$\begin{array}{r} 28 \\ + 39 \\ \hline \end{array}$$

you can add down (28 + 39) and then check by adding up (39 + 28). Similarly, it is easier to multiply

$$\begin{array}{r} 48 \\ \times 9 \\ \hline \end{array} \quad \text{instead of} \quad \begin{array}{r} 9 \\ \times 48 \\ \hline \end{array}$$

Answers

7. a. $-\dfrac{21}{20}$ **b.** $\dfrac{10}{7}$ **c.** $-\dfrac{1}{2}$

but by the commutative property, the result is the same. On the other hand, the **associative property** tells us that the *grouping* of the numbers does not affect the final answer. Thus if you wish to add $3 + 24 + 6$ by adding $24 + 6$ first—that is, by finding $3 + (24 + 6)$—you will get the same answer as if you had added $(3 + 24) + 6$.

Table 1 Properties of Real Numbers (a, b, and c represent real numbers)

Property	Addition	Multiplication
Closure	If a and b are real numbers, then $a + b$ is a real number. $-8 + \sqrt{5}$ is a real number.	If a and b are real numbers, then $a \cdot b$ is a real number. $-8 \cdot \sqrt{5}$ is a real number.
Commutative	$a + b = b + a$ $\sqrt{2} + 8 = 8 + \sqrt{2}$	$ab = ba$ $3 \cdot \dfrac{1}{5} = \dfrac{1}{5} \cdot 3$
Associative	$a + (b + c) = (a + b) + c$ $\dfrac{1}{3} + \left(\dfrac{2}{5} + \dfrac{3}{4}\right) = \left(\dfrac{1}{3} + \dfrac{2}{5}\right) + \dfrac{3}{4}$	$a(bc) = (ab)c$ $-3 \cdot \left(\sqrt{5} \cdot \dfrac{1}{8}\right) = (-3 \cdot \sqrt{5}) \cdot \dfrac{1}{8}$
Identity	$a + 0 = 0 + a = a$ $-\dfrac{4}{5} + 0 = 0 + \left(-\dfrac{4}{5}\right) = -\dfrac{4}{5}$ (0 is called the additive identity.)	$1 \cdot a = a \cdot 1 = a$ $0.\overline{38} \cdot 1 = 1 \cdot 0.\overline{38} = 0.\overline{38}$ (1 is called the multiplicative identity.)
Inverse	$a + (-a) = (-a) + a = 0$ $-a$ is the additive inverse (opposite) of a. $-\dfrac{1}{5}$ is the opposite of $\dfrac{1}{5}$. Thus $\dfrac{1}{5} + \left(-\dfrac{1}{5}\right) = 0$.	$a \cdot \dfrac{1}{a} = \dfrac{1}{a} \cdot a = 1 \qquad (a \neq 0)$ $\dfrac{1}{a}$ is the multiplicative inverse (reciprocal) of a. 8 is the reciprocal of $\dfrac{1}{8}$.
Distributive	$a(b + c) = ab + ac \quad 3(1 + 6) = 3 \cdot 1 + 3 \cdot 6$	$a(b - c) = ab - ac \quad 4(1 - 6) = 4 \cdot 1 - 4 \cdot 6$

EXAMPLE 8   Properties of real numbers

Which property is illustrated in the following statements?

a. $(-3) + \dfrac{7}{5} = \dfrac{7}{5} + (-3)$

b. $3 \cdot (4 \cdot 7) = (4 \cdot 7) \cdot 3$

c. $(-4) \cdot \left(\dfrac{1}{2} \cdot \dfrac{1}{8}\right) = \left(-4 \cdot \dfrac{1}{2}\right) \cdot \dfrac{1}{8}$

d. $2 + (3 \cdot 4) = 2 + (4 \cdot 3)$

SOLUTION

a. We changed the *order*, so the commutative property of addition applies.

b. Here again we changed the *order* (the 3 changed position). The commutative property of multiplication was used.

c. This time we *grouped* the numbers differently. The associative property of multiplication was used.

d. The commutative property of multiplication applies (the 3 and 4 changed position).

PROBLEM 8

Which multiplication property is illustrated in each statement?

a. $2 \cdot (3 \cdot 5) = (3 \cdot 5) \cdot 2$

b. $(-4) \cdot \dfrac{2}{5} = \dfrac{2}{5} \cdot (-4)$

c. $7 \cdot \left(\dfrac{1}{4} \cdot \dfrac{1}{2}\right) = \left(7 \cdot \dfrac{1}{4}\right) \cdot \dfrac{1}{2}$

d. $5 + (3 \cdot 9) = 5 + (9 \cdot 3)$

Answers

8. a. Commutative
b. Commutative   **c.** Associative
d. Commutative

Teaching Tip

One of the first properties to look for is the commutative. This can be done by ignoring grouping and operation symbols on either side of the equal sign and comparing the order of the left side to the right. If there is a change, then the commutative property is being illustrated. Thus, in Example 8(b), compare 3, 4, 7 to 4, 7, 3. The order is different, so the commutative property is being used.

The associative property can help you find the answer to a problem such as $(3)(-5)(6)$. To use this property, you can write either of the following:

1. $(3)(-5)(6) = (-15)(6)$ Multiply $(3)(-5)$ first and $(-15)(6)$ next.

$\qquad\qquad\quad = -90$

2. $(3)(-5)(6) = (3)(-30)$ Multiply $(-5)(6)$ first and $(3)(-30)$ next.

$\qquad\qquad\quad = -90$

The answer is the same in both cases. We shall practice with this type of problem in the exercises.

We have not yet added or subtracted expressions like $2\sqrt{3} + 5\sqrt{3}$ or $8\sqrt{2} - 12\sqrt{2}$. We can do this using the commutative and distributive properties as follows:

$$2\sqrt{3} + 5\sqrt{3} = (\sqrt{3} \cdot 2) + (\sqrt{3} \cdot 5) \qquad \text{By the commutative property of multiplication}$$

$$= \sqrt{3}(2 + 5) \qquad \text{By the distributive property}$$

$$= 7\sqrt{3} \qquad \text{By the commutative property of multiplication}$$

In practice you can see that $2\sqrt{3}$ and $5\sqrt{3}$ are **like** terms that can be combined. Using this idea, $8\sqrt{2} - 12\sqrt{2} = -4\sqrt{2}$. We shall study the distributive property and how to use it to combine like terms in Section 1.4.

Calculate It Exercises

1. What do you have to do so that the answers to Examples 1(d) and 1(e) are shown as fractions on your calculator screen?

2. What happens if you try dividing by zero on your calculator?

3. Do you get the same answer in Example 7(c) if you omit the grouping symbols? If not, what answer do you get?

Exercises 1.2

A In Problems 1–86, perform the indicated operations.

1. $\dfrac{3}{5} + \left(-\dfrac{1}{5}\right)$ $\dfrac{2}{5}$

2. $-0.4 + 0.9$ 0.5

3. $-0.3 + 0.2$ -0.1

4. $-8 + 5$ -3

5. $-4 + 6$ 2

6. $-0.2 + 0.3$ 0.1

7. $-0.5 + (-0.3)$ -0.8

8. $-7 + (-11)$ -18

9. $-\dfrac{1}{5} + \dfrac{1}{4} + \dfrac{3}{20}$ $\dfrac{1}{5}$

10. $-\dfrac{4}{7} + \dfrac{2}{9} + \dfrac{1}{63}$ $-\dfrac{1}{3}$

11. $6 - 13$ -7

12. $8 - 13$ -5

13. $0.6 - 0.9$ -0.3

14. $0.3 - 0.8$ -0.5

15. $\dfrac{1}{7} - \dfrac{3}{8}$ $-\dfrac{13}{56}$

16. $\dfrac{3}{8} - \dfrac{4}{8}$ $-\dfrac{1}{8}$

17. $-8 - 4 - 2$ -14

18. $-4 - 6 - 3$ -13

19. $-0.4 - 0.2$ -0.6

20. $-0.3 - 0.5$ -0.8

21. $-\dfrac{3}{7} - \dfrac{2}{9}$ $-\dfrac{41}{63}$

22. $-\dfrac{4}{9} - \dfrac{1}{9}$ $-\dfrac{5}{9}$

23. $-6 - (-5)$ -1

24. $-7 - (-9)$ 2

25. $-8 - (-4)$ -4

26. $-9 - (-2)$ -7

27. $-0.7 - (-0.6)$ -0.1

28. $-0.9 - (-0.3)$ -0.6

29. $-\dfrac{2}{7} - \left(-\dfrac{4}{3}\right)$ $\dfrac{22}{21}$

30. $-\dfrac{3}{4} - \left(-\dfrac{5}{3}\right)$ $\dfrac{11}{12}$

31. $-5(8)$ -40

32. $-9(6)$ -54

33. $4(-3)$ -12

34. $6(-8)$ -48

35. $-10(-5)$ 50

36. $-6(-9)$ 54

37. $-3(4)(-5)$ 60

38. $-5(2)(3)$ -30

39. $-4(-2)(5)$ 40

40. $-2(-5)(9)$ 90

41. $-3(5)(-2)$ 30

42. $-3(10)(-2)$ 60

43. $4(-5)(2)$ -40

44. $10(-3)(6)$ -180

45. $-2.2(3.3)$ -7.26

46. $-1.4(3.1)$ -4.34

47. $-1.3(-2.2)$ 2.86

48. $-1.5(-1.1)$ 1.65

49. $\dfrac{5}{6}\left(-\dfrac{5}{7}\right)$ $-\dfrac{25}{42}$

50. $\dfrac{3}{8}\left(-\dfrac{5}{7}\right)$ $-\dfrac{15}{56}$

51. $-\dfrac{3}{5}\left(-\dfrac{5}{12}\right)$ $\dfrac{1}{4}$

52. $-\dfrac{4}{7}\left(-\dfrac{21}{8}\right)$ $\dfrac{3}{2}$

53. $-\dfrac{6}{7}\left(\dfrac{35}{8}\right)$ $-\dfrac{15}{4}$

54. $-\dfrac{7}{5}\left(\dfrac{15}{28}\right)$ $-\dfrac{3}{4}$

55. $\dfrac{-18}{9}$ -2

56. $\dfrac{-32}{16}$ -2

57. $\dfrac{20}{-5}$ -4

58. $\dfrac{36}{-3}$ -12

59. $\dfrac{-14}{-7}$ 2

60. $\dfrac{-24}{-8}$ 3

61. $\dfrac{0}{-3}$ 0

62. $\dfrac{0}{-9}$ 0

63. $\dfrac{4}{0}$ Undefined

64. $\dfrac{-7}{0}$ Undefined

65. $-\left(\dfrac{-4}{-2}\right)$ -2

66. $-\left(\dfrac{-10}{-5}\right)$ -2

67. $-\left(\dfrac{-27}{3}\right)$ 9

68. $-\left(\dfrac{-9}{3}\right)$ 3

69. $-\left(\dfrac{15}{-5}\right)$ 3

70. $-\left(\dfrac{18}{-6}\right)$ 3

71. $\dfrac{-3}{-3}$ 1

72. $\dfrac{-18}{-9}$ 2

73. $\dfrac{-16}{4}$ -4

74. $\dfrac{-48}{6}$ -8

75. $\dfrac{-56}{8}$ -7

76. $\dfrac{-54}{6}$ -9

77. $\dfrac{3}{5} \div \left(-\dfrac{4}{7}\right)$ $-\dfrac{21}{20}$

78. $\dfrac{4}{9} \div \left(-\dfrac{1}{7}\right)$ $-\dfrac{28}{9}$

79. $-\dfrac{2}{3} \div \left(-\dfrac{7}{6}\right)$ $\dfrac{4}{7}$

80. $-\dfrac{5}{6} \div \left(-\dfrac{25}{18}\right)$ $\dfrac{3}{5}$

81. $-\dfrac{5}{8} \div \dfrac{7}{8}$ $-\dfrac{5}{7}$

82. $-\dfrac{4}{5} \div \dfrac{8}{15}$ $-\dfrac{3}{2}$

83. $\dfrac{-3.1}{6.2}$ $-\dfrac{1}{2}$

84. $\dfrac{1.2}{-4.8}$ $-\dfrac{1}{4}$

85. $\dfrac{-1.6}{-9.6}$ $\dfrac{1}{6}$

86. $\dfrac{-9.8}{-1.4}$ 7

B In Problems 87–94, indicate which property is illustrated in each statement (a and b represent real numbers).

87. $5.6 + 9.2 = 9.2 + 5.6$
Commutative property of addition

88. $-3 \cdot 4 = 4 \cdot (-3)$ Commutative property of multiplication

89. $5 \cdot (-2) = -2 \cdot 5$ Commutative property of multiplication

90. $\left(\dfrac{1}{5} + \dfrac{2}{7}\right) + \dfrac{1}{8} = \dfrac{1}{5} + \left(\dfrac{2}{7} + \dfrac{1}{8}\right)$
Associative property of addition

91. $(-3 \cdot a) \cdot 2 = 2 \cdot (-3 \cdot a)$
Commutative property of multiplication

92. $5(3a) = (3a)5$ Commutative property of multiplication

93. $1 \cdot (3 + b) = 3 + b$
Multiplicative identity

94. $(a + b) \cdot 1 = a + b$
Multiplicative identity

95. If the area of a rectangle is found by multiplying the length times the width, express the area of the rectangle in the figure in two ways to illustrate the distributive property for $a(b + c)$. $A = a(b + c) = ab + ac$

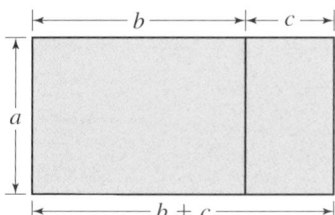

96. Express the shaded area of the rectangle in the figure in two ways to illustrate the distributive property for $a(b - c)$. $A = a(b - c) = ab - ac$

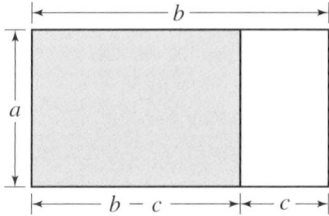

APPLICATIONS

97. The temperature in the center core of the Earth reaches $+5000°C$. In the thermosphere (a region in the upper atmosphere), the temperature is $+1500°C$. Find the difference in temperature between the center of the Earth and the thermosphere. $3500°C$

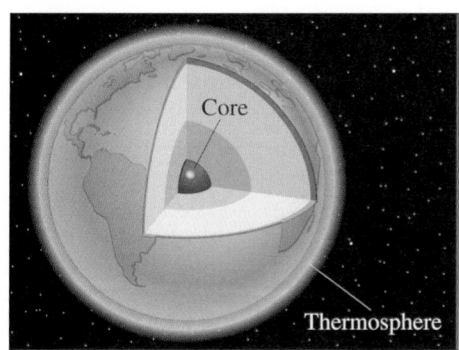

98. The record high temperature in Calgary, Alberta, is $+99°F$. The record low temperature is $-46°F$. Find the difference between these extremes. $145°F$

99. The price of a certain stock at the beginning of the week was $47. Here are the changes in price during the week: $+1, +2, -1, -2, -1$. What is the price of the stock at the end of the week? $46

100. The price of a stock at the beginning of the week was $37. On Monday, the price went up $2; on Tuesday, it went down $3; and on Wednesday, it went down another $1. What was the price of the stock then? $35

Here are the temperature changes (in degrees Celsius, C) by the hour in a certain city:

  1 P.M., $+2$ 2 P.M., $+1$ 3 P.M., -1 4 P.M., -3

101. If the temperature was initially $15°C$, what was it at 4 P.M.? $14°C$

102. If the temperature was initially $-15°C$, what was it at 4 P.M.? $-16°C$

SKILL CHECKER

Try the *Skill Checker Exercises* so you'll be ready for the next section.

In Problems 103–106, fill in the blanks.

	Number	Additive Inverse	Reciprocal
103.	7	-7	$\dfrac{1}{7}$
104.	-2.8	2.8	$-\dfrac{1}{2.8}$ or $-\dfrac{5}{14}$
105.	0	0	Undefined
106.	$-\dfrac{2}{3}$	$\dfrac{2}{3}$	$-\dfrac{3}{2}$

USING YOUR KNOWLEDGE

Would you rather be "very lovable" or "extremely good"? Use signed numbers to decide.

Have you met anybody *nice* today? Have you had an *unpleasant* experience? Perhaps the person you met was *very nice* or your experience *very unpleasant.* Psychologists and linguists have a numerical way to indicate the difference between nice and very nice or between unpleasant and very unpleasant. Suppose you assign a positive number ($+2$, for example) to the adjective *nice,* and a negative number (say, -2) to *unpleasant,* and a positive number greater than 1 (say, $+1.75$) to *very.* Then, by definition, "**very nice**" means

Adverbs		Adjectives	
Slightly	0.54	Wicked	-2.5
Rather	0.84	Disgusting	-2.1
Decidedly	0.16	Average	-0.8
Very	1.25	Good	3.1
Extremely	1.45	Lovable	2.4

very nice
$$\downarrow \quad \downarrow$$
$$(1.75) \cdot (2) = 3.50$$

and "very unpleasant" means

very unpleasant
$$\downarrow \quad \downarrow$$
$$(1.75) \cdot (-2) = -3.50$$

Here are some adverbs and adjectives and their average numerical values as rated by a panel of college students.

Find the value of:

107. Slightly wicked -1.35

108. Decidedly average -0.128

109. Extremely disgusting -3.045

110. Rather lovable 2.016

111. Very good 3.875

By the way, if you got all the answers correct, you are 4.495! (Extremely good)

WRITE ON

112. Explain why division by zero is not defined.
Answers may vary.

113. The distributive property $a(b + c) = ab + ac$ is called the distributive property of multiplication over addition. Is addition distributive over multiplication? Explain and give examples to support your answer.
Answers may vary.

114. We mentioned that the set of real numbers is *closed* under addition and multiplication. Are the natural numbers closed under addition? What about under subtraction? Explain and give examples to support your answer.
Answers may vary.

MASTERY TEST

If you know how to do these problems, you have learned your lesson!

Perform the indicated operation:

115. $-\dfrac{2}{5} \div \left(-\dfrac{5}{8}\right)$ $\dfrac{16}{25}$ **116.** $-\dfrac{9}{5} - \left(-\dfrac{3}{4}\right)$ $-\dfrac{21}{20}$ **117.** $-\left(\dfrac{3}{5}\right)\left(-\dfrac{10}{3}\right)$ 2   **118.** $23.4 + (-29.7)$ -6.3

119. $-\dfrac{5}{8} + \left(-\dfrac{3}{7}\right)$ $-\dfrac{59}{56}$ **120.** $18.7 - (-13.2)$ 31.9   **121.** $3.2(-4)$ -12.8 **122.** $-3.4 \div 1.7$ -2

123. $(-3 - 5)(-2) + 8(3 - 7 + 4)$ 16   **124.** $-4(5 - 7) + (-3 - 5)(-2)$ 24

Name the property used:

125. $(-8) + 8 = 0$
Inverse property of addition

126. $(a + 3) = 1 \cdot (a + 3)$
Multiplicative identity

127. $3(ab) = (3a)b$ Associative property of multiplication

128. $8 + (b + c) = (8 + b) + c$
Associative property of addition

129. $2 \cdot (b + 4) = (b + 4) \cdot 2$
Commutative property of multiplication

130. $(a + 1) \cdot \dfrac{1}{a + 1} = 1, a \neq -1$
Multiplicative inverse

1.3 PROPERTIES OF EXPONENTS

To Succeed, Review How To . . .

Perform the four fundamental operations (addition, subtraction, multiplication, and division) using signed numbers (pp. 17–23).

Objectives

A Evaluate expressions containing natural numbers as exponents.

B Write an expression containing negative exponents as a fraction.

C Multiply and divide expressions containing exponents.

D Raise a power to a power.

E Raise a quotient to a power.

F Convert between ordinary decimal notation and scientific notation, and use scientific notation in computations.

GETTING STARTED

Unlike the "prank" Rubik's cube shown in the picture, the side of each of the squares in a "real" Rubik's cube is 1 centimeter (cm) long. Can you find the **area** of the top of the cube? Can you find the **volume** of the whole cube? As you may recall, to find the area of a square, we multiply the length of two sides of the square. The top of the square is 3 cm long, therefore the area of the top is

$$(3 \text{ cm})(3 \text{ cm}) = 3^2 \text{ cm}^2 \text{ or } 9 \text{ \textbf{square} centimeters.}$$

Similarly, the volume of the cube is

$$(3 \text{ cm})(3 \text{ cm})(3 \text{ cm}) = 3^3 \text{ cm}^3 \text{ or } 27 \text{ \textbf{cubic} centimeters.}$$

$$\overbrace{3^2 = 3 \times 3}^{2 \text{ times}} \qquad \text{and} \qquad \overbrace{3^3 = 3 \times 3 \times 3}^{3 \text{ times}}$$

This is **exponential** notation, one of the topics we shall study in this section.

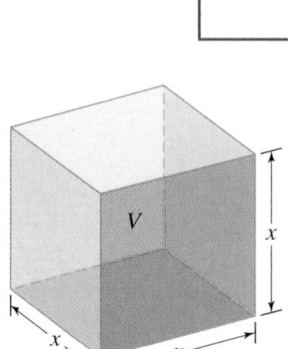

One of the most prominent landmarks in the East Village in Manhattan is a statue of a giant steel black cube. The cube was built at Astor Place in 1968, and has stood there ever since. One day a group of individuals decided to play a prank and turn it into the world's largest Rubik's cube. The cube measures 8 ft by 8 ft by 8 ft (2.44 m^3). It is standing on one vertex, such that the top reaches around 12 ft in the air. The cube remained painted like the Rubik's cube for an entire day before the New York City maintenance department cleaned it off.

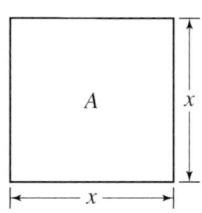

What is the area A of the square? The area is

$$A = x \cdot x = x^2 \qquad \text{Read "}x\text{ squared, or }x\text{ to the second power."}$$

In the expression x^2, the *exponent* 2 indicates that the *base* x is to be used as a factor twice. What about the volume V of the cube? It is

$$V = x \cdot x \cdot x = x^3 \qquad \text{Read "}x\text{ cubed, or }x\text{ to the third power."}$$

This time the *exponent* 3 indicates that the *base* x is used as a factor three times. In general, we have the following definition.

Exponent and Base	If a is a real number and n is a natural number,
	$$a^n = \underbrace{a \cdot a \cdot a \cdots a}_{n \text{ factors}}$$
	where n is called the **exponent** and a is the **base**.

When $n = 1$, the exponent is usually omitted. Thus, $a^1 = a$, $b^1 = b$, and $c^1 = c$.

Natural-Number Exponents

Calculate It

Verify $(-4)^2 \neq -4^2$

We can use a calculator to clarify the concepts in Example 1. Enter $(-4)^2$ and -4^2 in your calculator. Can you see that in $(-4)^2$ we are raising the base -4 to the second power, while in -4^2 we are finding the additive inverse of 4^2? Look at the way you enter each of the expressions. With a TI-83 Plus, $(-4)^2$ is entered as ⟨ (⟩ ⟨ (−) ⟩ 4 ⟨) ⟩ ⟨ x^2 ⟩, while -4^2 is entered as ⟨ (−) ⟩ 4 ⟨ x^2 ⟩. The answers are different. We can see that $(-4)^2 \neq -4^2$.

```
(-4)²
                    16
-4²
                   -16
```

We can **evaluate** (find the value of) expressions containing natural-number exponents. Thus,

$$4^2 = 4 \cdot 4 = 16$$

$$(-3)^2 = (-3)(-3) = 9$$

$$(-2)^3 = (-2)(-2)(-2)$$

$$= 4(-2) \qquad (-2)(-2) = 4$$

$$= -8$$

$$(-4)^4 = (-4)(-4)(-4)(-4)$$

$$= (16)(16)$$

$$= 256$$

For the **even** exponents 2 and 4 in $(-3)^2 = 9$ and $(-4)^4 = 256$, respectively, we get the **positive** answers 9 and 256, but for the **odd** exponent 3 in $(-2)^3 = -8$, the answer is **negative**. This suggests the following.

PRODUCTS OF NEGATIVE NUMBERS

The product of an **odd** number of **negative** factors is **negative**. The product of an **even** number of **negative** factors is **positive**.

EXAMPLE 1   Evaluating expressions with natural-number exponents

Evaluate:

a. $(-4)^2$ **b.** -4^2 **c.** $(-4)^3$ **d.** -4^3

SOLUTION

a. $(-4)^2 = (-4)(-4) = 16$ The exponent 2 is applied to the base (-4).

b. $-4^2 = -1 \cdot 4^2$ Here the exponent 2 applies to the base 4 only.

$\qquad = -1 \cdot 16$ Since $4^2 = 16$

$\qquad = -16$

c. $(-4)^3 = (-4)(-4)(-4) = -64$

d. $-4^3 = -1 \cdot 4^3$

$\qquad = -1 \cdot 64$ Since $4^3 = 4 \cdot 4 \cdot 4 = 64$

$\qquad = -64$

PROBLEM 1

Evaluate:

a. $(-5)^2$ **b.** -5^2

c. $(-6)^3$ **d.** -6^3

Web It

For a lesson on evaluating exponents, visit link 1-3-1 at mhhe.com/bello.

Answers

1. a. 25   **b.** -25 **c.** -216
d. -216

Teaching Tip

Stress that this definition works only when $n \in (1, 2, 3, \ldots)$. There will be new definitions to learn later for 2^{-3}, 5^0, and $16^{1/4}$.

CAUTION

$(-4)^2 \neq -4^2$ because $(-4)^2 = 16$, which is positive, while -4^2 is the additive inverse of 4^2, that is, $-4^2 = -16$. The placement of parentheses when using exponents is extremely important. Always interpret -4^2 as $-(4^2)$.

B Negative and Zero Exponents

Web It

For an elementary lesson dealing with 0 and negative exponents, including a motivation to define negative exponents, try link 1-3-2 at mhhe.com/bello.

In science and technology, negative numbers are used as exponents. For example, the diameter of a DNA molecule is 10^{-8} m, and the time it takes for an electron to go from source to screen in a TV tube is 10^{-6} sec. What do 10^{-8} and 10^{-6} mean? Look at the pattern obtained by dividing by 10 in each of the following:

$$10^3 = 1000$$
$$10^2 = 100$$
$$10^1 = 10$$
$$10^0 = 1$$

The exponents decrease by 1. ⟶   ⟵ The number decreases by a factor of 10.

$$10^{-1} = \frac{1}{10} = \frac{1}{10^1}$$
$$10^{-2} = \frac{1}{100} = \frac{1}{10^2}$$
$$10^{-3} = \frac{1}{1000} = \frac{1}{10^3}$$

Teaching Tip

Discuss ways you can show 0^0 is not defined. Can 0^{-1} be defined? Why or why not?

This procedure yields $10^0 = 1$. In general, we make the following definition.

Zero Exponent	If a is a nonzero real number, $$a^0 = 1$$ Moreover, 0^0 is *not* defined.

Thus $5^0 = 1$, $8^0 = 1$, and $9^0 = 1$. 0^0 is **not** defined. Now look again at the numbers in the first box. We obtained

$$10^{-1} = \frac{1}{10}, \qquad 10^{-2} = \frac{1}{10^2}, \qquad \text{and} \qquad 10^{-3} = \frac{1}{10^3}$$

We make the following definition.

Negative Exponents	If n is a positive integer, $$a^{-n} = \frac{1}{a^n} \quad (a \neq 0)$$

Web It

For a complete lesson on sim-plifying expressions involv-ing exponents, try link 1-3-3 at mhhe.com/bello.

This definition says that a^{-n} and a^n are reciprocals, provided $a \neq 0$, because

$$a^{-n} \cdot a^n = \frac{1}{a^n} \cdot a^n = 1$$

By definition

$$5^{-2} = \frac{1}{5^2} = \frac{1}{5 \cdot 5} = \frac{1}{25}$$

and

$$(-2)^{-3} = \frac{1}{(-2)^3} = \frac{1}{(-2) \cdot (-2) \cdot (-2)} = \frac{1}{-8} = -\frac{1}{8}$$

Similarly,

$$\frac{1}{4^2} = 4^{-2}$$

and

$$\frac{1}{3^4} = 3^{-4}$$

| EXAMPLE 2 | **From negative exponents to fractions** | PROBLEM 2 |

Write the following without negative exponents:

a. 6^{-2} **b.** $(-4)^{-3}$ **c.** x^{-4} **d.** $-a^{-4}$ **e.** $3x^{-2}$

SOLUTION

a. $6^{-2} = \frac{1}{6^2} = \frac{1}{6 \cdot 6} = \frac{1}{36}$

b. $(-4)^{-3} = \frac{1}{(-4)^3} = \frac{1}{(-4) \cdot (-4) \cdot (-4)} = \frac{1}{-64} = -\frac{1}{64}$

c. $x^{-4} = \frac{1}{x^4}$

d. $-a^{-4} = -\frac{1}{a^4}$

e. $3x^{-2} = 3 \cdot x^{-2} = 3 \cdot \frac{1}{x^2} = \frac{3}{x^2}$

Write the following without negative exponents:

a. 5^{-2} **b.** $(-3)^{-3}$ **c.** x^{-5}

d. $-b^{-5}$ **e.** $5x^{-3}$

C   Multiplication and Division with Exponents

Suppose we wish to multiply $x^2 \cdot x^3$. We write

$$\underbrace{x^2}_{} \cdot \underbrace{x^3}_{}$$
$$\underbrace{x \cdot x \cdot x \cdot x \cdot x}_{x^5}$$

We want to know the total number of factors, so we add the exponents 2 and 3 of x^2 and x^3 to find the exponent 5 of the result. Similarly,

$$a^3 \cdot a^4 = a^{3+4} = a^7$$

$$b^2 \cdot b^4 = b^{2+4} = b^6$$

$$c^5 \cdot c^0 = c^{5+0} = c^5$$

Answers

2. a. $\frac{1}{5^2} = \frac{1}{25}$ **b.** $\frac{1}{(-3)^3} = -\frac{1}{27}$

c. $\frac{1}{x^5}$ **d.** $-\frac{1}{b^5}$ **e.** $\frac{5}{x^3}$

Now, let's consider $x^5 \cdot x^{-3}$, where one of the exponents is negative. By the definition of exponents,

$$x^5 \cdot x^{-3} = x \cdot x \cdot x \cdot x \cdot x \cdot \frac{1}{x \cdot x \cdot x}$$

$$= \frac{x \cdot x \cdot \cancel{x} \cdot \cancel{x} \cdot \cancel{x}}{\cancel{x} \cdot \cancel{x} \cdot \cancel{x}}$$

$$= x^2$$

Adding exponents,

$$x^5 \cdot x^{-3} = x^{5+(-3)} = x^2 \qquad \text{Same answer}$$

Let us try $x^{-2} \cdot x^{-3}$. This time we have

$$x^{-2} \cdot x^{-3} = \frac{1}{x \cdot x} \cdot \frac{1}{x \cdot x \cdot x}$$

$$= \frac{1}{x \cdot x \cdot x \cdot x \cdot x} \qquad \text{Multiplying}$$

$$= x^{-5} \qquad \text{By the definition of negative exponents}$$

Adding exponents,

$$x^{-2} \cdot x^{-3} = x^{-2+(-3)} = x^{-5} \qquad \text{Same answer}$$

This suggests the following law.

First Law of Exponents (Product Rule)

If a is a real number and m and n are integers,

$$a^m \cdot a^n = a^{m+n} \quad (a \neq 0)$$

Web It

To study how to use the product rule, try link 1-3-4 at mhhe.com/bello.

Teaching Tip

Make clear that this law does *not* apply to $x^n \cdot y^m$ because the bases x and y are different. Thus, $2^3 \cdot 3^4 \neq 6^7$.

This law tells us that to *multiply* expressions with the *same* base, we *add* the exponents. This law *does not* apply to expressions such as $x^m \cdot y^n$ because the bases are *different*. $[3^2 \cdot 2^4 = 3 \cdot 3 \cdot 2 \cdot 2 \cdot 2 \cdot 2$, so we have different bases (factors) that cannot be combined.] If the expressions involved have numerical coefficients, we multiply numbers by numbers and letters (variables) by letters using the commutative and associative properties that we have studied. Thus, to multiply $(-3x^2)(5x^3)$, we write

$$(-3x^2)(5x^3) = (-3 \cdot 5)(x^2)(x^3)$$

$$= -15x^{2+3} = -15x^5$$

Similarly,

$$(-4x^{-2})(3x^5) = (-4 \cdot 3)(x^{-2+5})$$

$$= -12x^3 \qquad -2 + 5 = 3$$

and

$$(-2x^{-5})(-3x^2) = (-2)(-3)(x^{-5+2})$$

$$= 6x^{-3} \qquad -5 + 2 = -3$$

$$= 6 \cdot \frac{1}{x^3}$$

$$= \frac{6}{x^3}$$

NOTE

We write the answer without using negative exponents.

EXAMPLE 3 Using the product rule of exponents	**PROBLEM 3**

Multiply and simplify:

a. $(-4x^2)(5x^5)$ **b.** $(3x^7)(-2x^4)$ **c.** $(4x^3y)(-2x^{-8}y^6)$

SOLUTION

a. $(-4x^2)(5x^5) = (-4 \cdot 5)(x^2 \cdot x^5)$

$$= -20x^{2+5}$$

$$= -20x^7$$

b. $(3x^7)(-2x^4) = (3)(-2)(x^7 \cdot x^4)$

$$= -6x^{7+4}$$

$$= -6x^{11}$$

c. $(4x^3y)(-2x^{-8}y^6) = (4)(-2)(x^3 \cdot x^{-8})(y^1 \cdot y^6)$

$$= -8(x^{3+(-8)})(y^{1+6})$$

$$= -8(x^{-5})(y^7)$$

$$= -8\left(\frac{1}{x^5}\right)(y^7)$$

$$= -\frac{8y^7}{x^5}$$

Multiply and simplify:

a. $(-3x^{-3})(5x^5)$

b. $(3x^2)(-2x^3)$

c. $(5x^3y^4)(-2x^{-7}y)$

Calculate It   Verify the Product Rule

You can verify the answers in Example 3(b) by graphing the *original* expression $(3x^7)(-2x^4)$ and the *answer* $-6x^{11}$. If the resulting picture (graph) is the same, the original expression and the answer are also the same, and we have verified that the answer is correct. To graph $(3x^7)(-2x^4)$, first select a standard window (press ⬛ZOOM⬛ 6 on a TI-83 Plus) then graph $Y_1 = (3x^7)(-2x^4)$ and $Y_2 = -6x^{11}$. (With a TI-83 Plus, press $Y_1 =$ ⬛(⬛ 3 ⬛X,T,θ,n⬛ ⬛^⬛ 7 ⬛)⬛ ⬛(⬛ ⬛(−)⬛ 2 ⬛X,T,θ,n⬛ ⬛^⬛ 4 ⬛)⬛ ⬛GRAPH⬛.) The graph you have just keyed in appears in the screen.

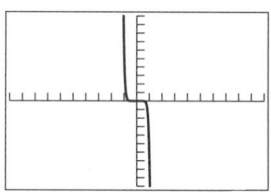

Now graph $Y_2 = -6x^{11}$ by entering $Y_2 =$ ⬛(−)⬛ 6 ⬛X,T,θ,n⬛ ⬛^⬛ 11 ⬛GRAPH⬛. We get the same graph, so our answer is correct. (Try graphing $-6x^{12}$ and you will see that you get a different graph.)

Now consider the following division:

$$\frac{x^5}{x^3} = \frac{x \cdot x \cdot x \cdot x \cdot x}{x \cdot x \cdot x}$$

$$= x \cdot x \cdot \frac{x \cdot x \cdot x}{x \cdot x \cdot x}$$

$$= x^2 \cdot 1$$

$$= x^2$$

$$= x^{5-3}$$

Answers

3. a. $-15x^2$ **b.** $-6x^5$

c. $-\dfrac{10y^5}{x^4}$

The exponent 2 in the answer can be obtained by *subtracting* $5 - 3 = 2$. Similarly,

$$\frac{x^5}{x^{-3}} = x^{5-(-3)} = x^8$$

This is because

$$\frac{x^5}{x^{-3}} = \frac{x^5}{\frac{1}{x^3}} = x^5 \div \frac{1}{x^3} = x^5 \cdot x^3 = x^8$$

This line of reasoning suggests the following law.

SECOND LAW OF EXPONENTS (QUOTIENT RULE)

If a is a real number and m and n are integers,

$$\frac{a^m}{a^n} = a^{m-n} \qquad (a \neq 0)$$

Web It

To evaluate expressions using the quotient rule for exponents, go to link 1-3-5 at mhhe.com/bello.

After that, you can try the quotient rule of exponents, the power rule, and raising a product to a power using variables at link 1-3-6 at mhhe.com/bello.

This law tells us that to *divide* expressions with the same base, we *subtract* the exponent of the denominator from that of the numerator. Thus

$$\frac{x^8}{x^5} = x^{8-5} = x^3$$

$$\frac{x^3}{x^8} = x^{3-8} = x^{-5} = \frac{1}{x^5}$$

$$\frac{x^{-3}}{x^{-8}} = x^{-3-(-8)} = x^{-3+8} = x^5$$

$$\frac{x^8}{x^8} = x^{8-8} = x^0 = 1$$

EXAMPLE 4 Using the quotient rule of exponents	**PROBLEM 4**
Divide and simplify:	Divide and simplify:

a. $\dfrac{4x^8}{2x^5}$ **b.** $\dfrac{-12x^4}{-3x^6}$ **c.** $\dfrac{5x^{-4}}{-15x^{-6}}$ **d.** $\dfrac{30x^3}{-15x^{-6}}$

SOLUTION

a. $\dfrac{4x^8}{2x^5} = \dfrac{4}{2} \cdot \dfrac{x^8}{x^5}$

$= 2x^{8-5}$

$= 2x^3$

b. $\dfrac{-12x^4}{-3x^6} = \dfrac{-12}{-3} \cdot \dfrac{x^4}{x^6}$

$= 4x^{4-6}$

$= 4x^{-2}$ $4 - 6 = -2$

$= \dfrac{4}{x^2}$

c. $\dfrac{5x^{-4}}{-15x^{-6}} = \dfrac{5}{-15} \cdot \dfrac{x^{-4}}{x^{-6}}$

$= -\dfrac{1}{3} \cdot x^{-4-(-6)}$

$= -\dfrac{1}{3} \cdot x^{-4+6}$ $-4 - (-6) =$
                                    $-4 + 6$

$= -\dfrac{1}{3} \cdot x^2$

$= -\dfrac{x^2}{3}$

d. $\dfrac{30x^3}{-15x^{-6}} = \dfrac{30}{-15} \cdot \dfrac{x^3}{x^{-6}}$

$= -2x^{3-(-6)}$

$= -2x^9$ $3 - (-6) =$
              $3 + 6 = 9$

PROBLEM 4

a. $\dfrac{6x^9}{12x^3}$ **b.** $\dfrac{-18x^3}{-9x^8}$

c. $\dfrac{10x^{-6}}{-20x^{-8}}$ **d.** $\dfrac{45x^4}{-15x^{-6}}$

Answers

4. **a.** $\dfrac{x^6}{2}$ **b.** $\dfrac{2}{x^5}$ **c.** $-\dfrac{x^2}{2}$

d. $-3x^{10}$

> **NOTE**
>
> A **negative coefficient,** like -15 in Example 4c, does not follow the same rule as the definition of a **negative exponent** like x^{-2} in Example 4(b). The -15 stays in the denominator, but the x^{-2} moves to the denominator and becomes x^2.

D Raising a Power to a Power

Now let us consider $(5^3)^2$. By definition,

$$(5^3)^2 = (5^3)(5^3) = 5^{3+3} = 5^6$$

The exponent is $6 = 3 \cdot 2$, so we could have obtained the answer by multiplying exponents in the expression $(5^3)^2$. Similarly,

$$(3^{-2})^3 = \left(\frac{1}{3^2}\right)\left(\frac{1}{3^2}\right)\left(\frac{1}{3^2}\right) = \frac{1}{3^6} = 3^{-6}$$

Again, the exponent -6 could be obtained by multiplying the original exponents -2 and 3. Generalizing this result, we have the following law.

Teaching Tip

To help students realize that this rule applies only to *products* raised to a power and not to *sums,* show the following examples:

Simplify $(3 + 5)^2$

$\begin{array}{ll} (3 + 5)^2 = (8)^2 & (3 + 5)^2 = 3^2 + 5^2 \\ \quad\quad = 64 & \quad\quad\quad = 9 + 25 \\ & \quad\quad\quad = 34 \\ \text{Correct} & \text{Incorrect} \end{array}$

Now simplify $(3 \cdot 5)^2$

$\begin{array}{ll} (3 \cdot 5)^2 = (15)^2 & (3 \cdot 5)^2 = 3^2 \cdot 5^2 \\ \quad\quad = 225 & \quad\quad\quad = 9 \cdot 25 \\ & \quad\quad\quad = 225 \end{array}$

Both are correct

Therefore, the power rule works only for *products* raised to a *power* (not sums).

THIRD LAW OF EXPONENTS (POWER RULE)

If a is a real number and m and n are integers,

$$(a^m)^n = a^{m \cdot n} \qquad (a \neq 0)$$

To raise a power to another power, we multiply the exponents,

$$(x^3)^4 = x^{3 \cdot 4} = x^{12}$$

$$(y^6)^{-3} = y^{6 \cdot (-3)} = y^{-18} = \frac{1}{y^{18}}$$

$$(z^{-5})^4 = z^{-5 \cdot 4} = z^{-20} = \frac{1}{z^{20}}$$

Let's consider $(5x^4)^3$. By definition,

$$\begin{aligned} (5x^4)^3 &= (5x^4)(5x^4)(5x^4) \\ &= (5 \cdot 5 \cdot 5)(x^4 \cdot x^4 \cdot x^4) \\ &= 5^3 \cdot x^{3 \cdot 4} \\ &= 5^3 \cdot x^{12} \end{aligned}$$

Web It

For a slide show on the laws of exponents with examples, try link 1-3-7 at mhhe.com/bello.

The exponent 3 is applied to the 5 and the x^4; that is, if we raise several factors inside parentheses to a power, we raise each factor to the given power. In general, we have the following law.

Raising a Product to a Power

If a and b are real numbers and m, n, and k are integers,

$$(a^m b^n)^k = a^{m \cdot k} b^{n \cdot k} \qquad (a \neq 0, b \neq 0)$$

This result can be generalized further to apply to any number of factors inside the parentheses.

EXAMPLE 5	**Using the power rule of exponents**

Simplify:

a. $(3x^5y^3)^{-2}$　　　　　　　　　　**b.** $(-2x^5y^{-4})^3$

SOLUTION

a. $(3x^5y^3)^{-2} = (3)^{-2}(x^5)^{-2}(y^3)^{-2}$

$$= \frac{1}{3^2} \cdot x^{5 \cdot (-2)} \cdot y^{3 \cdot (-2)} \qquad (3)^{-2} = \frac{1}{3^2}$$

$$= \frac{1}{3^2} \cdot x^{-10} \cdot y^{-6}$$

$$= \frac{1}{3^2} \cdot \frac{1}{x^{10}} \cdot \frac{1}{y^6} \qquad x^{-10} = \frac{1}{x^{10}}, \; y^{-6} = \frac{1}{y^6}$$

$$= \frac{1}{9x^{10}y^6}$$

b. $(-2x^5y^{-4})^3 = (-2)^3(x^5)^3(y^{-4})^3$

$$= -8x^{5 \cdot 3}y^{-4 \cdot 3}$$

$$= -8x^{15}y^{-12}$$

$$= -\frac{8x^{15}}{y^{12}} \qquad y^{-12} = \frac{1}{y^{12}}$$

PROBLEM 5

Simplify:

a. $(4x^3y^4)^{-2}$　　**b.** $(-3x^2y^{-5})^3$

Web It

Practice some examples using the laws of exponents at link 1-3-8 at mhhe.com/bello.

 **Raising a Quotient to a Power**

We have already raised a product to a power. Can we raise a quotient to a power? Let us try $(2^3/3^4)^2$. By the definition of exponents

$$\left(\frac{2^3}{3^4}\right)^2 = \frac{2^3}{3^4} \cdot \frac{2^3}{3^4} = \frac{2^{3+3}}{3^{4+4}} = \frac{2^6}{3^8} = \frac{2^{3 \cdot 2}}{3^{4 \cdot 2}}$$

The same answer is obtained by multiplying each of the exponents in the numerator and denominator by 2. Here is the general rule.

Raising a Quotient to a Power	If a and b are real numbers and m, n, and k are integers,
	$$\left(\frac{a^m}{b^n}\right)^k = \frac{a^{m \cdot k}}{b^{n \cdot k}} \qquad (a \neq 0, b \neq 0)$$

EXAMPLE 6	**Raising a quotient to a power**

Simplify:

a. $\left(\dfrac{x^4}{y^{-3}}\right)^{-2}$　　　　　　　　**b.** $\left(\dfrac{3x^{-3}y^2}{2y^3}\right)^3$

SOLUTION

a. $\left(\dfrac{x^4}{y^{-3}}\right)^{-2} = \dfrac{(x^4)^{-2}}{(y^{-3})^{-2}} = \dfrac{x^{4 \cdot (-2)}}{y^{-3 \cdot (-2)}} = \dfrac{x^{-8}}{y^6} \qquad x^{-8} = \dfrac{1}{x^8}$

$$= \frac{1}{x^8y^6}$$

PROBLEM 6

Simplify:

a. $\left(\dfrac{x^5}{y^{-3}}\right)^{-4}$　　**b.** $\left(\dfrac{2x^{-3}y^3}{3y^3}\right)^2$

Answers

5. a. $\dfrac{1}{16x^6y^8}$　**b.** $-\dfrac{27x^6}{y^{15}}$

6. a. $\dfrac{1}{x^{20}y^{12}}$　**b.** $\dfrac{4}{9x^6}$

b. In this case, it is easier to do the operations inside the parentheses first.

$$\left(\frac{3x^{-3}y^2}{2y^3}\right)^3 = \left(\frac{3}{2} \cdot x^{-3} \cdot y^{2-3}\right)^3 \qquad \frac{y^2}{y^3} = y^{2-3}$$

$$= \left(\frac{3y^{-1}}{2x^3}\right)^3 \qquad\qquad x^{-3} = \frac{1}{x^3}$$

$$= \left(\frac{3}{2x^3y}\right)^3 \qquad\qquad y^{-1} = \frac{1}{y}$$

$$= \frac{3^3}{(2x^3y)^3}$$

$$= \frac{27}{8x^9y^3}$$

Since the reciprocal of $\frac{a}{b}$ is $\frac{b}{a}$,

$$\left(\frac{a}{b}\right)^{-1} = \left(\frac{b}{a}\right)$$

Thus

$$\left[\left(\frac{a}{b}\right)^{-1}\right]^n = \left(\frac{b}{a}\right)^n$$

PROCEDURE

Raising a Quotient to a Negative Power

$$\left(\frac{a}{b}\right)^{-n} = \left(\frac{b}{a}\right)^n \qquad (a \neq 0, b \neq 0)$$

This means that a fraction raised to the $-n$th power is equivalent to its *reciprocal* raised to the nth power. In Example 6(a), we could write

$$\left(\frac{x^4}{y^{-3}}\right)^{-2} = \left(\frac{y^{-3}}{x^4}\right)^2 = \frac{y^{-6}}{x^8} = \frac{1}{x^8y^6}$$

You can use this method when you work the exercise set.

F   Scientific Notation

In science and other areas of endeavor, very large or very small numbers occur frequently. For example, a red cell of human blood contains 270,000,000 hemoglobin molecules, and the mass of a single carbon atom is 0.000 000 000 000 000 000 000 019 9 gram. Numbers in this form are difficult to write and to work with, so they are written in scientific notation for which we have the following definition.

Web It

For a good lesson explaining scientific notation and giving many applications, visit link 1-3-9 at mhhe.com/bello.

If you want to practice using a calculator, go to link 1-3-10 at mhhe.com/bello.

SCIENTIFIC NOTATION

A number is said to be in **scientific notation** if it is written in the form

$$m \times 10^n$$

where m is a number greater than or equal to 1 and less than 10 ($1 \leq m < 10$) and n is an integer.

Teaching Tip

Remind students that a positive number less than one is a fraction.

Example: $0.1 = \dfrac{1}{10}$

and $\dfrac{1}{10} = 10^{-1}$

Therefore, when writing a number less than one in scientific notation, we should expect a negative power of 10.

For any given number, the m is obtained by placing the decimal point so that there is exactly one nonzero digit to its left. The integer n is then the number of places that the decimal point must be moved from its position in m to its original position; it is positive if the point must be moved to the right and negative if the point must be moved to the left. Thus

$5.3 = 5.3 \times 10^0$ — Decimal point in 5.3 must be moved 0 places.

$87 = 8.7 \times 10^1 = 8.7 \times 10$ — Decimal point in 8.7 must be moved 1 place to the *right* to get 87.

$68{,}000 = 6.8 \times 10^4$ — Decimal point in 6.8 must be moved 4 places to the *right* to get 68,000.

$0.49 = 4.9 \times 10^{-1}$ — Decimal point in 4.9 must be moved 1 place to the *left* to get 0.49.

$0.072 = 7.2 \times 10^{-2}$ — Decimal point in 7.2 must be moved 2 places to the *left* to get 0.072.

$0.0003875 = 3.875 \times 10^{-4}$ — Decimal point in 3.875 must be moved 4 places to the *left* to get 0.0003875.

EXAMPLE 7   **Writing numbers in scientific notation**

Write in scientific notation:

a. 270,000,000

b. 0.000 000 000 000 000 000 000 019 9

SOLUTION

a. $270{,}000{,}000 = 2.7 \times 10^8$

b. $0.000\,000\,000\,000\,000\,000\,000\,019\,9 = 1.99 \times 10^{-23}$

PROBLEM 7

Write in scientific notation:

a. 350,000

b. 0.000000378

Calculate It   Scientific Notation

Consult the *Keystroke Guide* or your calculator manual to determine how to enter a number written in scientific notation. Thus, to do Example 7(a) with a TI-83 Plus, key in 270 000 000 **MODE** and use your ▸ button to select **Sci.** Press **ENTER**. Go back to the Home screen **2nd** **MODE** and press **ENTER**. The answer is 2.7E8, which means 2.7×10^8.

```
270000000
                2.7E8
```

EXAMPLE 8   **From scientific to standard decimal notation**

Write in standard decimal notation:

a. 2.5×10^{10} **b.** 7.4×10^{-6}

SOLUTION

a. $2.5 \times 10^{10} = 25{,}000{,}000{,}000$ **b.** $7.4 \times 10^{-6} = 0.0000074$

PROBLEM 8

Write in standard decimal notation:

a. 3.5×10^5 **b.** 8.2×10^{-3}

Answers

7. a. 3.5×10^5 **b.** 3.78×10^{-7}
8. a. 350,000 **b.** 0.0082

We can use the laws of exponents when working with numbers in scientific notation. The next example shows you how.

| EXAMPLE 9 | Calculations in scientific notation | PROBLEM 9 |

EXAMPLE 9 Calculations in scientific notation

Do the following calculations, and write the answers in scientific notation:

a. $(5 \times 10^4) \times (9 \times 10^{-7})$

b. $\dfrac{6 \times 10^5}{3 \times 10^{-4}}$

SOLUTION

a. $(5 \times 10^4) \times (9 \times 10^{-7}) = (5 \times 9) \times (10^4 \times 10^{-7})$

$$= 45 \times 10^{4-7}$$
$$= 45 \times 10^{-3}$$
$$= 4.5 \times 10^1 \times 10^{-3}$$
$$= 4.5 \times 10^{1-3}$$
$$= 4.5 \times 10^{-2}$$

b. $\dfrac{6 \times 10^5}{3 \times 10^{-4}} = \dfrac{6}{3} \times \dfrac{10^5}{10^{-4}}$

$$= 2 \times 10^{5-(-4)}$$
$$= 2 \times 10^9 \quad \text{or} \quad 2.0 \times 10^9$$

PROBLEM 9

Do the calculations and write the answers in scientific notation:

a. $(3 \times 10^3)(8 \times 10^{-7})$

b. $\dfrac{4 \times 10^6}{2 \times 10^{-5}}$

Calculate It Exercises

Answers

9. a. 2.4×10^{-3} **b.** 2.0×10^{11}
or 2×10^{11}

Write the given calculator answers in scientific notation.

1. 3.8 E 5 **2.** 6.4 E 2 **3.** 1.56 E −2 **4.** 2.45 E −5

Answers to Calculate It Exercises: **1.** 3.8×10^5 **2.** 6.4×10^2 **3.** 1.56×10^{-2} **4.** 2.45×10^{-5}

Exercises 1.3

**Boost *your* GRADE
at mathzone.com!**

MathZone

• Practice Problems
• Self-Tests
• Videos
• NetTutor
• e-Professors

A In Problems 1–10, evaluate.

1. $-4^2 = \underline{\quad -16 \quad}$

2. $(-4)^2 = \underline{\quad 16 \quad}$

3. $(-5)^2 = \underline{\quad 25 \quad}$

4. $-5^2 = \underline{\quad -25 \quad}$

5. $-5^3 = \underline{\quad -125 \quad}$

6. $(-5)^3 = \underline{\quad -125 \quad}$

7. $(-6)^4 = \underline{\quad 1296 \quad}$

8. $-6^4 = \underline{\quad -1296 \quad}$

9. $-2^5 = \underline{\quad -32 \quad}$

10. $(-2)^5 = \underline{\quad -32 \quad}$

B In Problems 11–20, write the expression given as a fraction in simplified form and without negative exponents.

11. $4^{-2} \;\; \frac{1}{16}$

12. $2^{-3} \;\; \frac{1}{8}$

13. $5^{-3} \;\; \frac{1}{125}$

14. $7^{-2} \;\; \frac{1}{49}$

15. $3^{-4} \;\; \frac{1}{81}$

16. $6^{-3} \;\; \frac{1}{216}$

17. $x^{-6} \;\; \frac{1}{x^6}$

18. $y^{-7} \;\; \frac{1}{y^7}$

19. $a^{-8} \;\; \frac{1}{a^8}$

20. $b^{-4} \;\; \frac{1}{b^4}$

C In Problems 21–50, perform the indicated operations and simplify.

21. $2^{-4} \cdot 2^{-2}$ $\dfrac{1}{64}$

22. $4^{-1} \cdot 4^{-2}$ $\dfrac{1}{64}$

23. $(3x^6) \cdot (4x^{-4})$ $12x^2$

24. $(4y^7) \cdot (5y^{-3})$ $20y^4$

25. $(-3y^{-3}) \cdot (5y^5)$ $-15y^2$

26. $(-5x^{-7}) \cdot (4x^8)$ $-20x$

27. $(-4a^3) \cdot (-5a^{-8})$ $\dfrac{20}{a^5}$

28. $(-2b^4) \cdot (-3b^{-7})$ $\dfrac{6}{b^3}$

29. $(3x^{-5}) \cdot (5x^2y)(-2xy^2)$ $-\dfrac{30y^3}{x^2}$

30. $(4y^{-6}) \cdot (5xy^4)(-2x^2y)$ $-\dfrac{40x^3}{y}$

31. $(-2x^{-3}y^2)(3x^{-2}y^3)(4xy)$ $-\dfrac{24y^6}{x^4}$

32. $(-3xy^{-5})(4x^2y)(2x^3y^2)$ $-\dfrac{24x^6}{y^2}$

33. $(4a^{-2} \cdot b^{-3})(5a^{-1}b^{-1})(-2ab)$ $-\dfrac{40}{a^2b^3}$

34. $(2a^{-5} \cdot b^{-2})(3a^{-1}b^{-1})(5ab)$ $\dfrac{30}{a^5b^2}$

35. $(6a^{-3} \cdot b^3)(5a^2b^2)(-ab^{-5})$ -30

36. $(7a^6 \cdot b^{-6})(2ab^5)(-a^{-7}b)$ -14

37. $\dfrac{8x^7}{4x^3}$ $2x^4$

38. $\dfrac{8a^3}{4a^2}$ $2a$

39. $\dfrac{-8a^4}{-16a^2}$ $\dfrac{a^2}{2}$

40. $\dfrac{-9y^5}{-18y^2}$ $\dfrac{y^3}{2}$

41. $\dfrac{12x^5y^3}{-6x^2y}$ $-2x^3y^2$

42. $\dfrac{18x^6y^2}{-9xy}$ $-2x^5y$

43. $\dfrac{-6x^{-4}}{12x^{-5}}$ $-\dfrac{x}{2}$

44. $\dfrac{8x^{-3}}{4x^{-4}}$ $2x$

45. $\dfrac{-14a^{-5}}{-21a^{-2}}$ $\dfrac{2}{3a^3}$

46. $\dfrac{-2a^{-6}}{-6a^{-3}}$ $\dfrac{1}{3a^3}$

47. $\dfrac{-27a^{-4}}{-36a^{-4}}$ $\dfrac{3}{4}$

48. $\dfrac{-5x^{-3}}{10x^{-3}}$ $-\dfrac{1}{2}$

49. $\dfrac{3a^{-2} \cdot b^5}{2a^4b^2}$ $\dfrac{3b^3}{2a^6}$

50. $\dfrac{x^{-3} \cdot y^6}{x^4 \cdot y^3}$ $\dfrac{y^3}{x^7}$

D In Problems 51–60, simplify the expression given and write your answer without negative exponents.

51. $(2x^3y^{-2})^3$ $\dfrac{8x^9}{y^6}$

52. $(3x^2y^{-3})^2$ $\dfrac{9x^4}{y^6}$

53. $(2x^{-2}y^3)^2$ $\dfrac{4y^6}{x^4}$

54. $(3x^{-4}y^4)^3$ $\dfrac{27y^{12}}{x^{12}}$

55. $(-3x^3y^2)^{-3}$ $-\dfrac{1}{27x^9y^6}$

56. $(-2x^5y^4)^{-4}$ $\dfrac{1}{16x^{20}y^{16}}$

57. $(x^{-6}y^{-3})^2$ $\dfrac{1}{x^{12}y^6}$

58. $(y^{-4}z^{-3})^5$ $\dfrac{1}{y^{20}z^{15}}$

59. $(x^{-4}y^{-4})^{-3}$ $x^{12}y^{12}$

60. $(y^{-5}z^{-3})^{-4}$ $y^{20}z^{12}$

E In Problems 61–70, simplify.

61. $\left(\dfrac{a}{b^3}\right)^2$ $\dfrac{a^2}{b^6}$

62. $\left(\dfrac{a^2}{b}\right)^3$ $\dfrac{a^6}{b^3}$

63. $\left(\dfrac{-3a}{2b^2}\right)^{-3}$ $-\dfrac{8b^6}{27a^3}$

64. $\left(\dfrac{-2a^2}{3b^0}\right)^{-2}$ $\dfrac{9}{4a^4}$

65. $\left(\dfrac{a^{-4}}{b^2}\right)^{-2}$ a^8b^4

66. $\left(\dfrac{a^{-2}}{b^3}\right)^{-3}$ a^6b^9

67. $\left(\dfrac{x^5}{y^{-2}}\right)^{-3}$ $\dfrac{1}{x^{15}y^6}$

68. $\left(\dfrac{x^6}{y^{-3}}\right)^{-2}$ $\dfrac{1}{x^{12}y^6}$

69. $\left(\dfrac{x^{-4}y^3}{x^5y^5}\right)^{-3}$ $x^{27}y^6$

70. $\left(\dfrac{x^{-2}y^0}{x^7y^2}\right)^{-2}$ $x^{18}y^4$

F In Problems 71–74, write the numbers in scientific notation.

71. 268,000,000 (U.S. population in the year 2000) 2.68×10^8

72. 1,900,000,000 (dollars spent on waterbeds and accessories in 1 year) 1.9×10^9

73. 0.00024 (probability of four of a kind in poker) 2.4×10^{-4}

74. 0.00000009 (wavelength in centimeters of an X-ray) 9×10^{-8}

In Problems 75–78, write the numbers in decimal notation.

75. 8×10^6 (bagels eaten per day in the United States) 8,000,000

76. $\$6.85 \times 10^9$ (estimated wealth of the five wealthiest women) $6,850,000,000

77. 2.3×10^{-1} (kilowatts per hour used by your TV) 0.23

78. 4×10^{-11} (joules of energy released by splitting one uranium atom) 0.000 000 000 04

APPLICATIONS

In Problems 79–81, write your answer in scientific notation.

79. The width of the asteroid belt is 2.8×10^8 kilometers (km). The speed of *Pioneer 10* in passing through this belt was 1.4×10^5 km/hr. Thus *Pioneer 10* took

$$\frac{2.8 \times 10^8}{1.4 \times 10^5} \text{ hr}$$

to go through the belt. How many hours was that?
2×10^3

81. The velocity of light can be measured by knowing the distance from the sun to Earth (1.47×10^{11} meter, m) and the time it takes for sunlight to reach Earth (490 sec). Thus the velocity of light is

$$\frac{1.47 \times 10^{11}}{490} \text{ m/sec}$$

How many meters per second is that? 3×10^8

83. The world's oil reserves are estimated to be 6.28×10^{11} barrels. Production is 2.0×10^{10} barrels per year. At this rate, how long would the world's oil reserves last? (Give your answer to the nearest year.)
31 yr

80. The mass of Earth is 6×10^{21} tons. The sun is about 300,000 times as massive. Thus the mass of the sun is $(6 \times 10^{21}) \times 300,000$ tons. How many tons is that?
1.8×10^{27}

82. Oil reserves in the United States are estimated to be 3.5×10^{10} barrels. Production amounts to 3.2×10^9 barrels per year. At this rate, how long would U.S. oil reserves last? (Give your answer to the nearest year.) 11 yr

84. Scientists have estimated that the total energy received from the sun each minute is 1.02×10^{19} calories. The area of Earth is 5.1×10^8 km^2 (square kilometers), so the amount of energy received per square centimeter of Earth's surface per minute (the solar constant) is

$$\frac{1.02 \times 10^{19}}{(5.1 \times 10^8) \times 10^{10}} \qquad (1 \text{ km}^2 = 10^{10} \text{ cm}^2)$$

How many calories per square centimeter is that?
2 cal/cm^2

SKILL CHECKER

Try the *Skill Checker Exercises* so you'll be ready for the next section.

Find:

85. a. $(-3.2)(-1.4)(-2.2)$ -9.856
    b. $(-1.1)(-1.2)(-2.1)$ -2.772

86. a. $\dfrac{-3.2}{1.6}$ -2 **b.** $\dfrac{-4.8}{-1.2}$ 4

USING YOUR KNOWLEDGE

If You Have a Scientific Calculator

If you have a scientific calculator, and you multiply 9,800,000 by 4,500,000, the display may show

4.41   13

This means that the answer is 4.41×10^{13}.

87. The display on a calculator shows

3.34   5

Write this number in scientific notation. 3.34×10^5

88. The display on a calculator shows

−9.97   −6

Write this number in scientific notation.
-9.97×10^{-6}

89. To enter large or small numbers in a calculator with scientific notation, you must write the number using this notation first. To enter the number 8,700,000,000 in the calculator, you must know that 8,700,000,000 is 8.7×10^9, *then* you can key in

(8) (·) (7) [EE↓] (9) or (8) (·) (7) [EXP] (9)

The calculator displays

8.7   09

a. What would the display read when you enter the number 73,000,000,000? | 7.3   10 |

b. What would the display read when you enter the number 0.000000123? | 1.23   −07 |

WRITE ON

Write an explanation for:

90. Why a^0 is defined as 1.
Answers may vary.

91. Why $0^n = 0$ for "n" a natural number.
Answers may vary.

92. Why 0^0 is not defined.
Answers may vary.

93. Why $x^m x^n = x^{m+n}$, using the concept of a factor.
Answers may vary.

94. Why $\dfrac{x^m}{x^n} = x^{m-n}$, using the concept of a factor.
Answers may vary.

95. Why $(x^m)^n = x^{m \cdot n}$, using the concept of a factor.
Answers may vary.

MASTERY TEST

If you know how to do these problems, you have learned your lesson!

Multiply and simplify:

96. $(-3x^4 y^2)(3x^{-7} y)$ $-\dfrac{9y^3}{x^3}$

97. $(-2x^{-5} y^{-3})(-4xy)$ $\dfrac{8}{x^4 y^2}$

Divide and simplify:

98. $\dfrac{45y^4}{-15y^{-7}}$ $-3y^{11}$

99. $\dfrac{-6x^{-8}}{30x^6}$ $-\dfrac{1}{5x^{14}}$

Evaluate:

100. $(-5)^{-4}$ $\dfrac{1}{625}$

101. -2^6 -64

Simplify:

102. $(-2x^4 y^{-5})^3$ $-\dfrac{8x^{12}}{y^{15}}$

103. $(-3x^{-4} y^5)^{-2}$ $\dfrac{x^8}{9y^{10}}$

Write as a fraction:

104. $(-3)^{-5}$ $-\dfrac{1}{243}$

105. $(-x)^{-5}$ $-\dfrac{1}{x^5}$

Simplify:

106. $\left(\dfrac{2x^{-4} y^3}{3y^5}\right)^{-2}$ $\dfrac{9x^8 y^4}{4}$

107. $\left(-\dfrac{5x^{-5} y^7}{7y^3}\right)^{-2}$ $\dfrac{49x^{10}}{25y^8}$

Write in scientific notation:

108. $387,000,000$ 3.87×10^8

109. $(4 \times 10^{-5}) \times (6 \times 10^2)$ 2.4×10^{-2}

1.4 ALGEBRAIC EXPRESSIONS AND THE ORDER OF OPERATIONS

To Succeed, Review How To . . .

1. Perform the four fundamental operations using signed numbers (pp. 17–23).

2. Calculate powers of integers (pp. 32–33).

3. Use the identity for multiplication (p. 26).

Objectives

A Evaluate numerical expressions with grouping symbols.

B Evaluate expressions using the correct order of operations.

C Evaluate algebraic expressions.

D Use the distributive property to simplify expressions.

E Simplify expressions by combining like terms.

F Simplify expressions by removing grouping symbols and combining like terms.

GETTING STARTED   ### Swimming and Your Heart Rate

How do you calculate your ideal heart rate when swimming? One way is to subtract your age from 205 and multiply by 0.70.

This means that if you are a years old, you would subtract your age a from 205 and multiply by 0.70.

Should you subtract your age from 205 first and then multiply by 0.70, or multiply your age by 0.70 first, and then subtract from 205? In algebra we can make our meaning clear by using parentheses. The formula is written as $(205 - a) \cdot 0.70$ to indicate that the subtraction should be done first. (Try it by substituting your age for a.)

In the "Getting Started" in 1.1, it was mentioned that in algebra we use letters or **variables** to represent different numbers, and when a letter stands for just one number it is called a **constant.** If an expression consists only of numbers, it is called a **numerical expression.**

$$4 + 3 - 2 \qquad 5(9 - 6) \qquad 12 - \frac{6}{2}$$

However, when an expression consists of variables, numbers, and operation symbols it is called an **algebraic expression.** Some examples are:

$$2x + 3 \qquad \frac{7 - 6z}{5} \qquad 4x^2 - 2x + 8 \qquad \sqrt{y} + 9$$

Terms in an algebraic expression are separated by $+$ or $-$ signs. In the algebraic expression "$4x^2 - 2x + 8$" there are three terms; $4x^2$, $-2x$, and the constant 8.

In Sections A and B we will learn how to evaluate *numerical expressions* with grouping symbols and the correct order of operations. In Section C we will use those skills to help us evaluate an *algebraic expression.*

A   ### Evaluate Numerical Expressions with Grouping Symbols

In algebra and arithmetic *parentheses* ( ) are grouping symbols used to indicate which operations are to be performed first. Square brackets [ ] and braces { } are also grouping symbols; they can be used in the same manner as parentheses.

Web It

To learn about evaluating expressions containing more than one operation, go to link 1-4-1 at mhhe.com/bello.

$$4 \cdot (3 + 2), \qquad 4 \cdot [3 + 2], \qquad \text{and} \qquad 4 \cdot \{3 + 2\}$$

all mean that we must first add 3 and 2 and then multiply this sum by 4. The expressions $4 \cdot (3 + 2)$ and $(4 \cdot 3) + 2$ have *different* meanings. In the first expression we add 3 and 2 first, while in the second we multiply 4 by 3 first. Thus $4 \cdot (3 + 2) = 4 \cdot (5) = 20$, but $(4 \cdot 3) + 2 = (12) + 2 = 14$. Hence, $4 \cdot (3 + 2) \neq (4 \cdot 3) + 2$.

EXAMPLE 1 Evaluating expressions containing parentheses	**PROBLEM 1**

Evaluate:

a. $(-4 \cdot 5) + 6$ **b.** $-4 \cdot (5 + 6)$

c. $-48 \div (4 \cdot 3)$ **d.** $(-48 \div 4) \cdot 3$

Evaluate:

a. $(-6 \cdot 4) + 7$ **b.** $-6 \cdot (4 + 7)$

c. $-36 \div (3 \cdot 4)$ **d.** $(-36 \div 3) \cdot 4$

SOLUTION   Perform the operations inside the parentheses *first*.

a. $(-4 \cdot 5) + 6 = -20 + 6$ Multiply -4 and 5 first.
$$= -14$$

b. $-4 \cdot (5 + 6) = -4 \cdot 11$ Add 5 and 6 first.
$$= -44$$

c. $-48 \div (4 \cdot 3) = -48 \div 12$ Multiply 4 and 3 first.
$$= -4$$

d. $(-48 \div 4) \cdot 3 = -12 \cdot 3$ Divide -48 by 4 first.
$$= -36$$

Web It

For a lesson covering simplifying either algebraic or arithmetic expressions containing grouping symbols, try link 1-4-2 at mhhe.com/bello.

If we have more than one set of grouping symbols, we first perform the operations inside the innermost grouping symbols.

$$[2 \cdot (88 + 14)] + 12 = [2 \cdot (102)] + 12 \quad \text{Add 88 and 14 inside the parentheses first.}$$
$$= 204 + 12$$
$$= 216$$

EXAMPLE 2 Evaluating expressions containing grouping symbols	**PROBLEM 2**

Evaluate:

a. $[-5 \cdot (6 + 4)] + 9$ **b.** $[-10 \cdot (8 - 3)] - 9$

Evaluate:

a. $[-8 \cdot (7 + 2)] + 8$

b. $[-10 \cdot (9 - 6)] - 9$

SOLUTION

a. $[-5 \cdot (6 + 4)] + 9 = [-5 \cdot 10] + 9$ Add 6 and 4 first.
$$= -50 + 9$$
$$= -41$$

b. $[-10 \cdot (8 - 3)] - 9 = [-10 \cdot 5] - 9$ Subtract 3 from 8 first.
$$= -50 - 9$$
$$= -59$$

Answers

1. a. -17 **b.** -66 **c.** -3
d. -48 **2. a.** -64 **b.** -39

Web It

In some cases we have to evaluate expressions in which a bar is used to indicate division. For instance, if we wish to convert a temperature given in degrees Fahrenheit to degrees Celsius, we have to evaluate the expression

$$\frac{5 \cdot (F - 32)}{9}$$

where F represents the temperature in degrees Fahrenheit. If the temperature is 77°F, the corresponding Celsius temperature is calculated as follows:

$$\frac{5 \cdot (77 - 32)}{9} = \frac{5 \cdot (45)}{9}$$

$$= \frac{225}{9}$$

You can also do this by dividing 45 by 9 first and then multiplying the result, 5, by 5, to obtain 25.

$$= 25$$

The corresponding temperature is 25°C.

EXAMPLE 3   An application using grouping symbols

In September 1933, a freak heat flash struck the city of Coimbra, in Portugal. On this day, the temperature rose to 158°F for 120 sec. How many degrees Celsius is that?

SOLUTION   In this case $F = 158$; thus the Celsius temperature is given by

$$\frac{5 \cdot (158 - 32)}{9} = \frac{5 \cdot (126)}{9}$$

$$= \frac{630}{9}$$

You can also do this by dividing 126 by 9 first and then multiplying the result, 14, by 5, to obtain 70.

$$= 70$$

The corresponding Celsius temperature is 70°C.

PROBLEM 3

Evaluate:

$$\frac{5 \cdot (149 - 32)}{9}$$

B   The Order of Operations

Web It

If an expression does not contain parentheses or brackets, we must establish the order in which operations are to be performed. For example, the expression $6^2 + 9 \div 3$ might be evaluated in two ways:

$6^2 + 9 \div 3$	Square 6 and divide 9 by 3.		$6^2 + 9 \div 3$	Square 6.
$36 + 3$	Add.		$36 + 9 \div 3$	Add 36 + 9.
39			$45 \div 3$	Divide by 3.
			15	

Answer

3. 65

To avoid this ambiguity, we agree to perform any sequence of operations from left to right and in the following order.

Calculate It

Calculator Does Order of Operations

Calculators automatically follow the order of operations. Thus if you key in 3 × 4 + 5, the calculator multiplies 3 by 4 *first,* and then adds 5 to obtain 17 as shown. On the other hand, if you enter 3 + 4 × 5, the calculator does **not** add 3 and 4 first, but rather *multiplies* 4 by 5 first and then adds 3 to obtain 23.

```
3*4+5
                        17
3+4*5
                        23
(3+4)*5
                        35
```

As before, if you want to add 3 and 4 first, you have to use parentheses and key in (3 + 4) × 5. The result is 35 as shown.

PROCEDURE

Order of Operations

P   **1.** Do the operations inside the parentheses (or other grouping symbols) starting with the innermost grouping symbols and operations above and below fraction bars.

E   **2.** Evaluate all exponential expressions.

(MD)   **3.** Perform multiplications and divisions as they occur from left to right.

(AS)   **4.** Perform additions and subtractions as they occur from left to right.

The letters to the left of the procedure are used to help remember this order of operations. The sentence used to remember the letters is, "**P**lease **E**xcuse **M**y **D**ear **A**unt **S**ally." The multiplications and divisions are equally important so they are done as they appear from left to right in the expression. Thus to evaluate $24 \div 6 \cdot 2 + (7 - 9) - 1 + 2^3$ we let **PE(MD)(AS)** guide us.

$$24 \div 6 \cdot 2 + (7 - 9) - 1 + 2^3 = 24 \div 6 \cdot 2 + (-2) - 1 + 2^3 \quad \text{Parentheses } (7-9) = -2$$

$$= 24 \div 6 \cdot 2 + (-2) - 1 + 8 \quad \text{Exponents } 2^3 = 8$$

$$= 4 \cdot 2 + (-2) - 1 + 8 \quad \text{(MD) Divide } 24 \div 6 = 4$$

$$= 8 + (-2) - 1 + 8 \quad \text{(MD) Multiply } 4 \cdot 2 = 8$$

$$= 6 - 1 + 8 \quad \text{(AS) Add } 8 + (-2) = 6$$

$$= 5 + 8 \quad \text{(AS) Subtract } 6 + (-1) = 5$$

$$= 13 \quad \text{(AS) Add } 5 + 8 = 13$$

EXAMPLE 4   Order of operations

Evaluate:

$$-6^2 + \frac{(4-8)}{2} + 10 \div 5$$

SOLUTION

$$-6^2 + \frac{(4-8)}{2} + 10 \div 5$$

$$= -6^2 + \frac{-4}{2} + 10 \div 5 \quad \text{P Perform the operation inside the parentheses.}$$

$$= -36 + \frac{-4}{2} + 10 \div 5 \quad \text{E Evaluate } -6^2 = -36.$$

$$= -36 + (-2) + 2 \quad \text{(MD) Perform multiplications and divisions as they occur from left to right.}$$

$$= -38 + 2 \quad \text{(AS) } -36 + (-2) = -38 \quad \text{Perform additions and subtractions as they occur from left to right.}$$

$$= -36 \quad -38 + 2 = -36$$

Answer

4. -48

PROBLEM 4

Evaluate:

$$-7^2 + \frac{(2-8)}{2} + 20 \div 5$$

Teaching Tip

Emphasize the importance of doing multiplication and division as they appear left to right by the following example:

Simplify $18 \div 6 \cdot 3$

Performing multiplication and division left to right, as they should be done:

$(18 \div 6) \cdot 3$

$3 \quad \cdot 3$

$9 \quad \text{Correct}$

Performing multiplication first and then division, as they should not be done:

$18 \div (6 \cdot 3)$

$18 \div \quad 18$

$1 \quad \text{Incorrect}$

C Evaluate Algebraic Expressions

Without explanation the algebraic expression $5x$ has no real application. However, if we know that we have to ride the bus to work for 5 days and the cost of the round trip will vary, (x), depending on whether we ride the regular bus or the express bus, then 5 times x or $5x$ could represent the cost of the bus fare for the 5 days.

Suppose we decide to ride the regular bus, which costs $2 for one round trip. Then we could evaluate the cost of the bus fare for the week by using the algebraic expression $5x$ and replacing the x with $2. Thus, the cost of the bus fare for the 5 days would be 5 ($2) or $10. $10 is called the **value** of the algebraic expression. What if we choose the express bus, which costs $3 for a round trip? Then the cost of the 5-day bus fare becomes 5($3) or $15. You can see that the value of the expression $5x$ will vary depending on the number used to replace x. This is a very simple example of how to **evaluate an algebraic expression.** To evaluate the expressions in Example 5, it will require some of the skills we learned in simplifying with grouping symbols and correct order of operations.

EXAMPLE 5 Evaluating an algebraic expression

a. $(n - 2)180$ gives the sum of the measures of the angles in a polygon of n sides. Find the sum of the measures of the angles if the polygon has 5 sides.

b. Evaluate the algebraic expression

$$2x - 3(y^2 + 1)$$

if $x = -4$ and $y = 5$.

SOLUTION

a. Replace n with 5.

$$(n - 2)180 = (5 - 2)180$$
$$= (3)180 \qquad \text{Simplify inside parentheses}$$
$$= 540 \qquad \text{Multiply}$$

Thus the sum of the measures of the angles in a polygon with 5 sides is 540°.

b. Replace x with -4 and y with 5

$$2x - 3(y^2 + 1) = 2(-4) - 3(5^2 + 1)$$
$$= 2(-4) - 3(25 + 1) \qquad \text{Simplify exponent inside parentheses}$$
$$= 2(-4) - 3(26) \qquad \text{Add inside parentheses}$$
$$= -8 - 78 \qquad \text{Multiply}$$
$$= -86 \qquad \text{Subtract}$$

The value of the algebraic expression is -86.

PROBLEM 5

a. $\dfrac{5(F - 32)}{9}$ gives the Celsius temperature when the degrees Fahrenheit, F, are known. Find the Celsius temperature when $F = 50$.

b. Evaluate the algebraic expression

$$3x^2 - (y + 11) + 10 \div 2$$

if $x = -1$ and $y = 7$.

Answers

5. a. 10°C   **b.** -10

D   Use the Distributive Property to Simplify

Web It

To practice simplifying expressions containing parentheses, go to link 1-4-5 at mhhe.com/bello.

In algebra the distributive property, $a(b + c) = ab + ac$, is used to remove parentheses in expressions such as $3(x + 5)$ or $4(x - 7)$, where x is a real number. Thus

$$3(x + 5) = 3x + 3 \cdot 5 = 3x + 15$$

and

$$4(x - 7) = 4x - 4 \cdot 7 = 4x - 28$$

EXAMPLE 6 **Removing parentheses**	**PROBLEM 6**
Remove the parentheses (simplify):	Simplify:
a. $-2(x + 8)$	**a.** $-3(a + b)$
b. $0.5(7 - y)$	**b.** $0.2(6 - b)$

SOLUTION

a. $-2(x + 8) = -2x + (-2 \cdot 8)$

$\qquad\qquad\quad = -2x + (-16)$

$\qquad\qquad\quad = -2x - 16$ Recall that $a - b = a + (-b)$.

b. $0.5(7 - y) = 0.5 \cdot 7 - 0.5y$

$\qquad\qquad\quad = 3.5 - 0.5y$

Expressions of the form $-(a + b)$ or $-(a - b)$, require special consideration. We first recall the following.

Identity for Multiplication
For any real number a, $a = 1 \cdot a$.

Any real number has an additive inverse and the additive inverse of a is $-a$, so the additive inverse of $1 \cdot a$ is $-1 \cdot a$.

Additive Inverse
For any real number a, $-a = -1 \cdot a$.

Hence,

$$-(a + b) = -1 \cdot (a + b)$$

$$= -1 \cdot a + (-1 \cdot b)$$

$$= -a - b$$

Answers

6. a. $-3a - 3b$ **b.** $1.2 - 0.2b$

Teaching Tip

Here again the "–" symbol is used
as the "opposite of" $(a + b)$.

PROCEDURE

Additive Inverse of a Sum

$$-(a + b) = -a - b$$

Similarly,

$$-(a - b) = -1 \cdot (a - b)$$
$$= -1 \cdot [a + (-b)]$$
$$= (-1)(a) + (-1)(-b)$$
$$= -a + b$$

PROCEDURE

Additive Inverse of a Difference

$$-(a - b) = -a + b$$

These rules tell us that to remove the parentheses in an expression preceded by a minus sign, we *change the sign of every term inside the parentheses* or, equivalently, *multiply each term inside the parentheses by* -1.

EXAMPLE 7   **Removing parentheses**

Remove the parentheses (simplify):

a. $-(x - 2)$ **b.** $-(ab + 3)$

SOLUTION

a. $-(x - 2) = -1 \cdot (x - 2)$
$$= -1 \cdot x + (-1)(-2)$$
$$= -x + 2$$

Changing the signs inside the parentheses in $-(x - 2)$ will immediately yield

change sign

$$-(x - 2) = -x + 2$$

change sign

b. $-(ab + 3) = -1 \cdot (ab + 3) = -1 \cdot ab + -1 \cdot (3)$
$$= -ab + (-3)$$
$$= -ab - 3$$

Changing signs inside the parentheses will immediately yield the answer $-ab - 3$.

PROBLEM 7

Simplify:

a. $-(y - 6)$ **b.** $-(xy + 7)$

Answers

7. a. $-y + 6$ **b.** $-xy - 7$

Web It

For an extensive lesson on removing parentheses, read the complete link 1-4-6 at mhhe.com/bello.

For another lesson with practice problems and answers, go to 1-4-7 at mhhe.com/bello.

We can summarize this discussion by the following two facts.

PROCEDURE

Removing Parentheses

1. If the factor in front of the parentheses has no written sign, multiply each term inside the parentheses by this factor; that is,

$$a(b - c + d - e) = ab - ac + ad - ae$$

2. If the factor in front of the parentheses is preceded by a minus sign, multiply this factor by each of the terms inside the parentheses and change the sign of each of these terms; that is,

$$-a(b - c + d - e) = -ab + ac - ad + ae$$

EXAMPLE 8   Removing parentheses

Remove the parentheses (simplify):

a. $4(x - 2y + 3)$ **b.** $-5(2x + y - z)$

c. $0.4(-3x + 2y - 7z - 8)$ **d.** $0.5x(y + 3z - 5)$

SOLUTION

a. $4(x - 2y + 3) = 4x - 8y + 12$

b. $-5(2x + y - z) = -10x - 5y + 5z$

c. $0.4(-3x + 2y - 7z - 8) = -1.2x + 0.8y - 2.8z - 3.2$

d. $0.5x(y + 3z - 5) = 0.5xy + 1.5xz - 2.5x$

PROBLEM 8

Simplify:

a. $5(a - 3b + 4)$

b. $-3(4x + y - z)$

c. $0.5(-2x + 3y - 6z - 4)$

d. $0.3x(y + 2z - 4)$

NOTE

There are four cases of removing parentheses to remember.

1. $(x + 2) = x + 2$ Drop ( ).

2. $-(x + 2) = -x - 2$ Change signs.

3. $3(x + 2) = 3x + 6$ Distribute 3.

4. $-3(x + 2) = -3x - 6$ Distribute -3.

 **Combining Like Terms**

Suppose we wish to simplify $3x + 2(x + 5)$. We start by simplifying $2(x + 5)$, to obtain

$$3x + 2(x + 5) = 3x + 2x + 10$$

The terms $3x$ and $2x$ are called *like terms*. They differ only in their numerical parts (coefficients). Similarly, $-3y$ and $5y$ are like terms, and $9z^2$ and $-3z^2$ are like terms. In general, we have the following definition.

Answers

8. a. $5a - 15b + 20$
b. $-12x - 3y + 3z$
c. $-x + 1.5y - 3z - 2$
d. $0.3xy + 0.6xz - 1.2x$

Web It

For an extensive lesson dealing with combining like terms, try 1-4-8 at mhhe.com/bello.

Teaching Tip

Mention that terms are separated by plus or minus signs. Give students the following examples to determine how many terms are in each expression, and identify each term:

$2xy$ (one)
$2x^3 + y$ (two; $2x^3$, y)
$\dfrac{2x}{y} + 3$ (two; $\dfrac{2x}{y}$, 3)
$4x^2 + 7x - 1$ (three; $4x^2$, $7x$, -1)

Calculate It

Verify Simplifications

We can check the results of Example 9(c) by making a picture (graph) of the expression $5x - 2(x + 1) + (x + 3)$ and the simplified version $4x + 1$. As before, use a standard window and graph

$Y_1 = 5x - 2(x + 1) + (x + 3)$ and $Y_2 = 4x + 1$

The two pictures (graphs) are the same, so

$\quad 5x - 2(x + 1) + (x + 3)$
$\quad\quad = 4x + 1$

[With a TI-83 Plus you can confirm that there are two graphs in the window (y_1 and y_2) by pressing TRACE and the ◁ and ▷ keys. The equations at the top should change to indicate which graph, Y_1 or Y_2, you are currently viewing.]

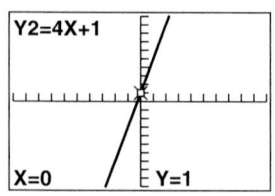

The graphs of Y_1 and Y_2 are identical, so our answer is correct.

LIKE TERMS

Constant terms or terms with exactly the same variable factors are called **similar** or **like** terms.

NOTE

Like terms differ only in their *numerical* coefficients (the numbers being multiplied by the variables).

We can *combine* like terms by using a variation of the distributive property. As you recall, the distributive property states that, for any real numbers a, b, and c,

$$a(b + c) = ab + ac$$

and

$$a(b - c) = ab - ac$$

Using the commutative property of multiplication, we can rewrite the two distributive properties as follows.

PROCEDURE

Distributive Properties for Like Terms

$$(b + c)a = ba + ca$$
$$(b - c)a = ba - ca$$

Now

$$3x + 2x = (3 + 2)x = 5x$$

Similarly,

$$7z^2 - 5z^2 = (7 - 5)z^2 = 2z^2$$

and

$$8xy + 3xy - 2xy = (8 + 3 - 2)xy$$
$$= 9xy$$

$x + x = 1 \cdot x + 1 \cdot x = (1 + 1)x = 2x$. The coefficient of x is understood to be 1. Also, if an expression within parentheses is preceded by a plus sign, we can remove the parentheses and combine any like terms. Using the commutative and associative properties,

$$3x + (2 + 5x) = 3x + 2 + 5x$$
$$= 8x + 2$$

NOTE

You can combine like terms by adding or subtracting their coefficients.

Use this idea in the next example.

EXAMPLE 9 **Combining like terms**	**PROBLEM 9**

Simplify:

a. $5x + 2(x - 4)$ **b.** $-3(x + 5) - 2x$ **c.** $5x - 2(x + 1) + (x + 3)$

SOLUTION

a. $5x + 2(x - 4) = 5x + 2x - 8$

$\qquad\qquad\qquad = 7x - 8$

b. $-3(x + 5) - 2x = -3x - 15 - 2x$

$\qquad\qquad\qquad = -5x - 15$ $-3x - 2x = (-3 - 2)x = -5x$

c. $5x - 2(x + 1) + (x + 3) = 5x - 2x - 2 + x + 3$

$\qquad\qquad\qquad\qquad = 4x - 2 + 3$ $5x - 2x + x = 4x$

$\qquad\qquad\qquad\qquad = 4x + 1$

PROBLEM 9

Simplify:

a. $8y + 3(y - 4)$

b. $-4(y + 2) - 3y$

c. $6y - 3(y + 2) + (y + 8)$

Removing Other Grouping Symbols

To avoid confusion when parentheses occur within other parentheses, we do not write $((x + 5) + 3)$. Instead, we may use a different grouping symbol, the brackets [ ], and write $[(x + 5) + 3]$. To simplify (combine like terms) in such expressions, the innermost grouping symbols are removed first. This procedure is illustrated in the next example.

EXAMPLE 10 **Removing other grouping symbols**	**PROBLEM 10**

Remove the grouping symbols and simplify:

$$[(4x^2 - 1) + (2x + 5)] - [(x - 2) + (3x^2 - 3)]$$

SOLUTION

We first remove the innermost parentheses and then add like terms. Thus

$[(4x^2 - 1) + (2x + 5)] - [(x - 2) + (3x^2 - 3)]$

$= [4x^2 - 1 + 2x + 5] - [x - 2 + 3x^2 - 3]$ Remove parentheses.

$= [4x^2 + 2x + 4] - [3x^2 + x - 5]$ Combine like terms.

$= 4x^2 + 2x + 4 - 3x^2 - x + 5$ Multiply by -1 and remove brackets.

$= x^2 + x + 9$ Combine like terms.

PROBLEM 10

Remove grouping symbols and simplify:

$[(2x^2 - 3) + (3x + 1)]$
$\quad - [(x + 1) + (x^2 - 2)]$

Calculate It Exercises

Check the results of the following examples with your calculator:

1. Example 4.

2. Example 5(a).

3. Examples 9(a) and 9(b) (see Calculate It p. 54)

4. Example 10.

5. Why can't we check the results of Example 7(b) with a calculator?

Answers

9. a. $11y - 12$ **b.** $-7y - 8$
c. $4y + 2$ **10.** $x^2 + 2x - 1$

Exercises 1.4

A In Problems 1–20, evaluate the expression.

1. a. $(-10 \cdot 3) + 4$ -26

   b. $-10 \cdot (3 + 4)$ -70

2. a. $(6 \cdot 4) + 6$ 30

   b. $6 \cdot (4 + 6)$ 60

3. a. $(36 \div 4) \cdot 3$ 27

   b. $36 \div (4 \cdot 3)$ 3

4. a. $(-28 \div 7) \cdot 2$ -8

   b. $-28 \div (7 \cdot 2)$ -2

5. $[-5 \cdot (8 + 2)] + 3$ -47

6. $[7 \cdot (4 + 3)] + 1$ 50

7. $-7 + [3 \cdot (4 + 5)]$ 20

8. $-8 + [3 \cdot (4 + 1)]$ 7

9. $[-6 \cdot (4 - 2)] - 3$ -15

10. $[-2(7 - 5)] - 8$ -12

11. $3 - [8 \cdot (5 - 3)]$ -13

12. $7 - [3(4 - 5)]$ 10

13. $-8[3 - 2(4 + 1)] + 1$ 57

14. $6[7 - 2(5 - 7)] - 2$ 64

15. $48 \div \{4(8 - 2[3 - 1])\}$ 3

16. $-96 \div \{4(8 - 2[1 - 3])\}$ -2

17. $\left[\dfrac{9 - (-3)}{8 - 6}\right]\left[\dfrac{3 + (-8)}{7 - 2}\right]$ -6

18. $\left[\dfrac{6 + (-2)}{3 + (-7)}\right]\left[\dfrac{8 + (-12)}{2 - 4}\right]$ -2

19. $\dfrac{3 - 5\left(\dfrac{4 + 2}{2 + 1}\right) - 2}{-4 + 3\left(\dfrac{4 - 2}{4 - 6}\right) - 2}$ 1

20. $\dfrac{8 + 2\left(\dfrac{9 - 15}{3 - 1}\right) - 2}{-4 + 8\left(\dfrac{6 - 3}{1 - 4}\right) + 12}$ Undefined

B In Problems 21–40, use the correct order of operations and simplify.

21. $-5 \cdot 6 - 6$ -36

22. $-5 \cdot 2 - 2$ -12

23. $-7 \cdot 3 \div 3 - 3$ -10

24. $-36 \cdot 2 \div 18 - 4$ -8

25. $(-20 - 5 + 3 \div 3) \div 6$ -4

26. $(-10 - 2 + 10 \div 5) \cdot 4$ -40

27. $\dfrac{8 + (-3)}{5} - 1$ 0

28. $\dfrac{7 + (-3)}{2} - 4$ -2

29. $\dfrac{4 \cdot (6 - 2)}{-8} - \dfrac{6}{-2}$ 1

30. $\dfrac{5 \cdot (6 - 2)}{-4} - \dfrac{16}{-4}$ -1

31. $4 \div 2 + 3 - 5^2$ -20

32. $8 \div 4 + 7 - 2^2$ 5

33. $4 + 6 \cdot 4 \div 2 - 2^3$ 8

34. $6 + 6 \div 3 - 3^3$ -19

35. $-5^2 + \dfrac{2 - 10}{4} + 12 \div 4$ -24

36. $-4^2 + \dfrac{3 - 7}{2} + 18 \div 9$ -16

37. $-3^3 + 4 - 6 \cdot 8 \div 4 - \dfrac{8 - 2}{-3}$ -33

38. $-2^3 + 6 - 6 \div 3 \cdot 2 - \dfrac{9 - 3}{-6}$ -5

39. $4 \cdot 9 \div 3 \cdot 10^3 - 2 \cdot 10^2$ $11{,}800$

40. $5 \cdot 8 \div 4 \cdot 10^3 - 2 \cdot 10^2$ 9800

C In Problems 41–44, evaluate the algebraic expressions.

41. $2(l + w)$ gives the perimeter of a rectangle. Find the value of the perimeter if $l = 12\frac{1}{2}$ and $w = 6$ 37

42. $1.8C + 32$ gives the temperature in degrees Fahrenheit. Evaluate the temperature when $C = 15$. 59

43. $P(1 + r)$ gives the amount compounded annually for one year. Find the value of the amount if $P = \$1000$ and $r = 5\%$. $\$1050$

44. Pe^{rt} gives the amount when continuous interest is computed. Find the amount if $P = \$2000$, $e = 3$, $r = 100\%$, and $t = 2$. $\$18{,}000$

In Problems 45–50, evaluate the algebraic expressions for the given values of the variables.

45. $2x - y$; for $x = 4$, $y = -5$ 13

46. $a - 3b$; for $a = -7$, $b = 20$ -67

47. $t^2 - 5t + 8$; for $t = -3$ 32

48. $-z^2 + z - 1$; for $z = 2$ -3

49. $5n^2 + 2(n - m)$; for $n = -4$, $m = 6$ 60

50. $5(p^2 - r) + p \div 5$; for $p = 10$, $r = 50$ 252

D In Problems 51–80, remove the parentheses (simplify).

51. $4(x - y)$ $4x - 4y$

52. $3(a - b)$ $3a - 3b$

53. $-9(a - b)$ $-9a + 9b$

54. $-6(x - y)$ $-6x + 6y$

55. $0.3(4x - 2)$ $1.2x - 0.6$

56. $0.2(3a - 9)$ $0.6a - 1.8$

57. $-\left(\dfrac{3a}{2} - \dfrac{6}{7}\right)$ $-\dfrac{3a}{2} + \dfrac{6}{7}$ or $\dfrac{-21a + 12}{14}$

58. $-\left(\dfrac{2x}{3} - \dfrac{1}{5}\right)$ $-\dfrac{2x}{3} + \dfrac{1}{5}$ or $\dfrac{-10x + 3}{15}$

59. $-(2x - 6y)$ $-2x + 6y$

60. $-(3a - 6b)$ $-3a + 6b$

61. $-(2.1 + 3y)$ $-2.1 - 3y$

62. $-(5.4 + 4b)$ $-5.4 - 4b$

63. $-4(a + 5)$ $-4a - 20$

64. $-6(x + 8)$ $-6x - 48$

65. $-x(6 + y)$ $-6x - xy$

66. $-y(2x + 3)$ $-2xy - 3y$

67. $-8(x - y)$ $-8x + 8y$

68. $-9(a - b)$ $-9a + 9b$

69. $-3(2a - 7b)$ $-6a + 21b$

70. $-4(3x - 9y)$ $-12x + 36y$

71. $0.5(x + y - 2)$
$0.5x + 0.5y - 1$

72. $0.8(a + b - 6)$ $0.8a + 0.8b - 4.8$

73. $-\dfrac{6}{5}(a - b + 5)$ $-\dfrac{6}{5}a + \dfrac{6}{5}b - 6$

74. $-\dfrac{2}{3}(x - y + 4)$ $-\dfrac{2}{3}x + \dfrac{2}{3}y - \dfrac{8}{3}$

75. $-2(x - y + 3z + 5)$
$-2x + 2y - 6z - 10$

76. $-4(a - b + 2c + 8)$
$-4a + 4b - 8c - 32$

77. $-0.3(x + y - 2z - 6)$
$-0.3x - 0.3y + 0.6z + 1.8$

78. $-0.2(a + b - 3c - 4)$
$-0.2a - 0.2b + 0.6c + 0.8$

79. $-\dfrac{5}{2}(a - 2b + c + 2d - 2)$
$-\dfrac{5}{2}a + 5b - \dfrac{5}{2}c - 5d + 5$

80. $-\dfrac{4}{7}(2a - b + 3c + 7d - 7)$
$-\dfrac{8}{7}a + \dfrac{4}{7}b - \dfrac{12}{7}c - 4d + 4$

E In Problems 81–95, remove the parentheses and combine like terms.

81. $6x + 3(x - 2)$ $9x - 6$

82. $8y + 6(y - 3)$ $14y - 18$

83. $-4(x + 2) - 5x$ $-9x - 8$

84. $-5(x + 3) - 6x$
$-11x - 15$

85. $(5L - 3W) - (W - 6L)$
$11L - 4W$

86. $(2ab - 2ac) - (ab - 4ac)$
$ab + 2ac$

87. $5x - (8x + 1) + (x + 1)$ $-2x$

88. $3x - (7x + 2) + (x + 2)$ $-3x$

89. $\dfrac{2x}{9} - \left(\dfrac{x}{9} - 2\right)$ $\dfrac{x}{9} + 2$

90. $\dfrac{5x}{7} - \left(\dfrac{2x}{7} - 3\right)$ $\dfrac{3x}{7} + 3$

91. $4a - (a + b) + 3(b + a)$ $6a + 2b$

92. $8x - 3(x + y) - (x - y)$ $4x - 2y$

93. $7x - 3(x + y) - (x + y)$ $3x - 4y$

94. $4(b - a) + 3(b + a) - 2(a + b)$ $-3a + 5b$

95. $-(x + y - 2) + 3(x - y + 6) - (x + y - 16)$ $x - 5y + 36$

F In Problems 96–105, remove the grouping symbols and simplify.

96. $[(a^2 - 4) + (2a^3 - 5)] + [(4a^3 + a) + (a^2 + 9)]$
$6a^3 + 2a^2 + a$

97. $(x^2 + 7 - x) + [-2x^3 + (8x^2 - 2x) + 5]$
$-2x^3 + 9x^2 - 3x + 12$

98. $[(0.4x - 7) + 0.6x^2] - [(0.3x^2 - 2) - 0.8x]$
$0.3x^2 + 1.2x - 5$

99. $\left[\left(\dfrac{5}{7}x^2 + \dfrac{1}{5}x\right) - \dfrac{1}{8}\right] - \left[\left(\dfrac{3}{7}x^2 - \dfrac{3}{5}x\right) + \dfrac{5}{8}\right]$
$\dfrac{2}{7}x^2 + \dfrac{4}{5}x - \dfrac{3}{4}$

100. $[3(x + 2) - 10] + [5 + 2(5 + x)]$ $5x + 11$

101. $[3(2a - 4) + 5] - [2(a - 1) + 6]$ $4a - 11$

102. $[6(a - b) + 2a] - [3b - 4(a - b)]$ $12a - 13b$

103. $[4a - (3 + 2b)] - [6(a - 2b) + 5a]$
$-7a + 10b - 3$

104. $-[-(x + y) + 3(x - y)] - [4(x + y) - (3x - 5y)]$
$-3x - 5y$

105. $-[-(0.2x + y) + 3(x - y)] - [2(x + 0.3y) - 5]$
$-4.8x + 3.4y + 5$

SKILL CHECKER

Try the *Skill Checker Exercises* so you'll be ready for the next section.

Find:

106. $[(-3)(-3)](-3)$ -27

107. $[(-2)(-2)][(-2)(-2)]$ 16

108. $(-2)[(-2)(-2)]$ -8

109. $\dfrac{(-2)(-2)(-2)}{(-2)(-2)}$ -2

110. $\dfrac{(-3)(-3)(-3)(-3)}{(-3)(-3)}$ 9

USING YOUR KNOWLEDGE

Average Velocity, Momentum, and Kinetic Energy

The distributive property is helpful in solving problems in many areas. Use your knowledge of the distributive property to remove the parentheses in Problems 111–114.

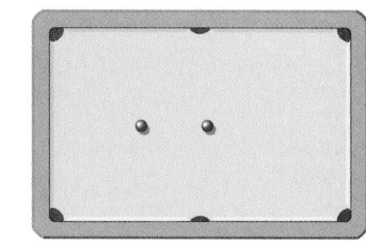

111. If your car is accelerating at a constant rate, and v_1 is the initial velocity and v_2 is the final velocity, the *average* velocity is

$$v_a = \frac{1}{2}(v_1 + v_2) \quad v_a = \frac{1}{2}v_1 + \frac{1}{2}v_2$$

112. The momentum M of a billiard ball is the product of its mass m and its velocity v. If two billiard balls of equal mass m and moving in the same straight line, collide with velocities v_1 and v_2, respectively, the total momentum M is given by

$$M = m(v_1 + v_2) \quad M = mv_1 + mv_2$$

113. The total kinetic energy (KE) of the billiard balls in Problem 112 is given by

$$\text{KE} = \frac{1}{2}m(v_1^2 + v_2^2) \quad \text{KE} = \frac{1}{2}mv_1^2 + \frac{1}{2}mv_2^2$$

114. The length of a belt L needed to connect two pulleys of radius r_1 and r_2, respectively, with centers d units apart is

$$L = \pi(r_1 + r_2) + 2d \quad L = \pi r_1 + \pi r_2 + 2d$$

WRITE ON

115. Explain why $(32 \div 4) \cdot 2$ is different from $32 \div (4 \cdot 2)$. Answers may vary.

116. Write in your words the definition for "like terms." Answers may vary.

MASTERY TEST

If you know how to do these problems, you have learned your lesson!

Evaluate:

117. $-3^2 + \left(\dfrac{4 - 8}{2}\right) + 48 \div 6$ -3

118. $\left(\dfrac{2 - 14}{6}\right) + 16 \div 4 - 5^2$ -23

Simplify:

119. $-2(x - 3y + 2z - 4)$ $-2x + 6y - 4z + 8$

120. $-\dfrac{2}{3}(a + 6b - 9c - 12)$ $-\dfrac{2}{3}a - 4b + 6c + 8$

121. $\dfrac{3}{8}x - \left(\dfrac{x}{8} - 5\right)$ $\dfrac{1}{4}x + 5$

122. $\dfrac{5}{7}x - \left(4 - \dfrac{5}{7}x\right)$ $\dfrac{10}{7}x - 4$

123. $[(a^2 - 5) + (2a^3 - 3)] - [(4a^3 + a) - (a^2 - 9)]$
   $-2a^3 + 2a^2 - a - 17$

124. $\left[\left(\dfrac{5}{9}x^2 + \dfrac{1}{5}x\right) - \dfrac{1}{3}\right] - \left[\left(\dfrac{2}{9}x^2 - \dfrac{2}{5}x\right) - \dfrac{4}{3}\right]$
   $\dfrac{1}{3}x^2 + \dfrac{3}{5}x + 1$

Evaluate:

125. $\left[\dfrac{8-(-4)}{8-10}\right]\left[\dfrac{5+(-9)}{7-3}\right]$ 6

126. $\left[\dfrac{7+(-4)}{4-7}\right]\left[\dfrac{9+(-14)}{3-5}\right]$ $-\dfrac{5}{2}$

COLLABORATIVE LEARNING 1A

The real numbers we studied in this chapter are used in many day-to-day applications. For instance, when deciding how much furniture to buy for a room and how to arrange it, it would be useful to be able to make a scale drawing. A scale drawing is a drawing that represents a real object. The scale of the drawing is the ratio of the size of the drawing to the actual size of the object. For example, you might let 1 inch = 1 foot. It would be helpful to use grid paper when making a scale drawing. Measurements can be in the English system or in the metric system. The United States has not totally changed to the metric system, so it is important to be able to measure in both systems.

Divide into groups and perform the following tasks. Have one person from each group present the results to the class.

1. Measure the dimensions of your classroom to the nearest inch and then measure it again to the nearest centimeter.

2. Measure each desk and cabinet in your classroom first to the nearest inch and then again to the nearest centimeter.

3. Using grid paper, make two scale drawings of your classroom, one in the English system and one in the metric system. The drawing should include the placement of the desks and any cabinets in the room.

After listening to each group's results and viewing their scale drawings, determine the following.

1. Were there any differences in the measurements when comparing each group's drawings? If so, discuss how the differences might occur.

2. Of the English and metric scales used, which scale did you like the best and why?

COLLABORATIVE LEARNING 1B

Even pranksters have to know their math. The giant Rubik's cube in the "Getting Started" from 1.3 took a lot of planning and math to accomplish. Let's see if you can do the math. Remember the cube is 8 ft by 8 ft by 8 ft. The colors of the Rubik's cube are red, orange, yellow, green, blue, and white. Each face of the cube should be one color and contain nine squares with a 2-inch black strip separating the squares. It was decided to use cardboard to cover each face of the cube and duct tape to make the black strip separating the squares. Working in groups, divide the following questions so that each group answers a share of the questions. Then the groups should collaborate to see how much the prank will cost.

1. Find the cost of the cardboard.

    a. How much cardboard is necessary to cover the six faces of the cube?

    b. Research the price and sizes of cardboard sheets. Make a table of the results.

    c. Use the results from **a** and **b** to choose the most cost-effective size cardboard sheets to buy and how many sheets will be required. Then, compute the total cost for the cardboard. Remember to include shipping if you decide to purchase online.

2. Find the cost of the paint.

    a. How many square feet of paint will be needed to cover each face of the cube?

    b. Research the price of paint by the various size containers (pints, quarts, gallons) and how many square feet each size will cover. Make a table of the results.

    c. Use the results from **a** and **b** to choose the most economical size to buy and how many containers of each paint color is necessary. Then, compute the total cost for the paint.

3. Find the cost of the duct tape.

    a. Two-inch duct tape can be used to make the strip separating the nine squares on each face. How many linear feet of duct tape are necessary for the strips on the entire cube?

    b. Research the price of 2-inch black duct tape and the various amounts contained in the roll. Make a table of the results.

    c. Use the results from **a** and **b** to decide what size roll of duct tape to buy and how many rolls are needed. Then, compute the total cost for the duct tape.

4. What is the most economical cost of the prank?

Research Questions

1. There is a charming story about the long-accumulated used wooden tally sticks mentioned in the chapter preview. Find out how their disposal literally resulted in the destruction of the old Houses of Parliament in England.

2. Write a paper detailing the Egyptian number system and the base and symbols used, and enumerate the similarities and differences between the Egyptian and our (Hindu-Arabic) system of numeration.

3. Write a paper detailing the Greek number system and the base and symbols used, and enumerate the similarities and differences between the Greek and our system of numeration.

4. Find out about the development of the symbols we use in our present numeration system. Where was the symbol for zero invented and by whom?

5. When were negative numbers introduced, by whom were they introduced, and what were they first called?

6. Write a short paper about the Rhind, or Ahmes, papyrus. What is the significance of the names *Rhind* and *Ahmes?* What is the content of the papyrus and who discovered it?

7. Find out what "gematria" (not geometry!) is, the significance of 666, and the reason why many old editions of the Bible substitute the number 99 for *amen* at the end of a prayer.

Summary

SECTION	ITEM	MEANING	EXAMPLE
1.1A	Empty or null set $\varnothing$	The set containing no elements	The set of natural numbers between 5 and 6 is $\varnothing$.
	Natural numbers	$N = \{1, 2, 3, \ldots\}$	2, 76, and 308 are natural numbers.
	Whole numbers	$W = \{0, 1, 2, \ldots\}$	0, 8, and 93 are whole numbers.
	Integers	$I = \{\ldots, -2, -1, 0, 1, 2, \ldots\}$	-7 and 23 are integers.
	Rational numbers	$Q = \left\{ r \mid r = \dfrac{a}{b}, a \text{ and } b \text{ are integers,} \text{ and } b \neq 0 \right\}$	$\dfrac{1}{5}, -\dfrac{2}{3}, 0, 9, 1.4,$ and $0.\overline{3}$ are rational numbers.
1.1B	Rational numbers	The set Q of rational numbers is the same as the set of terminating or repeating decimals.	0.345 and $0.\overline{3}$ are rational numbers.
	Irrational numbers	$H = \{x \mid x \text{ is not rational}\}$	$\sqrt{2}$ and π are irrational numbers.
	Real numbers (R)	The set of all rationals and irrationals	$\dfrac{2}{7}, -\dfrac{2}{3}, 0, 9, 1.4, 0.\overline{3}, \sqrt{2},$ and π are real numbers.
1.1C	$N \subseteq W \subseteq I \subseteq Q \subseteq R$	N is a subset of W, W is a subset of I, and so on.	Every natural number is a whole number, every whole number is an integer, and so on.
1.1D	Additive inverses (opposites)	a and $-a$ are additive inverses.	8 and -8 are additive inverses.
1.1E	Absolute value $\lvert a \rvert$	The distance from 0 to a on the number line. $\lvert a \rvert = \begin{cases} a & \text{when } a \geq 0 \\ -a & \text{when } a < 0 \end{cases}$	$\lvert -8 \rvert = 8, \left\lvert \dfrac{2}{3} \right\rvert = \dfrac{2}{3},$ and $\lvert -0.4 \rvert = 0.4$
1.1F	Trichotomy law	If a and b are real numbers, then 1. $a = b$, or 2. $a < b$, or 3. $a > b$	
1.2A	Adding signed numbers with the same sign	Add their absolute values and give the sum the common sign.	$-3 + (-7) = -(\lvert -3 \rvert + \lvert -7 \rvert)$ $= -10$
	Adding signed numbers with different signs	Subtract the smaller absolute value from the greater absolute value and use the sign of the number with the greater absolute value.	$3 + (-5) = -(5 - 3) = -2$ $-7 + 9 = +(9 - 7) = 2$
	Subtraction	If a and b are real numbers, $a - b = a + (-b)$.	$3 - (-4) = 3 + 4 = 7$
	$a \cdot b, ab, (a)(b), a(b), (a)b$	The product of a and b	$5 \cdot 2 = (5)(2) = 5(2) = (5)2 = 10$
	Multiplying signed numbers with different signs	Multiply their absolute values; the product is negative.	$3 \cdot (-4) = -12$ $-7 \cdot 2 = -14$

(Continued)

SECTION	ITEM	MEANING	EXAMPLE
1.2A	Multiplying signed numbers with the same sign	Multiply their absolute values; the product is positive.	$3 \cdot 8 = 24$ and $(-9)(-2) = 18$
	Multiplication of fractions	$\dfrac{a}{b} \cdot \dfrac{c}{d} = \dfrac{a \cdot c}{b \cdot d}$ $\quad (b \neq 0, d \neq 0)$	$\left(-\dfrac{3}{4}\right) \cdot \dfrac{2}{7} = -\dfrac{3}{14}$
	Division	If a, b, c are real numbers, $\dfrac{a}{b} = c$ means $a = bc$ $\quad (b \neq 0)$	$\dfrac{6}{3} = 2$ means $6 = 3 \cdot 2$.
	Dividing signed numbers	The quotient of two real numbers with the same sign is positive, with different signs, negative.	$\dfrac{6}{-3} = -2, \dfrac{-6}{3} = -2,$ and $\dfrac{-8}{-2} = 4$
	Zero in division problems	$\dfrac{0}{a} = 0 \quad (a \neq 0)$	$\dfrac{0}{9} = 0, \dfrac{0}{-8} = 0,$ and $\dfrac{0}{-2.4} = 0$
		$\dfrac{a}{0}$ is not defined.	$\dfrac{9}{0}, \dfrac{-8}{0},$ and $\dfrac{-2.4}{0}$ are not defined.
		$\dfrac{0}{0}$ is indeterminate.	
	Reciprocal	The reciprocal of a is $\dfrac{1}{a}$. $\quad (a \neq 0)$	The reciprocal of -7 is $-\dfrac{1}{7}$.
	Division of fractions	$\dfrac{a}{b} \div \dfrac{c}{d} = \dfrac{a}{b} \cdot \dfrac{d}{c}$ $\quad (b \neq 0, c \neq 0, d \neq 0)$	$\dfrac{3}{4} \div \dfrac{9}{2} = \dfrac{3}{4} \cdot \dfrac{2}{9} = \dfrac{1}{6}$

1.2B Real-Number Properties

If a, b, and c are real numbers:

Name	Addition	Multiplication
Closure	$a + b$ is a real number.	$a \cdot b$ is a real number.
Commutative	$a + b = b + a$	$a \cdot b = b \cdot a$
Associative	$a + (b + c) = (a + b) + c$	$a \cdot (b \cdot c) = (a \cdot b) \cdot c$
Identity	$a + 0 = 0 + a = a$	$1 \cdot a = a \cdot 1 = a$
	(0 is the identity.)	(1 is the identity.)
Inverse	For each real number a, there is a unique real number $-a$ such that $a + (-a) = -a + a = 0$.	For each nonzero real number a, there is a unique real number $\frac{1}{a}$ such that
	$(-a$ is called the opposite of a.)	$a \cdot \dfrac{1}{a} = \dfrac{1}{a} \cdot a = 1$
		$\left(\dfrac{1}{a}$ is called the reciprocal of a.$\right)$

Distributive property of multiplication over addition	$a(b + c) = ab + ac$	
Distributive property of multiplication over subtraction	$a(b - c) = ab - ac$	

1.3A	Exponent	$a^n = a \cdot a \cdot a \cdots a$ (n factors) n is called the exponent.	$3^4 = 3 \cdot 3 \cdot 3 \cdot 3$
	Base	In the expression a^n, a is called the base.	In the expression 2^5, 2 is the base.

SECTION	ITEM	MEANING	EXAMPLE
1.3B	Negative exponent	$a^{-n} = \dfrac{1}{a^n} \quad (a \neq 0)$	$5^{-2} = \dfrac{1}{5^2} = \dfrac{1}{25}$
	Zero exponent	$a^0 = 1 \quad (a \neq 0)$	$2^0 = 1$ and $\left(\dfrac{-1}{4}\right)^0 = 1$
1.3C	First law of exponents	$a^m \cdot a^n = a^{m+n} \quad (a \neq 0)$	$x^5 \cdot x^4 = x^9$
	Second law of exponents	$\dfrac{a^m}{a^n} = a^{m-n} \quad (a \neq 0)$	$\dfrac{x^8}{x^3} = x^5$
1.3D	Third law of exponents	$(a^m)^n = a^{m \cdot n} \quad (a \neq 0)$	$(x^2)^5 = x^{10}$
	Raising products to powers	$(a^m b^n)^k = a^{m \cdot k} b^{n \cdot k}$ $(a \neq 0, b \neq 0)$	$(x^3 y^5)^6 = x^{3 \cdot 6} y^{5 \cdot 6} = x^{18} y^{30}$
1.3E	Raising a quotient to a power	$\left(\dfrac{a^m}{b^n}\right)^k = \dfrac{a^{m \cdot k}}{b^{n \cdot k}} \quad (a \neq 0, b \neq 0)$	$\left(\dfrac{x^5}{y^3}\right)^4 = \dfrac{x^{5 \cdot 4}}{y^{3 \cdot 4}} = \dfrac{x^{20}}{y^{12}}$
1.3F	Scientific notation	A number is in scientific notation when it is written in the form $m \times 10^n$, where m is greater than or equal to 1 and less than 10 and n is an integer.	$352 = 3.52 \times 10^2$ is in scientific notation.
1.4B	Order of operations (from left to right)	**P** Parentheses (or other grouping symbols, including division bars) **E** Exponentiation **(MD)** Multiplication Division **(AS)** Addition Subtraction	$12 \div 6 \cdot 2 + 3^2 - (4 + 5)$ $= 12 \div 6 \cdot 2 + 3^2 - 9$ **Parentheses** $= 12 \div 6 \cdot 2 + 9 - 9$ **Exponents** $= 2 \cdot 2 + 9 - 9$ **(MD)** Divide $= 4 + 9 - 9$ Multiply $= 13 - 9$ **(AS)** Add $= 4$ Subtract
1.4D	$-1 \cdot a$ $-(a + b)$ $-(a - b)$	$-1 \cdot a = -a$ $-(a + b) = -a - b$ $-(a - b) = -a + b$	$-1 \cdot 4 = -4$ $-(x + 7) = -x - 7$ $-(x - 3) = -x + 3$
1.4E	Similar or like terms	Two or more terms that differ only in their numerical coefficients	$-3a$ and $7a$ are similar or like terms.

Review Exercises

(If you need help with these exercises, look in the section indicated in brackets.)

1. [1.1A] Use roster notation to list the natural numbers that fall between

   a. 3 and 9 $\{4, 5, 6, 7, 8\}$

   b. 4 and 8 $\{5, 6, 7\}$

2. [1.1B] Write as a decimal:

   a. $\dfrac{1}{5}$ 0.2   **b.** $\dfrac{2}{5}$ 0.4

3. [1.1B] Write as a decimal:

   a. $\dfrac{1}{9}$ $0.\overline{1}$ **b.** $\dfrac{2}{9}$ $0.\overline{2}$

4. [1.1C] Classify the given number by making a check mark (✓) in the appropriate row(s).

Set	0.3	0	$\frac{-3}{4}$	−5	$\sqrt{3}$
Natural numbers					
Whole numbers		✓			
Integers		✓		✓	
Rational numbers	✓	✓	✓	✓	
Irrational numbers					✓
Real numbers	✓	✓	✓	✓	✓

5. [1.1D] Find the additive inverse:

   a. -3.5 3.5   **b.** $\dfrac{3}{4}$ $-\dfrac{3}{4}$

6. [1.1E] Find:

   a. $|-9|$ 9   **b.** $|4.2|$ 4.2

7. [1.1E] Find:

   a. $\left|-\dfrac{1}{8}\right|$ $\dfrac{1}{8}$ **b.** $|0.\overline{4}|$ $0.\overline{4}$

8. [1.1F] Fill in the blank with $<$, $>$, or $=$ to make the resulting statement true:

   a. $-8 \underline{\ <\ } -7$ **b.** $-4 \underline{\ <\ } -3$

9. [1.1F] Fill in the blank with $<$, $>$, or $=$ to make the resulting statement true:

   a. $\dfrac{1}{4} \underline{\ >\ } \dfrac{1}{5}$ **b.** $\dfrac{3}{4} \underline{\ =\ } 0.75$

10. [1.1F] Fill in the blank with $<$, $>$, or $=$ to make the resulting statement true:

   a. $\dfrac{1}{5} \underline{\ <\ } 0.25$ **b.** $0.\overline{6} \underline{\ =\ } \dfrac{2}{3}$

11. [1.2A] Find:

   a. $-3 + (-8)$ -11 **b.** $-5 + 2$ -3

12. [1.2A] Find:

   a. $\dfrac{1}{7} - \dfrac{3}{7}$ $-\dfrac{2}{7}$ **b.** $-0.2 - 0.4$ -0.6

13. [1.2A] Find:

   a. $8 - (-4)$ 12   **b.** $-3 - (-7)$ 4

14. [1.2A] Find:

   a. $\dfrac{3}{4} - \left(-\dfrac{1}{5}\right)$ $\dfrac{19}{20}$ **b.** $\dfrac{5}{6} - \left(-\dfrac{1}{4}\right)$ $\dfrac{13}{12}$

15. [1.2A] Find:

   a. $9 \cdot (-4)$ -36 **b.** $-2.4 \cdot 6$ -14.4

16. [1.2A] Find:

   a. $\left(-\dfrac{3}{4}\right) \cdot \dfrac{7}{8}$ $-\dfrac{21}{32}$ **b.** $\left(-\dfrac{5}{6}\right) \cdot \left(-\dfrac{2}{7}\right)$ $\dfrac{5}{21}$

17. [1.2A] Find:

   a. $\dfrac{0}{7}$ 0   **b.** $\dfrac{8}{0}$ Undefined

18. [1.2A] Find the reciprocal:

   a. $-\dfrac{3}{5}$ $-\dfrac{5}{3}$ **b.** 0.3 $\dfrac{1}{0.3}$ or $\dfrac{10}{3}$

19. [1.2A] Find:

   a. $-\dfrac{3}{5} \div \dfrac{4}{15}$ $-\dfrac{9}{4}$ **b.** $\dfrac{3.6}{-1.2}$ -3

20. [1.2B] Name the property illustrated in the statement:

    a. $(4 + 9) + 5 = 5 + (4 + 9)$

      Commutative property of addition

    b. $(3 + 5) + 8 = 3 + (5 + 8)$

      Associative property of addition

21. [1.3A] Evaluate:

    a. $(-3)^4$ 81

    b. -3^4 -81

22. [1.3B] Evaluate:

    a. 9^0 1   **b.** $\left(\dfrac{1}{7}\right)^0$ 1

23. [1.3B] Write as a fraction:

    a. $(-8)^{-3}$ $-\dfrac{1}{512}$ **b.** x^{-10} $\dfrac{1}{x^{10}}$

24. [1.3C] Multiply and simplify:

    a. $(3x^4y)(-5x^{-8}y^9)$ $-\dfrac{15y^{10}}{x^4}$

    b. $(4x^{-3}y^{-1})(-6x^{-8}y^{-7})$ $-\dfrac{24}{x^{11}y^8}$

25. [1.3C] Divide and simplify:

    a. $\dfrac{48x^4}{16x^6}$ $\dfrac{3}{x^2}$ **b.** $\dfrac{8x^5}{-2x^{-6}}$ $-4x^{11}$

26. [1.3C] Divide and simplify:

    a. $\dfrac{-5x^{-3}}{15x^{-4}}$ $-\dfrac{x}{3}$ **b.** $\dfrac{8x^{-4}}{-4x^7}$ $-\dfrac{2}{x^{11}}$

27. [1.3D] Simplify:

    a. $(-2x^7y^{-6})^3$ $-\dfrac{8x^{21}}{y^{18}}$ **b.** $(-2x^{-6}y^{-6})^4$ $\dfrac{16}{x^{24}y^{24}}$

28. [1.3E] Simplify:

    a. $\left(\dfrac{x^6}{y^{-3}}\right)^{-4}$ $\dfrac{1}{x^{24}y^{12}}$ **b.** $\left(\dfrac{x^{-5}}{y^3}\right)^{-5}$ $x^{25}y^{15}$

29. [1.3F] Write in scientific notation:

    a. 340,000 3.4×10^5 **b.** 0.000047 4.7×10^{-5}

30. [1.3F] Write in decimal notation:

    a. 3.7×10^4 37,000   **b.** 7.8×10^{-3} 0.0078

31. [1.4A] Evaluate:

    a. $[-8 \cdot (9 + 2)] + 13$ -75

    b. $[-7(3 - 8)] + 15$ 50

32. [1.4B] Evaluate:

    a. $6^2 \div 3 - 9 \cdot 2 \div 3 + 3$ 9

    b. $\dfrac{5 \cdot (68 - 32)}{9}$ 20

33. [1.4B] Evaluate:

    a. $-3^2 + \dfrac{4 - 10}{2} + 15 \div 3$ -7

    b. $-4^3 + \dfrac{2 - 10}{2} - 25 \div 5$ -73

34. [1.4C] Evaluate the algebraic expression:

    a. $2(lw + lh + wh)$ gives the surface area of a rectangular box. Find the value of the area if $l = 12$ in., $w = 8$ in., and $h = 4$ in.   352 sq in.

    b. $-2x^2 - (4 - y)5$ if $x = -2$ and $y = -1$. -33

35. [1.4D] Remove the parentheses (simplify):

    a. $-3(x - 7)$ $-3x + 21$

    b. $3(x + 8) - (x + 7)$ $2x + 17$

36. [1.4E, F] Simplify:

    a. $3(x + 2y - 2) - 2(2x - 2y + 5)$ $-x + 10y - 16$

    b. $[(5x^2 - 3) + (4x + 5)] - [(x - 4) + (2x^2 - 2)]$

      $3x^2 + 3x + 8$

Practice Test 1

(Answers on pages 67–68)

1. Use roster notation to list the natural numbers between 5 and 9.

2. Write as a decimal:

   a. $\dfrac{3}{8}$ **b.** $\dfrac{2}{3}$

3. Classify the given number by making a check mark (✓) in the appropriate row(s).

Set	0.5	0	−6	$\frac{-2}{7}$	$\sqrt{5}$
Natural numbers					
Whole numbers					
Integers					
Rational numbers					
Irrational numbers					
Real numbers					

4. Find the additive inverse of $\dfrac{4}{5}$.

5. Find:

   a. $|-9|$ **b.** $|0.5|$

6. Fill in the blank with $<$, $>$, or $=$ to make the resulting statement true:

   a. $-\dfrac{1}{4}$ _____ $-\dfrac{1}{3}$ **b.** 0.4 _____ $\dfrac{2}{5}$

7. Find:

   a. $-9 + 5$ **b.** $-0.8 + (-0.7)$

8. Find:

   a. $-16 - 7$ **b.** $-0.6 - (-0.4)$

9. Find:

   a. $-\dfrac{1}{8} - \dfrac{3}{4}$ **b.** $-\dfrac{3}{4} - \left(-\dfrac{5}{6}\right)$

10. Find:

   a. $6 \cdot (-9)$ **b.** $-4 \cdot (-1.2)$

11. Find:

   a. $-\dfrac{1}{2} \cdot \dfrac{2}{9}$ **b.** $-\dfrac{3}{2} \div \dfrac{9}{8}$

12. Name the property illustrated in the statement:

   a. $(7 + 3) + 6 = (3 + 7) + 6$

   b. $(2 + 9) + 4 = 2 + (9 + 4)$

13. Name the property illustrated in the statement:

   a. $3 \cdot \dfrac{1}{3} = 1$ **b.** $0.3 + (-0.3) = 0$

14. Evaluate:

   a. $(-3)^4$ **b.** -3^4

15. Write without negative exponents:

   a. 7^{-2} **b.** x^{-8}

16. Perform the indicated operation and simplify:

   a. $(3x^4 y)(-4x^{-8} y^8)$ **b.** $\dfrac{48x^4}{16x^{-8}}$

17. Simplify:

   a. $(-2x^8 y^{-2})^3$ **b.** $\left(\dfrac{x^5}{y^{-3}}\right)^{-3}$

18. Write 6.5×10^{-3} in standard decimal notation.

19. Write 8.5×10^5 in standard decimal notation.

20. Perform the calculation and write your answer in scientific notation:

$$(7.1 \times 10^5) \times (4 \times 10^{-7})$$

21. Evaluate:

   a. $[-7(4 + 3)] + 9$ **b.** $\dfrac{5 \cdot (131 - 32)}{9}$

22. Evaluate: $-4^3 + \dfrac{6 - 12}{2} + 15 \div 3$.

23. Evaluate:

   a. $\frac{1}{2}(b_1 + b_2)h$ gives the area of a trapezoid. Find the area of the trapezoid if $b_1 = 8$, $b_2 = 3$, and $h = 6$.

   b. $7 - x^2 + (20 \div y) - xy$; if $x = -2$ and $y = 4$.

24. Simplify:

   a. $-5(x + 7)$ **b.** $7x - (3x + 1) + (2x + 2)$

25. Simplify:

   $[(5x^2 - 3) + (3x + 7)] - [(x - 3) + (2x^2 - 2)]$

Answers to Practice Test

ANSWER	IF YOU MISSED		REVIEW	
	QUESTION	SECTION	EXAMPLES	PAGE
1. $\{6, 7, 8\}$	1	1.1A	1	6
2. a. 0.375 **b.** $0.\overline{6}$	2	1.1B	2	7
3.	3	1.1C	3	8

Set	0.5	0	−6	$\frac{-2}{7}$	$\sqrt{5}$
N					
W		✓			
I		✓	✓		
Rat.	✓	✓	✓	✓	
Irr.					✓
R	✓	✓	✓	✓	✓

ANSWER	IF YOU MISSED		REVIEW	
4. $-\dfrac{4}{5}$	4	1.1D	4	10
5. a. 9 **b.** 0.5	5	1.1E	5	11
6. a. $>$ **b.** $=$	6	1.1F	6	12–13
7. a. -4 **b.** -1.5	7	1.2A	1	18
8. a. -23 **b.** -0.2	8	1.2A	2a, b	19
9. a. $-\dfrac{7}{8}$ **b.** $\dfrac{1}{12}$	9	1.2A	2c	19
10. a. -54 **b.** 4.8	10	1.2A	3	20
11. a. $-\dfrac{1}{9}$ **b.** $-\dfrac{4}{3}$	11	1.2A	4, 5, 6, 7	21–25
12. a. Commutative property of addition **b.** Associative property of addition	12	1.2B	8	26
13. a. Inverse (reciprocal) property for multiplication **b.** Inverse (opposite) property for addition	13	1.2B	8	26
14. a. 81 **b.** -81	14	1.3A	1	32
15. a. $\dfrac{1}{49}$ **b.** $\dfrac{1}{x^8}$	15	1.3B	2	34
16. a. $-\dfrac{12y^9}{x^4}$ **b.** $3x^{12}$	16	1.3C	3, 4	36–37
17. a. $-\dfrac{8x^{24}}{y^6}$ **b.** $\dfrac{1}{x^{15}y^9}$	17	1.3D, E	5, 6	39–40
18. 0.0065	18	1.3F	8	41

ANSWER		IF YOU MISSED	REVIEW		
		QUESTION	SECTION	EXAMPLES	PAGE
19. 850,000		19	1.3F	8	41
20. 2.84×10^{-1}		20	1.3F	7, 8, 9	41–42
21. a. -40	**b.** 55	21	1.4A	1, 2, 3	47–48
22. -62		22	1.4B	4	49
23. a. 33	**b.** 16	23	1.4C	5	50
24. a. $-5x - 35$	**b.** $6x + 1$	24	1.4D, E	6, 7, 8, 9	51–55
25. $3x^2 + 2x + 9$		25	1.4F	10	55

Linear Equations and Inequalities

2

The Human Side of Algebra

Who invented algebra, anyway? One of the earliest accounts of an algebra problem is in the Rhind papyrus, an ancient Egyptian document written by an Egyptian priest named Ahmes (ca. 1620 B.C.) and purchased by Henry Rhind. When Ahmes wished to find a number such that the number added to its seventh made 19, he symbolized the number by the sign we translate as "heap." Today, we write the problem as

$$x + \frac{x}{7} = 19$$

Can you find the answer? Ahmes says it is $16 + \frac{1}{2} + \frac{1}{8}$.

Western Europeans, however, learned their algebra from the works of Mohammed ibn Musa al-Khowarizmi (this translates as Mohammed the son of Moses of Khowarizmi; ca. A.D. 820), an astronomer and mathematician of Baghdad and author of the treatise *Hisab al-jabr w'al muqabalah,* the science of restoring (placing variables on one side of an equation) and reduction (collecting like terms). With the passage of time, the word *al-jabr* evolved into our present word *algebra,* a subject that we shall continue to study now.

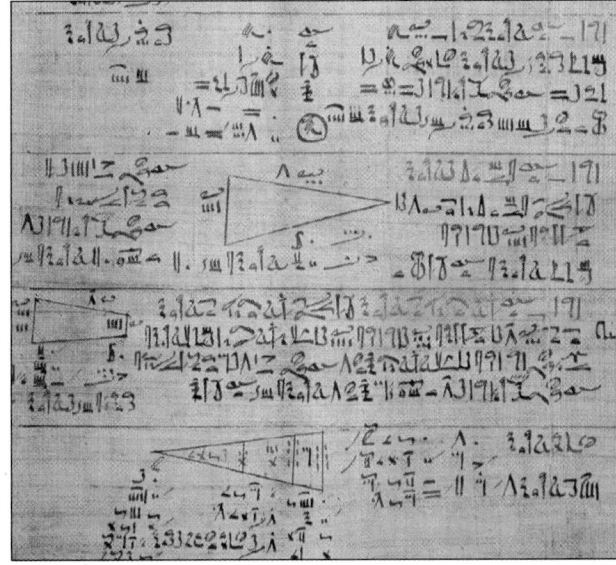

Rhind Papyrus
Source: Touregypt.net.

Pretest for Chapter 2

(Answers on page 71)

1. Does -2 satisfy the equation $7 - x = 5$?

2. Solve $5x + 7 = -3$

3. Solve $3x - 4 = -x + 6$

4. Solve $2(3 - x) + 7 = 2x - (x - 4)$

5. Solve $\dfrac{3}{4} + \dfrac{x}{2} = \dfrac{(x - 6)}{8}$

6. Solve $0.03x + 0.05(2500 - x) = 75$

7. Solve for y in $2x - y = 6$

8. If the perimeter of a rectangle is 120 cm and the length is 18 cm more than the width, find the dimensions.

9. If the lines are parallel, find x and the measure of the unknown angles.

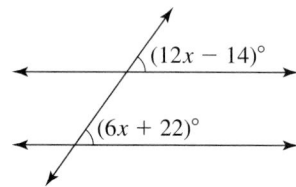

10. A car repair totals \$135. If the repair shop charges a flat fee of \$45 plus \$30 for each hour of labor, how many hours did the repair take?

11. The sum of three consecutive integers is 159. What are the integers?

12. A man's bonus increased by 15 percent to \$9200. What was his bonus before the increase?

13. A truck traveling at 55 mph leaves a rest stop going north on the interstate. One hour later a car leaves the same rest stop going north at 75 mph. How far from the rest stop does the car overtake the truck?

14. Solve, graph, and write the solution set in interval notation:

$$5(x - 2) < 7x + 8$$

15. Solve, graph, and write the solution set in interval notation:

$$\frac{x}{6} + \frac{x}{3} \le \frac{x - 4}{6}$$

16. Solve, graph, and write the solution set in interval notation:

$$-3 < -x + 2 < 2$$

17. Solve $\left| \dfrac{1}{2}x - 3 \right| - 5 = 4$

18. Solve $|2x + 4| = |x - 5|$

19. Solve, graph, and write the solution in interval notation:

$$|5x - 3| \ge 7$$

20. Solve, graph, and write the solution in interval notation:

$$|8x - 4| < 12$$

Answers to Pretest

	ANSWER	IF YOU MISSED	REVIEW		
		QUESTION	SECTION	EXAMPLES	PAGE
1.	no	1	2.1	1	74
2.	-2	2	2.1	2	75–76
3.	$\frac{5}{2}$	3	2.1	3	76–77
4.	3	4	2.1	4	77
5.	-4	5	2.1	5, 6	78–79
6.	2500	6	2.1	8, 9	82–83
7.	$y = 2x - 6$	7	2.2	1, 2	88–89
8.	21 cm by 39 cm	8	2.2	3	90
9.	$x = 6$; angles $= 58°$ each	9	2.2	7	94
10.	3 hr	10	2.3	2	102
11.	52; 53; 54	11	2.3	3	103
12.	$8000	12	2.4	1, 2	112–113
13.	206.25 miles	13	2.4	4	115
14.	$x > -9$; $(-9, \infty)$	14	2.5	2, 3	123–125
15.	$x \le -2$; $(-\infty, -2]$	15	2.5	4	126
16.	$0 < x < 5$; $(0, 5)$	16	2.5	8	131
17.	$x = -12$; $x = 24$	17	2.6	1, 2	138–139
18.	$x = -9$; $x = \frac{1}{3}$	18	2.6	3	140
19.	$x \le -\frac{4}{5}$ or $x \ge 2$; $(-\infty, -\frac{4}{5}] \cup [2, \infty)$	19	2.6	5	143
20.	$-1 < x < 2$; $(-1, 2)$	20	2.6	4	142

2.1   LINEAR EQUATIONS IN ONE VARIABLE

To Succeed, Review How To . . .

1. Add, subtract, multiply, and divide positive and negative numbers (pp. 17–23).

2. Use the commutative, associative, and distributive properties (p. 26).

3. Find the sum of opposites (additive inverses) (p. 10).

4. Find the product of two reciprocals (p. 24).

Objectives

A Determine whether a number is a solution of a given equation.

B Solve linear equations using the properties of equality.

C Solve linear equations in one variable using the CRAM procedure.

D Solve linear equations involving decimals.

GETTING STARTED

Crickets, Ants, and Temperatures

Does temperature affect animal behavior? You certainly know about bears hibernating and languid students in the spring. But what about crickets and ants? Farmers claim that they can tell the temperature F in degrees Fahrenheit by listening to the number of chirps N a certain type of cricket makes in 1 minute! How? By using the formula

$$F = \frac{N}{4} + 40$$

Thus, if a cricket is chirping 80 times a minute, the temperature is $F = \frac{80}{4} + 40 = 60°F$ (60 degrees Fahrenheit). Now suppose the temperature is 90°F. How fast is the cricket chirping? To find the answer, you must know how to solve the equation

$$90 = \frac{N}{4} + 40$$

Similarly, the speed S of an ant, in centimeters per second (cm/sec), is given by $S = \frac{1}{6}(C - 4)$, where C is the temperature in degrees Celsius. If an ant is moving at 4 cm/sec, what is the temperature? Here, you have to solve the equation $4 = \frac{1}{6}(C - 4)$. You will learn how to solve these and other similar equations in this chapter.

Finally, if you know the relationship between Fahrenheit and Celsius temperatures and you look at these two formulas, can you tell, as the weather gets cooler, whether the crickets stop chirping before the ants stop crawling?

How do we express our ideas in algebra? We start with a collection of letters (**variables**) and real numbers (**constants**) and then perform the basic operations of addition, subtraction, multiplication, and division. The result is an **algebraic expression** such as

$$2x + 7, \quad x^2 - 3x + 5, \quad \text{or} \quad \frac{x^5 y^7}{z^3}$$

CAUTION

In expressions such as

$$\frac{x^5 y^7}{z^3}$$

with variables in the denominator, the denominator cannot be zero.

An **equation** is a sentence stating that two algebraic expressions are equal. There are three important properties of equality: the reflexive, symmetric, and transitive properties.

Properties of Equality	For all real numbers a, b, and c,	
	1. $a = a$	Reflexive property
	2. If $a = b$, then $b = a$.	Symmetric property
	3. If $a = b$ and $b = c$, then $a = c$.	Transitive property

Study Skills Hint

Examples of the Reflexive Property

$$0.5 = 0.5$$
$$x + 7 = x + 7$$
$$x^2 + 5x - 7 = x^2 + 5x - 7$$

Examples of the Symmetric Property

If $x = 7$, then $7 = x$.

If $C = 2\pi r$, then $2\pi r = C$.

If $y = x^2 + 3x - 7$, then $x^2 + 3x - 7 = y$.

Examples of the Transitive Property

If $x = y$ and $y = 2a$, then $x = 2a$.

If $C = 2\pi r$ and $2\pi r = \pi d$, then $C = \pi d$.

If $x = \dfrac{a}{b}$ and $\dfrac{a}{b} = \dfrac{ac}{bc}$, then $x = \dfrac{ac}{bc}$.

In this section we consider *linear equations* involving only real numbers and *one* variable. Here are some examples of linear equations:

$$x = 8, \quad 2x - 5 = 7, \quad 3(y - 1) = 2y + 8, \quad \text{and} \quad 3k + 7 = 10$$

In general, we have the following definition.

Teaching Tip

Give students the following applied example of writing an equation to solve a real problem and then challenge them to work in groups and write others.

Example: What is the most miles you can drive if you can rent a car for $15 a day plus $0.10 for each mile, and you only have $100 to spend? Equation: $15 + \$0.10(x) = \$100$, where x stands for the number of miles driven.

LINEAR EQUATIONS

A **linear equation** in one variable is an equation that can be written in the form

$$ax + b = c$$

where a, b, and c are real numbers and $a \neq 0$.

The highest power of the variable in a linear equation is 1, so linear equations are also called **first-degree equations.**

Solutions of an Equation

Some equations are always *true* (identities: $2 + 2 = 4$, $5 - 3 = 2$), some are always *false* (contradictions: $2 + 2 = 22$, $5 - 3 = -2$), and some are neither true nor false. For example, the equation $x + 1 = 5$ is neither true nor false. It is a *conditional* equation, and its truth or falsity depends on the value of x.

In the equation $x + 1 = 5$, the **variable** x can be replaced by many numbers, but only one number will make the resulting statement true. This number is called the *solution* of the equation.

SOLUTIONS OF AN EQUATION

The **solutions (roots)** of an equation are the replacements of the variable that make the equation a true statement. To **solve** an equation is to find all its solutions.

Teaching Tip

Have students verify that $x = 850$ miles is the solution to the example in the previous Teaching Tip with equation $\$15 + \$0.10(x) = \$100$.

To determine whether a number is a solution of an equation, we replace the variable by the number. For example, 4 is a solution of $x + 1 = 5$ because replacing x with 4 in the equation yields $4 + 1 = 5$, a true statement, but -6 is *not* a solution because $-6 + 1 \neq 5$. Since 4 is the *only* number that yields a true statement, the solution set of $x + 1 = 5$ is $\{4\}$.

EXAMPLE 1 Determining when a number is a solution	**PROBLEM 1**

Determine whether:

a. 8 is a solution of $x - 5 = 3$

b. 5 is a solution of $3 = 2 - y$

c. 6 is a solution of $\frac{1}{3}z - 4 = 2z - 14$

Determine whether:

a. 6 is a solution of $x - 2 = 4$

b. 7 is a solution of $4 = 11 - y$

c. 4 is a solution of $\frac{1}{2}z - 2 = z - 2$

SOLUTION

a. Substituting 8 for x in $x - 5 = 3$, we have $8 - 5 = 3$, a true statement. Thus, 8 is a solution of $x - 5 = 3$.

b. Substituting 5 for y in $3 = 2 - y$, we obtain $3 = 2 - 5$, a false statement. Hence, 5 is not a solution of $3 = 2 - y$.

c. If we replace z by 6 in $\frac{1}{3}z - 4 = 2z - 14$, we obtain

$$\frac{1}{3}(6) - 4 = 2(6) - 14$$
$$2 - 4 = 12 - 14$$
$$-2 = -2$$

a true statement. Thus, 6 is a solution of $\frac{1}{3}z - 4 = 2z - 14$.

B Solving Equations Using the Properties of Equality

We have learned how to determine whether a number is a solution of an equation; now we will learn how to find these solutions—that is, how to *solve* the equation. The procedure is to find an *equivalent* equation whose solution is obvious. For example, the equations $x = 2$ and $x + 3 = 5$ are equivalent because $x = 2$, with the obvious solution 2, is the only solution of the equation $x + 3 = 5$.

Answers

1. a. Yes: $6 - 2 = 4$
b. Yes: $4 = 11 - 7$
c. No: $\frac{1}{2}(4) - 2 \neq 4 - 2$

EQUIVALENT EQUATIONS

Two or more equations are **equivalent** if they have the same solution set.

To solve equations, we use the idea that adding or subtracting the same number on both sides of the equation and multiplying or dividing both sides of an equation by the same nonzero number produces an equivalent equation. Here is the principle.

Properties of Equality	If C is a real number, then the following equations are equivalent.
	$A = B$
Add C.	$A + C = B + C$
Subtract C.	$A - C = B - C$
Multiply by C.	$A \cdot C = B \cdot C \qquad (C \neq 0)$
Divide by C.	$\dfrac{A}{C} = \dfrac{B}{C} \qquad (C \neq 0)$

Teaching Tip

Emphasize that "sides" of the equation are separated by the "=" sign. Also, only one equal sign should be written on each line.

We will use the properties of equality to write a series of equivalent equations until we reach the obvious solution. Suppose we want to solve $x + 5 = 7$. We are trying to find a value of x that will satisfy the equation, so we try to get x by itself on one side of the equation. To do this, we "undo" the addition of 5 by *adding* the inverse of 5 on both sides.

$$x + 5 = 7 \qquad \text{Given.}$$
$$x + 5 + (-5) = 7 + (-5) \qquad \text{Add } (-5) \text{ to both sides.}$$
$$x + 0 = 2 \qquad 5 + (-5) = 0, \text{ additive inverses}$$
$$x = 2 \qquad x + 0 = x, \text{ additive identity}$$

The solution is 2, and the solution set is $\{2\}$. We can also solve the equation by subtracting 5 from both sides:

$$x + 5 = 7 \qquad \text{Given.}$$
$$x + 5 - 5 = 7 - 5 \qquad \text{Subtract 5 from both sides.}$$
$$x + 0 = 2 \qquad 5 - 5 = 0$$
$$x = 2$$

Using the same idea, we can solve $8 = 3x - 7$ by first adding 7 to both sides.

$$8 = 3x - 7 \qquad \text{Given.}$$
$$8 + 7 = 3x - 7 + 7 \qquad \text{Add 7 to both sides.}$$
$$15 = 3x$$
$$\frac{15}{3} = \frac{3x}{3} \qquad \text{Divide both sides by 3.}$$
$$5 = x$$

The solution is 5, and the solution set is $\{5\}$.

EXAMPLE 2	**Solving linear equations using the equality properties**	**PROBLEM 2**

Solve:

a. $2x - 4 = 6$ **b.** $\dfrac{2}{3}y - 3 = 9$

Solve:

a. $3x - 8 = 4$ **b.** $\dfrac{3}{4}x - 5 = 10$

Answers

2. a. 4   **b.** 20

SOLUTION

a. To solve this equation, we want the variable x by itself on one side of the equation. We start by adding 4 to both sides.

$$2x - 4 = 6 \qquad \text{Given.}$$

$$2x - 4 + 4 = 6 + 4 \qquad \text{Add 4 to both sides.}$$

$$2x = 10$$

$$\frac{1}{2} \cdot 2x = \frac{1}{2} \cdot 10 \qquad \text{Multiply both sides by } \tfrac{1}{2}.$$

$$x = 5 \qquad \tfrac{1}{2} \cdot 2 = 1 \text{ and } \tfrac{1}{2} \cdot 10 = 5$$

The solution set is $\{5\}$.

To solve $2x = 10$, we could also divide both sides by 2

$$\frac{2x}{2} = \frac{10}{2}$$

to obtain the same result, $x = 5$. To check this solution, we substitute 5 for x in the original equation and use the following diagram:

$$
\begin{array}{c|c}
\multicolumn{2}{c}{2x - 4 \overset{?}{=} 6} \\
\hline
2(5) - 4 & 6 \\
10 - 4 & \\
6 & \\
\end{array}
$$

Both sides yield 6, so we have a true statement and our result is correct.

b.
$$\frac{2}{3}y - 3 = 9 \qquad \text{Given.}$$

$$\frac{2}{3}y - 3 + 3 = 9 + 3 \qquad \text{Add 3 to both sides.}$$

$$\frac{2}{3}y = 12$$

$$\frac{3}{2} \cdot \frac{2}{3}y = \frac{3}{2} \cdot 12 \qquad \text{Multiply both sides by } \tfrac{3}{2}.$$

$$y = 18 \qquad \tfrac{3}{2} \cdot 12 = \tfrac{3}{2} \cdot \tfrac{12}{1} = 18$$

The solution set is $\{18\}$. You can check this answer by substituting 18 for y in $\frac{2}{3}y - 3 = 9$ to obtain

$$
\begin{array}{c|c}
\multicolumn{2}{c}{\frac{2}{3}y - 3 \overset{?}{=} 9} \\
\hline
\frac{2}{3} \cdot 18 - 3 & 9 \\
12 - 3 & \\
9 & \\
\end{array}
$$

EXAMPLE 3	**Solving linear equations using the equality properties**

Solve: $4a - 7 = a + 4$

SOLUTION   We start by adding the inverse of a on both sides so that only variables are on the left. Here are the steps:

$$4a - 7 = a + 4 \qquad \text{Given.}$$

$$4a + (-a) - 7 = a + (-a) + 4 \qquad \text{Add } (-a) \text{ to both sides so that all variables are on the left.}$$

$$3a - 7 = 4$$

PROBLEM 3

Solve: $5a - 8 = a + 5$

Answer

3. $\frac{13}{4}$

$$3a - 7 + 7 = 4 + 7$$ Add 7 to both sides.

$$3a = 11$$

$$a = \frac{11}{3}$$ Divide both sides by 3 (or multiply by $\frac{1}{3}$).

The solution is $\frac{11}{3}$, and the solution set is $\left\{\frac{11}{3}\right\}$, as can be checked by substituting $\frac{11}{3}$ for a in $4a - 7 = a + 4$.

Teaching Tip

Remind students that $\frac{11}{3}$ may be written as $3\frac{2}{3}$ or $3.\overline{6}$.

Sometimes we need to simplify an equation before applying the properties of equality. For instance, to solve $x + 6 = 3(2x - 2)$, we use the distributive property to simplify the right-hand side of the equation and then solve for x. We do this next.

EXAMPLE 4 **Simplifying equations before solving**	**PROBLEM 4**

Solve: $x + 6 = 3(2x - 2)$

Solve: $3 + x = 2(2x - 3)$

SOLUTION

$$x + 6 = 3(2x - 2)$$ Given.

$$x + 6 = 6x - 6$$ Simplify the right-hand side (distributive property).

We have two choices; we can isolate the variables on the right or on the left. To avoid negative expressions, this time we keep them on the right by adding $(-x)$ to both sides. We have

$$x + (-x) + 6 = 6x + (-x) - 6$$ Add $(-x)$ so all variables are on the right.

$$6 = 5x - 6$$

$$6 + (6) = 5x - 6 + (6)$$ Add 6 to both sides.

$$12 = 5x$$

$$\frac{12}{5} = x$$ Divide by 5 (or multiply by $\frac{1}{5}$).

The solution is $\frac{12}{5}$, and the solution set is $\left\{\frac{12}{5}\right\}$.

If an equation involves fractions, we "clear" them by multiplying both sides of the equation by the *smallest* number that is a multiple of each denominator—by the least common denominator (LCD). Thus to solve

$$\frac{x}{6} + \frac{x}{4} = 10$$

Teaching Tip

Use the following two examples to contrast the solving of an equation with variables on the same side of the equal sign versus opposite sides.

Variable on same side	Variable on Opposite Sides
$2x + 3x = 5$	$2x + 5 = 3x$
$5x = 5$	$\underline{-2x} \quad \underline{-2x}$
$\dfrac{5x}{5} = \dfrac{5}{5}$	$5 = x$
$x = 1$	

we have to find the LCD of $\frac{x}{6}$ and $\frac{x}{4}$. One way is to pick the larger of the two denominators and double it, triple it, and so on, until the other number divides into the result. Using this idea, we find that 12 is the LCD. Multiplying both sides of the equation by 12, we have

$$12 \cdot \left(\frac{x}{6} + \frac{x}{4}\right) = 12 \cdot (10)$$

$$12 \cdot \frac{x}{6} + 12 \cdot \frac{x}{4} = 12 \cdot 10$$ Use the distributive property.

$$2x + 3x = 120$$ Simplify.

$$5x = 120$$ Combine like terms.

$$x = 24$$ Divide both sides by 5.

The solution is 24, and the solution set is $\{24\}$. Here is the check:

$$\frac{x}{6} + \frac{x}{4} \overset{?}{=} 10$$

$$\frac{24}{6} + \frac{24}{4} \quad \bigg| \quad 10$$

$$4 + 6$$

$$10$$

Answer

4. 3

Web It

If you would like to see more step-by-step examples of solving linear equations, try link 2-1-1 at mhhe.com/bello.

If you want more practice solving linear equations, try link 2-1-2 at mhhe.com/bello.

> **NOTE**
>
> When finding the LCD be sure to consider all denominators that appear in the equation (that includes looking on both sides of the equal sign).

| **EXAMPLE 5** Clearing fractions in linear equations | **PROBLEM 5** |

Solve:

a. $\dfrac{x+1}{3} + \dfrac{x-1}{10} = 5$ **b.** $\dfrac{x+1}{3} - \dfrac{x-1}{8} = 4$

Solve:

a. $\dfrac{x+2}{4} + \dfrac{x-1}{5} = 3$

b. $\dfrac{x+3}{2} - \dfrac{x-2}{3} = 5$

SOLUTION

a. The LCD of

$$\frac{x+1}{3} \quad \text{and} \quad \frac{x-1}{10}$$

is $3 \cdot 10 = 30$, since 3 and 10 do not have any common factors. Multiplying both sides by 30, we have

$$30\left(\frac{x+1}{3} + \frac{x-1}{10}\right) = 30 \cdot (5)$$

$$30\left(\frac{x+1}{3}\right) + 30\left(\frac{x-1}{10}\right) = 30 \cdot 5 \qquad \text{By the distributive property.}$$

$$10(x+1) + 3(x-1) = 150 \qquad \text{Since } \overset{10}{\cancel{30}}\left(\tfrac{x+1}{\underset{1}{\cancel{3}}}\right) = 10(x+1) \text{ and}$$
$$\overset{3}{\cancel{30}}\left(\tfrac{x-1}{\underset{1}{\cancel{10}}}\right) = 3(x-1).$$

$$10x + 10 + 3x - 3 = 150 \qquad \text{Use the distributive property.}$$

$$13x + 7 = 150 \qquad \text{Add like terms.}$$

$$13x = 143 \qquad \text{Subtract 7.}$$

$$x = 11 \qquad \text{Divide by 13.}$$

The solution set is $\{11\}$. The check is left to you.

b. Here the LCD is $3 \cdot 8 = 24$. Multiplying both sides by 24, we obtain

$$24\left(\frac{x+1}{3} - \frac{x-1}{8}\right) = 24 \cdot (4)$$

$$24\left(\frac{x+1}{3}\right) - 24\left(\frac{x-1}{8}\right) = 24 \cdot 4 \qquad \text{By the distributive property.}$$

$$8(x+1) - 3(x-1) = 96 \qquad \text{Since } \overset{8}{\cancel{24}}\left(\tfrac{x+1}{\underset{1}{\cancel{3}}}\right) = 8(x+1) \text{ and}$$
$$\overset{3}{\cancel{24}}\left(\tfrac{x-1}{\underset{1}{\cancel{8}}}\right) = 3(x-1).$$

$$8x + 8 - 3x + 3 = 96 \qquad \text{Use the distributive property.}$$

$$5x + 11 = 96 \qquad \text{Add like terms.}$$

$$5x = 85 \qquad \text{Subtract 11.}$$

$$x = 17 \qquad \text{Divide by 5.}$$

The solution set is $\{17\}$. Check this answer in the original equation.

Answers

5. a. 6 **b.** 17

C Solving Linear Equations

All the equations we have solved in this section can be written in the form $ax + b = c$ where a, b, and c are real numbers. Such equations are called **linear equations.** (You will see in Chapter 3 that the graph of $ax + b = y$ is a *straight line.*)

$$2x - 4 = 6 \quad \text{and} \quad \frac{x}{10} + \frac{x}{8} = 9$$

are linear equations. In the equation $2x - 4 = 6$, $2x$, -4, and 6 are called **terms:** $2x$ is a variable term and -4 and 6 are constant terms. We will use this terminology to give you a general procedure used to solve linear equations.

PROCEDURE

Procedure for Solving Linear Equations

1. If there are fractions or decimals, multiply both sides of the equation by the LCD of the fractions or decimals.

2. Remove parentheses and collect like terms (simplify) if necessary.

3. Add or subtract the same quantity on both sides of the equation so that one side has only the terms containing variables.

4. Add or subtract the same quantity on both sides of the equation so that the other side has only a constant.

5. If the coefficient of the variable is not 1, divide both sides of the equation by this coefficient (or, equivalently, multiply by the reciprocal of the coefficient of the variable).

6. Be sure to check your answer in the original equation.

You can remember these rules as "CRAM":

Clear fractions/decimals
Remove parentheses/simplify
Add/subtract to get variable isolated
Multiply/divide to make coefficient a 1

Teaching Tip

These steps have some flexibility regarding the order in which they are performed. Students should try to be consistent with their steps.

Calculate It Graph $ax + b = y$

The graph of any equation that can be written as $ax + b = y$ is a straight line. To illustrate this, let $a = 2$, $b = 3$, and graph $y = 2x + 3$. Begin with pressing the ⬛Y= key and inputting $2x + 3$. Press GRAPH and the result is the line shown. (Use a "standard" window, ZOOM 6, for the graph.)

Try other numbers for a and b and convince yourself that $ax + b = y$ always has a straight line for its graph.

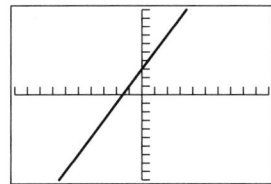

EXAMPLE 6 Solving linear equations using the CRAM procedure

Solve:

a. $\dfrac{7}{24} = \dfrac{x}{8} + \dfrac{1}{6}$

b. $\dfrac{1}{5} - \dfrac{x}{4} = \dfrac{7(x + 3)}{10}$

SOLUTION We use the six steps as follows.

a. Given: $\dfrac{7}{24} = \dfrac{x}{8} + \dfrac{1}{6}$ Remember CRAM.

1. $24 \cdot \dfrac{7}{24} = 24\left(\dfrac{x}{8} + \dfrac{1}{6}\right)$ Clear fractions; multiply by 24, the LCD.

$\overset{1}{\cancel{24}} \cdot \dfrac{7}{\cancel{24}} = \overset{3}{\cancel{24}} \cdot \dfrac{x}{8} + \overset{4}{\cancel{24}} \cdot \dfrac{1}{6}$ Remove parentheses/simplify.

PROBLEM 6

Solve:

a. $\dfrac{7}{12} = \dfrac{x}{4} + \dfrac{1}{3}$

b. $\dfrac{1}{3} - \dfrac{x}{5} = \dfrac{8(x + 2)}{15}$

Answers

6. a. 1   **b.** -1

2. $\qquad 7 = 3x + 4$

3. $\quad 7 - 4 = 3x + 4 - 4 \qquad$ Add -4 to both sides.

4. $\qquad 3 = 3x$

5. $\qquad \dfrac{3}{3} = \dfrac{3x}{3} \qquad\qquad$ Multiply by $\frac{1}{3}$ (or divide by 3).

$\qquad 1 = x$

$\qquad x = 1$

The solution set is $\{1\}$.

6. CHECK Since $\dfrac{1}{8} = \dfrac{3}{24}$ and $\dfrac{1}{6} = \dfrac{4}{24}$,

$$\frac{7}{24} = \frac{1}{8} + \frac{1}{6} = \frac{3}{24} + \frac{4}{24}$$

which is true.

b. Given: $\dfrac{1}{5} - \dfrac{x}{4} = \dfrac{7(x + 3)}{10}$

1. $\quad 20\left(\dfrac{1}{5} - \dfrac{x}{4}\right) = 20 \cdot \left[\dfrac{7(x + 3)}{10}\right] \qquad$ Clear fractions; multiply by 20, the LCD.

$\overset{4}{\cancel{20}} \cdot \dfrac{1}{5} - \overset{5}{\cancel{20}} \cdot \dfrac{x}{4} = \overset{2}{\cancel{20}} \cdot \dfrac{7(x + 3)}{\cancel{10}} \qquad$ Remove parentheses/simplify.

2. $\qquad 4 - 5x = 14(x + 3)$

$\qquad 4 - 5x = 14x + 42$

3. $\quad 4 - 5x + 5x = 14x + 5x + 42 \qquad$ Add $5x$ to both sides

$\qquad 4 = 19x + 42$

4. $\quad 4 - 42 = 19x + 42 - 42 \qquad$ and -42 to both sides.

$\qquad -38 = 19x$

5. $\qquad \dfrac{-38}{19} = \dfrac{19x}{19} \qquad\qquad$ Multiply by $\frac{1}{19}$ (or divide by 19).

$\qquad -2 = x$

$\qquad x = -2$

The solution set is $\{-2\}$.

6. CHECK $\qquad \dfrac{1}{5} - \dfrac{x}{4} \overset{?}{=} \dfrac{7(x + 3)}{10} \qquad$ for $x = -2$

$$\frac{1}{5} - \frac{(-2)}{4} \quad \bigg| \quad \frac{7(-2 + 3)}{10}$$

$$\frac{1}{5} + \frac{1}{2} \quad \bigg| \quad \frac{7(1)}{10}$$

$$\frac{7}{10} \quad \bigg| \quad \frac{7}{10}$$

Web It

For more practice solving equations with fractions, try link 2-1-3 at mhhe.com/bello.

Calculate It　Solve Linear Equations by Graphing

All the linear equations we have solved can be graphed by first writing an equivalent equation with all the variables on the left and zero on the right, that is, an equation of the form $A(x) = 0$ (read "A of x is 0"), and then graphing the line $y = A(x)$. The number at which the line crosses the x-axis (where $y = 0$) is the desired solution because at this point $y = A(x) = 0$.

Let's use a calculator to solve the equation in Example 6(b):

$$\frac{1}{5} - \frac{x}{4} = \frac{7(x + 3)}{10}$$

First, subtract

$$\frac{7(x + 3)}{10}$$

from both sides of the equation

$$\frac{1}{5} - \frac{x}{4} = \frac{7(x + 3)}{10}$$

to obtain the equivalent equation

$$\frac{1}{5} - \frac{x}{4} - \frac{7(x + 3)}{10} = 0$$

Next, graph

$$Y_1 = \frac{1}{5} - \frac{x}{4} - \frac{7(x + 3)}{10} \text{ and } Y_2 = 0$$

using a decimal or an integer window, [ZOOM] 4 or 8. With a decimal or integer window, the cursor moves in x-increments of 0.1 and 1, respectively. You should get a line crossing the x-axis at $x = -2$. Thus the solution set of the given equation is $\{-2\}$, as shown. Verify by [2nd] [TRACE] [5] [ENTER] [ENTER] [ENTER].

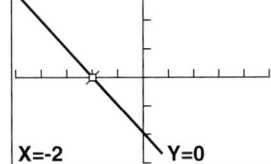

X=-2　　Y=0

So far all our equations have had a solution. They are called **conditional equations.** There are two other possibilities.

| **Equations with *No* Solutions and *Infinitely Many* Solutions** | **1.** Equations with *no* solution are *contradictions.* | $x + 4 = x - 2$ |
| | **2.** Equations with *infinitely many* solutions are *identities.* | $2x + 8 = 2(x + 4)$ |

Consider the following example.

$$-4x + 6 = 2(1 - 2x) + 3 \qquad \text{Clear fractions: none.}$$

$$-4x + 6 = 2 - 4x + 3 \qquad \text{Remove parentheses/simplify.}$$

$$-4x + 6 = -4x + 5$$

$$-4x + 4x + 6 = -4x + 4x + 5 \qquad \text{Add } 4x \text{ to both sides.}$$

$$6 = 5 \quad \text{False!}$$

We cannot find a replacement for x to satisfy this equation. This equation is a **contradiction;** it has no solution. Its solution set is $\varnothing$ (the empty set). On the other hand, consider the following equation.

$$-4x + 6 = 2(1 - 2x) + 4 \qquad \text{Clear fractions: none.}$$

$$-4x + 6 = 2 - 4x + 4 \qquad \text{Remove parentheses/simplify.}$$

$$-4x + 6 = -4x + 6 \qquad \text{Add } 4x \text{ to both sides.}$$

$$6 = 6 \quad \text{True!}$$

Here, any x will be a solution (try 0, 1, or any other number in the original equation). This equation is an **identity.** Any real number is a solution. Its solution set is ***R*, the set of real numbers.**

Teaching Tip

Do the following example to show that 0 can be a solution to an equation. This is not the same as "no solution," as in Example 7(b).

$$3x + 8 = 2(x + 1) + 6$$
$$3x + 8 = 2x + 2 + 6$$
$$3x + 8 = 2x + 8$$
$$\underline{-2x \qquad\quad -2x}$$
$$x + 8 = \quad 8$$
$$\underline{-8 \quad -8}$$
$$x = \quad 0$$

EXAMPLE 7	Solving linear equations with no solution or infinitely many solutions

Solve:

a. $3x + 8 = 3(x + 1) + 5$ **b.** $3x + 8 = 3(x + 1) + 2$

SOLUTION   We use the six-step procedure we studied.

a. Given: $3x + 8 = 3(x + 1) + 5$ Clear fractions: none.

$3x + 8 = 3x + 3 + 5$ Remove parentheses/simplify.

$3x + 8 = 3x + 8$

We can stop here. This equation is always true. It is an identity, and its solution set is R.

b. Given: $3x + 8 = 3(x + 1) + 2$ Clear fractions: none.

$3x + 8 = 3x + 3 + 2$ Remove parentheses/simplify.

$3x + 8 = 3x + 5$

We can stop here. This equation is a contradiction; it is always false. (Try subtracting $3x$ from both sides.) It has no solution, and its solution set is $\varnothing$.

PROBLEM 7

Solve:

a. $4(x + 1) + 3 = 4x + 7$

b. $5(x + 2) + 1 = 3 + 5x$

Teaching Tip

If the solution is $x = 0$, then the solution set is $\{0\}$ not $\{\varnothing\}$

Web It

For a tutorial practice on solving equations, try link 2-1-4 at mhhe.com/bello.

CAUTION

Do not write $\{\varnothing\}$ to represent the empty set because the set $\{\varnothing\}$ is **not** empty, it has one element, $\varnothing$! Always use $\varnothing$ or $\{\ \}$ to represent the empty set.

D   Linear Equations with Decimals

Remember that if an equation has fractions, we "clear" the fractions by multiplying both sides of the equation by the LCD. If an equation has decimals we will "clear" the decimals by multiplying both sides by a power of 10. The power of 10 should have as many zeros as the number of decimal places in the coefficient with the most decimal places. In Example 8, the coefficient 3.15 has the most decimal places (2), so we multiply both sides by 100.

EXAMPLE 8	Solving equations involving decimals

Solve: $14.5 - 3.15x = 5.5$ Remember CRAM.

SOLUTION

$$100 \cdot (14.5 - 3.15x) = 100 \cdot (5.5)$$ Clear decimals; multiply by 100.

$$100(14.5) - 100(3.15x) = 100(5.5)$$ Remove parentheses/simplify.

$$1450 - 315x = 550$$

$$1450 - 1450 - 315x = 550 - 1450$$ Add -1450 to both sides.

$$\frac{-315x}{-315} = \frac{-900}{-315}$$ Multiply by $-\frac{1}{315}$ (or divide by -315).

PROBLEM 8

Solve: $12.5 - 1.25x = 6.5$

Answers

7. a. R **b.** No solution or $\varnothing$

8. $\frac{24}{5}$ or 4.8

$$x = \frac{900}{315} \quad \text{(Reduce the fraction by dividing numerator and denominator by 45.)}$$

$$x = \frac{20}{7}$$

The solution is $\frac{20}{7}$, and the solution set is $\left\{\frac{20}{7}\right\}$, which you can check by substitution in the original equation. Also, the solution can be written as a repeating decimal or as 2.86 rounded to the nearest hundredth.

Teaching Tip

If students use a calculator to evaluate $\frac{20}{7}$, they will get the repeating decimal $2.\overline{857142}$.

Later in this chapter we translate word problems into equations such as $1.30P + 1.50(50 - P) = 72.50$. But, first let's solve this equation.

EXAMPLE 9   Solving equations involving decimals

Solve: $1.30P + 1.50(50 - P) = 72.50$

SOLUTION

$$1.3P + 1.5(50 - P) = 72.5 \qquad \text{Rewrite the equation by dropping the zero in the hundredths place for each coefficient.}$$

$$10[1.3P] + 10[1.5(50 - P)] = 10[72.5] \qquad \text{Clear decimals; multiply by 10.}$$

$$10(1.3P) + 10(1.5)(50 - P) = 10(72.5) \qquad \text{Remove parentheses/simplify.}$$

$$13P + 15(50 - P) = 725$$

$$13P + 750 - 15P = 725$$

$$-2P + 750 = 725$$

$$-2P + 750 - 750 = 725 - 750 \qquad \text{Add } -750 \text{ to both sides.}$$

$$\frac{-2P}{-2} = \frac{-25}{-2} \qquad \text{Multiply by } -\frac{1}{2} \text{ (or divide by } -2\text{).}$$

$$P = \frac{25}{2} \qquad \text{(Simplify the answer.)}$$

The solution is $\frac{25}{2}$, and the solution set is $\left\{\frac{25}{2}\right\}$. The solution set can also be written as $\{12.5\}$ or $\left\{12\frac{1}{2}\right\}$. Make sure you check this!

PROBLEM 9

Solve: $3.30d + 1.30(30 - d) = 42.50$

Teaching Tip

Usually, when the original equation has decimals, the solution is expressed in decimal form.

Calculate It

Compare the Graphs of Equations

What happens when you graph $Y_1 = 3x + 8$ and $Y_2 = 3(x + 1) + 5$? The *Calculate It Exercises* will explore this question.

Web It

For examples of solving more difficult linear equations, try link 2-1-5 at mhhe.com/bello.

Calculate It Exercises

Can we make a calculator show that the graph of the linear equation $y = ax + b$ is indeed a line? Graph $y = x$, $y = 3x$, and $y = 5x$. The TI-83 Plus will graph the **family** of lines $y = 1x$, $y = 3x$, $y = 5x$ by entering $Y_1 = \{1, 3, 5\}x$ and pressing **GRAPH**.

1. What happens to the graph of the line $y = ax$ as the coefficient a increases from 1 to 3 to 5?

2. What happens to the graph of the line $y = ax$ as the coefficient a decreases from -1 to -3 to -5?

3. Consider the family of lines $y = x + 1$, $y = x + 3$, $y = x + 5$.
   a. Do the lines intersect?
   b. What do you call lines that are in the same plane and don't intersect?

4. Graph $Y_1 = 3x + 8$ and $Y_2 = 3(x + 1) + 5$ from Example 7(a).
   a. How many lines do you see in your window?
   b. How many solutions do you get when you graph Y_1 and Y_2 (Y_1 and Y_2 linear equations) and you get only one line?

5. Graph $Y_1 = 3x + 8$ and $Y_2 = 3(x + 1) + 2$ from Example 7(b) using a square window.
   a. Do the lines intersect?
   b. How many solutions do you get when you graph Y_1 and Y_2 (Y_1 and Y_2 linear equations) and you get two lines that don't intersect?

Answer

9. $\frac{7}{4}$ or 1.75

Exercises 2.1

A In Problems 1–10, determine if the number in the box is a solution of the equation.

1. $2x + 8 = 14;$ $\boxed{3}$ Yes

2. $5x + 5 = 10;$ $\boxed{2}$ No

3. $-2x + 1 = 3;$ $\boxed{-1}$ Yes

4. $-3x + 4 = 10;$ $\boxed{-2}$ Yes

5. $2y - 5 = y - 2;$ $\boxed{3}$ Yes

6. $3y - 7 = y + 1;$ $\boxed{4}$ Yes

7. $\frac{4}{5}t - 1 = 5t;$ $\boxed{-\frac{1}{5}}$ No

8. $6t + 1 = t + \frac{2}{3};$ $\boxed{-\frac{1}{3}}$ No

9. $\frac{1}{2}x + 5 = 5 - \frac{1}{3}x;$ $\boxed{3}$ No

10. $-\frac{1}{3}x + 1 = x - 3;$ $\boxed{3}$ Yes

B In Problems 11–36, solve the equation.

11. $3x - 4 = 8$ 4

12. $5a + 16 = 6$ -2

13. $2y + 8 = 10$ 1

14. $4b - 6 = 2$ 2

15. $-3z - 6 = -12$ 2

16. $-4r - 3 = 5$ -2

17. $-5y + 2 = -8$ 2

18. $-3x + 2 = -10$ 4

19. $3x + 5 = x + 19$ 7

20. $4x + 6 = x + 9$ 1

21. $7(x - 1) - 3 + 5x$
$= 3(4x - 3) + x$ -1

22. $5(x + 1) + 3x + 2$
$= 8(x + 2) + 2x + 1$ -5

23. $6v - 8 = 8v + 8$ -8

24. $8t + 3 = 15t - 11$ 2

25. $7m - 4m + 12 = 0$ -4

26. $10k + 25 - 5k = 35$ 2

27. $4(2 - z) + 8 = 8(2 - z)$ 0

28. $4(3 - y) + 8 = 12(3 - y)$ 2

29. $5(x + 3) = 3(x + 3) + 6$ 0

30. $y - (5 - 2y) = 7(y - 1) - 2$ 1

31. $5(4 - 3a) = 7(3 - 4a)$ $\frac{1}{13}$

32. $\frac{3}{4}y - 4 = \frac{1}{4}y - 2$ 4

33. $-\frac{7}{8}c + 5 = -\frac{5}{8}c + 3$ 8

34. $x + \frac{2}{3}x = 10$ 6

35. $-2x + \frac{1}{4} = 2x + \frac{4}{5}$ $-\frac{11}{80}$

36. $6x + \frac{1}{7} = 2x - \frac{2}{7}$ $-\frac{3}{28}$

C In Problems 37–60, use the CRAM procedure given in the text to solve the equation.

37. $\frac{t}{6} + \frac{t}{8} = 7$ 24

38. $\frac{f}{9} + \frac{f}{12} = 14$ 72

39. $\frac{x}{2} + \frac{x}{5} = \frac{7}{10}$ 1

40. $\frac{a}{3} + \frac{a}{7} = \frac{20}{21}$ 2

41. $\frac{c}{3} - \frac{c}{5} = 2$ 15

42. $\frac{F}{4} - \frac{F}{7} = 3$ 28

43. $\frac{W}{6} - \frac{W}{8} = \frac{5}{12}$ 10

44. $\frac{m}{6} - \frac{m}{10} = \frac{4}{3}$ 20

45. $\frac{x}{5} - \frac{3}{10} = \frac{1}{2}$ 4

46. $\dfrac{3y}{7} - \dfrac{1}{14} = \dfrac{1}{14}$ $\dfrac{1}{3}$

47. $\dfrac{x+4}{4} - \dfrac{x+2}{3} = -\dfrac{1}{2}$ 10

48. $\dfrac{w-1}{2} + \dfrac{w}{8} = \dfrac{7w+1}{16}$ 3

49. $\dfrac{x+1}{4} - \dfrac{2x-2}{3} = 3$ -5

50. $\dfrac{z+4}{3} = \dfrac{z+6}{4}$ 2

51. $\dfrac{2h-1}{3} = \dfrac{h-4}{12}$ 0

52. $\dfrac{5-6y}{7} - \dfrac{-7-4y}{3} = 2$ $-\dfrac{11}{5}$

53. $\dfrac{2w+3}{2} - \dfrac{3w+1}{4} = 1$ -1

54. $\dfrac{8x-23}{6} + \dfrac{1}{3} = \dfrac{5}{2}x$ -3

55. $\dfrac{7r+2}{6} + \dfrac{1}{2} = \dfrac{r}{4}$ $-\dfrac{10}{11}$

56. $\dfrac{x+1}{2} + \dfrac{x+2}{3} + \dfrac{x+4}{4} = -8$ $-\dfrac{122}{13}$

57. $\dfrac{x-5}{2} - \dfrac{x-4}{3}$ $= \dfrac{x-3}{2} - (x-2)$ $\dfrac{5}{2}$

58. $\dfrac{x+1}{2} + \dfrac{x+2}{3} + \dfrac{x+3}{4} = 16$ 13

59. $4(x-2) + 4 = 4x - 4$ An identity

60. $8 - (2 - 3x) = 3(x + 2)$ An identity

D In Problems 61–76, solve the equation.

61. $6.3x - 8.4 = 16.8$ 4

62. $15.5a + 49.6 = 18.6$ -2

63. $-12.6y - 25.2 = 50.4$ -6

64. $6.4y - 19.2 = 32$ 8

65. $2.1y + 3.5 = 0.7y + 83.3$ 57

66. $2.4x + 3.6 = 0.6x + 5.4$ 1

67. $3.5(x + 3) = 2.1(x + 3) + 4.2$ 0

68. $7.2(3 - t) = 2.4(3 - t) + 4.8$ 2

69. $0.40y + 0.20(32 - y) = 9.60$ 16

70. $0.30x + 0.35(50 - x) = 16$ 30

71. $0.65x + 0.40(50 - x) = 25.375$ 21.5

72. $0.09y + 0.12(200 - y) = 19.20$ 160

73. $0.06P + 0.08(2000 - P) = 130$ 1500

74. $0.15P + 0.10(6000 - P) = 660$ 1200

75. $0.30y + 1.80 = 0.20(y + 12)$ 6

76. $0.10x + 4 = 0.30(x + 10)$ 5

SKILL CHECKER

Try the "Skill Checker" exercises so you'll be ready for the next section.

Simplify:

77. $3x - (4x + 7) - 2x$ $-3x - 7$

78. $5x - 3 - (7 + 4x)$ $x - 10$

79. $6x - 2(3x - 1)$ 2

80. $8x - 4(3 - 2x) + 5x$ $21x - 12$

Evaluate:

81. $\dfrac{400 - 2 \cdot 150}{2}$ 50

82. $\dfrac{300.48 - 2 \cdot 100}{2}$ 50.24

83. $\dfrac{5}{9}(F - 32)$ when $F = 41$ 5

84. $\dfrac{5}{9}(F - 32)$ when $F = 32$ 0

USING YOUR KNOWLEDGE

If the Shoe Fits

Use the knowledge you have gained to solve the following problems about shoes.

The relationship between your shoe size S and the length of your foot L (in inches) is given by

$$S = 3L - 22 \qquad \text{(for men)}$$
$$S = 3L - 21 \qquad \text{(for women)}$$

85. If Tyrone wears a size 11, what is the length L of his foot? 11 in.

86. If Maria wears a size 7, what is the length L of her foot? $9\frac{1}{3}$ in.

87. Sam's size 7 tennis shoes fit Sue perfectly! What size women's tennis shoe does Sue wear? 8

88. The largest shoes ever sold was a pair of size 42 built for the giant Harley Davidson of Avon Park, Florida. How long is Mr. Davidson's foot? $21\frac{1}{3}$ in.

89. How long of a foot requires a size 14, the largest standard shoe size for men? 12 in.

90. How long is your foot when your shoe size is the same as the length of your foot and you are

    a. A man? 11 in.

    b. A woman? $10\frac{1}{2}$ in.

91. In 1951, Eric Shipton photographed a 23-in. footprint believed to be that of the Abominable Snowman.

    a. What size shoe does the Abominable Snowman need? 47

    b. If the Abominable Snowman turned out to be a woman, what size shoe would she need? 48

WRITE ON

Write a paragraph:

92. Explaining the difference between a conditional equation and an identity.

93. Defining a contradictory equation.

94. Detailing the terminology used for equations that have no solution, exactly one solution, and infinitely many solutions, and give examples.

MASTERY TEST

If you know how to do these problems, you have learned your lesson!

Solve:

95. $\dfrac{x + 4}{8} - \dfrac{x + 2}{6} = -\dfrac{1}{4}$ 10

96. $\dfrac{y}{8} - \dfrac{1}{4} = \dfrac{7y + 2}{12}$ $-\dfrac{10}{11}$

97. $0.8t + 0.4(32 - t) = 19.2$ 16

98. $0.60x + 3.6 = 0.40(x + 12)$ 6

99. $7(x - 1) - x = 3 - 5x + 3(4x - 3)$ -1

100. $8(x + 2) + 4x + 3 = 5x + 4 + 5(x + 1)$ -5

101. Is -5 a solution of $-\dfrac{1}{5}x + 2 = -x + 4$? No

102. Is $-\dfrac{1}{4}$ a solution of $4t - 1 = t + \dfrac{1}{4}$? No

2.2   FORMULAS, GEOMETRY, AND PROBLEM SOLVING

To Succeed, Review How To . . .

1. Solve linear equations (p. 79).

2. Evaluate expressions using the correct order of operations (p. 49).

Objectives

A Solve a formula for a specified variable and then evaluate the answer for given values of the variables.

B Write a formula for a given situation that has been described in words.

C Solve problems about angle measures.

GETTING STARTED

Trucks, Axles, and Tolls

You are a truck driver traveling on a toll road in Florida. If your truck has 5 axles, how much do you have to pay? You can use the toll schedule and a linear equation to figure this out. The cost C for a truck with n axles is given by

$$C = 0.50 + 0.25(n - 3)$$

For a truck with 5 axles, $n = 5$ and the answer is

$$C = 0.50 + 0.25(5 - 3)$$
$$= 0.50 + 0.25(2)$$
$$= 0.50 + 0.50$$
$$= \$1.00$$

Many problems in algebra can be solved if the correct formula is used. Suppose a trucker pays $1.25 for the toll. How many axles does the truck have? Here, we are interested in n in the equation $1.25 = 0.50 + 0.25(n - 3)$. Thus we are *solving for a specified variable*. The steps that we use are similar to those we used in solving linear equations. To remind you that we are solving for n, the n is in color. Here is the procedure.

$1.25 = 0.50 + 0.25(n - 3)$	Given.
$125 = 50 + 25(n - 3)$	Clear decimals; multiply by 100.
$125 = 50 + 25n - 75$	Remove parentheses/simplify.
$125 = -25 + 25n$	
$25 + 125 = -25 + 25 + 25n$	Add 25 to both sides.
$150 = 25n$	
$\dfrac{150}{25} = n$	Multiply by $\frac{1}{25}$ (or divide by 25).
$6 = n$	

The truck has 6 axles.

A   Solving Formulas for a Specified Variable

Web It

To practice solving literal equations for a specific variable, try link 2-2-1 at mhhe.com/bello.

Sometimes we are asked to solve for one of the variables in an equation called a **literal equation**. For example, the equation $C = 0.50 + 0.25(n - 3)$ is a literal equation. The procedure used to solve $C = 0.50 + 0.25(n - 3)$ for n is similar to that used for solving linear equations (page 79). Here are some suggestions to follow when solving for a *specified variable*.

PROCEDURE

Procedure to Solve for a Specified Variable

1. Add or subtract the same quantity on both sides of the equation so that the terms containing the specified **variable** (the one for which you are solving) are isolated on one side. The terms that *do not* contain the specified variable are on the other side.

2. If necessary, use the distributive property to write the side containing the specified variable as a product of the variable and a sum (or difference) of terms.

3. In general, use the rules for solving linear equations. Use CRAM.

EXAMPLE 1   Solving for a variable in a literal equation

Anthropologists know how to estimate the height of a man (in centimeters) using only a bone. They use the formula

$$H = \underbrace{2.89h}_{\substack{\text{Height of} \\ \text{the man}}} + \underbrace{}_{\substack{\text{Length of} \\ \text{the humerus}}} 70.64$$

a. Solve for h.

b. A man is 157.34 cm tall. How long is his humerus?

SOLUTION

a. We have to solve for h; that is, we isolate h on one side of the equation. It is not necessary to clear decimals because we are rewriting the formula for "h."

$H = 2.89h + 70.64$ Clear decimals (not necessary).
Remove parentheses (not necessary).

$H - 70.64 = 2.89h + 70.64 - 70.64$ Add -70.64 to both sides.

$H - 70.64 = 2.89h$

$\dfrac{H - 70.64}{2.89} = h$ Multiply by $\dfrac{1}{2.89}$ or (divide by 2.89).

Now we can find h for any given value of H.

b. Substitute 157.34 for H to get

$$h = \frac{157.34 - 70.64}{2.89} = \frac{86.7}{2.89} = 30$$

The man's humerus is 30 cm long.

PROBLEM 1

The comparable formula for a woman is $H = 2.75h + 71.48$.

a. Solve for h.

b. A woman is 126.48 cm tall. How long is her humerus?

Answers

1. a. $h = \frac{H - 71.48}{2.75}$

b. 20 cm

| **EXAMPLE 2**   Solving for a specified variable | **PROBLEM 2** |

The formula for converting degrees Fahrenheit (°F) to degrees Celsius (°C) is

$$C = \frac{5}{9}(F - 32)$$

a. Solve for F.

b. If the temperature is 35°C, what is the equivalent Fahrenheit temperature?

a. Solve for a in the formula

$$A = \frac{1}{2}h(a + b).$$

b. If $A = 40$, $h = 10$, and $b = 5$, what is a?

SOLUTION

a. We use CRAM for solving linear equations.

Given: $C = \dfrac{5}{9}(F - 32)$

1. $9 \cdot C = \overset{1}{\cancel{9}} \cdot \dfrac{5}{\cancel{9}}(F - 32)$ Clear fractions; multiply by 9.

$9C = 5(F - 32)$ Remove parentheses.

2. $9C = 5F - 160$

3. $9C + 160 = 5F$ Add 160 to both sides.

4. $\dfrac{9C + 160}{5} = F$ Multiply by $\frac{1}{5}$ (or divide by 5).

b. Substitute 35 for C. $\dfrac{9 \cdot 35 + 160}{5} = \dfrac{475}{5} = 95$

The equivalent Fahrenheit temperature is 95°F.

Teaching Tip

Suggest that students use a highlighting marker to color the specified variable as they proceed through each step.

Calculate It   Rewrite a Linear Equation and Graph

Many students believe that if they have a calculator, they don't need to know any algebra. This is an appropriate time to dispel this notion. Suppose you want to graph the simple equation $2x + 5y = 10$ using a calculator. **You cannot do it unless you understand the concepts studied in this section!** Why? Because your calculator requires you to enter an expression representing y in $2x + 5y = 10$; that is, you have to *solve* for y.

How do we do it? The way we have suggested, except that, to remember we want to isolate y, we place the variable y in a box

$2x + 5y = 10$ Given.

$2x - 2x + 5\boxed{y} = 10 - 2x$ Subtract 2x.

$5\boxed{y} = 10 - 2x$ Simplify.

$\boxed{y} = \dfrac{10 - 2x}{5}$ Divide by 5.

Now, you can enter and graph

$$y_1 = \frac{(10 - 2x)}{5}$$

The result is shown.

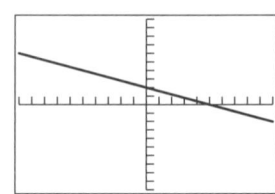

Answers

2. a. $a = \dfrac{2A}{h} - b$ or

$a = \dfrac{2A - bh}{h}$

b. $a = 3$

Many of the formulas we encounter in algebra come from geometry. In geometry, the distance around a polygon is called the **perimeter,** and the number of square units occupied by the figure is its **area.** Table 1 shows the formulas for the perimeters and areas of some common geometric figures.

Table 1 Perimeters and Areas

Name	Shape	Perimeter	Area
Square		$P = 4S$	$A = S^2$
Rectangle		$P = 2L + 2W$	$A = LW$
Triangle		$P = s_1 + s_2 + b$	$A = \dfrac{1}{2}bh$
Circle		$C = 2\pi r = \pi d$ — Circumference Radius Diameter	$A = \pi r^2$

$2r = d$ and $\pi \approx 3.14$

EXAMPLE 3   **Solving geometric shapes using a specified variable**

The perimeter P of a rectangle is given by

$$P = 2L + 2W$$

a. Solve for W.

b. One of the largest posters ever made was a rectangular greeting card 166 ft long and with a perimeter of 458.50 ft. How wide was this poster?

SOLUTION

a.
$$P = 2L + 2W$$
Clear fractions: none.
Remove parentheses: none.

$$P - 2L = 2L - 2L + 2W$$
Add $-2L$ to both sides.

$$P - 2L = 2W$$

$$\frac{P - 2L}{2} = W$$
Multiply by $\frac{1}{2}$ (or divide by 2).

b. The perimeter P was 458.50 ft and the length L was 166 ft, so we substitute 458.50 for P and 166 for L, to obtain

$$W = \frac{458.50 - 2 \cdot 166}{2}$$

$$= \frac{458.50 - 332}{2} = 63.25$$

The poster was 63.25 ft wide.

PROBLEM 3

a. Solve for L in the formula
$P = 2L + 2W$

b. The Mona Lisa, a painting by Leonardo da Vinci, has a perimeter of 102.8 in. and is 20.9 in. wide. What is the length of this painting?

Answers

3. a. $L = \frac{P - 2W}{2}$
b. 30.5 in.

| **EXAMPLE 4**   Solving interest rate problems | **PROBLEM 4** |

If you invest P dollars at the simple interest rate r, the amount of money A you will receive at the end of t years is $A = P + Prt$.

a. Solve for P.

b. At the end of 5 years, an investor receives $2250 on her investment at 10% simple interest. How much money did she invest?

PROBLEM 4

If $2A = ah + bh$,

a. Solve for h.

b. If A is 150, $a = 10$, and $b = 20$, what is h?

SOLUTION

a. This time, P appears twice on the right-hand side of the equation. Using the distributive property, we write $A = P(1 + rt)$. To get P by itself on the right, we need only divide by $(1 + rt)$. Here are all the steps:

$$A = P + Prt \qquad \text{Given.}$$

$$A = P(1 + rt) \qquad \text{Use the distributive property.}$$

$$\frac{A}{1 + rt} = \frac{P(1 + rt)}{1 + rt} \qquad \text{Divide by } 1 + rt.$$

$$P = \frac{A}{1 + rt}$$

b. We substitute 2250 for A, $10\% = 0.10$ for r, and 5 for t, to obtain

$$P = \frac{2250}{1 + 0.10 \cdot 5} = \frac{2250}{1.5} = 1500$$

She invested $1500.

| **EXAMPLE 5**   Solving for a specified variable | **PROBLEM 5** |

Do you know how to add $a_1 + a_2 + a_3 + \cdots + a_n$? If the difference between successive terms is a constant, the sum is

$$S_n = \frac{n(a_1 + a_n)}{2}$$

where n is the number of terms to be added. Note that a_1 is the first term and the subscript n in a_n indicates the number of the term. (We show you why in Appendix A.)

a. Solve for n.

b. $S_n = 2 + 4 + 6 + \cdots + 100 = 2550$. Thus $a_1 = 2$, $a_n = 100$, and $S_n = 2550$. Verify that we have added 50 terms.

PROBLEM 5

If $A = \dfrac{n(P_1 + P_n)}{2}$,

a. Solve for n.

b. If $A = 50$, $P_1 = 32$, $P_n = 18$, find n.

SOLUTION

a. We want n by itself on the right, so we multiply by 2 and divide by $a_1 + a_n$. Here are the steps:

$$S_n = \frac{n(a_1 + a_n)}{2}$$

$$2 \cdot S_n = \overset{1}{2} \cdot \frac{n(a_1 + a_n)}{2} \qquad \text{Clear fractions; multiply by 2.}$$

$$2S_n = n(a_1 + a_n)$$

$$\frac{2S_n}{a_1 + a_n} = \frac{n(a_1 + a_n)}{a_1 + a_n} \qquad \text{Multiply by } \frac{1}{a_1 + a_n} \text{ [or divide by } (a_1 + a_n)\text{].}$$

$$\frac{2S_n}{a_1 + a_n} = n$$

b. We substitute $a_1 = 2$, $a_n = 100$, and $S_n = 2550$ in the formula

$$n = \frac{2S_n}{a_1 + a_n} \qquad \text{to get} \qquad n = \frac{2 \cdot 2550}{2 + 100} = \frac{5100}{102} = 50$$

Web It

For a short quiz with answers on solving literal equations, try link 2-2-2 at mhhe.com/bello.

Answers

4. a. $h = \frac{2A}{a + b}$ **b.** 10

5. a. $n = \frac{2A}{P_1 + P_n}$ **b.** 2

B Writing and Evaluating Formulas

In the preceding examples you were asked to solve for a specified variable in a formula. There are many applications in which a formula is given in words and you have to translate it into algebra. For example, anthropologists know that the relationship between a female's height H (in centimeters) and the length f of her femur bone can be found by

$$\underbrace{\text{Multiplying 1.95 by } f}_{} \text{ and } \underbrace{\text{adding 28.68}}_{}$$

Algebraically this is $\quad\quad 1.95f \quad + \quad 28.68 = H$

See how your translation skills work in Example 6.

| EXAMPLE 6 | Writing formulas from words |

The cost of a long distance call anywhere, anytime in the United States is advertised as 39¢ for the first minute and 6¢ for each additional minute or fraction thereafter.

a. Write a formula for the cost C of a call from Miami to Dallas with time t minutes.

b. If C is the cost and t is the number of minutes for a call from Miami to Dallas, find t.

c. Roberto's telephone bill showed a $1.59 charge for a Miami-to-Dallas call. Use the formula found in part **b** to find out for how many minutes he was charged.

SOLUTION Let's look at the costs for 1, 2, 3, and 4 minutes and see whether we can notice a pattern.

For 1 minute: $\quad C = 0.39$
For 2 minutes: $\quad C = 0.39 + 0.06(1)$
For 3 minutes: $\quad C = 0.39 + 0.06(2)$
For 4 minutes: $\quad C = 0.39 + 0.06(3)$
For t minutes: $\quad C = 0.39 + 0.06(t - 1)$

You pay 6¢ for each minute after the first, so you have to multiply 0.06 by $(t - 1)$.

a. In general, the formula for C is $C = 0.39 + 0.06(t - 1)$.

b. To solve for t, think of t as the unknown and C as a number, then follow the procedure for solving a linear equation. (It is not necessary to clear the decimals.)

$C = 0.39 + 0.06(t - 1)$	Clear decimals; not necessary.
$C = 0.39 + 0.06t - 0.06$	Remove parentheses/simplify.
$C = 0.33 + 0.06t$	
$C - 0.33 = 0.06t$	Add -0.33 to both sides.
$\dfrac{C - 0.33}{0.06} = t$	Multiply by $\frac{1}{0.06}$ (or divide by 0.06).

c. Substituting 1.59 for C in the preceding equation gives us

$$\frac{1.59 - 0.33}{0.06} = \frac{1.26}{0.06} = 21$$

Roberto was charged for 21 minutes.

| PROBLEM 6 |

The cost of a long distance call is advertised as 25¢ for the first minute and 10¢ for each additional minute or fraction thereafter.

a. Write a formula for the cost C of a call with time t minutes.

b. Solve the formula in part **a** for the time t minutes.

c. If one long distance charge on Nancy's bill was $2.75, for how many minutes was she charged?

Answers

6. a. $C = 0.25 + 0.10(t - 1)$, $t > 1$
b. $t = \frac{C - 0.15}{0.1}$ **c.** 26 minutes

 **Angle Measures**

In geometry, angles are measured in **degrees**, and $1°$ is defined as $\frac{1}{360}$ of a complete revolution. Certain relationships between angles are shown in Table 2.

Table 2

Angle Relationships	Examples
If L_1 and L_2 are parallel lines crossed by a **transversal** as shown to the right, angles 1, 2, 7, and 8 are **exterior angles,** and angles 3, 4, 5, and 6 are **interior angles.** Angles 1 and 5 are **corresponding angles,** as are angles 3 and 7. The measures of angles 1 and 5 are equal; and the measures of angles 3 and 7 are equal.	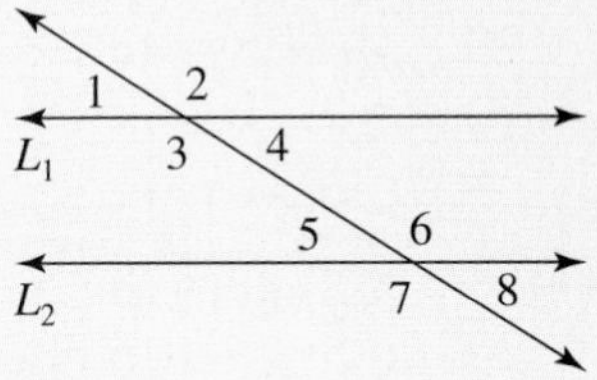
Alternate interior angles (the pairs of angles *between* the parallel lines and on *opposite* sides of the transversal) are equal; that is, the measures of angles 3 and 6 are equal. Similarly, the measures of angles 4 and 5 are equal.	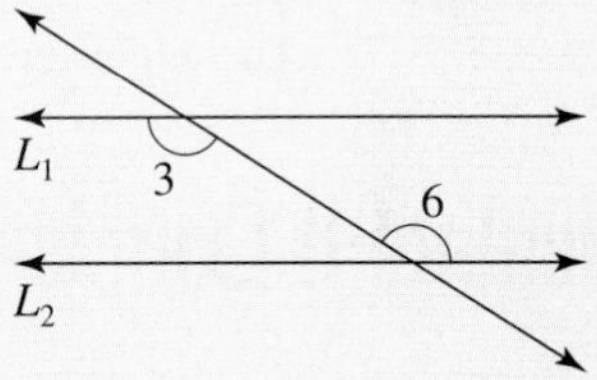 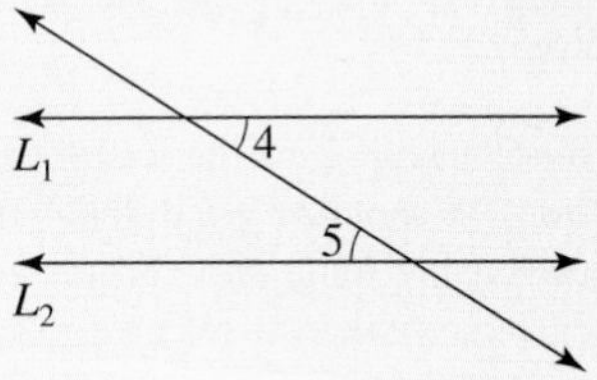
Vertical angles (angles *opposite* each other) are equal. That is, the following pairs of angles are equal: 1 and 4; 2 and 3; 5 and 8; and 6 and 7.	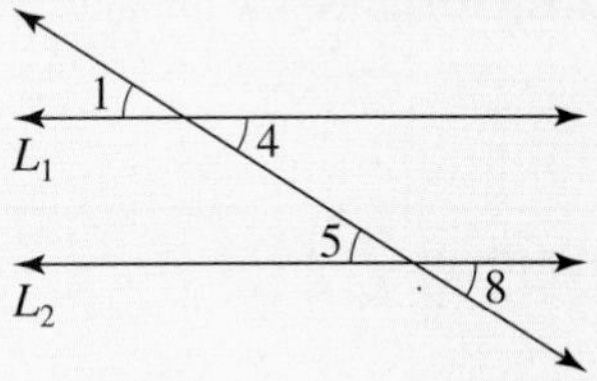 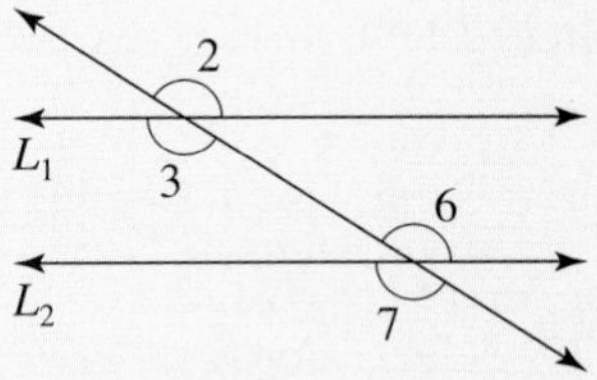

Here is quick way of remembering all these facts: Look at angles 1 and 2. Angle 1 is *acute* (less than $90°$), and angle 2 is *obtuse* (more than $90°$). Think of the measure of angle 1 as "small" and that of angle 2 as "big."

Remember:

All the acute ("small") angles in the figures have the same measure. Thus, angles 1, 4, 5, and 8 have the same measure.

All the obtuse ("big") angles in the figures have the same measure. Thus, angles 2, 3, 6, and 7 have the same measure.

EXAMPLE 7   **Finding the measure of angles**

If lines L_1 and L_2 are parallel, find x and the measure of the two unknown angles.

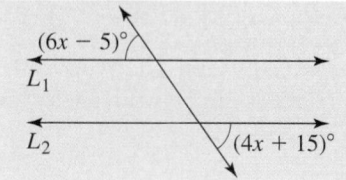

SOLUTION   The angles whose measures are given are both "small" (alternate exterior angles); hence, they have equal measures. Thus

$$6x - 5 = 4x + 15$$ Clear fractions: none.
                          Remove parentheses: none.

$$6x - 4x - 5 = 4x - 4x + 15$$ Add $-4x$ to both sides.

$$2x - 5 = 15$$

$$2x - 5 + 5 = 15 + 5$$ Add 5 to both sides.

$$2x = 20$$ Multiply by $\frac{1}{2}$ (or divide by 2).

$$x = 10$$

Substituting $x = 10$ in $6x - 5$, we find that the measure of one angle is $6 \cdot 10 - 5 = 55°$. The angles are equal, so the other angle also measures 55°.

PROBLEM 7

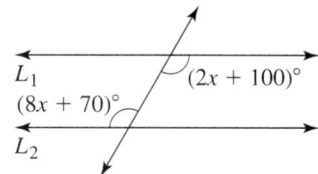

If lines L_1 and L_2 are parallel, find x and the measure of the two unknown angles.

Answer

7. $x = 5$; both angles measure 110°.

Calculate It Exercises

We can graph the equation $H = 2.89h + 70.64$ of Example 1 by using Y_1 instead of H, x instead of h, and entering $Y_1 = 2.89x + 70.64$.

1. Since x represents the length of the humerus (the bone going from the shoulder to the elbow) in centimeters, what would be a reasonable minimum and maximum for x? Press WINDOW and enter the values you selected as Xmin and Xmax with Xscl = 10.

2. Using the minimum and maximum values you selected, what would be the minimum and maximum values for Y_1?

Press WINDOW and enter the values you selected as Ymin and Ymax with YScl = 10.

3. Graph Y_1. Using your TRACE and the arrow keys, what value do you get when y is near 157.34?

4. A TI-83 Plus can evaluate expressions. You can check the result of Example 1(b) by entering 2nd TRACE 1 and, when the calculator shows X = , enter 30 and press ENTER .

Exercises 2.2

A In Problems 1–12, solve the given formula for the indicated letter.

1. $V = \pi r^2 h$ for h $h = \dfrac{V}{\pi r^2}$

2. $V = \dfrac{1}{3} \pi r^2 h$ for h $h = \dfrac{3V}{\pi r^2}$

3. $V = LWH$ for W $W = \dfrac{V}{LH}$

4. $V = LWH$ for H $H = \dfrac{V}{LW}$

5. $P = s_1 + s_2 + b$ for b $b = P - s_1 - s_2$

6. $P = s_1 + s_2 + b$ for s_2 $s_2 = P - s_1 - b$

7. $A = \pi(r^2 + rs)$ for s $s = \dfrac{A - \pi r^2}{\pi r}$ or $s = \dfrac{A}{\pi r} - r$

8. $T = 2\pi(r^2 + rh)$ for h $h = \dfrac{T - 2\pi r^2}{2\pi r}$ or $h = \dfrac{T}{2\pi r} - r$

9. $\dfrac{V_2}{V_1} = \dfrac{P_1}{P_2}$ for V_2 $V_2 = \dfrac{P_1 V_1}{P_2}$

10. $\dfrac{V_2}{V_1} = \dfrac{P_1}{P_2}$ for P_1 $P_1 = \dfrac{P_2 V_2}{V_1}$

11. $2x + 3y = 12$ for y $y = 4 - \dfrac{2}{3}x$ or $y = \dfrac{12 - 2x}{3}$

12. $2x + 3y = 12$ for x $x = 6 - \dfrac{3}{2}y$ or $x = \dfrac{12 - 3y}{2}$

13. The distance D traveled in time T by an object moving at rate R is given by $D = RT$.

    a. Solve for T. $T = \frac{D}{R}$

    b. The distance between two cities A and B is 220 miles (mi). How long would it take a driver traveling at 55 mi/hr to go from A to B? 4 hr

14. The ideal height H (in inches) of a man is related to his weight W (in pounds) by the formula $W = 5H - 190$.

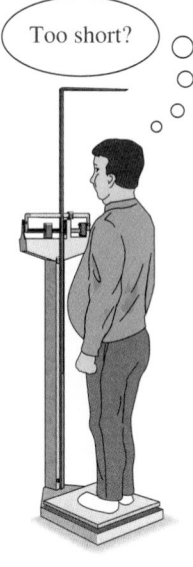

Too short?

    a. Solve for H. $H = \frac{W + 190}{5}$

    b. If a man weighs 160 lb, how tall should he be? 70 inches

15. The number of hours H a growing child should sleep is $H = 17 - \frac{A}{2}$, where A is the age of the child in years.

    a. Solve for A. $A = 34 - 2H$

    b. The parents of an infant cannot wait until the child sleeps just 8 hours a day. At what age will that happen? 18 yr

16. The efficiency energy rating (E) of an air conditioner is given by $E = \frac{B}{W}$, where (B) is the cooling capacity (per hour) in British thermal units and W is the watts of energy consumed.

    a. Solve for W. $W = \frac{B}{E}$

    b. How many watts of electricity would an air conditioner with $E = 9$ and rated at $9000 = B$ consume in 1 hr? 1000

17. The operating profit margin (O) for a business is

$$O = \frac{C + E}{N}$$

where C is the cost of goods sold, E is the operating expense, and N is the net sales.

    a. Solve for C. $C = (O)(N) - E$

    b. If the operating expenses of a business amounted to $18,500, the operating profit margin was 96% = 0.96, and the net sales were $50,000, what was the cost of the goods sold? $29,500

18. The acid-test (A) ratio for a business is given by

$$A = \frac{C + R}{L}$$

where C is the cash, R is the amount of receivables, and L is the current liability.

    a. Solve for R. $R = (A)(L) - C$

    b. If the A ratio for a business is 1, its current liability is $7800, and the business has $1200 cash on hand, what should the accounts receivable be? $6600

19. The sum of the measures of the three angles of a triangle is 180. The formula is

$$A + B + C = 180$$

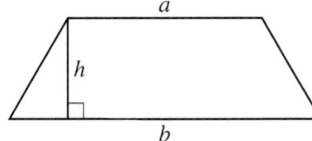

where A, B, and C represent the measures of the respective three angles.

    a. Solve for B. $B = 180 - A - C$

    b. If the measure of angle A is 47° and the measure of angle C is 119°, find the measure of angle B. 14°

20. The area A of a trapezoid is given by $A = \frac{1}{2}h(a + b)$.

    a. Solve for b. $b = \frac{2A - ha}{h}$

    b. If the area of a trapezoid is 60 square units, its height h is 10 units, and side a is 7 units, what is the length of side b? 5 units

Use the following information in Problems 21–22.

The relationship between the length L of the femur bone of a man and his height H is given by

$$H = 1.88L + 32 \tag{1}$$

where H and L are both measured in inches. The corresponding equation for a woman is

$$H = 1.95L + 29 \tag{2}$$

21. a. Solve equation (1) for L. $L = \dfrac{H - 32}{1.88}$

    b. Can a 20-in. femur belong to a man whose height was 6 ft? No

    c. Can the bone belong to a man whose height was 69.6 in.? Yes

    d. Solve equation (2) for L. $L = \dfrac{H - 29}{1.95}$

    e. Can a femur measuring 20 in. in length belong to a woman whose height was 5 ft, 8 in.? Yes

22. A police pathologist wants to check the accuracy of equations (1) and (2). Use your calculator to find L (to one decimal place) for

    a. A man 5 ft, 6 in. in height. 18.1

    b. A man 5 ft, 8 in. in height. 19.1

    c. A woman 5 ft, 6 in. in height. 19.0

    d. A woman 5 ft, 10 in. in height. 21.0

 Translate the following word problems into formulas.

23. A new Acura can be leased for $2000 down and $309 per month.

    a. Find a formula for the total cost C of the car after m months. $C = 2000 + 309m$

    b. Use the formula from part **a** to find the total cost C after leasing the car for the required 39 months. $14,051

24. A disc jockey charges $150 per hour with a set up fee of $75.

    a. Write a formula for the cost C of the disc jockey for h hours. $C = 75 + 150h$

    b. If the cost of the DJ for Cindy's birthday was $675, how many hours did he work? 4 hr

25. The cost C of a Tampa-to-Los Angeles call during business hours is 36¢ for the initial minute and 8¢ for each additional minute t or fraction thereof.

    a. Write a formula for the cost C of a Tampa-to-Los Angeles call. $C = 0.36 + 0.08(t - 1), t \geq 1$

    b. Use the formula from part **a** to find the time of a phone call from Tampa to Los Angeles if the cost was $4.04. 47 min

26. A plumber charges $25 per hour ($h$) plus $30 for the service call.

    a. Find a formula for C if C is the total cost for the call. $C = 30 + 25h$

    b. Use the formula from part **a** to find the number of hours worked by a plumber who charged $142.50. 4.5 hr

27. A finance company will lend you $1000 for a finance charge of $20 plus simple interest at 1% per month. This means that you will pay interest of 1% of $1000— that is, $10 per month in addition to the $20 finance charge.

    a. Find a formula for F if F is the finance charge and m is the number of months before you repay the loan. $F = 20 + 10m$

    b. Use the formula from part **a** to find the number of months outstanding on a $1000 loan that has already cost $140 in finance charges. 12

 In Problems 28–35, find x and then find the measure of each marked angle. In Problems 32–35, assume that L_1 and L_2 are parallel.

28.

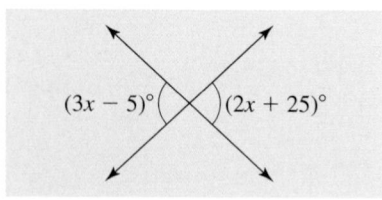

$x = 30$; angles $= 85°$ each

29.

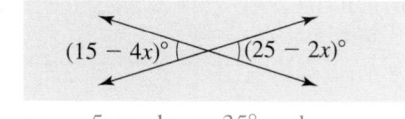

$x = -5$; angles $= 35°$ each

30.

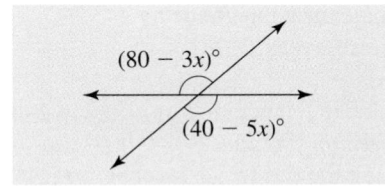

$x = -20$; angles $= 140°$ each

31.

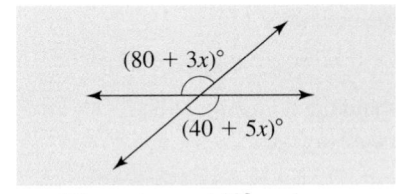

$x = 20$; angles $= 140°$ each

32.

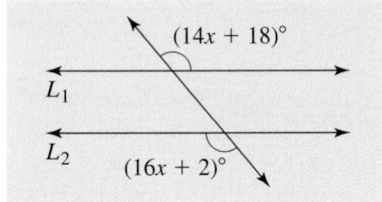

$x = 8$; angles $= 130°$ each

33.

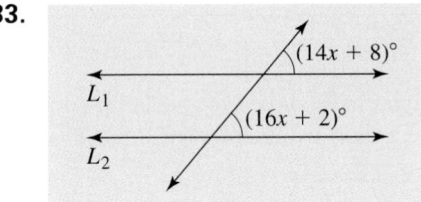

$x = 3$; angles $= 50°$ each

34.

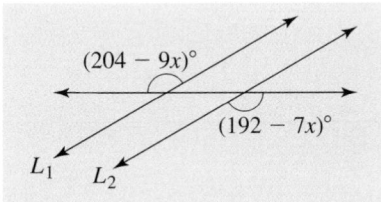

$x = 6$; angles $= 150°$ each

35.

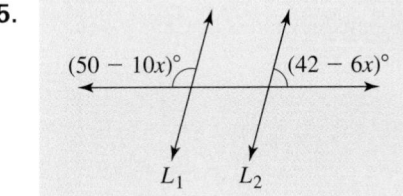

Hint: The two marked angles are supplementary. Their sum is 180°.

$x = -5.5$; $(42 - 6x)° = 75°$; $(50 - 10x)° = 105°$

SKILL CHECKER

Try the "Skill Checker" exercises so you'll be ready for the next section.

Simplify:

36. $n + (n + 1) + (n + 2)$ $3n + 3$

37. $n + (n + 2) + (n + 4)$ $3n + 6$

Solve:

38. $(p + 50) + p = 110$ $p = 30$

39. $p + (p + 60{,}000) = 110{,}000$ $p = 25{,}000$

40. $0.30m + 25 = 52$ $m = 90$

41. $0.25m + 20 = 37.50$ $m = 70$

42. $x + (90 - x) + 3x = 180$ $x = 30$

43. $(90 - x) + x + 5x = 180$ $x = 18$

USING YOUR KNOWLEDGE

Global Warming and Sea Levels

Do you know what global warming is? It is the theory that over the past century, the earth has begun to warm significantly due to factors like pollution (from your car and power plants); deforestation (fewer trees, less shade, less rain, more heat); and the depletion of the ozone layer (ozone absorbs ultraviolet radiation; less ozone, more sun and more heat). The average global temperature C (in degrees Celsius) can be approximated by

$$C = 14.84 + 0.08t$$

where t is the number of years after 1960.

44. Solve for t in the equation. $t = \frac{C - 14.84}{0.08}$

45. In what year would the temperature be 20°C? Give your answer to the nearest year. 2025

46. In what year would the temperature be 25°C? Give your answer to the nearest year. 2087

47. The average global temperature in 1962 was about 15°C. An increase of 1°C causes a 1-ft rise in the sea level. What would the rise in sea level be from 1962 to the year 2000? 3.04 ft

WRITE ON

48. In this section you have solved formulas for specified variables. Write a short paragraph explaining what that means. Answers may vary.

49. A linear equation in one variable is an equation that can be written as $ax + b = c$ (a, b, and c are real numbers and $a \neq 0$). Write an explanation of how you would solve for x in this equation. Use your explanation to solve the equation $-3x + 10 = 16$ for x. Answers may vary.

50. The definition of linear equation in one variable (Problem 49) states that a cannot be zero ($a \neq 0$). Explain what happens when a is 0. Can $c = 0$? Answers may vary.

MASTERY TEST

If you know how to do these problems, you have learned your lesson!

In Problems 51 and 52, find x and the measures of the marked angles:

51.

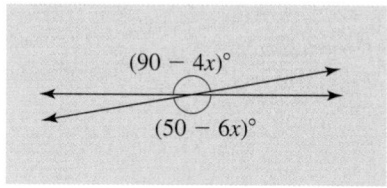

$(90 - 4x)°$
$(50 - 6x)°$

$x = -20$; angles $= 170°$ each

52. Assume L_1 and L_2 are parallel lines.

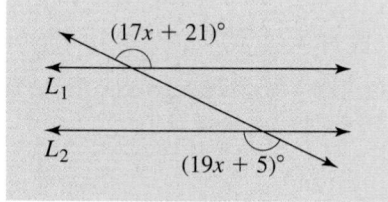

$(17x + 21)°$
L_1
L_2
$(19x + 5)°$

$x = 8$; angles $= 157°$ each

53. The cost C of mailing a first class letter is 37¢ for the first ounce (oz) and 23¢ for each additional ounce or fraction thereof. If w is the weight of the letter in ounces:

a. Write a formula for the cost C in dollars.
$C = 0.37 + 0.23(w - 1)$, $w > 1$
b. Solve for w in the formula. $w = \frac{C - 0.14}{0.23}$

c. What is the cost of mailing a letter weighing 10 oz? $2.44

54. The annual number N of cigarettes per person consumed in the United States is the difference between 4200 and the product of 70 and t, where t is the number of years after 1960.

a. Write a formula for N. $N = 4200 - 70t$

b. In what year would consumption drop to 2100 cigarettes per person? 1990

55. The speed S (in centimeters per second) for a certain type of ant is $S = \frac{1}{6}(C - 4)$, where C is the temperature in degrees Celsius.

a. Solve for C. $C = 6S + 4$

b. If the ant is moving at a rate of 2 cm/sec, what is the temperature? 16°C

56. The number of chirps N a certain type of cricket makes per minute is $N = 4(F - 40)$, where F is the temperature in degrees Fahrenheit.

a. Solve for F. $F = \frac{N + 160}{4}$

b. If the cricket is chirping 80 times per minute, what is the temperature? 60°F

2.3 # PROBLEM SOLVING: INTEGERS AND GEOMETRY

To Succeed, Review How To ...

1. Simplify expressions (pp. 51–55).

2. Solve linear equations involving decimals (p. 82).

Objectives

A Translate a word expression into a mathematical expression.

B Solve word problems of a general nature.

C Solve word problems about integers.

D Solve word problems about geometric formulas and angles.

GETTING STARTED

RSTUV Procedure

In the preceding sections we learned how to solve certain types of equations. Now we are ready to apply this knowledge to solve problems. These problems will be stated in words and are consequently called **word,** or **story, problems.** Word problems frighten many students, but do not panic. We have a surefire method for tackling *any* word problem. Here is our five-step procedure:

1. **R**ead the problem. Not once or twice, but until you understand it.

2. **S**elect the unknown; that is, find out what the problem asks for.

3. **T**hink of a plan to solve the problem.

4. **U**se the techniques you are studying to carry out the plan.

5. **V**erify the answer.

Look at the first letter in each sentence. To help you remember the steps, we call it the **RSTUV** method. Later in this section we have additional tips on how to use this method, but first, we have to discuss the terminology we shall use.

 **A** **Translating Words into Mathematical Expressions**

You have probably noticed the frequent occurrence of certain words in the statements of word problems. Table 3 presents a brief mathematics dictionary to help you translate these words properly and consistently.

We are now ready to use our mathematics dictionary to translate words into mathematical expressions.

Teaching Tip

Ask students if $(y - x)$ means the same as $(x - y)$. Have them provide examples that dispute it. See if they can remember for which operations changing the order of the variables does work and name the property used.

Table 3 Mathematics Dictionary

Words	Translation	Example	Translation
Add More than Sum Increased by Added to	+	Add n to 7 7 more than n The sum of n and 7 n increased by 7 7 added to n	$n + 7$
Subtract Less than Minus Difference Decreased by Subtracted from	−	Subtract 9 from x 9 less than x x minus 9 Difference of x and 9 x decreased by 9 9 subtracted from x	$x - 9$
Of The product Times Multiply by	×	$\frac{1}{2}$ of a number x The product of $\frac{1}{2}$ and x $\frac{1}{2}$ times a number x Multiply $\frac{1}{2}$ by x	$\frac{1}{2}x$
Divide Ratio Divided by Divide into The quotient	÷	Divide 10 by x Ratio of 10 to x 10 divided by x x divided into 10 The quotient of 10 and x	$\frac{10}{x}$
The same, yields, gives, is, equals	=	The sum of x and 10 is the same as the sum of 10 and x.	$x + 10 = 10 + x$

Study Skills Hint

To learn the terms in this mathematics dictionary, write them and their translation on an index card. These cards should be reviewed before and after each class and each homework assignment.

CAUTION

Don't confuse "7 less than x," that is, $x - 7$, with "7 is less than x," which is $7 < x$. Moreover, *do not* write "7 less than x" as $7 - x$, which means "x less than 7." A number x subtracted *from* y is always written as $y - x$, *not* $x - y$. For division, remember that the numerator is divided *by* the denominator. Thus x divided by y means

$$\frac{x}{y} \leftarrow \begin{array}{l}\text{numerator} \\ \leftarrow \text{denominator}\end{array}$$

y divided *into* x also means $\frac{x}{y}$.

EXAMPLE 1 Translating words into mathematical equations

Cover the "Translation" column and see if you can translate the sentences into equations.

Verbal Sentence	Translation
A number increased by 7 yields 25.	$x + 7 = 25$
The quotient of a number and the number minus 8 equals 22.	$\frac{x}{x - 8} = 22$
Twice a number decreased by 9 equals the number divided into 9.	$2x - 9 = \frac{9}{x}$
If the product of a number and 10 is subtracted from 15, the result is 45.	$15 - 10x = 45$
The difference of the consecutive integers n and $n + 1$ is always -1.	$n - (n + 1) = -1$

PROBLEM 1

Translate the sentences into equations:

a. A number increased by 5 yields 8.

b. The quotient of a number and the number minus 4 equals 16.

c. Twice a number decreased by 7 equals the number divided into 14.

d. If the product of a number and 6 is subtracted from 9, the result is 21.

e. The difference of the two consecutive even integers n and $n + 2$ is always -2.

General Word Problems

The mathematics dictionary we have presented plays an important role in the solution of word problems. You will find that the words contained in the dictionary are often key words when you translate a problem. But there are other details that can help you. Here is a restatement of the **RSTUV** method with hints and tips you can use when you solve word problems.

HINTS AND TIPS

Mathematics is a language. As such you have to learn how to read it. You may not understand or even get through reading the problem the first time. Read it again and as you do, pay attention to key words or instructions such as *compute, draw, write, construct, make, show, identify, state, simplify, solve,* and *graph.* (Can you think of others?)

1. Read the problem.
Using a complete sentence, write a "Find" statement. "Find the sales tax."

How can you answer a question if you don't know what the question is? One good way to look for the unknown is to look for the question mark (?) and read the material to its left. Try to determine what is given and what is missing.

2. Select the unknown.
Using a "Let" statement, represent the unknown(s) as a variable expression. "Let x = the sales tax."

Problem solving requires many skills and strategies: Some of them are *look for a pattern; examine a related problem; make tables, pictures, diagrams; write an equation; work backwards;* and *make a guess.* (Can you think of others?)

3. Think of a plan.
Using the information, translate it first to a word equation cost + tax = total and then to an algebra equation

$$\$2.98 + x = \$3.16$$

If you are studying a mathematical technique, it is almost certain that you will have to use it to solve the given problem. Look for specific procedures given to solve certain problems.

4. Use the techniques you are studying to carry out the plan.
Using your algebra skills, solve the equation.

$$
\begin{aligned}
2.98 + x &= 3.16 \\
-2.98 &\; -2.98 \\
\hline
x &= \$0.18
\end{aligned}
$$

Look back and check the results of the original problem. Is your answer reasonable? Can you find it some other way?

5. Verify the answer.
Replace the result into the word problem and check.

$$\$2.98 + \$0.18 = \$3.16$$

Now let's try a word problem. Did you know a man in England balanced a car on his head for 33 seconds? His name is John Evans and we are ready to solve a problem about it. This solution is presented in a two-column format. Cover the left-hand column (a 3-by-5 index card will do), and write *your answers* so you can practice the RSTUV method. After that, uncover the answers and check to see whether you are right. We will then give you another example and provide you with its solution.

Answers

1. a. $x + 5 = 8$ **b.** $\frac{x}{x-4} = 16$
c. $2x - 7 = \frac{14}{x}$ **d.** $9 - 6x = 21$
e. $n - (n + 2) = -2$

According to the *Guinness Book of World Records* professional "Head Balancer" John Evans balanced a Mini car on his head for 33 seconds at The London Studios, England, on May 24, 1999. The car weighed 3.9 kg more than Mr. Evans. The combined weight of the car and Mr. Evans was 315.3 kg. What was the weight of each?

1. Read and write a "Find" statement.
Find Mr. Evan's weight and the car's weight.

2. Select the unknown using a "Let" statement.
Let E = Mr. Evan's weight and
(E + 3.9) = the car's weight.

3. Think of a word equation and translate it to an algebra equation.

$$\underbrace{\text{The car's weight}}_{(E + 3.9)} \quad \text{plus} \quad \underbrace{\text{Mr. Evan's weight}}_{E} \quad = \quad \underbrace{\text{combined weight}}_{315.3}$$

4. Use your algebra skills to solve the equation.

$$2E + 3.9 = 315.3$$

$$2E = 311.4$$

$$E = 155.7$$

$$(E + 3.9) = 159.6$$

5. Verify the result.

$$159.6 + 155.7 = 315.3$$

Thus, Mr. Evan's weighed 155.7 kg and the car weighed 159.6 kg.

The problem asks for the weight of each—that is, the weight of Mr. Evans and the weight of the car.

Let E represent the weight of Mr. Evans in kg. Since the car weighed 3.9 kg, more than Mr. Evans, the car weighed (E + 3.9).

We use the sentence, "The combined weight of the car and Mr. Evans was 315.3 kg" to write the word equation.

Use the variable expressions from the "Let" statement as replacements in the word equation.

Simplify by dropping parenthesis and combining like terms.

Subtract 3.9.

Divide by 2 to get Mr. Evan's weight.

Find the car's weight by using (E + 3.9).

EXAMPLE 2 **Car loan costs**

J. R. Clementi buys a $20,000 car. He puts $2500 down in cash and trade-in value. His monthly payment is $335. He is told that when the car loan is paid off he will have paid a total of $26,620 for the car. How many months will it take J. R. to pay off the car?

SOLUTION We use the RSTUV method.

1. Read and write a "Find" statement.
Find how many months it will take J. R. to pay off the car.

2. Select the unknown using a "Let" statement.
Let m = the number of months to pay off the car.

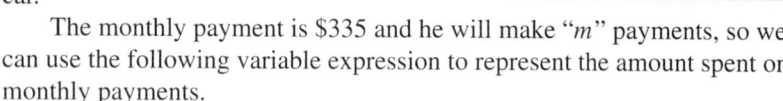

The monthly payment is $335 and he will make "$m$" payments, so we can use the following variable expression to represent the amount spent on monthly payments.

$$\$335 \, (m) = \text{money spent in monthly payments}$$

Hint: Often numbers that are not needed are mentioned in a problem, like the "33 seconds" in the car balance problem or the "$20,000 car" in the car loan cost problem.

PROBLEM 2

Kyra rented an intermediate sedan at $25 per day plus $0.20 per mile. How many miles can Kyra travel in one day for $50?

Answer

2. Kyra can travel 125 miles.

3. Think of a word equation and translate it to an algebra equation.

Down payment plus the monthly payments = total paid for the car.

$$\$2500 \quad + \quad \$335\,(m) \quad = \quad \$26{,}620$$

4. Use your algebra skills to solve the equation.

$$335\,m = 24{,}120 \qquad \text{Subtract 2500.}$$
$$m = 72 \qquad \text{Divide by 335.}$$

5. Verify the results.

$$2500 + 335(72) = 26{,}620.$$

It took J. R. 72 months (6 years) to pay off the car.

C Integer Word Problems

Many popular algebra problems deal with integers. If you are given an integer, can you find the integer that follows? For example, the integer that follows 7 is $7 + 1 = 8$, and the one that follows -6 is $-6 + 1 = -5$.

Consecutive Integers	If n is any integer, the next **consecutive** integer is $n + 1$.

On the other hand, if you are given an *even* integer such as 6, the next *even* integer is 8 (add 2 this time). The next even integer after 34 is $34 + 2 = 36$.

Consecutive Even (or Odd) Integers	If n is an *even* (or *odd*) integer, the next *even* (or *odd*) integer is $n + 2$.

Web It

For more examples of integer word problems, try link 2-3-1 at mhhe.com/bello.

If $n = 34$, the next even integer is $n + 2 = 34 + 2 = 36$. Similarly, if $n = 21$, the next odd integer is $n + 2 = 21 + 2 = 23$.

We use this idea in the next example. (Don't forget to consult the mathematics dictionary when necessary.)

EXAMPLE 3 Integer problems	**PROBLEM 3**

The sum of three consecutive odd integers is 129. Find the integers.

The sum of three consecutive odd integers is 249. Find the integers.

SOLUTION   Use the RSTUV method. We are asking for three consecutive odd integers. Look at the following number line and notice how far apart the odd integers are.

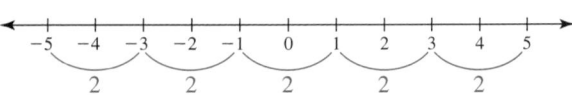

1. Read and write a "Find" statement.

Find three consecutive odd integers.

Answer

3. The three consecutive odd integers are 81, 83, and 85.

2. Select the unknown using a "Let" statement.

If we let n be the first of the odd integers, then we can represent the other two with variable expressions by using n and the spacing on the number line. We want three consecutive odd integers, so we need to find the next two consecutive odd integers. After noting on the graph that the odd integers are "2" apart, the next odd integer after n is $(n + 2)$ and the one after $(n + 2)$ is $(n + 4)$.

$$\text{Let } n = \text{the first odd integer}$$
$$(n + 2) = \text{the second odd integer}$$
$$(n + 4) = \text{the third odd integer}$$

3. Think of a word equation and translate it to an algebra equation.

First odd integer	plus	second	plus	third	=	the sum of the three
n	$+$	$(n + 2)$	$+$	$(n + 4)$	$=$	129

4. Use your algebra skills to solve the equation.

$$3n + 6 = 129 \qquad \text{Combine like terms.}$$
$$3n = 123 \qquad \text{Subtract 6.}$$
$$n = 41 \qquad \text{Divide by 3 for the first.}$$
$$(n + 2) = 43 \qquad \text{The second odd integer}$$
$$(n + 4) = 45 \qquad \text{The third odd integer}$$

5. Verify the results.

$$41 + 43 + 45 = 129$$

Thus, the three consecutive odd integers are 41, 43, and 45.

D Geometry Problems

As you recall, the *perimeter* (distance around) of a rectangle is $P = 2L + 2W$, where L is the length and W is the width of the rectangle. Here's a problem using this formula.

EXAMPLE 4 **A geometry word problem**

The students at Osaka Gakun University made a rectangular poster whose length was 130 ft more than its width. If the perimeter of the poster was 416 ft, give its dimensions.

SOLUTION Use the RSTUV method. The key words are "rectangular" and "perimeter." Drawing the geometric shape and labeling it will help.

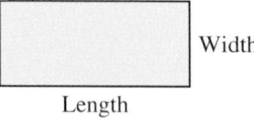

Width

Length

1. Read and write a "Find" statement.

Find the dimensions, which are the length and width.

2. Select the unknown using a "Let" statement.

If we let "w" represent the measure of the width, we can represent the measure of the length with a variable expression using "w."

The length is 130 feet more than the width so we can represent the length as $(w + 130)$.

$$\text{Let } w = \text{the measure of the width}$$
$$(w + 130) = \text{the measure of the length}$$

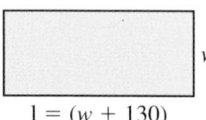

w

$1 = (w + 130)$

PROBLEM 4

A poster is 100 in. longer than it is wide. If its perimeter is 360 in., give its dimensions.

Answer

4. The width is 40 in. and the length is 140 in.

3. Think of a word equation and translate it to an algebra equation.

We use the formula for the perimeter of the rectangle to write the word equation and translate it to an algebra equation.

$$2L + 2W = P$$

$$2(\text{length}) + 2(\text{width}) = \text{the perimeter of the rectangle.}$$

$$2(w + 130) + 2(w) = 416$$

4. Use your algebra skills to solve the equation.

$2w + 260 + 2w = 416$	Distribute.
$4w + 260 = 416$	Combine like terms.
$4w = 156$	Subtract 260.
$w = 39$	Divide by 4 for the width.
$(w + 130) = 169$	The length

5. Verify the results.

$$2(169) + 2(39) = 416$$

Thus, the dimensions of the rectangle are width of 39 ft and length of 169 ft.

Web It

For more geometry word problems, try link 2-3-2 at mhhe.com/bello.

In Section 2.2, we studied vertical angles and angles formed by transversals that intersect parallel lines. Table 4 shows some other types of angles and their relationships.

Table 4

Angle	Relationship
A **right angle** is an angle whose measure is 90°. If the sum of the measures of two angles is 90°, the angles are **complementary angles** and they are **complements** of each other.	Right angle Complementary angles If the measure of one of the angles is $x°$, the measure of its complement is $(90 - x)°$.
A **straight angle** is an angle whose measure is 180°. If the sum of the measures of two angles is 180°, the angles are **supplementary angles** and they are **supplements** of each other.	Straight angle Supplementary angles If the measure of one of the angles is $x°$, the measure of its supplement is $(180 - x)°$.
The **sum** of the measures of the angles of a triangle is 180°. You can use this fact to find the measure of the third angle when the measures of the other two angles are given.	 The measure of the other angle, call it $x°$, is such that $$60 + 35 + x = 180$$ $$95 + x = 180$$ $$x = 85$$ The third angle is 85°.

EXAMPLE 5   **A geometry problem involving angles**

PROBLEM 5

Two of the angles in a triangle are complementary. The third angle is twice the measure of one of the complementary angles. What is the measure of each of the angles?

If the measures of the three angles of a triangle are related such that the second angle is twice the first angle and the third angle is 20 more than the first angle, find the measure of each of the angles.

SOLUTION   Use the RSTUV method. The key words are "complementary angles" and "angles in a triangle." You can use Table 4 to help with this problem and draw the geometric shape.

1. Read and write a "Find" statement.
Find the measures of the three angles of the triangle.

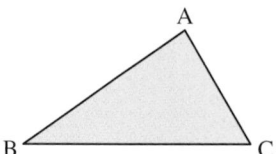

2. Select the unknown using a "Let" statement.
If we let x be the measure of one of the complementary angles, then we can represent the measure of the other complementary angle as $(90 - x)$. Well, that takes care of two of the three angles in the triangle, but what about the third angle? It says that the third angle is twice the measure of one of the complementary angles, so we represent its measure with $2(x)$.

Let x = the measure of one of the complementary angles

$(90 - x)$ = the measure of the other complementary angle

$2(x)$ = the measure of the third angle of the triangle

3. Think of a word equation and translate it to an algebra equation.
In Table 4 it states that the sum of the measures of the angles of a triangle is 180. We use this fact to write a word equation that we will translate into an algebra equation.

Measure of the first angle	+	measure of the second angle	+	measure of the third angle	is 180
x	+	$(90 - x)$	+	$2(x)$	= 180

4. Use your algebra skills to solve the equation.

$2x + 90 = 180$ Combine like terms.

$2x = 90$ Subtract 90.

$x = 45$ Divide by 2 for the first angle measure.

$(90 - x) = 45$ For the second angle measure

$2(x) = 90$ For the third angle measure

5. Verify the results.

$$45 + 45 + 90 = 180$$

The measures of the three angles of the triangle are 45, 45, and 90.

Answer

5. The three angle measures are 40, 80, and 60.

Calculate It   Suggestions for Solving Word Problems

How do you solve word problems with a calculator? The same way as without one! The calculator is an additional tool. Here are some suggestions for solving word problems with a calculator:

1. Use the RSTUV method until you reach step 3.

2. Graph the equation or inequality obtained in step 3 using your calculator. To do this, write the equation in the form $A(x) = 0$ and graph $Y_1 = A(x)$.

We must warn you, however, that in many cases it is *easier* to solve the problems *algebraically* than it is to solve them *graphically*.

Calculate It Exercises

1. The graph for Example 3, given in Window 1, uses a $[-5, 45]$ with Xscl = 5 by $[-5, 5]$ with Yscl = 5 window. The equations are $y_1 = 3n - 123$ and $y_2 = 0$. What window would you use so that the values of x are near the point at which the line crosses the horizontal axis?

2. In Example 4, show the key strokes to find the solution (root) of the given equation.

3. Check the results in Examples 2 and 5 by using a calculator to graph the equations involved.

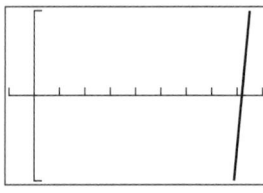

Window 1

Exercises 2.3

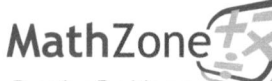
A In Problems 1–4, translate the sentences into equations.

Write an equation equivalent to the description. (Do not solve.)

1. The product of 4 and a number m is the number increased by 18. $4m = m + 18$

2. The quotient of a number and 3 yields 5 more than twice the number. $\frac{x}{3} = 2x + 5$

3. One half of a number less 12 is the same as two times the number divided by 3. $\frac{1}{2}x - 12 = \frac{2x}{3}$

4. The square of a number x, decreased by twice the number itself, is 10 more than the number. $x^2 - 2x = x + 10$

B In Problems 5–14, write the given statement as an equation and then solve it.

5. If 4 times a number is increased by 5, the result is 29. Find the number. $4x + 5 = 29; x = 6$

6. Eleven more than twice a number is 19. Find the number. $2x + 11 = 19; x = 4$

7. The sum of 3 times a number and 8 is 35. Find the number. $3x + 8 = 35; x = 9$

8. If 6 is added to 7 times a number, the result is 76. Find the number. $7x + 6 = 76; x = 10$

9. If the product of 3 and a number is decreased by 2, the result is 16. Find the number. $3x - 2 = 16; x = 6$

10. Five times a certain number is 9 less than twice the number. What is the number? $5x = 2x - 9; x = -3$

11. Five times a certain number is the same as 12 increased by twice the number. What is the number? $5x = 12 + 2x; x = 4$

12. If 5 is subtracted from half a number, the result is 1 less than the number itself. Find the number. $\frac{1}{2}x - 5 = x - 1; x = -8$

13. One-third of a number decreased by 2 yields 10. Find the number. $\frac{1}{3}x - 2 = 10; x = 36$

14. One-fifth of a certain number plus 2 times the number is 11. What is the number? $\frac{1}{5}x + 2x = 11; x = 5$

In Problems 15–20, use the RSTUV method to obtain the solution.

15. According to a *Sports Illustrated* 2003 survey of major league baseball players, golf was voted the most favorite nonbaseball activity. The other four on the list were hunting, fishing, time with the family, and the movies, in that order. Golf received approximately three times as many votes as the other four combined. Using 100% as the total amount of votes, find the percent of the votes received by golf. 75%

16. According to that same *Sports Illustrated* 2003 survey, Barry Bonds was voted the greatest living baseball player. The other five on the list were Alex Rodriguez, Willie Mays, Nolan Ryan, Hank Aaron, and Pete Rose, in that order. Bonds received approximately $1\frac{1}{2}$ times as many votes as the other five. Using 100% as the total amount of votes, find the percent of the votes received by Barry Bonds. 60%

17. Alex is 12 years old and his sister Ramie is 2 years old. In how many years will Alex be exactly twice as old as Ramie? 8 yr

18. Morgan is 19 years old and her sister Erin is 3 years old. In how many years will Morgan be exactly three times as old as Erin? 5 yr

19. The cost of renting a car is $18 per day plus $0.20 per mile traveled. Margie rented a car and paid $44 at the end of the day. How many miles did Margie travel? 130 mi

20. The Zone LD company offers a long distance phone rate plan that requires a monthly fee of $2 and an interstate rate of $0.04 per minute. If Kyra's long distance phone bill for one month is $13.32, how many long distance minutes did she talk? 283 min

C In Problems 21–32, consult your mathematics dictionary, if necessary, to solve the integer problems.

21. The sum of three consecutive even integers is 138. Find the integers. 44; 46; 48

22. The sum of three consecutive odd integers is 135. Find the integers. 43; 45; 47

23. The sum of three consecutive odd integers is -27. Find the integers. -11; -9; -7

24. The sum of three consecutive even integers is -24. Find the integers. -10; -8; -6

25. The sum of two numbers is 179, and one of them is 5 more than the other. Find the numbers. 87; 92

26. The larger of two numbers is 6 times the smaller. Their sum is 147. Find the numbers. 21; 126

27. Would you believe Florida gets more tornadoes than Kansas or Oklahoma? However, the tornadoes in Florida are rated F-0 and F-1 as compared to the F-5 rated storms of the Midwest. The Fujita scale uses the damage a tornado causes to gauge the storm's wind speed in mi/hr. An F-5 rated (incredible) tornado can have winds that exceed three times an F-0 rated (minor damage) tornado by 100 mi/hr. If the two wind rates were added they would total 388.
   a. Find the wind rate for an F-0 rated tornado. 72 mi/hr
   b. Find the wind rate for an F-5 rated tornado.
      316 mi/hr

28. The Saffir-Simpson scale is used to measure the intensity of a hurricane. They are measured in categories from 1 to 5 according to their wind rates in mi/hr. The wind speed of a category 4 hurricane exceeds that of the wind speed of a category 1 hurricane by 60 mi/hr. If the two wind rates were added it would total 250.
   a. Find the wind rate for a category 1 hurricane.
      95 mi/hr
   b. Find the wind rate for a category 4 hurricane.
      155 mi/hr

29. To find the weight of an object on the moon, you can divide its weight on Earth by 6. The crew of Apollo 16 collected lunar rocks and soil weighing 35.5 lb on the moon. What is their weight on Earth? 213 lb

30. The weight of an object on the moon is obtained by dividing its earth weight by 6. If an astronaut weighs 28 lb on the moon, what is the corresponding earth weight?
   168 lb

31. The Beatles have 20 more Recording Industry Association of America awards than Paul McCartney. If the number of awards received by the Beatles and McCartney total 74, how many awards do the Beatles have? 47 awards

32. The greatest weight difference ever recorded in a major boxing bout was 140 lb in a match between John Fitzsimmons and Ed Punkhorst. If the combined weight of the contestants was 484 lb, find Fitzsimmons' weight. (He was the lighter of the two.) 172 lb

D In Problems 33–40, solve the geometry problems.

33. The largest painting in the world used to be the *Panorama of the Mississippi,* by John Banvard. If the length of this painting was 4988 ft more than its width and its perimeter was 10,024 ft, find the dimensions of this rectangular painting. 12 ft by 5000 ft

34. In 2001, *Hero: The World's Largest Painting by One Artist,* painted by Eric Waugh, was recorded in the *Guinness Book of World Records* as having a perimeter of 820 feet. If the length of this painting exceeds its width by 50 feet, find its dimensions. 180 ft by 230 ft

35. The scientific building with the greatest capacity is the Vehicle Assembly Building at the John F. Kennedy Space Center. The width of this building is 198 ft less than its length. If the perimeter of the rectangular building is 2468 ft, find its dimensions. 518 ft by 716 ft

36. The largest fair hall is located in Hanover, Germany. The length of this hall exceeds its width by 295 ft. If the perimeter of this rectangular hall is 4130 ft, find its dimensions. 885 ft by 1180 ft

37. An angle has four times the measure of its complement. What is the measure of the angle? 72°

38. The sum of the measures of an angle and one third a second angle is 32°. If the angles are complementary, what are their measures? 3°; 87°

39. An angle has three times the measure of its supplement. What is the measure of the angle? 135°

40. An angle is 5° less than twice another angle. Find their measures if they are supplementary angles. $61\frac{2}{3}°$; $118\frac{1}{3}°$

41. Find x and the measure of each complementary angle.

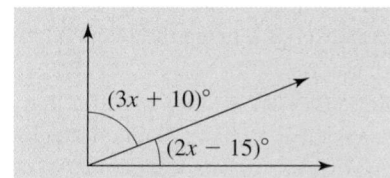

$x = 19$; angles: 67°, 23°

42. Find x and the measure of each complementary angle.

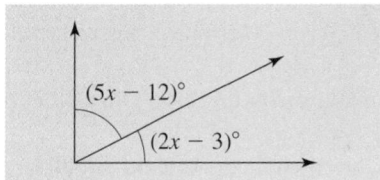

$x = 15$; angles: 27°, 63°

43. Find x and the measure of each supplementary angle.

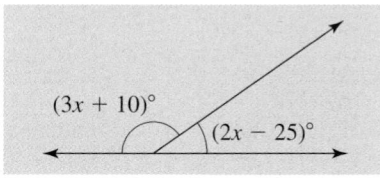

$x = 39$; angles: 127°, 53°

44. Find x and the measure of each supplementary angle.

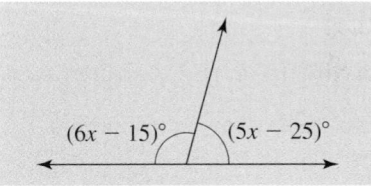

$x = 20$; angles: 75°, 105°

45. A triangle has two complementary angles. If the measure of the third angle is three times that of the first, what are the measures of the angles in this triangle? 30°; 60°; 90°

46. A triangle has two complementary angles. If the measure of the third angle is five times that of the first, what are the measures of the angles in this triangle? 18°; 72°; 90°

SKILL CHECKER

Try the "Skill Checker" exercises so you'll be ready for the next section.

Solve:

47. $n + 0.30n = 26$ $n = 20$

48. $n - 0.60n = 80$ $n = 200$

49. $220 = 2(W + 70) + 2W$ $W = 20$

50. $300 = 2(W + 60) + 2W$ $W = 45$

51. $0.05x + 0.10(800 - x) = 20$ $x = 1200$

52. $0.10s + 0.20(6000 - s) = 1000$ $s = 2000$

53. $110T = 80(T + 3)$ $T = 8$

54. $80T = 50(T + 3)$ $T = 5$

USING YOUR KNOWLEDGE

Puzzles and Riddles

Here is a puzzle and a riddle that you can solve using the knowledge you've gained in these sections.

55. "What number," asked Professor Bourbaki, absent-mindedly trying to find his algebra book, "must be added to the numerator and to the denominator of the fraction $\frac{1}{4}$ to obtain the fraction $\frac{2}{3}$?" "Well," he said after he found his book. Well? 5

56. Not much is known about Diophantus, sometimes called the father of algebra, except that he lived between A.D. 100 and 400. His age at death, however, is well known, because one of his admirers described his life using this algebraic riddle:

> Diophantus' youth lasted $\frac{1}{6}$ of his life. He grew a beard after $\frac{1}{12}$ more of his life. After $\frac{1}{7}$ more of his life, he married. Five years later, he had a son. The son lived exactly $\frac{1}{2}$ as long as his father, and Diophantus died just 4 years after his son's death.

How many years did Diophantus live? 84

WRITE ON

57. When reading a word problem, what is the first thing you should try to determine? Answers may vary.

58. How can you verify the answer in a word problem? Answers may vary.

59. Make your own integer word problem and show a detailed solution of it using the RSTUV method. Answers may vary.

60. Make your own geometry word problem and show a detailed solution of it using the RSTUV method. Answers may vary.

61. Step 3 of the RSTUV method calls for you to "**T**hink of a plan" to solve the problem. Some strategies you can use include *looking for a pattern* and *making a picture*. Can you think of three other strategies? Answers may vary.

MASTERY TEST

If you know how to do these problems, you have learned your lesson!

62. The sum of three consecutive integers is 27. What are the integers? 8; 9; 10

63. The sum of three consecutive odd integers is 99. What are the three integers? 31; 33; 35

64. The square of a number x, increased by 4 times the number itself, is 2 less than the number. Write an equation expressing this statement. $x^2 + 4x = x - 2$

65. If a number is decreased by 12 the result is the same as the product of 7 and the number. Write an equation expressing this statement. $x - 12 = 7x$

66. The measure of an angle is 8 times that of another.

  a. Find their measures if the angles are complementary. 10°; 80°

  b. Find their measures if the angles are supplementary. 20°; 160°

67. Find x and the measure of each marked angle.

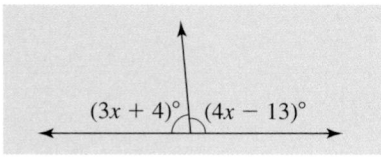

$x = 27$; angles: 85°, 95°

68. Denmark's cigarette tax tops the world. The tax on a pack of 20 cigarettes exceeds the price of the cigarettes by $3.03! If the total price of a pack of cigarettes in Denmark is $4.33, what is the tax? $3.68

69. The total number of students in U.S. schools in 1995 was 65 million. The number in secondary schools exceeded that of students in college by 2 million, while in elementary schools there were 3 times as many students as in college. What is the number (in millions) of students enrolled in college in 1995? Give your answer to two decimal places. 12.60 million

2.4   PROBLEM SOLVING: PERCENT, INVESTMENT, MOTION, AND MIXTURE PROBLEMS

To Succeed, Review How To . . .

1. Perform the four fundamental operations using decimals (pp. 17–23).

2. Solve linear equations involving decimals (p. 82).

Objectives

A Solve percent problems.

B Solve investment problems.

C Solve uniform motion problems.

D Solve mixture problems.

GETTING STARTED   **Web Shopping Costs**

List Price: ~~$1,299.00~~

Sale Price: ?

You Save: 63%

List Price: ~~$399.00~~

Sale Price: $139.65

You Save: ?%

List Price: ?

Sale Price: $9.99

You Save: $20.00 (66%)

These items can be purchased on the Internet. Underneath the items are the savings that can be realized by doing so. To supply the answers for the question marks, it would require knowing how to solve percent problems. Suppose you wanted to find out what the sale price of the laptop computer would be with a 63% discount. Knowing the percent of discount allows you to find the dollar amount of the discount.

$$\text{percent of discount times list price} = \text{discount dollar amount}$$
$$(0.63) \quad \times \quad \$1299 \quad = \$818.37$$

Remember $818.37 is the discount. We use the discount to find the sale price.

$$\text{list price minus discount} = \text{sale price}$$
$$\$1299 \quad - \quad \$818.37 = \$480.63$$

Those two equations could be combined to make one formula:

$$\text{list price} - (\text{percent of discount times list price}) = \text{sale price}$$
$$\$1299 - \quad (0.63)(\$1299) \quad = \$480.63$$

To find the percent when buying the camera, solve $399.00 - (x\%)(399.00) = 139.65$ to get 65%. To find the list price for the vacuum solve $x - 0.66(x) = 9.99$ to get $29. These types of equations are explained on p. 112.

Don't forget to read the fine print for the cost of shipping and handling. Some companies charge by the shipping weight and some charge a flat fee like $2.95. These were an example of solving percent problems. In this section we will also study how to solve investment, motion, and mixture problems.

Some students may already be familiar with the proportion

$$\frac{\text{"}is\text{"}}{\text{"}of\text{"}} = \frac{\text{"}p\text{" }(percent)}{100}$$

or $\dfrac{a}{b} = \dfrac{p}{100}$

and they cross multiply to get the equation that will solve:

Type 1: $\dfrac{a}{80} = \dfrac{30}{100}$

Type 2: $\dfrac{8}{40} = \dfrac{p}{100}$

Type 3: $\dfrac{10}{b} = \dfrac{40}{100}$

In algebra, rates, increases, decreases, and discounts are often written as percents (%). *Percent* means "by the hundred." Thus 28% means 28 parts of 100 or $\frac{28}{100}$. In most applications, however, percents are written as decimals. Thus

$$28\% = \frac{28}{100} = 0.28$$

$$30\% = \frac{30}{100} = 0.30$$

$$120\% = \frac{120}{100} = 1.2$$

and

$$34.8\% = \frac{34.8}{100} = \frac{348}{1000} = 0.348$$

There are three basic types of percent problems. All of these can be translated into algebra equations.

TYPE 1 asks to find a percent of a number.

PROBLEM   30% of 80 is what number?

TRANSLATION $0.30 \times 80 = n$

SOLUTION $24 = n$

TYPE 2 asks what percent of a number is another given number.

PROBLEM   What percent of 40 is 8? Or 8 is what percent of 40?

TRANSLATION $n \times 40 = 8$ or $8 = n \times 40$

SOLUTION $n = \dfrac{8}{40} = \dfrac{1}{5} = 20\%$

TYPE 3 asks to find a number when a percent of the number is given.

PROBLEM   10 is 40% of what number?

TRANSLATION $10 = 0.40 \times n$

SOLUTION $\dfrac{10}{0.40} = \dfrac{1000}{40} = 25 = n$

 **A**   **Percent Problems**

EXAMPLE 1   **Percent problems**

When purchasing their first home the Gonzalez' were prequalified to be able to purchase a home up to $110,000. They had saved $6000 for the down payment. When they found their dream home it was priced at $102,651 with a 5% down payment. Obviously the house was in their price range.

a. Did they have enough for the down payment?

b. If they did purchase the house what would be the amount of the mortgage?

SOLUTION   Use the RSTUV method.

a. Is $6000 enough for their down payment?

 1. Read and write a "Find" statement.
 Find the amount of the down payment and compare that to $6000.

 2. Select the unknown using a "Let" statement.
 Let x = the amount of the down payment.

PROBLEM 1

The Tylers' want to purchase a home for $122,852 with a 10 percent down payment.

a. What is the down payment (to the nearest dollar)?

b. What is the amount of the mortgage?

Answers

1. a. $12,285   **b.** $110,567

3. Think of a word equation and translate it to an algebra equation.

Percent times cost of house = down payment
(0.05)   ($102,651) = x

4. Use your algebra skills to solve the equation.

$5133 = x$ (Rounded to nearest dollar)

Comparison: $5133 is less than $6000.

5. Verify the results.

Verification left to you.

The Gonzalez' have enough for the down payment.

b. What is the amount of the mortgage?

Cost of the house − the down payment = amount of mortgage
$102,651   − $5133   = $97,518

The Gonzalez' will have a mortgage of $97,518.

EXAMPLE 2   Solving percent increase problems

According to O'Hare Airport statistics, the total number of passengers using the airport in January 2003 was approximately 5 million. This was a 12% increase over the total for January 2002. Approximately how many passengers used O'Hare Airport in January 2002? (Round the answer to the nearest thousand.)

SOLUTION   Use the RSTUV method. The key words are "12% increase." The problem will be solved like one in *Getting Started*.

1. Read and write a "Find" statement.

Find the number of passengers using O'Hare Airport in January 2002.

2. Select the unknown using a "Let" statement.

Let x = the number of passengers in January 2002.

3. Think of a word equation and translate it to an algebra equation.

If we add the increase to the number of passengers in January 2002, that should equal the number of passengers in January 2003. To find the increase we have to multiply 12% times the number in January 2002 [(0.12)(x)].

number in January 2002   +   increase   =   number in January 2003
x +   [(0.12)(x)]   =   5,000,000

4. Use your algebra skills to solve the equation.

$1x + [(0.12)(x)] = 5,000,000$ Remember $x = 1x$.

$1.12x = 5,000,000$ Simplify.

$x = 4,464,286$ Divide by 1.12.

$x = 4,464,000$ (Rounded to nearest thousand)

5. Verify the results.

Verification is left for you.

The number of passengers using O'Hare Airport in January 2002 was approximately 4,464,000.

PROBLEM 2

The number of passengers using a certain airport for the year 2002 was approximately 67 million. If that was down by 1.3% from 2001, how many passengers used the airport in 2001? (Round the answer to the nearest thousand.)

Web It

For practice with percent increase and percent decrease problems, try link 2-4-1 at mhhe.com/bello.

Teaching Tip

Remind students that the coefficient of n is understood to be "1," which can be written as 1.00. Thus, $0.50n + 1.00n = 1.50n$.

Answer

2. The number of passengers for the year 2001 was approximately 67,882,000.

 **Investment Problems**

Investment problems also use percents. If you invest P dollars at a rate r, your **annual interest** is

$$I = Pr$$

When working this type of problem, it's helpful to enter all the information in a table. We do this in Example 3.

EXAMPLE 3	Solving investment problems

A woman has some stocks yielding 5% annually and some bonds that yield 10%. If her investment totals $6000 and her annual income from the investment is $500, how much does she have invested in stocks and how much in bonds?

SOLUTION   Use the RSTUV method.

1. Read and write a "Find" statement.
Find how much she has invested in stocks and how much in bonds.

2. Select the unknown using a "Let" statement.
Let s = the amount she has invested in stocks. This makes the amount invested in bonds $(6000 - s)$.

3. Think of a word equation and translate it to an algebra equation.
Use the following to help visualize the problem. Enter the information in the chart:

	Principal	×	Rate	=	Interest
Stocks	s		0.05		$0.05s$
Bonds	$6000 - s$		0.10		$0.10(6000 - s)$

The total interest is the sum of the entries in the last column, that is, $0.05s$ and $0.10(6000 - s)$. This amount must be $500

$$\underbrace{0.05s}_{\text{interest in stocks}} + \underbrace{0.10(6000 - s)}_{\text{interest in bonds}} = \underbrace{500}_{\text{total interest}}$$

4. Use your algebra skills to solve the equation.

$0.05s + 0.10(6000 - s) = 500$	Given.
$0.05s + 600 - 0.10s = 500$	Remove parentheses; simplify.
$600 - 0.05s = 500$	
$600 - 600 - 0.05s = 500 - 600$	Add -600.
$-0.05s = -100$	
$\dfrac{-0.05s}{-0.05} = \dfrac{-100}{-0.05}$	Divide by -0.05.
$s = 2000$	

The woman has $2000 in stocks and the rest, $4000, in bonds.

5. Verify the results.
Five % of $2000 = $100 and 10% of $4000 = $400, so the total interest is $500.

PROBLEM 3

A man has two investments totaling $8000. One investment yields 7% and the other 10%. If the total annual interest is $710, how much is invested at each rate?

Web It

For more examples of investment problems, try link 2-4-2 at mhhe.com/bello.

Answer

3. $3000 at 7%; $5000 at 10%

 **Uniform Motion Problems**

When traveling at a constant rate R, the distance D traveled in time T is given by

$$D = RT$$

This formula is similar to the one used for interest problems, so we again use a table to enter the information. Here's how we do it.

EXAMPLE 4	Solving distance problems

A Supercruiser bus leaves Miami for San Francisco traveling at the rate of 40 mi/hr. Three hours later a car leaves Miami for San Francisco traveling at 55 mi/hr on the same route as the bus. How far from Miami does the car overtake the bus (assume they both travel at constant rates)?

SOLUTION   Use the RSTUV method.

1. Read and write a "Find" statement.
To find the distance, we need to know the time it takes the car to overtake the bus.

2. Select the unknown using a "Let" statement.
Let T = time it takes the car to overtake the bus.

3. Think of a word equation and translate it to an algebra equation.
We translate the given information and write it in a table. If the car travels for T hr, the bus travels for $(T + 3)$ hr because it left 3 hr earlier.

	Rate	×	**Time**	=	**Distance**
Car	55		T		$55T$
Bus	40		$(T + 3)$		$40(T + 3)$

When the car overtakes the bus, they will have traveled the same distance. According to the table, the car has traveled $55T$ miles in distance and the bus $40(T + 3)$ miles. Thus

$$\underbrace{\text{distance of car}} = \underbrace{\text{distance of bus}}$$
$$55T = 40(T + 3)$$

4. Use your algebra skills to solve the equation.

$$55T = 40(T + 3) \qquad \text{Given.}$$
$$55T = 40T + 120 \qquad \text{Remove parentheses.}$$
$$55T - 40T = 40T - 40T + 120 \qquad \text{Add } -40T.$$
$$15T = 120 \qquad \text{Simplify.}$$
$$\frac{15T}{15} = \frac{120}{15} \qquad \text{Divide by 15.}$$
$$T = 8$$

It takes the car 8 hr to overtake the bus. Since $R \cdot T = D$, in 8 hr the car travels $8 \times 55 = 440$ mi, and overtakes the bus 440 mi from Miami.

5. Verify the results.
The distances are equal so we can use either one.
   The car has traveled for 8 hr at 55 mi/hr; thus it travels $55 \times 8 = 440$ mi, whereas the bus has traveled 11 hr at 40 mi/hr, a total of $40 \times 11 = 440$ mi.

PROBLEM 4

A bus leaves Los Angeles traveling at 50 mi/hr. An hour later a car leaves at 60 mi/hr to try and catch the bus. How far from Los Angeles does the car catch the bus?

Teaching Tip

Sometimes a picture helps visualize the distances as being the same.

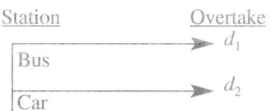

Therefore, $d_1 = d_2$.

Answer

4. 300 miles

 **Mixture Problems**

The last type of problem we discuss is the **mixture problem,** a type of problem in which two or more things are put together to form a mixture. Again, we use a table to enter the information.

EXAMPLE 5 Solving mixture problems

How many ounces of a 50% acetic acid solution should a photographer add to 32 oz of a 5% acetic acid solution to obtain a 10% acetic acid solution?

SOLUTION Use the RSTUV method.

1. Read and write a "Find" statement.
Find the number of ounces of the 50% solution that should be added.

2. Select the unknown using a "Let" statement.
Let x = the number of ounces of 50% solution to be added.

 +  =

5%   50%   10%

3. Think of a word equation and translate it to an algebra equation.
To translate the problem, we use a table. In this case, the headings for the table contain the percent of acetic acid and the amount to be mixed. The product of these two numbers will give us the amount of pure acetic acid.

Percent of Acetic Acid	×	Amount to be mixed	=	Amount of Pure Acid
50% solution or 0.50		x oz		$0.50x = 0.5x$
5% solution or 0.05		32 oz		$0.05(32) = 1.6$
10% solution or 0.10 (Final mixture)		$(x + 32)$ oz		$0.10(x + 32) = 0.1(x + 32)$

The percents have been converted to decimals. You *should not* add the percents in this column.

We have x oz of one and 32 oz of the other, so we have $(x + 32)$ ounces of the mixture.

The sum of the total amounts of pure acetic acid should be the same as the amount of pure acetic acid in the final mixture, so we have

amt. in 50% + amt. in 5% = amt. of acid in final mixture

$$0.5x + 1.6 = 0.1(x + 32)$$

4. Use your algebra skills to solve the equation.

$5x + 16 = x + 32$ Clear decimals; multiply by 10.

$5x + 16 - 16 = x + 32 - 16$ Add −16.

$5x = x + 16$

$5x - x = x - x + 16$ Add −x.

$4x = 16$

$\dfrac{4x}{4} = \dfrac{16}{4}$ Divide by 4.

$x = 4$

The photographer must add 4 oz of the 50% solution.

5. Verify the results.
The verification of this fact is left to you.

PROBLEM 5

How many gallons of a 10% salt solution should be added to 15 gallons of a 20% salt solution to obtain an 18% solution?

Web It

For detailed examples of integer, coin, mixture, and distance word problems, try link 2-4-3 at mhhe.com/bello.

Calculate It

Graphing Equations

1. Graph the equation in Example 4.

2. Graph the equation in Example 5.

Answer

5. 3.75 gallons

Calculate It   **Learning to Adjust the "Window" When Graphing**

Suppose you want to use your calculator to solve Example 3. First, do steps 1 and 2 in the RSTUV procedure in the usual manner, then go to step 3, write the equation

$$0.05s + 0.10(6000 - s) = 500$$

as

$$0.05s + 0.10(6000 - s) - 500 = 0$$

Then use x instead of s and graph

$$Y_1 = 0.05x + 0.10(6000 - x) - 500 \quad \text{and} \quad Y_2 = 0$$

What happens when you press the GRAPH button? Nothing, unless you have set your WINDOW correctly. If you look at step 4 in the solution of the problem, you will note that the answer is $s = 2000$; thus *before* you graph the equation, you have to set your window

at $[-5000, 5000]$ by $[-5000, 5000]$ with a scale of 1000. You will then see that the resulting line intersects the x-axis at $x = 2000$ as shown in Window 1. For a better view, adjust the y-scale to $[-1000, 1000]$ as shown in Window 2. To verify the answer, press  .

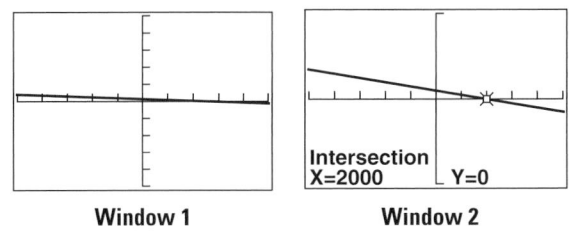

Window 1   **Window 2**

Exercises 2.4

A In Problems 1–14, use the RSTUV method to solve the percent problems.

1. By weight, the average adult is composed of 43% muscle, 26% skin, 17.5% bone, 7% blood, and 6.5% organs. Suppose a person weighs 150 lb.

   a. How many pounds of muscle does the person have? 64.5

   b. How many pounds of organs does the person have? 9.75

2. Refer to Problem 1.

   a. How many pounds of bone does the person have? 26.25

   b. How many pounds of skin does the person have? 39

3. A bicycle is priced at $196.50. If the sales tax rate is 5.5%, what is the tax? $10.81

4. The highest recorded shorthand speed was 300 words per minute for 5 min with 99.64% accuracy. How many errors were made? (Answer to the nearest whole number.) 5

5. In a recent year 41.2 million households with televisions watched the Super Bowl. This represents 41.7% of homes with televisions, in major cities. How many homes with televisions are there in major cities? 98.8 million

6. On February 28, 1983, 125 million people watched the final episode of M*A*S*H. If the 125 million represents a 77% share of the viewing audience, what was the viewing audience that day? 162.3 million

In Problems 7–10, use the following formula:

$$\text{Selling price} = \text{cost} + \text{markup}$$

7. The cost of an article is $18.50. If the markup is 25% of the cost, what is the selling price of the article? $23.13

8. The selling price of an article is $30. If the markup is 20% of the cost, what is the cost of the article? $25

9. An article costing $30 is sold for $54. What is the markup and the percent of markup on selling price? Markup: $24; 44.4% of selling price

10. An article costing $60 is sold for $90. What is the markup and the percent of markup on cost? Markup: $30; 50% of cost

11. If 25 is increased to 35, what is the percent increase? 40%

12. If 32 is decreased to 24, what is the percent decrease? 25%

13. If 40 is increased by 25% of itself, what is the result? 50

14. If 35 is decreased by 20% of itself, what is the result? 28

15. Luisa is a computer programmer. Her salary was increased from $25,000 to $30,000. What was her percent increase? 20%

16. The number of comedy shows on network TV went from 62 in 1997 to 35 in 2003. What percent of decrease does this represent? 43.5%

17. Tran was told that if he would switch to the next higher grade of gasoline, his mileage would increase to 120% of his present mileage of 25 mi/gal. What would be his increased mileage? 30 mi/gal

18. The average salary for pediatric doctors is $110,000. Surgeons make an astounding 250% more. On average, how much more do surgeons earn? $275,000

19. The number of beverages introduced in the market during two consecutive years decreased by 6% to 611 new beverages. How many beverages were introduced during the first year? 650

20. The total amount spent in corporate travel and entertainment increased by 21% in a 2-yr period. If the amount spent reached $115 billion at the end of the period, how much was spent at the beginning? 95.04 billion

21. The annual amount spent on entertainment by the average person will increase by 33% to $1463. How much is the average person now spending on entertainment? $1100

B In Problems 22–25, use the RSTUV method to solve the investment problems.

22. Two sums of money totaling $15,000 earn, respectively, 5% and 7% annual interest. If the interest from both investments amounts to $870, how much is invested at each rate? $9000 at 5%; $6000 at 7%

23. An investor invested $20,000, part at 6% and the rest at 8%. Find the amount invested at each rate if the annual income from the two investments is $1500. $5000 at 6%; $15,000 at 8%

24. A woman invested $25,000, part at 7.5% and the rest at 6%. If her annual interest from these two investments amounted to $1620, how much money did she have invested at each rate? $8000 at 7.5%; $17,000 at 6%

25. A man has a savings account that pays 5% annual interest and some certificates of deposit paying 7% annually. His total interest from the two investments is $1100, and the total amount of money in the two investments is $18,000. How much money does he have in the savings account? $8000

C In Problems 26–31, use the RSTUV method to solve the motion problems.

26. A car leaves a town traveling at an average speed of 60 km/hr. Two hours later a highway patrol officer leaves from the same starting point to overtake the car. If the average speed of the officer is 90 km/hr, how far from town does the officer overtake the car? 360 km

27. A group of smugglers cross the border in a car traveling in a straight line at 96 km/hr. An hour later, the border patrol starts after them in a light plane traveling 144 km/hr.
a. How long will it be before the border patrol reaches the smugglers? 2 hr
b. At what distance from the border will the border patrol overtake the smugglers? 288 km

28. A freight train leaves the station traveling at 30 mi/hr. One hour later, a passenger train leaves the same station on a parallel track traveling at 60 mi/hr. How far from the station does the passenger train overtake the freight train? 60 mi

29. A bus leaves the station traveling at 60 km/hr. Two hours later, the wife of one of the passengers shows up at the station with a briefcase belonging to an absent-minded professor riding the bus. If she immediately starts after the bus at 90 km/hr, how far from the station is the briefcase reunited with the professor? 360 km

30. An accountant and her boss have to travel to a nearby town. The accountant catches a train traveling at 50 mi/hr while the boss leaves 1 hr later in a car traveling at 60 mi/hr. They have decided to meet at the train station and, strangely enough, they get there at exactly the same time! If the train and the car traveled in straight lines on parallel paths, how far is it from one town to the other? 300 mi

31. The basketball coach at a local high school left for work on her bicycle, traveling at 15 mi/hr. Thirty minutes later, her husband noticed that she had left her lunch. He got in his car and traveled 60 mi/hr to take her lunch to her. Luckily, he got to school at exactly the same time as his wife. How far is it from the house to the school? 10 mi

D In Problems 32–36, use the RSTUV method to solve the mixture problems.

32. How many liters (L) of a 40% glycerin solution must be mixed with 10 L of an 80% glycerin solution to obtain a 65% solution? 6 L

33. How many parts of glacial acetic acid (99.5%) must be added to 100 parts of a 10% solution of acetic acid to give a 28% solution? 25.2

34. If the price of copper is 65¢ per pound and the price of zinc is 30¢ per pound, how many pounds of copper and zinc should be mixed to make 70 lb of brass selling for 45¢ per pound? 30 lb copper; 40 lb zinc

35. Oolong tea sells for $19 per pound. How many pounds of Oolong should be mixed with regular tea selling at $4 per pound to produce 50 lb of tea selling for $7 per pound? 10

36. You think the prices of coffee are high? You haven't seen anything yet! Jamaican Blue coffee sells for about $20 per pound! How many pounds of Jamaican Blue should be mixed with 80 lb of regular coffee selling at $8 per pound so that the result is a mixture selling for $10.40 per pound? (You can cleverly advertise this as "Containing the incomparable Jamaican Blue coffee.") 20 lb

SKILL CHECKER

Try the "Skill Checker" exercises so you'll be ready for the next section.

Fill in the blank with $<$ or $>$ to make the resulting statement true.

37. a. -3 __ $<$ __ -1 **b.** -1.3 __ $>$ __ -1.4 **38. a.** $-\dfrac{1}{3}$ __ $>$ __ $-\dfrac{1}{2}$ **b.** $-\dfrac{1}{5}$ __ $>$ __ $-\dfrac{1}{2}$

  c. $\dfrac{1}{3}$ __ $<$ __ $\dfrac{1}{2}$ **c.** $-\dfrac{1}{4}$ __ $>$ __ $-\dfrac{1}{2}$

USING YOUR KNOWLEDGE

We Owe, We Owe, and to the Deficit We Go.

The last page of one year's IRS instructions for Form 1040 indicated that the federal income that year was $1090.5 billion, while outlays (expenses) amounted to $1380.9 billion. (By the way, 1090.5 billion = 1.0905 *trillion*.)

39. What is the difference between income and expenses?
−$290.4 billion

40. The pie chart indicates that 21% of the federal income was borrowed to cover the deficit. How much money was borrowed to cover the deficit?
$229.005 billion

41. The pie chart indicates that 35% of the federal income comes from personal taxes. How much money comes from personal taxes? $381.675 billion

Social Security, Medicare, unemployment, and other retirement taxes 30%

Personal income taxes 35%

Borrowing to cover deficit 21%

Corporate income taxes 7%

Excise, customs, estate, gift, and miscellaneous taxes 7%

WRITE ON

42. Formulate three different problems involving percents, and show a detailed solution using the RSTUV method.
Answers may vary.

43. Formulate your own investment word problem, and show a detailed solution using the RSTUV method.
Answers may vary.

44. Formulate your own motion word problem, and show a detailed solution using the RSTUV method.
Answers may vary.

45. Ask your pharmacist or your chemistry instructor if he or she mixes products of different concentrations to make new mixtures. Write a paragraph on your findings.
Answers may vary.

46. All the problems in this text have precisely the information you need to solve them. In real life, however, some of the information given may be irrelevant. Such irrelevant information is called a *red herring*. Find some problems with red herrings and point them out.
Answers may vary.

MASTERY TEST

If you know how to do these problems, you have learned your lesson!

47. A man has two investments totaling $8000. One investment yields 5% and the other 10%. If the total annual interest is $650, how much is invested at each rate? $3000 at 5%; $5000 at 10%

48. How many gallons of a 10% salt solution must be added to 15 gallons of a 20% salt solution to obtain a 16% solution? 10 gallons

49. The number of passengers using the Atlanta airport in the year 2000 was 31.2 million, a 30% increase over the 1999 figure. How many passengers used the Atlanta airport in 1999? 24 million

50. A bus leaves Los Angeles traveling at 50 mi/hr. An hour later a car traveling on the same road leaves at 70 mi/hr to try and overtake the bus. How far from Los Angeles does the car overtake the bus? 175 mi

51. The number of students enrolled in higher education in the United States increased from 12 million in 1980 to 14.4 million in 1990. What was the percent increase? 20%

2.5 LINEAR AND COMPOUND INEQUALITIES

To Succeed, Review How To ...

1. Solve linear equations (p. 79).

2. Add, subtract, multiply, and divide integers (pp. 17–23).

3. Properly use the symbols $>$ and $<$ when comparing numbers (p. 12).

Objectives

A Graph linear inequalities.

B Solve and graph linear inequalities.

C Solve and graph compound inequalities.

D Use the inequality symbols to translate sentences into inequalities.

GETTING STARTED Inequalities in Rental Cars

Suppose a rental car costs $25 each day and $0.20 per mile. If you drive m miles, the cost C for the day is

$$C = 0.20m + 25$$

$\underrightarrow{\text{Cost for } m \text{ miles}}$ $\underrightarrow{\text{Cost per day}}$

Now suppose your daily cost must be under $50. This means that $0.20m + 25 < 50$. How many miles can you drive? The expression $0.20m + 25 < 50$ is an example of a linear inequality and can be solved using the same techniques as those used to solve linear equations. Thus

$0.20m + 25 < 50$	Given.
$0.20m + 25 - 25 < 50 - 25$	Add -25.
$0.20m < 25$	
$m < \dfrac{25}{0.20}$	Divide by 0.20.
$m < 125$	

Hence if you drive less than 125 mi, your cost will be less than $50 per day.

An **inequality** is a statement that two expressions are *not* equal. Thus $x + 2 < 5$, $x - 7 < 3$, and $3x - 8 \leq 15$ are *linear inequalities*.

Linear Inequalities	A **linear inequality** in one variable is an inequality that can be written in the form $$ax + b < c$$ where a, b, and c are real numbers, and $a \neq 0$.

The inequality $ax + b < c$ is still a linear inequality if the $<$ (less than) symbol is replaced by $>$ (greater than), $\leq$ (is less than or equal to), or $\geq$ (is greater than or equal to). Thus, $2x + 5 > 8$, $-3x - 7 \geq 9$, $-\frac{1}{2}x + 8 < -\frac{3}{4}$, and $0.20m + 25 \leq 50$ are all linear inequalities. As with linear equations, we *solve* an inequality by finding all the replacements of the variable that make the inequality a true statement. This is done by finding an *equivalent* inequality whose solution is obvious. For example, the inequalities $x + 2 > 5$ and $x > 3$ are equivalent. They are both satisfied by all real numbers greater than 3. We write the *solution set* of $x + 2 > 5$ in set-builder notation as

$$\{x \mid x > 3\}$$

This is read "the set of all x's such that x is greater than 3."

Graphing Linear Inequalities

There are infinitely many numbers in the solution set of the inequality $x > 3$ (4, 7.5, $\sqrt{10}$, and $\frac{10}{3}$ are a few of them). We cannot list all these numbers, so we show all solutions of $x > 3$ **graphically** by using a number line. This type of representation is called the **graph** of the inequality. The heavy line in the figure is the graph of $x > 3$.

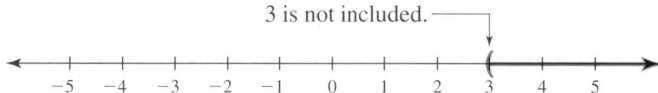

The number 3 is *excluded* from the graph. This is shown by drawing a parenthesis at the point 3.

This graph can also be drawn with an open circle at 3.

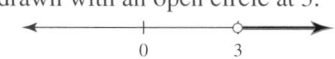

Teaching Tip

Encourage the students to rewrite statements of inequality from $3 < x$ to $x > 3$ before attempting to graph. Then the inequality arrow will be a reminder as to which direction of the number line should be shaded.

EXAMPLE 1 Graphing linear inequalities	**PROBLEM 1**

Graph:

a. $x \geq -1$ **b.** $x < -2$

SOLUTION

a. The numbers that satisfy the inequality $x \geq -1$ are the numbers *greater than or equal to* -1, that is, the number -1 and all the numbers to the *right* of -1 ($\geq$ points to the right). The graph is shown in the following figure. That -1 is included is shown by drawing a bracket at the point -1. (*Hint:* When graphing a linear inequality, the arrow will point the direction to shade on the number line as long as the variable is on the left side of the inequality symbol.)

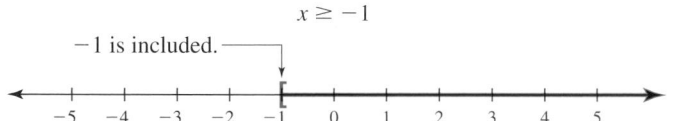

Graph:

a. $x \geq -2$ **b.** $x < 1$

Answers

1. a.

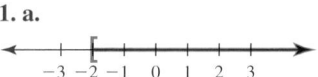

b.

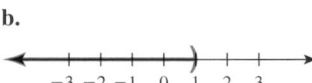

This graph can also be drawn with a closed circle at −1.

b. The numbers that satisfy the inequality $x < -2$ are the numbers *less than* −2, that is, the numbers to the *left* of but *not including* −2 (< points to the left). The graph of these points is

$$x < -2$$

−2 is not included.

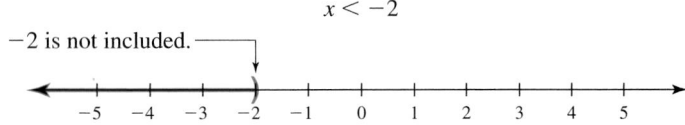

The inequalities $x > 3$, $x \geq -1$, and $x < -2$ resulted in graphs that did *not* have a finite length; they are called **unbounded** (or **infinite**) **intervals.** The four basic types of unbounded intervals are summarized in Table 5, where we use a parenthesis to denote the inequality symbol < or > and a bracket to indicate ≤ or ≥. An **open** interval does not contain either of its end points.

Study Skills Hint

To practice how to write your answer in interval notation, it would be a good idea to write your own examples of the four cases in Table 5 on an index card. Practice reading and writing this notation before and after each class and each homework assignment. For example you might begin with:

1. $\{x \mid x > 2\}$ *The set of all "x" such that x is greater than 2.*

2. $(2, \infty)$ *The open interval from 2 to positive infinity.*

Table 5 Unbounded Interval Notation

Set Notation	Interval Notation	Type of Interval	Graph	Inequality
$\{x \mid x > a\}$	$(a, +\infty)$	Open		$x > a$
$\{x \mid x < b\}$	$(-\infty, b)$	Open		$x < b$
$\{x \mid x \geq a\}$	$[a, +\infty)$	Half-open		$x \geq a$
$\{x \mid x \leq b\}$	$(-\infty, b]$	Half-open		$x \leq b$

When writing interval notation, the first number or symbol represents the left end of the interval and the second number or symbol represents the right end of the interval. For example, to name the interval on the graph

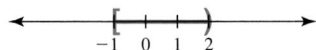

we write the left-end number, −1, and then the right-end number, 2, with the appropriate enclosures. The interval is written $[-1, 2)$.

In all these intervals a parenthesis is always used with the symbols ∞ (**positive** infinity) and $-\infty$ (**negative** infinity). The symbols ∞ and $-\infty$ do not represent real numbers; they are used to indicate that the interval is unbounded.

B   Solving Linear Inequalities

The inequalities $x + 2 > 5$ and $x > 3$ are equivalent. This is because we added −2 to both sides of $x + 2 > 5$ to obtain the equivalent inequality

$$x + 2 + (-2) > 5 + (-2) \quad \text{or} \quad x > 3$$

We used the first of the following properties.

Properties of Inequalities: **Addition and Subtraction**	If C is a real number, then the following inequalities are all *equivalent:*
	$A < B$
Add C.	$A + C < B + C$
Subtract C.	$A - C < B - C$

Now consider the inequality $2x - 3 < x + 1$. To solve this inequality, we need all the variables by themselves on one side. Thus we proceed as follows:

$$2x - 3 < x + 1$$ Clear fractions: none.
                                            Remove parentheses: none.

$$2x - x - 3 < x - x + 1$$ Add $-x$ to both sides.

$$x - 3 < 1$$

$$x - 3 + 3 < 1 + 3$$ Add 3 to both sides.

$$x < 4$$

The solution set is $\{x \mid x < 4\}$ or, in interval notation, $(-\infty, 4)$.

The graph of this inequality is shown in the figure. You can check that this solution is correct by selecting a number from the graph (say 0) and replacing x with the number 0 in the original inequality to obtain $2(0) - 3 < 0 + 1$ or $-3 < 1$, a true statement. This is only a "partial" check because we did not try *all* the numbers in the graph.

$$2x - 3 < x + 1$$
$$x < 4$$

↤———————)——→
−3 −2 −1 0 1 2 3 4 5

$(-\infty, 4)$

EXAMPLE 2 **Solving and graphing inequalities**	**PROBLEM 2**

Solve, graph, and write the solution set in interval notation.

a. $3x - 2 < 2(x - 2)$ **b.** $4(x + 1) \geq 3x + 7$

SOLUTION

a.
$$3x - 2 < 2(x - 2)$$ Clear fractions: none.

$$3x - 2 < 2x - 4$$ Remove parentheses.

$$3x - 2x - 2 < 2x - 2x - 4$$ Add $-2x$ to both sides.

$$x - 2 < -4$$

$$x - 2 + 2 < -4 + 2$$ Add 2 to both sides.

$$x < -2$$

The solution set is $\{x \mid x < -2\}$ or, in interval notation, $(-\infty, -2)$. The graph is

$$3x - 2 < 2(x - 2)$$
$$x < -2$$

↤————————)————————————————→
−5 −4 −3 −2 −1 0 1 2 3 4 5

b.
$$4(x + 1) \geq 3x + 7$$ Clear fractions: none.

$$4x + 4 \geq 3x + 7$$ Remove parentheses.

$$4x - 3x + 4 \geq 3x - 3x + 7$$ Add $-3x$ to both sides.

$$x + 4 \geq 7$$

$$x + 4 - 4 \geq 7 - 4$$ Add -4 to both sides.

$$x \geq 3$$

PROBLEM 2

Solve, graph, and write the solution set in interval notation.

a. $4x - 7 < 3(x - 2)$

b. $3(x + 1) \geq 2x + 5$

Answers

2. a. $x < 1$; $(-\infty, 1)$

↤——————)————→
−3 −2 −1 0 1 2 3

b. $x \geq 2$; $[2, \infty)$

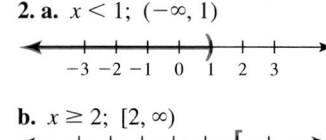

The solution set is $\{x | x \geq 3\}$ or, in interval notation, $[3, \infty)$. The graph of this inequality is

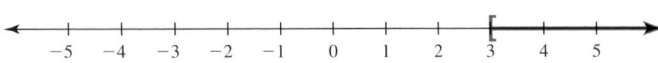

Before we state the multiplication and division properties of inequalities, let's see what happens if we multiply both sides of an inequality by a positive number. Consider these true inequalities:

$$1 < 3 \qquad\qquad -1 < 3 \qquad\qquad -3 < -1$$
$$4 \cdot 1 \; ? \; 4 \cdot 3 \qquad 4 \cdot (-1) \; ? \; 4 \cdot 3 \qquad 4 \cdot (-3) \; ? \; 4 \cdot (-1) \qquad \text{Multiply by 4.}$$
$$4 < 12 \qquad\qquad -4 < 12 \qquad\qquad -12 < -4$$

The resulting inequalities are all *true,* and the inequality symbol points in the *same* direction as the original. Multiplying both sides of an inequality by a *positive* number *preserves the sense,* or direction, of the inequality.

Now, let's multiply both sides of the original inequality by a *negative* number, say -4. We have

$$1 < 3 \qquad\qquad -1 < 3 \qquad\qquad -3 < -1$$
$$-4 \cdot 1 \; ? \; -4 \cdot 3 \qquad -4 \cdot (-1) \; ? \; -4 \cdot 3 \qquad -4 \cdot (-3) \; ? \; -4 \cdot (-1) \qquad \text{Multiply by } -4.$$
$$-4 > -12 \qquad\qquad 4 > -12 \qquad\qquad 12 > 4$$

This time, however, we had to reverse the direction of the inequalities to maintain a true statement. Multiplying both sides of an inequality by a *negative* number *reverses the sense* of the inequality. Division is defined in terms of multiplication, so these two properties apply to division as well. They can be stated as follows.

Properties of Inequalities: Multiplication and Division	If C is a real number, then the following inequalities are all equivalent:
	$A < B$
	$A \cdot C < B \cdot C \qquad$ if C is positive $(C > 0)$
	$\dfrac{A}{C} < \dfrac{B}{C} \qquad$ if C is positive $(C > 0)$
	$A \cdot C > B \cdot C \qquad$ if C is negative $(C < 0)$
	$\dfrac{A}{C} > \dfrac{B}{C} \qquad$ if C is negative $(C < 0)$

Calculate It   Solve Inequalities by Graphing

To solve the inequality in Example 2(a), $3x - 2 < 2(x - 2)$, subtract $2(x - 2)$ and write:

$$3x - 2 - 2(x - 2) < 0$$

Now graph $Y_1 = 3x - 2 - 2(x - 2)$ using a decimal window, ⟨ZOOM⟩ 4 (Window 1).

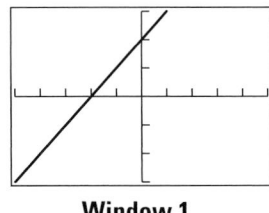

Window 1

The resulting line is *below* the x-axis for values of x less than -2, so the solution set is $\{x | x < -2\}$, as before.

Similarly, to solve the inequality in Example 2(b), $4(x + 1) \geq 3x + 7$, subtract $3x + 7$ and write

$$4(x + 1) - (3x + 7) \geq 0$$

Then graph

$$Y_1 = 4(x + 1) - (3x + 7)$$

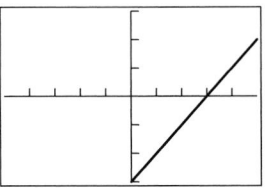

Window 2

This time (see Window 2), the resulting line is above or on the x-axis for values of x greater than or equal to 3. The solution set is $\{x | x \geq 3\}$.

Web It

If you would like more examples to practice solving linear inequalities, try link 2-5-1 at mhhe.com/bello.

If you still want more explanation and practice solving linear inequalities, try link 2-5-2 at mhhe.com/bello.

CAUTION

We can still multiply or divide both sides of an inequality by any nonzero number *as long as we remember to reverse the sense (direction) of the inequality if the number is negative.*

To solve $-2x > 4$, we must divide both sides by -2 (or multiply by $-\frac{1}{2}$). When doing this, remember to reverse the sense (direction) of the inequality.

$$-2x > 4 \qquad \text{Given.}$$
$$\frac{-2x}{-2} < \frac{4}{-2} \qquad \text{Multiply by } -\tfrac{1}{2} \text{ (or divide by } -2\text{) and reverse the sign.}$$
$$x < -2 \qquad \text{Simplify.}$$

EXAMPLE 3	**Solving and graphing inequalities**

Solve, graph, and write the solution set in interval notation.

a. $4(x - 1) \le 6x + 2$ **b.** $2x + 9 \ge 5x + 3$

SOLUTION

a. We follow the same procedure (CRAM) as that for solving linear equations given on page 79.

Given:	$4(x - 1) \le 6x + 2$	Clear fractions: none.
	$4x - 4 \le 6x + 2$	Remove parentheses.
	$4x - 6x - 4 \le 6x - 6x + 2$	Add $-6x$ to both sides.
	$-2x - 4 \le 2$	
	$-2x - 4 + 4 \le 2 + 4$	Add 4 to both sides.
	$-2x \le 6$	
	$\dfrac{-2x}{-2} \ge \dfrac{6}{-2}$	Multiply by $-\tfrac{1}{2}$ (or divide by -2).
	$x \ge -3$	(Remember to reverse the sense of the inequality.)

The solution set is $\{x \mid x \ge -3\}$ or, in interval notation, $[-3, \infty)$. The graph is

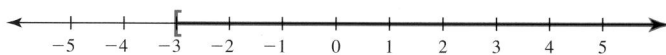

If we wish to avoid multiplying (or dividing) by negative numbers, we can add $-4x$ in step 3 and -2 in step 4 to obtain $-6 \le 2x$. Then, dividing by 2, we get $-3 \le x$, equivalent to $x \ge -3$.

b. Again, we isolate the x's on one side. This time, avoid multiplying or dividing by negative numbers. Do this by noting that there are more x's on the right-hand side of the inequality. Isolate the x's on the right.

Given:	$2x + 9 \ge 5x + 3$	Clear fractions: none. Remove parentheses: none.
	$2x - 2x + 9 \ge 5x - 2x + 3$	Add $-2x$ to both sides.
	$9 \ge 3x + 3$	
	$9 - 3 \ge 3x + 3 - 3$	Add -3 to both sides.
	$6 \ge 3x$	
	$\dfrac{6}{3} \ge \dfrac{3x}{3}$	Multiply by $\tfrac{1}{3}$ (or divide by 3).
	$2 \ge x$	

PROBLEM 3

Solve, graph, and write the solution set in interval notation:

a. $3(x - 1) \le 5x + 1$

b. $3x + 11 \ge 5x + 5$

Teaching Tip

Ask students to give some examples that would support why $-3 \le x$ is equivalent to $x \ge -3$.

Answers

3. a. $x \ge -2$; $[-2, \infty)$

b. $x \le 3$; $(-\infty, 3]$

The inequality $x \le 2$ is equivalent to $2 \ge x$. So the solution set is $\{x \mid x \le 2\}$ or, in interval notation $(-\infty, 2]$. The graph is

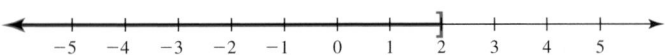

This could also be solved as follows:

$$2x + 9 \ge 5x + 3$$
$$2x - 5x + 9 \ge 5x - 5x + 3$$
$$-3x + 9 \ge 3$$
$$-3x + 9 - 9 \ge 3 - 9$$
$$\frac{-3x}{-3} \le \frac{-6}{-3}$$
$$x \le 2$$

This indicates $2 \ge x$ means the same as $x \le 2$. If you exchange sides in an inequality you must reverse the inequality sign. The last two inequalities contained no fractions. If fractions are present, we clear them by multiplying both sides of the inequality by the LCD of the fractions involved, as Example 4 shows.

EXAMPLE 4 **Solving and graphing inequalities with fractions**

Solve, graph, and write the solution set in interval notation:

$$\frac{x}{4} - \frac{x}{6} > \frac{x - 3}{6}$$

SOLUTION

Given: $\qquad \dfrac{x}{4} - \dfrac{x}{6} > \dfrac{x - 3}{6}$

$$12\left(\frac{x}{4} - \frac{x}{6}\right) > 12\left(\frac{x - 3}{6}\right) \qquad \text{Clear fractions: multiply by 12.}$$

$$\overset{3}{\cancel{12}}\left(\frac{x}{4}\right) - \overset{2}{\cancel{12}}\left(\frac{x}{6}\right) > \overset{2}{\cancel{12}}\left(\frac{x - 3}{6}\right)$$

$$3x - 2x > 2(x - 3) \qquad \text{Remove parentheses/simplify.}$$

$$x > 2x - 6$$

$$x - 2x > 2x - 2x - 6 \qquad \text{Add } -2x \text{ to both sides.}$$

$$-x > -6$$

$$\frac{-x}{-1} < \frac{-6}{-1} \qquad \text{Multiply by } -1 \text{ (or divide by } -1\text{).}$$

$$x < 6 \qquad \text{(Remember to reverse sense of inequality.)}$$

The solution set is $\{x \mid x < 6\}$ or, in interval notation, $(-\infty, 6)$, and the graph is

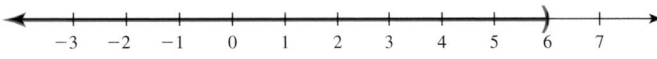

PROBLEM 4

Solve, graph, and write the solution set in interval notation:

$$\frac{x}{3} - \frac{x}{4} > \frac{x - 4}{4}$$

Answer

4. $x < 6$; $(-\infty, 6)$

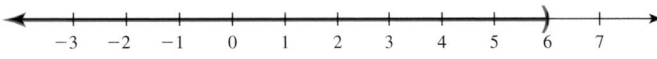

Calculate It Solve Inequalities with Fractions

If the inequality contains fractions, you **do not** have to clear the fractions. To solve Example 4,

$$\frac{x}{4} - \frac{x}{6} > \frac{x-3}{6}$$

subtract

$$\frac{x-3}{6}$$

to obtain

$$\frac{x}{4} - \frac{x}{6} - \frac{x-3}{6} > 0$$

Now graph

$$Y_1 = \frac{x}{4} - \frac{x}{6} - \frac{(x-3)}{6}$$

With a decimal window, you don't get the complete graph. Use an integer window (ZOOM) 8 (ENTER) and the (TRACE) key to get a better view of the graph, as shown. To find the x-intercept go to (Y=) and enter $y_2 = 0$, then press (GRAPH).

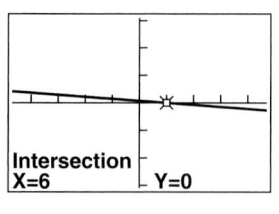
Intersection
X=6   Y=0

Then enter (2nd) (TRACE) **5** (ENTER) (ENTER) (ENTER). The resulting line crosses the x-axis at $x = 6$ and is *above* the x-axis ($y > 0$) for values less than 6. Thus the solution set is $\{x|x < 6\}$.

C Solving Compound Inequalities

Sometimes we connect inequalities by using the word *or* as in

$$x < -2 \text{ or } x > 1$$

The resulting inequality is a **compound** inequality and its graph is based on the idea of the *union* of two sets.

Union of Two Sets (A or B) **($A \cup B$)**	If A and B are sets, the **union** of A and B, denoted by $A \cup B$, is the set of elements in either A or B.

To graph $\{x|x < -2 \text{ or } x > 1\}$, first graph $x < -2$, then graph $x > 1$, and finally graph the *union*. The graph of $x < -2$ is

$(-\infty, -2)$

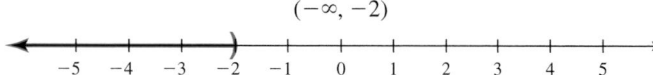

The graph of $x > 1$ is

$(1, \infty)$

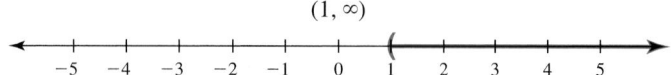

The union is

$(-\infty, -2) \cup (1, \infty)$

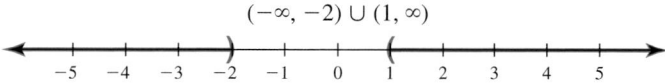

The solution set consists of all numbers less than -2 **or** greater than 1 or, in interval notation, $(-\infty, -2) \cup (1, \infty)$.

EXAMPLE 5   Solving a compound inequality with "or"

Graph and write the solution set in interval notation:

$$\{x|x < -3 \text{ or } x \ge 1\}$$

SOLUTION   We do each of the graphs separately and then take the union of the two graphs. The graph of $x < -3$ is

$(-\infty, -3)$

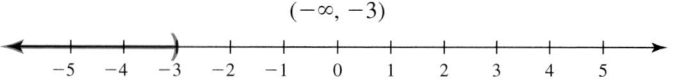

PROBLEM 5

Graph and write the solution set in interval notation:

$$\{x \mid x \le 1 \text{ or } x > 4\}$$

Answer

5. $(-\infty, 1] \cup (4, \infty)$

-3 -2 -1 0 1 2 3 4

The graph of $x \geq 1$ is

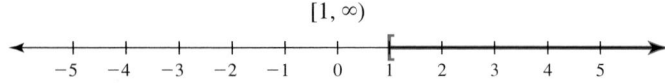

The union is

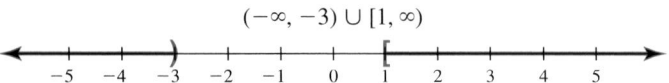

The inequality $3 \leq s \leq 5$ is also a compound inequality because it is equivalent to two other inequalities:

$$3 \leq s \quad \text{and} \quad s \leq 5$$

These inequalities use "and" as their connective. We can graph these inequalities using the idea of *intersection*.

Intersection of Two Sets **(A and B) (A ∩ B)**	If A and B are two sets, the intersection of A and B, denoted by A ∩ B, is the set of elements in both A and B.

The graph of $3 \leq s$ is (Remember $3 \leq s$ is the same as $s \geq 3$)

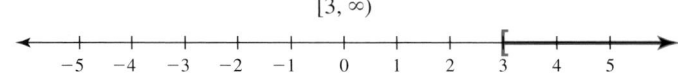

The graph of $s \leq 5$ is

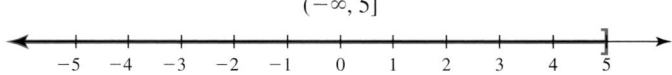

The intersection $A \cap B$ is

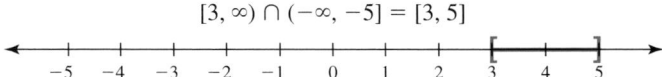

The intersection is the line segment that the two graphs have in common. It is the graph of $\{s \mid 3 \leq s \text{ and } s \leq 5\}$.

The interval $[3, 5]$ has a finite length, $5 - 3 = 2$. Because of this, $[3, 5]$ is called a **bounded** interval. In general, if a and b are real numbers and $a < b$, the intervals shown in Table 6 are bounded intervals and a and b are called the **endpoints** of each interval.

Table 6 Intervals

Set Notation	Interval Notation	Type of Interval	Graph	Inequality
$\{x \mid a \leq x \leq b\}$	$[a, b]$	Closed	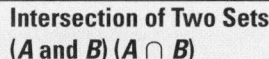	$a \leq x \leq b$
$\{x \mid a < x < b\}$	(a, b)	Open		$a < x < b$
$\{x \mid a \leq x < b\}$	$[a, b)$	Half-open		$a \leq x < b$
$\{x \mid a < x \leq b\}$	$(a, b]$	Half-open		$a < x \leq b$

EXAMPLE 6   Solving a compound inequality with "and"

Graph and write the solution set in interval notation:

$$\{x \mid x > -2 \ \text{and} \ x < 1\}$$

SOLUTION   The graph of $x > -2$ is

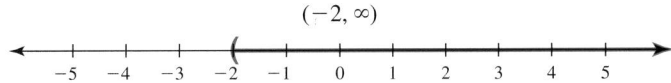

$$(-2, \infty)$$

The graph of $x < 1$ is

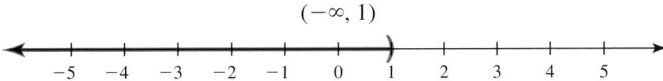

$$(-\infty, 1)$$

The numbers common to both graphs are in the open interval from -2 to 1. Thus, the intersection is $(-2, 1)$ or

$$(-2, \infty) \cap (-\infty, 1) = (-2, 1)$$

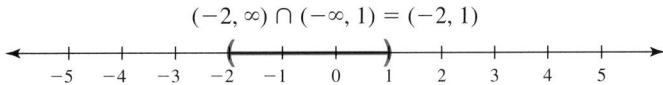

PROBLEM 6

Graph and write the solution set in interval notation:

$$\{x \mid x > -1 \ \text{and} \ x < 2\}$$

Now suppose you wish to solve the compound inequality $2x + 7 < 9$ *and* $x + 3 \geq -1$. We solve each inequality, obtaining

$2x + 7 < 9$	Given.	and	$x + 3 \geq -1$	
$2x < 2$	Add -7.		$x \geq -4$	Add -3.
$x < 1$	Divide by 2.			

The graph of $x < 1$ is

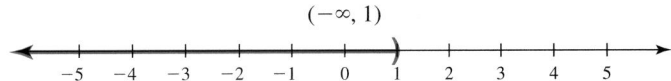

$$(-\infty, 1)$$

The graph of $x \geq -4$ is

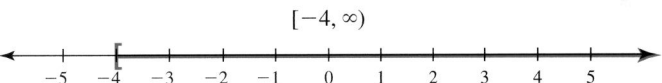

$$[-4, \infty)$$

Teaching Tip

Remind students that when we write $-4 \leq x < 1$, x is "in between" -4 and 1 including the -4 but not the 1.

The numbers common to both graphs are in the half-open interval from -4 to 1. The intersection is $[-4, 1)$ or

$$(-\infty, 1) \cap [-4, \infty) = [-4, 1)$$

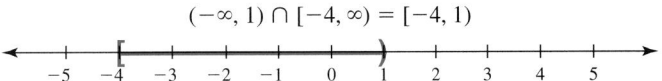

The intersection consists of all points such that $-4 \leq x$ *and* $x < 1$, which we can write as $-4 \leq x < 1$. A compound inequality that can be written in the form $a < x$ and $x < b$ (a and b real numbers) can be expressed more concisely as $a < x < b$.

Equivalent Statements for "AND"	$a < x$ and $x < b$ is equivalent to $a < x < b$ (where a and b are real numbers with $a < b$).

Answer

6. $(-1, 2)$

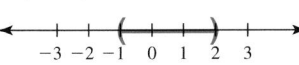

The graph of $a < x < b$ is the solution set of $a < x$ and $x < b$, that is, (a, b). Also be aware that inequality statements may be written with or without set builder notation. Solving $\{x \mid 3 < x + 2 \ \text{and} \ x < 3\}$ means the same as solving $3 < x + 2$ and $x < 3$. We use these ideas to solve compound inequalities in Example 7.

EXAMPLE 7   **Solving a compound inequality with "and"**

Solve, graph, and write the solution set in interval notation:

a. $3 < x + 2$ and $x < 3$ **b.** $5 \geq -x$ and $x - 1 \leq -4$

c. $x + 1 \leq 5$ and $-2x < 6$

SOLUTION

a. We wish to write this inequality in the form $a < x < b$, so we add -2 to both sides of the first inequality to obtain

$$1 < x \quad \text{and} \quad x < 3$$

equivalent to

$$1 < x < 3 \quad \text{or} \quad (1, 3)$$

The graph is

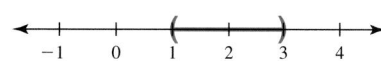

b. We wish to write this inequality in the form $a < x < b$, so we multiply the first inequality by -1 and add 1 to both sides of the second inequality to obtain

$$-5 \leq x \quad \text{and} \quad x \leq -3$$

equivalent to

$$-5 \leq x \leq -3 \quad \text{or} \quad [-5, -3]$$

The graph is

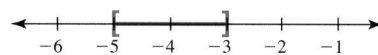

c. We solve the first inequality by subtracting 1 from both sides. We then have

$$x \leq 4 \quad \text{and} \quad -2x < 6$$

We then divide the second inequality by -2. (Remember to reverse the inequality sign.)

$$x \leq 4 \quad \text{and} \quad x > -3$$

Rearranging these inequalities gives

$$-3 < x \quad \text{and} \quad x \leq 4$$

equivalent to

$$-3 < x \leq 4 \quad \text{or} \quad (-3, 4]$$

The graph is

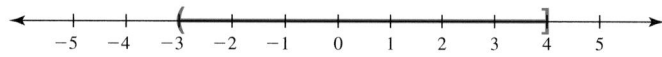

PROBLEM 7

Solve, graph, and write the solution set in interval notation.

a. $1 < x - 1$ and $x < 4$

b. $3 \geq -x$ and $x + 6 \leq 5$

c. $x + 2 \leq 6$ and $-3x < 12$

Web It

If you would like a slide show presentation on solving compound inequalities try link 2-5-3 at mhhe.com/bello.

Answers

7. a. $2 < x < 4$; $(2, 4)$

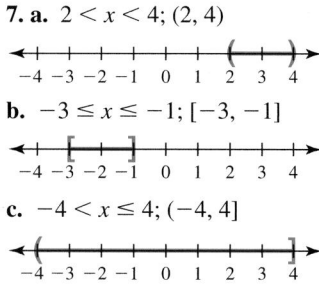

b. $-3 \leq x \leq -1$; $[-3, -1]$

c. $-4 < x \leq 4$; $(-4, 4]$

Suppose we want to solve $-5 < 2x + 3 \leq 9$. This inequality is equivalent to $-5 < 2x + 3$ and $2x + 3 \leq 9$, so it is also a compound inequality. We can solve this inequality by using the inequality properties we studied, keeping in mind that if we do any operations on the center expression ($2x + 3$), we must do the same operation on the outside expressions. As before, we wish to isolate x (this time in the middle). We start by adding

(−3) to all three parts. We then have

$$-5 < \quad 2x + 3 \le 9 \qquad \text{Given.}$$
$$-5 - 3 < 2x + 3 - 3 \le 9 - 3 \qquad \text{Add } -3.$$
$$-8 < \quad 2x \quad \le 6 \qquad \text{Simplify.}$$
$$\frac{-8}{2} < \quad \frac{2x}{2} \quad \le \frac{6}{2} \qquad \text{Multiply by } \tfrac{1}{2} \text{ (or divide by 2).}$$
$$-4 < \quad x \quad \le 3 \qquad \text{Simplify.}$$

The solution is $-4 < x \le 3$ or $(-4, 3]$
    The graph is

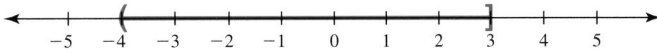

EXAMPLE 8   Solving and graphing compound inequalities

Solve, graph, and write the solution set in interval notation:

$$-2 \le -3x - 5 < 4$$

SOLUTION We start by adding 5 to each part.

$$-2 + 5 \le -3x - 5 + 5 < 4 + 5 \qquad \text{Add 5.}$$
$$3 \le \quad -3x \quad < 9$$
$$\frac{3}{-3} \ge \quad \frac{-3x}{-3} \quad > \frac{9}{-3} \qquad \text{Divide by } -3 \text{ and } reverse \text{ the sense (direction).}$$
$$-1 \ge \quad x \quad > -3$$
$$-3 < \quad x \quad \le -1 \qquad \text{Rewrite.}$$

The solution is $-3 < x \le -1$ or $(-3, -1]$
    The graph is

Check your answer by substituting a number from the interval $(-3, -1]$, say -2, into the original inequality.

PROBLEM 8

Solve, graph, and write the solution set in interval notation:

$$-4 \le -2x - 2 < 6$$

Teaching Tip

In Example 8, the original statement has the smaller number, -2, on the left and both arrows pointing left. Therefore, we should finish the problem in the same manner: smaller number, -3, on the left and both arrows pointing to the left. Thus, $-3 < x \le -1$.

CAUTION

When writing a compound inequality such as $-2 < -3x - 5 \le 4$, make sure that the numbers are in the correct position. If you write $4 \le -3x - 5 < -2$, you are implying that $4 < -2$, which is *wrong*. Compound inequalities must be written with the symbols pointing in the same direction toward the smaller number.

    No matter how hard you try, the inequality $\{x \mid x > 5 \text{ and } x < -5\}$ has no solution. (If you rewrite it as $5 < x < -5$, you can see that its solution set is $\varnothing$, the empty set.)

Answer

8. $-4 < x \le 1$ $(-4, 1]$

You may have noticed that all linear inequalities we have solved have *unbounded* intervals for their graph, while all compound inequalities have *bounded* intervals or *unbounded* intervals for their graph. Table 7 shows what happens in general.

Table 7 Equations, Inequalities, and Their Solution Sets

Type	Solution Set	Graph
Linear equation $ax + b = c$	$\{p\}$	
Linear inequality $ax + b < c$	$(-\infty, p)$ or (p, ∞)	
Compound inequality "or" $x < p$ or $x > q$	$(-\infty, p) \cup (q, \infty)$	
Compound inequality "and" $c < ax + b < d$ $c < ax + b$ and $ax + b < d$	(p, q)	

D   Translating Sentences into Inequalities

At the beginning of this section, we mentioned that the cost of renting a car must be under $50. We then translated this phrase by writing "< 50."

Here are some other phrases and their translations using inequalities.

Words	Translation	In Symbols
x is at least 10	x is 10 or more	$x \geq 10$
x is at most 20	x is 20 or less	$x \leq 20$
x is no more than 30	x is 30 or less	$x \leq 30$
x is no less than 40	x is 40 or more	$x \geq 40$

EXAMPLE 9   **Translating sentences involving inequalities**

Translate into an inequality:

a. The height h of a human (in feet) has never been known to exceed 9 ft.

b. The weight w of a human is at most 1400 lb.

c. The number n of puppies born in a single litter is no more than 23.

d. The cat population p in the United States is at least 74 million.

SOLUTION

a. $h \leq 9$ (Robert Wadlow was the tallest at 8 ft, 11.1 in.)

b. $w \leq 1400$ (Jon Browner Minnoch weighed 1400 lb)

c. $n \leq 23$ (Lena, a foxhound, had 23 live puppies June 9, 1944.)

d. $p \geq 74$ million (according to the American Pet Association, 1999)

PROBLEM 9

Translate into an inequality:

a. x does not exceed 23.

b. y is at most 180.

c. z is no more than 10.

d. p is at least 45.

Answers

9. a. $x \leq 23$ **b.** $y \leq 180$
c. $z \leq 10$ **d.** $p \geq 45$

Calculate It Exercises

There's another way to graphically solve linear inequalities. To solve Example 1(a)

$$3x - 2 < 2(x - 2)$$

Enter $Y_1 = 3x - 2$ and $Y_2 = 2(x - 2)$ and ZOOM 6 for a standard window. You have to check the points at which

$$Y_1 = 3x - 2 < Y_2 = 2(x - 2)$$

Find the point of intersection of Y_1 and Y_2 by entering 2nd

TRACE 5 ENTER ENTER ENTER . The point of intersection occurs when $X = -2$ and $Y = -8$ (see the window). $Y_1 < Y_2$; that is, the graph of the line Y_1 is below the graph of the line Y_2 when $x < -2$. The solution set is $\{x \mid x < -2\}$.

1. For Example 3(a),
   a. What window do you need to see the point of intersection for $Y_1 = 4(x - 1)$ and $Y_2 = 6x + 2$?
   b. What is the point of intersection for Y_1 and Y_2?
   c. When is $Y_1 = 4(x - 1)$ below $Y_2 = 6x + 2$?

2. Solve Example 4 by finding the points at which $Y_1 > Y_2$, where

$$Y_1 = \frac{x}{4} - \frac{x}{6} \qquad \text{and} \qquad Y_2 = \frac{x - 3}{6}$$

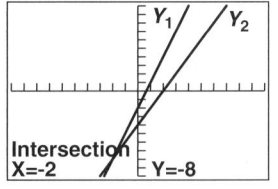

Intersection
X=-2   Y=-8

Exercises 2.5

A In Problems 1–4, graph the inequality and write the solution set using interval notation.

1. $x > 3$ $(3, \infty)$

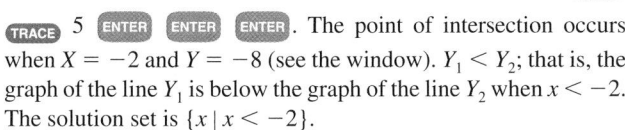

2. $x > 4$ $(4, \infty)$

3. $x \leq -3$ $(-\infty, -3]$

4. $x \leq -4$ $(-\infty, -4]$

B In Problems 5–30, solve the inequality, graph the solution set, and write the solution set in interval notation.

5. $2x \geq 6$ $[3, \infty)$; $x \geq 3$

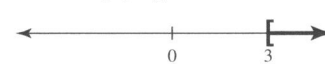

6. $3x \geq 8$ $\left[\frac{8}{3}, \infty\right)$; $x \geq \frac{8}{3}$

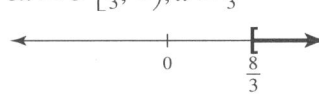

7. $-3x \leq 3$ $[-1, \infty)$; $x \geq -1$

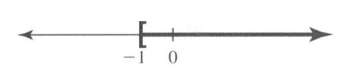

8. $-5x \leq 10$ $[-2, \infty)$; $x \geq -2$

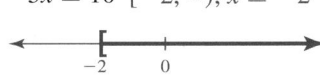

9. $-4x > -8$ $(-\infty, 2)$; $x < 2$

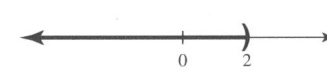

10. $-6x > -12$ $(-\infty, 2)$; $x < 2$

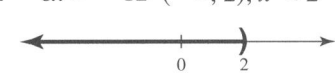

11. $3x + 6 \leq 9$ $(-\infty, 1]$; $x \leq 1$

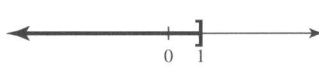

12. $4y - 9 \leq 3$ $(-\infty, 3]$; $y \leq 3$

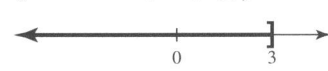

13. $-2y - 4 \geq -10$ $(-\infty, 3]$; $y \leq 3$

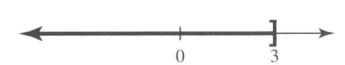

14. $-2z - 2 \geq -10$ $(-\infty, 4]$; $z \leq 4$

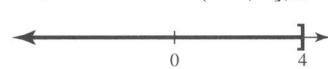

15. $-3x + 1 < -14$ $(5, \infty)$; $x > 5$

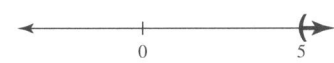

16. $-3x + 4 < -8$ $(4, \infty)$; $x > 4$

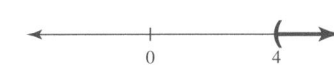

17. $3a + 6 \leq a + 10$ $(-\infty, 2]$; $a \leq 2$

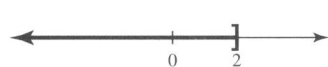

18. $4b + 4 \leq b + 7$ $(-\infty, 1]$; $b \leq 1$

19. $7z - 12 > 8z - 8$

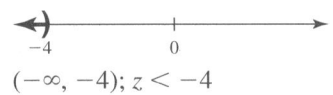

$(-\infty, -4)$; $z < -4$

20. $3z + 7 > 5z + 19$

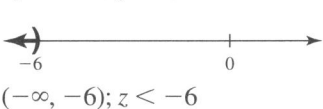

$(-\infty, -6)$; $z < -6$

21. $10 - 5x \leq 7 - 8x$

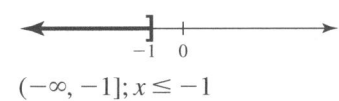

$(-\infty, -1]$; $x \leq -1$

22. $6 - 4y \leq -14 + 6y$

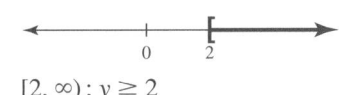

$[2, \infty)$; $y \geq 2$

23. $5(x + 2) \leq 3(x + 3) + 1$

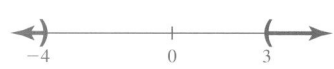

$(-\infty, 0]; x \leq 0$

24. $5(4 - 3x) < 7(3 - 4x) + 12$

$(-\infty, 1); x < 1$

25. $-4x + \dfrac{1}{2} > 4x + \dfrac{8}{5}$

$\left(-\infty, -\dfrac{11}{80}\right); x < -\dfrac{11}{80}$

26. $12x + \dfrac{2}{7} > 4x - \dfrac{4}{7}$

$\left(-\dfrac{3}{28}, \infty\right); x > -\dfrac{3}{28}$

27. $\dfrac{x}{5} - \dfrac{x}{4} \leq 1$

$[-20, \infty); x \geq -20$

28. $\dfrac{x}{3} - \dfrac{x}{2} \leq 1$

$[-6, \infty); x \geq -6$

29. $\dfrac{7x + 2}{6} + \dfrac{1}{2} \geq \dfrac{3}{4}x$

$[-2, \infty); x \geq -2$

30. $\dfrac{8x - 23}{6} + \dfrac{1}{3} \geq \dfrac{5}{2}x$

$(-\infty, -3]; x \leq -3$

C In Problems 31–70, solve, graph, and write the solution set in interval notation. If the solution set is empty, write $\varnothing$ for the answer. (Remember, inequality statements may be written with or without set builder notation.)

31. $\{x \mid x + 4 > 7 \text{ or } x - 2 < -6\}$

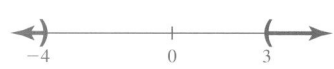

$x < -4 \text{ or } x > 3; (-\infty, -4) \cup (3, \infty)$

32. $\{x \mid x - 4 > 1 \text{ or } x + 2 < 1\}$

$x > 5 \text{ or } x < -1; (-\infty, -1) \cup (5, \infty)$

33. $3x + 4 > 10 \text{ or } 3x - 1 < 2$

$x < 1 \text{ or } x > 2; (-\infty, 1) \cup (2, \infty)$

34. $2x - 5 > 5 \text{ or } 2x < -4$

$x < -2 \text{ or } x > 5; (-\infty, -2) \cup (5, \infty)$

35. $2x \leq x + 4 \text{ or } x - 2 > 3$

$x \leq 4 \text{ or } x > 5; (-\infty, 4] \cup (5, \infty)$

36. $3x - 2 > 7 \text{ or } -5x \geq -5$

$x \leq 1 \text{ or } x > 3; (-\infty, 1] \cup (3, \infty)$

37. $\{x \mid -3x + 2 < -4 \text{ or } 2x - 1 < 3\}$

$x < 2 \text{ or } x > 2; (-\infty, 2) \cup (2, \infty)$

38. $\{x \mid -5x + 1 < -4 \text{ or } -2x + 1 > 5\}$

$x < -2 \text{ or } x > 1; (-\infty, -2) \cup (1, \infty)$

39. $-6x - 2 \geq -14 \text{ or } -7x + 2 < -19$

$x \leq 2 \text{ or } x > 3; (-\infty, 2] \cup (3, \infty)$

40. $-3x - 5 < 7 \text{ or } -2x + 1 > -5$

Real numbers; $(-\infty, \infty)$

41. $\{x \mid x \leq 4 \text{ and } x \geq -2\}$

$-2 \leq x \leq 4; [-2, 4]$

42. $\{x \mid x > 0 \text{ and } x \leq 5\}$

$0 < x \leq 5; (0, 5]$

43. $x + 1 \leq 7 \text{ and } x > 2$

$2 < x \leq 6; (2, 6]$

44. $x > -5 \text{ and } x - 1 < 0$

$-5 < x < 1; (-5, 1)$

45. $\{x \mid 2x - 1 > 1 \text{ and } x + 1 < 4\}$

$1 < x < 3; (1, 3)$

46. $\{x \mid x - 1 < 1 \text{ and } 3x - 1 > 11\}$
No solution; $\varnothing$

47. $x < -5 \text{ and } x > 5$
No solution; $\varnothing$

48. $x \geq 0 \text{ and } x < -2$
No solution; $\varnothing$

49. $\{x \mid x + 1 \geq 2 \text{ and } x \leq 4\}$

$1 \leq x \leq 4; [1, 4]$

50. $\{x \mid x \leq 5 \text{ and } x > -1\}$

$-1 < x \leq 5; (-1, 5]$

51. $\{x \mid x < 3 \text{ and } -x < -2\}$

$2 < x < 3; (2, 3)$

52. $\{x|-x < 5 \text{ and } x < 2\}$

$-5 < x < 2; (-5, 2)$

53. $\{x|x + 1 < 4 \text{ and } -x < -1\}$

$1 < x < 3; (1, 3)$

54. $\{x|x - 2 < 1 \text{ and } -x < 2\}$

$-2 < x < 3; (-2, 3)$

55. $\{x|x - 2 < 3 \text{ and } 2 > -x\}$

$-2 < x < 5; (-2, 5)$

56. $\{x|x - 3 < 1 \text{ and } 1 > -x\}$

$-1 < x < 4; (-1, 4)$

57. $\{x|x + 2 < 3 \text{ and } -4 < x + 1\}$

$-5 < x < 1; (-5, 1)$

58. $\{x|x + 4 < 5 \text{ and } -1 < x + 2\}$

$-3 < x < 1; (-3, 1)$

59. $\{x|x - 1 > 2 \text{ and } x + 7 < 11\}$

$3 < x < 4; (3, 4)$

60. $\{x|x - 2 > 1 \text{ and } -x > -5\}$

$3 < x < 5; (3, 5)$

61. $-3 < x - 1 < 3$

$-2 < x < 4; (-2, 4)$

62. $-4 < x + 1 < 4$

$-5 < x < 3; (-5, 3)$

63. $-8 < 2y + 4 < 6$

$-6 < y < 1; (-6, 1)$

64. $-2 \le 3x + 1 \le 7$

$-1 \le x \le 2; [-1, 2]$

65. $4 \le 3y - 8 \le 10$

$4 \le y \le 6; [4, 6]$

66. $3 \le 4z + 3 \le 5$

$0 \le z \le \frac{1}{2}; [0, \frac{1}{2}]$

67. $-1 < \frac{x}{2} < 2$

$-2 < x < 4; (-2, 4)$

68. $-2 < \frac{y}{2} < 1$

$-4 < y < 2; (-4, 2)$

69. $2 < 4 + \frac{2}{3}a < 6$

$-3 < a < 3; (-3, 3)$

70. $1 < 5 + \frac{4}{5}b < 9$

$-5 < b < 5; (-5, 5)$

D In Problems 71–76, translate the statement into an inequality.

71. The height h (in feet) of any mountain does not exceed that of Mt. Everest, 29,028 ft. $h \le 29,028$

72. The number e of possible eclipses in a year is at most 7. $e \le 7$

73. The number e of possible eclipses in a year is at least 2. $e \ge 2$

74. The altitude h (in feet) attained by the first liquid-fueled rocket was no more than 41 ft. $h \le 41$

75. There are no less than 4×10^{25} nematode sea worms in the world. (Let n be the number of nematodes.) $n \ge 4 \times 10^{25}$

76. There are at least 713 million people (p) that speak Mandarin Chinese. $p \ge 713$ million

77. When the variable cost per unit is \$12 and the fixed cost is \$160,000, the total cost for a certain product is $C = 12n + 160,000$ (n is the number of units sold). If the unit price is \$20, the revenue R is $20n$. What is the minimum number of units that must be sold to make a profit? (You need $R > C$ to make a profit.) 20,001

78. The cost of first-class mail is 37¢ for the first ounce and 25¢ for each additional ounce. A delivery company will charge \$4.00 for delivering a package weighing up to 2 lb (32 oz). When would the U.S. Post Office price, $P = 0.37 + 0.25(x - 1)$ (where x is the weight of the package in ounces), be cheaper than the delivery company's price? when $x < 15.52$ oz

79. The parking cost at a garage is $C = 1 + 0.75(h - 1)$, where h is the number of hours you park and C is the cost in dollars. When is the cost C less than \$10?

when $h < 13$ hr

SKILL CHECKER

Try the "Skill Checker" exercises so you'll be ready for the next section.

Fill in the blank with $<$ or $>$ to make the result a true statement.

80. $-7 \underline{>} -8$ **81.** $\dfrac{1}{3} \underline{\phantom{<}<\phantom{<}} \dfrac{1}{2}$ **82.** $0.34 \underline{\phantom{<}<\phantom{<}} 0.342$ **83.** $-0.234 \underline{\phantom{<}<\phantom{<}} -0.233$

Find:

84. $|-9|$ 9 **85.** $\left|-\dfrac{1}{5}\right|$ $\dfrac{1}{5}$ **86.** $|-0.34|$ 0.34 **87.** $|\sqrt{2}|$ $\sqrt{2}$

USING YOUR KNOWLEDGE

Inequalities and the Environment

88. Do you know why spray bottles use a pump rather than propellants? Because some of the propellants have chlorofluorocarbons (CFCs) and they deplete the ozone layer. In an international meeting in London, the countries represented agreed to stop producing CFCs by the year 2000. The production of CFCs (in thousands of tons) is given by $P = 1260 - 110x$, where x is the number of years after 1980. How many years after 1980 did it take for the production of CFCs to be less than 0? (Answer to the nearest whole number.) What year did that take place? $x > 11$; 1992

89. Cigarette smoking also produces air pollution. The annual number of cigarettes per person consumed in the United States is given by $N = 4200 - 70x$, where x is the number of years after 1960. How many years after 1960 will consumption be less than 1000 cigarettes per person annually? (Answer to the nearest whole year.) What year will that take place? $x > 46$; 2006

WRITE ON

90. What is the main difference in the techniques used when solving equations as opposed to inequalities?
Answers may vary.

How would you describe in words the real numbers in the following intervals?

91. $(1, \infty)$ All real numbers greater than 1.   **92.** $(-\infty, 3]$ All real numbers less than or equal to 3.

93. $(4, \infty)$ All real numbers greater than 4.   **94.** $[-2, \infty)$ All real numbers greater than or equal to -2.

95. Look at the following argument, where it is assumed that $a > b$:

$2 > 1$	Known.
$2(b - a) > 1(b - a)$	Multiply by $(b - a)$.
$2b - 2a > b - a$	Simplify.
$2b > b + a$	Add $2a$.
$b > a$	Subtract b.

But we assumed that $a > b$. Write a paragraph explaining what went wrong. Since $a > b$, $b - a < 0$.

MASTERY TEST

If you know how to do these problems, you have learned your lesson!

Translate into an inequality:

96. x does not exceed 23. $x \le 23$ **97.** p is at least 45. $p \ge 45$

Graph:

98. $x \le -3$

99. $x > 4$

Solve and graph:

100. $-3 < -3x - 6 < 3$

$-3 < x < -1$; $(-3, -1)$

101. $-2 \le 2 + \dfrac{4}{5}x < 6$

$-5 \le x < 5$; $[-5, 5)$

102. $2(x - 1) \geq 3x + 1$

$x \leq -3; \; (-\infty, -3]$

103. $3x + 10 < 6x + 4$

$x > 2; \; (2, \infty)$

104. $\dfrac{1}{4}(x - 5) < \dfrac{1}{3}(x - 4)$

$x > 1; \; (1, \infty)$

105. $2x + 3 > 1$ or $5x - 3 < -13$

$x < -2$ or $x > -1$;
$(-\infty, -2) \cup (-1, \infty)$

106. $\{x \mid 2x + 9 > 11 \text{ or } x + 3 > -1\}$

$x > -4; \; (-4, \infty)$

107. $\{x \mid 2x + 3 < 5x - 3 \text{ and } 3x - 2 < 3 + 2x\}$

$2 < x < 5; \; (2, 5)$

108. $\{x \mid -3x + 5 < 9 \text{ and } 3x - 7 < -3 + 2x\}$

$-\dfrac{4}{3} < x < 4; \; \left(-\dfrac{4}{3}, 4\right)$

2.6   ABSOLUTE-VALUE EQUATIONS AND INEQUALITIES

To Succeed, Review How To . . .

1. Find the absolute value of a number (p. 11).

2. Graph an inequality (p. 53).

Objectives

A Solve absolute-value equations.

B Solve absolute-value inequalities of the form $|ax + b| < c$ or $|ax + b| > c$, where $c > 0$.

GETTING STARTED

Budget Variances

Businesses and individuals usually try to predict how much money will be spent on certain items over a certain time. If you budget $120 for a month's utilities and a heat wave or cold snap makes your actual expenses jump to $150, the $30 difference might be an acceptable **variance.** Suppose b represents the budgeted amount for an item, a represents the actual expense, and you want to be within $10 of your estimate. The item will pass the variance test if the actual expenses a are within $10 of the budgeted amount b; that is, $b - a$ is between -10 and 10. In symbols, $-10 \leq b - a \leq 10$ or equivalently $|b - a| \leq 10$.

In general, if a and b are as before, a certain item will pass the variance test if $|b - a| \leq c$, where c is the variance. The quantity c can be a definite amount or a percent of the budget. For example, if you budget $50 for gas and you want to be within 10% of your budget, how much gas money can you spend and still be within your variance? Since 10% of $50 = (0.10)(50) = 5$, you can see that if you spend between $45 and $55, you will be within your 10% variance. Here $|b - a| \leq c$ becomes $|50 - a| \leq 5$ or equivalently $-5 \leq 50 - a \leq 5$.

$$-55 \leq -a \leq -45 \qquad \text{Add } -50.$$

$$55 \geq a \geq 45 \qquad \text{Multiply by } -1.$$

$$45 \leq a \leq 55 \text{ (as expected)} \qquad \text{Rewrite.}$$

$|50 - a| \leq 5$ is equivalent to $-5 \leq 50 - a \leq 5$; we shall develop the rule for this concept in this section.

 **Absolute-Value Equations**

The absolute value of a, denoted by $|a|$, is the distance between a and zero on a number line. Thus, $|3| = 3$, $|-5| = 5$, and $|-0.7| = 0.7$. The equation $|x| = 2$ is read as "the distance between x and zero on the number line is 2." x is 2 units from zero, so x can only be $+2$ or -2; that is, $x = \pm 2$.

To solve absolute-value equations, we need the following definition.

Web It

For a tutorial and practice in solving absolute value equations, try link 2-6-1 at mhhe.com/bello.

THE SOLUTIONS OF $|x| = a$ WHERE $a \geq 0$

If $a \geq 0$, the solutions of $|x| = a$ are $x = a$ and $x = -a$.

CAUTION

If $a < 0$, $|x| = a$ has no solution. An example of an absolute value equation with no solution is $|3x - 4| = -7$.

EXAMPLE 1 **Solving absolute-value equations**

Solve:

a. $|x| = 8$　　**b.** $|y| = 2.5$　　**c.** $|z| = -6$　　**d.** $|w| = 0$

SOLUTION

a. Using a number line we can locate those points on the line that are 8 units in distance from 0.

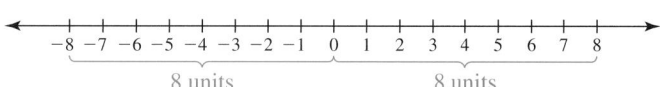

8 and -8 are 8 units from 0, so the solutions of $|x| = 8$ are 8 and -8. The solution set of $|x| = 8$ is $\{-8, 8\}$.

b. Using a number line we can locate those points on the line 2.5 units in distance from 0.

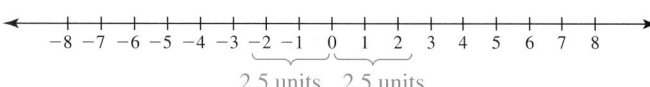

2.5 and -2.5 are 2.5 units from 0, so the solutions of $|y| = 2.5$ are 2.5 and -2.5. The solution set of $|y| = 2.5$ is $\{-2.5, 2.5\}$.

c. A number line can't help us solve $|z| = -6$ because there is no such thing as a -6 distance from 0. Distance is always expressed as a positive number or 0. Thus, the equation $|z| = -6$ has no solution. Its solution set is $\varnothing$ (the empty set).

d. Using a number line we can see that there is only one point 0 units from 0 and that is the number 0.

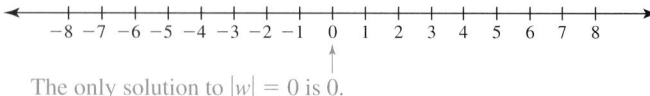

The only solution to $|w| = 0$ is 0.

PROBLEM 1

Solve:

a. $|x| = 7$　　**b.** $|y| = 3.4$

c. $|z| = -9$

Answers

1. a. -7; 7　**b.** -3.4; 3.4
c. No solution

Calculate It

Graph $|A(x)| = c$

To graph the solution set of $|A(x)| = c$, where $A(x)$ is a linear expression, you must find an equivalent equation of the form $|A(x)| - c = 0$ and then graph by pressing   . After entering the absolute value expression, close the parentheses and enter the constant. The x-values at which the graph crosses the x-axis are the solutions of the equation.

The ideas of Example 1 can be generalized to more complicated equations. Look at the pattern:

$|x| = 8$ has solutions	$x = 8$	and	$x = -8$
$|y| = 2.5$ has solutions	$y = 2.5$	and	$y = -2.5$
$|x + 1| = 8$ has solutions	$x + 1 = 8$	and	$x + 1 = -8$
(Solve each equation by adding (-1) on each side.)	$x = 7$	and	$x = -9$

We can check the solutions by substituting them into the equation $|x + 1| = 8$. Substituting 7 for x, we obtain $|7 + 1| = |8| = 8$, a true statement. If we substitute -9 for x, we have $|-9 + 1| = |-8| = 8$, also true.

EXAMPLE 2	**Solving more complicated absolute-value equations**

Solve:

a. $|2x + 1| = 6$ **b.** $\left|\dfrac{3}{4}x + 1\right| + 5 = 11$

SOLUTION

a. The solutions of $|2x + 1| = 6$ are

$$2x + 1 = 6 \quad \text{and} \quad 2x + 1 = -6$$
$$2x = 5 \quad \text{and} \quad 2x = -7 \qquad \text{Add } (-1) \text{ to both sides.}$$
$$x = \frac{5}{2} \quad \text{and} \quad x = -\frac{7}{2} \qquad \text{Multiply both sides by } \tfrac{1}{2}.$$

The solution set of $|2x + 1| = 6$ is $\left\{-\frac{7}{2}, \frac{5}{2}\right\}$.

b. To use the definition of absolute value, we must isolate $\left|\frac{3}{4}x + 1\right|$. We first add (-5) to both sides.

$$\left|\frac{3}{4}x + 1\right| + 5 = 11 \qquad \text{Given.}$$

$$\left|\frac{3}{4}x + 1\right| + 5 + (-5) = 11 + (-5) \qquad \text{Add } -5.$$

$$\left|\frac{3}{4}x + 1\right| = 6 \qquad \text{Simplify.}$$

The solutions of this equation are

$$\frac{3}{4}x + 1 = 6 \qquad \text{and} \qquad \frac{3}{4}x + 1 = -6$$

$$\frac{3}{4}x = 5 \qquad \text{and} \qquad \frac{3}{4}x = -7 \qquad \text{Add } -1.$$

$$\frac{4}{3} \cdot \frac{3}{4}x = \frac{4}{3} \cdot 5 \qquad \text{and} \qquad \frac{4}{3} \cdot \frac{3}{4}x = \frac{4}{3} \cdot (-7) \qquad \text{Multiply by } \tfrac{4}{3}.$$

$$x = \frac{20}{3} \qquad \text{and} \qquad x = -\frac{28}{3} \qquad \text{Simplify.}$$

The solution set of $\left|\frac{3}{4}x + 1\right| + 5 = 11$ is $\left\{-\frac{28}{3}, \frac{20}{3}\right\}$.

PROBLEM 2

Solve:

a. $|3x + 1| = 5$

b. $\left|\dfrac{3}{2}x + 5\right| - 4 = 6$

Answers

2. a. $-2; \frac{4}{3}$ **b.** $-10; \frac{10}{3}$

Calculate It   Graph $|A(x)| + B = C$

To solve $\left|\frac{3}{4}x + 1\right| + 5 = 11$, write an equivalent equation with zero on the right-hand side by subtracting 11 from both sides to obtain $\left|\frac{3}{4}x + 1\right| - 6 = 0$. (It's easier to enter $\frac{3}{4}x$ as $0.75x$.) Now graph $Y_1 = abs\,(0.75 + 1) - 6$ and $Y_2 = 0$. Use a standard window, ZOOM 6. The graph is V-shaped, and it crosses the x-axis at $x = -\frac{28}{3} = -9\frac{1}{3}$ and $x = \frac{20}{3} = 6\frac{2}{3}$. You can verify this by pressing 2nd TRACE 5 ENTER ENTER ENTER to get $X = 6.6666667$, an approximation for $\frac{20}{3}$ as shown in the window.

You can get the other root by using TRACE and ◀ repeatedly until the cursor is on the other point of intersection. Then press 2nd TRACE 5 ENTER ENTER ENTER .

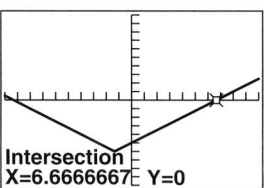

Intersection
X=6.6666667 Y=0

EXAMPLE 3   Solving absolute-value equations

Solve: $|x - 3| = |x - 5|$

SOLUTION   The equation $|x - 3| = |x - 5|$ is true if $x - 3$ and $x - 5$ are equal to each other or if they are additive inverses. The additive inverse of $(x - 5)$ is $-(x - 5)$. Here are the two cases.

Equal	**Opposites**	
$x - 3 = x - 5$	$x - 3 = -(x - 5)$	
$-x + x - 3 = -x + x - 5$ Add $-x$.	$x - 3 = -x + 5$	Remove parentheses.
$-3 = -5$	$x = -x + 8$	Add 3.
This is a contradiction. There is *no* solution for this case.	$2x = 8$	Add x.
	$x = 4$	Divide by 2.

The solution set is $\{4\}$. We can verify this by replacing x by 4 in $|x - 3| = |x - 5|$ to obtain

$$|4 - 3| = |4 - 5|$$
$$|1| = |-1|$$
$$1 = 1$$

This is a true statement, so our solution is correct.

PROBLEM 3

Solve:

$$|x - 5| = |x - 8|$$

Teaching Tip

Be sure that students identify whether the example is in the correct form, $|expression| = $ positive number or 0, before proceeding.

Statement Translation for Absolute Value Equations	If $|expression| = a$, where $a \geq 0$, then

If $|expression| = a$, where $a \geq 0$, then
$$expression = a \quad \text{or} \quad expression = -a$$

If $|expression| = |expression|$, then

Equal	**Opposites**
$expression = expression$	$expression = -(expression)$

Be sure $|expression|$ is isolated on one side before applying the translation statement. Example: $|x + 3| - 4 = 6$ cannot be translated until written as $|x + 3| = 10$.

Remember if the $|expression|$ equals a negative number, the equation has no solution.

Calculate It   Graph $|A(x)| = |B(x)|$

To solve $|x - 3| = |x - 5|$, first obtain the equivalent equation $|x - 3| - |x - 5| = 0$ and then graph $Y_1 = abs(x - 3) - abs(x - 5)$ and $Y_2 = 0$ and use a standard window. This time the graph is **not** V-shaped [$y = |A(x) - B(x)|$ is V-shaped when $A(x)$ and $B(x)$ are linear expressions, but this equation is of the form $y = |A(x)| - |B(x)|$]. By pressing 2nd TRACE 5 ENTER ENTER ENTER , the answer is $x = 4$.

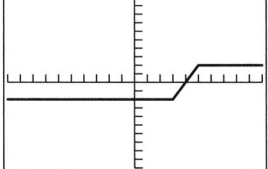

Answer

3. $\frac{13}{2}$

B   Absolute-Value Inequalities

As you recall, $|x| = 2$ means that the distance from 0 to x on the number line is 2 units. What would $|x| < 2$ and $|x| > 2$ mean? Here are the statements, their translations, their graphs, and the meaning in terms of the compound inequalities we just studied.

Statement Translation

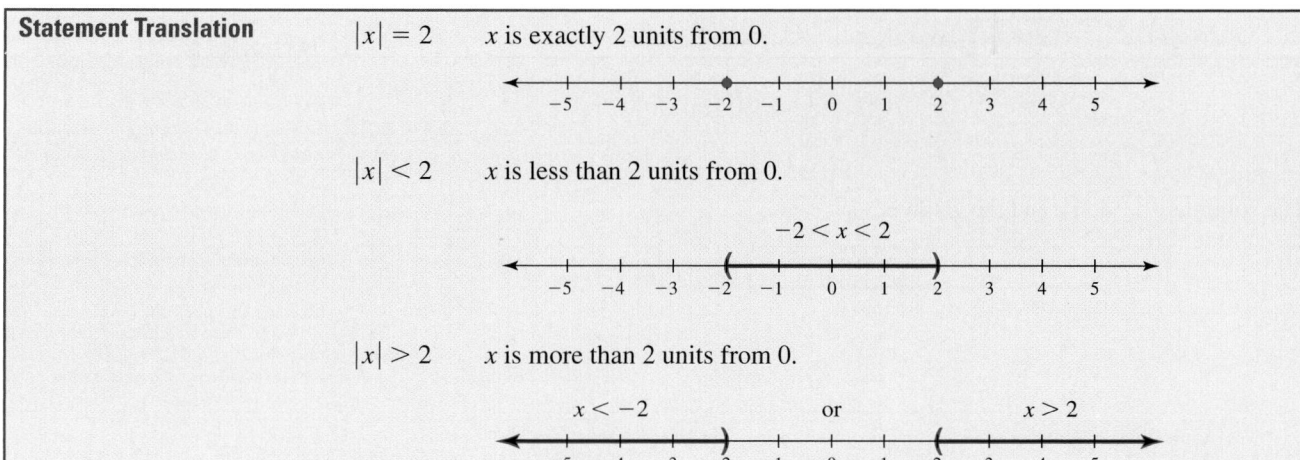

In general, the following holds.

$|x| < a$ WHERE $a > 0$ IS EQUIVALENT TO $-a < x < a$

The expression $|x| < a$ is equivalent to $-a < x < a$, where $a > 0$. A similar relationship exists if $\leq$ replaces $<$.

CAUTION

If $a < 0$, $|x| < a$ has no solution. An example of an absolute value inequality with no solution is $|2x - 5| < -3$.

$$|x| < 4 \qquad \text{is equivalent to} \qquad -4 < x < 4$$

$$|y| \leq 2.5 \qquad \text{is equivalent to} \qquad -2.5 \leq y \leq 2.5$$

$$|z| < \frac{3}{4} \qquad \text{is equivalent to} \qquad -\frac{3}{4} < z < \frac{3}{4}$$

What does $|x + 1| < 2$ mean? $|x| < a$ is equivalent to $-a < x < a$, so $|x + 1| < 2$ is equivalent to $-2 < x + 1 < 2$.

To solve $-2 < x + 1 < 2$, add -1, and simplify like this:

$$-2 - 1 < x + 1 - 1 < 2 - 1 \qquad \text{Add } -1.$$

$$-3 < \quad x \quad < 1 \qquad \text{Simplify.}$$

The solution is $(-3, 1)$, and the graph is

In general, $|ax + b| < c$, where $c > 0$, is equivalent to $-c < ax + b < c$.

EXAMPLE 4	Absolute-value inequalities of the form $|ax + b| \le c$

Solve, graph, and write the solution set in interval notation.

$$|3x + 4| \le 8$$

SOLUTION $|3x + 4| \le 8$ is equivalent to $-8 \le 3x + 4 \le 8$.   Translation.

$$-8 - 4 \le 3x + 4 - 4 \le 8 - 4 \quad \text{Add } -4.$$

$$-12 \le \quad 3x \quad \le 4$$

$$\frac{-12}{3} \le \quad \frac{3x}{3} \quad \le \frac{4}{3} \quad \text{Divide by 3.}$$

$$-4 \le \quad x \quad \le \frac{4}{3}$$

The solution is $[-4, \frac{4}{3}]$, and the graph is

PROBLEM 4

Solve, graph, and write the solution set in interval notation:

$$|3x - 2| \le 5$$

Teaching Tip

Point out to students that it is not mathematically correct to write the compound statement with "or" as a double inequality. Thus, "$x < 2$ or $x > 6$" may not be written as $2 < x > 6$, nor as $2 > x > 6$. The only correct notation is "$x < 2$ or $x > 6$."

Calculate It

Graph $|A(x)| \le C$

To solve Example 4, $|3x + 4| \le 8$ is equivalent to $|3x + 4| - 8 \le 0$ (subtract 8 from both sides of $|3x + 4| \le 8$). Thus we graph $Y_1 = abs(3x + 4) - 8$ and $Y_2 = 0$. The values that make $Y_1 \le 0$ are *below* the x-axis and have x-coordinates between -4 and $1\frac{1}{3} = \frac{4}{3}$, as shown in the window.

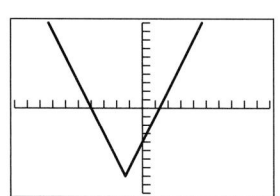

You can confirm that $x = -4$ and $x = \frac{4}{3}$ are the points at which the graph crosses the x-axis by pressing [2nd] [TRACE] [5] [ENTER] [ENTER] [ENTER]. The solution set is $[-4, \frac{4}{3}]$.

Answer

4. $-1 \le x \le \frac{7}{3}; \left[-1, \frac{7}{3}\right]$

Let's look at $|x| > 2$. Its graph consists of all points such that $x < -2$ or $x > 2$. $|x| > 2$ is equivalent to $x < -2$ or $x > 2$. If we generalize this result, we obtain the following.

$|x| > a$ WHERE $a > 0$ IS EQUIVALENT TO $x < -a$ OR $x > a$

The expression $|x| > a$ is equivalent to $x < -a$ or $x > a$, where $a > 0$. A similar relationship applies if $\ge$ replaces $>$.

$|x| > 5$	is equivalent to	$x < -5$ or $x > 5$
$|x| > 3.4$	is equivalent to	$x < -3.4$ or $x > 3.4$
$|x| \ge \frac{1}{5}$	is equivalent to	$x \le -\frac{1}{5}$ or $x \ge \frac{1}{5}$

What does $|x - 1| > 3$ mean? $|x| > a$ is equivalent to $x < -a$ or $x > a$, so $|x - 1| > 3$ means

$$x - 1 < -3 \quad \text{or} \quad x - 1 > 3$$

$$x < -2 \quad \text{or} \quad x > 4 \quad \text{Solve each inequality by adding 1 to both sides.}$$

The solution is $(-\infty, -2) \cup (4, \infty)$, and the graph of the solution set is

In general, $|ax + b| > c$, where $c > 0$, is equivalent to $ax + b < -c$ or $ax + b > c$.

Web It

For more examples of solving absolute value inequalities and writing the solutions in interval notation, try link 2-6-2 at mhhe.com/bello.

Statement Translation for Absolute Value Inequalities

If $|\text{expression}| > a$, where $a > 0$, then

$$\text{expression} < -a \quad \text{or} \quad \text{expression} > a$$

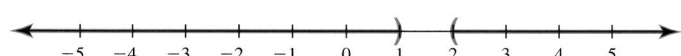

If $|\text{expression}| < a$, where $a > 0$, then

$$-a < \text{expression} < a$$

In the previous translations the $|\text{expression}|$ must be isolated on the left side of the inequality sign before the equivalent translation can be made.

EXAMPLE 5 Absolute-value inequalities of the form $|ax + b| > c$

Solve, graph, and write the solution set in interval notation:

$$|-2x + 3| > 1$$

SOLUTION The inequality $|-2x + 3| > 1$ is equivalent to

$$-2x + 3 < -1 \quad \text{or} \quad -2x + 3 > 1 \qquad \text{Translation.}$$

Add (-3) on both sides.

$$-2x + 3 + (-3) < -1 + (-3) \quad \text{or} \quad -2x + 3 + (-3) > 1 + (-3)$$

$$-2x < -4 \quad \text{or} \quad -2x > -2$$

Divide by -2 and reverse the inequality signs.

$$\frac{-2x}{-2} > \frac{-4}{-2} \quad \text{or} \quad \frac{-2x}{-2} < \frac{-2}{-2}$$

$$x > 2 \quad \text{or} \quad x < 1$$

The solution is $(2, \infty) \cup (-\infty, 1)$, and the graph is

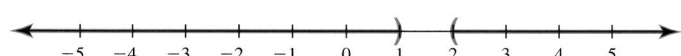

EXAMPLE 6 Complicated absolute-value inequalities

Solve, graph, and write the solution set in interval notation:

$$10 \geq 4|5x - 3| + 2$$

SOLUTION Before making the equivalent translation statement the $|\text{expression}|$ must be isolated on the left side of the inequality sign. We will subtract 2, divide by 4, and rewrite with $|\text{expression}|$ on the left side.

$$10 \geq 4|5x - 3| + 2$$

$$8 \geq 4|5x - 3| \qquad \text{Subtract 2.}$$

$$2 \geq |5x - 3| \qquad \text{Divide by 4.}$$

$$|5x - 3| \leq 2 \qquad \text{Rewrite with } |\text{expression}| \text{ on left side.}$$

PROBLEM 5

Solve, graph, and write the solution set in interval notation:

$$|-2x + 1| > 5$$

Teaching Tip

Have students discuss the similarities and differences between solving the statements:

$$|x + 2| \leq 0 \quad \text{and} \quad |x + 2| < 0$$

PROBLEM 6

Solve, graph, and write the solution set in interval notation:

$$5 \geq 2|2x - 6| - 3$$

Answers

5. $x < -2$ or $x > 3$;
$(-\infty, -2) \cup (3, \infty)$

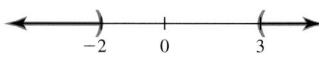

6. $1 \leq x \leq 5$; $[1, 5]$

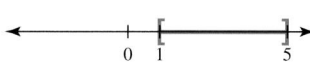

The statement is ready for the equivalent translation.

$$-2 \leq 5x - 3 \leq 2 \qquad \text{Translation.}$$

$$1 \leq 5x \leq 5 \qquad \text{Add 3.}$$

$$\frac{1}{5} \leq x \leq 1 \qquad \text{Divide by 5.}$$

The solution is $[\frac{1}{5}, 1]$, and the graph is

To find the solution sets to absolute value equations and inequalities you must be able to write the appropriate translation statement. Table 8 is a summary of this information.

Table 8 Summary for Solving Absolute-Value Equations and Inequalities

---

1. The solutions of the equation $|ax + b| = c$, $c \geq 0$, are obtained by solving the two equations

$$ax + b = c \qquad \text{and} \qquad ax + b = -c$$

Note: If c is negative ($c < 0$), there is no solution.

2. The solutions of the equation $|ax + b| = |cx + d|$ are obtained by solving the equations

$$ax + b = cx + d \qquad \text{and} \qquad ax + b = -(cx + d)$$

3. The solution set of the inequality $|ax + b| < c$, $c > 0$, is obtained by solving the inequality $-c < ax + b < c$. The graph is *bounded.* (If c is negative, there is no solution.)

4. The solution set of the inequality $|ax + b| > c$, $c > 0$, is obtained by solving the inequalities $ax + b > c$ or $ax + b < -c$. The graph is the union of two *unbounded* intervals. (If c is negative, $|ax + b| > c$ is always true, and the solution set is the set of real numbers, R.)

Calculate It Exercises

If you are proficient in finding the coordinates of the intersection of two graphs (using [2nd] [TRACE] [5]), there's another way to solve absolute-value inequalities. To do Example 2(b), $|\frac{3}{4}x + 1| + 5 = 11$, graph $Y_1 = |\frac{3}{4}x + 1| + 5$ and $Y_2 = 11$. The points of intersection at which $Y_1 = Y_2$ will satisfy both equations. The x-values at these points are the desired solution. The solution $X = 6.6666667$ or $\frac{20}{3}$ is shown in the window. Use this method to do:

1. Example 3   **2.** Example 4   **3.** Example 5

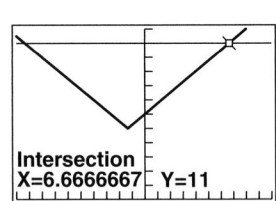

$Y_1 = |0.75x + 1| + 5$
and $Y_2 = 11$

Exercises 2.6

A In Problems 1–30, solve the equation.

1. $|x| = 13$ $-13; 13$

2. $|y| = 17$ $-17; 17$

3. $|y| - 2.3 = 0$ $-2.3; 2.3$

4. $|x| - 3.7 = 0$ $-3.7; 3.7$

5. $|x| = 0$ 0

6. $|y| = -3$ No solution

7. $|z| = -4$ No solution

8. $|x + 1| = 10$ $-11; 9$

9. $|x + 7| = 2$ $-9; -5$

10. $|x + 9| = 3$ $-12; -6$

11. $|2x - 4| = 8$ $-2; 6$

12. $|3x - 6| = 9$ $-1; 5$

13. $|5a - 2| - 8 = 0$ $-\frac{6}{5}; 2$

14. $|6b - 3| - 9 = 0$ $-1; 2$

15. $\left|\frac{1}{2}x + 4\right| = 6$ $-20; 4$

16. $\left|\frac{1}{3}x + 2\right| = 7$ $-27; 15$

17. $\left|\frac{2}{3}z - 3\right| = 9$ $-9; 18$

18. $\left|\frac{2}{5}x - 6\right| = 4$ $5; 25$

19. $|x + 2| = |x + 4|$ -3

20. $|y + 6| = |y + 2|$ -4

21. $|2y - 4| = |4y + 6|$ $-5; -\frac{1}{3}$

22. $|3x - 2| = |6x + 4|$ $-\frac{2}{9}; -2$

23. $2|a + 1| - 3 = 9$ $-7; 5$

24. $2|3a + 1| + 5 = 13$ $-\frac{5}{3}; 1$

25. $3|2x + 1| - 4 = -6$
No solution

26. $5|x - 1| - 6 = -8$
No solution

27. $|x - 4| = |4 - x|$
All real numbers

28. $|2x - 2| = |2 - 2x|$
All real numbers

29. $|5x - 10| = |10 - 5x|$
All real numbers

30. $|6x - 3| = |3 - 6x|$
All real numbers

B In Problems 31–68 solve, graph, and write the solution set in interval notation.

31. $|x| < 4$ $-4 < x < 4; (-4, 4)$

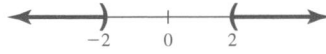

32. $|y| \leq 1.5$ $-1.5 \leq y \leq 1.5; [-1.5, 1.5]$

33. $|z| + 4 \leq 6$ $-2 \leq z \leq 2; [-2, 2]$

34. $|x| + 3 < 7$ $-4 < x < 4; (-4, 4)$

35. $|a| - 2 \leq 2$ $-4 \leq a \leq 4; [-4, 4]$

36. $|b| + 2 \leq 5$ $-3 \leq b \leq 3; [-3, 3]$

37. $|x - 1| < 2$ $-1 < x < 3; (-1, 3)$

38. $|x - 3| \leq 1$ $2 \leq x \leq 4; [2, 4]$

39. $|x + 3| < -2$ No solution

40. $|2x - 3| < -4$ No solution

41. $|2x + 3| \leq 1$ $-2 \leq x \leq -1; [-2, -1]$

42. $|3x + 2| < 5$ $-\frac{7}{3} < x < 1; \left(-\frac{7}{3}, 1\right)$

43. $|4x + 2| - 4 < 2$ $-2 < x < 1; (-2, 1)$

44. $|3x + 3| - 2 < 4$ $-3 < x < 1; (-3, 1)$

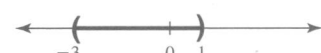

45. $|x| > 2$ $x < -2$ or $x > 2; (-\infty, -2) \cup (2, \infty)$

46. $|y| \geq 2.5$ $y \leq -2.5$ or $y \geq 2.5; (-\infty, -2.5] \cup [2.5, \infty)$

47. $|z| + 5 \geq 6$ $z \leq -1$ or $z \geq 1$; $(-\infty, -1] \cup [1, \infty)$

48. $|x| + 2 > 5$ $x < -3$ or $x > 3$; $(-\infty, -3) \cup (3, \infty)$

49. $|a| - 1 \geq 2$ $a \leq -3$ or $a \geq 3$; $(-\infty, -3] \cup [3, \infty)$

50. $|b| + 1 \geq 5$ $b \leq -4$ or $b \geq 4$; $(-\infty, -4] \cup [4, \infty)$

51. $|x - 1| > 1$ $x < 0$ or $x > 2$; $(-\infty, 0) \cup (2, \infty)$

52. $|x - 3| \geq 2$ $x \leq 1$ or $x \geq 5$; $(-\infty, 1] \cup [5, \infty)$

53. $|x + 3| > -1$ All real numbers; $(-\infty, \infty)$

54. $|2x - 3| > -4$ All real numbers; $(-\infty, \infty)$

55. $|2x + 3| \geq 1$ $x \leq -2$ or $x \geq -1$; $(-\infty, -2] \cup [-1, \infty)$

56. $|3x + 2| > 5$ $x < -\frac{7}{3}$ or $x > 1$; $(-\infty, -\frac{7}{3}) \cup (1, \infty)$

57. $|3 - 4x| > 7$ $x < -1$ or $x > \frac{5}{2}$; $(-\infty, -1) \cup (\frac{5}{2}, \infty)$

58. $|5 - 3x| > 11$ $x < -2$ or $x > \frac{16}{3}$; $(-\infty, -2) \cup (\frac{16}{3}, \infty)$

59. $|4x + 2| - 4 \geq 2$ $x \leq -2$ or $x \geq 1$; $(-\infty, -2] \cup [1, \infty)$

60. $|3x + 3| - 2 \geq 4$ $x \leq -3$ or $x \geq 1$; $(-\infty, -3] \cup [1, \infty)$

61. $\left|2 - \frac{1}{2}a\right| > 1$ $a < 2$ or $a > 6$; $(-\infty, 2) \cup (6, \infty)$

62. $\left|1 + \frac{2}{3}b\right| > 2$ $b < -\frac{9}{2}$ or $b > \frac{3}{2}$; $(-\infty, -\frac{9}{2}) \cup (\frac{3}{2}, \infty)$

63. $-2 < |2x + 4|$ All real numbers; $(-\infty, \infty)$

64. $-3 \leq |-2x - 4|$ All real numbers; $(-\infty, \infty)$

65. $-3|x - 5| > 6$ No solution; $\varnothing$

66. $-4|x + 3| > 12$ No solution; $\varnothing$

67. $9 \geq 5|3x - 2| + 4$ $\frac{1}{3} \leq x \leq 1$; $[\frac{1}{3}, 1]$

68. $3 < 4|-2x + 3| - 5$ $x < \frac{1}{2}$ or $x > \frac{5}{2}$; $(-\infty, \frac{1}{2}) \cup (\frac{5}{2}, \infty)$

APPLICATIONS

In Problems 69–72, use $|b - a| \leq c$, where b is the budgeted amount, a is the actual expense, and c is the variance.

69. A company budgets $500 for office supplies. If their variance is $50, write an inequality giving the amounts between which the actual expense a must fall.
$450 \leq a \leq $550

70. A company budgets $800 for maintenance. If their acceptable variance is 5% of their budgeted amount, write an inequality giving the amounts between which the actual expense a must fall.
$760 \leq a \leq $840

71. George budgets $300 for miscellaneous monthly expenses. His actual expenses for 1 month amounted to $290. Was he within a 5% budget variance? Yes

72. If George from Problem 71 spent $310, was he within a 5% budget variance? Yes

SKILL CHECKER

Try the "Skill Checker" exercises so you'll be ready for the next section.

Evaluate:

73. $-16t^2 + 10t - 15$ when $t = 2$ -59

74. $x^2 - 3x$ when $x = -3$ 18

Simplify:

75. $(-6x^2 + 3x + 5) + (7x^2 + 8x + 2)$ $\;x^2 + 11x + 7$ **76.** $(3x - 2x^2 + 4) + (5 - 7x + 4x^2)$ $\;2x^2 - 4x + 9$

77. $(3x - 7x^2 + 4) - (3x^2 + 5x + 1)$ $\;-10x^2 - 2x + 3$ **78.** $(5 - 3x^2 - 2x) - (8x^2 + 4x - 2)$ $\;-11x^2 - 6x + 7$

USING YOUR KNOWLEDGE

The Boundaries of Inequalities

The inequality $\left|-2x + 3\right| > 1$ can be solved by finding its *boundary numbers,* the solutions resulting when the inequality sign ($>$ in this case) is replaced by an equal sign. We write $\left|-2x + 3\right| = 1$ with solutions

$$-2x + 3 = 1 \quad \text{or} \quad -2x + 3 = -1$$
$$-2x = -2 \qquad\qquad -2x = -4 \quad \text{Add } -3.$$
$$x = 1 \qquad\qquad\quad x = 2 \quad \text{Multiply by } -\tfrac{1}{2}.$$

The boundary numbers are 1 and 2.

The graph of an inequality of the form $\left|ax + b\right| > c$ is the union of two unbounded line segments, so the graph is

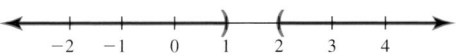

as in Example 5. Try doing Problems 37–42 and 51–60 using this technique.

WRITE ON

79. Using the idea of distance between two numbers, write your own definition for the absolute value of a number. Answers may vary.

80. Write your own explanation of why the equation $\left|x\right| = a$ has no solution if a is negative. Answers may vary.

81. Explain how to solve the equation $\left|A(x)\right| = a$, where $A(x)$ is a linear equation. When is the solution set of this equation empty? Answers may vary.

82. Write in your own words the reasons why
 a. The inequality $\left|x\right| < a$ ($a < 0$) has no solution. Answers may vary.
 b. The inequality $\left|x\right| > a$ ($a < 0$) is always true. Answers may vary.

MASTERY TEST

If you know how to do these problems, you have learned your lesson!

83. $\left|x - 1\right| = \left|x - 8\right|$ $\;x = \dfrac{9}{2}$ **84.** $\left|4 - x\right| = \left|7 + 2x\right|$ $\;x = -1; x = -11$

85. $\left|3x + 1\right| = 5$ $\;x = \dfrac{4}{3}; x = -2$ **86.** $\left|z\right| - \dfrac{3}{4} = 0$ $\;z = \dfrac{3}{4}; z = -\dfrac{3}{4}$

87. $\left|\dfrac{2}{3}x + 1\right| + 3 = 9$ $\;x = -\dfrac{21}{2}; x = \dfrac{15}{2}$ **88.** $\left|y\right| = 3.4$ $\;y = 3.4; y = -3.4$

Solve, graph, and write the solution set in interval notation.

89. $\left|-2x + 1\right| - 1 > 4$ $\;x < -2$ or $x > 3;$ $(-\infty, -2) \cup (3, \infty)$

90. $\left|x - \dfrac{2}{3}\right| \le 5$ $\;-\dfrac{13}{3} \le x \le \dfrac{17}{3};$ $[-\dfrac{13}{3}, \dfrac{17}{3}]$

91. $\left|-2 + 3x\right| \ge 5$ $\;x \le -1$ or $x \ge \dfrac{7}{3};$ $(-\infty, -1] \cup [\dfrac{7}{3}, \infty)$

92. $\left|2x - 3\right| - 5 < 0$ $\;-1 < x < 4;$ $(-1, 4)$

COLLABORATIVE LEARNING 2A

How much is "too much" when it comes to calories from fat in your daily diet? Most dieticians recommend that the proportion of calories from fat in your daily diet should be 30% or less, although many people consume more than 50% of their total calories as fat. Usually nutrition facts are based on a daily 2000 calorie diet and that fat contains 9 calories per gram of food energy.

To take a closer look at this we will divide into groups, plan two menus for a day, and then collaborate the findings. The first menu will be any items of your choosing as long as it totals 2000 calories for the day. The second menu will not only have to total 2000 calories, but it should not exceed 30% in fat calories. To plan the menus you may use the following charts or any other resource, including books or Internet sites.

1. Use two poster boards, one for each menu, to make a table that includes columns with all the food items for the day, their calorie content, fat grams, and fat percents. To find the percent of fat calories you will have to answer the following questions:

   a. What is the total amount of fat calories allowed in a 2000 calorie diet if 30% is allowed for fat calories?

   b. If fat contains 9 calories per gram and you know how many fat grams there are in a food, what would you do to find the amount of fat calories in that food?

   c. Knowing the amount of fat calories in a food and the total amount of fat calories allowed, what would you do to find the percent of fat calories?

2. After the groups have completed their menus on poster board have a class discussion about your findings. Be sure to include the following questions.

Item	Calorie content	Fat grams
Bread and Bakery		
Bagel (85g)	216	1.4
Biscuit (15g)	74	3.3
Bread, white (slice, 37g)	84	0.6
Danish pastry (67g)	287	17.4
Doughnut (49g)	140	2
Dairy		
Butter (10g)	74	8.2
Cheddar cheese (40g)	172	14.8
Cream cheese (34g)	58	4.8
Eggs, 3 (57g)	84	6.2
Milk, semi-skim (200ml)	96	3.2
Fruit		
Apple (112g)	53	0.1
Banana (150g)	143	0.5
Grapes (50g)	30	0.1
Melon (1 oz/28g)	7	0.1
Orange (160g)	59	0
Strawberries (1 oz/28g)	7	0
Drinks		
Coffee (1 cup/220ml)	15.4	0.9
Coke, can (330ml)	139	0
Orange juice (1 glass/200ml)	88	0
Tea (1 mug/270ml)	29	0.5

Item	Calorie content	Fat grams
Vegetables		
Carrots (60g)	13	0.2
Celery (40g)	2	0.1
Chips (100g)	253	9.9
Peas (60g)	32	0.4
Potato, baked (180g)	245	0.4
Salad (100g)	19	0.3
Meats		
Bacon (25g)	64	4
Beef in gravy (83 ml)	45	2.7
Chicken breast (200g)	342	13
Ham slice (30g)	35	1
Kebob (168g)	429	28.6
Fast Food		
Big Mac (215g)	492	23
Cheeseburger (148 g)	379	18.9
Chicken, KFC (67g)	195	12
Hamburger (108g)	254	7.7
Pizza (1/2 pizza/135g)	263	4.9
Pizza Deluxe (1 slice/66g)	171	6.7
Potato Wedges (135g)	279	13

a. What kinds of food items have a higher percent of fat calories? What kinds have a lower percent?

b. How does the percent of fat calories from fast food compare with the other types of food?

c. Which meal (breakfast, lunch, or dinner) contains most of the percent of fat calories?

COLLABORATIVE LEARNING 2B

Do you know the 10 hottest careers for college grads? Government economists estimated which occupations would grow fastest between 2002 and 2012, and they are listed in the following table. The numbers are in thousands. It seems these careers could be listed in one of two categories—the technology field or the medical/health field. Maybe you are trying to decide on one of these two areas to pursue as a career. Knowing which ones will have more opportunities could help make that choice. It is said that a picture is worth a thousand words, so that is what we will use to help you consider a career choice. Divide into groups and do the following.

1. Separate these 10 careers into two categories—Technical or Medical/Health. Make a chart for each field that will include a column for the percent of increase for each career. Percent increase is found by dividing the difference of the two numbers by the original number.

2. Make a bar graph for each of the two categories so that each bar represents the percent of increase for a career in that field.

3. Have each group present their bar graphs and discuss the following.

   a. Of the two categories, which field seems to have a higher percentage increase?

   b. Which career has the highest percent of increase in the technology field? The lowest?

   c. Which career has the highest percent of increase in the medical/health field? The lowest?

   d. What other information would you want to know about a career besides the availability of jobs?

10 Fastest Growing Occupations for College Grads

Occupation	2002	2012
Network systems and data communications analysts	186	292
Physician assistants	63	94
Medical records and health information technicians	147	216
Computer software engineers, applications	394	573
Computer software engineers, systems software	281	409
Physical therapist assistants	50	73
Fitness trainers and aerobics instructors	183	264
Database administrators	110	159
Veterinary technologists and technicians	53	76
Dental hygienists	148	212

Source: United States Bureau of Labor Statistics.

1. Write a short paragraph about the contents of the Rhind papyrus with special emphasis on the algebraic method called "the method of false position."

2. Try to find three problems that appeared on the Rhind papyrus and explain the techniques used to solve them. Then try to solve them yourself using the techniques you've learned in this chapter. (*Hint:* Problems 25, 26, and 27 are easy to solve.)

3. Find out about al-Khowarizmi and then write a few paragraphs about him and his life. (Include what academy he belonged to, the titles of the books he wrote, and the types of problems that appeared in these books.)

4. A Latin corruption of the name "al-Khowarizmi" meant "the art of computing with Hindu Arabic numerals." What is this word and what does it mean today?

5. The traditional explanation of the word *jabr* is "the setting of a broken bone." Write a paragraph relating how this word reached Spain as *algebrista* and the context in which it was used.

6. Diophantus of Alexandria wrote a book described "as the earliest treatise devoted to algebra." Write a few paragraphs about Diophantus and, in particular, the problems contained in his book.

7. Write a short paragraph detailing the types of symbols used in Diophantus' book.

Summary

SECTION	ITEM	MEANING	EXAMPLE
2.1	Equation	A sentence using $=$ as its verb	$x + 1 = 5$ and $3 - y = 7$ are equations.
2.1A	Solutions of an equation	The replacements of the variable that make the equation a true statement	5 is a solution of $x + 3 = 8$.
2.1B	Equivalent equations	Two or more equations are equivalent if they have the same solution set.	$x + 2 = 5$ and $x = 3$ are equivalent.
	Properties of equality	You can *add* or *subtract* the same quantity on both sides and *multiply* or *divide* both sides of an equation by the same nonzero quantity, and the result will be an equivalent equation.	If C is a real number and $A = B$, $A + C = B + C$ $A - C = B - C$ $A \cdot C = B \cdot C \ (C \neq 0)$ $\dfrac{A}{C} = \dfrac{B}{C} \ (C \neq 0)$ are equivalent equations.
2.1C	Linear equation	An equation that can be written in the form $ax + b = c$, where a, b, and c are real numbers and $a \neq 0$.	$3x + 7 = 9$ and $-x - 3 = \frac{2}{3}$ are linear equations.
	CRAM Procedure	Clear fractions/decimals Remove parentheses/simplify Add/subtract to get variable isolated Multiply/divide to make coefficient a 1.	
2.2A	Perimeter	The distance around a polygon (figure)	The perimeter P of a rectangle is $P = 2L + 2W$, where L is the length and W is the width.

SECTION	ITEM	MEANING	EXAMPLE
2.2C	Angles and transversals	All the acute angles (1, 4, 5, and 8) are equal. All the obtuse angles (2, 3, 6, and 7) are equal.	 L_1 and L_2 are parallel.
2.3B	Word problem	The RSTUV procedure to solve a word problem is **R**ead **S**elect a variable **T**hink of a plan **U**se algebra **V**erify	
2.3C	Consecutive integers	n and $n + 1$ are consecutive integers.	8 and 9; 28 and 29 are consecutive integers.
2.3D	Right angle	An angle whose measure is 90°	
	Complementary angles	Two angles whose measures add to 90° (A and B are complementary.)	
	Straight angle	An angle whose measure is 180°	
	Supplementary angles	Two angles whose measures add to 180° (A and B are supplementary.)	
2.4B	$I = Pr$	The annual interest I is the product of the principal P and the rate r.	The annual interest on a $3000 principal at a 5% rate is $I = 3000 \cdot 0.05 = \$150$.
2.4C	$D = RT$	When moving at a constant rate R, the distance D traveled in time T is $D = RT$.	A car moving at 55 mi/hr for 3 hr will travel $D = 55 \cdot 3 = 165$ mi.
2.5	Linear inequality	An inequality that can be written in the form $ax + b > c$, where a, b, and c are real numbers	$-2x + 3 > 5$ and $3 - 2x > 7$ are linear inequalities.
2.5B	Properties of inequalities	If $A < B$, then $A + C < B + C$.	The number C can be added to both sides to obtain an equivalent inequality.
		If $A < B$, then $A - C < B - C$.	The number C can be *subtracted* from both sides to obtain an equivalent inequality.

(Continued)

SECTION	ITEM	MEANING	EXAMPLE
		If $A < B$, then $A \cdot C < B \cdot C$ when $C > 0$.	Both sides can be *multiplied* by $C > 0$ to obtain an equivalent inequality.
		If $A < B$, then $A \cdot C > B \cdot C$ when $C < 0$.	Both sides can be *multiplied* by $C < 0$ to obtain an equivalent inequality provided you change the sense (direction) of the inequality.
		If $A < B$, then $\dfrac{A}{C} < \dfrac{B}{C}$ when $C > 0$.	Both sides can be *divided* by $C > 0$ to obtain an equivalent inequality.
		If $A < B$, then $\dfrac{A}{C} > \dfrac{B}{C}$ when $C < 0$.	Both sides can be *divided* by $C < 0$ to obtain an equivalent inequality provided you change the sense (direction) of the inequality.
2.5C	Union of sets	If A and B are sets, the union of A and B, denoted by $A \cup B$, is the set of elements that are in either A or B.	
	Intersection of sets	If A and B are sets, the intersection of A and B, denoted by $A \cap B$, is the set of elements that are in both A and B.	
2.6A	Absolute-value equation	The solutions of $|x| = a$ are $x = a$ or $x = -a$ for $a \geq 0$.	The solutions of $|x| = 2$ are 2 or -2.
2.6B	$|x| < a$	For $a \geq 0$, $|x| < a$ is equivalent to $-a < x < a$.	$|x| < 3$ is equivalent to $-3 < x < 3$.
	$|x| > a$	For $a \geq 0$, $|x| > a$ is equivalent to $x > a$ or $x < -a$.	$|x| > 3$ is equivalent to $x > 3$ or $x < -3$.

Review Exercises

(If you need help with these exercises, look in the section indicated in brackets.)

1. [2.1A] Does -3 satisfy the equation?

   a. $7 = 8 - x$ No   **b.** $9 = 8 + x$ No
   c. $4 = 1 - x$ Yes

2. [2.1B] Solve.

   a. $\dfrac{2}{3}y - 3 = 5$ 12   **b.** $\dfrac{2}{3}y - 5 = 5$ 15

   c. $\dfrac{2}{3}y - 7 = 5$ 18

3. [2.1B] Solve.

   a. $x + 2 = 2(2x - 2)$ 2
   b. $x + 3 = 3(2x - 4)$ 3
   c. $x + 4 = 4(2x - 6)$ 4

4. [2.1B] Solve.

   a. $\dfrac{x + 4}{3} - \dfrac{x - 4}{5} = 4$ 14   **b.** $\dfrac{x + 9}{4} - \dfrac{x - 12}{5} = 6$ 27

   c. $\dfrac{x + 8}{3} - \dfrac{x - 8}{5} = 8$ 28

5. [2.1C] Solve.

   a. $\dfrac{x}{4} - \dfrac{x}{3} = \dfrac{x - 4}{4}$ 3   **b.** $\dfrac{x}{5} - \dfrac{x}{4} = \dfrac{x - 5}{5}$ 4

   c. $\dfrac{x}{7} - \dfrac{x}{3} = \dfrac{x - 7}{7}$ 3

6. [2.1D] Solve.

   a. $0.05P + 0.10(2000 - P) = 175$ $P = 500$
   b. $0.08P + 0.10(5000 - P) = 460$ $P = 2000$
   c. $0.06P + 0.10(10{,}000 - P) = 840$ $P = 4000$

7. [2.2A] Solve for h and evaluate when $H = 82.48$.

   a. $H = 2.5h + 72.48$ $h = \dfrac{H - 72.48}{2.5}$; $h = 4$

   b. $H = 2.5h + 77.48$ $h = \dfrac{H - 77.48}{2.5}$; $h = 2$

   c. $H = 2.5h + 84.98$ $h = \dfrac{H - 84.98}{2.5}$; $h = -1$

8. [2.2A] Solve for A.

   a. $B = \dfrac{2}{7}(A - 7)$ $A = \dfrac{7B + 14}{2}$

   b. $B = \dfrac{3}{7}(A - 5)$ $A = \dfrac{7B + 15}{3}$

   c. $B = \dfrac{4}{5}(A - 7)$ $A = \dfrac{5B + 28}{4}$

9. [2.2A] The perimeter of a rectangle is $P = 2L + 2W$, where L is the length and W is the width. Solve for L and give the dimensions when

   a. The perimeter is 180 ft and the length is 10 ft more than the width. $L = \frac{P - 2W}{2}$; 40 ft by 50 ft

   b. The perimeter is 220 ft and the length is 10 ft more than the width. $L = \frac{P - 2W}{2}$; 50 ft by 60 ft

   c. The perimeter is 260 ft and the length is 10 ft more than the width. $L = \frac{P - 2W}{2}$; 60 ft by 70 ft

10. [2.2C]

   a. If L_1 and L_2 are parallel lines, find x and the measure of the unknown angles. $x = 10$; angles: $50°$ each

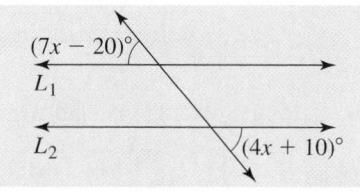

   b. Find x and the measure of the unknown angles if the measure of one angle is $(3x + 12)°$ and the measure of the other angle is $(4x - 3)°$. $x = 15$; angles: $57°$ each

   c. Find x and the measure of the unknown angles if the measure of one angle is $(2x + 18)°$ and the measure of the other angle is $(3x - 2)°$. $x = 20$; angles: $58°$ each

11. [2.3B] Juan rented a car for \$30 per day plus 15¢ per mile. How many miles did Juan travel if he paid

   a. \$45 for 1 day?   100 mi   **b.** \$52.50 for 1 day?   150 mi

   c. \$60 for 1 day?   200 mi

12. [2.3C] Find three consecutive odd integers whose sum is

   a. 153   49; 51; 53   **b.** 159   51; 53; 55

   c. 207   67; 69; 71

13. [2.4A] A woman's salary is increased by 20%. What was her salary before the increase if her new salary is

   a. \$24,000   \$20,000   **b.** \$36,000   \$30,000

   c. \$18,000   \$15,000

14. [2.4B] An investor bought some municipal bonds yielding 5% annually and some certificates of deposit (CDs) yielding 10% annually. If his total investment amounts to \$20,000, find how much money is invested in bonds and how much in certificates of deposit if his annual interest is

   a. \$1750   \$5000 in bonds; \$15,000 in CDs

   b. \$1150   \$17,000 in bonds; \$3000 in CDs

   c. \$1500   \$10,000 in bonds; \$10,000 in CDs

15. [2.4C] A car leaves a town traveling at 40 mi/hr. An hour later another car leaves the same town in the same direction traveling at 50 mi/hr.

   a. How far from town does the second car overtake the first?   200 mi

   b. Repeat the problem where the first car is traveling at 50 mi/hr and the second one, 60 mi/hr.   300 mi

   c. Repeat the problem where the first car is traveling at 40 mi/hr and the second one, 60 mi/hr.   120 mi

16. [2.4D] How many liters of a 40% salt solution must be mixed with 50 L of a 10% salt solution to obtain

   a. A 30% solution?   100 L

   b. A 20% solution?   25 L

   c. A 10% solution?   0 L

17. [2.5A] Graph:

   a. $x \geq -2$

   b. $x \geq -3$

   c. $x \geq -4$

18. [2.5B] Solve, graph, and write in interval notation:

a. $2(x - 1) \leq 4x + 4$ $x \geq -3$; $[-3, \infty)$

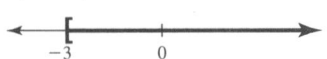

b. $3(x - 1) \leq 6x + 3$ $x \geq -2$; $[-2, \infty)$

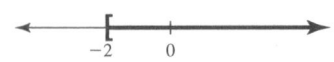

c. $4(x - 1) \leq 8x + 4$ $x \geq -2$; $[-2, \infty)$

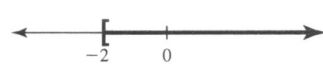

20. [2.5C] Solve, graph, and write in interval notation:

a. $\{x \mid x > -1 \text{ and } x < 2\}$ $-1 < x < 2$; $(-1, 2)$

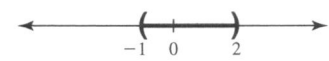

b. $\{x \mid x > -2 \text{ and } x < 3\}$ $-2 < x < 3$; $(-2, 3)$

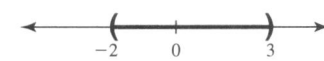

c. $\{x \mid x > -3 \text{ and } x < 4\}$ $-3 < x < 4$; $(-3, 4)$

22. [2.5C] Solve, graph, and write in interval notation:

a. $x + 1 \leq 3$ and $-4x < 8$ $-2 < x \leq 2$; $(-2, 2]$

b. $x + 1 \leq 4$ and $-3x < 9$ $-3 < x \leq 3$; $(-3, 3]$

c. $x + 1 \leq 5$ and $-2x < 4$ $-2 < x \leq 4$; $(-2, 4]$

24. [2.6A] Solve:

a. $\left| \frac{2}{7}x + 2 \right| + 5 = 9$ $x = 7$; $x = -21$

b. $\left| \frac{2}{7}x + 2 \right| + 3 = 9$ $x = 14$; $x = -28$

c. $\left| \frac{2}{7}x + 2 \right| + 1 = 9$ $x = 21$; $x = -35$

26. [2.6B] Solve, graph, and write in interval notation:

a. $|3x - 1| \leq 2$ $-\frac{1}{3} \leq x \leq 1$; $\left[-\frac{1}{3}, 1 \right]$

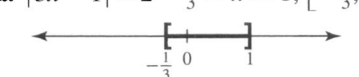

b. $|4x - 1| \leq 3$ $-\frac{1}{2} \leq x \leq 1$; $\left[-\frac{1}{2}, 1 \right]$

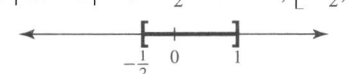

c. $|5x - 1| \leq 4$ $-\frac{3}{5} \leq x \leq 1$; $\left[-\frac{3}{5}, 1 \right]$

19. [2.5B] Solve, graph, and write in interval notation:

a. $\frac{x}{4} - \frac{x}{3} < \frac{x - 4}{4}$ $x > 3$; $(3, \infty)$

b. $\frac{x}{5} - \frac{x}{3} < \frac{x - 5}{5}$ $x > 3$; $(3, \infty)$

c. $\frac{x}{7} - \frac{x}{3} < \frac{x - 7}{7}$ $x > 3$; $(3, \infty)$

21. [2.5C] Solve, graph, and write in interval notation:

a. $\{x \mid x < -2 \text{ or } x \geq 3\}$ $(-\infty, -2) \cup [3, \infty)$

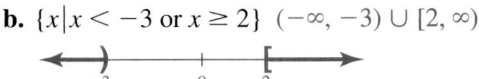

b. $\{x \mid x < -3 \text{ or } x \geq 2\}$ $(-\infty, -3) \cup [2, \infty)$

c. $\{x \mid x < -4 \text{ or } x \geq 1\}$ $(-\infty, -4) \cup [1, \infty)$

23. [2.5C] Solve, graph, and write in interval notation:

a. $-4 \leq -2x - 6 < 4$ $-5 < x \leq -1$; $(-5, -1]$

b. $-3 \leq -2x - 5 < 3$ $-4 < x \leq -1$; $(-4, -1]$

c. $-6 \leq -2x - 4 < 2$ $-3 < x \leq 1$; $(-3, -1]$

25. [2.6A] Solve:

a. $|x - 1| = |x - 3|$ $x = 2$

b. $|x - 3| = |x - 5|$ $x = 4$

c. $|x - 5| = |x - 7|$ $x = 6$

27. [2.6B] Solve, graph, and write in interval notation:

a. $|3x - 1| \geq 2$

$x \leq -\frac{1}{3}$ or $x \geq 1$; $\left(-\infty, -\frac{1}{3} \right] \cup [1, \infty)$

b. $|4x - 1| \geq 3$ $x \leq -\frac{1}{2}$ or $x \geq 1$; $\left(-\infty, -\frac{1}{2} \right] \cup [1, \infty)$

c. $|5x - 1| \geq 4$ $x \leq -\frac{3}{5}$ or $x \geq 1$; $\left(-\infty, -\frac{3}{5} \right] \cup [1, \infty)$

Practice Test 2

(Answers on pages 156–157)

1. Does 4 satisfy the equation $5 = 9 - x$?

2. Solve $4y + 3 = -9$.

3. Solve $-z + 8 = 5z - 6$.

4. Solve $4(5 - x) + 6 = 2x - (x + 4)$.

5. Solve $\dfrac{6}{5} + \dfrac{3x}{15} = \dfrac{x + 4}{10}$.

6. Solve $0.06P + 0.07(1500 - P) = 96$.

7. $H = 2.75h + 71.48$

   a. Solve for h.   **b.** Find h if $H = 140.23$.

8. Solve for A in $B = \dfrac{3}{4}(A - 8)$.

9. The perimeter of a rectangle is $P = 2L + 2W$, where L is the length and W is the width.

   a. Solve for L.

   b. If the perimeter is 100 ft and the length is 20 ft more than the width, what are the dimensions of the rectangle?

10. If L_1 and L_2 are parallel lines, find x and the measure of the unknown angles.

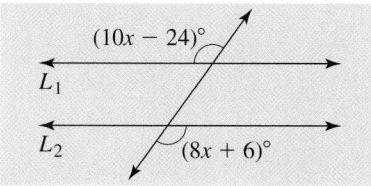

11. The bill for repairing an appliance totaled $72.50. If the repair shop charges $35 for the service call, plus $25 for each hour of labor, how many hours of labor did the repair take?

12. The sum of three consecutive odd integers is 117. What are the three integers?

13. A woman's salary is increased by 20% to $48,000. What was her salary before the increase?

14. An investor bought some municipal bonds yielding 5% annually and some certificates of deposit yielding 7%. If his total investment amounts to $20,000 and his annual interest is $1100, how much money is invested in bonds and how much in certificates of deposit?

15. A freight train leaves a station traveling at 40 mi/hr. Two hours later, a passenger train leaves the same station traveling in the same direction at 60 mi/hr. How far from the station does the passenger train overtake the freight train?

16. How many gallons of a 30% salt solution must be mixed with 40 gal of a 12% salt solution to obtain a 20% solution?

17. Solve, graph, and write the solution set in interval notation of $4(x - 1) \le 8x + 4$.

18. Solve, graph, and write the solution set in interval notation of $\dfrac{x}{8} - \dfrac{x}{3} < \dfrac{x - 8}{8}$.

19. Solve, graph, and write the solution set in interval notation of $\{x \mid x < -1 \text{ or } x \ge 2\}$.

20. Solve, graph, and write the solution set in interval notation of $x + 1 \le 4$ and $-2x < 6$.

21. Solve, graph, and write the solution set in interval notation of $-4 \le -2x - 6 < 0$.

22. Solve $\left|\dfrac{3}{4}x + 2\right| + 4 = 9$.

23. Solve $|x - 3| = |x - 7|$.

24. Solve and graph $|2x - 1| \le 5$.

25. Solve and graph $|2x + 1| > 3$.

Answers to Practice Test

ANSWER	IF YOU MISSED	REVIEW		
	QUESTION	SECTION	EXAMPLES	PAGE
1. Yes	1	2.1A	1	74
2. -3	2	2.1B	2	75–76
3. $\dfrac{7}{3}$	3	2.1B	3	76–77
4. 6	4	2.1B	4	77
5. -8	5	2.1B, C	5, 6	78–80
6. 900	6	2.1D	8, 9	82–83
7. a. $h = \dfrac{H - 71.48}{2.75}$ **b.** 25	7	2.2A	1	88
8. $A = \dfrac{4B + 24}{3}$	8	2.2A	2	89
9. a. $L = \dfrac{P - 2W}{2}$ **b.** 15 ft by 35 ft	9	2.2A	3	90
10. $x = 15$; $126°$	10	2.2C	7	94
11. 1.5 hr	11	2.3B	2	102–103
12. 37; 39; 41	12	2.3C	3	103–104
13. $40,000	13	2.4A	2	113
14. $15,000 bonds; $5000 CD	14	2.4B	3	114
15. 240 mi	15	2.4C	4	115
16. 32	16	2.4D	5	116
17. $x \geq -2$; $[-2, \infty)$ $-3\ -2\ -1\ 0\ 1\ 2\ 3$	17	2.5B	2, 3	123–125
18. $x > 3$; $(3, \infty)$ $0\ 1\ 2\ 3\ 4\ 5\ 6$	18	2.5B	4	126–127
19. $x < -1$ or $x \geq 2$; $(-\infty, -1) \cup [2, \infty)$ $-3\ -2\ -1\ 0\ 1\ 2\ 3$	19	2.5C	5	127–128
20. $-3 < x \leq 3$; $(-3, 3]$ $-4\ -3\ -2\ -1\ 0\ 1\ 2\ 3\ 4$	20	2.5C	6, 7	129–131

ANSWER	IF YOU MISSED	REVIEW		
	QUESTION	SECTION	EXAMPLES	PAGE
21. $-3 < x \le -1; (-3, -1]$	21	2.5C	8	131
22. $4; -\dfrac{28}{3}$	22	2.6A	1, 2	138–139
23. 5	23	2.6A	3	140
24. $-2 \le x \le 3$	24	2.6B	4	142
25. $-2 < x$ or $x > 1$	25	2.6B	5	143

Cumulative Review Chapters 1–2

1. Use braces to list the elements of the set of even natural numbers less than 8. $\{2, 4, 6\}$

2. Write 0.19 as a fraction. $\frac{19}{100}$

3. Find: $-5 + (-6)$. -11

4. Find: $5 \cdot (-4)$. -20

5. Simplify:

$[(3x^2 - 3) + (7x + 1)] - [(x - 2) + (8x^2 - 5)]$.
$-5x^2 + 6x + 5$

6. Perform the indicated operation and simplify: $\dfrac{80x^8}{16x^{-6}}$. $5x^{14}$

7. Evaluate: $[-2(4 + 4)] + 7$. -9

8. Does the number -4 satisfy the equation $5 = 9 - x$? No

9. Solve: $x + 5 = 3(2x - 1)$. $\frac{8}{5}$

10. Solve: $\dfrac{x + 2}{5} - \dfrac{x - 2}{7} = 4$. 58

11. Solve: $0.04P + 0.06(1700 - P) = 65$. 1850

12. Solve: $\left|\frac{3}{7}x + 8\right| + 3 = 9$. $-\frac{14}{3}; -\frac{98}{3}$

13. Graph: $3(x + 1) \le 4x - 3$.

14. Graph: $\{x \mid x < -1 \text{ or } x \ge 2\}$.

15. Graph: $-4 \le -4x - 8 < 4$.

16. Graph: $|2x + 3| > 5$.

17. Solve for A in $B = \frac{6}{7}(A - 11)$. $A = \frac{7B + 66}{6}$

18. The perimeter of a rectangle is $P = 2L + 2W$, where L is the length and W is the width. If the perimeter is 100 ft and the length is 10 ft more than the width, what are the dimensions? 20 ft by 30 ft

19. A woman's salary was increased by 30% to $29,900. What was her salary before the increase? $23,000

20. How many gallons of a 60% salt solution must be mixed with 60 gallons of a 21% salt solution to obtain a solution that is 50% salt? 174

Graphs and Functions

The Human Side of Algebra

The inventor of the Cartesian coordinate system, Rene Descartes, was born March 31, 1596, near Tours, France. Trained as a gentleman and educated in Latin, Greek, and rhetoric, he maintained a healthy skepticism toward all he was taught. Disenchanted with these studies, he took an interlude of pleasure in Paris, later moving to the suburb of St. Germain, where he worked on mathematics for 2 years. In 1637, at age 41, his friends persuaded him to print his masterpiece, known as the *Discourse on Method,* which included an essay on geometry that is probably the most important thing he ever did. His *Analytic Geometry,* a combination of algebra and geometry, revolutionized the study of geometry and made much of modern mathematics possible.

In 1649, Queen Christine of Sweden chose Descartes as her private philosophy teacher. Unfortunately, she insisted that her lessons begin promptly at five o'clock in the morning in the ice-cold library of her palace. This proved to be too much for the frail Descartes, who soon caught "inflammation of the lungs," from which he died on February 11, 1650, at the age of 54. What price for fame!

Pretest for Chapter 3

(Answers on pages 161–162)

1. Graph the solutions to $2x + y = 2$.

2. Find the x- and y-intercepts of $y = -x + 3$.

3. Graph the solutions to $-5x = -10$.

4. Find the domain and range of the relation

   $\{(2, -1), (4, 0), (1, 1), (-2, 1)\}$

5. **a.** Find the domain and range of the relation.

   b. Is the relation a function?

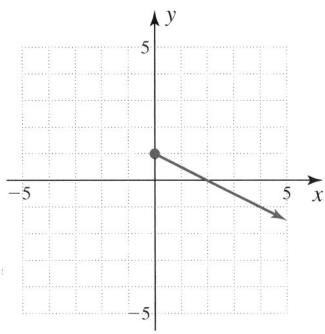

6. Given the function, $f(x) = \dfrac{3x}{x - 5}$.

   a. Find the domain of the function.

   b. Find $f(2) - f(6)$.

7. Given $f = \{(-3, 5), (7, 2), (5, -5)\}$, find $f(7) - f(5)$.

8. Find the slope of the line through the two given points.

   a. $(4, -3)$ and $(-6, -1)$

   b. $(-7, 1)$ and $(-7, 0)$

9. A line L_1 has slope -2. Determine whether the line through the two points $A\ (2, 5)$ and $B\ (4, 6)$ is parallel or perpendicular to L_1.

10. A line goes through the point $(-1, 2)$ and has slope $\frac{-1}{3}$. Graph this line.

11. Find the equation of the line through $(1, 3)$ and $(4, 2)$. Then write the equation in standard form.

12. Find the equation of the line with slope -4 and passing through the point $(-2, 3)$. Write the equation in slope-intercept form.

13. Find the slope and the y-intercept of the line $-4x + y = 6$.

14. A line has slope 5 and y-intercept -1. Find the slope-intercept equation of this line.

15. Find an equation of the line through the point $(8, 1)$ and parallel to the line with equation $-2y = 6x - 2$. Write the equation in standard form.

16. Write the equation of the line passing through the point $(-3, 5)$ with a slope of 0.

17. When graphed, the data in the table represents a linear model. Plot the points on a graph, draw the "line of best fit," and write the equation of that line in slope-intercept form.

X	2	4	6	8	10
Y	3	5	4	5	8

For the following questions, graph the solutions.

18. $x - 3y > 6$

19. $2x \leq -4$

20. $|y| \leq 3$

Answers to Pretest

ANSWER	IF YOU MISSED		REVIEW	
	QUESTION	SECTION	EXAMPLES	PAGE
1.	1	3.1	3	166
2. x: $(3, 0)$ and y: $(0, 3)$	2	3.1	4	168
3.	3	3.1	5	169
4. $D = \{-2, 1, 2, 4\}; R = \{-1, 0, 1\}$	4	3.2	1	177
5. a. $D = \{x \mid x \geq 0\}; R = \{y \mid y \leq 1\}$ **b.** Yes	5	3.2	2, 3, 4	177–179
6. a. All real numbers except 5 **b.** -20	6	3.2	5, 6	180, 182
7. 7	7	3.2	7	182
8. a. $\frac{-1}{5}$ **b.** Undefined	8	3.3	1	193–194
9. Perpendicular	9	3.3	3	197
10.	10	3.3	5	199
11. $x + 3y = 10$	11	3.4	1	209
12. $y = -4x - 5$	12	3.4	2	209

ANSWER	IF YOU MISSED		REVIEW	
	QUESTION	SECTION	EXAMPLES	PAGE
13. Slope = 4; y-intercept = 6	13	3.3	6	200–201
14. $y = 5x - 1$	14	3.4	3	210
15. $3x + y = 25$	15	3.4	4	210
16. $y = 5$	16	3.4	5	211–212
17. $y = 0.5x + 2$	17	3.4	6	212

(Answers will vary.)

18.	18	3.5	1	222
19.	19	3.5	3	224
20.	20	3.5	4	224

3.1 GRAPHS

To Succeed, Review How To ...

1. Evaluate an expression (p. 50).

2. Solve linear equations (p. 79).

Objectives

A Given an ordered pair of numbers, find its graph, and vice versa.

B Graph lines by finding two or more points satisfying the equation of the line.

C Graph lines by finding the x- and y-intercepts.

D Graph horizontal and vertical lines.

GETTING STARTED

Hurricane Preparedness

The map shows the **coordinates** of the hurricane to be near the vertical line indicating 90° of longitude and the horizontal line indicating 25° of latitude. If we agree to list the longitude first, we can identify this point by using the ordered pair

$$(90, \quad 25)$$

This is the longitude. ⟶   ⟵ This is the latitude.

So if we know the *coordinates* of the hurricane, we can find it on the map, and if we know the *point* where the hurricane is located on the map, we can find its coordinates. This is an example of a rectangular coordinate system, which we shall study in this section.

A   Graphing and Finding Ordered Pairs

In algebra we use a similar system to locate points by using **ordered pairs** of real numbers. We draw two perpendicular number lines called the x-axis and the y-axis, intersecting at a point O called the **origin.** (See Figure 1.) On the x-axis, *positive* is to the right, whereas on the y-axis, *positive* is up. The two axes divide the plane into four regions called **quadrants.** These quadrants are numbered counterclockwise using Roman numerals and starting in the upper right-hand region, as shown. The whole arrangement is called a **Cartesian coordinate system,** a **rectangular coordinate system,** or a **coordinate plane.**

Every point P in the plane can be associated with an ordered pair of real numbers (x, y), and every ordered pair (x, y) can be associated with a point P in the plane. For example, in Figure 2 the point A can be associated with the ordered pair $(1, 3)$, the point B with the ordered pair $(-3, 2)$, and the point C with the ordered pair $(0, -2)$. The **graphs** of the points $A(1, 3)$, $B(-3, 2)$, and $C(0, -2)$ are indicated by the blue dots.

"Beadwork and graphs"
Patterns of beauty require knowing about coordinates and the four quadrants. (See Section 3.1 Exercises 9–12)

Figure 1

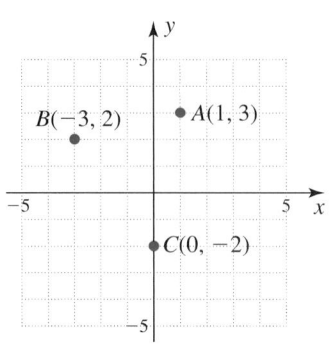

Figure 2

A NOTE ABOUT COORDINATES

If the *x*-coordinate is *positive,* the point is to the *right* of the vertical axis.

If the *x*-coordinate is *negative,* the point is to the *left* of the vertical axis.

If the *x*-coordinate is *zero,* the point is *on* the vertical axis.

If the *y*-coordinate is *positive,* the point is *above* the horizontal axis.

If the *y*-coordinate is *negative,* the point is *below* the horizontal axis.

If the *y*-coordinate is *zero,* the point is *on* the horizontal axis.

Teaching Tip

Have a discussion about these being real numbers and have the students plot some points with fractions and decimals. Pose the thought that there could be other real numbers not studied yet (irrationals) and maybe some not real numbers (imaginary).

| **EXAMPLE 1** | **Graphing ordered pairs** |

Graph the ordered pairs $A(2, 3)$, $B(-1, 2)$, $C(-2, -1)$, and $D(0, -3)$.

SOLUTION To graph the ordered pair $(2, 3)$, we start at the origin and move 2 units to the *right* and then 3 units *up,* reaching the point whose coordinates are $(2, 3)$. The other three pairs are graphed in a similar manner and are shown in Figure 3.

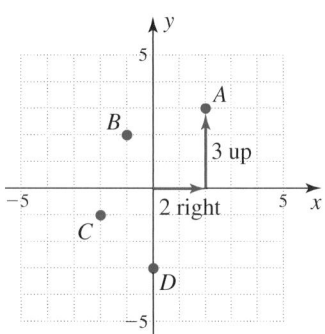

Figure 3

| **PROBLEM 1** |

Graph the ordered pairs.
$A(3, 2)$, $B(-1, 4)$, $C(-1, 2)$, $D(3, -1)$

Teaching Tip

To demonstrate the placement of a point according to the signs of the ordered pair, use the following:

$$\begin{array}{c|c} (-, +) & (+, +) \\ \hline (-, -) & (+, -) \end{array}$$

Answer

1.

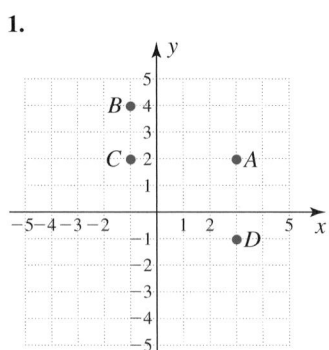

Calculate It   Entering Lists and Making a Scatter Plot

In Example 1, we are going to plot data using two features: plotting data points and making a scatter plot *or* scattergram. The procedure for doing this varies among calculators.

First, erase any old data points. With a TI-83 Plus, clear any old lists by pressing [2nd] [+] [4] [ENTER] to clear all the lists; then press [STAT] [1] . Enter the *x*-coordinates of points *A, B, C,* and *D* under L1, $(2, -1, -2, 0)$, by entering each number followed by [ENTER]. To enter the *y*-coordinates press the right arrow to get to L2 and enter 3, 2, -1, and -3. Now press [2nd] [Y=] [ENTER] [ENTER] to turn the plot feature ON. Go to "**TYPE**" (line 2) and select the first type of graph, called a scattergram, by pressing [ENTER]. Finally, press [ZOOM] 9 and the scattergram of the points appears, as shown in the window. Pressing [ZOOM] 9 selects the correct window for you!

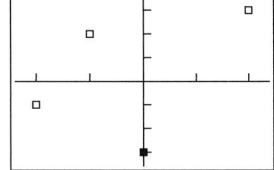

Every point P in the plane is associated with an ordered pair (x, y), so we should be able to find the coordinates of any point, as shown in Example 2.

EXAMPLE 2 **Finding coordinates**

Determine the coordinates of each of the points shown in Figure 4.

SOLUTION Point A is 4 units to the *right* of the origin and 1 unit *above* the horizontal axis. Thus the ordered pair corresponding to A is $(4, 1)$. The coordinates of the other four points can be found in a similar manner. The summary appears below.

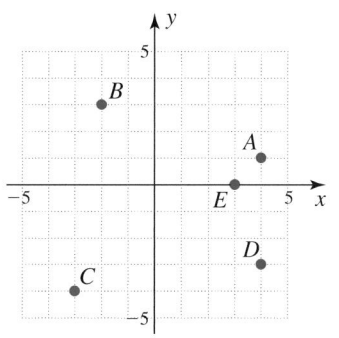

Figure 4

Point	Start at the origin, move	Coordinates
A	4 units *right*, 1 unit *up*	$(4, 1)$
B	2 units *left*, 3 units *up*	$(-2, 3)$
C	3 units *left*, 4 units *down*	$(-3, -4)$
D	4 units *right*, 3 units *down*	$(4, -3)$
E	3 units *right*, 0 units *up*	$(3, 0)$

PROBLEM 2

Determine the coordinates of each of the points shown.

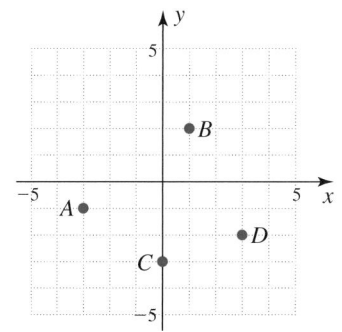

Web It

To verify that you are plotting points correctly, go to link 3-1-1 at mhhe.com/bello.

B **Graphing Lines**

Web It

For a step-by-step explanation of graphing a linear equation, go to link 3-1-2 at mhhe. com/bello.

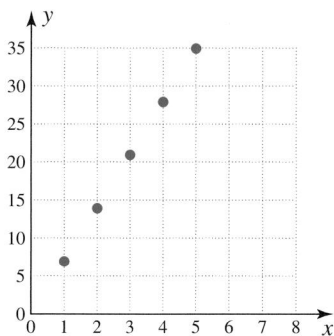

Figure 6

Answer

2. $A(-3, -1)$; $B(1, 2)$; $C(0, -3)$; $D(3, -2)$

Some information is best described graphically by displaying it as ordered pairs. The total cost of a long distance call with a flat rate of 7 cents per minute or portion thereof and ranging from 1 to 5 minutes is shown in Figure 5.

Figure 6 represents these points when plotted.

Consider the pattern involved for finding the total cost C.

$C = 1$ minute times (7)

$C = 2$ minutes times (7)

$C = 3$ minutes times (7)

$C = 4$ minutes times (7)

$C = 5$ minutes times (7)

Minutes	Total Cost	Ordered Pair
1	7	$(1, 7)$
2	14	$(2, 14)$
3	21	$(3, 21)$
4	28	$(4, 28)$
5	35	$(5, 35)$

Figure 5

In general, for x minutes, the total cost can be found by $C = x$ times (7) or $C = 7x$.

We used the variable C because that related to the context of the problem, but we could replace the variable C with a y and write the equation as $y = 7x$. We call this an equation in two variables, x and y. If in this equation x is replaced with 3 and y by 21, we obtain the true statement $21 = 7(3)$. Thus we say that the ordered pair $(3, 21)$ is a **solution** of the equation $y = 7x$ or that it satisfies the equation. On the other hand, $(21, 3)$ is *not* a solution of $y = 7x$ because $3 \neq 7(21)$. All the ordered pairs in the table are solutions to this equation. There are infinitely many ordered pairs of real numbers that will solve this equation and when graphed (see Figure 7) will appear to be on the same straight line.

It can be proved that every solution of $y = 7x$ corresponds to a point on the line, and vice versa. The line obtained by joining the points is the **graph** of all the solutions to the equation $y = 7x$.

The **solution set** of an equation in two variables can be written using set notation. For example, the solution set of $y = 7x$ can be written as $\{(x, y) \mid y = 7x\}$. The solution set consists of infinitely many points, so it would be impossible to list all these points. However, we can find some of these points like we did in the table, graph them, and then connect the points with a line. The line in Figure 7 represents only a *part* of the complete graph, which continues without end in both directions, as indicated by the arrows in the figure. Although it takes only *two* points to determine a straight line, we shall include a third point here in our solutions to make sure that we have drawn the correct line.

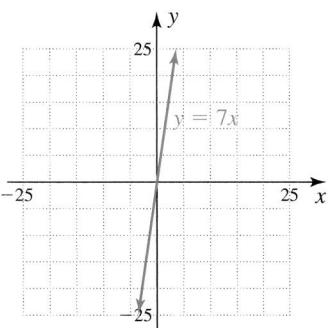

Figure 7

Calculate It Make a Table of Values for a Given Equation

You can use your TI-83 Plus to produce a table of values for any given linear equation and sets of *x* values. Suppose in the example of the long distance phone call the company offers the same 7 cents per minute but will time the call to the nearest half-minute. Then we would want a table of values that would display the minutes (*x*) as 1, 1.5, 2, 2.5, . . . , and so on. Remember to clear any old lists.

| MODE | ◇ ◇ ◇ | ENTER | (function mode) |

| Y= | 7 | X,T,θ,*n* | (enters equation) |

| 2nd | WINDOW | 1 | ENTER | .5 | ENTER |

| 2nd | GRAPH |

X	Y
1	7
1.5	10.5
2	14
2.5	17.5
3	21
3.5	24.5
4	28

$X=1$

So for a $2\frac{1}{2}$ (2.5) minute phone call we would owe 17.5 cents. By using the down arrow ⊙ you can scroll through a continuation of the table. How much would a $20\frac{1}{2}$ minute phone call cost?

EXAMPLE 3 **Graphing lines**

Graph the solutions to: $x + 2y = 8$

SOLUTION

For *x* = −2,	For *x* = 0,	For *x* = 2,
$-2 + 2y = 8$	$0 + 2y = 8$	$2 + 2y = 8$
$2y = 10$	$2y = 8$	$2y = 6$
$y = 5$	$y = 4$	$y = 3$
We have $(-2, 5)$.	We have $(0, 4)$.	We have $(2, 3)$.

We graph the points $(-2, 5)$, $(0, 4)$, and $(2, 3)$ and then draw a line through them, as shown in Figure 8.

Teaching Tip

Have a discussion emphasizing that though we use three points to draw the line, the points on the line represent *all* the solutions to the equation.

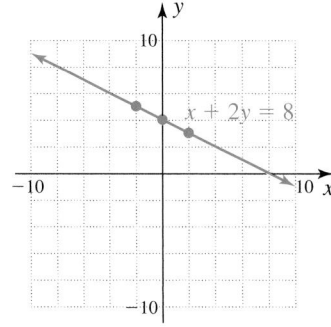

Figure 8

PROBLEM 3

Graph the solutions to: $3x - y = 3$

Answer

3.

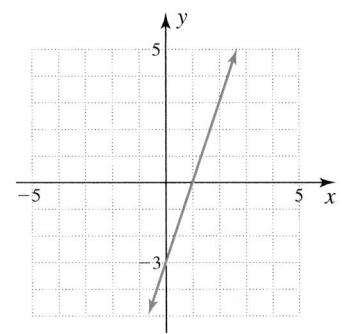

Calculate It   **Creating a Split Screen for a Table and the Graph**

You can graph the equation of a line on your TI-83 Plus and have it show a table of values on a split screen so we see both the graph and the table at the same time. We will use Example 3 to demonstrate this. The calculator will graph the equation only when it is solved for y, so the first step is to solve for y.

$$x + 2y = 8$$
$$2y = -x + 8 \qquad \text{Subtract } x.$$
$$y = -0.5x + 4 \qquad \text{Divide by 2.}$$

Be sure the STATPLOT is turned off by pressing [2nd] [Y=]
[ENTER] [◊] [ENTER] .

Press [Y=] and enter [(−)] 0.5 [X,T,θ,n] [+] 4. To set the table to integer values press [2nd] [WINDOW] [(−)] 8 [⌄] 1. To set the split screen press [MODE] ; continue to use the [⌄] to arrow

down to the line "**Full Horiz G-T**"; press the [◊] to get to G-T; and then press [ENTER] . To set a standard window for the graph press [ZOOM] 6. This should show the graph and a table of values for the graph. To scroll up and down the table of values press [2nd] [GRAPH] and use the **UP** and **DOWN** arrow keys to scroll the values in the table. Scroll down to $x = -2$ and verify that $y = 5$ and $x = 2$ and verify that $y = 3$ as we indicated in our solution to Example 3.

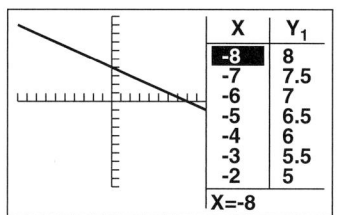

C   **Graphing Lines Using Intercepts**

The equation $x + 2y = 8$ has a *straight* line as its graph of solutions. It can be shown that any equation in two variables x and y of the form $Ax + By = C$ (where A and B are not both zero) has a straight line for its graph. This is why $Ax + By = C$ is called a *linear equation*. Here is the definition.

Standard Form of Linear Equations	Any equation that can be written in the form
	$$Ax + By = C$$
	is a **linear equation** in two variables, and the graph of its solutions is a *straight line*. The form $Ax + By = C$ is called the **standard form** of the equation, and A, B, and C are integers, where A and B are not both 0, and A is positive.

> **NOTE**
>
> We prefer to write an equation in the form $3x + 2y = 6$ instead of $\frac{1}{2}x + \frac{1}{3}y = 1$; that is, we use the standard form with integer coefficients for A, B, and C, and A is positive.

Calculate It   **Finding the x- and y-intercepts**

You can find the x- and y-intercepts of a line on your TI-83 Plus by graphing it using an integer window and the trace feature. To find the x- and y-intercepts of $x + 2y = 8$, solve for y.

$$y = -0.5x + 4$$

Then use the following steps:

[Y=] [(−)] 0.5 [X,T,θ,n] [+] 4

[ZOOM] 8 [ENTER] [TRACE] 0 [ENTER]

That should indicate "$y = 4$," the y-intercept.

To find the x-intercept, continue to press the right arrow [◊] until the screen displays "$y = 0$" on the right. On the left it says "$x = 8$."

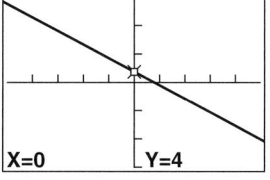

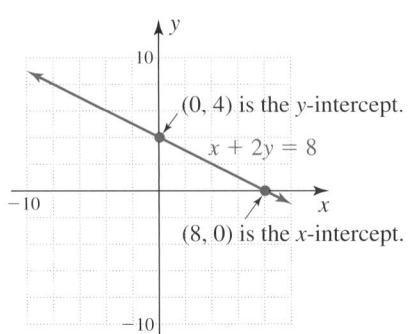

Figure 9

Two points determine a straight line, so it is sufficient to locate two points to graph all solutions to a linear equation. The easiest points to compute are those involving zeros—points of the form $(x, 0)$ and $(0, y)$. For example, if we let $x = 0$ in the equation $x + 2y = 8$, we obtain $2y = 8$, or $y = 4$. Thus $(0, 4)$ is a point on the graph. Similarly, if we let $y = 0$ in the equation $x + 2y = 8$, we have $x = 8$. Therefore $(8, 0)$ is also on the graph. The points $(8, 0)$ and $(0, 4)$ are the points at which the line crosses the x- and y-axes, respectively, so they are called the **x-** and **y-intercepts.** These two intercepts, as well as the line determined by them (the graph of $x + 2y = 8$), are shown in Figure 9. Here is the procedure we used to find the intercepts.

PROCEDURE

Finding the Intercepts

To find the x-intercept, let $y = 0$ and find x: $(x, 0)$ is the x-intercept.

To find the y-intercept, let $x = 0$ and find y: $(0, y)$ is the y-intercept.

EXAMPLE 4   **Finding intercepts**

Find the x- and y-intercepts and then graph the lines:

a. $y = 3x + 6$ **b.** $2x + 3y = 0$

SOLUTION

a. We first find the x- and y-intercepts. For $x = 0$, $y = 3x + 6$ becomes

$$y = 3(0) + 6 = 6$$

$(0, 6)$ is the y-intercept. For $y = 0$, $y = 3x + 6$ becomes

$$0 = 3x + 6$$
$$-6 = 3x \qquad \text{Subtract 6.}$$
$$x = -2 \qquad \text{Divide by 3.}$$

$(-2, 0)$ is the x-intercept. We join the points $(-2, 0)$ and $(0, 6)$ with a line and obtain the graph of all solutions to $y = 3x + 6$ shown in Figure 10.

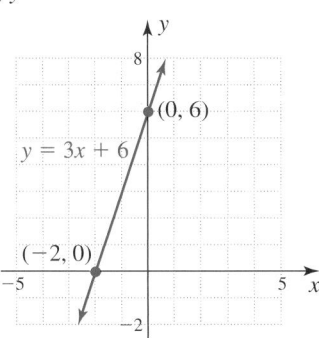

Figure 10

b. As before, we try to find the x- and y-intercepts. For $x = 0$, $2x + 3y = 0$ becomes

$$2(0) + 3y = 0 \quad \text{or} \quad y = 0$$

$(0, 0)$ is the y-intercept. If we now let $y = 0$, we get the ordered pair $(0, 0)$ again. This is because any line of the form $Ax + By = 0$ (where A and B are not both zero) goes through the origin. In these cases, we select a different value for x, say $x = 3$. Then $2x + 3y = 0$ becomes

$$2(3) + 3y = 0$$
$$3y = -6 \qquad \text{Subtract 6.}$$
$$y = -2 \qquad \text{Divide by 3.}$$

$(3, -2)$ is another point on the line. We join the points $(0, 0)$ and $(3, -2)$ with a line and obtain the graph of all solutions to $2x + 3y = 0$ shown in Figure 11.

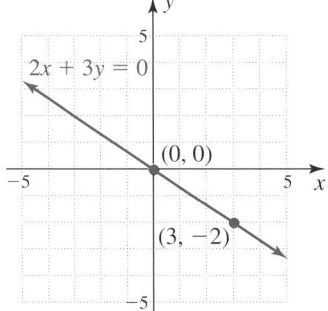

Figure 11

PROBLEM 4

Find the x- and y-intercepts and then graph the lines:

a. $y = x - 2$ **b.** $3x + y = 0$

Answers

4. a. x: $(2, 0)$ and y: $(0, -2)$

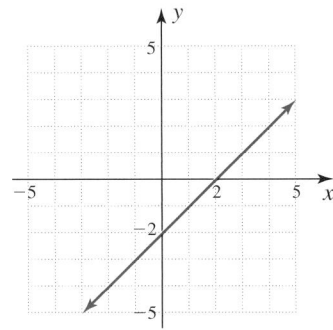

b. $(0, 0)$ is both the x- and y-intercept.

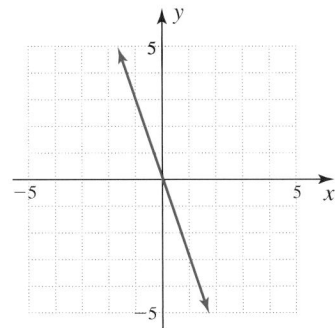

D   Graphing Horizontal and Vertical Lines

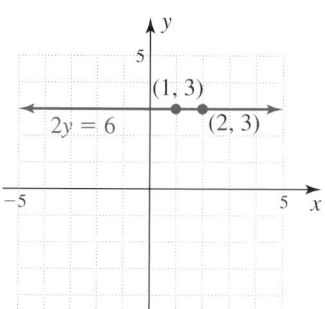

Figure 12

The procedure we have discussed works *only* for equations that can be written in the form $Ax + By = C$, where both A and B are not zero. The equation $2y = 6$ can be written in the form $0 \cdot x + 2y = 6$, so we *cannot* graph $2y = 6$ by finding its intercepts. The equation $2y = 6$ assigns to every value of x a y-value of 3. If we solve for $2y = 6$, we get $y = 3$, which has no specific x-coordinate. Thus for $x = 1$, $y = 3$, and for $x = 2$, $y = 3$ (y is always 3). If we graph the points $(1, 3)$ and $(2, 3)$ and connect them with a straight line, we see that the result is a horizontal line, as shown in Figure 12. Similarly, the equation $2x = 6$ assigns an x-value of 3 to every y. For $y = 1$, $x = 3$, and for $y = 5$, $x = 3$. If we graph the points $(3, 1)$ and $(3, 5)$ and draw a straight line through them, we see that the result is a vertical line, as shown in Figure 13. In general, we have the following.

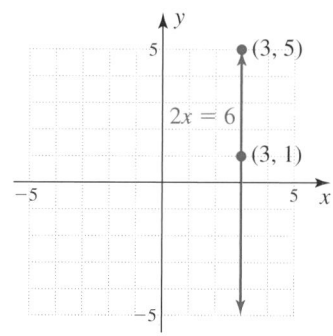

Figure 13

Horizontal and Vertical Lines

The graph of all solutions to $y = C$ is a *horizontal* line.

The graph of all solutions to $x = C$ is a *vertical* line.

EXAMPLE 5   **Graphing horizontal and vertical lines**

Graph the solutions to:

a. $2x = 8$

b. $5y = -10$

SOLUTION

a. $2x = 8$ is equivalent to $x = 4$, so the graph of $2x = 8$ is a vertical line for which $x = 4$. If we choose the solutions $(4, 1)$ and $(4, 2)$ and draw a straight line through them, we obtain the graph of all solutions to $2x = 8$, shown in Figure 14.

b. $5y = -10$ is equivalent to $y = -2$, so $5y = -10$ is a horizontal line for which $y = -2$. If we choose the solutions $(1, -2)$ and $(2, -2)$ and draw a straight line through them, we obtain the graph of all solutions to $5y = -10$, shown in Figure 15.

Teaching Tip

Remind students that they can make a table to determine whether $x = 3$ is horizontal or vertical where all x's are 3 and y's can be any real number. Plotting the points will produce the correct line.

x	y
3	0
3	-1
3	2

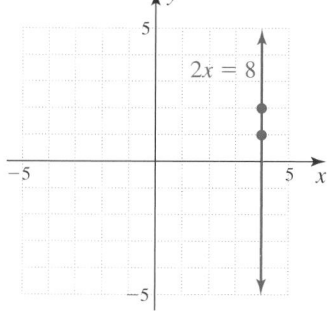

Figure 14

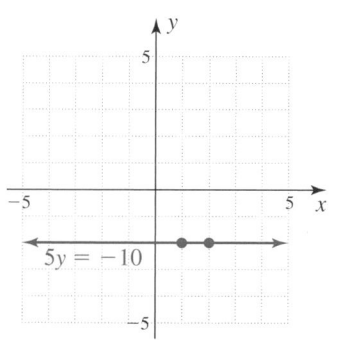

Figure 15

PROBLEM 5

Graph the solutions to:

a. $3x = -9$

b. $-2y = -4$

Answers

5. a.

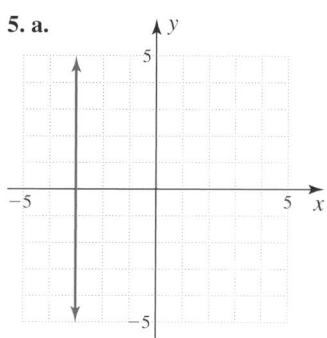

b.

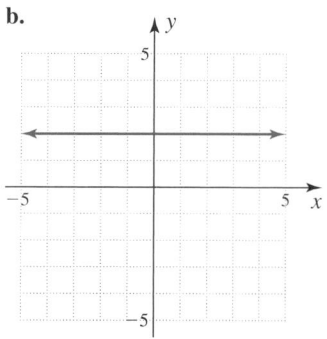

Calculate It Exercises

1. Can you graph the vertical line $x = 4$ using the ⬭ Y= key?

2. The TI-83 Plus has a draw feature to graph horizontal and vertical lines. Press ⬭ 2nd ▮ PRGM to access the feature. What keys do you have to press to draw the graph of $x = 4$?

3. What keys do you have to press to draw the line $y = -2$?

Exercises 3.1

A In Problems 1–8, graph the ordered pair.

1. $(-4, 3)$

2. $(-3, 4)$

3. $(-3, -2)$

4. $(-2, -3)$

5. $(0, -2)$

6. $(2, 0)$

7. $\left(\dfrac{1}{2}, 3\right)$

8. $\left(3, \dfrac{-1}{2}\right)$

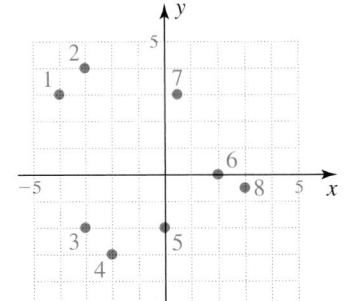

In Problems 9–12, Figure 16 depicts the colors of a beadwork pattern. Use the graph to answer the following questions.

9. What quadrant(s) will contain red beads? QI; QII

10. What quadrant(s) will contain blue beads? QIII; QIV

11. What quadrant(s) will not contain red beads? QIII; QIV

12. What quadrant(s) will not contain pink beads? QI; QII

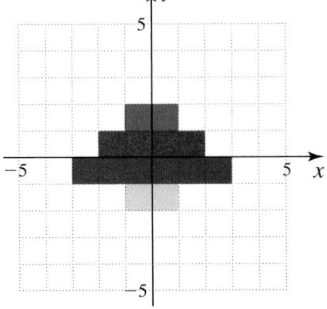

Figure 16

In Problems 13–20, determine the coordinates of each of the points in Figure 17.

13. C $(3, 5)$

14. D $(0, 4)$

15. E $(-2, 3)$

16. F $(-4, 1)$

17. G $(-4, 0)$

18. H $(-2, -2)$

19. I $(0, -3)$

20. J $(4, -2)$

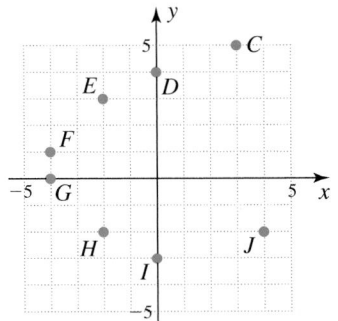

Figure 17

B In Problems 21–26, complete the ordered pairs to satisfy the equation and then graph the line of all solutions.

21. $y = x + 3$ $(-2,\ 1), (-1,\ 2),$
$(0,\ 3), (1,\ 4), (2,\ 5)$

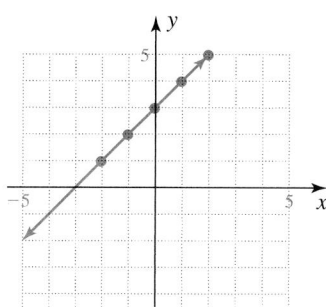

22. $y = 2x + 1$
$(-1,\ -1), (0,\ 1), (1,\ 3)$

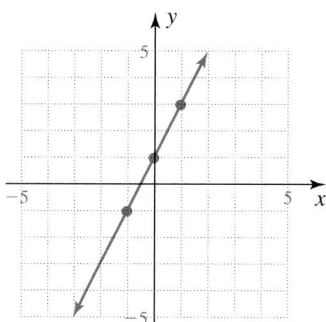

23. $x - y = 4$
$(-1,\ -5), (0,\ -4), (1,\ -3)$

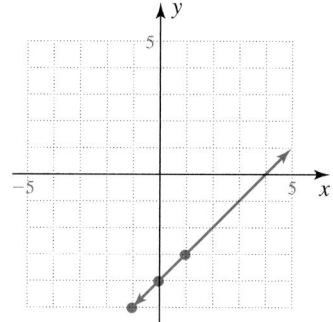

24. $x - 3y = 6$
$(0,\ -2), (3,\ -1), (-3,\ -3)$

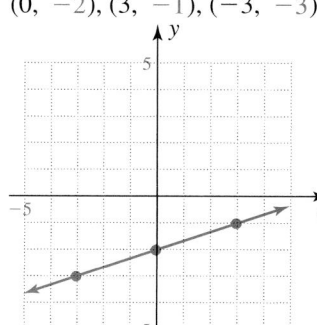

25. $2x - y - 3 = 0$
$(-1,\ -5), (0,\ -3), (1,\ -1)$

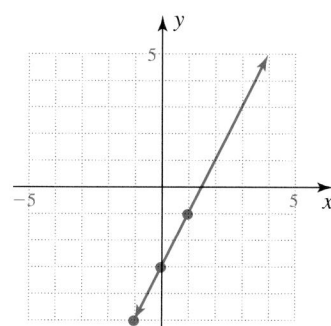

26. $2x + y + 2 = 0$
$(-1,\ 0), (0,\ -2), (1,\ -4)$

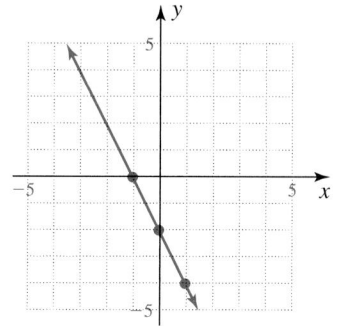

C In Problems 27–40, find the x- and y-intercepts and then graph all solutions to the equation.

27. $y = x - 5$ $(5, 0); (0, -5)$

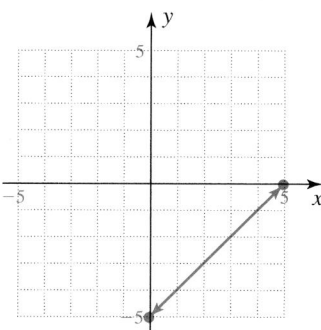

28. $2y = 4x - 2$ $(\frac{1}{2}, 0); (0, -1)$

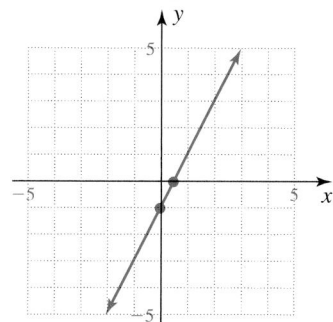

29. $2x + 3y = 6$ $(3, 0); (0, 2)$

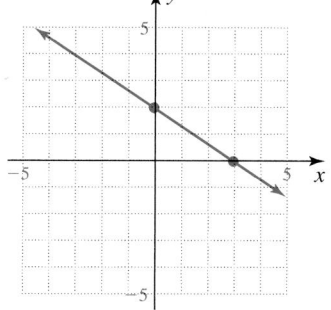

30. $3x + 2y = 6$ $(2, 0); (0, 3)$

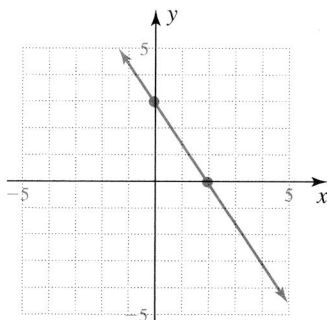

31. $2x - y = 4$ $(2, 0); (0, -4)$

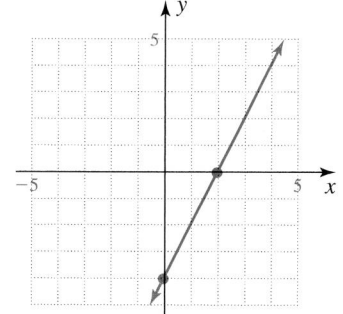

32. $3x - y = 3$ $(1, 0); (0, -3)$

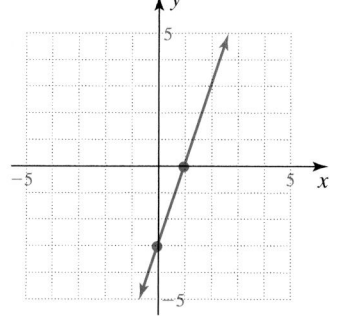

33. $2x + y - 4 = 0$ $(2, 0)$; $(0, 4)$

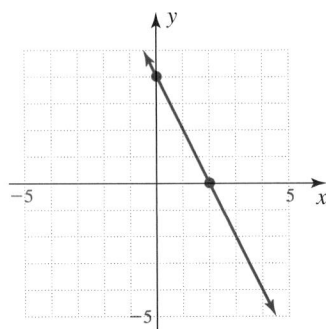

34. $3x + y - 3 = 0$ $(1, 0)$; $(0, 3)$

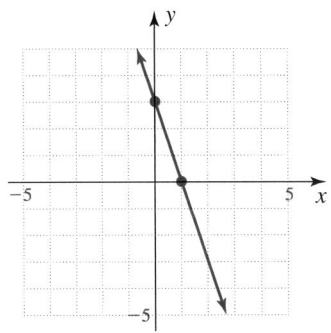

35. $y + 4x = 0$ $(0, 0)$

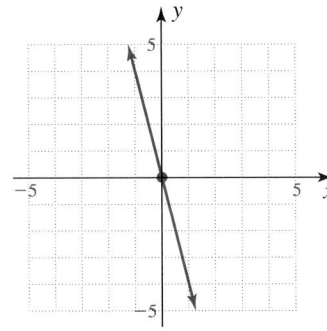

36. $y + 3x = 0$ $(0, 0)$

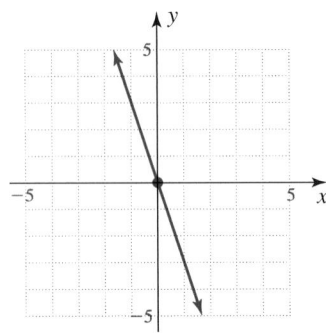

37. $2x - 5y = -10$ $(-5, 0)$; $(0, 2)$

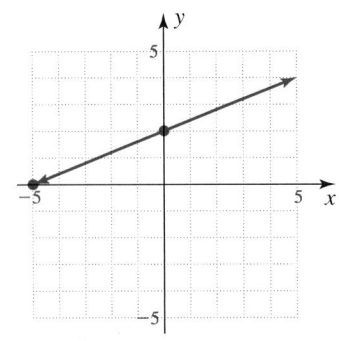

38. $2x - 3y = -6$ $(-3, 0)$; $(0, 2)$

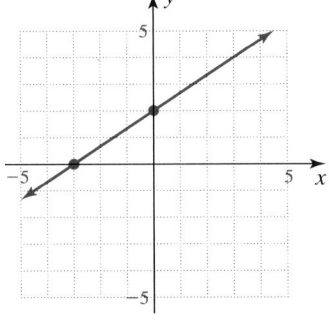

39. $x - y - 5 = 0$ $(5, 0)$; $(0, -5)$

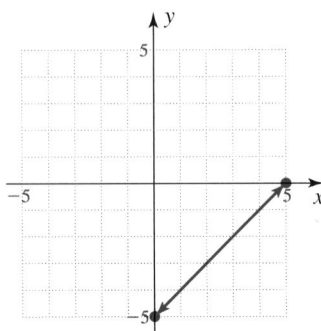

40. $2x - 3y - 12 = 0$ $(6, 0)$; $(0, -4)$

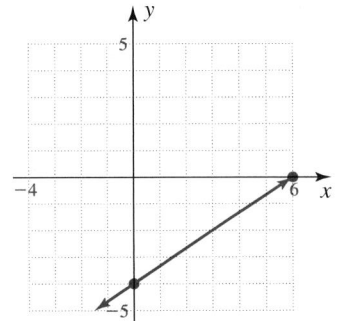

D In Problems 41–50, determine whether the given line is horizontal or vertical and then graph all the solutions.

41. $\dfrac{-7}{2}x = 14$ Vertical; $x = -4$

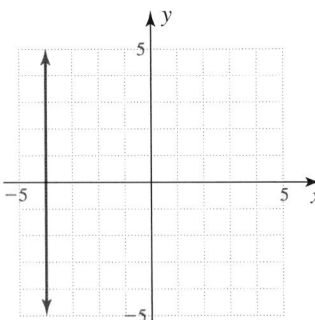

42. $\dfrac{-1}{2}x = 2$ Vertical; $x = -4$

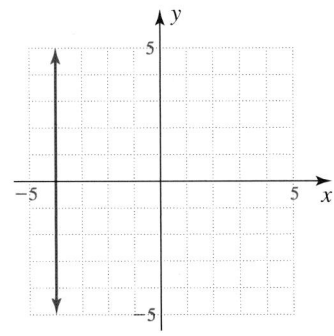

43. $\dfrac{3}{2}x = 6$ Vertical; $x = 4$

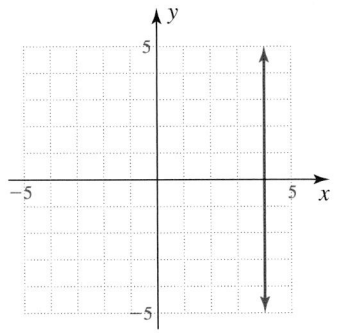

44. $\dfrac{-5}{2}y = 10$ Horizontal; $y = -4$

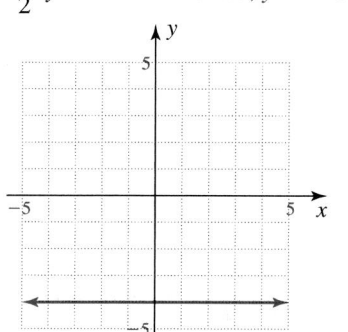

45. $\dfrac{-3}{4}x = 3$ Vertical; $x = -4$

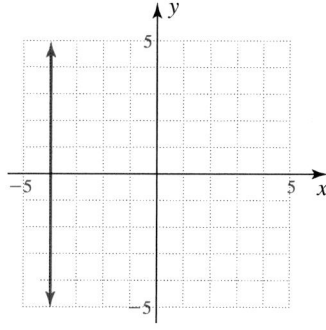

46. $\dfrac{-3}{7}y = \dfrac{-6}{7}$ Horizontal; $y = 2$

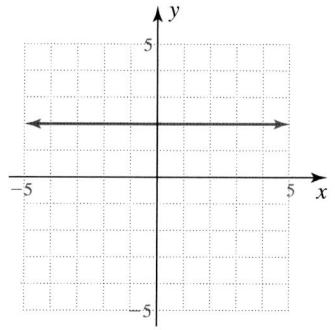

47. $\dfrac{-1}{3} + y = \dfrac{2}{3}$ Horizontal; $y = 1$

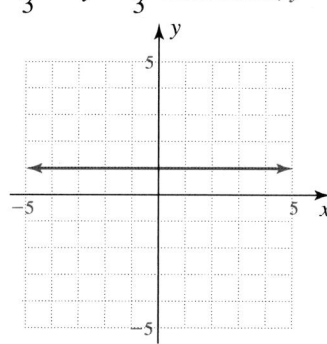

48. $\dfrac{-1}{5} + y = \dfrac{4}{5}$ Horizontal; $y = 1$

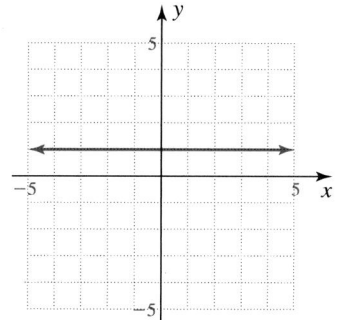

49. $\dfrac{2}{3} = x - \dfrac{4}{3}$ Vertical; $x = 2$

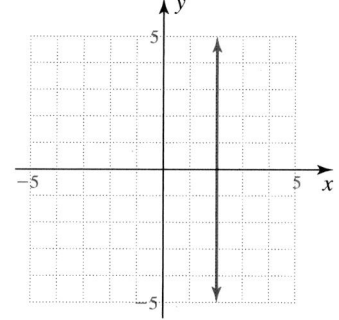

50. $\dfrac{3}{4} = x - \dfrac{5}{4}$ Vertical; $x = 2$

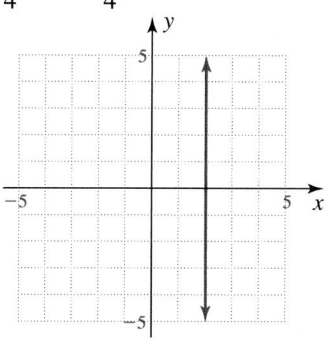

APPLICATIONS

51. *Cricket chirps* The number N of chirps a certain type of cricket makes per minute is given by

$$N = 4(T - 40)$$

where T is the temperature in degrees Fahrenheit. The ordered pairs corresponding to this equation are of the form (T, N).

a. Does the ordered pair $(60, 80)$ satisfy the equation? Yes

b. Based on your answer to part **a**, how many chirps does a cricket make when the temperature is 60°F? 80

c. How many chirps does a cricket make when the temperature is 80°F? 160

52. Based on the formula in Problem 51, crickets stop chirping when $N = 0$. The corresponding ordered pair will be $(T, 0)$.

a. Find T. $T = 40$

b. At what temperature will crickets stop chirping? 40°F

Are you exercising too hard? Your target zone can tell you. It works like this. Take your pulse after exercising, find your age on the *x*-axis in Figure 18, and follow a vertical line up to the lower edge of the shaded area; then go across to the number on the *y*-axis at the left. That pulse rate is the lower limit for your target zone. To find the upper limit, continue vertically on your age line to the top of the shaded area and then go across to the *y*-axis to locate your pulse rate. Use this information and Figure 18 to solve Problems 53–56.

53. What is the lower limit pulse rate for a 20-year-old person? Write the answer as an ordered pair (age, limit). (20, 140)

54. What is the upper limit pulse rate for a 20-year-old person? Write the answer as an ordered pair. (20, 170)

55. What is the upper limit pulse rate for a 45-year-old person? Write the answer as an ordered pair. (45, 148)

56. What is the lower limit pulse rate for a 50-year-old person? Write the answer as an ordered pair. (50, 118)

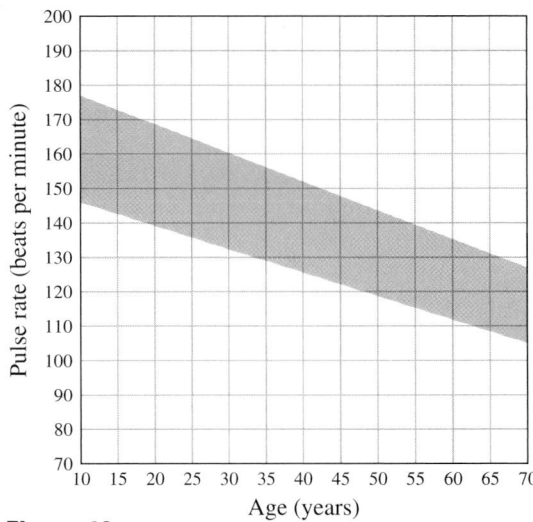

Figure 18

SKILL CHECKER

Try the "Skill Checker" exercises so you'll be ready for the next section.

Find *y* for the indicated value of *x*:

57. $y = 3x + 6, x = 2$ 12

58. $7y = 3x + 6, x = -2$ 0

59. $y = \dfrac{-2}{3}x + 4, x = 3$ 2

60. $y = \dfrac{-2}{3}x + 4, x = -3$ 6

Find *x*:

61. If $y = 3x + 6$ and $y = 0$. -2

62. If $y = 2x + 8$ and $y = 0$. -4

USING YOUR KNOWLEDGE

The Long, Hot Summer Exercise

The ideas presented in this section are vital for understanding graphs. For example, do you exercise in the summer? To determine the risk of exercising in the heat, you must know how to read the graph. To do this, first find the temperature on the *y*-axis, then read across to the right, stopping at the vertical line representing the relative humidity. For example, on a 90°F day, if the humidity is less than 30% the weather is in the safe zone.

63. If the humidity is 50%, how high can the temperature rise and still be in the safe zone for exercising? (Answer to the nearest degree.)
    86°F

64. If the humidity is 70%, at what temperature will the danger zone start?
    95°F

65. If the temperature is 100°F, what does the humidity have to be so that it is safe to exercise? Less than 10%

66. Between what temperatures should you use caution when exercising if the humidity is 80%? 81°–93°F

67. If you start jogging at 1 P.M. when the temperature is 86°F and the humidity is 60%, how many degrees can the temperature rise before you get to the danger zone? 11°F

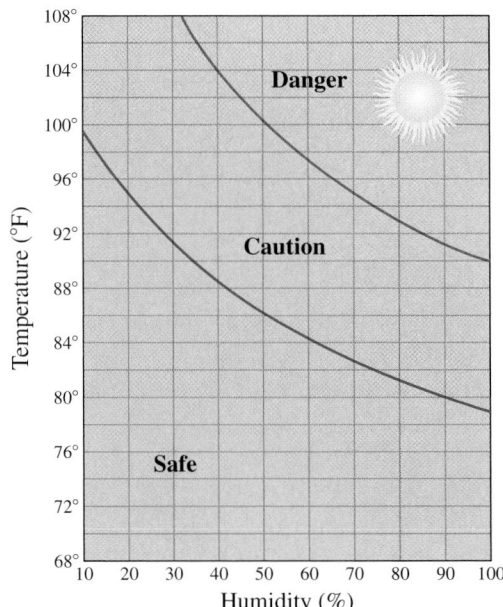

WRITE ON

68. What does the graph of an equation represent to you? Can you ever draw the *complete* graph of a line? Answers may vary.

69. Why are *linear equations* named that way? Answers may vary.

70. What happens when $A = 0$ in the equation $Ax + By = C$? Answers may vary.

71. What happens when $B = 0$ in the equation $Ax + By = C$? Answers may vary.

72. Why does setting $x = 0$ give the *y*-intercept for a line? Answers may vary.

73. Why does setting $y = 0$ give the *x*-intercept for a line? Answers may vary.

74. If two points determine a line, why did we use three points when graphing lines? Answers may vary.

75. Explain in your words the procedure you use to graph lines by (a) graphing points and (b) using the intercepts. Answers may vary.

MASTERY TEST

If you know how to do these problems, you have learned your lesson!

Graph the solutions to:

76. $3x = -9$

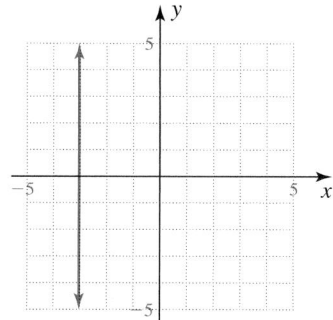

77. $2y = -6$

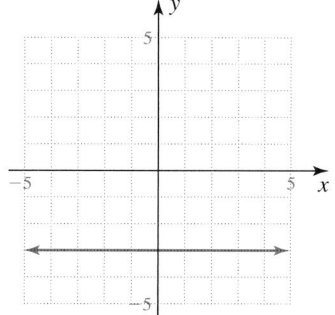

78. $2x - y = 6$

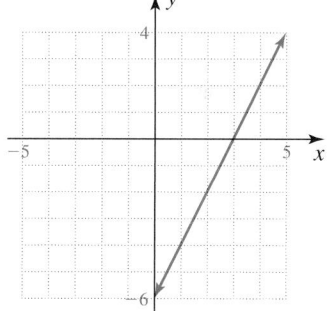

79. $-3x - y = -6$

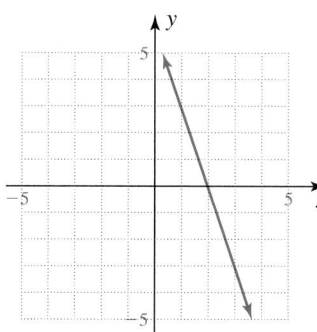

80. Find the *x*- and *y*-intercepts of $2x - 3y = 6$ and then graph all the solutions. $(3, 0); (0, -2)$

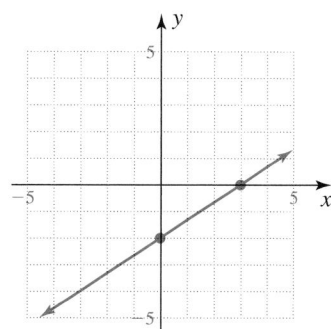

81. Find the *x*- and *y*-intercepts of $3x - 2y = -6$ and then graph all the solutions. $(-2, 0); (0, 3)$

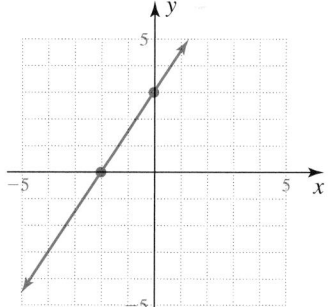

82. Graph the points $A\ (3, 2)$, $B\ (-1, 4)$, $C\ (0, -2)$, $D\ (-2, 1)$, and $E\ (2, -1)$.

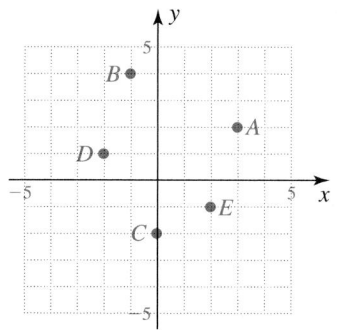

83. What are the coordinates of the points A, B, and C in Figure 19?

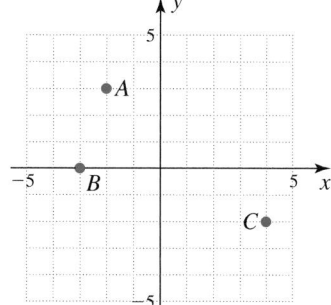

$A(-2, 3); B(-3, 0); C\ (4, -2)$

Figure 19

3.2 INTRODUCTION TO FUNCTIONS

To Succeed, Review How To ...

1. Graph a line (pp. 165–168).
2. Evaluate an expression (p. 50).

Objectives

A Find the domain and range of a relation.

B Use the vertical line test to determine if a relation is a function.

C Find the domain of a function defined by an equation.

D Find the value of a function.

GETTING STARTED

Traveling Ants

Did you know that there is a linear relationship between the temperature and the rate of travel (speed) of certain ants? If y is the speed in centimeters per second and x is the temperature in degrees Celsius, the relationship is

$$y = \frac{1}{6}(x - 4)$$

Thus if the temperature is 10°C, the speed is

$$y = \frac{1}{6}(10 - 4) = \frac{1}{6} \cdot 6 = 1 \text{ cm/sec}$$

If the temperature is 16°C, the speed is

$$y = \frac{1}{6}(16 - 4) = \frac{1}{6} \cdot 12 = 2 \text{ cm/sec}$$

We can make a table showing two related sets of numbers, one for the temperature x and the other for the speed y, as shown. (By the way, x has to be greater than 4°C and less than 35°C. Why?)

The numbers in the first column are called values of the **independent variable** because they are chosen *independently* of the second number. The numbers in the second column are called values of the **dependent variable** because they *depend* on the values of the numbers in the first column. The numbers in our table are written as ordered pairs, and these ordered pairs can also be written as the set

$$\{(4, 0), (10, 1), (16, 2), (22, 3)\}$$

The concept of ordered pairs begins our investigation into an important idea in algebra: the concept of a function.

Temperature (in degrees Celsius)	Rate of Travel (Speed)
4	0
10	1
16	2
22	3

(Temperature, Speed)

(4, 0)
(10, 1)
(16, 2)
(22, 3)

A Finding the Domain and Range

Teaching Tip

A relation is a correspondence between two sets, such as the relation between the books in a library and the number of pages in each book. Have students discuss other examples of relations.

Any set of ordered pairs is a relation, which we define as follows.

RELATION, DOMAIN, AND RANGE

A **relation** is a set of ordered pairs. The set of all first coordinates is the **domain** of the relation, and the set of all second coordinates is the **range** of the relation.

Now suppose $S = \{(1, 5), (2, 7), (3, 9)\}$. The domain D is the set of all first coordinates—$D = \{1, 2, 3\}$—and the range R is the set of all second coordinates—$R = \{5, 7, 9\}$.

The relation S can also be written by giving the rule used to obtain the ordered pairs. Thus, when $x = 1, 2,$ or 3, the corresponding range value can be found by

$$S = \{(x, y) \mid y = 2x + 3\}$$

EXAMPLE 1	**Finding the domain and range of a relation in roster form**

Find the domain and range of the relation $A = \{(1, 2), (2, 3), (3, 4)\}$.

SOLUTION   The domain of A is the set of first coordinates, so $D = \{1, 2, 3\}$. The range of A is the set of second coordinates, so $R = \{2, 3, 4\}$. The graph of the relation A is shown in Figure 20.

Teaching Tip

Have students make a list of operations of mathematics: add, subtract, multiply, divide, raise to a power, take the root, absolute value, and so on. Then discuss which operations have restrictions. For example,

Division: $\frac{5}{0}$ has no real number answer

Square root: $\sqrt{-9}$ has no real number answer

Therefore, when given $y = 2x - 4$, have the students discuss whether there are any values of x that must be excluded, depending on the operations being used in the equation.

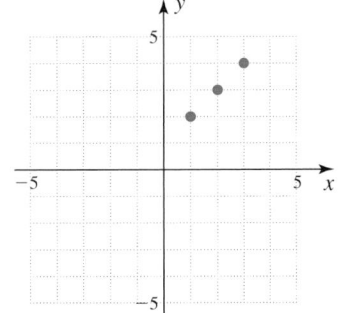

Figure 20

PROBLEM 1

Find the domain and range of the relation:

$$A = \{(2, 1), (7, 5), (9, 8)\}$$

A relation is a set of ordered pairs, so relations can be graphed in the Cartesian plane. We can then identify the domain and range by examining the graph of the relation.

> **NOTE**
>
> When no domain is specified, the domain is assumed to be the set of all real numbers for which the relation is defined.

EXAMPLE 2	**Finding the graph and the domain and range of a relation given an equation**

Find the graph, domain, and range of the relation $\{(x, y) \mid y = 2x - 4\}$.

SOLUTION   The graph of the relation is the graph of the equation $y = 2x - 4$, shown in Figure 21. The domain of this relation is the set of all real numbers since any real number x can be used as the first coordinate. Similarly, the range of y is the set of all real numbers. Some of the ordered pairs in the relation are $(0, -4)$, $(1, -2)$, and $(2, 0)$. The graph of the relation $\{(x, y) \mid y = 2x - 4\}$ is the graph of its ordered pairs and the graph is a "picture" of the relation.

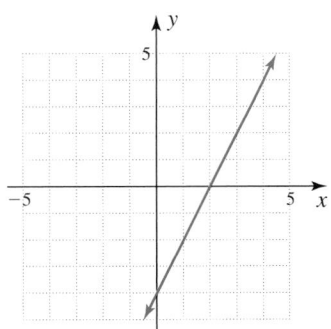

Figure 21

PROBLEM 2

Find the graph, domain, and range of the relation:

$$\{(x, y) \mid y = -2x + 1\}$$

Answers

1. $D = \{2, 7, 9\}$ and $R = \{1, 5, 8\}$
2. $D = $ Reals and $R = $ Reals

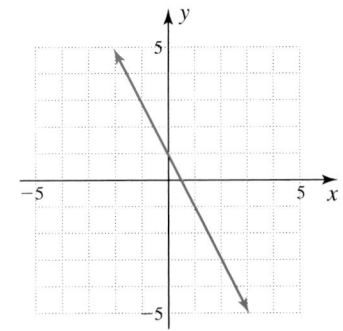

EXAMPLE 3	**More practice finding the domain and range given a graph**

Find the domain and range of:

a.

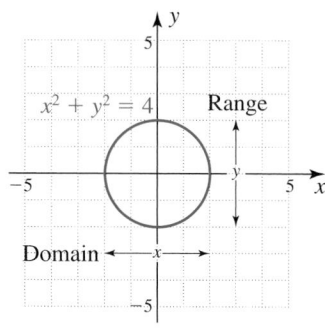

b.

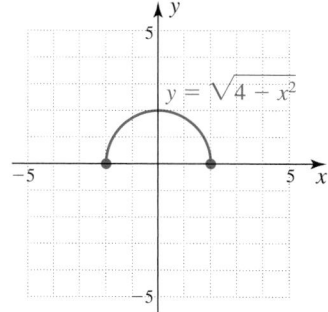

c.

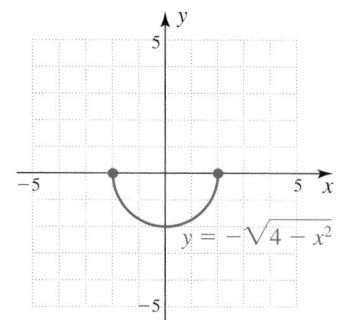

Teaching Tip

It may be helpful to use two different colors of chalk or some similar method to indicate the interval on the x-axis corresponding to the domain and the interval on the y-axis corresponding to the range of the relation. Students can visualize the actual intervals on the axes, not the graph itself.

SOLUTION

a. The graph is a circle of radius 2 centered at the origin. From the graph, it is clear that x and y can be any real numbers between -2 and 2, inclusive. The domain is $D = \{x \mid -2 \le x \le 2\}$, and the range is $R = \{y \mid -2 \le y \le 2\}$.

b. The graph is the top half of the circle from part **a**. The domain is $D = \{x \mid -2 \le x \le 2\}$, and the range is $R = \{y \mid 0 \le y \le 2\}$.

c. The graph is the bottom half of the circle from part **a**. The domain is $D = \{x \mid -2 \le x \le 2\}$, and the range is $R = \{y \mid -2 \le y \le 0\}$.

PROBLEM 3

Find the domain and range of:

a.

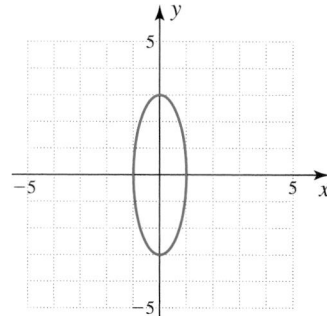

b.

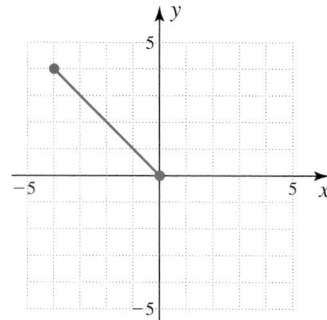

c.

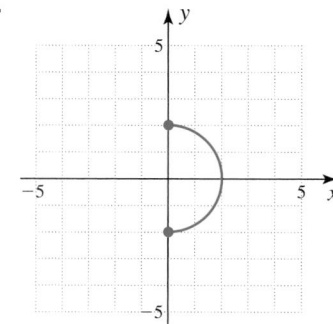

B	**Functions and the Vertical Line Test**

The relation in Example 3, part **a**, allows *two* values of y for some values of x. For instance, if $x = 0$, then $y = \pm 2$. On the other hand, the relations in parts **b** and **c** allow only *one* value of y for each value of x. These two relations are *functions*. Here is the definition.

FUNCTION

A **function** is a relation in which no two different ordered pairs have the same first coordinate.

Answers

3. a. $D = \{-1 \le x \le 1\}$ and $R = \{-3 \le y \le 3\}$
b. $D = \{-4 \le x \le 0\}$ and $R = \{0 \le y \le 4\}$
c. $D = \{0 \le x \le 2\}$ and $R = \{-2 \le y \le 2\}$

The relation $\{(1, 2), (2, 3), (3, 4)\}$ is a function because no two ordered pairs have the same first coordinate. On the other hand, the relation $\{(1, 2), (2, 3), (1, 3)\}$ is not a function because $(1, 2)$ and $(1, 3)$ have the same first coordinate.

The graph of a relation can be used to determine whether the relation is a function. Any two points with the same first coordinate will be on a vertical line parallel to the y-axis, so if any vertical line intersects the graph more than once, the relation is *not* a function. Testing to see if a relation is a function by determining whether a vertical line crosses the graph more than once is called the **vertical line test.**

VERTICAL LINE TEST

If a vertical line parallel to the y-axis intersects the graph of a relation more than once, the relation is *not* a function.

Using this test, we can see that the relation in Example 3, part **a**, is *not* a function. (A vertical line crosses the graph in more than one place.) On the other hand, the graphs in parts **b** and **c** are functions.

EXAMPLE 4   **Using the vertical line test**

Use the vertical line test to determine whether the graph of the relation defines a function:

a.

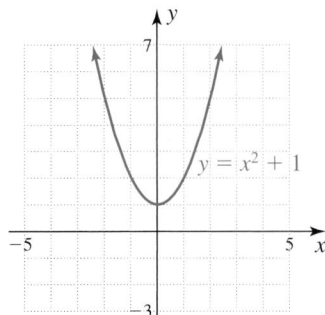

$y = x^2 + 1$

b.

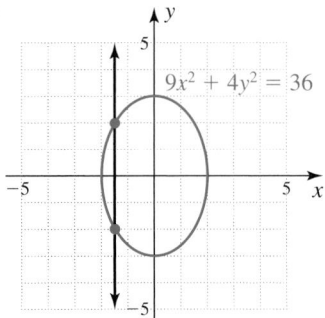

$9x^2 + 4y^2 = 36$

Teaching Tip

Show students that if a vertical line intersects the graph more than once, there is an x that has two y-values; thus, the graph is *not* that of a function

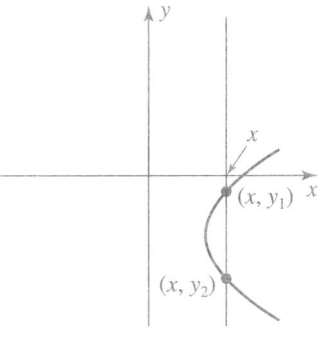

This vertical line crosses the graph in *two* points.

Not a function

PROBLEM 4

Use the vertical line test to determine whether the graph of the relation defines a function:

a.

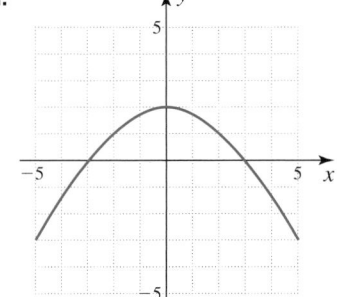

b.

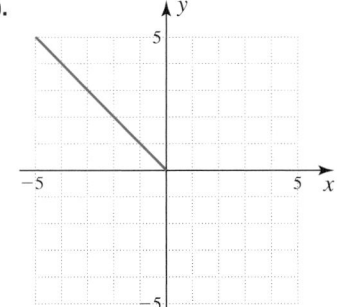

c.

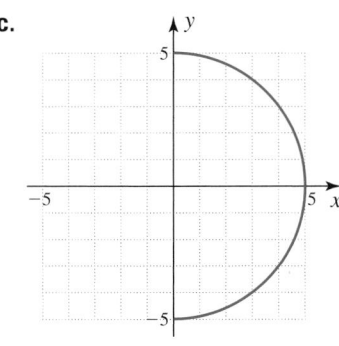

SOLUTION

a. The graph is a **parabola.** The relation is a function because no vertical line crosses the graph more than once.

b. The graph is an **ellipse.** We can draw a vertical line that crosses the graph in more than one point, so the relation is not a function.

Web It

For an interactive practice in using the vertical line test for deciding if the graph is a function, go to link 3-2-1 at mhhe.com/bello.

Roll the dice 5 times for 5 dots then connect them one at a time by clicking on them in order. Is the graph a function? It will tell you if you are correct.

Answers

4. a. Yes   **b.** Yes   **c.** No

 C   Finding the Domain of a Function

When relations are defined by means of an equation, the domain is the set of all possible replacements for the variable x that result in real numbers for y. We cannot replace x by values that will produce zero in the denominator of a fraction or the square root of a negative number. For example, the domain of $y = \dfrac{1}{x}$ is the set of all real numbers *except* 0 and the domain of $y = \sqrt{x}$ is the set of all real numbers $x \geq 0$.

EXAMPLE 5   Finding the domain of a function given an equation

Find the domain of the function defined by:

a. $y = \dfrac{1}{(x - 2)(x + 3)}$

b. $y = \sqrt{x - 3}$

SOLUTION

a. We cannot replace x by values that will produce zero in the denominator, so we must avoid the case in which

$$x - 2 = 0 \quad \text{or} \quad x + 3 = 0$$
$$x = 2 \qquad\qquad x = -3$$

The domain of

$$y = \dfrac{1}{(x - 2)(x + 3)}$$

is the set of all real numbers except 2 and -3 or $\{x \mid x \in \text{Reals, where } x \neq 2, -3\}$.

b. The square root of a negative number is not a real number, thus we must make the expression under the radical, $x - 3$, nonnegative, that is,

$$x - 3 \geq 0$$

or

$$x \geq 3$$

The domain is $\{x \mid x \geq 3\}$.

PROBLEM 5

Find the domain of the function defined by:

a. $f(x) = \dfrac{1}{(x + 1)(x - 4)}$

b. $y = \sqrt{x + 2}$

Teaching Tip

When division occurs, set the denominator equal to zero and solve. The solution is what is *excluded* from the domain. When square root occurs, set the radicand ≥ 0 and solve. The solution set *is* the domain. If the radical expression is in the denominator, remind students to set the radicand > 0 to find the domain.

Calculate It   Finding Domain and Range

To find the domain and range of

$$y = \dfrac{1}{(x - 2)(x + 3)}$$

using your TI-83 Plus: First graph Y using a decimal window:

[Y=] 1 [÷] (([X,T,θ,n] [−] 2) ([X,T,θ,n] [+] 3)) [ZOOM] 4

The dot mode of your calculator gives a better view of points that may not be in the domain.

press [MODE] , use arrows to DOT, then press [ENTER] [GRAPH]

Do you see why $x = -3$ and $x = 2$ are not part of the graph?

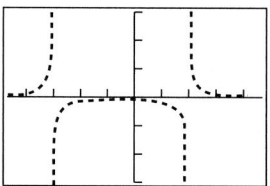

If you use the [TRACE] key to move your cursor to $x = 2$, what value do you get for y?
Try graphing $y = \sqrt{x - 3}$. What do you get for the domain?

Answers

5. a. All Reals except -1 and 4
b. $\{x \mid x \geq -2\}$

D　Finding the Value of a Function

Water pressure is a *function* of the depth, as you can see in the photo. The higher pressure at the lower holes of the bucket makes water squirt out in a flat trajectory, whereas the lower pressure at the upper holes produces only a weak stream. The pressure y of the water (in pounds per square foot) at a depth of x feet (x is a positive number) is

$$y = 62.5x$$

If we wish to emphasize that the pressure y is a function of the depth x, we can use the **function notation** $f(x)$ (read "f of x") and write

$$f(x) = 62.5x$$

To find the pressure at 2 ft below the surface, we find the value of y when $x = 2$. If $x = 2$, then $y = (62.5)(2) = 125$. Using functional notation, we would write $f(2) = (62.5)(2) = 125$. This table illustrates both notations:

y in terms of x $y = 62.5x$	Function Notation $f(x) = 62.5x$
If $x = 2$, then $y = (62.5)(2) = 125$.	$f(2) = (62.5)(2) = 125$
If $x = 4$, then $y = (62.5)(4) = 250$.	$f(4) = (62.5)(4) = 250$
If $x = 5$, then $y = (62.5)(5) = 312.5$.	$f(5) = (62.5)(5) = 312.5$

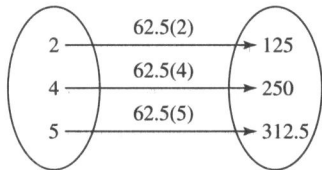

If $y = f(x)$, the symbols $f(x)$ and y are interchangeable; they both represent the value of the function for the given value of x. If

$$y = f(x) = 2x + 3$$

then

$$f(1) = 2(1) + 3 = 5$$
$$f(0) = 2(0) + 3 = 3$$
$$f(-6) = 2(-6) + 3 = -9$$
$$f(4) = 2(4) + 3 = 11$$
$$f(a) = 2(a) + 3 = 2a + 3$$
$$f(w + 2) = 2(w + 2) + 3 = 2w + 7$$

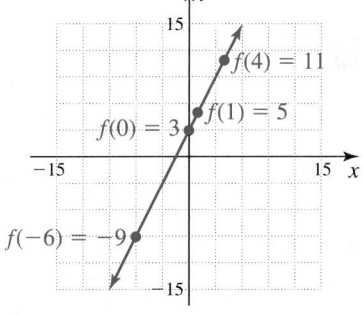

Figure 22

(shown in Figure 22).

Whatever appears between the parentheses in $f(\ )$ is to be substituted for x in the rule that defines $f(x)$.

Instead of describing a function in set notation, we frequently say "the function defined by $f(x) = \ldots$," where the ellipsis dots are to be replaced by the expression for the value of the function. For instance, "the function defined by $f(x) = 2x + 3$" has the same meaning as "the function $f = \{(x, y) \mid y = 2x + 3\}$." The function $f(x) = 2x + 3$ is an example of a linear function because its graph is a straight line. In general, we have the following definition.

Calculate It

Evaluate a Linear Function by Storing a Value

Use your TI-83 Plus to evaluate the function of Example 6:

Store a value for x, say, 4

Now tell the calculator you want to evaluate by entering:

Try evaluating the function for $f(-4)$.

> ### LINEAR FUNCTION
>
> A **linear function** is a function that can be written in the form
> $$f(x) = mx + b$$
> where m and b are real numbers.

EXAMPLE 6   **Finding the values of a function given the rule**

Let $f(x) = 3x + 5$ and find:

a. $f(4)$ **b.** $f(2)$ **c.** $f(2) + f(4)$ **d.** $f(x + 1)$

SOLUTION

a. $f(x) = 3x + 5$, so $f(4) = 3 \cdot 4 + 5 = 12 + 5 = 17$

b. $f(2) = 3 \cdot 2 + 5 = 6 + 5 = 11$

c. $f(2) = 11$ and $f(4) = 17$, so $f(2) + f(4) = 11 + 17 = 28$

d. $f(x + 1) = 3(x + 1) + 5 = 3x + 8$

PROBLEM 6

Let $f(x) = 2x - 3$ and find:

a. $f(5)$

b. $f(1)$

c. $f(1) - f(5)$

d. $f(x - 1)$

EXAMPLE 7   **Finding the values of a function given ordered pairs**

Let $f = \{(3, 2), (4, -2), (5, 0)\}$ and find:

a. $f(3)$ **b.** $f(4)$ **c.** $f(3) + f(4)$

SOLUTION

a. In the ordered pair $(3, 2)$, $x = 3$ and $y = 2$. Thus $f(3) = 2$.

b. In the ordered pair $(4, -2)$, $x = 4$ and $y = -2$. Thus $f(4) = -2$.

c. $f(3) + f(4) = 2 + (-2) = 0$

PROBLEM 7

Let $f = \{(2, 3), (-1, 4), (0, 6)\}$ and find:

a. $f(2)$ **b.** $f(-1)$ **c.** $f(2) + f(-1)$

Teaching Tip

For Example 7, point out that the function here is defined with ordered pairs instead of a rule such as $y = 2x + 3$. Therefore, all that is necessary is to identify the y-value that corresponds to the appropriate x-value.

Web It

If you need more practice evaluating functions, go to link 3-2-2 at mhhe.com/bello.

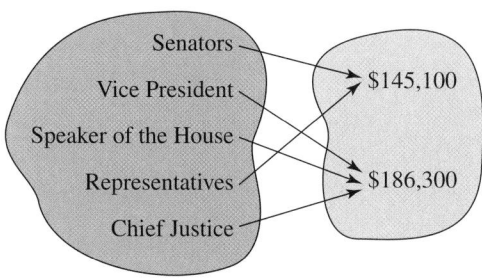

Figure 23

The rule or correspondence that assigns a range value to each domain value can also be shown by a method called **mapping**. For example, see Figure 23 where the mapping shown is a function that assigns to each federal government title its corresponding salary. The value of the function can be found by inspecting the "map." In this case,

$$f(\text{Senator}) = \$145{,}100$$
$$f(\text{Vice President}) = \$186{,}300$$
$$f(\text{Speaker}) = \$186{,}300$$
$$f(\text{Representative}) = \$145{,}100$$
$$f(\text{Chief Justice}) = \$186{,}300$$

The idea of the function is of great importance in algebra, so make sure you understand this concept and its notation. Here are three ways in which a function can be viewed.

DEFINITIONS OF A FUNCTION

1. A **function** is a rule or a correspondence that assigns exactly **one** *range* value to each *domain* value (as in Example 6).

2. A **function** is a relation in which **no two** ordered pairs have the same first coordinate (as in Example 7).

3. A **function** is a *mapping* that assigns exactly **one** range element to each domain element.

Answers

6. a. 7   **b.** -1 **c.** -8
d. $2x - 5$ **7. a.** 3   **b.** 4
c. 7

Calculate It Exercises

The ideas of domain and range of a linear function can be used to set up the [WINDOW] for a graph. Suppose we specify a [3, 7] by [−2, 6] [WINDOW] for a graph:

1. What is the domain of the function we are graphing?

2. What is the range of the function we are graphing?

3. In Problem 73, Exercise 3.2, what is a reasonable domain for the function $m(L)$?

4. In Problem 73, Exercise 3.2, what is a reasonable range for the function $f(L)$?

5. If you use the notation [a, b] by [c, d] to indicate a [WINDOW] that extends from a to b in the horizontal direction and from c to d in the vertical direction, write the domain and range of $m(L)$ using this notation.

6. Graph $m(L)$ using a [5, 15] by [1, 12] [WINDOW]. Can you see the x- and the y-axes? How would you modify the [WINDOW] so you can see the x- and y-axes?

If you have a TI-83 Plus, you can use the CALC feature to evaluate functions. To do Example 6, enter $Y_1 = 3x + 5$ and press [2nd] [TRACE] [1]. Enter the value $x = 4$ [ENTER]. The calculator gives $y = 17$ [remember $f(x) = y$]. You can also use the Table feature. Press [2nd] [WINDOW] and leave TblStart = 1, △Tbl = 1, and Indpnt, Depend on "AUTO." To enter the function you want, press [Y=] and enter $3x + 5$ [ENTER]. Now press [2nd] [GRAPH] and the calculator shows you the independent variable x in the first column and the values of $Y_1 = 3x + 5$ in the second column, and it's done automatically. Check to see if $f(4) = 17$ and $f(2) = 11$ as before.

Exercises 3.2

A In Problems 1–10, find the domain and range, and determine whether the relation is a function.

1. $\{(-3, 0), (-2, 1), (-1, 2)\}$
   $D = \{-3, -2, -1\}; R = \{0, 1, 2\};$
   a function

2. $\{(-1, -2), (0, -1), (1, 0)\}$
   $D = \{-1, 0, 1\}; R = \{-2, -1, 0\};$
   a function

3. $\{(3, 0), (4, 0), (5, 0)\}$
   $D = \{3, 4, 5\}; R = \{0\};$
   a function

4. $\{(0, 1), (0, 2), (0, 3)\}$
   $D = \{0\}; R = \{1, 2, 3\};$
   not a function

5. $\{(1, 2), (1, 3), (2, 2), (2, 3)\}$
   $D = \{1, 2\}; R = \{2, 3\};$
   not a function

6. $\{(2, 1), (1, 2), (3, 4), (4, 3)\}$
   $D = \{1, 2, 3, 4\}; R = \{1, 2, 3, 4\};$
   a function

7.
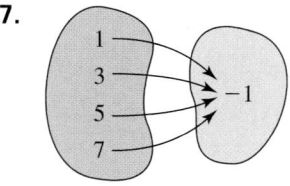
$D = \{1, 3, 5, 7\}; R = \{-1\};$
a function

8.
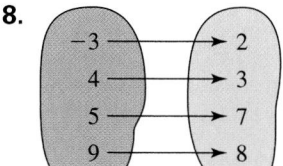
$D = \{-3, 4, 5, 9\};$
$R = \{2, 3, 7, 8\};$ a function

9.
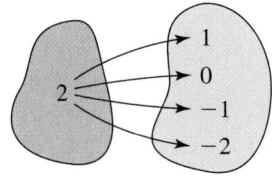
$D = \{2\}; R = \{1, 0, -1, -2\};$
not a function

10.
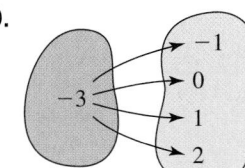
$D = \{-3\}; R = \{-1, 0, 1, 2\};$
not a function

A B In Problems 11–30, give the domain and range, and use the vertical line test to determine whether the relation is a function.

11.

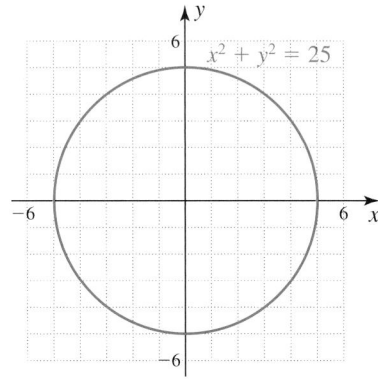

$D = \{x \mid -5 \le x \le 5\}$;
$R = \{y \mid -5 \le y \le 5\}$;
not a function

12.

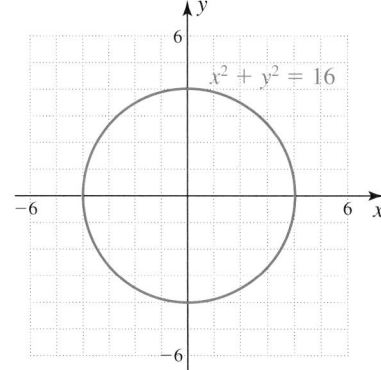

$D = \{x \mid -4 \le x \le 4\}$;
$R = \{y \mid -4 \le y \le 4\}$;
not a function

13.

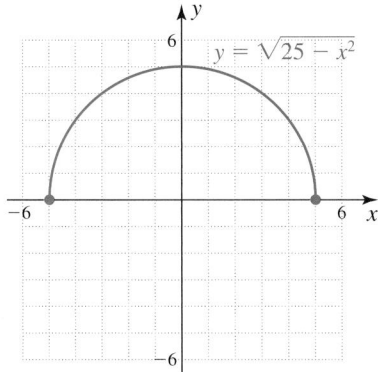

$D = \{x \mid -5 \le x \le 5\}$;
$R = \{y \mid 0 \le y \le 5\}$;
a function

14.

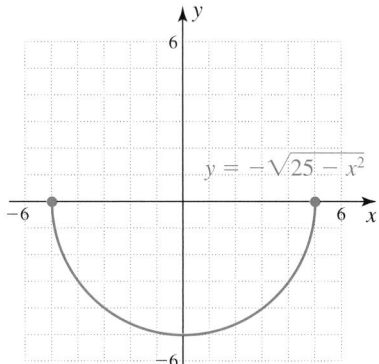

$D = \{x \mid -5 \le x \le 5\}$;
$R = \{y \mid -5 \le y \le 0\}$;
a function

15.

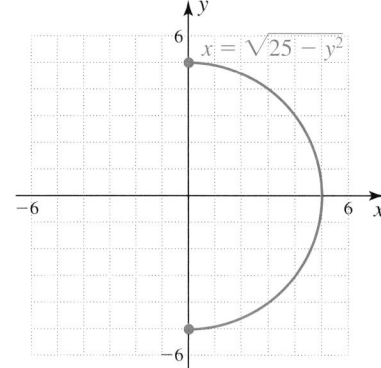

$D = \{x \mid 0 \le x \le 5\}$;
$R = \{y \mid -5 \le y \le 5\}$;
not a function

16.

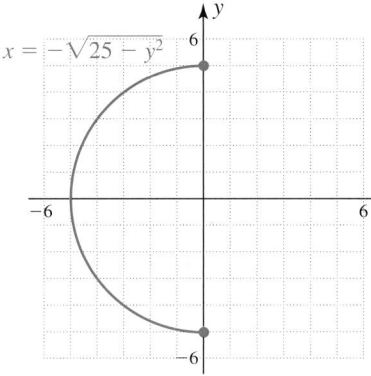

$D = \{x \mid -5 \le x \le 0\}$;
$R = \{y \mid -5 \le y \le 5\}$;
not a function

17.

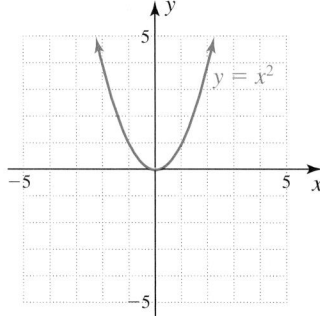

$D = \{x \mid x \text{ is a real number}\}$;
$R = \{y \mid y \ge 0\}$;
a function

18.

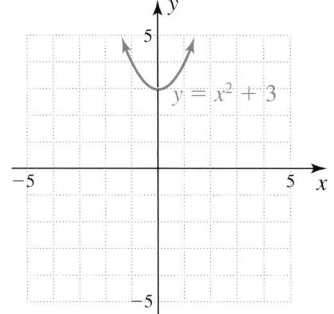

$D = \{x \mid x \text{ is a real number}\}$;
$R = \{y \mid y \ge 3\}$;
a function

19.

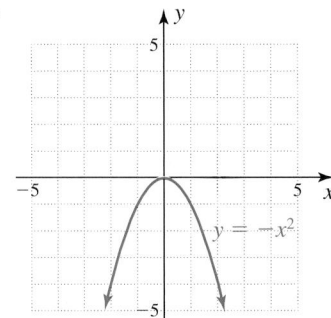

$D = \{x \mid x \text{ is a real number}\}$;
$R = \{y \mid y \le 0\}$;
a function

20.

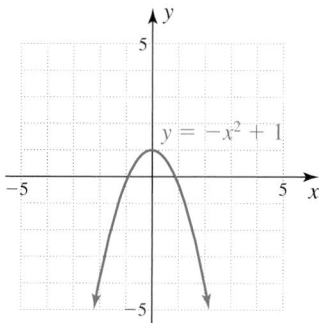

$D = \{x \mid x \text{ is a real number}\};$
$R = \{y \mid y \leq 1\};$
a function

21.

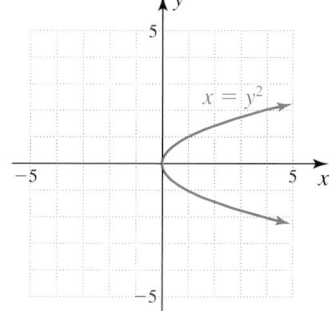

$D = \{x \mid x \geq 0\};$
$R = \{y \mid y \text{ is a real number}\};$
not a function

22.

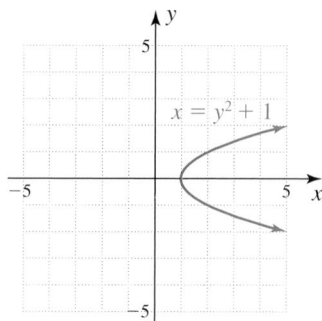

$D = \{x \mid x \geq 1\};$
$R = \{y \mid y \text{ is a real number}\};$
not a function

23.

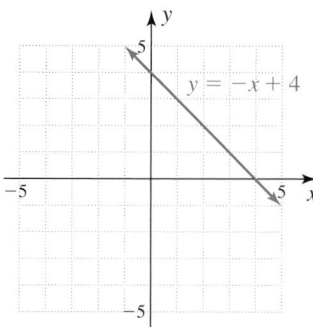

$D = \{x \mid x \text{ is a real number}\};$
$R = \{y \mid y \text{ is a real number}\};$
a function

24.

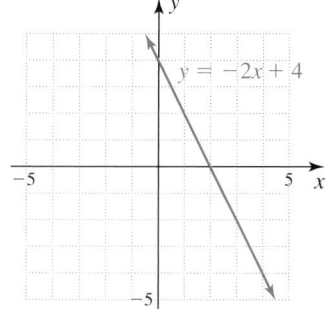

$D = \{x \mid x \text{ is a real number}\};$
$R = \{y \mid y \text{ is a real number}\};$
a function

25.

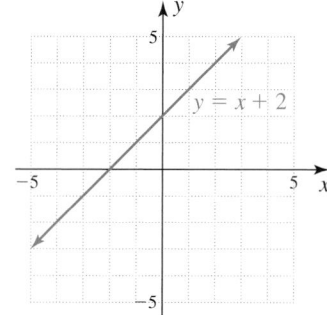

$D = \{x \mid x \text{ is a real number}\};$
$R = \{y \mid y \text{ is a real number}\};$
a function

26.

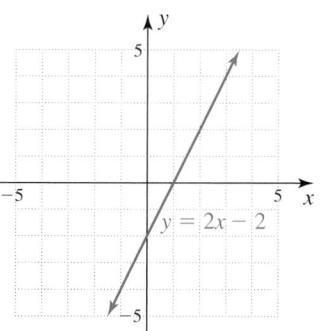

$D = \{x \mid x \text{ is a real number}\};$
$R = \{y \mid y \text{ is a real number}\};$
a function

27.

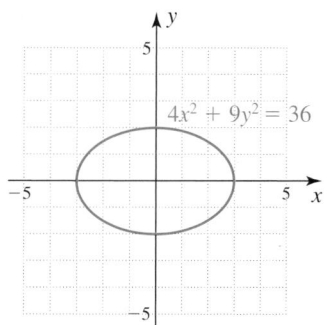

$D = \{x \mid -3 \leq x \leq 3\};$
$R = \{y \mid -2 \leq y \leq 2\};$
not a function

28.

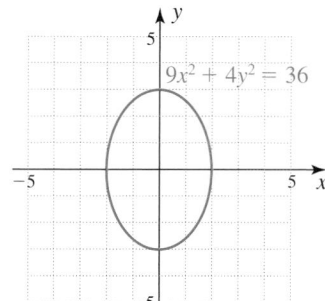

$D = \{x \mid -2 \leq x \leq 2\};$
$R = \{y \mid -3 \leq y \leq 3\};$
not a function

29.

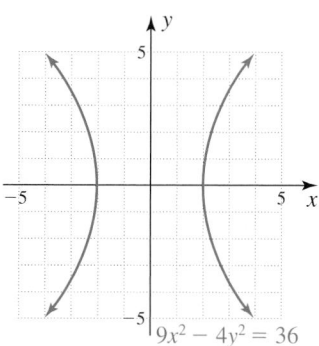

$D = \{x \mid x \geq 2 \text{ or } x \leq -2\};$
$R = \{y \mid y \text{ is a real number}\};$
not a function

30.

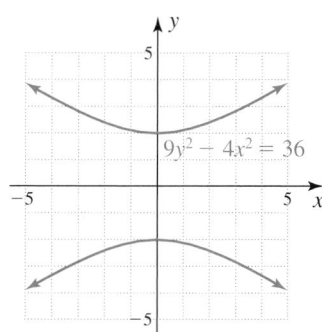

$D = \{x \mid x \text{ is a real number}\};$
$R = \{y \mid y \geq 2 \text{ or } y \leq -2\};$
not a function

B In Problems 31–36, use the vertical line test to determine whether the graphs represent functions.

31.

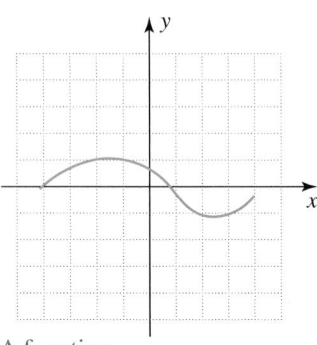

A function

32.

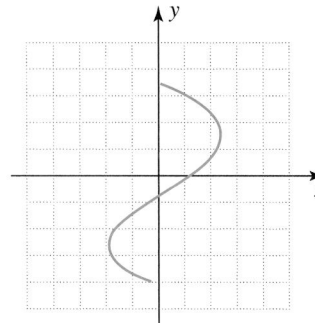

Not a function

33.

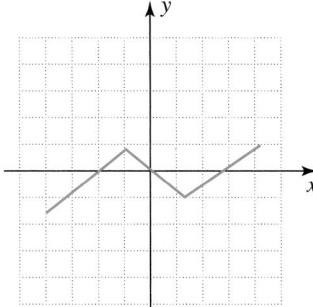

A function

34.

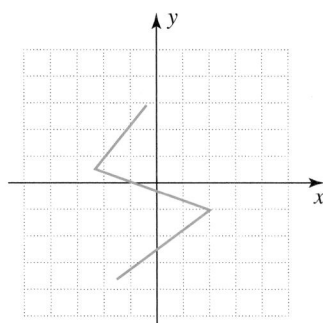

Not a function

35.

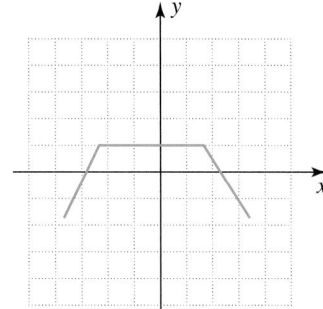

A function

36.

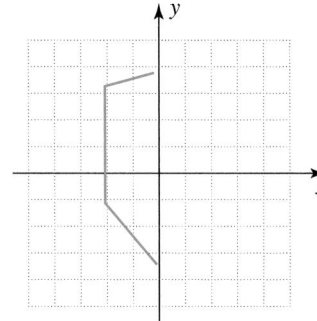

Not a function

C In Problems 37–50, find the domain of the function defined by the given equation.

37. $y = \sqrt{x - 5}$
$D = \{x \mid x \geq 5\}$

38. $y = \sqrt{x + 5}$
$D = \{x \mid x \geq -5\}$

39. $y = \sqrt{4 - 2x}$
$D = \{x \mid x \leq 2\}$

40. $y = \sqrt{6 + 3x}$
$D = \{x \mid x \geq -2\}$

41. $y = \sqrt{x^2 + 1}$

$D = \{x \mid x$ is a real number$\}$

42. $y = \sqrt{x^2 + 2}$

$D = \{x \mid x$ is a real number$\}$

43. $y = \dfrac{1}{x - 5}$

$D = \{x \mid x$ is a real number and $x \neq 5\}$

44. $y = \dfrac{1}{x + 5}$

$D = \{x \mid x$ is a real number and $x \neq -5\}$

45. $y = \dfrac{x + 2}{x + 5}$

$D = \{x \mid x$ is a real number and $x \neq -5\}$

46. $y = \dfrac{x + 3}{x - 6}$

$D = \{x \mid x$ is a real number and $x \neq 6\}$

47. $y = \dfrac{x}{(x + 2)(x + 1)}$

$D = \{x \mid x$ is a real number and $x \neq -2$ or $x \neq -1\}$

48. $y = \dfrac{x}{(x + 3)(x + 2)}$

$D = \{x \mid x$ is a real number and $x \neq -3$ or $x \neq -2\}$

49. $y = \dfrac{x}{(x + 4)(x - 4)}$

$D = \{x \mid x$ is a real number and $x \neq \pm 4\}$

50. $y = \dfrac{x + 2}{(x + 3)(x - 3)}$

$D = \{x \mid x$ is a real number and $x \neq \pm 3\}$

D In Problems 51–56, the definition of a function is given.

51. Given: $f(x) = 3x + 1$. Find:
  a. $f(0)$ 1   **b.** $f(2)$ 7   **c.** $f(-2)$ -5

52. Given: $g(x) = -2x + 1$. Find:
  a. $g(0)$ 1   **b.** $g(1)$ -1 **c.** $g(-1)$ 3

53. Given: $F(x) = \sqrt{x - 1}$. Find:
  a. $F(1)$ 0   **b.** $F(5)$ 2   **c.** $F(26)$ 5

54. Given: $G(x) = x^2 + 2x - 1$. Find:
  a. $G(0)$ -1 **b.** $G(2)$ 7   **c.** $G(-2)$ -1

55. Given: $f(x) = \dfrac{1}{3x + 1}$
Find:

   a. $f(1)$ $\frac{1}{4}$ **b.** $f(1) - f(2)$ $\frac{3}{28}$

   c. $\dfrac{f(1) - f(2)}{3}$ $\frac{1}{28}$

56. Given: $f(x) = \dfrac{x - 2}{x + 3}$
Find:

   a. $f(2)$ 0 **b.** $f(3)$ $\frac{1}{6}$

   c. $f(3) - f(2)$ $\frac{1}{6}$

The functions defined by $f(x) = 3x - 4$ and $g(x) = x^2 + 2x + 4$ are used in Problems 57–60.

57. Find:

   a. $f(3)$ 5
   b. $g(3)$ 19

   c. $f(3) + g(3)$ 24

58. Find:

   a. $f(4)$ 8
   b. $g(4)$ 28

   c. $f(4) - g(4)$ -20

59. Find:

   a. $f(-2)$ -10
   b. $g(-3)$ 7

   c. $f(-2) \cdot g(-3)$ -70

60. Find:

   a. $f(-1)$ -7
   b. $g(-2)$ 4

   c. $\dfrac{f(-1)}{g(-2)}$ $\frac{-7}{4}$

The functions $f = \{(1, 3), (-1, 5), (-3, 7), (-5, 9)\}$ and $g = \{(-2, 4), (0, 6), (2, 8), (4, 10)\}$ are used in Problems 61–64.

61. Find:

   a. $f(1)$ 3 **b.** $g(-2)$ 4 **c.** $f(1) + g(-2)$ 7

62. Find:

   a. $f(-1)$ 5 **b.** $g(0)$ 6 **c.** $f(-1) - g(0)$ -1

63. Find:

   a. $f(-3)$ 7 **b.** $g(2)$ 8 **c.** $f(-3) \cdot g(2)$ 56

64. Find:

   a. $f(-5)$ 9 **b.** $g(4)$ 10 **c.** $\dfrac{f(-5)}{g(4)}$ $\frac{9}{10}$

APPLICATIONS

Practicing Math with Real Word Problems

65. The revenue obtained from selling x textbooks is given by $R(x) = 30x - 0.0005x^2$. The cost of producing the books is $C(x) = 100,000 + 6x$.

   a. Find the profit function $P(x) = R(x) - C(x)$.
   $P(x) = -0.0005x^2 + 24x - 100,000$

   b. Find the profit when 10,000 books are sold.
   $90,000

66. The Fahrenheit temperature reading F is a function of the Celsius temperature reading C. This function is given by

$$F(C) = \frac{9}{5}C + 32$$

   a. If the temperature is 15°C, what is the Fahrenheit temperature?   59°F

   b. Water boils at 100°C. What is the corresponding Fahrenheit temperature?   212°F

   c. The freezing point of water is 0°C or 32°F. How many Fahrenheit degrees below freezing is a temperature of -10°C? $F = 14$, so 18° below freezing

   d. The lowest temperature attainable is -273°C; this is the zero point on the absolute temperature scale. What is the corresponding Fahrenheit temperature? -459.4°F

67. When you exercise, your pulse rate should be within a certain *target zone*. The *upper limit U* of your target zone when exercising is a function of age a (in years) and is given by

   $U(a) = -a + 190$ (your pulse or heart rate)

Find the highest safe heart rate for a person who is

   a. 50 years old $U(50) = 140$

   b. 60 years old $U(60) = 130$

68. The lower limit L of your target zone when exercising is a function of a (in years) and is given by

$$L(a) = \frac{-2}{3}a + 150$$

The target zone for a person a years old consists of all the heart rates between $L(a)$ and $U(a)$, inclusive. (See Problem 67.) If a person's heart rate is R, that person's target zone is described by $L(a) \le R \le U(a)$. Find the target zone for a person who is

   a. 30 years old $130 \le R \le 160$

   b. 45 years old $120 \le R \le 145$

69. The ideal weight w (in pounds) of a man is a function of his height h (in inches). This function is defined by

$$w(h) = 5h - 190$$

   a. If a man is 70 in. tall, what should his weight be?
   160 lb

   b. If a man weighs 200 lb, what should his height be?
   78 in.

70. The cost C in dollars of renting a car for 1 day is a function of the number of miles traveled, m. For a car renting for $20 per day and 20¢ per mile, this function is given by

$$C(m) = 0.20m + 20$$

   a. Find the cost of renting a car for 1 day and driving 290 mi. $78

   b. An executive is given a $60-a-day expense budget. How many miles can she drive? 200 miles

72. If a ball is dropped from a point above the surface of the earth, the distance s (in meters) that the ball falls in t seconds is a function of t. This function is given by

$$s(t) = 4.9t^2$$

   Find the distance that the ball falls in

   a. 2 sec 19.6 m

   b. 5 sec 122.5 m

74. The speed S (in centimeters per second) of an ant is a *linear* function of the temperature C (in degrees Celsius) and is given by

$$S = f(C) = \frac{1}{6}(C - 4)$$

   a. What is the independent variable? C

   b. What is the dependent variable? S

   c. What is the speed of an ant on a hot day when the temperature is 28°C? 4 cm/sec

   d. What is the speed of an ant on a cold day when the temperature is 10°C? 1 cm/sec

76. Do you recycle bottles, newspaper, and plastic? The waste recovered (in millions of tons) is a *quadratic* function of t, the number of years after 1960 and is given by

$$G(t) = 0.04t^2 - 0.59t + 7.42$$

   a. How many million tons of waste were recovered in 1960? 7.42

   b. How many million tons of waste would you expect to be recovered in the year 2010? 77.92

71. The pressure P (in pounds per square foot) at a depth of d feet below the surface of the ocean is a function of the depth. This function is given by

$$P(d) = 63.9d$$

   What is the pressure on a submarine at a depth of

   a. 10 ft? 639 lb/ft^2

   b. 100 ft? 6390 lb/ft^2

73. Your shoe size S is a *linear* function of the length L (in inches) of your foot and is given by

$$S = m(L) = 3L - 22 \quad \text{(for men)}$$
$$S = f(L) = 3L - 21 \quad \text{(for women)}$$

   a. What is the independent variable? L

   b. What is the dependent variable? S

   c. If the length of a man's foot is 11 in., what is his shoe size? Size 11

   d. If the length of a woman's foot is 11 in., what is her shoe size? Size 12

75. According to FBI data, the number of robberies (per 100,000 population) is a *quadratic* function of t, the number of years after 1980 and is given by

$$R(t) = 1.85t^2 - 19.14t + 262$$

   a. What was the number of robberies (per 100,000) in 2000? 619.2

   b. What do you predict that the number of robberies would be in the year 2010? 1352.8

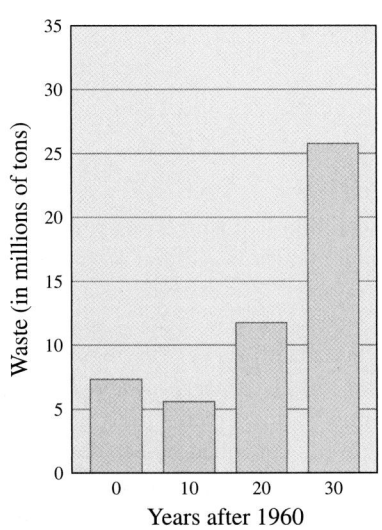

77. The graph shows that your blood alcohol level L is related to the number of drinks D you consume and your weight, sex, and the time period in which the drinks were consumed.

 a. Do the two graphs represent functions? Yes

 b. What are the domain and range for the graph representing males?
 $D = \{x \mid 1 \leq x \leq 6\}; R = \{y \mid 0 \leq y \leq 0.11\}$

 c. What are the domain and range for the graph representing females?
 $D = \{x \mid 1 \leq x \leq 4\}; R = \{y \mid 0 \leq y \leq 0.09\}$

Blood alcohol levels

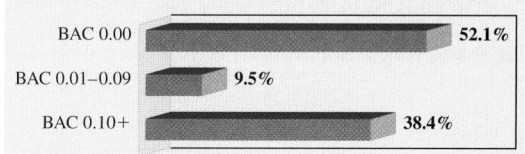

Fatalities

Most states consider a driver drunk with a blood alcohol content (BAC) of 0.10%. Nearly half of all fatal accidents involving drivers and pedestrians in 1991 were alcohol related. Here is a breakdown.

Effects of alcohol

How blood alcohol levels are affected by each drink over a 2-hour period:

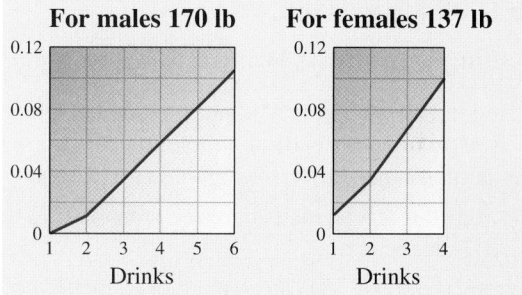

Source: Data from USA Today, 1993.

78. The graphs show the worldwide emissions (in parts per trillion, ppt) of two ozone-destroying chemicals: CFC-12 and CFC-11.

 a. Do both graphs represent functions? Yes

 b. What is the domain of the functions? $D = \{x \mid x \text{ is a year after } 1977\}$

 c. What is the range for CFC-11? $R = \{y \mid y \geq 150\}$

 d. What is the range for CFC-12? $R = \{y \mid y \geq 250\}$

 e. From 1980 to 1992, CFC-12 emissions can be modeled by the linear function $f(t) = 20t + 300$, where t is the number of years elapsed after 1980. What is $f(0)$, and is it close to the value for 1980 shown on the graph? $f(0) = 300$; Yes

 f. What is $f(10)$, and is it close to the corresponding value on the graph? $f(10) = 500$; Yes

 g. If you use this same linear model, what would be the CFC-12 emissions in the year 2000? Is this prediction consistent with the predictions in the graph? What is the difference in parts per trillion between the linear model and the observed values?
 $f(20) = 700$; 700 is 150 ppt greater than 550

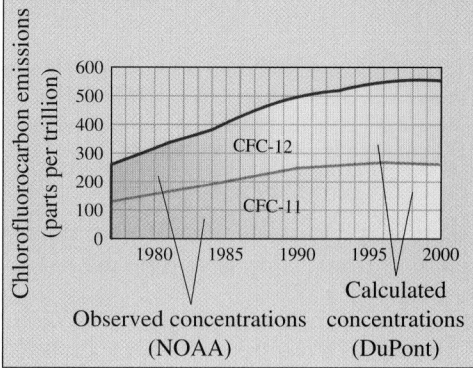

Decline in ozone-destroying chemicals seen

Worldwide emissions of two ozone-destroying chemicals are slowing sooner than expected. The gradual repair of the Earth's ozone layer should begin by 2000. Ozone molecules absorb some of the Sun's harmful ultraviolet radiation before it reaches the Earth's surface, making life on Earth possible.

Source: Data from AP/Wide World Photos.

WRITE ON

What is the difference between:

79. A function and a relation when they are written as a set of ordered pairs? Answers may vary.

80. The graph of a function and the graph of a relation? Answers may vary.

Write your own definition of:

81. The domain of a function. Answers may vary.

82. The range of a function. Answers may vary.

True or false? Explain your answer.

83. Every relation is a function. False; answers may vary.

84. Every function is a relation. True; answers may vary.

USING YOUR KNOWLEDGE

Everything You Always Wanted to Know about Relations (in Mathematics)

A special kind of relation that's important in mathematics is called an **equivalence relation.** A relation R is an equivalence relation if it has the following three properties:

a. Reflexive: If a is an element of the domain of R, then (a, a) is an element of R.

b. Symmetric: If (a, b) is an element of R, then (b, a) is an element of R.

c. Transitive: If (a, b) and (b, c) are both elements of R, then (a, c) is an element of R.

A simple example of an equivalence relation is

$$R = \{(x, y) \mid y = x, x \text{ an integer}\}$$

To show that R is an equivalence relation, we check the three properties:

a. Reflexive: If a is an integer, then $a = a$, so (a, a) is an element of R.

b. Symmetric: Suppose that (a, b) belongs to R. Then, by the definition of R, a and b are integers and $b = a$. But if $b = a$, then $a = b$, so (b, a) also belongs to R.

c. Transitive: Suppose that (a, b) and (b, c) both belong to R. Then $b = a$ and $c = b$, so $c = a$. Hence (a, c) also belongs to R.

Because R has all three properties, it is an equivalence relation.

The pairs in a relation don't have to be numbers, and some interesting relations occur outside the field of numbers. For example,

$$R = \{(x, y) \mid y \text{ is a member of the same immediate family as } x, x \text{ is a person}\}$$

is a relation. Is R an equivalence relation?

We check the three properties as before:

a. Reflexive: Given a person A, is (A, A) an element of R? Yes, A is obviously a member of the same immediate family as A.

b. Symmetric: Suppose that (A, B) is an element of R. Then B is a member of the same immediate family as A. But then A is a member of the same immediate family as B, so (B, A) is an element of R.

c. Transitive: Suppose that (A, B) and (B, C) both belong to R. Then A, B, and C are all members of the same immediate family. Thus (A, C) is an element of R.

Again, we see that R has all three properties, so it is an equivalence relation.

Can you determine which of the following are equivalence relations?

85. $R = \{(x, y) \mid x \text{ and } y \text{ are triangles and } y \text{ is similar to } x\}$ (*Note:* Here, "is similar to" means "has the same shape as.") Yes

86. $R = \{(x, y) \mid x \text{ and } y \text{ are integers and } y > x\}$
No, $x \not> x$, $x \not> y$.

87. $R = \{(x, y) \mid x \text{ and } y \text{ are positive integers and } y \text{ has the same parity as } x\}$; that is, y is odd if x is odd and y is even if x is even Yes

88. $R = \{(x, y) \mid x \text{ and } y \text{ are boys and } y \text{ is the brother of } x\}$
No, x is not the brother of x.

89. $R = \{(x, y) \mid x \text{ and } y \text{ are positive integers and when } x \text{ and } y \text{ are divided by 3, } y \text{ leaves the same remainder as } x\}$ Yes

90. $R = \{(x, y) \mid x \text{ is a fraction } \frac{a}{b} \text{ where } a \text{ and } b \text{ are integers } (b \neq 0), \text{ and } y \text{ is an equivalent fraction } \frac{ma}{mb} \text{ where } m \text{ is a nonzero integer}\}$ Yes

MASTERY TEST

If you know how to do these problems, you have learned your lesson!

91. If $f = \{(4, 3), (5, -1), (6, 0)\}$, find:

    a. $f(4)$ 3   **b.** $f(5)$ -1 **c.** $f(4) - f(5)$ 4

92. If $f(x) = 2x - 3$, find:

    a. $f(4)$ 5   **b.** $f(2)$ 1   **c.** $f(x + 1)$ $2x - 1$

Find the domain of:

93. $f(x) = \dfrac{1}{x - 2}$ $D = \{x \mid x \text{ is a real number and } x \neq 2\}$

94. $y = \sqrt{x - 3}$ $D = \{x \mid x \text{ is a real number and } x \geq 3\}$

Use the vertical line test to determine whether the relation shown is a function:

95.

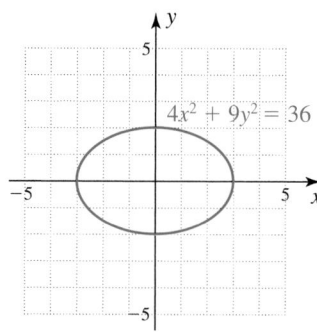

Not a function

96.

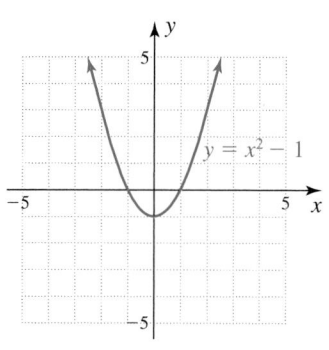

A function

Find the domain and range of the relations:

97.

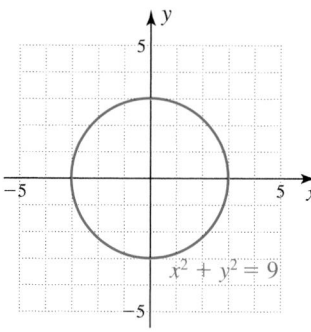

$D = \{x \mid -3 \leq x \leq 3\};$
$R = \{y \mid -3 \leq y \leq 3\}$

98.

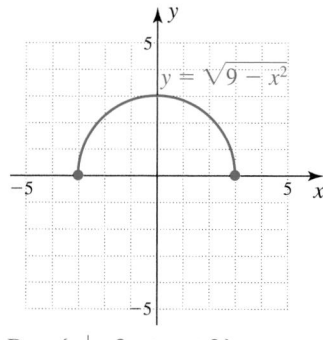

$D = \{x \mid -3 \leq x \leq 3\};$
$R = \{y \mid 0 \leq y \leq 3\}$

99.

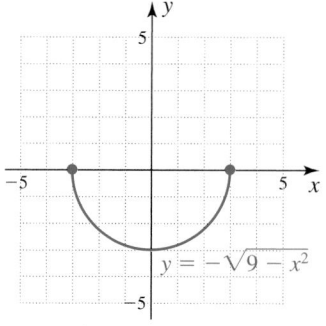

$D = \{x \mid -3 \leq x \leq 3\};$
$R = \{y \mid -3 \leq y \leq 0\}$

100. Find the domain and range of the relation
$\{(x, y) \mid y = 2x + 4\}$. $D = \{x \mid x \text{ is a real number}\};$
$R = \{y \mid y \text{ is a real number}\}$

101. Find the domain and range of the relation
$B = \{(7, 8), (8, 9), (9, 10)\}$.
$D = \{7, 8, 9\}; R = \{8, 9, 10\}$

| **3.3** | # USING SLOPES TO GRAPH LINES |

To Succeed, Review How To ...

1. Add, subtract, multiply, and divide integers (pp. 17–23).

2. Find the reciprocal of a number (p. 24).

Objectives

A Find the slope of a line passing through two given points.

B Use the definition of slope to decide whether two lines are perpendicular, parallel, or neither.

C Graph a line given its slope and a point on the line.

D Find the slope and *y*-intercept given the equation.

GETTING STARTED   **Sloping Salaries**

Beginning Teacher Salaries and Job Offers for New Graduates

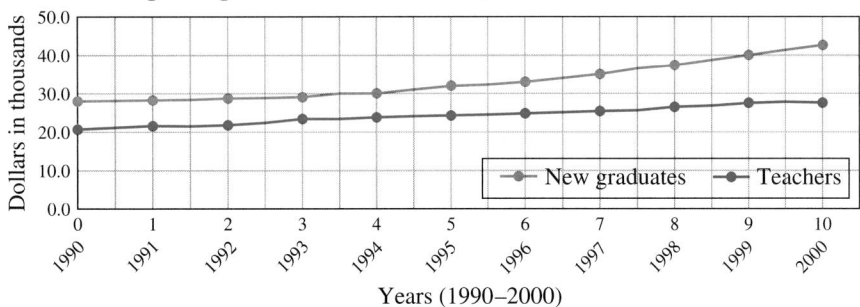

Years (1990–2000)

A line segment joining two points in a plane has a certain "steepness" or inclination. For example, suppose you want to find the average yearly raise in beginning teacher's average salaries. According to the graph, the salary in 1990 was about $22 (thousand); in 2000, it was about $28 (thousand). The change is

$$28 - 22 = 6 \text{ (thousand)}$$

during the 10-year period, and the average yearly raise for beginning teacher salaries is given by the ratio

$$\frac{\text{Increase in 10 yr}}{\text{Number of yr}} = \frac{28 - 22}{10} = \frac{6}{10} = 0.6$$

Thus, the teachers' average beginning salary went up about $600, (0.6 thousand), each year.

Now, how does that compare to the average yearly raise with the starting salaries for other job offers for new graduates? Using the ratio we get

$$\frac{\text{Increase in 10 yr}}{\text{Number of yr}} = \frac{43 - 28}{10} = \frac{15}{10} = 1.5$$

Starting salaries for other job offers for new graduates went up $1500 (1.5 thousand) each year. So not only were the other job offers for new graduates in 1990 higher than the beginning teacher salary, but the average yearly increase over 10 years was higher. The steepness of the two lines we just described algebraically is called the slope of the line, and we shall now show you how to determine it in various situations.

Teaching Tip

See if students can give other applications of "steepness."

Examples:

1. The pitch (steepness) of a roof.
2. The incline (steepness) of a mountain.

| **A** | **Finding the Slope of a Line** |

In mathematics the ratio of the change (difference) in *x* to the change in *x* is called the *slope* of the line segment and is denoted by the letter *m*. Thus

$$m = \frac{\text{change in } y}{\text{change in } x} = \frac{\text{difference in } y\text{'s}}{\text{difference in } x\text{'s}} = \frac{\text{rise } (\uparrow)}{\text{run } (\rightarrow)}$$

We summarize this discussion with the following definition and formula.

DEFINITION OF SLOPE

If $A(x_1, y_1)$ and $B(x_2, y_2)$ are any two distinct points on a line L (that is not parallel to the y-axis), then the **slope** of L, denoted by m, is

$$m = \frac{y_2 - y_1}{x_2 - x_1}$$

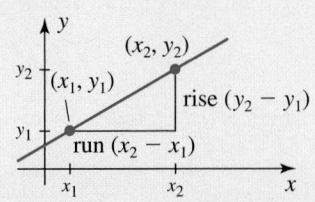

NOTE

The slope is a number that measures the "steepness" of a line. For positive numbers, the larger the number, the steeper the line.

Web It

For another explanation of finding slopes along with an interactive practice, go to link 3-3-1 at mhhe.com/bello.

When using this definition, it doesn't matter which point is taken for A and which for B. For example, the slope of the line passing through the points $A(0, -6)$ and $B(3, 3)$ is

$$m = \frac{3 - (-6)}{3 - 0} = \frac{9}{3} = 3$$

If we choose A to be $(3, 3)$ and B to be $(0, -6)$, the slope is the same.

$$m = \frac{-6 - 3}{0 - 3} = \frac{-9}{-3} = 3$$

Thus A and B can be interchanged without changing the value of the resulting slope.

On the left side (x_2, y_2) is the first point and (x_1, y_1) is the second point.   On the right side (x_1, y_1) is the first point and (x_2, y_2) is the second point.

$$\frac{y_2 - y_1}{x_2 - x_1} = \frac{y_1 - y_2}{x_1 - x_2}$$

EXAMPLE 1   Finding slopes given two points on the line

Find the slope of the line passing through the given points:

a. $A(-3, 1), B(-1, -2)$ **b.** $A(-1, 5), B(-1, 6)$

c. $A(1, 1), B(2, 3)$ **d.** $A(3, 4), B(1, 4)$

SOLUTION

a. The slope, as shown in Figure 24, is

$$m = \frac{-2 - 1}{-1 - (-3)} = \frac{-3}{2} \quad \begin{array}{l} \leftarrow \text{rise} \\ \leftarrow \text{run} \end{array}$$

$\frac{-3}{2}$ is in standard form, so if we start at A, we move 3 units *down* (the change in y) and 2 units *right* (the change in x) ending at B.

b. The line passing through $A(-1, 5)$ and $B(-1, 6)$ in Figure 25 has as its equation $x = -1$ and is a line *parallel* to the y-axis. The fact that the change in x is zero in lines parallel to the y-axis (that is, *vertical* lines)

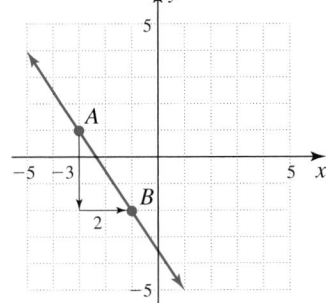

Figure 24

PROBLEM 1

Find the slope of the line passing through the given points:

a. $A(-2, 3), B(1, 5)$

b. $A(4, -5), B(4, 3)$

c. $A(2, 2), B(4, -4)$

d. $A(6, -1), B(7, -1)$

Answers

1. a. $\frac{2}{3}$ **b.** Undefined **c.** -3
d. 0

requires that they be excluded from the definition; their slope is *undefined*. If we tried to apply the definition we would get

$$m = \frac{6 - 5}{-1 - (-1)} = \frac{1}{0}$$

which is *undefined*.

> **NOTE**
>
> If the slope is a number that measures the "steepness" of a line, then the vertical line, being the steepest, should have a slope that names "the largest number." Is there a "largest number"? Of course not, so we can say there is "no slope" because there is not a "largest number." *Beware that "no slope" is not the same as "0" slope.*

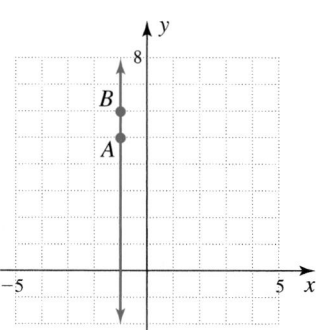

Figure 25

c. The slope, as shown in Figure 26, is

$$m = \frac{3 - 1}{2 - 1} = \frac{2}{1} = 2$$

The change in *y* (rise) is 2 units up, the change in *x* (run) is 1 unit right.

d. The slope, as shown in Figure 27, is

$$m = \frac{4 - 4}{1 - 3} = \frac{0}{-2}$$

Since

$$\frac{0}{-2} = 0$$

the slope is zero. The line is parallel to the *x*-axis and has equation $y = 4$, a horizontal line. Horizontal lines have zero slope.

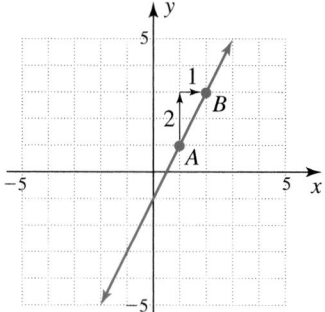

Figure 26

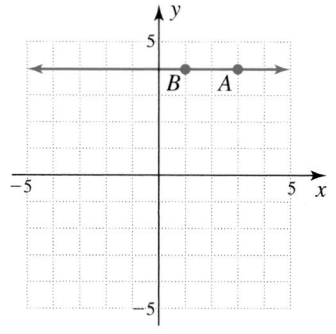

Figure 27

Table 1 summarizes the information in Example 1.

Table 1 Slope Summary

A line that *falls* from left to right has a *negative* slope.	The slope of a *vertical* line is *undefined*. Since $x_2 - x_1 = 0$,
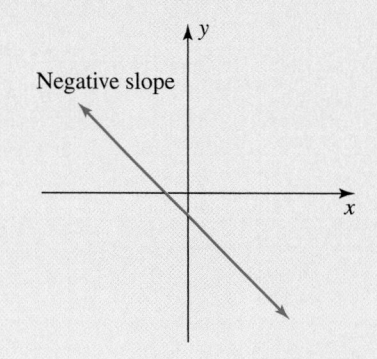	$$m = \frac{y_2 - y_1}{x_2 - x_1} = \frac{y_2 - y_1}{0}$$ so *m* is *undefined*. 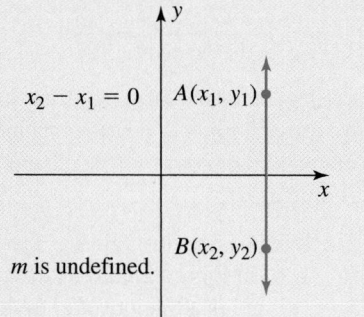

Table 1 (Continued)

A line that *rises* from left to right has a *positive* slope.

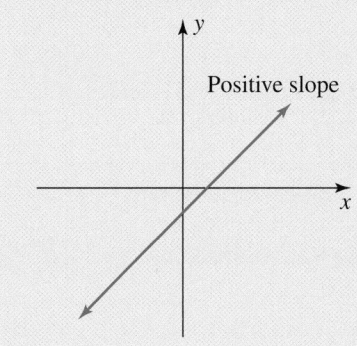

Positive slope

A *horizontal* line has zero slope. Since $y_2 - y_1 = 0$,

$$m = \frac{y_2 - y_1}{x_2 - x_1} = \frac{0}{x_2 - x_1}$$

$$m = 0$$

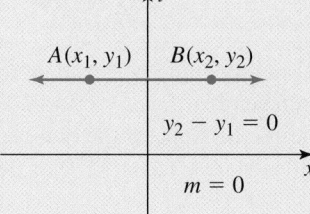

$A(x_1, y_1)$ $B(x_2, y_2)$

$y_2 - y_1 = 0$

$m = 0$

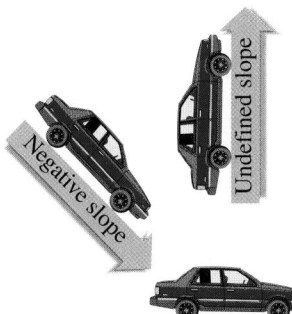

Negative slope

Undefined slope

Positive slope

Zero slope

EXAMPLE 2 **Application using the slope**

Fairfax County Public Safety Spending per Resident, 1975–2004 Adjusted for Inflation

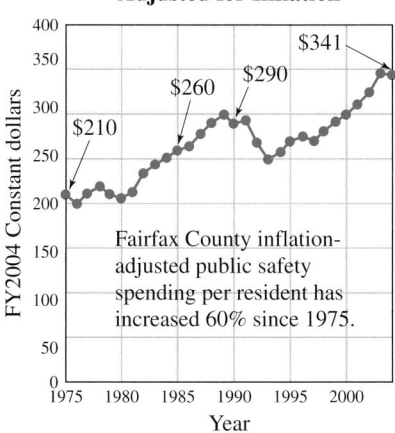

Fairfax County inflation-adjusted public safety spending per resident has increased 60% since 1975.

Fairfax County Taxpayers Alliance Analysis based on Fairfax County Adopted budgets, FY1977–2003, and FY04 Advertised budget 3/10/03

Source: Fairfax County Taxpayer Alliance.

Fairfax County Health and Welfare Spending per Resident, 1975–2004 Adjusted for Inflation

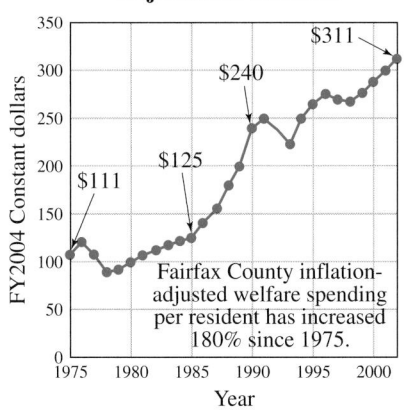

Fairfax County inflation-adjusted welfare spending per resident has increased 180% since 1975.

Fairfax County Taxpayers Alliance Analysis based on Fairfax County Adopted budgets, FY1977–2003, and FY04 Advertised budget 3/10/03

Each graph indicates Fairfax County spending per resident—one for public safety and the other for health and welfare. Which of the two had a higher rate of spending from 1998 to 2002?

PROBLEM 2

Using the graphs from Example 2, what was the rate of increase for each of the two from 1985 to 1990?

Answer

2. $6 per year increase for public safety spending and $23 per year for health and welfare spending.

SOLUTION It would be difficult to discover the answer just by looking at the graph as it is not obvious which line segment is steeper. However, now that we know how to find the slope of a line when given two points we can find the slope of each line segment and compare them. To find the slope of the line segment for public safety spending per resident, we will use $260 for 1998 and $325 for 2002.

$$m = \frac{y_2 - y_1}{x_2 - x_1} = \frac{325 - 260}{2002 - 1998} = \frac{65}{4} = \frac{16.25}{1} = 16.25$$

This means spending per resident for public safety has increased at the rate of $16.25 per year from 1998 to 2002.

To find the slope of the line segment for health and welfare spending per resident, we will use $270 for 1998 and $311 for 2002.

$$m = \frac{y_2 - y_1}{x_2 - x_1} = \frac{311 - 270}{2002 - 1998} = \frac{41}{4} = \frac{10.25}{1} = 10.25$$

This means spending per resident for health and welfare has increased at the rate of $10.25 per year from 1998 to 2002.

Thus, comparing the two we see that public safety spending per resident had a higher rate of increase from 1998 to 2002.

B Parallel and Perpendicular Lines

The definition of slope can be used to determine when two line segments are parallel. Two parallel lines have the same steepness (inclination) and thus the same slope, so we have the following.

Web It

For a good visual display of parallel and perpendicular lines, go to link 3-3-2 at mhhe.com/bello.

> **Slopes of Parallel Lines**
>
> Two nonvertical lines L_1 and L_2 with slopes m_1 and m_2 are parallel if and only if $m_1 = m_2$.

> **NOTE**
>
> If L_1 and L_2 are vertical lines, they are parallel.

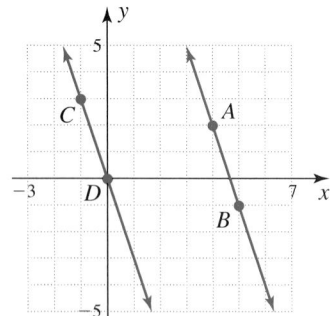

Figure 28

For example, consider the two lines, one passing through points $A(4, 2)$ and $B(5, -1)$ and the other through points $C(-1, 3)$ and $D(0, 0)$, as shown in Figure 28. The slope of AB is

$$m_1 = \frac{-1 - 2}{5 - 4} = \frac{-3}{1} = -3$$

and the slope of CD is

$$m_2 = \frac{0 - 3}{0 - (-1)} = \frac{-3}{1} = -3$$

Thus $m_1 = m_2$, so both lines have the same slope and are parallel.

Web It

For an interactive graph with practice finding parallel and perpendicular lines, go to link 3-3-3 at mhhe.com/bello.

It is shown in more advanced courses that two lines with slopes m_1 and m_2 are perpendicular (meet at a 90° angle) if

$$m_1 = \frac{-1}{m_2}$$

that is, if their slopes are *negative reciprocals*. This means that

$$\text{if} \quad m_1 = \frac{-1}{m_2}, \quad \text{then} \quad m_1 \cdot m_2 = \frac{-1}{m_2} \cdot m_2 = -1$$

Slopes of Perpendicular Lines	The lines L_1 and L_2 with slopes m_1 and m_2, respectively, are perpendicular if and only if the slopes are **negative reciprocals**; that is, $m_1 \cdot m_2 = -1$ $(m_1, m_2 \neq 0)$.

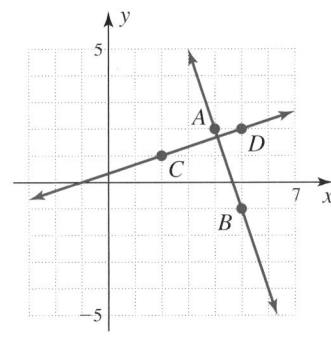

Figure 29

We can show that the line passing through $A(4, 2)$ and $B(5, -1)$ is perpendicular to the line passing through $C(2, 1)$ and $D(5, 2)$. Since the slope of AB is

$$m_1 = \frac{-1 - 2}{5 - 4} = \frac{-3}{1} = -3$$

and that of CD is

$$m_2 = \frac{2 - 1}{5 - 2} = \frac{1}{3}$$

we have $m_1 \cdot m_2 = -3 \cdot \frac{1}{3} = -1$. These perpendicular lines can be graphed as shown in Figure 29.

Teaching Tip

Another way to remember slopes of perpendicular lines:

1. They are opposite in sign
2. They are reciprocals

EXAMPLE 3	**Finding whether lines are parallel or perpendicular**

A line L_1 has slope $\frac{2}{3}$; find:

a. Whether another line passing through $A(5, 1)$ and $B(8, 3)$ is parallel or perpendicular to L_1

b. Whether the line passing through $C(-1, -4)$ and $D(-3, -1)$ is parallel or perpendicular to L_1

SOLUTION

a. The slope of AB is

$$m = \frac{3 - 1}{8 - 5} = \frac{2}{3}$$

The slope of L_1 is also $\frac{2}{3}$, so line AB is parallel to L_1.

b. The slope of CD is

$$m = \frac{-1 - (-4)}{-3 - (-1)} = \frac{3}{-2} = \frac{-3}{2}$$

The slope of L_1 is $\frac{2}{3}$ and

$$\frac{2}{3} \cdot \left(\frac{-3}{2} \right) = -1$$

so line CD is perpendicular to L_1.

PROBLEM 3

A line L_1 has slope $\frac{3}{4}$; find:

a. Whether another line passing through $A(2, 6)$ and $B(6, 9)$ is parallel or perpendicular to L_1

b. Whether the line passing through $C(-9, -1)$ and $D(-6, -5)$ is parallel or perpendicular to L_1

Answers

3. a. Parallel   **b.** Perpendicular

EXAMPLE 4	**Finding y when lines are perpendicular**	PROBLEM 4

If the line through $A(4, y)$ and $B(-2, -5)$ is perpendicular to a line whose slope is $\frac{-2}{3}$; find y.

SOLUTION   The slope of the line through points A and B is

$$\frac{y - (-5)}{4 - (-2)} = \frac{y + 5}{6}$$

If the line through A and B is perpendicular to a line whose slope is $\frac{-2}{3}$, the slope of the line through A and B must be $\frac{3}{2}$ (the negative reciprocal of $\frac{-2}{3}$). Thus

$$\frac{y + 5}{6} = \frac{3}{2}$$

$$\frac{6 \cdot (y + 5)}{6} = \frac{6 \cdot 3}{2} \qquad \text{Multiply both sides by 6.}$$

$$y + 5 = 9 \qquad \text{Simplify.}$$

$$y = 4 \qquad \text{Subtract 5 from both sides.}$$

The line through $A(4, 4)$ and $B(-2, -5)$ has slope

$$\frac{4 - (-5)}{4 - (-2)} = \frac{9}{6} = \frac{3}{2}$$

making it perpendicular to a line whose slope is $\frac{-2}{3}$, and our result is correct.

PROBLEM 4

The line through $A(5, y)$ and $B(-4, 8)$ is perpendicular to a line whose slope is $\frac{-3}{4}$. Find y.

Answer

4. 20

Calculate It   Comparing Slopes and y-intercepts

You can use your TI-83 Plus and the knowledge that parallel lines have the same slope to explore the graph of a line when the equation is solved for "y." Let's graph the following equations all on the same graph using the [Y=] for equations 1–3 and then compare the slopes and y-intercepts.

(Remember to turn off the Stat Plot by pressing [2nd]

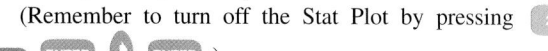 [Y=] [ENTER] ◊ [ENTER] .)

Begin by pressing [Y=] and then enter the right side of each equation on the appropriate line. Use the down arrow key ⬇ to move between lines.

$$Y_1 = 2x + 3$$
$$Y_2 = 2x$$
$$Y_3 = 2x - 3$$

Once the equations are entered, set the graph window by pressing [WINDOW] and entering $[-5, 5]$ and $[-5, 5]$. Press [GRAPH] and you should see the following screen.

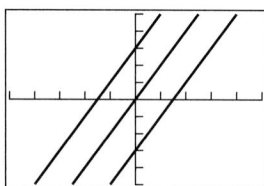

You can find the y-intercept by pressing [TRACE] , and notice that the blinking cursor is on the y-intercept of the first line and

at the bottom of the screen it says "$x=0$" and "$y=3$." Thus the first y-intercept is 3. Note that the constant in the first equation is 3. To locate the y-intercept of the 2nd line, press ⬇ and the blinking cursor should be on the 2nd y-intercept of "0." The constant in the 2nd equation is "0." Press ⬇ again and you will see the third y-intercept "-3," and the constant in the third equation is "-3." We can generalize that when the equation is solved for "y," the constant is the y-intercept.

To find the slope we can take two ordered pairs of values off the table on the calculator for these lines and use the slope formula,

$$m = \frac{y_2 - y_1}{x_2 - x_1}$$

To get to the table, we first set the table by pressing [2nd] [WINDOW]

[5] ⬇ [1] and then [2nd] [GRAPH] . Use the X and Y_1 columns to get $(5, 13)$ and $(6, 15)$ and then find the slope. Do the same with the X and Y_2 to get $(5, 10)$ and $(6, 12)$ and X and Y_3 to get $(5, 7)$ and $(6, 9)$. You will have to press ◊ three times to get to the Y_3 column. The slope obtained for all three should be 2, the coefficient of x in all three equations. We can generalize that when the equation is solved for "y," the coefficient of x is the slope of the line.

When the equation is written $y = mx + b$, "m" is the slope of the line and "b" is the y-intercept.

C   Graphing Lines Using the Slope and the *y*-intercept

Web It

For another visual demonstration of graphing a line using the slope and a point on the line, go to link 3-3-4 at mhhe.com/bello.

We can graph a line if we know its slope and a point on the line. For convenience we will use the *y*-intercept as the point but the procedure could be done using any point on the line. For example, suppose a line has a *y*-intercept of -2 and has a slope of $\frac{2}{3}$. The *y*-intercept of -2 means that one point on the line is $(0, -2)$. To graph the line, we recall that the slope of a line is the ratio

$$m = \frac{change\ in\ y}{change\ in\ x} = \frac{2}{3}$$

We start at the point $(0, -2)$ and go 2 units up (change in *y* is 2) and 3 units right (change in *x* is 3), ending at $(3, 0)$. We then draw a line through the points $(0, -2)$ and $(3, 0)$ to obtain the graph shown in blue in Figure 30. If the slope of the line had been $\frac{-2}{3}$, we would move 2 units down (change in *y*) and then go 3 units right (change in *x*), ending at $(3, -4)$. This line with the negative slope is shown in red.

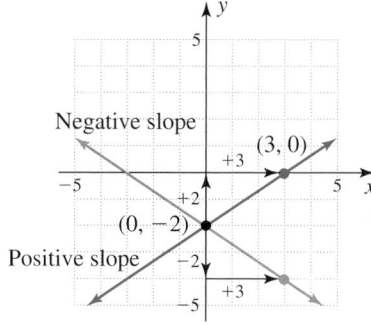

Figure 30

EXAMPLE 5   **Graphing lines using the slope and *y*-intercept**

Graph a line that has *y*-intercept 2 and has slope $-\frac{3}{4}$.

SOLUTION $-\frac{3}{4}$ can be written as $\frac{-3}{4}$.

We start at the *y*-intercept point $(0, 2)$ and go 3 units down (the change in *y*). We then go right 4 units (the change in *x*), ending at $(4, -1)$, as shown in Figure 31. The graph is obtained by drawing a line through the points $(0, 2)$ and $(4, -1)$.

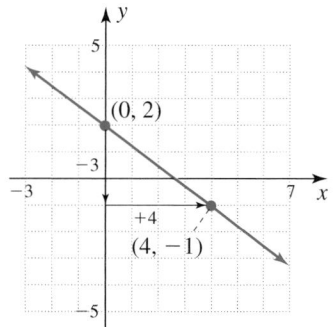

Figure 31

PROBLEM 5

Graph a line that has *y*-intercept -1 and has slope $-\frac{1}{2}$.

Teaching Tip

Emphasize in Example 5 that the slope is plotted from the given point and *not* from the origin.

D   Slope-Intercept Form for a Linear Equation

Can the coefficients and constant in a linear equation be related to the graph of the line? Let's graph the equation $-x + 2y = 4$. Using a table we get the following points to plot.

Answer

5.

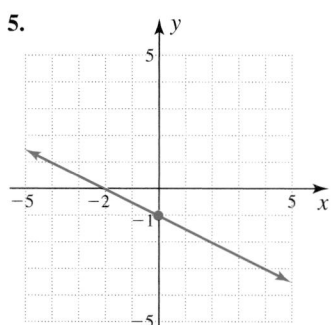

x	*y*
0	2
-4	0
-2	1

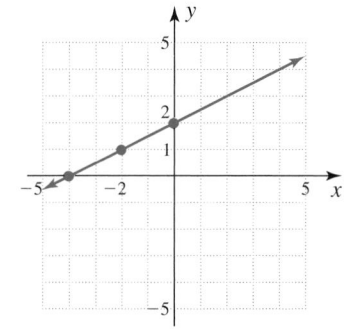

The y-intercept is 2 and the slope is $m = \frac{2-0}{0-(-4)} = \frac{2}{4} = \frac{1}{2}$.

The numbers in the equation $-x + 2y = 4$ don't reflect that information but solve the equation for "y" and see what happens.

$$-x + 2y = 4$$
$$2y = x + 4 \qquad \text{Add } x \text{ to both sides.}$$
$$y = \frac{1}{2}x + 2 \qquad \text{Divide both sides by 2.}$$

The coefficient of x, $\frac{1}{2}$, is the same value we obtained when we calculated the slope of the line. Also, the constant, 2, in the equation is the same as the y-intercept we observed on our graph. We can use this example to generalize that when the equation is solved for y, then the equation, $y = mx + b$, has m as its **slope** and b as its **y-intercept.**

The Slope-Intercept Form of a Line	$y = mx + b$
	slope ⟶↑ ↑⟶ y-intercept

The slope-intercept form of a line is also the form we used to define a linear function in 3.2 as $f(x) = mx + b$.

Calculate It   Finding the Slope Given 2 Points

Use your TI-83 Plus to find the slope of a line. To do this, you need two points. If you have an equation written in the form $y = ax + b$, a is the slope of the line. (The calculator uses a instead of m.) Let's do Example 1(a). The two given points are $A(-3, 1)$ and $B(-1, -2)$. We will make two lists on the calculator, but we must first be sure the lists are clear. Press 2nd + 4 ENTER to ClrAllLists. Press STAT 1 , and under L_1 press (−) 3 ENTER and (−) 1 ENTER . Press the right arrow to get to L_2 and then press 1 ENTER and (−) 2 ENTER . The strokes needed to give you the equation of a line of the form $y = ax + b$ using a technique called "linear regression" are STAT ◇ 4 ENTER . We see that the slope $a = -1.5 = \frac{-3}{2}$, as shown in the screen. Try Example 1(b) and see what happens.

```
LinReg
y=ax+b
a=-1.5
b=-3.5
```

EXAMPLE 6 Finding the slope and y-intercept given the equation of a line	**PROBLEM 6**

Given the following linear equations find the slope and y-intercept of each.

a. $y = 5x - 3$ **b.** $-y = \frac{1}{4}x + 2$ **c.** $-3x + 2y = 7$

SOLUTION

a. The equation is solved for y so the coefficient of x and the constant are the slope and y-intercept, respectively.

$$y = 5x - 3$$
$$\text{slope} = 5 \text{⟶↑}   \text{↑⟶ } y\text{-intercept} = -3$$

b. We start by solving the equation for y.

$$-y = \frac{1}{4}x + 2$$
$$y = \frac{-1}{4}x - 2 \qquad \text{Divide both sides by } -1.$$

slope is $\frac{-1}{4}$ and y-intercept is -2

PROBLEM 6

Given the following linear equations, find the slope and y-intercept of each:

a. $y = -4x + 1$

b. $-y = \frac{1}{2}x - 3$

c. $2x - 3y = 5$

Answers

6. a. Slope -4; y-intercept 1
b. Slope $\frac{-1}{2}$; y-intercept 3
c. Slope $\frac{2}{3}$; y-intercept $-\frac{5}{3}$

c. We solve the equation for y.

$$-3x + 2y = 7$$
$$2y = 3x + 7 \qquad \text{Add } 3x \text{ to both sides.}$$
$$y = \frac{3}{2}x + \frac{7}{2} \qquad \text{Divide both sides by 2.}$$

slope is $\frac{3}{2}$ and y-intercept is $\frac{7}{2}$.

Exercises 3.3

A In Problems 1–7, find the slope of the line passing through the points and graph the line.

1. $A(2, 4), B(-1, 0)$ $m = \frac{4}{3}$

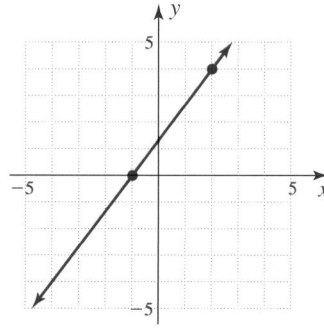

2. $A(3, -2), B(8, 4)$ $m = \frac{6}{5}$

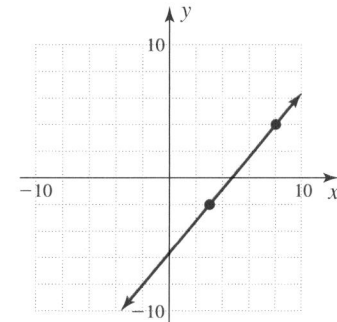

3. $C(-4, -5), D(-1, 3)$ $m = \frac{8}{3}$

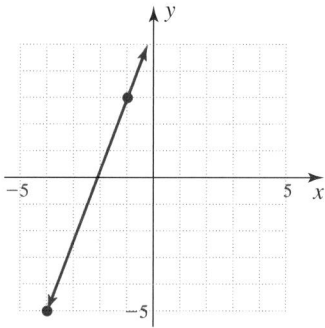

4. $C(5, 7), D(-2, 3)$ $m = \frac{4}{7}$

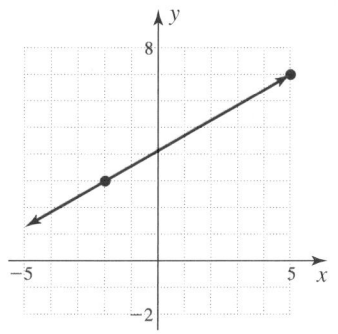

5. $E(4, 8), G(1, -1)$ $m = 3$

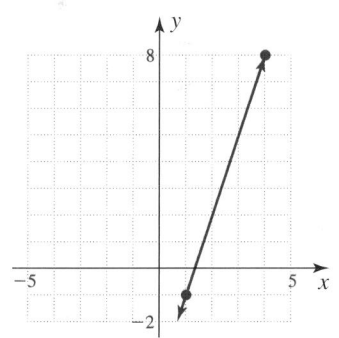

6. $H(-2, -2), I(-2, 4)$ $m = $ undefined

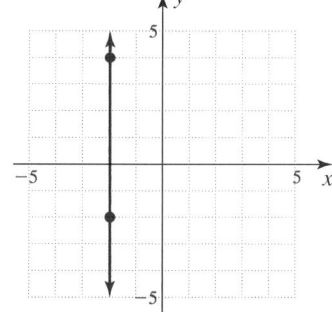

7. $A(3, -1), B(-2, -1)$ $m = 0$

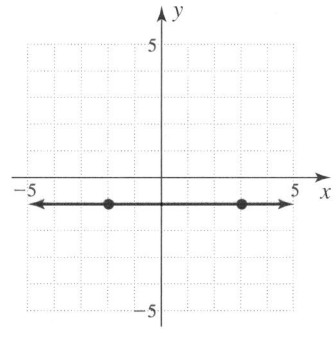

In Problems 8–10, use the graph.

The graph indicates the salary trends for the Fairfax County government and school system. Both the schools average salaries and the county average salaries show a steady increase from 1998 to 2004.

8. a. Find the slope of the line segment for the county average salary from 1998 to 2004. Use $50,000 for the 1998 salary and $58,000 for 2004. $m = \frac{4000}{3}$

b. Using the slope, describe the rate of change of the salary increase per year for the county average salary (to the nearest dollar). *The salary increased approximately $1333 per year.*

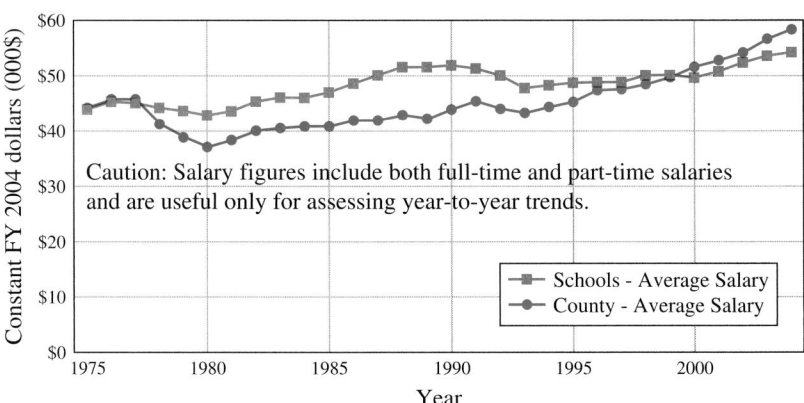

Fairfax County Government and School System Salary Trends, FY 1975–2004 (Inflation Adjusted)

Caution: Salary figures include both full-time and part-time salaries and are useful only for assessing year-to-year trends.

Fairfax County Taxpayers Alliance analysis based on Fairfax County and Fairfax County Public Schools budgets. FY 1975–FY 2004 and data from the Fairfax County Office of Management and Budget. 4/6/03

Source: Fairfax County Taxpayers Alliance.

9. a. Find the slope of the line segment for the schools average salary from 1998 to 2004. Use $50,000 for the 1998 salary and $54,000 for 2004. $m = \frac{2000}{3}$

b. Using the slope, describe the rate of change of the salary increase per year for the schools average salary (to the nearest dollar). *The salary increased approximately $667 per year.*

10. Comparing the results from Problems 8 and 9, which average salary is increasing at a higher rate? Explain how the graph confirms the answer. *The graph indicates the county average salary is a steeper line segment from 1998 to 2004 than the schools average salary. The comparisons of the salary increases per year (slopes) reflect the same result.*

B In Problems 11–20, determine whether lines *AB* and *CD* are parallel, perpendicular, or neither.

11. $A(1, 6)$, $B(-1, 4)$ and $C(1, -2)$, $D(2, -1)$ Parallel

12. $A(0, 4)$, $B(1, -1)$ and $C(0, 1)$, $D(-5, -1)$ Neither

13. $A(2, 0)$, $B(4, 5)$ and $C(8, 0)$, $D(3, 2)$ Perpendicular

14. $A(1, 1)$, $B(-1, 2)$ and $C(1, -1)$, $D(0, -3)$ Perpendicular

15. $A(-1, 1)$, $B(1, 2)$ and $C(1, -1)$, $D(0, -1)$ Neither

16. $A(1, 1)$, $B(3, 3)$ and $C(1, -1)$, $D(0, 2)$ Neither

17. $A(1, -1)$, $B\left(2, \frac{-1}{2}\right)$ and $C(2, -2)$, $D(1, 0)$ Perpendicular

18. $A(1, 1)$, $B\left(\frac{1}{5}, 0\right)$ and $C(1, 1)$, $D\left(0, \frac{9}{5}\right)$ Perpendicular

19. $A(0, 1)$, $B(14, -1)$ and $C(0, 2)$, $D(7, 1)$ Parallel

20. $A(2, -2)$, $B(1, -7)$ and $C(1, -3)$, $D(0, -8)$ Parallel

In Problems 21–30, find *x* or *y*.

21. The line through $A(x, 4)$ and $B(6, 8)$ is parallel to a line whose slope is 1. $x = 2$

22. The line through $A(x, 5)$ and $B(-2, 3)$ is parallel to a line whose slope is $\frac{2}{5}$. $x = 3$

23. The line through $A(x, 2)$ and $B(2, 6)$ is perpendicular to a line whose slope is $\frac{1}{2}$. $x = 4$

24. The line through $A(x, 4)$ and $B(-3, -4)$ is perpendicular to a line whose slope is $\frac{2}{5}$. $x = \frac{-31}{5}$

25. The line through $A(x, -6)$ and $B(-2, -1)$ is perpendicular to a line whose slope is $\frac{-2}{3}$. $x = \frac{-16}{3}$

26. The line through $A(2, y)$ and $B(3, 4)$ is parallel to a line whose slope is 2. $y = 2$

27. The line through $A(3, y)$ and $B(1, -2)$ is parallel to a line whose slope is -3. $y = -8$

28. The line through $A(2, y)$ and $B(1, -4)$ is perpendicular to a line whose slope is $\frac{1}{3}$. $y = -7$

29. The line through $A(x, 4)$ and $(3, 5)$ is perpendicular to the horizontal line $y = 5$. $x = 3$

30. The line through $A(-4, 2)$ and $(x, 7)$ is perpendicular to the horizontal line $y = -3$. $x = -4$

C   In Problems 31–40, graph the line with the indicated slope and passing through the given point.

31. Slope 2 through $(1, 1)$

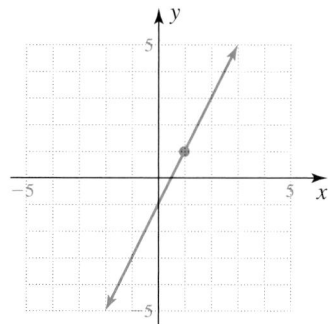

32. Slope $\frac{2}{3}$ through $(1, 2)$

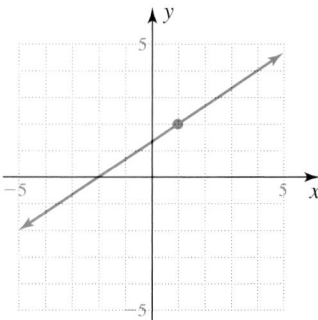

33. Slope $\frac{-2}{3}$ through $(1, 1)$

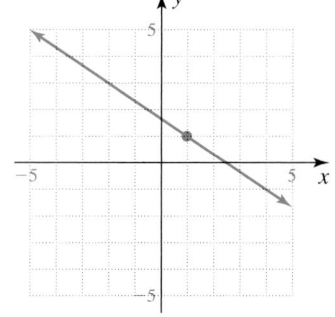

34. Slope $\frac{-3}{4}$ through $(-1, -1)$

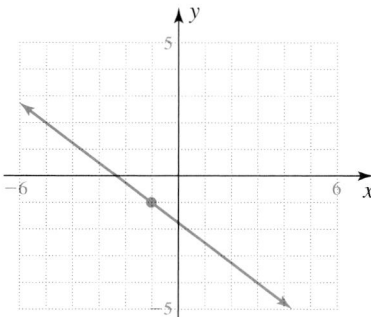

35. Slope 0 through $(2, 3)$

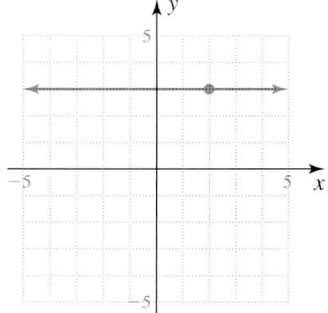

36. Slope 0 through $(3, 2)$

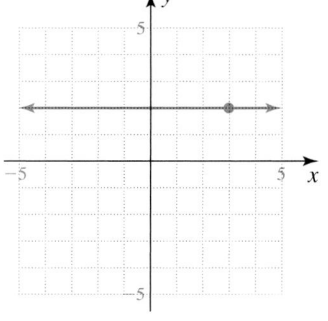

37. Slope 0 through $(0, 0)$

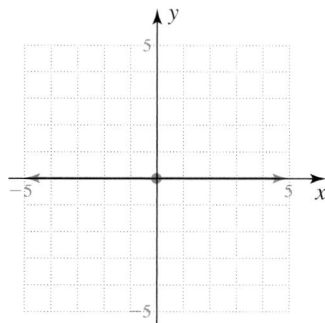

38. Slope undefined through $(-1, 2)$

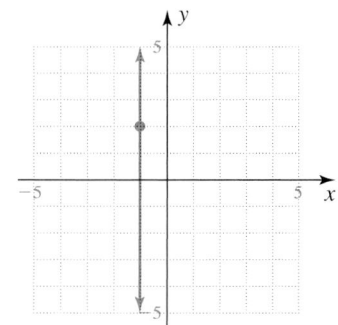

39. Slope undefined through $(2, -1)$

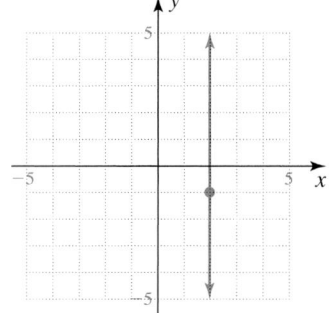

40. Slope undefined through $(0, 0)$

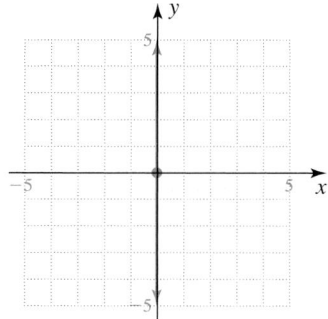

The midpoint of the line segment AB joining the points $A(x_1, y_1)$ and $B(x_2, y_2)$ is the point (x_m, y_m), where

$$x_m = \frac{x_1 + x_2}{2} \quad \text{and} \quad y_m = \frac{y_1 + y_2}{2}$$

In Problems 41–46, find the midpoint of the line segment AB.

41. $A(3, 4)$ and $B(7, 2)$ $(5, 3)$

42. $A(2, 5)$ and $B(-6, 3)$ $(-2, 4)$

43. $A(0, -8)$ and $B(-8, 0)$ $(-4, -4)$

44. $A(-3, -5)$ and $B(1, 3)$ $(-1, -1)$

45. $A(-5, -2)$ and $B(-6, 2)$ $\left(\frac{-11}{2}, 0\right)$

46. $A(0, -4)$ and $B(-5, -2)$ $\left(\frac{-5}{2}, -3\right)$

D In Problems 47–52, find the slope and y-intercept for the given linear equation.

47. $y = x - 4$ $m = 1$; y-int: -4

48. $y = -x + 2$ $m = -1$; y-int: 2

49. $-3y = x + 6$ $m = \frac{-1}{3}$; y-int: -2

50. $-2y = -x - 8$ $m = \frac{1}{2}$; y-int: 4

51. $4x + 2y = 8$ $m = -2$; y-int: 4

52. $15x + 3y = 9$ $m = -5$; y-int: 3

APPLICATIONS

Flying.

53. A pilot is flying from New Orleans, with coordinates (90, 30), to Philadelphia, with coordinates (76, 40). What are the coordinates of the point exactly halfway between the two cities? (83, 35)

54. A pilot is flying from Buenos Aires, with coordinates (58, 36), to Sao Paulo, with coordinates (46, 24). What are the coordinates of the point exactly halfway between the two cities? (52, 30)

In Problems 55–60, determine whether the three given points form the vertices (corners) of a right triangle. (*Hint:* A triangle is a right triangle if two of the sides are perpendicular. What do you know about the slopes of two perpendicular lines?)

55. $A(2, 2)$, $B(0, 5)$, $C(-20, 12)$ No

56. $A(0, 6)$, $B(-3, 0)$, $C(9, -6)$ Yes

57. $A(2, 2)$, $B(0, 5)$, $C(-19, -12)$ Yes

58. $A(0, 0)$, $B(6, 0)$, $C(3, 3)$ Yes

59. $A(2, 2)$, $B(-4, -14)$, $C(-20, -8)$ Yes

60. $A(3, 2)$, $B(0, -4)$, $C(12, -10)$ Yes

---

61. The U.S. population has been growing according to the equation

$$y = 2.2x + 180$$

where y is the approximate population (in millions) and x is the number of years after 1960. For example, in 1970, $x = 10$ and $y = 2.2 \cdot 10 + 180 = 202$. This means that the U.S. population was about 202 million in 1970. Use the equation $y = 2.2x + 180$ to find the approximate population in each of the following years:

a. 1990 246 million

b. 2000 268 million

c. 2010 290 million

d. In what year will the U.S. population reach approximately 312 million? year 2020

e. Graph the equation $y = 2.2x + 180$

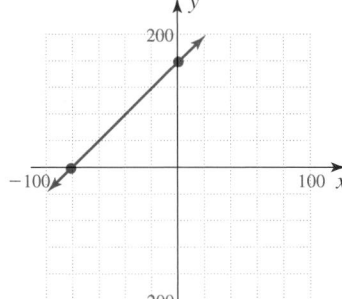

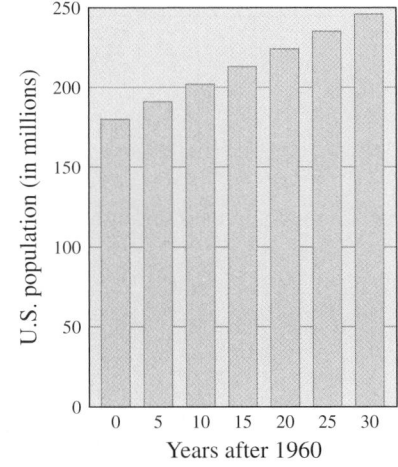

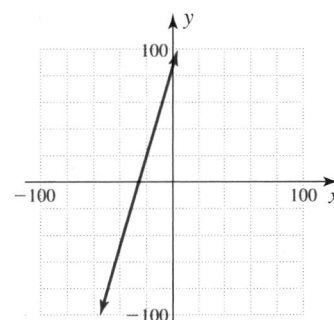

a. 1990 190 million

b. 2000 224 million

c. 2010 258 million

d. In what year will the amount of waste generated reach approximately 241 million tons? year 2005

e. Graph the equation $y = 3.4x + 88$.

62. The municipal waste y (in millions of tons) generated in the United States is approximately $y = 3.4x + 88$, where x is the number of years after 1960. Use the equation $y = 3.4x + 88$ to find the approximate amount of waste generated in each of the following years:

SKILL CHECKER

Try the "Skill Checker" exercises so you'll be ready for the next section.

Solve for y:

63. $6x + 3y = 12$ $y = -2x + 4$

64. $2x + 3y = 6$ $y = \frac{-2}{3}x + 2$

65. $3y - 2x = 12$ $y = \frac{2}{3}x + 4$

66. $5y - 2x = 10$ $y = \frac{2}{5}x + 2$

USING YOUR KNOWLEDGE

Distances and Slopes

The following problems require using your knowledge about a topic previously discussed.

67. A line has x-intercept 2 and y-intercept -3. What is the slope of the line? $\frac{3}{2}$

68. A line has x-intercept -3 and y-intercept 2. What is the slope of the line? $\frac{2}{3}$

69. Find the slope of a line parallel to the line through the points $(3, 1.3)$ and $(2, 3.6)$. -2.3

70. Find the slope of a line parallel to the line through the points $(5, 1.8)$ and $(6, \frac{-3}{4})$. -2.55

71. Find the slope of a line perpendicular to a second line passing through the points $(3, 0.\overline{6})$ and $(2, 0.\overline{3})$. -3

72. Find the slope of a line perpendicular to a second line passing through the points $(5, 2\pi)$ and $(4, -\pi)$. $\frac{-1}{3\pi}$

73. The line through the points $(1, c)$ and $(2, 3c)$ is parallel to another line whose slope is 2. What are the possible values for c? 1

74. The line through the points $(6, 2c)$ and $(3, \frac{1}{2}c)$ is perpendicular to another line whose slope is 2. What is the value of c? -1

WRITE ON

Explain in your own words:

75. Why the slope of a horizontal line is zero. Answers may vary.

76. Why the slope of a vertical line is undefined. Answers may vary.

77. Why two parallel lines have the same slope. Answers may vary.

What does it mean graphically:

78. If the slope of a line is positive?
The line rises from left to right.

79. If the slope of a line is negative?
The line falls from left to right.

MASTERY TEST

If you know how to do these problems, you have learned your lesson!

Graph:

80. The line through the point (2, 1) with slope $\frac{-1}{3}$.

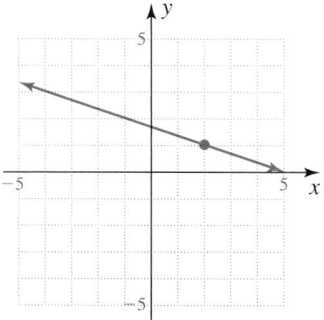

81. The line through the point $(-2, 1)$ with slope $\frac{1}{4}$.

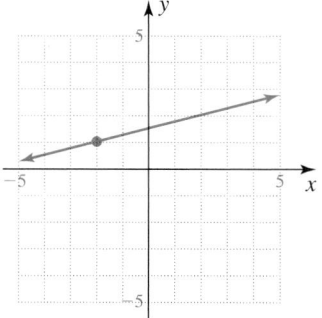

Find x or y:

82. The line through $A(5, y)$ and $B(-1, -4)$ is perpendicular to a line whose slope is $\frac{-3}{4}$. $y = 4$

83. The line through $A(x, 2)$ and $B(3, 4)$ is parallel to a line whose slope is 1. $x = 1$

84. Line L_1 has slope $\frac{3}{2}$. Find whether the line passing

    a. through $A(6, 3)$ and $B(3, 5)$ is parallel or perpendicular to L_1. Perpendicular

    b. through $A(-6, -8)$ and $B(-4, -5)$ is parallel or perpendicular to L_1. Parallel

85. Find the slope of the line passing through the points

    a. $A(-4, 2)$ and $B(-2, -3)$ $m = \frac{-5}{2}$

    b. $A(-2, 4)$ and $B(-2, 7)$ m undefined

    c. $A(2, 1)$ and $B(4, 5)$ $m = 2$

    d. $A(2, 3)$ and $B(1, 3)$ $m = 0$

3.4   EQUATIONS OF LINES

To Succeed, Review How To . . .

1. Graph lines when two points are given (pp. 165–166).

2. Write a linear equation in standard form (p. 167).

3. Solve an equation for a specified variable (p. 88).

Objectives

Find the equation and the graph of a line given

A Two points.

B One point and the slope.

C The slope and the y-intercept.

D One point and the fact that the line is parallel or perpendicular to a given line.

E The slope is that of a horizontal or vertical line.

F Mathematical modeling application.

G The graph of the line.

GETTING STARTED   How's the Profit?

Suppose a company wants to examine its profit over six years beginning in 2000. They might start with a table indicating those numbers.

Year	2000	2002	2004
Profit in Millions	$4.2	$5.6	$7.0

After comparing the profits, you see that for each two-year period, profit has increased by 1.4 million dollars. So the annual rate of change is 0.7 million dollars, which can be found using the ratio

$$\frac{\text{change in profit}}{\text{change in time}} = \frac{5.6 - 4.2}{2} = 0.7$$

Here's another way to create the table.

Suppose we let $t = 0$ in 2000. We are now creating a new time line. It is *relative* to the start of our information. Our table looks like this:

Time since 2000	0	2	4
Profit in Millions	$4.2	$5.6	$7.0

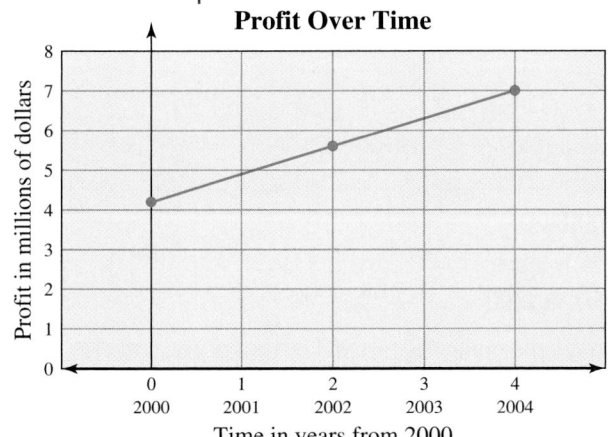

Profit Over Time

Now we can describe the data set with the ordered pairs (0, 4.2), (2, 5.6), and (4, 7.0). Let's use the x-y coordinate system to graph this data where the x-axis will represent the time in years since 2000 and the y-axis will represent the profit in millions of dollars. Here the annual rate is the *slope* of the linear relationship between profit and time. In the last section we learned that when an equation is written as $y = mx + b$, the m is the slope of the line and the b is the y-intercept. The y-intercept is (0, 4.2) and the slope is the annual rate of change, 0.7, so we can use the *slope-intercept* form, $y = mx + b$, to write the equation of this profit line. Thus the equation of the profit line is $y = 0.7x + 4.2$ where y is the profit in millions of dollars and x is the time in years since 2000.

Assuming the profit continues at the same rate, we can use this equation to predict our future profits. For example, let's calculate the year 2010 profit. The year 2010 is $x = 10$. This gives us $y = 0.7(10) + 4.2 = 11.2$ million dollars. There are no guarantees in the business world, but having information like this profit equation could help a business when planning for their future. As we study more about writing equations of lines in this section, remember it could mean a profit for your future.

A   Finding Equations Given Two Points

We are now ready to work with the graph of a line to find its equation. In Section 3.1, we defined standard form of an equation as $Ax + By = C$ where A, B, and C are integers and A is positive. We worked from an equation of a line to draw its graph. In general, if a line goes through two points $P_1(x_1, y_1)$ and $P_2(x_2, y_2)$, as shown in Figure 32, an equation for this line can be found as follows:

Teaching Tip

Equations of lines with slope zero or that are undefined should not be done with these formulas. Instead, they are done as follows:

1. For $m = 0$, the equation is $y = c$.
2. For m undefined, the equation is $x = c$.

1. Select a general point $P(x, y)$ on the line.

2. The slope of the line P_1P_2 is

$$m = \frac{y_2 - y_1}{x_2 - x_1} \quad (x_2 \neq x_1)$$

3. The slope of the line P_1P is

$$m = \frac{y - y_1}{x - x_1} \quad (x \neq x_1)$$

4. Since the slopes are equal,

$$\frac{y - y_1}{x - x_1} = \frac{y_2 - y_1}{x_2 - x_1}$$

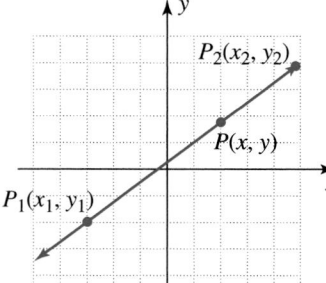

Figure 32

Teaching Tip

Remind students that standard form is $Ax + By = C$ and that they must first use the point-slope form of the line to get the equation before rearranging into standard form.

or $y - y_1 = \dfrac{y_2 - y_1}{x_2 - x_1} \cdot (x - x_1)$ Multiply both sides by $(x - x_1)$.

NOTE

Since

$$\frac{y_2 - y_1}{x_2 - x_1} = m$$

it's easier to remember this equation as $y - y_1 = m(x - x_1)$.

Teaching Tip

If the slope is a fraction, using the format

$$m = \frac{y - y_1}{x - x_1}$$

might be helpful. For example, if $m = \frac{2}{3}$, and the point is (3, 5) then

$$\frac{2}{3} = \frac{y - 5}{x - 3}$$

which can be readily simplified with cross multiplication

$$2(x - 3) = 3(y - 5)$$

and then written in standard form.

We summarize this discussion as follows.

THE POINT-SLOPE FORM OF A LINE

An equation of a line going through the points (x_1, y_1) and (x_2, y_2) is given by

$$y - y_1 = m(x - x_1) \tag{1}$$

where $m = \dfrac{y_2 - y_1}{x_2 - x_1}$ and $(x_2 \neq x_1)$

Calculate It   Find the Equation Given Two Points

As you recall from Section 3.3 we can use the TI-83 Plus to find the equation of a line given two points. Before we begin Example 1, let's clear any old lists by pressing [2nd] [+] [4] [ENTER]. Also, let's set the window to a standard window by pressing [ZOOM] [6]. Now, press [STAT] [1] and enter 5 and 6 under L1 and enter 2 and 4 under L2. Press [STAT] [▷] [4] [ENTER] to get the equation

$$y = ax + b, \text{ with } a = 2 \text{ and } b = -8, \text{ that is, the line}$$
$$y = 2x - 8, \text{ equivalent to } 2x - y = 8.$$

Now graph the two points we entered in the table by using [2nd] [Y=] [ENTER]. Remember to turn on the stat plot by highlighting [ON] and pressing [ENTER]; then choose the type by highlighting the scattergram and pressing [ENTER]. Pressing [GRAPH] will show the two points.

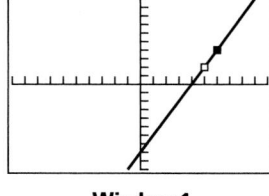

Window 1

To verify that the line, $y = 2x - 8$ does contain these two points press [Y=] and enter $2x - 8$ on the first line and then press [GRAPH]. Does the line pass through the two points? (See Window 1.)

EXAMPLE 1   Finding an equation given two points

Find, write in standard form, and graph an equation of the line going through the points $(5, 2)$ and $(6, 4)$.

SOLUTION   To use the point-slope form of the line we must first find the slope m. Letting $(x_1, y_1) = (5, 2)$ and $(x_2, y_2) = (6, 4)$ we substitute into the slope formula.

$$m = \frac{4 - 2}{6 - 5} = \frac{2}{1} = 2$$

Now we use the point $(5, 2)$ and $m = 2$ to substitute into the point-slope form of the line.

$y - 2 = 2(x - 5)$	Point-slope form.
$y - 2 = 2x - 10$	Distributive property.
$8 = 2x - y$	Subtract y and add 10.

In standard form, the equation of this line is $2x - y = 8$. The graph is shown in Figure 33.

Figure 33

PROBLEM 1

Find, write in standard form, and graph an equation of the line going through the points $(3, 1)$ and $(4, 3)$.

Teaching Tip

For Example 1, to graph $2x - y = 8$ use the x- and y-intercept method. Plot $(0, -8)$ and $(4, 0)$; then connect with a straight line.

B   Finding Equations Given a Point and the Slope

The point-slope form of a line enables us to find an equation of a line when a point $P(x_1, y_1)$ and the slope m are given. We use this equation in Example 2.

> **Using the Point-Slope Form of the Line**
>
> $$y - y_1 = m(x - x_1) \tag{1}$$

EXAMPLE 2   Finding an equation given a point and the slope

Find, write in standard form, and graph an equation of the line with slope $m = -2$ passing through the point $(3, 5)$.

SOLUTION   Here $m = -2$, $(x_1, y_1) = (3, 5)$. Substituting in equation (1), we get

$y - 5 = -2(x - 3)$	
$y - 5 = -2x + 6$	Distributive property.
$2x + y = 11$	Add $2x$ and 5 to both sides.

To graph the equation, we start at $(3, 5)$. Since

$$m = -2 = \frac{-2}{1}$$

go 2 units down (the change in y) and 1 unit right (the change in x), ending at $(4, 3)$. The graph is the line through the points $(3, 5)$ and $(4, 3)$, as shown in Figure 34.

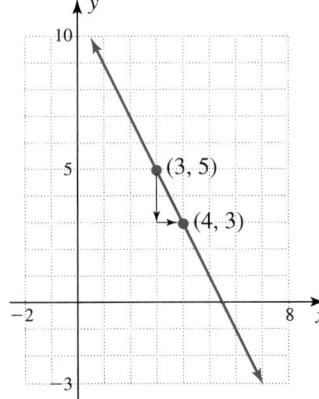

Figure 34

PROBLEM 2

Find, write in standard form, and graph an equation of the line with slope $m = -3$ passing through the point $(1, 2)$.

(Answer on page 211.)

Answer

1. $2x - y = 5$

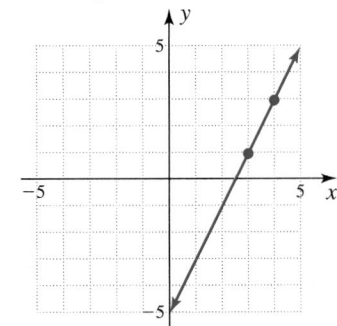

C   Finding Equations Given the Slope and the y-Intercept

Web It

For practice examples on how to write the equation of a line in standard form given certain information about the line, go to link 3-4-1 at mhhe.com/bello.

If, in $y - y_1 = m(x - x_1)$, the point $P(x_1, y_1)$ is on the y-axis, then $x_1 = 0$. If we let $y_1 = b$, then $P(x_1, y_1) = P(0, b)$. The point-slope form of the equation is $y - b = m(x - 0) = mx$. Solving for y by adding b to both sides, we have the following.

The Slope-Intercept Form of a Line

$$y = mx + b \qquad \qquad (2)$$

slope ⬏   ⬑ y-intercept

EXAMPLE 3   **Finding an equation given the slope and the y-intercept**

Find the slope-intercept form and graph an equation of the line with slope 5 and y-intercept 3.

SOLUTION   Using equation (2) with $m = 5$ and $b = 3$, we find the required equation to be $y = 5x + 3$. To graph this line, we start at the y-intercept, (0, 3). The slope is $5 = \frac{5}{1}$, so we go 5 units up and 1 unit right, ending at (1, 8). The graph is the line drawn through the points (0, 3) and (1, 8), as shown in Figure 35.

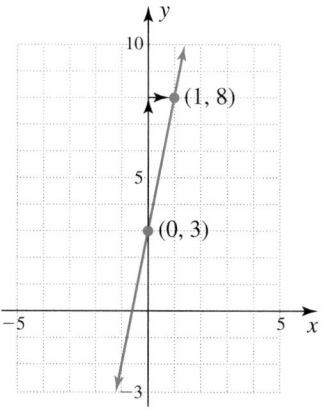

Figure 35

PROBLEM 3

Find the slope-intercept form and graph an equation of the line with slope 3 and y-intercept 2.

(Answer on page 211.)

Teaching Tip

Use the equation and graph from Example 1 to verify $y = mx + b$. By finding the intercepts $(0, -8)$ and $(4, 0)$ to graph, a student can use them to find the slope.

$$m = \frac{-8 - 0}{0 - 4} = \frac{-8}{-4} = 2$$

By taking the equation $2x - y = 8$ and solving for y we get $y = 2x - 8$. It is not a coincidence that 2, the coefficient of x, is the slope of the line and the constant (-8) is the y-intercept.

D   Finding the Equation of a Line Through a Given Point and Parallel or Perpendicular to a Given Line

We are already familiar with particular aspects of the slopes of parallel lines and perpendicular lines. For example, we can use the fact that two parallel lines have identical slopes to find an equation of a line parallel to a given line. Likewise, the fact that two perpendicular lines have slopes that are negative reciprocals of each other can be used to find an equation of a line perpendicular to a given line. We illustrate these ideas next.

EXAMPLE 4   **Finding an equation of a line through a given point and parallel or perpendicular to a given line**

Find an equation of the line passing through the point (6, 1) and

a. Parallel to the line $y - 3x = 1$. Write the final equation in slope-intercept form.

b. Perpendicular to the line $y - 3x = 1$. Write the final equation in slope-intercept form.

PROBLEM 4

Find an equation of the line passing through the point (4, 3) and

a. Parallel to the line $y + 2x = 7$. Write the final equation in slope-intercept form.

b. Perpendicular to the line $y + 2x = 7$. Write the final equation in slope-intercept form.

(Answer on page 211.)

SOLUTION

a. We first write the equation $y - 3x = 1$ in the slope-intercept form:
$y = 3x + 1$. The slope of this line is 3. If we wish to construct another
line parallel to $y = 3x + 1$ passing through $(6, 1)$, we use the
point-slope form, with $(x_1, y_1) = (6, 1)$ and $m = 3$, the same slope as
that of $y = 3x + 1$. Thus

$$y - 1 = 3(x - 6)$$
$$y - 1 = 3x - 18$$
$$y = 3x - 17$$

b. The line $y - 3x = 1$ has
slope 3. A line perpendicular
to this line must have a slope
$\frac{-1}{3}$. Using the point-slope
form again, we find that an
equation of the line perpendicular to $y - 3x = 1$ and
passing through $(6, 1)$ is

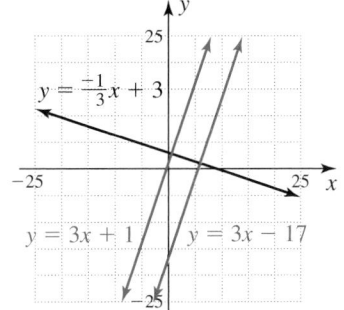

Figure 36

$$y - 1 = \frac{-1}{3}(x - 6)$$
$$y - 1 = \frac{-1}{3}x + 2$$
$$y = \frac{-1}{3}x + 3$$

The graphs of all three lines are shown in Figure 36.

Answers

2. $3x + y = 5$

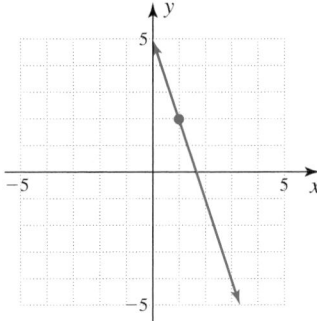

3. $y = 3x + 2$

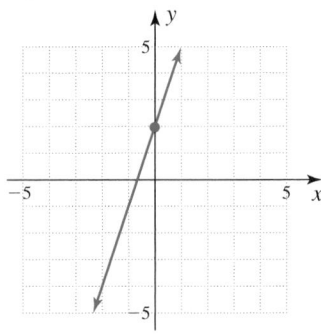

4. a. $y = -2x + 11$
b. $y = \frac{1}{2}x + 1$

E   Finding the Equations of Horizontal and Vertical Lines

Web It

For a visual example of how
to write the equation of a horizontal or vertical line, see
Examples 4 and 5 at link
3-4-2 at mhhe.com/bello.

In Section 3.3 we studied lines that have slope = 0, which are horizontal lines with equation $y = C$ (where C is some constant). Lines whose slope is undefined are vertical lines
and have equation $x = C$ (where C is some constant). The next example will use this information to illustrate how to write the equations of horizontal and vertical lines when
given a point on the line.

EXAMPLE 5   **Finding the equations of horizontal and
vertical lines**

Given that a line passes through the point $(-4, 2)$, find and graph the
equation of the line if:

a. the slope is 0.   **b.** the slope is undefined.

SOLUTION

a. Given the slope = 0, we know that it must be a horizontal line with
equation $y = C$. To find "C," the line passes through the point $(-4, 2)$,
which indicates we use $C = 2$ because that is the y-value of the point on the
line. Thus, the equation of the line is $y = 2$. The graph is the blue horizontal
line shown in Figure 37.

PROBLEM 5

Given that a line passes through the
point $(1, -3)$, find and graph the
equation of the line if:

a. the slope is 0.

b. the slope is undefined.

(Answers on page 212.)

b. Given the slope is undefined we know that it must be a vertical line with equation $x = C$. To find "C," the line passes through the point $(-4, 2)$, which indicates we use $C = -4$ because that is the x-value of the point on the line. Thus, the equation of the line is $x = -4$. The graph is the red vertical line shown in Figure 37.

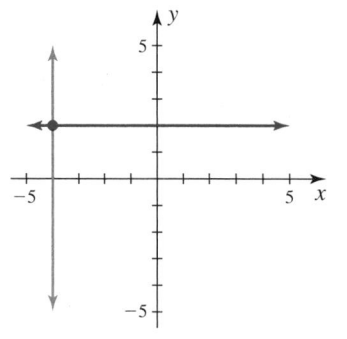

Figure 37

F | Mathematical Modeling Using Linear Equations

Mathematical modeling is the process of taking a verbal description of a problem, assigning labels to the unknown quantities, and forming a mathematical model or algebraic equation. We will use this idea of mathematical modeling to find the answer to Example 6.

| **EXAMPLE 6** | **How long can you expect to live?** |

According to the U.S. National Center for Health Statistics, life expectancy has been steadily increasing since 1950. Figure 38 indicates the life expectancy from birth of a person born in each of the years listed.

How can we use this information to find the approximate life expectancy for the year 2010?

Figure 38

Year	Life Expectancy
1950	68.2
1960	69.7
1970	70.8
1980	73.7
1990	75.4
2000	76.9
2010	**?**

SOLUTION If we could develop an algebraic equation using this information that would model the life expectancy based on the year the person was born after 1950, then the solution would be easy to obtain.

| **PROBLEM 6** |

Using the table in Example 6, find the equation of the line using $t = 0$, $y = 68.2$ and $t = 50$, $y = 76.9$.

How does that equation compare to the result in Example 6?

Answers

5. **a.** $y = -3$ **b.** $x = 1$

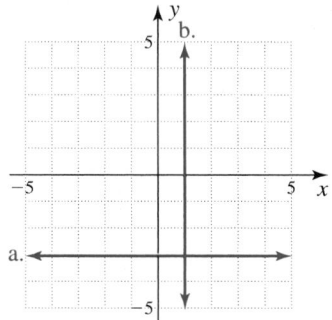

6. $y = 0.174t + 68.2$
This equation is similar to the solution of Example 6.

Calculate It | Write the Equation of a Line

As you recall from Section 3.3 we can use the TI-83 Plus to find the equation of a line given two points. Before we begin Example 6, let's clear any old lists by pressing 〔2nd〕 〔+〕 〔4〕

〔ENTER〕. Now, press 〔STAT〕 〔1〕 and enter 0, 10, 20, 30, 40, and 50 under L1 and enter the y-values under L2. Press 〔STAT〕 〔▶〕 〔4〕

〔ENTER〕 to get the equation $y = ax + b$, with $a \approx 0.18$ and $b \approx 67.9$, that is, the line with equation
$$y = 0.18x + 67.9$$
As you can see we came close to this equation when we approximated the line of best fit in Example 6, where we got
$$y = 0.19t + 67.8$$

Teaching Tip

Have students discuss why the resulting equations in Example 6 and Problem 6 are similar.

We will try to learn more about the information in the table by displaying it on a graph. The time column is in 10-year periods, so we will rename the numbers 0, 10, 20, 30, 40, and 50 as our t values, representing the 10-year periods after 1950. For 1950, $t = 0$, for 1960, $t = 10$, and so on. The second column, "Life Expectancy," will be labeled y.

The new table and its graph.

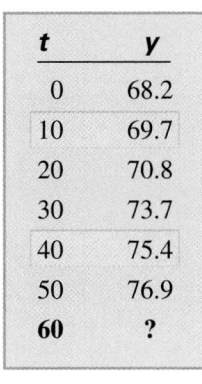

t	y
0	68.2
10	69.7
20	70.8
30	73.7
40	75.4
50	76.9
60	**?**

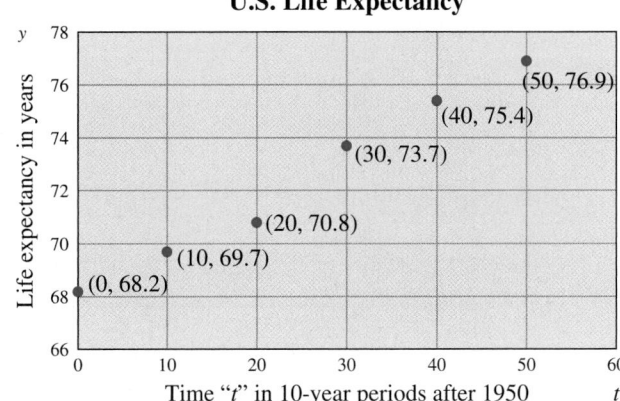

The points on the graph closely resemble a line. When this occurs we say this data represents a linear model. In this section we learned how to write the equation of a line given two points on the line. We can approximate the best line that will fit the data by connecting the two points that will have most of the points on the line or close to it. To be more precise we could use a graphing utility to find the line of best fit, or line of **least squares**, by finding the **linear regression equation, $y = ax + b$.** (See *Calculate It* on the previous page.)

The two points we have chosen to write the equation of the line are highlighted in the table. Find the slope first.

$$m = \frac{75.4 - 69.7}{40 - 10} = \frac{5.7}{30} = 0.19$$

Using the point-slope formula,

$$(y - y_1) = m(x - x_1)$$

with $m = 0.19$ and the point $(10, 69.7)$ we get

$$y - 69.7 = 0.19(t - 10)$$
$$y - 69.7 = 0.19t - 1.9$$
$$y = 0.19t + 67.8$$

The equation that will approximate the life expectancy for the year 2010 is $y = 0.19t + 67.8$. Since $t = 60$ for the year 2010, we substitute for t in the equation and $y = 0.19(60) + 67.8 = 79.2$. A person born in 2010 can expect to live about 79.2 years.

Finding the Equation of a Line from a Graph

In the *Getting Started* section we demonstrated how being able to write the equation of a line for a given data set could be helpful in predicting future values. Now that we have studied in detail how to write the equation of a line given certain information, we will apply it in the next example.

EXAMPLE 7   **Writing the equation of a line in an application**

The following line graph describes the number C of calories burned when running t minutes.

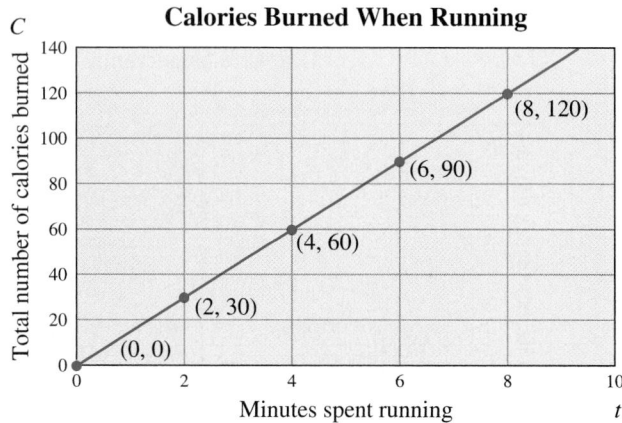

Calories Burned When Running

a. Write the equation of the line where C is the calories burned and t is time in minutes.

b. What is the rate of change of the number of calories burned?

c. How many calories would the runner burn if he ran 1 hour?

d. How long would it take the runner to burn 630 calories?

SOLUTION

a. To write the equation of the line, use two points, $(0, 0)$ and $(2, 30)$, and find the slope $m = \frac{30 - 0}{2 - 0} = \frac{30}{2} = 15$. The graph indicates the y-intercept is 0. So we use the slope and y-intercept form of the equation, $y = mx + b$, to write $C = 15t + 0$. Thus the equation of the line is $C = 15t$.

b. The rate of change can be found by comparing the change in calories to the change in minutes. We use the time interval from 0 to 2 minutes and get the ratio

$$\frac{\text{Change in calories}}{\text{Change in minutes}} = \frac{30 - 0}{2 - 0} = \frac{30}{2} = \frac{15}{1}$$

The rate of change is $\frac{15}{1}$ or we say the runner will burn 15 calories per minute. (The rate of change is the same as the slope of the line obtained in part **a**.)

c. The equation describing this linear relationship is written for t *minutes*, so we rewrite 1 hour as 60 minutes. Using the equation $C = 15t$, we replace t with 60.

$$C = 15(60) = 900$$

The runner will burn 900 calories if he runs for one hour.

d. The equation describing this linear relationship is $C = 15t$, so we replace C with 630 and solve for t.

$$630 = 15t \quad \text{Divide both sides by 15.}$$

$$42 = t$$

It will take the runner 42 minutes to burn 630 calories.

PROBLEM 7

The following line graph describes the possible linear relationship a salesman has between his monthly gross pay and his sales for his first 2 months.

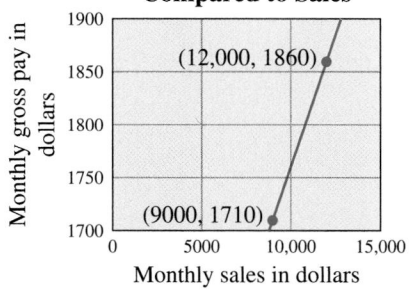

Monthly Gross Pay Compared to Sales

a. Write the equation of the line describing this linear relationship where P is the monthly gross pay and s is the monthly sales.

b. What is the rate of change of the monthly gross pay?

c. Assuming his sales continue to increase at this same rate, what would his monthly gross pay be if his monthly sales were \$20,000?

Web It

For an explanation of how writing the equation of a line can be applied to time and temperature, go to link 3-4-3 at mhhe.com/bello.

Answers

7. a. $P = 0.05s + 1260$
b. \$1 gross pay per \$20 in sales or $\frac{1}{20}$ **c.** \$2260

Study Skills Hint

Web It

Try the practice quiz on equations of lines at link 3-4-4 at mhhe.com/bello.

Table 2 gives you a summary of most of the formulas dealing with linear equations to which you can refer.

Table 2 Linear Equations

If You Want an *Equation*	Write
In standard form	$Ax + By = C$
For a vertical line	$x = C$ (C a constant)
For a horizontal line	$y = C$ (C a constant)
For a line going through (x_1, y_1) and (x_2, y_2)	$y - y_1 = m(x - x_1)$, where $m = \dfrac{y_2 - y_1}{x_2 - x_1}$
For a line with slope m going through (x_1, y_1)	$y - y_1 = m(x - x_1)$
For a line with slope m and y-intercept b	$y = mx + b$

If You Want the *Formula*	Write
For the slope m of a line passing through (x_1, y_1) and (x_2, y_2)	$m = \dfrac{y_2 - y_1}{x_2 - x_1}$
For the slope of a vertical line	Undefined
For the slope of a horizontal line	$m = 0$
For the slope of a line parallel to the line $y = mx + b$	m
For the slope of a line perpendicular to the line $y = mx + b$	$\dfrac{-1}{m}$

Calculate It

Be sure to clear all lists and clear the equation in (Y=) before you begin. Given the point (2, 1) we want a line parallel to $y - x = 1$, that is, $y = x + 1$, which has slope $1 = \frac{1}{1}$. This means that the change in x is 1 and the change in y is 1. Starting at the point (2, 1), go 1 unit right and 1 unit up, ending at (3, 2). We now have two points, (2, 1) and (3, 2), so we can follow our previous steps to make the two lists and get an equation of the form $y = ax + b$. For (2, 1) and (3, 2), $a = 1$ and $b = -1$, that is $y = x - 1$. Now graph the line perpendicular to $y - x = 1$ that goes through the point (2, 1).

A warning about your calculator. Look closely at the graph of the two lines. The lines $y = x - 1$ and $y = -x + 3$ are perpendicular. Turn off the Stat Plot, and graph these two lines using your calculator and a standard window (see Window 1). They don't look perpendicular. Why? Because the standard window is not square (the units on the x-axis are larger than the units on the y-axis). To remedy this, change to a square window by pressing (ZOOM) 5 (see Window 2). Your faith in your calculator should now be restored.

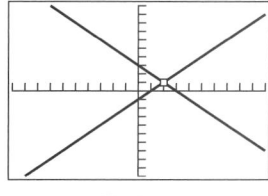

Window 1

$Y_1 = x - 1$
$Y_2 = -x + 3$

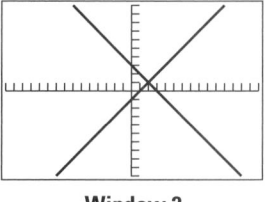

Window 2

Exercises 3.4

A In Problems 1–6, find the equation of the line passing through the given points. Write the final equation in standard form.

1. $(1, -1)$ and $(2, 2)$ $3x - y = 4$

2. $(-3, -4)$ and $(-2, 0)$ $4x - y = -8$

3. $(3, 2)$ and $(2, 3)$ $x + y = 5$

4. $(3, 0)$ and $(0, 5)$ $5x + 3y = 15$

5. The line with x-intercept 2 and y-intercept 4. $2x + y = 4$

6. The line with x-intercept -3 and y-intercept -1. $x + 3y = -3$

B In Problems 7–10, find the equation of the line with the given slope and passing through the given point. Write the final equation in standard form.

7. Slope 2, point $(-3, 5)$
$2x - y = -11$

8. Slope $\frac{1}{2}$, point $(2, 3)$
$x - 2y = -4$

9. Slope -3, point $(-1, -2)$
$3x + y = -5$

10. Slope $\frac{-1}{3}$, point $(2, -4)$
$x + 3y = -10$

C In Problems 11–14, find the slope-intercept form of the equation of the line with the given slope and intercept. Then graph the line.

11. Slope 5, y-intercept 2

$y = 5x + 2$

12. Slope $\frac{1}{4}$, y-intercept 2

$y = \frac{1}{4}x + 2$

13. Slope $\frac{-1}{5}$, y-intercept $\frac{-1}{3}$

$y = \frac{-1}{5}x - \frac{1}{3}$

14. Slope $\frac{2}{3}$, y-intercept $\frac{1}{2}$

$y = \frac{2}{3}x + \frac{1}{2}$

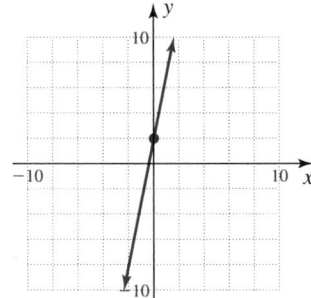

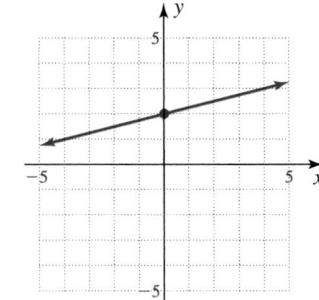

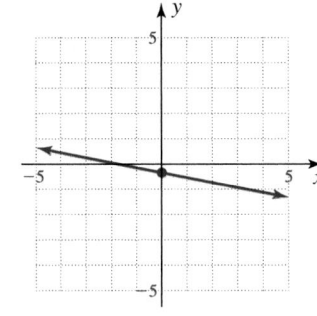

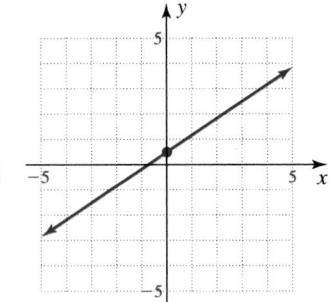

D In Problems 15–22, find the slope-intercept form of the equation of the line passing:

15. Through $(1, -2)$ and parallel to $y = 2x + 1$
$y = 2x - 4$

16. Through $(-1, -2)$ and parallel to $2y = -4x + 5$
$y = -2x - 4$

17. Through $(-5, 3)$ and parallel to $2y + 6x = 8$
$y = -3x - 12$

18. Through $(-3, -5)$ and parallel to $3y - 6x = 12$
$y = 2x + 1$

19. Through $(1, 1)$ and perpendicular to $2y = x + 6$
$y = -2x + 3$

20. Through $(2, 3)$ and perpendicular to $3y = -x + 5$
$y = 3x - 3$

21. Through $(-2, -4)$ and perpendicular to $y - x = 3$
$y = -x - 6$

22. Through $(-3, 5)$ and perpendicular to $2y - x = 5$
$y = -2x - 1$

E In Problems 23–30, find an equation of the line described.

23. A line passing through the point $(3, 4)$ and with slope 0.
$y = 4$

24. A line passing through the point $(-2, -4)$ and with slope 0. $y = -4$

25. The slope of the line is undefined, and it passes through the point $(-2, 4)$. $x = -2$

26. The slope of the line is undefined, and it passes through the point $(-4, -5)$. $x = -4$

27. A vertical line passing through $(-2, 3)$. $x = -2$

28. A vertical line passing through $(-3, -1)$. $x = -3$

29. A horizontal line passing through $(3, 2)$. $y = 2$

30. A horizontal line passing through $(-3, -4)$. $y = -4$

F In Problems 31–32, use mathematical modeling.

31. According to the U.S. Energy Information Administration, the energy prices in the following table indicate the amount of money spent on electricity in dollars per million Btu by the residential consumer in the United States for each of the years listed in the first column.

Year	Dollars per Million Btu Spent on Electricity
2000	$24.50
2001	$25.40
2002	$23.90
2003	$23.10
2004	$22.90
2008	?

a. Rename the table by assigning variable names to the columns, adjusting the numbers in the first column so 2000 corresponds to $x = 0$, and then graph the data.

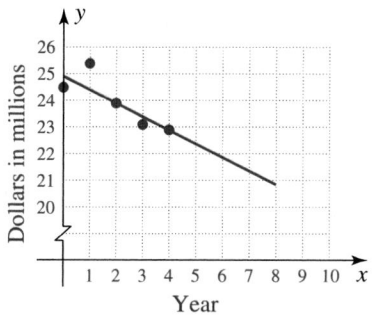

b. Draw the line of best fit and name the coordinates of the two points used to draw the line.
$(2, 23.9); (4, 22.9)$

c. Use the two points named in part **b** to write the equation of the line of best fit. $y = \frac{-1}{2}x + 24.9$

d. Use the equation in part **c** to predict the amount of money that will be spent on electricity in dollars per million Btu in the year 2008. $20.9 million

e. What seems to be the trend in the dollars per million Btu spent on electricity for the future in the United States? Each year it decreases by $0.5 million.

32. Have you ever had a great idea you thought should be patented? Thousands of people in the United States get patents each year. The data in the following table is from the U.S. Patent Office, and it indicates the number of patents in millions for each of the years listed.

Year	Number of Patents in Thousands
1998	90.7
1999	94.1
2000	97.0
2001	98.7

a. Rename the table by assigning variable names to the columns, adjust the numbers in the first column so 1998 corresponds to $x = 0$, and then graph the data.

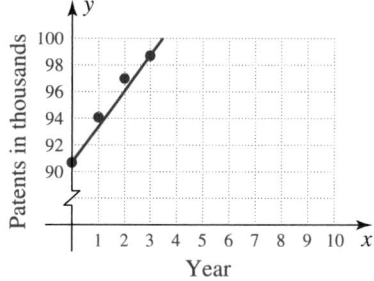

b. Draw the line of best fit, and name the coordinates of the two points used to draw the line.
$(0, 90.7); (3, 98.7)$

c. Use the two points named in part **b** to write the equation of the line of best fit. $y = \frac{8}{3}x + 90.7$

d. Use the equation in part **c** to predict the number of patents there will be in the year 2008.
117.4 thousand patents

G In Problems 33–36, find an equation for the specified line shown in the graph.

33. L_1 $y = -4$

34. L_2 $x - y = 0$

35. L_3 $2x - 3y = -6$

36. L_4 $x + y = 0$

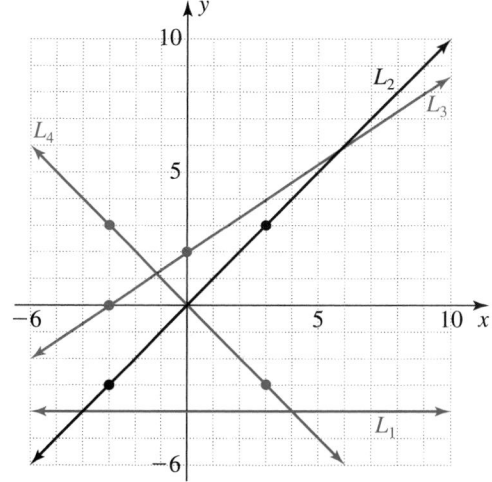

In Problems 37–40, use each graph to solve the application.

37. The line graph to the right describes the number N, in millions, of cases of diabetes in the country in years x since 1983.

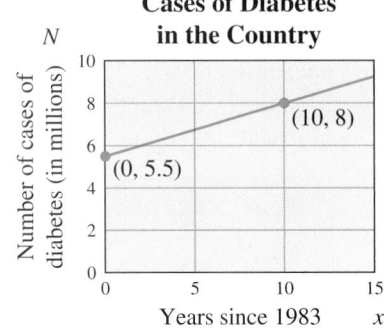

Cases of Diabetes in the Country

(0, 5.5)
(10, 8)

Number of cases of diabetes (in millions)
Years since 1983

a. Find the equation of the line where N is the number of cases and x is the time in years since 1983. $N = 0.25x + 5.5$

b. What is the rate of change of the number of cases of diabetes? **Each year there is a 0.25 million increase in cases of diabetes.**

c. What is x when the year is 2008? $x = 25$

d. Use the equation to predict the number of cases of diabetes in 2008. **11.75 million**

38. After walking into a cold room (45°F), the heat is turned on high. The line graph to the right describes the change in temperature F over the time t in minutes.

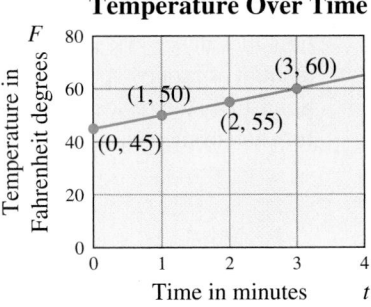

Temperature Over Time

(0, 45)
(1, 50)
(2, 55)
(3, 60)

Temperature in Fahrenheit degrees
Time in minutes

a. Find the equation of the line where F is Fahrenheit degrees and t is the time in minutes. $F = 5t + 45$

b. What is the rate of change of the temperature over time? **Each minute there is a 5°F rise in temperature.**

c. Use the equation to predict the temperature in 10 minutes. **95°F**

d. The heat should be turned down after the room reaches 80°F. How many minutes will that take? **7 minutes**

39. The line graph to the right describes the cost of homes in a new housing development. The cost is based on the square feet of living space in the home.

a. Find the equation of the line where C is the cost and x is the square feet. $C = 85x$

b. What is the cost of a home per square foot? **$85 per sq foot**

c. Use the equation to predict the cost of a home with 1475 square feet. **$125,375**

d. If your bank approved you for a home mortgage loan of $138,000, how big of a house could you buy? (Round the answer to the nearest square foot.) **1624 sq feet**

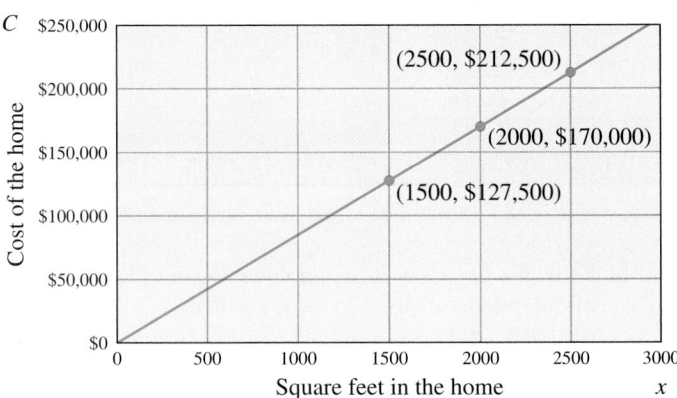

Cost of Home Based on Square Feet

(1500, $127,500)
(2000, $170,000)
(2500, $212,500)

Cost of the home
Square feet in the home

40. Studies show that college students are taking longer to graduate from college. The line graph to the right describes the percentage of college students graduating in less than 5 years in 1983 and in 1997.

a. Find the equation of the line where P is the percentage and x is the years after 1983. $P = \frac{-5}{14}x + 58$

b. What is the rate of change in the percentage of college students graduating in less than 5 years? **Each year there is about a 0.4% decrease in college students graduating in less than 5 years.**

c. Use the equation to estimate the percentage of college students that graduated in 1990 after less than 5 years in school. **55.5%**

d. Use the equation to predict the percentage of college students that will graduate in 2008 after less than 5 years in school. **49%**

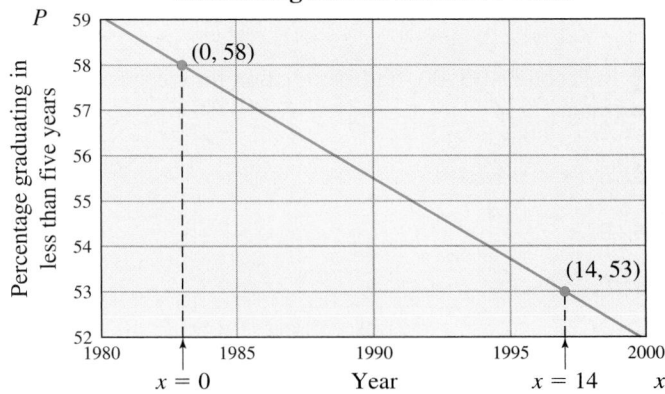

Percentage of College Students Graduating in Less than Five Years

(0, 58)
(14, 53)

Percentage graduating in less than five years
Year
$x = 0$
$x = 14$

APPLICATIONS

Supply and Demand.

41. When sunglasses are sold at the regular price of $6, a student purchases two pairs. Sold for $4 at the flea market, the student purchases six pairs. Let the ordered pair (d, p) represent the demand and price for sunglasses.

    a. Find the demand equation. $d + 2p = 14$

    b. How many pairs would the student buy if sunglasses are selling for $2? 10 pairs

42. When skateboards are sold for $80, 5 of them are sold each day. When they are on sale for $40, 13 of them are sold daily. Let (d, p) represent the demand and price for skateboards.

    a. Find the demand equation. $5d + p = 105$

    b. For 10 skateboards to be sold on a given day, what should the price be? $55

43. A wholesaler will supply 50 sets of video games at $35. If the price drops to $20, she will provide only 20. Let (s, p) represent the supply and the price for video games.

    a. Find the supply equation. $s - 2p = -20$

    b. At what price would the wholesaler stop selling the video games; in the other words, when is the supply zero? $10

    c. If the price went up to $40 per game, how many games would she be willing to supply? 60

44. When T-shirts are selling for $6, a retailer is willing to supply 6 of them each day. If the price increases to $12, he's willing to supply 12 T-shirts.

    a. Find the supply equation. $s = p$

    b. If the T-shirts were free, how many would he supply? None

45. The supply s of a product is given by $s = 3p - 6$, while the demand d is $d = -2p + 14$. What will the price p be when the supply s equals the demand d? (This price is called the *equilibrium price*.) $4

46. The supply and demand for a product are given by $s = 3p - 10$ and $d = -2p + 40$, respectively. At what price p will the supply equal the demand? $10

SKILL CHECKER

Try the "Skill Checker" exercises so you'll be ready for the next section.

Find the x- and y-intercepts of the line:

47. $2x - 4y = 8$ x: 4; y: -2 **48.** $3x - 2y = 6$ x: 2; y: -3 **49.** $y = 2x + 6$ x: -3; y: 6 **50.** $y = -4x + 8$ x: 2; y: 8

Solve:

51. $3(x - 1) = 6x + 3$ $x = -2$

52. $4(x - 1) = 8x + 4$ $x = -2$

USING YOUR KNOWLEDGE

Business by Candle Lights

In economics and business, the slope, m, and the y-intercept, b, of an equation play an important role. Let's see how.

    Suppose you wish to go into the business of manufacturing fancy candles. First, you have to buy some ingredients such as wax, paint, and so on. Assume these ingredients cost you $100. This is the *fixed cost*. Now suppose it costs $2 to manufacture each candle. This is the *marginal cost*. What would be the total cost y if the marginal cost is $2, x units are produced, and the fixed cost is $100? The answer is

$$\underset{\substack{\text{Total}\\\text{cost}}}{y} = \underset{\substack{\text{Cost for}\\x\text{ units}}}{2x} + \underset{\substack{\text{Fixed}\\\text{cost}}}{100}$$

In general, an equation of the form

$$y = mx + b$$

gives the total cost y of producing x units, where m is the cost of producing 1 unit and b is the fixed cost.

53. Find the total cost y of producing x units of a product costing $2 per unit if the fixed cost is $50. $y = 2x + 50$

54. Find the total cost y of producing x units of a product whose production cost is $7 per unit if the fixed cost is $300. $y = 7x + 300$

55. The total cost y of producing x units of a certain product is given by

$$y = 2x + 75$$

    a. What is the production cost for each unit? $2

    b. What is the fixed cost? $75

WRITE ON

56. How do you decide what formula to use when you're asked to find an equation of a line? Answers may vary.

Write the procedure you use to draw the graph of a line:

57. When a point and the slope are given.
Answers may vary.

58. When the *y*-intercept and the slope are given.
Answers may vary.

59. When the *x*- and *y*-intercepts are given. Answers may vary.

MASTERY TEST

If you know how to do these problems, you have learned your lesson!

60. Find an equation of the line passing through the point (1, 1) and

    a. Parallel to $2y - 6x = 5$. $3x - y = 2$

    b. Perpendicular to $2y - 6x = 5$. $x + 3y = 4$

61. Find the slope and the *y*-intercept of the line $8x + 4y = 16$.
$m = -2; b = 4$

62. Find the slope and the *y*-intercept of the line $y = 3$.
$m = 0; b = 3$

63. A line has slope 3 and *y*-intercept 2. Find the slope-intercept form of the equation of the line and then graph it.
$y = 3x + 2$

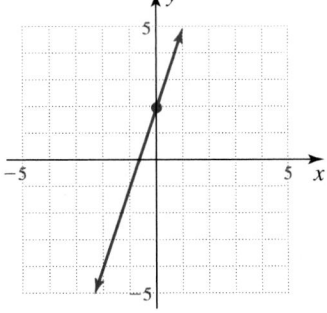

64. The slope of a line is undefined and it passes through the point $(-2, 0)$. What is the equation of this line?
$x = -2$

65. Find an equation of the line with slope -3 and passing through the point (1, 2) and then graph it. $y = -3x + 5$

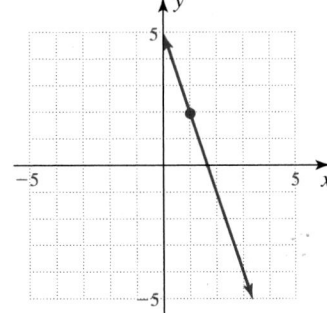

66. Find an equation of the line passing through points (3, 1) and (4, 3), write it in standard form, and then graph it. $2x - y = 5$

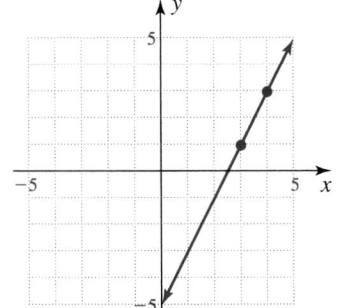

67. Find an equation of the line passing through points (3, 2) and (4, 2) and then graph it. $y = 2$

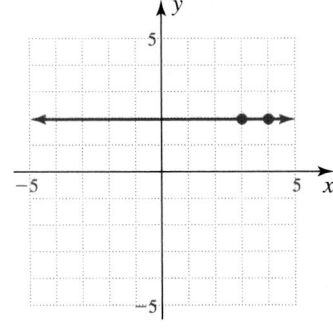

3.5 LINEAR INEQUALITIES IN TWO VARIABLES

To Succeed, Review How To ...

1. Find the *x*- and *y*-intercepts of a line (pp. 167–168).

2. Solve linear equations (p. 79).

3. Graph lines (pp. 165–169).

4. Solve inequalities involving absolute values (pp. 141–144).

Objectives

A Graph linear inequalities.

B Graph inequalities involving absolute values.

GETTING STARTED

Renting Cars

Suppose you want to rent a car for a few days. Here are some prices for an intermediate car rental.

Rental A:	$36 per day, $0.15 per mile
Rental B:	$49 per day, $0.33 per mile

The total cost *C* for the Rental A car is

$$C = \underbrace{36d}_{\text{Cost for } d \text{ days}} + \underbrace{0.15m}_{\text{Cost for } m \text{ miles}}$$

Now suppose you want the cost *C* to be $180. Then $180 = 36d + 0.15m$. We graph this equation by finding the intercepts. When $d = 0$, $180 = 0.15m$, or

$$m = \frac{180}{0.15} = 1200$$

When $m = 0$, $180 = 36d$, or

$$d = \frac{180}{36} = 5$$

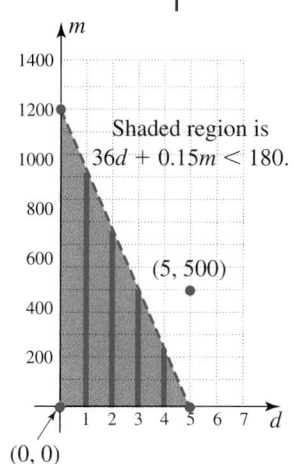

Shaded region is $36d + 0.15m < 180$.

(5, 500)

(0, 0)
Test point

We join (0, 1200) and (5, 0) with a line and then graph the discrete points corresponding to 1, 2, 3, or 4 days.

But what if you want the cost to be less than $180? We then have

$$36d + 0.15m < 180$$

We have graphed the points on the line $36d + 0.15m = 180$. Where are the points for which $36d + 0.15m < 180$? As the graph shows, the line $36d + 0.15m = 180$ divides the plane into three parts:

1. The points *below* the line

2. The points *on* the line

3. The points *above* the line

It can be shown that if any point on one side of the line $Ax + By = C$ satisfies the inequality $Ax + By < C$, then all points on that side satisfy the inequality and no point on the other side of the line does. Let's select (0, 0) as a test point. Since

$$36 \cdot 0 + 0.15 \cdot 0 < 180$$
$$0 < 180$$

is true, all points below the line (shown shaded) satisfy the inequality. [As a check, note that (5, 500), above the line, doesn't satisfy the inequality, because $36 \cdot 5 + 0.15 \cdot 500 < 180$ is a false statement.] The line $36d + 0.15m = 180$ is *not* part of the graph and is shown dashed. To apply this result to our rental-car problem, *d* must be an integer. The solution to our problem consists of the points on the heavy-line segments at $d = 1, 2, 3,$ and 4. The graph shows, for instance, that you can rent a car for 2 days and go about 700 miles at a cost less than $180. What we have just set up is called a *linear inequality,* which we shall now examine in more detail.

 **Graphing Linear Inequalities**

LINEAR INEQUALITY IN TWO VARIABLES

A **linear inequality** is a statement that can be written in the form

$$Ax + By \leq C, \ Ax + By \geq C, \ Ax + By < C, \text{ or } Ax + By > C$$

where A and B are not both zero.

Here is a summary of the procedure we used to graph the linear inequality in the *Getting Started*.

PROCEDURE

Graphing a Linear Inequality

1. Graph the line associated with the inequality. This line is called the boundary line. If the inequality involves $\leq$ or $\geq$, draw a solid line; this means the line is included in the solution. If the inequality involves $<$ or $>$, draw the line dashed, which means the line is not part of the solution.

2. Choose a test point not on the line. [$(0, 0)$ if possible.]

3. If the test point satisfies the inequality, shade the region containing the test point; otherwise, shade the region on the other side of the line. The shaded region represents all the solutions to the inequality.

EXAMPLE 1 **Graphing a linear inequality**

Graph all the solutions to $x - 2y < -4$

SOLUTION We follow the three-step procedure.

1. We first graph the boundary line $x - 2y = -4$ by finding the intercepts.

x	y	
0	2	When $x = 0$, $-2y = -4$ and $y = 2$
−4	0	When $y = 0$, $x - 2 \cdot 0 = -4$ and $x = -4$

The boundary line is shown dashed to indicate that it isn't part of the solution (Figure 39).

2. We select an easy test point and see if it satisfies the inequality. If it does, the solution lies on the same side of the boundary line as the test point; otherwise, the solution is on the other side of the line. An easy point is $(0, 0)$, *below* the line. If we substitute $x = 0$ and $y = 0$ in the inequality $x - 2y < -4$, we obtain

$$0 - 2 \cdot 0 < -4$$
$$0 < -4$$

which is false.

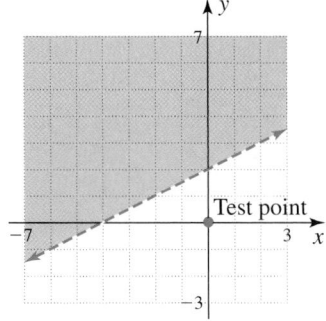

Test point

Figure 39

3. The point $(0, 0)$ is not part of the solution. Because of this, the solution consists of the points *above* (on the other side of) the boundary line $x - 2y = -4$ and is shown shaded in Figure 39.

PROBLEM 1

Graph all the solutions to

$$3x - 2y < -6$$

Answer

1.

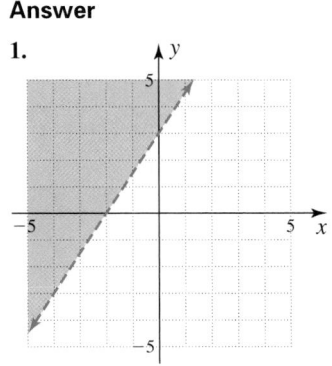

In Example 1, the test point was not part of the solution for the given inequality. Now we give an example in which the test point is part of the solution for the inequality.

Web It

For another step-by-step example of graphing linear inequalities, try link 3-5-1 at mhhe.com/bello.

EXAMPLE 2	**Graphing an inequality where the test point is part of the solution**

Graph all the solutions to: $y \leq -2x + 4$

SOLUTION   We use our three-step procedure.

1. As usual, we first graph the boundary line $y = -2x + 4$.

x	y	
0	4	When $x = 0$, $y = 4$.
2	0	When $y = 0$, $0 = -2x + 4$, or $x = 2$.

The graph of the boundary line is shown in Figure 40.

2. Now we select the point $(0, 0)$ as a test point. When $x = 0$ and $y = 0$, we have

$$y \leq -2x + 4$$
$$0 \leq -2 \cdot 0 + 4$$
$$0 \leq 4$$

which is true.

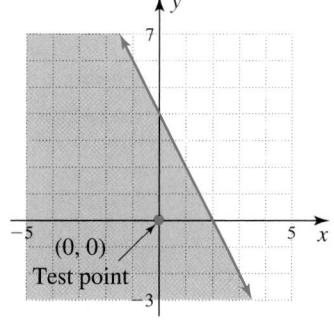

Figure 40

3. Thus all the points on the same side of the boundary line as $(0, 0)$—that is, the points *below* the line—are solutions of $y \leq -2x + 4$. These solutions are shown shaded in Figure 40. This time, the line is *solid* because it's part of the solution.

PROBLEM 2

Graph all the solutions to: $y \geq -x - 3$

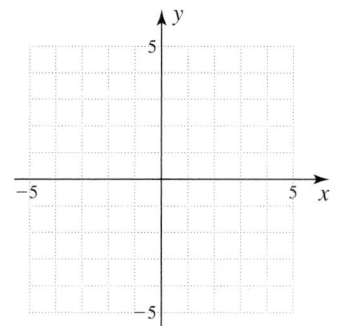

NOTE

If you solve an inequality for *y* obtaining:

(1) $y > ax + b$, the solution set consists of all the points *above* $y = ax + b$

(2) $y < ax + b$, the solution set consists of all the points *below* $y = ax + b$

Answer

2.

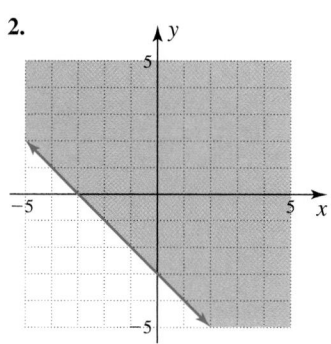

Using the information from this note, rework Example 1. First, solve for *y* so that it is on the left of the inequality sign.

$$x - 2y < -4$$
$$-2y < -x - 4 \qquad \text{Subtract } x.$$
$$y > \frac{1}{2}x + 2 \qquad \text{Divide by } -2 \text{ (reverse the inequality sign when dividing by a negative number).}$$

1. This time to graph the boundary line we use the y-intercept, 2, and the slope, $\frac{1}{2}$. We will get the same line as shown in Figure 39.

2. We don't have to use a test point when the inequality is in this format. Instead, we know the solutions are above the boundary line because the inequality sign is "greater than."

3. Thus, we shade above the boundary line.

EXAMPLE 3 **Graphing an inequality without using a test point**	**PROBLEM 3**

Graph all the solutions to: $x \geq -1$

SOLUTION   We first graph the vertical boundary line $x = -1$. This time, we don't even need a test point! All points to the *right* of this line have x-coordinates greater than -1 (points to the left have x-coordinates less than -1). The graph of all the solutions, which includes the boundary line $x = -1$, is shown in Figure 41.

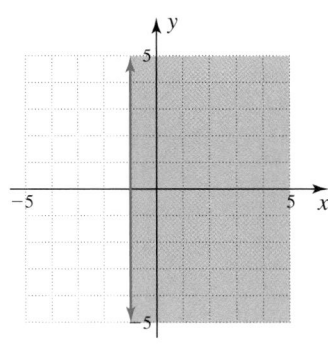

Figure 41

Graph all the solutions to: $x < 2$

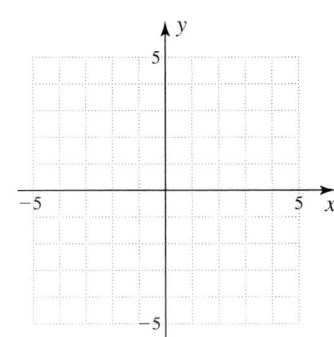

(Answer on page 225.)

B   **Graphing Absolute-Value Inequalities**

As you recall from Section 2.6,

$$|x| \leq a \quad \text{is equivalent to} \quad -a \leq x \leq a$$

Teaching Tip

Think of this double inequality statement as x being between the numbers on the ends. Graph the vertical lines corresponding to the numbers on the ends and shade between them.

If we graph the solutions to $|x| \leq 1$, we must graph all the points satisfying the inequality $-1 \leq x \leq 1$ (that is, the points between -1 and 1) as well as the boundary lines $x = -1$ and $x = 1$. These points are shown in Figure 42.

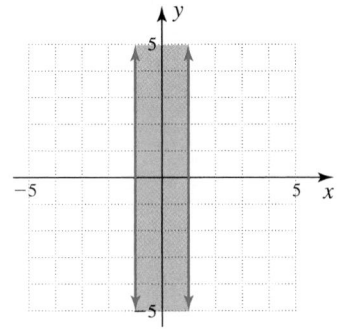

Figure 42

EXAMPLE 4 **Graphing an absolute-value inequality**	**PROBLEM 4**

Graph all the solutions to: $|y| \leq 2$

SOLUTION $|y| \leq 2$ is equivalent to $-2 \leq y \leq 2$, so the graph of all the solutions, shown in Figure 43, consists of all points bounded by the horizontal lines $y = -2$ and $y = 2$, as well as these two boundary lines; these lines are therefore shown as solid lines.

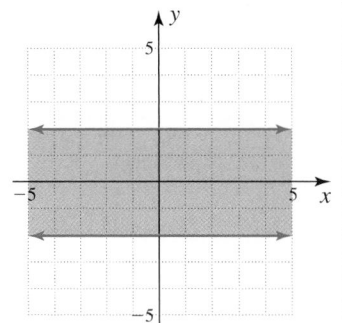

Figure 43

Graph all the solutions to: $|x - 3| \leq 4$

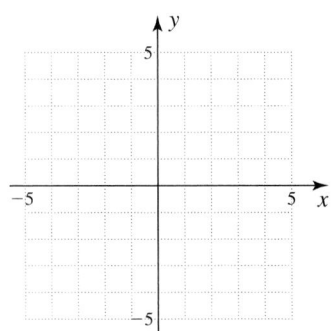

(Answer on page 225.)

EXAMPLE 5   **Graphing an absolute-value inequality**

Graph all the solutions to: $|x + 1| > 2$

SOLUTION   The inequality $|x + 1| > 2$ is equivalent to

$$x + 1 > 2 \quad \text{or} \quad x + 1 < -2$$
$$x > 1 \quad \text{or} \quad x < -3$$

The graph of all the solutions to $|x + 1| > 2$ consists of all points to the *right* of the vertical boundary line $x = 1$ and all points to the *left* of the boundary line $x = -3$. The boundary lines $x = 1$ and $x = -3$ are *not* part of the graph. They are therefore shown as dashed lines in Figure 44.

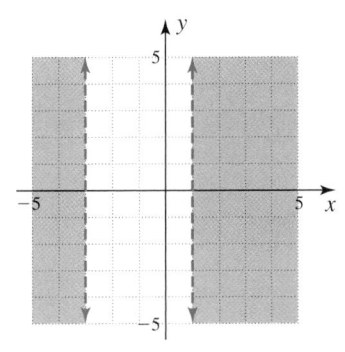

Figure 44

PROBLEM 5

Graph all the solutions to: $|y| \geq 1$

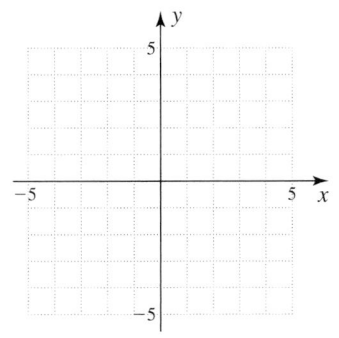

Calculate It   Graphing Linear Inequalities

Your calculator will be a tremendous help when you graph linear inequalities, but before you can use it properly, you need to do some work on your own. In Example 1, you graphed $x - 2y = -4$ and used a test point to find the region that satisfied $x - 2y < -4$. If you want to use your calculator to do this problem, you *first* have to solve for y in $x - 2y < -4$ (add $2y + 4$ to both sides and then divide each term by 2) to obtain

$$\frac{x}{2} + 2 < y \quad \text{or equivalently,} \quad y > \frac{x}{2} + 2$$

Now graph the associated equation,

$$y = \frac{x}{2} + 2$$

shown in Window 1.
   The solution set of

$$y > \frac{x}{2} + 2$$

(equivalent to the original inequality $x - 2y < -4$) consists of the points *above* the graph of the line, as shown in Window 2. (We want the y's that are strictly "greater than" so we do not include the points on the line itself in the solution set.) To shade the region above, press $Y_1 =$, two left arrows, and toggle the enter key until the appropriate shading appears, then press **GRAPH** (see Window 2). You try Example 2 (see Window 3).

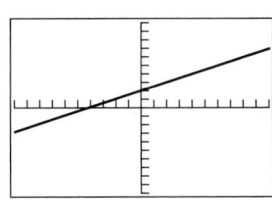

Window 1

$x - 2y < -4$ or $y > \dfrac{x}{2} + 2$

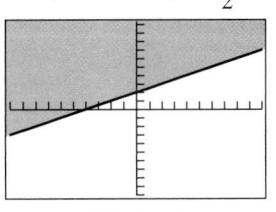

Window 2

$y \leq -2x + 4$

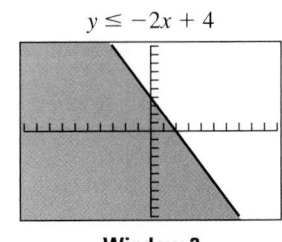

Window 3

Answers

3.

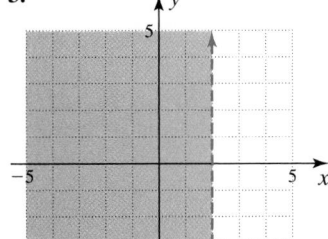

4.

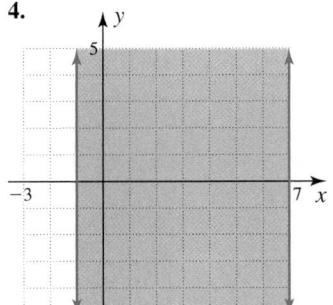

5.

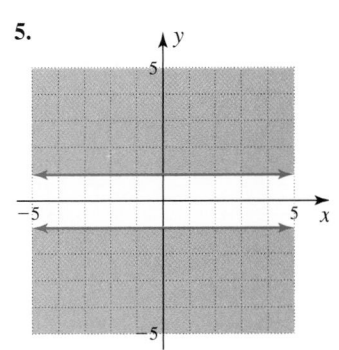

Exercises 3.5

A In Problems 1–20, graph the solutions to the inequalities.

1. $x + 2y > 4$

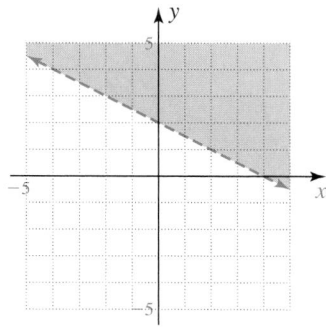

2. $x + 3y > 3$

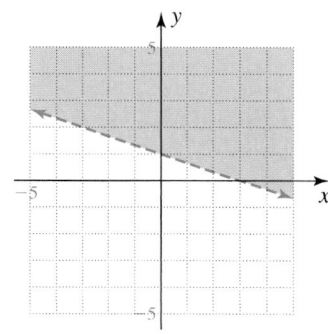

3. $-2x - 5y \leq -10$

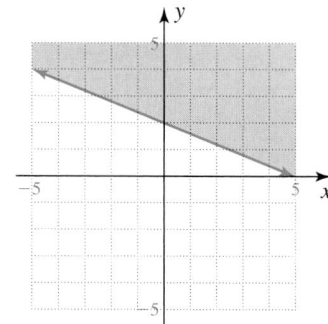

4. $-3x - 2y \leq -6$

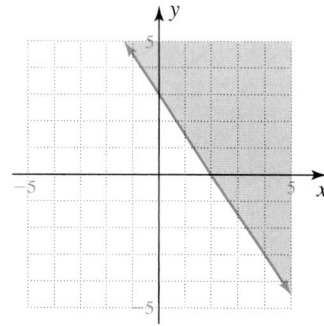

5. $y \geq 2x - 2$

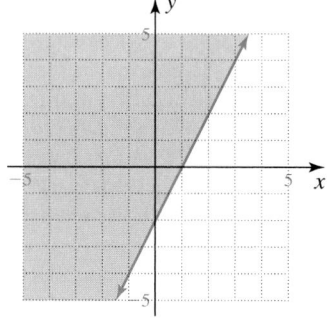

6. $y \geq -2x + 4$

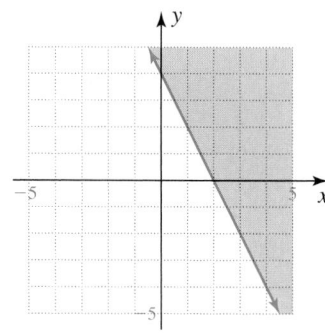

7. $6 < 3x - 2y$

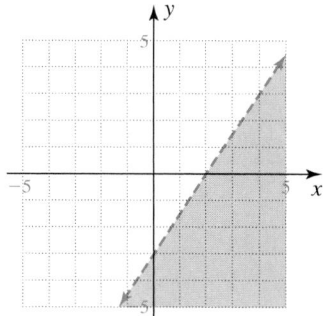

8. $6 < 2x - 3y$

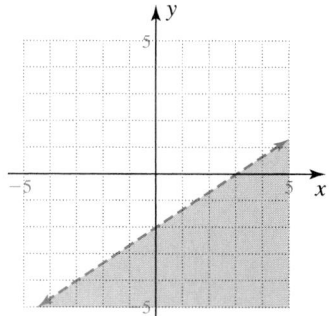

9. $4x + 3y \geq 12$

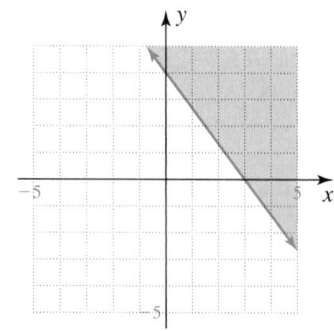

10. $-3y \geq 6x + 6$

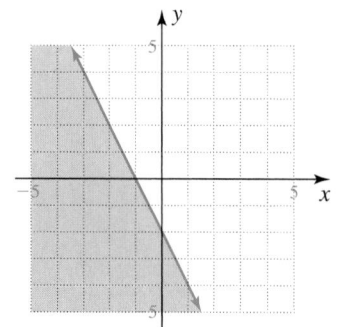

11. $10 < -5x + 2y$

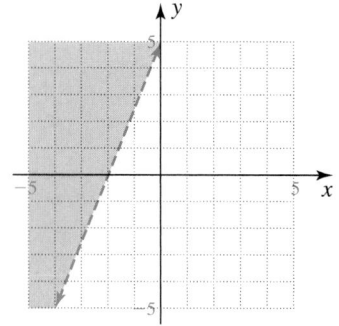

12. $4 < -2x - 4y$

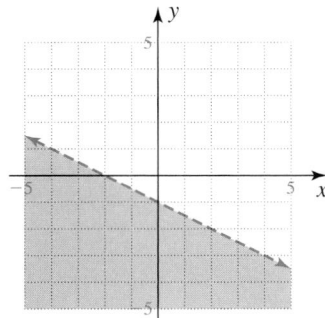

13. $2x \geq 2y - 4$

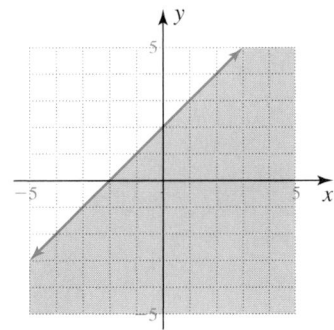

14. $2x \geq 4y + 2$

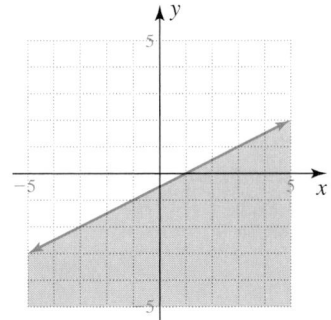

15. $2y < -4x + 8$

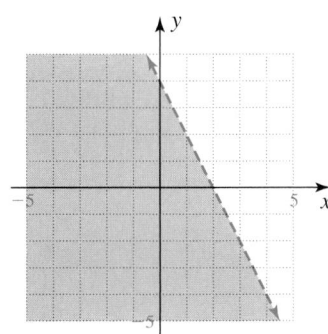

16. $3y < -6x + 9$

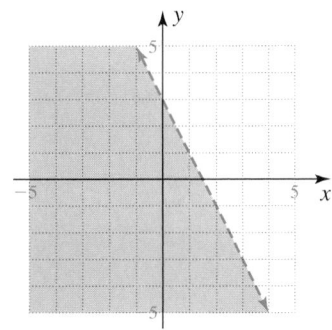

17. $x \geq -3$

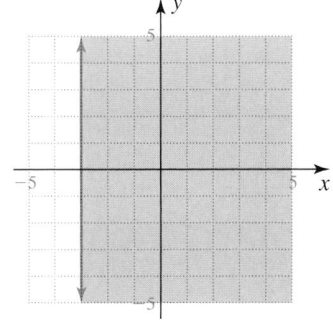

18. $x \geq -4$

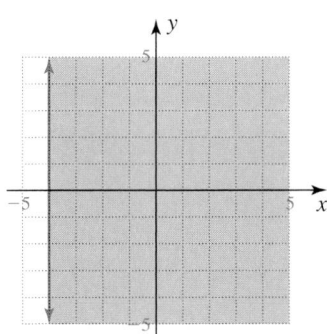

19. $y < 3$

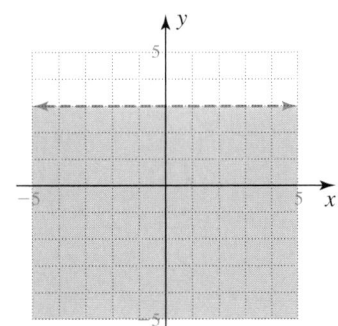

20. $y < -2$

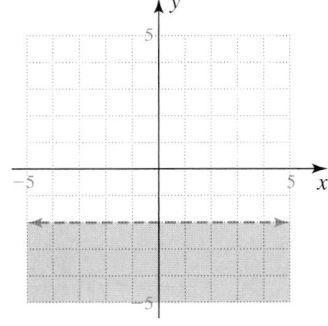

B In Problems 21–38, graph the solutions to the inequalities.

21. $|x| < 1$

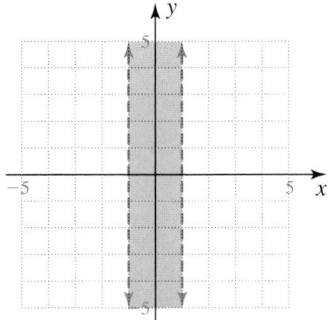

22. $|x| < 3$

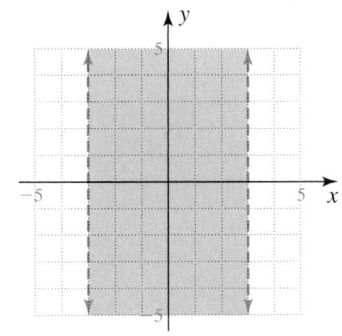

23. $|y| < 4$

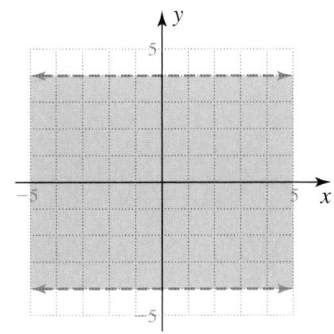

24. $|y| < 3$

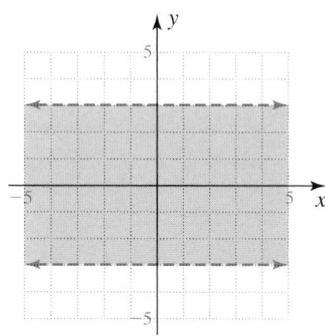

25. $|x| \geq 1$

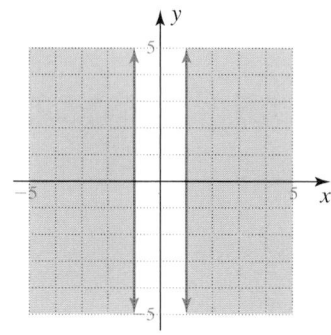

26. $|x| \geq 2$

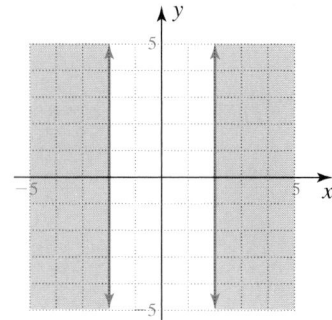

27. $|y| \geq 2$

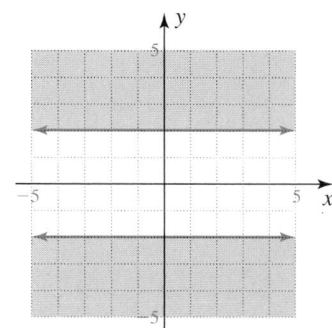

28. $|y| \geq 4$

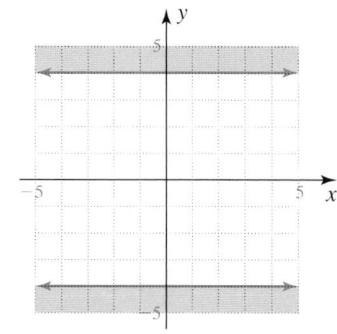

29. $|x + 2| < 1$

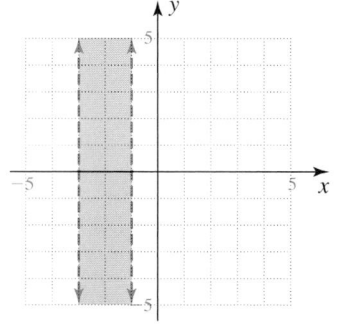

30. $|x + 3| < 1$

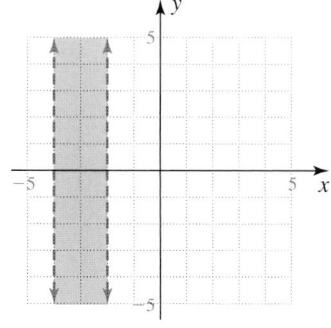

31. $|y + 2| < 1$

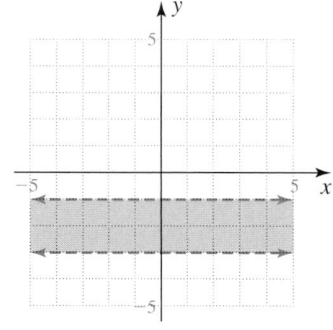

32. $|y + 2| < 2$

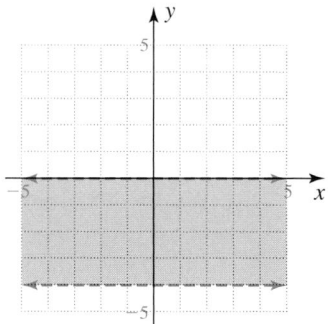

33. $|x + 1| \geq 3$

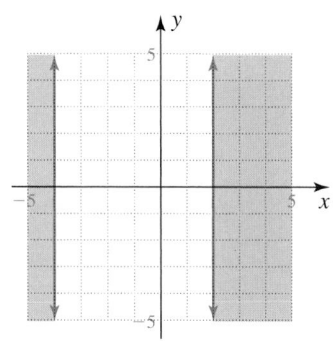

34. $|x + 2| \geq 1$

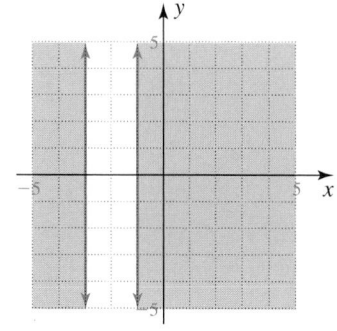

35. $|x - 1| \leq 2$

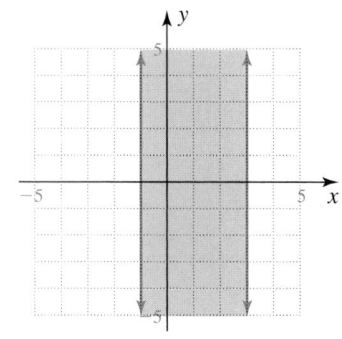

36. $|x - 2| \leq 1$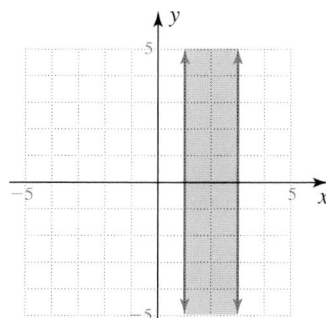

37. $|y - 2| < 1$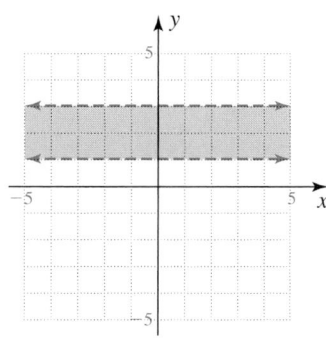

38. $|y - 3| < 1$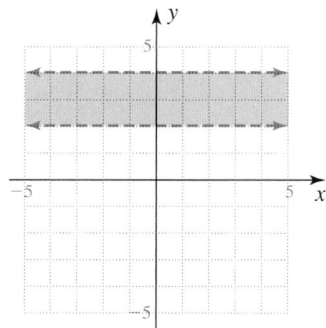

SKILL CHECKER

Try the "Skill Checker" exercises so you'll be ready for the next section.

Solve:

39. $56 = k \cdot 14$ 4

40. $39 = k \cdot 13$ 3

41. $60 = \dfrac{k}{3}$ 180

42. $40 = \dfrac{k}{5}$ 200

USING YOUR KNOWLEDGE

The More You Drive, the Costlier It Is.

You can use what you know about linear inequalities to save money when you rent a car. Here are the rental prices for an intermediate car obtained from a telephone survey

> Rental A: $41 a day, unlimited mileage
>
> Rental B: $36 a day, $0.15 per mile

43. If you compare the Rental A price ($41) with the Rental B price, you see that Rental B appears to be cheaper.

  a. How far can you drive a Rental B car in one day if you wish to spend exactly $41? (Answer to the nearest mile.) 33 mi

  b. If you are planning on driving 100 mi in one day, which car would you rent, Rental B or Rental A? Rental A

44. How far can you drive a Rental B car in 1 day if you wish to spend exactly $42? (Answer to the nearest mile.) 40 mi

45. Based on your answers to Problems 43 and 44, which is the cheaper rental price? (*Hint:* It has to do with the miles you drive.) If you plan to drive more than 33 mi, Rental A is cheaper.

WRITE ON

46. Write the procedure you would use to graph the solution set of the inequality $ax + by > c$. Answers may vary.

47. How would you describe the graph of the inequality $x \geq k$ (where k is a constant)? Answers may vary.

48. Explain what it means when a dashed line is used as the boundary in the graph of the solution set of a linear inequality. Answers may vary.

49. Explain what it means when a solid line is used as the boundary in the graph of the solution set of a linear inequality. Answers may vary.

MASTERY TEST

If you know how to do these problems, you have learned your lesson!

Graph the solutions to:

50. $|x + 2| > 3$

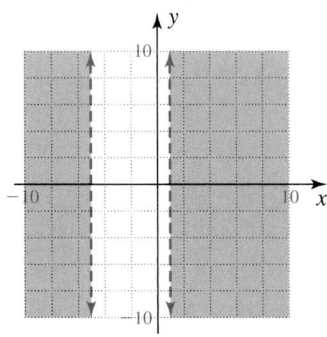

51. $|x - 1| < 4$

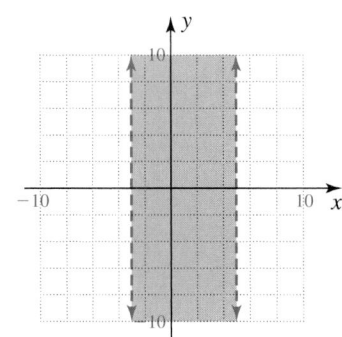

52. $|y| \leq 1$

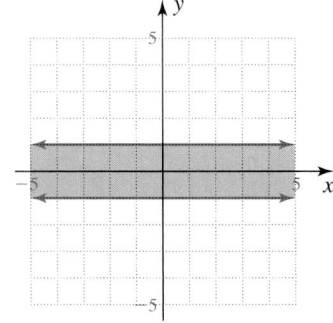

53. $|y| > 3$

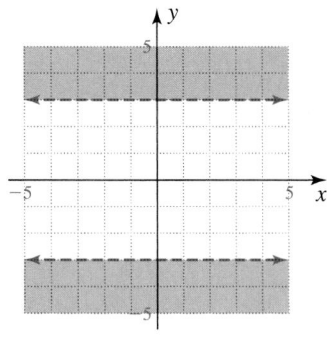

54. $x \geq -2$

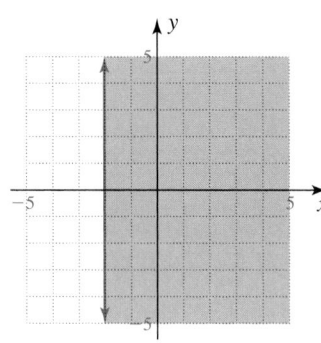

55. $x < 3$

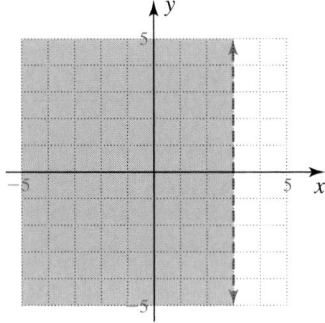

56. $3x - 2y < -6$

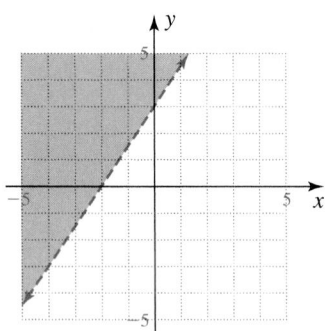

57. $-2x - 3y \geq -6$

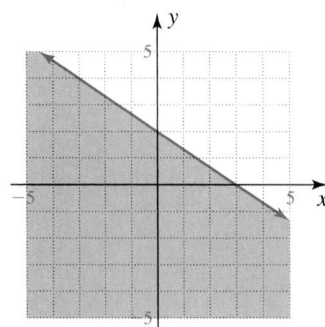

COLLABORATIVE LEARNING 3A

Suppose you interviewed for a job with the two companies shown here, and received a job offer from each of the companies. Should you be greedy and take the job with the higher base salary or is there more to it? Let's do some collaboration to help with your decision.

Divide into two groups and have each group select one of the two companies to investigate. To conduct the investigations do the following:

1. Make a table of values for potential sales (*x*-values) and their respective salary (*y*-values) based on that company's compensation plan. Let the potential sales begin with $0 and increase by increments of $50,000 up to $500,000.

2. Make a graph plotting the pairs of values from the table.

After each group has completed their company's graph, have someone from each group put their group's graph on the same large poster board graph for all to view. Have a group discussion about the comparison of these two graphs including the following questions:

1. Does the salary from the $80,000/year job always yield more income?

2. Is there a point where the amount of sales will yield the same income for both jobs? If so, when?

3. Does the income from the $60,000/year job ever yield more income? If so, when?

4. Which job should you take? Explain.

Company:	Sales Consultants of Tacoma	**Job Type:**	Banking
Location:	US-Washington		Business Development
Base Pay:	$80,000.00/Year		Finance
Other Pay:	10% of all sales		Marketing
Employee Type:	Full-Time Employee		Sales
Industry:	Banking–Financial Services	**Req'd Education:**	BS degree
	Consulting	**Req'd Experience:**	At Least 3 Years
	Sales–Marketing	**Req'd Travel:**	Negligible
		Relocation Covered:	No

Company:	Sales Reps of Syracuse	**Job Type:**	Banking
Location:	US-New York		Business Development
Base Pay:	$60,000.00/Year		Finance
Other Pay:	20% of all sales		Marketing
Employee Type:	Full-Time Employee		Sales
Industry:	Banking–Financial Services	**Req'd Education:**	BS degree
	Consulting	**Req'd Experience:**	At Least 3 Years
	Sales–Marketing	**Req'd Travel:**	Negligible
		Relocation Covered:	No

COLLABORATIVE LEARNING 3B

Do you know the annual average salary offered by U.S. employers to graduates of your college or university? The following tables indicate the annual salaries offered to MIT classes of 2001, 2002, 2003, and 2004. We will use some collaborative learning to study these numbers so we can predict the annual salary for a specific year in the future. We will assume this data is linearly related. Divide into groups and complete the following.

1. Make a graph using the average base salary for a bachelors degree graduate from each of the four years. Let the x-axis represent the year (for 2001 let $x = 1$, for 2002 let $x = 2$, etc.) and the y-axis represent the average salary.

2. Name the two points on the graph that can be connected to make the line of best fit. Use those ordered pairs to write the equation of the line in slope-intercept form. (This could also be done using a graphing calculator.)

3. This equation should be a predictor of salaries for the future. Let's see how close the equation value comes to the actual value for the first 4 years ($x = 1$, $x = 2$, $x = 3$, and $x = 4$). Make a table to display those comparisons and then add $x = 5$, $x = 6$, and $x = 7$.

4. Research you own college or university for this data, and answer the first three questions again. How does your school's annual salaries offered to bachelors degree graduates compare with those at MIT? Discuss some of the things that might make a difference in comparing these salaries.

Annual Salaries Offered By U.S. Employers to MIT Class of 2004 Graduates

| Degree | Base Yearly Salary | | | |
	Low	Average	High	# of Reports
Bachelors	$30,000	$56,211	$90,000	73
Masters	$35,000	$71,632	$105,000	55
M. Eng	$43,000	$71,010	$90,000	49
Doctoral	$50,000	$84,081	$110,000	31
			Total	208

Annual Salaries Offered By U.S. Employers to MIT Class of 2003 Graduates

| Degree | Base Yearly Salary | | | |
	Low	Average	High	# of Reports
Bachelors	$30,000	$54,761	$94,000	111
Masters	$40,000	$66,367	$94,000	68
M. Eng	$40,000	$73,514	$150,000	88
Doctoral	$42,000	$81,867	$105,000	45
			Total	312

Annual Salaries Offered By U.S. Employers to MIT Class of 2002 Graduates

| Degree | Base Yearly Salary | | | |
	Low	Average	High	# of Reports
Bachelors	$30,000	$53,497	$90,000	145
Masters	$38,000	$73,914	$187,000	99
M. Eng	$40,000	$72,032	$110,000	78
Doctoral	$42,000	$84,020	$120,000	94
			Total	416

Annual Salaries Offered By U.S. Employers to MIT Class of 2001 Graduates

| Degree | Base Yearly Salary | | | |
	Low	Average	High	# of Reports
Bachelors	$40,000	$53,000	$100,000	64
Masters	$38,000	$66,500	$120,000	39
M. Eng	$30,000	$74,000	$102,500	44
Doctoral	$30,000	$79,900	$110,000	30
			Total	177

Source: Data from Massachusetts Institute of Technology (MIT), Cambridge, MA.

Research Questions

1. Some historians claim that the official birthday of analytic geometry is November 10, 1619. Investigate and write a report on why this is so and the events that led Descartes to the discovery of analytic geometry.

2. Find out what led Descartes to make his famous pronouncement, *"Je pense, donc je suis"* (I think, therefore I am), and write a report about the contents of one of his works, *La Geometrie*.

3. She was 19, a capable ruler, a good classicist, a remarkable athlete, and an expert hunter and horsewoman. Find out who this queen was and what connections she had with Descartes.

4. Upon her arrival at the University of Stockholm, one newspaper reporter wrote "Today we do not herald the arrival of some vulgar, insignificant prince of noble blood. No, the Princess of Science has honored our city with her arrival." Write a report identifying this woman and discussing the circumstances leading to her arrival in Sweden.

5. When she was 6 years old, the "princess's" room was decorated with a unique type of wallpaper. Write a paragraph about this wallpaper and its influence on her career.

6. In 1888, the "princess" won the Prix Bordin offered by the French Academy of Sciences. Write a report about the contents of her prizewinning essay and the motto that accompanied it.

Summary

SECTION	ITEM	MEANING	EXAMPLE
3.1A	Origin	The origin is where the x- and y-axes intersect.	The coordinates of the origin are $(0, 0)$.
	Quadrant	The x- and y-axes divide the plane into four quadrants.	
3.1B	Solution	An ordered pair (a, b) is a solution of an equation if a true statement results when a and b are substituted for x and y.	$(3, 4)$ is a solution of $x + y = 7$.
3.1C	Linear equation in standard form	Any equation that can be written in the form $Ax + By = C$ (A, B, C are integers, A and B not both zero, and A is positive)	$7x + 8y = 3$ and $2x = 3y - 2$ are linear equations.
	x-intercept	The x-coordinate of the point at which the graph crosses the x-axis	The x-intercept of $2x + y = 6$ is $x = 3$.
	y-intercept	The y-coordinate of the point at which the graph crosses the y-axis	The y-intercept of $2x + y = 6$ is $y = 6$.
3.1D	Horizontal line	A line whose equation can be written in the form $y = C$, C a constant	$y = 4$, $y = -3$, $y = 0.25$, and $2y = 6$ are equations of horizontal lines.
	Vertical line	A line whose equation can be written in the form $x = C$, C a constant	$x = 4$, $x = -3$, $x = 0.25$, and $2x = 6$ are equations of vertical lines.

(Continued)

SECTION	ITEM	MEANING	EXAMPLE
3.2A	Relation	A set of ordered pairs	$\{(2, -1), (4, 6), (5, 2)\}$
	Domain	The set of first coordinates of a relation	$\{2, 4, 5\}$ is the domain of the preceding relation.
	Range	The set of second coordinates of a relation	$\{-1, 2, 6\}$ is the range of the preceding relation.
3.2B	Function	A relation in which no two different ordered pairs have the same first coordinate	$\{(1, 2), (2, 4), (3, 6)\}$ is a function.
	Vertical line test	If any vertical line intersects the graph of a relation more than once, the relation is *not* a function.	$\{(1, 3), (1, 4)\}$ is *not* a function.
3.2D	Function notation	Use of a letter such as f to denote a function and $f(x)$ to mean the value of the function for the given value of x.	$f = \{(x, y) \mid y = x^2\}$. For this function, $f(x) = x^2$, $f(2) = 4$, and $f(-3) = 9$.
	Linear function	A function that can be written as $f(x) = mx + b$, where m and b are real numbers.	$f(x) = 3x + 5$ and $g(x) = -2x - 7$ are linear functions, but $h(x) = x^2 - 1$ and $i(x) = \frac{1}{x}$ are not.
3.3A	Slope	The slope of the line through (x_1, y_1) and (x_2, y_2) is $m = \frac{y_2 - y_1}{x_2 - x_1}$.	The slope of the line through $(3, 5)$ and $(9, 7)$ is $m = \frac{7 - 5}{9 - 3} = \frac{1}{3}$.
	Slope of a vertical line	The slope of a vertical line is undefined.	The slope of the line $x = 2$ is undefined.
	Slope of a horizontal line	The slope of a horizontal line is zero.	The slope of the line $y = -3$ is zero.
3.3B	Slopes of parallel lines	Two different lines L_1 and L_2 with slopes m_1 and m_2 are parallel if and only if $m_1 = m_2$.	The lines $y = -2x + 3$ and $y = -2x + 9$ are parallel.
	Slopes of perpendicular lines	Two lines L_1 and L_2 with slopes m_1 and m_2 are perpendicular if and only if $m_1 = \frac{-1}{m_2}$.	A line perpendicular to the line $y = -2x + 3$ will have a slope of $\frac{1}{2}$, because the slope of $y = -2x + 3$ is -2.
3.3D	Slope-intercept form of a line	Given the equation $y = mx + b$, $m = $ slope, and $b = y$-intercept	Given the equation $y = 2x + 3$, the slope $= 2$ and the y-intercept is 3.
3.4A	Given two points	An equation of the line going through the points (x_1, y_1) and (x_2, y_2) is $y - y_1 = m(x - x_1)$ where $m = \frac{y_2 - y_1}{x_2 - x_1}$.	An equation of the line through the points $(3, 1)$ and $(4, 3)$ is $y - 1 = 2(x - 3)$ since $m = \frac{3 - 1}{4 - 3} = \frac{2}{1} = 2$.
3.4B	Point-slope form of a line	An equation of the line going through the point (x_1, y_1) and with slope m is $y - y_1 = m(x - x_1)$.	An equation of the line going through the point $(2, 5)$ and with slope -3 is $y - 5 = -3(x - 2)$.
3.4C	Slope-intercept form of a line	An equation of the line with slope m and y-intercept b is $y = mx + b$.	An equation of the line with slope 3 and y-intercept -4 is $y = 3x - 4$.

SECTION	ITEM	MEANING	EXAMPLE
3.4E	Equations of horizontal and vertical lines	$y = c$, $m = 0$, horizontal $x = c$, m is undefined, vertical	An equation of the line with $m = 0$ and passing through the point $(2, 5)$ is $y = 5$.
3.5A	Linear inequality	A statement that can be written in the form $Ax + By \leq C$ or $Ax + By \geq C$ where A and B are not both zero	$5x - 2y \geq 10$ and $y \leq 3x + 1$ are linear inequalities. Graph the boundary line and then shade the side where the solutions occur.

Review Exercises

(If you need help with these exercises, look in the section indicated in brackets.)

1. [3.1A] Graph.

   a. $A(1, 4)$, $B(-3, 1)$, $C(-3, -2)$, and $D(3, -1)$

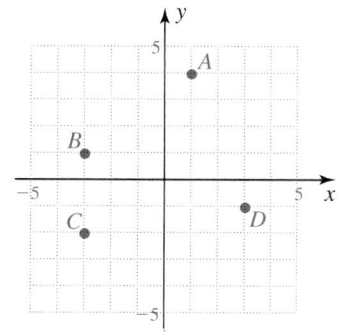

   b. $A(4, 1)$, $B(-1, 3)$, $C(-2, -3)$, and $D(1, -3)$

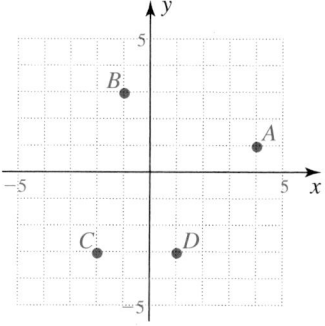

2. [3.1A] Determine the coordinates of each of the points.

   $A\,(2, 0)$;

   $B\,(-1, 1)$;

   $C\,(-3, -3)$;

   $D\,(4, -4)$

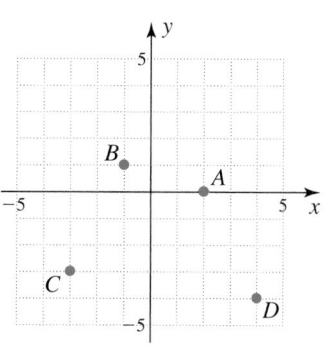

3. [3.1A] Determine the coordinates of each of the points.

   $A\,(2, 1)$;

   $B\,(0, 3)$;

   $C\,(-3, -1)$;

   $D\,(3, -3)$

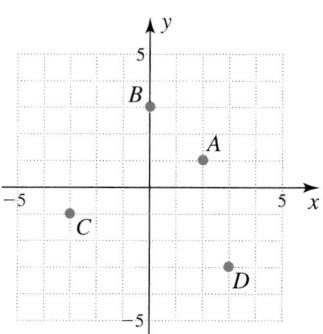

4. [3.1B] Graph the solutions to:

   a. $x + 2y = 4$

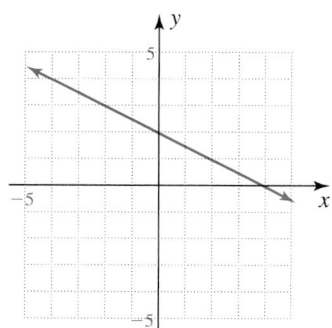

   b. $2x - y = 2$

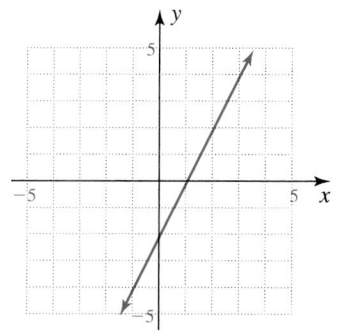

5. [3.1C] Find the *x*- and *y*-intercepts and graph the solutions to the line.

a. $y = 3x + 3$

$(-1, 0); (0,3)$

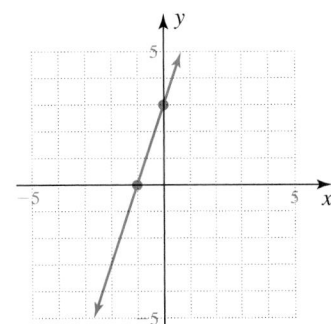

b. $y = 2x - 4$

$(2, 0); (0,-4)$

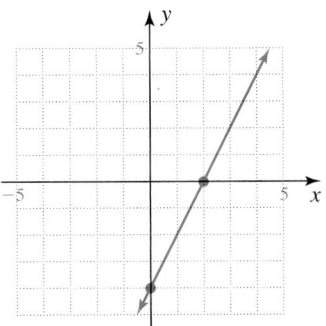

6. [3.1D] Graph the solutions to each equation on the same coordinate system.

a. $2x = 6$ **b.** $3y = 6$

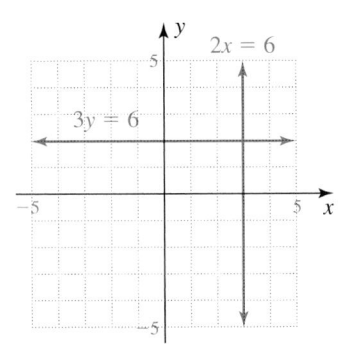

7. [3.1D] Graph the solutions to each equation on the same coordinate system.

a. $2x = -6$ **b.** $3y = -6$

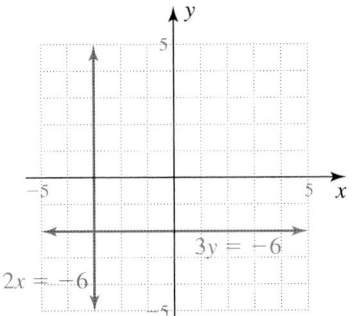

8. [3.2A] Find the domain and the range of each relation.

a. $\{(0, 5), (2, 9), (3, 10), (5, 8)\}$ $D = \{0, 2, 3, 5\}; R = \{5, 8, 9, 10\}$

b. $\{(0, 6), (2, 10), (3, 11), (5, 9)\}$ $D = \{0, 2, 3, 5\}; R = \{6, 9, 10, 11\}$

9. [3.2A] Find the domain and the range and graph the relation.

a. $\{(x, y) \mid y = 1 - x\}$

$D = \{x \mid x \text{ is a real number}\};$

$R = \{y \mid y \text{ is a real number}\}$

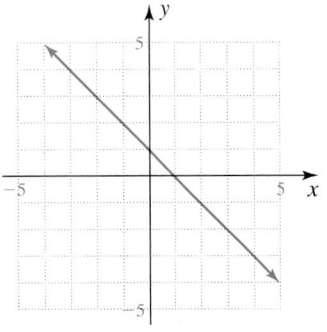

b. $\{(x, y) \mid y = 2 - x\}$

$D = \{x \mid x \text{ is a real number}\};$

$R = \{y \mid y \text{ is a real number}\}$

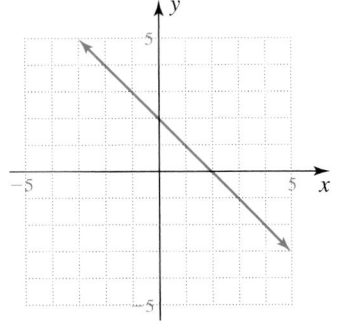

10. [3.2A] Find the domain and the range.

a.

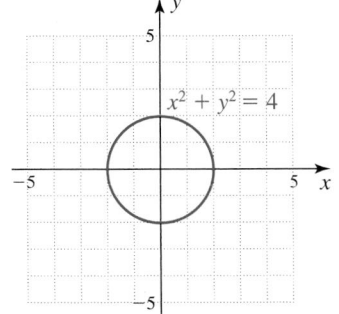

$D = \{x \mid -2 \le x \le 2\}; R = \{y \mid -2 \le y \le 2\}$

b.

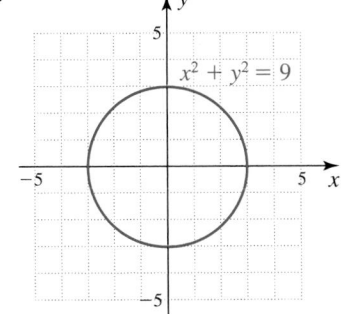

$D = \{x \mid -3 \le x \le 3\}; R = \{y \mid -3 \le y \le 3\}$

11. [3.2A] Find the domain and the range.

a.

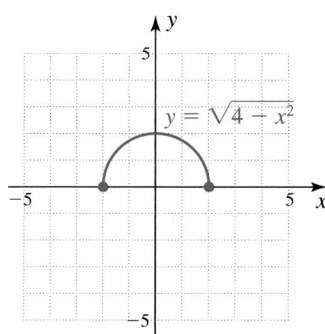

$D = \{x \mid -2 \le x \le 2\}; R = \{y \mid 0 \le y \le 2\}$

b.

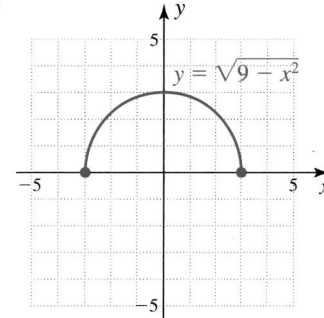

$D = \{x \mid -3 \le x \le 3\}; R = \{y \mid 0 \le y \le 3\}$

12. [3.2A] Find the domain and the range.

a.

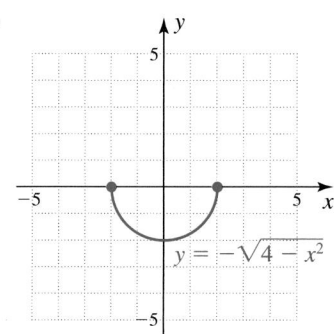

$D = \{x \mid -2 \le x \le 2\}; R = \{y \mid -2 \le y \le 0\}$

b.

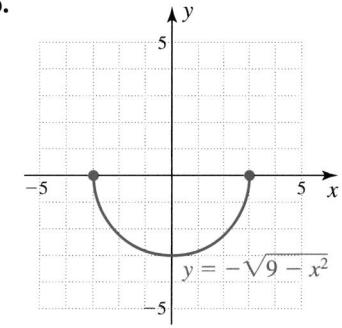

$D = \{x \mid -3 \le x \le 3\}; R = \{y \mid -3 \le y \le 0\}$

13. [3.2B] Use the vertical line test to determine whether the relation defines a function.

a.   Yes

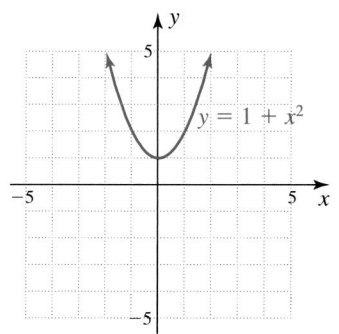

b.   No

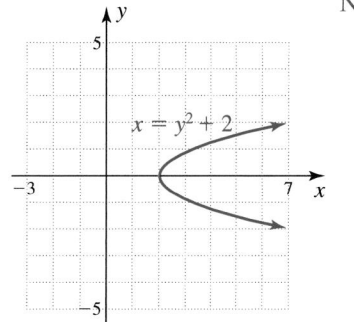

14. [3.2C] Find the domain of the function.

  a. $y = \dfrac{2}{x - 1}$ $D = \{x \mid x \text{ is a real number and } x \ne 1\}$

  b. $y = \dfrac{2}{x - 2}$ $D = \{x \mid x \text{ is a real number and } x \ne 2\}$

15. [3.2C] Find the domain of the function.

  a. $y = \sqrt{x - 3}$ $D = \{x \mid x \ge 3\}$

  b. $y = \sqrt{x - 4}$ $D = \{x \mid x \ge 4\}$

16. [3.2D] Let $f(x) = x - 4$. Find:

  a. $f(2)$ -2 **b.** $f(1)$ -3 **c.** $f(2) - f(1)$ 1

17. [3.2D] Let $f(x) = x^2 - 3$. Find:

  a. $f(2)$ 1 **b.** $f(1)$ -2 **c.** $f(2) - f(1)$ 3

18. [3.2D] Let $f = \{(2, 0), (3, 3), (1, -1)\}$. Find:

  a. $f(2)$ 0 **b.** $f(1)$ -1 **c.** $f(2) - f(1)$ 1

19. [3.2D] Let $f = \{(2, 1), (3, 4), (1, 0)\}$. Find:

  a. $f(2)$ 1 **b.** $f(1)$ 0 **c.** $f(2) - f(1)$ 1

20. [3.3A] Find the slope of the line through the given points.

   a. $A(-3, 2)$ and $B(1, 0)$ $m = \frac{-1}{2}$

   b. $A(4, -2)$ and $B(4, -7)$ Undefined

21. [3.3A] Find the slope of the line through the given points.

   a. $A(3, 4)$ and $B(4, 3)$ $m = -1$

   b. $A(1, -1)$ and $B(-3, 5)$ $m = \frac{-3}{2}$

22. [3.3B] A line L has slope $\frac{3}{4}$. Determine whether the line through the two given points is parallel or perpendicular to L.

   a. $A(1, 3)$ and $B(-2, 7)$ Perpendicular

   b. $A(1, 3)$ and $B(5, 6)$ Parallel

23. [3.3B] A line L has slope -2. Determine whether the line through the two given points is parallel or perpendicular to L.

   a. $A(2, -1)$ and $B(1, 1)$ Parallel

   b. $A(3, -1)$ and $B(2, -3)$ Neither

24. [3.3B] The line through $(2, 4)$ and $(5, y)$ is perpendicular to a line with the given slope. Find y.

   a. $m = \frac{3}{2}$ $y = 2$ **b.** $m = -2$ $y = 5\frac{1}{2}$ or $\frac{11}{2}$

25. [3.3C] A line passes through the point $(-1, 2)$ and has the given slope. Graph the line.

   a. $m = \frac{-1}{2}$ **b.** $m = \frac{2}{3}$

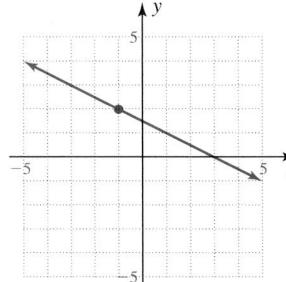   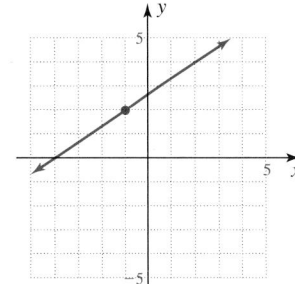

26. [3.3D] Find the slope and y-intercept for the given linear equation.

   a. $6x - y = 2$ $m = 6$; y-intercept $= -2$

   b. $4y = 2x - 8$ $m = \frac{1}{2}$; y-intercept $= -2$

27. [3.4A] Find an equation of the line through the two given points, and write the equation in standard form.

   a. $A(2, 5)$ and $B(-1, 2)$ $x - y = -3$

   b. $A(-4, 3)$ and $B(-2, -2)$ $5x + 2y = -14$

28. [3.4B] Find an equation of the line with slope 2 and passing through the given point, and write the equation in standard form.

   a. $A(-1, -3)$ $2x - y = 1$

   b. $A(5, 0)$ $2x - y = 10$

29. [3.4C]

   a. A line has slope 3 and y-intercept 2. Find the slope-intercept equation of this line. $y = 3x + 2$

   b. Repeat part **a** if the slope is -3 and the y-intercept is 4. $y = -3x + 4$

30. [3.4D] Find an equation of the line through the point $(2, 1)$ and parallel to the line

   a. $2x + y = 7$ $y = -2x + 5$

   b. $3x - y = 4$ $y = 3x - 5$

31. [3.4D] Find an equation of the line through the point $(2, 1)$ and perpendicular to the line.

   a. $2x + 3y = 7$ $y = \frac{3}{2}x - 2$

   b. $3x - 2y = 4$ $y = \frac{-2}{3}x + \frac{7}{3}$

32. [3.4E] Find the equation of the line passing through the point $(-3, 7)$ with

   a. slope 0 $y = 7$

   b. undefined slope $x = -3$

33. [3.4F] The points in the following table represent a linear model. Plot the points on a graph, draw the line of best fit, and write the equation of that line in slope-intercept form.

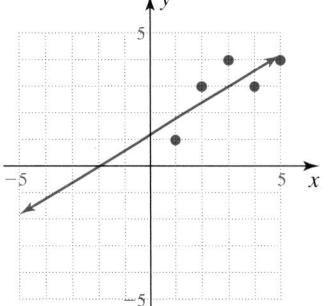

x	1	2	3	4	5
y	1	3	4	3	4

$y = \frac{3}{5}x + \frac{6}{5}$ (Answers may vary.)

34. [3.4G] Find the equation of the line shown in the graph.

$y = \frac{-1}{3}x + 1$

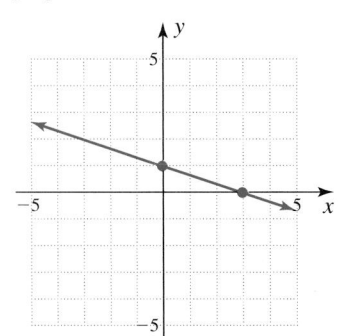

35. [3.5A] Graph:

   a. $2x - y < 1$ **b.** $y \geq x - 3$ **c.** $x \geq 4$ **d.** $2y < -6$

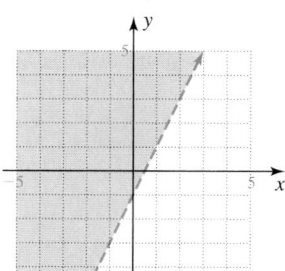 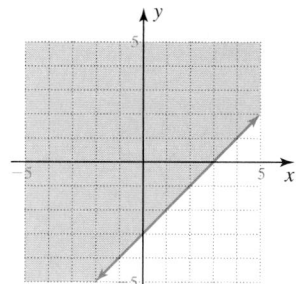 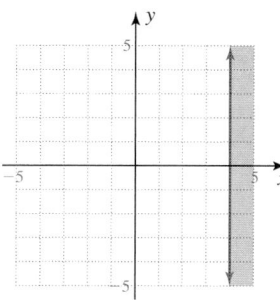 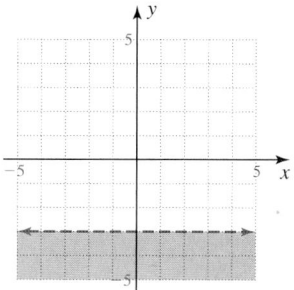

36. [3.5B] Graph:

   a. $|y| > 2$ **b.** $|x + 2| \leq 3$

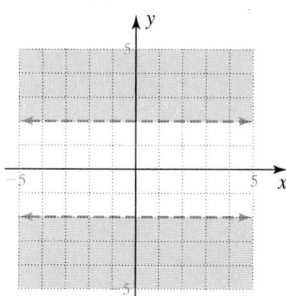 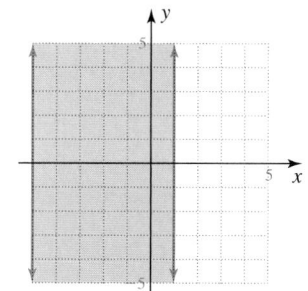

Practice Test 3

(Answers on pages 242–246)

1. a. Graph $A(3, 2)$, $B(4, -2)$, $C(-1, -2)$, and $D(-1, 3)$.

   b. Find the coordinates of the points in the figure.

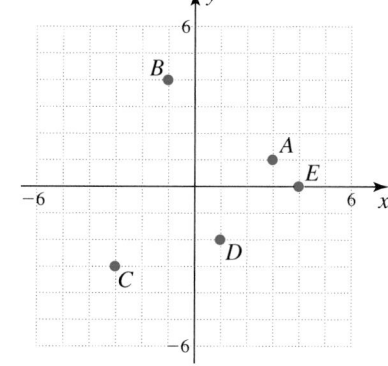

2. Graph the solutions to $3x - y = 3$.

3. Find the x- and y-intercepts of $y = 3x + 2$ and then graph the solutions to the equation.

4. Graph the solutions to:

   a. $3x = -9$ **b.** $2y = -4$

5. Find the domain and range of the relation $\{(1, 3), (2, 5), (3, 7), (4, 9)\}$.

6. Find the domain and the range of the relation $y = 4 + x$ shown in the graph.

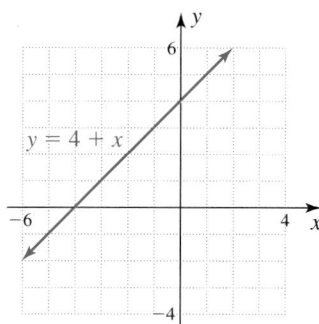

7. Find the domain and the range of the relation.

a.
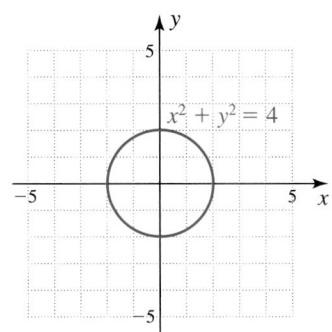

b.
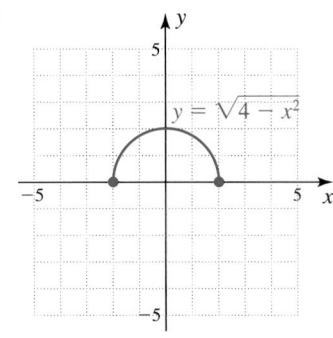

c.
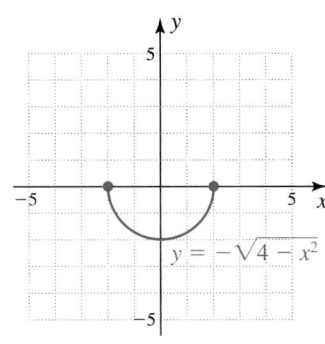

8. Use the vertical line test to determine whether the graph of the given relation defines a function.

a.
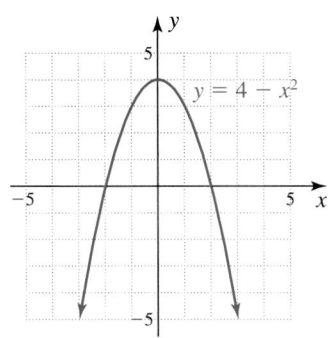

b.
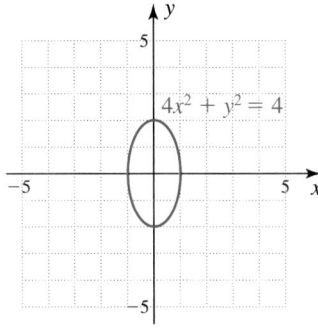

9. Find the domain of the function.

a. $y = \dfrac{2}{x + 3}$ **b.** $y = \sqrt{x + 2}$

10. Let $f(x) = 4x - 3$. Find:

a. $f(2)$ **b.** $f(1)$ **c.** $f(2) - f(1)$

11. Let $f = \{(1, 4), (2, -1), (3, 2)\}$. Find:

a. $f(2)$

b. $f(1)$

c. $f(2) - f(1)$

12. Find the slope of the line through the two given points.

a. $A(-2, 2)$ and $B(1, 1)$

b. $A(-2, 4)$ and $B(-2, 8)$

c. $A(5, 6)$ and $B(-5, -2)$

13. A line L_1 has slope $\frac{3}{2}$. Determine whether the line through the two given points is parallel or perpendicular to L_1.

a. $A(4, 2)$ and $B(1, 4)$

b. $A(-1, -4)$ and $B(-4, -6)$

14. The line through $A(1, -2)$ and $B(-1, y)$ is perpendicular to a line with slope $\frac{-2}{3}$. Find y.

15. A line has y-intercept 1 and has slope $\frac{-1}{2}$. Graph this line.

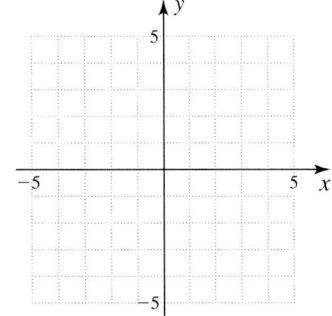

16. Find the slope and y-intercept of $4x - y = 3$.

17. Find an equation of the line through $(4, 3)$ and $(2, 4)$. Then write the equation in standard form and graph the line.

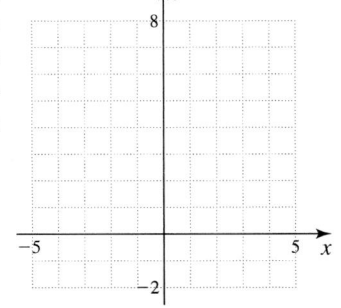

18. Find an equation of the line with slope -2 and passing through the point $(2, -3)$. Then graph the line.

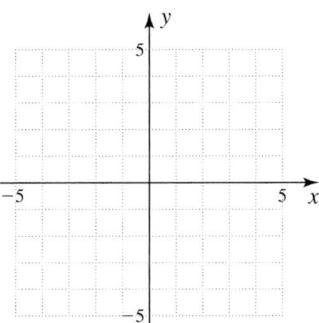

19. A line has slope 3 and y-intercept 2. Find the slope-intercept equation of this line and graph the line.

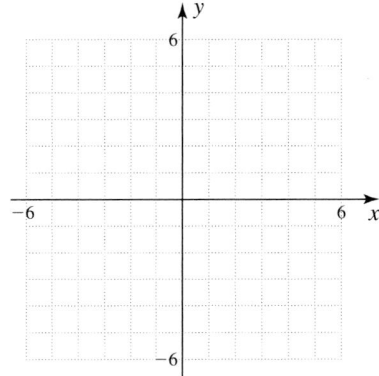

20. Find an equation of the line through the point $(1, 2)$ and

a. Parallel to the line $2x - 3y = 5$. Write the equation in standard form.

b. Perpendicular to the line $2x - 3y = 5$. Write the equation in standard form.

21. Given that a line passes through the point $(6, -5)$, find the equation of the line if the slope is 0.

22. Given that a line passes through the point $(-2, 4)$, find the equation of the line if it is a vertical line.

23. The points in the table represent a linear model. Plot the points on a graph, draw the line of best fit, and write the equation of that line in slope-intercept form.

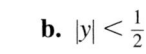

x	5	6	7	8	9
y	3	2	5	4	6

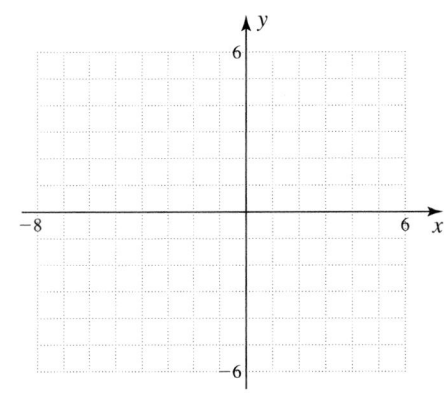

24. Graph:

a. $x + 4y < 4$ **b.** $y + 2 \geq 1$

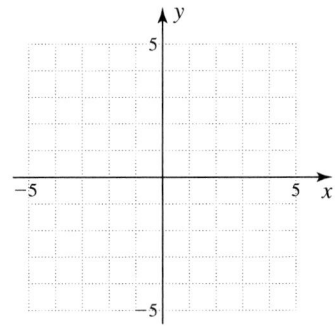

25. Graph:

a. $|x + 1| \geq 5$ **b.** $|y| < \frac{1}{2}$

Answers to Practice Test

ANSWER	IF YOU MISSED		REVIEW		
	QUESTION		**SECTION**	**EXAMPLES**	**PAGE**
1. a.	1a		3.1A	1	164
b. $A(3, 1)$; $B(-1, 4)$; $C(-3, -3)$; $D(1, -2)$; $E(4, 0)$	1b		3.1A	2	165
2.	2		3.1B	3	166
3. x: $(\frac{-2}{3}, 0)$; y: $(0, 2)$	3		3.1C	4	168
4. a.	4a		3.1D	5	169

ANSWER	IF YOU MISSED	REVIEW		
	QUESTION	SECTION	EXAMPLES	PAGE
b.	4b	3.1D	5	169
5. $D = \{1, 2, 3, 4\}$; $R = \{3, 5, 7, 9\}$	5	3.2A	1	177
6. The domain and range are the set of real numbers.	6	3.2A	2	177
7. a. $D = \{x \mid -2 \le x \le 2\}$; $R = \{y \mid -2 \le y \le 2\}$	7	3.2A	3	178
b. $D = \{x \mid -2 \le x \le 2\}$; $R = \{y \mid 0 \le y \le 2\}$				
c. $D = \{x \mid -2 \le x \le 2\}$; $R = \{y \mid -2 \le y \le 0\}$				
8. a. A function	8	3.2B	4	179
b. Not a function				
9. a. All real numbers except -3	9	3.2C	5	180
b. All real numbers greater than or equal to -2				
10. a. 5	10	3.2D	6	182
b. 1				
c. 4				
11. a. -1	11	3.2D	7	182
b. 4				
c. -5				
12. a. $\dfrac{-1}{3}$	12	3.3A	1	193–194
b. Undefined				
c. $\dfrac{4}{5}$				
13. a. Perpendicular	13	3.3B	3	197
b. Neither				
14. $y = -5$	14	3.3B	4	198

ANSWER	IF YOU MISSED	REVIEW		
	QUESTION	SECTION	EXAMPLES	PAGE
15. 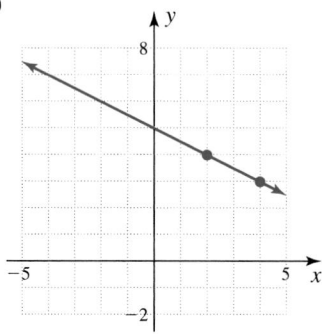	15	3.3C	5	199
16. Slope $= 4$; y-intercept $= -3$	16	3.3D	6	200–201
17. $x + 2y = 10$ 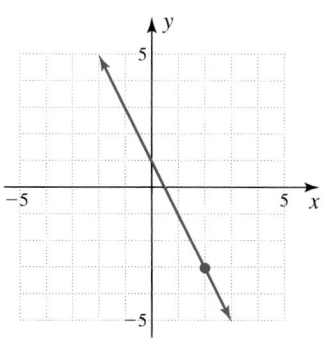	17	3.4A	1	209
18. $2x + y = 1$ 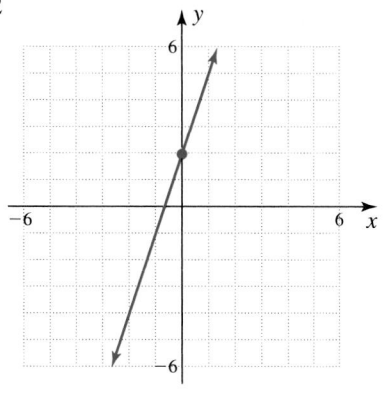	18	3.4B	2	209
19. $y = 3x + 2$	19	3.4C	3	210

ANSWER	IF YOU MISSED		REVIEW	
	QUESTION	SECTION	EXAMPLES	PAGE
20. a. $2x - 3y = -4$	20	3.4D	4	210–211
b. $3x + 2y = 7$				
21. $y = -5$	21	3.4E	5	211–212
22. $x = -2$	22	3.4E	5	211–212
23. $y = 0.8x - 1.6$ (Answers will vary.)	23	3.4F	6	212
24. a.	24a	3.5A	1	222
b.	24b	3.5A	3	224

	ANSWER	IF YOU MISSED		REVIEW	
		QUESTION	SECTION	EXAMPLES	PAGE
25. a.		25a	3.5B	5	225

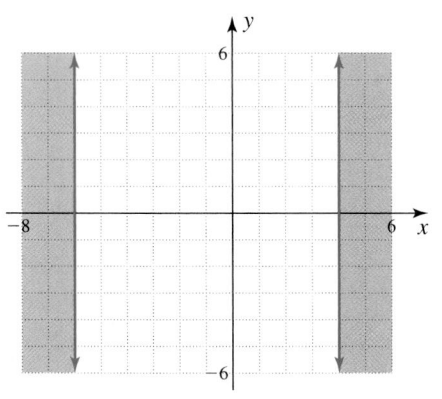

	ANSWER	IF YOU MISSED		REVIEW	
b.		25b	3.5B	4	224

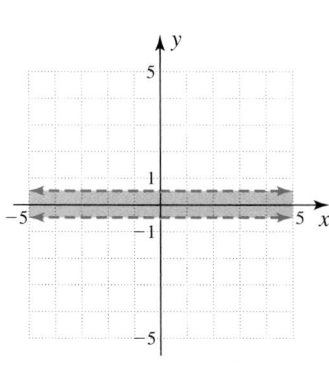

Cumulative Review Chapters 1–3

1. The number $\sqrt{13}$ belongs to which of these sets? Natural numbers, Whole numbers, Integers, Rational numbers, Irrational numbers, Real numbers. Name all that apply. Irrational numbers; Real numbers

3. Find: $\dfrac{-1}{6} \div \left(\dfrac{-7}{12}\right)$ $\dfrac{2}{7}$

5. Evaluate: $-7^3 + \dfrac{(14 - 10)}{2} + 35 \div 7$ -336

7. Solve: $|x - 5| = |x - 9|$ 7

9. Graph: $\{x \mid x > -4 \text{ and } x < 4\}$

11. The sum of three consecutive odd integers is 75. What are the three integers? 23; 25; 27

2. Graph the additive inverse of $\dfrac{5}{2}$ on the number line.

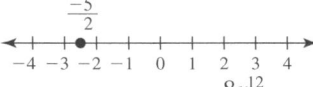

4. Simplify: $(2x^4y^{-4})^3$ $\dfrac{8x^{12}}{y^{12}}$

6. Solve: $\dfrac{3}{7}y - 4 = 2$ 14

8. Graph: $x \geq 1$

10. If $H = 2.45h + 72.98$, find h when $H = 136.68$. 26

12. A freight train leaves a station traveling at 30 mi/hr. Two hours later, a passenger train leaves the same station in the same direction at 40 mi/hr. How far from the station does the passenger train overtake the freight train? 240 mi

13. Find the range of the relation
$\{(-2, -3), (5, -2), (-3, -4)\}$. $\{-3, -2, -4\}$

14. Find the domain and range of $\{(x, y) | y = 6 + x\}$.
$D = \{\text{all real numbers}\};$
$R = \{\text{all real numbers}\}$

15. Find the domain of $y = \sqrt{x + 25}$. $\{x | x \geq -25\}$

16. Let $f = \{(-2, 1), (-3, 2), (1, 4)\}$.
Find $f(-3) - f(-2)$. 1

17. The lower limit L (heartbeats per minute) of your target zone is given by $L(a) = \frac{-2}{3}a + 150$, where a is your age in years. Find L for a person who is 24 years old. 134

18. Graph: $x - y = 2$

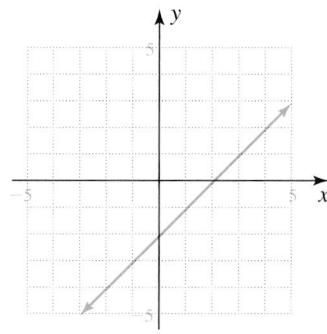

19. Find the x- and y-intercepts of $y = -5x + 2$.
x-intercept: $\frac{2}{5}$; y-intercept: 2

20. Graph: $2x = -2$

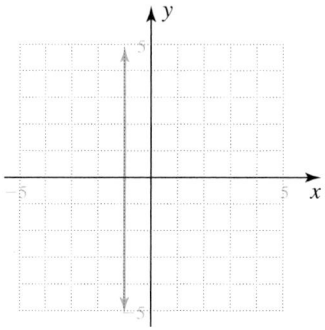

21. A line L_1 has slope 1. Find whether the line through $(4, 6)$ and $(7, 3)$ is parallel or perpendicular to line L_1.
Perpendicular

22. Find an equation of the line with slope -2 and passing through the point $(-2, 1)$. Write the answer in standard form. $2x + y = -3$

23. Find the slope and the y-intercept of the line $6x - 3y = 36$. $m = 2$; y-intercept: -12

24. Find an equation of the line that passes through the point $(-2, 6)$ and is parallel to the line $4x - 3y = -1$.
$4x - 3y = -26$

25. Graph: $y \leq 6x - 6$

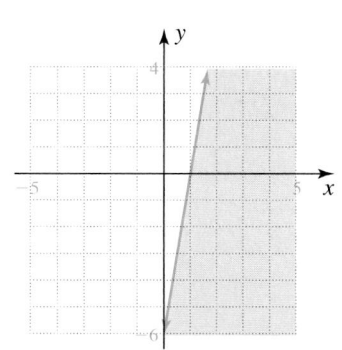

Solving Systems of Linear Equations and Inequalities

4

Who was the real author of Cramer's rule?

The Human Side of Algebra

The study of systems of linear equations in the Western world was initiated by Gottfried Wilhelm Leibniz. In 1693, Leibniz solved a system of three equations by eliminating two of the unknowns and using a determinant to obtain the solution. It was Colin Maclaurin, however, who used determinants to solve simultaneous linear equations in two, three, and four unknowns. Maclaurin was born in Scotland and educated at the University of Glasgow, which he entered at the incredible age of 11! He became professor of mathematics at Aberdeen at 19 and taught at the prestigious University of Edinburgh at 25.

Ironically, Maclaurin's name is associated with a portion of analysis (the Maclaurin series) discovered by Brook Taylor. The general Taylor series, in turn, had been known long before to James Gregory and Jean Bernoulli. If Maclaurin's name is recalled in connection with a series he did not first discover, this is compensated by a contribution he did make that bears the name of someone else who discovered and printed it later: Cramer's rule, published in 1750, was probably known to Maclaurin as early as 1729.

Pretest for Chapter 4

(Answers on page 251)

1. Use the graphical method to solve the system.

$$2x + y = -2$$
$$y = 2x - 6$$

2. Use the substitution method to solve the system.

$$3x - 2y = 2$$
$$2x + y = 6$$

3. Solve by the elimination method.

$$-5x + 3y = 6$$
$$2x - y = 1$$

4. Solve the system.

$$y + 3x = 4$$
$$4y + 12x = 8$$

5. Solve the system.

$$2x - 4y = -2$$
$$5x = 10y - 5$$

6. Solve the system.

$$3x - \frac{5}{2}y = \frac{13}{2}$$
$$\frac{3}{4}x = 3y - \frac{3}{4}$$

7. The demand function for a certain product is $D(p) = 203 - 4p$ (where p is the price in dollars), and the supply function is $S(p) = 7p + 170$. What will the price of this product be at the equilibrium point? What will the demand for the product be at the equilibrium?

8. Solve the system.

$$x + y + z = 8$$
$$-x + 2y + 4z = 6$$
$$3x - y + 2z = -6$$

9. Solve the system.

$$x + y + z = 14$$
$$2x - y + z = 4$$
$$x + y + z = 6$$

10. The total cars sold for GM, Ford, and Chrysler for the year 2002 was 20 million. The sales for GM and Ford combined were 11 million more than Chrysler. Chrysler sold 4 million less than GM. How many million did each of the companies sell?

11. Norma has $6.25 in quarters and dimes. She has 40 coins in all. How many quarters and how many dimes does she have?

12. A biologist estimates that the region she is studying can supply 50,000 kg of food for birds during the nesting season. The region has an area of 130,000 sq meters. There are two species of birds that compete for food and territory in that region. Each nesting pair of species X needs 30 kg of food and has a territory of 90 sq meters. Each nesting pair of species Y needs 50 kg of food and has a territory of 70 sq meters. How many nesting pairs of each species can the area support?

13. A boat travels 30 miles downstream for one hour. On the return trip the boat is going upstream and takes $1\frac{1}{2}$ hours. What is the speed of the boat in still water and the speed of the current?

14. Chip wants to invest $10,000—part at 4% and part at 6%. If the total annual interest income is to be $550, find the amount invested at each rate.

15. Write the augmented matrix in echelon form.

$$\begin{bmatrix} 1 & 1 & 1 & 8 \\ -1 & 2 & 4 & 6 \\ 3 & -1 & 2 & -4 \end{bmatrix}$$

16. Evaluate the determinant. $\begin{vmatrix} 3 & -2 \\ 4 & -1 \end{vmatrix}$

17. Solve by Cramer's rule.

$$3x - 4y = 14$$
$$5x - 7y = 18$$

18. Evaluate the determinant. $\begin{vmatrix} -1 & 0 & -2 \\ 1 & 4 & -3 \\ 0 & 2 & 5 \end{vmatrix}$

19. Solve by Cramer's rule.

$$x - 2y + 2z = 6$$
$$3x + y = 3$$
$$2x + y + z = 0$$

20. Graph the solution set of the system of linear inequalities.

$$2x - 3y < 6$$
$$x \leq 3$$
$$y \leq 1$$

Answers to Pretest

	ANSWER	IF YOU MISSED		REVIEW	
		QUESTION	SECTION	EXAMPLES	PAGE
1.	$(1, -4)$	1	4.1	1	253
2.	$(2, 2)$	2	4.1	4	257
3.	$(9, 17)$	3	4.1	7	259–260
4.	No solution; inconsistent system	4	4.1	8	261
5.	Infinitely many solutions; dependent $\{(x, y) \mid y = \frac{1}{2}x + \frac{1}{2}\}$	5	4.1	6	258
6.	$(3, 1)$	6	4.1	9	262
7.	Price is \$3; demand is 191	7	4.1	10	265
8.	$(2, 8, -2)$	8	4.2	1	275
9.	No solution; inconsistent system	9	4.2	2	275
10.	GM $= 8.5$, Ford $= 7$, Chrysler $= 4.5$ (in millions)	10	4.2	4	278–279
11.	15 quarters; 25 dimes	11	4.3	1	284–285
12.	$X = 1250$; $Y = 250$	12	4.3	2	285–286
13.	Boat $= 25$ mi/hr; current $= 5$ mi/hr	13	4.3	3	286–287
14.	\$7500 at 6%; \$2500 at 4%	14	4.3	4	287–288
15.	$\begin{bmatrix} 1 & 1 & 1 & 8 \\ 0 & 3 & 5 & 14 \\ 0 & 0 & 17 & -28 \end{bmatrix}$	15	4.4	1, 2, 3	298–300
16.	5	16	4.5	1	308
17.	$(26, 16)$	17	4.5	2	309–310
18.	-30	18	4.5	3	311–312
19.	$(2, -3, -1)$	19	4.5	5	313–314
20.		20	4.6	3	323

4.1 SYSTEMS WITH TWO VARIABLES

To Succeed, Review How To . . .

1. Find the *x*- and *y*-intercepts of a line (pp. 167–168).

2. Graph a line (pp. 165–169).

3. Find the slope of a line given its equation (pp. 199–200).

Objectives

Find the solution of a system of two linear equations using:

A The graphical method

B The substitution method

C The elimination method

D Solve applications involving systems of equations

GETTING STARTED **How Do Yearly Sales of SUVs Compare to Passenger Car Sales?**

Projected Yearly SUV Sales Versus Passenger Car Sales in North America

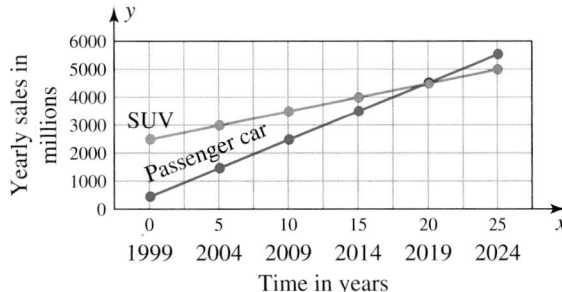

Sport utility vehicles (SUVs) are becoming more popular among car buyers. From 1999 to 2000, the number of yearly sales in North America in millions of SUVs sold and the number of yearly sales in millions of passenger cars sold can be approximately represented by

$$\text{SUVs sold: } y = 100x + 2500$$
$$\text{Passenger cars sold: } y = 200x + 500$$

where *x* represents the year, with *x* = 0 corresponding to 1999. This *system* of *two* equations can be graphed as shown.

If this rate continues, in which year will there be as many SUVs sold as passenger cars? The graph indicates it will occur in 2019—when *x* = 20.

Using the first equation from the system and replacing *x* with 20, the number of vehicles (in millions) sold was

$$y = 100(20) + 2500 = 4500$$

If we denote the year by *x* and the number of vehicles sold by *y*, the point $(x, y) = (20, 4500)$ represents the point at which the lines *intersect*. This point, (20, 4500), is the *solution* of the system of equations. The point (20, 4500) *satisfies* both equations.

We can also solve this system of equations by substitution. The variable *y* is the same in both equations. We substitute $100x + 2500$ for *y* in the second equation to obtain

$$100x + 2500 = 200x + 500 \qquad$$ There are no fractions to *C*lear or parentheses to *R*emove, so we can solve this by

$$2500 = 100x + 500 \qquad$$ *A*dding $(-100x)$ on both sides and then (-500) to both sides

$$2000 = 100x$$

$$20 = x \qquad$$ Multiplying both sides by $\frac{1}{100}$

If you substitute 20 for *x* in the first equation (as we did before), you obtain 4500 for the value of *y*.

These are the graphical and substitution methods for solving a system of equations; we investigate these and other methods in this section.

Finding Solutions Using the Graphical Method

Teaching Tip

A simple introduction to systems of equations: Have students name two numbers whose sum is 6 and difference is 4. They soon will answer 5 and 1. Then see if they can write two equations using x and y to illustrate the two requirements for the solution, namely, $x + y = 6$ and $x - y = 4$. They have just written a system of equations. The solution $x = 5$ and $y = 1$ is the only pair that will satisfy both equations. Thus we represent the solution to the system as the ordered pair, (5, 1).

As we have seen, the solution set of a system of linear equations can be estimated by graphing the equations on the same axes and determining the coordinates of any points of intersection. This is called the **graphical method,** and we use it in Example 1. Remember, the *solution* of the system is a point (an ordered pair) that satisfies both equations.

Calculate It Finding a Solution of a Linear System by Graphing

In Example 1, rewrite the first equation, $2x - y = 2$, so that it is solved for y ($y = 2x - 2$). Then graph $Y_1 = 2x - 2$ and $Y_2 = x - 1$. To verify that the solution is (1, 0), use a decimal window by pressing ⓩⓄⓄⓂ **4** , use the ⓣⓇⒶⒸⒺ key and arrows until you find the point of intersection or, with a TI-83 Plus; use the intersect feature by pressing ②ₙ𝒹 ⓣⓇⒶⒸⒺ **5** ⒺⓃⓉⒺⓇ ⒺⓃⓉⒺⓇ ⒺⓃⓉⒺⓇ to find the intersection, as shown in the window.

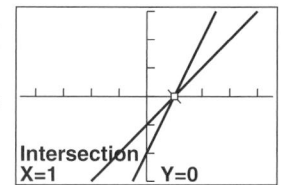

EXAMPLE 1 **Using the graphical method to solve a system with one solution**

Use the graphical method to find the solution of the system:

$$2x - y = 2$$
$$y = x - 1$$

SOLUTION We first graph the equation $2x - y = 2$ using the x- and y-intercepts shown in the table.

When $x = 0$, $2x - y = 2$ becomes $2(0) - y = 2$, or $y = -2$.

When $y = 0$, $2x - y = 2$ becomes $2x - 0 = 2$, or $x = 1$.

x	y
0	−2
1	0

Thus the points $(0, -2)$ and $(1, 0)$ are on the graph of the equation. We draw a line through these two points (the blue line in Figure 1). The two points and the complete graph are shown in color. We then graph $y = x - 1$.

The equation is already in slope-intercept form ($y = mx + b$). We can say the y-intercept is -1 (the constant) and the slope of the line is 1 (the coefficient of x). We use this information to graph the line like we did in Section 3.3. We plot -1 on the y-axis and use the slope ($1 = \frac{1}{1}$) to plot the point $(1, 0)$ so we can draw the line through the two points.

The graph for $y = x - 1$ is shown in red in Figure 1. The lines intersect at $(1, 0)$. The point $(1, 0)$ is the solution of the system of equations, and the system is said to be **consistent** because it has *one* solution. In fact, if a system has one *or more* solutions, it is called a consistent system.

CHECK For $x = 1$, $y = 0$, $2x - y = 2$ becomes $2(1) - 0 = 2$ (true). For $x = 1$, $y = 0$, $y = x - 1$ becomes $0 = 1 - 1$ (true). Thus $(1, 0)$ is the correct solution for the system; the solution set is $\{(1, 0)\}$.

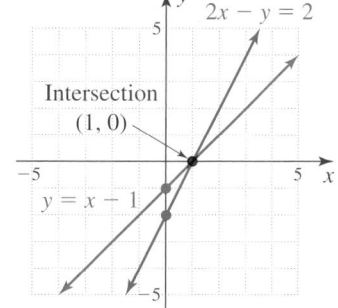

Figure 1

PROBLEM 1

Use the graphical method to find the solution of the system:

$$x - y = -1$$
$$y = -x - 1$$

Answer

1.

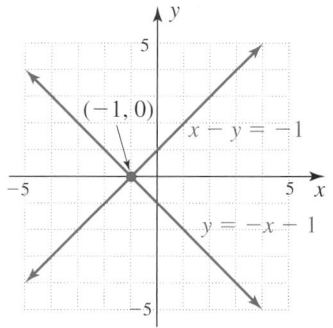

Solution: $(-1, 0)$

EXAMPLE 2 **Using the graphical method to solve a system with no solution**

Use the graphical method to find the solution of the system:

$$-2x + y = 4$$
$$-6x + 3y = 18$$

SOLUTION We first graph the equation $-2x + y = 4$ by solving the equation for y so it will be in slope-intercept form ($y = mx + b$). That will make it easy to name the y-intercept and slope of the line. We will use the y-intercept and slope to graph the line like we did in Example 1.

$$-2x + y = 4 \qquad \text{To solve for } y, \text{ add } 2x \text{ to both sides.}$$
$$y = 2x + 4 \qquad \text{The } y\text{-intercept is 4 and the slope is 2.}$$

We plot 4 on the y-axis and use the slope ($2 = \frac{2}{1}$) to plot the point $(1, 6)$ so we can draw the line through the two points. The two points, as well as the completed graph, are shown in red in Figure 2. We then graph $-6x + 3y = 18$.

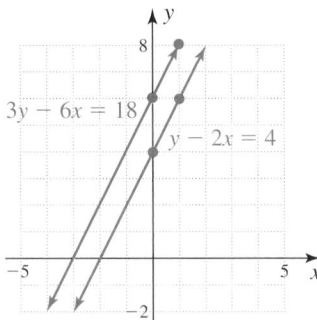

Figure 2

We solve the equation for y by adding $6x$ to both sides of the equation and then dividing by 3.

$$-6x + 3y = 18$$
$$3y = 6x + 18$$
$$y = 2x + 6 \qquad \text{The } y\text{-intercept is 6 and the slope is } \frac{2}{1}.$$

We plot 6 on the y-axis and use the slope, $\frac{2}{1}$, to plot the point $(1, 8)$ so we can draw the blue line in Figure 2 through the two points. The graph of the two lines we have drawn in Figure 2 look parallel. However, we must be mathematically sure that they are parallel lines. What do we notice when we compare the slopes of these two lines? Yes, they are the same—both have a slope of 2. In Chapter 3 we learned that if lines have the same slopes then they are parallel. So, not only do the lines look parallel, but we can conclude that they are parallel. Notice however that they have different y-intercepts. The lines are parallel so their graphs cannot intersect. There is *no solution* for this system. The solution set is $\varnothing$ and the system is said to be **inconsistent.** In fact, *any* system with no solution is called an inconsistent system.

PROBLEM 2

Use the graphical method to find the solution of the system:

$$y - 3x = 3$$
$$2y - 6x = 12$$

Web It

For another example of how to solve systems by the graphing method, try link 4-1-1 at mhhe.com/bello.

Teaching Tip

Before discussing the slope of these two lines, remind students of the slope-intercept form of equations. Have them rewrite $-2x + y = 4$ and $3y - 6x = 18$ in $y = mx + b$ form and name the slope.

Answer

2.

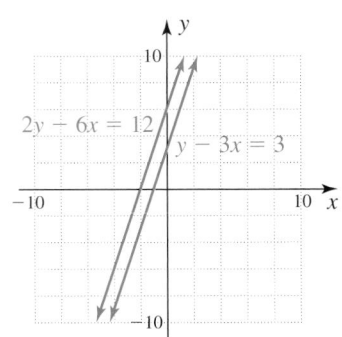

No solution; inconsistent

| **EXAMPLE 3** | **Using the graphical method to solve a system with infinitely many solutions** |

Use the graphical method to find the solution of the system:

$$2x + \frac{1}{2}y = 2$$
$$y = -4x + 4$$

SOLUTION   To write the first equation in slope-intercept form we solve for y by adding $(-2x)$ to both sides and then multiply both sides by 2.

$$2x + \frac{1}{2}y = 2$$

$$\frac{1}{2}y = -2x + 2$$

$$y = -4x + 4 \qquad \text{(}y\text{-intercept 4 and slope } -4\text{)}$$

The graph is shown in Figure 3. The second equation is already solved for y, and it is exactly the same as the equation we got for the first equation when we solved for y. What does this mean? It means that the graphs of the lines $2x + \frac{1}{2}y = 2$ and $y = -4x + 4$ coincide (are the same). Thus a solution of one equation is automatically a solution of the other. This solution set is $\{(x, y)\,|\,y = -4x + 4\}$. There are *infinitely* many solutions. Such a system is called **dependent**. Since the system has one or more solutions, it is also a consistent system.

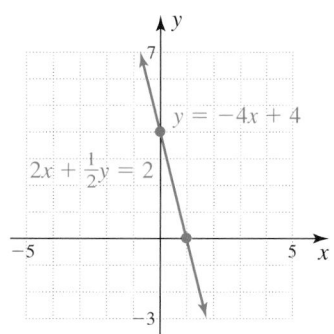

Figure 3

| **PROBLEM 3** |

Use the graphical method to find the solution of the system:

$$x + \frac{1}{2}y = -2$$
$$y = -2x - 4$$

Web It

For a step-by-step procedure for solving systems on the TI-83 Plus and some practice problems, try link 4-1-2 at mhhe.com/bello.

A system of two linear equations in two variables can have one of three possible solution sets. They are summarized in the following table followed by their graphs in Figure 4. For ease of classification, we will use the name "consistent" for a system with one solution, even though a dependent system is also consistent.

Answer

3.

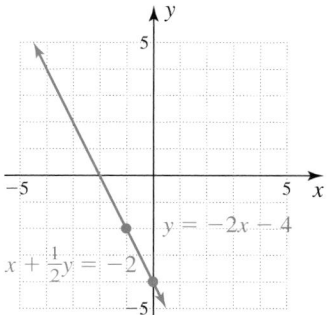

Infinitely many solutions;
dependent system

Summary Table for Solving a System by Graphing

Solution	Name of System	Slope Comparison	y-intercept Comparison
One Solution Lines intersect (Example 1)	**Consistent**	Different slopes	
No Solution Lines parallel (Example 2)	**Inconsistent**	Same slopes	Different y-intercepts
Infinitely Many Solutions Lines coincide (Example 3)	**Dependent**	Same slopes	Same y-intercepts

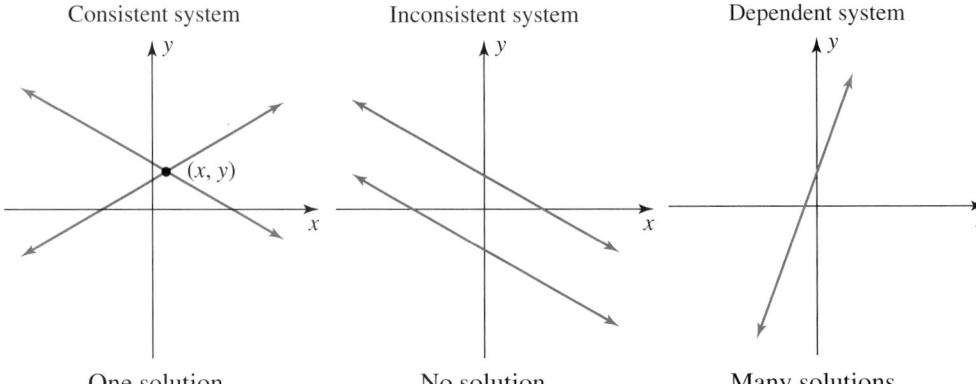

Consistent system	Inconsistent system	Dependent system
One solution	No solution	Many solutions
Two lines that *intersect* at the point (x, y); the solution is (x, y).	Two lines that are *parallel* and thus do not *intersect;* there is no solution, and the solution set is $\varnothing$.	Two lines that *coincide;* there are infinitely many solutions.

Figure 4

B   Finding Solutions Using the Substitution Method

Sometimes the graphical method isn't accurate enough to determine the exact solutions of a system of equations. For example, to solve the system

$$y = 2.5x + 72$$
$$y = 3x + 70$$

we can start by letting $x = 0$ in the first equation to obtain $y = 2.5(0) + 72$, or $y = 72$. To graph this point, we need a piece of graph paper with 72 units, or we have to make each division on the graph paper 10 units. But there is a way out. Keep in mind that we are looking for a point that satisfies *both* equations and notice that the second equation tells us that y is $3x + 70$. Thus we can *substitute* $3x + 70$ for y in the equation $y = 2.5x + 72$ to obtain

$$
\begin{array}{ll}
3x + 70 = 2.5x + 72 & \\
3x = 2.5x + 2 & \text{Subtract 70.} \\
0.5x = 2 & \text{Subtract } 2.5x. \\
x = \dfrac{2}{0.5} & \text{Divide by 0.5.} \\
x = 4 &
\end{array}
$$

Now substitute 4 for x in the equation $y = 3x + 70$ to obtain $y = 3(4) + 70 = 82$.

The solution of the system is $(4, 82)$, and the method we used is called the **substitution method.**

NOTE

The substitution method is most useful when both equations are solved for one of the variables in terms of the other (as in the system we just solved) or when at least one of the equations is in this form, as in Example 1.

To solve the system of Example 1 by substitution, we write

$$2x - y = 2$$
$$y = x - 1$$

Since $y = x - 1$, we substitute $x - 1$ for y in the first equation, $2x - y = 2$, which becomes

$$2x - \underline{y} = 2$$
$$2x - (x - 1) = 2$$
$$2x - x + 1 = 2$$
$$x + 1 = 2$$
$$x = 1$$

Substituting 1 for x in the second equation, $y = x - 1$, gives

$$y = 1 - 1 = 0$$

The solution is $(1, 0)$, and the solution set is $\{(1, 0)\}$ as shown in Example 1.

| **EXAMPLE 4** | **Using the substitution method to solve a system with one solution** | **PROBLEM 4** |

Use the substitution method to solve the system:

$$3y + x = 9 \qquad \textbf{(1)}$$
$$x - 3y = 0 \qquad \textbf{(2)}$$

PROBLEM 4

Use the substitution method to find the solution of the system:

$$x + 2y = 3$$
$$x - 3y = 0$$

SOLUTION   We solve equation (2) for x to obtain $x = 3y$. Substituting $3y$ for x in equation (1)

$$x = 3y$$
$$3y + x = 9 \quad \text{becomes} \quad 3y + 3y = 9$$
$$6y = 9$$
$$y = \frac{3}{2}$$

Since $x = 3y$ and $y = \frac{3}{2}$,

$$x = 3\left(\frac{3}{2}\right) = \frac{9}{2}$$

The solution of $3y + x = 9$ and $x - 3y = 0$ is $(\frac{9}{2}, \frac{3}{2})$, and the solution set is $\{(\frac{9}{2}, \frac{3}{2})\}$ or $\{(4\frac{1}{2}, 1\frac{1}{2})\}$.

CHECK $3\left(\frac{3}{2}\right) + \frac{9}{2} = 9$ and $\frac{9}{2} - 3 \cdot \left(\frac{3}{2}\right) = 0$

The system is consistent. You can verify this by graphing $3x + y = 9$ and $x - 3y = 0$.

How do we recognize that a system such as $x + y = 3$ and $y = -x - 3$ is inconsistent? We can do this with the substitution method as follows. The system is

$$x + y = 3 \qquad \textbf{(1)}$$
$$y = -x - 3 \qquad \textbf{(2)}$$

We substitute $-x - 3$ for y in equation (1), and

$$y = -x - 3$$
$$x + y = 3 \quad \text{becomes} \quad x + (-x - 3) = 3$$
$$-3 = 3$$

Since $-3 = 3$ is a contradiction, the system $x + y = 3$ and $y = -x - 3$ is inconsistent; it has *no* solution. Graphically, the lines are parallel, so the solution set is $\varnothing$.

EXAMPLE 5   **Using the substitution method to solve a system with no solution**

Use the substitution method to solve the system:

$$x + y = 5 \quad \textbf{(1)}$$

$$y = -x \quad \textbf{(2)}$$

SOLUTION   Substituting $y = -x$ in equation (1), we get

$$x + (-x) = 5$$

$$0 = 5$$

This is a contradiction, so the system $x + y = 5$ and $y = -x$ has *no* solution. The system is inconsistent and the solution set is $\varnothing$.

PROBLEM 5

Use the substitution method to solve the system:

$$x - y = 2$$

$$y = x$$

Calculate It   Find the Solution for an Inconsistent System

To verify the result of Example 5, solve each equation for y and graph

$$Y_1 = -x + 5$$

and

$$Y_2 = -x$$

The slopes are the same, so the lines are parallel, as shown in the window, and there is no solution. The system is **inconsistent** and the solution set is $\varnothing$.

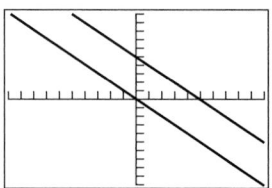

EXAMPLE 6   **Using the substitution method to solve a system with infinitely many solutions**

Use the substitution method to solve the system:

$$x + 2y = 4 \quad \textbf{(1)}$$

$$2x = 8 - 4y \quad \textbf{(2)}$$

SOLUTION   Solving equation (1) for x, we obtain $x = 4 - 2y$. We now substitute $4 - 2y$ for x in equation (2).

$$2\underset{\smile}{x} = 8 - 4y$$

$$2(4 - 2y) = 8 - 4y$$

$$8 - 4y = 8 - 4y$$

This equation is an identity, so any value of y will make it true. There are infinitely many solutions. The system is dependent. For example, if $x = 0$ in equation (1),

$$2y = 4 \quad \text{and} \quad y = 2$$

$(0, 2)$ is a solution of the system. If we let $y = 0$ in equation (2),

$$2x = 8 - 4(0) \quad \text{and} \quad x = 4$$

$(4, 0)$ is another solution. You can continue to obtain solutions by assigning numbers to one of the variables in either equation and solving for the other variable. The solution set is obtained by solving $x + 2y = 4$ for y and writing

$$\left\{ (x, y) \,\middle|\, y = -\frac{x}{2} + 2 \right\}$$

PROBLEM 6

Use the substitution method to solve the system:

$$x - 3y = 4$$

$$2x = 8 + 6y$$

Answers

5. There is no solution; the system is inconsistent and the solution set is $\varnothing$.

6. There are infinitely many solutions; the system is dependent.

$$\left\{ (x, y) \,\middle|\, y = \tfrac{1}{3}x - \tfrac{4}{3} \right\}$$

Calculate It   Find the Solution for a Dependent System

To verify the result of Example 6, solve each equation for y and graph

$$Y_1 = -\frac{x}{2} + 2$$

and

$$Y_2 = -\frac{x}{2} + 2$$

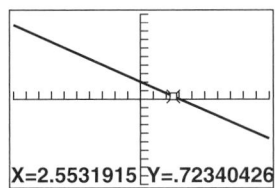

X=2.5531915 Y=.72340426

You can stop here; you don't need a calculator! The equations are equivalent, the system is dependent, and there are infinitely many solutions. You can find many of these solutions using the TRACE key. One such solution, point (2.5531915, .72340426), is shown in the window. Can you verify that this is a solution? If we had set the window with ZOOM 4, the TRACE key would have given solutions that would be easier to verify! You can also use 2nd GRAPH to find more solutions.

C Solving by Elimination

A third method for solving a system of linear equations is called the **elimination method.** The key step in this method is *to obtain coefficients that differ only in sign* for one of the variables so that this variable is eliminated by adding the two equations. The procedure is given next.

Web It

For a few more examples on solving by the elimination method, try link 4-1-4 at mhhe.com/bello.

PROCEDURE

Procedure for Solving a System of Two Equations in Two Unknowns by Elimination

1. Clear any fractions or decimals that may be present.
2. Multiply both sides of the equations (as needed) by numbers that make the coefficients of one of the variables additive inverses (opposites).
3. Add the two equations.
4. Solve for the remaining variable.
5. Substitute this solution into one of the given equations and solve for the second variable.
6. Check the solution.

We illustrate this method in Example 7.

EXAMPLE 7	**Using the elimination method to solve a system with one solution**

Use the elimination method to solve the system:

$$0.2x + 0.3y = 1 \qquad \textbf{(1)}$$

$$x - \tfrac{1}{2}y = 3 \qquad \textbf{(2)}$$

SOLUTION   We will eliminate x by the following steps.

1. Clear the fractions and decimals.

$0.2x + 0.3y = 1$	$2x + 3y = 10$	Multiply both sides of equation (1) by 10.	**(3)**
$x - \tfrac{1}{2}y = 3$	$2x - y = 6$	Multiply both sides of equation (2) by 2.	**(4)**

2. Make the x variables additive inverses.

$2x + 3y = 10$	$-2x - 3y = -10$	Multiply both sides of equation (3) by -1.	**(5)**
	$2x - y = \quad 6$	Copy equation (4).	**(6)**

PROBLEM 7

Use the elimination method to solve the system:

$$x + 0.2y = 4$$

$$4x + \tfrac{2}{3}y = 2$$

Answer

7. $(-17, 105)$

3. Add the two equations (5) and (6). $-4y = -4$

4. Solve for y by dividing by -4. $y = 1$

5. Substitute $y = 1$ into equation (4). $2x - y = 6$ **(4)**

$$2x - 1 = 6$$

$$2x = 7 \quad \text{Add 1 to both sides.}$$

$$x = 3.5 \quad \text{Divide both sides by 2.}$$

The solution is (3.5, 1) and the system is **consistent.**

6. The check is left for you.

Calculate It   Find the Solution for a Consistent System

We can verify the results of Example 7 by solving each equation for y and graphing

$$Y_1 = -\frac{2}{3}x + \frac{10}{3}$$

and

$$Y_2 = 2x - 6$$

Using an integer or decimal window and the ⬛TRACE key or the intersect (⬛2nd ⬛TRACE ⬛5) feature, we can verify that the system is
consistent and the solution is (3.5, 1) as shown in the window.

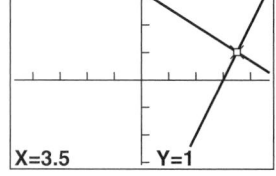

In many cases, it is not enough to multiply *one* of the equations by a number to eliminate
one of the variables. For example, to find the common solution of

$$2x + 3y = -1 \tag{1}$$

$$3x + 2y = -4 \tag{2}$$

we multiply both equations by numbers chosen so that the coefficients of one of the
variables become opposites in the resulting equations. We will eliminate the y variable by
using the 6 steps from the procedure we used in Example 7.

1. Clear the fractions and decimals.   There are none.

2. Make the y variables additive inverses by multiplying the top equation by the bottom
y coefficient and then multiplying the bottom equation by the negative of the top y
coefficient.

$$2x + 3y = -1 \qquad 4x + 6y = -2 \quad \text{Multiply both sides of equation (1) by 2.} \quad \textbf{(3)}$$

$$3x + 2y = -4 \qquad -9x - 6y = 12 \quad \text{Multiply both sides of equation (2) by } -3. \quad \textbf{(4)}$$

3. Add the two equations (3) and (4). $-5x = 10$

4. Solve for x by dividing by -5 $x = -2$

5. Substitute $x = -2$ into equation (1)

$$2x + 3y = -1$$

$$2(-2) + 3y = -1$$

$$-4 + 3y = -1 \quad \text{Simplify.}$$

$$3y = 3 \quad \text{Add 4 to both sides.}$$

$$y = 1 \quad \text{Divide both sides by 3.}$$

The solution is $(-2, 1)$ and the system is *consistent,* as shown in the *Calculate It.*

6. The check is left for you.

Calculate It   Find the Solution to a System

To find the solution of

$$2x + 3y = -1$$
$$3x + 2y = -4$$

solve each equation for y and graph

$$Y_1 = -\frac{2}{3}x - \frac{1}{3}$$

and

$$Y_2 = -\frac{3}{2}x - 2$$

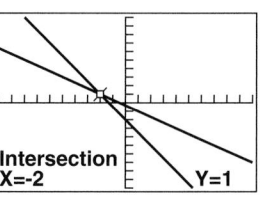

As shown in the window, the system is **consistent,** and the solution is $(-2, 1)$.

EXAMPLE 8	**Using the elimination method to solve a system with no solutions**

Solve the system:

$$3x - 6y = 2 \qquad \textbf{(1)}$$
$$-8x + 16y = 1 \qquad \textbf{(2)}$$

SOLUTION   We proceed as before.

1. Clear the fractions and decimals.   There are none.

2. Make the x variables additive inverses by multiplying the top equation by the negative of the bottom x coefficient (8) and then multiplying the bottom equation by the top x coefficient (3).

$$3x - 6y = 2 \qquad 24x - 48y = 16 \qquad$$ Multiply both sides of   **(3)**
equation (1) by 8.

$$-8x + 16y = 1 \qquad -24x + 48y = 3 \qquad$$ Multiply both sides of   **(4)**
equation (2) by 3.

3. Add the two equations (3) and (4). $0 = 43$

4. This is a contradiction, so the system has no solution.

The solution set is $\varnothing$ and the system is *inconsistent.*

    If we were to write the equations in slope-intercept form we would notice that the slopes are the same and the y-intercepts are different. This means the lines are parallel, as shown in the *Calculate It.*

PROBLEM 8

Use the elimination method to solve the system.

$$-2x + 6y = 1$$
$$5x - 15y = 2$$

Calculate It   Find the Solution to a System

In Example 8, solve each equation for y and graph the system

$$Y_1 = \frac{1}{2}x - \frac{1}{3}$$
$$Y_2 = \frac{1}{2}x + \frac{1}{16}$$

The slopes are the same and the lines are parallel, so there is no solution. The system is **inconsistent.** Looking at Window 1, however, it isn't clear that the lines are parallel.

    So we use a decimal or a "square" window by pressing ZOOM 4 , and we get two parallel lines, as shown in Window 2.

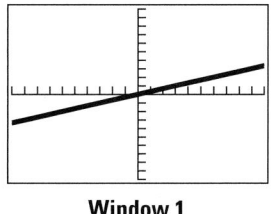

Window 1

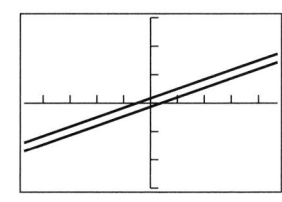

Window 2

Answer

8. No solution. The system is inconsistent and the solution set is $\varnothing$.

EXAMPLE 9 **Using the elimination method to solve a system with infinitely many solutions**

Solve the system:

$$\frac{x}{6} + \frac{y}{2} = 1 \qquad \textbf{(1)}$$

$$\frac{x}{9} + \frac{y}{3} = \frac{2}{3} \qquad \textbf{(2)}$$

SOLUTION We proceed as before.

1. Clear the fractions and decimals.

$x + 3y = 6$ Multiply both sides of equation (1) by 6.   **(3)**

$x + 3y = 6$ Multiply both sides of equation (2) by 9.   **(4)**

2. Make the x variables additive inverses.

$-x - 3y = -6$ Multiply both sides of equation (3) by -1.   **(5)**

$x + 3y = 6$ Copy equation (4).   **(6)**

3. Add the two equations (5) and (6). $0 = 0$

We obtain the true statement $0 = 0$, so the equations are dependent, and the system has infinitely many solutions. For example, if we substitute 6 for x in equation (1) we have

$$\frac{6}{6} + \frac{y}{2} = 1$$

$$1 + \frac{y}{2} = 1$$

$$\frac{y}{2} = 0$$

$$y = 0$$

Thus $(6, 0)$ is a solution. We can substitute 12 for x,

$$\frac{12}{6} + \frac{y}{2} = 1$$

$$2 + \frac{y}{2} = 1$$

$$\frac{y}{2} = -1$$

$$y = -2$$

and we have another solution, $(12, -2)$. The solution set is obtained by solving $x + 3y = 6$ for y and is written as $\{(x, y) \mid y = -\frac{x}{3} + 2\}$

NOTE

In step 1, we came up with identical equations. Any time we have two equivalent (or identical) equations, the system is dependent, and we write the solution set as $\{(x, y) \mid y = mx + b\}$.

PROBLEM 9

Use the elimination method to solve the system.

$$\frac{x}{20} + \frac{y}{12} = \frac{1}{2}$$

$$\frac{x}{5} + \frac{y}{3} = 2$$

Web It

For a slide presentation and practice problems on solving systems of equations by the elimination method, try link 4-1-5 at mhhe.com/bello.

Answer

9. Infinitely many solutions; the system is dependent; $\{(x, y) \mid y = \frac{-3}{5}x + 6\}$

Calculate It   Use the Table to Read the Solutions

In Example 9, the harder part is to solve for *y*. Multiply both sides of the first equation by the LCD 6 and both sides of the second equation by the LCD 9, then solve each equation for *y* to obtain the system

$$Y_1 = -\frac{x}{3} + 2$$

$$Y_2 = -\frac{x}{3} + 2$$

The lines are identical, so the system is **dependent** and there are infinitely many solutions. You can get some of them as we did in the *Calculate It* for Example 6 or you can press [2nd] [GRAPH] and read the solutions by scrolling through the table. The solution (18, −4) is shown in this window.

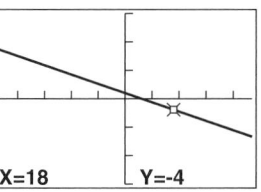

X=18   Y=-4

WHICH METHOD SHALL I USE?

We have discussed three methods for solving systems of equations: graphical, substitution, and elimination. If you are wondering how to decide which method to use, Table 1 gives you some guidelines.

Table 1 Solving Systems of Equations: A Summary

Method	Suggested Use	Disadvantages
Graphical	When the coefficients of the variables and the solutions are integers. You get a picture of the situation.	If the solutions are not integers, they are hard to read on the graph.
Substitution	When one of the variables is isolated (alone) on one side of the equation.	If fractions are involved, you may have much computation.
Elimination	When fractions, decimals, or variables with coefficients that are the same or additive inverses of each other (2*x* and −2*x*, for example) are present.	You may have lots of computations involving signed numbers.

D Solving Applications Involving Systems of Equations

Mercy Gonzalez is a buyer for the cosmetic counters of the BestDeal drug store chain. BestDeal has decided to have a fragrance promotion and sell one of their more popular fragrances at a premium price. But which one? She can choose from several different items that have been sold at different prices in past promotions. Her problem is to decide which selling price will best suit the needs of both the customers and the stores. Mercy has data from previous similar promotions to help her make a decision.

Supply Function, *S*

Selling Price of Each Item	Number Supplied Per Week Per Store
$15	55
$25	145
$35	235

Web It

To see a slide presentation on supply and demand with easy to follow animated graphs, try link 4-1-6 at mhhe.com/bello.

Based on this data, the equation of the supply function is $S = 9x - 80$, where x is the selling price of the item. The graph is shown in Figure 5.

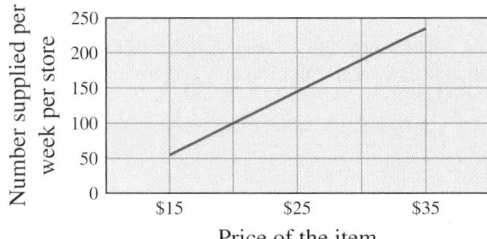

Figure 5

Demand Function, D

Selling Price of Each Item	Number Requested (Demand) Per Week Per Store
$15	295
$25	225
$35	155

The equation of the demand function is $D = -7x + 400$, where x is the selling price of the item. The graph is shown in Figure 6.

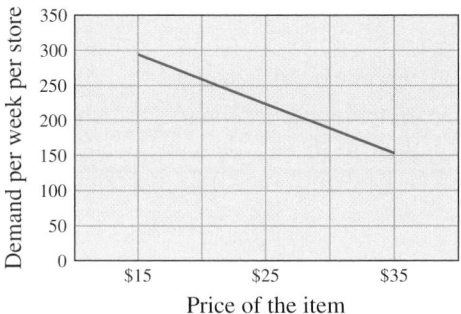

Figure 6

Now let's graph both functions on the same axes. As price increases, supply increases. As price increases, demand decreases. The point at which the graphs intersect is the **equilibrium point;** see Figure 7. That is, the price of the item at this point will indicate the number of items the customers are willing to buy (demand) is the same as the amount the store is willing to supply. By definition, if D is the demand function, S is the supply function, and p is the price, the equilibrium point is where the demand equals supply; $D(p) = S(p)$.

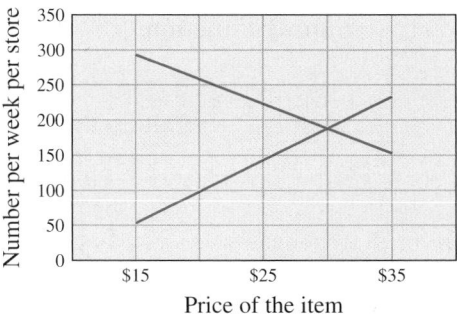

Figure 7

We can see by the graph in Figure 7 that Mercy should price the item at $30. Let's verify this by using the supply and demand equations for the equilibrium point.

$$D(p) = S(p)$$

$$-7x + 400 = 9x - 80 \qquad \text{Substitute } -7x + 400 \text{ for } D(p) \text{ and } 9x - 80 \text{ for } S(p).$$

$$-7x + 480 = 9x \qquad \text{Add 80 to both sides.}$$

$$480 = 16x \qquad \text{Add } 7x \text{ to both sides.}$$

$$\$30 = x \qquad \text{Divide both sides by 16.}$$

There is no doubt about it. Mercy makes the decision to go with the fragrance priced at $30.

EXAMPLE 10   **A supply and demand problem**

Booksellers supply fewer books when prices go down and more books when prices go up (because the profit margin is larger with the higher price). If a store knows the supply function is $S(p) = 20p + 200$ and the demand function is $D(p) = 740 - 100p$ for their art booklets, find the price of the book at the equilibrium point.

SOLUTION   We use the RSTUV method. The equilibrium point is the point at which

$$D(p) = S(p)$$

$$740 - 100p = 20p + 200 \qquad \text{Substitute } 740 - 100p \text{ for } D(p) \text{ and } 20p + 200 \text{ for } S(p).$$

$$540 - 100p = 20p \qquad \text{Subtract 200 from both sides.}$$

$$540 = 120p \qquad \text{Add } 100p \text{ to both sides.}$$

$$\frac{540}{120} = p \qquad \text{Divide both sides by 120.}$$

$$\$4.50 = p \qquad \text{Divide 540 by 120.}$$

The price at which the supply equals the demand is $4.50. At this point, the number of books suppliers are willing to offer is

$$S(4.5) = 20(4.5) + 200$$

$$= 90 + 200$$

$$= 290 \text{ books}$$

EXAMPLE 11   **Geometry problems**

Two angles are complementary. If the measure of one of the angles is 20° more than the other, what are the measures of these angles?

SOLUTION   We use the RSTUV method.

1. Read the problem.
We are asked to find the measures of these angles.

2. Select the unknown.
There are two angles involved. Let x be the measure of one of the angles and y the measure of the other angle.

3. Think of a plan. Do you know what complementary angles are? They are angles whose sum is 90°.
We have to translate the problem, but to do so, we must know that *complementary* angles are angles whose sum is 90°. With this

---

PROBLEM 10

Find the equilibrium point if the supply and demand functions are given by:

$$S(p) = 0.4p + 3$$

and

$$D(p) = -0.2p + 27$$

Calculate It

Graphing a System to Find the Equilibrium Point

You can verify the answer to Example 10 with your calculator. Using x instead of p, graph $Y_1 = 740 - 100x$ and $Y_2 = 20x + 200$ and find the point of intersection as shown in the window. (Use $[-1, 10]$ for x with a scale of 1 and $[0, 1000]$ for y with a scale of 100.)

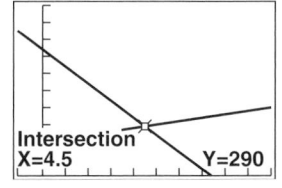

Intersection
X=4.5   Y=290

PROBLEM 11

Two angles are supplementary. If the measure of one of the angles is 40° less than the other, what are the measures of these angles?

Answers

10. $P = 40$ **11.** 70°, 110°

information, we have

$$x + y = 90 \qquad \text{The sum of the angles is } 90°.$$
$$y = x + 20 \qquad \text{One of the angles is } 20° \text{ more than the other.}$$

4. Use the substitution method to solve the system.

Substituting $y = x + 20$ in $x + y = 90$, we obtain

$$x + (x + 20) = 90$$
$$2x + 20 = 90 \qquad \text{Add like terms.}$$
$$2x = 70 \qquad \text{Subtract 20.}$$
$$x = 35 \qquad \text{Divide by 2.}$$

Thus one angle is $35°$ and the other $20°$ more, or $55°$.

5. Verify the answer.

Since the sum of the measures of the two angles is $35° + 55° = 90°$, the angles are complementary and our result is correct.

Calculate It   Summary of Solving Systems by Graphing

If you are using a calculator, you have to solve for y to graph each of the equations in a system. In doing so, you will know, even before graphing, whether the system is inconsistent (the lines will be parallel), dependent (the lines will coincide), or consistent [the lines will intersect at one point (x, y)]. To use the calculator, write the equation as

$$y = m_1 x + b_1$$
$$y = m_2 x + b_2$$

1. The system is *inconsistent* when $m_1 = m_2$ and $b_1 \neq b_2$.
2. The system is *dependent* when $m_1 = m_2$ and $b_1 = b_2$.
3. The system is *consistent* otherwise.

You can verify this:

a. Numerically. In Example 1, after entering $Y_1 = 2x - 2$ and $Y_2 = x - 1$, go to the home screen of a TI-83 Plus (2nd MODE) and store the value $x = 1$ by pressing 1 STO▸ X,T,θ,n ENTER .

Now let's evaluate Y_1. Press VARS ▸ 1 (the variable Y_1 should now be showing). Press ENTER ENTER . The answer is zero. [You can also evaluate $Y_1(1)$ by entering VARS ▸ 1 ENTER (1) ENTER .] Do the same for Y_2 and you should get zero again. Thus $(1, 0)$ satisfies both equations.

b. Graphically. By repeatedly using the TRACE and ZOOM keys of your calculator, you can find the intersection of two curves. With a TI-83 Plus, press 2nd TRACE 5 and ENTER three times to find the coordinates of the intersection of two functions.

In Examples 4, 5, and 6, remember that you can tell, even before you graph, the type of system you have by looking at the resulting equivalent equations

$$y = m_1 x + b_1$$
$$y = m_2 x + b_2$$

Calculate It Exercises

Problems 1–3 refer to Window 1.

1. How many solutions does the system have?

2. Do the lines have the same slope?

3. Is the system consistent or inconsistent?

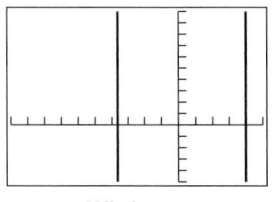

Window 1

Problems 4–6 refer to Window 2.

4. How many solutions does the system have?

5. Do the lines have the same slope?

6. Is the system consistent or inconsistent?

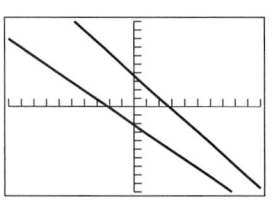

Window 2

Problems 7–9 refer to Window 3.

7. How many solutions does the system have? (Assume a pair of equations, Y_1 and Y_2, have been graphed.)

8. Is the system dependent or independent?

9. Do the lines have the same slope?

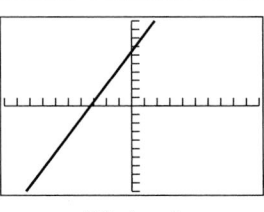

Window 3

Exercises 4.1

 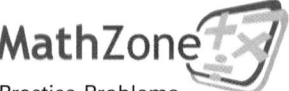
A In Problems 1–10, solve by the graphical method. Label each system as consistent (one solution), inconsistent (no solution), or dependent (many solutions).

1. $x - 2y = 6$
   $y = 2x$

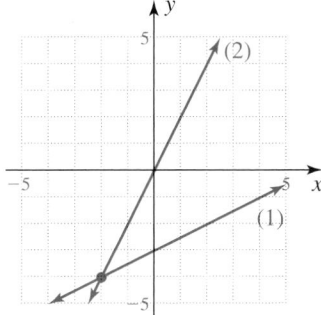

Solution: $(-2, -4)$; consistent

2. $2x + y = 4$
   $y = 2x$

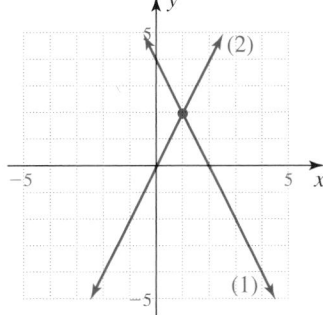

Solution: $(1, 2)$; consistent

3. $y = x - 3$
   $y = 2x - 4$

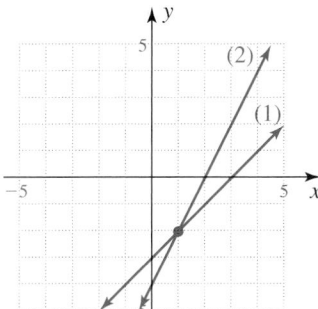

Solution: $(1, -2)$; consistent

4. $y = x - 1$
   $y = 3x - 3$

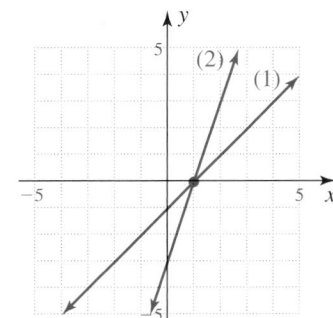

Solution: $(1, 0)$; consistent

5. $2y = -x + 4$
   $y = -2x + 4$

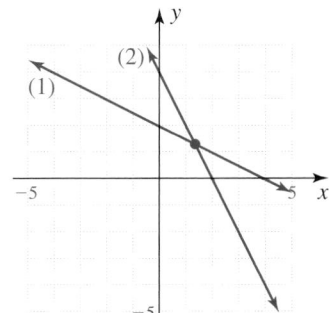

Solution: $(1\frac{1}{3}, 1\frac{1}{3})$; consistent

6. $2y = -x + 2$
   $y = -2x + 2$

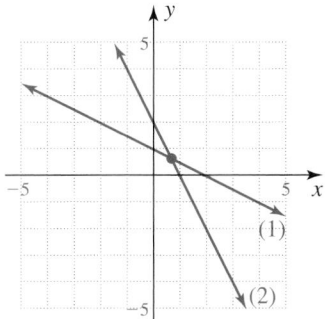

Solution $(\frac{2}{3}, \frac{2}{3})$; consistent

7. $2x - y = -2$
   $y = 2x + 4$

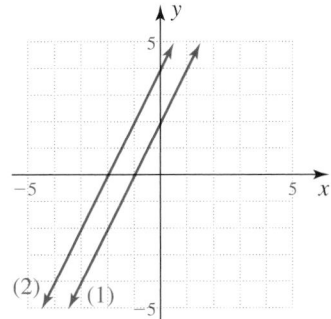

No solution; inconsistent

8. $2x + y = -2$
   $y = -2x + 4$

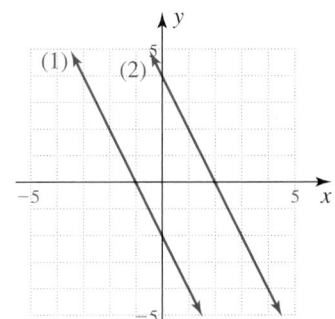

No solution; inconsistent

9. $3x + 4y = 12$
   $8y = 24 - 6x$

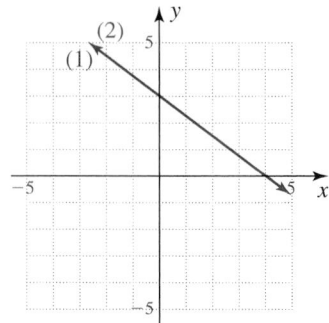

Infinitely many solutions; dependent

10. $2x - 3y = 6$
   $6x = 18 + 9y$

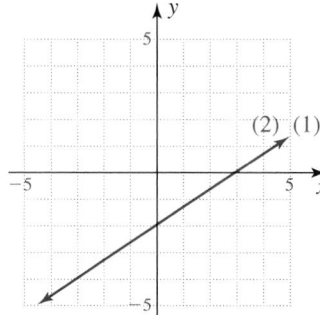

Infinitely many solutions; dependent

B In Problems 11–26, solve by the substitution method. Label each system as consistent (one solution), inconsistent (no solution), or dependent (many solutions).

11. $y = 2x - 4$
   $-2x = y - 4$
   $(2, 0)$; consistent

12. $y = 2x + 2$
   $-x = y + 1$
   $(-1, 0)$; consistent

13. $x + y = \dfrac{5}{2}$
   $3x + y = \dfrac{9}{2}$
   $(1, \frac{3}{2})$; consistent

14. $x + y = \dfrac{5}{2}$
   $3x + y = \dfrac{3}{2}$
   $(-\frac{1}{2}, 3)$; consistent

15. $y - 4 = 2x$
   $y = 2x + 2$
   No solution; inconsistent

16. $y + 5 = 4x$
   $y = 4x + 7$
   No solution; inconsistent

17. $x = 8 - 2y$
   $x + 2y = 4$
   No solution; inconsistent

18. $x = 4 - 2y$
   $x - 2y = 0$
   $(2, 1)$; consistent

19. $x + 2y = 4$
   $x = -2y + 4$
   Infinitely many solutions; dependent

20. $x + 3y = 6$
   $x = -3y + 6$
   Infinitely many solutions; dependent

21. $x = 2y + 1$
   $y = 2x + 1$
   $(-1, -1)$; consistent

22. $y = 3x + 2$
   $x = 3y + 2$
   $(-1, -1)$; consistent

23. $2x - y = -4$
   $4x = 4 + 2y$
   No solution; inconsistent

24. $5x + y = 5$
   $5x = 15 - 3y$
   $(0, 5)$; consistent

25. $x = 5 - y$
   $0 = x - 4y$
   $(4, 1)$; consistent

26. $x = 3 - y$
   $0 = 2x - y$
   $(1, 2)$; consistent

C In Problems 27–56, solve each system by elimination. Label the system consistent (one solution), inconsistent (no solution), or dependent (infinitely many solutions).

27. $x + y = 8$
   $x - y = 2$
   $(5, 3)$; consistent

28. $x + y = 3$
   $x - y = 1$
   $(2, 1)$; consistent

29. $x + 4y = 2$
   $x - 4y = -2$
   $(0, \frac{1}{2})$; consistent

30. $x - 5y = 15$
   $x + 5y = -5$
   $(5, -2)$; consistent

31. $-x - 2y = -2$
   $x - 2y = -2$
   $(0, 1)$; consistent

32. $x + 3y = -7$
   $-x + 2y = -3$
   $(-1, -2)$; consistent

33. $2x + y = 7$
   $3x - 2y = 0$
   $(2, 3)$; consistent

34. $2x + y = 4$
   $3x - 2y = -1$
   $(1, 2)$; consistent

35. $2x - 2y = 6$
   $x + y = 2$
   $(\frac{5}{2}, -\frac{1}{2})$; consistent

36. $3x - 3y = -15$
   $x - y = -5$
   Infinitely many solutions; dependent

37. $3x + 5y = 1$
   $-6x - 10y = 2$
   No solution; inconsistent

38. $5x - 2y = 4$
   $-10x + 4y = 1$
   No solution; inconsistent

39. $2x + y = 8$
   $3x - y = 7$
   $(3, 2)$; consistent

40. $x - 3y = -2$
   $x + 3y = 4$
   $(1, 1)$; consistent

41. $2x - 5y = 9$
   $4x - 10y = 18$
   Infinitely many solutions; dependent

42. $3x + 5y = 26$
   $5x + 3y = 22$
   $(2, 4)$; consistent

43. $6x + 5y = 12$
   $9x - 4y = -5$
   $(\frac{1}{3}, 2)$; consistent

44. $5x + 4y = 6$
   $4x - 3y = 11$
   $(2, -1)$; consistent

45. $2x - 3y = 16$
   $x - y = 7$
   $(5, -2)$; consistent

46. $3x - 2y = 35$
   $x - 5y = 42$
   $(7, -7)$; consistent

47. $18x - 15y = 1$
$10x - 12y = 3$
$(\frac{-1}{2}, \frac{-2}{3})$; consistent

48. $6x - 9y = -2$
$3x - 5y = -6$
$(\frac{44}{3}, 10)$; consistent

49. $\dfrac{x}{3} + \dfrac{y}{6} = \dfrac{2}{3}$
$\dfrac{2}{5}x + \dfrac{y}{4} = \dfrac{1}{5}$
$(8, -12)$; consistent

50. $\dfrac{x}{6} + \dfrac{y}{3} = \dfrac{1}{2}$
$\dfrac{3}{5}x + \dfrac{y}{4} = \dfrac{17}{20}$
$(1, 1)$; consistent

51. $\dfrac{5}{6}x + \dfrac{y}{4} = 7$
$\dfrac{2}{3}x - \dfrac{y}{8} = 3$
$(6, 8)$; consistent

52. $\dfrac{1}{5}x + \dfrac{2}{5}y = 1$
$\dfrac{1}{4}x - \dfrac{1}{3}y = \dfrac{-5}{12}$
$(1, 2)$; consistent

53. $\dfrac{2}{x} + \dfrac{3}{y} = \dfrac{-1}{2}$
$\dfrac{3}{x} - \dfrac{2}{y} = \dfrac{17}{12}$
$(4, -3)$; consistent

Hint: Multiply the first equation by 2, and the second by 3 and add, or let $a = \frac{1}{x}$, $b = \frac{1}{y}$ and solve the system

$$2a + 3b = \dfrac{-1}{2}$$

$$3a - 2b = \dfrac{17}{12}$$

54. $\dfrac{4}{x} + \dfrac{2}{y} = \dfrac{26}{21}$
$\dfrac{2}{x} - \dfrac{1}{y} = \dfrac{-1}{21}$
$(7, 3)$; consistent

Hint: Multiply the second equation by 2 and add, or let $a = \frac{1}{x}$, $b = \frac{1}{y}$ and solve the resulting system.

55. $\dfrac{2}{x} - \dfrac{1}{y} = 0$
$\dfrac{3}{x} + \dfrac{5}{y} = \dfrac{13}{4}$
$(4, 2)$; consistent

Hint: Multiply the first equation by 5 and add, or let $a = \frac{1}{x}$, $b = \frac{1}{y}$ and solve the resulting system.

56. $\dfrac{1}{x} - \dfrac{3}{y} = \dfrac{-13}{10}$
$\dfrac{5}{x} + \dfrac{2}{y} = 2$
$(5, 2)$; consistent

Hint: Multiply the first equation by -5 and add.

APPLICATIONS

Equilibrium, Angles, and More

In Problems 57–61, find the price p at the equilibrium point for each pair of demand and supply functions.

57. $D(p) = 620 - 10p$
$S(p) = 20p + 200$
$p = 14$

58. $D(p) = 200 - 2p$
$S(p) = 50 + p$
$p = 50$

59. $D(p) = 1500 - 5p$
$S(p) = 300 + p$
$p = 200$

60. $D(p) = 1100 - 8p$
$S(p) = 200 + 17p$
$p = 36$

61. $D(p) = 450 - 5p$
$S(p) = 2p + 170$
$p = 40$

62. The demand function for a certain product is $D(p) = 500 - 5p$, and the supply function is $S(p) = 8p + 110$. What will the price of this product be at the equilibrium point? What will the demand for the product be at the equilibrium point?
$p = 30$; $D(30) = 350$

63. The demand function for skateboards is $D(p) = 960 - 16p$, where p is the price in dollars. If the manufacturer is willing to supply $S(p) = 14p + 120$ boards when the price is p, what is the price of each skateboard when the equilibrium point is reached? What is the demand at that price; that is, how many skateboards sell at that price? $p = 28$; $D(28) = 512$

64. The demand function for sunglasses is $D(p) = 24 - 0.4p$, where p is the price in dollars. If the supply function is $S(p) = p + 10$, what will the price of sunglasses be when the equilibrium point is reached? How many sunglasses would suppliers be willing to offer at that price? $p = 10$; $S(10) = 20$

65. Find the price p (in dollars) and the supply and the demand at the equilibrium point when the demand function is given by $D(p) = 480 - 3p$ and the supply function is $S(p) = 25p + 60$. $p = 15$; $D(15) = S(15) = 435$

66. A store will buy 80 digital cellular phones if the price is $350 and 120 if the price is $300. The wholesaler is willing to supply 60 phones if the price is $280 and 140 if the price is $370. If the supply and demand functions are linear, write the equations of the functions, make a graph, and find the equilibrium point.

$$D(p) = 360 - \tfrac{4}{5}p$$

$$S(p) = \tfrac{8}{9}p - \tfrac{1700}{9}$$

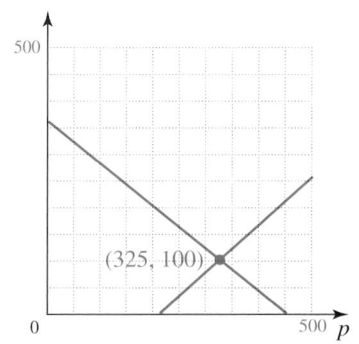

(325, 100)

In Problems 67–70, (a) write a system of two equations describing the situation, and (b) find the measure of each angle.

67. Two angles are complementary, and the measure of one of the angles is 15° more than the other.

(a) $x + y = 90$
$y = x + 15$

(b) $x = 37.5$
$y = 52.5$

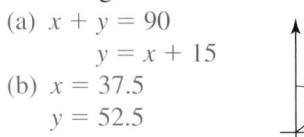

The sum of the measures of the two angles is 90°.

68. Two angles are complementary, and the measure of one of the angles is twice the measure of the other angle.

(a) $x + y = 90$
$y = 2x$

(b) $x = 30$
$y = 60$

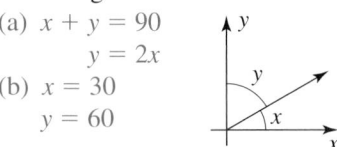

The sum of the measures of the two angles is 90°.

69. Two angles are supplementary, and the measure of one of the angles is four times the measure of the other.

(a) $x + y = 180$
$y = 4x$

(b) $x = 36$
$y = 144$

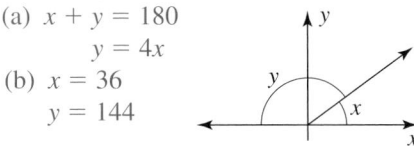

The sum of the measures of the two angles is 180°.

70. Two angles are supplementary, and the measure of one of the angles is 30° more than the other.

(a) $x + y = 180$
$y = x + 30$

(b) $x = 75$
$y = 105$

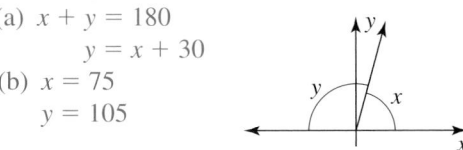

The sum of the measures of the two angles is 180°.

71. The heaviest pair of African elephant tusks on record weighed a total of 465 lb. One of the tusks weighed 15 lb more than the other.

a. Write a system of two equations representing this situation. $x + y = 465; x = y + 15$

b. What does each tusk weigh? 240 lb; 225 lb

72. Do you know who invented basketball? It was Dr. James Naismith who in 1921 created a game that could be played indoors in Massachusetts. Although the height of the players has changed dramatically, the size of the court hasn't. The perimeter (distance around) of the maximum-size basketball court is 288 ft, and the length L of the court is 44 ft more than the width W.

a. Write a system of two equations representing this situation. $2L + 2W = 288; L = W + 44$

b. What are the dimensions of the maximum-size basketball court? $L = 94$ ft; $W = 50$ ft

73. Do you smoke? How many cigarettes a year? Together, smokers in the United States and Japan consume 4637 cigarettes per person each year. If the average Japanese smoker consumes 437 cigarettes a year more than the average U.S. smoker, what is the annual consumption of cigarettes per person in each country?
Japan: 2537; United States: 2100

74. The average American drinks 29 more Cokes® per year than the average Mexican. If together they consume 555 Cokes per year, how many Cokes are consumed each year by the average American and how many by the average Mexican?
American: 292; Mexican: 263

75. Find the solution to this ancient Babylonian problem:

There are two silver rings; $\frac{1}{7}$ of the first ring and $\frac{1}{11}$ of the second ring are broken off so that what is broken off weighs 1 shekel. The first, diminished by $\frac{1}{7}$, weighs as much as the second diminished by $\frac{1}{11}$. What did the silver rings (together) originally weigh?

(*Hint:* Consider the system of equations

$$\frac{x}{7} + \frac{y}{11} = 1 \quad \text{and} \quad \frac{6x}{7} = \frac{10y}{11}$$

Clear denominators by multiplying by the LCD, then use the substitution method to solve the system.)
8.5 shekels

76. Is there a nuclear reactor near your city? The total number of reactors in the United States is 160, but not all of them are in operation. The number of operable reactors is four more than double the number of inoperable reactors. How many U.S. reactors are operable and how many are inoperable?
108 operable; 52 inoperable

77. According to the *Guinness Book of World Records*, the sum of the ages of the two oldest cats is 70 yr. The difference of their ages is 2 yr. What are their ages?
36 and 34

78. According to the *Guinness Book of World Records*, the sum of the heights of the shortest and tallest persons on record is 130 in. If the difference in their heights was 84 in., how tall was each? 107 in. and 23 in.

79. The height of the Empire State Building and its antenna is 1472 ft. The difference in height between the building and the antenna is 1028 ft. How tall is the antenna and how tall is the building? antenna: 222 ft; building: 1250 ft

80. At one time, the combined weight of the McCreary brothers was 1300 lb. Their weight difference was 20 lb. What was the weight of each brother? 660 lb and 640 lb

SKILL CHECKER

Try the "Skill Checker" exercises so you'll be ready for the next section.

Find the *x*- and *y*-intercept of each line.

81. $2x + y = 6$ x: 3; y: 6

82. $3x + 2y = 12$ x: 4; y: 6

83. $-2x + 3y = 9$ x: -4.5; y: 3

84. $-3x + 2y = 4$ x: $\frac{-4}{3}$; y: 2

85. $-2x - 3y = 7$ x: $\frac{-7}{2}$; y: $\frac{-7}{3}$

USING YOUR KNOWLEDGE

Tweedledee and Tweedledum

Have you read *Alice in Wonderland?* Do you know who the author of this book is? The answer is Lewis Carroll. Did you also know that he was an accomplished mathematician and logician? These interests show up occasionally in his children's books. Take, for example, *Through the Looking Glass,* where one of the characters, Tweedledee, is talking to Tweedledum:

Tweedledee: The sum of your weight and twice mine is 361 pounds.

Tweedledum: Contrariwise, the sum of your weight and twice mine is 360 pounds.

86. If Tweedledee weighs *x* pounds and Tweedledum weighs *y* pounds, can you use the knowledge gained in this section to find their weights?

Tweedledee: $120\frac{2}{3}$ lb; Tweedledum: $119\frac{2}{3}$ lb

WRITE ON

87. How would you know if a system of linear equations is consistent when you are using:

a. the graphical method?   **b.** the substitution method?

c. the elimination method? Answers may vary.

88. How would you know if a system of linear equations is inconsistent when you are using:

a. the graphical method?   **b.** the substitution method?

c. the elimination method? Answers may vary.

89. How would you know if a system of linear equations is dependent when you are using:

a. the graphical method?   **b.** the substitution method?

c. the elimination method? Answers may vary.

90. Write the possible advantages and disadvantages of each of the methods we have studied: graphical, substitution, and elimination. Answers may vary.

MASTERY TEST

If you know how to do these problems, you have learned your lesson!

Solve the system by elimination:

91. $5x = 6 - 4y$
$3y = 4x - 11$ $(2, -1)$

92. $3x = 6 - 3y$
$2y = 5x + 11$ $(-1, 3)$

93. $\dfrac{x}{3} + \dfrac{y}{9} = \dfrac{2}{3}$
$\dfrac{x}{2} + \dfrac{y}{6} = 1$
Infinitely many solutions

94. $\dfrac{x}{3} + \dfrac{y}{4} = \dfrac{11}{6}$
$\dfrac{x}{6} + \dfrac{y}{8} = \dfrac{7}{12}$ No solution

95. $3x + 2y = 8$
$x - y = 1$ $(2, 1)$

96. $2x + 3y = 9$
$x - y = 2$ $(3, 1)$

Solve the system by substitution:

97. $x - 3y = 4$
$2x = 8 + 6y$
Infinitely many solutions

98. $x - y = 2$
$y = x$
No solution

99. $2y + x = 8$
$x - 3y = 0$
$\left(\frac{24}{5}, \frac{8}{5}\right)$

Use the graphical method to solve the system and classify the system as consistent, inconsistent, or dependent:

100. $2x + 2y = 8$
$2x - y = 2$

101. $2x + y = 4$
$2y + 4x = 6$

102. $2x + y = 5$
$2x + 4y = 8$

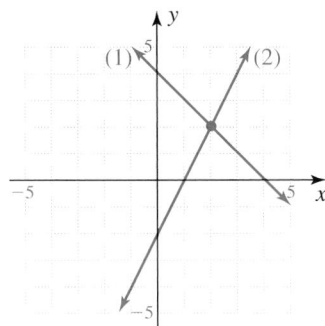

(2, 2); consistent

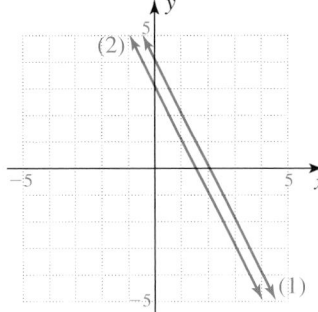

No solution; inconsistent

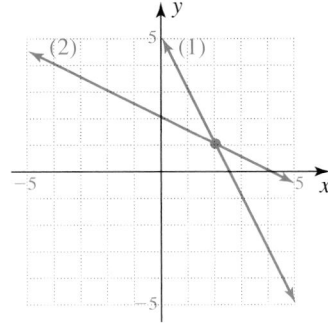

(2, 1); consistent

103. $x + \frac{1}{2}y = -2$
$y = -2x - 4$

104. $y - 3x = 3$
$2y - 6x = 12$

105. $x - y = -1$
$y = -x - 1$

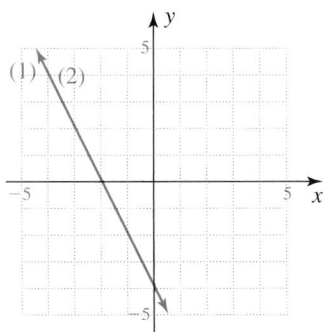

Infinitely many solutions; dependent

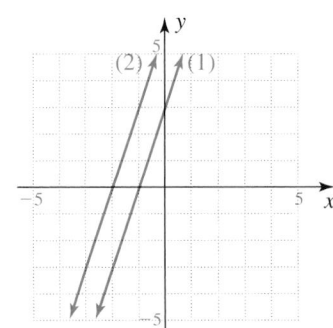

No solution; inconsistent

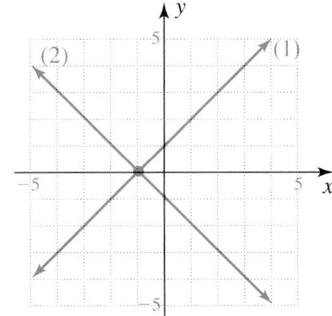

(-1, 0); consistent

106. The height of the Eiffel Tower and its antenna is 1052 ft. The difference in height between the tower and the antenna is 920 ft. How tall is the antenna and how tall is the tower? antenna: 66 ft; tower: 986 ft

4.2 SYSTEMS WITH THREE VARIABLES

To Succeed, Review How To ...

1. Solve linear equations (p. 79).

2. Evaluate an expression (p. 49).

3. Solve a system of two equations in two unknowns (pp. 256–263).

Objectives

A Solve a system of three equations and three unknowns by the elimination method.

B Determine whether a system of three equations in three unknowns is consistent, inconsistent, or dependent.

C Solve applications involving systems of three equations.

GETTING STARTED

Coffee Mixing: An Application

The man has three coffees that sell for \$13.00, \$14.00, and \$15.00 per pound, respectively. If he calls these coffees A, B, and C and decides to make 50 lb of a mixture containing x pounds of A, y pounds of B, and z pounds of C, then

$$x + y + z = 50 \qquad \textbf{(1)}$$

In this mixture he will have twice as much of brand B as of brand C. Thus

$$y = 2z \qquad \textbf{(2)}$$

Finally, if he sells the 50 lb at \$14.20 per pound, the total price will be \$710.00 so that

$$\$13x + \$14y + \$15z = \$710 \qquad \textbf{(3)}$$

Now we rewrite equations (1), (2), and (3) in standard form:

$$x + y + z = 50 \qquad \textbf{(4)}$$
$$y - 2z = 0 \qquad \text{Subtract } 2z \text{ from both sides of (2).} \qquad \textbf{(5)}$$
$$13x + 14y + 15z = 710 \qquad \textbf{(6)}$$

We have obtained a system of linear equations in three unknowns. There are many ways to solve this system. If we use the elimination method, we take two *different* pairs of equations and eliminate the same variable from each pair. Equation (5) *does not* contain x. Because of this, we select equations (4) and (6) and eliminate x by multiplying equation (4) by -13 and adding equation (6), as follows:

$$-13x - 13y - 13z = -650 \qquad \textbf{(7)}$$
$$\underline{13x + 14y + 15z = 710} \qquad \textbf{(6)}$$
$$y + 2z = 60 \qquad \text{Add.} \qquad \textbf{(8)}$$

The new system, consisting of equations (5) and (8), is a system of two equations in two unknowns. We solve it using the techniques of Section 4.1. To eliminate z from this system, we add equations (5) and (8), obtaining

$$y - 2z = 0 \qquad \textbf{(5)}$$
$$\underline{y + 2z = 60} \qquad \textbf{(8)}$$
$$2y = 60 \qquad \text{Add.} \qquad \textbf{(9)}$$
$$y = 30$$

(Continued)

Substituting $y = 30$ in equation (5) gives

$$y - 2z = 0$$
$$30 - 2z = 0$$
$$z = 15$$

Using $y = 30$ and $z = 15$ in equation (4) gives

$$x + 30 + 15 = 50$$
$$x + 45 = 50$$
$$x = 5$$

The solution for the system is the ordered triple (5, 30, 15). You can check this by substituting 5 for x, 30 for y, and 15 for z in each of the original equations. This means the man mixed 5 lb of the $13.00 coffee, 30 lb of the $14.00 coffee, and 15 lb of the $15.00 coffee.

Solving Equations by Elimination

In Section 4.1, we used the graphical, substitution, and elimination methods to solve systems of equations involving two variables in two equations. When we have three variables, the graphical method isn't used because a three-dimensional coordinate system is required. However, we can use the elimination method to solve a system of three linear equations in three unknowns, as we illustrated in the *Getting Started.* Now let's examine this method in more detail. To solve a system of three linear equations in three unknowns by elimination, we use the following procedure.

PROCEDURE

Solving a Three-Equation System with Three Unknowns by Elimination

1. Select a pair of equations and eliminate one variable from this pair.

2. Select a *different* pair of equations and eliminate the same variable as in step 1.

3. Solve the pair of equations resulting from steps 1 and 2. (Use the procedure outlined in Section 4.1.)

4. Substitute the values found in step 3 in the simplest of the original equations, and then solve for the third variable.

5. Check by substituting the values in each of the original equations.

Calculate It   Solving a System with Three Equations

To solve systems of three equations with your calculator, select two pairs of equations and eliminate the same variable from each pair; then graph the resulting pair of equations.

In Example 1, z was eliminated from each pair. We then graph

$$Y_1 = -\frac{2}{3}x + 6 \qquad \textbf{(4)}$$
$$Y_2 = -3x + 13 \qquad \textbf{(5)}$$

and find the point of intersection, as shown in the window, by using [2nd] [TRACE] [5] [ENTER] [ENTER] [ENTER].

The result is $x = 3$ and $y = 4$. Now substitute $x = 3$ and $y = 4$ in the first equation:

$$x + y + z = 12$$
$$3 + 4 + z = 12$$
$$z = 5$$

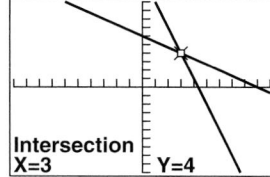

| **EXAMPLE 1** | **Using the elimination method to solve a three-equation system with one solution** |

Solve the system:

$$x + y + z = 12 \qquad (1)$$
$$2x - y + z = 7 \qquad (2)$$
$$x + 2y - z = 6 \qquad (3)$$

If we add these two, z is eliminated.
If we add these two, z is eliminated.

SOLUTION   It's easiest to eliminate z from the two pairs of equations, (1) and (3), and (2) and (3).

1. Adding (1) and (3), we obtain

$$2x + 3y = 18 \qquad (4)$$

2. Adding (2) and (3), we have

$$3x + y = 13 \qquad (5)$$

3. We now have the system

$$2x + 3y = 18 \qquad (4)$$
$$3x + y = 13 \qquad (5)$$

Multiplying equation (5) by -3, we have

$$2x + 3y = 18 \qquad (4)$$
$$-9x - 3y = -39 \qquad (6)$$
$$\overline{ -7x = -21} \qquad \text{Add.}$$
$$x = 3$$

Substituting 3 for x in equation (5), we get $9 + y = 13$, or $y = 4$.

4. We know now that x is 3 and y is 4. We can substitute these values in equation (1) to obtain $3 + 4 + z = 12$. Solving, we find $z = 5$.

5. The solution of the system is $(3, 4, 5)$, as can easily be verified:

$$3 + 4 + 5 = 12 \qquad (1)$$
$$2(3) - 4 + 5 = 7 \qquad (2)$$
$$3 + 2(4) - 5 = 6 \qquad (3)$$

The system is *consistent*.

Web It

For some more practice examples on solving systems of three equations, try link 4-2-1 at mhhe.com/bello.

| **EXAMPLE 2** | **Using the elimination method to solve a system of three equations with no solution** |

Solve the system:

$$x + 3y - z = 1 \qquad (1)$$
$$x - y + z = 4 \qquad (2)$$
$$3x + y + z = 3 \qquad (3)$$

If we add these two, z is eliminated.
If we add these two, z is eliminated.

| **PROBLEM 1** |

Solve the system:

$$x + y + z = 4$$
$$x - y + z = 2$$
$$2x + 2y - z = -4$$

Calculate It

Solving a System of Three Equations with No Solution

To verify Example 2, eliminate z by first adding equations (1) and (2), then (1) and (3). Solve each resulting equation for y to obtain

$$Y_1 = -x + \frac{5}{2}$$
$$Y_2 = -x + 1$$

You could stop here. The lines have the same slope and different y-intercepts, so they are parallel and the system is *inconsistent* as shown in the window.

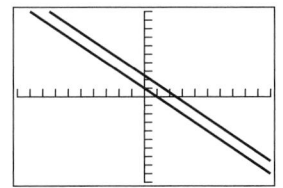

| **PROBLEM 2** |

Solve the system:

$$x + 8y - z = 8$$
$$-x + 2y - z = 4$$
$$2x + y + z = 2$$

Answers

1. $(-1, 1, 4)$; consistent system
2. No solution; inconsistent system

SOLUTION Adding first (1) and (2) and then (1) and (3), we obtain

$$2x + 2y = 5 \qquad \text{The sum of (1) and (2).} \qquad \textbf{(4)}$$

$$4x + 4y = 4 \qquad \text{The sum of (1) and (3).} \qquad \textbf{(5)}$$

Multiplying equation (4) by -2 to eliminate x, we have

$$-4x - 4y = -10 \qquad \text{This is } -2(2x + 2y) = -2 \cdot 5. \qquad \textbf{(6)}$$

$$\underline{4x + 4y = 4} \qquad\qquad\qquad\qquad\qquad \textbf{(5)}$$

$$0 = -6 \qquad \text{Add.} \qquad\qquad\qquad \textbf{(7)}$$

It's impossible for 0 to be equal to -6. This system has *no solution;* it is an *inconsistent system.*

| **EXAMPLE 3** | **Using the elimination method to solve a system of three equations with infinitely many solutions** |

Solve the system:

$$x - 2y + 3z = 4 \qquad \textbf{(1)}$$

$$2x - y + z = 1 \qquad \textbf{(2)}$$

$$x + y - 2z = -3 \qquad \textbf{(3)}$$

SOLUTION We use the five-step procedure.

1. Multiplying equation (2) by -2 and adding to equation (1) (to eliminate y), we get

$$x - 2y + 3z = 4 \qquad \textbf{(1)}$$

$$\underline{-4x + 2y - 2z = -2} \qquad \textbf{(4)}$$

$$-3x + z = 2 \qquad \textbf{(5)}$$

2. Adding equations (2) and (3) to eliminate y, we obtain

$$2x - y + z = 1 \qquad \textbf{(2)}$$

$$\underline{x + y - 2z = -3} \qquad \textbf{(3)}$$

$$3x - z = -2 \qquad \textbf{(6)}$$

3. We now have the system

$$-3x + z = 2 \qquad \textbf{(5)}$$

$$\underline{3x - z = -2} \qquad \textbf{(6)}$$

$$0 = 0 \qquad \text{Add.} \qquad \textbf{(7)}$$

The system is *dependent* and has infinitely many solutions of the form (x, y, z). One such solution is obtained if we let $x = 0$ in equation (6), which gives $z = 2$.

4. Substituting $x = 0$ and $z = 2$ in equation (2), we have

$$2 \cdot 0 - y + 2 = 1 \qquad \textbf{(2)}$$

$$-y + 2 = 1$$

$$y = 1$$

5. So $(0, 1, 2)$ is *one* of the solutions for the system, as can be easily checked by substituting $x = 0$, $y = 1$, and $z = 2$ in the original system. Here are some more solutions you can check: $(1, 6, 5)$ and $(-1, -4, -1)$.

| **PROBLEM 3** |

Solve the system:

$$x + 2y + z = -10$$

$$x + y - z = -3$$

$$5x + 7y - z = -29$$

Teaching Tip

Have students solve equation (6) for z ($z = 3x + 2$) and equation (3) for y ($y = 2z - x - 3$). This will make it easier to check possible solutions. Then, see if they can write the ordered triple in x, $(x, 5x + 1, 3x + 2)$.

Web It

For another explanation on solving a system of three equations with infinitely many solutions, go to link 4-2-2 at mhhe.com/bello.

Answer

3. Infinitely many solutions; dependent system

B   Consistent, Inconsistent, and Dependent Systems

As was the case with systems of two equations in two unknowns, the solution of three equations in three unknowns always produces one of three different possibilities.

Calculate It

**Solving a System of
Three Equations with
Infinitely Many Solutions**

To verify the results of Example 3, start by eliminating y to obtain

$$-3x + z = 2$$
$$3x - z = -2$$

Now solve each equation for z:

$$z = 3x + 2$$
$$z = 3x + 2$$

You get the same equation. The system is *dependent*.

POSSIBLE SOLUTIONS OF THREE EQUATIONS IN THREE UNKNOWNS

The system is *consistent* and *independent,* as in Example 1; it has one solution consisting of an ordered triple (x, y, z).

The system is *inconsistent,* as in Example 2. It has no solution.

The system is *consistent* and *dependent,* as in Example 3. It has infinitely many solutions.

In the case of two unknowns, that a system is *consistent* tells us that if we graph the lines associated with the system, the lines *intersect*. If we graph a linear equation in three unknowns, the graph is a plane. If three linear equations in three unknowns have a solution, it means that the three planes corresponding to their equations intersect at a point, as shown in Figure 8.

If the equations are *inconsistent,* the planes do not intersect at a common point, as shown in Figures 9 and 10. Finally, if the equations are *dependent,* the three planes can intersect in a line, as in Figure 11 (or the three planes can be coincident). Any point on the intersection is a solution; consequently, there are infinitely many solutions.

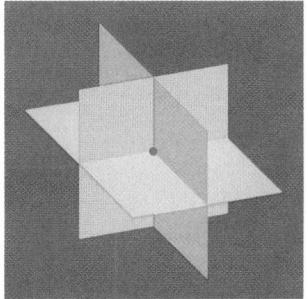

Figure 8
One solution: Planes intersect in exactly one point.

Figure 9
No solution: Planes intersect two at a time but have no common point of intersection. The equations are *inconsistent,* and the lines of intersection are parallel.

Figure 10
No solution: Three parallel planes with no point of intersection. The equations are *inconsistent.*

Figure 11
Infinitely many solutions: The three planes intersect along a common line. The equations are *dependent.*

C   Solving Applications Involving Systems of Three Equations

There are problems to solve that would be best accomplished by writing a system of three equations with three unknowns as we will demonstrate in Cindy's dilemma in Example 4.

| **EXAMPLE 4** | **Solve an application involving three equations** |

Cindy has been searching for a juicing recipe that will help her immune system and is told that it should contain vitamin C, vitamin B, and beta-carotene. She chooses three vegetables (V_1, V_2, V_3) known to contain these three vitamins.

   She finds a table that gives the amount of units per ounce that is in each of the three vegetables (see Table 2). Her dilemma: how many ounces of each vegetable should she juice to obtain 260 units of vitamin C, 140 units of vitamin B, and 90 units of beta-carotene in her juice drink?

Table 2

	V_1	V_2	V_3
Vitamin C	50	60	
Vitamin B	20		20
Beta-carotene	10	20	10

SOLUTION   We are asked to find how many ounces of each vegetable should be used. We use the following variables.

Let x = the number of ounces of V_1

   y = the number of ounces of V_2

   z = the number of ounces of V_3

Using Table 2 we notice that only V_1 and V_2 contain vitamin C and because the numbers given represent the units per ounce, we must multiply those numbers by the respective ounces for those two vegetables, namely x and y ounces.

This means: $50x + 60y$ represents the total vitamin C.

The total units for vitamin C is 260, so

$$50x + 60y = 260 \qquad \textbf{(1)}$$

The table also indicates that only V_1 and V_3 contain vitamin B. To represent the units of vitamin B from those vegetables we use the numbers from the table and the variables representing the ounces in those vegetables, x and z.

This means: $20x + 20z$ represents the total vitamin B.

The total units for vitamin B is 140, so

$$20x + 20z = 140 \qquad \textbf{(2)}$$

Beta-carotene is found in all three vegetables so we multiply those numbers times x, y, and z ounces.

This means: $10x + 20y + 10z$ represents the total beta-carotene.

The total units for beta-carotene is 90, so

$$10x + 20y + 10z = 90 \qquad \textbf{(3)}$$

Using equations (1), (2), and (3) we can solve the system to find the answer.
   By dividing each equation in the system by 10 we get,

$$5x + 6y \qquad = 26 \qquad \textbf{(1)}$$
$$2x \qquad + 2z = 14 \qquad \textbf{(2)}$$
$$x + 2y + \ z = \ 9 \qquad \textbf{(3)}$$

PROBLEM 4

Cindy is told that during the flu season she should increase the amount of vitamin C to 520 units, the amount of vitamin B to 260, and the amount of beta-carotene to 170 units. How many ounces of each vegetable should she juice for this recipe?

Web It

For more practice solving applications involving three equations, try link 4-2-3 at mhhe.com/bello.

Answer

4. 8 oz of V_1; 2 oz of V_2; 5 oz of V_3

Adding equation (2) with the result of multiplying equation (3) by -2 we get

$$2x \qquad\quad + 2z = \quad 14 \qquad\qquad\qquad\qquad\qquad \textbf{(2)}$$

$$\underline{-2x - 4y - 2z = -18} \quad \text{\scriptsize -2 times equation (3)} \qquad \textbf{(4)}$$

$$-4y \qquad\quad = -4$$

$$y = 1$$

We substitute $y = 1$ into equation (1).

$$5x + 6(1) = 26 \qquad\qquad\qquad\qquad\qquad \textbf{(1)}$$

$$5x + \quad 6 = 26$$

$$5x = 20 \quad \text{\scriptsize Subtract 6 from both sides.}$$

$$x = 4 \quad \text{\scriptsize Divide both sides by 5.}$$

Now we substitute $x = 4$ and $y = 1$ into equation (3) to find z.

$$4 + 2(1) + z = 9 \qquad\qquad\qquad\qquad \textbf{(3)}$$

$$6 + z = 9$$

$$z = 3 \quad \text{\scriptsize Subtract 6 from both sides.}$$

Cindy's dilemma is solved. To have the proper amount of each vitamin in her juice drink she must juice 4 oz of V_1, 1 oz of V_2, and 3 oz of V_3.

Calculate It   Summary of Solving Systems with Three Equations

The graphical method is not practical when solving a system of three equations in three unknowns because the resulting graphs would have to be in three dimensions; that is, you would need a coordinate system with three axes, x, y, and z. There are some 3D calculators available but with the TI-83 Plus the best you can do when solving systems of equations in three unknowns is to follow steps 1 and 2 of the procedure given in the text. At that point, you will have a linear system with two equations and two unknowns. To finish, you then solve each of the two equations for the same variable, graph the resulting equations, and find the point of intersection, if any. You will then have the values (answer) for two of the three variables. Substitute these two values in any of the original equations and solve that equation for the third value.

Exercises 4.2

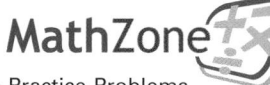
A B In Problems 1–20, solve the system. Label the system as consistent, inconsistent, or dependent.

1. $x + \quad y + z = 12$
$x - \quad y + z = \quad 6$
$x + 2y - z = \quad 7$
$(5, 3, 4)$; consistent

2. $x + \quad y + \quad z = 13$
$x - 2y + 4z = 10$
$3x + \quad y - 3z = \quad 5$
$(4, 5, 4)$; consistent

3. $x + \quad y + z = \quad 4$
$2x + 2y - z = -4$
$x - \quad y + z = \quad 2$
$(-1, 1, 4)$; consistent

4. $2x - \quad y + \quad z = \quad 3$
$x + 4y - \quad z = \quad 6$
$3x + 2y + 3z = 16$
$(1, 2, 3)$; consistent

5. $2x - \quad y + z = \quad 3$
$x + 2y + z = 12$
$4x - 3y + z = \quad 1$
$(3, 4, 1)$; consistent

6. $x - 3y - 2z = -12$
$2x + \quad y - 3z = \quad -1$
$3x - 2y - \quad z = \quad -5$
$(1, 3, 2)$; consistent

7. $x - 2y - \quad 3z = \quad 2$
$x - 4y - 13z = 14$
$-3x + 5y + \quad 4z = \quad 2$
No solution; inconsistent

8. $2x + 2y + z = 3$
$-x + \quad y - z = 5$
$3x + 5y + z = 8$
No solution; inconsistent

9. $2x + 4y + 3z = 3$
$10x - 8y - 9z = 0$
$4x + 4y - 3z = 2$
$(\frac{1}{2}, \frac{1}{4}, \frac{1}{3})$; consistent

10. $9x + 4y - 10z = 6$
$6x - 8y + 5z = -1$
$12x + 12y - 15z = 10$
$(\frac{1}{3}, \frac{1}{4}, -\frac{1}{5})$; consistent

11. $x - 2y - z = 3$
$2x - 5y + z = -1$
$x - 2y - z = -3$
No solution; inconsistent

12. $x - 3y + 6z = -8$
$3x - 2y - 10z = 11$
$5x - 6y - 2z = 7$
$(10, 7, \frac{1}{2})$; consistent

13. $2x + y + z = 5$
$-x + 2y - z = 3$
$3x + 4y + z = 10$
No solution; inconsistent

14. $x - 3y + z = 2$
$x + 2y - z = 1$
$-7x + y + z = -10$
Infinitely many solutions; dependent

15. $x + y = 5$
$y + z = 3$
$x + z = 7$
$(\frac{9}{2}, \frac{1}{2}, \frac{5}{2})$; consistent

16. $x + 2y = -1$
$2y + z = 0$
$x + 2z = 11$
$(3, -2, 4)$; consistent

17. $x - 2y = 0$
$y - 2z = 5$
$x + y + z = 8$
$(6, 3, -1)$; consistent

18. $2y + z = 9$
$z - 2y = 1$
$x + y + z = 1$
$(-6, 2, 5)$; consistent

19. $5x - 3z = 2$
$2z - y = -5$
$x + 2y - 4z = 8$
$(-2, -3, -4)$; consistent

20. $5x - 2z = 1$
$3z - y = 6$
$x + 2y - z = -1$
$(1, 0, 2)$; consistent

21. Find a, b, and c so that the ordered pairs $(1, 0)$, $(-1, -4)$, and $(2, 5)$ satisfy the equation $y = ax^2 + bx + c$. [*Hint:* For $(1, 0)$, $x = 1$ and $y = 0$, and

$$y = ax^2 + bx + c \qquad (1)$$
$$0 = a + b + c$$

Do the same for $(-1, -4)$ and $(2, 5)$ and then solve the resulting system for a, b, and c.]
$a = 1; b = 2; c = -3$

22. Show that the system

$2x + 4z = 6$	**(1)**
$3x + y + z = -1$	**(2)**
$2y - z = -2$	**(3)**
$x - y + z = -5$	**(4)**

does not have a solution. [*Hint:* Solve the system consisting of equations (1), (2), and (4) and then show that the solution does not satisfy equation (3).]
$(-3, 5, 3)$ is not a solution of $2y - z = -2$

23. Find the solution set of

$$2x + 4z = 6$$
$$3x + y + z = -1$$
$$2y - z = -2$$
$$x - y - 2z = -5 \quad \{(-1, 0, 2)\}$$

24. Find a value of k so that the system

$5x - y + 2z = 2$	**(1)**
$3x + y - 3z = 7$	**(2)**
$x + 5y + z = 5$	**(3)**
$x + ky - z = 9$	**(4)**

has a solution and give the solution. [*Hint:* Solve the system consisting of equations (1), (2), and (3). Then substitute the values of x, y, and z in equation (4) and solve for k.] $k = 7; (1, 1, -1)$

25. Find a value of k so that the system

$2x + 4z = 6$	**(1)**
$3x + y + z = -1$	**(2)**
$2y - z = -2$	**(3)**
$x - y + kz = -5$	**(4)**

has a solution and give the solution. (See Problem 24.)
$k = -2; (-1, 0, 2)$

APPLICATIONS

Equations for Unknowns

In Problems 26–36, use a system of three equations in three unknowns to solve the problem.

26. The sum of three numbers is 48. The second number is double the first, and the third number is triple the first. What are the numbers? $8; 16; 24$

27. The sum of three numbers is 49. The second number is 8 more than the first, and the third is 9 more than the second. Find the numbers. $8; 16; 25$

28. The sum of the measures of the three angles in a triangle is 180°. The measure of one of the angles is twice that of the smallest angle, and the measure of the largest angle is three times that of the smallest. Find the measure of each angle.　30°; 60°; 90°

29. The sum of the measures of three angles in a triangle is 180°. The sum of the measures of the first and the second angle is 20° less than that of the third angle, and the measure of the second angle is 20° more than that of the first. Find the measure of each angle.　30°; 50°; 100°

30. In 2004, the sum of the salaries of the president, vice president, and chief justice of the United States was $805,800. The president made $197,100 more than the vice president. If the vice president and the chief justice made the same salary, how much did each of these people make?　Pres.: $400,000; vice pres.: $202,900; chief justice: $202,900

31. In a survey of 1000 people, the number of respondents complaining about corns, heel pain, or ingrown toenails was 150. Two more people suffered heel pain than ingrown toenails, and the number of the respondents with either heel pain or ingrown toenails exceeded the number of those with corns by 10. Write a system of three equations and three unknowns, solve it, and find the number of people complaining about corns only, heel pain only, and ingrown toenails only.　Corns: 70; heel pain: 41; ingrown toenails: 39

32. Do you take vitamins? The annual amount spent on vitamins C, E, and B in the United States reached $885 million. Vitamin C is the most popular of the three; $90 million more was spent on it than on vitamin B, and $15 million more was spent on vitamin E than on B. Find the amount of money spent on each of these vitamins.　B: $260 million; C: $350 million; E: $275 million

33. Who makes all the dough? Pizza Hut, Domino's, and Papa John's account for 72% of the total pizza market. Pizza Hut sells 40% more than Domino's, while Domino's sells 4% more than Papa John's. Find each company's market share (%).　Pizza Hut: 52%; Domino's: 12%; Papa John's: 8%

34. More than 57 million adults play a musical instrument. The number of piano, guitar, and organ players combined is 46 million. The number of guitar and organ players exceeds the number of piano players by 4 million, but the number of guitar players is only 2 million less than those playing piano. Find out how many adults play each of these instruments.　Piano: 21 million; guitar: 19 million; organ: 6 million

35. In a recent year, Bill Gates (computer software), John Kluge (media), and the Walton family (retailing) had combined fortunes totaling $16.9 billion. Kluge's fortune was $0.4 billion more than the Walton family's and $0.8 billion less than Gates's. Find out just how rich these people were during this particular year.　Gates: $6.3 billion; Kluge: $5.5 billion; Walton family: $5.1 billion

36. If you went to school in Japan, then moved to Israel, and finally ended up in the United States the third year, you would have attended school for a total of 639 days. In Israel, you would have attended 36 more days than in the United States. In Japan, you would have attended even longer: 27 more days than in Israel. How long is the school year in each of these countries?　Israel: 216 days; Japan: 243 days; United States: 180 days

---

SKILL CHECKER

Try the "Skill Checker" exercises so you'll be ready for the next section.

Solve:

37. A person bought some bonds yielding 5% annually and some certificates yielding 7%. If the total investment amounts to $20,000 and the interest received is $1160, how much is invested in bonds and how much in certificates?　$12,000 at 5%; $8000 at 7%

38. A car leaves a town going north at 30 mi/hr. Two hours later, another car leaves the same town traveling on the same road in the same direction at 40 mi/hr. How far from the town does the second car overtake the first one?　240 mi

39. How many gallons of a 30% solution must be mixed with 40 gal of a 12% solution to obtain a 20% solution?　32 gal

40. A person can complete a job in 3 hr. Another person does it in 2 hr. How long would it take to complete the job if both people work together?　$1\frac{1}{5}$ hr

41. A plane traveled 840 mi with a 30-mi/hr tail wind in the same time it took to travel 660 mi against the wind. What is the plane's speed in still air?　250 mi/hr

USING YOUR KNOWLEDGE

Crickets, Ants, and the Weather

Can you tell how hot it is by listening to the crickets? You can if you know the formula. The number of chirps n a certain cricket makes per minute is related to the temperature F, in degrees Fahrenheit, by

$$F = \frac{n}{4} + 40 \tag{1}$$

Ants are also affected by temperature changes. The crawling speed d (in centimeters per second) of a certain ant is related to the temperature C, in degrees Celsius, by

$$C = 6d + 4 \tag{2}$$

The relationship between C and F is given by

$$C = \frac{5}{9}(F - 32) \tag{3}$$

42. Use the substitution method with equations (2) and (3) to solve for F in terms of d.

$$F = \frac{54d + 196}{5}$$

43. Substitute the expression obtained for F in Problem 42 in equation (1) to find the relationship between d and n.

$$d = \frac{5n + 16}{216}$$

44. Now use your answer from Problem 43 to solve this: If the cricket is chirping 112 times a minute, how fast is the ant crawling?

$2\frac{2}{3}$ cm/sec

WRITE ON

45. Why is the graphical method difficult to use when solving a system of equations in three unknowns? Answers may vary.

46. When would you use the substitution method when solving a system of three equations in three unknowns? Answers may vary.

47. If you graph the linear equation $2x + y = 4$, what is the resulting graph? What do you think the graph will be if you graph the equation $2x + y + z = 4$? Answers may vary.

48. How do you know if a system of three equations in three unknowns is inconsistent? Answers may vary.

49. How do you know if a system of three equations in three unknowns is dependent? Answers may vary.

MASTERY TEST

If you know how to do these problems, you have learned your lesson!

Solve:

50. $x + 2y + z = -10$
$x + \ y - z = \ -3$
$5x + 7y - z = -29$
Infinitely many solutions; dependent

51. $x + \ y + \ z = \ 4$
$x - 2y - \ z = \ 1$
$2x - \ y - 2z = -1$
$(2, -1, 3)$; consistent

52. $x + 8y - z = 8$
$-x + 2y - z = 4$
$2x + \ y + z = 2$
No solution; inconsistent

53. $2x + 2y - 6z = 5$
$-x - \ y + 3z = 4$
$3x - \ y + \ z = 2$
No solution; inconsistent

54. $x + \ y + z = \ 4$
$x - \ y + z = \ 2$
$2x + 2y - z = -4$
$(-1, 1, 4)$; consistent

55. $x + 2y + \ z = \ 4$
$-3x + 4y - \ z = -4$
$-2x - 4y - 2z = -8$
Infinitely many solutions; dependent

4.3   COIN, DISTANCE-RATE-TIME, INVESTMENT, AND GEOMETRY PROBLEMS

To Succeed, Review How To . . .

1. Use the RSTUV method (pp. 99–106).

2. Solve a system of linear equations in two or three unknowns (pp. 256–263, 274–277).

Objectives

A Solve coin problems with two or more unknowns.

B Solve general problems with two or more unknowns.

C Solve rate, time, and distance (motion) problems with two or more unknowns.

D Solve investment problems with two or more unknowns.

E Solve geometry problems with two or more unknowns.

GETTING STARTED   ## Money Problems!

The pile of money contains $3.25 in nickels and dimes. There are five more nickels than dimes. How many nickels and how many dimes are in the pile?

We use the RSTUV method to solve this problem.

1. Read the problem.

We are asked to find the number of nickels and dimes.

2. Select the unknown.

Let n be the number of nickels and d the number of dimes.

3. Think of a plan.

We have to translate two statements, yielding two equations and two unknowns. First note that:

> If you have 1 nickel, you have $0.05(1)
>
> If you have 2 nickels, you have $0.05(2)
>
> If you have n nickels, you have $0.05(n)

Similarly,

> If you have d dimes, you have $0.10(d)

Now we can translate the statement.

The pile contains $3.25		in nickels	and	dimes.
$\left\{\begin{array}{l}\text{Total}\\\text{amount}\end{array}\right\}$	$=$	$\left\{\begin{array}{l}\text{amount}\\\text{in nickels}\end{array}\right\}$	$+$	$\left\{\begin{array}{l}\text{amount}\\\text{in dimes}\end{array}\right\}$
3.25	$=$	0.05n	$+$	0.10d	
325	$=$	5n	$+$	10d	Multiply by 100.
65	$=$	n	$+$	2d	Divide by 5.

The statement "there are five more nickels than dimes" means that

$\left\{\begin{array}{l}\text{The number}\\\text{of nickels}\end{array}\right\}$	$=$	$\left\{\begin{array}{l}\text{5 more}\\\text{than}\end{array}\right\}$	$+$	$\left\{\begin{array}{l}\text{the number}\\\text{of dimes}\end{array}\right\}$
n	$=$	5	$+$	d

(Continued)

The complete problem can be reduced to the system of equations

$$65 = n + 2d$$
$$n = 5 + d$$

4. Use the substitution method or your calculator to solve this system.
Substituting $5 + d$ for n in the first equation gives

$$65 = (5 + d) + 2d$$
$$65 = 5 + d + 2d \qquad \text{Remove parentheses.}$$
$$65 = 5 + 3d \qquad \text{Combine like terms.}$$
$$60 = 3d \qquad \text{Subtract 5.}$$
$$20 = d \qquad \text{Divide by 3.}$$

$n = 5 + d$, so we substitute 20 for d:

$$n = 5 + 20 = 25$$

Hence we have 25 nickels and 20 dimes.

5. Verify the answer.
We do have 5 more nickels (25 in total) than dimes (20 in total) and $3.25 ($1.25 in nickels and $2.00 in dimes).

A   Solving Coin Problems

Coin problems are a classic way to practice setting up systems of equations.

EXAMPLE 1   **Nickels and dimes**

Jack has $3 in nickels and dimes. He has twice as many nickels as he has dimes. How many nickels and how many dimes does he have?

2d

d

Dimes   Nickels

"Twice as many nickels as dimes" $2d = n$

SOLUTION   As usual, we use the RSTUV method.

1. Read the problem.
We are asked to find the number of nickels and dimes, so we need two variables.

2. Select the unknown.
Let n be the number of nickels and d the number of dimes.

3. Think of a plan.
Translate the problem. Jack has $3 (300 cents) in nickels and dimes:

$$300 = 5n + 10d$$

He has twice as many nickels as he has dimes:

$$n = 2d$$

We then have the system

$$5n + 10d = 300$$
$$n = 2d$$

PROBLEM 1

Jill has $9.90 in dimes and quarters. She has 3 times as many dimes as quarters. How many dimes and how many quarters does she have?

Answer

1. 54 dimes; 18 quarters

4. Use the substitution method or your calculator to solve the system.
This time it's easy to use the substitution method.

$$5n + 10d = 300 \rightarrow 5(2d) + 10d = 300 \qquad \text{Let } n = 2d.$$

$$10d + 10d = 300 \qquad \text{Simplify.}$$

$$20d = 300 \qquad \text{Combine like terms.}$$

$$d = 15 \qquad \text{Divide by 20.}$$

$$n = 2(15) = 30 \qquad \text{Substitute } d = 15 \text{ in } n = 2d.$$

Thus Jack has 15 dimes ($1.50) and 30 nickels ($1.50).

5. Verify the answer.
It's easy to verify this answer because we know Jack has $3 and twice as many nickels as dimes.

Web It

For some more information and practice on solving coin problems, try link 4-3-1 at mhhe.com/bello.

B Solving General Problems

We can also use systems of equations to solve other problems. Here is an interesting one.

EXAMPLE 2   A lengthy matter

Santa is not the only one that can grow a beard. Did you know the average beard grows 5.5 inches in a year? According to the *Guinness Book of World Records,* Shamsher Singh of India has the longest beard on a living male and Vivian Wheeler of the United States has the longest female beard. The length differential in their beards is 61 inches. If the combined length of their beards is 83 inches, find the length of each beard.

PROBLEM 2

The total combined height of a redwood tree and a silver maple is 436 feet. If the height differential is 292 feet and the redwood tree is taller, what is the height of each tree?

SOLUTION

1. Read the problem.
We are asked to find the lengths of the longest beards on a living male and a living female, so we need two variables.

2. Select the unknown.
Let $m =$ the length of the longest beard on a living male
and $f =$ the length of the longest beard on a living female

3. Think of a plan.
We need two facts to translate into two algebra equations.

Fact 1: The length differential is 61 in. $m - f = 61$

Fact 2 : The combined lengths are 83 in. $m + f = 83$

Now we use our skills to solve the system.

$$m - f = 61$$
$$m + f = 83$$

Answer

2. 364 ft and 72 ft

4. Use the elimination method or your calculator to solve the system.

We use the elimination method since the *f* variables are additive inverses (opposites) and by adding the equations together they will be eliminated.

$$
\begin{array}{rl}
m - f = 61 & \\
m + f = 83 & \\
\hline
2m \phantom{{}+f} = 144 & \text{Add equations.} \\
m = 72 & \text{Divide both sides by 2.}
\end{array}
$$

Substitute $m = 72$ into $m + f = 83$

$$
\begin{array}{rl}
72 + f = 83 & \\
f = 11 & \text{Subtract 72.}
\end{array}
$$

Thus Mr. Singh's beard is 72 in. long, and Ms. Wheeler's beard is 11 in. long.

5. Verify the answer.

You can verify this by going online to the *Guinness Book of World Records*.

C   Motion Problems

Remember the motion problems we solved in Chapter 2? They can also be solved using two variables. The procedure is almost the same. We write the given information in a chart labeled $R \times T = D$, and use the RSTUV method.

EXAMPLE 3   The current in Norway

The world's strongest current is the Saltstraumen in Norway. The current is so strong that a boat, which can go 48 mi downstream (with the current) in 1 hr, takes 4 hr to go the same 48 mi upstream (against the current). How fast is the current flowing?

SOLUTION

1. Read the problem.

We are asked to find the rate of speed of the current, so we need two variables: the rate of speed of the boat and the rate of speed of the current.

2. Select the unknown.

Let x be the rate of speed of the boat in still water and y be the rate of speed of the current. Then $(x + y)$ is the rate of speed of the boat downstream; $(x - y)$ is the rate of speed of the boat upstream.

3. Think of a plan.

We enter this information in a chart:

	Rate of Speed $\times$ R (mi/hr)	Time $\times$ T (hr) =	Distance D (mi)	
Downstream	$(x + y)$	1	48	$\rightarrow$ $x + y = 48$
Upstream	$(x - y)$	4	48	$\rightarrow$ $(x - y)4 = 48$

PROBLEM 3

A plane goes 1200 miles with a tail wind in 3 hr. It takes 4 hr to travel the same distance against the wind. Find the velocity of the wind and the velocity of the plane in still air.

Answer

3. Wind velocity 50 mi/hr; plane velocity 350 mi/hr

4. Use the elimination method or your calculator to solve the system.

Our system of equations is simplified as follows:

$$x + y = 48 \xrightarrow{\text{leave as is}} x + y = 48$$

$$(x - y)4 = 48 \xrightarrow{\text{divide by 4}} \underline{x - y = 12}$$

$$2x = 60 \qquad \text{Add.}$$

$$x = 30 \qquad \text{Divide by 2.}$$

$$30 + y = 48 \qquad \begin{array}{l}\text{Substitute } x = 30 \text{ in}\\ x + y = 48.\end{array}$$

$$y = 18 \qquad \text{Subtract 30.}$$

The speed of the boat in still water is $x = 30$ mi/hr and the speed of the current is 18 mi/hr.

5. Verify the answer.

The verification is left to you. You may wish to compare how we solved this type of problem using one variable in Section 2.4 with the method used here.

Investment Problems

The investment problems we studied in Section 2.4 are easier to solve if we use more than one variable. We illustrate their solution next.

EXAMPLE 4　　**Tracking investments**

An investor divides $20,000 among three investments at 6%, 8%, and 10%. If the total annual income is $1700 and the income of the 10% investment exceeds the income from the 6% and 8% investments by $300, how much was invested at each rate?

SOLUTION

1. Read the problem.

We are asked to find how much was invested at each rate, so we need three variables.

2. Select the unknown.

Let x be the amount invested at 6%, y the amount invested at 8%, and z the amount invested at 10%.

3. Think of a plan.

Let's use a chart with the heading:

$$\text{Principal} \times \text{Rate} = \text{Annual Interest}$$

We enter the information in the chart:

	Principal	×	**Rate**	=	**Annual Interest**
1st investment	x		6%		$0.06x$
2nd investment	y		8%		$0.08y$
3rd investment	z		10%		$0.10z$
Total	$20,000				$1700

PROBLEM 4

Spencer received an inheritance. He wants to invest in three accounts—the first at 4%, the second at 6%, and the third at 8%. The second investment is three times the first and the third investment is $7000 more than the second. If the annual interest from all three is $2860, find the amount of Spencer's inheritance.

Web It

To solve an investment interest problem interactively, try link 4-3-2 at mhhe.com/bello.

Answer

4. 1st = $5000; 2nd = $15,000; 3rd = $22,000; thus, the inheritance was $42,000.

Looking at the column labeled "Principal," we see that

$$x + y + z = 20{,}000 \qquad \textbf{(1)}$$

From the column labeled "Annual Interest," we can see that the total interest earned is $1700; thus

$$0.06x + 0.08y + 0.10z = 1700$$

Multiplying by 100 gives

$$6x + 8y + 10z = 170{,}000 \qquad \textbf{(2)}$$

We also know that

$$\left\{ \begin{array}{c} \text{The income} \\ \text{from the 10\%} \\ \text{investment} \end{array} \right\} \quad \left\{ \begin{array}{c} \text{exceeds the income} \\ \text{from the 6\% and} \\ \text{8\% investments} \end{array} \right\} \quad \{\text{by \$300}\}$$

$0.10z$	$=$	$0.06x + 0.08y$	$+$	300	
$10z$	$=$	$6x + 8y$	$+$	$30{,}000$	Multiply by 100.

$$-6x - 8y + 10z = 30{,}000 \qquad \textbf{(3)}$$

4. Use the elimination method or your calculator to solve the system.
We now have the system

$$x + y + z = 20{,}000 \qquad \textbf{(1)}$$

$$6x + 8y + 10z = 170{,}000 \qquad \textbf{(2)}$$

$$-6x - 8y + 10z = 30{,}000 \qquad \textbf{(3)}$$

Adding equations (2) and (3), we have

$$20z = 200{,}000 \qquad \textbf{(4)}$$

$$z = 10{,}000$$

Now multiply equation (1) by 6 and add it to equation (3):

$$\begin{array}{rcl} 6x + 6y + 6z & = & 120{,}000 \qquad \textbf{(5)} \\ -6x - 8y + 10z & = & 30{,}000 \qquad \textbf{(3)} \\ \hline -2y + 16z & = & 150{,}000 \qquad \textbf{(6)} \end{array}$$

Substitute 10,000 for z in equation (6):

$$-2y + 16(10{,}000) = 150{,}000 \qquad \textbf{(7)}$$

$$-2y + 160{,}000 = 150{,}000$$

$$-2y = -10{,}000$$

$$y = 5000$$

Finally, substitute 10,000 for z and 5000 for y in equation (1):

$$x + 5000 + 10{,}000 = 20{,}000$$

$$x + 15{,}000 = 20{,}000$$

$$x = 5000$$

Thus $x = 5000$, $y = 5000$, and $z = 10{,}000$.

5. Verify the answer.
You can verify that if we invest $5000 at 6%, $5000 at 8%, and $10,000 at 10%, the conditions of the problem are satisfied.

 **Solving Geometry Problems**

Problems involving the perimeter of a rectangle also involve two unknowns. Example 5 shows how to set up a system of two equations to solve such problems.

EXAMPLE 5 A "check" using geometry and perimeter

The check is in the mail? Certainly not one of the physically largest checks ever written, which had a perimeter of 202 ft and was 39 ft longer than wide! Can you find the dimensions of this check given by InterMortgage of Leeds, England, to a Yorkshire TV telephone appeal?

SOLUTION We continue to use the RSTUV procedure.

1. Read the problem.
We are asked to find the dimensions of the check. Even though the problem doesn't specify it, we assume that the check is rectangular.

2. Select the unknown.
We let W be its width and L its length.

3. Think of a plan.
First, translate the problem. A picture helps.
   The perimeter P is 202 ft and the perimeter is also $2W + 2L$, so we have the equation

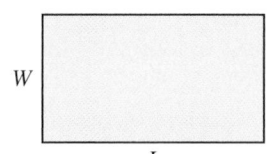

$$2W + 2L = 202 \qquad (1)$$

Also, "the check was 39 ft longer than wide" means

$$L = W + 39 \qquad (2)$$

Equation (1) can be simplified if we divide each of its terms by 2. Since $L = W + 39$, it will be easier to use the substitution method.
   We have the system

$$2W + 2L = 202 \qquad (1)$$
$$L = W + 39 \qquad (2)$$

First, divide each term in equation (1) by 2 to obtain

$$W + L = 101 \qquad (3)$$
$$L = W + 39 \qquad (2)$$

4. Use the substitution method or your calculator to solve the system.
Substituting $W + 39$ for L in equation (3), we have

$$W + (W + 39) = 101$$
$$2W + 39 = 101 \quad \text{Simplify.}$$
$$2W = 62 \quad \text{Subtract 39 from both sides.}$$
$$W = 31 \quad \text{Divide both sides by 2.}$$

From equation (2), $L = W + 39$ and we know that $W = 31$, so

$$L = 31 + 39 = 70$$

Thus the dimensions of the check are 31 ft wide by 70 ft long.

5. Verify the answer.
The verification that the perimeter is 202 ft and that the check is 39 ft longer than wide is left to you.

PROBLEM 5

The perimeter of the base of the Washington Monument is $183\frac{5}{6}$ ft. If the length is $18\frac{1}{4}$ ft longer than the width, find the dimensions of the base.

Answer

5. Width $= 36\frac{5}{6}$ ft or $36.8\overline{3}$ ft; length $= 55\frac{1}{12}$ ft or $55.08\overline{3}$ ft

Calculate It   Summary of Solving a Word Problem by Graphing a System

You can use your calculator as a tool to assist you in solving the problems in this section. There are two important considerations:

1. Which variable will you designate as the independent variable (the one that is easier to solve for)?

2. What size window will let you see the part of the graph at which the given lines intersect? (Use a window that contains the x- and y-intercepts of the lines involved.)

For instance, in the *Getting Started,* go through the first three steps of the RSTUV method (your calculator can't do this for you) to obtain the system

$$65 = n + 2d \quad \text{and} \quad n = 5 + d$$

It's easier to solve for n, so we have the equivalent system

$$n = 65 - 2d \quad \text{and} \quad n = 5 + d$$

Graph

$$Y_1 = 65 - 2X \quad \text{and} \quad Y_2 = 5 + X$$

In the first equation, when $X = 0$, $Y_1 = 65$ and when $Y_1 = 0$, $X = 32.5$, so we use a $[0, 35]$ by $[0, 65]$ window, with a scale of 5 for X and Y ($X\text{scl} = 5$, $Y\text{scl} = 5$). Using your intersection feature (2nd TRACE 5 and ENTER three times; see Window 1), we get $X = 20$ and $Y = 25$. Thus, $d = 20$ and $n = 25$ as before; that is, we have 20 dimes and 25 nickels. Now you can verify that these two numbers satisfy the conditions of the original problem.

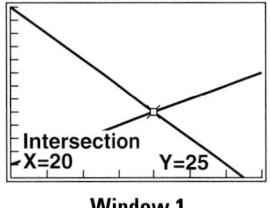

Window 1

In Example 2, we have the system $m - f = 61$ and $m + f = 83$, or

$$m = 61 + f \quad \text{and} \quad m = 83 - f$$

Graph

$$Y_1 = 61 + X \quad \text{and} \quad Y_2 = 83 - X$$

using a $[-70, 90]$ by $[0, 90]$ window with a scale of 10 for both the X and Y. Using your intersection feature, we obtain $X = 11$ and $Y = 72$ (Window 2). Thus $f = 11$ and $m = 72$ as before. Check this out with the conditions of the problem.

For Example 3, we graph the system

$$Y_1 = 48 - X \quad \text{and} \quad Y_2 = X - 12$$

with a $[0, 50]$ by $[0, 50]$ window and an X- and Y-scale of 5, which yields $X = 30$ and $Y = 18$ as the point of intersection (Window 3).

Now it's your turn to do Example 4. But wait, we have three variables. It's easier to do that problem algebraically. In other problems with three variables, reduce the system to two equations and two unknowns and graph. Then substitute your answers for x and y into any of the equations and solve for z. Now try Example 5.

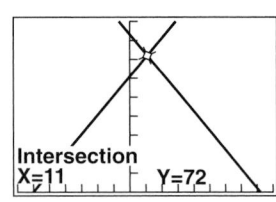

Window 2

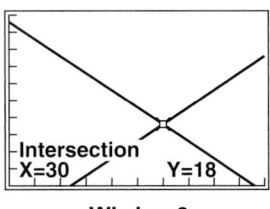

Window 3

Exercises 4.3

A In Problems 1–10, use two or more unknowns to solve these coin problems.

1. Natasha has $6.25 in nickels and dimes. If she has twice as many dimes as she has nickels, how many dimes and how many nickels does she have?
    50 dimes; 25 nickels

2. Mida has $2.25 in nickels and dimes. She has four times as many dimes as nickels. How many dimes and how many nickels does she have? 20 dimes; 5 nickels

3. Dora has $5.50 in nickels and quarters. She has twice as many quarters as she has nickels. How many of each coin does she have? 10 nickels; 20 quarters

4. Mongo has 20 coins consisting of nickels and dimes. If the nickels were dimes and the dimes were nickels, he would have 50¢ more than he now has. How many nickels and how many dimes does he have?
    15 nickels; 5 dimes

5. Desi has 10 coins consisting of pennies and nickels. Strangely enough, if the nickels were pennies and the pennies were nickels, she would have the same amount of money as she now has. How many pennies and nickels does she have? 5 pennies; 5 nickels

6. Don has $26 in his pocket. If he has only 1-dollar bills and 5-dollar bills, and he has a total of 10 bills, how many of each bill does he have? 4 fives; 6 ones

7. A person went to the bank to deposit $300. The money was in 10- and 20-dollar bills, 25 bills in all. How many of each did the person have? 20 tens; 5 twenties

8. A woman has $5.95 in nickels and dimes. If she has a total of 75 coins, how many nickels and how many dimes does she have? 31 nickels; 44 dimes

9. A man has $7.05 in nickels, dimes, and quarters. The quarters are worth $4.60 more than the dimes and the dimes are worth 25¢ more than the nickels. How many nickels, dimes, and quarters does the man have? 13 nickels; 9 dimes; 22 quarters

10. Amy has $2.50 consisting of nickels, dimes, and quarters in her piggy bank. She has the same amount in nickels and dimes, and twice as much in nickels as she has in quarters. How many nickels, dimes, and quarters does Amy have? 20 nickels; 10 dimes; 2 quarters

B In Problems 11–20, use two or more unknowns to solve these general problems.

11. The sum of two numbers is 102. Their difference is 16. What are the numbers? 43 and 59

12. The difference between two numbers is 28. Their sum is 82. What are the numbers? 55 and 27

13. The sum of two integers is 126. If one of the integers is 5 times the other, what are the integers? 21 and 105

14. The difference between two integers is 245. If one of the integers is 8 times the other, find the integers. 280 and 35

15. The difference between two numbers is 16. One of the numbers is 5 times the other. What are the numbers? 4 and 20

16. The sum of two numbers is 116. One of the numbers is 50 less than the other. What are the numbers? 83 and 33

17. Longs Peak is 145 ft higher than Pikes Peak. If you were to put these two peaks one on top of the other, you would still be 637 ft short of reaching the elevation of Mt. Everest, 29,002 ft. Find the elevations of Longs Peak and Pikes Peak.
Longs Peak: 14,255 ft; Pikes Peak: 14,110 ft

18. The height of the Empire State building and its antenna (Figure 12) is 1472 ft. The difference in height between the building and the antenna is 1028 ft. How tall is the antenna and how tall is the building? Antenna: 222 ft; building: 1250 ft

Figure 12
The Empire State building and its antenna.

19. One of the largest sundaes ever made contained about 6700 lb of topping. The topping flavors were chocolate, butterscotch, and caramel. There was the same amount of butterscotch as caramel but 600 lb more of chocolate than butterscotch. How many pounds of each were included in the topping? Butterscotch: $2033\frac{1}{3}$ lb; caramel: $2033\frac{1}{3}$ lb; chocolate: $2633\frac{1}{3}$ lb

20. One of the largest pancakes ever made used buckwheat flour, Puritan mix, and 15 gal of syrup. The flour and mix weighed 100 lb more than the 15 gal of syrup. What was the weight of the syrup if the whole pancake weighed 4100 lb? By the way, 68 lb of butter were added before it was consumed! 2000 lb

C In Problems 21–25, use two unknowns to solve these motion problems.

21. A plane flies 540 mi with a tail wind in $2\frac{1}{4}$ hr. The plane makes the return trip against the same wind and takes 3 hr. Find the speed of the plane in still air and the speed of the wind. Plane: 210 mi/hr; wind: 30 mi/hr

22. A motorboat runs 45 mi downstream in $2\frac{1}{2}$ hr and 39 mi upstream in $3\frac{1}{4}$ hr. Find the speed of the boat in still water and the speed of the current.
Boat: 15 mi/hr; current: 3 mi/hr

23. A motorboat can travel 15 mi/hr downstream and 9 mi/hr upstream on a certain river. Find the rate of the current and the rate at which the boat can travel in still water. Current: 3 mi/hr; boat: 12 mi/hr

24. It takes a motorboat $1\frac{1}{3}$ hr to go 20 mi downstream and $2\frac{2}{9}$ hr to return. Find the rate of the current and the rate at which the boat can travel in still water.
Current: 3 mi/hr; boat: 12 mi/hr

25. A plane flying with the wind took 2 hr for a 1000-mi flight and $2\frac{1}{2}$ hr for the return flight. Find the wind velocity and the speed of the plane in still air.
Wind: 50 mi/hr; plane: 450 mi/hr

D In Problems 26–30, use two or more unknowns to solve these investment problems.

26. Two sums of money totaling $20,000 earn 8% and 10% annual interest, respectively. If the interest from both investments amounts to $1900, how much is invested at each rate? $5000 at 8%; $15,000 at 10%

27. An investor invested $10,000, part at 6% and the rest at 8%. Find the amount invested at each rate if the annual income from the two investments is $720.
$6000 at 8%; $4000 at 6%

28. Andy Cabazos has $20,000 in three investments paying 6%, 8%, and 10%. The total interest on the 6% and 8% investments is $300 less than that obtained from the 10% investment. If his annual income from these investments is $1700, how much does he have invested at each rate? $5000 at 6%; $5000 at 8%; $10,000 at 10%

29. Marlene McGuire invested $25,000 in municipal bonds. The first investment paid 6%, the second 8%, and the third, 10%. If her annual income from these bonds was $2000 and the interest she received on the combined 6% and 8% investments equaled the interest on the 10% investment, how much money did she have in each category? $10,000 at 6%; $5000 at 8%; $10,000 at 10%

30. Marc Goldstein divided $20,000 into three parts. One part yielded 4%, another 8%, and the third one, 6%. If his total return was $1080 and he made $40 less on his 8% investment than on his 4% investment, what amount did he invest in each category? $11,000 at 4%; $5000 at 8%; $4000 at 6%

E In Problems 31–34, use two unknowns to solve these geometry problems.

31. The perimeter of the SuperFlag (Figure 13) is 1520 ft. If the length of the flag exceeds the width by 250 ft, what are the dimensions of the flag? $L = 505$ ft; $W = 255$ ft

32. One of the largest flags *actually flown* from a flagpole was 98 ft longer than it was wide. If its perimeter was 1016 ft, what were the dimensions of the flag? (It was a Brazilian flag, flown in Brazil.) $L = 303$ ft; $W = 205$ ft

Figure 13
The SuperFlag, shown hanging on Hoover Dam, is the Worlds Largest Flag according to the *Guinness Book of World Records.*

33. One of the world's largest quilts boasted a 438-ft perimeter with its length being 49 ft more than its width. What were the dimensions of the quilt? It took 7000 North Dakotans to make it. $L = 134$ ft; $W = 85$ ft

34. If you walked around one of the largest rectangular swimming pools in the world, located in Morocco, you would end up walking 3640 ft. If the pool is 1328 ft longer than it is wide, what are the dimensions of the pool?
$L = 1574$ ft; $W = 246$ ft

SKILL CHECKER

Try these "Skill Checker" exercises so you'll be ready for the next section.

Solve the system:

35.
$$2x - y + z = 3$$
$$x + y = -1$$
$$3x - y - 2z = 7$$
$(1, -2, -1)$

36.
$$2x - y + 2z = 3$$
$$2x + 2y - z = 0$$
$$-x + 2y + 2z = -12$$
$(2, -3, -2)$

USING YOUR KNOWLEDGE

The A, B, Cs of Vitamins

In this section we solved coin, general, distance, investment, and geometry problems. What type of problem is left out? Mixture problems! Use your knowledge to solve these mixture problems.

A dietician wants to arrange a diet composed of three basic foods A, B, and C. The diet must include 170 units of calcium, 90 units of iron, and 110 units of vitamin B. The table gives the number of units per ounce of each of the needed ingredients contained in each of the basic foods.

	Units per Ounce		
Nutrient	**Food A**	**Food B**	**Food C**
Calcium	15	5	20
Iron	5	5	10
Vitamin B	10	15	10

If a, b, and c are the number of ounces of basic foods A, B, and C taken by an individual, write an equation indicating:

37. The amount of calcium needed
$15a + 5b + 20c = 170$

38. The amount of iron needed
$5a + 5b + 10c = 90$

39. The amount of vitamin B needed
$10a + 15b + 10c = 110$

We shall come back to this problem and show you a new way to solve it in the next section.

WRITE ON

40. State the advantages and disadvantages of solving word problems using systems of equations instead of the techniques we studied in Chapter 2. Answers may vary.

42. Write the method you use to solve word problems.
Answers may vary.

41. Write a word problem that uses the system
$$x + y = 10$$
$$x - y = 2$$
to obtain its solution.
Answers may vary.

MASTERY TEST

If you know how to do these problems, you have learned your lesson!

43. The perimeter of a rectangle is 170 cm. If its length is 15 cm more than its width, what are the dimensions of the rectangle? $L = 50$ cm; $W = 35$ cm

44. An investor divides $20,000 among three investments at 6%, 8%, and 10%. If her total income is $1740 and the income from the 10% investment exceeds the income from the 6% and 8% investments by $260, how much did she invest at each rate? $3000 at 6%; $7000 at 8%; $10,000 at 10%

45. A plane flies 1200 mi with a tail wind in 3 hr. It takes 4 hr to fly the same distance against the wind. Find the wind velocity and the velocity of the plane in still air.
Wind: 50 mi/hr; plane: 350 mi/hr

46. A change machine gave Jill $2 in nickels and dimes. She had twice as many nickels as dimes. How many nickels and how many dimes did the machine give Jill?
20 nickels; 10 dimes

47. The McGuire twins weighed a total of 1466 lb. If their weight differential was 20 lb, what was the weight of each of the twins? 723 lb and 743 lb

48. The tallest president of the United States was Abraham Lincoln and the shortest was James Madison. Their height differential was $1\frac{1}{12}$ ft. If their combined height was $11\frac{7}{12}$ ft, how tall was each president?
Lincoln: $6\frac{1}{3}$ ft tall; Madison: $5\frac{1}{4}$ ft tall

4.4 MATRICES

To Succeed, Review How To ...

1. Solve a system of three equations in three unknowns (pp. 274–276).

2. Recognize whether a system is inconsistent or dependent (p. 277).

Objectives

A Perform elementary operations on systems of equations.

B Solve systems of linear equations using matrices.

C Solve applications using matrices.

GETTING STARTED

Alice meets Tweedledee and Tweedledum in *Through the Looking Glass.*

Tweedledee and Tweedledum

Do you remember Tweedledee and Tweedledum from Problem 86 in Exercises 4.1? Here's what they said:

Tweedledee: The sum of your weight and twice mine is 361 pounds.
Tweedledum: Contrariwise, the sum of your weight and twice mine is 360 pounds.

If Tweedledee weighs x pounds and Tweedledum weighs y pounds, the two sentences can be translated as

$$2x + y = 361$$
$$x + 2y = 360$$

We are going to solve this system again, but this time we shall also write the equivalent operations using *matrices.*

> **Matrix**
>
> A **matrix** (plural matrices) is a rectangular array of numbers enclosed in brackets.

For example,

$$\begin{bmatrix} 2 & 1 \\ 1 & 1 \end{bmatrix} \quad \text{and} \quad \begin{bmatrix} -5 & 3 & 2 \\ 4 & 0 & 1 \end{bmatrix}$$

are matrices. The first matrix has two rows and two columns (2×2), while the second one has two rows and three columns (2×3). A matrix derived from a system of linear equations (each written in standard form with the constant terms on the right) is called the **augmented matrix** of the system. For example, the augmented matrix of the system

$$\begin{matrix} 2x + y = 361 \\ x + 2y = 360 \end{matrix} \quad \text{is} \quad \begin{bmatrix} 2 & 1 & 361 \\ 1 & 2 & 360 \end{bmatrix}$$

Now compare the solution of the two-equation system with its solution using matrices:

$$\begin{matrix} 2x + y = 361 \\ x + 2y = 360 \end{matrix} \qquad \begin{bmatrix} 2 & 1 & 361 \\ 1 & 2 & 360 \end{bmatrix}$$

$$\begin{matrix} 2x + y = 361 \\ -2x - 4y = -720 \end{matrix} \qquad \begin{bmatrix} 2 & 1 & 361 \\ -2 & -4 & -720 \end{bmatrix}$$ Multiply the second equation by -2.

$$\begin{matrix} 2x + y = 361 \\ -3y = -359 \end{matrix} \qquad \begin{bmatrix} 2 & 1 & 361 \\ 0 & -3 & -359 \end{bmatrix}$$ Copy first equation.
Result of adding the two equations.

$$\begin{matrix} 2x + y = 361 \\ y = 119\frac{2}{3} \end{matrix} \qquad \begin{bmatrix} 2 & 1 & 361 \\ 0 & 1 & 119\frac{2}{3} \end{bmatrix}$$ Copy first equation.
Divide the second equation by -3.

Now you can substitute $119\frac{2}{3}$ for y in $2x + y = 361$ and solve for x. Did you note that the operations performed on the equations were identical to those performed on the matrices? In this section we shall solve systems of three equations using matrices.

 **Perform Elementary Operations on Systems of Equations**

Suppose we wish to solve a system of three linear equations in three unknowns. This means that we wish to find all sets of values of (x, y, z) that make all three equations true. We first need to consider what changes can be made in the system to yield an **equivalent system,** a system that has exactly the same solutions as the original system. We need to consider only three simple operations.

Teaching Tip

Have students solve the system

$$x + y + z = 6$$
$$x - y + z = 2$$
$$x + y - z = 0$$

by the elimination method and, side by side (as in the *Getting Started*), using matrices.

PROCEDURE

Elementary Operations on Systems of Equations

1. **The order of the equations may be changed.** This clearly cannot affect the solutions.

2. **Any of the equations may be multiplied by any nonzero real number.** If (m, n, p) is a solution of $ax + by + cz = d$, then, for $k \neq 0$, it is also a solution of $kax + kby + kcz = kd$, and conversely.

3. **Any equation of the system may be replaced by the sum (term by term) of itself and any other equation of the system.** (You can show this by doing Problem 11 of Exercises 4.4.)

These three **elementary operations** are used to simplify systems of equations and to find their solutions. Let's use them to solve the system

$$2x - y + z = 3$$

System I
(original system)
$$x + y \qquad = -1$$

$$3x - y - 2z = 7$$

Step 1. In the second equation of the system, x and y have unit coefficients, so we interchange the first two equations to get the following more convenient arrangement:

$$x + y \qquad = -1 \qquad \text{(1)}$$

System II
$$2x - y + z = 3 \qquad \text{(2)}$$

$$3x - y - 2z = 7 \qquad \text{(3)}$$

Step 2. To make the coefficients of x in equations (1) and (2) the same in absolute value but opposite in sign, multiply both sides of equation (1) by -2:

$$-2x - 2y = 2$$

Step 3. To eliminate x between equations (1) and (2), add the equation obtained in step 2 to equation (2) to get

$$-3y + z = 5 \qquad \text{(Replaces 2nd equation)} \qquad \text{(4)}$$

and restore equation (1) by dividing out the -2. The system now reads

$$x + y \qquad = -1 \qquad \text{(1)}$$

System III
$$-3y + z = 5 \qquad \text{(4)}$$

$$3x - y - 2z = 7 \qquad \text{(3)}$$

Now we proceed in a similar way to eliminate x between equations (1) and (3).

Step 4. Multiply both sides of equation (1) by -3:

$$-3x - 3y = 3$$

Step 5. Add this last equation to equation (3) to get

$$-4y - 2z = 10 \qquad \text{(Replaces 3rd equation)} \tag{5}$$

and restore equation (1) by dividing out the -3. The system now reads

System IV
$$
\begin{aligned}
x + y &= -1 \tag{1}\\
-3y + z &= 5 \tag{4}\\
-4y - 2z &= 10 \tag{5}
\end{aligned}
$$

We eliminate y between equations (4) and (5) in the following way:

Step 6. Multiply both sides of equation (4) by -4 and both sides of equation (5) by 3 to get

$$
\begin{aligned}
12y - 4z &= -20\\
-12y - 6z &= 30
\end{aligned}
$$

Step 7. Add the last two equations to get

$$-10z = 10 \qquad \text{(Replaces 3rd equation)} \tag{6}$$

and restore the second equation by dividing out the -4. The system is now

System V
$$
\begin{aligned}
x + y &= -1 \tag{1}\\
-3y + z &= 5 \tag{4}\\
-10z &= 10 \tag{6}
\end{aligned}
$$

Step 8. It's easy to solve system V: Equation (6) immediately gives $z = -1$. Then, by substitution into equation (4), we get

$$
\begin{aligned}
-3y - 1 &= 5 \tag{4}\\
-3y &= 6\\
y &= -2
\end{aligned}
$$

By substituting $y = -2$ into equation (1) we find

$$
\begin{aligned}
x - 2 &= -1 \tag{1}\\
x &= 1
\end{aligned}
$$

The solution of the system is $x = 1$, $y = -2$, $z = -1$. This is easily checked in the given set of equations.

B   Solve Systems of Linear Equations Using Matrices

A general system of the same form as system V may be written

System VI
$$
\begin{aligned}
ax + by + cz &= d\\
ey + fz &= g\\
hz &= k
\end{aligned}
$$

A system such as system VI, in which the first unknown, x, is missing from the second and third equations and the second unknown, y, is missing from the third equation, is said to be in **echelon form.** A system in echelon form is easy to solve. The third equation immediately yields the value of z. Back-substitution of this value into the second equation yields the value of y. Finally, back-substitution of the values of y and z into the first equation yields the value of x. Briefly, we say that we solve the system by **back-substitution.**

Every system of linear equations can be brought into echelon form by the use of the elementary operations. It remains for us only to organize and make the procedure more efficient by employing the augmented matrix.

Let's compare the augmented matrices of system I and system V:

System I
(original system)

$$\begin{aligned}\text{Main} &\to\\ \text{diagonal} &\end{aligned}\left[\begin{array}{ccc|c} 2 & -1 & 1 & 3 \\ 1 & 1 & 0 & -1 \\ 3 & -1 & -2 & 7 \end{array}\right]$$

System V

$$\begin{aligned}\text{Main} &\to\\ \text{diagonal} &\end{aligned}\left[\begin{array}{ccc|c} 1 & 1 & 0 & -1 \\ 0 & -3 & 1 & 5 \\ 0 & 0 & -10 & 10 \end{array}\right]$$

The augmented matrix for system V shows that the system is in echelon form because it has only 0's below the main diagonal (the diagonal left to right of the coefficient matrix). We should be able to obtain the second matrix from the first by performing operations corresponding to the elementary operations on the equations. These operations are called **elementary row operations,** and they always yield matrices of equivalent systems. Such matrices are called **row-equivalent.** If two matrices A and B are row-equivalent, we write $A \sim B$. The elementary row operations are as follows.

> **PROCEDURE**
>
> **Elementary Row Operations on Matrices**
>
> **1.** Change the order of the rows.
>
> **2.** Multiply all the elements of a row by any nonzero number.
>
> **3.** Replace any row by the element-by-element sum of itself and any other row.

These operations are performed as necessary to write the equivalent system in echelon form.

> **NOTE**
>
> Compare the elementary row operations with the elementary operations on a system. They are analogous!

We illustrate the procedure by showing the transition from the matrix of the original system I to that of system V. To explain what is happening at each step, we use the notation R_1, R_2, and R_3 for the respective rows of the matrix, along with the following typical abbreviations:

Step	Notation	Meaning
1	$R_1 \longleftrightarrow R_2$	Interchange R_1 and R_2.
2	$2 \times R_1$	Multiply each element of R_1 by 2.
3	$2 \times R_1 + R_2 \to R_2$	Replace R_2 by $2 \times R_1 + R_2$.

We write step 1 like this:

$$\left[\begin{array}{ccc|c} 2 & -1 & 1 & 3 \\ 1 & 1 & 0 & -1 \\ 3 & -1 & -2 & 7 \end{array}\right] \sim \left[\begin{array}{ccc|c} 1 & 1 & 0 & -1 \\ 2 & -1 & 1 & 3 \\ 3 & -1 & -2 & 7 \end{array}\right]$$
$$R_1 \longleftrightarrow R_2$$

Next, we proceed to get 0's in the second and third rows of the first column:

$$\left[\begin{array}{ccc|c} 1 & 1 & 0 & -1 \\ 2 & -1 & 1 & 3 \\ 3 & -1 & -2 & 7 \end{array}\right] \sim \left[\begin{array}{ccc|c} 1 & 1 & 0 & -1 \\ 0 & -3 & 1 & 5 \\ 0 & -4 & -2 & 10 \end{array}\right]$$
$$-2 \times R_1 + R_2 \to R_2$$
$$-3 \times R_1 + R_3 \to R_3$$

To complete the procedure, we get a 0 in the third row of the second column:

$$\begin{bmatrix} 1 & 1 & 0 & | & -1 \\ 0 & -3 & 1 & | & 5 \\ 0 & -4 & -2 & | & 10 \end{bmatrix} \sim \begin{bmatrix} 1 & 1 & 0 & | & -1 \\ 0 & -3 & 1 & | & 5 \\ 0 & 0 & -10 & | & 10 \end{bmatrix}$$

$$-4 \times R_2 + 3 \times R_3 \rightarrow R_3$$

Now we need to verify the result, the augmented matrix of system V. Once we obtain this last matrix, we can solve the system by back-substitution, as before.

We shall now provide additional illustrations of this procedure, but these will help you only if you take a pencil and paper and carry out the detailed row operations as they are indicated.

EXAMPLE 1 **Using matrices to solve a system of equations with one solution**

Use matrices to solve the system:

$$\begin{aligned} 2x - y + 2z &= 3 \\ 2x + 2y - z &= 0 \\ -x + 2y + 2z &= -12 \end{aligned}$$

SOLUTION The augmented matrix is

$$\begin{bmatrix} 2 & -1 & 2 & | & 3 \\ 2 & 2 & -1 & | & 0 \\ -1 & 2 & 2 & | & -12 \end{bmatrix} \sim \begin{bmatrix} 2 & -1 & 2 & | & 3 \\ 0 & 3 & -3 & | & -3 \\ 0 & 3 & 6 & | & -21 \end{bmatrix}$$

$$\begin{aligned} -R_1 + R_2 &\rightarrow R_2 \\ R_1 + 2 \times R_3 &\rightarrow R_3 \end{aligned}$$

$$\sim \begin{bmatrix} 2 & -1 & 2 & | & 3 \\ 0 & 1 & -1 & | & -1 \\ 0 & 0 & 9 & | & -18 \end{bmatrix}$$

$$\begin{aligned} -R_2 + R_3 &\rightarrow R_3 \\ \frac{1}{3} \times R_2 &\rightarrow R_2 \end{aligned}$$

This last matrix is in echelon form and corresponds to the system

$$2x - y + 2z = 3 \qquad \textbf{(1)}$$

$$y - z = -1 \qquad \textbf{(2)}$$

$$9z = -18 \qquad \textbf{(3)}$$

We solve this system by back-substitution. Equation (3) immediately yields $z = -2$. Equation (2) then becomes

$$y + 2 = -1$$

so that

$$y = -3$$

Equation (1) then becomes

$$2x + 3 - 4 = 3$$

so that

$$x = 2$$

The final answer, $x = 2$, $y = -3$, $z = -2$, can be checked in the given system.

PROBLEM 1

Use matrices to solve the system:

$$\begin{aligned} 2x + 3y + z &= 14 \\ x - y + 2z &= 8 \\ -x + 4y - z &= 2 \end{aligned}$$

Answer

1. (2, 2, 4)

Calculate It   Solve a System Using Matrices

Calculators use matrix operations (too advanced to discuss in detail here) to solve systems of equations. Let's illustrate the procedure to solve Example 1. Let A be the coefficient matrix and B the constant matrix defined by:

$$A = \begin{bmatrix} 2 & -1 & 2 \\ 2 & 2 & -1 \\ -1 & 2 & 2 \end{bmatrix}$$

$$B = \begin{bmatrix} 3 \\ 0 \\ -12 \end{bmatrix}$$

To enter A, press [2nd] [x^{-1}] [◄][►][►] [1] [3] [ENTER] [3] [ENTER]. This tells the calculator you are about to enter a 3×3 matrix. Now, enter the values for A by pressing [2] [ENTER] [(−)] [1]

[ENTER] [2] [ENTER] [2] [ENTER] [2] [ENTER] [(−)] [1] [ENTER] [(−)] [1] [ENTER] [2] [ENTER] [2] [ENTER] .

To enter B press [2nd] [x^{-1}] [◄][►] [2] [3] [ENTER] [1] [ENTER]. Next, enter the values for B by pressing [3] [ENTER] [0] [ENTER] [(−)] [1] [2] [ENTER]. Press [2nd] [MODE] to go to the home screen. Finally, press [2nd] [x^{-1}] [1] [x^{-1}] [2nd] [x^{-1}] [2] [ENTER]. The solution is as shown in the window.

This means that $x = 2$, $y = -3$, and $z = -2$, as obtained in Example 1. See if you can follow the same procedure to solve Example 2 but do not get alarmed if you get an error message. (There is no solution to the system in Example 2.)

```
[A]⁻¹[B]
              [[2 ]
               [-3]
               [-2]]
```

| **EXAMPLE 2** | **Using matrices to solve a system of equations with no solution** |

Use matrices to solve the system:

$$2x - y + 2z = 3$$
$$2x + 2y - z = 0$$
$$4x + y + z = 5$$

SOLUTION

$$\begin{bmatrix} 2 & -1 & 2 & | & 3 \\ 2 & 2 & -1 & | & 0 \\ 4 & 1 & 1 & | & 5 \end{bmatrix} \sim \begin{bmatrix} 2 & -1 & 2 & | & 3 \\ 0 & 3 & -3 & | & -3 \\ 0 & 3 & -3 & | & -1 \end{bmatrix}$$

$$-R_1 + R_2 \rightarrow R_2$$
$$-2 \times R_1 + R_3 \rightarrow R_3$$

$$\sim \begin{bmatrix} 2 & -1 & 2 & | & 3 \\ 0 & 3 & -3 & | & -3 \\ 0 & 0 & 0 & | & 2 \end{bmatrix}$$

$$-R_2 + R_3 \rightarrow R_3$$

The final matrix is in echelon form, and the last line corresponds to the equation

$$0x + 0y + 0z = 2$$

which is false for all values of x, y, z. Hence the given system has *no solution*.

| **PROBLEM 2** |

Use matrices to solve the system:

$$x + 2y - z = 4$$
$$5x - 3y + 2z = 1$$
$$6x - y + z = 3$$

Answer

2. No solution

> **NOTE**
>
> If reduction to echelon form introduces any row with all 0's to the left and a nonzero number to the right of the vertical line, then the system has no solution.

EXAMPLE 3 **Using matrices to solve a system of equations with infinitely many solutions**

Solve the system:

$$2x - y + 2z = 3$$
$$2x + 2y - z = 0$$
$$4x + y + z = 3$$

SOLUTION

$$\begin{bmatrix} 2 & -1 & 2 & | & 3 \\ 2 & 2 & -1 & | & 0 \\ 4 & 1 & 1 & | & 3 \end{bmatrix} \sim \begin{bmatrix} 2 & -1 & 2 & | & 3 \\ 0 & 3 & -3 & | & -3 \\ 0 & 3 & -3 & | & -3 \end{bmatrix}$$

$$-R_1 + R_2 \rightarrow R_2$$
$$-2 \times R_1 + R_3 \rightarrow R_3$$

$$\sim \begin{bmatrix} 2 & -1 & 2 & | & 3 \\ 0 & 3 & -3 & | & -3 \\ 0 & 0 & 0 & | & 0 \end{bmatrix}$$

$$-R_2 + R_3 \rightarrow R_3$$

The last matrix is in echelon form, and the last line corresponds to the equation

$$0x + 0y + 0z = 0$$

which is true for all values of x, y, and z. Any solution of the first two equations will be a solution of the system. The first two equations from the equivalent matrix in echelon form are

$$2x - y + 2z = 3 \qquad \textbf{(1)}$$
$$3y - 3z = -3 \qquad \textbf{(2)}$$

This system is equivalent to

$$2x - y = 3 - 2z \qquad \text{Subtract } 2z \text{ from both sides.} \qquad \textbf{(3)}$$
$$y = -1 + z \qquad \text{Solve equation (2) for } y. \qquad \textbf{(4)}$$

Suppose we let $z = k$, where k is any real number. Then equations (3) and (4) become

$$2x - y = 3 - 2k \qquad \textbf{(5)}$$
$$y = -1 + k \qquad \textbf{(6)}$$

Substitution of $y = -1 + k$ into equation (5) results in

$$2x - (-1 + k) = 3 - 2k$$

and we solve for x.

$$2x + 1 - k = 3 - 2k \qquad \text{Remove parentheses.}$$
$$2x = 2 - k \qquad \text{Subtract 1 and add } k.$$
$$x = 1 - \frac{1}{2}k \qquad \text{Divide by 2.}$$

Thus if k is any real number, then $x = 1 - \frac{1}{2}k$, $y = k - 1$, $z = k$ is a solution of the system. You can verify this by substitution in the original system. We see that the system in this example has infinitely many solutions, because the value of k may be arbitrarily chosen. For instance, if $k = 2$, then the solution is $x = 0$, $y = 1$, $z = 2$; if $k = 5$, then $x = -\frac{3}{2}$, $y = 4$, $z = 5$; if $k = -4$, then $x = 3$, $y = -5$, $z = -4$; and so on. In conclusion, we can say this system has infinitely many solutions and for any real number, k, solutions can be obtained from the ordered triple $(1 - \frac{1}{2}k, k - 1, k)$.

PROBLEM 3

Use matrices to solve the system of equations.

$$-4x + y + z = 4$$
$$2x - y + 3z = 5$$
$$6x - 2y + 2z = 1$$

Answer

3. Infinitely many solutions such that for any real number, k, solutions can be obtained from the ordered triple $(2k - 4\frac{1}{2}, 7k - 14, k)$.

In the final echelon form, the system has infinite solutions if there is no row with all 0's to the left and a nonzero number to the right of the vertical line but there *is* a row with all 0's to both the left and the right.

Examples 1–3 illustrate the three possibilities for three linear equations in three unknowns. The system may have **one unique solution** as in Example 1; the system may have **no solution** as in Example 2; or the system may have **infinitely many solutions** as in Example 3. The final echelon form of the matrix always shows which case is at hand.

Solving Applications Using Matrices

MATRICES AND NUTRITION

Here is a problem from *Using Your Knowledge,* Section 4.3. Now let's solve this problem using matrices.

EXAMPLE 4   **Where are the nutrients?**

A dietitian wants to arrange a diet composed of three basic foods A, B, and C. The diet must include 170 units of calcium, 90 units of iron, and 110 units of vitamin B. The table gives the number of units per ounce of each of the needed ingredients contained in each of the basic foods.

Nutrient	Units per Ounce		
	Food A	Food B	Food C
Calcium	15	5	20
Iron	5	5	10
Vitamin B	10	15	10

SOLUTION

1. Read the problem.

If a, b, and c are the number of ounces of basic foods A, B, and C taken by an individual, find the number of ounces of each of the basic foods needed to meet the diet requirements.

2. Select the unknown.

We want to find the values of a, b, and c, the number of ounces of basic foods A, B, and C taken by an individual.

3. Think of a plan.

We write a system of equations: What is the amount of calcium needed? The individual gets 15 units of calcium from A, 5 from B, and 20 from C, so the amount of calcium is

$$15a + 5b + 20c = 170$$

What is the amount of iron needed?

$$5a + 5b + 10c = 90$$

What is the amount of vitamin B needed?

$$10a + 15b + 10c = 110$$

4. Use matrices or your calculator to solve the system.

Write the equations obtained using matrices. The simplified system of three equations and three unknowns, obtained by dividing each term in each of the

PROBLEM 4

A landscaping company placed three orders with a nursery. The first order was for 24 bushes, 16 flowering plants, and 10 trees and totaled $374. The second order was for 36 bushes, 48 flowering plants, and 20 trees and totaled $828. The third order was for 50 bushes, 40 flowering plants, and 30 trees and totaled $970. The bill does not list the per item price. What is the cost of one bush, one flowering plant, and one tree?

Answer

4. One bush is $4, one flowering plant is $8, and one tree is $15.

equations by 5, is written as

$$3a + b + 4c = 34$$
$$a + b + 2c = 18$$
$$2a + 3b + 2c = 22$$

The corresponding augmented matrix is thus

$$\begin{bmatrix} 3 & 1 & 4 & | & 34 \\ 1 & 1 & 2 & | & 18 \\ 2 & 3 & 2 & | & 22 \end{bmatrix}$$

$$\begin{bmatrix} 3 & 1 & 4 & | & 34 \\ 1 & 1 & 2 & | & 18 \\ 2 & 3 & 2 & | & 22 \end{bmatrix} \sim \begin{bmatrix} 1 & 1 & 2 & | & 18 \\ 3 & 1 & 4 & | & 34 \\ 2 & 3 & 2 & | & 22 \end{bmatrix} \sim \begin{bmatrix} 1 & 1 & 2 & | & 18 \\ 0 & -2 & -2 & | & -20 \\ 0 & 1 & -2 & | & -14 \end{bmatrix}$$

$$R_1 \leftrightarrow R_2 \qquad \begin{array}{l} R_2 - 3R_1 \to R_2 \\ R_3 - 2R_1 \to R_3 \end{array}$$

$$\sim \begin{bmatrix} 1 & 1 & 2 & | & 18 \\ 0 & 1 & 1 & | & 10 \\ 0 & 1 & -2 & | & -14 \end{bmatrix} \sim \begin{bmatrix} 1 & 1 & 2 & | & 18 \\ 0 & 1 & 1 & | & 10 \\ 0 & 0 & -3 & | & -24 \end{bmatrix}$$

$$\frac{R_2}{-2} \to R_2 \qquad R_3 - R_2 \to R_3$$

From the third row, $-3c = -24$, or $c = 8$. Substituting in the second row, we have $b + 8 = 10$, or $b = 2$. Finally, substituting in the first row, we have $a + 2 + 2(8) = 18$, or $a = 0$.

5. Verify the solution.

Substitute $a = 0$, $b = 2$, and $c = 8$ into the first equation to obtain $15 \cdot 0 + 5 \cdot 2 + 20 \cdot 8 = 10 + 160 = 170$. Then use the same procedure to check the second and third equations. The required answer is 0 oz of food A, 2 oz of food B, and 8 oz of food C.

EXAMPLE 5 **Nailing down the problem**

Tom Jones, who was building a workshop, went to the hardware store and bought 1 lb each of three types of nails: small, medium, and large. After completing part of the work, Tom found that he had underestimated the number of small and large nails he needed. So he bought another pound of the small nails and 2 lb more of the large nails. After some more work, he again ran short of nails and had to buy another pound of each of the small and the medium nails. Upon looking over his bills, he found that the hardware store charged him $2.10 for nails the first time, $2.30 the second time, and $1.20 the third time. The prices for the various sizes of nails were not listed. Find these prices.

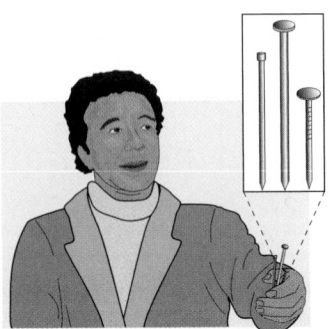

Web It

For another clear example of how to use the augmented matrix and echelon form to solve a system, try link 4-4-1 at mhhe.com/bello.

PROBLEM 5

After Tom built his workshop (see Example 5), he had another project requiring small, medium, and large nails. This time he went to a different hardware store. On the first trip he bought 1 lb of each size nail for $2.15. On the second trip, he bought 4 lb of small nails and 1 lb of medium nails for $2.85. He needed one more trip, where he bought 1 lb of medium nails and 2 lb of large nails and paid $2.55. How much did this hardware store charge for each size of nail?

Answer

5. Small: 55¢/lb; medium: $65¢/lb; large: 95¢/lb

SOLUTION

1. Read the problem.
There were three purchases of nails costing $2.10, $2.30, and $1.20.

2. Select the unknown.
We let x, y, and z be the prices in cents per pound for the small, medium, and large nails, respectively.

3. Think of a plan.
Then we know that

$$x + y + z = 210$$
$$x \quad\;\; + 2z = 230$$
$$x + y \quad\;\;\; = 120$$

4. Use matrices to solve the system.
We solve this system as follows:

$$\begin{bmatrix} 1 & 1 & 1 & | & 210 \\ 1 & 0 & 2 & | & 230 \\ 1 & 1 & 0 & | & 120 \end{bmatrix} \sim \begin{bmatrix} 1 & 1 & 1 & | & 210 \\ 0 & 1 & -1 & | & -20 \\ 0 & 0 & 1 & | & 90 \end{bmatrix}$$

$$R_1 - R_2 \to R_2$$
$$R_1 - R_3 \to R_3$$

The second matrix is in echelon form, and the solution of the system is easily found by back-substitution to be $x = 50$, $y = 70$, $z = 90$, giving the schedule of prices shown.

Tom's Schedule of Nail Prices	
Nail Size	**Price Per Pound**
Small	50¢
Medium	70¢
Large	90¢

5. Verify the solution.
Substitute $x = 50$, $y = 70$, and $z = 90$ in $x + y + z = 210$ to obtain $50 + 70 + 90 = 210$, which is true.

Exercises 4.4

A B In Problems 1–10, find all the solutions (if there are any).

1. $x + y - z = 3$
$x - 2y + z = -3$
$2x + y + z = 4$
$(1, 2, 0)$

2. $x + 2y - z = 5$
$2x + y + z = 1$
$x - y + z = -1$
$(2, 0, -3)$

3. $2x - y + 2z = 5$
$2x + y - z = -6$
$3x \quad\;\; + 2z = 3$
$(-1, -1, 3)$

4. $x + 2y - z = 0$
$2x + 3y \quad\;\; = 3$
$2y + z = -1$
$(3, -1, 1)$

5. $3x + 2y + z = -5$
$2x - y - z = -6$
$2x + y + 3z = 4$
$(-2, -1, 3)$

6. $4x + 3y - z = 12$
$2x - 3y - z = -10$
$x + y - 2z = -5$
$(2, 3, 5)$

7. $x + y + z = 3$
$x - 2y + z = -3$
$3x + 3z = 5$
No solution

8. $x + y + z = 3$
$x - 2y + 3z = 5$
$5x - 4y + 11z = 20$
No solution

9. $x + y + z = 3$
$x - 2y + z = -3$
$x + z = 1$
$(1 - k, 2, k)$, k is any real number

10. $x - y - 2z = -1$
$x + 2y + z = 5$
$5x + 4y - z = 13$
$(1 + k, 2 - k, k)$, k is any real number

APPLICATIONS

In Problems 11–16, use matrices to solve the system.

11. Nancy's change drawer contains $58 in dimes, quarters, and one-dollar coins. If twice the number of quarters is 80 more than the number of one-dollar coins, and the number of quarters plus twice the number of dimes is three times the number of one-dollar coins, find the number of each type of coin in Nancy's drawer.
30 dimes; 60 quarters; 40 one-dollar coins

12. The sum of $8.50 is made up of nickels, dimes, and quarters. The number of dimes is equal to the number of quarters plus twice the number of nickels. The value of the dimes exceeds the combined value of the nickels and the quarters by $1.50. How many of each coin are there? 20 nickels; 50 dimes; 10 quarters

13. The Mechano Distributing Company has three types of vending machines, which dispense snacks as listed in the table. Mechano fills all the machines once a day and finds them all sold out before the next day. The total daily sales are candy, 760; peanuts, 380; sandwiches, 660. How many of each type of machine does Mechano have?
Type I: 8; type II: 10; type III: 12

Mechano Distributing Company Data

Snack	Vending Machine Type		
	I	II	III
Candy	20	24	30
Peanuts	10	18	10
Sandwiches	0	30	30

14. Suppose the total daily income from the various types of machines in Problem 13 is as follows: type I, $32.00; type II, $159.00; type III, $192.00. What is the selling price for each type of snack? (Use your answers from Problem 13.) Candy: 15¢; peanuts: 10¢; sandwiches: 35¢

15. Gro-Kwik Garden Supply has three types of fertilizer, which contain chemicals A, B, and C in the percentages shown in the table. In what proportions must Gro-Kwik mix these three types to get an 8-8-8 fertilizer (one that has 8% of each of the three chemicals)?
50% type I; 25% type II; 25% type III

Gro-Kwik Garden Supply Data

Chemical	Type of Fertilizer		
	I	II	III
A	6%	8%	12%
B	6%	12%	8%
C	8%	4%	12%

16. Three water supply valves, A, B, and C, are connected to a tank. If all three valves are opened, the tank is filled in 8 hr. The tank can also be filled by opening A for 8 hr and B for 12 hr, while keeping C closed, or by opening B for 10 hr and C for 28 hr, while keeping A closed. Find the time needed by each valve to fill the tank by itself. (*Hint:* Let x, y, and z, respectively, be the fractions of the tank that valves A, B, and C can fill alone in 1 hr.) A: 16 hr; B: 24 hr; C: 48 hr

17. A 2 × 2 matrix (2 rows and 2 columns)
$$\begin{bmatrix} a & b \\ c & d \end{bmatrix}$$
is said to be **singular** if $ad - bc = 0$. Otherwise, it is nonsingular. Determine whether the following matrices are singular or nonsingular.

a. $\begin{bmatrix} 1 & 2 \\ 2 & 4 \end{bmatrix}$ Singular
b. $\begin{bmatrix} 2 & -3 \\ 3 & 5 \end{bmatrix}$ Nonsingular
c. $\begin{bmatrix} 0 & 2 \\ 2 & 4 \end{bmatrix}$ Nonsingular

18. The product of two matrices,

$$\begin{bmatrix} a & b \\ c & d \end{bmatrix} \text{ and } \begin{bmatrix} x & y \\ z & w \end{bmatrix}$$

is found by a row-column multiplication defined as follows:

$$\begin{bmatrix} a & b \\ c & d \end{bmatrix} \begin{bmatrix} x & y \\ z & w \end{bmatrix} \begin{bmatrix} ax + bz & ay + bw \\ cx + dz & cy + dw \end{bmatrix}$$

If this product equals

$$\begin{bmatrix} 1 & 0 \\ 0 & 1 \end{bmatrix}$$

each of the two matrices is said to be the **multiplicative inverse** of the other. To find the inverse of

$$\begin{bmatrix} a & b \\ c & d \end{bmatrix}$$

you have to find a matrix

$$\begin{bmatrix} x & y \\ z & w \end{bmatrix}$$

such that

$$\begin{bmatrix} a & b \\ c & d \end{bmatrix} \begin{bmatrix} x & y \\ z & w \end{bmatrix} = \begin{bmatrix} 1 & 0 \\ 0 & 1 \end{bmatrix}$$

This means that you must solve the systems

$$ax + bz = 1 \qquad ay + bw = 0$$
$$\text{and}$$
$$cx + dz = 0 \qquad cy + dw = 1$$

To solve the first system when $c \neq 0$, we can write

$$\begin{bmatrix} a & b & | & 1 \\ c & d & | & 0 \end{bmatrix} \sim \begin{bmatrix} ac & bc & | & c \\ ac & ad & | & 0 \end{bmatrix} \sim \begin{bmatrix} ac & bc & | & c \\ 0 & ad - bc & | & -c \end{bmatrix}$$

Now we see that the second equation has a unique solution if and only if $ad - bc \neq 0$.

Use these ideas to find the inverse of

a. $\begin{bmatrix} 3 & 2 \\ 2 & 1 \end{bmatrix}$ $\begin{bmatrix} -1 & 2 \\ 2 & -3 \end{bmatrix}$

b. $\begin{bmatrix} 1 & -2 \\ 2 & 1 \end{bmatrix}$ $\begin{bmatrix} \frac{1}{5} & \frac{2}{5} \\ -\frac{2}{5} & \frac{1}{5} \end{bmatrix}$

SKILL CHECKER

Try these "Skill Checker" exercises so you'll be ready for the next section.

Evaluate:

19. $(8)(-5) - (22)(5)$ -150

20. $(13)(-4) - (21)(3)$ -115

21. $(2)(-4) - (5)(93)$ -473

22. $\dfrac{(5)(-1) - (3)(1)}{(1)(-1) - (3)(1)}$ 2

23. $\dfrac{(1)(3) - (3)(5)}{(1)(-3) - (1)(93)}$ $\frac{1}{8}$

24. $\dfrac{(-3)(1) - (-2)(5)}{(2)(1) - (-1)(-3)}$ -7

USING YOUR KNOWLEDGE

Finding Your Identity

Instead of stopping with the echelon form of the matrix and then using back-substitution to solve a system of equations, many people prefer to transform the matrix of coefficients into the **identity matrix** (a square matrix with 1's along the diagonal and 0's everywhere else) and then read off the solution by inspection. If the final augmented matrix reads

$$\begin{bmatrix} 1 & 0 & 0 & | & a \\ 0 & 1 & 0 & | & b \\ 0 & 0 & 1 & | & c \end{bmatrix}$$

then the solution of the system is $x = a$, $y = b$, $z = c$. You need only read the column to the right of the vertical line.

Suppose, for example, that we have reduced the augmented matrix to the form

$$\begin{bmatrix} 2 & -1 & 2 & | & 3 \\ 0 & 3 & -3 & | & -3 \\ 0 & 0 & 2 & | & 5 \end{bmatrix}$$

We can now divide R_2 by 3 and R_3 by 2 to obtain

$$\begin{bmatrix} 2 & -1 & 2 & | & 3 \\ 0 & 1 & -1 & | & -1 \\ 0 & 0 & 1 & | & \frac{5}{2} \end{bmatrix}$$

To get 0's in the off-diagonal places of the matrix to the left of the vertical line, we first add R_2 to R_1 to get the new R_1:

$$\begin{bmatrix} 2 & 0 & 1 & | & 2 \\ 0 & 1 & -1 & | & -1 \\ 0 & 0 & 1 & | & \frac{5}{2} \end{bmatrix}$$

Next, we subtract R_3 from R_1 and add R_3 to R_2, with the result

$$\begin{bmatrix} 2 & 0 & 0 & | & -\frac{1}{2} \\ 0 & 1 & 0 & | & \frac{3}{2} \\ 0 & 0 & 1 & | & \frac{5}{2} \end{bmatrix}$$

Finally, we divide R_1 by 2 to obtain

$$\begin{bmatrix} 1 & 0 & 0 & \Big| & -\frac{1}{4} \\ 0 & 1 & 0 & \Big| & \frac{3}{2} \\ 0 & 0 & 1 & \Big| & \frac{5}{2} \end{bmatrix}$$

from which we can see by inspection that the solution of the system is $x = -\frac{1}{4}$, $y = \frac{3}{2}$, $z = \frac{5}{2}$.

After you bring the augmented matrix into echelon form, you perform additional elementary row operations to get 0's in all the off-diagonal places of the coefficient matrix and 1's on the diagonal. We assume that the system has a unique solution. If this is not the case, you will see what the situation is when you obtain the echelon form, and you will also see that it's impossible to transform the coefficient matrix into the identity matrix if the system doesn't have a unique solution. Use these ideas to solve Problems 1–10 in Exercises 4.4.

WRITE ON

When solving systems of equations using matrices, explain how you recognize:

25. A consistent system
Answers may vary.

26. An inconsistent system
Answers may vary.

27. A dependent system
Answers may vary.

28. Describe the relationship between elementary row operations on matrices and elementary operations on systems of equations. Answers may vary.

MASTERY TEST

If you know how to do these problems, you have learned your lesson!

Solve using matrices:

29. $2x + 3y - z = -1$
$3x + 4y + 2z = 14$
$x - 6y - 5z = 4$
$(6, -3, 4)$

30. $5x + 2y + 4z = -5$
$7x + 8y - 2z = 13$
$2x - 5y + 3z = 4$
$(3, -2, -4)$

31. $5x + 6y - 30z = 13$
$2x + 4y - 12z = 6$
$x + 2y - 6z = 8$
No solution

32. $x + y + z = 4$
$x - 2y + z = 7$
$2x - y + 2z = 11$
$(5 - k, -1, k)$, k is any real number

33. $3x - y + z = 3$
$x - 2y + 2z = 1$
$2x + y - z = 2$
$(1, k, k)$, k is any real number

34. $x - y + z = 3$
$x + 2y - z = 3$
$2x - 2y + 2z = 6$
$(k, 6 - 2k, 9 - 3k)$, k is any real number

35. The sum of $14.10 is made up of nickels, dimes, and quarters. The number of quarters is equal to the number of dimes plus three times the number of nickels. The value of the quarters exceeds the combined value of the dimes and nickels by $8.90. How many of each coin are there? Dimes: 22; nickels: 8; quarters: 46

4.5   DETERMINANTS AND CRAMER'S RULE

To Succeed, Review How To . . .

1. Evaluate expressions involving integers (p. 49).

Objectives

A Evaluate a 2 × 2 determinant.

B Use Cramer's rule to solve a system of two equations in two unknowns.

C Use minors to evaluate 3 × 3 determinants.

D Use Cramer's rule to solve a system of three equations.

GETTING STARTED

Leibniz and Matrices

Gottfried Wilhelm Leibniz developed the theory of determinants. In 1693, Leibniz studied and used *determinants* to solve systems of simultaneous equations. A **determinant** is a square array of numbers of the form

$$\begin{vmatrix} a_1 & b_1 \\ a_2 & b_2 \end{vmatrix}$$

The numbers a_1, a_2, b_1, and b_2 are called the *elements* of the determinant. As you can see, this determinant has *two* rows and *two* columns. For this reason,

$$\det A = \begin{vmatrix} a_1 & b_1 \\ a_2 & b_2 \end{vmatrix}$$

Gottfried Wilhelm Leibniz
(1646–1716)

is called a two-by-two (2 × 2) determinant. Every **square matrix** (a matrix with the same number of rows and columns) has a determinant associated with it.

A   Evaluating 2 × 2 Determinants

Each 2 × 2 determinant has a real number associated with it. The value of

$$\det A = \begin{vmatrix} a_1 & b_1 \\ a_2 & b_2 \end{vmatrix}$$

is defined to be

$$a_1 b_2 - a_2 b_1$$

which can be obtained by multiplying along the diagonals, as indicated in the following definition.

Determinant

The determinant of the matrix $\begin{bmatrix} a_1 & b_1 \\ a_2 & b_2 \end{bmatrix}$ is denoted by $\begin{vmatrix} a_1 & b_1 \\ a_2 & b_2 \end{vmatrix}$ and is defined as

$$\begin{vmatrix} a_1 & b_1 \\ a_2 & b_2 \end{vmatrix} = a_1 b_2 - a_2 b_1$$

For example,

$$\begin{vmatrix} 2 & 5 \\ -3 & -9 \end{vmatrix} = (2)(-9) - (-3)(5) = -18 + 15 = -3$$

EXAMPLE 1 **Evaluating determinants**	**PROBLEM 1**

Evaluate:

a. $\begin{vmatrix} -3 & 7 \\ -5 & 4 \end{vmatrix}$ **b.** $\begin{vmatrix} -3 & 6 \\ -5 & 10 \end{vmatrix}$

SOLUTION

a. $\begin{vmatrix} -3 & 7 \\ -5 & 4 \end{vmatrix} = (-3)(4) - (-5)(7) = -12 - (-35) = 23$

b. $\begin{vmatrix} -3 & 6 \\ -5 & 10 \end{vmatrix} = (-3)(10) - (-5)(6) = -30 - (-30) = 0$

Evaluate:

a. $\begin{vmatrix} -2 & 6 \\ -3 & -5 \end{vmatrix}$ **b.** $\begin{vmatrix} 2 & -8 \\ -3 & -12 \end{vmatrix}$

Calculate It

Looking Ahead

You can learn to evaluate determinants with your calculator by looking at the *Calculate It* at the end of this section.

In Example 1(b) the second-column elements of the determinant are both the same multiple of the corresponding first-column elements:

$$6 = (-2)(-3) \quad \text{and} \quad 10 = (-2)(-5)$$

In general, if the elements of one row are just some constant k times the elements of the other row (or if this is true of the columns), then the value of the determinant is zero. This is easy to see:

$$\begin{vmatrix} a & b \\ ka & kb \end{vmatrix} = akb - bka = 0$$

This result will be of use to us in *Using Your Knowledge* in Exercises 4.5.

B   Using Cramer's Rule to Solve a System of Two Equations

Web It

For a video demonstration of finding the value of 2 × 2 and 3 × 3 determinants, along with using Cramer's rule to solve systems of equations, try link 4-5-1 at mhhe.com/bello.

One of the important applications of determinants is in the solution of a system of linear equations. Let's look at the simplest case, two equations in two unknowns. Such a system can be written

$$a_1 x + b_1 y = d_1 \tag{1}$$

$$a_2 x + b_2 y = d_2 \tag{2}$$

We can eliminate y from this system by multiplying equation (1) by b_2 and equation (2) by b_1 and then subtracting, as follows.

$$a_1 b_2 x + b_1 b_2 y = d_1 b_2 \qquad \text{This is } b_2 \text{ times equation (1).}$$

$$\underline{(-)\, a_2 b_1 x + b_1 b_2 y = d_2 b_1} \qquad \text{This is } b_1 \text{ times equation (2).}$$

$$a_1 b_2 x - a_2 b_1 x = d_1 b_2 - d_2 b_1 \qquad \text{Subtract the second equation from the first.}$$

$$(a_1 b_2 - a_2 b_1) x = d_1 b_2 - d_2 b_1 \qquad \text{Factor out } x.$$

Now if the quantity $a_1 b_2 - a_2 b_1 \neq 0$, we can divide by this quantity to get

$$x = \frac{d_1 b_2 - d_2 b_1}{a_1 b_2 - a_2 b_1} \qquad \text{Solve for } x.$$

The denominator, $a_1 b_2 - a_2 b_1$, which we shall denote by D, can be written as the determinant

$$D = \begin{vmatrix} a_1 & b_1 \\ a_2 & b_2 \end{vmatrix} = a_1 b_2 - a_2 b_1$$

Answers

1. a. 28   **b.** −48

This determinant is naturally called the **determinant of the coefficients.** The numerator, $d_1b_2 - d_2b_1$, can be obtained from the denominator by replacing the coefficients of x (the a's) by the corresponding constant terms, the d's. We denote the numerator of x by D_x. Thus

Constant terms replace the a's. $D_x = \begin{vmatrix} d_1 & b_1 \\ d_2 & b_2 \end{vmatrix}$ y coefficients unchanged

We can now write the solution for x in the form

$$x = \frac{D_x}{D}$$

A similar procedure shows that the solution for y is

$$y = \frac{a_1d_2 - a_2d_1}{a_1b_2 - a_2b_1}$$

The denominator, $a_1b_2 - a_2b_1$, is again the determinant D. The numerator, $a_1d_2 - a_2d_1$, which we shall denote by D_y, can be formed from D by replacing the coefficients of y (the b's) by the corresponding d's. Hence

x coefficients unchanged $D_y = \begin{vmatrix} a_1 & d_1 \\ a_2 & d_2 \end{vmatrix}$ Constant terms replace the y coefficients.

and we can write the solution for y in the form

$$y = \frac{D_y}{D}$$

We summarize these results as follows.

Cramer's Rule for Solving a System of Two Equations in Two Unknowns	

The system

$$a_1x + b_1y = d_1$$
$$a_2x + b_2y = d_2$$

1.   Has the unique solution

$$x = \frac{D_x}{D} \quad \text{and} \quad y = \frac{D_y}{D}$$

where $D \neq 0$ and

$$D_x = \begin{vmatrix} d_1 & b_1 \\ d_2 & b_2 \end{vmatrix}, \qquad D_y = \begin{vmatrix} a_1 & d_1 \\ a_2 & d_2 \end{vmatrix}, \qquad D = \begin{vmatrix} a_1 & b_1 \\ a_2 & b_2 \end{vmatrix}$$

2.   Is inconsistent and has no solution if $D = 0$ and any one of D_x or D_y is different from 0.

3.   Has no unique solution if $D = 0$ and $D_x = D_y = 0$. (In this case, the system either has no solution or infinitely many solutions. This situation is studied more fully in advanced algebra.)

EXAMPLE 2	Using Cramer's rule to solve a system of two equations

Use Cramer's rule to solve the system:

$$2x + 3y = 7$$
$$5x + 9y = 11$$

Answer

2. (5, 2)

PROBLEM 2

Use Cramer's rule to solve the system:

$$2x + 3y = 16$$
$$5x - 4y = 17$$

SOLUTION

$$D = \begin{vmatrix} 2 & 3 \\ 5 & 9 \end{vmatrix} = 18 - 15 = 3$$ Use the coefficients of the variables.

$$D_x = \begin{vmatrix} 7 & 3 \\ 11 & 9 \end{vmatrix} = 63 - 33 = 30$$ These are the constant terms.
These are the coefficients of y.

$$D_y = \begin{vmatrix} 2 & 7 \\ 5 & 11 \end{vmatrix} = 22 - 35 = -13$$ These are the coefficients of x.
These are the constant terms.

Therefore,

$$x = \frac{D_x}{D} = \frac{30}{3} = 10 \qquad y = \frac{D_y}{D} = \frac{-13}{3} = -\frac{13}{3}$$

CHECK Substituting $x = 10$, $y = -\frac{13}{3}$ in the equations, we obtain

$$(2)(10) + (3)\left(-\frac{13}{3}\right) = 20 - 13 = 7$$

$$(5)(10) + (9)\left(-\frac{13}{3}\right) = 50 - 39 = 11$$

The answers do satisfy the given equations, and the solution is $x = 10$, $y = -\frac{13}{3}$, or $\left(10, -\frac{13}{3}\right)$.

Teaching Tip

It might be helpful to label the columns as follows:

$$D = \begin{vmatrix} x & y \\ & \end{vmatrix}$$

$$D_x = \begin{vmatrix} c & y \\ & \end{vmatrix}$$

$$D_y = \begin{vmatrix} x & c \\ & \end{vmatrix}$$

The x and y columns contain the respective coefficients and the c column contains the constants.

C Using Minors to Evaluate 3 × 3 Determinants

We can also solve a system of three linear equations in three unknowns by using determinants. A 3×3 determinant—a determinant with three rows and three columns—is written as

$$\det A = \begin{vmatrix} a_1 & b_1 & c_1 \\ a_2 & b_2 & c_2 \\ a_3 & b_3 & c_3 \end{vmatrix}$$

The value of a 3×3 determinant can be defined by using the idea of a *minor* of an element in a determinant. Here is the definition of minor.

Minor

In the determinant

$$\begin{vmatrix} a_1 & b_1 & c_1 \\ a_2 & b_2 & c_2 \\ a_3 & b_3 & c_3 \end{vmatrix}$$

the **minor** of an element is the determinant that remains after deleting the row and column in which the element appears. Thus the minor of a_1 is

$$\begin{vmatrix} b_2 & c_2 \\ b_3 & c_3 \end{vmatrix}$$

the minor of b_1 is

$$\begin{vmatrix} a_2 & c_2 \\ a_3 & c_3 \end{vmatrix}$$

and the minor of c_1 is

$$\begin{vmatrix} a_2 & b_2 \\ a_3 & b_3 \end{vmatrix}$$

The value of a 3×3 determinant is defined as follows:

$$\begin{vmatrix} a_1 & b_1 & c_1 \\ a_2 & b_2 & c_2 \\ a_3 & b_3 & c_3 \end{vmatrix} = a_1 \begin{vmatrix} b_2 & c_2 \\ b_3 & c_3 \end{vmatrix} - b_1 \begin{vmatrix} a_2 & c_2 \\ a_3 & c_3 \end{vmatrix} + c_1 \begin{vmatrix} a_2 & b_2 \\ a_3 & b_3 \end{vmatrix}$$

The 2×2 determinants in this equation are clearly minors of the elements in the first row of the 3×3 determinant on the left; the right-hand side is called the **expansion** of the 3×3 determinant by minors along the first row. An expansion with numbers is carried out in Example 3.

EXAMPLE 3   Using minors to evaluate a 3 × 3 determinant

Evaluate the determinant by expanding by minors along the first row:

$$\begin{vmatrix} 1 & 1 & 1 \\ 1 & 2 & 1 \\ 1 & 1 & 2 \end{vmatrix}$$

SOLUTION

$$\begin{vmatrix} 1 & 1 & 1 \\ 1 & 2 & 1 \\ 1 & 1 & 2 \end{vmatrix} = (1)\begin{vmatrix} 2 & 1 \\ 1 & 2 \end{vmatrix} - (1)\begin{vmatrix} 1 & 1 \\ 1 & 2 \end{vmatrix} + (1)\begin{vmatrix} 1 & 2 \\ 1 & 1 \end{vmatrix}$$

$$= (1)(3) - (1)(1) + (1)(-1)$$

$$= 3 - 1 - 1 = 1$$

PROBLEM 3

Evaluate the determinant by expanding by minors along the first row:

$$\begin{vmatrix} 2 & -3 & 4 \\ -1 & 2 & 1 \\ 1 & 1 & -2 \end{vmatrix}$$

Teaching Tip

Here is an alternate method of evaluating only applicable to 3×3 determinants. Recopy the first two columns on the right. Find the products of the diagonals as marked and evaluate the products as shown.

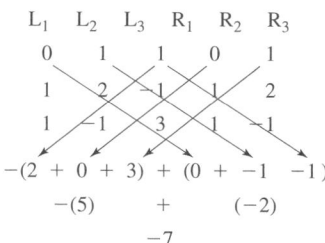

$$-(2 + 0 + 3) + (0 + -1 + -1)$$
$$-(5) \quad + \quad (-2)$$
$$-7$$

It's also possible to expand a determinant by the minors of *any row* or *any column*. To do this, it's necessary to define the *sign array* of a determinant.

> **Sign Array**
>
> For a 3×3 determinant, the **sign array** is the following arrangement of alternating signs:
>
> $$\begin{matrix} + & - & + \\ - & + & - \\ + & - & + \end{matrix}$$

To obtain the expansion of

$$\begin{vmatrix} a_1 & b_1 & c_1 \\ a_2 & b_2 & c_2 \\ a_3 & b_3 & c_3 \end{vmatrix}$$

Teaching Tip

Point out that not only must you remember the sign array but you must also multiply by the numbers in the row.

along a particular row or column, we write the corresponding sign from the array in front of each term in the expansion. For example, to expand

$$\begin{vmatrix} 1 & 1 & 1 \\ 1 & 2 & 1 \\ 1 & 1 & 2 \end{vmatrix}$$

Answer

3. -19

along the *second row*, we write

$$\begin{vmatrix} 1 & 1 & 1 \\ 1 & 2 & 1 \\ 1 & 1 & 2 \end{vmatrix} = -(1)\begin{vmatrix} 1 & 1 \\ 1 & 2 \end{vmatrix} + (2)\begin{vmatrix} 1 & 1 \\ 1 & 2 \end{vmatrix} - (1)\begin{vmatrix} 1 & 1 \\ 1 & 1 \end{vmatrix}$$

$$= -1(1) + 2(1) - 1(0) = 1$$

We used the signs of the *second row* of the sign array, $- + -$, for the first, second, and third terms, respectively.

| **EXAMPLE 4** | **Expanding determinants by minors** |

Expand the given determinant along the third column:

a. $\begin{vmatrix} 0 & 1 & 1 \\ 1 & 2 & -1 \\ 1 & -1 & 3 \end{vmatrix}$ **b.** $\begin{vmatrix} 1 & 1 & 0 \\ 0 & -1 & 1 \\ 2 & -1 & -3 \end{vmatrix}$

SOLUTION

a. $\begin{vmatrix} 0 & 1 & 1 \\ 1 & 2 & -1 \\ 1 & -1 & 3 \end{vmatrix} = +(1)\begin{vmatrix} 1 & 2 \\ 1 & -1 \end{vmatrix} - (-1)\begin{vmatrix} 0 & 1 \\ 1 & -1 \end{vmatrix} + (3)\begin{vmatrix} 0 & 1 \\ 1 & 2 \end{vmatrix}$

$$= (1)(-3) + (1)(-1) + (3)(-1) = -3 - 1 - 3 = -7$$

b. $\begin{vmatrix} 1 & 1 & 0 \\ 0 & -1 & 1 \\ 2 & -1 & -3 \end{vmatrix} = +(0)\begin{vmatrix} 0 & -1 \\ 2 & -1 \end{vmatrix} - (1)\begin{vmatrix} 1 & 1 \\ 2 & -1 \end{vmatrix} + (-3)\begin{vmatrix} 1 & 1 \\ 0 & -1 \end{vmatrix}$

$$= 0 - (1)(-3) + (-3)(-1) = 6$$

| **PROBLEM 4** |

Expand the given determinant along the third column:

a. $\begin{vmatrix} -2 & -1 & 4 \\ 0 & 1 & -1 \\ -1 & 2 & 1 \end{vmatrix}$

b. $\begin{vmatrix} -1 & 3 & -2 \\ -4 & 2 & 0 \\ 0 & -1 & 1 \end{vmatrix}$

NOTE

A last word before you do the problems. When expanding a 3×3 determinant, it's easier to expand along the row or column containing the most zeros, since the coefficient of the resulting minors would then be zero. This will save you time.

| **D** | **Cramer's Rule for Solving a System of Three Equations** |

We can solve the system

$$a_1x + b_1y + c_1z = d_1$$

$$a_2x + b_2y + c_2z = d_2$$

$$a_3x + b_3y + c_3z = d_3$$

in exactly the same manner as we solved the system of equations with two unknowns. The details are similar to those for a system of two equations. We state these results here.

Answers

4. **a.** -3 **b.** 2

Cramer's Rule for Solving a System of Three Equations in Three Unknowns	The system

$$a_1x + b_1y + c_1z = d_1$$
$$a_2x + b_2y + c_2z = d_2$$
$$a_3x + b_3y + c_3z = d_3$$

1. Has the unique solution

$$x = \frac{D_x}{D}, \quad y = \frac{D_y}{D}, \quad z = \frac{D_z}{D}$$

where $D \neq 0$ and

$$D = \begin{vmatrix} a_1 & b_1 & c_1 \\ a_2 & b_2 & c_2 \\ a_3 & b_3 & c_3 \end{vmatrix}, \qquad D_x = \begin{vmatrix} d_1 & b_1 & c_1 \\ d_2 & b_2 & c_2 \\ d_3 & b_3 & c_3 \end{vmatrix},$$

$$D_y = \begin{vmatrix} a_1 & d_1 & c_1 \\ a_2 & d_2 & c_2 \\ a_3 & d_3 & c_3 \end{vmatrix}, \quad \text{and} \quad D_z = \begin{vmatrix} a_1 & b_1 & d_1 \\ a_2 & b_2 & d_2 \\ a_3 & b_3 & d_3 \end{vmatrix}.$$

2. Is *inconsistent* and has no solution if $D = 0$ and any one of D_x, D_y, D_z is different from zero.

3. Has *no unique* solution if $D = 0$ and $D_x = D_y = D_z = 0$. (In this case, the system either has no solution or infinitely many solutions. This situation is studied more fully in advanced algebra.)

D is the determinant of the coefficients; D_x is formed from D by replacing the coefficients of x (the a's) by the corresponding d's; D_y is formed from D by replacing the coefficients of y (the b's) by the corresponding d's; and D_z is formed similarly.

EXAMPLE 5   **Using Cramer's rule to solve a system of three equations in three unknowns**

Use Cramer's rule to solve the system:

$$x + y + 2z = 7$$
$$x - y - 3z = -6$$
$$2x + 3y + z = 4$$

SOLUTION   To use Cramer's rule, we first have to evaluate D, the determinant of the coefficients. If D is not zero, then we calculate the other three determinants, D_x, D_y, and D_z. For the given system,

$$D = \begin{vmatrix} 1 & 1 & 2 \\ 1 & -1 & -3 \\ 2 & 3 & 1 \end{vmatrix} = (1)\begin{vmatrix} -1 & -3 \\ 3 & 1 \end{vmatrix} - (1)\begin{vmatrix} 1 & -3 \\ 2 & 1 \end{vmatrix} + (2)\begin{vmatrix} 1 & -1 \\ 2 & 3 \end{vmatrix}$$

$$= (1)(-1 + 9) - (1)(1 + 6) + (2)(3 + 2)$$

$$= 8 - 7 + 10 = 11$$

PROBLEM 5

Use Cramer's rule to solve the system:

$$x + y + z = 6$$
$$x - y + z = 2$$
$$x + y - 2z = -3$$

Answer

5. $(1, 2, 3)$

Since $D \neq 0$, we calculate the other three determinants.

$$D_x = \begin{vmatrix} 7 & 1 & 2 \\ -6 & -1 & -3 \\ 4 & 3 & 1 \end{vmatrix} = (7)\begin{vmatrix} -1 & -3 \\ 3 & 1 \end{vmatrix} - (1)\begin{vmatrix} -6 & -3 \\ 4 & 1 \end{vmatrix} + (2)\begin{vmatrix} -6 & -1 \\ 4 & 3 \end{vmatrix}$$

$$= (7)(-1 + 9) - (1)(-6 + 12) + (2)(-18 + 4)$$

$$= 56 - 6 - 28 = 22$$

$$D_y = \begin{vmatrix} 1 & 7 & 2 \\ 1 & -6 & -3 \\ 2 & 4 & 1 \end{vmatrix} = (1)\begin{vmatrix} -6 & -3 \\ 4 & 1 \end{vmatrix} - (7)\begin{vmatrix} 1 & -3 \\ 2 & 1 \end{vmatrix} + (2)\begin{vmatrix} 1 & -6 \\ 2 & 4 \end{vmatrix}$$

$$= (1)(-6 + 12) - (7)(1 + 6) + (2)(4 + 12)$$

$$= 6 - 49 + 32 = -11$$

$$D_z = \begin{vmatrix} 1 & 1 & 7 \\ 1 & -1 & -6 \\ 2 & 3 & 4 \end{vmatrix} = (1)\begin{vmatrix} -1 & -6 \\ 3 & 4 \end{vmatrix} - (1)\begin{vmatrix} 1 & -6 \\ 2 & 4 \end{vmatrix} + (7)\begin{vmatrix} 1 & -1 \\ 2 & 3 \end{vmatrix}$$

$$= (1)(-4 + 18) - (1)(4 + 12) + (7)(3 + 2)$$

$$= 14 - 16 + 35 = 33$$

By Cramer's rule,

$$x = \frac{D_x}{D} = \frac{22}{11} = 2, \qquad y = \frac{D_y}{D} = \frac{-11}{11} = -1, \qquad z = \frac{D_z}{D} = \frac{33}{11} = 3$$

You can check the solution $(2, -1, 3)$ by substituting in the given equations.

EXAMPLE 6 **More practice using Cramer's rule to solve a system of three equations in three unknowns**

Use Cramer's rule to solve the system:

$$x + y - z = 2$$
$$2x + y + z = 4$$
$$-x - y + z = 3$$

SOLUTION By Cramer's rule, if there is a unique solution, it is given by

$$x = \frac{D_x}{D}, \qquad y = \frac{D_y}{D}, \qquad \text{and} \qquad z = \frac{D_z}{D}, \text{ where } D \neq 0$$

However,

$$D = \begin{vmatrix} 1 & 1 & -1 \\ 2 & 1 & 1 \\ -1 & -1 & 1 \end{vmatrix} = (1)\begin{vmatrix} 1 & 1 \\ -1 & 1 \end{vmatrix} - (1)\begin{vmatrix} 2 & 1 \\ -1 & 1 \end{vmatrix} + (-1)\begin{vmatrix} 2 & 1 \\ -1 & -1 \end{vmatrix}$$

$$= 2 - 3 + 1 = 0$$

Since $D = 0$, we must try calculating each of the other three determinants to see if we get one of them to be different from zero.

$$D_x = \begin{vmatrix} 2 & 1 & -1 \\ 4 & 1 & 1 \\ 3 & -1 & 1 \end{vmatrix}$$

$$= (2)\begin{vmatrix} 1 & 1 \\ -1 & 1 \end{vmatrix} - (1)\begin{vmatrix} 4 & 1 \\ 3 & 1 \end{vmatrix} + (-1)\begin{vmatrix} 4 & 1 \\ 3 & -1 \end{vmatrix}$$

$$= 4 - 1 + 7 = 10 \neq 0$$

D_x is different from zero. The system is inconsistent and has no solution.

Teaching Tip

Suggest that students label the columns when setting up the determinants:

$$D = \begin{vmatrix} x & y & z \\ & & \\ & & \end{vmatrix}$$

$$D_x = \begin{vmatrix} c & y & z \\ & & \\ & & \end{vmatrix}$$

$$D_y = \begin{vmatrix} x & c & z \\ & & \\ & & \end{vmatrix}$$

$$D_z = \begin{vmatrix} x & y & c \\ & & \\ & & \end{vmatrix}$$

The x, y, and z columns contain the respective coefficients and the c column contains the constants.

PROBLEM 6

Use Cramer's rule to solve the system:

$$x - y - z = 2$$
$$x + 2y + z = 6$$
$$-x + y + z = 4$$

Answer

6. Since $D = 0$ and $D_x \neq 0$, the system is inconsistent. There is no solution.

Calculate It　Solving a System by Matrices

Most calculators have a feature that calculates the determinant of a matrix. Let's start by erasing any matrices you may have in memory by pressing 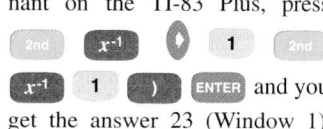 and DEL as many times as there are matrices listed. Let's enter the matrix A from Example 1(a) and find its determinant. As you recall,

$$A = \begin{bmatrix} -3 & 7 \\ -5 & 4 \end{bmatrix}$$

Press 2nd x⁻¹ ◄ ► ENTER to edit matrix A. Now enter **2** ENTER **2** ENTER for the number of rows and columns and then the values of the matrix A. Go to the home screen by pressing 2nd MODE. To find the determinant on the TI-83 Plus, press 2nd x⁻¹ ◄ **1** 2nd x⁻¹ **1** **)** ENTER and you get the answer 23 (Window 1). Now let's do Example 2. Since

the calculator uses the matrices A, B, C, D, and so on, let $A = D$, $B = D_x$, and $C = D_y$. Enter the values for A, B, and C as defined in Example 2. In your home screen, matrices A and B should look like those shown in Window 2. Since $x = \frac{B}{A}$, press 2nd x⁻¹ ► **1** 2nd x⁻¹ **2** **)** (this will enter det [B] on the home screen). Then press ÷ 2nd x⁻¹ ► **1** 2nd x⁻¹ **1** **)** ENTER (this will divide by det [A]). The result is 10, as shown in Window 3. You can use the same procedure to find y. Now that you know how to find determinants, you can do the rest of the examples.

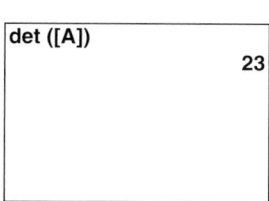
Window 1
Det [A] from Example 1

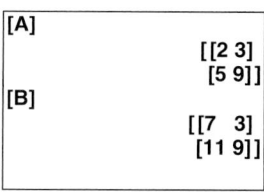
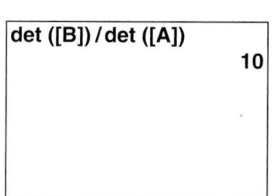
Window 2
Matrices A and B from Example 2

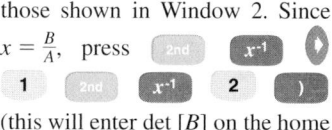
Window 3
Finding the solution for x in Example 2

Exercises 4.5

A In Problems 1–10, evaluate the determinant.

1. $\begin{vmatrix} 1 & 1 \\ 0 & 2 \end{vmatrix}$ 2

2. $\begin{vmatrix} 2 & -1 \\ 4 & 3 \end{vmatrix}$ 10

3. $\begin{vmatrix} -3 & -2 \\ 5 & 1 \end{vmatrix}$ 7

4. $\begin{vmatrix} 2 & -1 \\ -3 & 1 \end{vmatrix}$ −1

5. $\begin{vmatrix} -2 & 0 \\ 5 & -3 \end{vmatrix}$ 6

6. $\begin{vmatrix} 5 & 2 \\ -10 & -4 \end{vmatrix}$ 0

7. $\begin{vmatrix} \frac{1}{2} & \frac{-1}{4} \\ \frac{1}{2} & \frac{3}{4} \end{vmatrix}$ $\frac{1}{2}$

8. $\begin{vmatrix} \frac{1}{5} & \frac{1}{10} \\ \frac{1}{2} & \frac{1}{4} \end{vmatrix}$ 0

9. $\begin{vmatrix} \frac{3}{5} & \frac{1}{2} \\ -1 & -1 \\ \frac{}{4} & \frac{}{2} \end{vmatrix}$ $\frac{-7}{40}$

10. $\begin{vmatrix} \frac{4}{5} & \frac{-1}{3} \\ \frac{-1}{2} & \frac{1}{2} \end{vmatrix}$ $\frac{7}{30}$

B In Problems 11–30, solve the system using Cramer's rule. If the system is dependent or inconsistent, state that fact.

11. $x + y = 5$
$3x - y = 3$ (2, 3)

12. $x + y = 9$
$x - y = 3$ (6, 3)

13. $x + y = 9$
$x - y = -1$ (4, 5)

14. $2x + y = -1$
$x - 2y = -13$ (−3, 5)

15. $4x + 9y = 3$
$3x + 7y = 2$ (3, −1)

16. $5x + 2y = 32$
$3x + y = 18$ (4, 6)

17. $x - y = -1$
$x - 2y = -6$ (4, 5)

18. $x - 2y = -13$
$3x - 2y = -19$ (−3, 5)

19. $2x + 3y = -13$
$6x + 9y = -39$
$\left(x, \frac{-2x - 13}{3}\right)$; dependent

20. $4x + 5y = -2$
$12x + 15y = -6$
$\left(x, \frac{-4x - 2}{5}\right)$; dependent

21. $x - y = 1$
$x - 2y = 4$ (−2, −3)

22. $x - 2y = 4$
$4x - 5y = 7$ (−2, −3)

23. $x + 3y = 6$
$2x + 6y = 5$
No solution; inconsistent

24. $x - 2y = 3$
$-x + 2y = 6$
No solution; inconsistent

25. $x = 7y + 3$
$2x + 3y = 23$ (10, 1)

26. $x = 3y + 1$
$2x + 3y = 20$ (7, 2)

27. $y = -3x + 17$
$2x - y = 8$ (5, 2)

28. $y = -2x + 14$
$3x - y = 11$ (5, 4)

29. $\dfrac{x}{2} - \dfrac{y}{3} = \dfrac{-1}{6}$
$\dfrac{x}{3} + \dfrac{y}{4} = \dfrac{-7}{12}$ $(-1, -1)$

30. $\dfrac{x}{3} - \dfrac{y}{5} = \dfrac{4}{3}$
$\dfrac{x}{4} - \dfrac{y}{3} = \dfrac{1}{12}$ (7, 5)

C In Problems 31–40, evaluate the determinant.

31. $\begin{vmatrix} 1 & 3 & 2 \\ 2 & 4 & 1 \\ 3 & 6 & 5 \end{vmatrix}$ -7

32. $\begin{vmatrix} 1 & 3 & 5 \\ 2 & 0 & 10 \\ -3 & 1 & -15 \end{vmatrix}$ 0

33. $\begin{vmatrix} 1 & 2 & 3 \\ 4 & 5 & 6 \\ 7 & 8 & 9 \end{vmatrix}$ 0

34. $\begin{vmatrix} 1 & 1 & 1 \\ 2 & 3 & 1 \\ 2 & 4 & 1 \end{vmatrix}$ 1

35. $\begin{vmatrix} 2 & 1 & 3 \\ 1 & 2 & -1 \\ 3 & 1 & 5 \end{vmatrix}$ -1

36. $\begin{vmatrix} -1 & 1 & -1 \\ -2 & 2 & -6 \\ 3 & -3 & 4 \end{vmatrix}$ 0

37. $\begin{vmatrix} 1 & 1 & 6 \\ 1 & 1 & 4 \\ 1 & -1 & 2 \end{vmatrix}$ -4

38. $\begin{vmatrix} 1 & 4 & 0 \\ 1 & -3 & 1 \\ 0 & 8 & -1 \end{vmatrix}$ -1

39. $\begin{vmatrix} 0 & -1 & 2 \\ 2 & 1 & -3 \\ 1 & -3 & 1 \end{vmatrix}$ -9

40. $\begin{vmatrix} -3 & 2 & -4 \\ 1 & -1 & 3 \\ 1 & 2 & 10 \end{vmatrix}$ 22

D In Problems 41–50, solve the system using Cramer's rule. You can use minors to expand the resulting determinants.

41. $x + y + z = 6$
$2x - 3y + 3z = 5$
$3x - 2y - z = -4$
(1, 2, 3)

42. $x + y + z = 13$
$3x + y - 3z = 5$
$x - 2y + 4z = 10$
(4, 5, 4)

43. $6x + 5y + 4z = 5$
$5x + 4y + 3z = 5$
$4x + 3y + z = 7$
$(3, -1, -2)$

44. $3x + 2y + z = 4$
$4x + 3y + z = 5$
$5x + y + z = 9$
$(2, -1, 0)$

45. $x - 2y + 3z = 15$
$5x + 7y - 11z = -29$
$-13x + 17y + 19z = 37$
(3, 0, 4)

46. $2x - y + z = 3$
$x + 2y + z = 12$
$4x - 3y + z = 1$
(3, 4, 1)

47. $5x + 3y + 5z = 3$
$3x + 5y + z = -5$
$2x + 2y + 3z = 7$
$(-5, 1, 5)$

48. $x + y = 5$
$y + z = 3$
$x + z = 7$
$\left(\dfrac{9}{2}, \dfrac{1}{2}, \dfrac{5}{2}\right)$

49. $2y + z = 9$
$-2y + z = 1$
$x + y + z = 1$
$(-6, 2, 5)$

50. $x - y = 3$
$y - z = 3$
$x + z = 9$
$\left(\dfrac{15}{2}, \dfrac{9}{2}, \dfrac{3}{2}\right)$

In Problems 51–60, show that the statement is true.

51. $\begin{vmatrix} a & b & 0 \\ c & d & 0 \\ e & f & 0 \end{vmatrix} = 0$ $\begin{vmatrix} a & b & 0 \\ c & d & 0 \\ e & f & 0 \end{vmatrix} = a\begin{vmatrix} d & 0 \\ f & 0 \end{vmatrix} - b\begin{vmatrix} c & 0 \\ e & 0 \end{vmatrix} + 0\begin{vmatrix} c & d \\ e & f \end{vmatrix} = a(0) - b(0) + 0 = 0$

52. $\begin{vmatrix} a & b & c \\ d & e & f \\ 0 & 0 & 0 \end{vmatrix} = 0$ $\begin{vmatrix} a & b & c \\ d & e & f \\ 0 & 0 & 0 \end{vmatrix} = 0\begin{vmatrix} b & c \\ e & f \end{vmatrix} - 0\begin{vmatrix} a & c \\ d & f \end{vmatrix} + 0\begin{vmatrix} a & b \\ d & e \end{vmatrix} = 0 - 0 + 0 = 0$

53. $\begin{vmatrix} a & b & c \\ 1 & 2 & 3 \\ a & b & c \end{vmatrix} = 0 \quad \begin{vmatrix} a & b & c \\ 1 & 2 & 3 \\ a & b & c \end{vmatrix} = -1\begin{vmatrix} b & c \\ b & c \end{vmatrix} + 2\begin{vmatrix} a & c \\ a & c \end{vmatrix} - 3\begin{vmatrix} a & b \\ a & b \end{vmatrix} = -1(bc - bc) + 2(ac - ac) - 3(ab - ab)$

$$= -1(0) + 2(0) - 3(0)$$
$$= -0 + 0 - 0 = 0$$

54. $\begin{vmatrix} 1 & a & a \\ 2 & b & b \\ 3 & c & c \end{vmatrix} = 0 \quad \begin{vmatrix} 1 & a & a \\ 2 & b & b \\ 3 & c & c \end{vmatrix} = 1\begin{vmatrix} b & b \\ c & c \end{vmatrix} - 2\begin{vmatrix} a & a \\ c & c \end{vmatrix} + 3\begin{vmatrix} a & a \\ b & b \end{vmatrix} = 1(bc - bc) - 2(ac - ac) + 3(ab - ab)$

$$= 1(0) - 2(0) + 3(0) = 0$$

55. $\begin{vmatrix} 1 & 2 & 3 \\ 3 & 1 & 2 \\ 3k & 2k & k \end{vmatrix} = k\begin{vmatrix} 1 & 2 & 3 \\ 3 & 1 & 2 \\ 3 & 2 & 1 \end{vmatrix} \quad \begin{vmatrix} 1 & 2 & 3 \\ 3 & 1 & 2 \\ 3k & 2k & k \end{vmatrix} = 3k\begin{vmatrix} 2 & 3 \\ 1 & 2 \end{vmatrix} - 2k\begin{vmatrix} 1 & 3 \\ 3 & 2 \end{vmatrix} + k\begin{vmatrix} 1 & 2 \\ 3 & 1 \end{vmatrix} = 3k(1) - 2k(-7) + k(-5)$

$$= 3k + 14k - 5k$$

$$k\begin{vmatrix} 1 & 2 & 3 \\ 3 & 1 & 2 \\ 3 & 2 & 1 \end{vmatrix} = k\left[3\begin{vmatrix} 2 & 3 \\ 1 & 2 \end{vmatrix} - 2\begin{vmatrix} 1 & 3 \\ 3 & 2 \end{vmatrix} + 1\begin{vmatrix} 1 & 2 \\ 3 & 1 \end{vmatrix} \right] = k[3(1) - 2(-7) + 1(-5)]$$

$$= k(3 + 14 - 5)$$
$$= 3k + 14k - 5k$$

$$\therefore \begin{vmatrix} 1 & 2 & 3 \\ 3 & 1 & 2 \\ 3k & 2k & k \end{vmatrix} = k\begin{vmatrix} 1 & 2 & 3 \\ 3 & 1 & 2 \\ 3 & 2 & 1 \end{vmatrix}$$

56. $\begin{vmatrix} 1 & 2 & 3k \\ 3 & 2 & k \\ 3 & 1 & 2k \end{vmatrix} = k\begin{vmatrix} 1 & 2 & 3 \\ 3 & 2 & 1 \\ 3 & 1 & 2 \end{vmatrix} \quad \begin{vmatrix} 1 & 2 & 3k \\ 3 & 2 & k \\ 3 & 1 & 2k \end{vmatrix} = 3k\begin{vmatrix} 3 & 2 \\ 3 & 1 \end{vmatrix} - k\begin{vmatrix} 1 & 2 \\ 3 & 1 \end{vmatrix} + 2k\begin{vmatrix} 1 & 2 \\ 3 & 2 \end{vmatrix} = 3k(-3) - k(-5) + 2k(-4)$

$$= -9k + 5k - 8k$$

$$k\begin{vmatrix} 1 & 2 & 3 \\ 3 & 2 & 1 \\ 3 & 1 & 2 \end{vmatrix} = k\left[3\begin{vmatrix} 3 & 2 \\ 3 & 1 \end{vmatrix} - 1\begin{vmatrix} 1 & 2 \\ 3 & 1 \end{vmatrix} + 2\begin{vmatrix} 1 & 2 \\ 3 & 2 \end{vmatrix} \right] = k[3(-3) - 1(-5) + 2(-4)]$$

$$= k(-9 + 5 - 8)$$
$$= -9k + 5k - 8k$$

$$\therefore \begin{vmatrix} 1 & 2 & 3k \\ 3 & 2 & k \\ 3 & 1 & 2k \end{vmatrix} = k\begin{vmatrix} 1 & 2 & 3 \\ 3 & 2 & 1 \\ 3 & 1 & 2 \end{vmatrix}$$

57. $\begin{vmatrix} kb_1 & b_1 & 1 \\ kb_2 & b_2 & 2 \\ kb_3 & b_3 & 3 \end{vmatrix} = 0 \quad \begin{vmatrix} kb_1 & b_1 & 1 \\ kb_2 & b_2 & 2 \\ kb_3 & b_3 & 3 \end{vmatrix} = kb_1\begin{vmatrix} b_2 & 2 \\ b_3 & 3 \end{vmatrix} - kb_2\begin{vmatrix} b_1 & 1 \\ b_3 & 3 \end{vmatrix} + kb_3\begin{vmatrix} b_1 & 1 \\ b_2 & 2 \end{vmatrix}$

$$= kb_1(3b_2 - 2b_3) - kb_2(3b_1 - b_3) + kb_3(2b_1 - b_2)$$
$$= 3kb_1b_2 - 2kb_1b_3 - 3kb_1b_2 + kb_2b_3 + 2kb_1b_3 - kb_2b_3$$
$$= 3kb_1b_2 - 3kb_1b_2 - 2kb_1b_3 + 2kb_1b_3 + kb_2b_3 - kb_2b_3$$
$$= 0$$

58. $\begin{vmatrix} b_1 & b_2 & b_3 \\ kb_1 & kb_2 & kb_3 \\ 1 & 2 & 3 \end{vmatrix} = 0 \quad \begin{vmatrix} b_1 & b_2 & b_3 \\ kb_1 & kb_2 & kb_3 \\ 1 & 2 & 3 \end{vmatrix} = -kb_1\begin{vmatrix} b_2 & b_3 \\ 2 & 3 \end{vmatrix} + kb_2\begin{vmatrix} b_1 & b_3 \\ 1 & 3 \end{vmatrix} - kb_3\begin{vmatrix} b_1 & b_2 \\ 1 & 2 \end{vmatrix}$

$$= -kb_1(3b_2 - 2b_3) + kb_2(3b_1 - b_3) - kb_3(2b_1 - b_2)$$
$$= -3kb_1b_2 + 2kb_1b_3 + 3kb_1b_2 - kb_2b_3 - 2kb_1b_3 + kb_2b_3$$
$$= -3kb_1b_2 + 3kb_1b_2 + 2kb_1b_3 - 2kb_1b_3 - kb_2b_3 + kb_2b_3$$
$$= 0$$

59. $\begin{vmatrix} 1 & 1 & 1 \\ 2 & a & a \\ 3 & b & b \end{vmatrix} = 0$ $\begin{vmatrix} 1 & 1 & 1 \\ 2 & a & a \\ 3 & b & b \end{vmatrix} = 1\begin{vmatrix} a & a \\ b & b \end{vmatrix} - 2\begin{vmatrix} 1 & 1 \\ b & b \end{vmatrix} + 3\begin{vmatrix} 1 & 1 \\ a & a \end{vmatrix}$

$$= 1(ab - ab) - 2(b - b) + 3(a - a)$$
$$= 1(0) - 2(0) + 3(0) = 0$$

60. $\begin{vmatrix} 0 & 0 & 0 \\ a & b & c \\ d & e & f \end{vmatrix} = 0$ $\begin{vmatrix} 0 & 0 & 0 \\ a & b & c \\ d & e & f \end{vmatrix} = 0\begin{vmatrix} b & c \\ e & f \end{vmatrix} - 0\begin{vmatrix} a & c \\ d & f \end{vmatrix} + 0\begin{vmatrix} a & b \\ d & e \end{vmatrix}$

$$= 0 - 0 + 0 = 0$$

SKILL CHECKER

Try these "Skill Checker" exercises so you'll be ready for the next section.

Determine whether the point $(-3, 1)$ satisfies the inequality.

61. $3x + 2y > 6$ No

62. $2x + y \geq -2$ No

63. $-2y < 6$ Yes

64. $x + 3 \leq 5$ Yes

65. $y < x$ No

66. $y \geq -x$ No

USING YOUR KNOWLEDGE

Determining the Equations of Lines

Determinants provide a convenient and neat way of writing the equation of a line through two points. This is one of the problems that we studied in Section 3.4. Suppose the line is to pass through the points (x_1, y_1) and (x_2, y_2); then an equation of the line can be written in the form

$$\begin{vmatrix} x & y & 1 \\ x_1 & y_1 & 1 \\ x_2 & y_2 & 1 \end{vmatrix} = 0$$

It's easy to verify. First, think of expanding the determinant by minors along the first row. The coefficients of x and y will be constants, so the equation is linear. Next, you can see that (x_1, y_1) and (x_2, y_2) both satisfy the equation, because if you substitute either of these pairs for (x, y) in the first row of the determinant, the result is a determinant with two identical rows and this you know is zero. (See Problem 53.) The equation is that of the desired line.

As an illustration, let's find an equation of the line through $(1, 3)$ and $(-5, -2)$. The determinant form of this equation is

$$\begin{vmatrix} x & y & 1 \\ 1 & 3 & 1 \\ -5 & -2 & 1 \end{vmatrix} = 0$$

or

$$[3 - (-2)]x - [1 - (-5)]y + [-2 - (-15)](1) = 0$$

where the quantities in brackets are the expanded minors of the first-row elements. The final equation is

$$5x - 6y + 13 = 0$$

In Problems 67–72, use the method described above to find the equation of the line through the two given points.

67. $(2, 7)$ and $(0, 3)$ $2x - y + 3 = 0$

68. $(10, 12)$ and $(-7, 1)$ $11x - 17y + 94 = 0$

69. $(-1, 4)$ and $(8, 2)$ $2x + 9y - 34 = 0$

70. $(5, 0)$ and $(0, -3)$ $3x - 5y - 15 = 0$

71. $(a, 0)$ and $(0, b)$, $ab \neq 0$ $bx + ay - ab = 0$

72. (a, b) and $(0, 0)$, $(a, b) \neq (0, 0)$ $bx - ay = 0$

WRITE ON

How can you tell:

73. If a system of equations is consistent when using determinants to solve it? Answers may vary.

74. If a system of equations is inconsistent when using determinants to solve it? Answers may vary.

75. If a system of equations is dependent when using determinants to solve it? Answers may vary.

76. If a system of linear equations has a determinant $D = 0$, what do you know about the system? Answers may vary.

77. Explain why a determinant that has a column or row of zeros has a value of zero. (See Problems 51 and 52.) Answers may vary.

MASTERY TEST

If you know how to do these problems, you have learned your lesson!

Expand the determinant by minors:

78. Along the third row

$$\begin{vmatrix} 0 & 1 & 1 \\ 1 & 2 & -1 \\ 1 & -1 & 3 \end{vmatrix} \quad 1\begin{vmatrix} 1 & 1 \\ 2 & -1 \end{vmatrix} - (-1)\begin{vmatrix} 0 & 1 \\ 1 & -1 \end{vmatrix} + 3\begin{vmatrix} 0 & 1 \\ 1 & 2 \end{vmatrix} = 1(-3) + 1(-1) + 3(-1)$$
$$= -3 - 1 - 3$$
$$= -7$$

79. Along the third row

$$\begin{vmatrix} 1 & 1 & 0 \\ 0 & -1 & 1 \\ 2 & -1 & -3 \end{vmatrix} \quad 2\begin{vmatrix} 1 & 0 \\ -1 & 1 \end{vmatrix} - (-1)\begin{vmatrix} 1 & 0 \\ 0 & 1 \end{vmatrix} + (-3)\begin{vmatrix} 1 & 1 \\ 0 & -1 \end{vmatrix} = 2(1) + 1(1) - 3(-1)$$
$$= 2 + 1 + 3$$
$$= 6$$

80. Along the first column

$$\begin{vmatrix} 1 & 1 & 1 \\ 1 & 2 & 1 \\ 1 & 1 & 2 \end{vmatrix} \quad 1\begin{vmatrix} 2 & 1 \\ 1 & 2 \end{vmatrix} - 1\begin{vmatrix} 1 & 1 \\ 1 & 2 \end{vmatrix} + 1\begin{vmatrix} 1 & 1 \\ 2 & 1 \end{vmatrix} = 1(3) - 1(1) + 1(-1)$$
$$= 3 - 1 - 1$$
$$= 1$$

Use Cramer's rule to solve the system:

81.
$$\begin{aligned} x - y - z &= 2 \\ x + 2y + z &= 6 \\ -x + y + z &= 4 \end{aligned}$$

No solution

82.
$$\begin{aligned} x + y + z &= 6 \\ x - y + z &= 2 \\ x + y - 2z &= 4 \end{aligned}$$

$(\frac{10}{3}, 2, \frac{2}{3})$

83. Evaluate the determinant

$$\begin{vmatrix} 2 & -3 & 4 \\ -1 & 2 & 1 \\ 1 & 1 & -2 \end{vmatrix} \quad -19$$

84. Use Cramer's rule to solve the system:

$$\begin{aligned} 2x + 3y &= 13 \\ 5x - 4y &= 21 \end{aligned} \quad (5, 1)$$

Evaluate:

85. $\begin{vmatrix} -2 & 6 \\ -3 & 5 \end{vmatrix} \quad 8$

86. $\begin{vmatrix} -2 & 8 \\ -3 & 12 \end{vmatrix} \quad 0$

4.6 SYSTEMS OF LINEAR INEQUALITIES

To Succeed, Review How To ...

1. Evaluate functions (pp. 181–182).

2. Solve systems of equations (pp. 256–263, 274–277).

Objectives

A Graphing systems of two linear inequalities.

B Graphing systems of inequalities.

GETTING STARTED

Nutrition: You Want a Coke with That?

Are burgers and fries your menu of choice but you've got a budget problem and you're concerned about your fat and protein intake? A regular hamburger contains about 11 g of fat and 12 g of protein while French fries have 12 g of fat and 3 g of protein. Your daily consumption of fat and protein should be about 56 and 51 g, respectively. (You can check this out in the *Fast Food Guide* by Jacobson and Fritschner.) Suppose that a regular hamburger costs 39¢ and regular fries are 79¢. How many of each can you eat so that you meet the recommended fat and protein intake and, at the same time, minimize your cost? Let's approach this problem by constructing a table where h and f represent the number of hamburgers and fries, respectively.

	Hamburgers	French Fries	Total
Fat (g)	11	12	$11h + 12f$
Protein (g)	12	3	$12h + 3f$
Cost (¢)	39	79	$39h + 79f$

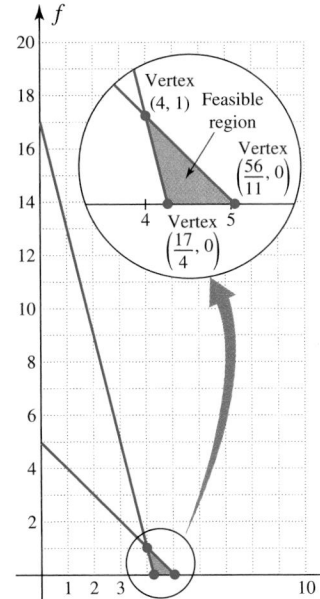

The recommended amounts of fat and protein are 56 and 51 g, respectively, so we have the following system of equations.

$$\text{Fat:} \quad 11h + 12f = 56$$
$$\text{Protein:} \quad 12h + 3f = 51$$

When $h = 0$ in the first equation, $f = \frac{56}{12} = \frac{14}{3}$, so we graph $(0, \frac{14}{3})$.
When $f = 0$, $h = \frac{56}{11}$, and we graph $(\frac{56}{11}, 0)$ and join the two points with a line. We graph the second equation accordingly.

Now suppose you want to consume no more than 56 g of fat and at least 51 g of protein *and* you want to do this as cheaply as possible. The total cost function C is given by $C = 39h + 79f$, where h is the number of hamburgers and f the number of fries you buy. The shaded triangle in the graph is called the **feasible region,** and the minimum or maximum for a linear function such as C occurs at the corner points (**vertices**) or $(\frac{17}{4}, 0)$, $(\frac{56}{11}, 0)$, or $(4, 1)$. To find the minimum or maximum of C, evaluate $C = 39h + 79f$ at each of the vertices.

For $(\frac{17}{4}, 0)$, $\quad C = 39 \cdot \frac{17}{4} + 79 \cdot 0 = \$1.66$

For $(\frac{56}{11}, 0)$ $\quad C = 39 \cdot \frac{56}{11} + 79 \cdot 0 = \$1.99$

For $(4, 1)$ $\quad C = 39 \cdot 4 + 79 \cdot 1 \ = \$2.35$

The ordered pair $(\frac{17}{4}, 0)$ means that you eat $4\frac{1}{4}$ hamburgers with no French fries. This produced the smallest cost, \$1.66. Thus, the most economical way to meet your daily fat and protein requirements is to eat approximately 4 hamburgers and no fries. We leave the decision about the Coke up to you!

This problem is an example of an optimization problem, and linear programming, which we shall discuss in *Using Your Knowledge.* First we learn how to graph solutions to systems with two linear inequalities.

A   Graphing Systems of Two Linear Inequalities

If we graph two or more linear inequalities **on the same coordinate system,** we have a *system of linear inequalities.* The solution set of the system is the set of points that satisfies *all* the inequalities in the system—the region that is *common* to every graph in the system. We illustrate how to find this solution set in Examples 1 and 2 for systems of two linear inequalities.

| **EXAMPLE 1**   **Graphing systems of two linear inequalities** | **PROBLEM 1** |

Graph the solution set of the system of inequalities:

$$x \leq 0 \quad \text{and} \quad y \geq 2$$

SOLUTION $x = 0$ is a vertical line corresponding to the y-axis, so $x \leq 0$ consists of the points to the *left* of or *on* the line $x = 0$, as shown in Figure 14. The condition $y \geq 2$ defines all points *above* or *on* the line $y = 2$, as shown in Figure 15. The solution set is the set satisfying both conditions $x \leq 0$ and $y \geq 2$; this is the **intersection** of the two solution sets and is the region common to both graphs, as shown by the darker region in Figure 16.

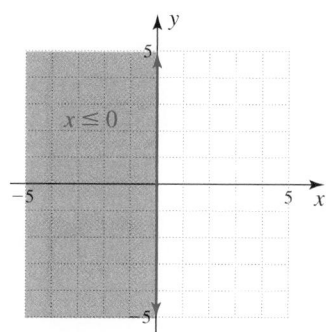

Figure 14

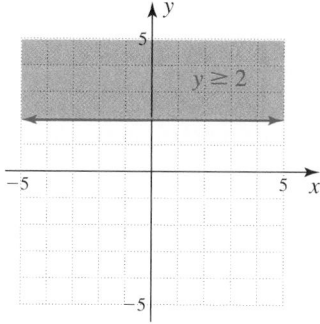

Figure 15

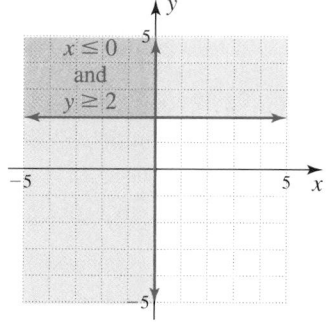

Figure 16

PROBLEM 1

Graph:

$$y \geq 0$$
$$x < 3$$

Answers

1.

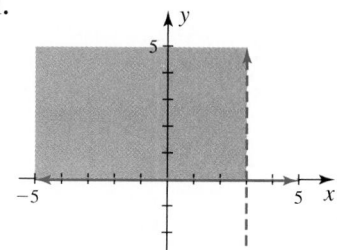

2.

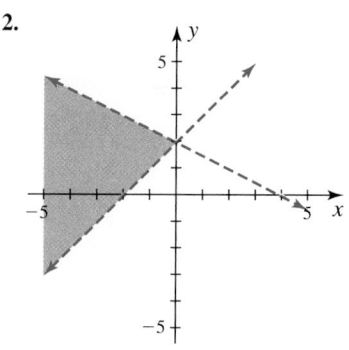

| **EXAMPLE 2**   **More practice graphing systems of two linear inequalities** | **PROBLEM 2** |

Graph the solution set of the system of inequalities:

$$y + x \geq 2 \quad \text{and} \quad y - x \leq 2$$

SOLUTION   First, we graph the line $y + x = 2$. When $x = 0$, $y = 2$, so we graph $(0, 2)$. When $y = 0$, $x = 2$, and we graph $(2, 0)$. Joining the points $(0, 2)$ and $(2, 0)$ with a line, we have the graph of $y + x = 2$. Use $(0, 0)$ as a test point. Does $(0, 0)$ satisfy $y + x \geq 2$? Letting $x = 0$ and $y = 0$ we obtain $0 + 0 \geq 2$, which is not true. The solution set consists of all points *above* or *on* the line $y + x = 2$, as shown in Figure 17.

PROBLEM 2

Graph:

$$y > x + 2$$
$$x + 2y < 4$$

Now graph the line $y - x = 2$ (Figure 18). When $x = 0$, $y = 2$, so we graph $(0, 2)$. When $y = 0$, $x = -2$, so we graph $(-2, 0)$ and join it to $(0, 2)$ with a line, the graph of $y - x = 2$. Using $(0, 0)$ as a test point for $y - x \le 2$, we let $x = 0$ and $y = 0$ to obtain $0 - 0 \le 2$, which is true, so we shade all the points *below* or *on* the line $y - x = 2$.

The solution set of the system is the **intersection** of the solution sets of $y + x \ge 2$ and $y - x \le 2$, as shown in Figure 19. We have shown the solution sets of $y + x \ge 2$ and $y - x \le 2$ separately to illustrate the procedure step by step. When you do the exercises, try to graph the solution set of *all* inequalities involved on the same coordinate axes and use different colors to distinguish between the different regions.

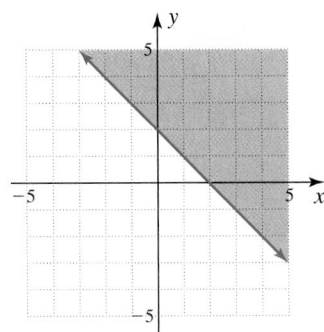

Figure 17

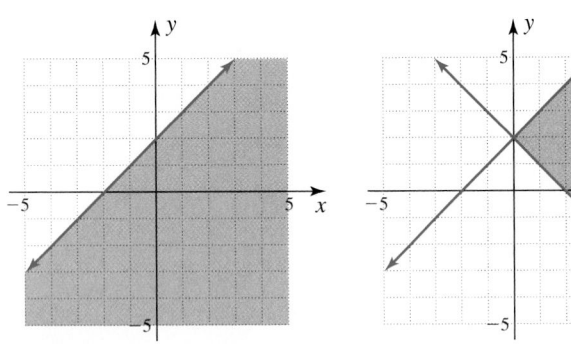

Figure 18   **Figure 19**

B Graphing Systems of Inequalities

Many problems in business, economics, and the social sciences involve solving systems of linear inequalities that have certain restrictions or conditions (*constraints*) such as "less than," "more than," "at least," "no more than," "a minimum of," and "a maximum of." As you recall, a **solution** of a system of linear inequalities is a point (x, y) that satisfies each inequality in the system. For example, in our fat/protein intake problem in *Getting Started*, the point $(4, 1)$ satisfies the system of inequalities:

$$11h + 12f \le 56 \qquad \text{No more than 56 g of fat.} \qquad \textbf{(1)}$$
$$12h + 3f \ge 51 \qquad \text{At least 51 g of protein.}$$
$$f \ge 0 \qquad \text{Some fries.}$$
$$h \ge 0 \qquad \text{Some hamburgers.}$$

We then use this information and the following procedure.

PROCEDURE

Graphing a System of Inequalities

1. Sketch the line corresponding to each inequality using dashed lines for inequalities involving $<$ or $>$ and solid lines for inequalities with $\le$ or $\ge$.

2. Use a test point to shade the half-plane that is the graph of each linear inequality. (If the test point satisfies the inequality, shade all points on the same side of the line as the test point.)

3. The graph of the system is the intersection of the half-planes, that is, the region consisting of the points satisfying *all* the inequalities.

EXAMPLE 3   **Graphing a system of inequalities**

Graph the system of inequalities and label the vertices:

$$x + y \leq 5$$

$$2x + 3y < 12$$

$$x \geq 0$$

$$y \geq 0$$

SOLUTION   Sketch the line $x + y = 5$, as shown in Figure 20. Select the test point $(0, 0)$. Since $0 + 0 \leq 5$ is true, select all the points *below* or *on* the line $x + y = 5$ as the solution set. Now sketch the line $2x + 3y = 12$, as shown in Figure 20. Again, select the test point $(0, 0)$. $2 \cdot 0 + 3 \cdot 0 < 12$, so the solution set consists of the points *below* the line $2x + 3y = 12$. (This line is shown dashed in Figure 20.) The graph of $x \geq 0$ consists of the points *on* and to the *right* of the

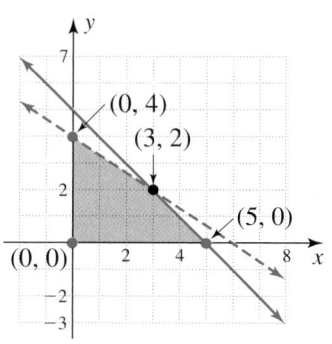

Figure 20

y-axis, and the graph of $y \geq 0$ consists of all points *on* or *above* the x-axis. The region common to all these graphs (shown shaded) is the solution set for the system of inequalities. The vertices are $(0, 0)$, $(0, 4)$, $(5, 0)$, and $(3, 2)$. (This last vertex is obtained by solving the system of equations $x + y = 5$ and $2x + 3y = 12$.)

PROBLEM 3

Graph:

$$x \leq 0$$

$$y \leq 0$$

$$y > -x - 2$$

Answer

3.

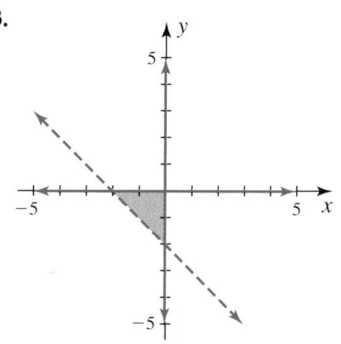

Exercises 4.6

A In Problems 1–10, graph the solution set of the given system of inequalities.

1. $x - y \geq 2$ and $x + y \leq 6$

2. $x + 2y \leq 3$ and $x \leq y$

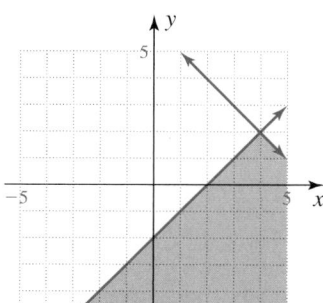

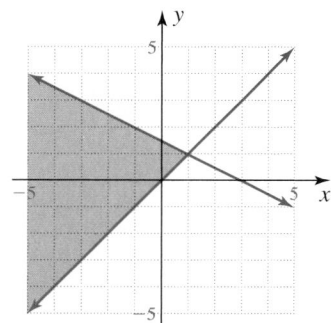

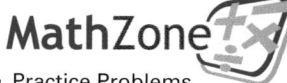

3. $2x - 3y \leq 6$ and $4x - 3y \geq 12$

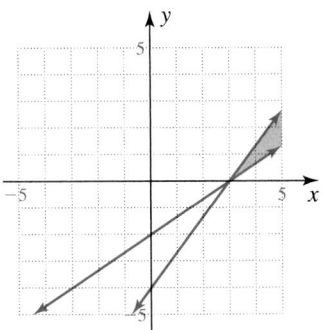

4. $2x - 5y \leq 10$ and $3x + 2y \leq 6$

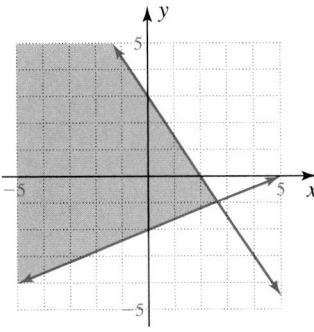

5. $2x - 3y \leq 5$ and $x \geq y$

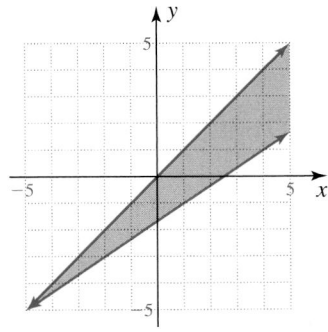

6. $x \leq 2y$ and $x + y < 4$

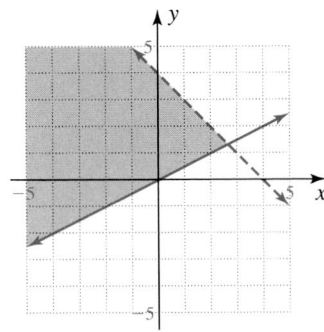

7. $x + 3y \leq 6$ and $x \geq y$

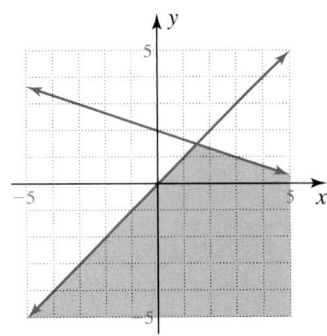

8. $2x - y \leq 2$ and $x \leq y$

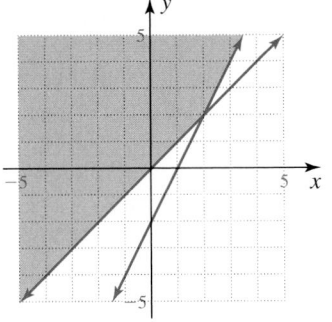

9. $x - y \leq 1$ and $3x - y < 3$

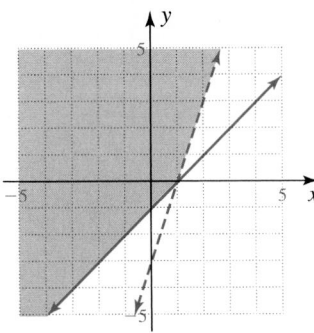

10. $x - y \geq -2$ and $x + y \leq 6$

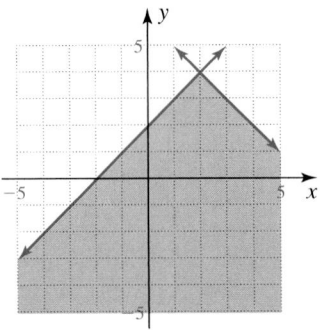

B In Problems 11–16, graph the solution set of the systems of inequalities.

11.
$$x \geq 1$$
$$x \leq 4$$
$$y \leq 4$$
$$x - 3y \leq -2$$

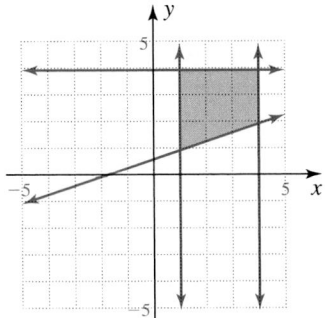

12.
$$y - x \leq 0$$
$$x \leq 4$$
$$y \geq 0$$
$$x + 2y \leq 6$$

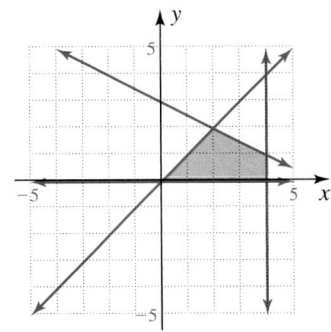

13.
$$x + y \geq 1$$
$$2y - x \leq 1$$
$$x \leq 1$$

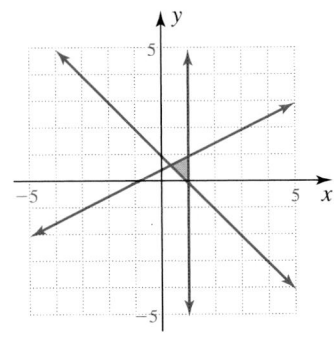

14. $2x + y \geq 18$
$\quad\ x + y \geq 12$
$\quad\quad\ \ 2y \leq 30$

15. $\quad\ x \geq 1$
$\quad\quad\ y \geq 2$
$\quad\ 4 \leq 2x + y$
$\quad\ 2x + y \leq 6$

16. $2x + y \geq 6$
$\quad\ 0 \leq y \leq 4$
$\quad\ 0 \leq x \leq 3$

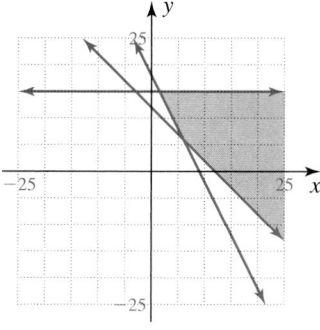
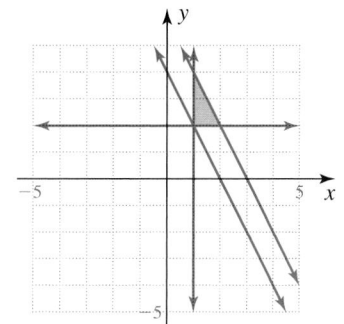
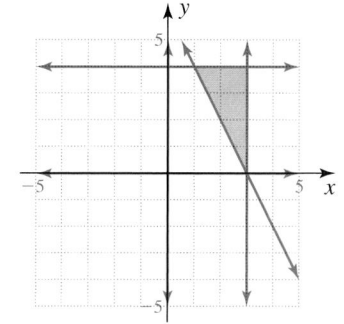

SKILL CHECKER

Try the "Skill Checker" exercises so you'll be ready for the next section.

For Problems 17–20, write the equations in slope-intercept form.

17. $2x - 4y = 12$ $y = \frac{1}{2}x - 3$

18. $-3x + 6y = 18$ $y = \frac{1}{2}x + 3$

19. $x - y = 5$ $y = x - 5$

20. $-4y = x - 8$ $y = -\frac{1}{4}x + 2$

USING YOUR KNOWLEDGE

Linear Programming

As we demonstrated in the *Getting Started,* linear inequalities can be used to solve **optimization problems,** problems in which we find the greatest or the least value of a function. The technique used to solve such problems is called **linear programming.** A two-variable linear programming problem consists of two parts:

1. An **objective function** giving the quantity we wish to maximize or minimize. (In the *Getting Started,* $C = 39h + 79f$ is the objective function.)

2. A system of **constraints** (linear inequalities) whose solution set is called the set of **feasible solutions.** (In the *Getting Started,* the system of inequalities (1) is the set of constraints, and the shaded triangle in the graph is the set of feasible solutions.) Sometimes there is a set of **implied constraints** that state that the variables *cannot* be negative.

As mentioned, if there is an optimal solution, it must occur at one of the vertices of the set of feasible solutions. This solution can be found by testing the function at each of the vertices. Let's clarify this with an example of finding the optimal solution.
    Find the maximum value of:

$$C = 2x + 3y \quad \text{Objective function.}$$
$$2x + y \leq 6$$
$$x - y \leq 3 \quad \text{Constraints on } C.$$
$$x \geq 0$$
$$y \geq 0 \quad \text{Implied constraints.}$$

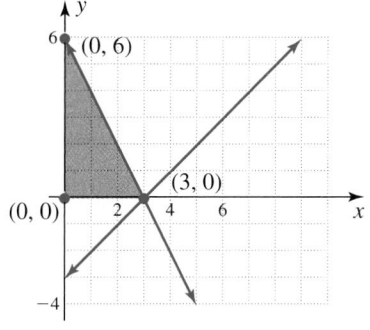

Figure 21

SOLUTION   As before, graph the line $2x + y = 6$ (Figure 21) and use the test point $(0, 0)$. Since $0 + 0 \leq 6$, the points *below* or *on* the line $2x + y = 6$ are in the solution set of $2x + y \leq 6$. Next, graph the line $x - y = 3$ and select the test point $(0, 0)$. Again, $0 - 0 \leq 3$ so the solution set consists of the points *above* or *on* the line $x - y = 3$. The lines $2x + y = 6$ and $x - y = 3$ intersect at the point $(3, 0)$.

The points satisfying $x \geq 0$ are the points *on* or to the *right* of the *y*-axis, and the points satisfying $y \geq 0$ are *on* or *above* the *x*-axis. The vertices of the feasible region are (0, 0), (0, 6), and (3, 0). At these three vertices, the objective function *C* has the following values:

At (0, 0), $C = 2 \cdot 0 + 3 \cdot 0 = 0$

At (0, 6), $C = 2 \cdot 0 + 3 \cdot 6 = 18$ ← Maximum

At (3, 0), $C = 2 \cdot 3 + 3 \cdot 0 = 6$

The maximum value of *C* is 18 and occurs when $x = 0$ and $y = 6$.

For Problems 21–22, use linear programming to solve.

21. The E-Z-Park storage lot can hold at most 100 cars and trucks. A car occupies 100 ft² and a truck 200 ft², and the lot has a usable area of 12,000 ft². The storage charge is $20 per month for a car and $35 per month for a truck. How many of each should be stored to bring E-Z-Park the maximum revenue? 80 cars; 20 trucks

22. The Zig-Zag Manufacturing Company produces two products, zigs and zags. Each of these products has to be processed through all three machines, as shown in the table. If Zig-Zag makes $12 profit on each zig and $8 profit on each zag, find the number of each that the company should produce to maximize its profit.
14 zigs; 1 zag

Zig-Zag Company data

Machine	Machine Time Available (hours)	Production Time (hours)	
		Zigs	Zags
I	Up to 100	4	12
II	Up to 120	8	8
III	Up to 84	6	0

WRITE ON

23. Is it possible for a system of linear inequalities to have no solution? If so, find an example of such a system.
Answers may vary.

24. Would it be possible to list all the solutions in the solution set of a system of linear inequalities? Explain.
Answers may vary.

MASTERY TEST

If you know how to do these problems, you have learned your lesson!

25. Graph the solution set to the system of linear inequalities:

$$y > 2x + 1$$
$$x \leq -1$$

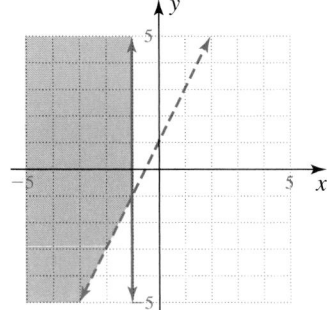

26. Graph the solution set to the system of linear inequalities:

$$y \leq -2$$
$$x - y < 4$$

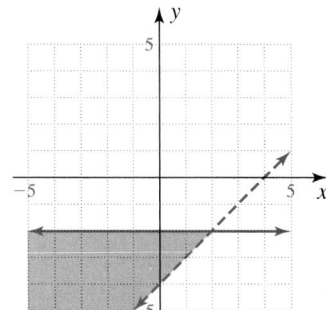

27. Graph the solution set to the system of linear inequalities:

$$x + y \leq 5$$
$$2x - y \leq 4$$
$$x \geq 0$$
$$y \geq 0$$

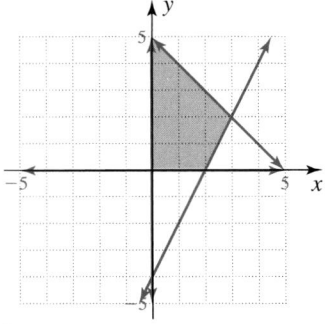

COLLABORATIVE LEARNING 4A

The transition to digital and high-definition television (HDTV) is driving up the sales for digital displays as shown in the first bar graph. How does that compare to the United States' investment in upgrading to HDTV as shown in the second bar graph? Let's do some collaboration and find out. Divide into two groups and answer the following questions.

Projected Sales of HDTVs

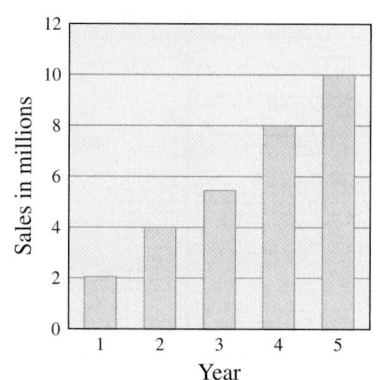

Group 1

1. Using the bar graph for the Projected Sales of HDTVs, write the equation of the line of best fit.

2. Graph the equation of the line found in number 1 on a coordinate system on a large poster board. Label it the "Supply" equation.

Group 2

3. Using the bar graph for the "United States Investment Needed for Upgrading to HDTV," write the equation of the line of best fit.

4. Graph the equation of the line found in number 3 on the same poster board graph as Group 1. Label it the "Demand" equation.

United States Investment Needed for Upgrading to HDTV

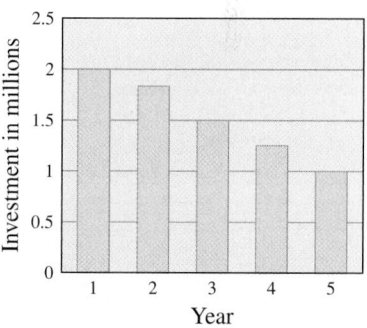

The two lines graphed on the poster board represent a system of equations where one line represents the supply (Projected Sales) of HDTVs and the other represents the demand for investment by the United States for upgrading to HDTVs. As a class, discuss the following questions.

5. Use the graph to estimate the year at which the projected sales of HDTVs will equal the United States investment for upgrading to HDTVs. What will the amount of sales for HDTVs be for that point?

6. What do you call the point at which the supply equation intersects the demand equation?

7. Discuss other real world data that could be considered as a supply and demand situation.

8. Choose one or two of the topics discussed in number 7, research the data, analyze it, and find the equilibrium point.

COLLABORATIVE LEARNING 4B

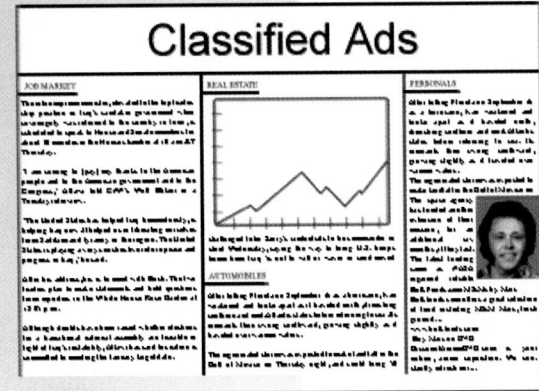

Classified Ads

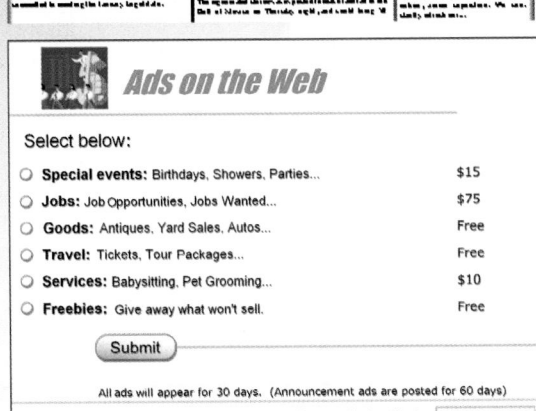

Ads on the Web

Select below:

○ **Special events:** Birthdays, Showers, Parties... $15

○ **Jobs:** Job Opportunities, Jobs Wanted... $75

○ **Goods:** Antiques, Yard Sales, Autos...   Free

○ **Travel:** Tickets, Tour Packages...   Free

○ **Services:** Babysitting, Pet Grooming... $10

○ **Freebies:** Give away what won't sell.   Free

[Submit]

All ads will appear for 30 days. (Announcement ads are posted for 60 days)

FAQ   Contact Us   Privacy Policy   Take the Tour   Members Sign In: [_____]

Have you ever considered selling an item by advertising it in the classified ads? One way to advertise is in your local newspaper classified ads while another might be to use the Web.

Here is one newspaper's pricing for classified ads.

Our classified line ads traditionally do not have artwork and are charged out by the number of lines of text. There are approximately 18 characters (including spaces) per line. The charge for an ad depends on the number of lines and the number of days the ad runs. The minimum number of times an ad can be scheduled is three days. We have price breaks for ads running in 6-, 12-, and 24-day increments. However, there will be a set fee of $2 for all 6-day ads plus line cost and a set fee of $1 for all 12-day ads plus line cost.

For example: A three-line ad for three days costs $8.74; a three-line ad for six days costs $15.04; and a three-line ad for 12 days costs $24.32.

Divide into three groups. Have the first group complete the table using the information from the ad above (based on a three-line ad). Then use the information from the table to graph each of the three rate lines (three-day, six-day, and 12-day rate) on a large poster board graph.

Days	Cost at 3-day Rate	Cost at 6-day Rate	Cost at 12-day Rate
3	$8.74		
6	17.48	$15.04 + $2 = ?	
9	26.22		
12	?	$30.08 + $2 = ?	$24.32 + $1 = ?
15	?		
18	?	?	
21	?		
24	?	?	?

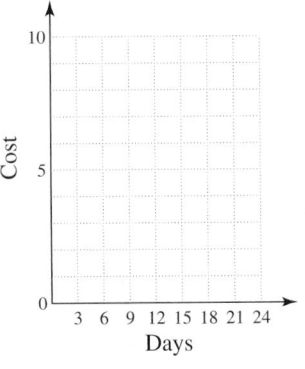

The second group should research the pricing to advertise in your local newspaper and the third group should research the pricing to advertise on the Web. These groups should construct tables and graphs as well. When all groups have completed their work discuss the following.

1. Using the graph from group 1, will the three lines ever intersect at the same point? What is the name of that type of system of equations?

2. In the graph from group 1, what is the cost at the point of intersection of the line representing the three-day rate and the line representing the six-day rate? Which of the two was the better rate before the intersection point and which is the better rate after that point?

3. How does the information from group 2 and group 3 compare to group 1? Discuss some of the reasons why one form of advertising may or may not be cheaper than another.

Research Questions

1. Leibniz was probably the first person to use elimination and determinants to solve a system of three equations in three unknowns. But the evidence of a systematic method of solving a system of three equations appeared in China probably in about 250 B.C. Here is the problem:

   There are three grades of corn. After threshing, three bundles of top grade, two bundles of medium grade, and one bundle of low grade make 39 dou (a measure of volume). Two bundles of top grade, three bundles of medium grade, and one bundle of low grade will produce 34 dou. The yield of one bundle of top grade, two bundles of medium grade, and three bundles of low grade is 26 dou. How many dou are contained in each bundle of each grade?

   a. Write a system of three equations in three unknowns representing this situation.

   b. Find out the name of the book where this problem originated and write a short paragraph detailing the contents of the book.

   c. Write a short paragraph explaining how the Chinese solved this problem.

2. Who introduced the term *matrix* into mathematical literature?

3. The mathematician viewed as the "creator of the theory of matrices" wrote a book called *A Memoir on the Theory of Matrices*. Who was this mathematician, what were his contributions to the study of matrices, and what was the name of the only theorem appearing in the book?

4. Who invented Cramer's rule? Write a paragraph supporting your findings.

5. Write a report detailing the uses of matrices in science with particular emphasis on the work of Werner Heisenberg, W. J. Duncan, and A. R. Collar.

6. Who received the 1973 Nobel prize for economics, and how did he use matrices?

Summary

SECTION	ITEM	MEANING	EXAMPLE
4.1A	Consistent system	Graphs intersect at one point. There is one solution.	$2x - y = 2$ and $y = x - 1$ form a consistent system intersecting at $(1, 0)$.
	Inconsistent system	Graphs are parallel lines. There is no solution.	$y - 2x = 4$ and $3y - 6x = 18$ form an inconsistent system.
	Dependent system	Graphs coincide. There are infinitely many solutions.	$2x + \frac{1}{2}y = 2$ and $y = -4x + 4$ form a dependent system.
4.1B	Substitution method (two variables)	A method where one equation is solved for a variable that is substituted into the other equation.	$y = 2x$ and $2x + y = 4$ can be solved by substituting $y = 2x$ into $2x + y = 4$ to obtain $$2x + 2x = 4 \text{ or}$$ $$4x = 4$$ $$x = 1$$ Thus $y = 2 \cdot 1 = 2$.

(Continued)

SECTION	ITEM	MEANING	EXAMPLE
4.1C	Elimination method (two variables)	A method where equations are multiplied by suitable numbers so that addition eliminates one of the variables.	For the system $$x - 2y = 4$$ $$x + y = 6$$ multiplying the second equation by 2 yields $$x - 2y = 4$$ $$2x + 2y = 12$$ so that addition eliminates the y, leaving $3x = 16$ or $x = \frac{16}{3}$.
4.2A	Elimination method (three variables)	A method in which two equations are selected and one variable is eliminated. Then a different pair of equations is selected, and the same variable is eliminated. The system is then solved as in 4.1C.	Consider the system $$2x - y + z = 4 \quad \textbf{(1)}$$ $$-x - y - z = 0 \quad \textbf{(2)}$$ $$-x + 2y - z = 2 \quad \textbf{(3)}$$ Add (1) and (2), then add (1) and (3). We get $$x - 2y = 4$$ $$x + y = 6$$ Solve this system as in 4.1C, then substitute the values of x and y into (1), (2), or (3) to find the value of z.
4.2B	Consistent system	A system with one solution consisting of an ordered triple of the form (x, y, z)	The system $$x + y + z = 6$$ $$x - y - z = -4$$ $$x + y - z = 0$$ is consistent. The solution is $(1, 2, 3)$.
	Inconsistent system	A system with no solution	The system $$x + y + z = 4$$ $$-x - y - z = 3$$ $$x + 2y + z = 5$$ is inconsistent. The solution set is $\varnothing$.
	Dependent system	A system with infinitely many solutions	The system $$x + y + z = 4$$ $$-x - y - z = -4$$ $$x + 2y - z = 5$$ is dependent.
4.4	Matrix	A rectangular array of numbers	$\begin{bmatrix} 2 & -3 \\ -1 & 0 \end{bmatrix}$ is a matrix.
	Augmented matrix	A matrix consisting of the coefficients of the variables and the constants in a system of equations	$\begin{bmatrix} 1 & 2 & 3 & -1 \\ 2 & 3 & 4 & 0 \\ -1 & -2 & 3 & 5 \end{bmatrix}$ is the augmented matrix for the system $$x + 2y + 3z = -1$$ $$2x + 3y + 4z = 0$$ $$-x - 2y + 3z = 5$$

SECTION	ITEM	MEANING	EXAMPLE
4.5	Determinant	$\det A = \begin{vmatrix} a_1 & b_1 \\ a_2 & b_2 \end{vmatrix}$	$\begin{vmatrix} 2 & -4 \\ -3 & 5 \end{vmatrix}$ is a 2×2 determinant.
4.5A	The value of det A	$\begin{vmatrix} a_1 & b_1 \\ a_2 & b_2 \end{vmatrix} = a_1 b_2 - a_2 b_1$	The value of $\begin{vmatrix} 2 & -4 \\ -3 & 5 \end{vmatrix}$ is $(2)(5) - (-3)(-4) = -2$.
4.5B	Cramer's rule for a 2×2 system	The solution to the system $$a_1 x + b_1 y = d_1$$ $$a_2 x + b_2 y = d_2$$ is given by $$x = \frac{D_x}{D} \quad \text{and} \quad y = \frac{D_y}{D} \quad (D \neq 0)$$ where $$D = \begin{vmatrix} a_1 & b_1 \\ a_2 & b_2 \end{vmatrix}, \quad D_x = \begin{vmatrix} d_1 & b_1 \\ d_2 & b_2 \end{vmatrix},$$ and $D_y = \begin{vmatrix} a_1 & d_1 \\ a_2 & d_2 \end{vmatrix}$	The solution of $$2x + 3y = 7$$ $$5x + 9y = 11 \text{ is}$$ $$x = \frac{D_x}{D} = \frac{\begin{vmatrix} 7 & 3 \\ 11 & 9 \end{vmatrix}}{\begin{vmatrix} 2 & 3 \\ 5 & 9 \end{vmatrix}} = \frac{30}{3} = 10$$ $$y = \frac{D_y}{D} = \frac{\begin{vmatrix} 2 & 7 \\ 5 & 11 \end{vmatrix}}{\begin{vmatrix} 2 & 3 \\ 5 & 9 \end{vmatrix}} = \frac{-13}{3}$$
4.5C	Minor	The minor of an element of a determinant is the determinant that remains after deleting the row and column in which the element appears.	The minor of 6 in $$\begin{vmatrix} 3 & 4 & ⑥ \\ 1 & 2 & 3 \\ 0 & 1 & 4 \end{vmatrix}$$ is $\begin{vmatrix} 1 & 2 \\ 0 & 1 \end{vmatrix}$
	The value of $\begin{vmatrix} a_1 & b_1 & c_1 \\ a_2 & b_2 & c_2 \\ a_3 & b_3 & c_3 \end{vmatrix}$	$a_1 \begin{vmatrix} b_2 & c_2 \\ b_3 & c_3 \end{vmatrix} - b_1 \begin{vmatrix} a_2 & c_2 \\ a_3 & c_3 \end{vmatrix} + c_1 \begin{vmatrix} a_2 & b_2 \\ a_3 & b_3 \end{vmatrix}$	
4.6B	System of linear inequalities	The solution set is the set of all points that satisfy all the inequalities of the system.	$$2x + y < 1$$ $$x \leq 0$$ $$y \geq 0$$ Solution:

Review Exercises

(If you need help with these exercises, look in the section indicated in brackets.)

1. [4.1A] Use the graphical method to solve the system.

    a. $2x - y = 2$
           $y = 3x - 4$

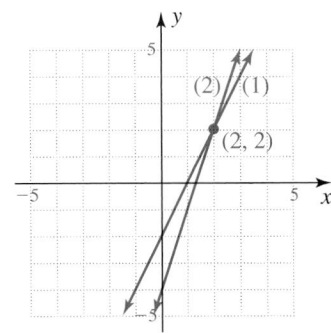

    b. $x - 2y = 0$
           $y = x - 2$

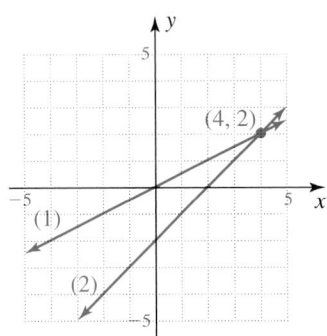

2. [4.1A] Use the graphical method to solve the system.

    a. $2y - x = 3$
           $4y = 2x + 8$

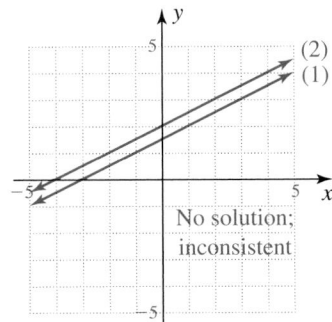

No solution; inconsistent

    b. $3y + x = 5$
           $2x = 8 - 6y$

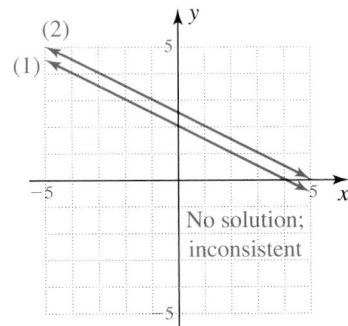

No solution; inconsistent

3. [4.1A] Use the graphical method to solve the system.

    a. $3x + 2y = 6$
           $y = 3 - \dfrac{3}{2}x$

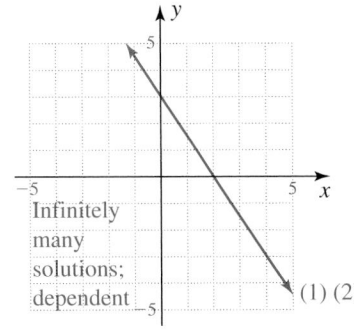

Infinitely many solutions; dependent

    b. $x + 2y = 4$
           $2x = 8 - 4y$

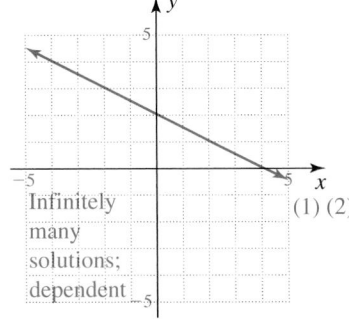

Infinitely many solutions; dependent

4. [4.1B] Solve by the substitution method.

    a. $2x - y = 4$
         $x + y = 5$
        $(3, 2)$

    b. $2x + 3y = 10$
         $x - y = -1$
        $\left(\dfrac{7}{5}, \dfrac{12}{5}\right)$

5. [4.1B] Solve by the substitution method.

    a. $2x + 4y = 7$
         $x = -2y - 1$
    No solution

    b. $2y + x = 5$
         $3x = 10 - 6y$
    No solution

6. [4.1B] Solve by the substitution method.

    a. $2y - x = 5$
         $2x = 4y - 10$
    $\left(x, \dfrac{1}{2}x + \dfrac{5}{2}\right)$; infinitely many solutions

    b. $x + 5y = 5$
         $y = 1 - \dfrac{x}{5}$
    $\left(x, 1 - \dfrac{1}{5}x\right)$; infinitely many solutions

7. [4.1C] Solve by the elimination method.

   a. $x - 3y = 7$ **b.** $2x + 3y = 4$
      $2x - y = 9$ $x + y = 1$
      $(4, -1)$ $(-1, 2)$

8. [4.1C] Solve by the elimination method.

   a. $2x + 3y = 7$ **b.** $3x - 4y = 5$
      $6x + 9y = 14$ $6x - 8y = 15$
      No solution   No solution

9. [4.1C] Solve by the elimination method.

   a. $\dfrac{x}{5} + \dfrac{y}{2} = \dfrac{1}{5}$ **b.** $\dfrac{x}{3} - \dfrac{y}{4} = 2$

      $2x + 5y = 2$ $\dfrac{x}{6} - \dfrac{y}{8} = 1$

      $\left(x, \dfrac{2-2x}{5}\right);$ $\left(x, \dfrac{4}{3}x - 8\right);$
      infinitely many solutions   infinitely many solutions

10. [4.1C] Solve by the elimination method.

   a. $4y = -5x - 2$ **b.** $2x = 3y - 1$
      $4x = -3y - 1$ $2y = 3x - 1$
      $(2, -3)$ $(1, 1)$

11. [4.2A] Solve the system.

   a. $x - y + z = 1$ **b.** $2x - 3y + z = -4$
      $x + y - z = 7$ $2x + y + z = 4$
      $2x + y + z = 9$ $4x + 9y + 3z = 10$
      $(4, 2, -1)$ $(7, 2, -12)$

12. [4.2B] Solve the system.

   a. $2x + y - 2z = 4$ **b.** $2x + 2y + z = 4$
     $-x + y + 2z = 2$ $-2x - y + z = 2$
        $3y + 2z = 12$ $-2x + 3z = 9$
      No solution   No solution

13. [4.2B] Solve the system.

   a. $x + 2y + 3z = 6$ **b.** $x + 2y = 4$
      $x - 2y - z = -2$ $y + 2z = 6$
      $x + z = 2$ $2x + 2y - 4z = -4$
      Infinitely many solutions;   Infinitely many solutions;
      $(2 - k, 2 - k, k)$, k any $(4k - 8, 6 - 2k, k)$, k
      real number   any real number

14. [4.3A] How many of each coin do Joey and Alice have?

   a. Joey has $4 in nickels and dimes, and he has five more nickels than dimes. 30 nickels; 25 dimes

   b. Alice has $2 in nickels and dimes, and she has five fewer nickels than dimes. 10 nickels; 15 dimes

15. [4.3B] How tall are these buildings?

   a. The total height of a building and a flagpole on the roof is 200 ft. The building is nine times as high as the flagpole. 180 ft

   b. The total height of a building and a flagpole on the roof is 180 ft. The building is eight times as high as the flagpole. 160 ft

16. [4.3C] Find the speed of each current.

   a. A motorboat can go 12 mi downstream on a river in 20 min. It takes this boat 30 min to go upstream the same 12 mi. 6 mi/hr

   b. A motorboat can go 6 mi downstream on a river in 15 min. It takes this boat 20 min to go upstream the same 6 mi. 3 mi/hr

17. [4.3D] Find out how much Bill and Betty have invested at each rate.

   a. Bill has three investments totaling $40,000. These investments earn interest at 4%, 6%, and 8%. Bill's annual income from these investments is $2600. The income from the 8% investment exceeds the total income from the other two investments by $600. $10,000 at 4%; $10,000 at 6%; $20,000 at 8%

   b. Betty has three investments totaling $45,000. These investments earn interest at 4%, 6%, and 8%. Betty's annual income from these investments is $2900. The income from the 8% investment exceeds the total income from the other two investments by $300. $10,000 at 4%; $15,000 at 6%; $20,000 at 8%

18. [4.3E] What are the dimensions of each rectangle?

   a. The perimeter of a rectangle is 100 in. and the length is 30 in. more than the width. 10 in. by 40 in.

   b. The perimeter of a rectangle is 80 in. and the length is three times the width. 10 in. by 30 in.

19. [4.4B] Use matrices to solve the system.

   a. $2x - y - z = 3$ **b.** $2x - 6y + 2z = -6$
      $x + y + z = 6$ $2x + y - 4z = -12$
      $3x + 2y + z = 10$ $-x + 3y + z = 5$
      $(3, -2, 5)$ $(-4, 0, 1)$

20. [4.4B] Use matrices to solve the system.

   a. $3x + y - 2z = 1$ **b.** $x + y + z = 2$
      $9x + 3y - 6z = 6$ $2x - y + z = -1$
     $-2x - y + 3z = -1$ $x - y - z = 0$
      No solution $(1, 2, -1)$

21. [4.5A] Evaluate.

   a. $\begin{vmatrix} 3 & 5 \\ 2 & -4 \end{vmatrix}$ -22 **b.** $\begin{vmatrix} -4 & 5 \\ -6 & 4 \end{vmatrix}$ 14

22. [4.5B] Solve by Cramer's rule.

   a. $2x + 5y = -8$ **b.** $4x + 2y = 1$
      $3x - 4y = 11$ $2x - 6y = 4$
      $(1, -2)$; $D = -23$; $\left(\frac{1}{2}, -\frac{1}{2}\right)$; $D = -28$;
      $D_x = -23$; $D_y = 46$ $D_x = -14$; $D_y = 14$

23. [4.5C] Evaluate.

a. $\begin{vmatrix} 1 & -2 & -2 \\ 3 & 0 & -1 \\ 4 & 1 & 2 \end{vmatrix}$ 15

b. $\begin{vmatrix} 0 & 2 & 4 \\ 1 & 2 & 0 \\ 2 & 1 & 3 \end{vmatrix}$ -18

24. [4.5C] Expand by minors along the first row.

a. $\begin{vmatrix} 1 & -1 & 1 \\ 2 & 3 & 1 \\ 1 & 3 & 2 \end{vmatrix}$ $1\begin{vmatrix} 3 & 1 \\ 3 & 2 \end{vmatrix} + 1\begin{vmatrix} 2 & 1 \\ 1 & 2 \end{vmatrix} + 1\begin{vmatrix} 2 & 3 \\ 1 & 3 \end{vmatrix} = 1(3) + 1(3) + 1(3)$
$= 3 + 3 + 3 = 9$

b. $\begin{vmatrix} 4 & -2 & -1 \\ 2 & 5 & -2 \\ 1 & -2 & 2 \end{vmatrix}$ $4\begin{vmatrix} 5 & -2 \\ -2 & 2 \end{vmatrix} + 2\begin{vmatrix} 2 & -2 \\ 1 & 2 \end{vmatrix} - 1\begin{vmatrix} 2 & 5 \\ 1 & -2 \end{vmatrix} = 4(6) + 2(6) - 1(-9)$
$= 24 + 12 + 9 = 45$

25. [4.5C] Expand by minors along the second column.

a. $\begin{vmatrix} 1 & 0 & 5 \\ 3 & 2 & 1 \\ 5 & 3 & -1 \end{vmatrix}$ $0\begin{vmatrix} 3 & 1 \\ 5 & -1 \end{vmatrix} + 2\begin{vmatrix} 1 & 5 \\ 5 & -1 \end{vmatrix} - 3\begin{vmatrix} 1 & 5 \\ 3 & 1 \end{vmatrix} = 0 + 2(-26) - 3(-14)$
$= 0 - 52 + 42 = -10$

b. $\begin{vmatrix} 1 & 2 & 1 \\ 0 & 4 & -2 \\ 3 & 6 & -2 \end{vmatrix}$ $-2\begin{vmatrix} 0 & -2 \\ 3 & -2 \end{vmatrix} + 4\begin{vmatrix} 1 & 1 \\ 3 & -2 \end{vmatrix} - 6\begin{vmatrix} 1 & 1 \\ 0 & -2 \end{vmatrix} = -2(6) + 4(-5) - 6(-2)$
$= -12 - 20 + 12 = -20$

26. [4.5C] Expand by minors along the third column.

a. $\begin{vmatrix} 1 & 3 & 0 \\ 0 & 1 & -2 \\ 2 & 4 & 3 \end{vmatrix}$ $0\begin{vmatrix} 0 & 1 \\ 2 & 4 \end{vmatrix} + 2\begin{vmatrix} 1 & 3 \\ 2 & 4 \end{vmatrix} + 3\begin{vmatrix} 1 & 3 \\ 0 & 1 \end{vmatrix} = 0 + 2(-2) + 3(1)$
$= 0 - 4 + 3 = -1$

b. $\begin{vmatrix} 3 & 1 & 5 \\ 1 & 0 & -2 \\ 6 & 1 & 3 \end{vmatrix}$ $5\begin{vmatrix} 1 & 0 \\ 6 & 1 \end{vmatrix} + 2\begin{vmatrix} 3 & 1 \\ 6 & 1 \end{vmatrix} + 3\begin{vmatrix} 3 & 1 \\ 1 & 0 \end{vmatrix} = 5(1) + 2(-3) + 3(-1)$
$= 5 - 6 - 3 = -4$

27. [4.5D] Solve by Cramer's rule.

a. $x + 2y + z = 6$
$x + y - z = 7$
$2x - y + 2z = -3$
$(2, 3, -2); D = -10;$
$D_x = -20; D_y = -30;$
$D_z = 20$

b. $x + y + 2z = -3$
$x - y + 2z = 1$
$x + 2y - z = -2$
$(1, -2, -1); D = 6;$
$D_x = 6; D_y = -12;$
$D_z = -6$

28. [4.6A, B] Graph the solution set.

a. $y \geq x$
$x > -2$

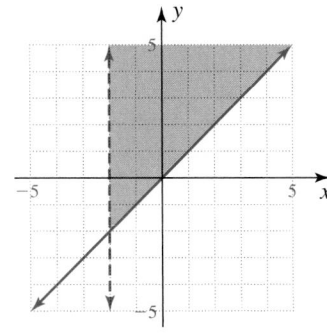

b. $3x - 4y \geq -12$
$x < 1$
$y \geq 0$

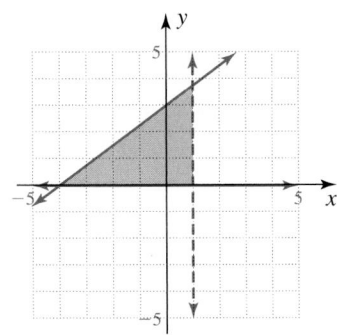

Practice Test 4

(Answers are on pages 337–338)

1. Use the graphical method to solve the system.

$$x - 3y = 3$$
$$y = x - 1$$

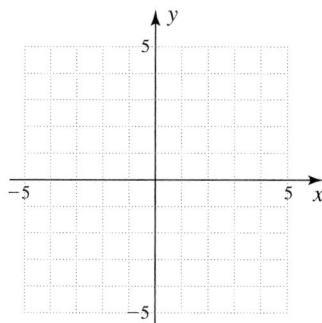

2. Use the graphical method to solve the system.

$$y - 3x = -3$$
$$3y = 9x + 9$$

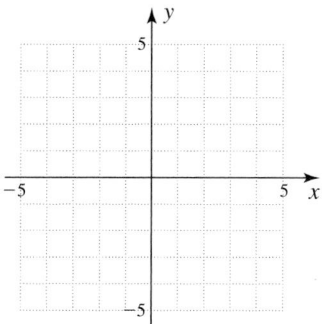

3. Use the graphical method to solve the system.

$$3x + 2y = 6$$
$$x = 2 - \frac{2}{3}y$$

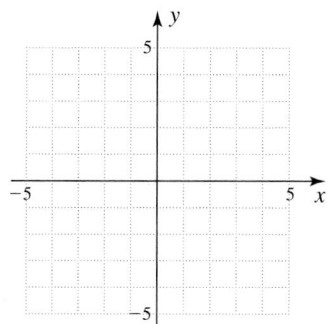

4. Use the substitution method to solve the system.

$$x - 2y = 4$$
$$x = 1 + y$$

5. Use the substitution method to solve the system.

$$2x - 3y = 6$$
$$4x = 6y + 7$$

6. Use the substitution method to solve the system.

$$2x - y = 6$$
$$2y = 4x - 12$$

7. Solve by the elimination method.

$$3x + 4y = 5$$
$$x + y = 1$$

8. Solve by the elimination method.

$$3x + 4y = 5$$
$$6x + 8y = 9$$

9. Solve the system.

$$\frac{x}{2} + \frac{y}{3} = 2$$
$$\frac{x}{4} + \frac{y}{8} = 1$$

10. Solve the system.

$$2x = 3y - 10$$
$$2y = 3x + 10$$

11. Solve the system.

$$x + y + z = 2$$
$$2x + y - z = 5$$
$$x + y - z = 4$$

12. Solve the system.

$$2x + y - 2z = 4$$
$$-x + y + 2z = 2$$
$$3y + 2z = 12$$

13. Solve the system.

$$2x + y + z = 4$$
$$-x + 2y + z = 3$$
$$-2x + 9y + 5z = 16$$

14. Pedro has $3.50 in nickels and dimes. He has 10 more nickels than dimes. How many of each coin does he have?

15. The total height of a building and a flagpole on the roof is 240 ft. The building is nine times as high as the flagpole. How high is the building?

16. A motorboat can go 10 mi downstream on a river in 20 min. It takes 30 min for this boat to go back upstream the same 10 mi. Find the speed of the current.

17. Annie has three investments totaling $60,000. These investments earn interest at 4%, 6%, and 8%. Annie's total annual income from these investments is $4000. The income from the 8% investment exceeds the total income from the other two investments by $800. Find how much she has invested at each rate.

18. Use matrices to solve the system.
$$\begin{aligned} x + y + z &= -2 \\ 2x + y - z &= 3 \\ -x + y + z &= 0 \end{aligned}$$

19. Evaluate.

a. $\begin{vmatrix} 4 & -2 \\ 5 & -6 \end{vmatrix}$ **b.** $\begin{vmatrix} 4 & -2 \\ 6 & -3 \end{vmatrix}$

20. Solve by Cramer's rule.
$$\begin{aligned} 2x - 3y &= 5 \\ 6x - 4y &= 7 \end{aligned}$$

21. Evaluate.
$$\begin{vmatrix} -2 & 4 & -2 \\ 1 & 0 & 3 \\ 4 & 5 & 2 \end{vmatrix}$$

22. Expand by minors along the first row.
$$\begin{vmatrix} 3 & 2 & -1 \\ 1 & 3 & 2 \\ 1 & 1 & -1 \end{vmatrix}$$

23. Solve by Cramer's rule.
$$\begin{aligned} x + 2y + z &= 6 \\ x + y - z &= 7 \\ 2x - y + 2z &= -3 \end{aligned}$$

24. Graph the solution set.
$$\begin{aligned} x &> -1 \\ y &\geq x \end{aligned}$$

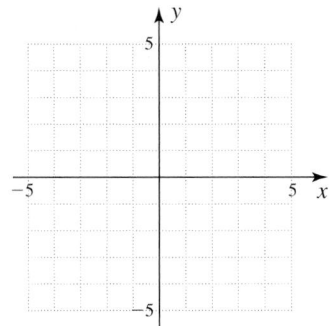

25. Graph the solution set.
$$\begin{aligned} x + y &\leq 2 \\ y &\leq 2 \\ x &\leq 2 \end{aligned}$$

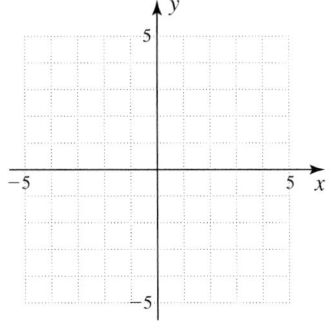

Answers to Practice Test

ANSWER	IF YOU MISSED		REVIEW		
	QUESTION		SECTION	EXAMPLES	PAGE
1. The solution is $(0, -1)$.	1		4.1A	1	253

2. Inconsistent; no solution	2		4.1A	2	254

3. Dependent, consistent; infinitely many solutions	3		4.1A	3	255

4. $(-2, -3)$	4		4.1B	4	257
5. Inconsistent; no solution	5		4.1B	5	258
6. Dependent, consistent; infinitely many solutions	6		4.1B	6	258
7. $(-1, 2)$	7		4.1C	7	259–260
8. Inconsistent; no solution	8		4.1C	8	261
9. $(4, 0)$	9		4.1C	7	259–260

ANSWER	IF YOU MISSED		REVIEW	
	QUESTION	SECTION	EXAMPLES	PAGE
10. $(-2, 2)$	10	4.1C	7–8	259–261
11. $(1, 2, -1)$; consistent	11	4.2A	1	275
12. Inconsistent; no solution	12	4.2A	2	275–276
13. Infinitely many solutions; dependent	13	4.2A	3	276
14. 30 nickels; 20 dimes	14	4.3A	1	284–285
15. 216 ft	15	4.3B	2	285–286
16. 5 mi/hr	16	4.3C	3	286–287
17. $10,000 at 4%; $20,000 at 6%; $30,000 at 8%	17	4.3D	4	287–288
18. $(-1, 2, -3)$	18	4.4A	1, 2, 3	298–300
19. a. -14 **b.** 0	19	4.5A	1	308
20. $\left(\frac{1}{10}, -\frac{8}{5}\right)$	20	4.5B	2	309–310
21. 60	21	4.5C	3	311–312
22. -7	22	4.5C	3	311–312
23. $(2, 3, -2)$	23	4.5D	5	313–314
24.	24	4.6A	1	321

25.	25	4.6B	3	323

Cumulative Review Chapters 1–4

1. The number $\sqrt{13}$ belongs to which of these sets? Natural numbers, whole numbers, integers, rational numbers, irrational numbers, and real numbers. Name all that apply. Irrational numbers; real numbers

2. Simplify:

$$[(2x^2 - 6) + (6x + 6)] - [(x - 2) + (4x^2 - 5)]$$
$-2x^2 + 5x + 7$

3. Perform the indicated operation and simplify:
$\dfrac{36x^6}{9x^{-7}}$ $4x^{13}$

4. Simplify: $(3x^3y^{-4})^2$ $\dfrac{9x^6}{y^8}$

5. Evaluate: $[-7(2 + 6)] + 6$ -50

6. Solve: $x + 6 = 3(5x - 2)$ $x = \frac{6}{7}$

7. Solve: $0.03P + 0.08(1300 - P) = 75$ $P = 580$

8. Solve: $|x - 6| = |x - 8|$ $x = 7$

9. Graph: $2(x - 3) \le 3x - 5$

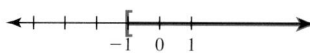

10. Graph: $-5 \le -5x - 15 < 5$

11. Graph: $|2x + 3| > 4$

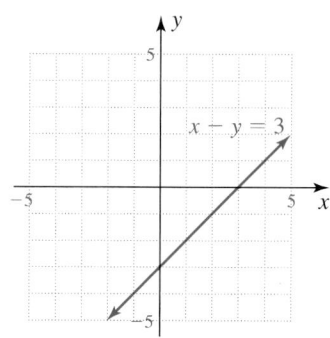

12. If $H = 2.85h + 73.82$, find h when $H = 139.37$. 23

13. A woman's salary was increased by 10% to $29,700. What was her salary before the increase? $27,000

14. How many gallons of a 40% acid solution must be mixed with 30 gallons of a 16% acid solution to obtain a solution that is 30% acid? 42 gallons

15. Graph: $x - y = 3$

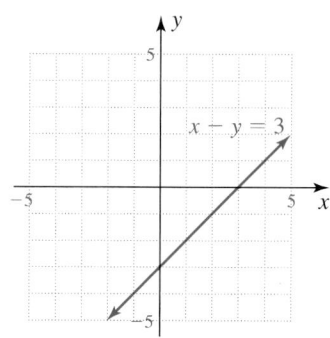

16. Find the slope of the line passing through the points $A(-7, -7)$ and $B(0, 8)$. $\frac{15}{7}$

17. A line goes through the point $(5, -5)$ and has slope 2. Graph this line.

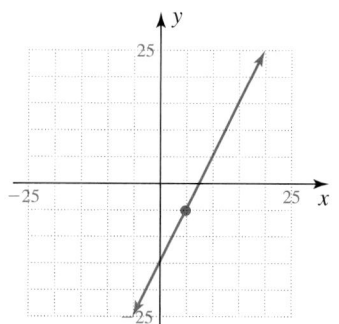

18. Find the slope and the y-intercept of the line $9x + 3y = -54$. $m = -3$; y-intercept $= -18$

19. Find an equation of the line that passes through the point (3, 6) and is parallel to the line $8x + 6y = -2$. Write the equation in standard form.
$4x + 3y = 30$

20. Graph: $y \leq x + 1$

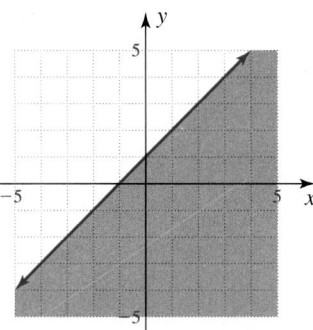

21. Graph: $|y| \leq 4$

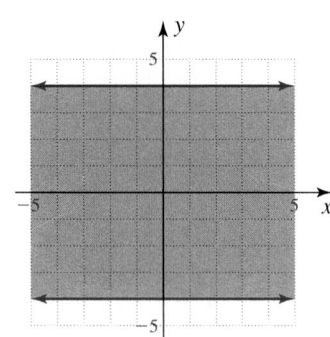

22. Find the domain and range of $\{(x, y) | y = 2 + x\}$.
$D = \{\text{all real numbers}\};$
$R = \{\text{all real numbers}\}$

23. Find $f(2)$ given $f(x) = 3x - 2$. 4

24. Use the graphical method to solve the system:
$$x - 3y = -3$$
$$x = -3 - 3y$$

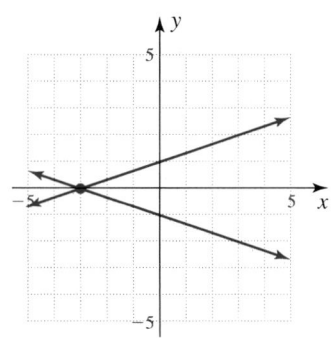

Solution: $(-3, 0)$

25. Use the substitution method to solve the system:
$$x - 2y = 3$$
$$6y = 3x - 9$$
Infinitely many solutions: $\{(x, y) | y = \frac{1}{2}x - 1\frac{1}{2}\}$

26. Solve the system:
$$\frac{x}{2} + \frac{y}{5} = 6$$
$$\frac{x}{6} + \frac{y}{3} = 6 \quad (6, 15)$$

27. Solve the system:
$$2x + y - 3z = 5$$
$$-4x + y - 2z = 4$$
$$3y - 8z = -6 \quad \text{No solution; inconsistent}$$

28. Evaluate:
$$\begin{vmatrix} 2 & -5 \\ 1 & 1 \end{vmatrix} \quad 7$$

29. Expand by minors along the first row:
$$\begin{vmatrix} 2 & 3 & -1 \\ 1 & 2 & 3 \\ 1 & 1 & -1 \end{vmatrix} \quad 3$$

30. A motorboat can go 12 mi downstream on a river in 20 min. It takes 30 min for this boat to go back upstream the same 12 mi. Find the speed of the boat.
30 mi/hr

Polynomials

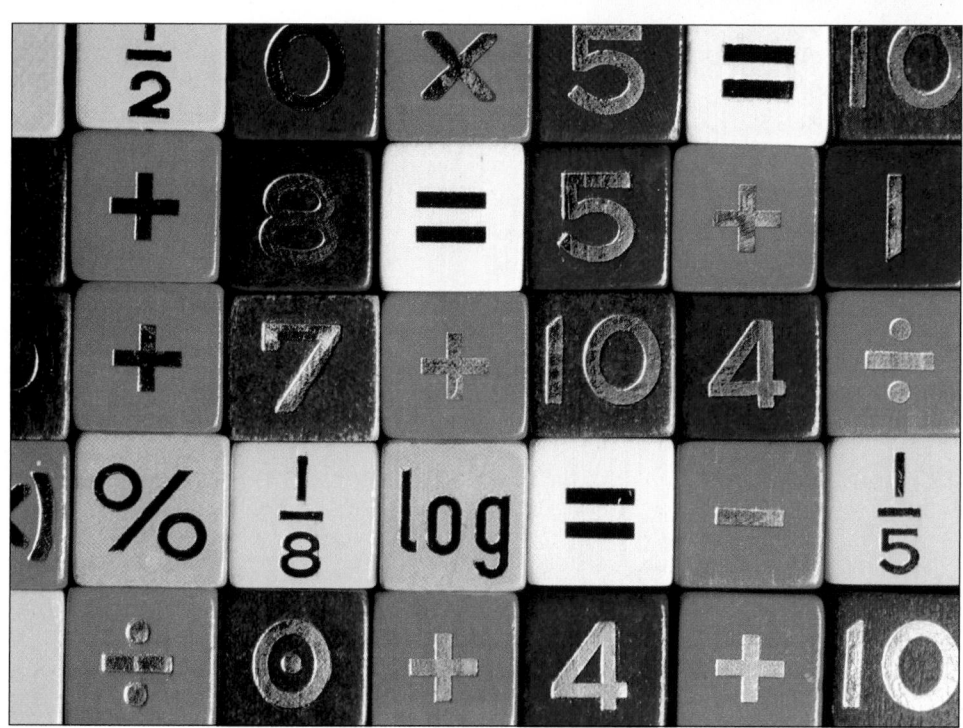

The Human Side of Algebra

Algebra has gone through three stages: rhetorical, in which statements and equations were written in ordinary language; syncopated, in which familiar terms were abbreviated; and symbolic, in which every part of an expression is written in symbols. At the time of Euclid, letters were used to represent quantities entered into equations until Diophantus introduced "the syncopation of algebra," using his own shorthand to express quantities and operations. (For example, Diophantus wrote the square of the unknown as Δ^Y.)

Francois Vieta (1540–1603), a French lawyer and member of parliament, used vowels to designate unknown quantities and consonants to represent constants. Vieta, however, retained part of the verbal algebra by writing *A quadratus* for x^2, *A cubus* for x^3, and so on. Vieta's contribution was a significant step toward a more abstract mathematics, but one of his most interesting contributions was his motto, *Leave no problem unsolved*. See if you can apply this motto to your study of algebra.

Pretest for Chapter 5

(Answers on page 343)

1. Classify as a monomial, binomial, or trinomial and give the degree of $-x^2 + xy^2z^4 - x^6$

2. Write the polynomial, $-4x + 2x^3 - x^2 + 7$ in descending order.

3. The height of an object thrown straight up with an initial velocity of 96 ft/sec after t seconds is given by $H(t) = -16t^2 + 96t$. What is the height of the object after 4 sec?

4. Let $P(x) = x^2 - 3x + 2$. Find $P(-1)$.

5. Add $(5x^3 + 8x^2 - 6x - 4)$ and $(6 - 3x + x^2 - 2x^3)$.

6. Subtract $(8x^3 - 6x^2 + 4x - 2)$ from $(5x^3 + 3x^2 + 2)$.

7. Multiply $-2x^2y(x^2 + 5xy - 3y^3)$.

8. Multiply $(x - 3)(x^2 - 4x - 5)$.

9. Multiply $(3x + 4y)(4x - 7y)$.

10. Multiply

    a. $(2x + 5y)^2$ **b.** $(5x - 4y)(5x + 4y)$

11. Factor completely $12x^5 - 16x^4 + 8x^3 + 20x^2$.

12. Factor completely $6x^6 - 4x^4 + 15x^3 - 10x$.

13. Factor completely

    a. $x^2 - 2xy - 24y^2$ **b.** $2x^2 - xy - 15y^2$

14. Factor completely $24x^4y + 4x^3y^2 - 8x^2y^3$.

15. Factor completely

    a. $16x^2 - 24xy + 9y^2$ **b.** $9x^2 + 12xy + 4y^2$

16. Factor completely $16x^4 - y^4$.

17. Factor completely $x^2 - 8x + 16 - y^2$.

18. Factor completely

    a. $8x^3 + 27y^3$ **b.** $27y^3 - 8x^3$

19. Factor completely

    a. $8x^6 - x^3y^3$

    b. $6x^6 + 24x^4$

20. Factor completely

    a. $18x^4 + 24x^3y + 8x^2y^2$

    b. $12x^2y^2 - 36xy^3 + 27y^4$

21. Factor completely $9x^3y - 33x^2y^2 - 12xy^3$.

22. Factor completely $12x^3 - 8x^2 - 3x + 2$.

23. Solve

    a. $x^2 = -2x + 8$ **b.** $6x^2 - x = 2$

24. Solve $x^3 + 4x^2 - x - 4 = 0$.

25. The sides of a right triangle are x, $x + 7$, and $x + 8$ units long. Find the dimensions of the triangle.

Answers to Pretest

ANSWER	IF YOU MISSED	REVIEW		
	QUESTION	SECTION	EXAMPLES	PAGE
1. Trinomial; 7	1	5.1	1, 2	345–346
2. $2x^3 - x^2 - 4x + 7$	2	5.1	2	346
3. 128 ft	3	5.1	3	347
4. 6	4	5.1	4	347
5. $3x^3 + 9x^2 - 9x + 2$	5	5.1	5, 6	349
6. $-3x^3 + 9x^2 - 4x + 4$	6	5.1	7, 8	350
7. $-2x^4y - 10x^3y^2 + 6x^2y^4$	7	5.2	1	357
8. $x^3 - 7x^2 + 7x + 15$	8	5.2	2	358
9. $12x^2 - 5xy - 28y^2$	9	5.2	3	360
10. a. $4x^2 + 20xy + 25y^2$ **b.** $25x^2 - 16y^2$	10	5.2	4, 5	361–362
11. $4x^2(3x^3 - 4x^2 + 2x + 5)$	11	5.3	1, 2, 3	369–370
12. $x(3x^2 - 2)(2x^3 + 5)$	12	5.3	4, 5	370–372
13. a. $(x + 4y)(x - 6y)$ **b.** $(2x + 5y)(x - 3y)$	13	5.4	1, 2, 3, 4, 5	376–377, 379
14. $4x^2y(3x + 2y)(2x - y)$	14	5.4	6	380
15. a. $(4x - 3y)^2$ **b.** $(3x + 2y)^2$	15	5.5	1, 2	385–386
16. $(4x^2 + y^2)(2x + y)(2x - y)$	16	5.5	3	386–387
17. $(x - 4 + y)(x - 4 - y)$	17	5.5	4	387
18. a. $(2x + 3y)(4x^2 - 6xy + 9y^2)$ **b.** $(3y - 2x)(9y^2 + 6xy + 4y^2)$	18	5.5	5	388
19. a. $x^3(2x - y)(4x^2 + 2xy + y^2)$ **b.** $6x^4(x^2 + 4)$	19	5.6	1	394
20. a. $2x^2(3x + 2y)^2$ **b.** $3y^2(2x - 3y)^2$	20	5.6	2	394
21. $3xy(3x + y)(x - 4y)$	21	5.6	3	395
22. $(3x - 2)(2x + 1)(2x - 1)$	22	5.6	5	395
23. a. $2, -4$ **b.** $\frac{2}{3}, -\frac{1}{2}$	23	5.7	1, 2, 3	400–403
24. $1, -1, -4$	24	5.7	4	404
25. 5, 12, 13 units	25	5.7	5	405

5.1 # POLYNOMIALS: ADDITION AND SUBTRACTION

To Succeed, Review How To . . .

1. Define base and exponent (p. 32).

2. Evaluate expressions involving exponents (pp. 32–40).

3. Use the properties of real numbers (pp. 25–27).

4. Collect like terms (pp. 53–55).

5. Remove parentheses in an expression preceded by a minus sign (p. 52).

Objectives

A Classify polynomials.

B Find the degree of a polynomial and write in descending order.

C Evaluate a polynomial.

D Add or subtract polynomials.

E Solve applications involving sums or differences of polynomials.

GETTING STARTED

Diving

A man dives from an altitude of 118 ft into water 12 ft deep. Do you know how high above sea level he will be after falling for t seconds? This height is given by

$$H = -16t^2 + 118$$

The right-hand side of this formula is an algebraic expression called a *polynomial*.

Polynomials have many applications. However, this section will begin by introducing some concepts and definitions before solving applications involving polynomials.

 **A**

Classify Polynomials—Monomials, Binomials, and Trinomials

An expression consisting of a constant or a constant times a product of variables with *whole-number exponents* is called a *monomial*. For example,

$$2x, \quad -5x^2y, \quad 10x^5yz^2, \quad -\frac{2}{3}x^4, \quad \text{and} \quad -7y^3$$

are all monomials.

A **polynomial** is a sum or difference of monomials. Thus

$$2x^2 + xy - 7y^3$$

is a polynomial. The individual monomials in a polynomial are called the **terms** of the polynomial. The terms are separated by $+$ and $-$ signs. In the term ax^k, where x is the only variable, a is the **coefficient** and k is the **degree** of the term. In the polynomial $2x^2 + xy - 7y^3$, the terms are $2x^2$, with coefficient 2; xy, with coefficient 1 (since $xy = 1xy$); and $-7y^3$, with coefficient -7.

Notice,

$$5x^{-1} \text{ or } \frac{5}{x}, \quad \frac{x+y}{z}, \quad \text{and} \quad 2 + \sqrt{x} \text{ or } 2 + x^{1/2}$$

are *not* polynomials. In each case not all variables have whole number exponents. (In the first two expressions we are dividing by a variable. The third expression involves taking the square root of the variable.)

Polynomials can be classified according to the number of terms they have. Polynomials with one term are called **monomials,** polynomials with two *unlike* terms are called **binomials,** and polynomials with three *unlike* terms are called **trinomials.** Polynomials with more than three unlike terms are also called polynomials, as shown in the table.

Teaching Tip

Point out that the coefficient of x, x^2y, and xy^3z^4 is understood to be 1. Also, -3, a number without a variable, has a coefficient of -3.

Type	Example
Monomials	$-8x$, $3x^2y$, $9m^{10}$, -3
Binomials	$x - y$, $-3x^2 + xy$, $-16t^2 + 118$
Trinomials	$x + y - z^5$, $2x^2 - 3x + \sqrt{2}$, $z^3 + 2z - 12$
Polynomials	$t^5 + 3t^3 - 2t^2 - 8$, $-3m^4 + 3n^2 + mn - 4m^3 + 2$

EXAMPLE 1 **Classifying polynomials**	**PROBLEM 1**

Classify each of the following polynomials as a monomial, binomial, trinomial, or polynomial:

a. $x + x^2$ **b.** $-9x$ **c.** $3x^2 - y + 3xyz$ **d.** $x^5 - 3x^3 + 5x^2 - 7$

SOLUTION

a. $x + x^2$ has two terms. It is a binomial.

b. $-9x$ has one term. It is a monomial.

c. $3x^2 - y + 3xyz$ has three terms. It is a trinomial.

d. $x^5 - 3x^3 + 5x^2 - 7$ has more than three terms. It is a polynomial.

Problem 1 column:

Classify each of the following polynomials as a monomial, binomial, trinomial, or polynomial:

a. $-12xy$

b. $7 + x + x^3$

c. $z^6 - 7z^4 - 8z^2 + 5z - 4$

d. $25x^2 - 16y^2$

B   Degree of a Polynomial and Writing in Descending Order

In Example 1, the polynomials $(x + x^2)$ and $(x^5 - 3x^3 + 5x^2 - 7)$ contain only one variable. The second polynomial is written in *descending order,* from the highest exponent on the variable to the lowest exponent on the variable. The first polynomial written in descending order would be, $x^2 + x$.

Single-variable polynomials can also be classified according to the *greatest exponent* of the variable. This number is called the *degree* of the polynomial. (Recall that in the monomial ax^k, k is the degree.) In general, we have the following definition.

Teaching Tip

Point out that the degree of $5x$, $7y$, and $-3z$ is 1 because the exponent of x, y, and z is understood to be 1.

DEGREE OF A POLYNOMIAL IN ONE VARIABLE

The **degree** of a polynomial in one variable is the *greatest* exponent of that variable.

Thus $-8x^5$ is of the *fifth* degree, $-3x^2 + 8x^4 - 2x$ is of the *fourth* degree, and $0.5x$ is of the *first* degree. (Note that $x = x^1$.) Since $x^0 = 1$, $-3 = -3 \cdot 1 = -3x^0$. Similarly, $9 = 9x^0$. Thus the degree of nonzero numbers such as -3 and 9 is 0. The number 0 is called the **zero polynomial** and is not assigned a degree. These ideas can be extended to include polynomials in more than one variable. For example, the expression $3x^2 - y$ is a polynomial in *two* variables (x and y), and $3x^2 - 2xy - 3xyz^2$ is a polynomial in *three* variables (x, y, and z). To find the degree of these polynomials, we look at the degree of each term. Here is the definition.

Answers

1. a. monomial   **b.** trinomial
c. polynomial   **d.** binomial

Web It

For a tutorial on finding the degree and practice on adding and subtracting polynomials, try link 5-1-1 at mhhe.com/bello.

DEGREE OF A POLYNOMIAL IN SEVERAL VARIABLES

The degree of a polynomial in several variables is the greatest sum of the exponents of the variables in any one term of the polynomial.

Thus the degree of $3x^2 - y$ is 2 (the degree of the first term, $3x^2$). To find the degree of $3x^2 - 2xy - 3xyz^2$, find the degree of each term:

$$\underbrace{3x^2}_{} \quad - \quad \underbrace{2x^1y^1}_{} \quad - \quad \underbrace{3x^1y^1z^2}_{}$$

Degree: 2   1 + 1 = 2   1 + 1 + 2 = 4

The greatest sum of exponents in any one term is 4, so the degree of $3x^2 - 2xy - 3xyz^2$ is 4.

EXAMPLE 2 **Find the degree of a polynomial and write in descending order**

Find the degree of the given polynomials:

a. $-5x^2 + 3x^5 + 9$ **b.** $-x^2 + xy^2z^3 - x^5$ **c.** 3   **d.** 0

e. Write Example 2(a) in descending order.

SOLUTION

a. The degree of $-5x^2 + 3x^5 + 9$ is 5.

b. The degree of $-x^2 + x^1y^2z^3 - x^5$ is $1 + 2 + 3 = 6$, the sum of the exponents in $x^1y^2z^3$.

c. The degree of 3 is 0.

d. 0 has no degree.

e. Descending order means highest exponent on the variable to lowest. Thus, $3x^5 - 5x^2 + 9$.

PROBLEM 2

Find the degree of the given polynomials:

a. $-2x$

b. $4 - 7x + 9x^4$

c. $3xz^6 - 7z^4 - z^2 + 5z - 8$

d. 25

e. Write Problem 2(b) in descending order.

C   Evaluating Polynomial Functions

In mathematics, polynomials in one variable are also called polynomial functions and are represented by using symbols such as $P(t)$ (read "P of t"), $Q(x)$, and $D(y)$, *where the symbol in parentheses indicates the variable being used.* For example, we may have

$$P(t) = -16t^2 + 10t - 15 \qquad \text{t is the variable.}$$

$$Q(x) = x^2 - 3x \qquad \text{x is the variable.}$$

$$D(y) = -3y - 9 \qquad \text{y is the variable.}$$

Teaching Tip

Have students note that all of the evaluation takes place on the right side of the equal sign. The left side is simply the name of the polynomial.

With this notation, it's easy to indicate the value of a polynomial for specific values of the variable as was done in Chapter 3. Thus $P(2)$ represents the value of the polynomial $P(t)$ when 2 is substituted for t in the polynomial. Similarly, $Q(3)$ represents the value of $Q(x)$ for $x = 3$. Thus

$$P(t) = -16t^2 + 10t - 15$$

$$P(2) = -16(2)^2 + 10(2) - 15$$

$$= -64 + 20 - 15 = -59$$

$P(2)$ represents the value of the polynomial function $P(t)$ when $t = 2$. We also say that we are *evaluating* $P(t)$ at $t = 2$.

Similarly, to evaluate $Q(x)$ at $x = 3$ in

$$Q(x) = x^2 - 3x$$

we find

$$Q(3) = 3^2 - 3(3) = 0$$

Answers

2. a. 1   **b.** 4   **c.** 7   **d.** 0

e. $9x^4 - 7x + 4$

Teaching Tip

To emphasize this caution do the following example:

Evaluate for $x = 3$.

$P(x) = -x^2$	$Q(x) = (-x)^2$
$P(3) = -(3)^2$	$Q(3) = (-(3))^2$
$= -9$	$= 9$
Exponents done first.	Inner parentheses done first.

CAUTION

Always enclose the substituted number in parentheses.

Using these ideas, we can find the height above sea level of the diver in the *Getting Started* section. As you recall, his altitude after t seconds was given as $H = -16t^2 + 118$. This can also be written as $H(t) = -16t^2 + 118$. Thus after 1 sec, his altitude will be

$$H(1) = -16(1)^2 + 118 = 102 \text{ ft}$$ Here we are evaluating $H(t)$ when $t = 1$.

After 2 sec, it will be

$$H(2) = -16(2)^2 + 118 = -64 + 118 = 54 \text{ ft}$$ Here we are evaluating $H(t)$ when $t = 2$.

EXAMPLE 3   **Evaluating polynomials using a formula**

Find the altitude of the diver in the *Getting Started* after 3 sec. Recall the height is given by $H(t) = -16t^2 + 118$. (Here $H = H(t)$.)

SOLUTION   After 3 sec, the formula tells us that his altitude will be $H(3) = -16(3)^2 + 118 = -26$ ft— that is, 26 ft below sea level. The water is only 12 ft deep at this point, so this is impossible. Moreover, divers cannot continue to free-fall after they hit the surface. In other words, the formula doesn't apply.

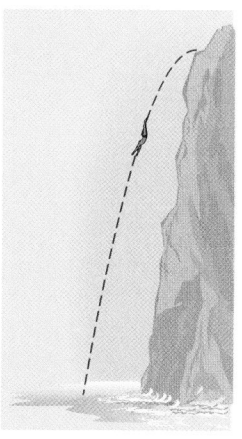

PROBLEM 3

Find the altitude of the diver in the *Getting Started* after 2.5 seconds.

EXAMPLE 4   **Evaluating polynomials**

Let $P(x) = x^2 - 2x + 3$ and $Q(x) = x^2 + 3x - 5$.

a. Find: $P(0)$ **b.** Find: $Q(-1)$ **c.** Find: $P(0) + Q(-1)$

SOLUTION

a. To find $P(0)$, we substitute 0 for x in $P(x)$.

$$P(x) = x^2 - 2x + 3$$
$$P(0) = 0^2 - 2 \cdot 0 + 3$$
$$= 0 - 0 + 3$$
$$= 3$$

Hence $P(0) = 3$.

b. To find $Q(-1)$, we substitute -1 for x in $Q(x)$.

$$Q(x) = x^2 + 3x - 5$$
$$Q(-1) = (-1)^2 + 3(-1) - 5$$
$$= 1 - 3 - 5$$
$$= -7$$

Hence $Q(-1) = -7$.

c. Since $P(0) = 3$ and $Q(-1) = -7$,

$$P(0) + Q(-1) = 3 + (-7) = -4$$

PROBLEM 4

If $P(x)$ and $Q(x)$ are as in Example 4, find:

a. $P(1)$

b. $Q(-2)$

c. $P(1) - Q(-2)$

Answers

3. 18 ft   **4. a.** 2   **b.** -7 **c.** 9

D · Adding and Subtracting Polynomials

Web It

For interactive practice on adding and subtracting polynomials, try link 5-1-2 at mhhe.com/bello.

The graph in Figure 1 illustrating the U.S. automobile fuel cost in cents per mile from 1970 to 2000 shows several peaks and valleys. For any given year during the five-year span when U.S. automobile fuel cost was at its all-time high (1978–1982), the cost in cents per mile could have been approximated using $H(y) = -0.9y^2 + 3y + 13$, where y represents the years (0, 1, 2, 3, or 4).

For any given year during the five-year span when U.S. automobile fuel cost was at its all-time low (1996–2000), the cost in cents per mile could have been approximated using $L(y) = 0.4y^2 - 2y + 7$, where y represents the years (0, 1, 2, 3, or 4).

We can find the difference in these two extremes by subtracting the two polynomials, which we will do in Example 7.

U.S. Automobile Fuel Cost

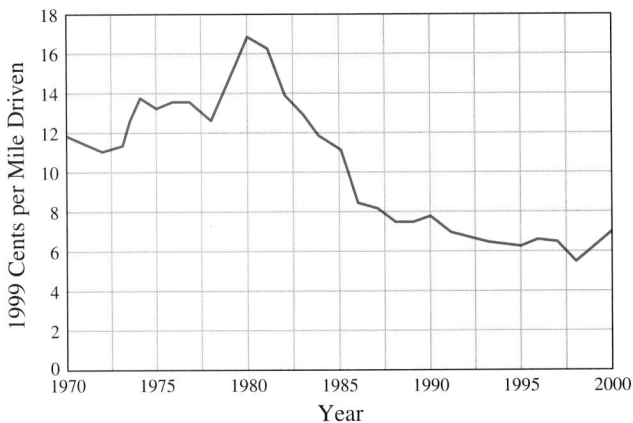

Figure 1

The procedure we use to add polynomials is dependent on the fact that the same properties used in the addition of numbers also apply to polynomials. We list these properties next.

Properties for Adding Polynomials	If P, Q, and R are polynomials,	
	$P + Q = Q + P$	Commutative property of addition
	$P + (Q + R) = (P + Q) + R$	Associative property of addition
	$P(Q + R) = PQ + PR$	Distributive property
	$(Q + R)P = QP + RP$	

Now suppose we wish to add $(5x^2 + 3x + 9) + (7x^2 + 2x + 1)$. For our solution, we use the commutative, associative, and distributive properties just mentioned. With these facts, the terms in the expression $(5x^2 + 3x + 9) + (7x^2 + 2x + 1)$ can be added as follows:

$$(5x^2 + 3x + 9) + (7x^2 + 2x + 1) \qquad \text{Given.}$$
$$= (5x^2 + 7x^2) + (3x + 2x) + (9 + 1) \qquad \text{Group like terms.}$$
$$= (5 + 7)x^2 + (3 + 2)x + (9 + 1) \qquad \text{Use the distributive property.}$$
$$= 12x^2 + 5x + 10$$

This addition is done more efficiently by writing the terms of the polynomials in order of descending (or ascending) degree and then placing like terms in a column:

$$
\begin{array}{r}
5x^2 + 3x + 9 \\
(+) \ 7x^2 + 2x + 1 \\
\hline
12x^2 + 5x + 10
\end{array}
$$

Teaching Tip

Students should note that coefficients are being added in each column but exponents remain the same.

As we have already seen, $a - b = a + (-b)$. The signs in a polynomial are always taken to indicate positive or negative coefficients, and the operation involved is assumed to be addition. Thus

$$(6x^2 - 9x - 8) + (x^2 + 3x - 1)$$
$$= [6x^2 + (-9x) + (-8)] + [x^2 + 3x + (-1)]$$
$$= (6x^2 + x^2) + (-9x + 3x) + [-8 + (-1)]$$
$$= (6 + 1)x^2 + (-9 + 3)x + [-8 + (-1)]$$
$$= 7x^2 + (-6)x + (-9)$$
$$= 7x^2 - 6x - 9$$

Using the column method, this problem can be done more efficiently by writing

$$
\begin{array}{r}
6x^2 - 9x - 8 \\
(+)\ \underline{x^2 + 3x - 1} \\
7x^2 - 6x - 9
\end{array}
$$

EXAMPLE 5   **Adding polynomials using the column method**

Add: $10x^3 + 8x^2 - 7x - 3$ and $9 - 4x + x^2 - 5x^3$

SOLUTION   We write $9 - 4x + x^2 - 5x^3$ in descending order: $-5x^3 + x^2 - 4x + 9$. We then place like terms in a column and add.

$$
\begin{array}{r}
10x^3 + 8x^2 - 7x - 3 \\
(+)\ \underline{-5x^3 + x^2 - 4x + 9} \\
5x^3 + 9x^2 - 11x + 6
\end{array}
$$

PROBLEM 5

Add the polynomials:
$(9z^4 - 6z^3 - z^2 + 12z - 8)$ and
$(5 + z - 7z^2 + 4z^3 - 8z^4)$

EXAMPLE 6   **Adding polynomials with missing terms**

Add: $P(x) = 2x^3 - 1.5x + 6.1$ and $H(x) = -x^2 + 4x - 3.8$

SOLUTION   As before, we place like terms in a column, leaving space for any missing terms, and then add as follows:

$$
\begin{array}{r}
2x^3 \quad\quad - 1.5x + 6.1 \\
(+)\ \underline{\quad\quad -x^2 + 4.0x - 3.8} \\
2x^3 - x^2 + 2.5x + 2.3
\end{array}
$$

PROBLEM 6

Add the polynomials:
$P(t) = -1.2t^3 - 4.5t^2 + 3.6t - 15$ and
$Q(t) = 9.18t^2 + 1.4t - 7.5$

Teaching Tip

Decimals need to be lined up as well as terms placed in columns with similar terms. Remind students that $4 = 4.0$.

To subtract polynomials, we first recall the following.

> **Subtracting Polynomials**
>
> $$a - (b + c) = a - b - c \quad\quad \text{By the distributive property}$$

For example, the difference between the revenue $R(p) = 60 - 0.3p^2$ and the cost $C(p) = 4000 - 20p$ is

$$(60p - 0.3p^2) - (4000 - 20p) = 60p - 0.3p^2 - 4000 + 20p$$
$$= -0.3p^2 + 80p - 4000$$

Similarly,

$$(3x^2 + 4x - 5) - (5x^2 + 2x + 3) = 3x^2 + 4x - 5 - 5x^2 - 2x - 3$$
$$= -2x^2 + 2x - 8$$

Answers

5. $z^4 - 2z^3 - 8z^2 + 13z - 3$
6. $-1.2t^3 + 4.68t^2 + 5t - 22.5$

To subtract $(5x^2 + 2x + 3)$ from $(3x^2 + 4x - 5)$, we changed the sign of each term in $(5x^2 + 2x + 3)$ and then added. This procedure can also be done in columns:

$$3x^2 + 4x - 5$$
$$(-)\ \underline{(5x^2 + 2x + 3)} \qquad \text{is written} \qquad (+)\ \underline{-5x^2 - 2x - 3}$$

$$3x^2 + 4x - 5$$
$$-2x^2 + 2x - 8$$

EXAMPLE 7   Subtracting polynomials

Subtract: $(0.4y^2 - 2y + 7)$ from $(-0.9y^2 + 3y + 13)$

SOLUTION

NOTE

Subtracting a *from* b means to find $b - a$. Thus we find $(-0.9y^2 + 3y + 13) - (0.4y^2 - 2y + 7)$ in column form.

$$-0.9y^2 + 3y + 13 \qquad\qquad -0.9y^2 + 3y + 13$$
$$(-)\ \underline{(0.4y^2 - 2y + \ \ 7)} \quad \text{is written} \quad (+)\ \underline{-0.4y^2 + 2y - \ \ 7}$$
$$-1.3y^2 + 5y + \ \ 6$$

The same result can be obtained by combining like terms, using the usual rules of signs. Thus

$$(-0.9y^2 + 3y + 13) - (0.4y^2 - 2y + 7)$$
$$= -0.9y^2 + 3y + 13 - 0.4y^2 + 2y - 7$$
$$= -1.3y^2 + 5y + 6$$

PROBLEM 7

Subtract: $(8y^3 - 6y^2 + 5y - 3)$ from $(4y^3 + 4y^2 - 3)$

E   Applications

The profit P derived from selling x units of a product is related to the cost C and the revenue R, and is given by $P = R - C$. (Thus the profit P is the revenue R minus the cost C.)

EXAMPLE 8   Solving a word problem with polynomials

A company produces x DVDs at a weekly cost $C = 2x + 1000$ (dollars). What is their weekly profit P if their revenue R is given by $R = 50x - 0.1x^2$ and they produce and sell 300 cassettes a week?

SOLUTION   We need to find

$$P = \qquad R \qquad - \qquad C$$
$$= (50x - 0.1x^2) - (2x + 1000)$$
$$= 50x - 0.1x^2 - 2x - 1000$$
$$= -0.1x^2 + 48x - 1000$$

If they produced and sold 300 DVDs, $x = 300$ and

$$P(300) = -0.1(300)^2 + 48(300) - 1000$$
$$= -9000 + 14{,}400 - 1000$$
$$= 4400 \text{ (dollars)}$$

Their profit P when they sell 300 DVDs is $4400.

PROBLEM 8

Find the profit if cost C and revenue R are as follows and they produce and sell 200 items, x:

$$C = 5x + 400 \text{ (dollars) and}$$
$$R = 50x - 0.2x^2 \text{ (dollars)}$$

Teaching Tip

Have students solve Example 8 by finding $R(300)$ and $C(300)$ and then subtracting the results. Have them compare answers. (The results are the same.) Ask them to comment on which is the shorter method and why.

Answers

7. $-4y^3 + 10y^2 - 5y$
8. $600

Calculate It Exercises

Finding the Intersection of Two Polynomial Functions and Finding the Maximum Values

According to data, the annual number of robberies (per 100,000 population) can be approximated by

$$R(t) = 194t^2 - 1430t + 9800$$

and the number of aggravated assaults can be approximated by

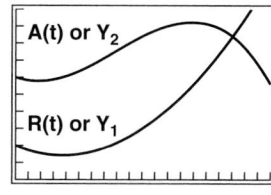

$$A(t) = -20t^3 + 470t^2 - 1500t + 30,000$$

where t is the number of years after 1996. Let's see how graphing these two polynomials can help answer questions about how to find the year in which one of these will reach its maximum or maybe the year in which they will both have the same number of occurrences. Using the $Y=$ key, enter the $R(t)$ polynomial as Y_1 and the $A(t)$ polynomial as Y_2.

We need to find a suitable window. Since t represents the number of years after 1996, set the minimum for t (or x if you prefer) at 0 and the maximum at 20 with a scale of 1. What do we need for the y-values representing the number of robberies and assaults? First, $R(0) = 9800$ and $A(0) = 30,000$ so we let Ymin = 0, Ymax = 50,000 and Yscl = 5000. We now have a [0, 20] by [0, 50000] window. Press $Y=$ and enter $Y_1 = 194t^2 - 1430t + 9800$ and $Y_2 = -20t^3 + 470t^2 - 1500t + 30,000$. (Remember, you can use x for the variable instead of t.) Press $GRAPH$.

You can now answer questions not evident when you only have the polynomials defined algebraically. For example, it seems that the number of robberies is increasing but the number of aggravated assaults peaks at a point and then begins to decrease. Can you find in what year the number of aggravated assaults $A(t)$ peaks? You can use the $TRACE$ key and decide, or better yet, you can let a TI-83 Plus decide for you. Press $2nd$

$TRACE$ and 4 for maximum. You will need to press ◄ or ► to get to the graph of $A(t)$ or Y_2. Now, use the ◄ key to trace around the curve $A(t)$ to a point to the *left* of the maximum and press $ENTER$. Use the ► key to trace along the curve to a point to the *right* of the maximum and press $ENTER$. When the calculator asks for a "Guess," press $ENTER$. The calculator gives the x- and y-values of the maximum point. Can you find when the number of aggravated assaults and robberies will be the same? Use the $TRACE$ key to find out or use the intersection feature on a TI-83 Plus by pressing $2nd$ $TRACE$ 5 $ENTER$ $ENTER$ $ENTER$.

Some calculators (TI-83 Plus) can even evaluate polynomials. If you have graphed $R(t)$ and $A(t)$, press $2nd$ $TRACE$ 1, which prompts you for an x-value by showing "X=" on your screen. Enter 16 and press $ENTER$. The calculator shows $y = 36,584$, so $R(16) = 36,584$. At the top of the screen, the Y_1 equation appears. This means that you have found the value of Y_1. To find the value of Y_2 at 16, press the ▼ key. The y-value is 44,400; thus $A(16) = 44,400$.

Refer to the graph.

1. During what time is the number of robberies less than the number of aggravated assaults?

2. In what year is the number of robberies the same as the number of aggravated assaults? (Answer to the nearest year.)

3. How many robberies and aggravated assaults occur at the point at which the graphs intersect?

4. In what year is the number of aggravated assaults a maximum? (Answer to the nearest year.)

5. What is the degree of $R(t)$?

6. What is the degree of $A(t)$?

Exercises 5.1

A **B** In Problems 1–10, classify the given polynomial as a monomial, binomial, or trinomial and give its degree.

1. xyz^2 Monomial; 4

2. u^2vw^3 Monomial; 6

3. $x^2 + yz^2$ Binomial; 3

4. $x^2y + z^3 - x^6$ Trinomial; 6

5. $x + y^2 + z^3$ Trinomial; 3

6. $xy + y^3$ Binomial; 3

7. $x^2yz - xy^3 - u^2v^3$ Trinomial; 5

8. 8 Monomial; 0

9. 0 Zero polynomial; no degree

10. $3xyz - uv^2 + v^7$ Trinomial; 7

B In Problems 11–16, write the polynomials in descending order.

11. $3x^2 + 5x - x^4 + 7$
$-x^4 + 3x^2 + 5x + 7$

12. $-6 + x - x^5 + x^3$
$-x^5 + x^3 + x - 6$

13. $9x - 8x^2 + 20$
$-8x^2 + 9x + 20$

14. $114x^2 + 12x^4 - 5 + x^6$
$x^6 + 12x^4 + 114x^2 - 5$

15. $2 + x - 5x^2$
$-5x^2 + x + 2$

16. $-11x + x^3 - 7 + x^4$
$x^4 + x^3 - 11x - 7$

C In Problems 17–20, evaluate the polynomial for the specified values of the variables.

17. z^3 for $z = -2$ -8

18. $(xy)^3$ for $x = 2, y = -1$ -8

19. $(x - 2y + z)^2$ for $x = 2, y = -1, z = -2$ 4

20. $(x - y)(x - z)$ for $x = 2, y = -1, z = -2$ 12

In Problems 21–30, find the specified function values.

21. If $P(x) = 4x^2 + 4x - 1$, find $P(-1)$. -1

22. If $P(x) = -3x^2 + 3x + 2$, find $P(0)$. 2

23. If $Q(y) = y^2 - 7y - 2$, find $Q(-2)$. 16

24. If $R(t) = t^2 - 2t + 7$, find $R(-2)$. 15

25. If $S(u) = -16u^2 + 120$, find $S(4)$. -136

26. If $V(t) = -16t^2 + 80t$, find $V(3)$. 96

27. $P(x) = 2x^2 + 3x$ and $Q(y) = -3y^2 - 7y + 1$.
  a. Find $P(0)$. 0
  b. Find $Q(1)$. -9
  c. Find $P(0) + Q(1)$. -9

28. $P(x) = 3x^2 - 2x + 5$ and $Q(y) = -2y^2 + 3y - 1$.
  a. Find $P(-2)$. 21
  b. Find $Q(0)$. -1
  c. Find $P(-2) + Q(0)$. 20

29. $P(x) = x^2 - 2x + 5$ and $Q(y) = -2y^2 + 5y - 1$.
  a. Find $P(-1)$. 8
  b. Find $Q(1)$. 2
  c. Find $P(-1) - Q(1)$. 6

30. $P(x) = x^2 - 3x + 5$ and $Q(y) = -2y^2 + 5y - 1$.
  a. Find $P(-2)$. 15
  b. Find $Q(2)$. 1
  c. Find $P(-2) - Q(2)$. 14

D In Problems 31–55, perform the indicated operations.

31. $(x^2 + 4x - 8) + (5x^2 - 4x + 3)$ $6x^2 - 5$

32. $(3x^2 + 2x + 1) + (8x^2 - 7x + 5)$ $11x^2 - 5x + 6$

33.
$$\begin{array}{r} 5x^2 + 3x + 4 \\ (+)\ -4x^2 - 5x - 8 \\ \hline x^2 - 2x - 4 \end{array}$$

34.
$$\begin{array}{r} -5x^2 + 4x - 3 \\ (+)\ \ \ 6x^2 + 4x - 7 \\ \hline x^2 + 8x - 10 \end{array}$$

35. $(4x^2 + 7x - 5) - (3x + x^2 + 4)$ $3x^2 + 4x - 9$

36. $(8x^2 - 6x + 3) - (4x + 2x^2 - 6)$ $6x^2 - 10x + 9$

37.
$$\begin{array}{r} -3y^2 + 6y - 5 \\ (-)\ \ \ 8y^2 + 7y - 2 \\ \hline -11y^2 - \ y - 3 \end{array}$$

38.
$$\begin{array}{r} -4y^2 + 5y - 2 \\ (-)\ \ \ 5y^2 - 3y + 6 \\ \hline -9y^2 + 8y - 8 \end{array}$$

39. $(x^3 - 6x^2 + 4x - 2) + (3x^3 - 6x^2 + 5x - 4)$
$4x^3 - 12x^2 + 9x - 6$

40. $(-6x^3 - 3x + 2x^2 + 2) + (2x^3 - 6x^2 + 8x - 4)$
$-4x^3 - 4x^2 + 5x - 2$

41. Add $(-8y^3 + 5y + 7y^2 - 5)$ and $(8y^3 + 7y - 6)$.
$7y^2 + 12y - 11$

42. Add $(5y^3 + 3y - 8)$ and $(-9y^3 - 6y + 2y^2 + 3)$.
$-4y^3 + 2y^2 - 3y - 5$

43. Subtract $(3v^3 + v - v^2 + 2)$ from $(6v^3 - 3v^2 + 2v - 5)$.
$3v^3 - 2v^2 + v - 7$

44. Subtract $(5v^3 + 3v^2 - 6v + 4)$ from $(3v^3 - 7v^2 + 3v - 1)$.
$-2v^3 - 10v^2 + 9v - 5$

45. $(4u^3 - 5u^2 - u + 3) - (2u + 9u^3 - 7)$
$-5u^3 - 5u^2 - 3u + 10$

46. $(x^3 + y^3 - 8xy + 3) + (10xy - y^3 + 2x^3 - 6)$
$3x^3 + 2xy - 3$

47. $(x^3 + y^3 - 6xy + 7) + (3x^3 - y^3 + 8xy - 8)$
$4x^3 + 2xy - 1$

48. $(2x^3 - y^3 + 3xy - 5) - (x^3 + y^3 - 3xy + 9)$
$x^3 + 6xy - 2y^3 - 14$

49. $(x^3 - y^3 + 5xy - 2) - (x^2 - y^3 + 5xy + 2)$
$x^3 - x^2 - 4$

50. $(4x^2 + y^2 - 3x^2y^2) - (x^3 + 3y^2 - 3x^2y^2)$
$-x^3 + 4x^2 - 2y^2$

51. $(a + a^2) + (9a - 4a^2) + (a^2 - 5a)$
$-2a^2 + 5a$

52. Subtract $(x + x^2)$ from $(2x - 5x^2) + (7x - x^2)$.
$8x - 7x^2$

53. $2y + (x + 3y) - (x + y)$ $4y$

54. $8y - (y + 3x) + 7y$ $14y - 3x$

55. $(3x^2 + y) - (x^2 - 3y) + (3y + x^2)$ $3x^2 + 7y$

In Problems 56–60, let $P(x) = x^2 - 2x + 3$ and $Q(x) = 2x^2 + 3x - 1$ and find the solution.

56. $P(x) - Q(x)$ $-x^2 - 5x + 4$

57. $P(0) + Q(0)$ 2

58. $P(1) - Q(-1)$ 4

59. $P(x) - P(x)$ 0

60. $[P(x) + Q(x)] + P(x)$ $4x^2 - x + 5$

In Problems 61–70, justify each of the equalities by using one of the three properties given in the text.

61. $3x^2 + 9x = 9x + 3x^2$
Commutative property of addition

62. $(-y^3) + 7y = 7y + (-y)^3$
Commutative property of addition

63. $(8x + 9)4 = 32x + 36$
Distributive property

64. $(7 - 2x)(-3) = -21 + 6x$
Distributive property

65. $x^2 + (x + 5) = (x^2 + x) + 5$
Associative property of addition

66. $(y^3 + y^2) + 4 = y^3 + (y^2 + 4)$
Associative property of addition

67. $x^7 + (3 + x) = x^7 + (x + 3)$
Commutative property of addition

68. $(y + 7) + y^3 = y^3 + (y + 7)$
Commutative property of addition

69. $3(x^2 + 5) = 3x^2 + 15$
Distributive property

70. $8(x^3 - 5x) = 8x^3 - 40x$
Distributive property

APPLICATIONS

71. The height $H(t)$ (in feet) of an object t seconds (sec) after being thrown straight up with an initial velocity of 64 ft/sec is given by

$$H(t) = -16t^2 + 64t$$

find the height of the object after

a. 1 sec.   48 ft

b. 2 sec.   64 ft

72. The total dollar cost $C(x)$ of manufacturing x units of a certain product each week is given by

$$C(x) = 10x + 400$$

find the cost of manufacturing

a. 500 units. $5400

b. 1000 units. $10,400

73. The dollar revenue obtained by selling x units of a certain product each week is given by

$$R(x) = 100x - 0.03x^2$$

find the revenue when

a. 500 units are sold. $42,500

b. 1000 units are sold. $70,000

74. In business, you can calculate your gross profit by subtracting the cost from the revenue. If the cost C and revenue R (in dollars) when x units are sold are given by

$$C(x) = 10x + 400 \quad \text{and} \quad R(x) = 100x - 0.03x^2$$

find the gross profit when

a. 500 units are sold. $37,100

b. 1000 units are sold. $59,600

75. If a \$100,000 computer depreciates 10% each year, its value $V(t)$ after t years is given by

$$V(t) = 100,000 - 0.10t(100,000)$$
$$= 100,000(1 - 0.10t)$$

find the value of the computer after

a. 5 yr. $50,000

b. 10 yr. $0 (no value)

76. The total number N of units of output per day when the number of employees m is given by $N = 20m - \frac{1}{2}m^2$. Find how many units are produced per day in a company with

a. 10 employees. 150

b. 20 employees. 200

In Problems 77–81, $P = R - C$, where P is the profit, R is the revenue, and C is the cost.

77. The cost C in dollars of producing x items is given by $C = 100 + 0.3x$. If the revenue is given by $R = 1.50x$, find the profit P when 100 items are produced and sold. $20

78. The cost C in dollars of producing x pairs of jogging shoes is given by $C = 2000 + 60x$. If the revenue is given by $R = 90x$, find the profit when 200 pairs are produced and sold. $4000

79. The cost C in dollars of producing x pairs of jeans is given by $C = 1500 + 20x$. If the revenue is given by

$$R = 50x - \frac{x^2}{20}$$

find the profit when 100 pairs of jeans are produced and sold. $1000

80. The cost C in dollars of producing x pairs of sunglasses is given by $C = 30,000 + 60x$. Find the profit when 300 pairs of sunglasses are manufactured and sold if the revenue is given by

$$R = 200x - \frac{x^2}{30} \quad \$9000$$

81. The cost C in dollars of producing x pairs of shoes is given by $C = 100,000 + 30x$. Find the profit when 500 pairs of shoes are manufactured and sold if the revenue is given by

$$R = 300x - \frac{x^2}{50} \quad \$30,000$$

SKILL CHECKER

Try these "Skill Checker" exercises so you'll be ready for the next section.

Multiply:

82. $6x^2y \cdot 2x$ $12x^3y$

83. $6x^2y \cdot 3xy$ $18x^3y^2$

84. $-2x^3y \cdot (-2y^2)$ $4x^3y^3$

85. $-2x^3y \cdot (-7xy)$ $14x^4y^2$

86. $(2x)^2$ $4x^2$

USING YOUR KNOWLEDGE

Sums of Areas

The addition of polynomials can be used to find the sum of the areas of several rectangles. To find the total area of the rectangles, add the individual areas as shown.

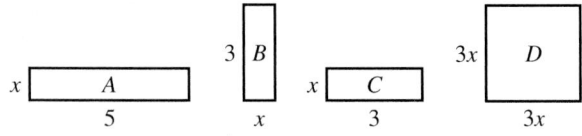

The total area is

$$\underbrace{\text{Area of A}}_{} + \underbrace{\text{area of B}}_{} + \underbrace{\text{area of C}}_{} + \underbrace{\text{area of D}}_{}$$
$$\underbrace{5x \quad + \quad 3x \quad + \quad 3x}_{11x} \quad + \quad \underbrace{(3x)^2}_{+ \quad 9x^2}$$

or, in descending order, $9x^2 + 11x$.

Find the sum of the areas of the rectangles.

87.

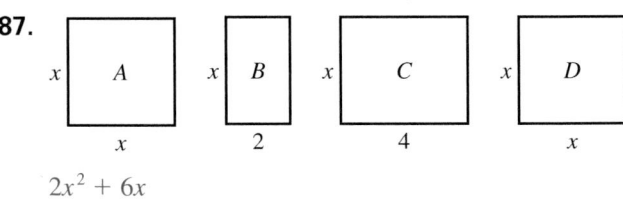

$2x^2 + 6x$

88.

$x^2 + 15x$

89.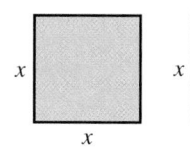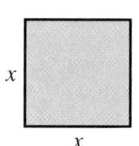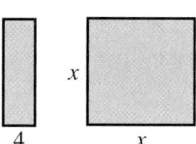

$3x^2 + 4x$

90.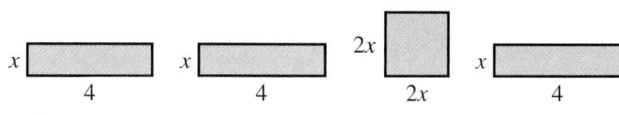

$4x^2 + 12x$

91.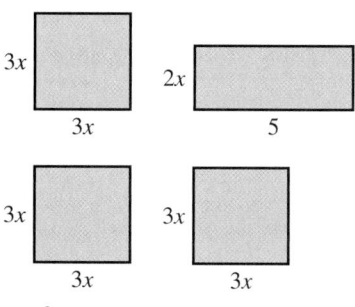

$27x^2 + 10x$

WRITE ON

92. Write your own definition of a polynomial.
Answers may vary.

93. Describe the procedure you use to find the degree of a polynomial. Answers may vary.

94. If $P(x)$ and $Q(x)$ are polynomials and you subtract $Q(x)$ from $P(x)$, what happens to the signs of all the terms of $Q(x)$? Answers may vary.

MASTERY TEST

If you know how to do these problems, you have learned your lesson!

95. A company produces x units of a product at a cost $C = 3x + 100$ (dollars).

  a. If their revenue R is given by $R = 50x - 0.2x^2$ and their profit is $P = R - C$, find P.
$P = -0.2x^2 + 47x - 100$

  b. What is their profit when they produce and sell 100 units? $2600

96. The height of an object t seconds after being thrown straight up with an initial velocity of 96 ft/sec is given by

$$H(t) = -16t^2 + 96t$$

What is the height of the object after 3 sec? 144 ft

Find the degree of:

97. $-2x^2 + 9 - 7x^4$ 4

98. $-3xy^2z + x^2 - x^3$ 4

99. Classify as monomials, binomials, trinomials, or polynomials:

  a. $x^2 - x$ Binomial

  b. $t^5 - 4t^2 + 3t - 2t^3 + 5$ Polynomial

  c. $x^2y^3z^5$ Monomial

  d. $\frac{2}{3}x^2 - 8 - 3x$ Trinomial

If $P(x) = x^2 - 3x + 2$ and $Q(x) = x^2 + 2x - 5$, find:

100. $P(2)$ and $Q(2)$ $P(2) = 0$; $Q(2) = 3$

101. $P(2) - Q(2)$ -3

102. $P(x) + Q(x)$ $2x^2 - x - 3$

103. $P(x) - Q(x)$ $-5x + 7$

104. Subtract $(8x^3 - 6x^2 + 5x - 3)$ from $(4x^2 + 4x + 3)$.
$-8x^3 + 10x^2 - x + 6$

5.2 MULTIPLICATION OF POLYNOMIALS

To Succeed, Review How To ...

1. Use the distributive property to simplify expressions (pp. 51–53).

2. Use the properties of exponents (pp. 32–40).

3. Combine like terms (pp. 53–55).

Objectives

A Multiply a monomial by a polynomial.

B Multiply two polynomials.

C Use the FOIL method to multiply two binomials.

D Square a binomial sum or difference.

E Find the product of the sum and the difference of two terms.

F Use the ideas discussed to solve applications.

GETTING STARTED

Building Bridges with Algebra

How much does a bridge beam bend (deflect) when a car or truck goes over the bridge? There's a formula that can tell us. For a certain beam of length L at a distance x from one end, the deflection is given by

$$(x - L)(x - 2L)$$

To multiply these two binomials, we first learn how to do several types of related multiplications. We show you how to multiply $(x - L)(x - 2L)$ in Example 3 of this section.

A Multiplying Monomials by Polynomials

When multiplying polynomials such as $6x^2y$ and $(2x + 3xy)$, we use the commutative, associative, and distributive properties of multiplication for polynomials. These properties are generalizations of the properties discussed in Chapter 1. We state them here for polynomials.

Multiplication Properties for Polynomials	If P, Q, and R are polynomials,	
	$P \cdot Q = Q \cdot P$	Commutative property of multiplication
	$P \cdot (Q \cdot R) = (P \cdot Q) \cdot R$	Associative property of multiplication
	$P(Q + R) = PQ + PR$	Distributive property
	$(Q + R)P = QP + RP$	

Calculate It Using Graphs to Check Polynomial Multiplication

A calculator is the perfect tool to verify products of polynomials in one variable. The procedure is as follows: Graph the original problem as Y_1 and the answer (the product) as Y_2. If the two graphs are identical, you have done the multiplication correctly.

Before you do this, however, find out how exponents are entered in your calculator. With a TI-83 Plus, press the exponent key  and then enter the exponent. (Some calculators also have $\boxed{x^2}$ and $\boxed{x^3}$ keys.)

To multiply $6x^2y(2x + 3xy)$, we proceed as follows:

$$6x^2y(2x + 3xy) = 6x^2y(2x) + 6x^2y(3xy) \qquad \text{Use the distributive property.}$$
$$= 12x^3y + 18x^3y^2 \qquad \text{Multiply and rearrange using the commutative and associative properties.}$$

We multiplied the coefficients but added the exponents. We can multiply $4x^2(3x^3 + 7x^2 - 4x + 8)$ in a similar way. Thus

$$4x^2(3x^3 + 7x^2 - 4x + 8) = 4x^2(3x^3) + 4x^2(7x^2) + 4x^2(-4x) + 4x^2(8)$$
$$= 12x^5 + 28x^4 - 16x^3 + 32x^2$$

EXAMPLE 1 Multiplying a monomial by a polynomial	**PROBLEM 1**

Multiply:

a. $5x^2(3x^3 + 3x^2 - 2x - 3)$ **b.** $-2x^3y(x^2 + 7xy - 2y^2)$

SOLUTION

a. $5x^2(3x^3 + 3x^2 - 2x - 3) = 15x^5 + 15x^4 - 10x^3 - 15x^2$

b. $-2x^3y(x^2 + 7xy - 2y^2) = -2x^5y - 14x^4y^2 + 4x^3y^3$

PROBLEM 1

Multiply the polynomials:

a. $4x^3(5x^4 - 7x^3 + x^2 + 2x - 1)$

b. $-3xy^2(9x^2 - 6xy + 10y^2)$

Calculate It Checking Multiplication of Polynomials

To verify the results in Example 1(a), use a standard window (**ZOOM** 6) and graph

$$Y_1 = 5x^2(3x^3 + 3x^2 - 2x - 3)$$
$$Y_2 = 15x^5 + 15x^4 - 10x^3 - 15x^2$$

The result is shown in Window 1. What happens when you encounter a wrong answer? For example, suppose you obtained $(x + 2)^2 = x^2 + 4$. If you let $Y_1 = (x + 2)^2$ and $Y_2 = x^2 + 4$, you get the two graphs in Window 2. Since the graphs are different, it indicates that the multiplication is incorrect.

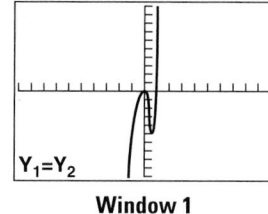

Window 1

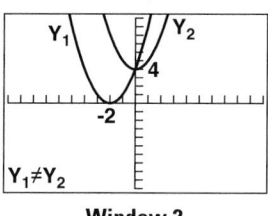

Window 2

B Multiplying Two Polynomials

To multiply $(3x + 2)(x + 3)$, we use the distributive property

$$a(b + c) = ab + ac$$

and multiply $(3x + 2)$ by each term of $(x + 3)$ to obtain

$$\underbrace{(3x + 2)}_{a}\ \underbrace{(x + 3)}_{(b + c)} = \underbrace{(3x + 2)}_{a}\underbrace{(x)}_{b} + \underbrace{(3x + 2)}_{a}\underbrace{(3)}_{c}$$
$$= 3x^2 + 2x + 9x + 6$$
$$= 3x^2 + 11x + 6$$

This multiplication can also be done by arranging the work as in ordinary multiplication and placing the resulting like terms of the products in the same column. The procedure looks like this:

$$
\begin{array}{r}
x + 3 \\
3x + 2 \\
\hline
2x + 6 \\
3x^2 + 9x \\
\hline
3x^2 + 11x + 6
\end{array}
$$

Multiply by 2: $2(x + 3) = 2x + 6$.
Multiply by $3x$: $3x(x + 3) = 3x^2 + 9x$.
Add like terms: $2x + 9x = 11x$.

Now let's use this same technique to multiply two polynomials like $(x + 5)$ and $(x^2 + x - 2)$. We call this technique the **vertical scheme**.

PROCEDURE

Vertical Scheme for Multiplying Polynomials

Step 1	Step 2	Step 3

Step 1

$$x^2 + x - 2$$
$$\underline{x + 5}$$
$$5x^2 + 5x - 10$$

Multiply.

$$5(x^2 + x - 2)$$
$$= 5x^2 + 5x - 10$$

Step 2

$$x^2 + x - 2$$
$$\underline{x + 5}$$
$$5x^2 + 5x - 10$$
$$\underline{x^3 + x^2 - 2x}$$

Multiply.

$$x(x^2 + x - 2)$$
$$= x^3 + x^2 - 2x$$

Step 3

$$x^2 + x - 2$$
$$\underline{x + 5}$$
$$5x^2 + 5x - 10$$
$$\underline{x^3 + x^2 - 2x}$$
$$x^3 + 6x^2 + 3x - 10$$

Add like terms.

We could have obtained the same result by using the distributive property and multiplying $(x + 5)$ by x^2, $(x + 5)$ by x, and $(x + 5)$ by -2. The result would look like this:

$$(x + 5)(x^2 + x - 2) = (x + 5)(x^2) + (x + 5)x + (x + 5)(-2)$$
$$= x^3 + 5x^2 + x^2 + 5x - 2x - 10$$
$$= x^3 + (5x^2 + x^2) + (5x - 2x) - 10$$
$$= x^3 + 6x^2 + 3x - 10$$

EXAMPLE 2   **Using the vertical scheme to multiply polynomials**

Multiply:

a. $(x - 3)(x^2 - 2x - 4)$ **b.** $(x + 2)(x^2 - 2x + 4)$

SOLUTION   We use the vertical scheme.

a.
$$x^2 - 2x - 4$$
$$\underline{x - 3}$$
$$-3x^2 + 6x + 12 \quad \text{Multiply by } -3.$$
$$\underline{x^3 - 2x^2 - 4x} \quad \text{Multiply by } x.$$
$$x^3 - 5x^2 + 2x + 12 \quad \text{Add like terms.}$$

b.
$$x^2 - 2x + 4$$
$$\underline{x + 2}$$
$$2x^2 - 4x + 8 \quad \text{Multiply by } 2.$$
$$\underline{x^3 - 2x^2 + 4x} \quad \text{Multiply by } x.$$
$$x^3 \qquad\quad + 8 \quad \text{Add like terms.}$$

PROBLEM 2

Multiply the polynomials:

a. $(x + 4)(3x^2 + x - 5)$

b. $(3x - 1)(9x^2 - 6x + 2)$

The result of multiplying $(x - 3)(x^2 - 2x - 4)$ is $x^3 - 5x^2 + 2x + 12$. Since $(b - c)a = ba - ca$, you can also do this problem by multiplying $x(x^2 - 2x - 4)$ first and then multiplying $-3(x^2 - 2x - 4)$ to obtain the following result:

$$\overbrace{(x - 3)}^{(b-c)}\overbrace{(x^2 - 2x - 4)}^{a} = \overbrace{x}^{b}\overbrace{(x^2 - 2x - 4)}^{a} \overbrace{- 3}^{-c} \overbrace{(x^2 - 2x - 4)}^{a}$$
$$= x^3 - 2x^2 - 4x - 3x^2 + 6x + 12$$
$$= x^3 + (-2x^2 - 3x^2) + (-4x + 6x) + 12$$
$$= x^3 - 5x^2 + 2x + 12$$

The same result is obtained in both cases. Here is the rule.

Answers

2. a. $3x^3 + 13x^2 - x - 20$
b. $27x^3 - 27x^2 + 12x - 2$

Rule to Multiply Any Two Polynomials	To multiply two polynomials, multiply each term of one by every term of the other and combine like terms.

C Multiplying Two Binomials

Let's use the rule we just stated and the vertical scheme to multiply $(x + 4)$ and $(x - 7)$.

$$
\begin{array}{r}
x + 4 \\
x - 7 \\
\hline
-7x \qquad - 28 \\
x^2 \qquad + 4x \\
\hline
x^2 - 7x + 4x - 28
\end{array}
$$

Multiply by -7.
Multiply by x.
Add.

If you look at the four terms in the last line before adding like terms, you can see the procedure for multiplying two binomials.

The first term, x^2, is the product of the first terms.

$$(x + 4)(x - 7) \qquad x^2 \qquad \text{First terms} \qquad \textbf{F}$$

The second term, $-7x$, is the product of the outside terms.

$$(x + 4)(x - 7) \qquad -7x \qquad \text{Outside terms} \qquad \textbf{O}$$

The third term, $+4x$, is the product of the inside terms.

$$(x + 4)(x - 7) \qquad 4x \qquad \text{Inside terms} \qquad \textbf{I}$$

The last term, -28, is the product of the last two terms.

$$(x + 4)(x - 7) \qquad -28 \qquad \text{Last terms} \qquad \textbf{L}$$

To multiply $(x + 4)(x - 7)$, we write

$$
\begin{array}{ccccc}
 & \text{First} & \text{Outside} & \text{Inside} & \text{Last} \\
 & \text{F} & \text{O} & \text{I} & \text{L} \\
(x + 4)(x - 7) = & x^2 & - 7x & + 4x & - 28 \\
 & = x^2 - 3x - 28
\end{array}
$$

Here is the general rule for multiplying two binomials using the FOIL method. This is the first of several special products.

Web It

For examples of how to multiple two binomials four different ways (FOIL, vertical, tiles, and grid), try link 5-2-1 at mhhe.com/bello.

PROCEDURE

Using FOIL to Multiply the Two Binomials $(x + a)(x + b)$

To find the product of two binomials, multiply the terms in this order:

$$
\begin{array}{ccccc}
 & \text{First} & \text{Outside} & \text{Inside} & \text{Last} \\
 & \text{F} & \text{O} & \text{I} & \text{L} \\
(x + a)(x + b) = & x^2 & + bx & + ax & + ab \qquad \text{Special product (1)} \\
 & = x^2 + (b + a)x + ab
\end{array}
$$

Let's do one more example, step by step, to give you more practice.

F $(x + 7)(x - 4)$ x^2

O $(x + 7)(x - 4)$ $x^2 - 4x$

I $(x + 7)(x - 4)$ $x^2 - 4x + 7x$

L $(x + 7)(x - 4) = x^2 - 4x + 7x - 28$

Thus

$$(x + 7)(x - 4) = x^2 + 3x - 28$$

EXAMPLE 3	Using the FOIL method

Multiply:

a. $(5x + 2y)(2x + 3y)$ **b.** $(3x - y)(4x - 3y)$ **c.** $(x - L)(x - 2L)$

SOLUTION

                F   O   I   L

a. $(5x + 2y)(2x + 3y) = (5x)(2x) + (5x)(3y) + (2y)(2x) + (2y)(3y)$

$\qquad\qquad\qquad\qquad = 10x^2 + 15xy + 4xy + 6y^2$

$\qquad\qquad\qquad\qquad = 10x^2 + 19xy + 6y^2$

                F   O   I   L

b. $(3x - y)(4x - 3y) = (3x)(4x) + (3x)(-3y) + (-y)(4x) + (-y)(-3y)$

$\qquad\qquad\qquad\qquad = 12x^2 - 9xy - 4xy + 3y^2$

$\qquad\qquad\qquad\qquad = 12x^2 - 13xy + 3y^2$

                F   O   I   L

c. $(x - L)(x - 2L) = x \cdot x + (x)(-2L) + (-L)(x) + (-L)(-2L)$

$\qquad\qquad\qquad\qquad = x^2 - 2Lx - Lx + 2L^2$

$\qquad\qquad\qquad\qquad = x^2 - 3Lx + 2L^2$

PROBLEM 3

Multiply:

a. $(x + 4)(7x - 5)$

b. $(3x - y)(9x + 2y)$

c. $(B - 3A)(B - 8A)$

Teaching Tip

Once students learn the vertical scheme for multiplying polynomials, they will hesitate to use these special products. Give them a sneak preview of factoring trinomials using the FOIL method to anticipate the binomials.

Example:

Factor $x^2 + 6x + 5$ into two binomials using the knowledge of "FOIL":

Last product is 5

(?   ?)(?   ?)

First product is x^2

D   Squaring Sums or Differences of Binomials

Teaching Tip

Before introducing the perfect square trinomial, refer students to the perfect square numbers 1, 4, 9, 16, 25, 36, 49, and so on and see if they can suggest the like factors that result in those products [1(1), 2(2), 3(3), 4(4), etc.]. If so, remind them that this is called squaring the number, hence the results are called perfect square numbers. When we square a binomial, the result is called a perfect square trinomial.

Answers

3. a. $7x^2 + 23x - 20$
b. $27x^2 - 3xy - 2y^2$
c. $B^2 - 11AB + 24A^2$

Now suppose we want to find $(x + 7)^2$. The exponent 2 means that we must multiply $(x + 7)(x + 7)$. Using FOIL, we write

$$(x + 7)(x + 7) = x^2 + 7x + 7x + 7 \cdot 7$$

$$= x^2 + 14x + 49$$

The result, $x^2 + 14x + 49$, is called a perfect square trinomial. Did you see how the middle term was calculated? It's the sum of $7x$ and $7x$, that is, $2 \cdot 7x$. Also, the last term is 7^2. In general, we have the following.

PROCEDURE

Square of a Binomial Sum $(x + a)^2$

                                  F   O   I   L

$$(x + a)^2 = (x + a)(x + a) = x^2 + ax + ax + a \cdot a$$

$$= x^2 + 2ax + a^2 \qquad \text{Special product (2)}$$

The same pattern applies to the difference of two binomials.

> ### Square of a Binomial Difference $(x - a)^2$
>
> $$\overset{\text{F} \quad\quad \text{O} \quad\quad \text{I} \quad\quad \text{L}}{(x - a)^2 = (x - a)(x - a) = x^2 - ax - ax + a \cdot a}$$
> $$= x^2 - 2ax + a^2 \qquad \text{Special product (3)}$$

To find the square of a binomial, add the square of the first term, twice the product of the two terms, and the square of the second term. The sign of the middle term is $+$ for binomial sums and $-$ for binomial differences. The sign of the last term is always $+$. You can use this arrow diagram to remember the steps.

EXAMPLE 4   Square of a binomial

Multiply:

a. $(2x + 3y)^2$

b. $(5x - 2y)^2$

c. $-3x(2x - 3y)^2$

d. $[(2x + 1) + y]^2$

SOLUTION

a. $(2x + 3y)^2 = (2x)^2 + 2 \cdot 3y \cdot 2x + (3y)^2$ or
$$= 4x^2 + 12xy + 9y^2$$

b. $(5x - 2y)^2 = (5x)^2 - 2 \cdot 2y \cdot 5x + (-2y)^2$ or
$$= 25x^2 - 20xy + 4y^2$$

c. $-3x(2x - 3y)^2 = -3x[(2x)^2 - 2 \cdot (3y)(2x) + (3y)^2]$
$$= -3x[4x^2 - 12xy + 9y^2]$$
$$= -12x^3 + 36x^2y - 27xy^2$$

d. $[(2x + 1) + y]^2 = (2x + 1)^2 + 2 \cdot y \cdot (2x + 1) + y^2$
$$= (4x^2 + 4x + 1) + 4xy + 2y + y^2$$
$$= 4x^2 + 4xy + y^2 + 4x + 2y + 1$$

PROBLEM 4

Multiply:

a. $(x + 4y)^2$

b. $(7x - 5y)^2$

c. $-6x(2x + 3)^2$

d. $[(x - 8) + y]^2$

Teaching Tip

Point out that the result of squaring a binomial is three terms, hence three steps. Use the following drawing:

$(x + a)^2 =$ _____ + _____ + _____
                   square the   multiply   square the
                   first term   and double   last term

Also, give a sneak preview of factoring a perfect square trinomial to see how this special product can help in factoring.

Example:

Factor $x^2 - 10x + 25$

E   Product of a Sum and a Difference

Suppose we multiply the sum of two terms by the difference of the same two terms; that is, suppose we want to find the product of

sum $\cdot$ difference

$(x + 7)(x - 7)$

Answers

4. a. $x^2 + 8xy + 16y^2$
b. $49x^2 - 70xy + 25y^2$
c. $-24x^3 - 72x^2 - 54x$
d.
$x^2 - 16x + 64 + 2xy - 16y + y^2$

Web It

For practice multiplying all types of polynomials, go to link 5-2-2 at mhhe.com/bello.

Using FOIL, we get

$$\overset{\text{F}\quad\text{O}\quad\text{I}\quad\text{L}}{(x + 7)(x - 7) = x^2 - 7x + 7x - 7^2}$$
$$= x^2 + 0x - 7^2$$
$$= x^2 - 49$$

Since multiplication is commutative, $(x - 7)(x + 7) = (x + 7)(x - 7)$; thus

$$(x - 7)(x + 7) = (x + 7)(x - 7) = x^2 - 49$$

Product of the Sum and Difference of the Same Two Monomials

$$(x - a)(x + a) = x^2 - a^2$$
$$(x + a)(x - a) = x^2 - a^2$$

Special product (4)

The FOIL method is still operative here.

EXAMPLE 5   Finding the product of the sum and difference of the same two monomials

Multiply:

a. $(x + 10)(x - 10)$

b. $(2x + y)(2x - y)$

c. $-(3x - 5y)(3x + 5y)$

d. $[2x - (3y + 1)][2x + (3y + 1)]$

SOLUTION

a. $(x + 10)(x - 10) = x^2 - 10^2$
$$= x^2 - 100$$

b. $(2x + y)(2x - y) = (2x)^2 - y^2$
$$= 4x^2 - y^2$$

c. $-(3x - 5y)(3x + 5y) = -[(3x)^2 - (5y)^2]$
$$= -9x^2 + 25y^2 \qquad \text{Since } -(a - b) = -a + b$$
$$= 25y^2 - 9x^2 \qquad \text{By the commutative property}$$

d. $[2x - (3y + 1)][2x + (3y + 1)] = (2x)^2 - (3y + 1)^2$
$$= 4x^2 - (9y^2 + 6y + 1)$$
$$= 4x^2 - 9y^2 - 6y - 1$$

PROBLEM 5

Multiply:

a. $(z - 9)(z + 9)$

b. $(y + 3x)(y - 3x)$

c. $-2(4x + 3y)(4x - 3y)$

d. $[6x + (2y - 5)][6x - (2y - 5)]$

F   Applications

The profit P is the revenue R minus the cost C; that is, $P = R - C$. Do you know how revenue is calculated? Suppose you have 10 skateboards and sell them for $50 each. Your revenue is $R = 10 \cdot 50 = \$500$.

$$R = \underbrace{\left(\begin{array}{c}\text{number of}\\\text{items sold}\end{array}\right)}_{x} \cdot \underbrace{\left(\begin{array}{c}\text{price of}\\\text{each item}\end{array}\right)}_{p}$$

or

$$R = xp$$

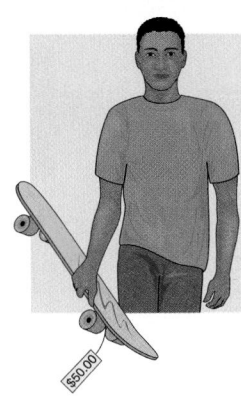

Answers

5. a. $z^2 - 81$ **b.** $y^2 - 9x^2$
c. $18y^2 - 32x^2$
d. $36x^2 - 4y^2 + 20y - 25$

<table>
<tr><td>

EXAMPLE 6   **Supply, demand, and skateboards**

The research department of a company determines that the number of skateboards that are selling (the demand) is given by $x = 1000 - 10p$, where p is the price of each skateboard.

a. Write a formula for the revenue R.

b. Find the revenue obtained from selling the skateboards for $50 each.

</td><td>

PROBLEM 6

Using the formula for Example 6(a) find the revenue obtained from selling the skateboards for $75 each.

</td></tr>
</table>

SOLUTION

a. The revenue is $R = \begin{pmatrix} \text{number of} \\ \text{items sold} \end{pmatrix} \cdot \begin{pmatrix} \text{price of} \\ \text{each item} \end{pmatrix}$

$$R = \qquad x \qquad \cdot p$$

$$R = (1000 - 10p)p \qquad \text{Substitute } x = 1000 - 10p.$$

$$= 1000p - 10p^2$$

b. When $p = 50$,

$$R = 1000(50) - 10(50)^2$$
$$= 50{,}000 - 10(2500)$$
$$= 50{,}000 - 25{,}000$$
$$= 25{,}000$$

The revenue is $25,000.

Calculate It   Equivalent Expressions

In Example 1(a), we discovered that:

$$5x^2(3x^3 + 3x^2 - 2x - 3) = 15x^5 + 15x^4 - 10x^3 - 15x^2$$

Thus $5x^2(3x^3 + 3x^2 - 2x - 3)$ and $15x^5 + 15x^4 - 10x^3 - 15x^2$ are equivalent expressions. This means that for any replacement of x, we obtain the same number for either $5x^2(3x^3 + 3x^2 - 2x - 3)$ or $15x^5 + 15x^4 - 10x^3 - 15x^2$. For example, when $x = 1$,

$$5x^2(3x^3 + 3x^2 - 2x - 3) = 5(1)^2[3(1)^3 + 3(1)^2 - 2(1) - 3]$$
$$= 5[3 + 3 - 2 - 3]$$
$$= 5[1]$$
$$= 5$$

and

$$15x^5 + 15x^4 - 10x^3 - 15x^2$$
$$= 15(1)^5 + 15(1)^4 - 10(1)^3 - 15(1)^2$$
$$= 15 + 15 - 10 - 15$$
$$= 5$$

If we view the polynomials as functions,

$$f(x) = 5x^2(3x^3 + 3x^2 - 2x - 3)$$

and

$$g(x) = 15x^5 + 15x^4 - 10x^3 - 15x^2$$

For any given x, $f(x)$ and $g(x)$ are identical and so their graphs are also identical. This can be verified by entering $f(x)$ as Y_1, $g(x)$ as Y_2. The graphs are the same. Most calculators can numerically verify these equivalencies. To show that $f(x) = g(x)$, store 1 into X by pressing [1] [STO▸] [ALPHA] [STO▸] [ENTER], and then enter $f(x)$ and [ENTER]. The answer will be 5 as shown in Window 1. Now, enter $g(x)$ and press [ENTER]. You get 5 again.

1. Verify the results of Example 2(a) graphically and numerically.

2. Verify the results of Example 5(a) graphically and numerically.

```
                          1
5X^2(3X^3+3X^2-2
X-3)
                          5
15X^5+15X^4-10X^
3-15X^2
                          5
```

Window 1

Answer

6. $18,750

Exercises 5.2

A In Problems 1–10, do the indicated multiplications.

1. $3x(4x - 2)$ $12x^2 - 6x$

2. $4x(x - 6)$ $4x^2 - 24x$

3. $-3x^2(x - 3)$ $-3x^3 + 9x^2$

4. $-5x^3(x^2 - 8)$ $-5x^5 + 40x^3$

5. $-8x(3x^2 - 2x + 1)$
$-24x^3 + 16x^2 - 8x$

6. $-4x^2(3x^2 - 5x - 1)$
$-12x^4 + 20x^3 + 4x^2$

7. $-3xy^2(6x^2 + 3y^2 - 7)$
$-18x^3y^2 - 9xy^4 + 21xy^2$

8. $-2x^2y^3(6xy^3 - 2x^2y + 9)$
$-12x^3y^6 + 4x^4y^4 - 18x^2y^3$

9. $2xy^3(3x^2y^3 - 5xy^2 + xy)$
$6x^3y^6 - 10x^2y^5 + 2x^2y^4$

10. $3x^4y(6x^3y^2 - 10x^2y + xy)$
$18x^7y^3 - 30x^6y^2 + 3x^5y^2$

B In Problems 11–20, do the indicated multiplications.

11. $(x + 3)(x^2 + x + 5)$ $x^3 + 4x^2 + 8x + 15$

12. $(x + 2)(x^2 + 5x + 6)$ $x^3 + 7x^2 + 16x + 12$

13. $(x + 4)(x^2 - x + 3)$ $x^3 + 3x^2 - x + 12$

14. $(x + 5)(x^2 - x + 2)$ $x^3 + 4x^2 - 3x + 10$

15. $x^2 - x - 2$
$\underline{\quad x + 3}$
$x^3 + 2x^2 - 5x - 6$

16. $x^2 - x - 3$
$\underline{\quad x + 4}$
$x^3 + 3x^2 - 7x - 12$

17. $(x - 2)(x^2 + 2x + 4)$ $x^3 - 8$

18. $(x - 3)(x^2 + x + 1)$ $x^3 - 2x^2 - 2x - 3$

19. $x^2 - x + 2$
$\underline{\quad x^2 - 1}$
$x^4 - x^3 + x^2 + x - 2$

20. $x^2 - 2x + 1$
$\underline{\quad x^2 - 2}$
$x^4 - 2x^3 - x^2 + 4x - 2$

C In Problems 21–38, do the indicated multiplications.

21. $(3x + 2)(3x + 1)$
$9x^2 + 9x + 2$

22. $(x + 5)(2x + 7)$
$2x^2 + 17x + 35$

23. $(5x - 4)(x + 3)$
$5x^2 + 11x - 12$

24. $(2x - 1)(x + 5)$
$2x^2 + 9x - 5$

25. $(3a - 1)(a + 5)$
$3a^2 + 14a - 5$

26. $(3a - 2)(a + 7)$
$3a^2 + 19a - 14$

27. $(y + 5)(2y - 3)$
$2y^2 + 7y - 15$

28. $(y + 1)(5y - 1)$
$5y^2 + 4y - 1$

29. $(x - 3)(x - 5)$
$x^2 - 8x + 15$

30. $(x - 6)(x - 1)$
$x^2 - 7x + 6$

31. $(2x - 1)(3x - 2)$
$6x^2 - 7x + 2$

32. $(3x - 5)(x - 1)$
$3x^2 - 8x + 5$

33. $(2x - 3a)(2x + 5a)$
$4x^2 + 4ax - 15a^2$

34. $(5x - 2a)(x + 5a)$
$5x^2 + 23ax - 10a^2$

35. $(x + 7)(x + 8)$
$x^2 + 15x + 56$

36. $(x + 1)(x + 9)$
$x^2 + 10x + 9$

37. $(2a + b)(2a + 4b)$
$4a^2 + 10ab + 4b^2$

38. $(3a + 2b)(3a + 5b)$
$9a^2 + 21ab + 10b^2$

D **E** In Problems 39–74, use the special products to do the indicated multiplications.

39. $(4u + v)^2$
$16u^2 + 8uv + v^2$

40. $(3u + 2v)^2$
$9u^2 + 12uv + 4v^2$

41. $(2y + z)^2$
$4y^2 + 4yz + z^2$

42. $(4y + 3z)^2$
$16y^2 + 24yz + 9z^2$

43. $(3a - b)^2$
$9a^2 - 6ab + b^2$

44. $(4a - 3b)^2$
$16a^2 - 24ab + 9b^2$

45. $(a + b)(a - b)$
$a^2 - b^2$

46. $(a + 4)(a - 4)$
$a^2 - 16$

47. $(5x - 2y)(5x + 2y)$
$25x^2 - 4y^2$

48. $(2x - 7y)(2x + 7y)$
$4x^2 - 49y^2$

49. $-(3a - b)(3a + b)$
$b^2 - 9a^2$

50. $-(2a - 5b)(2a + 5b)$
$25b^2 - 4a^2$

51. $3x(x + 1)(x + 2)$
$3x^3 + 9x^2 + 6x$

52. $3x(x + 2)(x + 3)$
$3x^3 + 15x^2 + 18x$

53. $-3x(x - 1)(x - 3)$
$-3x^3 + 12x^2 - 9x$

54. $-2x(x - 5)(x - 1)$
$-2x^3 + 12x^2 - 10x$

55. $x(x + 3)^2$
$x^3 + 6x^2 + 9x$

56. $3x(x + 7)^2$
$3x^3 + 42x^2 + 147x$

57. $-2x(x - 1)^2$
$-2x^3 + 4x^2 - 2x$

58. $-5x(x - 3)^2$
$-5x^3 + 30x^2 - 45x$

59. $(2x + y)(2x - y)y^2$
$4x^2y^2 - y^4$

60. $(3x + y)(3x - y)x^2$
$9x^4 - x^2y^2$

61. $\left(x + \dfrac{3}{4}\right)^2$
$x^2 + \frac{3}{2}x + \frac{9}{16}$

62. $\left(x + \dfrac{2}{5}\right)^2$
$x^2 + \frac{4}{5}x + \frac{4}{25}$

63. $\left(2y - \dfrac{1}{5}\right)^2$
$4y^2 - \frac{4}{5}y + \frac{1}{25}$

64. $\left(3y - \dfrac{3}{4}\right)^2$
$9y^2 - \frac{9}{2}y + \frac{9}{16}$

65. $\left(\dfrac{3}{4}p + \dfrac{1}{5}q\right)^2$
$\frac{9}{16}p^2 + \frac{3}{10}pq + \frac{1}{25}q^2$

66. $\left(\dfrac{2}{5}p + \dfrac{1}{4}q\right)^2$
$\frac{4}{25}p^2 + \frac{1}{5}pq + \frac{1}{16}q^2$

67. $[(3x + 1) + 4y]^2$
$9x^2 + 24xy + 16y^2 + 6x + 8y + 1$

68. $[(2x + 1) + 3y]^2$
$4x^2 + 12xy + 9y^2 + 4x + 6y + 1$

69. $[(3x - 1) - 4y]^2$
$9x^2 - 24xy + 16y^2 - 6x + 8y + 1$

70. $[(2x - 1) - 4y]^2$
$4x^2 - 16xy + 16y^2 - 4x + 8y + 1$

71. $[2y + (3x - 1)]^2$
$9x^2 + 12xy + 4y^2 - 6x - 4y + 1$

72. $[3y + (2x - 1)]^2$
$4x^2 + 12xy + 9y^2 - 4x - 6y + 1$

73. $[4p - (3q - 1)]^2$
$16p^2 - 24pq + 9q^2 + 8p - 6q + 1$

74. $[2p - (3q - 1)]^2$
$4p^2 - 12pq + 9q^2 + 4p - 6q + 1$

APPLICATIONS

In Problems 75 and 76, $R = xp$, where x is the number of items sold and p is the price of the item.

75. The demand x for a certain product is given by $x = 1000 - 30p$.

   a. Write a formula for the revenue R.
      $R = 1000p - 30p^2$

   b. What is the revenue when the price is $20? $8000

76. A company manufactures and sells x jogging suits at p dollars every day. If $x = 3000 - 30p$, write a formula for the daily revenue R and use it to find the revenue on a day in which the suits were selling for $40.
$R = 3000p - 30p^2$; $72,000

In Problems 77–79, multiply the expression given.

77. The heat transmission between two objects of temperature T_2 and T_1 involves the expression
$$(T_1^2 + T_2^2)(T_1^2 - T_2^2)$$
where T_1^2 means the square of T_1 and T_2^2 means the square of T_2.
$T_1^4 - T_2^4$

78. The deflection of a certain beam involves the expression
$$w(l^2 - x^2)^2$$
Multiply this expression.
$l^4w - 2l^2x^2w + x^4w$

79. The heat output from a natural draught convector is given by
$$K(t_n - t_a)^2$$
Multiply this expression.
$Kt_n^2 - 2Kt_nt_a + Kt_a^2$

SKILL CHECKER

Try these "Skill Checker" exercises so you'll be ready for the next section.

The distributive property can be written as $ab + ac = a(b + c)$. Use this rule to rewrite each expression:

80. $3x + 3y$ $3(x + y)$

81. $5x + 5y$ $5(x + y)$

82. $2xz + 2xy$ $2x(z + y)$

83. $3ab + 3ac$ $3a(b + c)$

84. $6bc + 6bd$ $6b(c + d)$

USING YOUR KNOWLEDGE

Avoiding Multiplication Mistakes

A common fallacy (mistake) when multiplying binomials is to assume that

$$(x + y)^2 = x^2 + y^2$$

Here are some arguments that should convince you otherwise.

85. Let $x = 1$, $y = 2$.

  a. What is $(x + y)^2$? 9

  b. What is $x^2 + y^2$? 5

  c. Does $(x + y)^2 = x^2 + y^2$? No

86. Let $x = 2$, $y = 1$.

  a. What is $(x - y)^2$? 1

  b. What is $x^2 - y^2$? 3

  c. Does $(x - y)^2 = x^2 - y^2$? No

87. Look at the large square. Its area is $(x + y)^2$. The square is divided into four smaller areas numbered 1, 2, 3, and 4.

  a. What is the area of square 1? x^2

  b. What is the area of rectangle 2? xy

  c. What is the area of square 4? y^2

  d. What is the area of rectangle 3? xy

	x	y
x	1	2
y	3	4

88. The total area of the square is $(x + y)^2$. It's also the sum of the areas of the regions numbered 1, 2, 3, and 4. What is the sum of these four areas? (Simplify your answer.) $x^2 + 2xy + y^2$

89. From your answer to Problem 88, what can you say about $x^2 + 2xy + y^2$ and $(x + y)^2$? $(x + y)^2 = x^2 + 2xy + y^2$

90. If $x^2 + y^2$ is the sum of the areas of squares 1 and 4, does $x^2 + y^2 = (x + y)^2$? No, $x^2 + y^2 \neq (x + y)^2$

WRITE ON

91. Write in your own words the procedure for multiplying two monomials with integer coefficients. Answers may vary.

92. Explain why $(x - y)^2$ is different from $x^2 - y^2$. Answers may vary.

93. Describe in your own words the procedure for multiplying two polynomials. Answers may vary.

MASTERY TEST

If you know how to do these problems, you have learned your lesson!

Multiply:

94. $(5x - y)(2x - 3y)$
$10x^2 - 17xy + 3y^2$

95. $(4x + 3y)(3x + 2y)$
$12x^2 + 17xy + 6y^2$

96. $(3x + 5y)^2$
$9x^2 + 30xy + 25y^2$

97. $(5x - 3y)^2$
$25x^2 - 30xy + 9y^2$

98. $(5x - 3y)(5x + 3y)$
$25x^2 - 9y^2$

99. $(3x + y)(3x - y)$
$9x^2 - y^2$

100. $(2x - 3)(x^2 - 3x + 5)$
$2x^3 - 9x^2 + 19x - 15$

101. $(x - 2)(x^2 - 4x - 3)$
$x^3 - 6x^2 + 5x + 6$

102. $-3xy^2(x^3 - 7xy - 2x^2)$
$-3x^4y^2 + 21x^2y^3 + 6x^3y^2$

103. $4x^3(5x^3 + 3x^2 - 2x - 5)$
$20x^6 + 12x^5 - 8x^4 - 20x^3$

104. $[3x - (2y - 1)]^2$
$9x^2 - 12xy + 4y^2 + 6x - 4y + 1$

105. $[(3x + 1) - 4y]^2$
$9x^2 - 24xy + 16y^2 + 6x - 8y + 1$

106. The research department of a company determined that the number of skateboards that will be sold is given by $x = 1000 - 20p$, where p is the price of each skateboard.

  a. If the revenue is the number of skateboards to be sold times the price of each skateboard, write a formula for the revenue R. $R = 1000p - 20p^2$

  b. Find the revenue obtained by selling the skateboards for $50 each. $0, no revenue

5.3   THE GREATEST COMMON FACTOR AND FACTORING BY GROUPING

To Succeed, Review How To . . .

1. Use the distributive property to multiply expressions (pp. 51–53).

2. Use the properties of exponents (pp. 32–40).

Objectives

A Factor out the greatest common factor in a polynomial.

B Factor a polynomial with four terms by grouping.

GETTING STARTED   Factoring with Interest

Fixed Rate Callable CDs				
Noncallable CD Term	**Possible CD Term**	**Current CD Rates**	**Theoretical APY**	**Minimum Deposit**
0.5 Yr	6.0 Yr	4.00%	4.07%	$25,000
0.5 Yr	10.0 Yr	5.25%	5.32%	$25,000
1.0 Yr	15.0 Yr	5.50%	5.64%	$25,000

Source: Federally Insured Savings Network.

If you buy a 1.0 yr, $25,000 certificate of deposit, what amount A would you get at the end of the year? Assuming the interest will be calculated once at the end of the year, you will get $25,000 plus interest. The annual interest is $25,000 \cdot 0.055$, so the amount will be $A = 25,000 + 25,000 \cdot 0.055$.

If you invest P dollars at an annual rate r, the amount A dollars that you get is

$$A = P + Pr$$

The expression $P + Pr$ can be written in a simpler way if we **factor** it, that is, if we write it as the product of its factors. Factoring is the reverse of multiplication. If you multiply P by $1 + r$, you will write $P(1 + r) = P + Pr$. If you factor $P + Pr$, you should write

$$P + Pr = P(1 + r) \qquad \text{Use the distributive property to factor } P + Pr.$$

Using the distributive property to factor an expression changes a *sum of terms* into a *product of factors*. Here are some examples of factoring. Check the results by multiplying the *factors* on the right-hand side of the equation to obtain the *terms* on the left-hand side.

$$5x + 15 = 5(x + 3)$$
$$6x^2 + 6xy = 6x(x + y)$$
$$-12x^3 - 42x^2y = -6x^2(2x + 7y)$$

A   The Greatest Common Factor

To factor the preceding expressions, you must know how to find the **greatest common factor (GCF)** of their terms. In arithmetic, we factor integers into a product of **primes.** (A **prime number** is a number that has exactly two factors: itself and 1.) The prime

Teaching Tip

Remind students that *factors* are multiplied and because $12 = 3(4)$, 3 and 4 are *factors* of 12 and 3(4) is a way to factor 12.

numbers are 2, 3, 5, 7, 11, 13, 17, and so on. We can write 18 and 12 as a product of primes by using successive divisions by primes:

$$
\begin{array}{ll}
2\ \lfloor 18 & 2\ \lfloor 12 \\
3\ \lfloor 9 & 2\ \lfloor 6 \\
3\ \lfloor 3 & 3\ \lfloor 3 \\
\quad 1 & \quad 1
\end{array}
$$

Thus $18 = 2 \cdot 3 \cdot 3 = 2 \cdot 3^2$ and $12 = 2 \cdot 2 \cdot 3 = 2^2 \cdot 3$. To obtain the GCF, we write with all the primes in columns as shown, and then we select the prime with the smallest exponent from each column.

1. In the first column, we have 2 and 2^2; select 2.

2. In the second column, we have 3^2 and 3; select 3.

3. The GCF is the product of the numbers selected: $2 \cdot 3 = 6$.

Pick the factor with the smallest exponent in each column.

$$
\begin{array}{l}
18 = \boxed{2} \cdot \boxed{3^2} \\
12 = \boxed{2^2} \cdot \boxed{3} \\
\text{GCF} = 2 \cdot 3 = 6
\end{array}
$$

The common factors in $18 = 2 \cdot 3 \cdot 3$ and $12 = 2 \cdot 2 \cdot 3$ are 2 and 3. The GCF of 18 and 12 is $2 \cdot 3 = 6$.

Similarly, to find the greatest common factor of two terms, such as xy^2 and x^2y, we compare the factors of each term to see if they have any in common. If they do, then the GCF is the product of all the common factors with the smallest exponent.

$$
\begin{array}{l}
xy^2 = \boxed{x} \cdot \boxed{y^2} \\
x^2y = \boxed{x^2} \cdot \boxed{y} \\
\text{GCF} = \quad x \cdot y = xy
\end{array}
$$

Web It

For another introduction to factoring out the GCF, go to link 5-3-1 at mhhe.com/bello.

To find the GCF of x^3y^5 and x^2y^7

$$
\begin{array}{l}
x^3y^5 = \boxed{x^3} \cdot \boxed{y^5} \\
x^2y^7 = \boxed{x^2} \cdot \boxed{y^7} \\
\text{GCF} = \quad x^2 \cdot y^5 = x^2y^5
\end{array}
$$

In general, we have the following.

Greatest Common Factor

The term ax^n is the **greatest common monomial factor** (GCF) of a polynomial in x (with integer coefficients) if

1. a is the *greatest* integer that divides each of the coefficients in the polynomial.

2. n is the *smallest* exponent of x in all the terms of the polynomial.

Teaching Tip

Sometimes it's easy for a student to find the GCF, but not to complete the factorization. Once the GCF is found, write it down with a set of parentheses next to it. Proceed to divide each term by the GCF and place results in the parentheses.

Example:

$$
\begin{array}{ccc}
\text{Factor} & 12x^3 & + & 18x^5 \\
& \text{Divide} & & \text{Divide} \\
\text{GCF} & \dfrac{12x^3}{6x^3} & & \dfrac{18x^5}{6x^3} \\
& \downarrow & & \downarrow \\
6x^3 & (2 & + & 3x^2)
\end{array}
$$

To factor $12x^3 + 18x^5$, we could write

$$12x^3 + 18x^5 = 2x(6x^2 + 9x^4) \qquad \text{Check by multiplying}$$

But this is *not* completely factored. The greatest number that divides 12 and 18 is 6, and the power of x with the smallest exponent in all the terms of the polynomial is x^3. Thus the GCF is $6x^3$. The complete factorization is

$$12x^3 + 18x^5 = 6x^3(2 + 3x^2)$$

It may help your accuracy and understanding if you write an intermediate step showing the greatest common factor present in each term. When factoring $12x^3 + 18x^5$, you can write

$$12x^3 + 18x^5 = 6x^3 \cdot 2 + 6x^3 \cdot 3x^2$$
$$= 6x^3(2 + 3x^2)$$

In addition, when factoring expressions such as $-6x + 18$, there are two choices:

$$-6(x - 3) \quad \text{and} \quad 6(-x + 3)$$

Both are correct, but the factorization $-6(x - 3)$ is the *preferred* choice because the first term of the binomial $x - 3$ has a positive sign.

EXAMPLE 1 **Factoring binomials**	**PROBLEM 1**

Factor completely:

a. $8x - 24$ **b.** $-6y^2 + 12y$ **c.** $10x^2 - 25x^3$

SOLUTION

a. $8x - 24 = 8 \cdot x - 8 \cdot 3$ 8 is the GCF of $8x$ and 24.
$\quad\quad = 8(x - 3)$

b. $-6y^2 + 12y = -6y \cdot y - 6y \cdot (-2)$ $-6y$ is the GCF of $-6y^2$ and $12y$.
$\quad\quad\quad\quad = -6y(y - 2)$

c. $10x^2 - 25x^3 = 5x^2 \cdot 2 - 5x^2 \cdot 5x$ $5x^2$ is the GCF of $10x^2$ and $25x^3$.
$\quad\quad\quad\quad = 5x^2(2 - 5x)$

Check your answers by multiplying the expressions to the right of the equals sign.

PROBLEM 1

Factor completely:

a. $6x + 48$

b. $-3y^2 + 21y$

c. $4x^2 - 32x^3$

You can also factor polynomials with more than two terms, as shown in Example 2.

EXAMPLE 2   **Factoring polynomials**

Factor completely:

a. $6x^3 + 12x^4 + 18x^2$ **b.** $10x^6 - 15x^5 + 20x^7 + 30x^2$

SOLUTION

a. $6x^3 + 12x^4 + 18x^2 = 6x^2 \cdot x + 6x^2 \cdot 2x^2 + 6x^2 \cdot 3$
$\quad\quad\quad\quad = 6x^2(x + 2x^2 + 3) = 6x^2(2x^2 + x + 3)$

b. $10x^6 - 15x^5 + 20x^7 + 30x^2 = 5x^2 \cdot 2x^4 - 5x^2 \cdot 3x^3 + 5x^2 \cdot 4x^5 + 5x^2 \cdot 6$
$\quad\quad\quad\quad\quad = 5x^2(2x^4 - 3x^3 + 4x^5 + 6)$
$\quad\quad\quad\quad\quad = 5x^2(4x^5 + 2x^4 - 3x^3 + 6)$

Check by multiplying.

PROBLEM 2

Factor completely:

a. $7x^3 + 14x^4 - 49x^2$

b. $3x^6 - 6x^5 + 12x^7 + 27x^2$

Answers

1. a. $6(x + 8)$ **b.** $-3y(y - 7)$
c. $4x^2(1 - 8x)$
2. a. $7x^2(2x^2 + x - 7)$
b. $3x^2(4x^5 + x^4 - 2x^3 + 9)$

Calculate It   Adjusting the Window to Check Factoring Polynomials

To check the factorization in Example 2(b), graph

$$Y_1 = 10x^6 - 15x^5 + 20x^7 + 30x^2$$

and

$$Y_2 = 5x^2(4x^5 + 2x^4 - 3x^3 + 6)$$

Window 1 shows the same graph for Y_1 and Y_2.

But how do we know that portions of the graph that may not be shown are the same? What we need is a more complete graph. To do this, adjust the vertical scale by selecting Ymin $= -100$, Ymax $= 100$ and Xscl $= 10$. A more complete graph is shown in

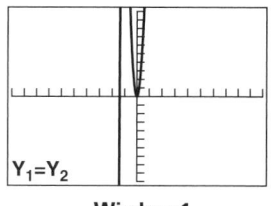
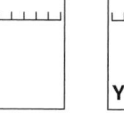
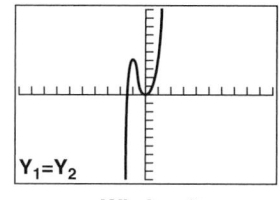

Window 1   **Window 2**

Window 2 and Y_1 and Y_2 appear to be identical, so the factorization is correct.

EXAMPLE 3 **Factoring polynomials with fractional coefficients**

Factor completely:

$$\frac{3}{4}x^2 - \frac{1}{4}x^4 + \frac{5}{4}x^5$$

SOLUTION The procedure for finding the GCF only works with polynomials with integer coefficients. However, since all the coefficients have a denominator of 4, the greatest common factor is $\frac{1}{4}x^2$.

$$\frac{3}{4}x^2 - \frac{1}{4}x^4 + \frac{5}{4}x^5 = \frac{1}{4}x^2(3 - x^2 + 5x^3)$$

Check by multiplying.

PROBLEM 3

Factor completely:

$$\frac{2}{5}x^2 - \frac{3}{5}x^4 + \frac{4}{5}x^5$$

B Factoring by Grouping

How do we factor expressions with four terms when there seems to be no common factor except 1? One way is to group and factor the first two terms and also the last two terms and then use the distributive property. Here is the three-step procedure.

PROCEDURE

Factoring by Grouping (Four Terms)

1. Group terms with common factors using the associative property.
2. Factor each resulting group.
3. Factor out the common binomial (GCF), by the distributive property.

Let's factor $x^3 + 2x^2 + 3x + 6$. As you can see, this expression doesn't appear to have a common factor other than 1. We use our three-step procedure:

1. Group terms. $\quad x^3 + 2x^2 + 3x + 6 = (x^3 + 2x^2) + (3x + 6)$
2. Factor each group. $\quad = x^2(x + 2) + 3(x + 2)$
3. Factor out the (binomial) GCF. $\quad = (x + 2)(x^2 + 3)$

Thus $x^3 + 2x^2 + 3x + 6 = (x + 2)(x^2 + 3)$. By the commutative property of multiplication, either $(x + 2)(x^2 + 3)$ or $(x^2 + 3)(x + 2)$ is a correct factorization for $x^3 + 2x^2 + 3x + 6$. Check by multiplying.

EXAMPLE 4 **Factoring by grouping**

Factor completely:

a. $3x^3 + 6x^2 + 4x + 8$ **b.** $6x^3 - 3x^2 - 4x + 2$

SOLUTION

a. Proceed by steps.

1. Group terms. $\quad 3x^3 + 6x^2 + 4x + 8 = (3x^3 + 6x^2) + (4x + 8)$
2. Factor each group. $\quad = 3x^2(x + 2) + 4(x + 2)$
3. Factor out the (binomial) GCF. $\quad = (x + 2)(3x^2 + 4)$

You can write $3x^3 + 6x^2 + 4x + 8$ as $(3x^2 + 4)(x + 2)$ in step 3. Since $(x + 2)(3x^2 + 4) = (3x^2 + 4)(x + 2)$, both answers are correct! Do you know why?

PROBLEM 4

Factor completely:

a. $2x^3 - 2x^2 + 3x - 3$

b. $6x^3 - 9x^2 - 2x + 3$

Answers

3. $\frac{1}{5}x^2(2 - 3x^2 + 4x^3)$
4. a. $(x - 1)(2x^2 + 3)$
b. $(2x - 3)(3x^2 - 1)$

b. Proceed by steps.

1. Group terms. $6x^3 - 3x^2 - 4x + 2 = (6x^3 - 3x^2) + (-4x + 2)$

2. Factor each group. $= 3x^2(2x - 1) + (-2)(2x - 1)$

3. Factor out the (binomial) GCF. $= (2x - 1)(3x^2 - 2)$

Thus, $6x^3 - 3x^2 - 4x + 2 = (2x - 1)(3x^2 - 2)$. We leave the verification to you.

     You might save some time if in step 1 you realize that $-4x + 2 = -(4x - 2)$ and write

$$6x^3 - 3x^2 - 4x + 2 = (6x^3 - 3x^2) - (4x - 2)$$
$$= 3x^2(2x - 1) - 2(2x - 1)$$
$$= (2x - 1)(3x^2 - 2)$$

If the third term is preceded by a minus sign, write $-(\quad)$ with the appropriate signs for the expression inside the parentheses.

Calculate It   Adjusting the Window to Check Factoring Polynomials

To verify the results of Example 4(a), we graph

$$Y_1 = 3x^3 + 6x^2 + 4x + 8$$

and

$$Y_2 = (x + 2)(3x^2 + 4)$$

Using a standard window (ZOOM 6), we get part of the graph as shown in Window 1.

     But we need a more complete graph, so in WINDOW we change the vertical (Y) scale to $[-100, 100]$ with Yscl = 10.

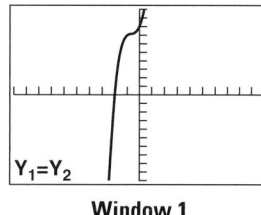

Window 1

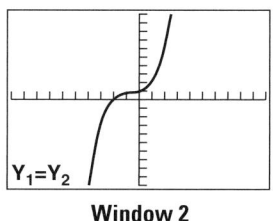

Window 2

Window 2 shows that both graphs are identical and our factorization is probably correct.

Web It

For more examples and practice factoring out the GCF and factoring by grouping, go to link 5-3-2 at mhhe.com/bello.

The next example will require two types of factoring. This will be done in two steps where the first step should be to factor out the GCF and the second step to factor by grouping. The polynomial would not be completely factored if you just did one or the other of these steps. For this reason we shall continue to use the term "factor completely" when giving factoring directions.

EXAMPLE 5   **Factoring out the common factor and factor by grouping**

Factor completely:

a. $2x^4 - 4x^3 - x^2 + 2x$ **b.** $6x^6 - 9x^4 + 4x^3 - 6x$

SOLUTION

a. First, factor out the common factor x and write

$2x^4 - 4x^3 - x^2 + 2x = x(2x^3 - 4x^2 - x + 2)$ Then factor $2x^3 - 4x^2 - x + 2$.

Use the three-step procedure. Can you tell what is happening without the words? If not, check the procedure box on the preceding page.

1. $2x^4 - 4x^3 - x^2 = x[(2x^3 - 4x^2) - (x - 2)]$

2. $= x[2x^2(x - 2) - 1(x - 2)]$

3. $= x[(x - 2)(2x^2 - 1)]$

Thus $2x^4 - 4x^3 - x^2 + 2x = x(x - 2)(2x^2 - 1)$.

PROBLEM 5

Factor completely:

a. $3x^4 - 6x^3 - x^2 + 2x$

b. $6x^6 - 9x^4 + 2x^3 - 3x$

Teaching Tip

Remind students that when the GCF is 1 in this type of factoring, it is important to write the 1.

Example:

    $-(x - 2)$ $-1(x - 2)$

Answers

5. a. $x(x - 2)(3x^2 - 1)$
b. $x(3x^3 + 1)(2x^2 - 3)$

b. Factor out x to obtain

$$6x^6 - 9x^4 + 4x^3 - 6x = x(6x^5 - 9x^3 + 4x^2 - 6)$$ Then factor
$6x^5 - 9x^3 + 4x^2 - 6$.

Now use the procedure.

1. $6x^6 - 9x^4 + 4x^3 - 6x = x[(6x^5 - 9x^3) + (4x^2 - 6)]$

2. $\qquad\qquad\qquad = x[3x^3(2x^2 - 3) + 2(2x^2 - 3)]$

3. $\qquad\qquad\qquad = x(2x^2 - 3)(3x^3 + 2)$

Thus

$$6x^6 - 9x^4 + 4x^3 - 6x = x(2x^2 - 3)(3x^3 + 2)$$
$\qquad\qquad\qquad\qquad\qquad\qquad \uparrow$
$\qquad\qquad\qquad\qquad$ Don't forget the x!

We leave the verification to you.

Calculate It Exercises

1. Use your calculator to verify the results of Example 1(b).

2. Use your calculator to verify the results of Example 2(a).

3. Use your calculator to verify the results of Example 5.

Exercises 5.3

Boost *your* **GRADE**
at mathzone.com!

MathZone

• Practice Problems
• Self-Tests
• Videos
• NetTutor
• e-Professors

A In Problems 1–26, factor completely.

1. $8x + 16$ $\;8(x + 2)$

2. $15x + 45$ $\;15(x + 3)$

3. $9y - 18$ $\;9(y - 2)$

4. $11y - 88$ $\;11(y - 8)$

5. $-5y + 25$ $\;-5(y - 5)$

6. $-4y + 28$ $\;-4(y - 7)$

7. $-8x - 24$ $\;-8(x + 3)$

8. $-6x - 36$ $\;-6(x + 6)$

9. $4x^2 + 36x$ $\;4x(x + 9)$

10. $5x^3 + 20x$ $\;5x(x^2 + 4)$

11. $6x - 42x^3$ $\;6x(1 - 7x^2)$

12. $7x - 14x^5$ $\;7x(1 - 2x^4)$

13. $-5x^2 - 35x^4$ $\;-5x^2(1 + 7x^2)$

14. $-8x^3 - 16x^6$ $\;-8x^3(1 + 2x^3)$

15. $3x^3 + 6x^2 + 39x$
$\quad 3x(x^2 + 2x + 13)$

16. $8x^3 + 4x^2 - 36x$
$\quad 4x(2x^2 + x - 9)$

17. $63y^3 - 18y^2 + 27y$
$\quad 9y(7y^2 - 2y + 3)$

18. $10y^3 - 5y^2 + 20y$
$\quad 5y(2y^2 - y + 4)$

19. $36x^6 + 12x^5 - 18x^4 + 30x^2$
$\quad 6x^2(6x^4 + 2x^3 - 3x^2 + 5)$

20. $15x^7 - 15x^6 + 30x^3 - 20x^2$
$\quad 5x^2(3x^5 - 3x^4 + 6x - 4)$

21. $48y^8 + 16y^5 - 24y^4 + 16y^3$
$\quad 8y^3(6y^5 + 2y^2 - 3y + 2)$

22. $12y^9 - 4y^6 + 6y^5 + 8y^4$
$\quad 2y^4(6y^5 - 2y^2 + 3y + 4)$

23. $\dfrac{4}{7}x^3 + \dfrac{3}{7}x^2 - \dfrac{9}{7}x + \dfrac{3}{7}$
$\quad \frac{1}{7}(4x^3 + 3x^2 - 9x + 3)$

24. $\dfrac{2}{5}x^3 + \dfrac{3}{5}x^2 - \dfrac{2}{5}x + \dfrac{4}{5}$
$\quad \frac{1}{5}(2x^3 + 3x^2 - 2x + 4)$

25. $\dfrac{7}{8}y^9 + \dfrac{3}{8}y^6 - \dfrac{5}{8}y^4 + \dfrac{5}{8}y^2$
$\quad \frac{1}{8}y^2(7y^7 + 3y^4 - 5y^2 + 5)$

26. $\dfrac{4}{3}y^7 - \dfrac{1}{3}y^5 + \dfrac{2}{3}y^4 - \dfrac{7}{3}y^3$
$\quad \frac{1}{3}y^3(4y^4 - y^2 + 2y - 7)$

B In Problems 27–52, factor completely.

27. $x^3 + 2x^2 + x + 2$
$\quad (x + 2)(x^2 + 1)$

28. $x^3 + 3x^2 + x + 3$
$\quad (x + 3)(x^2 + 1)$

29. $y^3 - 3y^2 + y - 3$
$\quad (y - 3)(y^2 + 1)$

30. $y^3 - 5y^2 + y - 5$
$\quad (y - 5)(y^2 + 1)$

31. $4x^3 + 6x^2 + 2x + 3$
$\quad (2x + 3)(2x^2 + 1)$

32. $6x^3 + 3x^2 + 2x + 1$
$\quad (2x + 1)(3x^2 + 1)$

33. $6x^3 - 2x^2 + 3x - 1$
$(3x - 1)(2x^2 + 1)$

34. $6x^3 - 9x^2 + 2x - 3$
$(2x - 3)(3x^2 + 1)$

35. $4y^3 + 8y^2 + y + 2$
$(y + 2)(4y^2 + 1)$

36. $2y^3 + 6y^2 + y + 3$
$(y + 3)(2y^2 + 1)$

37. $2a^6 + 3a^4 + 2a^2 + 3$
$(2a^2 + 3)(a^4 + 1)$

38. $3a^6 + 2a^4 + 3a^2 + 2$
$(3a^2 + 2)(a^4 + 1)$

39. $3x^5 + 12x^3 + x^2 + 4$
$(x^2 + 4)(3x^3 + 1)$

40. $2x^5 + 2x^3 + x^2 + 1$
$(x^2 + 1)(2x^3 + 1)$

41. $6y^5 + 9y^3 + 2y^2 + 3$
$(2y^2 + 3)(3y^3 + 1)$

42. $12y^5 + 8y^3 + 3y^2 + 2$
$(3y^2 + 2)(4y^3 + 1)$

43. $4y^7 + 12y^5 + y^4 + 3y^2$
$y^2(y^2 + 3)(4y^3 + 1)$

44. $2y^7 + 2y^5 + y^4 + y^2$
$y^2(y^2 + 1)(2y^3 + 1)$

45. $3a^7 - 6a^5 - 2a^4 + 4a^2$
$a^2(a^2 - 2)(3a^3 - 2)$

46. $4a^7 - 12a^5 - 3a^4 + 9a^2$
$a^2(a^2 - 3)(4a^3 - 3)$

47. $8a^5 - 12a^4 - 10a^3 + 15a^2$
$a^2(2a - 3)(4a^2 - 5)$

48. $3y^7 - 21y^4 + y^5 - 7y^2$
$y^2(y^3 - 7)(3y^2 + 1)$

49. $x^6 - 2x^5 + 2x^4 - 4x^3$
$x^3(x - 2)(x^2 + 2)$

50. $v^6 + 3v^5 - 7v^4 - 21v^3$
$v^3(v + 3)(v^2 - 7)$

51. $(x - 4)(x + 2) + (x - 4)(x + 3)$
$(x - 4)(2x + 5)$

52. $(y + 2)(y - 7) + (2y + 3)(y - 7)$
$(y - 7)(3y + 5)$

53. Factor completely $\alpha Lt_2 - \alpha Lt_1$, where α is the coefficient of linear expansion, L is the length of the material, and t_2 and t_1 are the temperatures in degrees Celsius.
$\alpha L(t_2 - t_1)$

54. Factor completely the expression $-kx - kl$, which represents the restoring force of a spring stretched an amount l from its equilibrium position and then an additional x units. $-k(x + l)$

55. When solving for the equivalent resistance of two circuits, we have to factor the expression $R^2 - R - R + 1$. Factor completely this expression by grouping.
$(R - 1)(R - 1)$ or $(R - 1)^2$

56. The bending moment of a cantilever beam of length L, at x inches from its support, involves the expression $L^2 - Lx - Lx + x^2$. Factor completely this expression by grouping. $(L - x)(L - x)$ or $(L - x)^2$

SKILL CHECKER

Try these "Skill Checker" exercises so you'll be ready for the next section.

Multiply:

57. $(x + 3)(x + 4)$ $x^2 + 7x + 12$

58. $(x + 7)(x + 2)$ $x^2 + 9x + 14$

59. $(x + 5)(x - 2)$ $x^2 + 3x - 10$

60. $(x - 5)(x + 3)$ $x^2 - 2x - 15$

61. $(5x + 2y)^2$ $25x^2 + 20xy + 4y^2$

62. $(3x + 4y)^2$ $9x^2 + 24xy + 16y^2$

63. $(5x - 2y)^2$ $25x^2 - 20xy + 4y^2$

64. $(3x - 4y)^2$ $9x^2 - 24xy + 16y^2$

65. $(u + 6)(u - 6)$ $u^2 - 36$

66. $(2a + 7b)(2a - 7b)$ $4a^2 - 49b^2$

USING YOUR KNOWLEDGE

General Formulas

There are many formulas that can be simplified by factoring completely. Here are a few.

67. The vertical shear at any section of a cantilever beam of uniform cross section is

$$-wl + wz$$

Factor completely this expression. $-w(l - z)$

68. The bending moment of any section of a cantilever beam of uniform cross section is

$$-Pl + Px$$

Factor completely this expression. $-P(l - x)$

69. The surface area of a square pyramid is
$$a^2 + 2as$$
Factor completely this expression. $a(a + 2s)$

71. The height (in feet after t seconds) of a rock thrown from the roof of a certain building is given by
$$-16t^2 + 80t + 240$$
Factor completely this expression. (*Hint:* -16 is a common factor.) $-16(t^2 - 5t - 15)$

70. The energy of a moving object is given by
$$800m - mv^2$$
Factor completely this expression. $m(800 - v^2)$

WRITE ON

72. What should the first step be in factoring a polynomial? Answers may vary.

74. The procedure we outlined to find the greatest common monomial factor of a polynomial doesn't apply to Example 3. Explain why. Answers may vary.

73. Write your own definition for the greatest common factor of a list of integers. Answers may vary.

75. Describe the relationship between multiplying and factoring a polynomial. Answers may vary.

MASTERY TEST

If you know how to do these problems, you have learned your lesson!

Factor completely:

76. $3x^4 - 6x^3 - x^2 + 2x$
$x(x - 2)(3x^2 - 1)$

77. $6x^6 - 9x^4 + 2x^3 - 3x$
$x(2x^2 - 3)(3x^3 + 1)$

78. $2x^3 + 2x^2 + 3x + 3$
$(x + 1)(2x^2 + 3)$

79. $6x^3 - 9x^2 - 2x + 3$
$(2x - 3)(3x^2 - 1)$

80. $7x^3 + 14x^4 - 49x^2$
$7x^2(2x^2 + x - 7)$

81. $3x^6 - 6x^5 + 12x^7 + 27x^2$
$3x^2(4x^5 + x^4 - 2x^3 + 9)$

82. $4x^2 - 32x^3$ $4x^2(1 - 8x)$

83. $6x - 48$ $6(x - 8)$

84. $-3y^2 + 21y$ $-3y(y - 7)$

85. $\frac{2}{5}x^2 - \frac{4}{5}x^4 - \frac{1}{5}x^5$ $-\frac{1}{5}x^2(x^3 + 4x^2 - 2)$ or $\frac{1}{5}x^2(2 - 4x^2 - x^3)$

| **5.4** | **FACTORING TRINOMIALS** |

To Succeed, Review How To . . .

1. Multiply two binomials using the FOIL method (pp. 359–360).

2. Add and multiply signed numbers (pp. 18–22).

Objectives

A Factor a trinomial of the form $x^2 + bx + c$ (b and c are integers).

B Factor a trinomial of the form $ax^2 + bx + c$ using trial and error.

C Factor a trinomial of the form $ax^2 + bx + c$ using the ac test.

GETTING STARTED

Applying Factoring to Fire Fighting

How many hundred gallons of water per minute can this fire truck pump? If it has a hose 100 ft long, you can find the answer by solving $2g^2 + g - 36 = 0$. The expression $2g^2 + g - 36$ is factorable. How do we factor it? By using reverse multiplication, but let's start with a simpler problem.

Factoring Trinomials of the Form $x^2 + bx + c$ (b and c integers)

In Section 5.2, we multiplied two binomials using the FOIL method:

$$\overset{\mathsf{F \qquad O \quad I \qquad L}}{(x + 4)(x - 7) = x^2 - 7x + 4x - 28}$$

and

$$(x + 7)(x - 4) = x^2 - 4x + 7x - 28$$

To factor $x^2 - 3x - 28$, we recall that

$$(x + a)(x + b) = x^2 + bx + ax + ab$$
$$= x^2 + (b + a)x + ab$$

Rewriting this trinomial, we have the following factoring form.

Teaching Tip

As in Section 5.3, we want to write a sum of terms as a product of factors. In this case the sum of 3 terms is written as the product of two binomials. To help students recognize the correct three-term model to be able to factor ($x^2 + 5x + 6$), have them give some examples that do not look like this model ($x^3 + 5x + 6$).

PROCEDURE

Factoring Trinomials of the Form $x^2 + (b + a)x + ab$

$$x^2 + (b + a)x + ab = (x + a)(x + b)$$

How to Determine the Signs When Factoring Trinomials	
If ab is $+$ then both have same sign	**If ab is $-$ then one is $+$ and one is $-$**
if middle term is $+$ then both are $+$ $+a, +b$	if middle term is $-$ then number with larger absolute value is $-$ $-a, +b$
if middle term is $-$ then both are $-$ $-a, -b$	if middle term is $+$ then number with larger absolute value is $+$ $+a, -b$

Teaching Tip

Remind students that at this level factoring answers should be done with integers, so we wouldn't expect a factoring answer to be $(x + \frac{1}{2})(x - 4)$.

This means that to factor a trinomial with a leading coefficient of 1, we need two integers a and b whose product is the last term and whose sum is the coefficient of the middle term. Since

$$x^2 + \underbrace{(b + a)}x + \underbrace{ab} = (x + a)(x + b)$$

we write

$$x^2 - \quad 3x \quad - 28 = (x + a)(x + b)$$

We need two integers a and b whose product ab is -28 and whose sum is -3. The product is negative, so one number must be positive and the other negative, with the larger number being negative, since the sum is -3. The numbers are -7 and 4. [*Check:* $-7 \cdot 4 = -28$ and $-7 + 4 = -3$, the coefficient of the middle term.] Thus

$$x^2 - 3x - 28 = (x - 7)(x + 4)$$

Since the multiplication of polynomials is commutative,

$$x^2 - 3x - 28 = (x + 4)(x - 7)$$

Now suppose we want to factor $x^2 - 8x + 12$. This time, we need two integers whose product is 12 and whose sum is -8 (the coefficient of the middle term). The numbers are -6 and -2. [*Check:* $(-6)(-2) = 12$ and $(-6) + (-2) = -8$.] Thus

$$x^2 - 8x + 12 = (x - 6)(x - 2)$$

You can check this by multiplying $(x - 6)$ by $(x - 2)$.

Web It

For a practice quiz with answers on factoring simple trinomials, go to link 5-4-1 at mhhe.com/bello.

EXAMPLE 1 **Factoring trinomials of the form $x^2 + bx + c$**

Factor completely:

a. $x^2 + 7x + 12$ **b.** $x^2 - 6x + 8$ **c.** $x^2 - 4 - 3x$

SOLUTION

a. To factor $x^2 + 7x + 12$, we need two integers whose product is 12 and whose sum is 7. The numbers are 3 and 4. Thus

$$x^2 + 7x + 12 = (x + 3)(x + 4)$$

b. To factor $x^2 - 6x + 8$, we need two integers whose product is 8 and whose sum is -6. The product is positive, so both numbers must be negative. They are -2 and -4. [*Check:* $(-2)(-4) = 8$ and $(-2) + (-4) = -6$.] Thus

$$x^2 - 6x + 8 = (x - 2)(x - 4)$$

c. First, rewrite $x^2 - 4 - 3x$ in descending order as $x^2 - 3x - 4$. This time, we need integers with product -4 and sum -3. The numbers are 1 and -4. Thus

$$x^2 - 4 - 3x = x^2 - 3x - 4 = (x + 1)(x - 4)$$

You can check all these results by multiplying. For example,
$(x + 1)(x - 4) = x^2 - 4x + x - 4 = x^2 - 3x - 4$.

EXAMPLE 2 **Factoring trinomials with two variables of the form $x^2 + bxy + cy^2$**

Factor completely:

a. $x^2 - 8xy + 7y^2$ **b.** $x^2 + 3xy + 7y^2$

SOLUTION

a. We need two integers whose product is 7 and whose sum is -8. The product is positive, so both numbers must be negative. The numbers are -1 and -7. Thus

$$x^2 - 8xy + 7y^2 = (x - 1y)(x - 7y)$$ (The y variable is in the 2nd term of each binomial.)

$$= (x - y)(x - 7y)$$ Check by multiplication.

b. We need two integers whose product is 7 and whose sum is 3. There are no such numbers. The polynomial $x^2 + 3xy + 7y^2$ is not factorable using integer coefficients.

   A polynomial not factorable using integer coefficients is called a **prime polynomial.**

PROBLEM 1

Factor completely:

a. $x^2 + 7x + 10$

b. $x^2 - 3x - 10$

c. $x^2 - 6 - 5x$

PROBLEM 2

Factor completely:

a. $x^2 - 2xy + 5y^2$

b. $x^2 - 6xy - 16y^2$

Teaching Tip

Be sure to note the change in model from $(x^2 + 5x + 6)$ to $(x^2 + 5xy + 6y^2)$. The procedure used in Example 1 can still be used; just add the variable y to the second term of each binomial factor.

B **Factoring Trinomials of the Form $ax^2 + bx + c$**

To factor the polynomial $2g^2 + g - 36$ mentioned in the *Getting Started* at the beginning of this section, we can rely on our experience with FOIL. To obtain $2g^2 + 1g - 36$, multiply

$$(2g + \underline{\quad\quad})(g - \underline{\quad\quad})$$

Answers

1. a. $(x + 2)(x + 5)$
b. $(x + 2)(x - 5)$
c. $(x + 1)(x - 6)$
2. a. Prime   **b.** $(x + 2y)(x - 8y)$

Fill the blanks with two integers that have a product of -36 and that give a middle term of $1g$. The possibilities are as follows.

Trial Factors	Middle Term
$(2g + \underline{1})(g - \underline{36})$	$-71g$
$(2g + \underline{2})(g - \underline{18})$	$-34g$
$(2g + 2) = 2(g + 1)$, so we won't use even numbers in both terms of a factor.	
$(2g + \underline{3})(g - \underline{12})$	$-21g$
$(2g + \underline{9})(g - \underline{4})$	$1g$

Thus $2g^2 + g - 36 = (2g + 9)(g - 4)$.

EXAMPLE 3   **Factoring trinomials of the form $ax^2 + bx + c$**

Factor completely: $6x^2 + 17x + 12$

SOLUTION   The factors of 6 are 6 and 1 or 3 and 2. The possible combinations are

$$(6x + \underline{\hspace{1cm}})(x + \underline{\hspace{1cm}}) \quad \text{or} \quad (3x + \underline{\hspace{1cm}})(2x + \underline{\hspace{1cm}})$$

Trial Factors	
$(6x + 12)(x + 1)$	$(6x + 1)(x + 12)$
$(6x + 6)(x + 2)$	$(6x + 2)(x + 6)$
$(6x + 4)(x + 3)$	$(6x + 3)(x + 4)$
$(3x + 12)(2x + 1)$	$(3x + 1)(2x + 12)$
$(3x + 6)(2x + 2)$	$(3x + 2)(2x + 6)$
$(3x + 4)(2x + 3)$	$(3x + 3)(2x + 4)$

The only combination yielding the correct middle term is

$$(3x + 4)(2x + 3) = 6x^2 + 9x + 8x + 12 = 6x^2 + 17x + 12$$

PROBLEM 3

Factor completely:

$$6x^2 + 13x + 6$$

C   The *ac* Test

The method used in Example 3 is not efficient when the coefficients are large. Moreover, we don't even know whether the polynomial is factorable. We remedy this situation with the following test.

THE *ac* TEST

The polynomial $ax^2 + bx + c$ is factorable only if there are two integers whose product is ac and whose sum is b.

Answer

3. $(3x + 2)(2x + 3)$

To find if $3x^2 + 2x + 5$ is factorable, we first look at $3 \cdot 5 = 15$. If we can find two integers whose product is 15 and whose sum is 2, we can factor the polynomial. No such integers exist, so $3x^2 + 2x + 5$ is *prime*. On the other hand, the polynomial $5x^2 + 11x + 2$ is factorable because we can find two integers (10 and 1) whose product, ac, is $5 \cdot 2 = 10$ and whose sum is 11. The number ac plays such an important part in the procedure used to factor $ax^2 + bx + c$ that we call it the **key number.**

We now give you a method that uses the key number to factor polynomials. For example, suppose we want to factor $2x^2 - 7x - 4$. We proceed as follows.

$$\overset{a \cdot c}{}$$

1. Find the key number, ac, $2 \cdot (-4) = -8$. $\quad 2x^2 - 7x - 4 \quad \boxed{-8}$

2. Find the factors of the key number and use the appropriate ones to rewrite the middle term so that it equals $-7x$. $\quad 2x^2 - 8x + 1x - 4 \quad -8, 1$

3. Group the terms (as in Section 5.3). $\quad (2x^2 - 8x) + (1x - 4)$

4. Factor each group. $\quad 2x(x - 4) + 1(x - 4)$

5. The binomial $(x - 4)$ is the GCF. $\quad (x - 4)(2x + 1)$

Thus $2x^2 - 7x - 4 = (x - 4)(2x + 1)$. You can check that this is the correct factorization by multiplying $(x - 4)$ by $(2x + 1)$.

Suppose you want to factor the trinomial $5x^2 + 7x + 2$. Here is one way of doing it.

1. Find the key number ($5 \cdot 2 = 10$). $\quad 5x^2 + 7x + 2 \quad \boxed{10}$

2. Find the factors of the key number and use them to rewrite the middle term. $\quad 5x^2 + 5x + 2x + 2 \quad 5, 2$

3. Group the terms. $\quad (5x^2 + 5x) + (2x + 2)$

4. Factor each group. $\quad 5x(x + 1) + 2(x + 1)$

5. $(x + 1)$ is the GCF. $\quad (x + 1)(5x + 2)$

Thus $5x^2 + 7x + 2 = (x + 1)(5x + 2)$.
   Here is another way to proceed.

Web It

For more examples and practice factoring trinomials, go to link 5-4-2 at mhhe.com/bello.

1. Find the key number. $\quad 5x^2 + 7x + 2 \quad \boxed{10}$

2. Find the factors of the key number and use them to rewrite the middle term. $\quad 5x^2 + 2x + 5x + 2 \quad 2, 5$

3. Group the terms. $\quad (5x^2 + 2x) + (5x + 2)$

4. Factor each group. $\quad x(5x + 2) + 1(5x + 2)$

5. $(5x + 2)$ is the GCF. $\quad (5x + 2)(x + 1)$

In this case, we found that

$$5x^2 + 7x + 2 = (5x + 2)(x + 1)$$

Which is the correct factorization, $(5x + 2)(x + 1)$ or $(x + 1)(5x + 2)$? Both are correct. The multiplication of polynomials is commutative, and the order in which the product is written makes no difference. You can write the factorization of $ax^2 + bx + c$ in *two* ways.

| **EXAMPLE 4**   **Factoring trinomials using the key number** | **PROBLEM 4** |

Factor completely:

a. $6x^2 - 3x + 4$ **b.** $4x^2 - 3 - 4x$

SOLUTION

a. We proceed by steps:

   1. Find the key number ($6 \cdot 4 = 24$). $6x^2 - 3x + 4$ (24)

   2. Find the factors of the key numbers and use them to rewrite the middle term. Unfortunately, it is impossible to find two integers with product 24 and sum -3. This trinomial is *not* factorable. It is a prime polynomial.

b. We first rewrite the polynomial (in descending order) as $4x^2 - 4x - 3$ and then proceed by steps.

   1. Find the key number $4x^2 - \underline{4x} - 3$ (−12)
      $[4 \cdot (-3) = -12]$.

   2. Find the factors of the key number and use them to rewrite the middle term. $4x^2 - \underline{6x + 2x} - 3$ $-6, 2$

   3. Group the terms. $(4x^2 - 6x) + (2x - 3)$

   4. Factor each group. $2x(2x - 3) + 1(2x - 3)$

   5. $(2x - 3)$ is the GCF. $(2x - 3)(2x + 1)$

$4x^2 - 4x - 3 = (2x - 3)(2x + 1)$ can be easily verified by multiplication.

PROBLEM 4

Factor completely:

a. $5x^2 - 2x + 2$

b. $3x^2 - 4 - 4x$

Teaching Tip

When factoring trinomials, $x^2 + bx + c$, we noted the importance of the correct model and order of terms. This is also true for the trinomial $ax^2 + bx + c$, as in Example 4(b).

| **EXAMPLE 5**   **Factoring trinomials with two variables using the key number** | **PROBLEM 5** |

Factor completely: $6x^2 + xy - y^2$

SOLUTION

1. Find the key number $[6 \cdot (-1) = -6]$. $6x^2 + \underline{xy} - y^2$ (−6)

2. Find the factors of the key number and use them to rewrite the middle term. $6x^2 + \underline{3xy - 2xy} - y^2$ $3, -2$

3. Group the terms. $(6x^2 + 3xy) - (2xy + y^2)$ $-(2xy + y^2) = -2xy - y^2$

4. Factor each group. $3x(2x + y) - y(2x + y)$

5. $(2x + y)$ is the GCF. $(2x + y)(3x - y)$

Thus $6x^2 + xy - y^2 = (2x + y)(3x - y)$.

PROBLEM 5

Factor completely:

$$2x^2 + xy - 3y^2$$

Answers

4. a. Prime   **b.** $(3x + 2)(x - 2)$
5. $(2x + 3y)(x - y)$

If the terms of the trinomial have a common factor, we factor it out first, as in the next example.

EXAMPLE 6   Factoring out common factors

Factor completely: $12x^3y^2 + 14x^2y^3 - 6xy^4$

SOLUTION   The greatest common factor of these three terms is $2xy^2$. Thus $12x^3y^2 + 14x^2y^3 - 6xy^4 = 2xy^2(6x^2 + 7xy - 3y^2)$. We then factor $6x^2 + 7xy - 3y^2$.

1. The key number is -18. $6x^2 + \underline{7xy} - 3y^2$

2. The factors of -18 with a sum of
7 are 9 and -2. Rewrite the middle term. $6x^2 + 9xy - 2xy - 3y^2$

3. Group the terms. $(6x^2 + 9xy) - (2xy + 3y^2)$

4. Factor each group. $3x(2x + 3y) - y(2x + 3y)$

5. $(2x + 3y)$ is the GCF. $(2x + 3y)(3x - y)$

Thus

$$12x^3y^2 + 14x^2y^3 - 6xy^4 = 2xy^2(6x^2 + 7xy - 3y^2)$$
$$= 2xy^2(2x + 3y)(3x - y)$$

You can check this by multiplying all the factors on the right-hand side of the equation.

PROBLEM 6

Factor completely:

$$12x^4y + 2x^3y^2 - 4x^2y^3$$

We have factored polynomials of the form $ax^2 + bx + c$, where $a > 0$. If a is negative, it's helpful to factor out -1 first and proceed as before.

EXAMPLE 7   Factoring $ax^2 + bx + c$ where a is negative

Factor completely: $-2x^2 - 7x - 5$

SOLUTION

$$-2x^2 - 7x - 5 = -1(2x^2 + 7x + 5)   \text{Factor out } -1.$$

1. Find the key number $[2 \cdot (5) = 10]$. $2x^2 + \underline{7x} + 5$

2. Find the factors of the key number
(5 and 2) that add up to the middle
term and rewrite the middle term. $-1[2x^2 + 5x + 2x + 5]$

3. Group the terms. $-1[(2x^2 + 5x) + (2x + 5)]$

4. Factor each group. $-1[x(2x + 5) + 1(2x + 5)]$

5. $(2x + 5)$ is the GCF. $-1[(2x + 5)(x + 1)]$

Since $-1 \cdot a = -a$,

$$-1(2x^2 + 7x + 5) = -(2x + 5)(x + 1)$$

PROBLEM 7

Factor completely:

$$-3x^2 + 14x + 24$$

Answers

6. $2x^2y(3x + 2y)(2x - y)$
7. $-(3x + 4)(x - 6)$

Finally, some polynomials that look complicated can be factored if we make a substitution. For example, to factor $6(y - 1)^2 - 5(y - 1) - 4$, we can think of $(y - 1)$ as A, and write $6A^2 - 5A - 4$, factor this polynomial, and substitute $(y - 1)$ for A in the final answer. We illustrate the procedure in the next example.

EXAMPLE 8 **Factoring by substitution**

Factor completely: $6(y - 1)^2 - 5(y - 1) - 4$

SOLUTION Let $A = (y - 1)$; that is,
$$6(y - 1)^2 - 5(y - 1) - 4 = 6A^2 - 5A - 4.$$

1. The key number is $6(-4) = -24$. $6A^2 - \underline{5A} - 4$

2. The factors of -24 adding to
   -5 are -8 and 3. $6A^2 - \underline{8A + 3A} - 4$

3. Group the terms. $(6A^2 - 8A) + (3A - 4)$

4. Factor each group. $2A(3A - 4) + 1(3A - 4)$

5. Factor out the GCF, $(3A - 4)$. $(3A - 4)(2A + 1)$

Substitute $(y - 1)$ for A and simplify:

$$6A^2 - 5A - 4 = (3A - 4)(2A + 1)$$
$$= [3(y - 1) - 4][2(y - 1) + 1]$$
$$= [3y - 3 - 4][2y - 2 + 1]$$
$$= (3y - 7)(2y - 1)$$

PROBLEM 8

Factor completely:

$$7(z - 3)^2 - 2(z - 3) - 5$$

_____

Answer

8. $(7z - 16)(z - 4)$

Calculate It Exercises

1. Can you factor $x^2 - 2$? The graph of $Y_1 = x^2 - 2$ is shown in Window 1. It crosses the horizontal axis, but $x^2 - 2$ is *not* factorable using integer coefficients. Can you explain what is wrong?

   Y₁=X²-2

   Window 1

2. To find the positive x-intercept of the points at which $x^2 - 2$ crosses the x-axis, let $Y_1 = x^2 - 2$ and $Y_2 = 0$, then press `2nd` `TRACE` `5` and `ENTER` three times to find the point

   of intersection. To find the negative x-intercept, use the `TRACE` key and arrows to move the cursor near the negative intercept. Press `2nd` `TRACE` `5` , and `ENTER` three times.

3. One of the x-intercepts of $x^2 - 2$ is $x = 1.4142136$. Do you recognize this number? If not, go to the home screen and press $\sqrt{2}$.

4. The x-intercepts of $x^2 - 4$ are $x = 2$ and $x = -2$. Also, $x^2 - 4 = (x + 2)(x - 2)$. If the x-intercepts of $x^2 - 2$ are $x = -\sqrt{2}$ and $x = \sqrt{2}$, what is the factorization of $x^2 - 2$?

5. Reexamine Exercise 1. Do you know what is wrong now?

Exercises 5.4

A In Problems 1–16, factor completely.

1. $x^2 + 5x + 6$
   $(x + 2)(x + 3)$

2. $x^2 + 15x + 56$
   $(x + 8)(x + 7)$

3. $a^2 + 7a + 10$
   $(a + 2)(a + 5)$

4. $a^2 + 10a + 24$
   $(a + 4)(a + 6)$

5. $x^2 + x - 12$
   $(x + 4)(x - 3)$

6. $x^2 + 5x - 6$
   $(x + 6)(x - 1)$

7. $x^2 - 2 + x$
   $(x + 2)(x - 1)$

8. $x^2 - 18 + 7x$
   $(x + 9)(x - 2)$

9. $x^2 - x - 2$
   $(x + 1)(x - 2)$

10. $x^2 - 5x - 14$
    $(x - 7)(x + 2)$

**Boost *your* GRADE
at mathzone.com!**

MathZone

- Practice Problems
- Self-Tests
- Videos
- NetTutor
- e-Professors

11. $x^2 - 3x - 10$
$(x + 2)(x - 5)$

12. $x^2 - 4x - 21$
$(x - 7)(x + 3)$

13. $a^2 - 16a + 63$
$(a - 7)(a - 9)$

14. $a^2 - 4a + 3$
$(a - 3)(a - 1)$

15. $y^2 + 22 - 13y$
$(y - 2)(y - 11)$

16. $y^2 + 11 - 12y$
$(y - 11)(y - 1)$

B **C** In Problems 17–46, factor completely.

17. $9x^2 + 37x + 4$
$(9x + 1)(x + 4)$

18. $2x^2 + 5x + 2$
$(2x + 1)(x + 2)$

19. $3a^2 - 5a - 2$
$(3a + 1)(a - 2)$

20. $8a^2 - 2a - 21$
$(4a - 7)(2a + 3)$

21. $2y^2 - 3y - 20$
$(2y + 5)(y - 4)$

22. $6y^2 - 13y - 5$
$(3y + 1)(2y - 5)$

23. $4x^2 - 11x + 6$
$(4x - 3)(x - 2)$

24. $16x^2 - 16x + 3$
$(4x - 3)(4x - 1)$

25. $6x^2 + x - 12$
$(2x + 3)(3x - 4)$

26. $20y^2 + y - 1$
$(5y - 1)(4y + 1)$

27. $21a^2 + 11a - 2$
$(3a + 2)(7a - 1)$

28. $18x^2 - 3x - 10$
$(6x - 5)(3x + 2)$

29. $6x^2 + 7xy - 3y^2$
$(2x + 3y)(3x - y)$

30. $3x^2 + 13xy - 10y^2$
$(3x - 2y)(x + 5y)$

31. $7x^4 - 10x^3y + 3x^2y^2$
$x^2(7x - 3y)(x - y)$

32. $6x^4 - 17x^3y + 5x^2y^2$
$x^2(3x - y)(2x - 5y)$

33. $15x^2y^3 - xy^4 - 2y^5$
$y^3(3x + y)(5x - 2y)$

34. $5x^2y^3 - 6xy^4 - 8y^5$
$y^3(5x + 4y)(x - 2y)$

35. $15x^3y^2 - 2x^2y^3 - 2xy^4$
$xy^2(15x^2 - 2xy - 2y^2)$

36. $4x^3y^2 - 13x^2y^3 - 3xy^4$
$xy^2(4x^2 - 13xy - 3y^2)$

37. $-2b^2 + 13b - 20$
$-(2b - 5)(b - 4)$

38. $-3b^2 - 16b + 12$
$-(3b - 2)(b + 6)$

39. $-12y^2 - 7y + 12$
$-(4y - 3)(3y + 4)$

40. $-12y^2 - 8y + 15$
$-(2y + 3)(6y - 5)$

41. $2(y + 2)^2 + (y + 2) - 3$
$(2y + 7)(y + 1)$

42. $3(y + 3)^2 - 11(y + 3) + 6$
$y(3y + 7)$

43. $2(x + 1)^2 - 13(x + 1) + 20$
$(2x - 3)(x - 3)$

44. $3(u - 1)^2 + 16(u - 1) - 12$
$(3u - 5)(u + 5)$

45. $-(a^2 + 2a)^2 - 2(a^2 + 2a) - 1$
$-(a + 1)^4$

46. $-(y^2 - 6y)^2 - 18(y^2 - 6y) - 81$
$-(y - 3)^4$

In Problems 47–50, factor completely the given expression.

47. The flow g (in hundreds of gallons per minute) in 100 ft of $2\frac{1}{2}$-in. rubber-lined hose when the friction loss is 21 lb/in.^{2} is given by

$$2g^2 + g - 21$$
$$(2g + 7)(g - 3)$$

48. The flow g (in hundreds of gallons per minute) in 100 ft of $2\frac{1}{2}$-in. rubber-lined hose when the friction loss is 55 lb/in.^{2} is given by

$$2g^2 + g - 55$$
$$(2g + 11)(g - 5)$$

49. The equivalent resistance R of two electric circuits is given by

$$2R^2 - 3R + 1$$
$$(2R - 1)(R - 1)$$

50. The time t at which an object thrown upward at 12 m/sec will be 4 m above the ground is given by

$$5t^2 - 12t + 4$$
$$(5t - 2)(t - 2)$$

SKILL CHECKER

Try the "Skill Checker" exercises so you'll be ready for the next section.

Multiply:

51. $(2a + b)^2$
$4a^2 + 4ab + b^2$

52. $(3a + 2b)^2$
$9a^2 + 12ab + 4b^2$

53. $(a - 2b)^2$
$a^2 - 4ab + 4b^2$

54. $(2a - 3b)^2$
$4a^2 - 12ab + 9b^2$

55. $(a + b)(a - b)$
$a^2 - b^2$

56. $(9a - 2b)(9a + 2b)$
$81a^2 - 4b^2$

57. $(2x - 3y)(2x + 3y)$
$4x^2 - 9y^2$

58. $(5x + 7y)(5x - 7y)$
$25x^2 - 49y^2$

USING YOUR KNOWLEDGE

Factoring Applications

The ideas presented in this section are important in many fields. Use your knowledge to factor completely the expressions given in Problems 59 and 60.

59. The deflection of a beam of length L at a distance of 3 ft from its end is given by

$$2L^2 - 9L + 9 \quad (L - 3)(2L - 3)$$

60. If the distance from the end of the beam in Problem 59 is x feet, then the deflection is given by

$$2L^2 - 3xL + x^2 \quad (2L - x)(L - x)$$

61. The distance (in meters) traveled in t seconds by an object thrown upward at 12 m/sec is

$$-5t^2 + 12t$$

To determine the time at which the object will be 7 m above ground, we must solve the equation

$$5t^2 - 12t + 7 = 0$$

Factor completely the trinomial on the left-hand side of this equation. $(5t - 7)(t - 1)$

WRITE ON

62. The factors of a polynomial are $3x^3$, $x + 5$, and $2x - 1$. Write a procedure that can be used to find the polynomial. Answers may vary.

63. If you multiply $(2x - 2)(x - 1)$, you get $2x^2 - 4x + 2$. However, $(2x - 2)(x - 1)$ is *not* the complete factorization for $2x^2 - 4x + 2$. Why not? Answers may vary.

64. Without using trial and error, explain why the polynomial $x^2 + 3x - 17$ is prime. Answers may vary.

MASTERY TEST

If you know how to do these problems, you have learned your lesson!

Factor completely if possible:

65. $6x^2 + 13x + 6$
$(3x + 2)(2x + 3)$

66. $6x^2 - 17x + 5$
$(3x - 1)(2x - 5)$

67. $x^2 - 3x - 10$
$(x - 5)(x + 2)$

68. $x^2 - 5x - 6$
$(x - 6)(x + 1)$

69. $x^2 - 2xy + 5y^2$
Not factorable

70. $x^2 - 7xy + 10y^2$
$(x - 5y)(x - 2y)$

71. $12x^4y + 2x^3y^2 - 4x^2y^3$
$2x^2y(2x - y)(3x + 2y)$

72. $2x^2 + xy - 3y^2$
$(2x + 3y)(x - y)$

73. $5x^2 - 2x + 2$
Not factorable

74. $3x^2 - 4 - 4x$
$(3x + 2)(x - 2)$

75. $-8x^2 + 2x + 21$
$-(4x - 7)(2x + 3)$

76. $-21x^2 - 11x + 2$
$-(7x - 1)(3x + 2)$

77. $2(y - 3)^2 + 5(y - 3) + 2$
$(2y - 5)(y - 1)$

78. $-9(y - 2)^2 - 37(y - 2) - 4$
$-(9y - 17)(y + 2)$

5.5 # SPECIAL FACTORING

To Succeed, Review How To ...

1. Square a binomial (pp. 360–361).

2. Multiply a binomial sum by a binomial difference (pp. 359–362).

3. Multiply a binomial and a trinomial (pp. 357–359).

Objectives

A Factor a perfect square trinomial.

B Factor the difference of two squares.

C Factor the sum or difference of two cubes.

GETTING STARTED

Applying Factoring to Bending Moments

At x feet from its support, the bending moment (the product of a quantity and the distance from its perpendicular axis) for the crane in the photograph involves the expression

$$\frac{w}{2}(x^2 - 20x + 100)$$

where w is the weight of the crane in pounds per foot. The expression $x^2 - 20x + 100$ is the result of expanding the binomial $(x - 10)^2$ and is called a *perfect square trinomial.* We learned that

$$(a + b)^2 = a^2 + 2ab + b^2 \quad \text{and} \quad (a - b)^2 = a^2 - 2ab + b^2$$

Perfect square trinomial   Perfect square trinomial

The trinomials on the right-hand side of the equations are perfect square trinomials.

The expression $x^2 - 20x + 100$ is also a perfect square trinomial. Let's factor it using the fact that $(a - b)^2 = a^2 - 2ab + b^2$. Write $x^2 - 20x + 100$ as $x^2 - 2 \cdot x \cdot 10 + 10^2$, where $a = x$ and $b = 10$. We then have $x^2 - 2 \cdot x \cdot 10 + 10^2 = (x - 10)^2$.

A Factoring Perfect Square Trinomials

To factor perfect square trinomials, we rewrite $(x + a)^2 = x^2 + 2ax + a^2$ and $(x - a)^2 = x^2 - 2ax + a^2$ as follows.

PROCEDURE

Factoring Perfect Square Trinomials

$$x^2 + 2ax + a^2 = (x + a)^2$$
$$x^2 - 2ax + a^2 = (x - a)^2$$

In a perfect square trinomial the following applies:

1. The first and last terms (x^2 and a^2) are perfect squares and positive.

2. The middle term is twice the product of the two terms in the binomial being squared ($2ax$), or it is the additive inverse of this product ($-2ax$).

If you can write the trinomial in the form shown on the left, then you may factor it as shown on the right. To factor $x^2 + 6x + 9$, note that the first and last terms are perfect squares [$(x)^2 = x^2$ and $3^2 = 9$] and positive and the middle term is $2 \cdot 3 \cdot x = 6x$. Hence

$$x^2 + 6x + 9 = x^2 + 2 \cdot 3x + 3^2 = (x + 3)^2$$

Teaching Tip

Have students square the following and make a list of their results:

$(x^1)^2$	x^2
$(x^2)^2$	x^4
$(x^3)^2$	x^6
$(x^4)^2$	x^8
$(x^5)^2$	x^{10}

They should be able to identify them as perfect squares and in general, know the requirements for the exponent if x^n is a perfect square.

In trying to factor $x^2 - 8x + 16$ we notice the first and last terms are perfect squares and positive.

$$x^2 - 8x + 16$$

The additive inverse of twice the two terms of the binomial

$(x)^2 \qquad (4)^2$ Middle term check: $-2 \cdot x \cdot 4 = -8x$

$(x - 4)^2$

The sign between the terms in the binomial is the same as the middle sign of the trinomial. You can also factor $x^2 - 8x + 16$ by finding two factors whose product is 16 and whose sum is -8, as we did in the previous section. The factors are -4 and -4. Thus $x^2 - 8x + 16 = (x - 4)(x - 4) = (x - 4)^2$.

EXAMPLE 1   Factoring perfect square trinomials in one variable

Factor completely:

a. $x^2 - 10x + 25$ **b.** $x^2 + 12x + 36$ **c.** $x^2 + 7x + 49$

SOLUTION

a. $x^2 - 10x + 25 = x^2 - 2 \cdot 5 \cdot x + 5^2 = (x - 5)^2$

b. $x^2 + 12x + 36 = x^2 + 2 \cdot 6 \cdot x + 6^2 = (x + 6)^2$

You can verify that parts **a** and **b** are correct by expanding $(x - 5)^2$ and $(x + 6)^2$.

c. $x^2 + 7x + 49$ has positive perfect squares for the first (x^2) and last (7^2) terms. However, the middle term is *not* $2 \cdot 7 \cdot x$. Thus $x^2 + 7x + 49$ is not a perfect square; it isn't even factorable. We cannot find two integers whose product is 49 and whose sum is 7, so, by the *ac* test, $x^2 + 7x + 49$ is prime.

PROBLEM 1

Factor completely:

a. $x^2 - 16x + 64$

b. $x^2 + 8x + 64$

c. $x^2 + 18x + 81$

We can use the same idea to factor trinomials in two variables. To factor $25x^2 + 20xy + 4y^2$, we write

$$25x^2 + 20xy + 4y^2 = (5x)^2 + 2 \cdot (2y)(5x) + (2y)^2$$
$$= (5x + 2y)^2$$

Calculate It   Factor Perfect Square Trinomials by Graphing

The graph of $x^2 - 10x + 25$ in Example 1(a) touches the horizontal axis at $x = 5$ (Window 1), and its factorization is $(x - 5)^2$. Graph $x^2 + 10x + 25$. Where does it touch the horizontal axis?

Thus if the graph of a perfect square trinomial just touches the horizontal axis at $x = a$, the factorization is $(x - a)^2$. If you know your algebra, you noticed that when the middle term of a perfect square trinomial is preceded by a *minus* sign, the factorization is of the form $(x - a)^2$.

Can you use your calculator to help you factor $x^2 + 7x + 49$ in Example 1(c)? No! The graph does not touch or cut the horizontal axis (Window 2).

To graph this polynomial, you need to set your viewing window so that Ymin $= 0$ and Ymax $= 100$. Then you have to know the *ac* test to discover that $x^2 + 7x + 49$ is *prime*.

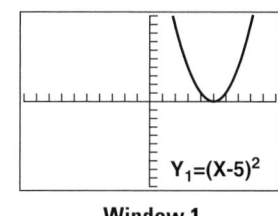

Window 1

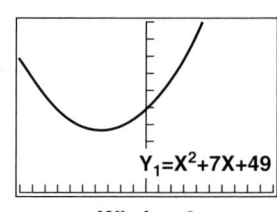

Window 2

Answers

1. a. $(x - 8)^2$ **b.** Prime
c. $(x + 9)^2$

EXAMPLE 2 **Factoring perfect square trinomials in two variables**

Factor completely:

a. $9x^2 - 12xy + 4y^2$ **b.** $4x^2 - 10xy + 25y^2$ **c.** $16x^2 + 24xy + 9y^2$

SOLUTION

a. $9x^2 - 12xy + 4y^2 = (3x)^2 - 2 \cdot 3x \cdot 2y + (2y)^2 = (3x - 2y)^2$

b. Even though the first term, $(2x)^2$, and last term, $(5y)^2$, are perfect squares, the middle term is *not* $2 \cdot 2x \cdot 5y$. Thus $4x^2 - 10xy + 25y^2$ is not a perfect square. The *ac* test shows that it is prime.

c. $16x^2 + 24xy + 9y^2 = (4x)^2 + 2 \cdot 4x \cdot 3y + (3y)^2 = (4x + 3y)^2$

PROBLEM 2

Factor completely:

a. $6x^2 + 30xy + 25y^2$

b. $4x^2 - 12xy + 9y^2$

B **Factoring the Difference of Two Squares**

Web It

For more details about special factoring, go to link 5-5-1 at mhhe.com/bello.

For a quiz on special factoring with answers, go to link 5-5-2 at mhhe.com/bello.

Teaching Tip

$x^2 + a^2$ is not factorable over the reals, but we will learn later that it can be factored over the complex numbers.

As you recall from Section 5.2,

$$(x + a)(x - a) = x^2 - a^2$$

where $x^2 - a^2$ is called the **difference of two squares.** The corresponding factoring formula is as follows.

PROCEDURE

Factoring the Difference of Two Squares

$$x^2 - a^2 = (x + a)(x - a)$$

NOTE

$x^2 + a^2$ is not factorable using real numbers.

To factor $x^2 - 9$, we write

$$x^2 - 9 = x^2 - 3^2 = (x + 3)(x - 3)$$

Similarly,

$$25x^2 - 16y^2 = (5x)^2 - (4y)^2 = (5x + 4y)(5x - 4y)$$

EXAMPLE 3 **Factoring the difference of two squares**

Factor completely:

a. $9x^2 - 1$ **b.** $81x^4 - 16y^4$ **c.** $x^2 + 49$

SOLUTION

a. $9x^2 - 1 = (3x)^2 - 1^2 = (3x + 1)(3x - 1)$

b. $81x^4 - 16y^4 = (9x^2)^2 - (4y^2)^2$ — Difference of two squares

$$= (9x^2 + 4y^2)(9x^2 - 4y^2)$$
$$= (9x^2 + 4y^2)(3x + 2y)(3x - 2y)$$

Not factorable   Factored

c. Since $x^2 + 49$ is not a difference of squares, it is prime.

PROBLEM 3

Factor completely:

a. $4x^2 - 25$

b. $x^2 + 64$

c. $16x^4 - 81y^4$

Answers

2. a. Prime   **b.** $(2x - 3y)^2$
3. a. $(2x + 5)(2x - 5)$ **b.** Prime
c. $(4x^2 + 9y^2)(2x + 3y)(2x - 3y)$

The polynomial $(x + y)^2 - 4$ is also the difference of two squares. If you think of $(x + y)$ as A and 4 as 2^2, you are factoring

$$A^2 - 2^2 = (A + 2)(A - 2)$$

or equivalently

$$(x + y)^2 - 2^2 = [(x + y) + 2][(x + y) - 2]$$
$$= (x + y + 2)(x + y - 2)$$

If $(x + y)^2$ appears as $x^2 + 2xy + y^2$, factor it first. The procedure is

$x^2 + 2xy + y^2 - 4$	Given.
$= (x^2 + 2xy + y^2) - 4$	Group the first three terms.
$= (x + y)^2 - 2^2$	Factor the trinomial.
$= [(x + y) + 2][(x + y) - 2]$	Factor the difference of two squares.
$= (x + y + 2)(x + y - 2)$	

EXAMPLE 4 **Factoring by grouping**	**PROBLEM 4**

Factor completely:

a. $x^2 - 6x + 9 - y^2$

b. $x^2 + 6xy + 9y^2 - 4$

SOLUTION

a. If you try to factor by grouping the four terms into pairs as $(x^2 - 6x) + (9 - y^2)$ and factor each binomial, you get $x(x - 6) + (3 + y)(3 - y)$ but there is no common binomial factor. However, the first three terms are a perfect square trinomial. Thus

$$x^2 - 6x + 9 = (x - 3)^2$$

and

$x^2 - 6x + 9 - y^2 = (x^2 - 6x + 9) - y^2$	Group the first three terms.
$= (x - 3)^2 - y^2$	The difference of two squares
$= [(x - 3) + y][(x - 3) - y]$	
$= (x + y - 3)(x - y - 3)$	

b. Since $x^2 + 6xy + 9y^2 = (x + 3y)^2$,

$x^2 + 6xy + 9y^2 - 4 = (x^2 + 6xy + 9y^2) - 4$	Group the first three terms.
$= (x + 3y)^2 - 2^2$	The difference of two squares
$= [(x + 3y) + 2][(x + 3y) - 2]$	
$= (x + 3y + 2)(x + 3y - 2)$	

PROBLEM 4

Factor completely:

a. $x^2 + 2xy + y^2 - z^2$

b. $x^2 + 10xy + 25y^2 - 16$

C The Sum and Difference of Two Cubes

We know how to factor the difference of two squares. Can we factor the difference of two cubes—that is, $x^3 - a^3$? Yes! We can also factor $x^3 + a^3$, the sum of two cubes. Here are the formulas.

Answers

4. a. $(x + y + z)(x + y - z)$
b. $(x + 5y + 4)(x + 5y - 4)$

Web It

For more details on factoring cubes, go to link 5-5-3 at mhhe.com/bello.

For more details on special factoring, go to link 5-5-4 at mhhe.com/bello.

PROCEDURE

Factoring the Sum and Difference of Two Cubes

$$x^3 + a^3 = (x + a)(x^2 - ax + a^2)$$

$$x^3 - a^3 = (x - a)(x^2 + ax + a^2)$$

We did not give corresponding product formulas, so we verify these results:

$$\begin{array}{r} x^2 - ax + a^2 \\ x + a \\ \hline ax^2 - a^2x + a^3 \\ x^3 - ax^2 + a^2x \\ \hline x^3 \qquad\qquad + a^3 \end{array}$$

Multiply $a(x^2 - ax + a^2)$.
Multiply $x(x^2 - ax + a^2)$.

Calculate It

Using Graphs to Check the Factorization of Cubes

The graph of $x^3 + 125$ cuts the horizontal axis at $x = -5$, so you know that $x - (-5) = x + 5$ is a factor of $x^3 + 125$.

You can verify that $x^3 + 125 = (x + 5)(x^2 - 5x + 25)$ by graphing $x^3 + 125$ and $(x + 5)(x^2 - 5x + 25)$ and making sure you get the same graph. However, only algebra can help you find the other factor, $x^2 - 5x + 25$. If you divide $x^3 + 125$ by $x + 5$, you get $x^2 - 5x + 25$. The moral continues to be: Even with a calculator, you need algebra!

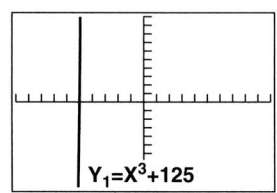
$Y_1 = X^3 + 125$

Similarly,

$$\begin{array}{r} x^2 + ax + a^2 \\ x - a \\ \hline -ax^2 - a^2x - a^3 \\ x^3 + ax^2 + a^2x \\ \hline x^3 \qquad\qquad - a^3 \end{array}$$

Multiply $-a(x^2 + ax + a^2)$.
Multiply $x(x^2 + ax + a^2)$.

Now that we have verified the formulas, we can factor sums and differences of cubes. For example,

$$\begin{aligned} 8 - x^3 &= 2^3 - x^3 \\ &= (2 - x)(2^2 + 2x + x^2) \\ &= (2 - x)\underbrace{(4 + 2x + x^2)}_{\text{Not factorable}} \end{aligned}$$

Middle term of the trinomial factor is the product of the two terms of the binomial factor with the opposite sign.

and

$$\begin{aligned} x^3 + 125 &= x^3 + 5^3 \\ &= (x + 5)(x^2 - 5x + 5^2) \\ &= (x + 5)\underbrace{(x^2 - 5x + 25)}_{\text{Not factorable}} \end{aligned}$$

EXAMPLE 5   **Factoring the sum and difference of two cubes**

Factor completely:

a. $27 - 8x^3$ **b.** $125x^3 + 64y^3$ **c.** $(x + y)^3 - 8$

SOLUTION

a. $27 - 8x^3 = 3^3 - (2x)^3$

These will always be opposite signs.

$$= (3 - 2x)[3^2 + 3 \cdot 2x + (2x)^2]$$
$$= (3 - 2x)(9 + 6x + 4x^2)$$

b. $125x^3 + 64y^3 = (5x)^3 + (4y)^3$

$$= (5x + 4y)[(5x)^2 - 5x \cdot 4y + (4y)^2]$$
$$= (5x + 4y)(25x^2 - 20xy + 16y^2)$$

PROBLEM 5

Factor completely:

a. $64 + \dfrac{1}{27}x^3$

b. $8x^3 + y^3$

c. $(a - b)^3 + 1$

Answers

5. a. $(4 + \frac{1}{3}x)(16 - \frac{4}{3}x + \frac{1}{9}x^2)$
b. $(2x + y)(4x^2 - 2xy + y^2)$
c. $(a - b + 1)(a^2 - 2ab + b^2 - a + b + 1)$

c. $(x + y)^3 - 8$ is the difference of two cubes. As before, think of $(x + y)$ as A and 8 as 2^3. Thus you are factoring

$$A^3 - 2^3 = (A - 2)(A^2 + 2A + 2^2)$$

then replace A with $(x + y)$ to get

$$[(x + y)^3 - 2^3] = [(x + y) - 2][(x + y)^2 + 2(x + y) + 4] \quad \text{Simplify.}$$
$$= (x + y - 2)(x^2 + 2xy + y^2 + 2x + 2y + 4)$$

EXAMPLE 6 **Factoring the difference of squares and then cubes**

Factor completely: $x^6 - 64$

SOLUTION x^6 and 64 are both perfect squares and perfect cubes. Factor first as the difference of two squares and then as the sum and difference of cubes.

$$x^6 - 64 = x^6 - 2^6$$
$$= (x^3)^2 - (2^3)^2 \quad \text{Difference of squares}$$
$$= (x^3 + 2^3)(x^3 - 2^3) \quad \text{Sum/difference of cubes}$$
$$= [(x + 2)(x^2 - 2x + 4)][(x - 2)(x^2 + 2x + 4)]$$
$$= (x + 2)(x - 2)(x^2 + 2x + 4)(x^2 - 2x + 4)$$

You can also factor $x^6 - 64$ by writing it as the difference of two cubes;

$$x^6 - 64 = (x^2)^3 - (2^2)^3$$
$$= (x^2 - 2^2)[(x^2)^2 + 2^2 x^2 + (2^2)^2]$$
$$= (x^2 - 4)(x^4 + 4x^2 + 16)$$
$$= (x + 2)(x - 2)(x^4 + 4x^2 + 16)$$

But the expression $(x^4 + 4x^2 + 16)$ is **not** completely factored. You can factor it by first *adding* and *subtracting* $4x^2$ to obtain

$$x^4 + 4x^2 + 16 = x^4 + 4x^2 + 4x^2 + 16 - 4x^2$$
$$= (x^4 + 8x^2 + 16) - 4x^2 \quad \text{By the associative property}$$
$$= (x^2 + 4)^2 - 4x^2 \quad \text{Factor } x^4 + 8x^2 + 16.$$
$$= [(x^2 + 4) + 2x][(x^2 + 4) - 2x] \quad \text{Factor the difference of two squares.}$$
$$= (x^2 + 2x + 4)(x^2 - 2x + 4) \quad \text{By the commutative property}$$

Thus, as before,

$$x^6 - 64 = (x^2)^3 - (2^2)^3 = (x + 2)(x - 2)(x^2 + 2x + 4)(x^2 - 2x + 4)$$

PROBLEM 6

Factor completely: $x^6 - 1$

Answer

6. $(x + 1)(x - 1)(x^2 + x + 1)(x^2 - x + 1)$

NOTE

As you can see, it's easier to factor $x^6 - 64$ when it is *first* written as the difference of two squares as we did here.

Calculate It Exercises

1. The graph of the quadratic function $f(x)$ in Window 1 touches the x-axis at $x = 3$. What is the y-intercept? What is $f(x)$?

2. If the graph of the quadratic function $g(x)$ touches the x-axis at $x = -2$ only and intersects the y-axis at a distance of 4 units from the origin, there are two possible definitions for $g(x)$. What are these two definitions?

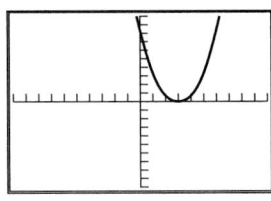

Window 1

3. The function $h(x)$ in Window 2 is a cubic function and has an x-intercept of 2 only and a y-intercept of -8. What is $h(x)$?

4. If there is another cubic function $i(x)$ with an x-intercept of -2 only and a y-intercept of -8, what is $i(x)$?

5. Use your calculator to show that $x^2 + 9 \neq (x + 3)^2$.

6. Use your calculator to show that $x^2 - 8x - 16 \neq (x - 4)^2$.

7. Given a polynomial function $f(x)$, how can you tell if $f(x)$ cannot be factored as a product of linear factors with integer coefficients?

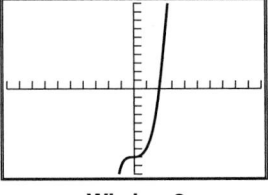

Window 2

Exercises 5.5

Boost *your* GRADE at mathzone.com!

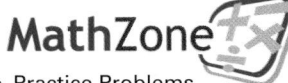

- Practice Problems
- Self-Tests
- Videos
- NetTutor
- e-Professors

A In Problems 1–24, factor completely.

1. $x^2 + 2x + 1$
$(x + 1)^2$

2. $x^2 + 20x + 100$
$(x + 10)^2$

3. $y^2 + 22y + 121$
$(y + 11)^2$

4. $y^2 + 14x + 49$
$(y + 7)^2$

5. $1 + 4x + 4x^2$
$(1 + 2x)^2$

6. $1 + 6x + 9x^2$
$(1 + 3x)^2$

7. $9x^2 + 30xy + 25y^2$
$(3x + 5y)^2$

8. $25x^2 + 30xy + 9y^2$
$(5x + 3y)^2$

9. $36a^2 + 48a + 16$
$4(3a + 2)^2$

10. $9a^2 + 60a + 100$
$(3a + 10)^2$

11. $y^2 - 2y + 1$
$(y - 1)^2$

12. $25 - 10y + y^2$
$(5 - y)^2$

13. $49 - 14x + x^2$
$(7 - x)^2$

14. $x^2 - 100x + 2500$
$(x - 50)^2$

15. $49a^2 - 28ax + 4x^2$
$(7a - 2x)^2$

16. $4a^2 - 12ax + 9x^2$
$(2a - 3x)^2$

17. $16x^2 - 24xy + 9y^2$
$(4x - 3y)^2$

18. $9x^2 - 42xy + 49y^2$
$(3x - 7y)^2$

19. $9x^4 + 12x^2 + 4$
$(3x^2 + 2)^2$

20. $25y^4 + 20y^2 + 4$
$(5y^2 + 2)^2$

21. $16x^4 - 24x^2 + 9$
$(4x^2 - 3)^2$

22. $4y^4 - 20y^2 + 25$
$(2y^2 - 5)^2$

23. $1 + 2x^2 + x^4$
$(1 + x^2)^2$

24. $4 + 12x^2 + 9x^4$
$(2 + 3x^2)^2$

B In Problems 25–44, factor completely.

25. $y^2 - 64$ $(y + 8)(y - 8)$

26. $y^2 - 121$ $(y + 11)(y - 11)$

27. $a^2 - \dfrac{1}{9}$ $\left(a + \dfrac{1}{3}\right)\left(a - \dfrac{1}{3}\right)$

28. $x^2 - \dfrac{1}{16}$ $\left(x + \dfrac{1}{4}\right)\left(x - \dfrac{1}{4}\right)$

29. $64 - b^2$ $(8 + b)(8 - b)$

30. $81 - b^2$ $(9 + b)(9 - b)$

31. $36a^2 - 49b^2$ $(6a + 7b)(6a - 7b)$

32. $36a^2 - 25b^2$ $(6a + 5b)(6a - 5b)$

33. $\dfrac{x^2}{9} - \dfrac{y^2}{16}$ $\left(\dfrac{x}{3} + \dfrac{y}{4}\right)\left(\dfrac{x}{3} - \dfrac{y}{4}\right)$

34. $\dfrac{x^2}{16} - \dfrac{9y^2}{25}$ $\left(\dfrac{x}{4} + \dfrac{3y}{5}\right)\left(\dfrac{x}{4} - \dfrac{3y}{5}\right)$

35. $a^2 + 4ab + 4b^2 - c^2$
$(a + 2b + c)(a + 2b - c)$

36. $9a^2 + 6ab + b^2 - 1$
$(3a + b + 1)(3a + b - 1)$

37. $4x^2 - 4xy + y^2 - 1$
$(2x - y + 1)(2x - y - 1)$

38. $9x^2 - 30xy + 25y^2 - 9$
$(3x - 5y + 3)(3x - 5y - 3)$

39. $9y^2 - 12xy + 4x^2 - 25$
$(3y - 2x + 5)(3y - 2x - 5)$

40. $16y^2 - 40xy + 25x^2 - 36$
$(4y - 5x + 6)(4y - 5x - 6)$

41. $16a^2 - (x^2 + 6xy + 9y^2)$
$(4a + x + 3y)(4a - x - 3y)$

42. $25a^2 - (4x^2 - 4xy + y^2)$
$(5a + 2x - y)(5a - 2x + y)$

43. $y^2 - a^2 + 2ab - b^2$
$(y + a - b)(y - a + b)$

44. $9y^2 - 9x^2 + 6xz - z^2$
$(3y + 3x - z)(3y - 3x + z)$

C In Problems 45–70, factor completely.

45. $x^3 + 125$
$(x + 5)(x^2 - 5x + 25)$

46. $x^3 + 64$
$(x + 4)(x^2 - 4x + 16)$

47. $1 + a^3$
$(1 + a)(1 - a + a^2)$

48. $343 + a^3$
$(7 + a)(49 - 7a + a^2)$

49. $8x^3 + y^3$
$(2x + y)(4x^2 - 2xy + y^2)$

50. $125x^3 + 8y^3$
$(5x + 2y)(25x^2 - 10xy + 4y^2)$

51. $x^3 - 1$
$(x - 1)(x^2 + x + 1)$

52. $x^3 - 216$
$(x - 6)(x^2 + 6x + 36)$

53. $125a^3 - 8b^3$
$(5a - 2b)(25a^2 + 10ab + 4b^2)$

54. $216a^3 - 125b^3$ $(6a - 5b)(36a^2 + 30ab + 25b^2)$

55. $x^6 - 64$ $(x + 2)(x - 2)(x^2 - 2x + 4)(x^2 + 2x + 4)$

56. $y^6 - 1$ $(y + 1)(y - 1)(y^2 + y + 1)(y^2 - y + 1)$

57. $x^6 - \dfrac{1}{64}$ $\left(x + \dfrac{1}{2}\right)\left(x - \dfrac{1}{2}\right)\left(x^2 - \dfrac{1}{2}x + \dfrac{1}{4}\right)\left(x^2 + \dfrac{1}{2}x + \dfrac{1}{4}\right)$

58. $y^6 - 729$ $(y + 3)(y - 3)(y^2 - 3y + 9)(y^2 + 3y + 9)$

59. $\dfrac{x^6}{64} - 1$ $\left(\dfrac{x}{2} + 1\right)\left(\dfrac{x}{2} - 1\right)\left(\dfrac{x^2}{4} - \dfrac{x}{2} + 1\right)\left(\dfrac{x^2}{4} + \dfrac{x}{2} + 1\right)$

60. $\dfrac{y^6}{729} - 1$ $\left(\dfrac{y}{3} + 1\right)\left(\dfrac{y}{3} - 1\right)\left(\dfrac{y^2}{9} - \dfrac{y}{3} + 1\right)\left(\dfrac{y^2}{9} + \dfrac{y}{3} + 1\right)$

61. $(x - y)^3 + 1$ $(x - y + 1)(x^2 - 2xy + y^2 - x + y + 1)$

62. $(x + 2y)^3 + 8$
$(x + 2y + 2)(x^2 + 4xy + 4y^2 - 2x - 4y + 4)$

63. $1 + (x + 2y)^3$
$(1 + x + 2y)(1 - x - 2y + x^2 + 4xy + 4y^2)$

64. $27 + (x + y)^3$
$(3 + x + y)(9 - 3x - 3y + x^2 + 2xy + y^2)$

65. $(y - 2x)^3 - 1$
$(y - 2x - 1)(y^2 - 4xy + 4x^2 + y - 2x + 1)$

66. $(y - 4x)^3 - 1$
$(y - 4x - 1)(y^2 - 8xy + 16x^2 + y - 4x + 1)$

67. $27 - (x + 2y)^3$
$(3 - x - 2y)(9 + 3x + 6y + x^2 + 4xy + 4y^2)$

68. $8 - (y - 4x)^3$
$(2 - y + 4x)(4 + 2y - 8x + y^2 - 8xy + 16x^2)$

69. $64 + (x^2 - y^2)^3$
$(4 + x^2 - y^2)(16 - 4x^2 + 4y^2 + x^4 - 2x^2y^2 + y^4)$

70. $27 + (y^2 - x^2)^3$
$(3 + y^2 - x^2)(9 - 3y^2 + 3x^2 + y^4 - 2x^2y^2 + x^4)$

SKILL CHECKER

Try the "Skill Checker" exercises so you'll be ready for the next section.

Multiply:

71. $(x + 3)(x - 5)$
$x^2 - 2x - 15$

72. $(x - 5)(x + 7)$
$x^2 + 2x - 35$

73. $(x - 8)(x + 2)$
$x^2 - 6x - 16$

74. $(x + 2y)^2$
$x^2 + 4xy + 4y^2$

75. $(2x - 3y)^2$
$4x^2 - 12xy + 9y^2$

76. $(x + 2y)(x - 2y)$
$x^2 - 4y^2$

USING YOUR KNOWLEDGE

Is There Demand for the Supply?

Have you heard of supply and demand? In business the supply and demand of a product can be expressed by using a polynomial. In Problems 77–79, factor completely the given expression.

77. A business finds that when x units of an item are demanded by consumers, the price per unit is given by

$$D(x) = 100 - x^2$$

Factor completely $100 - x^2$. $(10 + x)(10 - x)$

78. When x units are supplied by sellers, the price per unit of an item is given by

$$S(x) = x^3 + 216$$

Factor completely $x^3 + 216$. $(x + 6)(x^2 - 6x + 36)$

79. When x units of a certain item are produced, the cost is given by

$$C(x) = 8x^3 + 1$$

Factor completely $8x^3 + 1$. $(2x + 1)(4x^2 - 2x + 1)$

WRITE ON

80. In Example 6, $(x + 2)(x - 2)(x^2 - 2x + 4)(x^2 + 2x + 4)$ is the "preferred" factorization of $x^6 - 64$. Explain why you think this is true. Answers may vary.

81. Write a procedure to determine whether a quadratic trinomial is a perfect square trinomial. Answers may vary.

82. Find two different values of k that will make $9x^2 + kx + 4$ a perfect square trinomial. Explain the procedure you used to find your answer. 12, −12; answers may vary.

83. Find a value for k that will make $4x^2 + 12x + k$ a perfect square trinomial. Describe the procedure you used to find your answer. 9; answers may vary.

84. Explain why $a^2 + b^2$ cannot be factored. Answers may vary.

85. Can $a^4 + 64$ be factored? (*Hint:* Add and subtract $16a^2$.)
$$a^4 + 64 = a^4 + 16a^2 + 64 - 16a^2$$
$$= (a^4 + 16a^2 + 64) - 16a^2$$
Can you factor it now? Yes; yes

MASTERY TEST

If you know how to do these problems, you have learned your lesson!

Factor completely if possible:

86. $64 - \dfrac{1}{27}x^3$
$(4 - \frac{1}{3}x)(16 + \frac{4}{3}x + \frac{1}{9}x^2)$

87. $8x^3 + 27y^3$
$(2x + 3y)(4x^2 - 6xy + 9y^2)$

88. $x^6 - 1$
$(x + 1)(x - 1)(x^2 - x + 1)(x^2 + x + 1)$

89. $x^6 + 1$
$(x^2 + 1)(x^4 - x^2 + 1)$

90. $16x^2 - 1$
$(4x + 1)(4x - 1)$

91. $16x^4 - 81y^4$
$(4x^2 + 9y^2)(2x + 3y)(2x - 3y)$

92. $x^2 + 2xy + y^2 + 16$
Not factorable

93. $x^2 + 10xy + 25y^2 - 16$
$(x + 5y + 4)(x + 5y - 4)$

94. $4x^2 - 4xy + 9y^2$
Not factorable

95. $4x^2 - 12xy + 9y^2$
$(2x - 3y)^2$

96. $x^2 - 16x + 64$
$(x - 8)^2$

97. $x^2 + 18x + 81$
$(x + 9)^2$

98. $(x - y)^4 - (x - y)^2$
$(x - y)^2(x - y + 1)(x - y - 1)$

99. $(x + y)^2 - (y - x)^2$
$4xy$

100. $(x - y)^3 + 125$
$(x - y + 5)(x^2 - 2xy + y^2 - 5x + 5y + 25)$

101. $125 + (x + y)^3$
$(5 + x + y)(25 - 5x - 5y + x^2 + 2xy + y^2)$

5.6   GENERAL METHODS OF FACTORING

To Succeed, Review How To . . .

1. Use the factoring formulas (pp. 368, 370, 375, 377, 384, 386, 388).

Objective

A Factor a polynomial using the procedure given in the text.

GETTING STARTED   Applying Factoring to Your Arteries

Have you heard of blocked arteries? The velocity of the blood inside a blocked artery (see photo) depends on the thickness of the inside wall (r) and the diameter of the artery to the outside wall (R) and is given by

$$CR^2 - Cr^2$$

where C is a constant. How do we factor this expression? We shall follow a general pattern that uses one or more of the techniques we've learned.

Large Artery

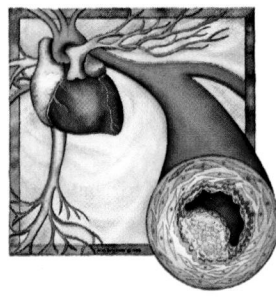

A   Factoring Polynomials

One of the most important skills when factoring polynomials is to know when the polynomial is completely factored. Here are the guidelines you need.

Completely Factored Polynomial	A polynomial is **completely factored** when
	1. The polynomial is written as the product of prime polynomials with **integer** coefficients.
	2. All the polynomial factors are prime, *except* that monomial factors need not be factored completely. $6x^3(2x + 1)$ is the factored form of $12x^4 + 6x^3$. You don't need to write $6x^3$ as $2 \cdot 3 \cdot x \cdot x \cdot x$.

Web It

For some practice with the general factoring strategy, go to link 5-6-1 at mhhe.com/bello.

PROCEDURE

A General Factoring Strategy

1. Factor out the GCF, if there is one.

2. Count the number of terms in the given polynomial (or inside the parentheses if the GCF was factored out).

    A. If there are *two terms,* check for

- Difference of two squares
$$x^2 - a^2 = (x + a)(x - a)$$

- Difference of two cubes
$$x^3 - a^3 = (x - a)(x^2 + ax + a^2)$$

- Sum of two cubes
$$x^3 + a^3 = (x + a)(x^2 - ax + a^2)$$

      The sum of two squares, $x^2 + a^2$, is not factorable.

    B. If there are *three terms,* check for

- Perfect square trinomial
$$x^2 + 2ab + b^2 = (x + a)^2$$
$$x^2 - 2ab + b^2 = (x - a)^2$$

- Trinomials of the form
$$ax^2 + bx + c \qquad (a > 0)$$

      Use the *ac* method or trial and error. If $a < 0$, factor out -1 first.

    C. If there are *four terms,* factor by grouping.

3. Check the result by multiplying the factors.

Factoring Strategy Flow Chart

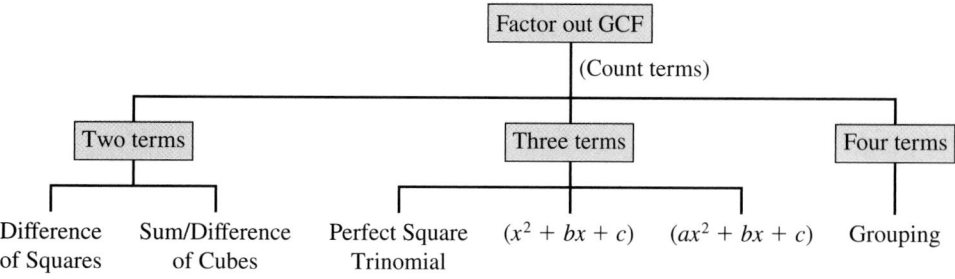

Teaching Tip

Students should be reminded that capital R and lower case r stand for different numbers as stated in *Getting Started*. This is often done in formulas.

To factor $CR^2 - Cr^2$ (see *Getting Started*) we follow these steps.

1. Factor out the GCF, C.

$$CR^2 - Cr^2 = C(R^2 - r^2)$$

2. Factor the difference of two squares inside the parentheses.

$$= C(R + r)(R - r)$$

3. Now check.

If you were the doctor you could tell the patient that when R and r are very close ($R - r$ is close to 0) the artery will be almost blocked.

| **EXAMPLE 1** | **Using the general factoring strategy to factor a binomial** |

Factor completely:

a. $8x^5 - x^2y^3$ **b.** $6x^5 + 24x^3$

SOLUTION   We use the steps in our general factoring strategy.

a. $8x^5 - x^2y^3$

1. Factor out the GCF, x^2. $8x^5 - x^2y^3 = x^2(8x^3 - y^3)$

2. Factor the difference of two cubes inside the parentheses. $= x^2(2x - y)(4x^2 + 2xy + y^2)$

3. Now check.

b. $6x^5 + 24x^3$

1. Factor out $6x^3$, the GCF: $6x^5 + 24x^3 = 6x^3(x^2 + 4)$.

2. Since $x^2 + 4$ is the sum of two squares, it is not factorable.

3. Check this.

The complete factorization of $6x^5 + 24x^3$ is $6x^3(x^2 + 4)$.

PROBLEM 1

Factor completely:

a. $27x^5 - x^2y^3$

b. $8x^5 + 72x^3$

| **EXAMPLE 2** | **Using the general factoring strategy to factor a trinomial** |

Factor completely:

a. $12x^5 + 12x^4y + 3x^3y^2$ **b.** $36x^2y^2 - 24xy^3 + 4y^4$

SOLUTION   As usual, factor out the GCFs, $3x^3$ in part **a** and $4y^2$ in part **b**, first. We then have perfect square trinomials inside the parentheses. Here are the steps.

a. $12x^5 + 12x^4y + 3x^3y^2$

1. Factor out the GCF, $3x^3$.

$$12x^5 + 12x^4y + 3x^3y^2 = 3x^3\underbrace{(4x^2 + 4xy + y^2)}$$

2. Factor the perfect square trinomial. $= 3x^3 \quad (2x + y)^2$

3. Now check.

b. $36x^2y^2 - 24xy^3 + 4y^4$

1. Factor out the GCF, $4y^2$.

$$36x^2y^2 - 24xy^3 + 4y^4 = 4y^2\underbrace{(9x^2 - 6xy + y^2)}$$

2. Factor the perfect square trinomial inside the parentheses. $= 4y^2 \quad (3x - y)^2$

3. Don't forget to check this.

PROBLEM 2

Factor completely:

a. $36x^5 + 24x^4y + 4x^3y^2$

b. $12x^2y^2 - 12xy^3 + 3y^4$

Answers

1. a. $x^2(3x - y)(9x^2 + 3xy + y^2)$
b. $8x^3(x^2 + 9)$
2. a. $4x^3(3x + y)^2$
b. $3y^2(2x - y)^2$

| **EXAMPLE 3** | **Using the general factoring strategy to factor a polynomial** |

Factor completely:

a. $4x^3y - 10x^2y^2 - 6xy^3$

b. $2x^5 + x^4y + x^3y^2$

SOLUTION

a. The GCF is $2xy$. After factoring out this GCF, we have three terms inside the parentheses. We can use the *ac* method or trial and error to finish the problem. The steps are as follows.

1. Factor out the GCF, $2xy$.

$$4x^3y - 10x^2y^2 - 6xy^3 = 2xy(2x^2 - 5xy - 3y^2)$$

2. Use the *ac* method or trial and error to factor $2x^2 - 5xy - 3y^2$. $= 2xy(2x + y)(x - 3y)$

3. Check the answer by multiplying the factors. $2xy(2x + y)(x - 3y) = 4x^3y - 10x^2y^2 - 6xy^3$

b. Factor out the GCF, x^3. $2x^5 + x^4y + x^3y^2 = x^3(2x^2 + xy + y^2)$
The expression inside the parentheses is *not* factorable. According to the *ac* method, the key number is 2. We need two integers whose product is 2 and whose sum is 1; no such integers exist.

| **PROBLEM 3** |

Factor completely:

a. $9x^3y - 15x^2y^2 - 6xy^3$

b. $24x^4 + x^3y + x^2y^2$

| **EXAMPLE 4** | **Finding the complete factorization** |

Factor completely: $2x^5 - x^4y + 4x^3y^2$

SOLUTION   We start by factoring out the GCF, x^3.

$$2x^5 - x^4y + 4x^3y^2 = x^3(2x^2 - xy + 4y^2)$$

Although $2x^2 - xy + 4y^2$ is a trinomial, it is *not* factorable. (The *ac* method gives $2 \cdot 4 = 8$, and there are no factors whose product is 8 and whose sum is -1.) Thus the factorization shown is the complete factorization.

| **PROBLEM 4** |

Factor completely:

$$2x^4 + x^3y + 2x^2y^2$$

Teaching Tip

The constant reminder in a general factoring strategy is to *always* look for a GCF *first*, then count terms (inside the parentheses if a GCF was factored out) and proceed with the appropriate type of factoring for that many terms.

| **EXAMPLE 5** | **Factoring by grouping** |

Factor completely: $4x^3 - 12x^2 - x + 3$

SOLUTION   In this case, there is no common factor. The polynomial has four terms, so we factor by grouping.

$4x^3 - 12x^2 - x + 3 = (4x^3 - 12x^2) - (x - 3)$

                $= 4x^2(x - 3) - 1(x - 3)$ Group by twos and factor.

                $= (x - 3) \quad (4x^2 - 1)$ Factor out the GCF, $(x - 3)$.

                $= (x - 3)(2x + 1)(2x - 1)$ Factor the difference of two squares, $(4x^2 - 1)$.

| **PROBLEM 5** |

Factor completely:

$$9x^3 + 18x^2 - 25x - 50$$

Answers

3. a. $3xy(3x + y)(x - 2y)$
   b. $x^2(24x^2 + xy + y^2)$
4. $x^2(2x^2 + xy + 2y^2)$
5. $(x + 2)(3x + 5)(3x - 5)$

EXAMPLE 6 **Factoring by grouping**

Factor completely: $x^2 - 6x + 9 - 9y^2$

SOLUTION The polynomial has four terms, so you may try to factor it by grouping. If we try groups of two, we have

$$x^2 - 6x + 9 - 9y^2 = (x^2 - 6x) + (9 - 9y^2)$$
$$= x(x - 6) + 9(1 - y^2)$$

But there is no GCF, so we can't factor any farther using this technique. However, the first three terms form a perfect square trinomial, so we write

$$x^2 - 6x + 9 - 9y^2 = (x^2 - 6x + 9) - 9y^2 \quad \text{Group the first 3 terms.}$$
$$= (x - 3)^2 - (3y)^2 \quad \text{The difference of two squares}$$
$$= [(x - 3) + 3y][(x - 3) - 3y] \quad \text{Simplify.}$$
$$= (x + 3y - 3)(x - 3y - 3)$$

PROBLEM 6

Factor completely:

$$x^2 + 8x + 16 - 81y^2$$

Web It

For a summary on general factoring strategy and more practice, go to link 5-6-2 at mhhe.com/bello.

EXAMPLE 7 **General factoring strategies**

Factor completely: $-2x^2 - 9x + 5$

SOLUTION This is a polynomial of the form $ax^2 + bx + c$ with $a = -2$, which is negative, so we first factor out -1 to obtain

Factor out -1. $-2x^2 - 9x + 5 = -1(2x^2 \underline{+ 9x} - 5)$

1. The key number is $2(-5) = -10$.

2. We write the middle term as $10x - 1x$. $= -1(2x^2 + \underline{10x - 1x} - 5)$

3. Grouping by twos. $= -1[(2x^2 + 10x) + (-1x - 5)]$

4. Factoring each group. $= -1[2x(x + 5) - 1(x + 5)]$

5. Factoring the common binomial. $= -1(x + 5)(2x - 1)$

Since $-1 \cdot a = -a$, $-2x^2 - 9x + 5 = -(x + 5)(2x - 1)$

PROBLEM 7

Factor completely:

$$-3x^2 + 17x + 28$$

Answers

6. $(x + 9y + 4)(x - 9y + 4)$
or $(x - 9y + 4)(x + 9y + 4)$
7. $-(x - 7)(3x + 4)$

Calculate It Factoring Strategy for the Graphing Calculator

If you are using a calculator, you need a general factoring strategy too. Here is one.

A General Factoring Strategy

1. Factor out the GCF if there is one. After you do this algebraically, graph the original polynomial and the factored one. You must get the same graph.

2. Look at the number of terms in the given polynomial (or inside the parentheses if the GCF was factored out).

- If there are *two* terms, graph the polynomial and determine the points at which the graph cuts the horizontal axis.

   If you have the difference of two squares and the points are $x = \pm p$, the factorization is

   $$(x + p)(x - p)$$

   If you have the sum or difference of two cubes and the point is $x = \pm p$, the factorization is

   $$(x \mp p)(x^2 \pm px + p^2)$$

- If there are *three* terms and the degree is 2, graph the polynomial and determine the point at which the graph crosses or touches the horizontal axis.

   If the point is $x = p$, the factorization is

   $$(x - p)^2$$

   If the points are

   $$x = \frac{p}{q} \quad \text{and} \quad x = \frac{r}{s}$$

   the factorization is $(qx - p)(sx - r)$. (If the leading coefficient is negative, factor out a -1 first.)

- If there are *four* terms, factor by grouping and check by graphing the original and the factored polynomial. The graphs should be identical.

- If you graph the polynomial and nothing shows, try adjusting the windows. Press TRACE to give you an idea of the x- and y-values so you can adjust your window. If the graph does not touch or cross the horizontal axis and you graphed correctly, the polynomial is not factorable. (Try the ac test for polynomials of the form $ax^2 + bx + c$.)

Exercises 5.6

A In Problems 1–78, factor completely.

1. $3x^4 - 3x^3 - 18x^2$
$3x^2(x + 2)(x - 3)$

2. $4x^5 - 12x^4 - 16x^3$
$4x^3(x + 1)(x - 4)$

3. $5x^4 + 10x^3y - 40x^2y^2$
$5x^2(x + 4y)(x - 2y)$

4. $6x^7 + 18x^6y - 60x^5y^2$
$6x^5(x + 5y)(x - 2y)$

5. $-3x^6 - 6x^5 - 21x^4$
$-3x^4(x^2 + 2x + 7)$

6. $-6x^5 - 18x^4 - 12x^3$
$-6x^3(x + 2)(x + 1)$

7. $2x^6y - 4x^5y^2 - 10x^4y^3$
$2x^4y(x^2 - 2xy - 5y^2)$

8. $3x^8y - 12x^7y^2 - 9x^6y^3$
$3x^6y(x^2 - 4xy - 3y^2)$

9. $-4x^6 - 12x^5y - 18x^4y^2$
$-2x^4(2x^2 + 6xy + 9y^2)$

10. $-5x^6 - 25x^5 - 30x^4$
$-5x^4(x + 3)(x + 2)$

11. $6x^3y^2 + 12x^2y^2 + 2xy^2 + 4y^2$
$2y^2(x + 2)(3x^2 + 1)$

12. $6x^3y^2 + 24x^2y^2 + 3xy^2 + 12y^2$
$3y^2(x + 4)(2x^2 + 1)$

13. $-9x^4y - 9x^3y - 6x^2y - 6xy$
$-3xy(x + 1)(3x^2 + 2)$

14. $-8x^4y - 16x^3y - 6x^2y - 12xy$
$-2xy(x + 2)(4x^2 + 3)$

15. $-4x^4 - 4x^3y + 2x^2y + 2xy^2$
$-2x(x + y)(2x^2 - y)$

16. $-9x^4 - 18x^3y + 3x^2y + 6xy^2$
$-3x(x + 2y)(3x^2 - y)$

17. $3x^2y^2 + 24xy^3 + 48y^4$
$3y^2(x + 4y)^2$

18. $8x^2y^2 + 24xy^3 + 18y^4$
$2y^2(2x + 3y)^2$

19. $-18kx^2 - 24kxy - 8ky^2$
$-2k(3x + 2y)^2$

20. $-12kx^2 - 60kxy - 75ky^2$
$-3k(2x + 5y)^2$

21. $16x^3y^2 - 48x^2y^3 + 36xy^4$
$4xy^2(2x - 3y)^2$

22. $45x^3y^2 - 60x^2y^3 + 20xy^4$
$5xy^2(3x - 2y)^2$

23. $kx^2 - 12kx + 36$
Not factorable

24. $kx^2 - 20kx + 25$
Not factorable

25. $3x^5 + 12x^4y + 12x^3y^2$
$3x^3(x + 2y)^2$

26. $2x^5 + 16x^4y + 32x^3y^2$
$2x^3(x + 4y)^2$

27. $18x^6 + 12x^5y + 2x^4y^2$
$2x^4(3x + y)^2$

28. $12x^6 + 12x^5y + 3x^4y^2$
$3x^4(2x + y)^2$

29. $12x^4y^2 - 36x^3y^3 + 27x^2y^4$
$3x^2y^2(2x - 3y)^2$

30. $18x^4y^2 - 24x^3y^3 + 8x^2y^4$
$2x^2y^2(3x - 2y)^2$

31. $6x^3 + 12x^2 - 6x - 12$
$6(x + 2)(x + 1)(x - 1)$

32. $4x^3 + 16x^2 - 16x - 64$
$4(x + 4)(x + 2)(x - 2)$

33. $7x^4 - 7y^4$
$7(x^2 + y^2)(x + y)(x - y)$

34. $9x^4 - 9z^4$
$9(x^2 + z^2)(x + z)(x - z)$

35. $2x^6 - 32x^2y^4$
$2x^2(x^2 + 4y^2)(x + 2y)(x - 2y)$

36. $x^7 - 81x^3y^4$
$x^3(x^2 + 9y^2)(x + 3y)(x - 3y)$

37. $-2x^2 - 12x - 18$
$-2(x + 3)^2$

38. $-2x^2 - 20x - 50$
$-2(x + 5)^2$

39. $-3x^2 - 12x - 12$
$-3(x + 2)^2$

40. $-4x^2 - 24x - 36$
$-4(x + 3)^2$

41. $-4x^4 - 4x^3y - x^2y^2$
$-x^2(2x + y)^2$

42. $-9x^4 - 6x^3y - x^2y^2$
$-x^2(3x + y)^2$

43. $-9x^2y^2 - 12xy^3 - 4y^4$
$-y^2(3x + 2y)^2$

44. $-4x^2y^2 - 12xy^3 - 9y^4$
$-y^2(2x + 3y)^2$

45. $-8x^2y^2 + 24xy^3 - 18y^4$
$-2y^2(2x - 3y)^2$

46. $-18x^4 + 24x^3y - 8x^2y^2$
$-2x^2(3x - 2y)^2$

47. $-18x^3 - 24x^2y - 8xy^2$
$-2x(3x + 2y)^2$

48. $-12x^3 - 36x^2y - 27xy^2$
$-3x(2x + 3y)^2$

49. $-18x^3 - 60x^2y - 50xy^2$
$-2x(3x + 5y)^2$

50. $-12x^3 - 60x^2y - 75xy^2$
$-3x(2x + 5y)^2$

51. $-x^3 + xy^2$
$-x(x + y)(x - y)$

52. $-x^3 + 9xy^2$
$-x(x + 3y)(x - 3y)$

53. $-x^4 + 4x^2y^2$
$-x^2(x + 2y)(x - 2y)$

54. $-x^4 + 16x^2y^2$
$-x^2(x + 4y)(x - 4y)$

55. $-4x^4 + 9x^2y^2$
$-x^2(2x + 3y)(2x - 3y)$

56. $-9x^4 + 4x^2y^2$
$-x^2(3x + 2y)(3x - 2y)$

57. $-8x^3 + 18xy^2$
$-2x(2x + 3y)(2x - 3y)$

58. $-12x^3 + 3x$
$-3x(2x + 1)(2x - 1)$

59. $-18x^4 + 8x^2y^2$
$-2x^2(3x + 2y)(3x - 2y)$

60. $-12x^4 + 27x^2y^2$
$-3x^2(2x + 3y)(2x - 3y)$

61. $27x^2 - x^5$
$x^2(3 - x)(9 + 3x + x^2)$

62. $64x^3 - x^6$
$x^3(4 - x)(16 + 4x + x^2)$

63. $x^7 - 8x^4$
$x^4(x - 2)(x^2 + 2x + 4)$

64. $8x^{10} - \frac{1}{27}x^7$
$x^7(2x - \frac{1}{3})(4x^2 + \frac{2}{3}x + \frac{1}{9})$

65. $27x^4 + 8x^7$
$x^4(3 + 2x)(9 - 6x + 4x^2)$

66. $8x^5 + 27x^8$
$x^5(2 + 3x)(4 - 6x + 9x^2)$

67. $27x^7 + 64x^4y^3$
$x^4(3x + 4y)(9x^2 - 12xy + 16y^2)$

68. $8x^8 + 27x^5y^3$
$x^5(2x + 3y)(4x^2 - 6xy + 9y^2)$

69. $x^2 + 4x + 4 - y^2$
$(x + y + 2)(x - y + 2)$

70. $x^2 + 8x + 16 - y^2$
$(x + y + 4)(x - y + 4)$

71. $x^2 + y^2 - 6y + 9$
Not factorable

72. $x^2 + y^2 - 8x + 16$
Not factorable

73. $x^2 - y^2 - 4y - 4$
$(x + y + 2)(x - y - 2)$

74. $x^2 - y^2 - 8y - 16$
$(x + y + 4)(x - y - 4)$

75. $-9x^2 + 30xy - 25y^2$
$-(3x - 5y)^2$

76. $-9x^2 + 12xy - 4y^2$
$-(3x - 2y)^2$

77. $18x^3 - 60x^2y + 50xy^2$
$2x(3x - 5y)^2$

78. $-12x^3 + 60x^2y - 72xy^2$
$-12(x - 3y)(x - 2y)$

SKILL CHECKER

Try these "Skill Checker" exercises so you'll be ready for the next section.

Factor completely.

79. $6x^2 - x - 2$
$(2x + 1)(3x - 2)$

80. $6x^2 - 7x - 3$
$(3x + 1)(2x - 3)$

81. $12x^2 - x - 1$
$(3x - 1)(4x + 1)$

82. $10x^2 - 17x + 3$
$(5x - 1)(2x - 3)$

USING YOUR KNOWLEDGE

Technical Applications

Many of the ideas presented in this section are used by engineers and technicians. Use your knowledge to factor completely the expressions in Problems 83–86.

83. The bend allowance needed to bend a piece of metal of thickness t through an angle A when the inside radius of the bend is R is given by

$$\frac{2\pi A}{360}R + \frac{2\pi A}{360}Kt \qquad K \text{ is a constant.}$$

Factor completely this expression. $\frac{2\pi A}{360}(R + Kt)$

84. The change in kinetic energy of a moving object of mass m with initial velocity v_1 and terminal velocity v_2 is given by

$$\frac{1}{2}mv_1^2 - \frac{1}{2}mv_2^2$$

Factor completely this expression.
$\frac{1}{2}m(v_1 + v_2)(v_1 - v_2)$

85. The parabolic distribution of shear stress on the cross section of a certain beam is given by

$$\frac{3Sd^2}{2bd^3} - \frac{12Sz^2}{2bd^3}$$

Factor completely this expression.
$\frac{3S}{2bd^3}(d + 2z)(d - 2z)$

86. The polar moment of inertia, J, of a hollow round shaft of inner diameter d_1 and outer diameter d is given by

$$\frac{\pi d^4}{32} - \frac{\pi d_1^4}{32}$$

Factor completely this expression.
$\frac{\pi}{32}(d^2 + d_1^2)(d + d_1)(d - d_1)$

WRITE ON

87. Explain why the expression $x^2(x^2 - 5) + y^2(x^2 - 5)$ is not completely factored. Answers may vary.

88. Explain the difference between the statements "a polynomial is factored" and "a polynomial is completely factored." Give examples of polynomials that are factored but are not completely factored.
Answers may vary.

89. Is the statement "$9(x^2 - x)$ is factored" true or false? Explain. False; answers may vary.

90. Suppose you factor $x^2 - 2$ as $(x + \sqrt{2})(x - \sqrt{2})$. Is $x^2 - 2$ completely factored according to the guidelines we mentioned at the beginning of this section? Explain.
No; answers may vary.

MASTERY TEST

If you know how to do these problems, you have learned your lesson!

Factor completely if possible:

91. $9x^3 - 18x^2 - x + 2$
$(x - 2)(3x + 1)(3x - 1)$

92. $4x^3 - 12x^2 - x + 3$
$(x - 3)(2x + 1)(2x - 1)$

93. $2x^4 + x^3y + 2x^2y^2$
$x^2(2x^2 + xy + 2y^2)$

94. $3x^5 + x^4y + x^3y^2$
$x^3(3x^2 + xy + y^2)$

95. $x^2 - 10x + 25 - y^2$
$(x + y - 5)(x - y - 5)$

96. $y^2 - 9x^2 + 12x - 4$
$(y + 3x - 2)(y - 3x + 2)$

97. $-8y^2 + 2y + 21$
$-(4y - 7)(2y + 3)$

98. $-2 - 5y - 2y^2$
$-(2 + y)(1 + 2y)$

99. $36x^5 + 24x^4y + 4x^3y^2$
$4x^3(3x + y)^2$

100. $12x^2y^2 - 12xy^3 + 3y^4$
$3y^2(2x - y)^2$

101. $8x^5 + 72x^3$
$8x^3(x^2 + 9)$

102. $27x^5 - x^2y^3$
$x^2(3x - y)(9x^2 + 3xy + y^2)$

| **5.7** | # SOLVING EQUATIONS BY FACTORING: APPLICATIONS |

To Succeed, Review How To . . .

1. Factor polynomials (pp. 393–396).
2. Solve linear equations (pp. 79–82).
3. Evaluate expressions (pp. 46–48).

Objectives

A Solve equations by factoring.

B Use the Pythagorean theorem to find the length of one side of a right triangle when the lengths of the other two sides are given.

C Solve applications involving quadratic equations.

GETTING STARTED

Quadratic Equations in Fire Fighting

How much water is the fire truck pumping if the friction loss is 36 lb/in.^{2}? You can find out by solving the equation

$$2g^2 + g - 36 = 0$$ (*g* in hundreds of gallons per minute)

This equation is a *quadratic equation in standard form*. In this section we shall study how to solve equations like this one.

We have already studied linear equations, equations that can be written in the form $ax + b = c$, where *a*, *b*, and *c* are real numbers and $a \neq 0$. We are now ready to study *quadratic equations*. Some of these equations can be written in standard form and then solved by the factoring methods we have studied. Here is the definition we need.

Standard Form of a Quadratic Equation

An equation that can be written in the **standard form**

$$ax^2 + bx + c = 0$$

where *a*, *b*, and *c* are constants and $a \neq 0$ is a **quadratic equation.**

Here are some quadratic equations:

$$x^2 = 5, \quad 3x^2 - 8x + 7 = 0, \quad \text{and} \quad x^2 - 2x = 4$$

Of these, only $3x^2 - 8x + 7 = 0$ is in standard form, with $a = 3$, $b = -8$, and $c = 7$. *Note:* In a quadratic equation, the highest exponent on the variable is 2.

| **A** | ## Solving Equations by Factoring |

Web It

For another explanation and examples of solving quadratic equations by factoring, go to link 5-7-1 at mhhe.com/bello.

The equation $2g^2 + g - 36 = 0$ can be solved by factoring. As you recall from Section 5.4, $2g^2 + g - 36$ can be factored as $(2g + 9)(g - 4)$. We then write

$$2g^2 + g - 36 = 0 \qquad \text{Given.}$$
$$(2g + 9)(g - 4) = 0 \qquad \text{Factor.}$$

The product of the factors is zero. The only way this can happen is if at least one of the factors is zero. (Try getting zero for a product without having any zero factors.) Here is the property we need.

Teaching Tip

To emphasize the importance of the zero-factor property—the equation being set equal to 0—do the following example to illustrate a common mistake made by students.

Solve: $(x + 2)(x - 3) = -4$

$x + 2 = -4$ $x - 3 = -4$

$x = -6$ $x = -1$

Check:

$(-6 + 2)(-6 - 3) = -4$

$(-4)(-9) = -4$

$36 = -4$

$(-1 + 2)(-1 - 3) = -4$

$(1)(-4) = -4$

$-4 = -4$

Since both solutions do not check we have to rework the problem, this time setting the equation equal to 0. This would be a good lead into Example 3(c). (The correct solutions are 2 and -1.)

Zero-Factor Property

For all real numbers a and b, $a \cdot b = 0$ means that $a = 0$ or $b = 0$ (or both).

Thus

$$(2g + 9)(g - 4) = 0$$

means that

$2g + 9 = 0$ or $g - 4 = 0$

$2g = -9$ or $g = 4$ Solve the linear equations.

$g = -\dfrac{9}{2}$ or $g = 4$

The two possible solutions are $g = -\frac{9}{2}$ and $g = 4$. Since g is the flow of water in hundreds of gallons per minute, g must be positive, so we discard the negative solution $g = -\frac{9}{2}$. Thus the fire truck can pump 400 gal/min.

PROCEDURE

Solving a Quadratic Equation

To solve a quadratic equation, follow these three steps using the suggested acronym, "OFF" to help remember the steps.

1. Set the equation equal to **0**.   **0**

2. Factor completely.   **F**

3. Set each linear Factor equal to 0 and solve each.   **F**

EXAMPLE 1   **Solving equations using the zero-factor property**

Solve by factoring:

a. $x^2 - 9 = 0$ **b.** $x^2 + 8x = 0$

SOLUTION

a. $x^2 - 9 = 0$ 0

$(x + 3)(x - 3) = 0$ F   Factor.

$x + 3 = 0$ or $x - 3 = 0$ F   Factors = 0. Use the zero-factor property, with $a = x + 3$ and $b = x - 3$.

$x = -3$ or $x = 3$ Solve the equations $x + 3 = 0$ and $x - 3 = 0$.

We can check the solutions by substituting in the original equation, $x^2 - 9 = 0$.

CHECK $\dfrac{x^2 - 9 = 0}{}$ $\dfrac{x^2 - 9 = 0}{}$

$\begin{array}{c|c} (-3)^2 - 9 & 0 \\ 9 - 9 & \\ 0 & \end{array}$ $\begin{array}{c|c} 3^2 - 9 & 0 \\ 9 - 9 & \\ 0 & \end{array}$

In both cases the result is true. The solution set is $\{3, -3\}$.

PROBLEM 1

Solve by factoring:

a. $x^2 - 36 = 0$

b. $2x^2 - 50x = 0$

Answers

1. a. $6, -6$ **b.** $0, 25$

b. $x^2 + 8x = 0$ 0

$x(x + 8) = 0$ F   Factor.

$x = 0$ or $x + 8 = 0$ F   Factors = 0. Use the zero-factor property,
                                                with $a = x$ and $b = x + 8$.

$x = 0$ or $x = -8$ Solve the equation $x + 8 = 0$.

CHECK $\dfrac{x^2 + 8x = 0}{\begin{array}{c|c}(0)^2 + 8(0) & 0 \\ 0 + 0 \\ 0 \end{array}}$ $\dfrac{x^2 + 8x = 0}{\begin{array}{c|c}(-8)^2 + 8(-8) & 0 \\ 64 - 64 \\ 0 \end{array}}$

Since both results check, the solution set is $\{0, -8\}$.

Calculate It

Using Graphs to Find Solutions to a Quadratic Equation

To solve $x^2 - 9 = 0$ in Example l(a), graph $Y_1 = x^2 - 9$ with a standard window. The points at which the graph cuts the horizontal axis, 3 and -3, are the points at which $Y_1 = x^2 - 9 = 0$. Thus 3 and -3 are the solutions of $x^2 - 9 = 0$.

You can verify this by letting $Y_2 = 0$ and pressing [2nd] [TRACE] 5, then [ENTER] three times.

The equation $x^2 = -x + 6$ is *not* in standard form. To write it in standard form, we add x and subtract 6 on both sides to obtain $x^2 + x - 6 = 0$, which can be solved by factoring. It is easier to factor a quadratic expression when the ax^2 term is positive. That is why we left the x^2 on the left side and moved the other two terms to the left side by using the addition property of equality. As you recall, to factor $x^2 + x - 6$, we need to find integers whose sum is 1 and whose product is -6. These integers are 3 and -2. Thus we have

$x^2 = -x + 6$ Given.

$x^2 + x - 6 = 0$ 0   Write in standard form (add x and subtract 6).

$(x + 3)(x - 2) = 0$ F   Factor.

$x + 3 = 0$ or $x - 2 = 0$ F   Factors = 0. Use the zero-factor property.

$x = -3$ or $x = 2$ Solve $x + 3 = 0$ and $x - 2 = 0$.

CHECK $\dfrac{x^2 = -x + 6}{\begin{array}{c|c}(2)^2 & -(2) + 6 \\ 4 & 4 \end{array}}$ $\dfrac{x^2 = -x + 6}{\begin{array}{c|c}(-3)^2 & -(-3) + 6 \\ 9 & 3 + 6 \\ & 9 \end{array}}$

The solution set is $\{2, -3\}$.

EXAMPLE 2 Writing and solving equations in standard form	**PROBLEM 2**

Solve by factoring: Solve by factoring:

a. $x^2 = 6x - 8$ **b.** $x^2 + x = 2$ **a.** $x^2 = 6x - 5$

                                                              b. $x^2 = 12 - x$

SOLUTION

a. We first write the equation in standard form by subtracting $6x$ and adding 8 to obtain $x^2 - 6x + 8 = 0$. To factor $x^2 - 6x + 8$, we must find two integers whose sum is -6 and whose product is 8. These numbers are -4 and -2. Here is the procedure,

$x^2 = 6x - 8$ Given.

$x^2 - 6x + 8 = 0$ 0   Subtract $6x$ and add 8.

$(x - 4)(x - 2) = 0$ F   Factor.

$x - 4 = 0$ or $x - 2 = 0$ F   Factors = 0. Use the zero-factor property.

$x = 4$ or $x = 2$ Solve $x - 4 = 0$ and $x - 2 = 0$.

The solution set is $\{2, 4\}$. Check this.

Answers

2. a. 1, 5   **b.** $-4, 3$

b. The equation $x^2 + x = 2$ is *not* in standard form. To solve an equation by factoring, we must first write the equation in the standard form $ax^2 + bx + c = 0$. Thus, subtracting 2 from both sides of $x^2 + x = 2$, we have

$$x^2 + x - 2 = 0 \qquad\qquad 0$$

$$(x + 2)(x - 1) = 0 \qquad\qquad \text{F Factor.}$$

$$x + 2 = 0 \quad \text{or} \quad x - 1 = 0 \qquad \text{F Factors} = 0. \text{ Use the zero-factor property.}$$

$$x = -2 \quad \text{or} \qquad x = 1 \qquad \text{Solve } x + 2 = 0 \text{ and } x - 1 = 0.$$

The solution set is $\{1, -2\}$. The check is left to you.

Web It

For more practice on solving equations by factoring, go to link 5-7-2 at mhhe.com/bello.

Calculate It

Using Graphs to Find Solutions to a Quadratic Equation

To do Example 2(a), write the equation in the standard form $x^2 - 6x + 8 = 0$, and graph. Since the graph cuts the horizontal axis at $x = 2$ and $x = 4$, these are the solutions. Graph $x^2 + x - 2 = 0$ to verify the solutions to Example 2(b).

To solve the equation $6x^2 - x - 2 = 0$, we first factor $6x^2 - x - 2$ by trial and error or by the *ac* method (shown). For $6x^2 - x - 2 = 0$, the key number is $6 \cdot (-2) = -12$. Thus we have to find integers whose product is -12 and whose sum is -1 (that is, -4 and 3) and use these numbers to rewrite the middle term $-x$. We have

$$6x^2 - x - 2 = 0 \qquad 0 \quad \text{Given.}$$

$$6x^2 - 4x + 3x - 2 = 0 \qquad \text{Write the middle term } -x \text{ as } -4x + 3x.$$

$$2x(3x - 2) + 1(3x - 2) = 0 \qquad \text{Factor the first and last pair of terms.}$$

$$(3x - 2)(2x + 1) = 0 \qquad \text{F Factor out the common factor, } 3x - 2.$$

$$3x - 2 = 0 \quad \text{or} \quad 2x + 1 = 0 \qquad \text{F Factors} = 0. \text{ Use the zero-factor property.}$$

$$3x = 2 \quad \text{or} \qquad 2x = -1 \qquad \text{Solve } 3x - 2 = 0 \text{ and } 2x + 1 = 0.$$

$$x = \frac{2}{3} \quad \text{or} \qquad x = \frac{-1}{2}$$

The solution set is $\{\frac{2}{3}, -\frac{1}{2}\}$. We check this for $x = \frac{2}{3}$.

Web It

For more practice, practice, and practice on solving quadratic equations by factoring, go to link 5-7-3 at mhhe.com/bello.

CHECK

$$6x^2 - x - 2 = 0$$

$$6\left(\frac{2}{3}\right)^2 - \frac{2}{3} - 2 \quad \Big| \quad 0$$

$$6\left(\frac{4}{9}\right) - \frac{2}{3} - 2$$

$$\frac{8}{3} - \frac{2}{3} - 2$$

$$2 - 2$$

$$0$$

The check that $x = -\frac{1}{2}$ is a solution is left to you.

EXAMPLE 3	Solving quadratic equations by factoring using the *ac* method

a. $12x^2 + 5x - 3 = 0$

b. $6x^2 - x = 1$

c. $x(x + 2) = (3x - 1)x + 1$

PROBLEM 3

Solve by factoring:

a. $12x^2 + 13x - 4 = 0$

b. $8x^2 - 2x = 1$

c. $x(x - 3) = (4x + 7)x - 8$

Answers

3. **a.** $-\frac{4}{3}, \frac{1}{4}$ **b.** $-\frac{1}{4}, \frac{1}{2}$ **c.** $-4, \frac{2}{3}$

SOLUTION

a. You can factor $12x^2 + 5x - 3$ by trial and error or note that the key number for $12x^2 + 5x - 3 = 0$ is $(12)(-3) = -36$. Thus

$12x^2 + 5x - 3 = 0$	**0** Given.
$12x^2 + 9x - 4x - 3 = 0$	Write $5x$ using coefficients whose sum is 5 and whose product is -36; that is, $5x = 9x - 4x$.
$3x(4x + 3) - 1(4x + 3) = 0$	**F** Factor the first and last pairs of terms.
$(4x + 3)(3x - 1) = 0$	Factor out the common binomial factor, $4x + 3$.
$4x + 3 = 0$ or $3x - 1 = 0$	**F** Factors $= 0$. Use the zero-factor property.
$4x = -3$ or $3x = 1$	Solve $4x + 3 = 0$ and $3x - 1 = 0$.
$x = \dfrac{-3}{4}$ or $x = \dfrac{1}{3}$	

The solution set is $\{\frac{1}{3}, -\frac{3}{4}\}$. Check this.

b. $6x^2 - x = 1$ is *not* in standard form. Subtracting 1 from both sides of the equation, we have

$6x^2 - x - 1 = 0$	**0** The key number is -6.
$6x^2 - 3x + 2x - 1 = 0$	The integers whose product is -6 with sum -1 are -3 and 2.
$3x(2x - 1) + (2x - 1) = 0$	**F** Factor the first and last pairs of terms.
$(2x - 1)(3x + 1) = 0$	Factor out $(2x - 1)$.
$2x - 1 = 0$ or $3x + 1 = 0$	**F** Factors $= 0$. Use the zero-factor property.
$2x = 1$ or $3x = -1$	Solve the equations $2x - 1 = 0$ and $3x + 1 = 0$.
$x = \dfrac{1}{2}$ or $x = \dfrac{-1}{3}$	

The solution set is $\{\frac{1}{2}, -\frac{1}{3}\}$. Check this.

c. $x(x + 2) = (3x - 1)x + 1$ is *not* in standard form. First simplify both sides and then write the result in standard form.

$x(x + 2) = (3x - 1)x + 1$	Given.
$x^2 + 2x = 3x^2 - x + 1$	Distributive property
$0 = 3x^2 - x^2 - 2x - x + 1$	Subtract $x^2 + 2x$.
$0 = 2x^2 - 3x + 1$	**0** Simplify.
$2x^2 - 3x + 1 = 0$	Symmetric property
$2x^2 - 2x - x + 1 = 0$	Since the key number is 2, rewrite $-3x = -2x - x$.
$(2x^2 - 2x) + (-x + 1) = 0$	Group in pairs.
$2x(x - 1) - 1 \cdot (x - 1) = 0$	**F** Factor the GCF from each pair.
$(x - 1)(2x - 1) = 0$	The GCF is $(x - 1)$.
$x - 1 = 0$ or $2x - 1 = 0$	**F** Factors $= 0$. Zero-factor property
$x = 1$ or $x = \dfrac{1}{2}$	Solve $x - 1 = 0$ and $2x - 1 = 0$.

The solution set is $\{1, \frac{1}{2}\}$. Check this.

Calculate It   Using Graphs to Find Solutions to a Quadratic Equation

To do Example 3(c), first rewrite the equation as

$$Y_1 = x(x + 2) - (3x - 1)x - 1,$$
$$Y_2 = 0$$

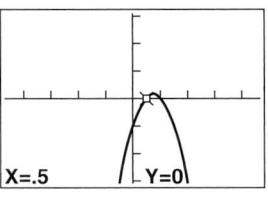

X=.5   Y=0

and graph (you don't have to do the multiplication). Using a decimal window, [ZOOM] [4], and* [2nd] [TRACE] 5 with [ENTER] three times, you will find the solution 0.5 as shown. (Verify that the other solution is 1 by tracing to the other x-intercept and using [2nd] [TRACE] 5 with [ENTER] three times.)

*Some calculators have an **intersect** and a **solve** feature. To find the point at which the curve intersects the horizontal axis, graph the horizontal axis $Y_2 = 0$. Get the cursor to a point near the intersection you want to find. Press [2nd] [TRACE] 5. The calculator asks "First Curve?" "Second Curve?" and "Guess?" Press [ENTER] each time, and the calculator gives the intersection $x = 0.5$, $y = 0$.

EXAMPLE 4   **Solving equations by grouping and the zero-factor property**

Solve: $x^3 + 3x^2 - 4x - 12 = 0$

SOLUTION   Since the polynomial has four terms, try to factor it by grouping. Here are the steps.

$x^3 + 3x^2 - 4x - 12 = 0$	0 Given.
$x^2(x + 3) - 4(x + 3) = 0$	Group for factoring.
$(x + 3)(x^2 - 4) = 0$	F Factor out the GCF.
$(x + 3)(x + 2)(x - 2) = 0$	Factor $x^2 - 4$.
	F Factors = 0.

$$x + 3 = 0 \quad \text{or} \quad x + 2 = 0 \quad \text{or} \quad x - 2 = 0$$
$$x = -3 \quad \text{or} \quad x = -2 \quad \text{or} \quad x = 2$$

The solution set is $\{2, -2, -3\}$. Check this.

PROBLEM 4

Solve:

$$x^3 + 2x^2 - 9x - 18 = 0$$

Calculate It   Using Graphs to Find Solutions to a Quadratic Equation

Let's find the solutions of $x^3 + 3x^2 - 4x - 12 = 0$ in Example 4 using a calculator. First graph $Y_1 = x^3 + 3x^2 - 4x - 12$ and $Y_2 = 0$ in a standard window. The points at which the graph cuts the horizontal axis, $x = -3$, $x = -2$, and $x = 2$ are the solutions of the equation as shown in Window 1 and can be verified using [2nd] [TRACE] 5 with [ENTER] three times.

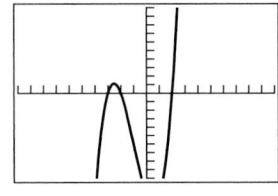

Window 1

B   **Using the Pythagorean Theorem**

Quadratic equations can be used to find the lengths of the sides of right triangles using the **Pythagorean theorem.**

Answer

4. $-2, -3, 3$

The Pythagorean Theorem

In any right triangle (a triangle with a 90° angle), the square of the longest side (hypotenuse) is equal to the sum of the squares of the other two sides (the legs). In symbols, this is

$$c^2 = a^2 + b^2$$

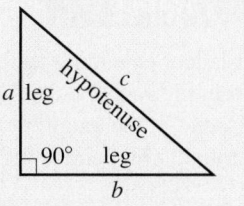

EXAMPLE 5   **Using the Pythagorean theorem to solve word problems**

The length of the three sides of a right triangle are consecutive integers. What are these lengths?

SOLUTION   We use the RSTUV method.

1. Read the problem.
We need to find the lengths of the sides of the right triangle.

2. Select the unknown.
If x is an integer, what are the next two consecutive integers?

Let the length of the shortest side be x. Since the lengths of the sides are consecutive integers, we have

x	Length of the shortest side
$x + 1$	Length of the next side
$x + 2$	Length of the hypotenuse (the longest side)

3. Think of a plan.
Make a diagram and use the Pythagorean theorem to obtain the equation

$$c^2 \;\; = \;\; a^2 \;\; + b^2$$
$$(x + 2)^2 = (x + 1)^2 + x^2$$

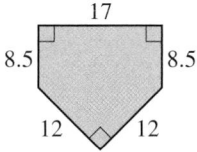

4. Use algebra to solve the resulting equation.

$x^2 + 4x + 4 = x^2 + 2x + 1 + x^2$	Multiply.
$x^2 + 4x + 4 = 2x^2 + 2x + 1$	Simplify.
$0 = x^2 - 2x - 3$	0 Subtract x^2, $4x$, and 4 from both sides.
$x^2 - 2x - 3 = 0$	Write in standard form.
$(x - 3)(x + 1) = 0$	F Factor.
$x - 3 = 0 \;\;$ or $\;\; x + 1 = 0$	F Factors = 0. Use the zero-factor property.
$x = 3 \;\;$ or $\;\;\;\;\;\; x = -1$	Solve $x - 3 = 0$ and $x + 1 = 0$.

The lengths of the sides must be positive, so we discard the negative answer, -1. Thus the shortest side is 3 units, so the other two sides are 4 and 5 units.

5. Verify the solution.
The verification is left to you.

Answer

5. No, this is not possible because

$$12^2 + 12^2 = 17^2$$
$$288 \neq 289$$

PROBLEM 5

According to Kreutzer and Kerley (1990) the Little League rule book's specification of the shape of home plate is a pentagon and looks like this:

Is this physically possible? Explain.

Web It

To verify if Problem 5 can be answered using the Pythagorean theorem, go to link 5-7-4 at mhhe.com/bello.

EXAMPLE 6   An application of the Pythagorean theorem

A ladder is leaning against the side of a house with its bottom x feet away from the wall and resting against the wall of the house $x + 7$ feet above the ground. If the length of the ladder is 1 ft more than its height above the ground, how long is the ladder?

SOLUTION   As before, we use the RSTUV method.

1. Read the problem.
Form a mental picture of the dimensions.

2. Select the unknown.
Let x feet be the distance from the foot of the ladder to the wall.

3. Think of a plan.
Make a diagram of the situation.

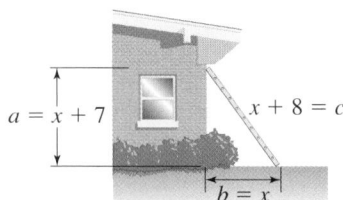

The distance from the bottom of the ladder to the house is x feet. The ladder is resting on the side of the house $x + 7$ feet from the ground, and the length of the ladder is 1 ft more than the height from the ground. Thus the length of the ladder is $x + 7 + 1$ or $x + 8$ feet. By the Pythagorean theorem, we have the equation

$$c^2 \quad = \quad a^2 \quad + b^2$$
$$(x + 8)^2 = (x + 7)^2 + x^2$$

4. Use algebra to solve the resulting equation.

$(x + 8)^2 = (x + 7)^2 + x^2$	Given.
$x^2 + 16x + 64 = x^2 + 14x + 49 + x^2$	Simplify.
$0 = x^2 - 2x - 15$	0 Subtract $x^2 + 16x + 64$ from both sides.
$x^2 - 2x - 15 = 0$	By the symmetric property
$(x - 5)(x + 3) = 0$	F Factor. We need two numbers whose product is -15 and whose sum is -2 (-5 and 3).
$x - 5 = 0$ or $x + 3 = 0$	F Factors $= 0$. Use the zero-factor property.
$x = 5$ or $x = -3$	Solve each equation.

x represents a length, so $x = -3$ must be discarded as an answer. The foot of the ladder is 5 ft from the wall and leans against the wall $x + 7 = 5 + 7 = 12$ ft from the ground. Its length is $x + 8 = 5 + 8 = 13$ ft.

5. Verify the solution.
We note that $13^2 = 5^2 + 12^2$.

PROBLEM 6

Gina swims diagonally across her rectangular pool every day. If her pool is 7 m longer than it is wide, and it is 17 m diagonally across, how long is her pool?

Answer

6. 15 m

C   Applications

Many problems can be studied using quadratic equations. For example, do you use hair spray containing chlorofluorocarbons (CFCs) for propellants? A U.N.-sponsored conference negotiated an agreement to stop producing CFCs by the year 2000 because they harm the ozone layer. In Example 7 we will check to see if the goal was reached.

EXAMPLE 7   An application of quadratic equations

The production of CFCs for use as aerosol propellants (in thousands of tons) can be approximately represented by $P(t) = -0.4t^2 + 22t + 120$, where t is the number of years after 1960. When will production be stopped?

SOLUTION   Production will be stopped when $P(t) = 0$; we need to solve the equation

$$P(t) = -0.4t^2 + 22t + 120 = 0$$

$$-4t^2 + 220t + 1200 = 0 \qquad \text{O} \quad \text{Multiply by 10 (to clear decimals).}$$

$$t^2 - 55t - 300 = 0 \qquad \text{Divide by } -4 \text{ (to obtain } t^2\text{).}$$

$$(t - 60)(t + 5) = 0 \qquad \text{F} \quad \text{Factor. We need two numbers whose product is } -300 \text{ and whose sum is } -55 \ (-60 \text{ and } 5).$$

$$t - 60 = 0 \quad \text{or} \quad t + 5 = 0 \qquad \text{F} \quad \text{Factors} = 0. \text{ By the zero-factor property}$$

$$t = 60 \quad \text{or} \qquad t = -5 \qquad \text{Solve each equation.}$$

Since t represents the number of years *after* 1960, production will be zero (stopped) 60 years after 1960 or in 2020 (not in the year 2000 as promised!). The answer $t = -5$ has to be discarded because it represents 5 years *before* 1960, but the equation applies only to years *after* 1960.

PROBLEM 7

John's courtyard measures 6 m by 8 m. He plans a garden inside, with a grass border of uniform width on all four sides. The garden must contain exactly 15 sq m of area to meet the plans. How wide a border should John create? He comes up with the equation $(6 - 2x)(8 - 2x) = 15$ to solve the problem. (*To see how he arrived at that equation go to http://math.uww. edu/~mcfarlat/141/story11j.htm.*) Solve the quadratic equation to find the width of the border.

Answer

7. 1.5 m

Calculate It Exercises

1. A polynomial $P(x)$ has the graph shown in Window 1. It has two roots (solutions).

   a. What is the degree of $P(x)$?

   b. What is the equation for $P(x)$?

   [*Hint:* Look at Example 2(a).]

2. A polynomial $Q(x)$ has the graph shown in Window 2. It has three roots (solutions).

   a. What is the degree of $Q(x)$?

   b. What is the equation for $Q(x)$?

   (*Hint:* Look at Example 4.)

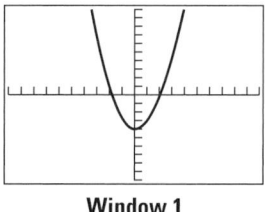

Window 1

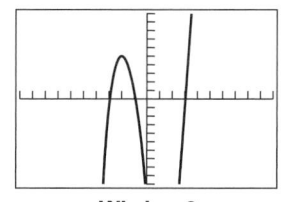

Window 2

3. Suppose a polynomial $R(x)$ crosses the x-axis n times.

   a. How many roots (solutions) does $R(x)$ have?

   b. What is the degree of $R(x)$?

   c. If the roots (solutions) of $R(x)$ are $r_1, r_2, r_3, \ldots r_n$, can you write an equation for $R(x)$?

Exercises 5.7

A In Problems 1–46, solve the equations.

1. $(x + 1)(x + 2) = 0$
$-1, -2$

2. $(x + 3)(x + 4) = 0$
$-3, -4$

3. $(x - 1)(x + 4)(x + 3) = 0$
$1, -4, -3$

4. $(x + 5)(x - 3)(x + 2) = 0$
$-5, 3, -2$

5. $\left(x - \dfrac{1}{2}\right)\left(x - \dfrac{1}{3}\right) = 0$ $\dfrac{1}{2}, \dfrac{1}{3}$

6. $\left(x - \dfrac{1}{4}\right)\left(x - \dfrac{1}{7}\right) = 0$ $\dfrac{1}{4}, \dfrac{1}{7}$

7. $y(y - 3) = 0$ $0, 3$

8. $y(y - 4) = 0$ $0, 4$

9. $y^2 - 64 = 0$ $8, -8$

10. $y^2 - 1 = 0$ $1, -1$

11. $y^2 - 81 = 0$ $9, -9$

12. $y^2 - 100 = 0$ $10, -10$

13. $x^2 + 6x = 0$ $0, -6$

14. $x^2 + 2x = 0$ $0, -2$

15. $x^2 - 3x = 0$ $0, 3$

16. $x^2 - 8x = 0$ $0, 8$

17. $y^2 - 12y = -27$ $3, 9$

18. $y^2 - 10y = -21$ $7, 3$

19. $y^2 = -6y - 5$ $-1, -5$

20. $y^2 = -3y - 2$ $-1, -2$

21. $x^2 = 2x + 15$ $5, -3$

22. $x^2 = 4x + 12$ $6, -2$

23. $3y^2 + 5y + 2 = 0$
$-\dfrac{2}{3}, -1$

24. $3y^2 + 7y + 2 = 0$
$-\dfrac{1}{3}, -2$

25. $2y^2 - 3y + 1 = 0$
$1, \dfrac{1}{2}$

26. $2y^2 - 3y - 20 = 0$
$-\dfrac{5}{2}, 4$

27. $2y^2 - y - 1 = 0$ $1, -\dfrac{1}{2}$

28. $2y^2 - y - 15 = 0$ $-\dfrac{5}{2}, 3$

Hint for Problems 29–34: multiply each term by the LCD *first*.

29. $\dfrac{x^2}{12} + \dfrac{x}{3} - 1 = 0$
$2, -6$

30. $\dfrac{x^2}{2} - \dfrac{x}{12} - 1 = 0$
$-\dfrac{4}{3}, \dfrac{3}{2}$

31. $\dfrac{x^2}{3} - \dfrac{x}{2} = -\dfrac{1}{6}$
$1, \dfrac{1}{2}$

32. $\dfrac{x^2}{6} + \dfrac{x}{3} = \dfrac{1}{2}$
$-3, 1$

33. $\dfrac{x^2}{12} + \dfrac{x}{2} = -\dfrac{2}{3}$ $-2, -4$

34. $\dfrac{x^2}{3} + \dfrac{x}{3} = \dfrac{1}{4}$ $-\dfrac{3}{2}, \dfrac{1}{2}$

35. $(2x - 1)(x - 3) = 3x - 5$
$4, 1$

36. $(3x + 1)(x - 2) = x + 7$
$-1, 3$

37. $(2x + 3)(x + 4) = 2(x - 1) + 4$ $-2, -\dfrac{5}{2}$

38. $(5x - 2)(x + 2) = 3(x + 1) - 7$ $0, -1$

39. $(2x - 1)(x - 1) = x - 1$ 1

40. $(3x - 2)(3x - 1) = 1 - 3x$ $\dfrac{1}{3}$

41. $x^3 + 4x^2 - 4x - 16 = 0$ $2, -2, -4$

42. $x^3 - 4x^2 - 4x + 16 = 0$ $-2, 2, 4$

43. $x^3 - 5x^2 - 9x + 45 = 0$ $5, 3, -3$

44. $x^3 + 5x^2 - 9x - 45 = 0$ $-5, -3, 3$

45. $3x^3 + 3x^2 = 12x + 12$ $2, -2, -1$

46. $2x^3 - 2x^2 - 18x + 18 = 0$ $1, 3, -3$

B Use the Pythagorean theorem to solve Problems 47–50.

47. The sides of a right triangle are consecutive even integers. Find their lengths. $6, 8, 10$

48. The hypotenuse of a right triangle is 4 cm longer than the shortest side and 2 cm longer than the remaining side. Find the dimensions of the triangle.
6 cm, 8 cm, 10 cm

49. The hypotenuse of a right triangle is 16 in. longer than the shortest side and 2 in. longer than the remaining side. Find the dimensions of the triangle.
10 in., 24 in., 26 in.

50. The hypotenuse of a right triangle is 8 in. longer than the shortest side and 1 in. longer than the remaining side. Find the dimensions of the triangle.
5 in., 12 in., 13 in.

APPLICATIONS

In Problems 51–54, use

$$d = 5t^2 + V_0 t$$

where d is the distance (in meters) traveled in t seconds by an object thrown downward with an initial velocity V_0 (in meters per second).

51. An object is thrown downward with an initial velocity of 5 m/sec from a height of 10 m. How long does it take the object to hit the ground? 1 sec

52. An object is thrown downward from a height of 28 m with an initial velocity of 4 m/sec. How long does it take the object to reach the ground? 2 sec

53. An object is thrown downward from a building 15 m tall with an initial velocity of 10 m/sec. How long does it take the object to hit the ground? 1 sec

54. How long would it take a package thrown downward from a plane with an initial velocity of 10 m/sec to hit the ground 175 m below? 5 sec

55. It costs a business $(0.1x^2 + x + 50)$ dollars to serve x customers. How many customers can be served if $250 is the cost? 40

56. The cost of serving x customers is given by $(x^2 + 10x + 100)$ dollars. If $1300 is spent serving customers, how many customers are served? 30

57. A manufacturer will produce x units of a product when its price is $(x^2 + 25x)$ dollars per unit. How many units will be produced when the price is $350 per unit? 10

58. When the price of a ton of raw materials is $(0.01x^2 + 5x)$ dollars, a supplier will produce x tons of it. How many tons will be produced when the price is $5000 per ton? 500

59. To attract more students, the campus theater decides to reduce ticket prices by x dollars from the current $5.50 price.

    a. What is the new price after the x dollar reduction? $5.50 - x$

    b. If the number of tickets sold is $100 + 100x$ and the revenue is the number of tickets sold times the price of each ticket, what is the revenue? $550 + 450x - 100x^2$

    c. If the theater wishes to have $750 in revenue, how much is the price reduction? $0.50 or $4

    d. If the reduction must be less than $1, what is the reduction? $0.50

60. An apartment owner wants to increase the monthly rent from the current $250 in n increases of $10.

    a. What is the new price after the n increases of $10? $250 + 10n$

    b. If the number of apartments rented is $70 - 2n$ and the revenue is the number of apartments rented times the rent, what is the revenue? $17{,}500 + 200n - 20n^2$

    c. If the owner wants to receive $17,980 per month, how many $10 increases can the owner make? 4 or 6

    d. What will be the monthly rent? $290 (if 62 apts), $310 (if 58 apts)

SKILL CHECKER

Try these "Skill Checker" exercises so you'll be ready for the next section.

Simplify:

61. $\dfrac{3x^5}{15x^7}$ $\dfrac{1}{5x^2}$

62. $\dfrac{10x^{-3}}{5x^6}$ $\dfrac{2}{x^9}$

63. $\dfrac{20x^7}{10x^{-3}}$ $2x^{10}$

64. $\dfrac{8x^{-4}}{16x^{-5}}$ $\dfrac{x}{2}$

65. $\dfrac{18x^{-10}}{9x^2}$ $\dfrac{2}{x^{12}}$

66. $\dfrac{4x^{-5}}{2x^{-5}}$ 2

USING YOUR KNOWLEDGE

Play Ball!

Have you been to a baseball game lately? Did anybody hit a home run? The trajectory of a baseball is usually complicated, but we can get help from *The Physics of Baseball,* by Robert Adair. According to Mr. Adair, after t seconds, starting 1 second after the ball leaves the bat, the height of a ball hit at a 35° angle rotating with an initial backspin of 2000 revolutions per minute (rpm) and hit at about 110 mi/hr is given by

$$H(t) = -80t^2 + 239t + 3 \text{ (in feet)}$$

67. How many seconds will it be before the ball hits the ground? 4 sec $(3 + 1)$

The distance traveled by the ball is given by

$$D(t) = -5t^2 + 115t - 110 \text{ (in feet)}$$

68. How far will the ball travel before it hits the ground? Approximately 288 ft

69. How far will the ball travel in 6 sec, the time it takes a high fly ball to hit the ground? 400 ft

WRITE ON

70. Write an explanation of the difference between quadratic and linear equations. Answers may vary.

71. Explain why the zero-factor property works for more than two numbers whose product is zero. Answers may vary.

72. Explain the differences in the procedure for solving $3(x - 1)(x + 4) = 0$ and $3x(x - 1)(x + 4) = 0$. Answers may vary.

73. Write a word problem that uses the Pythagorean theorem in its solution. Answers may vary.

MASTERY TEST

If you know how to do these problems, you have learned your lesson!

Solve:

74. $x^3 + 2x^2 - 9x - 18 = 0$
$-2, 3, -3$

75. $8x^2 - 2x = 1$ $-\frac{1}{4}, \frac{1}{2}$

76. $12x^2 + 13x - 4 = 0$ $\frac{1}{4}, -\frac{4}{3}$

77. $x^2 = 6x - 5$ $1, 5$

78. $x^2 + x = 6$ $-3, 2$

79. $x^2 - 16 = 0$ $-4, 4$

80. $x^2 + 9x = 0$ $0, -9$

81. The lengths of one leg of a right triangle and its hypotenuse are consecutive integers. If the shortest leg is 7 units shorter than the longer leg, find the lengths of the three sides. 5 units, 12 units, 13 units

COLLABORATIVE LEARNING 5A

How high could the heel of a shoe measure before you risk taking a tumble to the floor? At the Institute of Physics in London, scientists have come up with a polynomial formula that will tell you how high the heels can be without the risk of falling.

$$h = Q\left(12 + \frac{3s}{8}\right)$$

In this formula h is the maximum height of the heel in centimeters, Q is a sociological factor with a value between 0 and 1 (explained later), and s is the size of the shoe in British measurement. According to the physicist at the University of Surrey who led the research, Paul Stevenson, it is based on the Pythagorean formula. Using this formula it was calculated that, when sober, Carrie Bradshaw of the television show "Sex and the City," could wear a heel of just over five inches.

The formula for Q:

$$Q = \frac{p(y + 9)L}{(t + 1)(A + 1)(y + 10)(L + £20)}$$

The variables for this formula are as follows: p is the probability—from 0 to 1, with 1 being very probable—that the shoes will attract a mate for their wearer; y is the number of years' experience in high heels; L is the shoes' price in British pounds; t is the time in months since the shoes were fashionable (with 0 being right now); and A is the units of alcohol that the wearer plans to consume while in the shoes.

To use this formula in the United States, some of the measurements must be converted to U.S. measurements.

1. Find the conversion equivalences for changing from British women shoe sizes to U.S. women sizes.

2. Find the conversion equivalence for changing from the British pound to the U.S. dollar.

3. Find the conversion equivalence for changing from centimeters to inches.

4. Complete the table below to find the height of the heel in inches.

s (U.S.)	s (Brit)	p (0–1)	y (yr)	L (US \$)	L (Brit £)	A (units)	t (mo)	Height of heel (cm)	Height of heel (in.)
6	?	0.5	10	\$60	?	2	4		
7	?	1	20	\$100	?	0	0		
8	?	0.8	15	\$30	?	1	10		

5. Interview five people to get the necessary measurements so you can determine the appropriate height of heel. Study the results and have a discussion to see if there are any conclusions that can be drawn about what variables may or may not lead to a "safe" height of the heel.

6. Rewrite the formula so that the conversions are in the formula.

COLLABORATIVE LEARNING 5B

width

length

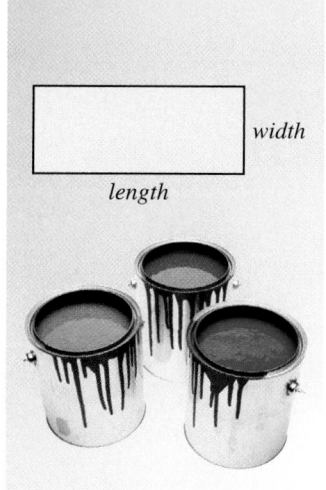

Did Picasso know about polynomials? Who knows, but if he did, it would have helped him paint his home! Here is how: suppose there are four walls and opposite sides have the same dimensions, then the total area can be found by the polynomial, $A = 2(l_1w_1) + 2(l_2w_2)$, where l_1w_1 are the dimensions of one wall and l_2w_2 are the dimensions of the other wall. Divide into groups and prepare a bid to paint your classroom. Along with this polynomial for finding the total area, you will need paint store information (from fliers or Internet research), measuring tape, calculators, and a chart (provided here) to itemize your bid.

Start by measuring the walls to find how many square feet you will have to paint. Remember to take into consideration things like whether you will paint around the bulletin boards and whether you will subtract the space taken up by the windows and doors when calculating total area.

Each gallon of paint could cover approximately 400–500 sq ft of wall space depending on the brand and grade of paint. Supplies will include brushes, rollers, drop cloths, and so on.

Determine how many painters will be painting, the amount of hours they will work, and the hourly wage. Then complete the information on the chart. Transfer the information to a poster board to present to the class. Have a class discussion about your results addressing the following questions.

Items			Total
Number of Painters			
Hours to Complete the Job			
Area to be Painted			
	How Many	**Cost Each**	
Cost of Paint			
Cost of Total Hours of Wages			
Cost of Supplies			
Profit			
Final Bid			

1. Was the estimate low enough to be in the running for the job, but high enough to give you some profit?

2. How important was it to prepare a precise estimate?

3. Does the size of the job have anything to do with how precise an estimate should be?

4. What happens if the estimate is too low or too high?

5. What other kinds of work would involve an understanding of area?

6. After listening to all the presenters, what would you do differently if you were asked to make another painting bid?

1. Who was the first person to use the notation x, xx, x^3, x^4, ... for exponents, and when did this notation first occur?

2. In Section 5.1, we mentioned that the height of a diver is given by $H(t) = -16t^2 + 118$. An eminent mathematician, born in Pisa in 1564, performed experiments from the Leaning Tower of Pisa. Find out the name of this mathematician, the nature of his experiments, and his conclusions.

3. How old do you think quadratic equations are? There are clay tablets indicating that the Babylonians of 2000 B.C. were familiar with formulas for solving quadratic equations. Write a paper about Babylonian mathematics with special emphasis on the solution of quadratic equations.

4. About 4000 years ago, the Egyptians used trained surveyors, the *harpedonaptae*. Find out what the word *harpedonaptae* means. Then write a paragraph explaining what these surveyors did and the ways in which they used the Pythagorean theorem in their work.

5. If you are looking for multicultural discovery, the Pythagorean theorem is it! Write a paper about the proofs of the Pythagorean theorem by

   a. The ancient Chinese

   b. Bhaskara

   c. The early Greeks

   d. Euclid

   e. Garfield (not the cat, the twentieth U.S. president!)

   f. Pappus

Summary

SECTION	ITEM	MEANING	EXAMPLE
5.1A	Monomial	A constant times a product of variables with whole-number exponents	$3x^2y$, $-7x$, $0.5x^3$, $\frac{2}{3}x^2yz^4$
	Polynomial	A sum or difference of monomials	$3x^2 - 7x + 8$, $x^2y + y^3$
	Terms	The individual monomials in a polynomial	The terms of $3x^2 - 7x + 8$ are $3x^2$, $-7x$, and 8.
	Coefficient	The numerical factor of a term	The coefficient of $3x^2$ is 3.
	Binomial	A polynomial with two terms	$5x^2 - 7$ is a binomial.
	Trinomial	A polynomial with three terms	$-3 + x^2 + x$ is a trinomial.
5.1B	Degree of a polynomial	Largest sum of the exponents in any term	The degree of $x^3 + 7x$ is 3. The degree of $-2x^3yz^2$ is 6.
	Descending order	Polynomial in one variable ordered from highest exponent to lowest	$7x^3 - 5x^2 + x - 3$ is in descending order.

SECTION	ITEM	MEANING	EXAMPLE
5.1C	Evaluate a polynomial	Indicate the value of a polynomial by replacing specific values for the variable.	If $P(x) = x^2 + 3x - 1$ then $P(2) = (2)^2 + 3(2) - 1$ $= 4 + 6 - 1$ $= 9$
5.1D	Commutative property of addition	If P and Q are polynomials, $P + Q = Q + P$.	$x + 3x^2 = 3x^2 + x$
	Associative property of addition	If P, Q, and R are polynomials, $P + (Q + R) = (P + Q) + R$.	$x^2 + (3 + 7x) = (x^2 + 3) + 7x$
	Distributive property	If P, Q, and R are polynomials, $P(Q + R) = PQ + PR$ $(Q + R)P = QP + RP$.	$2x(x^2 + 3) = 2x^3 + 6x$
5.2A	Commutative property of multiplication	If P and Q are polynomials, $P \cdot Q = Q \cdot P$.	$x \cdot 3x^2 = 3x^2 \cdot x$
	Associative property of multiplication	If P, Q, and R are polynomials, $P \cdot (Q \cdot R) = (P \cdot Q) \cdot R$.	$x^2 \cdot (3 \cdot 7x) = (x^2 \cdot 3) \cdot 7x$
5.2C	Product of two binomials	$(x + a)(x + b) = x^2 + (b + a)x + ab$	$(x + 5)(x - 7) = x^2 - 2x - 35$
5.2D	Square of a binomial sum	$(x + a)^2 = x^2 + 2ax + a^2$	$(x + 5y)^2 = x^2 + 10xy + 25y^2$
	Square of a binomial difference	$(x - a)^2 = x^2 - 2ax + a^2$	$(x - 5y)^2 = x^2 - 10xy + 25y^2$
5.2E	Product of a sum and a difference	$(x + a)(x - a) = x^2 - a^2$	$(x + 2y)(x - 2y) = x^2 - 4y^2$
5.3A	Greatest common factor (GCF) of a polynomial in x (ax^n)	1. a is the greatest integer that divides each of the coefficients in the polynomial. 2. n is the smallest exponent of x in all terms of the polynomial.	The GCF of $3x^6 + 6x^3$ is $3x^3$.
5.3B	Factoring by grouping (four terms)	1. Group terms with common factors. 2. Factor each group. 3. Factor out the (binomial) GCF.	$2x^3 + 4x^2 - 3x - 6$ $= (2x^3 + 4x^2) - (3x + 6)$ $= 2x^2(x + 2) - 3(x + 2)$ $= (x + 2)(2x^2 - 3)$
5.4A	Factoring $x^2 + (b + a)x + ab$	$x^2 + (b + a)x + ab = (x + a)(x + b)$	$x^2 + 5x + 4 = (x + 1)(x + 4)$
5.4C	The ac test	$ax^2 + bx + c$ is factorable only if there are two integers whose product is ac and whose sum is b.	$3x^2 + 5x + 2$ is factorable, since there are two integers with product 6 and sum 5.
5.5A	Factoring perfect square trinomials	$x^2 + 2ax + a^2 = (x + a)^2$ $x^2 - 2ax + a^2 = (x - a)^2$	$x^2 + 10x + 25 = (x + 5)^2$ $x^2 - 14x + 49 = (x - 7)^2$
5.5B	Factoring the difference of two squares	$x^2 - a^2 = (x + a)(x - a)$	$16x^2 - 9y^2 = (4x + 3y)(4x - 3y)$
5.5C	Factoring the sum or difference of two cubes	$x^3 + a^3 = (x + a)(x^2 - ax + a^2)$ $x^3 - a^3 = (x - a)(x^2 + ax + a^2)$	$8x^3 + 27 = (2x + 3)(4x^2 - 6x + 9)$ $8x^3 - 27 = (2x - 3)(4x^2 + 6x + 9)$

(Continued)

SECTION	ITEM	MEANING	EXAMPLE

5.6A	General factoring strategy	1. Factor out the GCF.
		2. Check for:
		  Difference of two squares
		  Difference of two cubes
		  Sum of two cubes
		  Perfect square trinomials
		  Trinomials of the form
		    $ax^2 + bx + c$
		  Four terms (grouping)
		3. Check by multiplying factors.	

5.7	Quadratic equation	An equation that can be written in the form $ax^2 + bx + c = 0 \ (a \neq 0)$	$3x^2 + 5x = -6$ is a quadratic equation.
5.7A	Zero-factor property	For all real numbers a and b, $a \cdot b = 0$ means that $a = 0$ or $b = 0$, or both.	$(x + 1)(x + 2) = 0$ means $x + 1 = 0$ or $x + 2 = 0$.
5.7B	Pythagorean theorem	In any right triangle, the square of the longest side is equal to the sum of the squares of the other two sides: $c^2 = a^2 + b^2$.	If the sides of a right triangle are of lengths 3 and 4 and the hypotenuse is 5, $3^2 + 4^2 = 5^2$.

Review Exercises

(If you need help with these exercises, look in the section indicated in brackets.)

1. [5.1A, B] Classify as a monomial, binomial, or trinomial and give the degree.

a. $x^3 + x^2y^3z$
   Binomial; degree 6
b. $x^3y^2z^3$
   Monomial; degree 8
c. $x^4 - 5x^2y^3 + xyz$
   Trinomial; degree 5

2. [5.1B] Write the polynomials in descending order.

a. $-x^2 + 3x^4 - 5x + 2$
   $3x^4 - x^2 - 5x + 2$
b. $3x - x^2 + 4x^3$
   $4x^3 - x^2 + 3x$
c. $6x^2 - 2 + x$
   $6x^2 + x - 2$

3. [5.1C, E] If a \$50,000 computer depreciates 20% each year and its value $v(t)$ after t years is given by $v(t) = 50,000(1 - 0.20t)$.

a. Find the value of the computer after 3 yr. \$20,000
b. Find the value of the computer after 5 yr. \$0

4. [5.1C] Let $P(x) = x^2 - 3x + 5$ and find.

a. $P(-1)$ 9
b. $P(2)$ 3
c. $P(-3)$ 23

5. [5.1D] Add $2x^3 + 5x^2 - 3x - 1$ and

a. $8 - 7x + 3x^2 - x^3$
   $x^3 + 8x^2 - 10x + 7$
b. $9 - 8x^2 + 3x^3$
   $5x^3 - 3x^2 - 3x + 8$
c. $7 - 4x + x^3$
   $3x^3 + 5x^2 - 7x + 6$

6. [5.2D] Subtract $7x^3 - 5x^2 + 3x - 1$ from

a. $4x^3 + 2x^2 + 2$
   $-3x^3 + 7x^2 - 3x + 3$
b. $7x^3 + 5x^2 + 4x - 7$
   $10x^2 + x - 6$
c. $8x^2 - 9x + 3$
   $-7x^3 + 13x^2 - 12x + 4$

7. [5.2A] Multiply.

   a. $-2x^2y(x^2 + 3xy - 2y^3)$
     $-2x^4y - 6x^3y^2 + 4x^2y^4$

   b. $-3x^2y^2(x^2 + 3xy - 2y^3)$
     $-3x^4y^2 - 9x^3y^3 + 6x^2y^5$

   c. $-4xy^2(x^2 + 3xy - 2y^3)$
     $-4x^3y^2 - 12x^2y^3 + 8xy^5$

8. [5.2B] Multiply.

   a. $(x - 1)(x^2 - 3x - 2)$
     $x^3 - 4x^2 + x + 2$

   b. $(x - 2)(x^2 - 3x - 2)$
     $x^3 - 5x^2 + 4x + 4$

   c. $(x + 3)(x^2 - 3x - 2)$
     $x^3 - 11x - 6$

9. [5.2C] Multiply.

   a. $(2x + 3y)(4x - 5y)$
     $8x^2 + 2xy - 15y^2$

   b. $(2x + 3y)(3x + 2y)$
     $6x^2 + 13xy + 6y^2$

   c. $(2x + 3y)(5x - 3y)$
     $10x^2 + 9xy - 9y^2$

10. [5.2D] Multiply.

   a. $(2x + 5y)^2$ $4x^2 + 20xy + 25y^2$

   b. $(3x + 7y)^2$ $9x^2 + 42xy + 49y^2$

   c. $(4x + 9y)^2$ $16x^2 + 72xy + 81y^2$

11. [5.2D] Multiply.

   a. $(3x - 2y)^2$ $9x^2 - 12xy + 4y^2$

   b. $(4x - 7y)^2$ $16x^2 - 56xy + 49y^2$

   c. $(5x - 6y)^2$ $25x^2 - 60xy + 36y^2$

12. [5.2E] Multiply.

   a. $(3x + 2y)(3x - 2y)$ $9x^2 - 4y^2$

   b. $(4x + 3y)(4x - 3y)$ $16x^2 - 9y^2$

   c. $(5x + 3y)(5x - 3y)$
     $25x^2 - 9y^2$

13. [5.3A] Factor completely.

   a. $15x^5 - 20x^4 + 10x^3 + 25x^2$
     $5x^2(3x^3 - 4x^2 + 2x + 5)$

   b. $9x^5 - 12x^4 + 6x^3 + 15x^2$
     $3x^2(3x^3 - 4x^2 + 2x + 5)$

   c. $6x^5 - 8x^4 + 4x^3 + 10x^2$
     $2x^2(3x^3 - 4x^2 + 2x + 5)$

14. [5.3B] Factor completely.

   a. $6x^6 - 2x^4 + 15x^3 - 5x$
     $x(3x^2 - 1)(2x^3 + 5)$

   b. $6x^6 - 8x^4 + 15x^3 - 20x$
     $x(3x^2 - 4)(2x^3 + 5)$

   c. $6x^6 - 4x^4 + 9x^3 - 6x$
     $x(3x^2 - 2)(2x^3 + 3)$

15. [5.4A] Factor completely.

   a. $x^2 - 3xy - 18y^2$
     $(x - 6y)(x + 3y)$

   b. $x^2 - 4xy - 12y^2$
     $(x - 6y)(x + 2y)$

   c. $x^2 - 5xy - 6y^2$
     $(x - 6y)(x + y)$

16. [5.4B] Factor completely.

   a. $2x^2 - 7xy - 30y^2$
     $(2x + 5y)(x - 6y)$

   b. $2x^2 - 3xy - 20y^2$
     $(2x + 5y)(x - 4y)$

   c. $2x^2 - 5xy - 25y^2$
     $(2x + 5y)(x - 5y)$

17. [5.4C] Factor completely.

   a. $-18x^4y - 3x^3y^2 + 6x^2y^3$
     $-3x^2y(3x + 2y)(2x - y)$

   b. $-30x^4y - 35x^3y^2 - 10x^2y^3$
     $-5x^2y(3x + 2y)(2x + y)$

   c. $36x^4y - 30x^3y^2 - 36x^2y^3$
     $6x^2y(3x + 2y)(2x - 3y)$

18. [5.5A] Factor completely.

   a. $4x^2 - 28xy + 49y^2$
     $(2x - 7y)^2$

   b. $9x^2 - 42xy + 49y^2$
     $(3x - 7y)^2$

   c. $16x^2 - 56xy + 49y^2$
     $(4x - 7y)^2$

19. [5.5A] Factor completely.

   a. $9x^2 + 24xy + 16y^2$
     $(3x + 4y)^2$

   b. $9x^2 + 30xy + 25y^2$
     $(3x + 5y)^2$

   c. $9x^2 + 36xy + 36y^2$
     $9(x + 2y)^2$

20. [5.5B] Factor completely.

   a. $81x^4 - y^4$
     $(9x^2 + y^2)(3x + y)(3x - y)$

   b. $x^4 - 16y^4$
     $(x^2 + 4y^2)(x + 2y)(x - 2y)$

   c. $81x^4 - 16y^4$
     $(9x^2 + 4y^2)(3x + 2y)(3x - 2y)$

21. [5.5B] Factor completely.

   a. $x^2 - 4x + 4 - y^2$
     $(x + y - 2)(x - y - 2)$

   b. $x^2 - 6x + 9 - y^2$
     $(x + y - 3)(x - y - 3)$

   c. $x^2 + 8x + 16 - y^2$
     $(x + y + 4)(x - y + 4)$

22. [5.5C] Factor completely.

   a. $27x^3 + 8y^3$
     $(3x + 2y)(9x^2 - 6xy + 4y^2)$

   b. $27x^3 + 64y^3$
     $(3x + 4y)(9x^2 - 12xy + 16y^2)$

   c. $64x^3 + 27y^3$
     $(4x + 3y)(16x^2 - 12xy + 9y^2)$

23. [5.5C] Factor completely.

   a. $27x^3 - 8y^3$
     $(3x - 2y)(9x^2 + 6xy + 4y^2)$

   b. $27x^3 - 64y^3$
     $(3x - 4y)(9x^2 + 12xy + 16y^2)$

   c. $64x^3 - 27y^3$
     $(4x - 3y)(16x^2 + 12xy + 9y^2)$

24. [5.6A] Factor completely.

   a. $27x^6 - 8x^3y^3$
     $x^3(3x - 2y)(9x^2 + 6xy + 4y^2)$

   b. $27x^7 - 64x^4y^3$
     $x^4(3x - 4y)(9x^2 + 12xy + 16y^2)$

   c. $64x^8 - 27x^5y^3$
     $x^5(4x - 3y)(16x^2 + 12xy + 9y^2)$

25. [5.6A] Factor completely.

   a. $27x^6 + 3x^4$
     $3x^4(9x^2 + 1)$

   b. $4x^6 + 64x^4$
     $4x^4(x^2 + 16)$

   c. $2x^6 - 18x^4$
     $2x^4(x + 3)(x - 3)$

26. [5.6A] Factor completely.

   a. $27x^4 + 36x^3y + 12x^2y^2$
     $3x^2(3x + 2y)^2$

   b. $36x^4 - 48x^3y + 16x^2y^2$
     $4x^2(3x - 2y)^2$

   c. $45x^4 + 30x^3y + 5x^2y^2$
     $5x^2(3x + y)^2$

27. [5.6A] Factor completely.

   a. $27x^4 - 18x^3y + 3x^2y^2$
     $3x^2(3x - y)^2$

   b. $36x^4 - 48x^3y + 16x^2y^2$
     $4x^2(3x - 2y)^2$

   c. $45x^4 + 60x^3y + 20x^2y^2$
     $5x^2(3x + 2y)^2$

28. [5.6A] Factor completely.

   a. $12x^3y - 44x^2y^2 - 16xy^3$
      $4xy(3x + y)(x - 4y)$

   b. $15x^3y + 65x^2y^2 + 20xy^3$
      $5xy(3x + y)(x + 4y)$

   c. $18x^3y - 60x^2y^2 - 48xy^3$
      $6xy(3x + 2y)(x - 4y)$

29. [5.6A] Factor completely.

   a. $2x^3 - x^2 - 2x + 1$
      $(2x - 1)(x + 1)(x - 1)$

   b. $18x^3 - 9x^2 - 2x + 1$
      $(2x - 1)(3x + 1)(3x - 1)$

   c. $32x^3 - 16x^2 - 2x + 1$
      $(2x - 1)(4x + 1)(4x - 1)$

30. [5.7A] Solve.

   a. $x^2 = -x + 12$
      $3, -4$

   b. $x^2 = -x + 20$
      $4, -5$

   c. $x^2 = -2x + 24$
      $4, -6$

31. [5.7A] Solve.

   a. $6x^2 + x = 1$ $\frac{1}{3}, -\frac{1}{2}$

   b. $8x^2 + 2x = 1$ $\frac{1}{4}, -\frac{1}{2}$

   c. $10x^2 + 3x = 1$ $\frac{1}{5}, -\frac{1}{2}$

32. [5.7A] Solve.

   a. $x^3 + 2x^2 - x - 2 = 0$
      $1, -1, -2$

   b. $x^3 + 4x^2 - x - 4 = 0$
      $1, -1, -4$

   c. $x^3 + 2x^2 - 9x - 18 = 0$
      $3, -2, -3$

33. [5.7B] Find the dimensions of a right triangle whose sides are

   a. $x, x + 7$, and $x + 8$ units long
      5 units, 12 units, 13 units

   b. $x, x + 3$, and $x + 6$ units long
      9 units, 12 units, 15 units

   c. $x, x + 4$, and $x + 8$ units long
      12 units, 16 units, 20 units

Practice Test 5

(Answers on page 417)

1. Classify as a monomial, binomial, or trinomial and give the degree of $xy^3z^4 - x^7$.

2. Write in descending order: $-4 + 3x^2 - x^4 + 2x^3$

3. The total dollar cost $C(x)$ of manufacturing x units of a product each week is given by $C(x) = 15x + 300$.

   a. Find the cost of manufacturing 400 units.

   b. Find the cost of manufacturing 1500 units.

4. Let $P(x) = x^2 - 3x + 2$. Find $P(-2)$.

5. Add $6x^3 + 8x^2 - 6x - 4$ and $6 - 3x + x^2 - 3x^3$.

6. Subtract $8x^3 - 6x^2 + 5x - 3$ from $5x^3 + 3x^2 + 3$.

7. Multiply $-3x^2y(x^2 + 5xy - 3y^3)$.

8. Multiply.

   a. $(x - 2)(x^2 - 4x - 5)$ **b.** $(3x + 5y)(4x - 7y)$

9. Multiply.

   a. $(2x + 3y)^2$ **b.** $(3x - 4y)^2$

10. Multiply $(3x + 4y)(3x - 4y)$.

11. Factor completely $12x^6 - 16x^5 + 8x^4 + 20x^3$.

12. Factor completely $6x^7 + 6x^5 + 15x^4 + 15x^2$.

13. Factor completely.

   a. $x^2 - 3xy - 18y^2$ **b.** $2x^2 + xy - 10y^2$

14. Factor completely $36x^4y + 12x^3y^2 - 8x^2y^3$.

15. Factor completely.

   a. $16x^2 - 24xy + 9y^2$ **b.** $9x^2 + 30xy + 25y^2$

16. Factor completely $x^4 - 16y^4$.

17. Factor completely $x^2 - 10x + 25 - y^2$.

18. Factor completely.

   a. $27x^3 + 8y^3$ **b.** $8y^3 - 27x^3$

19. Factor completely.

   a. $8x^7 - x^4y^3$ **b.** $6x^8 + 24x^6$

20. Factor completely.

   a. $8x^4 + 24x^3y + 18x^2y^2$

   b. $48x^2y^2 - 72xy^3 + 27y^4$

21. Factor completely $9x^3y - 33x^2y^2 - 12xy^3$.

22. Factor completely $16x^3 - 12x^2 - 4x + 3$.

23. Solve.

   a. $x^2 = -3x + 10$ **b.** $6x^2 + 7x = 3$

24. Solve $x^3 - x^2 - 4x + 4 = 0$.

25. The sides of a right triangle are $x, x + 2$, and $x + 4$ units long. Find the dimensions of the triangle.

Answers to Practice Test

ANSWER	IF YOU MISSED	REVIEW		
	QUESTION	SECTION	EXAMPLES	PAGE
1. Binomial; 8	1	5.1A, B	1, 2	345 & 346
2. $-x^4 + 2x^3 + 3x^2 - 4$	2	5.1B	2	346
3. **a.** 6300 **b.** $22,800$	3	5.1C, E	3, 8	347 & 350
4. 12	4	5.1C	3	347
5. $3x^3 + 9x^2 - 9x + 2$	5	5.1D	5, 6	349
6. $-3x^3 + 9x^2 - 5x + 6$	6	5.1D	7	350
7. $-3x^4y - 15x^3y^2 + 9x^2y^4$	7	5.2A	1	357
8. **a.** $x^3 - 6x^2 + 3x + 10$	8	5.2B, C	2, 3	358 & 360
b. $12x^2 - xy - 35y^2$				
9. **a.** $4x^2 + 12xy + 9y^2$	9	5.2D	4	361
b. $9x^2 - 24xy + 16y^2$				
10. $9x^2 - 16y^2$	10	5.2E	5	362
11. $4x^3(3x^3 - 4x^2 + 2x + 5)$	11	5.3A	1, 2, 3	369 & 370
12. $3x^2(x^2 + 1)(2x^3 + 5)$	12	5.3B	4, 5	370–372
13. **a.** $(x + 3y)(x - 6y)$	13	5.4A, B, C	1, 2, 3, 4, 5	376–377, 379
b. $(x - 2y)(2x + 5y)$				
14. $4x^2y(3x + 2y)(3x - y)$	14	5.4C	6	380
15. **a.** $(4x - 3y)^2$ **b.** $(3x + 5y)^2$	15	5.5A	1, 2	385 & 386
16. $(x^2 + 4y^2)(x + 2y)(x - 2y)$	16	5.5B	3	386
17. $(x - 5 + y)(x - 5 - y)$ or $(x + y - 5)(x - y - 5)$	17	5.5B	4	387
18. **a.** $(3x + 2y)(9x^2 - 6xy + 4y^2)$	18	5.5C	5	388
b. $(2y - 3x)(4y^2 + 6xy + 9x^2)$				
19. **a.** $x^4(2x - y)(4x^2 + 2xy + y^2)$	19	5.6A	1	394
b. $6x^6(x^2 + 4)$				
20. **a.** $2x^2(2x + 3y)^2$	20	5.6A	2	394
b. $3y^2(4x - 3y)^2$				
21. $3xy(3x + y)(x - 4y)$	21	5.6A	3	395
22. $(4x - 3)(2x + 1)(2x - 1)$	22	5.6A	5	395
23. **a.** $2, -5$ **b.** $\dfrac{1}{3}, -\dfrac{3}{2}$	23	5.7A	1, 2, 3	400–403
24. $1, 2, -2$	24	5.7A	4	404
25. 6, 8, and 10 units	25	5.7B	5	405

Cumulative Review Chapters 1–5

1. Use braces to list the elements of the set of even natural numbers less than 8. $\{2, 4, 6\}$

2. Write 0.19 as a fraction. $\frac{19}{100}$

3. Find: $-\dfrac{5}{9} \div \left(-\dfrac{1}{27}\right)$ 15

4. Do the calculation and write the answer in scientific notation: $(7.5 \times 10^3) \times (8 \times 10^{-7})$ 6×10^{-3}

5. Evaluate: $-4^3 + \dfrac{(4 - 12)}{2} + 6 \div 3$ -66

6. Solve: $\dfrac{3}{7}y - 4 = 2$ 14

7. Solve: $\left|\dfrac{4}{3}x + 6\right| + 8 = 15$ $\dfrac{3}{4}; -\dfrac{39}{4}$

8. Graph: $\dfrac{x}{8} - \dfrac{x}{3} < \dfrac{x - 8}{8}$

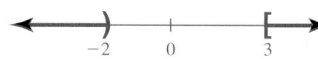

9. Graph: $\{x \mid x < -2 \text{ or } x \geq 3\}$

10. Graph: $|2x - 3| \leq 1$

11. Solve for A in $B = \dfrac{2}{5}(A - 11)$. $A = \dfrac{5B + 22}{2}$

12. A freight train leaves a station traveling at 45 mi/hr. Two hours later, a passenger train leaves the same station in the same direction at 55 mi/hr. How far from the station does the passenger train overtake the freight train? 495 mi

13. Find the x- and y-intercepts of $y = -3x - 5$. x-intercept $= -\dfrac{5}{3}$; y-intercept $= -5$

14. Graph: $4x = -16$

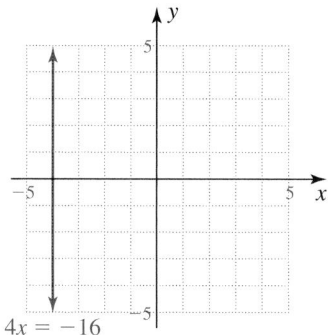

15. A line L_1 has slope $\dfrac{1}{2}$. Find whether the line through $(-3, -7)$ and $(3, -4)$ is parallel or perpendicular to line L_1. Parallel

16. Find the standard form of the equation of the line that passes through the points $(3, -1)$ and $(7, 4)$. $5x - 4y = 19$

17. A line has slope -3 and y-intercept -1. Find the slope-intercept equation of the line. $y = -3x - 1$

18. Graph: $x \geq 3$

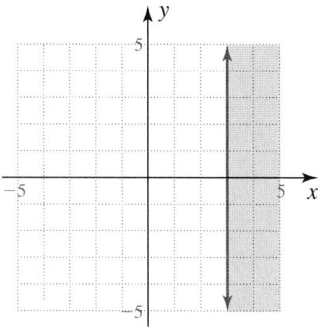

19. Solve by the elimination method:

$3x + 4y = -13$
$x + \ y = -4 \ (-3, -1)$

20. Solve the system:

$3x - 2y + \ z = 5$
$2x - \ y - 3z = 0$
$3x - \ y - \ z = 4 \ (2, 1, 1)$

21. Evaluate: $\begin{vmatrix} -1 & 2 & 4 \\ 3 & 0 & 0 \\ 3 & 3 & -4 \end{vmatrix}$ 60

22. Use Cramer's rule to solve for z:

$2x + \ y - 3z = -1$
$x - 2y - 2z = \ 7$
$3x + \ y + 3z = \ 0 \ z = 0$

23. Bert has \$3.75 in nickels and dimes. He has 12 more nickels than dimes. How many of each coin does he have? 33 nickels; 21 dimes

24. Classify as a monomial, binomial, or trinomial: $2x^4 + x^2$ Binomial

25. If $R(x) = x^2 - 5x - 5$, find $R(-4)$. 31

26. Subtract $8x^3 - 4x^2 + 7x - 7$ from $7x^3 + 3x^2 + 7$.
$-x^3 + 7x^2 - 7x + 14$

27. Multiply: $(2x - 7y)^2$ $4x^2 - 28xy + 49y^2$

28. Factor completely: $625x^4 - y^4$
$(25x^2 + y^2) \ (5x + y)(5x - y)$

29. Factor completely: $8c^3 + 125$
$(2c + 5)(4c^2 - 10c + 25)$

30. Solve for x: $2x^2 + 9x = -9$ $-\frac{3}{2}; -3$

Rational Expressions

The Human Side of Algebra

One of the most spectacular mathematical achievements of the sixteenth century was the discovery of an algebraic solution to cubic and quartic equations. The story of the discovery rivals the plots of contemporary novels. It starts with Scipione del Ferro's formula for solving the cubic $x^3 + mx = n$, a formula he passed on to his pupil Antonio Maria Fior. Enter Nicolo de Brescia, cruelly known as Tartaglia, "the stammerer," because of a speech impediment acquired during the sacking of Brescia by the French, which left his father dead and Nicolo with a sabre cut that cleft his jaw and palate. So poor was his mother that she could only pay his tutor for a meager 15 days, and even then he was relegated to using tombstones as slates on which to work his exercises. Tartaglia announced in 1535 the discovery of a more general formula to solve $x^3 + mx^2 = n$. Fior, believing the announcement to be a bluff, challenged him to a problem-solving contest. Who won? Tartaglia and his two formulas handily defeated Fior, who failed to solve a single problem, thus entering the annals of mathematical history in ignominious defeat.

Pretest for Chapter 6

(Answers on page 423)

1. a. For what value(s) is $\frac{3x}{2x+1}$ undefined?

b. Write an equivalent fraction with the indicated denominator.

$$\frac{2x^2}{9y^4}, \quad \text{denominator } 27y^6$$

2. Write in standard form.

a. $-\dfrac{-3}{y}$ **b.** $-\dfrac{x-y}{7}$

3. Reduce to lowest terms.

a. $\dfrac{x^4 y^5}{xy^2}$ **b.** $\dfrac{x^2 + xy}{x^2 - y^2}$

4. Reduce $\dfrac{x^2 - y^2}{y^3 - x^3}$ to lowest terms.

5. Multiply: $\dfrac{x-2}{3x-2y} \cdot \dfrac{9x^2 - 4y^2}{2x^2 - 3x - 2}$

6. Divide: $\dfrac{x+2}{x-2} \div (x^2 + 4x + 4)$

7. Perform the indicated operations:

$$\frac{2-x}{x+3} \div \frac{x^3 - 8}{x+6} \cdot \frac{x^3 + 27}{x+6}$$

8. Perform the indicated operations.

a. $\dfrac{x}{x^2 - 9} + \dfrac{3}{x^2 - 9}$ **b.** $\dfrac{x}{x^2 - 9} - \dfrac{3}{x^2 - 9}$

9. Perform the indicated operations.

a. $\dfrac{x+2}{x^2 + 2x - 3} + \dfrac{x+4}{x^2 - 9}$ **b.** $\dfrac{x-4}{x^2 - x - 6} - \dfrac{x+1}{x^2 - 9}$

10. Simplify: $\dfrac{x + \dfrac{1}{x^2}}{x - \dfrac{1}{x^3}}$

11. Simplify: $3 + \dfrac{a}{3 + \dfrac{3}{3+a}}$

12. Divide: $\dfrac{28x^5 - 14x^3 + 7x^2}{7x^4}$

13. Divide: $2x^3 - 4 - 4x$ by $2 + 2x$

14. Factor $2x^3 + 3x^2 - 23x - 12$ if $x + 4$ is one of its factors.

15. Use synthetic division to divide $x^4 + 2x^3 - 7x^2 - 30x + 18$ by $x + 2$

16. Use synthetic division to show that 3 is a solution of $x^4 + 2x^3 - 7x^2 - 30x + 18 = 0$.

17. Solve: $\dfrac{3}{x+2} + \dfrac{3}{4} = 1$

18. Solve: $1 + \dfrac{2}{x-3} = \dfrac{12}{x^2 - 9}$

19. Solve: $15x^{-2} + 7x^{-1} = 2$

20. Kyle can mow 3 acres of grass in 2 hr. At that rate, how long will it take him to mow 7 acres?

21. Find two consecutive even integers such that the sum of their reciprocals is $\frac{9}{40}$.

22. Jack can paint a room in 3 hr and Jill can paint it in 2 hr. How long would it take to paint the room if both work together?

23. A plane traveled 840 mi with a 30-mi/hr tail wind in the same time it took to travel 660 mi against the wind. What is the plane's speed in still air?

24. Pressure can be defined using the formula $P = \dfrac{F}{A}$, where F is force in newtons and A is area in meters squared.

a. Solve the formula for F.

b. If $P = 30$ and $A = 25$, find F.

25. The force, F, with which the earth attracts an object above the earth's surface varies inversely as the square of the distance, d, from the center of the earth.

a. Write an equation of variation with k as the constant of variation.

b. A meteorite weighs 50 lb on the earth's surface. If the radius of the earth is about 4000 mi, find k.

Answers to Pretest

	ANSWER		IF YOU MISSED		REVIEW	
			QUESTION	SECTION	EXAMPLES	PAGE
1. a. $-\dfrac{1}{2}$		**b.** $\dfrac{6x^2y^2}{27y^6}$	1	6.1	1, 2	425, 427
2. a. $\dfrac{3}{y}$		**b.** $\dfrac{y-x}{7}$	2	6.1	3	428
3. a. x^3y^3		**b.** $\dfrac{x}{x-y}$	3	6.1	4	430
4. $\dfrac{-(x+y)}{y^2+xy+x^2}$			4	6.1	5	431
5. $\dfrac{3x+2y}{2x+1}$			5	6.2	1, 2	436–437
6. $\dfrac{1}{x^2-4}$			6	6.2	3	438
7. $\dfrac{-(x^2-3x+9)}{x^2+2x+4}$ or $\dfrac{-x^2+3x-9}{x^2+2x+4}$			7	6.2	4, 5	439
8. a. $\dfrac{1}{x-3}$		**b.** $\dfrac{1}{x+3}$	8	6.3	1	446–447
9. a. $\dfrac{2x^2+2x-10}{(x+3)(x-1)(x-3)}$		**b.** $\dfrac{-4x-14}{(x+2)(x+3)(x-3)}$	9	6.3	2, 3	448–450
10. $\dfrac{x(x^2-x+1)}{(x^2+1)(x-1)}$			10	6.4	1, 2	456
11. $\dfrac{a^2+12a+36}{3a+12}$			11	6.4	3, 4	457–459
12. $4x-\dfrac{2}{x}+\dfrac{1}{x^2}$			12	6.5	1	466
13. $(x^2-x-1)\quad R-2$			13	6.5	2, 3	467–468
14. $(x+4)(2x+1)(x-3)$			14	6.5	4	469–470
15. $x^3-7x-16+\dfrac{50}{x+2}$			15	6.5	5	471
16. $3\overline{)1+2-\ 7-30-18}$ $\underline{\ \ \ 3+15+24-18}$ $1+5+\ \ 8-\ \ 6\quad\ \ 0$			16	6.5	6	472
17. 10			17	6.6	1	477–478
18. -5			18	6.6	2	478–480
19. $-\dfrac{3}{2},5$			19	6.6	5	482
20. $4\dfrac{2}{3}$ hr			20	6.6	6	483
21. 8 and 10			21	6.7	1	489–490
22. $1\dfrac{1}{5}$ hr			22	6.7	2	490
23. 250 mi/hr			23	6.7	4	492–493
24. a. $F=PA$		**b.** 750 newtons	24	6.7	5	493–494
25. a. $F=\dfrac{k}{d^2}$		**b.** 800,000,000	25	6.8	2	500

<table>
<tr><td>**6.1**</td><td colspan="2"># RATIONAL EXPRESSIONS</td></tr>
</table>

To Succeed, Review How To . . .

1. Find the GCF of two expressions (pp. 367–370).

2. Use the properties of exponents (pp. 33–40).

3. Factor polynomials (pp. 370–396).

Objectives

A Find the values that make a rational expression undefined.

B Write an equivalent rational expression with the indicated denominator.

C Write a rational expression in the standard forms.

D Reduce a rational expression to lowest terms.

GETTING STARTED   **An Application to Recycling and Waste**

Waste Recycling Rates 1960–2001

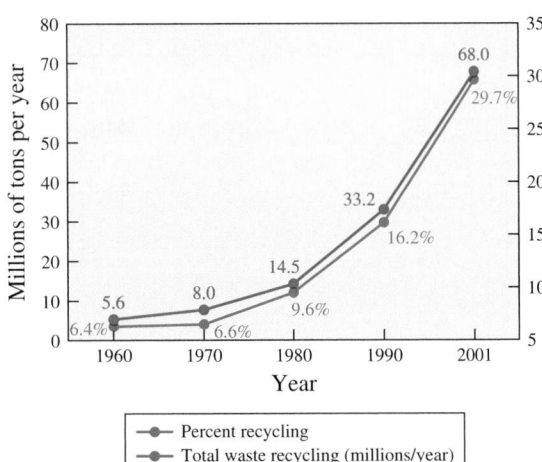

Source: Data from EPA.

Do you recycle bottles, newspaper, and plastic? What fraction of the waste generated in the United States is recycled? According to the EPA, the approximate waste *generated* in millions of tons is: $G(t) = 0.004t^2 + 3.52t + 86.5$, where t is the number of years after 1960. On the other hand, the approximate waste *recycled* in millions of tons is: $R(t) = 0.055t^2 - 0.7t + 6.86$. The fraction of the waste that gets recycled is

$$P(t) = \frac{R(t)}{G(t)} = \frac{0.055t^2 - 0.7t + 6.86}{0.004t^2 + 3.52t + 86.5}$$

The expression

$$\frac{R(t)}{G(t)}$$

is called a rational expression or rational function. Rational expressions are the topic of this section. To approximate the fraction of our waste expected to be recycled in the year 2010, 50 years after 1960, find

$$P(50) = \frac{R(50)}{G(50)} = \frac{109.36}{272.5} \approx 40.1\%$$

Can you compare this with the percent recycled in 1960?

The word *fraction* is derived from the Latin word *fractio,* which means "to break" or "to divide." Any **fraction** of the form $\frac{a}{b}$, where a and b are integers and $b \neq 0$, is a rational number. We extend this idea to polynomial expressions as follows:

Rational Expression	If P and Q are polynomials, the algebraic expression
	$$\frac{P}{Q} \quad (Q \neq 0)$$
	is a **rational expression.** If $Q = 0$, the expression is **undefined.**

Thus

$$\frac{x^2 - 2x + 1}{x + 2}$$

is undefined when $x + 2 = 0$, that is, when $x = -2$. Similarly,

$$\frac{1}{y^2 - 1}$$

is undefined when $y^2 - 1 = (y + 1)(y - 1) = 0$, that is, when $y = 1$ or $y = -1$. On the other hand,

$$\frac{x - 2}{3}$$

is *never* undefined, because the denominator 3 is *never* zero.

A Finding the Values That Make a Rational Expression Undefined

The values that make a rational expression, or rational function, undefined are the values that make the denominator of the fraction zero. We are looking for the domain of the function. For instance,

$$\frac{1}{x + 3} \quad \text{is undefined for} \quad x = -3$$

The domain is all real numbers where $x \neq -3$.

EXAMPLE 1   Finding values that make a rational expression undefined

For what values are the following rational expressions undefined?

a. $\dfrac{x + 1}{3x - 4}$ **b.** $\dfrac{8x}{x^2 + 9}$ **c.** $\dfrac{x^2 + 3x + 8}{x^2 + 2x - 3}$

SOLUTION

A rational expression is undefined for values of x that make the denominator zero. We find these domain values by setting the denominator equal to zero and solving the resulting equation.

a.
$$3x - 4 = 0 \qquad \text{Set the denominator equal to zero.}$$
$$3x = 4 \qquad \text{Add 4 to both sides.}$$
$$x = \frac{4}{3} \qquad \text{Divide both sides by 3.}$$

The number $\frac{4}{3}$ will make the rational expression undefined.

b.
$$x^2 + 9 = 0 \qquad \text{Set the denominator equal to zero.}$$

Since $x^2 = -9$ has no real number solution we conclude there aren't any real numbers that will make the rational expression undefined.

c.
$$x^2 + 2x - 3 = 0 \qquad \text{Set the denominator equal to zero.}$$
$$(x - 1)(x + 3) = 0 \qquad \text{Factor.}$$
$$x - 1 = 0 \quad \text{or} \quad x + 3 = 0 \qquad \text{Factors} = 0. \text{ Use the zero-factor property.}$$
$$x = 1 \quad \text{or} \qquad x = -3 \qquad \text{Solve each equation.}$$

The numbers 1 and -3 make the rational expression undefined.

PROBLEM 1

For what values are the following rational expressions undefined?

a. $\dfrac{-5}{2x}$

b. $\dfrac{x - 5}{2x + 3}$

c. $\dfrac{x^2 - 6x}{x^2 - 6x - 7}$

Answers

1. a. 0   **b.** $-\frac{3}{2}$ **c.** $-1; 7$

To avoid mentioning over and over that the denominators of algebraic fractions must not be zero, we make the following rule.

Undefined Rational Expressions

The variables in a rational expression may not be replaced by values that will make the denominator zero.

B   Writing Equivalent Rational Expressions

The fraction $\frac{4}{5}$ can be written as an equivalent fraction with a denominator of 10 by multiplying both numerator and denominator by 1 in the form of $\frac{2}{2}$ to obtain

$$\frac{4}{5} \cdot 1 = \frac{4 \cdot 2}{5 \cdot 2} = \frac{8}{10}$$

We can generalize this idea to rational expressions as follows.

Fundamental Property of Rational Expressions

If P, Q, and K are polynomials,

$$\frac{P}{Q} = \frac{P \cdot K}{Q \cdot K}$$

for all values for which the denominator is not zero.

To write

$$\frac{5x^3}{3y^2}$$

with a denominator of $6y^7$, first write the new equivalent rational expression with the old denominator factored out:

$$\frac{5x^3}{3y^2} \qquad = \qquad \frac{?}{3y^2(2y^5)}$$

$$\underset{\substack{\text{Multiply by } 2y^5 \\ \text{to obtain } 6y^7}}{\underline{\qquad\qquad\qquad}}$$

The multiplier is $2y^5$. We have

$$\frac{5x^3}{3y^2} = \frac{5x^3(2y^5)}{3y^2(2y^5)} = \frac{10x^3y^5}{6y^7}$$

(Multiply by $2y^5$)

We are multiplying the denominator by $2y^5$, so we have to multiply the numerator by $2y^5$, since $\frac{2y^5}{2y^5} = 1$.

Thus

$$\frac{5x^3}{3y^2} = \frac{10x^3y^5}{6y^7}$$

This skill will be used when adding or subtracting rational expressions with unlike denominators.

EXAMPLE 2 **Writing equivalent rational expressions with a specified denominator**

Write:

a. $\dfrac{5}{8a^2}$ with a denominator of $16a^5$

b. $\dfrac{x+7}{9y^3}$ with a denominator of $18y^4$

c. $\dfrac{3x+1}{x-1}$ with a denominator of $x^2 + 2x - 3$

SOLUTION

a. Since $16a^5 = 8a^2(2a^3)$, $\dfrac{5}{8a^2} = \dfrac{?}{8a^2(2a^3)}$

Multiply by $2a^3$

Multiply by $2a^3$

$\dfrac{5}{8a^2} = \dfrac{5(2a^3)}{8a^2(2a^3)}$

Thus $\dfrac{5}{8a^2} = \dfrac{10a^3}{16a^5}$

b. Since $18y^4 = 9y^3(2y)$,

$\dfrac{x+7}{9y^3} = \dfrac{?}{9y^3(2y)}$

Multiply by $2y$

Multiply by $2y$

$\dfrac{x+7}{9y^3} = \dfrac{(x+7)(2y)}{9y^3(2y)} = \dfrac{2xy + 14y}{18y^4}$

c. We first note that $x^2 + 2x - 3 = (x-1)(x+3)$. Thus

$\dfrac{3x+1}{x-1} = \dfrac{?}{(x-1)(x+3)}$

Multiply by $(x+3)$

Multiply by $(x+3)$

$\dfrac{3x+1}{x-1} = \dfrac{(3x+1)(x+3)}{(x-1)(x+3)} = \dfrac{3x^2 + 10x + 3}{x^2 + 2x - 3}$

PROBLEM 2

Write:

a. $\dfrac{3}{7x}$ with a denominator of $28x^4$

b. $\dfrac{4-x}{5y^2}$ with a denominator of $40y^2$

c. $\dfrac{x+1}{2x+1}$ with a denominator of $2x^2 - 5x - 3$

Web It

Do you know the difference between a rational expression and a rational equation? For some examples go to link 6-1-2 at mhhe.com/bello.

Teaching Tip

The students should be made aware that

$$\dfrac{3x+1}{x-1} = \dfrac{3x^2 + 10x + 3}{x^2 + 2x - 3}$$

are not equivalent for the value of -3 or 1. Have a class discussion about why this occurs.

C **Writing a Rational Expression in the Standard Form**

There are three signs associated with a fraction:

1. The sign before the fraction.

2. The sign of the numerator.

3. The sign of the denominator.

Answers

2. a. $\dfrac{12x^3}{28x^4}$ **b.** $\dfrac{32 - 8x}{40y^2}$

c. $\dfrac{x^2 - 2x - 3}{(2x+1)(x-3)}$ or $\dfrac{x^2 - 2x - 3}{2x^2 - 5x - 3}$

Using our definition of a quotient and the fundamental property of fractions, we can conclude that

$$\frac{-a}{b} = \frac{a}{-b} = -\frac{a}{b} = -\frac{-a}{-b} \quad \text{and} \quad \frac{a}{b} = \frac{-a}{-b} = -\frac{a}{-b} = -\frac{-a}{b}$$

The forms $\frac{-a}{b}$ and $\frac{a}{b}$, in which the sign of the fraction and that of the denominator are positive, are called the **standard forms** of the fractions. Thus $\frac{-2}{9}$ and $\frac{4}{7}$ are in standard form, but $\frac{2}{-9}$ and $\frac{-4}{-7}$ are not. In expressions with more than one term in the numerator or denominator, there are alternative standard forms. For example,

$$\frac{-1}{x-y} = \frac{-1}{-(y-x)} = \frac{1}{y-x}$$

Recall that $x - y = -(y - x)$, since
$-(y - x) = -y + x = x - y$

$\underbrace{\qquad}_{\text{same}}$

Either

$$\frac{-1}{x-y} \quad \text{or} \quad \frac{1}{y-x}$$

can be used as the standard form. We prefer

$$\frac{1}{y-x}$$

because it has only one minus sign, whereas

$$\frac{-1}{x-y}$$

has two.

EXAMPLE 3 **Writing rational expressions in standard form**

Write in standard form:

a. $\dfrac{x}{-2}$ **b.** $-\dfrac{-3}{y}$ **c.** $-\dfrac{x-y}{5}$

SOLUTION

a. $\dfrac{x}{-2} = \dfrac{-x}{2}$ **b.** $-\dfrac{-3}{y} = \dfrac{3}{y}$ **c.** $-\dfrac{x-y}{5} = \dfrac{-(x-y)}{5}$, or $\dfrac{y-x}{5}$

PROBLEM 3

Write in standard form:

a. $-\dfrac{7}{y}$ **b.** $-\dfrac{-x}{4}$

c. $-\dfrac{3a-b}{8}$

D **Reducing Rational Expressions to Lowest Terms**

The fundamental property of fractions can also be used to **simplify (reduce)** fractions—to write fractions as equivalent ones in which no integers other than 1 can be divided exactly into both the numerator and denominator. For example, the fraction $\frac{14}{21}$ can be simplified by writing the numerator and denominator in factored form and using the fundamental principle of fractions. Thus

$$\frac{14}{21} = \frac{2 \cdot \overset{1}{\cancel{7}}}{3 \cdot \underset{1}{\cancel{7}}} = \frac{2}{3}$$

Here we are dividing the numerator and denominator by the common factor 7. We usually write

$$\frac{\overset{2}{\cancel{14}}}{\underset{3}{\cancel{21}}}$$

and say that the fraction $\frac{2}{3}$ is in **lowest terms.**

Teaching Tip

Students should be reminded that we can rearrange due to the commutative property of multiplication. Also, we can divide out common factors without commuting.

Answers

3. a. $\frac{-7}{y}$ b. $\frac{x}{4}$ c. $\frac{b-3a}{8}$

The rational expression

$$\frac{(x + 3)(x^2 - 4)}{3(x + 2)(x^2 + x - 6)}$$

can also be written in lowest terms by using the fundamental property of fractions. We do it by steps.

PROCEDURE

Procedure for Reducing Rational Expressions

1. Write the numerator and denominator of the rational expression in completely factored form.

2. Find the greatest common factor (GCF) of the numerator and denominator.

3. Replace the quotient of the common factors by the number 1, since $\frac{a}{a} = 1$.

4. Rewrite the rational expression in lowest terms.

We are now ready to simplify

$$\frac{(x + 3)(x^2 - 4)}{3(x + 2)(x^2 + x - 6)}$$

Here are the steps:

1. Write the numerator and denominator in completely factored form.

$$\frac{(x + 3)(x + 2)(x - 2)}{3(x + 2)(x + 3)(x - 2)}$$

2. Find the GCF of the numerator and denominator (factors are rearranged to be in the same order).

$$\frac{(x + 2)(x + 3)(x - 2)}{3(x + 2)(x + 3)(x - 2)}$$

3. Replace the quotient of the common factors by the number 1.

$$\frac{\overset{1}{\cancel{(x + 2)(x + 3)(x - 2)}}}{3\cancel{(x + 2)(x + 3)(x - 2)}}$$

4. Rewrite the rational expression in lowest terms.

$$\frac{1}{3}$$

The whole procedure can be written as

$$\frac{(x + 3)(x^2 - 4)}{3(x + 2)(x^2 + x - 6)} = \frac{\overset{1}{\cancel{(x + 3)}}\overset{1}{\cancel{(x + 2)}}\overset{1}{\cancel{(x - 2)}}}{3\cancel{(x + 2)}\cancel{(x + 3)}\cancel{(x - 2)}} = \frac{1}{3}$$

Web It

For more examples on simplifying rational expressions, go to link 6-1-3 at mhhe.com/bello.

Teaching Tip

Contrast

$$\frac{x + 8}{x + 4} \quad \text{with} \quad \frac{8x}{4x}$$

where the first numerator and denominator represent sums and the second numerator and denominator represent factors. The factors can be reduced, the sums cannot. Thus, the second fraction reduces to 2.

CAUTION

Only common factors can be divided out. It is incorrect to write

$$\frac{\cancel{x} + 8}{\cancel{x} + 4} = 2$$

The x is *not* a factor.

EXAMPLE 4   **Reducing rational expressions to lowest terms**

Reduce each rational expression to lowest terms:

a. $\dfrac{x^3y^4}{xy^6}$ **b.** $\dfrac{xy - y^2}{x^2 - y^2}$ **c.** $\dfrac{2x + xy}{x}$

SOLUTION

a. $\dfrac{x^3y^4}{xy^6} = \dfrac{x^2 \cdot \overset{1}{\cancel{xy^4}}}{y^2 \cdot \cancel{xy^4}}$ Factor the numerator and denominator using the GCF as a factor.

$= \dfrac{x^2}{y^2}$ Divide out xy^4, the GCF.

b. $\dfrac{xy - y^2}{x^2 - y^2} = \dfrac{y \cdot \overset{1}{\cancel{(x - y)}}}{(x + y)\cancel{(x - y)}}$ Factor the numerator and denominator using the GCF as a factor.

$= \dfrac{y}{x + y}$ Divide out $(x - y)$, the GCF.

c. $\dfrac{2x + xy}{x} = \dfrac{\overset{1}{\cancel{x}}(2 + y)}{\cancel{x}}$ Factor the numerator and denominator using the GCF as a factor.

$= 2 + y$ Divide out x, the GCF.

PROBLEM 4

Reduce to lowest terms:

a. $\dfrac{3x^5y^7}{6x^2y^3}$

b. $\dfrac{2x^2 + 4xy}{x^2 - 4y^2}$

c. $\dfrac{3y + xy}{y^2}$

Teaching Tip

Do a simple arithmetic example to illustrate:

$$\frac{2}{3 + 2} = \frac{2}{5}$$

However,

$$\frac{\overset{1}{\cancel{2}}}{3 + \underset{1}{\cancel{2}}} = \frac{1}{4} \neq \frac{2}{5}$$

Since we didn't get the same answer, it must not be correct.

CAUTION

In Example 4(b), the final answer is

$$\frac{y}{x + y}$$

The answer cannot be reduced further. A common mistake is to try to divide out the y. This is not correct. You can only divide out *factors,* and y is not a factor of both the numerator and denominator.

Now, consider the case where additive inverses are divided, such as $\frac{2}{-2}$ or $\frac{-9}{9}$. We can see the result will always be -1. That is easy to recognize in arithmetic but what about in algebra?

Using the expression $\frac{x - 5}{5 - x}$, if $x = 7$, the expression becomes

$$\frac{7 - 5}{5 - 7} = \frac{2}{-2} = -1$$

In the same expression, if $x = -4$, the expression becomes

$$\frac{-4 - 5}{5 - (-4)} = \frac{-9}{9} = -1$$

In general, $\frac{a - b}{b - a}$ will yield the quotient of additive inverses. The result will always equal -1.

Thus

Quotient of Additive Inverses

$$\frac{a - b}{b - a} = -1$$

Answers

4. a. $\dfrac{x^3y^4}{2}$ **b.** $\dfrac{2x}{x - 2y}$ **c.** $\dfrac{3 + x}{y}$

We use this idea in Example 5.

EXAMPLE 5   **Quotients involving additive inverses**

Simplify:

a. $\dfrac{x^3 - y^3}{y - x}$ **b.** $\dfrac{x^2 - y^2}{y^3 - x^3}$

SOLUTION

a. $\dfrac{x^3 - y^3}{y - x} = \dfrac{\overset{-1}{\cancel{(x - y)}}(x^2 + xy + y^2)}{\cancel{y - x}}$ Factor the numerator and denominator.

$\qquad\quad = -1(x^2 + xy + y^2)$ Quotient of additive inverses: $\frac{x-y}{y-x} = -1$

$\qquad\quad = -(x^2 + xy + y^2)$

b. $\dfrac{x^2 - y^2}{y^3 - x^3} = \dfrac{(x + y)(x - y)}{(y - x)(y^2 + xy + x^2)}$ Factor the numerator and denominator.

$\qquad\quad = \dfrac{\overset{-1}{\cancel{(x - y)}}(x + y)}{\cancel{(y - x)}(y^2 + xy + x^2)}$ Quotient of additive inverses: $\frac{x-y}{y-x} = -1$

$\qquad\quad = \dfrac{-(x + y)}{y^2 + xy + x^2}$

PROBLEM 5

Reduce to lowest terms:

a. $\dfrac{x^3 - 8}{2 - x}$ **b.** $\dfrac{1 - 9x^2}{27x^3 - 1}$

Answers

5. a. $-(x^2 + 2x + 4)$ or

$-x^2 - 2x - 4$

b. $\dfrac{-(1 + 3x)}{9x^2 + 3x + 1}$

Calculate It   Graphing Rational Expressions

Your calculator can help you determine the points at which a rational expression is undefined. For instance, in Example 1(c), enter

$$Y_1 = \frac{(x^2 + 3x + 8)}{(x^2 + 2x - 3)}$$

and graph using a standard window. (Don't forget to enter the parentheses in the numerator and denominator.) You will notice two vertical lines crossing the horizontal axis at $x = -3$ and $x = 1$, the two points at which the fraction is undefined (see Window 1). The reason for this is that the calculator is trying to connect all the dots to produce a smoother picture. The graph of a rational expression with a zero denominator at $x = a$ will show a vertical line at $x = a$.

We can verify this by using the dot mode. With a TI-83 Plus, press [MODE], go to line 5, press [▶] to move right and select DOT. Now press [ENTER] [GRAPH]. The new graph, using dots that are *not* connected, shows that there is *nothing* at $x = 1$ and $x = -3$ (see Window 2). To further confirm your suspicions, press [TRACE]. As you approach $x = -3$ from the left, the value of y increases. If you continue to [TRACE] the graph, x

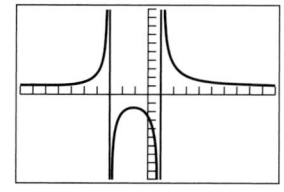

Window 1

$Y_1 = \frac{x^2 + 3x + 8}{x^2 + 2x - 3}$

using the connected mode.

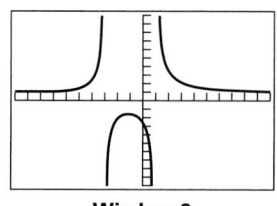

Window 2

$Y_1 = \frac{x^2 + 3x + 8}{x^2 + 2x - 3}$

using the dot mode.

becomes greater than -3, and y becomes negative. However, x is never -3. You can do the same for $x = 1$.

Another way to verify the undefined point is to evaluate the function at $x = -3$. To do this with a TI-83 Plus, go to the home screen ([2nd] [MODE]) store the -3 in memory by pressing [(−)] [3] [STO▶] [X,T,θ,n] [ENTER]. Since Y_1 is already entered, evaluate Y_1 at -3—that is, $Y_1(-3)$—by pressing [VARS] [▶] [ENTER] [ENTER].When you press [ENTER] again, the calculator gives you an error message (see Window 3).

Finally, the calculator can also help with reducing fractions. In Objective D, the fraction $\frac{(x + 3)(x^2 - 4)}{3(x + 2)(x^2 + x - 6)}$ was reduced to $\frac{1}{3}$. Enter

$$Y_1 = \frac{((x + 3)(x^2 - 4))}{(3(x + 2)(x^2 + x - 6))}$$

You will get a horizontal line. To see it better, change the window to $[-1, 1]$ by $[-1, 1]$. The trace cursor will verify that $y = 0.33333333$, an approximation for $y = \frac{1}{3}$, the simplified value of the given rational expression (see Window 4).

```
ERR:DIVIDE BY 0
1:Goto
2:Quit
```

Window 3

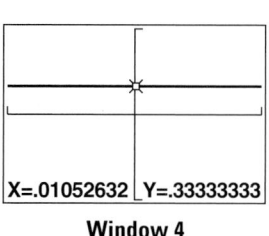

X=.01052632   Y=.33333333

Window 4

$\frac{(x + 3)(x^2 - 4)}{3(x + 2)(x^2 + x - 6)}$

Exercises 6.1

A In Problems 1–12, find the values (if any exist) that make the rational expression undefined.

1. $\dfrac{x}{x+3}$ $x=-3$

2. $\dfrac{y}{y-4}$ $y=4$

3. $\dfrac{5m-5}{5m+10}$ $m=-2$

4. $\dfrac{4q+4}{4q-4}$ $q=1$

5. $\dfrac{m+7}{m^2-m-2}$ $m=-1;2$

6. $\dfrac{u-9}{u^2+6u+5}$ $u=-5;-1$

7. $\dfrac{p^2-2p}{p^2-9}$ $p=-3;3$

8. $\dfrac{z^2-3z-4}{9z^2-4}$ $z=\dfrac{2}{3};\dfrac{-2}{3}$

9. $\dfrac{a^2+4}{2a^2-11a-6}$ $a=\dfrac{-1}{2};6$

10. $\dfrac{b^2+9}{6a^2-5a-6}$ $a=\dfrac{-2}{3};\dfrac{3}{2}$

11. $\dfrac{4v^2-9}{4v^2+9}$ None, defined for all values of v

12. $\dfrac{9y^2-16}{16+y^2}$ None, defined for all values of y

B In Problems 13–30, write the given rational expression with the indicated denominator.

13. $\dfrac{2x}{3y}$; denominator $6y^3$ $\dfrac{4xy^2}{6y^3}$

14. $\dfrac{-3y}{2x}$; denominator $8x^2$ $\dfrac{-12xy}{8x^2}$

15. $\dfrac{x}{x+y}$; denominator x^2-y^2
$\dfrac{x(x-y)}{x^2-y^2}$ or $\dfrac{x^2-xy}{x^2-y^2}$

16. $\dfrac{-y}{x-y}$; denominator x^2-y^2
$\dfrac{-y(x+y)}{x^2-y^2}=\dfrac{-xy-y^2}{x^2-y^2}$

17. $\dfrac{-x}{y-x}=\dfrac{?}{y^2-x^2}$
$\dfrac{-x(y+x)}{y^2-x^2}$ or $\dfrac{-xy-x^2}{y^2-x^2}$

18. $\dfrac{-4x}{y-x}=\dfrac{?}{y^2-x^2}$
$\dfrac{-4x(y+x)}{y^2-x^2}=\dfrac{-4xy-4x^2}{y^2-x^2}$

19. $\dfrac{-x}{2x-3y}$; denominator $4x^2-9y^2$
$\dfrac{-x(2x+3y)}{4x^2-9y^2}$ or $\dfrac{-2x^2-3xy}{4x^2-9y^2}$

20. $\dfrac{-x}{2x-y}$; denominator $4x^2-y^2$
$\dfrac{-x(2x+y)}{4x^2-y^2}$ or $\dfrac{-2x^2-xy}{4x^2-y^2}$

21. $\dfrac{4x}{x+1}=\dfrac{?}{x^2-x-2}$ $\dfrac{4x(x-2)}{x^2-x-2}$ or $\dfrac{4x^2-8x}{x^2-x-2}$

22. $\dfrac{5y}{y-1}=\dfrac{?}{y^2+2y-3}$ $\dfrac{5y(y+3)}{y^2+2y-3}$ or $\dfrac{5y^2+15y}{y^2+2y-3}$

23. $\dfrac{-5x}{x+3}$; denominator x^2+x-6
$\dfrac{-5x(x-2)}{x^2+x-6}$ or $\dfrac{-5x^2+10x}{x^2+x-6}$

24. $\dfrac{-3y}{y-4}$; denominator y^2-2y-8
$\dfrac{-3y(y+2)}{y^2-2y-8}$ or $\dfrac{-3y^2-6y}{y^2-2y-8}$

25. $\dfrac{3}{x+y}$; denominator x^3+y^3
$\dfrac{3(x^2-xy+y^2)}{x^3+y^3}$ or $\dfrac{3x^2-3xy+3y^2}{x^3+y^3}$

26. $\dfrac{-4}{x+y}$; denominator x^3+y^3
$\dfrac{-4(x^2-xy+y^2)}{x^3+y^3}$ or $\dfrac{-4x^2+4xy-4y^2}{x^3+y^3}$

27. $\dfrac{x}{x-y}=\dfrac{?}{x^3-y^3}$ $\dfrac{x(x^2+xy+y^2)}{x^3-y^3}$ or $\dfrac{x^3+x^2y+xy^2}{x^3-y^3}$

28. $\dfrac{-y}{x-y}=\dfrac{?}{x^3-y^3}$ $\dfrac{-y(x^2+xy+y^2)}{x^3-y^3}$ or $\dfrac{-x^2y-xy^2-y^3}{x^3-y^3}$

29. $\dfrac{x}{x^2-xy+y^2}=\dfrac{?}{x^3+y^3}$ $\dfrac{x(x+y)}{x^3+y^3}$ or $\dfrac{x^2+xy}{x^3+y^3}$

30. $\dfrac{x}{x^2+xy+y^2}=\dfrac{?}{x^3-y^3}$ $\dfrac{x(x-y)}{x^3-y^3}$ or $\dfrac{x^2-xy}{x^3-y^3}$

C In Problems 31–40, write each fraction in standard form.

31. $-\dfrac{y}{-2}$ $\dfrac{y}{2}$

32. $-\dfrac{x-3}{y}$ $\dfrac{3-x}{y}$

33. $-\dfrac{x}{x-5}$ $\dfrac{x}{5-x}$

34. $\dfrac{2x-y}{-x}$ $\dfrac{y-2x}{x}$

35. $-\dfrac{-2x}{-5y}$ $\dfrac{-2x}{5y}$

36. $\dfrac{-x}{-y}$ $\dfrac{x}{y}$

37. $\dfrac{-(x+y)}{-(x-y)}$ $\dfrac{x+y}{x-y}$

38. $-\dfrac{-(3x+y)}{-(x-5y)}$ $\dfrac{-(3x+y)}{x-5y}$ or $\dfrac{3x+y}{5y-x}$

39. $\dfrac{-1}{-(x-2)}$ $\dfrac{1}{x-2}$

40. $\dfrac{-y}{-(x+1)}$ $\dfrac{y}{x+1}$

D In Problems 41–70, simplify each fraction.

41. $\dfrac{x^4 y^2}{xy^5}$ $\dfrac{x^3}{y^3}$

42. $\dfrac{x^5 y^3 z^2}{x^2 y^6 z^4}$ $\dfrac{x^3}{y^3 z^2}$

43. $\dfrac{3x-3y}{x-y}$ 3

44. $\dfrac{4x^2}{4x-4y}$ $\dfrac{x^2}{x-y}$

45. $\dfrac{3x-2y}{9x^2-4y^2}$ $\dfrac{1}{3x+2y}$

46. $\dfrac{4x^2-9y^2}{2x+3y}$ $2x-3y$

47. $\dfrac{(x-y)^3}{x^2-y^2}$ $\dfrac{(x-y)^2}{x+y}$

48. $\dfrac{x^2-y^2}{(x+y)^3}$ $\dfrac{x-y}{(x+y)^2}$

49. $\dfrac{ay^2-ay}{ay}$ $y-1$

50. $\dfrac{a^3+2a^2+a}{a}$ a^2+2a+1

51. $\dfrac{x^2+2xy+y^2}{x^2-y^2}$ $\dfrac{x+y}{x-y}$

52. $\dfrac{x^2+3x+2}{x^2+2x+1}$ $\dfrac{x+2}{x+1}$

53. $\dfrac{y^2-8y+15}{y^2+3y-18}$ $\dfrac{y-5}{y+6}$

54. $\dfrac{y^2+7y-18}{y^2-3y+2}$ $\dfrac{y+9}{y-1}$

55. $\dfrac{2-y}{y-2}$ -1

56. $\dfrac{3(x-y)}{4(y-x)}$ $\dfrac{-3}{4}$

57. $\dfrac{9-x^2}{x-3}$ $-(x+3)$

58. $\dfrac{25-9x^2}{3x-5}$ $-(3x+5)$

59. $\dfrac{y^3-8}{2-y}$ $-(y^2+2y+4)$

60. $\dfrac{2+x}{x^3+8}$ $\dfrac{1}{x^2-2x+4}$

61. $\dfrac{3x-2y}{2y-3x}$ -1

62. $\dfrac{5y-2x}{2x-5y}$ -1

63. $\dfrac{x^2+4x-5}{1-x}$ $-(x+5)$

64. $\dfrac{x^2-2x-15}{5-x}$ $-(x+3)$

65. $\dfrac{x^2-6x+8}{4-x}$ $2-x$

66. $\dfrac{x^2-8x+15}{3-x}$ $5-x$

67. $\dfrac{2-x}{x^2+4x-12}$ $\dfrac{-1}{x+6}$

68. $\dfrac{3-x}{x^2+3x-18}$ $\dfrac{-1}{x+6}$

69. $-\dfrac{3-x}{x^2-5x+6}$ $\dfrac{1}{x-2}$

70. $-\dfrac{4-x}{x^2-3x-4}$ $\dfrac{1}{x+1}$

SKILL CHECKER

Try the "Skill Checker" exercises so you'll be ready for the next section.

Simplify:

71. $\dfrac{4x^3}{2x}$ $2x^2$

72. $\dfrac{8x^4}{2x}$ $4x^3$

73. $\dfrac{-16x^4}{8x^2}$ $-2x^2$

74. $\dfrac{-48x^5}{12x^3}$ $-4x^2$

Factor completely:

75. x^2+x-12
$(x+4)(x-3)$

76. $x^2-2x-15$
$(x-5)(x+3)$

77. $9x^2-4y^2$
$(3x+2y)(3x-2y)$

78. $4x^2-9y^2$
$(2x+3y)(2x-3y)$

79. x^3-1
$(x-1)(x^2+x+1)$

80. x^3-8
$(x-2)(x^2+2x+4)$

81. $8x^3+1$
$(2x+1)(4x^2-2x+1)$

82. $27x^3+8$
$(3x+2)(9x^2-6x+4)$

Multiply:

83. $\dfrac{3}{7}\cdot\dfrac{14}{9}$ $\dfrac{2}{3}$

84. $-\dfrac{5}{8}\cdot\dfrac{16}{15}$ $\dfrac{-2}{3}$

85. $\left(-\dfrac{4}{9}\right)\left(-\dfrac{27}{8}\right)$ $\dfrac{3}{2}$

Divide:

86. $-\dfrac{3}{4} \div \left(-\dfrac{3}{8}\right)$ 2

87. $-\dfrac{4}{5} \div \dfrac{8}{15}$ $\dfrac{-3}{2}$

88. $\dfrac{6}{7} \div \left(-\dfrac{3}{14}\right)$ -4

USING YOUR KNOWLEDGE

Buying Pollution Permits

89. Can you buy a permit to pollute the air? Unfortunately, yes. On March 29, 1993, the Chicago Board of Trade sold a permit for about $21 million to emit 150,000 tons of sulfur dioxide. (This amount represents 1% of the yearly sulfur dioxide emissions.) If the price (in millions of dollars) for removing $p\%$ of the sulfur dioxide is given by

$$\frac{2100p}{100 - p}$$

find the price that should be paid to remove

a. 20% $525 million

b. 40% $1400 million or $1.4 billion

c. 60% $3150 million or $3.15 billion

d. Can we afford to remove 100% of the sulfur dioxide? Explain. No; Denominator = 0 for $p = 100$. As p increases to 100, price increases without bound.

90. The demand for a product is given by

$$N(p) = \frac{2p + 100}{10p + 10}$$

where $N(p)$ represents the number of units people are willing to buy when the price is p dollars ($1 \le p \le 10$).

a. Reduce this expression to lowest terms. $N(p) = \frac{p + 50}{5(p + 1)}$

b. Find the demand when the price is $3. 2.65 units

c. What happens to the demand as the price increases? The demand decreases as the price increases.

91. A vendor's profit (in dollars) for the sale of x sunglasses is

$$\frac{5x^2 - 5}{x + 1} \quad (1 \le x \le 100)$$

a. Reduce this expression to lowest terms. $5(x - 1)$

b. What is the profit when 10 sunglasses are sold? $45

c. What is the maximum profit possible? $495 (when $x = 100$)

WRITE ON

We have already mentioned that when reducing rational expressions, you may divide out only factors. Explain what is wrong with the simplifications in Problems 92 and 93.

92. $\dfrac{x + y}{x} = 1 + y$ x is a term in the numerator, not a factor.

93. $\dfrac{y}{x + y} = \dfrac{1}{x + 1}$ y is a term in the numerator, not a factor.

94. In this section, we used the fundamental property of rational expressions in two different ways. Write a paragraph explaining these two ways.
Answers may vary.

95. Just before Example 4, we showed that

$$\frac{(x + 3)(x^2 - 4)}{3(x + 2)(x^2 + x - 6)} = \frac{1}{3}$$

Is this equation always true? Explain why or why not.
No; answers may vary.

96. Explain that since

$$\frac{a - b}{b - a} = -1, \quad \text{then} \quad \frac{a - b}{-a + b} = -1$$

Answers may vary.

97. Explain how you would reduce

$$\frac{4x^2 - 7x + 1}{-4x^2 + 7x - 1}$$

and then explain how to reduce a rational expression where the numerator and denominator differ only in sign.
Answers may vary.

MASTERY TEST

If you know how to do these problems, you have learned your lesson!

Reduce to lowest terms:

98. $\dfrac{y^3 - x^3}{y - x}$ $y^2 + xy + x^2$

99. $\dfrac{x^5y^7}{xy^3}$ x^4y^4

100. $\dfrac{3y + xy}{y}$ $3 + x$

101. $\dfrac{x^2 - y^2}{y^3 - x^3}$ $\dfrac{-(x + y)}{x^2 + xy + y^2}$

102. $\dfrac{x^2 - xy}{x^2 - y^2}$ $\dfrac{x}{x + y}$

103. $\dfrac{4y^2 - xy^2}{y^2}$ $4 - x$

104. For what values of x is $\dfrac{x-1}{x^2+2x-3}$ undefined? $x = -3; 1$

105. For what values of x is $\dfrac{x+1}{x^2-1}$ undefined? $x = -1; 1$

Write in standard form:

106. $\dfrac{x}{-7} \quad \dfrac{-x}{7}$

107. $-\dfrac{-4}{x} \quad \dfrac{4}{x}$

108. $-\dfrac{a-b}{8} \quad \dfrac{b-a}{8}$

Write:

109. $\dfrac{7}{8}$ with a denominator of 16. $\dfrac{14}{16}$

110. $\dfrac{3x^2}{8y^3}$ with a denominator of $24y^6$. $\dfrac{9x^2y^3}{24y^6}$

111. $\dfrac{4x+1}{x+2}$ with a denominator of $x^2 - x - 6$. $\dfrac{4x^2-11x-3}{x^2-x-6}$

6.2   MULTIPLICATION AND DIVISION OF RATIONAL EXPRESSIONS

To Succeed, Review How To . . .

1. Simplify quotients using the properties of exponents (pp. 39–40).
2. Multiply polynomials (pp. 356–362).

Objectives

A Multiply rational expressions.

B Divide rational expressions.

C Use multiplication and division together.

GETTING STARTED   It's All in the Cards

A regular deck of 52 cards has four kings. The probability that you pick two kings when drawing two cards is

$$\frac{4}{52} \cdot \frac{3}{51} = \frac{4}{4 \cdot 13} \cdot \frac{3}{3 \cdot 17} \qquad \text{Factor the denominators.}$$

$$= \frac{\overset{1}{\cancel{4}}}{\cancel{4} \cdot 13} \cdot \frac{\overset{1}{\cancel{3}}}{\cancel{3} \cdot 17} \qquad \text{Simplify.}$$

$$= \frac{1}{221} \qquad \text{Multiply.}$$

Save time by simplifying in the first step:

$$\frac{4}{52} \cdot \frac{3}{51} = \frac{\overset{1}{\cancel{4}}}{\underset{13}{\cancel{52}}} \cdot \frac{\overset{1}{\cancel{3}}}{\underset{17}{\cancel{51}}}$$

In this section, you will learn how to multiply rational expressions using rules involving these three steps: factor, simplify, and multiply.

A   Multiplying Rational Expressions

The multiplication of rational expressions follows the same procedure as the multiplication of rational numbers. Here is the definition from Section 1.2.

Multiplication of Rational Expressions	If a, b, c, and d are real numbers, $$\frac{a}{b} \cdot \frac{c}{d} = \frac{a \cdot c}{b \cdot d} \quad (b \neq 0, d \neq 0)$$

PROCEDURE

Procedure to Multiply Rational Expressions

1. **Factor** the numerators and denominators completely.

2. **Simplify** each rational expression completely.

3. **Multiply** the remaining factors in the numerator and denominator.

4. Make sure the final product is in **lowest terms.**

It is not necessary to factor if the numerators and denominators are monomials as in the case of this next example.

$$\frac{6x^3}{2y} \cdot \frac{4y^5}{9x^2}$$

Both numerators and denominators are to be simplified and reduced; proceed as follows:

$$\frac{6x^3}{2y} \cdot \frac{4y^5}{9x^2} = \frac{24x^3y^5}{18x^2y} = \frac{4xy^4 \cdot 6x^2y}{3 \cdot 6x^2y} = \frac{4xy^4}{3}$$

As in *Getting Started*, the common factors in the numerator and denominator can be divided out *before* doing the multiplications in the numerator and denominator. The procedure goes like this:

$$\frac{\overset{2}{6x^3}}{\underset{1}{2y}} \cdot \frac{\overset{2\ y^4}{4y^5}}{\underset{3\ 1}{9x^2}} = \frac{2 \cdot x \cdot 2 \cdot y^4}{1 \cdot 1 \cdot 3 \cdot 1} = \frac{4xy^4}{3}$$

Now let's multiply when the numerators and denominators are not monomials. Remember to factor and then divide out common factors before you do the multiplications.

$$\frac{x^2 + 2x - 8}{x^2 + 5x + 6} \cdot \frac{x + 2}{x + 4} = \frac{(x - 2)(x + 4)}{(x + 2)(x + 3)} \cdot \frac{(x + 2)}{(x + 4)} \quad \text{Factor.}$$

$$= \frac{(x - 2)(x + 4)}{(x + 2)(x + 3)} \cdot \frac{(x + 2)}{(x + 4)} \quad \text{Simplify.}$$

$$= \frac{x - 2}{x + 3}$$

EXAMPLE 1   Multiplying rational expressions

Multiply:

a. $\dfrac{x - 1}{2x - 3y} \cdot \dfrac{4x^2 - 9y^2}{2x^2 - x - 1}$

b. $\dfrac{x^2 - 4}{4x^2 - 9y^2} \cdot \dfrac{2x^2 - 3xy}{2x + 4}$

SOLUTION

a. $\dfrac{x - 1}{2x - 3y} \cdot \dfrac{4x^2 - 9y^2}{2x^2 - x - 1} = \dfrac{(x - 1)}{(2x - 3y)} \cdot \dfrac{(2x + 3y)(2x - 3y)}{(x - 1)(2x + 1)}$ Factor.

$= \dfrac{(x - 1)}{(2x - 3y)} \cdot \dfrac{(2x + 3y)(2x - 3y)}{(x - 1)(2x + 1)}$ Simplify.

$= \dfrac{2x + 3y}{2x + 1}$

PROBLEM 1

Multiply:

a. $\dfrac{x - 2}{3x - 2y} \cdot \dfrac{9x^2 - 4y^2}{2x^2 - 3x - 2}$

b. $\dfrac{x^2 - 9}{16x^2 - 25} \cdot \dfrac{4x^2 + 5x}{3x + 9}$

Answers

1. a. $\frac{3x + 2y}{2x + 1}$

b. $\frac{x(x - 3)}{3(4x - 5)}$ or $\frac{x^2 - 3x}{12x - 15}$

b. $\dfrac{x^2 - 4}{4x^2 - 9y^2} \cdot \dfrac{2x^2 - 3xy}{2x + 4} = \dfrac{(x + 2)(x - 2)}{(2x + 3y)(2x - 3y)} \cdot \dfrac{x(2x - 3y)}{2(x + 2)}$ Factor.

$= \dfrac{(x + 2)(x - 2)}{(2x + 3y)(2x - 3y)} \cdot \dfrac{x(2x - 3y)}{2(x + 2)}$ Simplify.

$= \dfrac{x(x - 2)}{2(2x + 3y)}$ Multiply.

$= \dfrac{x^2 - 2x}{4x + 6y}$

> ## Web It
>
> For more examples and practice on multiplying rational expressions, go to link 6-2-1 at mhhe.com/bello.

EXAMPLE 2   More practice at multiplying rational expressions

Multiply:

a. $\dfrac{2 - x}{x + 1} \cdot \dfrac{x^2 + 3x + 2}{x^2 - 4}$

b. $\dfrac{x^2 + 3x + 2}{x^2 + 5x + 4} \cdot \dfrac{x^2 + 2x - 3}{x^2 + x - 2}$

SOLUTION

a. $\dfrac{2 - x}{x + 1} \cdot \dfrac{x^2 + 3x + 2}{x^2 - 4} = \dfrac{(2 - x)}{(x + 1)} \cdot \dfrac{(x + 1)(x + 2)}{(x + 2)(x - 2)}$ Factor.

$= \dfrac{\overset{-1}{(2 - x)}}{(x + 1)} \cdot \dfrac{(x + 1)(x + 2)}{(x + 2)(x - 2)}$ Simplify. Recall that $\frac{2 - x}{x - 2} = -1$.

$= -1$

b. $\dfrac{x^2 + 3x + 2}{x^2 + 5x + 4} \cdot \dfrac{x^2 + 2x - 3}{x^2 + x - 2} = \dfrac{(x + 2)(x + 1)}{(x + 4)(x + 1)} \cdot \dfrac{(x - 1)(x + 3)}{(x - 1)(x + 2)}$ Factor.

$= \dfrac{(x + 2)(x + 1)}{(x + 4)(x + 1)} \cdot \dfrac{(x - 1)(x + 3)}{(x - 1)(x + 2)}$ Simplify.

$= \dfrac{x + 3}{x + 4}$

PROBLEM 2

Multiply:

a. $\dfrac{3 - x}{2x + 4} \cdot \dfrac{x^2 + 5x + 6}{x^2 - 9}$

b. $\dfrac{x^2 + 4x + 3}{x^2 + 5x + 6} \cdot \dfrac{x^2 - 2x - 8}{x^2 - 2x - 3}$

B   Dividing Rational Expressions

To divide one rational expression by another, we use the definition of division from Section 1.2:

Division of Real Numbers	If a, b, c, and d are real numbers,
	$$\dfrac{a}{b} \div \dfrac{c}{d} = \dfrac{a}{b} \cdot \dfrac{d}{c} \quad (b, d, \text{ and } c \neq 0)$$

The quotient of two rational expressions is the product of the first and the reciprocal of the second. For example,

$$\dfrac{2}{5} \div \dfrac{3}{8} = \dfrac{2}{5} \cdot \dfrac{8}{3} = \dfrac{16}{15}$$ The reciprocal of $\frac{3}{8}$ is $\frac{8}{3}$.

Reciprocal

and

$$\dfrac{3x^2}{2y} \div \dfrac{6x}{4y^2} = \dfrac{3x^2}{2y} \cdot \dfrac{4y^2}{6x}$$

Reciprocal

$$= \dfrac{\overset{1xy}{12x^2y^2}}{12xy}$$ Answers should be given in simplified form.

$$= xy$$

Answers

2. **a.** $\frac{-1}{2}$ **b.** $\frac{x - 4}{x - 3}$

EXAMPLE 3   **Dividing rational expressions**

Divide:

a. $\dfrac{x^3 y}{z} \div \dfrac{x^2 y}{z^4}$ **b.** $\dfrac{5x^2 - 5}{3x + 6} \div \dfrac{x + 1}{3}$ **c.** $\dfrac{x + 3}{x - 3} \div (x^2 + 6x + 9)$

SOLUTION

a. $\dfrac{x^3 y}{z} \div \dfrac{x^2 y}{z^4} = \dfrac{x^3 y}{z} \cdot \dfrac{z^4}{x^2 y}$

$\underbrace{\qquad}_{\text{Reciprocal}}$

$= \dfrac{\overset{xz^3}{\cancel{x^3} \cancel{y} \cancel{z^4}}}{\cancel{x^2} \cancel{y} \cancel{z}}$

$= xz^3$

b. $\dfrac{5x^2 - 5}{3x + 6} \div \dfrac{x + 1}{3} = \dfrac{5(x + 1)(x - 1)}{3(x + 2)} \cdot \dfrac{3}{x + 1}$ Factor.

$\underbrace{\qquad}_{\text{Reciprocal}}$

$= \dfrac{5(\cancel{x + 1})(x - 1)}{\cancel{3}(x + 2)} \cdot \dfrac{\cancel{3}}{\cancel{x + 1}}$ Simplify.

$= \dfrac{5(x - 1)}{x + 2}$ or $\dfrac{5x - 5}{x + 2}$ Multiply.

c. We first write $x^2 + 6x + 9$ as $\dfrac{x^2 + 6x + 9}{1}$.

$\dfrac{x + 3}{x - 3} \div (x^2 + 6x + 9) = \dfrac{x + 3}{x - 3} \div \dfrac{x^2 + 6x + 9}{1}$

$= \dfrac{x + 3}{x - 3} \cdot \dfrac{1}{x^2 + 6x + 9}$ Multiply by reciprocal.

$= \dfrac{(x + 3)}{(x - 3)} \cdot \dfrac{1}{(x + 3)(x + 3)}$ Factor.

$= \dfrac{(\cancel{x + 3})}{(x - 3)} \cdot \dfrac{1}{(\cancel{x + 3})(x + 3)}$ Simplify.

$= \dfrac{1}{(x - 3)(x + 3)}$ or $\dfrac{1}{x^2 - 9}$ Multiply.

PROBLEM 3

Divide:

a. $\dfrac{x^5 y}{z} \div \dfrac{x^3 y}{z^6}$

b. $\dfrac{3x^2 - 3}{5x + 10} \div \dfrac{x + 1}{5}$

c. $\dfrac{x + 2}{x - 2} \div (x^2 + 4x + 4)$

Web It

For more examples and practice on multiplying and dividing rational expressions, go to link 6-2-2 at mhhe.com/bello.

C   **Using Multiplication and Division**

Sometimes both multiplications and divisions are involved, as shown in Example 4.

Answers

3. a. $x^2 z^5$ **b.** $\dfrac{3(x - 1)}{x + 2}$ or $\dfrac{3x - 3}{x + 2}$

c. $\dfrac{1}{(x + 2)(x - 2)}$ or $\dfrac{1}{x^2 - 4}$

EXAMPLE 4   **Operations involving multiplications and divisions**

Perform the indicated operations:

$$\frac{3-x}{x+2} \div \frac{x^3-27}{x+4} \cdot \frac{x^3+8}{x+4}$$

SOLUTION   First rewrite the division as multiplication by the reciprocal:

$$\frac{3-x}{x+2} \div \frac{x^3-27}{x+4} \cdot \frac{x^3+8}{x+4}$$

$$= \frac{3-x}{x+2} \cdot \frac{x+4}{x^3-27} \cdot \frac{x^3+8}{x+4}$$ Change the division to multiplication by the reciprocal.

$$= \frac{(3-x)}{(x+2)} \cdot \frac{(x+4)}{(x-3)(x^2+3x+9)} \cdot \frac{(x+2)(x^2-2x+4)}{x+4}$$ Factor.

$$= \frac{\overset{-1}{\cancel{(3-x)}}}{\cancel{(x+2)}} \cdot \frac{\cancel{(x+4)}}{\cancel{(x-3)}(x^2+3x+9)} \cdot \frac{\cancel{(x+2)}(x^2-2x+4)}{\cancel{(x+4)}}$$ Simplify. Recall that $\frac{3-x}{x-3} = -1$.

$$= \frac{-(x^2-2x+4)}{x^2+3x+9} \quad \text{or} \quad \frac{-x^2+2x-4}{x^2+3x+9}$$ Multiply $(-1 \cdot a = -a)$.

PROBLEM 4

Perform the indicated operations:

$$\frac{2-x}{x+3} \div \frac{x^3-8}{x-5} \cdot \frac{x^3+27}{x-5}$$

Web It

For a review of rational expressions, go to link 6-2-3 at mhhe.com/bello.

EXAMPLE 5   **More practice at using multiplication and division**

Perform the indicated operations:

$$\frac{x^3+2x^2-x-2}{x+3} \div \frac{x^3+8}{x^2-9} \cdot \frac{1}{x^2-1}$$

SOLUTION   First rewrite the division as multiplication by the reciprocal to obtain

$$\frac{x^3+2x^2-x-2}{x+3} \div \frac{x^3+8}{x^2-9} \cdot \frac{1}{x^2-1}$$

$$= \frac{x^3+2x^2-x-2}{x+3} \cdot \frac{x^2-9}{x^3+8} \cdot \frac{1}{x^2-1}$$

Now we have the polynomial x^3+2x^2-x-2 with four terms in the first numerator. Factor it by grouping

$$x^3+2x^2-x-2 = x^2(x+2) - 1 \cdot (x+2)$$

$$= (x+2)(x^2-1)$$

$$= (x+2)(x+1)(x-1)$$

Hence, $(x+2)(x+1)(x-1)$ replaces the first numerator. Factor the remaining numerators and denominators and simplify.

$$\frac{(x+2)(x+1)(x-1)}{(x+3)} \cdot \frac{(x+3)(x-3)}{(x+2)(x^2-2x+4)} \cdot \frac{1}{(x+1)(x-1)}$$ Factor.

$$= \frac{\cancel{(x+2)}\cancel{(x+1)}\cancel{(x-1)}}{\cancel{(x+3)}} \cdot \frac{\cancel{(x+3)}(x-3)}{\cancel{(x+2)}(x^2-2x+4)} \cdot \frac{1}{\cancel{(x+1)}\cancel{(x-1)}}$$ Simplify.

$$= \frac{x-3}{x^2-2x+4}$$ Multiply.

PROBLEM 5

Perform the indicated operations:

$$\frac{x^3+x^2-x-1}{x+4} \div \frac{x^3+1}{x^2-16} \cdot \frac{1}{x^2-1}$$

Answers

4. $\frac{-(x^2-3x+9)}{x^2+2x+4}$ or $\frac{-x^2+3x-9}{x^2+2x+4}$

5. $\frac{x-4}{x^2-x+1}$

Calculate It Exercises
Using Graphs to Check Multiplication and Division of Rational Expressions

A calculator can be used to check the multiplication of rational expressions involving one variable. In Example 2(a), use a $[-2, 2]$ by $[-2, 2]$ window and enter

$$Y_1 = \left(\frac{(2 - x)}{(x + 1)}\right) \cdot \left(\frac{(x^2 + 3x + 2)}{(x^2 - 4)}\right)$$

Note the parentheses. The result is $Y_1 = -1$ as before (see Window 1). For part 2(b), use **ZOOM** **6** , to change to a standard window and enter

$$Y_1 = \left(\frac{(x^2 + 3x + 2)}{(x^2 + 5x + 4)}\right) \cdot \left(\frac{(x^2 + 2x - 3)}{(x^2 + x - 2)}\right)$$

and

$$Y_2 = \frac{(x + 3)}{(x + 4)}$$

Note the parentheses again. The graphs of Y_1 and Y_2 are identical (see Window 2). Some calculators (TI-83 Plus) give additional support for this fact by using the **TRACE** and ◄ ► keys. When you press ◄ and then ►, the equation at the top of your window changes from Y_1 to Y_2, telling you that there are really two curves being graphed. The **TRACE** key gives the same x- and y-coordinates for both graphs. Division problems can be verified similarly. For Example 3(b), you need to be extra careful with the parentheses. Note the extra set when entering Y_1:

$$Y_1 = \left(\frac{(5x^2 - 5)}{(3x + 6)}\right) \div \left(\frac{(x + 1)}{3}\right) \quad \text{and} \quad Y_2 = \frac{(5x - 5)}{(x + 2)}$$

The graphs for Y_1 and Y_2 are identical (Window 3). You can use the trace feature to confirm this.

To do Example 5, we use a new technique to avoid multiple sets of parentheses. Enter

$$Y_1 = \frac{(x^3 + 2x^2 - x - 2)}{(x + 3)}, \quad Y_2 = \frac{(x^3 + 8)}{(x^2 - 9)} \quad \text{and} \quad Y_3 = \frac{1}{(x^2 - 1)}$$

Now for Y_4, enter $Y_1 \div Y_2 \cdot Y_3$ by pressing **VARS** ► **1** **1** **÷**
VARS ► **1** **2** **×** **VARS** ► **1** **3** . Turn Y_1, Y_2 and Y_3 **off** (so their graphs don't clutter the window) by pressing **Y=**, going to Y_1 and placing the cursor on top of the = sign. Then press **ENTER**. Y_1 is now off. Do the same for Y_2 and Y_3. Finally, let

$$Y_5 = \frac{(x - 3)}{(x^2 - 2x + 4)}$$

The graphs of Y_4 and Y_5 are identical (Window 4). (If you want to see this better, make the vertical scale go from -2 to 2.)

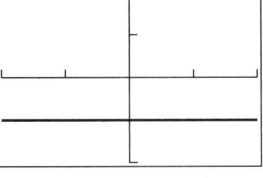

Window 1
$$Y_1 = \frac{2 - x}{x + 1} \cdot \frac{x^2 + 3x + 2}{x^2 - 4}$$

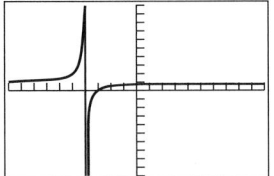

Window 2
$$Y_2 = \frac{x + 3}{x + 4}$$

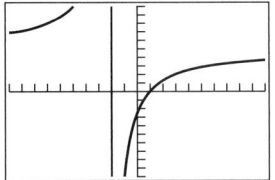

Window 3
$$Y_2 = \frac{5x - 5}{x + 2}$$

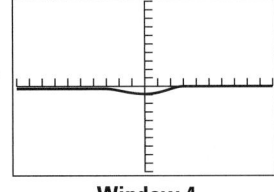

Window 4
$$Y_5 = \frac{x - 3}{x^2 - 2x + 4}$$

1. What happens if you try to evaluate Y_4 at $x = -3$? What about Y_5? (With a TI-83 Plus, press **2nd** **GRAPH** and scroll to $x = -3$.)

2. At what other values is Y_4 (above) undefined? Based on your answer, can $Y_4 = Y_5$?

3. How would you check the results of Example 4 using the technique above?

Exercises 6.2

A In Problems 1–20, perform the indicated multiplications.

1. $\dfrac{3}{4} \cdot \dfrac{2}{5} \quad \dfrac{3}{10}$

2. $\dfrac{-9}{10} \cdot \dfrac{2}{3} \quad \dfrac{-3}{5}$

3. $\dfrac{14x^2}{15} \cdot \dfrac{5}{7x} \quad \dfrac{2x}{3}$

4. $\dfrac{-5x^3}{7y} \cdot \dfrac{4y^3}{9x^6} \quad \dfrac{-20y^2}{63x^3}$

5. $\dfrac{-2xy^4}{9z^5} \cdot \dfrac{-3z}{7x^3y^3} \quad \dfrac{2y}{21x^2z^4}$

6. $\dfrac{-35x^5z}{24x^3y^9} \cdot \dfrac{84x^3y^8}{15x^4y^7z} \quad \dfrac{-49x}{6y^8}$

7. $\dfrac{10x + 50}{6x + 6} \cdot \dfrac{12}{5x + 25} \quad \dfrac{4}{x + 1}$

8. $\dfrac{x + y}{xy - y^2} \cdot \dfrac{y^2}{x^2 - y^2} \quad \dfrac{y}{(x - y)^2}$

9. $\dfrac{6y + 3}{2y^2 - 3y - 2} \cdot \dfrac{y^2 - 4}{3y + 6} \quad 1$

10. $\dfrac{y^2 + 9y + 18}{y - 2} \cdot \dfrac{2y - 4}{5y + 15} \quad \dfrac{2(y + 6)}{5}$

11. $\dfrac{y - x}{x^2 + 2xy} \cdot \dfrac{5x + 10y}{x^2 - y^2} \quad \dfrac{-5}{x(x + y)}$

12. $\dfrac{2 - 2x}{9x^2 - 25} \cdot \dfrac{6x - 10}{x^2 - 1} \quad \dfrac{-4}{(3x + 5)(x + 1)}$

13. $\dfrac{3y^2 - 17y + 10}{y^2 - 4y - 5} \cdot \dfrac{y^2 + 3y + 2}{y^2 + y - 2} \quad \dfrac{3y - 2}{y - 1}$

14. $\dfrac{y^2 + 2y - 3}{y^2 - 4y - 5} \cdot \dfrac{y^2 - 3y - 10}{y^2 + 5y - 6}$

$\dfrac{(y + 3)(y + 2)}{(y + 1)(y + 6)}$ or $\dfrac{y^2 + 5y + 6}{y^2 + 7y + 6}$

15. $\dfrac{y^2 + 2y - 8}{y^2 + 7y + 12} \cdot \dfrac{y^2 + 2y - 3}{y^2 - 3y + 2} \quad 1$

16. $\dfrac{y^2 + 2y - 15}{y^2 - 7y + 10} \cdot \dfrac{y^2 - 6y + 8}{y^2 - y - 12}$

$\dfrac{(y + 5)(y - 3)}{(y - 5)(y + 3)}$ or $\dfrac{y^2 + 2y - 15}{y^2 - 2y - 15}$

17. $\dfrac{x^3 - 8}{4 - x^2} \cdot \dfrac{x^2 + x - 2}{x^2 + 2x + 4} \quad 1 - x$

18. $\dfrac{x^3 + y^3}{y^2 - x^2} \cdot \dfrac{x - y}{x^2 - xy + y^2} \quad -1$

19. $\dfrac{a^3 + b^3}{a^3 - b^3} \cdot \dfrac{a^2 + ab + b^2}{a^2 - ab + b^2} \quad \dfrac{a + b}{a - b}$

20. $\dfrac{a^3 - 8}{a^2 + 2a + 4} \cdot \dfrac{a^2 + 3a + 9}{a^3 - 27} \quad \dfrac{a - 2}{a - 3}$

B In Problems 21–40, perform the indicated division.

21. $\dfrac{3}{5} \div \dfrac{10}{9} \quad \dfrac{27}{50}$

22. $\dfrac{3}{7} \div \dfrac{-9}{14} \quad \dfrac{-2}{3}$

23. $\dfrac{4}{5x^2} \div \dfrac{12}{25x^3} \quad \dfrac{5x}{3}$

24. $\dfrac{6x^2}{7} \div \dfrac{30x}{28} \quad \dfrac{4x}{5}$

25. $\dfrac{24a^2b}{7c^2d} \div \dfrac{8ab}{21cd^2} \quad \dfrac{9ad}{c}$

26. $\dfrac{16a^3b}{15a} \div \dfrac{12ab^2}{20b^4} \quad \dfrac{16ab^3}{9}$

27. $\dfrac{3x - 3}{x} \div \dfrac{x^2 - 1}{x^2} \quad \dfrac{3x}{x + 1}$

28. $\dfrac{5x^2 - 45}{x^3} \div \dfrac{x + 3}{x}$

$\dfrac{5(x - 3)}{x^2}$ or $\dfrac{5x - 15}{x^2}$

29. $\dfrac{y^2 - 25}{y^2 - 4} \div \dfrac{3y - 15}{4y - 8}$

$\dfrac{4(y + 5)}{3(y + 2)}$ or $\dfrac{4y + 20}{3y + 6}$

30. $\dfrac{y^2 + y - 12}{y^2 - 1} \div \dfrac{3y + 12}{4y^2 + 4y}$

$\dfrac{4y(y - 3)}{3(y - 1)}$ or $\dfrac{4y^2 - 12y}{3y - 3}$

31. $\dfrac{a^3 + b^3}{a^3 - b^3} \div \dfrac{a^2 - ab + b^2}{a^2 + ab + b^2}$

$\dfrac{a + b}{a - b}$

32. $\dfrac{a^2 + ab + b^2}{a^3 + b^3} \div \dfrac{a^2 + ab + b^2}{a^2 - ab + b^2}$

$\dfrac{1}{a + b}$

33. $\dfrac{8a^3 - 1}{6u^4w^3} \div \dfrac{1 - 2a}{3u^2w}$ $\dfrac{-(4a^2 + 2a + 1)}{2u^2w^2}$

34. $\dfrac{-b^2c}{27a^3 - 1} \div \dfrac{b^3c^2}{3a - 1}$ $\dfrac{-1}{bc(9a^2 + 3a + 1)}$

35. $\dfrac{x - x^3}{2x^2 + 6x} \div \dfrac{5x^2 - 5x}{2x + 6}$ $\dfrac{-(1 + x)}{5x}$ or $\dfrac{-1 - x}{5x}$

36. $\dfrac{121y - y^3}{y^2 - 49} \div \dfrac{y^2 - 11y}{y + 7}$ $\dfrac{-(y + 11)}{y - 7}$ or $\dfrac{y + 11}{7 - y}$

37. $\dfrac{y^2 + y - 12}{y^2 - 8y + 15} \div \dfrac{3y^2 + 7y - 20}{2y^2 - 7y - 15}$ $\dfrac{2y + 3}{3y - 5}$

38. $\dfrac{3y^2 + 11y + 6}{4y^2 + 16y + 7} \div \dfrac{3y^2 - y - 2}{2y^2 - y - 28}$ $\dfrac{(y + 3)(y - 4)}{(2y + 1)(y - 1)}$ or $\dfrac{y^2 - y - 12}{2y^2 - y - 1}$

39. $\dfrac{4x^2 - 12x + 9}{25 - 4x^2} \div \dfrac{6x^2 - 5x - 6}{6x^2 + 19x + 10}$ $\dfrac{2x - 3}{5 - 2x}$

40. $\dfrac{4x^2 + 12x + 9}{9 - 4x^2} \div \dfrac{10x^2 + 27x + 18}{8x^2 - 2x - 15}$ $\dfrac{-(4x + 5)}{5x + 6}$ or $\dfrac{-4x - 5}{5x + 6}$

C In Problems 41–54, perform the indicated operations.

41. $\dfrac{x^2 + 2x - 3}{x - 5} \div \dfrac{x^2 + 6x + 9}{x^2 - 2x - 15} \cdot \dfrac{1}{x^2 - 1}$ $\dfrac{1}{x + 1}$

42. $\dfrac{x^2 - 3x + 2}{x^2 - 5x + 6} \div \dfrac{x^2 - 5x + 4}{x^2 - 7x + 12} \cdot \dfrac{x^3 + 1}{x^2 - 1}$ $\dfrac{x^2 - x + 1}{x - 1}$

43. $\dfrac{x^2 - 1}{x^2 + 3x - 10} \div \dfrac{x^2 - 3x - 4}{x^2 - 25} \cdot \dfrac{x - 2}{x - 5}$ $\dfrac{x - 1}{x - 4}$

44. $\dfrac{x^2 - 25}{x^2 - 49} \cdot \dfrac{x^2 - 4x - 21}{x^2 - 10x + 25} \div \dfrac{x^2 + 2x - 3}{x^2 - 6x + 5}$ $\dfrac{x + 5}{x + 7}$

45. $\dfrac{x - 3}{3 - x} \cdot \dfrac{x^2 + 3x - 4}{x^2 + 7x + 12} \div \dfrac{x^2 + x - 2}{x^2 + 5x + 6}$ -1

46. $\dfrac{x^3 - 125}{x^3 - 8} \cdot \dfrac{x^2 + x - 2}{x^2 + 6x - 7} \div \dfrac{x^2 - 3x - 10}{x^2 + 5x - 14}$ $\dfrac{x^2 + 5x + 25}{x^2 + 2x + 4}$

47. $\dfrac{x^2 - y^2}{x^2 - 2xy} \div \dfrac{x^2 + xy - 2y^2}{x^2 - 4y^2} \cdot \dfrac{x^2}{(x + y)^2}$ $\dfrac{x}{x + y}$

48. $\dfrac{x^2 + xy - 2y^2}{x^2 - 4y^2} \div \dfrac{x^2 - y^2}{x^2 - 2xy} \cdot \dfrac{(x + y)^2}{x^2}$ $\dfrac{x + y}{x}$

49. $\dfrac{x^2 + 2xy - 3y^2}{y^2 - 7y + 10} \div \dfrac{x^2 - 3xy + 2y^2}{y^2 - 3y - 10} \cdot \dfrac{x^2 - 4y^2}{x^2 - 9y^2}$ $\dfrac{(x + 2y)(y + 2)}{(x - 3y)(y - 2)}$ or $\dfrac{xy + 2x + 2y^2 + 4y}{xy - 2x - 3y^2 + 6y}$

50. $\dfrac{x^2 + 2xy - 8y^2}{x^2 + 7xy + 12y^2} \div \dfrac{x^2 - 3xy + 2y^2}{x^2 + 2xy - 3y^2} \cdot \dfrac{x^2 - 9}{9 - x^2}$ -1

51. $\dfrac{x^3 + x^2 - x - 1}{x + 2} \div \dfrac{x^3 + 1}{x + 3} \cdot \dfrac{x + 2}{x - 1}$ $\dfrac{(x + 1)(x + 3)}{x^2 - x + 1}$ or $\dfrac{x^2 + 4x + 3}{x^2 - x + 1}$

52. $\dfrac{x^3 + 3x^2 - 9x - 27}{x + 4} \div \dfrac{x^3 + 27}{x + 4} \cdot \dfrac{x + 1}{x - 3}$ $\dfrac{(x + 3)(x + 1)}{x^2 - 3x + 9}$ or $\dfrac{x^2 + 4x + 3}{x^2 - 3x + 9}$

53. $\dfrac{x - 2}{x^2 - 9} \div \dfrac{x^3 - 8}{x + 3} \cdot \dfrac{x - 3}{x}$ $\dfrac{1}{x(x^2 + 2x + 4)}$ or $\dfrac{1}{x^3 + 2x^2 + 4x}$

54. $\dfrac{x - 1}{x^2 - 25} \div \dfrac{x - 3}{x^3 + 125} \cdot \dfrac{(x - 5)^2}{x - 1}$ $\dfrac{(x - 5)(x^2 - 5x + 25)}{x - 3}$ or $\dfrac{x^3 - 10x^2 + 50x - 125}{x - 3}$

APPLICATIONS

Prices and Currents

55. If the price for x units of a product is

$$\dfrac{3x + 9}{4}$$

and the demand for these is

$$\dfrac{600}{x^2 + 3x}$$

find the product of the price and the demand. $\dfrac{450}{x}$

56. If the price for x units of a product is given by

$$\dfrac{4x + 8}{5}$$

and the demand is

$$\dfrac{200}{x^2 + 2x}$$

what is the product of the price and the demand? $\dfrac{160}{x}$

57. In a simple electrical circuit, the current I is the quotient of the voltage E and the resistance R. If the resistance changes with the time t according to

$$R = \frac{t^2 + 9}{t^2 + 6t + 9}$$

and the voltage changes according to the formula

$$E = \frac{4t}{t + 3}$$

find the current I. $\frac{4t(t + 3)}{t^2 + 9}$ or $\frac{4t^2 + 12t}{t^2 + 9}$

58. If in Problem 57 the resistance is

$$R = \frac{t^2 + 5}{t^2 + 4t + 4}$$

and the voltage is

$$E = \frac{5t}{t + 2}$$

find the current I. $\frac{5t(t + 2)}{t^2 + 5}$ or $\frac{5t^2 + 10t}{t^2 + 5}$

SKILL CHECKER

Write the rational expression with the indicated denominator:

59. $\dfrac{x - 1}{x - 3}$; denominator $x^2 - x - 6$ $\dfrac{x^2 + x - 2}{x^2 - x - 6}$

60. $\dfrac{x + 4}{x - 3}$; denominator $x^2 - 9$ $\dfrac{x^2 + 7x + 12}{x^2 - 9}$

61. $\dfrac{1}{x^2 + 2x + 4}$; denominator $x^3 - 8$ $\dfrac{x - 2}{x^3 - 8}$

62. $\dfrac{1}{x^2 - 3x + 9}$; denominator $x^3 + 27$ $\dfrac{x + 3}{x^3 + 27}$

USING YOUR KNOWLEDGE

More Applications

63. When studying parallel resistors, the expression

$$R \cdot \frac{R_T}{R - R_T}$$

occurs, where R is a known resistance and R_T is a required resistance. Perform the indicated multiplication. $\frac{RR_T}{R - R_T}$

64. The molecular model predicts that the pressure of a gas is given by

$$\frac{2}{3} \cdot \frac{mv^2}{2} \cdot \frac{N}{v}$$

where m is the mass, N is a constant, and v is the velocity. Perform the indicated multiplication. $\frac{mvN}{3}$

65. Suppose a store orders 3000 items each year. If it orders x units at a time, the number N of reorders is

$$N = \frac{3000}{x}$$

If there is a fixed \$20 reorder fee and a \$3 charge per item, the cost of each order is

$$C = 20 + 3x$$

The yearly reorder cost R is then given by

$$R = N \cdot C$$

Find R. $R = \frac{60{,}000 + 9000x}{x}$

The formula for the area A of a rectangle is $A = LW$, where L is the length and W is the width of the rectangle. In Problems 66 and 67, find the area of the shaded rectangle.

66. $\dfrac{3x + 2}{3}$ $\dfrac{3x + 2}{3}$ $\dfrac{3x + 2}{3}$ $\dfrac{3x^2 + 2x}{6}$

$\dfrac{x}{2}$, $\dfrac{x}{2}$

67. $\dfrac{2w - L}{2}$ $\dfrac{2w - L}{2}$ $\dfrac{2w^2 - Lw}{6}$

$\dfrac{w}{3}$, $\dfrac{w}{3}$, $\dfrac{w}{3}$

WRITE ON

68. Write in your own words the procedure you use to multiply two rational expressions.
Answers may vary.

69. Write in your own words the procedure you use to divide two rational expressions.
Answers may vary.

70. When multiplying $(x + 2)(x + 3)$, you get $x^2 + 5x + 6$. When multiplying

$$\frac{x + 2}{2} \cdot \frac{x + 3}{3}$$

most textbooks write the answer as

$$\frac{(x + 2)(x + 3)}{6}$$

Do you agree? Why or why not?
Answers may vary.

MASTERY TEST

If you know how to do these problems, you have learned your lesson!

Perform the indicated operations:

71. $\dfrac{2 - x}{x + 3} \div \dfrac{x^3 - 8}{x - 5} \cdot \dfrac{x^3 + 27}{x - 5}$ $\dfrac{-(x^2 - 3x + 9)}{x^2 + 2x + 4}$

72. $\dfrac{x^5y}{z} \div \dfrac{x^3y}{z^6}$ x^2z^5

73. $\dfrac{3x^2 - 3}{5x + 10} \div \dfrac{x + 1}{5}$ $\dfrac{3(x - 1)}{x + 2}$ or $\dfrac{3x - 3}{x + 2}$

74. $\dfrac{x + 2}{x - 2} \div (x^2 + 4x + 4)$ $\dfrac{1}{(x - 2)(x + 2)}$ or $\dfrac{1}{x^2 - 4}$

75. $\dfrac{3 - x}{x + 2} \cdot \dfrac{x^2 + 5x + 6}{x^2 - 9}$ -1

76. $\dfrac{x^2 + 4x + 3}{x^2 + 5x + 6} \cdot \dfrac{x^2 - 2x - 8}{x^2 - 2x - 3}$ $\dfrac{x - 4}{x - 3}$

77. $\dfrac{x^3 + x^2 - x - 1}{x + 4} \div \dfrac{x^3 + 1}{x^2 - 16} \cdot \dfrac{1}{x^2 - 1}$ $\dfrac{x - 4}{x^2 - x + 1}$

78. $\dfrac{x - 2}{3x - 2y} \cdot \dfrac{9x^2 - 4y^2}{2x^2 - 3x - 2}$ $\dfrac{3x + 2y}{2x + 1}$

79. $\dfrac{x^2 - 9}{9x^2 - 4y^2} \cdot \dfrac{3x^2 - 2xy}{3x + 9}$ $\dfrac{x(x - 3)}{3(3x + 2y)}$ or $\dfrac{x^2 - 3x}{9x + 6y}$

6.3   ADDITION AND SUBTRACTION OF RATIONAL EXPRESSIONS

To Succeed, Review How To . . .

1. Add and subtract real numbers (pp. 17–19).

2. Write a rational expression with a given denominator (pp. 426–427).

3. Factor polynomials (pp. 370–396).

Objectives

A   Add and subtract rational expressions with the same denominator.

B   Add and subtract rational expressions with different denominators.

GETTING STARTED

Paper—don't throw it away, recycle it!

What Type of Materials Are More Likely To Be Recycled?

In the *Getting Started* for Section 6.1, we mentioned that the fraction of the waste recycled is

$$P(t) = \frac{R(t)}{G(t)} = \frac{0.055t^2 - 0.7t + 6.86}{0.004t^2 + 3.52t + 86.5}$$

where t is the number of years after 1960. Much of this waste is paper and paperboard. The approximate fraction of paper and paperboard recycled is given by

$$Q(t) = \frac{0.004t^2 - 1.25t + 30.28}{0.004t^2 + 3.52t + 86.5}$$

Can you find the fraction of the waste recycled that is not paper and paperboard? To do this, you need to find $P(t) - Q(t)$. Fortunately, $P(t)$ and $Q(t)$ have the same denominator, so you only need to subtract the numerators and keep the denominator. Make sure you understand that the subtraction sign must be distributed to every term in the numerator of the fraction that follows it, like this:

$$P(t) - Q(t) = \frac{0.055t^2 - 0.7t + 6.86}{0.004t^2 + 3.52t + 86.5} - \frac{0.004t^2 - 1.25t + 30.28}{0.004t^2 + 3.52t + 86.5}$$

$$= \frac{0.055t^2 - 0.7t + 6.86 - (0.004t^2 - 1.25t + 30.28)}{0.004t^2 + 3.52t + 86.5}$$

$$= \frac{0.055t^2 - 0.7t + 6.86 - 0.004t^2 + 1.25t - 30.28}{0.004t^2 + 3.52t + 86.5}$$

$$= \frac{0.051t^2 + 0.55t - 23.42}{0.004t^2 + 3.52t + 86.5}$$

If you let $t = 50$ in this expression, you can find, approximately, what fraction of the waste will not be paper or paperboard in the year 2010. You should come out with 48%. More of the material recycled (52%) will be paper or paperboard.

A   Adding and Subtracting Rational Expressions with the Same Denominator

In general, if a, b, and c are real numbers ($b \neq 0$),

$$\frac{a}{b} + \frac{c}{b} = \frac{a + c}{b} \quad \longleftarrow \text{Add numerators.}$$
$$\quad\quad\quad\quad\quad \longleftarrow \text{Keep the denominator.}$$

$$\frac{a}{b} - \frac{c}{b} = \frac{a - c}{b} \quad \longleftarrow \text{Subtract numerators.}$$
$$\quad\quad\quad\quad\quad \longleftarrow \text{Keep the denominator.}$$

Web It

For more examples on how to add rational expressions with the same denominator, go to link 6-3-1 at mhhe.com/bello.

To add (or subtract) rational expressions with the *same* denominators, add (or subtract) the numerators and *keep* the denominator. Thus

$$\frac{2}{x} + \frac{6}{x} = \frac{2+6}{x} = \frac{8}{x} \quad \longleftarrow \text{ Add numerators.}$$
$$\qquad\qquad\qquad\qquad \longleftarrow \text{ Keep denominator.}$$

Similarly,

$$\frac{5x}{x^2+1} + \frac{2x}{x^2+1} = \frac{5x+2x}{x^2+1} = \frac{7x}{x^2+1}$$

$$\frac{3x}{7(x-1)^2} + \frac{4x}{7(x-1)^2} = \frac{3x+4x}{7(x-1)^2} = \frac{7x}{7(x-1)^2} = \frac{x}{(x-1)^2}$$

and

$$\frac{8x}{x^2+5} - \frac{2x}{x^2+5} = \frac{8x-2x}{x^2+5} = \frac{6x}{x^2+5}$$

$$\frac{10x}{9(x-3)^2} - \frac{x}{9(x-3)^2} = \frac{10x-x}{9(x-3)^2} = \frac{9x}{9(x-3)^2} = \frac{x}{(x-3)^2}$$

The answer is written in simplified form.

EXAMPLE 1 Adding and subtracting with the same denominator

Perform the indicated operations:

a. $\dfrac{8x}{3(x-2)} + \dfrac{x}{3(x-2)}$

b. $\dfrac{7x}{5(x+4)^2} + \dfrac{3x}{5(x+4)^2}$

c. $\dfrac{x}{x^2-1} - \dfrac{1}{x^2-1}$

d. $\dfrac{4x}{x+2} - \dfrac{3x-2}{x+2}$

SOLUTION

a. $\dfrac{8x}{3(x-2)} + \dfrac{x}{3(x-2)} = \dfrac{8x+x}{3(x-2)} = \dfrac{\overset{3}{9x}}{3(x-2)} = \dfrac{3x}{x-2}$

b. $\dfrac{7x}{5(x+4)^2} + \dfrac{3x}{5(x+4)^2} = \dfrac{7x+3x}{5(x+4)^2} = \dfrac{\overset{2}{10x}}{5(x+4)^2} = \dfrac{2x}{(x+4)^2}$

c. Both expressions have the same denominator, so we subtract numerators and use the same denominator. However, the answer can be simplified. You can see that only when the denominator is completely factored. Make sure your denominators are in completely factored form so you can determine all the possible ways to simplify the answer.

$$\frac{x}{x^2-1} - \frac{1}{x^2-1} = \frac{x-1}{x^2-1} \quad \longleftarrow \text{ Subtract numerators.}$$
$$\qquad\qquad\qquad\qquad\qquad \longleftarrow \text{ Keep denominator.}$$

$$= \frac{\overset{1}{\cancel{x-1}}}{(x+1)\underset{1}{\cancel{(x-1)}}} \qquad \text{Factor the denominator and divide out the common factor } (x-1).$$

$$= \frac{1}{x+1}$$

PROBLEM 1

Perform the indicated operations:

a. $\dfrac{4x}{5(x+3)} + \dfrac{6x}{5(x+3)}$

b. $\dfrac{11x}{24(7+x)} + \dfrac{x}{24(7+x)}$

c. $\dfrac{3x}{9x^2-16} - \dfrac{4}{9x^2-16}$

d. $\dfrac{5x}{x+8} - \dfrac{4x-8}{x+8}$

Teaching Tip

Remind the students that in algebra adding and subtracting can be called combining like terms.

Answers

1. a. $\frac{2x}{x+3}$ **b.** $\frac{x}{2(7+x)}$ or $\frac{x}{14+2x}$

c. $\frac{1}{3x+4}$ **d.** 1

d. We indicate the subtraction of the *numerators* by using parentheses. Be careful with the signs when removing the parentheses.

Teaching Tip

Parentheses are necessary when the second numerator has more than one term.

$$\frac{4x}{x+2} - \frac{3x-2}{x+2} = \frac{4x - (3x-2)}{x+2}$$

$$= \frac{4x - 3x + 2}{x+2}$$

Recall that $-(3x-2) = -1(3x-2)$
$$= -3x + 2$$

$$= \frac{\overset{1}{\cancel{x+2}}}{\cancel{x+2}} = 1$$

Combine like terms and simplify.

B Adding and Subtracting Rational Expressions with Different Denominators

To add or subtract fractions with different denominators, first find a common denominator. It is most convenient to use the smallest one available, called the **Least Common Denominator (LCD)**—the smallest multiple of the denominators. To add

$$\frac{5}{12} + \frac{7}{18}$$

we start by writing 12 and 18 as products of *primes*. We have

$$12 = 2 \cdot 2 \cdot 3 \quad = 2^2 \cdot 3$$

The 2's are in one column and the 3's are written in another column.

$$18 = \quad 2 \cdot 3 \cdot 3 = 2 \cdot 3^2$$

Web It

For a lesson on adding and subtracting rational expressions, go to link 6-3-2 at mhhe.com/bello.

For more practice on adding and subtracting rational expressions, go to link 6-3-3.

Since we need the *smallest* number that is a multiple of 12 and 18, we select the factors raised to the *greatest* power in each column. The product of these factors is the LCD. The LCD of 12 and 18 is $2^2 \cdot 3^2 = 4 \cdot 9 = 36$. We then write each fraction with a denominator of 36 and add:

$$\frac{5}{12} = \frac{5 \cdot 3}{12 \cdot 3} = \frac{15}{36}$$

Multiply the denominator of $\frac{5}{12}$ by 3 (to get 36), and do the same to the numerator.

$$\frac{7}{18} = \frac{7 \cdot 2}{18 \cdot 2} = \frac{14}{36}$$

Multiply the numerator and denominator of $\frac{7}{18}$ by 2 to get 36 as the denominator.

$$\frac{5}{12} + \frac{7}{18} = \frac{15}{36} + \frac{14}{36} = \frac{29}{36}$$

The procedure used to find the LCD of two or more rational expressions is similar to that used to find the LCD of two numbers.

PROCEDURE

Finding the LCD of Two or More Rational Expressions

1. Factor each denominator. Place identical factors in columns. (Not necessary to factor monomials.)

2. From each column, select the factor with the greatest exponent.

3. The product of all the factors obtained in step 2 is the LCD.

If the denominators involved have no common factors, the LCD is the product of the denominators. For example, the denominators in $\frac{3}{5}$ and $\frac{1}{7}$ have no common factors. The LCD is $5 \cdot 7$. To subtract $\frac{1}{7}$ from $\frac{3}{5}$, we first write each fraction with a denominator of 35 and then subtract. Here are the steps.

1. The LCD is $5 \cdot 7 = 35$.

2. Write each fraction with 35 as the denominator.
$$\frac{3}{5} = \frac{3 \cdot 7}{5 \cdot 7} = \frac{21}{35} \quad \text{and} \quad \frac{1}{7} = \frac{1 \cdot 5}{7 \cdot 5} = \frac{5}{35}$$

3. Subtract: $\frac{3}{5} - \frac{1}{7} = \frac{21}{35} - \frac{5}{35} = \frac{16}{35}$

EXAMPLE 2 **Adding and subtracting with different denominators**

Perform the indicated operation:
$$\frac{2x}{x+1} - \frac{x}{x+2}$$

SOLUTION $(x+1)$ and $(x+2)$ do not have any common factors so we do not need to place the factors in columns. The LCD in
$$\frac{2x}{x+1} - \frac{x}{x+2} \quad \text{is} \quad (x+1)(x+2)$$

Now rewrite
$$\frac{2x}{x+1} \quad \text{and} \quad \frac{x}{x+2}$$

with the LCD, $(x+1)(x+2)$, as the denominator.
$$\frac{2x}{x+1} = \frac{2x(x+2)}{(x+1)(x+2)} \quad \text{and} \quad \frac{x}{x+2} = \frac{x(x+1)}{(x+2)(x+1)}$$

$$\frac{2x}{x+1} - \frac{x}{x+2} = \frac{2x(x+2)}{(x+1)(x+2)} - \frac{x(x+1)}{(x+1)(x+2)}$$

$$= \frac{2x(x+2) - x(x+1)}{(x+1)(x+2)} \qquad \text{Subtract.}$$

$$= \frac{2x^2 + 4x - x^2 - x}{(x+1)(x+2)} \qquad \text{Use the distributive property,} \\ -x(x+1) = -x^2 - x.$$

$$= \frac{x^2 + 3x}{(x+1)(x+2)} \qquad \text{Combine like terms in the numerator.}$$

$$\text{or} \quad \frac{x(x+3)}{(x+1)(x+2)}$$

PROBLEM 2

Perform the indicated operation:
$$\frac{5x}{x+1} - \frac{3x}{x+3}$$

In general, to add or subtract fractions with different denominators, use the following procedure.

PROCEDURE

Procedure to Add (or Subtract) Fractions with Different Denominators

1. Find the LCD.

2. Write all fractions as equivalent ones with the LCD as the denominator.

3. Add (or subtract) numerators; keep the LCD as denominator.

4. Simplify if possible.

Answer

2. $\frac{2x^2 + 12x}{(x+1)(x+3)}$ or $\frac{2x(x+6)}{(x+1)(x+3)}$

EXAMPLE 3	Using the LCD to add and subtract with different denominators

Perform the indicated operations:

a. $\dfrac{x + 1}{x^2 + x - 2} + \dfrac{x + 3}{x^2 - 1}$ **b.** $\dfrac{x - 1}{x^2 - x - 6} - \dfrac{x + 4}{x^2 - 9}$

SOLUTION

a. Use the four-step procedure to add fractions.

1. First find the LCD of the denominators. Write the denominators in factored form with the same factors in a column.

$$x^2 + x - 2 = \;\; (x + 2) \;\; (x - 1)$$
$$x^2 - 1 = \;\;\;\;\;\;\;\;\;\;\; (x - 1) \;\; (x + 1)$$

Select the factors with the greatest exponents in each column, $(x + 2)$ in column 1, $(x - 1)$ in column 2, and $(x + 1)$ in column 3. The LCD is the product of these factors,

$$(x + 2)(x - 1)(x + 1)$$

2. Then write each fraction as an equivalent one with the LCD as the denominator.

$$\frac{x + 1}{x^2 + x - 2} = \frac{x + 1}{(x + 2)(x - 1)} = \frac{(x + 1)(x + 1)}{(x + 2)(x - 1)(x + 1)} = \frac{x^2 + 2x + 1}{(x + 2)(x - 1)(x + 1)}$$

$$\frac{x + 3}{x^2 - 1} = \frac{x + 3}{(x + 1)(x - 1)} = \frac{(x + 3)(x + 2)}{(x + 1)(x - 1)(x + 2)} = \frac{x^2 + 5x + 6}{(x + 1)(x - 1)(x + 2)}$$

3. $\dfrac{x + 1}{x^2 + x - 2} + \dfrac{x + 3}{x^2 - 1} = \dfrac{x^2 + 2x + 1}{(x + 2)(x - 1)(x + 1)} + \dfrac{x^2 + 5x + 6}{(x + 2)(x - 1)(x + 1)}$

$$= \frac{(x^2 + 2x + 1) + (x^2 + 5x + 6)}{(x + 2)(x - 1)(x + 1)}$$ Add the numerators; keep the denominator.

$$= \frac{2x^2 + 7x + 7}{(x + 2)(x - 1)(x + 1)}$$ Combine like terms in the numerator.

4. The answer cannot be simplified; $2x^2 + 7x + 7$ is *not* factorable.

b. To subtract fractions again use the four-step procedure.

1. To find the LCD, factor the denominators keeping the same factors in a column:

$$x^2 - x - 6 = \;\;\;\;\;\;\;\;\;\;\; (x - 3) \;\; (x + 2)$$
$$x^2 - 9 = \;\; (x + 3) \;\; (x - 3)$$

The LCD is $(x + 3)(x - 3)(x + 2)$.

2. Write each fraction as an equivalent one with the LCD as the denominator:

$$\frac{x - 1}{x^2 - x - 6} = \frac{(x - 1)(x + 3)}{(x - 3)(x + 2)(x + 3)} = \frac{x^2 + 2x - 3}{(x + 3)(x - 3)(x + 2)}$$

$$\frac{x + 4}{x^2 - 9} = \frac{(x + 4)(x + 2)}{(x + 3)(x - 3)(x + 2)} = \frac{x^2 + 6x + 8}{(x + 3)(x - 3)(x + 2)}$$

PROBLEM 3

Perform the indicated operations:

a. $\dfrac{x + 1}{(x + 3)(x - 1)} + \dfrac{x + 4}{x^2 - 1}$

b. $\dfrac{x - 3}{x^2 - x - 2} - \dfrac{x + 3}{x^2 - 4}$

Web It

For a review on rational expressions and the operations, go to link 6-3-4 at mhhe.com/bello.

Answers

3. a. $\dfrac{2x^2 + 9x + 13}{(x + 3)(x - 1)(x + 1)}$

b. $\dfrac{-5x - 9}{(x + 1)(x - 2)(x + 2)}$

3. $\dfrac{x-1}{x^2-x-6} - \dfrac{x+4}{x^2-9} = \dfrac{x^2+2x-3}{(x+3)(x-3)(x+2)} - \dfrac{x^2+6x+8}{(x+3)(x-3)(x+2)}$

$= \dfrac{(x^2+2x-3)-(x^2+6x+8)}{(x+3)(x-3)(x+2)}$ Subtract the numerators; keep the denominator.

$= \dfrac{x^2+2x-3-x^2-6x-8}{(x+3)(x-3)(x+2)}$ Remember that $-(x^2+6x+8) = -x^2-6x-8$.

$= \dfrac{-4x-11}{(x+3)(x-3)(x+2)}$ Combine like terms in the numerator.

4. The answer is not reducible.

Teaching Tip

Use a simple arithmetic example to introduce Example 4:

$$\frac{4}{6} + \frac{4}{12} = \frac{2}{3} + \frac{1}{3} = \frac{3}{3} = 1$$

How would you add $\frac{6}{12} + \frac{1}{8}$? You can start by finding the LCD, 24. However, it is easier to reduce $\frac{6}{12}$ to $\frac{1}{2}$ first. Then add $\frac{1}{2} + \frac{1}{8} = \frac{4}{8} + \frac{1}{8} = \frac{5}{8}$. We illustrate a similar problem in Example 4.

EXAMPLE 4   **Simplifying before adding and subtracting**

Perform the indicated operations:

a. $\dfrac{x+y}{x^2+2xy+y^2} + \dfrac{x-y}{x^2-2xy+y^2}$ **b.** $\dfrac{x}{(x+2)(x-2)} - \dfrac{2}{(2-x)(x+2)}$

SOLUTION

a. $\dfrac{x+y}{x^2+2xy+y^2} = \dfrac{x+y}{(x+y)^2} = \dfrac{1}{x+y}$ Factor and simplify the first fraction.

$\dfrac{x-y}{x^2-2xy+y^2} = \dfrac{x-y}{(x-y)^2} = \dfrac{1}{x-y}$ Factor and simplify the second fraction.

$(x+y)$ and $(x-y)$ have no common factors, so the LCD is $(x+y)(x-y)$. We have

$\dfrac{x+y}{x^2+2xy+y^2} + \dfrac{x-y}{x^2-2xy+y^2} = \dfrac{1}{x+y} + \dfrac{1}{x-y}$

$= \dfrac{(x-y)}{(x+y)(x-y)} + \dfrac{(x+y)}{(x+y)(x-y)}$ Write each fraction with $(x+y)(x-y)$ as the denominator.

$= \dfrac{x-y+x+y}{(x+y)(x-y)}$ Combine like terms in the numerator.

$= \dfrac{2x}{(x+y)(x-y)}$

b. $x-2 = -(2-x)$, so start by multiplying the numerator and denominator of the second fraction by -1.

$\dfrac{x}{(x+2)(x-2)} - \dfrac{2}{(2-x)(x+2)}$

$= \dfrac{x}{(x+2)(x-2)} - \dfrac{-1\cdot(2)}{-1\cdot(2-x)(x+2)}$ Multiply the numerator and denominator by -1.

$= \dfrac{x}{(x+2)(x-2)} - \dfrac{-2}{(x-2)(x+2)}$ $-1\cdot2 = -2$ and $-1(2-x) = -2+x = x-2$

$= \dfrac{x-(-2)}{(x+2)(x-2)}$

$= \dfrac{(x+2)}{(x+2)(x-2)}$ Divide out $x+2$.

$= \dfrac{1}{x-2}$

PROBLEM 4

Perform the indicated operations:

a. $\dfrac{2x-2y}{x^2-2xy+y^2} + \dfrac{x-y}{x^2-y^2}$

b. $\dfrac{x}{(x+3)(x-3)} + \dfrac{3}{(3-x)(x+3)}$

Answers

4. a. $\dfrac{3x+y}{(x-y)(x+y)}$ or $\dfrac{3x+y}{x^2-y^2}$

b. $\dfrac{1}{x+3}$

Calculate It Exercises   Using Graphs to Verify Adding and Subtracting of Rational Expressions

The results of adding or subtracting rational expressions can be verified by graphing the original problem and the answer. Thus to verify Example 4(b), use a decimal window to graph

$$Y_1 = \left(\frac{x}{(x+2)(x-2)}\right) - \left(\frac{2}{(2-x)(x+2)}\right)$$

and $Y_2 = \frac{1}{(x-2)}$

and make sure that both graphs are identical.

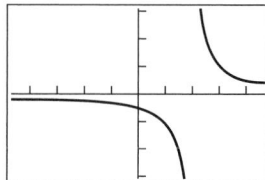

However, Y_1 is not defined for $x = -2$ but Y_2 is. The calculator didn't indicate this. You have to know some algebra to recognize it.

1. Verify the results of Example 1(d) using your calculator.

2. Can you verify the results in Example 4(a)? Explain.

3. Look at the graph in the window. Can you find the integer values of x for which the denominator is zero?

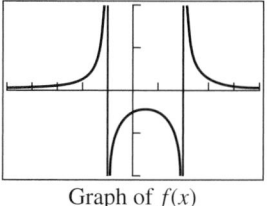

Graph of $f(x)$

4. If it is known that the numerator of the function $f(x)$ whose graph is shown in the window is 1 and the denominator is a quadratic function, what is $f(x)$? (*Note:* We have used a $[-5, 5]$ by $[-2, 2]$ window.)

Exercises 6.3

A In Problems 1–10, perform the indicated operations.

1. $\dfrac{x}{5} + \dfrac{2x}{5} - \dfrac{3x}{5}$

2. $\dfrac{x+1}{3x} + \dfrac{2x+7}{3x} - \dfrac{3x+8}{3x}$

3. $\dfrac{7x}{3} - \dfrac{2x}{3} - \dfrac{5x}{3}$

4. $\dfrac{2x-1}{5x} - \dfrac{x+1}{5x} - \dfrac{x-2}{5x}$

5. $\dfrac{3}{5x+10} + \dfrac{2x}{5(x+2)} - \dfrac{3+2x}{5(x+2)}$

6. $\dfrac{2x+1}{3(x+2)} + \dfrac{3x+1}{3x+6} - \dfrac{5x+2}{3(x+2)}$

7. $\dfrac{2x+1}{2(x+1)} - \dfrac{x-1}{2x+2} - \dfrac{x+2}{2(x+1)}$

8. $\dfrac{3x-1}{4(x-1)} - \dfrac{4x-1}{4x-4} - \dfrac{-x}{4(x-1)}$

9. $\dfrac{2x+1}{3(x-1)} + \dfrac{x+3}{3x-3} - \dfrac{x-1}{3(x-1)} - \dfrac{2x+5}{3(x-1)}$

10. $\dfrac{3x-1}{5(x+1)} - \dfrac{x+1}{5x+5} + \dfrac{2x-5}{5(x+1)} - \dfrac{4x-7}{5(x+1)}$

B In Problems 11–50, perform the indicated operations.

11. $\dfrac{x}{x^2+3x-4} + \dfrac{x}{x^2-16} - \dfrac{2x^2-5x}{(x-1)(x+4)(x-4)}$

12. $\dfrac{x-2}{x^2-9} + \dfrac{x+1}{x^2-x-12} - \dfrac{2x^2-8x+5}{(x+3)(x-3)(x-4)}$

13. $\dfrac{3x}{x^2+3x-10} + \dfrac{2x}{x^2+x-6} - \dfrac{5x^2+19x}{(x+5)(x-2)(x+3)}$

14. $\dfrac{x+3}{x^2-x-2} + \dfrac{x-1}{x^2+2x+1} - \dfrac{2x^2+x+5}{(x-2)(x+1)^2}$

15. $\dfrac{1}{x^2-y^2} + \dfrac{5}{(x+y)^2} - \dfrac{6x-4y}{(x+y)^2(x-y)}$

16. $\dfrac{3}{(x+y)^2} + \dfrac{5}{x+y} - \dfrac{5x+5y+3}{(x+y)^2}$

17. $\dfrac{2}{x-5} - \dfrac{3x}{x^2-25} - \dfrac{10-x}{(x-5)(x+5)}$

18. $\dfrac{x+3}{x^2-x-2} - \dfrac{x-1}{x^2+2x+1} - \dfrac{7x+1}{(x-2)(x+1)^2}$

19. $\dfrac{x-1}{x^2+3x+2} - \dfrac{x+7}{x^2+5x+6} - \dfrac{-6x-10}{(x+1)(x+2)(x+3)}$

20. $\dfrac{2}{x^2+3xy+2y^2} - \dfrac{1}{x^2-xy-2y^2} - \dfrac{x-6y}{(x+2y)(x+y)(x-2y)}$

Hint: For Problems 21–29, first simplify the fractions.

21. $\dfrac{x+2}{x^2-4} + \dfrac{x+3}{x^2-9} \quad \dfrac{2x-5}{(x-2)(x-3)}$

22. $\dfrac{x-4}{x^2-16} + \dfrac{x+3}{x^2-9} \quad \dfrac{2x+1}{(x+4)(x-3)}$

23. $\dfrac{x-3}{x^2-9} + \dfrac{x+3}{x^2+6x+9} \quad \dfrac{2}{x+3}$

24. $\dfrac{a-4}{a^2-16} + \dfrac{a+3}{a^2+5a+6} \quad \dfrac{2a+6}{(a+4)(a+2)}$

25. $\dfrac{a+3}{a^2+5a+6} + \dfrac{a+2}{a^2+6a+8} \quad \dfrac{2a+6}{(a+2)(a+4)}$

26. $\dfrac{a+3}{a^2+5a+6} - \dfrac{a-4}{a^2-16} \quad \dfrac{2}{(a+2)(a+4)}$

27. $\dfrac{3a+3}{a^2+5a+4} - \dfrac{a-3}{a^2+a-12} \quad \dfrac{2}{a+4}$

28. $\dfrac{2a}{5a-7b} - \dfrac{5a+7b}{25a^2-49b^2} \quad \dfrac{2a-1}{5a-7b}$

29. $\dfrac{5a-15}{a^2+2a-15} - \dfrac{a^2+5a}{a^2+8a+15} \quad \dfrac{15-a^2}{(a+5)(a+3)}$

30. $\dfrac{3}{y^2-9} + \dfrac{2y}{y-3} \quad \dfrac{2y^2+6y+3}{(y+3)(y-3)}$

31. $\dfrac{y}{y^2-1} + \dfrac{y}{y-1} \quad \dfrac{y^2+2y}{(y+1)(y-1)}$

32. $\dfrac{3y}{y^2-4} - \dfrac{y}{y+2} \quad \dfrac{5y-y^2}{(y+2)(y-2)}$

33. $\dfrac{3y+1}{y^2-16} - \dfrac{2y-1}{y-4} \quad \dfrac{-2y^2-4y+5}{(y+4)(y-4)}$

34. $\dfrac{3x-5y}{2x-3y} + \dfrac{2x-3y}{2x+3y} \quad \dfrac{10x^2-13xy-6y^2}{(2x-3y)(2x+3y)}$

35. $\dfrac{5x+2y}{5x-2y} + \dfrac{5x-2y}{5x+2y} \quad \dfrac{50x^2+8y^2}{(5x-2y)(5x+2y)}$

36. $\dfrac{x+3y}{x-5y} - \dfrac{x+5y}{x-3y} \quad \dfrac{16y^2}{(x-5y)(x-3y)}$

37. $\dfrac{3x-y}{2x-y} - \dfrac{2x+y}{3x+y} \quad \dfrac{5x^2}{(2x-y)(3x+y)}$

38. $\dfrac{a+3}{a^2+a-6} + \dfrac{a-2}{a^2+3a-10} \quad \dfrac{2a+3}{(a-2)(a+5)}$

39. $\dfrac{x+3}{x^2-x-2} + \dfrac{x-1}{x^2+2x+1} \quad \dfrac{2x^2+x+5}{(x-2)(x+1)^2}$

40. $\dfrac{8x}{x^2-4y^2} - \dfrac{2x}{x^2-5xy+6y^2} \quad \dfrac{6x^2-28xy}{(x+2y)(x-2y)(x-3y)}$

41. $\dfrac{x+1}{x^2-x-2} - \dfrac{x}{x^2-5x+4} \quad \dfrac{4-3x}{(x-2)(x-4)(x-1)}$

42. $\dfrac{3}{x^2-4} + \dfrac{1}{2-x} - \dfrac{1}{2+x} \quad \dfrac{3-2x}{(x+2)(x-2)}$

43. $\dfrac{2}{5+x} + \dfrac{5x}{x^2-25} + \dfrac{7}{5-x} \quad \dfrac{-45}{(x+5)(x-5)}$

44. $\dfrac{1}{x^2+x-12} + \dfrac{2}{x^2+2x-15} + \dfrac{3}{x^2+9x+20}$
$\dfrac{6x+4}{(x+4)(x-3)(x+5)}$

45. $\dfrac{x}{(x-y)(2-x)} - \dfrac{y}{(y-x)(2-x)} + \dfrac{y}{(x-y)(x-2)}$
$\dfrac{x}{(x-y)(2-x)}$

46. $\dfrac{a}{(b-a)(c-a)} - \dfrac{b}{(b-c)(a-b)} + \dfrac{c}{(a-c)(b-c)} \quad 0$

47. $\dfrac{4a^2-9b^2}{4a^2-12ab+9b^2} + \dfrac{12a+18b}{4a^2+12ab+9b^2} - \dfrac{2a+3b}{2a+3b}$
$\dfrac{18b^2+12ab+12a-18b}{(2a+3b)(2a-3b)}$

48. $\dfrac{x+2y}{x^3+8y^3} + \dfrac{5}{x+2y} + \dfrac{2x-3y}{x^2-2xy+4y^2}$
$\dfrac{7x^2-9xy+14y^2+x+2y}{(x+2y)(x^2-2xy+y^2)}$

49. $\dfrac{x+5}{x^3+125} + \dfrac{x-5}{x^2-25} - \dfrac{1}{x+5} \quad \dfrac{1}{x^2-5x+25}$

50. $\dfrac{a}{a+3} + \dfrac{a-2}{a^2-3a+9} + \dfrac{5a^2-13a}{a^3+27} \quad \dfrac{a^3+3a^2-3a-6}{(a+3)(a^2-3a+9)}$

APPLICATIONS

Single Rational Expressions

51. The moment M of a cantilever beam of length L, x units from the end is given by

$$-\frac{w_0 x^3}{6L} + \frac{w_0 Lx}{2} - \frac{w_0 L^2}{3}$$

Write this expression as a single rational expression in reduced form. $\quad \dfrac{-w_0 x^3 + 3w_0 L^2 x - 2w_0 L^3}{6L}$

52. The deflection d of the beam of Problem 51 involves the expression

$$\frac{-x^4}{24L} + \frac{Lx^2}{4} - \frac{L^2 x}{3}$$

Write this expression as a single rational expression in reduced form. $\quad \dfrac{-x^4 + 6L^2 x^2 - 8L^3 x}{24L}$

53. In astronomy, planetary motion is given by

$$\frac{p^2}{2mr^2} - \frac{gmM}{r}$$

Write this expression as a single rational expression in reduced form. $\frac{p^2 - 2gm^2rM}{2mr^2}$

54. The motion of a pendulum is given by

$$\frac{P_1^2 + P_2^2}{2(h_1 + h_2)} + \frac{P_1^2 - P_2^2}{2(h_1 - h_2)}$$

Write this expression as a single rational expression in reduced form. $\frac{P_1^2 h_1 - P_2^2 h_2}{(h_1 + h_2)(h_1 - h_2)}$

SKILL CHECKER

Try the "Skill Checker" exercises so you'll be ready for the next section.

Multiply:

55. $9\left(2 + \frac{2}{9}\right)$ 20

56. $4\left(60 - \frac{15}{2}\right)$ 210

57. $12xy\left(\frac{2}{y} + \frac{3}{2x}\right)$ $24x + 18y$

58. $6ab\left(\frac{3}{a} - \frac{4}{b}\right)$ $18b - 24a$

59. $x^2\left(1 - \frac{1}{x^2}\right)$ $x^2 - 1$

60. $x^3\left(1 - \frac{1}{x^3}\right)$ $x^3 - 1$

USING YOUR KNOWLEDGE

Looking Ahead to Calculus

In calculus, the derivative of a polynomial P is defined as the limiting value of

$$\frac{P(x + h) - P(x)}{h}$$

as h approaches zero. Let $P(x) = x^2$.

61. Find $P(x + h)$ and write it in expanded form.

$P(x + h) = x^2 + 2xh + h^2$

62. Find $P(x + h) - P(x)$ and simplify it.

$P(x + h) - P(x) = 2xh + h^2$

63. Find $\dfrac{P(x + h) - P(x)}{h}$ in simplified form.

$\frac{P(x + h) - P(x)}{h} = 2x + h$

64. Find $\dfrac{P(x + h) - P(x)}{h}$ for $P(x) = x^2 + x$.

$\frac{P(x + h) - P(x)}{h} = 2x + h + 1$

WRITE ON

65. Write in your own words the procedure you use to find the LCD of two or more rational expressions.

Answers may vary.

66. Write in your own words the procedure you use to add or subtract two rational expressions.

Answers may vary.

67. When adding two or more rational expressions, do you *always* have to find the LCD or can you use *any* common denominator? What is the advantage of using the LCD?

Answers may vary.

MASTERY TEST

If you know how to do these problems, you have learned your lesson!

68. $\dfrac{x + 1}{(x + 3)(x - 1)} + \dfrac{x + 4}{x^2 - 1}$ $\frac{2x^2 + 9x + 13}{(x + 3)(x - 1)(x + 1)}$

69. $\dfrac{x - 3}{x^2 - x - 2} - \dfrac{x + 3}{x^2 - 4}$ $\frac{-(5x + 9)}{(x + 1)(x - 2)(x + 2)}$

70. $\dfrac{x - y}{x^2 - 2xy + y^2} + \dfrac{x - y}{x^2 - y^2}$ $\frac{2x}{(x - y)(x + y)}$

71. $\dfrac{x}{(x + 3)(x - 3)} + \dfrac{3}{(3 - x)(x + 3)}$ $\frac{1}{x + 3}$

72. $\dfrac{5x}{x + 1} - \dfrac{3x}{x + 3}$ $\frac{2x^2 + 12x}{(x + 1)(x + 3)}$

73. $\dfrac{4x}{5(x - 2)} + \dfrac{6x}{5(x - 2)}$ $\frac{2x}{x - 2}$

74. $\dfrac{11x}{3(x + 2)^2} + \dfrac{4x}{3(x + 2)^2}$ $\frac{5x}{(x + 2)^2}$

75. $\dfrac{x}{x^2 - 9} - \dfrac{3}{x^2 - 9}$ $\frac{1}{x + 3}$

76. $\dfrac{5x}{x + 3} - \dfrac{4x - 3}{x + 3}$ 1

6.4 COMPLEX FRACTIONS

To Succeed, Review How To ...

1. Find the LCD of two or more rational expressions (pp. 447–450).

2. Remove parentheses using the distributive property (pp. 51–53).

3. Add, subtract, multiply, and divide rational expressions (pp. 445–450, 436–439).

Objective

A Write a complex fraction as a simple fraction in reduced form.

GETTING STARTED

Rocking Along with Fractions

Suppose a disc jockey devotes $7\frac{1}{2}$ min each hour to commercials, leaving $60 - 7\frac{1}{2}$ min for music. If the songs she plays last an average of $3\frac{1}{4}$ min and it takes her about $\frac{1}{2}$ min to get a song going, how many songs can she play each hour? The answer is

$$\frac{60 - 7\dfrac{1}{2}}{3\dfrac{1}{4} + \dfrac{1}{2}} \quad \begin{matrix}\leftarrow \text{ Time allowed for music}\\[1em] \leftarrow \text{ Time it takes to play each song}\end{matrix}$$

but you have to know how to simplify it to find out how many songs she plays each hour.

A Complex Fractions

The fraction

$$\frac{60 - 7\dfrac{1}{2}}{3\dfrac{1}{4} + \dfrac{1}{2}} \quad \begin{matrix}\leftarrow \text{ Numerator fraction}\\ \leftarrow \text{ Main fraction bar}\\ \leftarrow \text{ Denominator fraction}\end{matrix}$$

contains other fractions in its numerator and denominator. A fraction whose numerator or denominator (or both) contains other fractions is called a **complex fraction.**

A fraction that is not complex is called a **simple fraction.** Thus,

$$\frac{\dfrac{1}{2}}{\dfrac{3}{4} + \dfrac{1}{5}}, \quad \frac{\dfrac{3x}{5} - \dfrac{1}{8}}{\dfrac{x}{7}}, \quad \frac{-\dfrac{1}{3}}{\dfrac{1}{9}}, \quad \text{and} \quad \frac{x}{\dfrac{7}{8}}$$

are all complex fractions, but $\frac{1}{7}$, $\frac{3}{5}$, and $\frac{x}{9}$ are simple fractions.

To simplify a complex fraction, it's necessary to recall that the main fraction bar indicates that the numerator of the fraction is to be divided by the denominator of the fraction. Thus

$$\frac{60 - 7\dfrac{1}{2}}{3\dfrac{1}{4} + \dfrac{1}{2}} \quad \text{means} \quad \left(60 - 7\frac{1}{2}\right) \div \left(3\frac{1}{4} + \frac{1}{2}\right)$$

Use the ÷ sign instead of the bar.

Teaching Tip

Have students discuss applications that might involve using complex fractions.

Example: Finding how many $1\frac{1}{4}$ yd pieces of fabric you can cut from a piece of fabric that is $5\frac{1}{2}$ yd long.

$$\frac{5\frac{1}{2}}{1\frac{1}{4}}$$

Here are the procedures we use to simplify complex fractions.

> **PROCEDURE**
>
> **Procedures for Simplifying Complex Fractions**
>
> **Method 1.** Multiply the numerator and denominator of the complex fraction by the LCD of all the simple fractions appearing; or
>
> **Method 2.** Perform the operations indicated in the numerator and denominator of the given complex fraction, and then divide the numerator by the denominator.

Web It

For practice recognizing and simplifying complex rational expressions by both methods, go to link 6-4-1 at mhhe.com/bello.

Now simplify

$$\frac{60 - 7\frac{1}{2}}{3\frac{1}{4} + \frac{1}{2}} = \frac{60 - \frac{15}{2}}{\frac{13}{4} + \frac{1}{2}}$$

using each of these methods.

Method 1. The LCD of $\frac{15}{2}$, $\frac{13}{4}$, and $\frac{1}{2}$ is 4, so we multiply by 1 in the form of $\frac{4}{4}$ to obtain

$$\frac{60 - \frac{15}{2}}{\frac{13}{4} + \frac{1}{2}} = \frac{4 \cdot \left(60 - \frac{15}{2}\right)}{4 \cdot \left(\frac{13}{4} + \frac{1}{2}\right)} \quad \text{Multiply numerator and denominator by 4, the LCD of } \frac{15}{2}, \frac{13}{4}, \text{ and } \frac{1}{2}.$$

$$= \frac{240 - 30}{13 + 2} \quad \begin{array}{l} \leftarrow 4\left(60 - \frac{15}{2}\right) = 4 \cdot 60 - \overset{2}{\cancel{4}} \cdot \frac{15}{\cancel{2}} = 240 - 30 \\ \leftarrow 4\left(\frac{13}{4} + \frac{1}{2}\right) = \cancel{4} \cdot \frac{13}{\cancel{4}} + \overset{2}{\cancel{4}} \cdot \frac{1}{\cancel{2}} = 13 + 2 \end{array}$$

$$= \frac{210}{15} \quad \begin{array}{l} \leftarrow 240 - 30 = 210 \\ \leftarrow 13 + 2 = 15 \end{array}$$

$$= 14 \quad \text{Divide.}$$

The disc jockey plays 14 songs each hour.

Method 2.

$$\frac{60 - \frac{15}{2}}{\frac{13}{4} + \frac{1}{2}} = \frac{\frac{120}{2} - \frac{15}{2}}{\frac{13}{4} + \frac{2}{4}} \quad \begin{array}{l} \leftarrow \text{Write 60 as } \frac{120}{2}, \text{ so we can subtract } \frac{15}{2}. \\ \\ \leftarrow \text{Write } \frac{1}{2} \text{ as } \frac{2}{4}, \text{ so we can add it to } \frac{13}{4}. \end{array}$$

$$= \frac{\frac{105}{2}}{\frac{15}{4}} \quad \begin{array}{l} \leftarrow \frac{120}{2} - \frac{15}{2} = \frac{105}{2} \\ \\ \leftarrow \frac{13}{4} + \frac{2}{4} = \frac{15}{4} \end{array}$$

$$= \frac{105}{2} \div \frac{15}{4} \quad \text{Replace the bar by the division sign, } \div.$$

$$= \frac{\overset{7}{\cancel{105}}}{\underset{1}{\cancel{2}}} \cdot \frac{\overset{2}{\cancel{4}}}{\underset{1}{\cancel{15}}} \quad \text{Multiply by the reciprocal of } \frac{15}{4}, \text{ which is } \frac{4}{15}, \text{ and simplify.}$$

$$= 14 \quad \text{As before}$$

EXAMPLE 1 **Simplifying complex fractions using method 1**

Use method 1 to write the following as a simple fraction in simplified form:

$$\frac{\dfrac{3}{a} - \dfrac{4}{b}}{\dfrac{1}{2a} + \dfrac{2}{3b}}$$

SOLUTION The LCD of

$$\frac{3}{a}, \quad \frac{4}{b}, \quad \frac{1}{2a}, \quad \text{and} \quad \frac{2}{3b}$$

is $6ab$. Therefore, we multiply the numerator and denominator of the given fraction by $6ab$ to obtain

$$\frac{6ab \cdot \left(\dfrac{3}{a} - \dfrac{4}{b}\right)}{6ab \cdot \left(\dfrac{1}{2a} + \dfrac{2}{3b}\right)} = \frac{6ab \cdot \dfrac{3}{a} - 6ab \cdot \dfrac{4}{b}}{\overset{3}{6ab} \cdot \dfrac{1}{2a} + \overset{2}{6ab} \cdot \dfrac{2}{3b}} \quad \text{Use the distributive property and simplify.}$$

$$= \frac{18b - 24a}{3b + 4a} \quad \begin{array}{l} \leftarrow 6b \cdot 3 = 18b \text{ and } 6a \cdot 4 = 24a \\ \leftarrow 3b \cdot 1 = 3b \text{ and } 2a \cdot 2 = 4a \end{array}$$

or

$$\frac{6(3b - 4a)}{3b + 4a}$$

EXAMPLE 2 **More practice using method 1**

Use method 1 to simplify:

$$\frac{x - \dfrac{1}{x^3}}{x + \dfrac{1}{x^2}}$$

SOLUTION Here the LCD is x^3. Thus

$$\frac{x - \dfrac{1}{x^3}}{x + \dfrac{1}{x^2}} = \frac{x^3 \cdot \left(x - \dfrac{1}{x^3}\right)}{x^3 \cdot \left(x + \dfrac{1}{x^2}\right)}$$

$$= \frac{x^3 \cdot x - \overset{1}{\cancel{x^3}} \cdot \dfrac{1}{x^3}}{x^3 \cdot x + \overset{x}{\cancel{x^3}} \cdot \dfrac{1}{x^2}}$$

$$= \frac{x^4 - 1}{x^4 + x} = \frac{(x^2 + 1)(x^2 - 1)}{x(x^3 + 1)} \quad \text{Completely factor the numerator and denominator.}$$

$$= \frac{(x^2 + 1)\cancel{(x + 1)}(x - 1)}{x\cancel{(x + 1)}(x^2 - x + 1)} \quad \text{Divide out } x + 1.$$

$$= \frac{(x^2 + 1)(x - 1)}{x(x^2 - x + 1)}$$

or

$$\frac{x^3 - x^2 + x - 1}{x^3 - x^2 + x}$$

PROBLEM 1

Simplify the complex rational expression using method 1:

$$\frac{\dfrac{2}{b} - \dfrac{3}{a}}{\dfrac{1}{2b} - \dfrac{3}{4a}}$$

PROBLEM 2

Simplify:

$$\frac{2x^2 + \dfrac{1}{4x}}{4 - \dfrac{1}{x^2}}$$

Answers

1. 4   **2.** $\frac{x(4x^2 - 2x + 1)}{4(2x - 1)}$ or $\frac{4x^3 - 2x^2 + x}{8x - 4}$

EXAMPLE 3   **Comparing method 1 with method 2**

Simplify:

$$\frac{\dfrac{x}{x-2}+x}{1+\dfrac{3}{x^2-4}}$$

a. Use method 1.

b. Use method 2.

SOLUTION

a. First write x^2-4 as $(x+2)(x-2)$ to obtain

$$\frac{\dfrac{x}{x-2}+x}{1+\dfrac{3}{x^2-4}}=\frac{\dfrac{x}{x-2}+x}{1+\dfrac{3}{(x+2)(x-2)}}$$

The LCD of the fractions is $(x+2)(x-2)$. Multiply numerator and denominator by this LCD.

$$\frac{\dfrac{(x+2)(x-2)}{1}\cdot\left(\dfrac{x}{x-2}+x\right)}{\dfrac{(x+2)(x-2)}{1}\cdot\left[1+\dfrac{3}{(x+2)(x-2)}\right]}$$

$$=\frac{x(x+2)+x(x+2)(x-2)}{(x+2)(x-2)+3}$$ Distribute the LCD.

$$=\frac{x^2+2x+x^3-4x}{x^2-4+3}$$ Remove parentheses.

$$=\frac{x^3+x^2-2x}{x^2-1}$$ Collect like terms.

$$=\frac{x(x^2+x-2)}{(x+1)(x-1)}$$ Factor out x in numerator. Factor $x^2-1=(x+1)(x-1)$ in denominator.

$$=\frac{x(x-1)(x+2)}{(x+1)(x-1)}$$ Factor $x^2+x-2=(x-1)(x+2)$. Divide out $(x-1)$.

$$=\frac{x(x+2)}{x+1}\text{ or }\frac{x^2+2x}{x+1}$$

NOTE

$\dfrac{x^2+2x}{x+1}$ is also an acceptable answer.

Web It

For more examples on simplifying complex rational expressions, go to link 6-4-2 at mhhe.com/bello.

PROBLEM 3

Simplify:

$$\frac{\dfrac{x}{x+3}+x}{1-\dfrac{7}{x^2-9}}$$

Calculate It

We do Example 3 in the *Calculate It* on page 461.

Answer

3. $\frac{x(x-3)}{x-4}$ or $\frac{x^2-3x}{x-4}$

b. Using method 2, first perform the operations in the numerator and denominator, and then divide.

$$\frac{\dfrac{x}{x-2}+x}{1+\dfrac{3}{x^2-4}} = \frac{\dfrac{x}{x-2}+\dfrac{x(x-2)}{x-2}}{\dfrac{1(x^2-4)}{x^2-4}+\dfrac{3}{x^2-4}} \quad \longleftarrow \text{Rewrite } x \text{ as } \tfrac{x(x-2)}{x-2}.$$
$$\longleftarrow \text{Rewrite 1 as } \tfrac{1(x^2-4)}{x^2-4}.$$

$$= \frac{\dfrac{x+x(x-2)}{x-2}}{\dfrac{1(x^2-4)+3}{x^2-4}} \qquad \text{Add.}$$

$$= \frac{\dfrac{x+x^2-2x}{x-2}}{\dfrac{x^2-4+3}{x^2-4}} \qquad \text{Remove parentheses.}$$

$$= \frac{\dfrac{x^2-x}{x-2}}{\dfrac{x^2-1}{x^2-4}} \qquad \text{Simplify.}$$

$$= \frac{x^2-x}{x-2} \div \frac{x^2-1}{x^2-4} \qquad \begin{array}{l}\text{Use the division sign, } \div, \\ \text{instead of the bar.}\end{array}$$

$$= \frac{x(x-1)}{x-2} \cdot \frac{(x+2)(x-2)}{(x+1)(x-1)} \qquad \begin{array}{l}\text{Multiply by the reciprocal} \\ \text{of } \tfrac{x^2-1}{x^2-4} \text{ and factor } x^2-x, \\ x^2-4, \text{ and } x^2-1.\end{array}$$

$$= \frac{x\overset{1}{\cancel{(x-1)}}(x+2) \cdot \overset{}{\cancel{(x-2)}}}{\underset{1}{\cancel{(x-2)}}(x+1)\underset{1}{\cancel{(x-1)}}} \qquad \begin{array}{l}\text{Multiply. Divide out } x-2 \\ \text{and } x-1.\end{array}$$

$$= \frac{x(x+2)}{x+1}$$

or

$$\frac{x^2+2x}{x+1}$$

In comparing method 1 with method 2, the results are the same. Which is the easiest method? That will depend on the problem and the knowledge you bring to the problem.

EXAMPLE 4 **Simplifying a complex fraction using method 2**

Simplify:

$$1+\frac{a}{1+\dfrac{1}{1+a}}$$

SOLUTION Start by working on the denominator of the fraction

$$\frac{a}{1+\dfrac{1}{1+a}}$$

PROBLEM 4

Simplify:

$$2+\frac{a}{2+\dfrac{2}{2+a}}$$

Answer

4. $\dfrac{a^2+6a+12}{2a+6}$ or $\dfrac{a^2+6a+12}{2(a+3)}$

Since we have to add 1 to

$$\frac{1}{1 + a}$$

we rewrite 1 as

$$\frac{1 + a}{1 + a}$$

$$1 + \cfrac{a}{\boxed{1 + \cfrac{1}{1 + a}}} = 1 + \cfrac{a}{\boxed{\dfrac{1 + a}{1 + a} + \dfrac{1}{1 + a}}}$$

Work on the denominator.

$$= 1 + \cfrac{a}{\boxed{\dfrac{2 + a}{1 + a}}}$$

$$\frac{1 + a}{1 + a} + \frac{1}{1 + a} = \frac{1 + a + 1}{1 + a}$$
$$= \frac{2 + a}{1 + a}$$

$$= 1 + a \div \boxed{\dfrac{2 + a}{1 + a}}$$

Use the ÷ sign instead of the bar.

$$= 1 + a \cdot \frac{1 + a}{2 + a}$$

To divide a by $\frac{2+a}{1+a}$, multiply by the reciprocal $\frac{1+a}{2+a}$.

$$= 1 + \frac{a(1 + a)}{2 + a}$$

$$= \frac{2 + a}{2 + a} + \frac{a(1 + a)}{2 + a}$$

Write 1 as $\frac{2+a}{2+a}$.

$$= \frac{2 + a + a + a^2}{2 + a}$$

Add the numerator and keep the denominator.

$$= \frac{2 + 2a + a^2}{2 + a}$$

$$= \frac{a^2 + 2a + 2}{a + 2}$$

Write the numerator and denominator in descending order.

Sometimes complex fractions are written using negative exponents. To simplify such expressions, begin by rewriting them using the definition of a negative exponent, that is,

$$a^{-n} = \frac{1}{a^n}$$

Using this definition,

$$(x - 1)^{-1} = \frac{1}{x - 1}, \quad (x + 2)^{-3} = \frac{1}{(x + 2)^3}, \quad \text{and} \quad x(x + y)^{-5} = \frac{x}{(x + y)^5}$$

Notice the x in $x(x + y)^{-5}$ does not have a negative exponent. It remains as x in the numerator.

We shall use these ideas in Example 5. Pay close attention to what we do in each step of the solution.

EXAMPLE 5 **Simplifying a complex fraction involving negative exponents**

Simplify: $\dfrac{x(x - 3)^{-1} + x}{x(3 - x)^{-1} - x}$

PROBLEM 5

Simplify:

$$\frac{x + (x - 2)^{-1}}{1 - x(2 - x)^{-1}}$$

Answer

5. $\frac{x - 1}{2}$

SOLUTION

$$\frac{x(x-3)^{-1}+x}{x(3-x)^{-1}-x} = \frac{\dfrac{x}{x-3}+\dfrac{x}{1}}{\dfrac{x}{3-x}-\dfrac{x}{1}}$$

Rewrite $x(x-3)^{-1}$ as $\frac{x}{x-3}$, $x(3-x)^{-1}$ as $\frac{x}{3-x}$, and x as $\frac{x}{1}$.

$$= \frac{\dfrac{x}{x-3}+\dfrac{x}{1}}{\dfrac{x}{-(x-3)}-\dfrac{x}{1}}$$

$3-x$ and $x-3$ are opposites, that is, $3-x = -(x-3)$, so we substitute $-(x-3)$ for $3-x$.

$$= \frac{(x-3)\left[\dfrac{x}{x-3}+\dfrac{x}{1}\right]}{(x-3)\left[\dfrac{x}{-(x-3)}-\dfrac{x}{1}\right]}$$

The LCD of all the denominators is $x-3$. Multiply the numerator and denominator by $(x-3)$.

$$= \frac{\dfrac{(x-3)x}{(x-3)}+\dfrac{(x-3)x}{1}}{\dfrac{(x-3)x}{-(x-3)}-\dfrac{(x-3)x}{1}}$$

Use the distributive property to distribute the $(x-3)$.

$$= \frac{x+(x-3)x}{-x-(x-3)x}$$

Divide out the $(x-3)$ terms and note that $\frac{(x-3)x}{1} = (x-3)x$.

$$= \frac{x+x^2-3x}{-x-(x^2-3x)}$$

Multiply $(x-3)x$ in the numerator and denominator.

$$= \frac{x+x^2-3x}{-x-x^2+3x}$$

Use the distributive property, $-(x^2-3x) = -x^2+3x$.

$$= \frac{x^2-2x}{-x^2+2x}$$

Collect like terms in the numerator and denominator.

$$= \frac{x^2-2x}{-(x^2-2x)}$$

Rewrite the denominator $-x^2+2x$ as $-(x^2-2x)$.

$$= -1$$

Since $\frac{a}{-a} = -1$, the answer is -1.

Teaching Tip

Remember to identify the correct base with the negative exponent.

Examples:

$$2x^{-3} = \frac{2}{x^3}$$

$$(2x)^{-3} = \frac{1}{(2x)^3}$$

$$x(x-3)^{-1} = \frac{x}{x-3}$$

Web It

For a tutorial and more examples on simplifying complex rational expressions, go to link 6-4-3 at mhhe.com/bello.

NOTE

In step 2, the numerator is

$$\frac{x}{x-3}+\frac{x}{1}$$

and the denominator is

$$\frac{x}{-(x-3)}-\frac{x}{1} = -\left(\frac{x}{x-3}+\frac{x}{1}\right)$$

which is the additive inverse of the numerator. Thus

$$\frac{\dfrac{x}{x-3}+\dfrac{x}{1}}{\dfrac{x}{-(x-3)}-\dfrac{x}{1}} = \frac{\dfrac{x}{x-3}+\dfrac{x}{1}}{-\left(\dfrac{x}{x-3}+\dfrac{x}{1}\right)} = -1$$

These two expressions are additive inverses of each other.

Calculate It Exercises   **Verifying Results to Simplifying Complex Fractions**

In the last few sections we have been verifying the results of multiplying, dividing, adding, and subtracting polynomials by making sure that the graph of the original problem and the graph of the final answer are identical. This can also be done with complex fractions. However, we want to explore the numerical capabilities of your calculator. In Example 3, we found that

```
                                              4
( X / ( X - 2 ) + X ) / ( 1 + 3
/ ( X ^ 2 - 4 ) )
                                            4.8
X ( X + 2 ) / ( X + 1 )

                                            4.8
```

$$\frac{\left(\dfrac{x}{(x-2)}+x\right)}{\left(1+\dfrac{3}{(x^2-4)}\right)}=\frac{x(x+2)}{(x+1)}$$

To check the answer, substitute any convenient number for x and see if both sides are equal. You cannot choose numbers that will give you a zero denominator (2 and -2 will yield zero denominators). A simple number to use is $x=4$. Now, go to the home screen and store the value 4 as x. With a TI-83 Plus press **4** **STO▶** **X,T,θ,n** **ENTER** . The calculator confirms the entry by showing $4{\to}X$ and the answer 4. Now enter the original complex

fraction as

$$\left(\frac{x}{(x-2)}+x\right)\div\left(1+\frac{3}{(x^2-4)}\right)$$

Be extremely careful with the parentheses. Press **ENTER** . The answer is 4.8, as shown in the window. Next, enter the simplified complex fraction as $x(x+2)/(x+1)$. Press **ENTER** again. You should get the same answer as before, 4.8. The final confirmation is shown on the screen. You can check the rest of the problems involving one variable using this method. If you are checking Example 4, you can enter an x instead of an a when writing the complex fraction involved, but there are so many parentheses involved that it's probably easier to do the problem without the calculator. Try it if you don't believe this!

1. Verify the results of Example 2 numerically using the techniques shown.

2. Verify the results of Example 5 numerically using the techniques shown.

3. Can you verify the results of Example 1
   a. graphically? Explain.
   b. numerically? Explain.

Exercises 6.4

A In Problems 1–42, perform the indicated operation and give the answer in simplified form.

1. $\dfrac{50-5\frac{1}{2}}{7\frac{3}{4}+\frac{1}{2}}$ $\dfrac{178}{33}$

2. $\dfrac{70-17\frac{1}{2}}{2\frac{1}{4}+1\frac{1}{2}}$ 14

3. $\dfrac{\frac{a}{b}}{\frac{c}{b}}$ $\dfrac{a}{c}$

4. $\dfrac{\frac{-a^2}{c}}{\frac{-b^2}{c}}$ $\dfrac{a^2}{b^2}$

5. $\dfrac{\frac{x}{y}}{\frac{x^2}{z}}$ $\dfrac{z}{xy}$

6. $\dfrac{\frac{x^2}{y^2}}{\frac{x}{z}}$ $\dfrac{xz}{y^2}$

7. $\dfrac{\frac{3x}{5y}}{\frac{3x}{2z}}$ $\dfrac{2z}{5y}$

8. $\dfrac{\frac{7x}{3y}}{\frac{14x}{5y}}$ $\dfrac{5}{6}$

9. $\dfrac{\frac{1}{2}}{2-\frac{1}{2}}$ $\dfrac{1}{3}$

10. $\dfrac{\frac{1}{4}}{3-\frac{1}{4}}$ $\dfrac{1}{11}$

11. $\dfrac{a-\frac{a}{b}}{1+\frac{a}{b}}$ $\dfrac{ab-a}{b+a}$

12. $\dfrac{1-\frac{1}{a}}{1+\frac{1}{a}}$ $\dfrac{a-1}{a+1}$

13. $\dfrac{\frac{1}{x}+2}{2-\frac{1}{x}}$ $\dfrac{1+2x}{2x-1}$

14. $\dfrac{3-\frac{2}{y}}{\frac{1}{y}+4}$ $\dfrac{3y-2}{1+4y}$

15. $\dfrac{\frac{2}{3}+x}{x-\frac{1}{2}}$ $\dfrac{4+6x}{6x-3}$

16. $\dfrac{x+\frac{1}{3}}{\frac{3}{4}-x}$ $\dfrac{12x+4}{9-12x}$

17. $\dfrac{y+\frac{2}{x}}{y^2-\frac{4}{x^2}}$ $\dfrac{x}{xy-2}$

18. $\dfrac{y-\frac{3}{x}}{y^2-\frac{9}{x^2}}$ $\dfrac{x}{xy+3}$

19. $\dfrac{\frac{x}{y^2}-\frac{y}{x^2}}{x^2+xy+y^2}$ $\dfrac{x-y}{x^2y^2}$

20. $\dfrac{\frac{x}{y^2}+\frac{y}{x^2}}{x^2-xy+y^2}$ $\dfrac{x+y}{x^2y^2}$

21. $3-\dfrac{3}{3-\frac{1}{2}}$ $\dfrac{9}{5}$

22. $2-\dfrac{2}{2-\frac{1}{2}}$ $\dfrac{2}{3}$

23. $a-\dfrac{a}{a+\frac{1}{2}}$ $\dfrac{2a^2-a}{2a+1}$

24. $a + \cfrac{a}{a + \cfrac{1}{2}}$ $\dfrac{2a^2 + 3a}{2a + 1}$

25. $x - \cfrac{x}{1 - \cfrac{x}{1 - x}}$ $\dfrac{x^2}{2x - 1}$

26. $2x - \cfrac{x}{2 - \cfrac{x}{2 - x}}$ $\dfrac{6x - 5x^2}{4 - 3x}$

27. $\dfrac{\dfrac{x-1}{x+1} + \dfrac{x+1}{x-1}}{\dfrac{x-1}{x+1} - \dfrac{x+1}{x-1}}$ $\dfrac{-(x^2 + 1)}{2x}$

28. $\dfrac{\dfrac{x-1}{x+1} - \dfrac{x+1}{x-1}}{\dfrac{x-1}{x+1} + \dfrac{x+1}{x-1}}$ $\dfrac{-2x}{x^2 + 1}$

29. $\dfrac{(x-y)^{-1} + (x+y)^{-1}}{(x-y)^{-1} - (x+y)^{-1}}$ $\dfrac{x}{y}$

30. $\dfrac{a(a+b)^{-1} + 1}{1 - b(a+b)^{-1}}$ $\dfrac{2a + b}{a}$

31. $\dfrac{x(x-2)^{-1} - x}{x(2-x)^{-1} + x} - 1$

32. $\dfrac{x + x(4-x)^{-1}}{x(x-4)^{-1}}$ $x - 5$

33. $\dfrac{\dfrac{1}{x^2} + \dfrac{3}{x} - 4}{\dfrac{1}{x^2} + \dfrac{5}{x} + 4}$ $\dfrac{1 - x}{1 + x}$

34. $\dfrac{\dfrac{6}{v^2} - \dfrac{11}{v} - 10}{\dfrac{2}{v^2} + \dfrac{1}{v} - 15}$ $\dfrac{3 + 2v}{1 + 3v}$

35. $\dfrac{y + 3 - \dfrac{16}{y+3}}{y - 6 + \dfrac{20}{y+3}}$ $\dfrac{y + 7}{y - 2}$

36. $\dfrac{w + 2 - \dfrac{18}{w-5}}{w - 1 - \dfrac{12}{w-5}}$ $\dfrac{w + 4}{w + 1}$

37. $\dfrac{\dfrac{8x}{3x+1} - \dfrac{3x-1}{x}}{\dfrac{x}{3x+1} - \dfrac{2x-2}{x}}$ $\dfrac{x^2 - 1}{5x^2 - 4x - 2}$

38. $\dfrac{\dfrac{3}{m-4} - \dfrac{16}{m-3}}{\dfrac{2}{m-3} - \dfrac{15}{m+5}}$ $\dfrac{m + 5}{m - 4}$

39. $\dfrac{\dfrac{c}{d} - \dfrac{d}{c}}{\dfrac{c}{d} - 2 + \dfrac{d}{c}}$ $\dfrac{c + d}{c - d}$

40. $\dfrac{\dfrac{a^2}{b^2} + 4 + \dfrac{4b^2}{a^2}}{\dfrac{a}{b} + \dfrac{2b}{a}}$ $\dfrac{a^2 + 2b^2}{ab}$

41. $\dfrac{\dfrac{a^2 - b^2}{a^2 + b^2} - \dfrac{a^2 + b^2}{a^2 - b^2}}{\dfrac{a - b}{a + b} - \dfrac{a + b}{a - b}}$ $\dfrac{ab}{a^2 + b^2}$

42. $\dfrac{1 + \dfrac{4uv}{(u - v)^2}}{1 + \dfrac{uv - 3v^2}{(u - v)^2}}$ $\dfrac{u + v}{u - 2v}$

APPLICATIONS

Simplifying Expressions

43. When connected in parallel, the combined resistance R of two electrical resistances R_1 and R_2 is given by

$$R = \cfrac{1}{\dfrac{1}{R_1} + \dfrac{1}{R_2}}$$

Simplify the expression for R. $R = \dfrac{R_1 R_2}{R_2 + R_1}$

44. When connected in parallel, the combined resistance R of three electrical resistances R_1, R_2, and R_3 is given by

$$R = \cfrac{1}{\dfrac{1}{R_1} + \dfrac{1}{R_2} + \dfrac{1}{R_3}}$$

Simplify the expression for R.

$$R = \dfrac{R_1 R_2 R_3}{R_2 R_3 + R_1 R_3 + R_1 R_2}$$

45. The formula for the Doppler effect in light is

$$f = f_{static} \sqrt{\cfrac{1 + \dfrac{v}{c}}{1 - \dfrac{v}{c}}}$$

where f and f_{static} are frequencies, v is the velocity of the moving body, and c is the speed of light.

Simplify the expression for f. $f = f_{static} \sqrt{\dfrac{c + v}{c - v}}$

46. Balmer's formula for the wavelength λ (lambda) of the hydrogen spectrum light is given by

$$\lambda = \cfrac{1}{\dfrac{1}{m^2} - \dfrac{1}{n^2}}$$

Simplify the expression for λ. $\lambda = \dfrac{m^2 n^2}{n^2 - m^2}$

SKILL CHECKER

Try the "Skill Checker" exercises so you'll be ready for the next section.

Simplify:

47. $\dfrac{8x^4}{2x^3}$ $4x$ **48.** $\dfrac{28x^7}{7x^5}$ $4x^2$ **49.** $\dfrac{-30x^6}{6x^4}$ $-5x^2$ **50.** $\dfrac{-10x^2}{20x^4}$ $\dfrac{-1}{2x^2}$ **51.** $\dfrac{-30x}{10x^3}$ $\dfrac{-3}{x^2}$ **52.** $\dfrac{-50}{10x^3}$ $\dfrac{-5}{x^3}$

Multiply:

53. $x^2(5x + 5)$ $5x^3 + 5x^2$ **54.** $6x^2(x - 2)$ $6x^3 - 12x^2$ **55.** $3x^4(3x - 5)$ $9x^5 - 15x^4$

Factor completely:

56. $6x^2 + x - 2$ $(3x + 2)(2x - 1)$ **57.** $6x^2 + 7x - 3$ $(3x - 1)(2x + 3)$ **58.** $20x^2 - 7x - 6$ $(5x + 2)(4x - 3)$

59. Subtract $6x^3 + 24x^2$ from $6x^3 + 25x^2 + 2x - 8$
    $x^2 + 2x - 8$

60. Subtract $5x^2 + 15x$ from $5x^2 + 9x - 18$
    $-6x - 18$

USING YOUR KNOWLEDGE

Interest Rates and Planetary Orbits

Do you have monthly payments on any type of loan? Do you know what your **A**nnual **P**ercentage **R**ate (**APR**) is? If you financed P dollars to be paid in N monthly payments of M dollars, your APR is

$$\dfrac{\dfrac{24(NM - P)}{N}}{P + \dfrac{NM}{12}}$$

61. Simplify the APR formula. $\text{APR} = \dfrac{288(NM - P)}{N(12P + NM)}$

62. Use the simplified version of the APR formula to determine the APR on a $20,000 loan with payments of:

   a. $500 a month for 4 years. 9.09%

   b. $400 a month for 5 years. 7.27%

In the seventeenth century, the Dutch mathematician and astronomer Christian Huygens made a model of the solar system and found out that Saturn takes

$$29 + \dfrac{1}{2 + \dfrac{2}{9}}$$

years to go around the Sun. Now,

$$\dfrac{1}{2 + \dfrac{2}{9}} = \dfrac{9 \cdot 1}{9 \cdot \left(2 + \dfrac{2}{9}\right)}$$

$$= \dfrac{9}{18 + 2}$$

$$= \dfrac{9}{20}$$

Thus it takes Saturn $29 + \dfrac{9}{20} = 29\dfrac{9}{20}$ yr to go around the Sun. Use your knowledge to find the number of years it takes each of the following planets to go around the Sun by simplifying the fraction. (Write your answers as mixed numbers when necessary.)

63. Mercury: $\dfrac{1}{4+\dfrac{1}{6}}$ yr $\dfrac{6}{25}$ yr

64. Venus: $\dfrac{1}{1+\dfrac{2}{3}}$ yr $\dfrac{3}{5}$ yr

65. Jupiter: $11+\dfrac{1}{1+\dfrac{7}{43}}$ yr $11\dfrac{43}{50}$ yr

66. Mars: $1+\dfrac{1}{1+\dfrac{3}{22}}$ yr $1\dfrac{22}{25}$ yr

WRITE ON

67. Write in your own words the definition of a complex fraction. Answers may vary.

68. We have given two methods to simplify a complex fraction. Which method do you think is simpler and why? Answers may vary.

69. Can you explain when one should use method 1 to simplify a complex fraction and what types of fractions are easier to simplify using this method? Answers may vary.

70. Can you explain when one should use method 2 to simplify a complex fraction and what types of fractions are easier to simplify using this method? Answers may vary.

MASTERY TEST

If you know how to do these problems, you have learned your lesson!

Simplify:

71. $2+\dfrac{a}{2+\dfrac{2}{2+a}}$ $\dfrac{a^2+6a+12}{2(a+3)}$

72. $3+\dfrac{a}{3+\dfrac{3}{3+a}}$ $\dfrac{(a+6)^2}{3(a+4)}$

73. $\dfrac{\dfrac{x}{x+2}+x}{1-\dfrac{5}{x^2-4}}$ $\dfrac{x(x-2)}{x-3}$

74. $\dfrac{x+\dfrac{x}{x+3}}{\dfrac{1}{x^2-9}+1}$ $\dfrac{x(x-3)(x+4)}{x^2-8}$

75. $\dfrac{\dfrac{2}{b}-\dfrac{3}{a}}{\dfrac{1}{2b}+\dfrac{3}{4a}}$ $\dfrac{4(2a-3b)}{2a+3b}$

76. $\dfrac{\dfrac{3}{b}+\dfrac{2}{a}}{\dfrac{1}{2a}-\dfrac{3}{4b}}$ $\dfrac{4(2b+3a)}{2b-3a}$

77. $\dfrac{x-\dfrac{1}{x^3}}{x-\dfrac{1}{x^2}}$ $\dfrac{(x+1)(x^2+1)}{x(x^2+x+1)}$

78. $\dfrac{\dfrac{1}{x^2}-x}{\dfrac{1}{x^3}-x}$ $\dfrac{x(x^2+x+1)}{(x+1)(x^2+1)}$

79. $\dfrac{x(x-4)^{-1}+x}{x(4-x)^{-1}-x}$ -1

80. $\dfrac{y-y(y-4)^{-1}}{y+y(4-y)^{-1}}$ 1

6.5   DIVISION OF POLYNOMIALS AND SYNTHETIC DIVISION

To Succeed, Review How To . . .

1. Simplify quotients using properties of exponents (pp. 39–40).

2. Multiply polynomials (pp. 356–362).

3. Evaluate expressions using the order of operations (p. 49).

Objectives

A Divide a polynomial by a monomial.

B Use long division to divide one polynomial by another.

C Completely factor a polynomial when one of the factors is known.

D Use synthetic division to divide one polynomial by a binomial.

E Use the remainder theorem to verify that a number is a solution of a given equation.

GETTING STARTED

Efficiency Quotients

How efficient is your car engine? The efficiency E of an engine is given by

$$E = \frac{Q_1 - Q_2}{Q_1}$$

where Q_1 is the horsepower rating of the engine and Q_2 is the horsepower delivered to the transmission. Can you do the indicated division in the rational expression

$$\frac{Q_1 - Q_2}{Q_1}$$

Follow the steps in the procedure.

$$\frac{Q_1 - Q_2}{Q_1} = (Q_1 - Q_2)\left(\frac{1}{Q_1}\right) \qquad \text{Dividing by } Q_1 \text{ is the same as multiplying by the reciprocal of } Q_1, \text{ that is, } \frac{1}{Q_1}.$$

$$= Q_1\left(\frac{1}{Q_1}\right) - Q_2\left(\frac{1}{Q_1}\right) \qquad \text{Use the distributive property.}$$

$$= \frac{Q_1}{Q_1} - \frac{Q_2}{Q_1} \qquad \text{Multiply.}$$

$$= 1 - \frac{Q_2}{Q_1} \qquad \frac{Q_1}{Q_1} = 1$$

Next, how to divide a polynomial by a monomial.

A   Dividing a Polynomial by a Monomial

To divide the trinomial $4x^4 - 8x^3 + 12x^2$ by $2x^2$, we proceed as in the *Getting Started*.

$$\frac{4x^4 - 8x^3 + 12x^2}{2x^2} = (4x^4 - 8x^3 + 12x^2)\left(\frac{1}{2x^2}\right)$$

$$= 4x^4\left(\frac{1}{2x^2}\right) - 8x^3\left(\frac{1}{2x^2}\right) + 12x^2\left(\frac{1}{2x^2}\right)$$

$$= \frac{4x^4}{2x^2} - \frac{8x^3}{2x^2} + \frac{12x^2}{2x^2}$$

$$= 2x^2 - 4x + 6$$

Teaching Tip

This method is used when dividing by a monomial. Remember that a monomial is one term, and terms are separated by plus or minus signs.

Or, more simply $\dfrac{4x^4 - 8x^3 + 12x^2}{2x^2} = \dfrac{4x^4}{2x^2} - \dfrac{8x^3}{2x^2} + \dfrac{12x^2}{2x^2} = 2x^2 - 4x + 6$

Web It

For some more examples of dividing a polynomial by a monomial, go to link 6-5-1 at mhhe.com/bello.

To avoid doing all these steps, we can state the following rule.

PROCEDURE

Rule for Dividing a Polynomial by a Monomial

To divide a polynomial by a monomial, divide each term in the polynomial by the monomial.

EXAMPLE 1   Dividing a polynomial by a monomial

Divide:

a. $\dfrac{28x^5 - 14x^4 + 7x^3}{7x^2}$

b. $\dfrac{20x^4 - 15x^3 + 10x^2 - 30x + 50}{10x^3}$

SOLUTION

a. $\dfrac{28x^5 - 14x^4 + 7x^3}{7x^2} = \dfrac{28x^5}{7x^2} - \dfrac{14x^4}{7x^2} + \dfrac{7x^3}{7x^2}$

$= 4x^3 - 2x^2 + x$

b. $\dfrac{20x^4 - 15x^3 + 10x^2 - 30x + 50}{10x^3} = \dfrac{20x^4}{10x^3} - \dfrac{15x^3}{10x^3} + \dfrac{10x^2}{10x^3} - \dfrac{30x}{10x^3} + \dfrac{50}{10x^3}$

$= 2x - \dfrac{3}{2} + \dfrac{1}{x} - \dfrac{3}{x^2} + \dfrac{5}{x^3}$

In this case, the answer is *not* a polynomial.

PROBLEM 1

Divide:

a. $\dfrac{24x^7 - 18x^5 - 12x^3}{-6x^3}$

b. $\dfrac{16x^4 - 4x^3 + 8x^2 - 16x + 40}{8x^2}$

B   **Dividing One Polynomial by Another Polynomial**

Teaching Tip

This method is used when the divisor is other than a monomial.

If we wish to divide a polynomial (called the **dividend**) by another polynomial (called the **divisor**), we proceed very much as we did in long division in arithmetic. To show you that this is so, we shall perform the division of 337 by 16 and $(x^3 + 3x^2 + 3x + 1)$ by $(x^2 + x + 1)$ side by side.

Think $\dfrac{x^3}{x^2} = x$.   ←——1st term inside
←——1st term outside

1. $16\overline{)337}^{\,2}$ Divide 33 by 16. It goes twice. Write 2 over the 33.

$(x^2) + x + 1\overline{)(x^3) + 3x^2 + 3x + 1}^{\quad x}$ Write x over the 3x.

2. $16\overline{)\,337}^{\;2}$
$\underline{-32}$
1

Multiply 16 by 2 and subtract the product 32 from 33, obtaining 1.

$x^2 + x + 1\overline{)\,x^3 + 3x^2 + 3x + 1}^{\quad x}$
$\underline{-(x^3 + x^2 + x)}$
$0 + 2x^2 + 2x$

Multiply $x(x^2 + x + 1)$. Subtract $(x^3 + x^2 + x)$ from $x^3 + 3x^2 + 3x + 1$. You can omit the zero.

Think $\dfrac{2x^2}{x^2} = 2$.   ←——1st term inside
←——1st term outside

3. $16\overline{)\,337}^{\;21}$
$\underline{-32}$
17

"Bring down" the 7. Now, divide 17 by 16. It goes once. Write 1 after the 2.

$(x^2) + x + 1\overline{)\,x^3 + 3x^2 + 3x + 1}^{\quad x + 2}$
$\underline{-(x^3 + x^2 + x)}$
$0 \;(+\; 2x^2) + 2x + 1$

"Bring down" the 1. Write + 2 after the x.

Answers

1. a. $-4x^4 + 3x^2 + 2$
b. $2x^2 - \dfrac{x}{2} + 1 - \dfrac{2}{x} + \dfrac{5}{x^2}$

$$\begin{array}{r} 21 \\ 16\overline{)\;337} \\ -32 \\ \hline 17 \\ -16 \\ \hline 1 \end{array}$$

4. Multiply 16 by 1 and subtract the result from 17. The remainder is 1.

$$\begin{array}{r} x + 2 \\ x^2 + x + 1\overline{)\;x^3 + 3x^2 + 3x + 1} \\ -(x^3 + x^2 + x) \\ \hline 0 + 2x^2 + 2x + 1 \\ -(2x^2 + 2x + 2) \\ \hline -1 \end{array}$$

Multiply $2(x^2 + x + 1)$. Subtract $(2x^2 + 2x + 2)$ from $2x^2 + 2x + 1$. The remainder is -1.

5. The answer (**quotient**) can be written as 21 R 1 (read "21 remainder 1") or as $21 + \frac{1}{16}$, which is $21\frac{1}{16}$.

The answer (**quotient**) can be written as $(x + 2)$ R -1 (read "$x + 2$ remainder -1"), or you can write the result more completely as

$$\underbrace{\frac{\overbrace{x^3 + 3x^2 + 3x + 1}^{\text{Dividend}}}{\underbrace{x^2 + x + 1}_{\text{Divisor}}}}_{} = \overbrace{x + 2}^{\text{Quotient}} - \overbrace{\frac{1}{\underbrace{x^2 + x + 1}_{\text{Divisor}}}}^{\text{Remainder}}$$

6. You can check this answer by multiplying 21 by 16 (336) and adding the remainder 1 to obtain 337, the dividend.

You can check the answer by multiplying $(x + 2)(x^2 + x + 1) = x^3 + 3x^2 + 3x + 2$ and adding the remainder -1 to get the dividend $x^3 + 3x^2 + 3x + 1$.

EXAMPLE 2	**Dividing polynomials using long division**

Divide: $(x^3 + 2x^2 - 17x) \div (x^2 + x - 3)$

SOLUTION

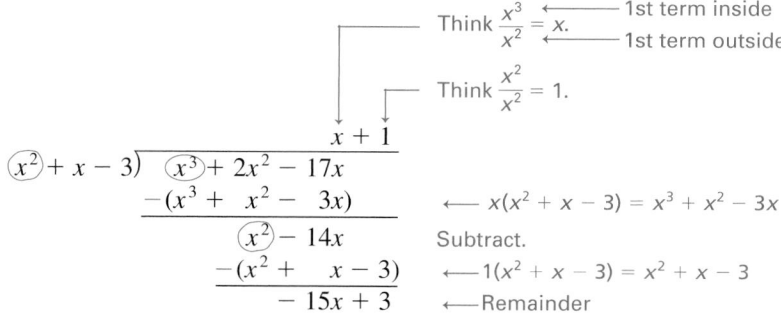

Think $\dfrac{x^3}{x^2} = x.$ ← 1st term inside / ← 1st term outside

Think $\dfrac{x^2}{x^2} = 1.$

$$\begin{array}{r} x + 1 \\ x^2 + x - 3\overline{)\;x^3 + 2x^2 - 17x} \\ -(x^3 + x^2 - 3x) \\ \hline x^2 - 14x \\ -(x^2 + x - 3) \\ \hline -15x + 3 \end{array}$$

← $x(x^2 + x - 3) = x^3 + x^2 - 3x$

Subtract.

← $1(x^2 + x - 3) = x^2 + x - 3$

← Remainder

Thus

$$\frac{x^3 + 2x^2 - 17x}{x^2 + x - 3} = x + 1 + \frac{-15x + 3}{x^2 + x - 3}$$

PROBLEM 2

Divide:

$$(x^3 + 4x^2 - 15x) \div (x^2 - 2x + 3)$$

Teaching Tip

Use the "Daddy, Mother, Sister, Brother" method to remember these steps.

(Daddy) **D**ivide first term inside by first term outside.

(Mother) **M**ultiply that result times the divisor and place answers under like terms of the dividend.

(Sister) **S**ubtract, remembering to change *all* signs of the subtrahend.

(Brother) **B**ring down the remaining terms and start over.

If there are missing terms in the polynomial being divided, we insert zero coefficients, as shown in the next example.

Calculate It   Using Graphs to Verify Long Division with Polynomials

Division problems can be checked with your calculator. To check the result of Example 2, graph the original problem

$$Y_1 = \frac{(x^3 + 2x^2 - 17x)}{(x^2 + x - 3)}$$

and the answer

$$Y_2 = x + 1 + \frac{(-15x + 3)}{(x^2 + x - 3)}$$

using a standard window. (Note the parentheses when entering the expression in your calculator.) The same graph is obtained in both cases, as shown in the window.

Answer

2. $(x + 6)$ R$(-6x - 18)$

EXAMPLE 3 Using long division when there are missing terms

Divide: $(4x^3 - 4 - 8x) \div (4 + 4x)$

SOLUTION We write the polynomials in **descending** order, inserting $0x^2$ in the dividend.

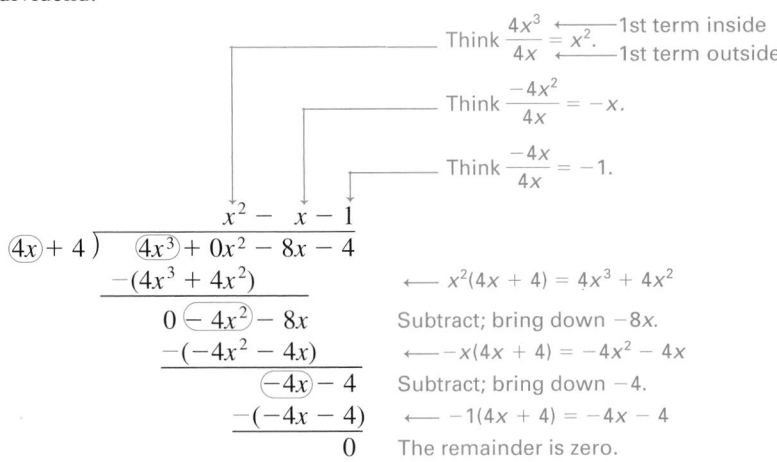

Think $\frac{4x^3}{4x} = x^2$. ←1st term inside
←1st term outside

Think $\frac{-4x^2}{4x} = -x$.

Think $\frac{-4x}{4x} = -1$.

$$
\begin{array}{r}
x^2 - x - 1 \\
(4x) + 4) \overline{\quad (4x^3) + 0x^2 - 8x - 4} \\
-(4x^3 + 4x^2) \\
\hline
0 \overline{-4x^2} - 8x \\
-(-4x^2 - 4x) \\
\hline
\overline{(-4x)} - 4 \\
-(-4x - 4) \\
\hline
0
\end{array}
$$

← $x^2(4x + 4) = 4x^3 + 4x^2$

Subtract; bring down $-8x$.

← $-x(4x + 4) = -4x^2 - 4x$

Subtract; bring down -4.

← $-1(4x + 4) = -4x - 4$

The remainder is zero.

NOTE

When the remainder is zero, the denominator divides *exactly* into the numerator. We say $(4x + 4)$ is a factor of $4x^3 - 8x - 4$.

Thus

$$\frac{4x^3 - 4 - 8x}{4 + 4x} = x^2 - x - 1$$

You can check this by multiplying $(4 + 4x)(x^2 - x - 1)$, obtaining $4x^3 - 4 - 8x$.

PROBLEM 3

Divide:

$$(6x^3 + 3x - 9) \div (3x - 3)$$

Teaching Tip

Remember that the denominator is said to be a *factor* of the numerator when the remainder is zero.

C Factoring When One of the Factors Is Known

Teaching Tip

Use a simple arithmetic example to demonstrate.

Example:

Factor 30 if we know that 2 is a factor.

$$\frac{30}{2} = P$$

$$\frac{30}{2} = 15$$

$$\frac{30}{2} = 3 \cdot 5$$

Thus, $30 = 2 \cdot 3 \cdot 5$

Answer

3. $2x^2 + 2x + 3$

Suppose we wish to factor the polynomial

$$6x^3 + 23x^2 + 9x - 18$$

None of the methods we have studied so far will work here (try them!), so we need some more information. If we know that $(x + 3)$ is one of the factors, we can write $6x^3 + 23x^2 + 9x - 18 = (x + 3)P$, where P is a polynomial. Dividing both sides by $x + 3$ gives

$$\frac{6x^3 + 23x^2 + 9x - 18}{x + 3} = P$$

$(x + 3)$ is a factor, so we should anticipate zero as the remainder when we divide (as in Example 3).

Web It

For more practice dividing polynomials, go to link 6-5-2 at mhhe.com/bello.

Let's do the division:

Think $\dfrac{6x^3}{x} = 6x^2$. ←1st term inside ←1st term outside

Think $\dfrac{5x^2}{x} = 5x$.

Think $\dfrac{-6x}{x} = -6$.

$$
\begin{array}{r}
6x^2 + 5x - 6 \\
x + 3 \overline{)\; 6x^3 + 23x^2 + 9x - 18} \\
-(6x^3 + 18x^2) \\
\hline
0 + 5x^2 + 9x \\
-(5x^2 + 15x) \\
\hline
0 - 6x - 18 \\
-(-6x - 18) \\
\hline
0
\end{array}
$$

← $6x^2(x + 3) = 6x^3 + 18x^2$
Subtract; bring down the $9x$.
← $5x(x + 3) = 5x^2 + 15x$
Subtract; bring down the -18.
← $-6(x + 3) = -6x - 18$
Subtract.

The polynomial P is $6x^2 + 5x - 6$, and we have

$$\frac{6x^3 + 23x^2 + 9x - 18}{x + 3} = 6x^2 + 5x - 6$$

Now $6x^2 + 5x - 6 = (3x - 2)(2x + 3)$, which gives

$$\frac{6x^3 + 23x^2 + 9x - 18}{x + 3} = (3x - 2)(2x + 3)$$

Multiplying both sides by $x + 3$ yields

$$6x^3 + 23x^2 + 9x - 18 = (x + 3)(3x - 2)(2x + 3)$$

If we wish to factor a polynomial and one of the factors is given, we can divide by this factor. The product of the quotient obtained (factored, if possible) and the given factor gives the complete factorization of the polynomial.

EXAMPLE 4   **Factoring when one factor is known**

Factor $6x^3 + 25x^2 + 2x - 8$ if $(x + 4)$ is one of its factors.

SOLUTION   We start by dividing $6x^3 + 25x^2 + 2x - 8$ by $x + 4$.

Think $\dfrac{6x^3}{x} = 6x^2$. ←1st term inside ←1st term outside

Think $\dfrac{x^2}{x} = x$.

Think $\dfrac{-2x}{x} = -2$.

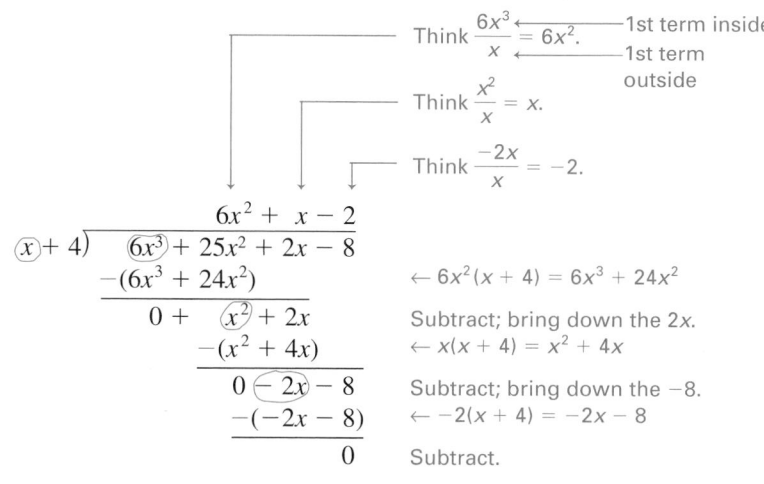

$$
\begin{array}{r}
6x^2 + x - 2 \\
x + 4 \overline{)\; 6x^3 + 25x^2 + 2x - 8} \\
-(6x^3 + 24x^2) \\
\hline
0 + x^2 + 2x \\
-(x^2 + 4x) \\
\hline
0 - 2x - 8 \\
-(-2x - 8) \\
\hline
0
\end{array}
$$

← $6x^2(x + 4) = 6x^3 + 24x^2$
Subtract; bring down the $2x$.
← $x(x + 4) = x^2 + 4x$
Subtract; bring down the -8.
← $-2(x + 4) = -2x - 8$
Subtract.

PROBLEM 4

Factor $2x^3 - x^2 - 18x + 9$ if $(x + 3)$ is one of its factors.

Calculate It

Using Graphs to Verify Factorization of Polynomials

Example 4 can be checked by graphing the original expression $Y_1 = 6x^3 + 25x^2 + 2x - 8$ and the factored answer $Y_2 = (x + 4)(3x + 2)(2x - 1)$.

Answer

4. $(x + 3)(2x - 1)(x - 3)$

Now we know that the polynomial we are trying to factor is the product of $(x + 4)$ and the quotient we obtained, $6x^2 + x - 2$. To factor completely, we have to factor $6x^2 + x - 2$ as $(3x + 2)(2x - 1)$.

$$6x^3 + 25x^2 + 2x - 8 = (x + 4)(6x^2 + x - 2)$$
$$= (x + 4)(3x + 2)(2x - 1)$$

D Using Synthetic Division to Divide a Polynomial by a Binomial

Web It

For another explanation of how to divide polynomials using synthetic division, go to link 6-5-3 at mhhe.com/bello.

The factorization process we have discussed can be made faster and more efficient if we can find a simpler method for doing the division. Look at the long division (left) and its simplified version (right):

$$
\begin{array}{r}
3x^2 + 11x + 15 \\
x - 2 \overline{\smash{)}\ 3x^3 + 5x^2 - 7x + 10} \\
\underline{3x^3 - 6x^2} \\
11x^2 \\
\underline{11x^2 - 22x} \\
15x \\
\underline{15x - 30} \\
+ 40
\end{array}
$$

Quotient
Dividend
Divisor
Remainder

$$
\begin{array}{r}
3 + 11 + 15 \\
1 - 2 \overline{\smash{)}\ 3 + 5 - 7 + 10} \\
\boxed{3}\,\boxed{-6} \\
\boxed{+11} \\
\boxed{+11}\,\boxed{-22} \\
\boxed{+15} \\
\boxed{+15}\,\boxed{-30} \\
+ 40
\end{array}
$$

Quotient
Dividend
Divisor
Remainder

First, the diagram on the right omits the variables. Moreover, all the numbers in the boxes are repeated. Finally, the only numbers that we have to use to do the division are the circled ones and the ones in the answer (the quotient). We use these facts to write the division in an even shorter version using only three lines:

$$
\begin{array}{r}
2\,\overline{)\ 3 \quad 5 \quad -7 \quad 10} \\
\downarrow \quad \text{Add} \\
\hline
3
\end{array}
$$

1st line:
To make it easier, we replace the subtraction with addition by changing the sign of the constant in the indicated divisor (in this case -2 to 2) so that we can add at each step. Thus, the first line has the additive inverse of the constant of the divisor, followed by the coefficients of the dividend.

$$
\begin{array}{r}
2\,\overline{)\ 3 \quad 5 \quad -7 \quad 10} \\
\text{Multiply} \quad 6 \\
\hline
3
\end{array}
$$

$$
\begin{array}{r}
2\,\overline{)\ 3 \quad 5 \quad -7 \quad 10} \\
6 \quad \text{Add} \\
\hline
3 \quad 11
\end{array}
$$

$$
\begin{array}{r}
2\,\overline{)\ 3 \quad 5 \quad -7 \quad 10} \\
\text{Multiply} \quad 6 \quad 22 \\
\hline
3 \quad 11
\end{array}
$$

$$
\begin{array}{r}
2\,\overline{)\ 3 \quad 5 \quad -7 \quad 10} \\
6 \quad 22 \quad \text{Add} \\
\hline
3 \quad 11 \quad 15
\end{array}
$$

$$
\begin{array}{r}
2\,\overline{)\ 3 \quad 5 \quad -7 \quad 10} \\
\text{Multiply} \quad 6 \quad 22 \quad 30 \\
\hline
3 \quad 11 \quad 15
\end{array}
$$

2nd line:
The second line shows the results of multiplying each of the sums by 2, the additive inverse of the constant of the divisor. These numbers are the additive inverses of the numbers we circled above.

$$
\begin{array}{r}
2\,\overline{)\ 3 \quad 5 \quad -7 \quad 10} \\
6 \quad 22 \quad 30 \quad \text{Add} \\
\hline
3 \quad 11 \quad 15 \quad 40
\end{array}
$$

3rd line:
Results of adding line 1 with line 2.

Quotient: $3x^2 + 11x + 15$ remainder 40.
The degree (2) of the quotient is one less than the degree (3) of the dividend.

SYNTHETIC DIVISION

Synthetic division is a procedure used when dividing a polynomial by a binomial of the form $x - k$. The degree of the quotient is one less than the degree of the dividend.

EXAMPLE 5	**Using synthetic division to divide a polynomial by a binomial**	**PROBLEM 5**

Use synthetic division to divide: $(2x^4 - 3x^2 + 5x - 7) \div (x + 3)$

PROBLEM 5

Use synthetic division to divide:

$$(3x^4 - 2x^3 - 9x + 1) \div (x - 2)$$

SOLUTION   Since synthetic division works only when dividing by binomials of the form $x - k$, write $x + 3$ as $x - (-3)$ to obtain the indicated divisor. The zero in the first line of the division is in place of the missing x^3 term.

$$-3 \overline{)\begin{array}{ccccc} 2 & 0 & -3 & +5 & -7 \\ & -6 & +18 & -45 & +120 \\ \hline 2 & -6 & +15 & -40 & +113 \end{array}}$$

Quotient: $(2x^3 - 6x^2 + 15x - 40)$ R 113

The answer is read from the bottom row, $2x^3 - 6x^2 + 15x - 40$ with 113 as the remainder,

$$\frac{2x^4 - 3x^2 + 5x - 7}{x + 3} = 2x^3 - 6x^2 + 15x - 40 + \frac{113}{x + 3}$$

E   Using the Remainder and the Factor Theorems

Is there a quick way of checking at least part of the division process in Example 5? Amazingly enough, we can find the remainder in the division by using the *remainder theorem*, whose proof is given in more advanced courses.

The Remainder Theorem	If the polynomial $P(x)$ is divided by $x - k$, then the remainder is $P(k)$.

This theorem says that when $P(x)$ is divided by $x - k$, the remainder can be found by evaluating P at k, that is, by finding $P(k)$.

Thus we can find the remainder in Example 5 by finding

$$P(-3) = 2(-3)^4 - 3(-3)^2 + 5(-3) - 7$$
$$= 2(81) - 3(9) - 15 - 7$$
$$= 162 - 27 - 15 - 7$$
$$= 113 \qquad \text{Remainder}$$

But there is a more important application of the remainder theorem. If you divide a polynomial $P(x)$ by $x - k$ and the remainder is zero, then $P(k) = 0$, which means that k is a solution of the equation $P(k) = 0$. Thus one way to show that $k = -3$ is a solution of $x^4 - 4x^3 - 5x^2 + 36x - 36 = 0$ is to substitute -3 for x in the equation. Another way is to use synthetic division and show that the remainder is zero, that is, $P(k) = 0$. We do this in Example 6.

Answer

5. $3x^3 + 4x^2 + 8x + 7 + \frac{15}{x - 2}$

EXAMPLE 6 **Using the remainder theorem with synthetic division**

Use synthetic division to show that -3 is a solution of

$$P(x) = x^4 - 4x^3 - 5x^2 + 36x - 36 = 0$$

SOLUTION We divide $x^4 - 4x^3 - 5x^2 + 36x - 36$ by $x - (-3)$.

$$
\begin{array}{r}
-3\,\overline{)\,1 - 4 - 5 + 36 - 36} \\
\underline{-3 + 21 - 48 + 36} \\
1 - 7 + 16 - 12 \quad\;\; 0
\end{array}
$$

The remainder is zero, $P(-3) = 0$, so -3 is a solution of the given equation.

PROBLEM 6

Use synthetic division to show that 1 is a solution of

$$P(x) = (2x^4 + x^3 - 35x^2 - 16x + 48)$$
$$= 0$$

The remainder is zero when the given polynomial is divided by $x + 3$, so $x + 3$ must be a factor of the polynomial. Thus

$$x^4 - 4x^3 - 5x^2 + 36x - 36 = (x + 3)(x^3 - 7x^2 + 16x - 12)$$

where the second factor is the quotient polynomial found in the last row of the synthetic division. Thus the remainder theorem, used together with synthetic division, can help us evaluate a polynomial (by finding the remainder) and, if the remainder is zero, can even help us factor the polynomial. This last important result, which can be derived from the remainder theorem, is called the *factor theorem*.

Answer

6.
$$
\begin{array}{r}
1\,\overline{)\,2 + 1 - 35 - 16 + 48} \\
\underline{2 + 3 - 32 - 48} \\
2 + 3 - 32 - 48 \quad\;\; 0
\end{array}
$$

The Factor Theorem

When a polynomial $P(x)$ has a factor $(x - k)$, it means that $P(k) = 0$.

Calculate It Exercises Exploring the Remainder Theorem and Factor Theorem

We can do some exploring with the remainder theorem. In Example 6, we found out that -3 is a root (solution) of $x^4 - 4x^3 - 5x^2 + 36x - 36 = 0$. How can we verify this graphically? We graph $Y_1 = x^4 - 4x^3 - 5x^2 + 36x - 36$. If we use a standard window, it seems that the curve cuts or touches the horizontal axis at $x = -3$, $x = 2$, and $x = 3$, but we cannot see the complete graph. To see more of the vertical axis, we let $Y\min = -100$ and $Y\max = 100$. We can clearly see that $x = -3$ is a solution (see Window 1).

If you have a TI-83 Plus, you can check the two other possible solutions, $x = 2$ and $x = 3$. Start with $x = 2$. Press `2nd` `TRACE` `2` and move the cursor to the left of 2. Press `ENTER`. Now move the cursor to the right of 2 (but less than 3) and press `ENTER`. When the calculator asks "Guess?" press `ENTER`. The root (solution) is $x = 2.0000012$. Do the same for 3. You can also verify that -3, 2, and 3 are the solutions of the equation by using the remainder theorem and checking that $P(-3) = 0$, $P(2) = 0$, and $P(3) = 0$. Here the calculator gives you the hint and the algebra confirms it.

Can we use all this information to factor the polynomial? Since -3, 2, and 3 are solutions of the equation, $x - (-3)$, $x - 2$, and $x - 3$ are factors of the polynomial; thus $x^4 - 4x^3 - 5x^2 + 36x - 36 = (x + 3)(x - 2)(x - 3)Q(x)$. To find $Q(x)$, divide $x^4 - 4x^3 - 5x^2 + 36x - 36$ by $(x + 3)(x - 2)(x + 3)$, that is, by $x^3 - 2x^2 - 9x + 18$. Using long division, we get the answer, $x - 2$. Substituting $x - 2$ for $Q(x)$, we have $x^4 - 4x^3 - 5x^2 + 36x - 36 = (x + 3)(x - 2)(x - 3)(x - 2)$,

which means that $x - 2$ is a solution twice. The number 2 is called a **double root** of the equation. In the following exercises, we show you how to factor a polynomial using your calculator and the factor theorem.

1. Graph the polynomial $P(x) = x^3 - 3x^2 - 6x + 8$.
   a. Find the values of c for which $P(c) = 0$.
   b. According to the factor theorem, if $P(c) = 0$, then $x - c$ is a factor of $P(x)$. Name the factors of $P(x)$.
   c. Write $P(x)$ as a product of its factors.
   d. Graph the $P(x)$ obtained in part **c**. Is the graph the same as that of $P(x) = x^3 - 3x^2 - 6x + 8$?

2. Use the procedure of Exercise 1 to factor the polynomial $Q(x) = x^3 + 3x^2 - 6x - 8$.

3. Look at the graph of $R(x)$ shown in Window 2.
   a. What are the solutions (roots) of $R(x) = 0$?
   b. Write $R(x)$ as a product of its factors.
   c. Verify your work by graphing the $R(x)$ you obtained and comparing with the graph shown in the window.

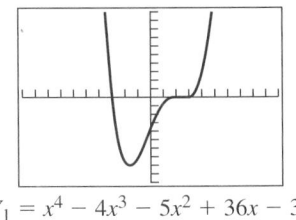

$Y_1 = x^4 - 4x^3 - 5x^2 + 36x - 36$

Window 1

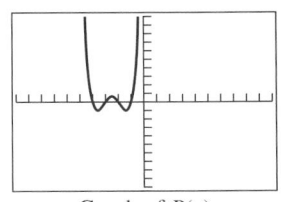

Graph of $R(x)$

Window 2

Exercises 6.5

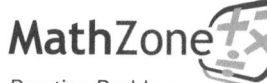
A In Problems 1–10, perform the indicated divisions.

1. $\dfrac{3x^3 + 9x^2 - 6x}{3x}$ $x^2 + 3x - 2$

2. $\dfrac{6x^3 + 8x^2 - 4x}{2x}$ $3x^2 + 4x - 2$

3. $\dfrac{10x^3 - 5x^2 + 15x}{-5x}$ $-2x^2 + x - 3$

4. $\dfrac{24x^3 - 12x^2 + 6x}{-6x}$ $-4x^2 + 2x - 1$

5. $\dfrac{8y^4 - 32y^3 + 12y^2}{-4y^2}$
$-2y^2 + 8y - 3$

6. $\dfrac{9y^4 - 45y^3 + 18y^2}{-3y^2}$
$-3y^2 + 15y - 6$

7. $\dfrac{10x^5 + 8x^4 - 16x^3 + 6x^2}{2x^3}$
$5x^2 + 4x - 8 + \frac{3}{x}$

8. $\dfrac{12x^4 + 18x^3 + 16x^2}{4x^3}$
$3x + \frac{9}{2} + \frac{4}{x}$

9. $\dfrac{15x^3y^2 - 10x^2y + 15x}{5x^2y}$
$3xy - 2 + \frac{3}{xy}$

10. $\dfrac{18x^4y^4 - 24x^2y^3 + 6xy^2}{3x^2y^2}$
$6x^2y^2 - 8y + \frac{2}{x}$

B In Problems 11–36, divide using long division.

11. $x^2 + 5x + 6$ by $x + 2$
$x + 3$

12. $x^2 + 9x + 20$ by $x + 5$
$x + 4$

13. $y^2 + 3y - 10$ by $y - 2$
$y + 5$

14. $y^2 + 2y - 15$ by $y - 3$
$y + 5$

15. $2x^3 - 4x - 2$ by $2x + 2$
$x^2 - x - 1$

16. $2x^3 + 5x^2 - x - 2$ by $2x - 1$
$(x^2 + 3x + 1)\,\text{R}\,-1$

17. $3x^3 + 14x^2 + 13x - 6$ by $3x - 1$
$x^2 + 5x + 6$

18. $2x^3 - 5x^2 - 14x + 3$ by $2x - 1$
$(x^2 - 2x - 8)\,\text{R}\,-5$

19. $2x^3 - 10x - 7x^2 + 24$ by $2x - 3$
$x^2 - 2x - 8$

20. $3x^3 + 8x + 13x^2 - 12$ by $3x - 2$
$x^2 + 5x + 6$

21. $2x^3 + 2x + 7x^2 - 2$ by $-1 + 2x$
$(x^2 + 4x + 3)\,\text{R}\,1$

22. $3x^3 + 3x + 8x^2 - 2$ by $-1 + 3x$
$x^2 + 3x + 2$

23. $y^4 - y^2 - 2y - 1$ by $y^2 + y + 1$
$y^2 - y - 1$

24. $y^4 - y^2 - 4y - 4$ by $y^2 + y + 2$
$y^2 - y - 2$

25. $8x^3 - 6x^2 + 5x - 9$ by $2x - 3$
$(4x^2 + 3x + 7)\,\text{R}\,12$

26. $2x^4 - x^3 + 7x - 2$ by $2x + 3$
$(x^3 - 2x^2 + 3x - 1)\,\text{R}\,1$

27. $x^3 + 8$ by $x + 2$
$x^2 - 2x + 4$

28. $x^3 + 64$ by $x + 4$
$x^2 - 4x + 16$

29. $8y^3 - 64$ by $2y - 4$
$4y^2 + 8y + 16$

30. $27x^3 - 8$ by $3x - 2$
$9x^2 + 6x + 4$

31. $a^4 - 4a^2 - 4a - 1$ by $a^2 + 2a + 1$
$a^2 - 2a - 1$

32. $b^4 - b^2 - 2b - 1$ by $b^2 - b - 1$
$b^2 + b + 1$

33. $x^5 - 5x + 12x^2$ by $x^2 + 5 - 2x$
$x^3 + 2x^2 - x$

34. $y^5 - y^4 + 10 - 27y + 7y^2$ by
$y^2 + 5 - y$ $y^3 - 5y + 2$

35. $4x^4 - 13x^2 + 4x^3 - 3x - 21$ by
$2x + 5$ $(2x^3 - 3x^2 + x - 4)\,\text{R}\,-1$

36. $8y^4 - 75y^2 - 18y^3 + 46y + 121$ by
$4y + 5$ $(2y^3 - 7y^2 - 10y + 24)\,\text{R}\,1$

C In Problems 37–42, factor completely.

37. $x^3 - 4x^2 + x + 6$ if $x + 1$ is one of the factors
$(x + 1)(x - 3)(x - 2)$

38. $x^3 - 4x^2 + x + 6$ if $x - 3$ is one of the factors
$(x - 3)(x - 2)(x + 1)$

39. $x^4 - 4x^3 + 3x^2 + 4x - 4$ if $x^2 - 4x + 4$ is one of the
factors $(x - 2)(x - 2)(x + 1)(x - 1)$

40. $x^4 - 2x^3 - 13x^2 + 14x + 24$ if $x^2 - 6x + 8$ is one of
the factors $(x - 4)(x - 2)(x + 3)(x + 1)$

41. $x^4 + 6x^3 + 3x + 140$ if $x^2 - 3x + 7$ is one of the
factors $(x^2 - 3x + 7)(x + 5)(x + 4)$

42. $x^4 - 22x^2 - 75$ if $x^2 + 3$ is one of the factors
$(x^2 + 3)(x + 5)(x - 5)$

D In Problems 43–52, use synthetic division to find the quotient and the remainder.

43. $(v^3 - 8v - 3) \div (v - 3)$
   $v^2 + 3v + 1$

44. $(y^3 - 4y^2 - 25) \div (y - 5)$
   $y^2 + y + 5$

45. $(x^3 + 4x^2 - 7x + 5) \div (x - 2)$
   $(x^2 + 6x + 5)$ R 15

46. $(4w^3 - w^2 + 92) \div (w + 3)$
   $(4w^2 - 13w + 39)$ R -25

47. $(z^3 - 32z + 24) \div (z + 6)$
   $z^2 - 6z + 4$

48. $(2y^4 - 3y^3 + y^2 - 3y) \div (y - 2)$
   $(2y^3 + y^2 + 3y + 3)$ R 6

49. $(3y^4 - 41y^2 - 13y - 8) \div (y - 4)$
   $(3y^3 + 12y^2 + 7y + 15)$ R 52

50. $(v^5 - 4v^3 + 5v^2 - 5) \div (v + 1)$
   $(v^4 - v^3 - 3v^2 + 8v - 8)$ R 3

51. $(2y^4 - 13y^3 + 6y^2 + 5y - 30) \div (y - 6)$
   $2y^3 - y^2 + 5$

52. $(4w^4 + 20w^3 - w^2 - 2w + 15) \div (w + 5)$
   $4w^3 - w + 3$

E In Problems 53–60, use synthetic division to show that the given number is a solution of the equation.

53. 4; $z^3 + 6z^2 - 6z - 136 = 0$
   Rem. = 0; 4 is a solution

54. -7; $3y^3 + 13y^2 - 57y - 7 = 0$
   Rem. = 0; -7 is a solution

55. -4; $5y^3 + 18y^2 - y + 28 = 0$
   Rem. = 0; -4 is a solution

56. 6; $7x^3 - 39x^2 - 26x + 48 = 0$
   Rem. = 0; 6 is a solution

57. 5; $3v^4 - 14v^3 - 7v^2 + 21v - 55 = 0$
   Rem. = 0; 5 is a solution

58. -8; $8w^4 + 62w^3 - 15w^2 + 10w + 16 = 0$
   Rem. = 0; -8 is a solution

59. -1; $y^5 + y^4 + 2y^3 + 5y^2 - 2y - 5 = 0$
   Rem. = 0; -1 is a solution

60. 1; $3z^6 - z^5 + 5z^4 - 3z - 4 = 0$
   Rem. = 0; 1 is a solution

APPLICATIONS

Finding the Average Cost

In business, the average cost per unit, $\overline{C}$, is given by

$$\overline{C} = \frac{C}{x}$$

where C is the total cost and x is the number of units.

61. Find the average cost when $C = 500 + 4x$. $\frac{500}{x} + 4$

62. Find the average cost when $C = 200 + 2x^2$. $\frac{200}{x} + 2x$

SKILL CHECKER

Try the "Skill Checker" exercises so you'll be ready for the next section.

Solve:

63. $4x + 8 = 6x$ $x = 4$

64. $5x + 10 = 7x$ $x = 5$

65. $x(x + 2) - (x - 3)(x - 4) = 4x + 3$ $x = 3$

66. $x(x + 1) - (x - 1)(x - 2) = 2x + 2$ $x = 2$

USING YOUR KNOWLEDGE

Finding Possible Factors

In this section we factored some polynomials by using a given first-degree factor. How did we find that one factor? Here is one way to do it.

If a polynomial of the form $c_n x^n + c_{n-1} x^{n-1} + \cdots + c_0$ with integer coefficients has a factor $ax + b$, where a and b are integers and $a > 0$, then a must divide c_n and b must divide c_0. For example, to find a and b for the polynomial $2x^3 + 3x^2 - 8x + 3$, we consider the positive divisors of 2 and all the divisors of 3. Thus the possibilities for a are 1 and 2 and for b are 1, -1, 3, and -3. This means that only the binomials $x + 1$, $x - 1$, $x + 3$, $x - 3$, $2x + 1$, $2x - 1$, $2x + 3$, and $2x - 3$ have to be checked. It turns out that $x - 1$ is a

factor and that division gives

$$2x^3 + 3x^2 - 8x + 3 = (x - 1)(2x^2 + 5x - 3)$$

Since $2x^2 + 5x - 3 = (2x - 1)(x + 3)$, we have

$$2x^3 + 3x^2 - 8x + 3 = (x - 1)(2x - 1)(x + 3)$$

67. What binomials should be checked as possible factors for the polynomial $x^3 + 3x^2 + 5x + 6$ if a and b are positive? $x + 1, x + 2, x + 3, x + 6$

68. What binomials should be checked as possible factors for the polynomial $3x^3 + 5x^2 - 3x - 2$ if a and b are positive? $x + 1, x + 2, 3x + 1, 3x + 2$

WRITE ON

69. Describe the procedure you use to divide a polynomial by a binomial. Answers may vary.

70. How can you check the result when dividing one polynomial by another? Use your answer to check that $(2x^3 + 5x^2 - 8x + 6) \div (x + 2) = (2x^2 + x - 10)$ R 26. Answers may vary.

71. Describe two different procedures that can be used to determine that there is a zero remainder when dividing a polynomial by a binomial of the form $x - k$. Answers may vary.

72. When dividing a polynomial by a monomial will the result always be a polynomial? Give an example and explain. Answers may vary.

MASTERY TEST

If you know how to do these problems, you have learned your lesson!

Use synthetic division to find the quotient and remainder when dividing:

73. $x^3 - 3x^2 + 3$ by $x - 3$ x^2 R 3

74. $2x^4 - 13x^3 + 16x^2 - 9x + 20$ by $x - 5$
$2x^3 - 3x^2 + x - 4$

Use synthetic division to show that the given number is a solution of the equation:

75. **2**; $2x^3 + 5x^2 - 8x - 20 = 0$
Rem. $= 0$; 2 is a solution

76. **-2**; $3x^4 + 5x^3 - x^2 + x - 2 = 0$
Rem. $= 0$; -2 is a solution

Factor completely:

77. $2x^3 - x^2 - 18x + 9$ if $x + 3$ is one of the factors
$(x + 3)(x - 3)(2x - 1)$

78. $z^3 - 3z^2 - 4z + 12$ if $z + 2$ is one of the factors
$(z + 2)(z - 2)(z - 3)$

Use long division to divide:

79. $6x^3 + 3x - 9$ by $x - 2$
$(6x^2 + 12x + 27)$ R 45

80. $3x^3 - 2x^2 - x - 6$ by $x + 2$
$(3x^2 - 8x + 15)$ R -36

Divide:

81. $\dfrac{24x^5 - 18x^4 + 12x^3}{6x^2}$

$4x^3 - 3x^2 + 2x$

82. $\dfrac{16x^4 - 4x^3 + 8x^2 - 16x + 40}{8x^3}$

$2x - \dfrac{1}{2} + \dfrac{1}{x} - \dfrac{2}{x^2} + \dfrac{5}{x^3}$

6.6   EQUATIONS INVOLVING RATIONAL EXPRESSIONS

To Succeed, Review How To . . .

1. Find the LCD of two or more fractions (pp. 447–450).

2. Factor polynomials (pp. 370–396).

3. Solve linear and quadratic equations (pp. 79, 399–403).

Objectives

A Solve equations involving rational expressions.

B Solve applications using proportions.

GETTING STARTED   **Play Ball with Equations Involving Rational Expressions**

The baseball player in this picture wants to improve his batting average. His batting average near the beginning of the season (to three decimal places) is found by using the formula:

$$\text{Average} = \frac{\text{number of hits}}{\text{number of times at bat}} = \frac{6}{25} = 0.240$$

How many consecutive hits h would he need to bring his average to 0.320? Here is the information we have:

	Actual	New
Number of hits	6	$6 + h$
Times at bat	25	$25 + h$
New average $= 0.320 = \dfrac{32}{100} =$		$\dfrac{6 + h}{25 + h}$

To find the answer, we must solve the equation. The first step in solving equations containing rational expressions is to multiply both sides of the equation by the LCD to clear the rational expressions; then we use the distributive property to clear parentheses and solve as usual. Here is the solution.

$$\frac{32}{100} = \frac{6 + h}{25 + h}$$

1. Multiply by the LCD, $100(25 + h)$.

$$\cancel{100}(25 + h)\frac{32}{\cancel{100}} = 100\cancel{(25 + h)}\frac{6 + h}{\cancel{(25 + h)}}$$

2. Reduce and use the distributive property.

$$800 + 32h = 600 + 100h$$

3. Subtract 600 from both sides.

$$200 + 32h = 100h$$

4. Subtract $32h$ from both sides.

$$200 = 68h$$

5. Divide by 68.

$$h = \frac{200}{68} \approx 2.941$$

He would need approximately three consecutive hits to obtain a very respectable 0.320 average.

 **Solving Equations Involving Rational Expressions**

What is the difference between the equations we have solved up to now and an equation such as the following?

$$\frac{32}{100} = \frac{6 + h}{25 + h}$$

This equation has a variable in the *denominator,* so we have to avoid values of the variable that make the denominator zero. Such equations are solved with a procedure similar to the one given on page 79, except that when multiplying by the LCD of the fractions involved, we may end up with a quadratic equation, an equation that can be written in the form $ax^2 + bx + c = 0$. Here is the procedure we need.

Teaching Tip

Remember the King "LEAR" method for adding rational expressions? Now we will learn the Queen "LISA" method for solving rational equations.

L Find **LCD.**

I **Improve** the equation by multiplying through by the LCD.

S **Solve** the resulting equation.

A **Adjust** the solution set, if necessary.

PROCEDURE

Procedure for Solving Equations Containing Rational Expressions

1. Factor all denominators and multiply both sides of the equation by the LCD of all rational expressions in the equation.

2. Write the result in reduced form and use the distributive property to remove parentheses.

3. Determine whether the equation is *linear* (can be written in the form $ax + b = 0$) or *quadratic* ($ax^2 + bx + c = 0$) and solve accordingly. (For linear equations see page 79; for quadratic equations, see page 400.)

4. Check that the *proposed* or *trial* solution satisfies the original equation. If it does not, discard it as an **extraneous solution.** (Denominators may not equal zero.)

Note: The guide to solving linear equations is CRAM. The guide to solving quadratic equations is OFF.

EXAMPLE 1 Solving equations containing rational expressions

Solve:

a. $\dfrac{4}{x} = \dfrac{6}{x + 2}$

b. $\dfrac{1}{x + 1} = \dfrac{2}{x + 2}$

SOLUTION

a. We use the four-step procedure.

1. The LCD of

$$\frac{4}{x} \quad \text{and} \quad \frac{6}{x + 2}$$

is $x(x + 2)$. Multiplying both sides of the equation by the LCD,

$$x(x + 2) = \frac{x(x + 2)}{1}$$

gives

$$\frac{\cancel{x}(x + 2)}{1} \cdot \frac{4}{\cancel{x}} = \frac{x\cancel{(x + 2)}}{1} \cdot \frac{6}{\cancel{x + 2}}$$

PROBLEM 1

Solve:

a. $\dfrac{-10}{2x + 1} = \dfrac{5}{x}$

b. $\dfrac{6}{x + 4} = \dfrac{5}{x - 3}$

Answers

1. a. $-\frac{1}{4}$ **b.** 38

2. Reduce and remove parentheses.

$$(x + 2) \cdot 4 = x \cdot 6 \qquad \text{Divide out } x \text{ and } x + 2.$$

$$4x + 8 = 6x \qquad \text{Remove parentheses.}$$

3. Solve.

$$8 = 2x \qquad \text{Subtract } 4x \text{ from both sides.}$$

$$4 = x \qquad \text{Divide by 2.}$$

4. The proposed solution is 4. To check the answer, we substitute 4 for x in the original equation to obtain

$$\frac{4}{4} = \frac{6}{4 + 2}$$

or $1 = 1$. Therefore, the solution 4 is correct, and the solution set is $\{4\}$.

b. Again, we use the four-step procedure.

1. The LCD of

$$\frac{1}{x + 1} \quad \text{and} \quad \frac{2}{x + 2}$$

is $(x + 1)(x + 2)$. Multiplying both sides of the equation by

$$\frac{(x + 1)(x + 2)}{1}$$

we obtain

$$\frac{(x + 1)(x + 2)}{1} \cdot \frac{1}{x + 1} = \frac{(x + 1)(x + 2)}{1} \cdot \frac{2}{x + 2}$$

2.

$$x + 2 = (x + 1) \cdot 2 \qquad \text{Simplify.}$$

$$x + 2 = 2x + 2 \qquad \text{Use the distributive property.}$$

3.

$$x = 2x \qquad \text{Subtract 2.}$$

$$0 = x \qquad \text{Subtract } x.$$

4. The proposed solution is zero. The verification that zero is the actual solution is left to you.

Teaching Tip

The equations in Example 1 are like proportions and can be solved by cross multiplication. Thus

$$\frac{4}{x} = \frac{6}{x + 2}$$

will have the same solution set as the equation

$$4(x + 2) = 6x$$

providing $x \neq 0$ and $x \neq -2$.

Teaching Tip

This is where "A" of the Queen "LISA" method of solving rational equations comes in. We adjust the solution set, if necessary.

As we mentioned, when variables occur in the denominator, it's possible to multiply both sides of the equation by the LCD of the fractions involved and obtain a solution of the resulting equation that does *not* satisfy the original equation. This points out the necessity of *checking*, by direct substitution in the original equation, any proposed solutions obtained after multiplying both sides of an equation by factors containing the unknown. If the proposed solution does not satisfy the original equation, it is called an *extraneous solution*.

CAUTION

By definition an extraneous solution does *not* satisfy the given equation and must *not* be listed as a solution.

EXAMPLE 2	**More practice solving equations containing rational expressions**

Solve:

a. $\dfrac{1}{x - 4} - \dfrac{1}{x - 2} = \dfrac{2x}{x^2 - 6x + 8}$

b. $\dfrac{x}{x - 3} - \dfrac{x - 4}{x + 2} = \dfrac{4x + 3}{x^2 - x - 6}$

PROBLEM 2

Solve:

a. $\dfrac{1}{x - 6} - \dfrac{1}{x - 4} = \dfrac{6}{(x - 6)(x - 2)}$

b. $\dfrac{x}{x - 4} - \dfrac{x - 5}{x + 3} = \dfrac{12x + 16}{x^2 - x - 12}$

Answers

2. a. 5 **b.** No solution

SOLUTION

a. We use our four-step procedure.

1. First factor the denominator of the right-hand side of the equation to obtain

$$\frac{1}{x-4} - \frac{1}{x-2} = \frac{2x}{(x-4)(x-2)}$$

Since the LCD of the fractions involved is

$$\frac{(x-4)(x-2)}{1}$$

we multiply each side of the equation by this LCD to get

$$\frac{(x-4)(x-2)}{1} \cdot \left[\frac{1}{x-4} - \frac{1}{x-2}\right] = \frac{(x-4)(x-2)}{1} \cdot \left[\frac{2x}{(x-4)(x-2)}\right]$$

2. Reduce and remove parentheses.

$$\frac{\cancel{(x-4)}(x-2)}{1} \cdot \frac{1}{\cancel{x-4}} - \frac{(x-4)\cancel{(x-2)}}{1} \cdot \frac{1}{\cancel{x-2}}$$

$$= \frac{\cancel{(x-4)(x-2)}}{1} \cdot \frac{2x}{\cancel{(x-4)(x-2)}}$$

3. Solve.
$$(x-2) - (x-4) = 2x$$
$$x - 2 - x + 4 = 2x$$
$$2 = 2x \qquad \text{Collect like terms.}$$
$$x = 1 \qquad \text{Divide by 2.}$$

4. The proposed solution is 1. Substituting 1 for x in the original equation gives

$$\frac{1}{1-4} - \frac{1}{1-2} \overset{?}{=} \frac{2 \cdot 1}{1^2 - 6 \cdot 1 + 8}$$

$$\frac{1}{-3} - \frac{1}{-1} \overset{?}{=} \frac{2}{3}$$

$$-\frac{1}{3} + 1 = \frac{2}{3}$$

which is a true statement. Thus our result is correct. The actual solution is 1 and the solution set is $\{1\}$.

b. We use our four-step procedure.

1. We first write the right-hand side of the equation with the denominator factored.

$$\frac{x}{x-3} - \frac{x-4}{x+2} = \frac{4x+3}{(x-3)(x+2)}$$

The LCD is

$$\frac{(x-3)(x+2)}{1}$$

We multiply both sides by the LCD.

$$\frac{(x-3)(x+2)}{1}\left[\frac{x}{x-3} - \frac{x-4}{x+2}\right] = \frac{(x-3)(x+2)}{1}\left[\frac{4x+3}{(x-3)(x+2)}\right]$$

Web It

For another explanation of how to solve equations with rational expressions, go to link 6-6-1 at mhhe.com/bello.

2. Reduce and remove parentheses.

$$\frac{(x-3)(x+2)}{1} \cdot \frac{x}{x-3} - \frac{(x-3)(x+2)}{1} \cdot \frac{x-4}{x+2}$$

$$= \frac{(x-3)(x+2)}{1} \cdot \frac{4x+3}{(x-3)(x+2)}$$

$$(x+2) \cdot x - (x-3)(x-4) = 4x+3$$

$$x^2 + 2x - (x^2 - 7x + 12) = 4x + 3$$

3. Solve.

$$9x - 12 = 4x + 3 \qquad \text{Combine like terms.}$$
$$5x = 15 \qquad \text{Add 12 and subtract } 4x.$$
$$x = 3 \qquad \text{Divide by 5.}$$

4. The proposed solution is 3. However, if x is replaced by 3 in the original equation, the term

$$\frac{x}{x-3}$$

yields $\frac{3}{0}$, which is meaningless. Consequently, the equation

$$\frac{x}{x-3} - \frac{x-4}{x+2} = \frac{4x+3}{x^2-x-6}$$

has no solution. Its solution set is $\varnothing$. 3 is an extraneous solution.

Web It

For a tutorial and more practice on how to solve equations with rational expressions, go to link 6-6-2 at mhhe.com/bello.

Teaching Tip

Remind students that the standard form for a quadratic equation is

$$ax^2 + bx + c = 0$$

Finally, the equations that result from clearing denominators are not *always* linear equations. For example, to solve the equation

$$\frac{x^2}{x+3} = \frac{9}{x+3}$$

we first multiply by the LCD $(x+3)$ to obtain

$$(x+3) \frac{x^2}{(x+3)} = (x+3) \frac{9}{(x+3)} \quad \text{or} \quad x^2 = 9$$

In this equation, the variable x has 2 as an exponent; thus it is a quadratic equation and can be solved when written in standard form—by writing the equation as

$$x^2 - 9 = 0 \qquad 0 \quad \text{Subtract 9 to set } = 0.$$

$$(x+3)(x-3) = 0 \qquad \text{F} \quad \text{Factor.}$$

$$x + 3 = 0 \quad \text{or} \quad x - 3 = 0 \qquad \text{F} \quad \text{Factors} = 0. \text{ Use the zero-factor property.}$$

$$x = -3 \quad \text{or} \qquad x = 3 \qquad \text{Solve each equation.}$$

3 is a solution, since

$$\frac{3^2}{3+3} = \frac{9}{3+3}$$

However, for -3, the denominator $x+3$ becomes zero. Thus -3 is an extraneous solution. The only actual solution is 3, and the solution set is $\{3\}$.

| EXAMPLE 3 | Solving equations having extraneous solutions | PROBLEM 3 |

Solve:

$$1 + \frac{3}{x-2} = \frac{12}{x^2-4}$$

Solve:

$$3 - \frac{4}{x^2-1} = \frac{-2}{x-1}$$

Answer

3. $-\frac{5}{3}$

SOLUTION   Since $x^2 - 4 = (x + 2)(x - 2)$, the LCD is $(x + 2)(x - 2)$. We then write the equation with the denominator $x^2 - 4$ in factored form and multiply each term by the LCD, as before. Here are the steps.

1. Multiply each term by the LCD.

$$(x + 2)(x - 2) \cdot 1 + (x + 2)(x - 2) \cdot \frac{3}{(x - 2)} = (x + 2)(x - 2) \cdot \frac{12}{(x + 2)(x - 2)}$$

2. Reduce and remove parentheses.

$$(x^2 - 4) + 3(x + 2) = 12$$
$$x^2 - 4 + 3x + 6 = 12$$

3. Solve the resulting quadratic equation.

$$x^2 + 3x + 2 = 12$$

$$x^2 + 3x - 10 = 0 \qquad \text{0 Subtract 12 to set } = 0.$$

$$(x + 5)(x - 2) = 0 \qquad \text{F Factor.}$$

$$x + 5 = 0 \quad \text{or} \quad x - 2 = 0 \qquad \text{F Factors} = 0. \text{ Use the zero-factor property.}$$

$$x = -5 \quad \text{or} \qquad x = 2 \qquad \text{Solve each equation.}$$

4. 2 makes the denominator $x - 2$ equal to zero, so it is an extraneous solution. The only actual solution is -5, and the solution set is $\{-5\}$. This solution can be checked in the original equation.

Teaching Tip

Students must remember to set the equation equal to zero *before* factoring.

EXAMPLE 4   More practice with extraneous solutions

Solve:

$$\frac{x - 3}{x^2 - 4x} = \frac{2}{x^2 - 16}$$

SOLUTION   As usual, we solve by steps.

1. Write all expressions in factored form and then multiply by the LCD,

$$\frac{x(x - 4)(x + 4)}{1}$$

2. Reduce and remove parentheses.

$$\frac{x(x - 4)(x + 4)}{1} \cdot \frac{x - 3}{x(x - 4)} = \frac{x(x - 4)(x + 4)}{1} \cdot \frac{2}{(x + 4)(x - 4)}$$

$$(x + 4)(x - 3) = 2x$$

3. Solve the resulting quadratic equation.

$$x^2 + x - 12 = 2x$$

$$x^2 - x - 12 = 0 \qquad \text{0 Subtract } 2x \text{ to set } = 0.$$

$$(x + 3)(x - 4) = 0 \qquad \text{F Factor.}$$

$$x + 3 = 0 \quad \text{or} \quad x - 4 = 0 \qquad \text{F Factors} = 0. \text{ Use the zero-factor property.}$$

$$x = -3 \quad \text{or} \qquad x = 4 \qquad \text{Solve.}$$

4. The proposed solutions are -3 or 4. If we substitute -3 for x in the original equation, we get a true statement. On the other hand, if we substitute 4 for x in the original equation, the denominators on both sides are zero, so the fractions are not defined. The only actual solution is -3, and 4 is an extraneous solution. The solution set is $\{-3\}$.

PROBLEM 4

Solve:

$$\frac{x - 7}{x^2 - 8x} = \frac{2}{x^2 - 64}$$

Answer

4. -7

EXAMPLE 5　　Solving equations involving negative exponents	**PROBLEM 5**

Solve: $3x(x+2)^{-1} + 8(x-3)^{-1} = 4$

Solve: $2(x+1)^{-1} + 4(x-4)^{-1} = 1$

SOLUTION　At first, it seems that there are *no* rational expressions involved. However, since $a^{-1} = \frac{1}{a}$,

$$3x(x+2)^{-1} = 3x \cdot \frac{1}{x+2} = \frac{3x}{x+2}$$

and

$$8(x-3)^{-1} = 8 \cdot \frac{1}{x-3} = \frac{8}{x-3}$$

Thus $3x(x+2)^{-1} + 8(x-3)^{-1} = 4$ becomes

$$\frac{3x}{x+2} + \frac{8}{x-3} = 4$$

Using our four-step procedure, we have the following:

1. The only denominators are $x+2$ and $x-3$, so the LCD is $(x+2)(x-3)$. Multiply both sides of the equation by the LCD.

$$(x+2)(x-3)\left(\frac{3x}{x+2} + \frac{8}{x-3}\right) = (x+2)(x-3)4$$

2. Reduce and remove parentheses.

$$(x+2)(x-3) \cdot \frac{3x}{(x+2)} + (x+2)(x-3) \cdot \frac{8}{(x-3)} = (x+2)(x-3)4$$

$$(x-3)3x + (x+2)8 = (x^2 - x - 6)4$$

$$3x^2 - 9x + 8x + 16 = 4x^2 - 4x - 24$$

3. Solve the resulting quadratic equation.

$$3x^2 - x + 16 = 4x^2 - 4x - 24$$

Subtract $3x^2 - x + 16$ from both sides.　　$0 = x^2 - 3x - 40$

Factor.　　$0 = (x+5)(x-8)$

Factors = 0.

$$x + 5 = 0 \quad \text{or} \quad x - 8 = 0$$

$$x = -5 \quad \text{or} \quad x = 8 \quad \text{Solve each equation.}$$

4. Neither of these proposed solutions makes a denominator zero. The solutions are -5 and 8 and the solution set is $\{-5, 8\}$. Check this.

B　　Solving Applications Using Proportions

Some equations containing rational expressions are special and can be solved with a different method. The equation used to solve the *Getting Started* problem, $\frac{32}{100} = \frac{6+h}{25+h}$, is an example of one. To be able to recognize these special cases you must know a little about ratios and proportions. A **ratio** is the comparison of two numbers or quantities and is usually expressed as a fraction such as $\frac{2}{3}$ or $\frac{x}{x+4}$. When two ratios are equal the statement is called a **proportion.** Two examples of proportions are

$$\frac{5}{10} = \frac{1}{2} \quad \text{and} \quad \frac{2}{3} = \frac{x}{x+4}$$

Answer

5. $0, 9$

Property of Proportions	If $\dfrac{a}{b} = \dfrac{c}{d}$ (where $b, d \neq 0$), then $a \cdot d = b \cdot c$
	A proportion is true if the cross products are equal.

Web It

For more practice on solving applications using a proportion, go to link 6-6-3 at mhhe.com/bello.

In the first example, we use this property to verify that the proportion is true.

$$\frac{5}{10} = \frac{1}{2} \qquad \text{Given.}$$

$$10 \cdot 1 = 5 \cdot 2 \qquad \text{Cross products are equal.}$$

$$10 = 10 \qquad \text{Simplify.}$$

10 does equal 10, so the proportion is true.

In the second example, we use this property to find the value of x that will make it a true proportion.

$$\frac{2}{3} = \frac{x}{x + 4} \qquad \text{Given.}$$

$$3 \cdot x = 2 \cdot (x + 4) \qquad \text{Cross products are equal.}$$

$$3x = 2x + 8 \qquad \text{Simplify.}$$

$$x = 8$$

As long as the variable, x, is replaced with 8, the proportion is true.

A **rate** is a ratio of two different quantities such as

$$\frac{5 \text{ ft}}{2 \text{ min}} \quad \text{or} \quad \frac{30 \text{ mi}}{1 \text{ hr}}$$

When writing a proportion with rates, be sure the units are written in the same order in each of the ratios.

There are many real-life applications for proportions such as proportionally decreasing the size of a large object in a scale drawing, increasing the dosage of a medicine proportional to the weight of a person, or decreasing the proportion of the ingredients in a recipe.

Now that we have reviewed proportions and how to solve them, in Example 6 we will solve an application using a proportion.

EXAMPLE 6 Solve an application using a proportion	**PROBLEM 6**

One of the ingredients in a recipe for braised steak with vegetables is 14.5 ounces of stewed tomatoes. The recipe will feed four people but you will be having six guests for dinner. Find the amount of stewed tomatoes needed to increase the recipe to feed six people.

If a map uses a scale of 1 in. = 25 mi, find the number of miles between two cities if the map indicates they are 3.5 in. apart.

SOLUTION   To solve this problem we write a proportion in which the ratios compare ounces of stewed tomatoes with serving size.

$$\frac{14.5 \text{ oz}}{4 \text{ people}} \qquad \text{First ratio}$$

$$\frac{x \text{ oz}}{6 \text{ people}} \qquad \text{Second ratio}$$

$$\frac{14.5 \text{ oz}}{4 \text{ people}} = \frac{x \text{ oz}}{6 \text{ people}} \qquad \text{Proportion}$$

To solve the proportion, we use the property of proportions, which allows us to set the cross products equal.

$$14.5(6) = 4(x) \qquad \text{Cross products are equal.}$$

$$87 = 4x \qquad \text{Multiply.}$$

$$21.75 = x \qquad \text{Divide by 4.}$$

To increase the recipe to feed 6 people, you will need 21.75 ounces of stewed tomatoes.

Answer

6. 87.5 mi

Calculate It Techniques for Solving Rational Equations

Now we shall discuss four techniques for solving rational equations: **graphing, finding roots (solutions), finding intersections,** and **using solve.** Let's select a typical equation, the one from Example 3:

$$1 + \frac{3}{x - 2} = \frac{12}{x^2 - 4}$$

To solve this equation by graphing, first subtract $\frac{12}{x^2 - 4}$ from both sides to obtain the equivalent rational expression $R(x) = 0$. To find the values of x for which $R(x) = 0$, that is, the values at which $R(x)$ crosses the horizontal axis, we graph $Y_1 = 1 + \frac{3}{(x-2)} - \frac{12}{(x^2-4)}$ (Window 1). The graph seems to cross the horizontal axis at $x = -5$. You can confirm this by using the TRACE and ZOOM keys as shown in Window 2. At $x = -5$, $y = -3E - 10 = -3 \times 10^{-10}$, which is nearly zero.

A second way of solving the equation is to find the roots (solutions) of $1 + \frac{3}{x-2} = \frac{12}{x^2-4}$ or equivalently, the roots of $1 + \frac{3}{x-2} - \frac{12}{x^2-4} = 0$. To do this on a TI-83 Plus, graph $Y_1 = 1 + \frac{3}{(x-2)} - \frac{12}{(x^2-4)}$ and press 2nd TRACE 2. Since the graph crosses the horizontal axis at about -5, move the cursor to the left of $x = -5$. When the calculator asks "Left Bound?" press ENTER. Now move the cursor to the right of $x = -5$. When the calculator asks "Right Bound?" and "Guess?" press ENTER each time. The calculator responds with "Zero X=−5" as shown in Window 3.

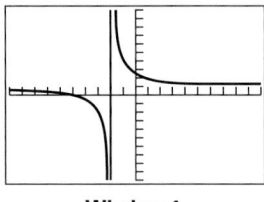

Window 1
$Y_1 = 1 + \frac{3}{(x-2)} - \frac{12}{(x^2-4)}$

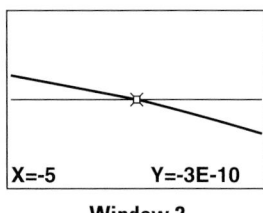

X=-5   Y=-3E-10
Window 2

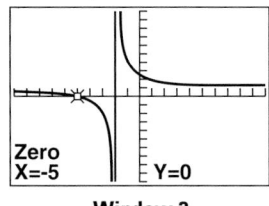

Zero
X=-5   Y=0
Window 3

The third technique is to graph the left-hand side as $Y_1 = 1 + \frac{3}{(x-2)}$ and the right-hand side as $Y_2 = \frac{12}{(x^2-4)}$ then find the point at which the two graphs intersect. The x-value is the solution of the equation. Look for solutions in three regions: points to the left of $x = -2$, points between -2 and 2, and points to the right of $x = 2$. First, move the cursor to the left of $x = -2$ and press 2nd TRACE 5. Press ENTER after the questions "First Curve?" "Second Curve?" and "Guess?" The calculator tells you "Intersection X= −5" (Window 4).

Finally, if your calculator has a **solve** feature, press MATH 0 and enter $1 + \frac{3}{(x-2)} - \frac{12}{(x^2-4)}$ (Window 5). Again, note the parentheses in the denominators. Press ENTER, then put in a guess for the answer. To have the calculator find the answer, press ALPHA ENTER. If you enter a number greater than 2 for your guess (say, 3), you will get an error message. If you enter a guess between -2 and 2 (say, 0), you will also get an error message. Finally, try a number less than -2 for your guess, say, -3. After you press ALPHA ENTER, the calculator shows the answer -5. To get to the final answer, it is essential that you understand what the graph looks like and where it is likely to have a solution. (This is your "guess.")

Now you can practice by verifying the rest of the examples using any one of the four techniques available on your calculator.

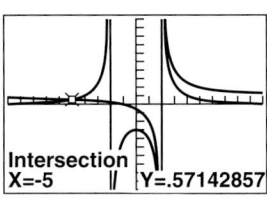

Intersection
X=-5   Y=.57142857
Window 4
$Y_1 = 1 + \frac{3}{(x-2)}$
$Y_2 = \frac{12}{(x^2-4)}$

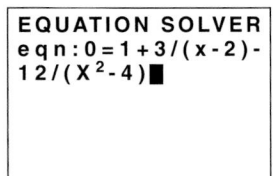

EQUATION SOLVER
eqn:0 = 1 + 3/(x-2)-
12/(X²-4)■

Window 5
Note the parentheses when entering the expression.

Exercises 6.6

A In Problems 1–36, solve the given equation.

1. $\frac{x}{3} + \frac{x}{6} = 3$ 6

2. $\frac{x}{2} + \frac{x}{4} = \frac{3}{8}$ $\frac{1}{2}$

3. $\frac{x}{5} - \frac{3x}{10} = \frac{1}{2}$ -5

4. $\frac{x}{6} - \frac{x}{5} = \frac{1}{15}$ -2

5. $\frac{1}{y} + \frac{4}{3y} = 7$ $\frac{1}{3}$

6. $\frac{10}{3y} - \frac{9}{2y} = \frac{7}{30}$ -5

7. $\frac{2}{y-8} = \frac{1}{y-2}$ -4

8. $\frac{2}{y-4} = \frac{3}{y-2}$ 8

9. $\frac{3}{3z+4} = \frac{2}{5z-6}$ $\frac{26}{9}$

10. $\frac{2}{4z-1} = \frac{3}{2z+1}$ $\frac{5}{8}$

11. $\dfrac{-2}{2x+1}=\dfrac{3}{3x-1}$ $\dfrac{-1}{12}$

12. $\dfrac{-5}{2x+3}=\dfrac{2}{3x-1}$ $\dfrac{-1}{19}$

13. $\dfrac{-1}{x+1}=\dfrac{-2}{2x-1}$ No solution

14. $\dfrac{-5}{5x-2}=\dfrac{-3}{3x+1}$ No solution

15. $\dfrac{2}{3x+1}=\dfrac{5}{6x+2}$ No solution

16. $\dfrac{3}{2x-1}=\dfrac{6}{4x-5}$ No solution

17. $\dfrac{2}{x^2-4}+\dfrac{5}{x+2}=\dfrac{7}{x-2}$ -11

18. $\dfrac{3}{x^2-9}+\dfrac{5}{x+3}=\dfrac{8}{x-3}$ -12

19. $\dfrac{t+2}{t^2-3t+2}=\dfrac{3}{t-1}-\dfrac{1}{t-2}$ 7

20. $\dfrac{t+3}{t^2+4t+3}=\dfrac{4}{t+3}-\dfrac{1}{t+1}$ 1

21. $\dfrac{x^2}{x^2-1}=1+\dfrac{1}{x+1}$ 2

22. $\dfrac{x^2}{x^2-9}=1+\dfrac{1}{x-3}$ 6

23. $\dfrac{1}{x^2-4x+3}+\dfrac{1}{x^2-2x-3}=\dfrac{1}{x^2-1}$ -3

24. $\dfrac{1}{x^2+3x+2}+\dfrac{1}{x^2+x-2}=\dfrac{1}{x^2-1}$ 2

25. $\dfrac{x+2}{3x^2+4x+1}=\dfrac{x+1}{3x^2+7x+2}$ $\dfrac{-3}{2}$

26. $\dfrac{x+2}{2x^2+x-1}=\dfrac{x-2}{2x^2+x-1}$ No solution

27. $\dfrac{2z+13}{2z^2+5z-3}+\dfrac{3}{z+3}=\dfrac{4}{2z-1}$ No solution

28. $\dfrac{z-14}{2z^2-3z-2}+\dfrac{3}{z-2}=\dfrac{4}{2z+1}$ 1

29. $\dfrac{3-x}{5x^2-4x-1}+\dfrac{2}{5x+1}=\dfrac{1}{x-1}$ 0

30. $\dfrac{16-x}{4x^2-11x-3}+\dfrac{5}{4x+1}=\dfrac{2}{x-3}$ No solution

31. $4x^{-1}+2=7$ (Recall that $x^{-1}=\tfrac{1}{x}$.) $\dfrac{4}{5}$

32. $3+6x^{-1}=5$ 3

33. $4x^{-1}+6x^{-1}=15(x+1)^{-1}$ 2

34. $6x^{-1}+9x^{-1}=25(x+2)^{-1}$ 3

35. $2(x-8)^{-1}=(x-2)^{-1}$ -4

36. $3(3y+4)^{-1}=2(5y-6)^{-1}$ $\dfrac{26}{9}$

A **B** In Problems 37–46, solve the given equation by using the cross products property of proportions.

37. $\dfrac{2}{y-8}=\dfrac{1}{y-2}$ -4

38. $\dfrac{2}{y-4}=\dfrac{3}{y-2}$ 8

39. $\dfrac{3}{3z+4}=\dfrac{2}{5z-6}$ $\dfrac{26}{9}$

40. $\dfrac{2}{4z-1}=\dfrac{3}{2z+1}$ $\dfrac{5}{8}$

41. $\dfrac{-2}{2x+1}=\dfrac{3}{3x-1}$ $\dfrac{-1}{12}$

42. $\dfrac{-5}{2x+3}=\dfrac{2}{3x-1}$ $\dfrac{-1}{19}$

43. $\dfrac{-1}{x+1}=\dfrac{-2}{2x-1}$ No solution

44. $\dfrac{-5}{5x-2}=\dfrac{-3}{3x+1}$ No solution

45. $\dfrac{2}{3x+1}=\dfrac{5}{6x+2}$ No solution

46. $\dfrac{3}{2x-1}=\dfrac{6}{4x-5}$ No solution

APPLICATIONS

Figure how long or how many

47. A film processing department can process nine rolls of film in 2 hours. At that rate, how long will it take them to process 20 rolls of film? $4\tfrac{4}{9}$ hr

48. If a blueprint uses a scale of $\tfrac{1}{2}$ in. = 3 ft, find the scaled-down dimensions of a room that is 9 ft × 12 ft.   1.5 in. × 2 in.

49. Latrice and Bob want to drive from Miramar, Florida, to Pittsburgh, Pennsylvania, a distance of approximately 1000 miles. Each day they plan to drive 7 hr and cover 425 mi. At that rate,

  a. How many driving hours should the trip take? Approximately $16\frac{1}{2}$ hr
  b. How many days should the trip take? Approximately $2\frac{1}{2}$ days

50. Marquel is an avid bicyclist and averages 16 mi/hr when he rides. He likes to ride 80 mi per day. He plans to take a 510-mile, round trip, between New Orleans and Panama City. At that rate,

  a. How many hours of riding will the trip take? Approximately 32 hr
  b. How many days should the trip take? Approximately $6\frac{1}{2}$ days

51. Early in the season a softball player has a 0.250 batting average (5 hits for 20 times at bat). How many consecutive hits would she need to bring her average to 0.350? (*Hint:* See the *Getting Started.*)
Approximately 3 consecutive hits

52. Norma takes a medicine that requires a dosage of 20 ml as a baseline amount plus 5 ml for every 30 lbs of body weight. Norma weighs 150 lbs.

  a. How many milliliters should each dosage be? 45 ml
  b. If a teaspoon is 5 ml, how many teaspoons would the dosage be? 9 teaspoons

SKILL CHECKER

Try the "Skill Checker" exercises so you'll be ready for the next section.

Solve using the RSTUV method:

53. The sum of three consecutive odd integers is 69. What are the integers? 21, 23, 25

54. An investor bought some municipal bonds yielding 5% annually and some certificates of deposit yielding 8%. If the total investment amounts to $10,000 and the annual interest is $680, how much is invested in bonds and how much in certificates of deposit? $4000 at 5%; $6000 at 8%

55. How many gallons of a 20% salt solution must be mixed with 40 gal of a 15% solution to obtain an 18% solution? 60 gallons

56. A car leaves a town traveling at 50 mi/hr. Two hours later, another car traveling at 60 mi/hr leaves on the same road in the same direction. How far from the town does the second car overtake the first? 600 mi from town

USING YOUR KNOWLEDGE

Solving for Letters

There are many instances in which a given formula must be changed to an equivalent form. For example, the formula

$$\frac{P}{R} = \frac{T}{V}$$

is frequently discussed in chemistry. Suppose you know P, R, and T. Can you find V? As before, we proceed by steps to solve for V.

Step 1. Since the LCD is RV, we multiply each term by RV to obtain

$$RV \cdot \frac{P}{R} = \frac{T}{V} \cdot RV$$

Step 2. Simplify.

$$VP = TR$$

Step 3. Divide by P.

$$V = \frac{TR}{P}$$

Thus

$$V = \frac{TR}{P}$$

In Problems 57–61, use your knowledge of rational equations to solve for the indicated variable.

57. The area A of a trapezoid is

$$A = \frac{h(b_1 + b_2)}{2}$$

Solve for h. $h = \dfrac{2A}{b_1 + b_2}$

58. In an electrical circuit we have

$$\frac{1}{R} = \frac{1}{R_1} + \frac{1}{R_2}$$

Solve for R. $R = \dfrac{R_1 R_2}{R_1 + R_2}$

59. In refrigeration we find the formula

$$\frac{Q_1}{Q_2 - Q_1} = P$$

Solve for Q_1. $Q_1 = \dfrac{PQ_2}{1 + P}$

60. When studying the expansion of metals, we use the formula

$$\frac{L}{1 + at} = L_0$$

Solve for t. $t = \dfrac{L - L_0}{aL_0}$

61. Students of photography use the formula

$$\frac{1}{f} = \frac{1}{a} + \frac{1}{b}$$

Solve for f. $f = \dfrac{ab}{a + b}$

WRITE ON

62. Consider the expression

$$\frac{x}{2} + \frac{x}{3}$$

and the equation

$$\frac{x}{2} + \frac{x}{3} = 5$$

a. What is the first step in simplifying the expression?

b. What is the first step in solving the equation?

c. What is the difference in the use of the LCD in adding two rational expressions as contrasted with solving an equation containing rational expressions? Answers may vary.

63. Write your definition of an extraneous root. Answers may vary.

64. Do you have to check the solutions to

$$\frac{x}{2} + \frac{x}{3} = 5$$

for extraneous solutions? Why or why not? No; answers may vary.

65. Do you have to check the solutions of

$$\frac{6}{x} + \frac{3}{x} = 4$$

for extraneous solutions? Why or why not? Yes; answers may vary.

66. In general, when would you check the solutions of an equation for extraneous solutions? Answers may vary.

MASTERY TEST

If you know how to do these problems, you have learned your lesson!

Solve:

67. $\dfrac{x - 7}{x^2 - 8x} = \dfrac{2}{x^2 - 64}$ $x = -7$

68. $1 - \dfrac{4}{x^2 - 1} = \dfrac{-2}{x - 1}$ $x = -3$

69. $\dfrac{3}{x} = \dfrac{5}{x + 2}$ $x = 3$

70. $\dfrac{2}{x + 1} = \dfrac{3}{x + 2}$ $x = 1$

71. $4(x + 3)^{-1} + 3x(x - 2)^{-1} = -2$ $x = -4$ and $x = 1$

72. $2x(x - 3)^{-1} + 6(x + 1)^{-1} = 6$ $x = 0$ and $x = 5$

73. $\dfrac{1}{x - 6} - \dfrac{1}{x - 4} = \dfrac{6}{(x - 6)(x - 2)}$ $x = 5$

74. $\dfrac{x}{x - 4} - \dfrac{x - 5}{x + 3} = \dfrac{3x + 16}{x^2 - x - 12}$ No solution

APPLICATIONS: PROBLEM SOLVING

To Succeed, Review How To . . .

1. Solve equations involving rational expressions (pp. 477–482).

2. Use the RSTUV procedure to solve word problems (pp. 101–106).

Objectives

A Solve integer problems.

B Solve work problems.

C Solve distance problems.

D Solve for a specified variable.

GETTING STARTED

Getting the Golden

The unfinished canvas pictured here was painted by Leonardo da Vinci and is entitled *St. Jerome.* A Golden Rectangle fits so neatly around St. Jerome that experts conjecture that da Vinci painted the figure to conform to those proportions. For many years it has been said that the Golden Rectangle is one of the most visually satisfying of all geometric forms. Do you know how to construct a Golden Rectangle? Such a rectangle has a special ratio of length to width of about 8 to 5 and can be described by writing

$$\frac{\text{Length of rectangle}}{\text{Width of rectangle}} = \frac{8}{5}$$

Now suppose you want to make a Golden Rectangle of your own, but you want the length to be 6 in. longer than the width. What are the dimensions of your rectangle?

To solve this problem, you need to review the RSTUV procedure we used in Section 2.3.

The Golden Rectangle

1. Read the problem.

You have to find the dimensions of the rectangle.

2. Select the unknown.

Let w be the width.

3. Think of a plan.

In Section 6.6 we learned how to write proportions that would solve a problem in which the ratios are the same.

Since you want the new dimensions,

$$\frac{\text{Length}}{\text{Width}} = \frac{w + 6}{w} \quad \begin{array}{l} \leftarrow \text{ Length is 6 in. more than the width} \\ \leftarrow \text{ Width is } w \end{array}$$

to be in the same proportion as the Golden Rectangle,

$$\frac{\text{Length}}{\text{Width}} = \frac{8}{5}$$

we set the ratios equal.

$$\frac{w + 6}{w} = \frac{8}{5}$$

4. Use the property for proportions and solve the resulting linear equation.
To solve the proportion, set the cross products equal and simplify.

$$\frac{w + 6}{w} = \frac{8}{5}$$

$$5(w + 6) = 8(w)$$

$$5w + 30 = 8w \qquad \text{Simplify.}$$

$$30 = 3w \qquad \text{Subtract } 5w.$$

$$10 = w \qquad \text{Divide by 3.}$$

Thus the width is 10 in. and the length is 6 in. more than that, or 16 in.

5. Verify the answer.
The rectangle is 10 in. by 16 in., so the ratio of length (16) to width (10) is $\frac{16}{10}$, or $\frac{8}{5}$, as desired.

A   Solving Integer Problems

We now discuss problems involving consecutive integers and other number properties. You can solve these problems using the RSTUV procedure.

EXAMPLE 1 Consecutive integers	**PROBLEM 1**

There are two consecutive even integers such that the reciprocal of the first added to the reciprocal of the second is $\frac{3}{4}$. What are the integers?

SOLUTION   We use the RSTUV method.

1. Read the problem.
Make sure you understand the meaning of "consecutive even integers." For example, 6, 8 are consecutive even integers, as are 78, 80.

2. Select the unknown.
Let n be the first integer. The next even integer is $n + 2$.

3. Think of a plan.
First, translate the problem:

The reciprocal of the first	added to	the reciprocal of the second	is $\frac{3}{4}$.
$\dfrac{1}{n}$	$+$	$\dfrac{1}{n + 2}$	$= \dfrac{3}{4}$

4. Use the procedure for solving equations with rational expressions.
Start by finding the LCD, $4n(n + 2)$. Multiply both sides by this LCD.

$$4n(n + 2)\left(\frac{1}{n} + \frac{1}{n + 2}\right) = 4n(n + 2) \cdot \frac{3}{4}$$

$$4\overset{1}{\cancel{n}}(n + 2) \cdot \frac{1}{\cancel{n}} + 4n\overset{1}{(\cancel{n + 2})} \cdot \frac{1}{\cancel{n + 2}} = \overset{1}{\cancel{4}}n(n + 2) \cdot \frac{3}{\cancel{4}}$$

$$4n + 8 + 4n = 3n^2 + 6n \qquad \text{Remove parentheses.}$$

$$0 = 3n^2 - 2n - 8 \qquad \begin{array}{l}\text{Subtract } 8n \text{ and } 8 \\ \text{from both sides.}\end{array}$$

$$0 = (3n + 4)(n - 2) \qquad \text{Factor.}$$

The sum of the reciprocals of two consecutive odd integers is $\frac{-8}{15}$. What are the integers?

Answer

1. -5 and -3

By the zero-factor property,

$$3n + 4 = 0 \quad \text{or} \quad n - 2 = 0$$
$$4n = \frac{-4}{3} \quad \text{or} \quad n = 2 \quad \text{Solve } 3n + 4 = 0, \, n - 2 = 0.$$

Since n was assumed to be an integer, we discard the answer $\frac{-4}{3}$. Thus the first even integer is 2 and the next one is 4.

5. Verify the answer.

Verify that the sum of the reciprocals is $\frac{3}{4}$. Since $\frac{1}{2} + \frac{1}{4} = \frac{3}{4}$, our result is correct.

B   Solving Work Problems

Have you ever wished for help with your taxes? The Internal Revenue Service estimates that it takes about 7 hr to file your 1040A form. (This includes record-keeping, familiarizing yourself with the form, and preparing and sending it.)

EXAMPLE 2   Work problems

A couple is about to file their form 1040A. One of them can complete it in 8 hr, and the other can do it in 6 hr. How long would it take if they work on it together?

PROBLEM 2

An accountant can finish form 1040A in 5 hr. An assistant can do it in 10 hr. How long would it take if they work together?

SOLUTION   Again, we use the RSTUV method.

1. Read the problem.

We need to find the total time it takes them when they work together.

2. Select the unknown.

Let t be the time it takes the couple to complete the form working together.

3. Think of a plan.

We concentrate on what happens each hour. Since one person can fill the form in 8 hr and the second can do it in 6 hr, the first person will complete $\frac{1}{8}$ of the form and the second will complete $\frac{1}{6}$ of the form each hour. They are working together and it takes t hours to do the whole thing, so they complete $\frac{1}{t}$ of the form each hour. Here is what we have in one hour:

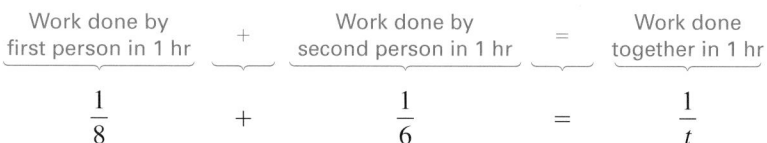

Work done by first person in 1 hr	+	Work done by second person in 1 hr	=	Work done together in 1 hr
$\dfrac{1}{8}$	+	$\dfrac{1}{6}$	=	$\dfrac{1}{t}$

4. Use the procedure for solving equations with rational expressions.

First, find the LCD of $\frac{1}{8}$, $\frac{1}{6}$ and $\frac{1}{t}$. The LCD of these fractions is $24t$.

$$24t\left(\frac{1}{8} + \frac{1}{6}\right) = 24t \cdot \frac{1}{t} \qquad \text{Multiply by } 24t.$$

$$\overset{3}{\cancel{24t}} \cdot \frac{1}{\underset{}{8}} + \overset{4}{\cancel{24t}} \cdot \frac{1}{\underset{}{6}} = 24\cancel{t} \cdot \frac{1}{\cancel{t}}$$

$$3t + 4t = 24 \qquad \text{Simplify.}$$

$$7t = 24$$

$$t = \frac{24}{7} = 3\frac{3}{7} \text{ hr}$$

It takes $3\frac{3}{7}$ hr (about 3 hr 26 min) to complete the job.

5. Verify the answer.

The verification is left to you.

Answer

2. $3\frac{1}{3}$ hr or 3 hr 20 min

Another type of problem can also be thought of as a work problem; this is the tank or pool problem. The idea is that the pipes filling or emptying a tank or pool are doing the *work* to fill or empty the pool. Here is how we solve these problems.

EXAMPLE 3 Pool problems	**PROBLEM 3**

A pool is filled by an intake pipe in 4 hr and is emptied by a drain pipe in 5 hr. How long will it take to fill the pool with both pipes open?

SOLUTION   As before, we use the RSTUV method.

1. Read the problem.
We are asked for the time it takes to fill the pool.

2. Select the unknown.
Let this time be T hours.

3. Think of a plan.
In 1 hr, the intake pipe fills $\frac{1}{4}$ of the pool, the drain pipe empties $\frac{1}{5}$, and together they fill $\frac{1}{T}$ of the pool. Thus, in 1 hr

Amount filled by intake pipe in 1 hr	−	Amount emptied by drain pipe in 1 hr	=	Amount filled by both in 1 hr
$\frac{1}{4}$	−	$\frac{1}{5}$	=	$\frac{1}{T}$

4. Use algebra to solve the equation.
The LCD is $20T$.

$$\overset{5}{\cancel{20T}}\cdot\frac{1}{4} - \overset{4}{\cancel{20T}}\cdot\frac{1}{5} = \cancel{20T}\cdot\frac{1}{T}$$
$$5T - 4T = 20$$
$$T = 20$$

5. Verify the answer.
It takes 20 hr to fill the pool if the intake and drain pipes are both open.

The intake pipe can fill the pool in 4 hr. It can then fill the pool five times in 20 hr. The drain pipe can empty the pool in 5 hr, so it can empty the pool four times in 20 hr. Since the intake can fill the pool five times and the drain can empty it four times in 20 hr, the pool would be filled once at the end of 20 hr.

Repeat Example 3 if the intake pipe can fill the pool in 6 hr and the drain pipe can empty it in 7 hr.

Web It

To see three examples of an alternate way to work applications involving "work," go to link 6-7-1 at mhhe.com/bello.

C   Solving Distance Problems

The ideas we have studied can be used to solve uniform motion problems like the ones discussed in Section 2.4. As you recall, when traveling at a constant rate R, the distance D traveled in time T is given by $D = RT$. We use this information in Example 4.

Answer

3. 42 hr

EXAMPLE 4   Uniform motion problems

One of the world's strongest currents is the Saltstraumen in Norway, reaching as much as 18 mi/hr. A speed boat can travel 48 mi downstream in this current in the same time it takes to go 12 mi upstream. What is the speed of the boat in still water?

SOLUTION   Once again, we use the RSTUV method.

1. Read the problem.
We want to find the speed (rate) of the boat in still water.

2. Select the unknown.
Let R be the rate of the boat in still water.

3. Think of a plan.
Make a chart with D, R, and T as headings.

	D (mi)	R (mi/hr)	T (hr)
Downstream			
Upstream			

Speed downstream: $R + 18$ Current helps, add 18 to R.

Speed upstream: $R - 18$ Current hinders, subtract 18 from R.

The time T is given by

$$T = \frac{D}{R}$$

(solving for T in $D = RT$). We then have

Time downstream: $\dfrac{48}{R + 18}$

Time upstream: $\dfrac{12}{R - 18}$

Enter this information in the chart.

	D (mi)	R (mi/hr)	$\left(T = \dfrac{D}{R}\right)$ (hr)
Downstream	48	$R + 18$	$\dfrac{48}{R + 18}$
Upstream	12	$R - 18$	$\dfrac{12}{R - 18}$

Since it takes the same time to go upstream as downstream, we have

$$T_{\text{up}} = T_{\text{down}}$$

$$\frac{48}{R + 18} = \frac{12}{R - 18}$$

PROBLEM 4

Repeat Example 4 if the boat travels 36 mi downstream in the same time it takes it to go 12 mi upstream.

Teaching Tip

Students should be reminded that an adjustment needs to be made on the rate due to the current helping or hindering the speed. Sometimes we call $(R + 18)$ and $(R - 18)$ the actual rates of speed.

Web It

For applications involving sociology and aviation using rational equations to solve, go to link 6-7-2 at mhhe.com/bello.

Answer

4. 36 mi/hr

4. Use algebra to solve the resulting equation.
The LCD is $\frac{(R + 18)(R - 18)}{1}$ so we multiply both sides of the equation by this LCD

$$\frac{\cancel{(R + 18)}(R - 18)}{1} \cdot \frac{48}{\cancel{R + 18}} = \frac{(R + 18)\cancel{(R - 18)}}{1} \cdot \frac{12}{\cancel{R - 18}}$$

$$48(R - 18) = 12(R + 18)$$

$$4(R - 18) = R + 18 \qquad \text{Divide by 12.}$$

$$4R - 72 = R + 18 \qquad \text{Simplify.}$$

$$4R = R + 90 \qquad \text{Add 72.}$$

$$3R = 90 \qquad \text{Subtract } R.$$

$$R = 30$$

The speed of the boat in still water is 30 mi/hr.

5. Verify the answer.
The verification is left to you.

D   Solving for Specified Variables

Many students get the impression that the RSTUV procedure should only be used when solving word problems. This procedure, however, is very general and can be used in other situations, as we shall see in Example 5.

EXAMPLE 5   **Medicine dosages**

There are many formulas that determine the medicine dosage c for children of age A when the adult dosage d is known. One rule, applicable to children aged 3–12 yr, is Young's rule:

$$c = \frac{A}{A + 12} d$$

Solve for d in Young's rule and find out what the adult dose is if a 6-yr-old child is supposed to take 2 tablets every 12 hr.

SOLUTION   Solve for d in

$$c = \frac{A}{A + 12} d$$

1. Read the problem.
We are asked to solve for d and find the adult dose if a 6-yr-old child is supposed to take $c = 2$ tablets every 12 hr.

2. Select the unknown.
The unknown is d.

3. Think of a plan.
Since we want to solve for d, we place the d in a box to isolate it:

$$c = \frac{A}{A + 12} \boxed{d}$$

PROBLEM 5

One of the trigonometry identities is

$$\sin^2 u = \frac{1 - \cos(2u)}{2}$$

Solve this identity for $\cos(2u)$.

Teaching Tip

It is sometimes helpful to do an example where numbers are substituted for all variables except the specific one for which we are solving.

Example: Solve

$$c = \frac{A}{A + 12} d \quad \text{for } d$$

First, go through the steps that would solve

$$6 = \frac{5}{5 + 12} d$$

Answer

5. $\cos(2u) = 1 - 2 \sin^2 u$

4. Use the procedure for solving equations involving rational expressions.

$$(A + 12)c = \cancel{(A + 12)} \dfrac{A}{\cancel{(A + 12)}} \boxed{d} \qquad \text{Multiply both sides by } (A + 12) \text{ and reduce.}$$

$$(A + 12)c = A\boxed{d}$$

$$\dfrac{(A + 12)c}{A} = \boxed{d} \qquad \text{Divide by } A.$$

$$d = \dfrac{(A + 12)c}{A} \qquad \text{By the symmetric property.}$$

When the child is 6 years old, $A = 6$, and we know that $c = 2$. Substituting in the equation, we get

$$d = \dfrac{(6 + 12)2}{6} = 6$$

The adult dosage is 6 tablets every 12 hr.

5. Verify the solution.
The verification is left to you.

There are more problems in the *Using Your Knowledge* section of the Exercises in which you are asked to solve for a variable.

Calculate It Exercises　Using Tables to Solve Rational Equations

By this time you should be convinced that graphing is a powerful tool in solving algebra problems. But is that all a calculator does? No, we haven't even touched on an important feature of some calculators—the construction of tables listing values that may be solutions. This would be useful in the problem in the *Getting Started,* which required us to construct a Golden Rectangle whose length is 6 in. more than its width. As before, solve the equation

$$\dfrac{w + 6}{w} = \dfrac{8}{5} = 1.6$$

Now we will show you how to set up a table using a TI-83 Plus that will solve the problem for us. First, let's agree to use x instead of w in the solution. Start by pressing [2nd] [WINDOW]. The calculator wants to know the minimum at which you wish to start x. Let TblStart = 1. It then asks for ΔTbl=. This sets up the increments for x in your table. Let ΔTbl = 1. This means that the first x is 1, the next one is increased by 1 to 2, the next one is 3, and so on. For the "Indpnt" and "Depend," leave the setting in auto mode (Window 1). Now, press [Y=] and enter

$$Y_1 = \dfrac{(x + 6)}{x}$$

```
TABLE SETUP
 TblStart=1
 ΔTbl=1
Indpnt:  AUTO  Ask
Depend:  AUTO  Ask
```
Window 1

then press [2nd] [GRAPH]. The first column shows successive values for x, while the second shows the values of

$$Y_1 = \dfrac{(x + 6)}{x}$$

(see Window 2). Press the ⊙ key while you are in the first

X	Y1	
1	7	
2	4	
3	3	
4	2.5	
5	2.2	
6	2	
7	1.8571	
X=1		

Window 2

column. It will show successive values of x and corresponding Y_1 values. You only need to move down until

$$Y_1 = \dfrac{8}{5} = 1.6$$

(see Window 3). This occurs when $x = 10$, the same answer as before, but it is more fun letting the calculator do the work for you.

X	Y1	
4	2.5	
5	2.2	
6	2	
7	1.8571	
8	1.75	
9	1.6667	
10	1.6	
X=10		

Window 3

Let's see if you can do Example 4 using this technique. (Remember, we are using x instead of R.) This time, let TblStart = 0 and ΔTbl = 10. Let

$$\text{Time downstream: } Y_1 = \dfrac{48}{x + 18}$$

$$\text{Time upstream: } Y_2 = \dfrac{12}{x - 18}$$

Now, look in your table. For what value of x is $Y_1 = Y_2$? That is your answer. Try solving some other problems using calculator tables.

1. Use the table feature to solve Example 1.

2. Example 2 can also be solved using the table feature of your calculator. First, assume that the whole job takes T hours. The first person can do $\frac{1}{8}T = \frac{T}{8}$ of the job in 1 hr, while the second person can do $\frac{T}{6}$ of the job in 1 hr. Together, they can do $\frac{T}{8} + \frac{T}{6}$ of the job in 1 hr. When will the job be complete? When $\frac{T}{8} + \frac{T}{6} = 1$. Now set up a table and see what T is. (*Hint:* Let Tblstart = 1 and ΔTbl = 0.1.) The answer you get will be pretty close to the one in Example 2.

3. Do Example 3 using the ideas discussed in Exercise 2. Is it a good idea to set ΔTbl = 0.1 in this case? Explain.

Exercises 6.7

A In Problems 1–10, solve the integer problems.

1. The sum of an integer and its reciprocal is $\frac{65}{8}$. Find the integer. 8

2. The sum of an integer and its reciprocal is $\frac{50}{7}$. What is the integer? 7

3. One number is twice another. The sum of their reciprocals is $\frac{3}{10}$. Find the numbers. 5 and 10

4. One number is three times another. The sum of their reciprocals is $\frac{1}{3}$. Find the numbers. 4 and 12

5. Find two consecutive even integers the sum of whose reciprocals is $\frac{7}{24}$. 6 and 8

6. Find two consecutive odd integers such that the sum of their reciprocals is $\frac{16}{63}$. 7 and 9

7. The denominator of a fraction is 5 more than the numerator. If 3 is added to both numerator and denominator, the resulting fraction is $\frac{1}{2}$. Find the fraction. $\frac{2}{7}$

8. The numerator of a certain fraction is 4 less than the denominator. If the numerator is increased by 8 and the denominator by 35, the resulting fraction is $\frac{1}{2}$. Find the fraction. $\frac{23}{27}$

9. The current ratio of your business is defined by

$$\text{Current ratio} = \frac{\text{current assets}}{\text{current liabilities}}$$

By how much should you increase your current liabilities if they are \$40,000 right now, your current assets amount to \$90,000, and you wish to maintain the current ratio at $\frac{3}{2}$? \$20,000

10. Repeat Problem 9 where you want the current ratio to be 2. \$5000

B In Problems 11–23, solve the work problems.

11. If one word processor can finish a job in 3 hr while another word processor can finish in 5 hr, how long will it take both of them working together to finish the job? $1\frac{7}{8}$ hr

12. A carpenter can finish a job in 8 hr, and another one can do it in 10 hr. How long will it take them to finish the job working together? $4\frac{4}{9}$ hr

13. The world record for riveting is 11,209 rivets in 9 hr, by J. Mair of Ireland. If another person can rivet 11,209 rivets in 10 hr, how long will it take both of them working together to rivet the 11,209 rivets? $4\frac{14}{19}$ hr

14. Mr. Gerry Harley of England shaved 130 men in 60 min. If another barber can shave all these men in 5 hr, how long will it take both of them working together to shave the 130 men? $\frac{5}{6}$ hr

15. A printing press can print the evening paper in half the time another press takes to print it. Together, the presses can print the paper in 2 hr. How long will it take each of them to print the paper? 6 hr; 3 hr

16. A computer can do a job in 4 hr. With the help of a newer computer, the job can be completed in 1 hr. How long will it take the newer computer to do the job alone? $1\frac{1}{3}$ hr

17. A tank can be filled by an intake pipe in 9 hr and drained by another pipe in 21 hr. If both pipes are open, how long will it take to fill the tank? $15\frac{3}{4}$ hr

18. A faucet fills a tank in 12 hr, and the drain pipe empties it in 18 hr. If the faucet and the drain pipe are both open, how long does it take to fill the tank? 36 hr

19. A pipe fills a pool in 7 hr, and another one fills it in 21 hr. How long will it take to fill the pool using both pipes? $5\frac{1}{4}$ hr

20. One pipe fills a tank in 6 hr, and another fills it in 4 hr. How long will it take both pipes together to fill the tank? $2\frac{2}{5}$ hr

21. The main engine of a rocket burns for 60 sec on the fuel in the rocket's tank, while the auxiliary engine burns for 90 sec on the same amount of fuel. How long do both engines burn if they are operated together on the single tank of fuel? 36 sec

22. An in-flow pipe fills a pool in 12 hr, and another pipe drains it in 4 hr. How long does it take to empty the pool if both pipes are open simultaneously? (Assume that the pool is full at the start.) 6 hr

23. A pipe fills a tank in 9 hr, but the drain empties it in 6 hr. How long does it take to empty the tank if both pipes are open simultaneously? (Assume that the tank is full at the start.) 18 hr

C In Problems 24–30, solve the distance problems.

24. A water skier travels 30 mi downstream in the same time it takes him to go 20 mi upstream. If the river current flows at 5 mi/hr, what is the skier's speed in still water? 25 mi/hr

25. A small plane flies 240 mi against the wind in the same time it takes it to fly 360 mi with a tail wind. If the wind velocity is 30 mi/hr, find the plane's speed in still air. 150 mi/hr

26. A jet plane flies 700 mi against the wind in the same time it takes it to fly 900 mi with a tail wind. If the wind velocity is 50 mi/hr, what is the plane's speed in still air? 400 mi/hr

27. A small plane cruises at 120 mi/hr in still air. It takes this plane the same time to travel 270 mi against the wind as it does to travel 450 mi with a tail wind. What is the wind velocity? 30 mi/hr

28. A small plane can travel 200 mi against the wind in the same time it takes it to travel 260 mi with a tail wind. If the plane's speed in still air is 115 mi/hr, find the wind velocity. 15 mi/hr

29. An automobile travels 200 mi in the same time in which a small plane travels 1000 mi. Find their rates of speed if the airplane is 100 mi/hr faster than the automobile. Auto 25 mi/hr; plane 125 mi/hr

30. Janice ran 1000 m in the same time that Paula ran 950 m. If Paula's speed was $\frac{1}{4}$ m/sec less than Janice's, what was Janice's speed? 5 m/sec

SKILL CHECKER

Try the "Skill Checker" exercises so you'll be ready for the next section.

Simplify:

31. $\dfrac{x^{-5}}{x^3}$ $\dfrac{1}{x^8}$

32. $\dfrac{x^5}{x^{-3}}$ x^8

33. $(x^{-4})^{-5}$ x^{20}

34. $(x^{-4})^5$ $\dfrac{1}{x^{20}}$

35. $(-2xy^2)^3$ $-8x^3y^6$

36. $(-2x^2y)^{-3}$ $\dfrac{-1}{8x^6y^3}$

37. $x^{-9} \cdot x^7$ $\dfrac{1}{x^2}$

38. $x^9 \cdot x^{-11}$ $\dfrac{1}{x^2}$

39. $\left(\dfrac{a^{-4}}{b^3}\right)^2$ $\dfrac{1}{a^8b^6}$

40. $\left(\dfrac{a^4}{b^{-3}}\right)^{-2}$ $\dfrac{1}{a^8b^6}$

USING YOUR KNOWLEDGE

Formulas, Formulas, and More Formulas

In Section 2.2, we learned how to solve a formula for a specified variable. You will find that many formulas involve rational expressions. Use your knowledge of solving equations involving rational expressions to solve for specified variables in the following equations.

41. To find the focal length of lenses, lens makers use the formula
$$\frac{1}{F} = \frac{1}{f_1} + \frac{1}{f_2}$$
Solve for F. $F = \dfrac{f_1 f_2}{f_1 + f_2}$

42. To find the radius of curvature R of a sphere, we use the formula
$$R = \frac{2AS}{L - 2S}$$
Solve for A. $A = \dfrac{RL - 2RS}{2S}$

43. The electric current i in a simple series circuit is given by
$$i = \frac{2E}{R + 2r}$$
Solve for R. $R = \dfrac{2E - 2ri}{i}$

44. Cowling's rule states that a child's dose c for a child A years old, where A is between 2 and 13, is given by
$$c = \frac{A + 1}{24}d$$
where d is the adult dose. Solve for d. $d = \dfrac{24c}{A + 1}$

45. In Problem 44, what would the adult dose be if a 5-year-old child's dose for aspirin is 3 tablets a day?

12 tablets a day

46. Is there an integer A for which the dosages are the same for Cowling's rule and Young's rule?

$$\text{Young's rule: } c = \frac{A}{A + 12}d$$

No, A is close to 10

47. Given this trigonometry identity, solve it for $\cos(2u)$.

$$\cos^2 u = \frac{1 + \cos(2u)}{2}$$

$\cos(2u) = 2\cos^2 u - 1$

48. Given this trigonometry identity, solve it for the sum, $\tan u + \tan v$

$$\tan(u + v) = \frac{\tan u + \tan v}{1 - \tan u \tan v}$$

$\tan u + \tan v = \tan(u + v)(1 - \tan u \tan v)$

WRITE ON

49. There's another way to solve Example 2. If you assume that t is the time it takes for the couple to complete the form working together, then in t hr the first person will do $\frac{t}{6}$ of the job and the second person will do $\frac{t}{8}$ of the job. Working together, they will do $\frac{t}{6} + \frac{t}{8}$ of the job. To what should this sum be equal? Explain.

Answers may vary.

50. Using the method in Problem 49, what equation would you use to solve Example 3?

$\frac{T}{4} - \frac{T}{5} = 1$

MASTERY TEST

If you know how to do these problems, you have learned your lesson!

51. A speed boat can travel 36 mi downstream in the same time it takes it to go 12 mi upstream. If the current is moving at 18 mi/hr, what is the speed of the boat in still water? 36 mi/hr

52. A pool is filled by an intake pipe in 5 hr and emptied by a drain pipe in 6 hr. How long would it take to fill the pool with both pipes open? 30 hr

53. The sum of the reciprocals of two consecutive even integers is $\frac{5}{12}$. What are the integers? 4 and 6

54. According to Clark's rule, the dose c for a child weighing W pounds is

$$c = \frac{W}{150}d$$

where d is the adult dose. Solve for d. $d = \frac{150c}{W}$

6.8 VARIATION

To Succeed, Review How To ...

1. Evaluate an expression (p. 49).

2. Solve linear equations (p. 79).

Objectives

Write an equation expressing:

A Direct variation.

B Inverse variation.

C Joint variation.

D Solve applications involving direct, inverse, and joint variation.

GETTING STARTED

Pendulums and Gas Mileage

What could pendulums and gas mileage possibly have in common? Each have measures associated with them that can be expressed in a formula referred to as a **variation.**

As the length L of the string of the pendulum increases, the time T it takes the pendulum to make a full back and forth swing increases. What is the formula relating the length L and the time T? Galileo Galilei discovered that the time T (in seconds) it takes for one swing of the pendulum varies directly as the square root of the length L of the pendulum.

$$k = \frac{T}{\sqrt{L}}$$

In the same manner, the number m of miles you drive a car is *proportional to,* or *varies directly as,* the number g of gallons of gas used. This means that the ratio

$$\frac{m}{g} \text{ is a constant: } \frac{m}{g} = k \quad \text{or} \quad m = kg$$

In 2002 inventor Doug Malewicki, with a team of fellow enthusiasts, began construction of a "C2C" car that could go from California to New York (about 3500 mi) on one 25 gal tank of standard gasoline. Using $m/g = k$ as before, what is k? Can you explain what k means?

A Direct Variation

Do you get higher grades when you study more hours? If this is the case, your grades vary directly or are directly proportional to the number of hours you study. Here is the definition for direct variation.

Direct Variation

y **varies directly as** x if there is a constant k such that

$$y = kx$$

(k is usually called the constant of variation.)

Other words can be used to indicate direct variation. Here's a list of some of these words and how they translate into an equation.

English Phrase	Translation
y varies with x	$y = kx$
y varies directly as t	$y = kt$
y is proportional to v	$y = kv$
v varies as the square of t	$v = kt^2$
p varies as the cube of r	$p = kr^3$
T varies as the square root of L	$T = k\sqrt{L}$

EXAMPLE 1   Mustaches and variations

The length L of a mustache varies directly as the time t that it takes to grow.

a. Write an equation of variation.

b. One of the longest mustaches on record was grown by Masuriya Din. His mustache grew 56 in. (on each side) over a 14-yr period. Find k and explain what it represents.

SOLUTION

a. Since the length L varies directly as the time t,

$$L = kt$$

b. We know that when $L = 56$, $t = 14$. Thus

$$56 = k \cdot 14$$
$$4 = k$$

This means that Mr. Din's mustache grew 4 in. each year.

PROBLEM 1

Hair length L is proportional to time t.

a. Write an equation of variation.

b. If your hair grew 6 in. in 2 mo find k.

B   Inverse Variation

Sometimes, as one quantity increases, a related quantity decreases proportionately. For example, the *more* time you spend practicing a task, the *less* time it will take you to do the task. In such cases, we say that the quantities *vary inversely as* each other.

Inverse Variation

y **varies inversely as** x if there is a constant k such that

$$y = \frac{k}{x}$$

Here are some other words that also mean "vary inversely."

English Phrase	Translation
y varies inversely with x	$y = \frac{k}{x}$
y is inversely proportional to x	$y = \frac{k}{x}$
v varies inversely as the square of t	$v = \frac{k}{t^2}$
p varies inversely as the cube of r	$p = \frac{k}{r^3}$
T varies inversely as the square root of L	$T = \frac{k}{\sqrt{L}}$

Answers

1. **a.** $L = kt$ **b.** 3 (in./mo)

EXAMPLE 2 Speeds and distances	**PROBLEM 2**

The speed s that a car travels is inversely proportional to the time t it takes to travel a given distance.

a. Write the equation of variation.

b. If a car travels at 60 mi/hr for 3 hr, what is k and what does it represent?

SOLUTION

a. The equation is

$$s = \frac{k}{t}$$

b. We know that $s = 60$ when $t = 3$. Substituting 60 for s and 3 for t,

$$60 = \frac{k}{3}$$

$$k = 180$$

In this case, k represents the total distance traveled, 180 mi.

PROBLEM 2

The principal P invested is inversely proportional to the annual rate of interest r.

a. Write an equation of variation.

b. Find k if $r = 8\%$ and $P = \$100$.

EXAMPLE 3 Deafening sound	**PROBLEM 3**

Have you ever heard one of those loud boom boxes or a car sound system that makes your bones vibrate? The loudness L of sound is inversely proportional to the square of your distance d from the source.

a. Write an equation of variation.

b. The loudness of rock music coming from a boom box 5 ft away is 100 dB (decibels). Find k.

c. If you move to 10 ft away from the boom box, how loud is the sound?

SOLUTION

a. The equation is

$$L = \frac{k}{d^2}$$

b. We know that $L = 100$ for $d = 5$ so that

$$100 = \frac{k}{5^2} = \frac{k}{25}$$

Multiplying both sides by 25, we find that $k = 2500$.

c. Since $k = 2500$,

$$L = \frac{2500}{d^2}$$ Substitute 2500 for k.

When $d = 10$,

$$L = \frac{2500}{10^2} = 25 \text{ dB}$$

100 dB is only 20 dB from the threshold of pain, which causes immediate and permanent hearing loss.

PROBLEM 3

The f-number on a camera varies inversely as the diameter a of the aperture when the distance is set at infinity.

a. Write an equation of variation.

b. Find k when the f-number is 6 and $a = \frac{1}{2}$.

c. Find a if the f-number is 18.

C Expressing Joint Variation

Besides the direct and inverse variations we have discussed, there can be variation involving a third variable. A variable z can vary *jointly* with the variables x and y. For example, labor costs c vary jointly with the number of workers w used and the number of hours h that they work. The formal expression of joint variation is given here.

Answers

2. **a.** $P = \frac{k}{r}$ **b.** \$8

3. **a.** $f = \frac{k}{a}$ **b.** 3 **c.** $\frac{1}{6}$

Web It

For more on writing and solving variation equations, go to link 6-8-1 at mhhe.com/bello.

Joint Variation

z **varies jointly** with x and y if there is a constant k such that

$$z = kxy$$

The statement *z is proportional to x and y* is sometimes used to mean *z varies jointly with variables x and y*. Thus the fact that labor costs c vary jointly with the number w of workers used and the number h of hours worked can be expressed as $c = kwh$, where k is a constant.

EXAMPLE 4   **The lifting force**

The lifting force P exerted by the atmosphere on the wings of an airplane varies jointly with the wing area A in square feet and the square of the plane's speed V in miles per hour. Suppose the lift is 1200 lb for a wing area of 100 ft² and a speed of 75 mi/hr.

a. Find an equation of variation.

b. Find k.

c. Find the lifting force on a wing area of 60 ft² when $V = 125$.

SOLUTION

a. Since P varies jointly with the area A and the square of the velocity V, we have $P = kAV^2$.

b. When the lift $P = 1200$, we know that $A = 100$ and $V = 75$. Substituting these values in the equation $P = kAV^2$, we obtain

$$1200 = k \cdot 100 \cdot (75)^2$$

Dividing both sides by $100 \cdot 75^2$, we find

$$k = \frac{1200}{100 \cdot 75^2} = \frac{12}{75^2} = \frac{4}{1875}$$

Thus $P = kAV^2$ becomes

$$P = \frac{4}{1875} AV^2$$

c. $P = \frac{4}{1875} AV^2$, where $A = 60$ and $V = 125$.

$$P = \frac{4}{1875} (60)(125^2)$$

$$= 2000 \text{ lb}$$

The lifting force is 2000 lb.

PROBLEM 4

The wind force F on a vertical surface varies jointly with the area A of the surface and the square of the wind velocity V. Suppose the wind force on 1 ft² of surface is 2.2 lb when $V = 20$ mi/hr.

a. Find an equation of variation.

b. Find k.

c. Find the force on a 2-ft² vertical surface when $V = 60$ mi/hr.

 **Solving an Application**

EXAMPLE 5   **An awful lot of snow**

Figure 1 (next page) shows the number of gallons of water, g (in millions), produced by an inch of snow in different cities. The larger the area of the city, the more gallons of water are produced, so g is directly proportional to A, the area of the city (in square miles).

a. Write an equation of variation.

b. If the area of St. Louis is about 62 mi², what is k?

PROBLEM 5

Using the graph in Example 5 and the formula found in part **a**, find the approximate area of Chicago. (Round your answer to the nearest square mile.)

Answers

4. a. $F = kAV^2$ **b.** 0.0055
c. 39.6 lb   **5.** 233 mi²

c. Find the amount of water produced by 1 in. of snow falling in Anchorage, Alaska, with an area of 1700 mi².

SOLUTION

a. Since g is directly proportional to A, $g = kA$.

b. From Figure 1, we can see that $g = 100$ (million) is the number of gallons of water produced by 1 in. of snow in St. Louis. Since it is given that $A = 62$, $g = kA$ becomes

$$100 = k \cdot 62 \quad \text{or} \quad k = \frac{100}{62} = \frac{50}{31}$$

c. For Anchorage, $A = 1700$, thus $g = \frac{50}{31} \cdot 1700 \approx 2742$ million gallons of water (rounded to the nearest million).

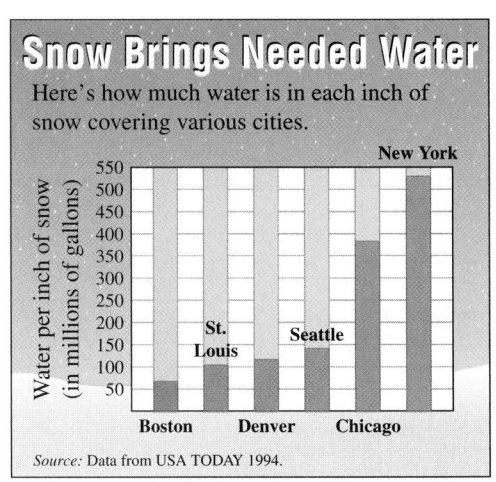

Snow Brings Needed Water

Here's how much water is in each inch of snow covering various cities.

Source: Data from USA TODAY 1994.

Figure 1

Calculate It   Using StatPlot and Lists to Verify Direct and Inverse Variation Equations

How do you recognize different types of variation with your calculator? If you have several points, you can graph them and examine the result. In Example 1, we concluded that Mr. Din's mustache grew 4 in. each year. This means that at the end of the first year, his mustache was 4 in. long; at the end of the second year, it was 8 in. long; and at the end of 14 yr, it was 56 in. long. You now have three ordered pairs: (1, 4), (2, 8), and (14, 56). You can make a list (clear lists by pressing [2nd] [+] [4] [ENTER]) by pressing [STAT] [1] and then entering 1, 2, and 14 under L_1 and 4, 8, and 56 under L_2. Press [2nd] [Y=] [ENTER] [ENTER] to turn plot 1 on; select the first type of graph, L_1, L_2, and ■. So that we don't have to contend with deciding what type of window to use, press [ZOOM] [9]. As you can see in Window 1, the points are on a line. (Remember that the length varies directly as the time.)

If you want to find the equation of the line passing through the three points in the graph, press [STAT] [◊] [4] [ENTER]. The calculator will tell you that $y = ax + b$, where $a = 4$ and $b = 0$— $y = 4x$ as before.

To get a feel for inverse variation, in Example 2, select a $[-1, 200]$ by $[-1, 5]$ window with Xscl = 10 and Yscl = 1. Using Y_1 instead of x and x instead of t, graph

$$Y_1 = \frac{180}{x}$$

The result is shown in Window 2. Can you see that as x increases, Y_1 decreases? Try looking at the graph

$$L = \frac{2500}{d^2}$$

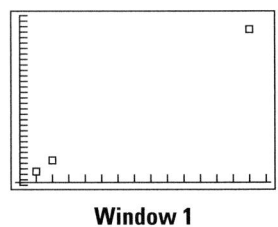

Window 1

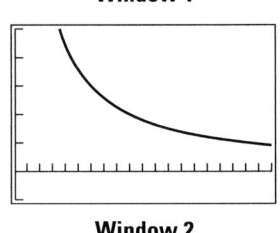

Window 2

What window do you need? What does the graph look like?

Exercises 6.8

A In Problems 1–5, write an equation of variation using k as the constant.

1. The tension T on a spring varies directly with the distance s it is stretched. (This is usually called Hooke's law.) $T = ks$

2. The distance s a body falls in t seconds is directly proportional to the square of t. $s = kt^2$

3. The weight W of a dam varies directly with the cube of its height h. $W = kh^3$

4. The kinetic energy KE of a moving body is proportional to the square of its velocity v. $KE = kv^2$

5. The weight W of a human brain is directly proportional to the body weight B. $W = kB$

B In Problems 6–8, write an equation of variation using k as the constant.

6. In a circuit with constant voltage, the current I varies inversely with the resistance R of the circuit. $I = \frac{k}{R}$

7. For a wire of fixed length, the resistance R varies inversely with the square of its diameter D. $R = \frac{k}{D^2}$

8. The intensity of illumination I from a source of light varies inversely with the square of the distance d from the source. $I = \frac{k}{d^2}$

C In Problems 9–20, write an equation of variation using k as the constant.

9. The annual interest I received on a savings account varies jointly with the principal P (the amount in the account) and the interest rate r paid by the bank. $I = kPr$

10. The cost C of a building varies jointly as the number w of workers used to build it and the cost of materials m. $C = kwm$

11. The amount of oil A used by a ship traveling at a uniform speed varies jointly with the distance s and the square of the speed v. $A = ksv^2$

12. The power P in an electric circuit varies jointly with the resistance R and the square of the current I. $P = kRI^2$

13. The volume V of a rectangular container of fixed length varies jointly with its depth d and width w. $V = kdw$

14. The force of attraction F between two spheres of mass m_1 and m_2 varies directly as the product of the masses and inversely as the square of the distance d between their centers. $F = \frac{km_1 m_2}{d^2}$

15. The illumination I in foot-candles upon a wall varies directly with the intensity i in candlepower of the source of light and inversely with the square of the distance d from the light. $I = \frac{ki}{d^2}$

16. The strength S of a horizontal beam of rectangular cross section and of length L varies jointly as the breadth b and the square of the depth d and inversely as the length L. $S = \frac{kbd^2}{L}$

17. The electrical resistance R of a wire of uniform cross section varies directly as its length L and inversely as its cross-sectional area A. $R = \frac{kL}{A}$

18. The electrical resistance R of a wire varies directly as the length L and inversely as the square of its diameter d. $I = \frac{kL}{d^2}$

19. The weight W of a body varies inversely as the square of its distance d from the center of the earth. $W = \frac{k}{d^2}$

20. z varies directly as the cube of x and inversely as the square of y. $z = \frac{kx^3}{y^2}$

APPLICATIONS

Real Life Variation Problems

21. The amount of annual interest I you receive on a savings account is directly proportional to the amount of money m you have in the account.

    a. Write an equation of variation. $I = km$

    b. If \$480 produces \$26.40 in interest, what is k? $k = 0.055$ or 5.5%

    c. How much annual interest would you receive if the account had \$750? \$41.25

22. The number of revolutions, R (rev), a record makes as it is being played varies directly as the time t that it is on the turntable.

    a. Write an equation of variation. $R = kt$

    b. A record that lasted $2\frac{1}{2}$ min made 112.5 rev. What is k? $k = 45$

    c. If a record makes 108 rev, how long does it take to play it? 2.4 min

23. The distance d an automobile travels after the brakes have been applied varies directly as the square of its speed s.

    a. Write an equation of variation. $d = ks^2$

    b. If the stopping distance for a car going 30 mi/hr is 54 ft, what is k? $k = 0.06$

    c. What is the stopping distance for a car going 60 mi/hr? 216 ft

24. The weight of a person varies directly as the cube of the person's height h (in inches). The **threshold weight** T (in pounds) for a person is defined as "the crucial weight, above which the mortality (risk) for the patient rises astronomically."

    a. Write an equation of variation relating T and h. $T = kh^3$

    b. If $T = 196$ when $h = 70$, find k written as a fraction. $\frac{1}{1750}$

    c. To the nearest pound, what is the threshold weight T for a person 75 in. tall? 241 lb

25. The number S of new songs a rock band needs to stay on top each year is inversely proportional to the number of years y the band has been in the business.

   a. Write an equation of variation. $S = \frac{k}{y}$

   b. If, after 3 yr in the business, the band needs 50 new songs, how many songs will it need after 5 yr? 30 new songs

26. When the distance is set at infinity, the f-number on a camera lens varies inversely as the diameter d of the aperture (opening).

   a. Write an equation of variation. $f = \frac{k}{d}$

   b. If the f-number on a camera is 8 when the aperture is $\frac{1}{2}$ in., what is k? $k = 4$

   c. Find the f-number when the aperture is $\frac{1}{4}$ in. 16

27. The weight W of an object varies inversely as the square of its distance d from the center of the earth.

   a. Write an equation of variation. $W = \frac{k}{d^2}$

   b. An astronaut weighs 121 lb on the surface of the earth. If the radius of the earth is 3960 mi, find the value of k for this astronaut. (Do not multiply out your answer.) $k = 121(3960)^2$

   c. What will this astronaut weigh when she is 880 mi above the surface of the earth? 81 lb

28. The number of miles m you can drive in your car is directly proportional to the amount of fuel g in your gas tank.

   a. Write an equation of variation. $m = kg$

   b. The greatest distance yet driven without refueling on a single fill in a standard vehicle is 1691.6 mi. If the twin tanks used to do this carried a total of 38.2 gal of fuel, what is k? (Round the answer to the nearest tenth.) $k = 44.3$

   c. How many mi/gal is this? 44.3 mi/gal

29. The distance d (in miles) traveled by a car is directly proportional to the average speed s (in mi/hr) of the car, even when driving in reverse.

   a. Write an equation of variation. $d = ks$

   b. The highest average speed attained in any nonstop reverse drive of more than 500 mi is 28.41 mi/hr. If the distance traveled was 501 mi, find k. (Round the answer to the nearest hundredth.) $k = 17.63$

   c. What does k represent? The number of hours needed to travel d distance at s speed.

30. Have you called in on a radio contest lately? According to Don Burley, a radio talk-show host in Kansas City, the listener response to a radio call-in contest is directly proportional to the size of the prize.

   a. If 40 listeners call when the prize is $100, write an equation of variation using N for the number of listeners and P for the prize in dollars. $N = 0.4P$

   b. How many calls would you expect for a $5000 prize? 2000

31. The number of chirps C a cricket makes each minute is directly proportional to 37 less than the temperature F in degrees Fahrenheit.

   a. If a cricket chirps 80 times when the temperature is 57°F, what is the equation of variation?
$C = 4(F - 37)$

   b. How many chirps per minute would the cricket make when the temperature is 90°F? 212 chirps

32. According to George Flick, the ship's surgeon of the SS *Constitution,* the number of hours H your life is shortened by smoking cigarettes varies jointly as N and $t + 10$, where N is the number of cigarettes you smoke and t is the time in minutes it takes you to smoke each cigarette. If it takes 5 min to smoke a cigarette and smoking 100 of them shortens your lifespan by 25 hr, how long would smoking 2 packs a day for a year (360 days) shorten your lifespan? (*Note:* There are 20 cigarettes in a pack.) 3600 hr (about 150 days)

33. The concentration of carbon dioxide (CO_2) in the atmosphere has been increasing due to human activities such as automobile emissions, electricity generation, and deforestation. In 1965, CO_2 concentration was 319.9 parts per million (ppm), and 23 years later, it increased to 351.3 ppm. The *increase* of carbon dioxide concentration, I, in the atmosphere is directly proportional to number of years n elapsed since 1965.

   a. Write an equation of variation for I. $I = kn$

   b. Find k. (Round to the nearest thousandth.)
$k = 1.365$

   c. What would you predict the CO_2 concentration to be in the year 2000? 367.675 ppm

34. The increase I, in the percent of college graduates in the United States among persons 25 years and older between 1930 and 1990 is proportional to the square of the number of years after 1930. In 1940, the increase was about 5%.

   a. Write an equation of variation for I if n is the number of years elapsed since 1930. $I = kn^2$

   b. Find k. $k = 0.0005$

   c. What would you predict the percent increase to be in the year 2000? 245%

35. The simple interest I on an account varies jointly as the time t and the principal P. After one quarter (3 months), an $8000 principal earned $100 in interest. How much would a $10,000 principal earn in 5 months? $208.33

36. At depths of more than 1000 m (a kilometer), water temperature T (in degrees Celsius) in the Pacific Ocean varies inversely as the water depth d (in meters). If the water temperature at 4000 m is 1°C, what would it be at 8000 m? $\frac{1}{2}$°C

37. Anthropologists use the cephalic index C in the study of human races and groupings. This index is directly proportional to the width w and inversely proportional to the length L of the head. The width of the head in a skull found in 1921 and named Rhodesian man was 15 cm, and its length was 21 cm. If the cephalic index of Rhodesian man was 98, what would the cephalic index of Cro-Magnon man be, whose head was 20 cm long and 15 cm wide? $C = 102.9$

SKILL CHECKER

Try the "Skill Checker" exercises so you'll be ready for the next section.

Graph:

38. $x + y = 3$

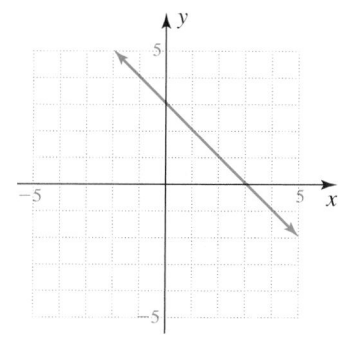

39. $2x - y = 2$

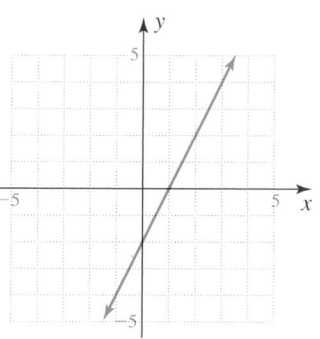

40. $2x + \frac{1}{2}y = 2$

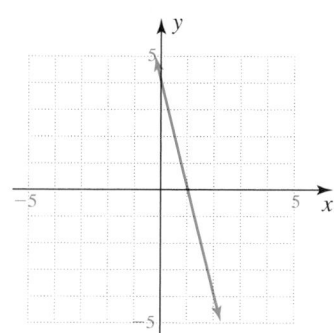

41. $y = -x - 3$

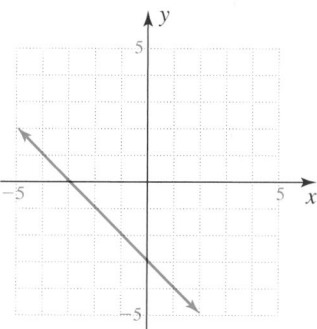

42. $y = -4x + 4$

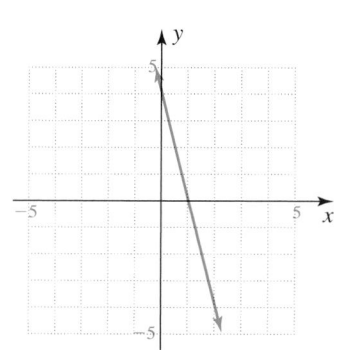

USING YOUR KNOWLEDGE

The Pressure of Diving

The equation for direct variation between x and y ($y = kx$) and the equation of a line of slope m passing through the origin ($y = mx$) are very similar. Look at the following table, which gives the water pressure in pounds per square inch exerted on a diver.

Depth of diver (ft)	10	25	40	55
Pressure on diver (lb/in.^{2})	4.2	10.5	16.8	23.1

43. Graph these points using x as the depth and y as the pressure.

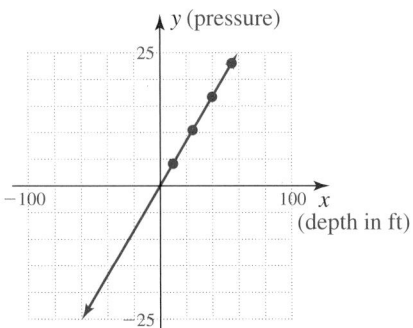

44. What is the slope of the resulting line? 0.42

45. As it turns out, the pressure p on a diver is directly proportional to the depth d. Write an equation of variation. $p = kd$

46. Use one of the points in the table to find k. $k = 0.42$

47. What is the relationship between k and the slope found in Problem 44? They are equal.

48. Predict the pressure on a diver at a depth of 125 ft. 52.5 lb/in.^{2}

WRITE ON

49. Explain the difference between direct variation and inverse variation. Answers may vary.

50. Find two different ways of expressing the idea that "y is directly proportional to x." Answers may vary.

51. Find two different ways of expressing the idea that "y is inversely proportional to x." Answers may vary.

52. Find two different ways of expressing the idea that y and z are directly proportional to x. Answers may vary.

MASTERY TEST

If you know how to do these problems, you have learned your lesson!

53. The wind force F on a vertical surface varies jointly with the area A of the surface and the square of the wind velocity V. Suppose the wind force on 1 ft^2 of surface is 1.8 lb when $V = 20$ mi/hr.

  a. Find an equation of variation. $F = kAV^2$

  b. Find k. $k = 0.0045$

  c. Find the force on a 2-ft^2 vertical surface when $V = 60$ mi/hr. 32.4 lb

54. The f-number on a camera varies inversely as the diameter a of the aperture when the distance is set at infinity.

  a. Write an equation of variation. $f = \frac{k}{a}$

  b. Find k when the f-number is 8 and $a = \frac{1}{2}$. $k = 4$

  c. Find a if the f-number is 16. $a = \frac{1}{4}$

55. The principal P invested is inversely proportional to the annual rate of interest r.

  a. Write an equation of variation. $P = \frac{k}{r}$

  b. Find k when $r = 10\%$ and $P = \$100$. $k = 10$

56. Hair length L is proportional to time t.

  a. Write an equation of variation. $L = kt$

  b. Find k if your hair grew 0.6 in. in 2 mo. $k = 0.3$

COLLABORATIVE LEARNING 6A

A bar graph is a chart that uses either horizontal or vertical bars to show comparisons among categories. One axis shows the categories being compared and the other axis represents the values. Some bar graphs show the bars being divided into subparts so that cumulative effects can be shown and are called *stacked bar graphs* as shown in Figure 2.

When making a stacked bar graph, examine the data to find:

1. the bar with the largest value that will help determine the range and the increments on the vertical axis.

2. how many bars will be needed for the horizontal axis.

3. whether the bars will be arranged from largest to smallest or chronologically.

4. the sequence of order to be used on the stacking of the subparts to each bar.

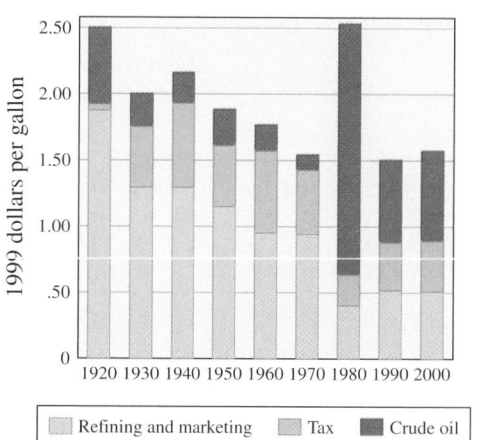

Components of U.S. Retail Gasoline Prices

Refining and marketing — Tax — Crude oil

Figure 2

Source: Data from Cambridge Energy Research Associates.

Make a stacked bar graph using the following information.

The University of Guelph in Ontario, Canada has published some statistics indicating the amount of tons of various recyclable waste materials during the years of 1999 to 2002. They are listed in the following table.

Waste Material	1999 Percentage of Total Tons	Tons	2000 Percentage of Total Tons	Tons	2001 Percentage of Total Tons	Tons	2002 Percentage of Total Tons	Tons
Landfill	66		57		58		72	
Mixed	12		12.5		11.5		0.5	
Manure	21		30		30		27.25	
Fine paper	0.5		0.25		0.2		0.15	
Misc.	0.5		0.25		0.3		0.1	
Total Tons	3346		3935		3891		3608	

1. Complete the table by finding the amount of tons recycled for each type of waste material for each year.

2. Which will be the largest bar, and what will be the increments used for the vertical axis?

3. How will the horizontal axis be arranged?

4. What order will be used to "stack" the sequence of different types of waste materials on each bar?

5. Make a stacked bar graph for these statistics using a poster board.

6. Discuss the trend in the waste recycling at the University of Guelph over the 4-yr period, and what are some possible reasons for that change in the recycling of the waste.

7. Divide into groups, research waste recycling statistics in your area, and present the information to the class using a stacked bar graph.

COLLABORATIVE LEARNING 6B

The Family Catering Company advertises they can cater events for 30 to 50 people. Recently they were asked to bid on a dinner for 45 people. Some of the company data is indicated in the following table. You will use it to help figure out what price to quote. Divide into groups and answer the questions.

Banquet Size	Cost of Food	Rate Per Person	Cost of Beverage	Rate Per Person	Cost of Labor	Rate Per Person	Total Cost	Rate Per Person	Amount Charged	Rate Per Person
30	$189		$33		$159		$381		$660	
35	$210		$35		$164.50		$409.50		$728	
40	$228		$36		$170		$434		$784	

1. Complete the table by computing the following rates:

   a. Cost of food per person

   b. Cost of beverage per person

   c. Cost of labor per person

   d. Total cost per person

   e. Amount charged per person

2. If the rates remain the same, what is the expected cost for the dinner?

3. How much should the company bid for the dinner?

4. Sometimes the customer wants a quote that indicates the cost per person. What cost per person would you quote?

5. Discuss other applications for rates, research some data, and then present the results in a similar table.

Research Questions

1. As we mentioned in *The Human Side of Algebra,* Tartaglia discovered new methods for solving cubics. But there is more to the story. Unfortunately, "an unprincipled genius who taught mathematics and practiced medicine in Milan, upon giving a solemn pledge of secrecy, wheedled the key to the cubic from Tartaglia." What was the name of this dastardly man, what publication did Tartaglia's work appear in, and what was the subject of the publication?

2. This "dastardly man" had a pupil who argued that his teacher received his information from del Ferro through a third party; the pupil then accused Tartaglia of plagiarism. What was the name of the pupil? Write a short paragraph about this pupil's contributions to mathematics.

3. Quadratic, cubic, and quartic equations were solved by formulas formed from the coefficients of the equation by using the four fundamental operations and taking radicals (which you shall study in the next chapter) of various sorts. Write a short paper detailing the struggles to solve quintic equations by these methods, which have culminated in the proof of the impossibility of obtaining such solutions by algebraic methods.

4. In the book *La Geometrie,* the author obtains the factor theorem as a major result. Who is the author of *La Geometrie,* and according to historians, what are the mathematical implications of this theorem?

Summary

SECTION	ITEM	MEANING	EXAMPLE
6.1	Fraction Rational expression	An expression denoting a division An expression of the form $\frac{P}{Q}$, where P and Q are polynomials, $Q \neq 0$	$\frac{3}{4}, \frac{-8}{7}$, and $\frac{1}{2}$ are fractions. $\frac{1}{x}, \frac{x}{x+y}, \frac{x+y}{z}$, and $\frac{x^2 + 21x - 1}{x^3 + 3}$ are rational expressions.
6.1B	Fundamental property of rational expressions	If P, Q, and K are polynomials, $$\frac{P}{Q} = \frac{P \cdot K}{Q \cdot K}$$ for all values for which the denominator is not zero.	$\frac{3}{4} = \frac{3 \cdot 8}{4 \cdot 8}, \frac{x}{2} = \frac{x \cdot 5}{2 \cdot 5}$, and $$\frac{3}{x+2} = \frac{3(x+5)}{(x+2)(x+5)}$$
6.1C	Standard forms of a fraction	The forms $\frac{-a}{b}$ and $\frac{a}{b}$ are the standard forms of a fraction.	$-\frac{-5}{4}$ is written as $\frac{5}{4}, -\frac{-8}{x}$ is written as $\frac{-8}{x}$, and $\frac{5}{-4}$ is written as $\frac{-5}{4}$.
6.1D	Simplified (reduced) fraction	A fraction is simplified (reduced) if the numerator and denominator have no common factor.	$\frac{3}{4}, \frac{9}{8}$, and $\frac{x}{7}$ are reduced, but $\frac{3}{6}$ and $\frac{x+y}{x^2 - y^2}$ are not.
6.2A	Multiplication of rational expressions	If a, b, c, and d are rational expressions ($b \neq 0$, $d \neq 0$), $$\frac{a}{b} \cdot \frac{c}{d} = \frac{a \cdot c}{b \cdot d}$$	$\frac{3}{4} \cdot \frac{5}{7} = \frac{3 \cdot 5}{4 \cdot 7} = \frac{15}{28}$ $\frac{x}{x+y} \cdot \frac{3}{x-y} = \frac{3x}{(x+y)(x-y)}$ $= \frac{3x}{x^2 - y^2}$
6.2B	Division of rational expressions	If a, b, c, and d are rational expressions ($b \neq 0$, $c \neq 0$, $d \neq 0$), $$\frac{a}{b} \div \frac{c}{d} = \frac{a}{b} \cdot \frac{d}{c}$$	$\frac{3}{4} \div \frac{7}{5} = \frac{3}{4} \cdot \frac{5}{7}$ $\frac{x}{x+y} \div \frac{x-y}{3} = \frac{x}{x+y} \cdot \frac{3}{x-y}$
6.3A	Addition and subtraction of rational expressions with the same denominator	If a, b, and c are rational expressions and $b \neq 0$, $$\frac{a}{b} + \frac{c}{b} = \frac{a+c}{b} \text{ and } \frac{a}{b} - \frac{c}{b} = \frac{a-c}{b}$$	$\frac{3}{5} + \frac{1}{5} = \frac{4}{5}$ and $\frac{2}{x} + \frac{1}{x} = \frac{3}{x}$ $\frac{3}{5} - \frac{1}{5} = \frac{2}{5}$ and $\frac{2}{x} - \frac{1}{x} = \frac{1}{x}$
6.3B	Addition and subtraction of rational expressions with different denominators	$$\frac{a}{b} + \frac{c}{d} = \frac{ad + bc}{bd}$$ $$\frac{a}{b} - \frac{c}{d} = \frac{ad - bc}{bd}$$	$\frac{3}{4} + \frac{1}{7} = \frac{21 + 4}{28} = \frac{25}{28}$ $\frac{4}{5} - \frac{1}{3} = \frac{12 - 5}{15} = \frac{7}{15}$

(Continued)

SECTION	ITEM	MEANING	EXAMPLE
6.4A	Complex fraction	A fraction whose numerator or denominator (or both) contain other fractions	$\dfrac{\frac{1}{2}}{x+1}$, $\dfrac{3}{\frac{1}{5}+x}$, and $\dfrac{\frac{x}{2}}{x+\frac{1}{2}}$ are complex fractions.
	Simple fraction	A fraction that is not complex	$\frac{1}{2}$, $\frac{2x}{4}$, and $\frac{x+y}{x-y}$ are simple fractions.
6.5E	Remainder theorem	If the polynomial $P(x)$ is divided by $x-k$, then the remainder is $P(k)$.	If $P(x) = x^2 + 2x + 5$ is divided by $x-3$, the remainder is $P(3) = 3^2 + 2(3) + 5 = 20$.
	Factor theorem	When $P(x)$ has a factor $(x-k)$, $P(k) = 0$.	$P(x) = x^2 + x - 6$ has $x-2$ as a factor, thus $P(2) = 0$.
6.6A	Extraneous solution	A trial solution that does not satisfy the equation	3 is an extraneous solution of $3 + \dfrac{1}{x-3} = \dfrac{1}{x-3}$.
6.6B	Proportion	Two equal ratios; cross products are equal	$\dfrac{6}{8} = \dfrac{9}{12}$ is a proportion and $6 \cdot 12 = 8 \cdot 9$.
6.7	RSTUV method for solving word problems	**R**ead the problem. **S**elect a variable for the unknown. **T**hink of a plan. **U**se algebra to solve. **V**erify the answer.	
6.8A	Direct variation	y varies directly as x if there is a constant k, such that $y = k \cdot x$	"b varies directly as c" means $b = k \cdot c$
6.8B	Inverse variation	y varies inversely as x if there is a constant k, such that $y = \frac{k}{x}$	"y varies inversely as d" means $y = \frac{k}{d}$
6.8C	Joint variation	z varies jointly with x and y if there is a constant k such that $z = kxy$	"m varies jointly as p and q" means $m = kpq$

Review Exercises

(If you need help with these exercises, look in the section indicated in brackets.)

1. [6.1A] For what value(s) are the following rational expressions undefined?

a. $\dfrac{-3x}{x-4}$ $x = 4$

b. $\dfrac{x^2 + 9}{x^2 + 10x + 9}$ $x = -1; -9$

c. $\dfrac{8 + x}{x^2 + 25}$

None, defined for all values

2. [6.1B] Write the fraction with the indicated denominator.

a. $\dfrac{2x^2}{9y^4}$; denominator $36y^7$ $\dfrac{8x^2y^3}{36y^7}$

b. $\dfrac{2x^2}{9y^4}$; denominator $45y^8$ $\dfrac{10x^2y^4}{45y^8}$

c. $\dfrac{2x+1}{x+1}$; denominator $x^2 + 6x + 5$ $\dfrac{2x^2 + 11x + 5}{x^2 + 6x + 5}$

d. $\dfrac{2x+1}{x+1}$; denominator $x^2 + 7x + 6$ $\dfrac{2x^2 + 13x + 6}{x^2 + 7x + 6}$

3. [6.1C] Write in standard form.

a. $-\dfrac{-6}{y}$ $\dfrac{6}{y}$

b. $\dfrac{-7}{-y}$ $\dfrac{7}{y}$

c. $-\dfrac{-8}{-y}$ $\dfrac{-8}{y}$

4. [6.1C] Write in standard form.

a. $-\dfrac{x-y}{6} \quad \dfrac{y-x}{6}$

b. $-\dfrac{7}{x-y} \quad \dfrac{7}{y-x}$

c. $-\dfrac{x-y}{8} \quad \dfrac{y-x}{8}$

5. [6.1D] Reduce to lowest terms.

a. $\dfrac{x^4y^7}{xy^2} \quad x^3y^5$

b. $\dfrac{x^4y^8}{xy^2} \quad x^3y^6$

c. $\dfrac{x^4y^9}{xy^2} \quad x^3y^7$

6. [6.1D] Simplify.

a. $\dfrac{xy^2+y^3}{x^2-y^2} \quad \dfrac{y^2}{x-y}$

b. $\dfrac{xy^3+y^4}{x^2-y^2} \quad \dfrac{y^3}{x-y}$

c. $\dfrac{xy^4+y^5}{x^2-y^2} \quad \dfrac{y^4}{x-y}$

7. [6.1D] Simplify.

a. $\dfrac{4y^2-x^2}{x^3+8y^3} \quad \dfrac{2y-x}{x^2-2xy+4y^2}$

b. $\dfrac{4y^2-x^2}{x^3-8y^3} \quad \dfrac{-(x+2y)}{x^2+2xy+4y^2}$

c. $\dfrac{9y^2-x^2}{x^3-27y^3} \quad \dfrac{-(x+3y)}{x^2+3xy+9y^2}$

8. [6.2A] Multiply.

a. $\dfrac{x-2}{3x-2y} \cdot \dfrac{9x^2-4y^2}{3x^2-5x-2}$
   $\dfrac{3x+2y}{3x+1}$

b. $\dfrac{x-2}{3x-2y} \cdot \dfrac{9x^2-4y^2}{4x^2-11x+6}$
   $\dfrac{3x+2y}{4x-3}$

c. $\dfrac{x-2}{3x-2y} \cdot \dfrac{9x^2-4y^2}{5x^2-8x-4}$
   $\dfrac{3x+2y}{5x+2}$

9. [6.2B] Divide.

a. $\dfrac{x+4}{x-2} \div (x^2+8x+16)$
   $\dfrac{1}{(x-2)(x+4)}$

b. $\dfrac{x+5}{x-2} \div (x^2+10x+25)$
   $\dfrac{1}{(x-2)(x+5)}$

c. $\dfrac{x+6}{x-2} \div (x^2+12x+36)$
   $\dfrac{1}{(x-2)(x+6)}$

10. [6.2C] Perform the indicated operations.

a. $\dfrac{2-x}{x+3} \div \dfrac{x^3-8}{x+6} \cdot \dfrac{x^3+27}{x+6}$
   $\dfrac{-(x^2-3x+9)}{x^2+2x+4}$

b. $\dfrac{4-x}{x+3} \div \dfrac{x^3-64}{x+7} \cdot \dfrac{x^3+27}{x+7}$
   $\dfrac{-(x^2-3x+9)}{x^2+4x+16}$

c. $\dfrac{2-x}{x+5} \div \dfrac{x^3-8}{x+8} \cdot \dfrac{x^3+125}{x+8}$
   $\dfrac{-(x^2-5x+25)}{x^2+2x+4}$

11. [6.3A] Perform the indicated operations.

a. $\dfrac{x}{x^2-4} + \dfrac{2}{x^2-4} \quad \dfrac{1}{x-2}$

b. $\dfrac{x}{x^2-9} + \dfrac{3}{x^2-9} \quad \dfrac{1}{x-3}$

c. $\dfrac{x}{x^2-16} + \dfrac{4}{x^2-16} \quad \dfrac{1}{x-4}$

12. [6.3A] Perform the indicated operations.

a. $\dfrac{x}{x^2-9} - \dfrac{3}{x^2-9} \quad \dfrac{1}{x+3}$

b. $\dfrac{x}{x^2-16} - \dfrac{4}{x^2-16} \quad \dfrac{1}{x+4}$

c. $\dfrac{x}{x^2-25} - \dfrac{5}{x^2-25} \quad \dfrac{1}{x+5}$

13. [6.3B] Perform the indicated operations.

a. $\dfrac{x+1}{x^2+x-2} + \dfrac{x+5}{x^2-1}$
   $\dfrac{2x^2+9x+11}{(x-1)(x+2)(x+1)}$

b. $\dfrac{x+1}{x^2+x-2} + \dfrac{x+6}{x^2-1}$
   $\dfrac{2x^2+10x+13}{(x-1)(x+2)(x+1)}$

c. $\dfrac{x+1}{x^2+x-2} + \dfrac{x+7}{x^2-1}$
   $\dfrac{2x^2+11x+15}{(x-1)(x+2)(x+1)}$

14. [6.3B] Perform the indicated operations.

a. $\dfrac{x-4}{x^2-x-6} - \dfrac{x+1}{x^2-9}$
   $\dfrac{-4x-14}{(x-3)(x+2)(x+3)}$

b. $\dfrac{x-3}{x^2-x-6} - \dfrac{x+1}{x^2-9}$
   $\dfrac{-3x-11}{(x-3)(x+2)(x+3)}$

c. $\dfrac{x-1}{x^2-x-6} - \dfrac{x+1}{x^2-9}$
   $\dfrac{-x-5}{(x-3)(x+2)(x+3)}$

15. [6.4A] Simplify.

a. $\dfrac{\frac{1}{x}+\frac{1}{x^4}}{\frac{1}{x}-\frac{1}{x^5}} \quad \dfrac{x(x^2-x+1)}{(x^2+1)(x-1)}$

b. $\dfrac{\frac{1}{x^2}+\frac{1}{x^5}}{\frac{1}{x^2}-\frac{1}{x^6}} \quad \dfrac{x(x^2-x+1)}{(x^2+1)(x-1)}$

c. $\dfrac{\frac{1}{x^3}+\frac{1}{x^6}}{\frac{1}{x^3}-\frac{1}{x^7}} \quad \dfrac{x(x^2-x+1)}{(x^2+1)(x-1)}$

16. [6.4A] Simplify.

a. $4+\dfrac{a}{4+\frac{4}{4+a}} \quad \dfrac{a^2+20a+80}{4a+20}$

b. $5+\dfrac{a}{5+\frac{5}{5+a}} \quad \dfrac{a^2+30a+150}{5a+30}$

c. $6+\dfrac{a}{6+\frac{6}{6+a}} \quad \dfrac{a^2+42a+252}{6a+42}$

17. [6.5A] Divide.

a. $\dfrac{18x^5-12x^3+6x^2}{6x^2}$
   $3x^3-2x+1$

b. $\dfrac{18x^5-12x^3+6x^2}{6x^3}$
   $3x^2-2+\frac{1}{x}$

c. $\dfrac{18x^5-12x^3+6x^2}{6x^4}$
   $3x-\frac{2}{x}+\frac{1}{x^2}$

18. [6.5B] Divide.

a. $2x^3-8-4x$ by $2+2x$
   $(x^2-x-1)\,\text{R}-6$

b. $2x^3-9-4x$ by $2+2x$
   $(x^2-x-1)\,\text{R}-7$

c. $2x^3-10-4x$ by $2+2x$
   $(x^2-x-1)\,\text{R}-8$

19. [6.5C] Factor $x^3 - 6x^2 + 11x - 6$ if
  a. $x - 1$ is one of its factors
    $(x - 1)(x - 2)(x - 3)$
  b. $x - 2$ is one of its factors
    $(x - 2)(x - 1)(x - 3)$
  c. $x - 3$ is one of its factors
    $(x - 3)(x - 1)(x - 2)$

20. [6.5D] Use synthetic division to divide $x^4 + 10x^3 + 35x^2 + 50x + 28$ by
  a. $x + 1$
    $(x^3 + 9x^2 + 26x + 24)$ R 4
  b. $x + 2$
    $(x^3 + 8x^2 + 19x + 12)$ R 4
  c. $x + 3$
    $(x^3 + 7x^2 + 14x + 8)$ R 4

21. [6.5E] If $P(x) = x^4 + 10x^3 + 35x^2 + 50x + 24$, use synthetic division to show that
  a. -1 is a solution of $P(x) = 0$
    Rem. $= 0$, so -1 is a solution
  b. -2 is a solution of $P(x) = 0$
    Rem. $= 0$, so -2 is a solution
  c. -3 is a solution of $P(x) = 0$
    Rem. $= 0$, so -3 is a solution

22. [6.6A] Solve.
  a. $\dfrac{x}{x + 4} - \dfrac{x}{x - 4} = \dfrac{x^2 + 16}{x^2 - 16}$
    No solution
  b. $1 + \dfrac{4}{x - 5} = \dfrac{40}{x^2 - 25}$
    $x = -9$
  c. $1 + \dfrac{5}{x - 6} = \dfrac{60}{x^2 - 36}$
    $x = -11$

23. [6.6B] Solve using a proportion.
  a. The scale $\frac{1}{4}$ inch = 2 ft is used to draw a scaled-down drawing of a backyard. If the measurements on the drawing are 4 in. by 6 in., what are the dimensions of the backyard?
    32 ft by 48 ft
  b. The average NBA player makes 75% of his free throws. At that rate, if an NBA player takes 10 free throws during a game, how many should he be expected to make?
    Approximately 7 or 8

24. [6.7A] Find two consecutive even integers such that the sum of their reciprocals is
  a. $\dfrac{11}{60}$ 10 and 12
  b. $\dfrac{13}{84}$ 12 and 14
  c. $\dfrac{15}{112}$ 14 and 16

25. [6.7B] Jack can paint a room in 4 hr. Find how long it would take to paint the room if he is helped by Jill, who can paint the same room in
  a. 5 hr $2\frac{2}{9}$ hr
  b. 6 hr $2\frac{2}{5}$ hr
  c. 7 hr $2\frac{6}{11}$ hr

26. [6.7C] A plane traveled 1200 mi with a 25-mi/hr tail wind. What is the plane's speed in still air if it took the plane the same time to travel the given mileage against the wind?
  a. 960 mi 225 mi/hr
  b. 1000 mi 275 mi/hr
  c. 1040 mi 350 mi/hr

27. [6.7D] If $A = \dfrac{a + 2b + 3c}{2}$,
  a. solve for a.
    $a = 2A - 2b - 3c$
  b. solve for b.
    $b = \frac{2A - a - 3c}{2}$ or
    $b = A - \frac{a}{2} - \frac{3c}{2}$
  c. solve for c.
    $c = \frac{2A - a - 2b}{3}$ or
    $c = \frac{2A}{3} - \frac{a}{3} - \frac{2b}{3}$

28. [6.8A] The gas in a closed container exerts a pressure P on the walls of the container. This pressure varies directly as the temperature T of the gas.
  a. Write an equation of variation using k for the constant.
    $P = kT$
  b. If the pressure is 3 lb/in.^{2} when the temperature is 360°F, find k. $k = \frac{1}{120}$

29. [6.8B] If the temperature of a gas is held constant, the pressure P varies inversely as the volume V.
  a. Write an equation of variation using k for the constant.
    $P = \frac{k}{V}$
  b. A pressure of 1600 lb/in.^{2} is exerted by 2 ft^3 of air in a cylinder fitted with a piston. Find k. $k = 3200$

30. [6.8C] If g varies jointly as x and the square of t, write an equation of variation using k for the constant. $g = kxt^2$

Practice Test 6

(Answers on page 514)

1. **a.** For what value(s) is $\frac{x-2}{3x+4}$ undefined?

  b. Write the fraction with the indicated denominator.

  $$\frac{2x^2}{9y^4};\ \text{denominator } 36y^7$$

2. Write in standard form.

  a. $-\dfrac{-5}{y}$ **b.** $-\dfrac{x-y}{5}$

3. Reduce to lowest terms.

  a. $\dfrac{x^4 y^6}{xy^2}$ **b.** $\dfrac{xy+y^2}{x^2-y^2}$

4. Reduce $\dfrac{y^2-x^2}{x^3-y^3}$ to lowest terms.

5. Multiply: $\dfrac{x-2}{3x-2y} \cdot \dfrac{9x^2-4y^2}{2x^2-x-6}$

6. Divide: $\dfrac{x+3}{x-2}$ by (x^2+6x+9)

7. Perform the indicated operations.

  $$\frac{2-x}{x+3} \div \frac{x^3-8}{x+5} \cdot \frac{x^3+27}{x+5}$$

8. Perform the indicated operations.

  a. $\dfrac{x}{x^2-1}+\dfrac{1}{x^2-1}$ **b.** $\dfrac{x}{x^2-4}-\dfrac{2}{x^2-4}$

9. Perform the indicated operations.

  a. $\dfrac{x+1}{x^2+x-2}+\dfrac{x+4}{x^2-1}$

  b. $\dfrac{x-5}{x^2-x-6}-\dfrac{x+1}{x^2-9}$

10. Simplify: $\dfrac{x+\dfrac{1}{x^2}}{x-\dfrac{1}{x^3}}$

11. Simplify: $\dfrac{\dfrac{x}{x+3}+x}{1-\dfrac{7}{x^2-9}}$

12. Divide: $\dfrac{28x^5-14x^3+7x^2}{7x^3}$

13. Divide: $2x^3-6-4x$ by $2+2x$

14. Factor $2x^3+3x^2-23x-12$ if $x-3$ is one of its factors.

15. Use synthetic division to divide $x^4-4x^3-7x^2+22x+25$ by $x+1$.

16. Use synthetic division to show that -1 is a solution of $x^4-4x^3-7x^2+22x+24=0$.

17. Solve: $\dfrac{x}{x+3}-\dfrac{x}{x-3}=\dfrac{x^2+9}{x^2-9}$

18. Solve: $18x^{-2}-3x^{-1}=1$

19. A recipe for curried shrimp that normally serves four was once served to 200 guests at a wedding reception. One of the ingredients in the recipe is $1\frac{1}{2}$ cups of chicken broth.

  a. How much chicken broth was required to make the recipe for 200 people?

  b. If a medium-sized can of chicken broth contains 2 cups of broth, how many cans are necessary?

20. Find two consecutive even integers such that the sum of their reciprocals is $\frac{7}{24}$.

21. Jack can mow the lawn in 4 hr and Jill can mow it in 3. How long would it take them to mow the lawn if they work together?

22. A plane traveled 990 mi with a 30-mi/hr tail wind in the same time it took to travel 810 mi against the wind. What is the plane's speed in still air?

23. The area A of a trapezoid is given by

  $$A=\frac{(B+b)h}{2}$$

  where B is the length of one base, b is the length of the other base, and h is the height of the trapezoid. Solve for h.

24. C is directly proportional to m.

  a. Write an equation of variation with k as the constant.

  b. Find k when $C=12$ and $m=\frac{1}{3}$.

25. The force F with which the earth attracts an object above the earth's surface varies inversely as the square of the distance d from the center of the earth.

  a. Write an equation of variation with k as the constant.

  b. A meteorite weighs 40 lb on the earth's surface. If the radius of the earth is about 4000 mi, find k.

Answers to Practice Test

ANSWER		IF YOU MISSED	REVIEW		
		QUESTION	SECTION	EXAMPLES	PAGE
1. a. $\frac{-4}{3}$	**b.** $\frac{8x^2y^3}{36y^7}$	1	6.1A, B	1, 2	425, 427
2. a. $\frac{5}{y}$	**b.** $\frac{y-x}{5}$	2	6.1C	3	428
3. a. x^3y^4	**b.** $\frac{y}{x-y}$	3	6.1D	4	430
4. $\frac{-(x+y)}{x^2+xy+y^2}$		4	6.1D	5	431
5. $\frac{3x+2y}{2x+3}$		5	6.2A	1, 2	436–437
6. $\frac{1}{(x-2)(x+3)}$ or $\frac{1}{x^2+x-6}$		6	6.2B	3	438
7. $\frac{-(x^2-3x+9)}{x^2+2x+4}$		7	6.2C	4, 5	439
8. a. $\frac{1}{x-1}$	**b.** $\frac{1}{x+2}$	8	6.3A	1	446–447
9. a. $\frac{2x^2+8x+9}{(x+2)(x-1)(x+1)}$		9	6.3B	2, 3	448–450
b. $\frac{-5x-17}{(x+2)(x+3)(x-3)}$					
10. $\frac{x(x^2-x+1)}{(x^2+1)(x-1)}$		10	6.4A	1, 2	456
11. $\frac{x(x-3)}{x-4}$ or $\frac{x^2-3x}{x-4}$		11	6.4A	3	457–459
12. $4x^2-2+\frac{1}{x}$		12	6.5A	1	466
13. (x^2-x-1) R -4		13	6.5B	2, 3	467–468
14. $(x-3)(x+4)(2x+1)$		14	6.5C	4	469–470
15. $(x^3-5x^2-2x+24)$ R 1		15	6.5D	5	471
16. When the division is done, the remainder is zero.		16	6.5E	6	472
17. No solution		17	6.6A	3, 4	480–481
18. $-6, 3$		18	6.6A	5	482
19. a. 75 cups	**b.** 38 cans	19	6.6B	6	483
20. 6 and 8		20	6.7A	1	489–490
21. $1\frac{5}{7}$ hr		21	6.7B	2	490
22. 300 mi/hr		22	6.7C	4	492–493
23. $h=\frac{2A}{B+b}$		23	6.7D	5	493–494
24. a. $C=km$	**b.** $k=36$	24	6.8A	1	499
25. a. $F=\frac{k}{d^2}$	**b.** $k=640,000,000$	25	6.8B	2, 3	500

Cumulative Review Chapters 1–6

1. Find: $-8 - 6$ -14

2. Which law is illustrated by the following statement?
$5 \cdot (8 \cdot 7) = (5 \cdot 8) \cdot 7$
Associative law of multiplication

3. Simplify:
$[(5x^2 - 1) + (x + 2)] - [(x - 6) + (8x^2 - 8)]$
$-3x^2 + 15$

4. Solve: $\dfrac{x + 6}{5} - \dfrac{x - 6}{7} = 2$ -1

5. Graph: $x \geq -4$

6. Graph: $x + 6 \leq 7$ and $-4x < 16$

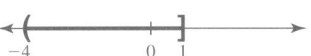

7. If $H = 2.75h + 73.49$, find h when $H = 136.74$. 23

8. The sum of three consecutive odd integers is 93. What are the three integers? 29, 31, 33

9. Find the slope of the line passing through the points A(4, 2) and B(−2, 4). $\dfrac{-1}{3}$

10. Find an equation of the line with slope -5 and passing through the point (1, 1). $5x + y = 6$

11. Find the slope and the y-intercept of the line $6x - 3y = 36$. $m = 2; y = -12$

12. Graph: $2x - 5y \leq -10$

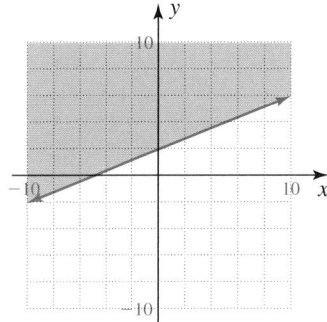

13. Graph: $|x + 4| > 2$

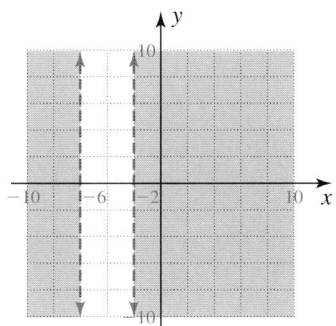

14. Use the graphical method to solve the system:
$y + x = -1$
$2y = -2x - 4$

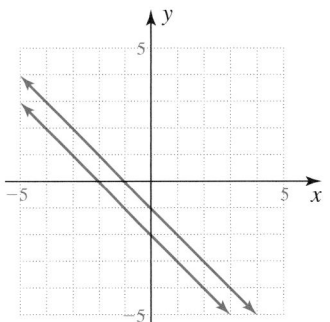

No solution

15. Use the substitution method to solve the system:
$x - 4y = 10$
$2x = 8y + 18$ No solution

16. Solve by the elimination method:
$3x + 4y = 18$
$x + y = 5$ (2, 3)

17. Evaluate: $\begin{vmatrix} -4 & -4 \\ 1 & -5 \end{vmatrix}$ 24

18. Solve by Cramer's rule:
$x + 2y + z = 5$
$x + y - z = 6$
$3x + 5y + z = 7$ No solution

19. Expand by minors along the third column:

$$\begin{vmatrix} -3 & -3 & 2 \\ -2 & 2 & -3 \\ 0 & -2 & -3 \end{vmatrix} \quad 62$$

20. The total height of a building and the flagpole on the roof is 210 ft. The building is 9 times as high as the flagpole. How high is the flagpole? 21 ft

21. Find the domain and range of $\{(x, y) \mid y = 4 + x\}$.
$D = \{\text{all real numbers}\}; R = \{\text{all real numbers}\}$

22. Add $2x^3 - 7x^2 - 2x - 3$ and $3 - 7x + 6x^2 - 7x^3$.
$-5x^3 - x^2 - 9x$

23. Multiply: $(2n - 1)(3n - 5)$ $6n^2 - 13n + 5$

24. Factor: $12t^2 - 7t - 10$ $(3t + 2)(4t - 5)$

25. Solve for x: $x^3 + 3x^2 - 16x - 48 = 0$
$-3; -4; 4$

26. Perform the indicated operations:

$$\frac{1 - x}{x + 3} \div \frac{x^3 - 1}{x + 2} \cdot \frac{x^3 + 27}{x + 2} \quad \frac{-(x^2 - 3x + 9)}{x^2 + x + 1}$$

27. Perform the indicated operations:

$$\frac{x - 1}{x^2 - 5x + 6} + \frac{x + 3}{x^2 - 4} \quad \frac{2x^2 + x - 11}{(x - 3)(x - 2)(x + 2)}$$

28. Solve for x: $1 + \dfrac{3}{x + 2} = \dfrac{-12}{x^2 - 4}$ -1

29. Sandra can paint a kitchen in 3 hr and Roger can paint the same kitchen in 5 hr. How long would it take for both working together to paint the kitchen? $1\frac{7}{8}$ hr

30. An enclosed gas exerts a pressure P on the walls of the container. This pressure is directly proportional to the temperature T of the gas. If the pressure is 8 lb/in.^{2} when the temperature is 480°F, find k. $\frac{1}{60}$

Rational Exponents and Radicals

7

The Human Side of Algebra

Imagine a universe in which only whole numbers, some classified as "perfect" and "amicable," existed. Even numbers were "feminine" and odd numbers "masculine," while 1 was the generator of all other numbers. This was the universe of the Pythagoreans, an ancient Greek secret society (ca. 540–500 B.C.). And then, disaster. In the midst of these charming fantasies, a new type of number was discovered, a type so unexpected that the brotherhood tried to suppress its discovery—the set of irrational numbers!

Pythagoras was born between 580 and 569 B.C., but accounts of his eventual demise differ. One claims that he died in about 501 B.C. when popular revolt erupted and the meeting house of the Pythagoreans was set afire, which resulted in his death. A more dramatic ending claims that his disciples made a bridge over the fire with their bodies enabling the master to escape to Metapontum. In the ensuing fight, Pythagoras was caught between freedom and a field of sacred beans. Rather than trampling the plants, he valiantly chose to die at the hands of his enemies.

Pretest for Chapter 7

(Answers on page 519)

1. Find, if possible.

    a. $\sqrt[3]{-27}$ **b.** $\sqrt{-16}$

2. Find, if possible.

    a. $(-8)^{1/3}$ **b.** $\left(\dfrac{1}{81}\right)^{1/4}$

3. Evaluate, if possible.

    a. $27^{2/3}$ **b.** $(-16)^{3/2}$

4. Evaluate, if possible.

    a. $(-8)^{-2/3}$ **b.** $27^{-2/3}$

5. If x and y are positive, simplify.

    a. $x^{1/2} \cdot x^{1/3}$ **b.** $\dfrac{x^{-1/3}}{x^{1/5}}$

6. If x and y are positive, simplify.

    a. $(x^{1/4}y^{2/5})^{-20}$ **b.** $x^{3/4}(x^{-1/4} + y^{2/5})$

7. Simplify.

    a. $\sqrt[4]{(-3)^4}$ **b.** $\sqrt[6]{(-x)^6}$

8. Simplify.

    a. $\sqrt[3]{128}$ **b.** $\sqrt[3]{54a^5b^9}$

9. Simplify.

    a. $\sqrt{\dfrac{5}{32}}$ **b.** $\sqrt[3]{\dfrac{6}{x^3}}$

10. Rationalize the denominator.

    a. $\dfrac{\sqrt{5}}{\sqrt{7}}$ **b.** $\dfrac{\sqrt{2}}{\sqrt{3x}}, x > 0$

11. Rationalize the denominator.

    a. $\dfrac{1}{\sqrt[3]{2x}}$ **b.** $\dfrac{\sqrt[4]{3}}{\sqrt[4]{16x^3}}$

12. Reduce the order (index).

    a. $\sqrt[4]{9x^2}$ **b.** $\sqrt[6]{16c^4d^4}$

13. Simplify. $\sqrt[6]{\dfrac{2a^2}{32c^{16}}}$

14. Perform the indicated operations.

    a. $\sqrt{98} + \sqrt{32}$ **b.** $2\sqrt{175} - \sqrt{28}$

15. Perform the indicated operations.

    a. $3\sqrt{\dfrac{1}{2}} - \sqrt{\dfrac{1}{8}}$ **b.** $\sqrt[3]{\dfrac{1}{2}} - \sqrt[3]{\dfrac{1}{54}}$

16. Perform the indicated operations.

    a. $\sqrt{2}(\sqrt{8} + \sqrt{3})$ **b.** $\sqrt[3]{2x}(\sqrt[3]{4x^2} - \sqrt[3]{16x})$

17. Find the product.

    a. $(\sqrt{27} + \sqrt{50})(\sqrt{12} + \sqrt{8})$

    b. $(\sqrt{3} + 2)(\sqrt{3} + 2)$

18. Find the product.

    a. $(2 - \sqrt{3})^2$ **b.** $(\sqrt{5} + \sqrt{3})(\sqrt{5} - \sqrt{3})$

19. Reduce. $\dfrac{10 - \sqrt{50}}{5}$

20. Rationalize the denominator. $\dfrac{3}{\sqrt{x} + 2}$

21. Solve. $\sqrt{x + 2} = -1$

22. Solve. $\sqrt{x + 2} = x - 4$

23. Solve. $\sqrt{x - 2} - x = -2$

24. Solve. $\sqrt{x - 7} - \sqrt{x} = -1$

25. Solve. $\sqrt[3]{x - 3} = 2$

26. Write the given expression in terms of i.

    a. $\sqrt{-49}$ **b.** $\sqrt{-160}$

27. Find.

    a. $(3 + 5i) + (6 - 14i)$ **b.** $(4 + 2i) - (7 - 8i)$

28. Multiply.

    a. $(3 + 2i)(4 - 3i)$ **b.** $\sqrt{-9}(4 - \sqrt{-8})$

29. Find.

    a. $\dfrac{4 + 5i}{2 + 3i}$ **b.** $\dfrac{4 - 2i}{3 - 5i}$

30. Write the answer as $1, -1, i,$ or $-i$.

    a. i^{55} **b.** i^{-7}

Answers to Pretest

ANSWER		IF YOU MISSED		REVIEW		
		QUESTION	SECTION	EXAMPLES	PAGE	
1. a. -3	**b.** Not a real number	1	7.1	1	522	
2. a. -2	**b.** $\frac{1}{3}$	2	7.1	2	523	
3. a. 9	**b.** Not a real number	3	7.1	3	523–524	
4. a. $\frac{1}{4}$	**b.** $\frac{1}{9}$	4	7.1	4	524	
5. a. $x^{5/6}$	**b.** $\frac{1}{x^{8/15}}$	5	7.1	5	525	
6. a. $\frac{1}{x^5 y^8}$	**b.** $x^{1/2} + x^{3/4} y^{2/5}$	6	7.1	6	525–526	
7. a. 3	**b.** $|-x|$	7	7.2	1	531	
8. a. $4\sqrt[3]{2}$	**b.** $3ab^3 \sqrt[3]{2a^2}$	8	7.2	2	532–533	
9. a. $\frac{\sqrt{10}}{8}$	**b.** $\frac{\sqrt[3]{6}}{x}$	9	7.2	3	533	
10. a. $\frac{\sqrt{35}}{7}$	**b.** $\frac{\sqrt{6x}}{3x}$	10	7.2	4	535	
11. a. $\frac{\sqrt[3]{4x^2}}{2x}$	**b.** $\frac{\sqrt[4]{3x}}{2x}$	11	7.2	5	535–536	
12. a. $\sqrt{3x}$	**b.** $\sqrt[3]{4c^2 d^2}$	12	7.2	6	536	
13. $\frac{\sqrt[3]{2ac}}{2c^3}$		13	7.2	7	537	
14. a. $11\sqrt{2}$	**b.** $8\sqrt{7}$	14	7.3	1	542	
15. a. $\frac{5\sqrt{2}}{4}$	**b.** $\frac{\sqrt[3]{4}}{3}$	15	7.3	2	543	
16. a. $4 + \sqrt{6}$	**b.** $2x - 2\sqrt[3]{4x^2}$	16	7.3	3	544	
17. a. $38 + 16\sqrt{6}$	**b.** $7 + 4\sqrt{3}$	17	7.3	4, 5	545	
18. a. $7 - 4\sqrt{3}$	**b.** 2	18	7.3	5	545	
19. $2 - \sqrt{2}$		19	7.3	6	546	
20. $\frac{3\sqrt{x} - 6}{x - 4}$		20	7.3	7	547–548	
21. No real-number solution		21	7.4	1	553–554	
22. 7		22	7.4	1	553–554	
23. 2 or 3		23	7.4	1	553–554	
24. 16		24	7.4	2	555	
25. 11		25	7.4	3, 4	555–556	
26. a. $7i$	**b.** $4i\sqrt{10}$	26	7.5	1	561	
27. a. $9 - 9i$	**b.** $-3 + 10i$	27	7.5	2	562	
28. a. $18 - i$	**b.** $6\sqrt{2} + 12i$	28	7.5	3, 4	563	
29. a. $\frac{23}{13} - \frac{2}{13}i$	**b.** $\frac{11}{17} + \frac{7}{17}i$	29	7.5	5	564–565	
30. a. $-i$	**b.** i	30	7.5	6	565–566	

7.1 **RATIONAL EXPONENTS AND RADICALS**

To Succeed, Review How To ...

1. Use the laws of exponents (pp. 33–40).

2. Do operations involving signed numbers (pp. 17–23).

Objectives

A Find the nth root of a number, if it exists.

B Evaluate expressions containing rational exponents.

C Simplify expressions involving rational exponents.

GETTING STARTED

Speeding Along to Exponents

Have you noticed that the speed limit on a curve is lower than on a straight road? The velocity v (in mi/hr) that a car can travel on a curved concrete highway of radius r feet without skidding is $v = \sqrt{9r}$. If the radius r of the curve is 100 ft, the velocity is $\sqrt{900}$ (read "the square root of 900"). A *square root* of 900 is a number whose square is 900. Since $(30)^2 = 900$, a square root of 900 is 30. Similarly, a square root of 25 is 5 (since $5^2 = 25$), and a square root of 36 is 6 (since $6^2 = 36$).

There are many numbers that are not square roots of perfect squares—for example, $\sqrt{2}$, $\sqrt{3}$, $\sqrt{10}$, and $\sqrt{24}$. These real numbers are irrational; they cannot be written as the quotient of two integers. The first irrational number was probably discovered by the Pythagoreans, an ancient Greek society of mathematicians who believed that everything was based on the whole numbers and that harmony consisted of numerical ratios. In this section we shall study rational exponents and radical expressions. Many of the important applications of mathematics use these two concepts. (See Problems 71–74.) We start by giving the definition of an nth root.

nth Root	If a and x are real numbers and n is a positive integer, then x is an nth root of a if $x^n = a$.

For example,

1. A square (second) root of 4 is 2 because $2^2 = 4$.

2. Another square root of 4 is -2 because $(-2)^2 = 4$.

3. A cube (third) root of 27 is 3 because $(3)^3 = 27$.

4. A cube (third) root of -64 is -4 because $(-4)^3 = -64$.

5. A fourth root of $\frac{16}{81}$ is $\frac{2}{3}$ because $(\frac{2}{3})^4 = \frac{16}{81}$.

6. Another fourth root of $\frac{16}{81}$ is $-\frac{2}{3}$ because $(-\frac{2}{3})^4 = \frac{16}{81}$.

Note that 4 has two square roots, 2 and -2, and $\frac{16}{81}$ has two real fourth roots, $-\frac{2}{3}$ and $\frac{2}{3}$. To avoid this situation and make our work more precise, we introduce the idea of the principal nth root.

A   Radicals

Principal *n*th Root	If *n* is a positive integer, then $\sqrt[n]{a}$ denotes the **principal *n*th root** of *a*, and
	1. If *a* is positive ($a > 0$), $\sqrt[n]{a}$ is the *positive n*th root of *a*.
	2. If *a* is negative ($a < 0$) and *n* is odd, $\sqrt[n]{a}$ is the *negative n*th root of *a*.
	3. If *a* is negative and *n* is even, there is no real *n*th root.
	4. If $a = 0$, then $\sqrt[n]{0} = 0$.

In this definition, $\sqrt[n]{a}$ is called a **radical expression,** $\sqrt{}$ is called the **radical sign,** *a* is the **radicand,** and *n* is the **index** that tells you what root is being considered. By convention, the index 2 for square root is understood but not written, $\sqrt{x}$ means $\sqrt[2]{x}$. Thus $\sqrt{9}$ means the principal square root of 9 and $\sqrt{25}$ means the principal square root of 25.

Now let's look at part 1 of the definition. It tells you that whenever the number under the radical sign is positive, the resulting root is also positive; "If $a > 0$, $\sqrt[n]{a} > 0$." Thus,

$$\sqrt{64} = 8 \quad \text{because} \quad 8^2 = 64$$

$$\sqrt[3]{8} = 2 \quad \text{because} \quad 2^3 = 8$$

$$\sqrt[4]{81} = 3 \quad \text{because} \quad 3^4 = 81$$

> **NOTE**
>
> A common mistake is to assume that $\sqrt{64}$ has two values. This is not correct. By our definition of principal root, $\sqrt{64} = 8$. If we wish to refer to the negative *n*th root, we write $-\sqrt[n]{a}$. Thus $-\sqrt{64} = -8$, $-\sqrt{16} = -4$, and $-\sqrt{25} = -5$.

Web It

For a lesson on how to evaluate square root and cube root functions using a calculator, go to link 7-1-1 at mhhe.com/bello.

Part 2 of the definition tells you that if *a* is *negative* and *n* is *odd*, $\sqrt[n]{a}$ is negative, that is, "If $a < 0$, $\sqrt[n]{a} < 0$." Thus,

negative

$$\text{odd} \longrightarrow \sqrt[3]{\boxed{-27}} = -3 \quad \text{because} \quad (-3)^3 = -27 \qquad \text{Negative answer}$$

$$\text{odd} \longrightarrow \sqrt[5]{\boxed{-32}} = -2 \quad \text{because} \quad (-2)^5 = -32 \qquad \text{Negative answer}$$

negative

To illustrate part 3 of the definition consider $\sqrt{-16}$. There is no real number whose square is -16 because the square of a nonzero number is positive. Thus $\sqrt{-16}$ is not a real number. Similarly, $\sqrt[4]{-81}$ is not a real number because there is no real number whose fourth power is -81. In both cases, the index *n* is even and the radicand is negative.

negative

$$\text{even} \longrightarrow \sqrt{-16} = \text{no real number} \quad \text{because} \quad (n)^2 \neq -16$$

$$\text{even} \longrightarrow \sqrt[4]{-81} = \text{no real number} \quad \text{because} \quad (n)^4 \neq -81$$

negative

EXAMPLE 1	**Finding the roots of numbers**

Find, if possible:

a. $\sqrt[3]{-64}$ **b.** $\sqrt{-64}$ **c.** $\sqrt[3]{\left(\dfrac{-1}{8}\right)}$

SOLUTION

a. $\sqrt[3]{-64} = -4$, because $(-4)^3 = -64$.

b. $\sqrt{-64}$ is not a real number. $\sqrt{-64} \ne -8$, because $(-8)(-8) = 64$ and not -64.

c. $\sqrt[3]{\left(\dfrac{-1}{8}\right)} = \dfrac{-1}{2}$, because $\left(\dfrac{-1}{2}\right)^3 = \dfrac{-1}{8}$.

PROBLEM 1

Find, if possible:

a. $\sqrt{-25}$

b. $\sqrt[3]{-125}$

c. $\sqrt[3]{\dfrac{-1}{27}}$

B	**From Rational Exponents to Radicals**

Calculate It

Two Ways to Find Roots

You can approximate the roots in Examples 1, 2, 3, and 4 in different ways, but be aware that different calculators have different procedures for this. To find $\sqrt[3]{-64}$ in Example 1(a) with a TI-83 Plus, press `MATH` `4` `(-)` `64` `)` `ENTER` and you get -4. You can also use the $\boxed{\sqrt[x]{\;}}$ feature to find roots. Thus to find $\sqrt[4]{16}$, press 4 `MATH` 5 16 `ENTER` and you get 2. The first entry is the index of the root you want (square, cube, fourth, fifth, and so on).

We have used radicals to define the nth root of a number. We can also define nth roots using rational exponents. For example, what do you think $a^{1/3}$ means? To find out,

let $\qquad\qquad x = a^{1/3}$

then $\qquad\qquad x^3 = (a^{1/3})^3$ $\qquad$ Cube both sides.

$\qquad\qquad\quad x^3 = a^{(1/3)\cdot(3)}$ $\qquad$ Assume $(a^{1/n})^n = a^{(1/n)\cdot(n)}$.

$\qquad\qquad\quad x^3 = a^1$

$\qquad\qquad\quad x^3 = a$

Since $x^3 = a$, x must be the cube root of a; $x = \sqrt[3]{a}$. But $x = a^{1/3}$, so

$$a^{1/3} = \sqrt[3]{a}$$

Similarly, if $\sqrt[4]{a}$ is defined,

$$a^{1/4} = \sqrt[4]{a}$$

In general, the following definition establishes the relationship between rational exponents and roots.

Rational Exponents and Their Roots	If n is a positive integer and $\sqrt[n]{a}$ is a real number, then $$a^{1/n} = \sqrt[n]{a}$$

When we write

$$a^{1/n} = \sqrt[n]{a} \quad \text{(same)}$$

the denominator of the rational exponent is the index of the radical:

$$16^{1/2} = \sqrt[2]{16} = 4 \qquad 4^2 = 16. \text{ Remember } \sqrt{16} \text{ means } \sqrt[2]{16}.$$

$$(-8)^{1/3} = \sqrt[3]{-8} = -2 \qquad (-2)^3 = -8$$

$$\left(\frac{1}{81}\right)^{1/4} = \sqrt[4]{\frac{1}{81}} = \frac{1}{3} \qquad \left(\frac{1}{3}\right)^4 = \frac{1}{81}$$

Answers

1. a. Not a real number

b. -5 **c.** $-\dfrac{1}{3}$

EXAMPLE 2   **Evaluating expressions containing rational exponents**

Evaluate, if possible:

a. $9^{1/2}$ **b.** $(-125)^{1/3}$ **c.** $\left(\dfrac{1}{16}\right)^{1/4}$

SOLUTION

a. $9^{1/2} = \sqrt{9} = 3$ $3^2 = 9$

b. $(-125)^{1/3} = \sqrt[3]{-125} = -5$ $(-5)^3 = -125$

c. $\left(\dfrac{1}{16}\right)^{1/4} = \sqrt[4]{\dfrac{1}{16}} = \dfrac{1}{2}$ $\left(\dfrac{1}{2}\right)^4 = \dfrac{1}{16}$

PROBLEM 2

Evaluate, if possible:

a. $49^{1/2}$ **b.** $(-216)^{1/3}$ **c.** $\left(\dfrac{1}{81}\right)^{1/4}$

Table 1 shows a summary of our work so far.

Table 1 Rational Exponents and Radicals

If a is a real number as specified and n is a positive integer		
	n even	**n odd**
For a positive a ($a > 0$)	$\sqrt[n]{a} = a^{1/n}$ is positive.	$\sqrt[n]{a} = a^{1/n}$ is positive.
For a negative a ($a < 0$)	$\sqrt[n]{a} = a^{1/n}$ is not a real number.	$\sqrt[n]{a} = a^{1/n}$ is negative.
For $a = 0$	$\sqrt[n]{a} = a^{1/n} = 0.$	$\sqrt[n]{a} = a^{1/n} = 0.$

We have already defined rational exponents of the form $1/n$. How do we define $a^{m/n}$, where m and n are positive integers when $n > 1$ and $\sqrt[n]{a}$ is a real number? If we assume that $(a^m)^n = a^{m \cdot n}$, then

$$a^{m/n} = (a^{1/n})^m = (a^m)^{1/n} = (\sqrt[n]{a})^m = \sqrt[n]{a^m}$$

From this we arrive at the following definition.

Radical Expression with an m/n Exponent	$a^{m/n} = (\sqrt[n]{a})^m = \sqrt[n]{a^m}$
	provided m and n are positive integers, $n > 1$, and $\sqrt[n]{a}$ is a real number.

The numerator of the exponent m/n is the exponent of the radical expression, and the denominator is the index of the radical

$$\overset{\text{exponent}}{\underset{\text{index}}{a^{m/n} = (\sqrt[n]{a})^m}}$$

For example,

$$a^{1/5} = \sqrt[5]{a} \quad \text{and} \quad a^{2/5} = (\sqrt[5]{a})^2 \quad \text{or} \quad \sqrt[5]{a^2}$$

EXAMPLE 3   **Evaluating expressions containing rational exponents**

Evaluate, if possible:

a. $8^{2/3}$ **b.** $(-27)^{2/3}$ **c.** $(-25)^{3/2}$

PROBLEM 3

Evaluate, if possible:

a. $27^{2/3}$ **b.** $(-64)^{2/3}$ **c.** $(-36)^{3/2}$

Answers

2. a. 7   **b.** -6 **c.** $\frac{1}{3}$ **3. a.** 9
b. 16   **c.** Not a real number

SOLUTION

a. $8^{2/3}$ can be evaluated in two ways.

1. $8^{2/3} = (\sqrt[3]{8})^2 = (2)^2 = 4$ $\qquad$ $\sqrt[3]{8} = 2$

2. $8^{2/3} = \sqrt[3]{8^2} = \sqrt[3]{64} = 4$

b. We can evaluate $(-27)^{2/3}$ in a similar manner.

1. $(-27)^{2/3} = (\sqrt[3]{-27})^2$

$\qquad = (-3)^2 \qquad \sqrt[3]{-27} = -3$

$\qquad = 9$

2. $(-27)^{2/3} = \sqrt[3]{(-27)^2} = \sqrt[3]{729} = 9$

c. $(-25)^{3/2} = \sqrt[2]{(-25)^3} = \sqrt{-15{,}625}$, which is not a real number.

Teaching Tip

Have students evaluate $32^{3/5}$ by using $\sqrt[5]{32^3}$, where they have to raise 32 to the third power and then take the fifth root. Have them evaluate $32^{3/5}$ again by first doing $\sqrt[5]{32} = 2$ and then $2^3 = 8$. Let them discuss which way is easier and why. The first way will require a calculator and the second way doesn't.

Calculate It

Two Ways to Find $a^{m/n}$

To do Example 3(b) with a TI-83 Plus, enter $((-27)^2)^{(1/3)}$ or $((-27)^{(1/3)})^2$ and you get 9. You can also enter $(-27)^{(2/3)}$ and get 9. These are all shown in the window.

```
((-27)^2)^(1/3)
                    9
((-27)^(1/3))^2
                    9
(-27)^(2/3)
                    9
```

Now try Example 3(c). Your screen should indicate "ERR." It will say "NONREAL ANS."

Finally, to define negative rational exponents, if m and n are positive integers with no common factors,

$$-\frac{m}{n} = \frac{-m}{n}$$

Thus if $(a^m)^n = a^{m \cdot n}$,

$$a^{-m/n} = (a^{1/n})^{-m}$$

$$= \frac{1}{(a^{1/n})^m} \qquad m \text{ is negative, by definition, } a^{-m} = \frac{1}{a^m}$$

We make the following definition for rational exponents.

> **Definition of $a^{-m/n}$**
>
> $$a^{-m/n} = \frac{1}{(a^{m/n})}$$
>
> where m and n are positive integers, $n > 1$, and $a^{1/n}$ a real number, $a \neq 0$.

Using this definition, we have the following.

1. $a^{-1/2} = \dfrac{1}{a^{1/2}} = \dfrac{1}{\sqrt{a}}$

2. $32^{-3/5} = \dfrac{1}{32^{3/5}} = \dfrac{1}{(\sqrt[5]{32})^3} = \dfrac{1}{2^3} = \dfrac{1}{8}$

3. $1000^{-2/3} = \dfrac{1}{1000^{2/3}} = \dfrac{1}{(\sqrt[3]{1000})^2} = \dfrac{1}{10^2} = \dfrac{1}{100}$

| **EXAMPLE 4** | **Evaluating expressions containing negative rational exponents** |

Evaluate, if possible:

a. $16^{-3/4}$ $\qquad$ **b.** $(-8)^{-4/3}$ $\qquad$ **c.** $125^{-2/3}$

SOLUTION

a. $16^{-3/4} = \dfrac{1}{16^{3/4}} = \dfrac{1}{(\sqrt[4]{16})^3} = \dfrac{1}{2^3} = \dfrac{1}{8}$

b. $(-8)^{-4/3} = \dfrac{1}{(-8)^{4/3}} = \dfrac{1}{(\sqrt[3]{-8})^4} = \dfrac{1}{(-2)^4} = \dfrac{1}{16}$

c. $125^{-2/3} = \dfrac{1}{125^{2/3}} = \dfrac{1}{(\sqrt[3]{125})^2} = \dfrac{1}{5^2} = \dfrac{1}{25}$

PROBLEM 4

Evaluate, if possible:

a. $81^{-3/4}$ $\quad$ **b.** $(-27)^{-4/3}$ $\quad$ **c.** $216^{-2/3}$

Answers

4. a. $\frac{1}{27}$ $\quad$ **b.** $\frac{1}{81}$ $\quad$ **c.** $\frac{1}{36}$

 **Operations with Rational Exponents**

The properties of exponents that we studied in Chapter 1 can be extended to rational exponents. If this is done, we have the following results.

Laws of Exponents

Let r, s, and t be rational numbers. If a and b are real numbers and the indicated expressions exist, then the following are true.

I. $a^r \cdot a^s = a^{r+s}$ **II.** $\dfrac{a^r}{a^s} = a^{r-s}$ **III.** $(a^r)^s = a^{r \cdot s}$

IV. $(a^r b^s)^t = a^{rt} b^{st}$ **V.** $\left(\dfrac{a^r}{b^s}\right)^t = \dfrac{a^{rt}}{b^{st}}$

EXAMPLE 5 Using the laws of exponents

If x and y are positive, simplify:

a. $x^{1/3} \cdot x^{1/4}$ **b.** $\dfrac{x^{-2/3}}{x^{1/5}}$ **c.** $(y^{3/5})^{-1/6}$

SOLUTION

a. $x^{1/3} \cdot x^{1/4} = x^{1/3+1/4}$ Law I
$= x^{4/12+3/12}$ The LCD is 12.
$= x^{7/12}$ Add exponents.

b. $\dfrac{x^{-2/3}}{x^{1/5}} = x^{-2/3-1/5}$ Law II
$= x^{-10/15-3/15}$ The LCD is 15.
$= x^{-13/15}$
$= \dfrac{1}{x^{13/15}}$ $a^{-n} = \dfrac{1}{a^n}$

c. $(y^{3/5})^{-1/6} = y^{(3/5)\cdot(-1/6)}$ Law III
$= y^{-1/10}$ $\frac{3}{5} \cdot \left(-\frac{1}{6}\right) = -\frac{1}{10}$
$= \dfrac{1}{y^{1/10}}$

EXAMPLE 6 More practice using the laws of exponents

If x and y are positive, simplify:

a. $(x^{1/5} y^{3/4})^{-20}$ **b.** $\dfrac{x^{1/3}y^{3/4}}{x^{2/3}y^{1/4}}$ **c.** $x^{2/3}(x^{-1/3} + y^{1/5})$

SOLUTION

a. $(x^{1/5}y^{3/4})^{-20} = (x^{1/5})^{-20} \cdot (y^{3/4})^{-20}$ Law IV
$= x^{(1/5)(-20)} \cdot y^{(3/4)(-20)}$ Law III
$= x^{-4} \cdot y^{-15}$
$= \dfrac{1}{x^4 y^{15}}$ Definition of negative exponent

b. $\dfrac{x^{1/3}y^{3/4}}{x^{2/3}y^{1/4}} = x^{1/3-2/3} \cdot y^{3/4-1/4}$ Law II
$= x^{-1/3} \cdot y^{1/2}$
$= \dfrac{y^{1/2}}{x^{1/3}}$

PROBLEM 5

If x and y are positive, simplify:

a. $x^{1/3} \cdot x^{2/5}$ **b.** $\dfrac{x^{-3/4}}{x^{-1/6}}$

c. $(y^{-3/2})^{-1/9}$

Teaching Tip

Give some integer exponent examples before doing Example 5.

Examples:
$x^2 \cdot x^3 = x^{2+3}$
$\dfrac{x^7}{x^4} = x^{7-4}$

Also remind students to never leave a final answer with a negative exponent.
$(x^{-3})^5 = x^{-3\cdot5} = x^{-15} = \dfrac{1}{x^{15}}$

PROBLEM 6

If x and y are positive, simplify:

a. $(x^{-1/4} y^{3/5})^{-20}$

b. $\dfrac{x^{1/4}y^{2/3}}{x^{3/4}y^{1/3}}$

c. $x^{2/5}(x^{-1/5} + y^{1/4})$

Answers

5. a. $x^{11/15}$ **b.** $x^{-7/12} = \dfrac{1}{x^{7/12}}$
c. $y^{1/6}$ **6. a.** $\dfrac{x^5}{y^{12}}$ **b.** $\dfrac{y^{1/3}}{x^{1/2}}$
c. $x^{1/5} + x^{2/5}y^{1/4}$

c. We first use the distributive property and then simplify.

$$x^{2/3}(x^{-1/3} + y^{1/5}) = x^{2/3} \cdot x^{-1/3} + x^{2/3} \cdot y^{1/5}$$
$$= x^{2/3+(-1/3)} + x^{2/3}y^{1/5}$$
$$= x^{1/3} + x^{2/3}y^{1/5}$$

Teaching Tip

In Example 6(b), you might add a step before Law II that shows:

$$\left[\frac{x^{1/3}}{x^{2/3}}\right] \cdot \left[\frac{y^{3/4}}{y^{1/4}}\right]$$

Web It

For practice on finding the *n*th root, go to link 7-1-2 at mhhe.com/bello.

Web It

For examples and practice on the laws of rational exponents, go to link 7-1-3 at mhhe.com/bello.

The laws of exponents for rational numbers provide a good way of simplifying some of the expressions we have studied. In Example 3, we evaluated $(8^{2/3})$ and $(-27)^{2/3}$. If we write 8 as 2^3, we can write

$$8^{2/3} = (2^3)^{2/3}$$
$$= 2^{(3) \cdot (2/3)} \quad \text{Law III}$$
$$= 2^2$$
$$= 4$$

Similarly, because $-27 = (-3)^3$,

$$(-27)^{2/3} = [(-3)^3]^{2/3} \quad \text{Law III}$$
$$= (-3)^2$$
$$= 9$$

Use these ideas when working the exercises. One word of caution, however. We cannot evaluate $[(-2)^2]^{1/2}$ using Law III. If we do, we obtain

$$[(-2)^2]^{1/2} = (-2)^{(2)(1/2)}$$
$$= (-2)^1$$
$$= -2$$

If we use the order of operations we have studied and square -2 first, we have

$$[(-2)^2]^{1/2} = [4]^{1/2}$$
$$= 2$$

The problem is that $a^{m/n}$ must be defined differently when m and n have common factors. In this case, $m = 2$ and $n = 2$, so they have a common factor. To remedy this situation, we make the following definition.

Definition of $(a^m)^{1/n}$	If m and n are positive even integers, then $$(a^m)^{1/n} = |a|^{m/n}$$

Thus $[(-2)^2]^{1/2} = |2|^{2/2} = |2|^1 = 2$. $[(-2)^{1/2}]^2$ is not defined since $(-2)^{1/2} = \sqrt{-2}$, not a real number. We shall discuss this further in Section 7.2.

Calculate It Exercises   Making Tables of Roots

A few years ago you had to find square and cube roots using a long and boring table. No more! You can make your own table with a TI-83 Plus. For example, to construct a table of square roots, press [2nd] [WINDOW] and set Tblstart = 1 and ΔTbl = 1. Now to tell the calculator you want a square root table, enter $Y_1 = x^{(1 \div 2)}$. Remember that to enter the exponent 1/2, you press [^] [(] [1] [÷] [2] [)] or the square root feature ([2nd] [x^2]). Now, press [2nd] [GRAPH]. You see two columns. The X column shows the number whose square root you want and the Y1 column shows the corresponding square root. In the window the second line shows a 2 for X and 1.4142 for Y1, meaning that the

square root of 2 is approximately 1.4142. You can make tables of cube, fourth, or *n*th roots by entering $Y_1 = x^{1/3}$ or $x^{1/4}$ or $x^{1/n}$.

There is at least one more way to approximate roots with a calculator. If you want the root of a specific number such as $\sqrt[5]{243}$, go to the home screen and enter 243^(1/5), then press [ENTER]. (Don't forget the parentheses when entering the exponent.) The answer is 3.

X	Y1	
1	1	
2	**1.4142**	
3	1.7321	
4	2	
5	2.2361	
6	2.4495	
7	2.6458	
X=2		

(continued)

1. What procedure would you use to verify the results in Example 5(a)?

2. Use the procedure to verify the results in Example 5(a).

3. Can you graphically verify the results of Example 6? Explain.

4. Local police and the highway patrol use the function $s(d) = \sqrt{30fd}$ to estimate the speed (in mi/hr) that a car was going when the skid mark left on the road is d feet long. Here, f is the coefficient of friction determined by the type of road (concrete, asphalt, gravel, tar) and whether the road is wet or dry. Some values for f are given as follows:

	Concrete	Tar
Wet road	0.4	0.5
Dry road	0.8	1.0

a. Graph $s(d)$ for a wet concrete road.

b. A car skidded 27 ft on a wet concrete road. How fast was the car going?

c. Graph $s(d)$ for a dry concrete road.

d. A car skidded 28 ft on a dry concrete road. How fast was the car going?

e. Graph $s(d)$ for a wet tar road and for a dry tar road.

f. In a court case the investigating officer testified that the skid mark left on the road by a car allegedly breaking the posted 25 mi/hr speed limit on a tar road was exactly 30 ft and this *proved* that the driver was speeding. No other information was provided by the officer. Based on the evidence presented, the judge dismissed the case. Why?

Exercises 7.1

Boost *your* **GRADE**
at mathzone.com!

- Practice Problems
- Self-Tests
- Videos
- NetTutor
- e-Professors

A In Problems 1–12, evaluate if possible.

1. $\sqrt{4}$ 2
2. $\sqrt{25}$ 5
3. $\sqrt[3]{8}$ 2
4. $\sqrt[3]{125}$ 5
5. $\sqrt[3]{-8}$ -2
6. $\sqrt[3]{-125}$ -5
7. $\sqrt[3]{\dfrac{-1}{64}}$ $\dfrac{-1}{4}$
8. $\sqrt[3]{\dfrac{-1}{27}}$ $\dfrac{-1}{3}$
9. $\sqrt[4]{16}$ 2
10. $\sqrt[4]{625}$ 5
11. $\sqrt[5]{32}$ 2
12. $\sqrt[5]{\dfrac{-1}{243}}$ $\dfrac{-1}{3}$

B In Problems 13–40, evaluate if possible.

13. $9^{1/2}$ 3
14. $16^{1/2}$ 4
15. $(-4)^{1/2}$ Not a real number
16. $-4^{1/2}$ -2
17. $27^{1/3}$ 3
18. $125^{1/3}$ 5
19. $81^{1/4}$ 3
20. $16^{1/4}$ 2
21. $\left(\dfrac{-1}{8}\right)^{1/3}$ $\dfrac{-1}{2}$
22. $\left(\dfrac{-1}{27}\right)^{1/3}$ $\dfrac{-1}{3}$
23. $\left(\dfrac{-1}{256}\right)^{1/4}$ Not a real number
24. $\left(\dfrac{1}{256}\right)^{1/4}$ $\dfrac{1}{4}$
25. $27^{2/3}$ 9
26. $(-27)^{2/3}$ 9
27. $125^{2/3}$ 25
28. $216^{2/3}$ 36
29. $\left(\dfrac{1}{8}\right)^{2/3}$ $\dfrac{1}{4}$
30. $\left(\dfrac{1}{81}\right)^{3/4}$ $\dfrac{1}{27}$
31. $(-8)^{4/3}$ 16
32. $(-27)^{4/3}$ 81
33. $(32)^{4/5}$ 16
34. $(-32)^{4/5}$ 16
35. $-32^{4/5}$ -16
36. $(-64)^{5/3}$ -1024
37. $64^{-2/3}$ $\dfrac{1}{16}$
38. $27^{-2/3}$ $\dfrac{1}{9}$
39. $[(-7)^4]^{1/4}$ 7
40. $[(-11)^6]^{1/6}$ 11

C In Problems 41–70, simplify and write the expression with positive exponents. All letters represent positive numbers.

41. $x^{1/7} \cdot x^{2/7}$ $x^{3/7}$
42. $y^{1/6} \cdot y^{1/6}$ $y^{1/3}$
43. $x^{-1/9} \cdot x^{-4/9}$ $\dfrac{1}{x^{5/9}}$
44. $y^{-5/2} \cdot y^{-3/2}$ $\dfrac{1}{y^4}$
45. $\dfrac{x^{4/5}}{x^{2/5}}$ $x^{2/5}$
46. $\dfrac{y^{5/7}}{y^{2/7}}$ $y^{3/7}$
47. $\dfrac{z^{2/3}}{z^{-1/3}}$ z
48. $\dfrac{a^{4/5}}{a^{-3/5}}$ $a^{7/5}$

49. $(x^{1/5})^{10}$ x^2

50. $(y^{1/3})^{12}$ y^4

51. $(z^{1/3})^{-6}$ $\dfrac{1}{z^2}$

52. $(a^{1/4})^{-8}$ $\dfrac{1}{a^2}$

53. $(b^{2/3})^{-6/5}$ $\dfrac{1}{b^{4/5}}$

54. $(c^{2/7})^{-7/8}$ $\dfrac{1}{c^{1/4}}$

55. $(a^{2/3}b^{3/4})^{-12}$ $\dfrac{1}{a^8 b^9}$

56. $(x^{1/8}y^{2/3})^{-24}$ $\dfrac{1}{x^3 y^{16}}$

57. $\left(\dfrac{a^{2/3}}{b^{3/5}}\right)^{-15}$ $\dfrac{b^9}{a^{10}}$

58. $\left(\dfrac{x^{1/2}}{y^{3/5}}\right)^{-20}$ $\dfrac{y^{12}}{x^{10}}$

59. $\left(\dfrac{x^{-2/5}}{y^{3/4}}\right)^{-40}$ $x^{16}y^{30}$

60. $\left(\dfrac{x^{-1/3}}{y^{3/8}}\right)^{-48}$ $x^{16}y^{18}$

61. $x^{1/3}(x^{2/3} + y^{1/2})$
$x + x^{1/3}y^{1/2}$

62. $x^{-4/5}(y^{1/3} + x^{-1/5})$
$\dfrac{y^{1/3}}{x^{4/5}} + \dfrac{1}{x}$

63. $y^{3/4}(x^{1/2} - y^{1/2})$
$x^{1/2}y^{3/4} - y^{5/4}$

64. $y^{2/3}(y^{1/2} - x^{2/3})$
$y^{7/6} - x^{2/3}y^{2/3}$

65. $\dfrac{x^{1/6} \cdot x^{-5/6}}{x^{1/3}}$
$\dfrac{1}{x}$

66. $\dfrac{(x^{1/3} \cdot x^{1/2})^2}{x^{1/2}}$
$x^{7/6}$

67. $\dfrac{(x^{1/3} \cdot y^{-1/2})^6}{(y^{1/2})^{-4}}$
$\dfrac{x^2}{y}$

68. $\left(\dfrac{x^{4/3} \cdot y^{1/2}}{x^{1/3}}\right)^{-1/2}$
$\dfrac{1}{x^{1/2}y^{1/4}}$

69. $\dfrac{(x^{1/4} \cdot y^2)^4}{(x^{2/3} \cdot y)^{-3}}$ $x^3 y^{11}$

70. $\left(\dfrac{-8a^{-3}b^{12}}{c^{15}}\right)^{-1/3}$ $\dfrac{-ac^5}{2b^4}$

APPLICATIONS

A Falling Body

If air resistance is ignored, the terminal velocity v of a falling body in meters per second is given by

$$v = (20h + v_0)^{1/2}$$

71. Find v if $h = 10$ and $v_0 = 25$ m/sec. $v = 15$ m/sec

72. Find v if a body is dropped ($v_0 = 0$) from a height of 45 m. $v = 30$ m/sec

73. If the velocity as measured in feet per second is
$$v = (64h + v_0)^{1/2}$$
find v if $h = 12$ ft and $v_0 = 16$ ft/sec. $v = 28$ ft/sec

74. Find v if a body is dropped ($v_0 = 0$) from a height of 25 ft (see Problem 73). $v = 40$ ft/sec

---

SKILL CHECKER

Try the "Skill Checker" exercises so you'll be ready for the next section.

In Problems 75–80, evaluate $\sqrt{b^2 - 4ac^2}$.

75. $a = 1, b = 5, c = 4$ 3

76. $a = 1, b = 3, c = 2$ 1

77. $a = 2, b = -3, c = -20$ 13

78. $a = \dfrac{1}{2}, b = -\dfrac{1}{12}, c = -1$ $\dfrac{17}{12}$

79. $a = \dfrac{1}{12}, b = \dfrac{1}{3}, c = -1$ $\dfrac{2}{3}$

80. $a = \dfrac{1}{12}, b = \dfrac{1}{2}, c = \dfrac{2}{3}$ $\dfrac{1}{6}$

Write:

81. $\dfrac{2x}{xy^2}$ with a denominator of $8x^3y^3$ $\dfrac{16x^3y}{8x^3y^3}$

82. $\dfrac{1}{5xy}$ with a denominator of $125x^3y^2$ $\dfrac{25x^2y}{125x^3y^2}$

83. $\dfrac{3xy}{x^2y^3}$ with a denominator of $16x^4y^4$ $\dfrac{48x^3y^2}{16x^4y^4}$

84. $\dfrac{2}{27x^5}$ with a denominator of $81x^8$ $\dfrac{6x^3}{81x^8}$

85. $\dfrac{5}{8x^4}$ with a denominator of $32x^5$ $\dfrac{20x}{32x^5}$

USING YOUR KNOWLEDGE

Using a Calculator to Find Rational Roots and Radicals

86. You already know that $\sqrt{x} = x^{1/2}$. Use your knowledge to write

$$\sqrt{\sqrt{x}}$$

using exponents. $x^{1/4}$

88. a. Write $\sqrt[3]{x}$ using exponents. $x^{1/3}$

    b. Write $\sqrt[3]{\sqrt{x}}$ using a single exponent. $x^{1/6}$

    c. If you have a calculator with a $\boxed{\sqrt{}}$ and a $\boxed{\sqrt[3]{}}$ key, find $\sqrt[6]{729}$. 3

87. If you have a calculator with a $\boxed{\sqrt{}}$ key, you can find $\sqrt{9}$ by pressing $\boxed{9}$ $\boxed{\sqrt{}}$. You can also find $\sqrt[4]{16}$ using the $\boxed{\sqrt{}}$ key and the result of Problem 86.

    a. Find $\sqrt[4]{16}$. 2   **b.** Find $\sqrt[4]{4096}$. 8

CALCULATOR CORNER

Many of the numerical evaluations in this section can be done with the $\boxed{y^x}$ key of a non-graphing calculator. This key will raise the number y to the x power. Thus to find 2^3, enter $\boxed{2}$ $\boxed{y^x}$ $\boxed{3}$ $\boxed{=}$. The answer is 8. To find $\sqrt[3]{-64}$, first recall that $\sqrt[3]{-64} = (-64)^{1/3}$. Thus we enter

Note that $\frac{1}{3}$ has to be entered in parentheses.

If your instructor permits, use the $\boxed{y^x}$ key on your calculator to find the following values.

89. $\sqrt[3]{-\dfrac{1}{8}}$ **90.** $(-125)^{1/3}$ **91.** $\sqrt[4]{\dfrac{1}{16}}$

    $\dfrac{-1}{2}$ -5 $\dfrac{1}{2}$

WRITE ON

92. Explain why a square root of 4 is -2 but $\sqrt{4} = 2$. Answers may vary.

94. Explain why $\sqrt[n]{a} = a^{1/n}$ is not a real number when a is negative and n is even. Answers may vary.

96. Explain why the even root of a negative number is not a real number. (For example, $\sqrt{-4}$ is not a real number.) Answers may vary.

93. How many square roots does every positive real number a have? Name them when $a = 36$. Answers may vary.

95. Explain what is meant by "the nth root of a real number a." Answers may vary.

MASTERY TEST

If you know how to do these problems, you have learned your lesson!

Simplify (x and y positive):

97. $(x^{1/4}y^{3/5})^{-20}$ $\dfrac{1}{x^5 y^{12}}$

98. $(x^{1/7}y^{3/14})^{-21}$ $\dfrac{1}{x^3 y^{9/2}}$

99. $\dfrac{x^{1/4}y^{2/3}}{x^{3/4}y^{1/3}}$ $\dfrac{y^{1/3}}{x^{1/2}}$

100. $\dfrac{x^{1/5}y^{3/7}}{x^{2/5}y^{2/7}}$ $\dfrac{y^{1/7}}{x^{1/5}}$

101. $x^{2/5}(x^{-1/5} + y^{1/4})$ $x^{1/5} + x^{2/5}y^{1/4}$

102. $x^{3/7}(x^{-3/7} + y^{3/4})$ $1 + x^{3/7}y^{3/4}$

103. $x^{1/3} \cdot x^{1/5}$ $x^{8/15}$

104. $y^{1/5} \cdot y^{1/4}$ $y^{9/20}$

105. $\dfrac{x^{-2/3}}{x^{1/4}}$ $\dfrac{1}{x^{11/12}}$

106. $\dfrac{y^{-3/4}}{y^{1/5}}$ $\dfrac{1}{y^{19/20}}$

107. $(y^{3/4})^{-1/5}$ $\dfrac{1}{y^{3/20}}$

108. $(x^{2/3})^{-2/5}$ $\dfrac{1}{x^{4/15}}$

Evaluate if possible:

109. $81^{-3/4}$ $\dfrac{1}{27}$

110. $(-27)^{-4/3}$ $\dfrac{1}{81}$

111. $216^{-2/3}$ $\dfrac{1}{36}$

112. $27^{2/3}$ 9

113. $(-64)^{-2/3}$ $\dfrac{1}{16}$

114. $(-36)^{3/2}$ Not a real number

115. $49^{1/2}$ 7

116. $(-216)^{1/3}$ -6

117. $\left(\dfrac{1}{81}\right)^{1/4}$ $\dfrac{1}{3}$

118. $\sqrt{-25}$ Not a real number

119. $\sqrt[3]{-125}$ -5

120. $\sqrt[3]{\dfrac{-1}{27}}$ $\dfrac{-1}{3}$

7.2 SIMPLIFYING RADICALS

To Succeed, Review How To...

1. Use the laws of exponents (pp. 33–40).

2. Factor perfect square trinomials (pp. 384–386).

Objectives

A Simplify radical expressions.

B Rationalize the denominator of a fraction.

C Reduce the order of a radical expression.

GETTING STARTED

Radical Speeding

Have you seen your local police measuring skid marks at the scene of an accident? The speed s (in mi/ hr) a car was traveling if it skidded d feet after the brakes were applied on a dry concrete road is given by $s = \sqrt{24d}$. A car leaving a 50-ft skid mark was traveling at a speed $s = \sqrt{24 \cdot 50} = \sqrt{1200}$. How can we simplify this? By factoring 1200 into factors that are perfect squares. Thus $\sqrt{1200} = \sqrt{100 \cdot 4 \cdot 3} = 20\sqrt{3}$ or about 35 mi/hr. We assumed $\sqrt{100 \cdot 4 \cdot 3} = \sqrt{100} \cdot \sqrt{4} \cdot \sqrt{3}$. Is this true?

In this section, we shall study how to simplify radical expressions by using properties analogous to the properties of rational exponents we studied in Section 7.1. First, let's recall the relationship between rational exponents and radicals. In general, the nth root of a number a has the following definition.

Web It

For a program to input on the TI-83 Plus that will simplify the square root of a number, go to link 7-2-1 at mhhe.com/bello.

---

nth Root

$$a^{1/n} = \sqrt[n]{a} \quad (a \geq 0)$$

where n is a positive integer.

From this definition, we can derive three important relationships involving radicals that can be proved using the properties of exponents discussed previously.

Properties of Radicals

Laws for Simplifying Radical Expressions

I. $\sqrt[n]{a^n} = a$ $(a \geq 0)$

II. $\sqrt[n]{ab} = \sqrt[n]{a}\sqrt[n]{b}$ $(a, b \geq 0)$ Product rule

III. $\sqrt[n]{\dfrac{a}{b}} = \dfrac{\sqrt[n]{a}}{\sqrt[n]{b}}$ $(a \geq 0, b > 0)$ Quotient rule

---

The first of these laws is equivalent to the definition of the principal nth root of a. Thus $\sqrt[n]{a^n} = (a^n)^{1/n} = a^{n \cdot 1/n} = a$. The other two laws are obtained as follows:

II. $\sqrt[n]{ab} = (ab)^{1/n} = a^{1/n} \cdot b^{1/n} = \sqrt[n]{a} \cdot \sqrt[n]{b}$

III. $\sqrt[n]{\dfrac{a}{b}} = \left(\dfrac{a}{b}\right)^{1/n} = \dfrac{a^{1/n}}{b^{1/n}} = \dfrac{\sqrt[n]{a}}{\sqrt[n]{b}}$

We have already mentioned that when m and n are even, $(a^m)^{1/n} = |a|^{m/n}$. For $m = n$, we have the following definition.

DEFINITION OF $\sqrt[n]{a^n}$

$\sqrt[n]{a^n} = |a|$ n is an **even** positive integer.

$\sqrt[n]{a^n} = a$ n is an **odd** positive integer.

Thus for even indices, $\sqrt{3^2} = |3| = 3$, $\sqrt{(-3)^2} = |-3| = 3$, and $\sqrt[4]{(-x)^4} = |-x|$. When the index is odd, absolute value is not necessary. Thus, $\sqrt[3]{-8} = -2$ and $\sqrt[5]{-x^5} = -x$.

EXAMPLE 1   Simplifying expressions containing radicals

Simplify:

a. $\sqrt[4]{(-2)^4}$ **b.** $\sqrt[8]{(-x)^8}$ **c.** $\sqrt[9]{(-x)^9}$ **d.** $\sqrt{x^2 + 8x + 16}$

SOLUTION

a. Since the index 4 is even, $\sqrt[4]{(-2)^4} = |-2| = 2$.

b. The index is even. Thus $\sqrt[8]{(-x)^8} = |-x| = |x|$.

c. $\sqrt[9]{(-x)^9} = -x$. (Absolute value is not necessary because the index 9 is not even.)

d. We start by factoring: $x^2 + 8x + 16 = (x + 4)^2$. Thus

$$\sqrt{x^2 + 8x + 16} = \sqrt{(x + 4)^2}$$
$$= |x + 4|$$

(Absolute value is needed, because the index 2 is even.)

PROBLEM 1

Simplify:

a. $\sqrt[4]{(-5)^4}$ **b.** $\sqrt[6]{(-x)^6}$

c. $\sqrt[7]{(-x)^7}$ **d.** $\sqrt{x^2 - 10x + 25}$

Calculate It   Finding the Numerical Root

Most of the numerical problems in this section can be verified with a calculator. Verifying Example 1(a) can be done two ways. The first way is to enter the information using rational exponents as shown in Window 1, press ENTER, and the answer "2" appears. Note how parentheses were used to enter the expression. Moreover, you have to know that taking the fourth root is the same as raising to the ¼ power.

The second way is to press 4 (for the index) MATH 5, the expression $(-2)^4$ ENTER, and the answer "2" appears (Window 2).

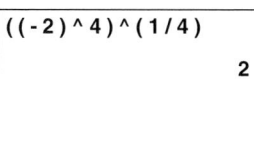

Window 1

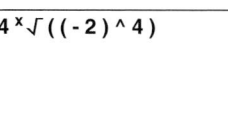

Window 2

Answers

1. a. $|-5| = 5$ **b.** $|-x| = |x|$
c. $-x$ **d.** $|x - 5|$

Web It

For a tutorial on simplifying radicals, go to link 7-2-2 at mhhe.com/bello.

Teaching Tip

Students should note the following:

$(x^1)^2 = x^2$ $(x^1)^3 = x^3$

$(x^2)^2 = x^4$ $(x^2)^3 = x^6$

$(x^3)^2 = x^6$ $(x^3)^3 = x^9$

$(x^4)^2 = x^8$ $(x^4)^3 = x^{12}$

and so on.   and so on.

Notice in the perfect squares the exponents are divisible by 2 and the perfect cube exponents are divisible by 3.

If the radicand does not have a perfect root, we can sometimes simplify it by factoring out any perfect roots. To help you, we list the first few square, cube, and fourth roots in Table 2.

Table 2 Partial List of Square, Cube, and Fourth Roots. (Assume variables are positive real numbers.)

Square Roots		Cube Roots		Fourth Roots	
$\sqrt{0} = 0$	$\sqrt{x^2} = x$	$\sqrt[3]{0} = 0$		$\sqrt[4]{0} = 0$	
$\sqrt{1} = 1$	$\sqrt{x^4} = x^2$	$\sqrt[3]{1} = 1$	$\sqrt[3]{x^3} = x$	$\sqrt[4]{1} = 1$	$\sqrt[4]{x^4} = x$
$\sqrt{4} = 2$	$\sqrt{x^6} = x^3$	$\sqrt[3]{8} = 2$	$\sqrt[3]{x^6} = x^2$	$\sqrt[4]{16} = 2$	$\sqrt[4]{x^8} = x^2$
$\sqrt{9} = 3$	$\sqrt{x^8} = x^4$	$\sqrt[3]{27} = 3$	$\sqrt[3]{x^9} = x^3$	$\sqrt[4]{81} = 3$	$\sqrt[4]{x^{12}} = x^3$
$\sqrt{16} = 4$	$\sqrt{x^{10}} = x^5$	$\sqrt[3]{64} = 4$	$\sqrt[3]{x^{12}} = x^4$	$\sqrt[4]{256} = 4$	$\sqrt[4]{x^{16}} = x^4$
$\sqrt{25} = 5$	$\sqrt{x^{12}} = x^6$	$\sqrt[3]{125} = 5$	$\sqrt[3]{x^{15}} = x^5$	$\sqrt[4]{625} = 5$	$\sqrt[4]{x^{20}} = x^5$

Variables that have perfect roots have exponents divisible by the index. To simplify $\sqrt[3]{40}$, we find a factor of 40 that has a perfect cube root. This factor is 8. Thus $40 = 8 \cdot 5$ and using Law II we can write

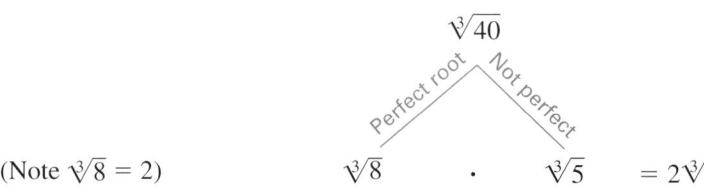

(Note $\sqrt[3]{8} = 2$) $\sqrt[3]{8} \quad \cdot \quad \sqrt[3]{5} \quad = 2\sqrt[3]{5}$

To simplify $\sqrt[3]{a^5}$ we can write

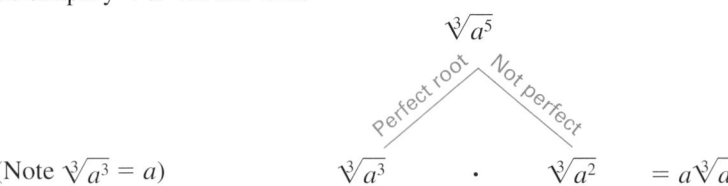

(Note $\sqrt[3]{a^3} = a$) $\sqrt[3]{a^3} \quad \cdot \quad \sqrt[3]{a^2} \quad = a\sqrt[3]{a^2}$

EXAMPLE 2   Simplifying square and cube roots

Simplify:

a. $\sqrt{48}$ **b.** $\sqrt[3]{54}$ **c.** $\sqrt[3]{128a^4b^6}$

SOLUTION

a. Since $48 = 16 \cdot 3$

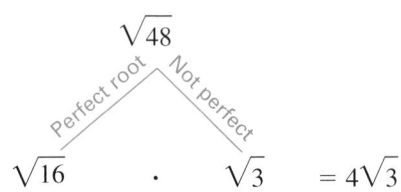

(Note $\sqrt{16} = 4$) $\sqrt{16} \quad \cdot \quad \sqrt{3} \quad = 4\sqrt{3}$

b. Since $54 = 27 \cdot 2$

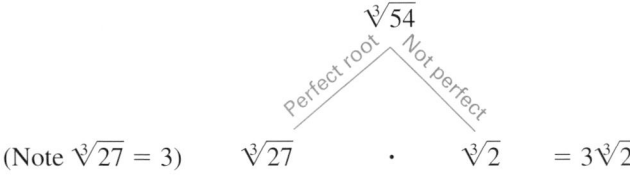

(Note $\sqrt[3]{27} = 3$) $\sqrt[3]{27} \quad \cdot \quad \sqrt[3]{2} \quad = 3\sqrt[3]{2}$

PROBLEM 2

Simplify:

a. $\sqrt{32}$ **b.** $\sqrt[3]{32}$ **c.** $\sqrt[3]{81a^6b^4}$

Answers

2. a. $4\sqrt{2}$ **b.** $2\sqrt[3]{4}$

c. $3a^2b\sqrt[3]{3b}$

c. We factor the radicand into factors that have perfect cube roots. 64 is a perfect cube and $128 = 64 \cdot 2$; a^3 is a perfect cube and $a^4 = a^3 \cdot a$, and b^6 is a perfect cube. Thus,

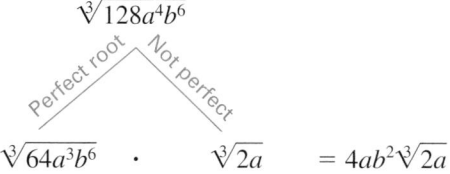

$$\sqrt[3]{128a^4b^6}$$

Perfect root　Not perfect

(Note $\sqrt[3]{64a^3b^6} = 4ab^2$)　$\sqrt[3]{64a^3b^6}$　$\cdot$　$\sqrt[3]{2a}$　$= 4ab^2\sqrt[3]{2a}$

Web It

The third law mentioned in this section can be used to change a radical into a form in which the radicand contains no fractions. For example,

$$\sqrt{\frac{3}{16}} = \frac{\sqrt{3}}{\sqrt{16}} = \frac{\sqrt{3}}{4}$$

$$\sqrt[3]{\frac{7}{8}} = \frac{\sqrt[3]{7}}{\sqrt[3]{8}} = \frac{\sqrt[3]{7}}{2}$$

If the denominator does not have a perfect root, multiply the numerator and denominator by a factor that will yield a perfect root. To simplify

$$\sqrt{\frac{3}{8}}$$

multiply the denominator by 2 to obtain 16, which has a perfect root. You must also multiply the numerator by 2 so that you get an equivalent fraction.

$$\sqrt{\frac{3}{8}} = \sqrt{\frac{3 \cdot 2}{8 \cdot 2}} = \sqrt{\frac{6}{16}} = \frac{\sqrt{6}}{4}$$

EXAMPLE 3　　**Simplifying radical expressions containing fractions**

Simplify:

a. $\sqrt{\dfrac{7}{32}}$　　**b.** $\sqrt[3]{\dfrac{9}{x^3}}$　　**c.** $\sqrt[4]{\dfrac{2}{27x^5}}$

SOLUTION

a. We multiply the denominator and numerator by 2 because $32 \cdot 2 = 64$ is a perfect square:

$$\sqrt{\frac{7}{32}} = \sqrt{\frac{7 \cdot 2}{32 \cdot 2}} = \sqrt{\frac{14}{64}} = \frac{\sqrt{14}}{\sqrt{64}} = \frac{\sqrt{14}}{8}$$

b. Since $\sqrt[3]{x^3} = x$, we simplify the denominator.

$$\sqrt[3]{\frac{9}{x^3}} = \frac{\sqrt[3]{9}}{\sqrt[3]{x^3}} = \frac{\sqrt[3]{9}}{x}$$

c. To obtain a perfect fourth power, multiply 27 by 3 to obtain $81 = 3^4$ and x^5 by x^3 to obtain x^8.

$$\sqrt[4]{\frac{2}{27x^5}} = \sqrt[4]{\frac{2 \cdot 3x^3}{27x^5 \cdot 3x^3}} = \sqrt[4]{\frac{6x^3}{81x^8}} = \frac{\sqrt[4]{6x^3}}{\sqrt[4]{81x^8}} = \frac{\sqrt[4]{6x^3}}{3x^2}$$

PROBLEM 3

Simplify:

a. $\sqrt{\dfrac{11}{12}}$　　**b.** $\sqrt[3]{\dfrac{6}{x^3}}$　　**c.** $\sqrt[4]{\dfrac{3}{8x^6}}$

Answers

3. **a.** $\dfrac{\sqrt{33}}{6}$　**b.** $\dfrac{\sqrt[3]{6}}{x}$　**c.** $\dfrac{\sqrt[4]{6x^2}}{2x^2}$

Calculate It Verifying Simplified Radical Expressions

To verify Example 3(a), enter the original problem,

$$\sqrt{\frac{7}{32}}$$

and the simplified version,

$$\frac{\sqrt{14}}{8}$$

You should get the approximation 0.4677071733 in both cases, as shown in Window 1.

```
√(7/32)
          .4677071733
√(14)/8
          .4677071733
```

Window 1

If the examples to be verified are not numerical, you can still graph the original problem and answer to make sure the graphs are the same. For example, to verify the results of Example 3(c), let

$$Y_1 = (2/(27x^5))^\wedge(1/4)$$

and

$$Y_2 = (\sqrt[4]{(6x^\wedge 3)})/(3x^2)$$

Graph both of these using a standard window first; then re-graph using a $[-1, 10]$ by $[-1, 10]$ window for a better view. The graph is shown in Window 2.

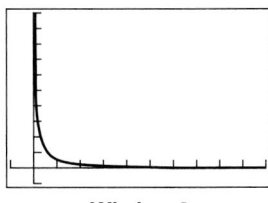

Window 2

B Rationalizing the Denominator

Web It

For more explanation and examples on rationalizing denominators, go to link 7-2-4 at mhhe.com/bello.

As we saw in Example 3, in some cases the denominators of fractions contain radical expressions. For example,

$$\frac{\sqrt{3}}{\sqrt{5}}$$

has $\sqrt{5}$ in the denominator. To simplify

$$\frac{\sqrt{3}}{\sqrt{5}}$$

we use the fundamental property of fractions to multiply numerator and denominator by $\sqrt{5}$, which causes the denominator to be rational. Thus

$$\frac{\sqrt{3}}{\sqrt{5}} = \frac{\sqrt{3} \cdot \sqrt{5}}{\sqrt{5} \cdot \sqrt{5}} \qquad \text{Fundamental property of fractions}$$

$$= \frac{\sqrt{15}}{\sqrt{5^2}} \qquad \text{Law II}$$

$$= \frac{\sqrt{15}}{5} \qquad \text{Law I}$$

This process is called **rationalizing the denominator.** To rationalize the denominator in the expression

$$\frac{\sqrt{7}}{\sqrt{3x}} \quad (x > 0)$$

we have to change the fraction to an equivalent one without a radical in the denominator. As before, we multiply the denominator and numerator by $\sqrt{3x}$ to obtain

$$\frac{\sqrt{7}}{\sqrt{3x}} = \frac{\sqrt{7} \cdot \sqrt{3x}}{\sqrt{3x} \cdot \sqrt{3x}} \qquad \text{Fundamental property of fractions}$$

$$= \frac{\sqrt{21x}}{\sqrt{3^2x^2}} \qquad \text{Law II}$$

$$= \frac{\sqrt{21x}}{3x} \qquad \text{Law I}$$

| **EXAMPLE 4**　　**Rationalizing the denominator** | **PROBLEM 4** |

Rationalize the denominator:

a. $\dfrac{\sqrt{11}}{\sqrt{6}}$　　　　**b.** $\dfrac{\sqrt{3}}{\sqrt{5x}}$ $(x>0)$　　　　**c.** $\dfrac{\sqrt{5}}{\sqrt{18x^2}}$

SOLUTION

a. To obtain a perfect square in the denominator, multiply by $\sqrt{6}$.

$$\frac{\sqrt{11}}{\sqrt{6}}=\frac{\sqrt{11}\cdot\sqrt{6}}{\sqrt{6}\cdot\sqrt{6}}=\frac{\sqrt{66}}{6}$$

b. To obtain a perfect square in the denominator, multiply by $\sqrt{5x}$.

$$\frac{\sqrt{3}}{\sqrt{5x}}=\frac{\sqrt{3}\cdot\sqrt{5x}}{\sqrt{5x}\cdot\sqrt{5x}}=\frac{\sqrt{15x}}{5x}$$

c. To convert $\sqrt{18}$ to a perfect square, multiply by $\sqrt{2}$.

$$\frac{\sqrt{5}}{\sqrt{18x^2}}=\frac{\sqrt{5}\cdot\sqrt{2}}{\sqrt{18x^2}\cdot\sqrt{2}}=\frac{\sqrt{10}}{\sqrt{36x^2}}=\frac{\sqrt{10}}{6x^2}$$

x^2 is *not* under the radical, so x^2 does not change.

Rationalize the denominator:

a. $\dfrac{\sqrt{7}}{\sqrt{3}}$　　　　**b.** $\dfrac{\sqrt{5}}{\sqrt{6x}}$, $x>0$

c. $\dfrac{\sqrt{11}}{\sqrt{32x^3}}$

Teaching Tip

In Example 4(c), as in Example 3(a), we can factor 18 to help rationalize the denominator.

$$\frac{\sqrt{5}}{\sqrt{18x^2}}=\frac{5}{\sqrt{3^2\cdot 2\,x^2}}\cdot\frac{\sqrt{2}}{\sqrt{2}}$$

Because the index is 2 we want the exponents of both primes to be divisible by 2. Hence we multiply by $\sqrt{2}$.

Web It

For a tutorial on rationalizing denominators, go to link 7-2-5 at mhhe.com/bello.

When the radical in the denominator is of index n, we must make the radicand an exact nth power. For example, to rationalize

$$\frac{\sqrt[3]{5}}{\sqrt[3]{3x}}$$

we convert $\sqrt[3]{3}$ to a perfect cube root by multiplying by $\sqrt[3]{3^2}$, and we convert $\sqrt[3]{x}$ to a perfect cube root by multiplying by $\sqrt[3]{x^2}$. We can combine these two steps and multiply by $\sqrt[3]{3^2 x^2}$ to obtain

$$\frac{\sqrt[3]{5}}{\sqrt[3]{3x}}=\frac{\sqrt[3]{5}\cdot\sqrt[3]{3^2\cdot x^2}}{\sqrt[3]{3x}\cdot\sqrt[3]{3^2\cdot x^2}}$$ 　Multiply the numerator and denominator by $\sqrt[3]{3^2\cdot x^2}$ to make the denominator a perfect cube root.

$$=\frac{\sqrt[3]{5\cdot 9\cdot x^2}}{\sqrt[3]{3^3 x^3}}$$ 　Law II

$$=\frac{\sqrt[3]{45x^2}}{3x}$$ 　Law I

Teaching Tip

An exact nth root must have an exponent divisible by n.

| **EXAMPLE 5**　　**Making the radicand an exact nth power** | **PROBLEM 5** |

Rationalize the denominator:

a. $\dfrac{1}{\sqrt[3]{5x}}$　　　　**b.** $\dfrac{\sqrt[4]{5}}{\sqrt[4]{8x}}$

SOLUTION

a. To convert $\sqrt[3]{5x}$ to a perfect cube root, multiply by $\sqrt[3]{5^2 x^2}$.

$$\frac{1}{\sqrt[3]{5x}}=\frac{1\cdot\sqrt[3]{5^2 x^2}}{\sqrt[3]{5x}\cdot\sqrt[3]{5^2 x^2}}$$

$$=\frac{\sqrt[3]{25x^2}}{\sqrt[3]{5^3 x^3}}$$

$$=\frac{\sqrt[3]{25x^2}}{5x}$$

Rationalize the denominator:

a. $\dfrac{1}{\sqrt[3]{54x}}$　　　　**b.** $\dfrac{\sqrt[5]{5}}{\sqrt[5]{27x^4}}$

Answers

4. a. $\dfrac{\sqrt{21}}{3}$　**b.** $\dfrac{\sqrt{30x}}{6x}$　**c.** $\dfrac{\sqrt{22x}}{8x^2}$

5. a. $\dfrac{\sqrt[3]{4x^2}}{6x}$　**b.** $\dfrac{\sqrt[5]{45x}}{3x}$

b. $\dfrac{\sqrt[4]{5}}{\sqrt[4]{8x}} = \dfrac{\sqrt[4]{5}}{\sqrt[4]{2^3 x}}$ Write $8x$ as $2^3 x$.

$= \dfrac{\sqrt[4]{5} \cdot \sqrt[4]{2 \cdot x^3}}{\sqrt[4]{2^3 \cdot x} \cdot \sqrt[4]{2 \cdot x^3}}$ Use the fundamental property of fractions to make the denominator a perfect 4th root.

$= \dfrac{\sqrt[4]{10x^3}}{\sqrt[4]{2^4 x^4}}$ Law II

$= \dfrac{\sqrt[4]{10x^3}}{2x}$ Law I

Web It

For a program to input on the TI-83 Plus that will reduce the index of a radical, go to link 7-2-6 at mhhe.com/bello.

C Reducing the Index of a Radical Expression

The index of a radical expression can sometimes be reduced by writing the radical as a power with a rational exponent and then reducing the exponent. For example, if $x \geq 0$,

Web It

For another example on how to reduce the index (change the order) of a radical, go to link 7-2-7 at mhhe.com/bello.

$$\sqrt[6]{x^3} = (x^3)^{1/6}$$
$$= x^{3 \cdot 1/6}$$
$$= x^{1/2}$$
$$= \sqrt{x}$$

Similarly, for $x \geq 0$ and $y \geq 0$,

$$\sqrt[4]{64x^2 y^2} = \sqrt[4]{(8xy)^2}$$
$$= [(8xy)^2]^{1/4}$$
$$= [8xy]^{2 \cdot 1/4}$$
$$= (8xy)^{1/2} = \sqrt{8xy}$$

Perfect root / Not perfect

$$\sqrt{4} \quad \cdot \quad \sqrt{2xy} \quad = 2\sqrt{2xy}$$

EXAMPLE 6	Reducing the index

Reduce the index (change the order):

a. $\sqrt[4]{\dfrac{4}{9}}$ **b.** $\sqrt[6]{27c^3 d^3}$ $(c \geq 0, d \geq 0)$

SOLUTION

a. $\sqrt[4]{\dfrac{4}{9}} = \left[\left(\dfrac{2}{3}\right)^2\right]^{1/4}$ $4 = 2^2$ and $9 = 3^2$

$= \left(\dfrac{2}{3}\right)^{1/2}$ Simplify exponents.

$= \sqrt{\dfrac{2}{3}}$ Write as a radical.

$= \dfrac{\sqrt{6}}{3}$ Rationalize the denominator. $\left(\dfrac{\sqrt{2} \cdot \sqrt{3}}{\sqrt{3} \cdot \sqrt{3}} = \dfrac{\sqrt{6}}{3}\right)$

b. $\sqrt[6]{27c^3 d^3} = \sqrt[6]{(3cd)^3}$ $27 = 3^3$

$= [(3cd)^3]^{1/6}$ Write with rational exponent.

$= [3cd]^{1/2}$ Simplify exponents.

$= \sqrt{3cd}$ Write as a radical.

PROBLEM 6

Reduce the index (order):

a. $\sqrt[4]{\dfrac{25}{x^2}}$

b. $\sqrt[6]{4c^2 d^2}$ $(c \geq 0, d \geq 0)$

Answers

6. a. $\sqrt{\dfrac{5}{x}} = \dfrac{\sqrt{5x}}{x}$ **b.** $\sqrt[3]{2cd}$

Calculate It Exercises

1. Verify the results of Example l(a) and l(b).

2. Verify the results of Example 3(b) and 3(c).

3. Graph $\sqrt[n]{a^n} = (a^n)^{(1/n)}$ for $n = 1, 2, 3,$ and 4.
   a. For what values of n is $(a^n)^{(1/n)} = a$?
   b. For what values of n is $(a^n)^{(1/n)} = |a|$?
   c. What can you conjecture from parts **a** and **b**?

We have used different techniques to *simplify* radical expressions. To make sure that the resulting radicals are simplified, use these rules.

Rules for Simplifying Radical Expressions

A radical expression is in **simplified** form if

1. All exponents in the radicand (the expression under the radical) are less than the index.

2. There are no fractions under the radical sign.

3. There are no radicals in the denominator.

4. The index is as low as possible.

EXAMPLE 7 **Simplifying radical expressions**	**PROBLEM 7**

Simplify:

$$\sqrt[6]{\frac{a^2}{16c^{10}}} \quad \text{where } a > 0, c > 0$$

Simplify:

$$\sqrt[6]{\frac{a^3}{8x^3}} \quad (a > 0, x > 0)$$

SOLUTION   To make the denominator a perfect sixth root, note that $16 = 2^4$, and then multiply numerator and denominator by $\sqrt[6]{2^2 c^2}$. We have

$$\sqrt[6]{\frac{a^2}{16c^{10}}} = \sqrt[6]{\frac{a^2}{2^4 c^{10}} \cdot \frac{\sqrt[6]{2^2 c^2}}{\sqrt[6]{2^2 c^2}}} = \sqrt[6]{\frac{2^2 a^2 c^2}{2^6 c^{12}}}$$

$$= \frac{\sqrt[6]{2^2 a^2 c^2}}{2c^2} \qquad \sqrt[6]{2^6 c^{12}} = 2c^2$$

$$= \frac{(2^2 a^2 c^2)^{1/6}}{2c^2} \qquad \text{Rewrite numerator.}$$

$$= \frac{[(2ac)^2]^{1/6}}{2c^2}$$

$$= \frac{(2ac)^{1/3}}{2c^2} \qquad \text{Reduce index.}$$

$$= \frac{\sqrt[3]{2ac}}{2c^2} \qquad \text{Write in radical form.}$$

Answer

7. $\dfrac{\sqrt{2ax}}{2x}$

Exercises 7.2

Boost *your* GRADE at mathzone.com!

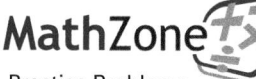

- Practice Problems
- Self-Tests
- Videos
- NetTutor
- e-Professors

A In Problems 1–24, simplify the expression given. (*Hint:* Some answers require absolute values.)

1. $\sqrt{(-5)^2}$ 5

2. $\sqrt{(5)^2}$ 5

3. $\sqrt[3]{-64}$ -4

4. $\sqrt[3]{-125}$ -5

5. $\sqrt[6]{(-x)^6}$ $|-x| = |x|$

6. $\sqrt[5]{(-x)^5}$ $-x$

7. $\sqrt{x^2 + 12x + 36}$ $|x + 6|$

8. $\sqrt{4x^2 + 12x + 9}$ $|2x + 3|$

9. $\sqrt{9x^2 - 12x + 4}$ $|3x - 2|$

10. $\sqrt{16x^2 + 8x + 1}$ $|4x + 1|$

11. $\sqrt{16x^3 y^3}$ $4|xy|\sqrt{xy}$

12. $\sqrt{81x^3 y^4}$ $9|xy^2|\sqrt{x}$

13. $\sqrt[3]{40x^4y}$ $2x\sqrt[3]{5xy}$

14. $\sqrt[3]{81x^3y^6}$ $3xy^2\sqrt[3]{3}$

15. $\sqrt[4]{x^5y^7}$ $|xy|\sqrt[4]{xy^3}$

16. $\sqrt[4]{162x^4y^7}$ $3|xy|\sqrt[4]{2y^3}$

17. $\sqrt[5]{-243a^{10}b^{17}}$ $-3a^2b^3\sqrt[5]{b^2}$

18. $\sqrt[5]{-32a^{15}b^{20}}$ $-2a^3b^4$

19. $\sqrt{\dfrac{13}{49}}$ $\dfrac{\sqrt{13}}{7}$

20. $\sqrt{\dfrac{17}{64}}$ $\dfrac{\sqrt{17}}{8}$

21. $\sqrt{\dfrac{17}{4x^2}}$ $\dfrac{\sqrt{17}}{2|x|}$

22. $\sqrt{\dfrac{19}{64x^4}}$ $\dfrac{\sqrt{19}}{8|x^2|}$

23. $\sqrt[3]{\dfrac{3}{64x^3}}$ $\dfrac{\sqrt[3]{3}}{4x}$

24. $\sqrt[3]{\dfrac{-7}{27x^6}}$ $\dfrac{-\sqrt[3]{7}}{3x^2}$

B In Problems 25–44, rationalize the denominator. (Assume all variables represent positive real numbers.)

25. $\sqrt{\dfrac{2}{3}}$ $\dfrac{\sqrt{6}}{3}$

26. $\sqrt{\dfrac{4}{5}}$ $\dfrac{2\sqrt{5}}{5}$

27. $\dfrac{-\sqrt{2}}{\sqrt{7}}$ $\dfrac{-\sqrt{14}}{7}$

28. $\dfrac{-\sqrt{3}}{\sqrt{11}}$ $\dfrac{-\sqrt{33}}{11}$

29. $\sqrt{\dfrac{5}{2a}}$ $\dfrac{\sqrt{10a}}{2a}$

30. $\sqrt{\dfrac{7}{36}}$ $\dfrac{\sqrt{7}}{6}$

31. $\sqrt{\dfrac{5}{32ab}}$ $\dfrac{\sqrt{10ab}}{8ab}$

32. $\sqrt{\dfrac{5}{8ab}}$ $\dfrac{\sqrt{10ab}}{4ab}$

33. $-\sqrt{\dfrac{3}{2a^3b^3}}$ $\dfrac{-\sqrt{6ab}}{2a^2b^2}$

34. $-\sqrt{\dfrac{3}{8ab^3}}$ $\dfrac{-\sqrt{6ab}}{4ab^2}$

35. $\dfrac{\sqrt{x}\sqrt{xy^3}}{\sqrt{y}}$ xy

36. $\dfrac{\sqrt{xy}\sqrt{xy^4}}{\sqrt{y}}$ xy^2

37. $-\sqrt[3]{\dfrac{7}{9}}$ $\dfrac{-\sqrt[3]{21}}{3}$

38. $-\sqrt[3]{\dfrac{3}{32}}$ $\dfrac{-\sqrt[3]{6}}{4}$

39. $\sqrt[3]{\dfrac{3}{16x^2}}$ $\dfrac{\sqrt[3]{12x}}{4x}$

40. $\sqrt[3]{\dfrac{5}{16x}}$ $\dfrac{\sqrt[3]{20x^2}}{4x}$

41. $\sqrt[3]{\dfrac{1}{8x^2}}$ $\dfrac{\sqrt[3]{x}}{2x}$

42. $\sqrt[3]{\dfrac{2}{4x^6}}$ $\dfrac{\sqrt[3]{4}}{2x^2}$

43. $\sqrt[4]{\dfrac{3}{2}}$ $\dfrac{\sqrt[4]{24}}{2}$

44. $\sqrt[4]{\dfrac{1}{2x^3}}$ $\dfrac{\sqrt[4]{8x}}{2x}$

C In Problems 45–54, reduce the index (order) of the given radical and simplify. (Assume the variables represent positive real numbers.)

45. $\sqrt[6]{9}$ $\sqrt[3]{3}$

46. $\sqrt[6]{4}$ $\sqrt[3]{2}$

47. $\sqrt[4]{4a^2}$ $\sqrt{2a}$

48. $\sqrt[4]{9a^2}$ $\sqrt{3a}$

49. $\sqrt[4]{25x^6y^2}$ $x\sqrt{5xy}$

50. $\sqrt[4]{36x^2y^6}$ $y\sqrt{6xy}$

51. $\sqrt[4]{49x^{10}y^6}$ $x^2y\sqrt{7xy}$

52. $\sqrt[4]{100x^{10}y^{10}}$ $x^2y^2\sqrt{10xy}$

53. $\sqrt[6]{8a^3b^3}$ $\sqrt{2ab}$

54. $\sqrt[6]{27a^3b^9}$ $b\sqrt{3ab}$

In Problems 55–59, simplify the radical expression. (Assume all variables represent positive real numbers.)

55. $\sqrt[6]{\dfrac{a^4}{b^8}}$ $\dfrac{\sqrt[3]{a^2b^2}}{b^2}$

56. $\sqrt[4]{\dfrac{c^6}{4b^2}}$ $\dfrac{c\sqrt{2bc}}{2b}$

57. $\sqrt[4]{\dfrac{64a^2}{9b^6}}$ $\dfrac{2\sqrt{6ab}}{3b^2}$

58. $\sqrt[4]{\dfrac{4c^2y^6}{9b^4}}$ $\dfrac{y\sqrt{6cy}}{3b}$

59. $\sqrt[6]{\dfrac{b^3a^3}{8x^3}}$ $\dfrac{\sqrt{2abx}}{2x}$

APPLICATIONS

Measurements

60. A body starting at rest takes t seconds to fall a distance of d feet, where

$$t = \sqrt{\dfrac{d}{16}}$$

   a. Simplify this expression. $t = \dfrac{\sqrt{d}}{4}$

   b. How long would it take an object starting at rest to fall 100 ft? 2.5 sec

61. The radius of a sphere is given by

$$r = \sqrt[3]{\dfrac{3V}{4\pi}}$$

where V is the volume of the sphere and π is about $\frac{22}{7}$.

   a. Simplify. $\sqrt[3]{\dfrac{3V}{4\pi}}$ $\dfrac{\sqrt[3]{6\pi^2V}}{2\pi}$

   b. If the volume of a sphere is 36π ft^3, what is its radius? 3 ft

62. The root-mean-square velocity, $\bar{v}$, of a gas particle is given by the formula

$$\bar{v} = \frac{\sqrt{3kT}}{\sqrt{m}}$$

where k is a constant, T is the temperature (in degrees Kelvin), and m is the mass of the particle. Rationalize the denominator of the expression on the right-hand side. $\bar{v} = \frac{\sqrt{3kTm}}{m}$

63. The mass m of an object depends on its speed v and the speed of light c. The relationship is given by the formula

$$m = \frac{m_0}{\sqrt{1 - \dfrac{v^2}{c^2}}} \qquad m = \frac{m_0 c \sqrt{c^2 - v^2}}{c^2 - v^2}$$

where m_0 is the *rest mass*, the mass when $v = 0$. Simplify the expression on the right-hand side and rationalize the denominator.

SKILL CHECKER

Try the "Skill Checker" exercises so you'll be ready for the next section.

64. Multiply $(a + b)(a - b)$. $a^2 - b^2$

65. Use the result of Problem 64 to multiply
$(\sqrt{x} + \sqrt{y})(\sqrt{x} - \sqrt{y})$. $x - y$

66. Use the result of Problem 64 to multiply
$(x^{3/2} + y^{3/2})(x^{3/2} - y^{3/2})$. $x^3 - y^3$

Divide:

67. $\dfrac{6 + 3y}{3}$ $2 + y$

68. $\dfrac{3xy + 6x^2 y}{3xy}$ $1 + 2x$

69. $\dfrac{12x^2 y^3 + 18xy^3}{6xy}$ $2xy^2 + 3y^2$

Evaluate and simplify $\sqrt{b^2 - 4ac}$ in each case:

70. $a = 2$, $b = -1$, and $c = -6$ 7

71. $a = 2$, $b = -5$, and $c = -12$ 11

72. $a = 6$, $b = -4$, and $c = -2$ 8

73. $a = -1$, $b = 1$, and $c = 12$ 7

USING YOUR KNOWLEDGE

From Radicals to Rational and Back

We've studied rational exponents and radical expressions. We will now use this knowledge to translate one notation to the other.

74. In Problem 63,

$$m = \frac{m_0}{\sqrt{1 - \dfrac{v^2}{c^2}}}$$

Rationalize the denominator and write the result using rational exponents. $m = \frac{m_0 c(c^2 - v^2)^{1/2}}{c^2 - v^2}$

75. Simplify the expression in Problem 62 defining $\bar{v}$ and write the result using rational exponents. $\bar{v} = \frac{(3kTm)^{1/2}}{m}$

76. The period T of a pendulum is

$$T = \sqrt{\frac{2\pi L}{g}} \qquad T = \frac{(2\pi Lg)^{1/2}}{g}$$

where L is the length and g is the gravity constant. Simplify this expression and write the result using rational exponents.

77. The pressure P of a gas is related to its volume V by the formula $P = kV^{-7/5}$, where k is the proportionality constant. Write this formula using radical notation.

$$P = \frac{k}{\sqrt[5]{V^7}} ; \; P = \frac{k\sqrt[5]{V^3}}{V^2}$$

78. The average speed $\bar{v}$ of oxygen molecules is given by the formula $\bar{v} = (3kT)^{1/2} m^{-1/2}$. Write this formula in simplified form using radicals. $\bar{v} = \frac{\sqrt{3kTm}}{m}$

CALCULATOR CORNER

Some of the simplifications we have made can be checked using a calculator with a $\boxed{\sqrt[x]{y}}$ key. To access this feature, you usually have to press the $\boxed{\text{2nd}}$ or $\boxed{\text{2ndF}}$ key first and then the $\boxed{\sqrt[x]{y}}$ key. The calculator will then find the xth root of y.

In Example 2, we learned that $\sqrt[3]{54} = 3\sqrt[3]{2}$. To check this, enter $\boxed{5}$ $\boxed{4}$ $\boxed{\text{2nd}}$ $\boxed{\sqrt[x]{y}}$ $\boxed{3}$ $\boxed{\text{ENTER}}$. The display

shows 3.77976315. Now enter $\boxed{3}$ $\boxed{\times}$ $\boxed{2}$ $\boxed{\sqrt[x]{y}}$ $\boxed{3}$ $\boxed{\text{ENTER}}$. The same result appears, so our answer is correct. If your instructor agrees, check the numerical problems in this section (Problems 25–28, for example) with a calculator.

WRITE ON

79. Law I states that $\sqrt[n]{a^n} = a$ for $a \geq 0$. What happens if $a < 0$? Explain and give examples. Answers may vary.

80. Write the procedure you use to rationalize a radical denominator in a quotient. Answers may vary.

81. State the conditions under which $\sqrt[n]{a^n} = (\sqrt[n]{a})^n = a$. Answers may vary.

82. Use some counterexamples to show that $(a^2 + b^2)^{1/2} \neq a + b$. Answers may vary.

83. Use some counterexamples to show that $(a^{1/2} + b^{1/2})^2 \neq a + b$. Answers may vary.

MASTERY TEST

If you know how to do these problems, you have learned your lesson!

Simplify (assume variables are positive):

84. $\sqrt{\dfrac{11}{12}}$ $\dfrac{\sqrt{33}}{6}$

85. $\sqrt[3]{\dfrac{6}{x^3}}$ $\dfrac{\sqrt[3]{6}}{x}$

86. $\sqrt[4]{\dfrac{3}{8x^6}}$ $\dfrac{\sqrt[4]{6x^2}}{2x^2}$

87. $\sqrt{32}$ $4\sqrt{2}$

88. $\sqrt[3]{32}$ $2\sqrt[3]{4}$

89. $\sqrt[3]{81a^6b^4}$ $3a^2b\sqrt[3]{3b}$

90. $\sqrt[6]{\dfrac{a^3}{8x^3}}$ $\dfrac{\sqrt{2ax}}{2x}$

91. $\sqrt[8]{\dfrac{x^4}{16a^4}}$ $\dfrac{\sqrt{2ax}}{2a}$

92. $\sqrt[4]{(-10)^4}$ 10

93. $\sqrt[6]{(-x)^6}$ x

94. $\sqrt[7]{(-x)^7}$ $-x$

95. $\sqrt{x^2 + 10x + 25}$ $x + 5$

Reduce the index (assume variables are positive):

96. $\sqrt[8]{\dfrac{81}{256}}$ $\dfrac{\sqrt{3}}{2}$

97. $\sqrt[6]{4c^2d^2}$ $\sqrt[3]{2cd}$

Rationalize the denominator (assume variables are positive):

98. $\dfrac{\sqrt{7}}{\sqrt{5}}$ $\dfrac{\sqrt{35}}{5}$

99. $\dfrac{\sqrt[3]{5}}{\sqrt[3]{4}}$ $\dfrac{\sqrt[3]{10}}{2}$

100. $\dfrac{\sqrt{5}}{\sqrt{6x}}$ $\dfrac{\sqrt{30x}}{6x}$

101. $\dfrac{\sqrt{11}}{\sqrt{32x^3}}$ $\dfrac{\sqrt{22x}}{8x^2}$

7.3 # OPERATIONS WITH RADICALS

To Succeed, Review How To ...

1. Combine like terms (pp. 53–55).

2. Remove parentheses using the distributive property (pp. 51–53).

3. Write a fraction with a specified denominator (pp. 426–427).

4. Reduce fractions (pp. 428–431).

Objectives

A Add and subtract similar radical expressions.

B Multiply and divide radical expressions.

C Rationalize the denominators of radical expressions involving sums or differences.

GETTING STARTED

Radical Flight

Have you heard of supersonic airplanes with a speed of Mach 2? The Mach number is named for the Austrian physicist **Ernst Mach** (1838–1916) and is used to indicate how fast one is going when compared to the speed of sound. Mach 2 means the speed of the plane is *twice* the speed of sound (747 mi/hr). How fast can the plane in this photo travel? The answer is classified information, but it is said that the plane's speed is more than Mach 2. The Mach number M can be found from the formula

$$M = \sqrt{\frac{2}{\gamma}}\ \sqrt{\frac{P_2 - P_1}{P_1}}$$

where P_1 and P_2 are air pressures and γ is the ratio of the specific heat at constant pressure to the specific heat at constant volume. This expression can be simplified by multiplying both radical expressions and rationalizing the denominator. In this section we will add, subtract, multiply, and divide radical expressions.

A ## Adding and Subtracting Similar Radical Expressions

In Section 1.4 we combined like terms using the distributive property. Thus

$$3x + 5x = (3 + 5)x = 8x$$

$$7x - 4x = (7 - 4)x = 3x$$

Similarly,

$$3\sqrt{2} + 5\sqrt{2} = (3 + 5)\sqrt{2} = 8\sqrt{2}$$

$$7\sqrt[3]{7} - 4\sqrt[3]{7} = (7 - 4)\sqrt[3]{7} = 3\sqrt[3]{7}$$

Thus we can combine *like* (*similar*) radical expressions. Here is the definition.

LIKE (SIMILAR) RADICAL EXPRESSIONS

Radical expressions with the same index and the same radicand are **like (similar) expressions.**

Web It

For more on operations with radical expressions, go to link 7-3-1 at mhhe.com/bello.

Teaching Tip

Remind students that terms are separated by plus or minus signs. Thus, $3 + 2\sqrt{5}$ has two terms and cannot be added because they aren't similar or "like" radical terms. Also, $\sqrt{5} + 2\sqrt{5}$ are similar or "like" radical terms and the first term is understood to have one as the coefficient. The result is $3\sqrt{5}$.

If the expressions don't appear to be similar or like, we must try to simplify them first. Thus to add $\sqrt{75} + \sqrt{27}$, we proceed as follows:

$$\sqrt{75} + \sqrt{27} = \sqrt{25 \cdot 3} + \sqrt{9 \cdot 3}$$
$$= \sqrt{25} \cdot \sqrt{3} + \sqrt{9} \cdot \sqrt{3} \qquad \sqrt{ab} = \sqrt{a}\sqrt{b}$$
$$= 5\sqrt{3} + 3\sqrt{3}$$
$$= (5 + 3)\sqrt{3} = 8\sqrt{3} \qquad \text{Add like radicals.}$$

The subtraction of similar (like) radicals is done in the same way. Thus

$$\sqrt{80} - \sqrt{20} = \sqrt{16 \cdot 5} - \sqrt{4 \cdot 5}$$
$$= \sqrt{16} \cdot \sqrt{5} - \sqrt{4} \cdot \sqrt{5}$$
$$= 4\sqrt{5} - 2\sqrt{5}$$
$$= (4 - 2) \cdot \sqrt{5} = 2\sqrt{5} \qquad \text{Subtract like radicals.}$$

EXAMPLE 1   Adding and subtracting radical expressions

Perform the indicated operations:

a. $\sqrt{175} + \sqrt{28}$

b. $\sqrt{98} - \sqrt{32}$

c. $3\sqrt{18x} - 5\sqrt{8x}$

d. $5\sqrt[3]{80x} - 3\sqrt[3]{270x}$

SOLUTION

a. $\sqrt{175} + \sqrt{28} = \sqrt{25 \cdot 7} + \sqrt{4 \cdot 7}$
$$= \sqrt{25} \cdot \sqrt{7} + \sqrt{4} \cdot \sqrt{7}$$
$$= 5\sqrt{7} + 2\sqrt{7} = 7\sqrt{7}$$

b. $\sqrt{98} - \sqrt{32} = \sqrt{49 \cdot 2} - \sqrt{16 \cdot 2}$
$$= \sqrt{49} \cdot \sqrt{2} - \sqrt{16} \cdot \sqrt{2}$$
$$= 7\sqrt{2} - 4\sqrt{2} = 3\sqrt{2}$$

c. $3\sqrt{18x} - 5\sqrt{8x} = 3 \cdot \sqrt{9 \cdot 2x} - 5 \cdot \sqrt{4 \cdot 2x}$
$$= 3 \cdot \sqrt{9} \cdot \sqrt{2x} - 5 \cdot \sqrt{4} \cdot \sqrt{2x}$$
$$= 3 \cdot 3 \cdot \sqrt{2x} - 5 \cdot 2 \cdot \sqrt{2x}$$
$$= 9\sqrt{2x} - 10\sqrt{2x}$$
$$= -\sqrt{2x}$$

d. This time we must find factors of 80 and 270 that are perfect cubes: $80 = 8 \cdot 10 = 2^3 \cdot 10$ and $270 = 27 \cdot 10 = 3^3 \cdot 10$. Thus

$$5\sqrt[3]{80x} - 3\sqrt[3]{270x} = 5 \cdot \sqrt[3]{2^3 \cdot 10x} - 3 \cdot \sqrt[3]{3^3 \cdot 10x}$$
$$= 5 \cdot \sqrt[3]{2^3} \cdot \sqrt[3]{10x} - 3 \cdot \sqrt[3]{3^3} \cdot \sqrt[3]{10x}$$
$$= 5 \cdot 2 \cdot \sqrt[3]{10x} - 3 \cdot 3\sqrt[3]{10x}$$
$$= 10\sqrt[3]{10x} - 9\sqrt[3]{10x}$$
$$= \sqrt[3]{10x}$$

PROBLEM 1

Perform the indicated operations:

a. $\sqrt{44} + \sqrt{99}$

b. $\sqrt{98} - \sqrt{50}$

c. $3\sqrt{20x} - 5\sqrt{45x}$

d. $2\sqrt[3]{250x} - 4\sqrt[3]{16x}$

Answers

1. a. $5\sqrt{11}$ **b.** $2\sqrt{2}$
c. $-9\sqrt{5x}$ **d.** $2\sqrt[3]{2x}$

We now show you how to combine like radicals that are fractions.

EXAMPLE 2   **Subtracting radical expressions containing fractions**

Perform the indicated operations:

a. $3\sqrt{\dfrac{1}{2}} - 5\sqrt{\dfrac{1}{8}}$

b. $\sqrt[3]{\dfrac{3}{16}} - \sqrt[3]{\dfrac{3}{2}}$

SOLUTION

a. We first simplify each of the radicals by making the denominator a perfect square.

$$\sqrt{\frac{1}{2}} = \sqrt{\frac{1 \cdot 2}{2 \cdot 2}} = \frac{\sqrt{2}}{2} \quad \text{and} \quad \sqrt{\frac{1}{8}} = \sqrt{\frac{1 \cdot 2}{8 \cdot 2}} = \frac{\sqrt{2}}{\sqrt{16}} = \frac{\sqrt{2}}{4}$$

Thus

$$3\sqrt{\frac{1}{2}} - 5\sqrt{\frac{1}{8}} = 3 \cdot \frac{\sqrt{2}}{2} - 5 \cdot \frac{\sqrt{2}}{4} \qquad \text{Substitute } \frac{\sqrt{2}}{2} \text{ for } \sqrt{\frac{1}{2}} \text{ and } \frac{\sqrt{2}}{4} \text{ for } \sqrt{\frac{1}{8}}.$$

$$= \frac{3\sqrt{2}}{2} - \frac{5\sqrt{2}}{4} \qquad \text{Multiply.}$$

$$= \frac{2 \cdot 3\sqrt{2}}{2 \cdot 2} - \frac{5\sqrt{2}}{4} \qquad \text{Since the LCD of 2 and 4 is 4, write the first fraction with 4 as the denominator.}$$

$$= \frac{6\sqrt{2} - 5\sqrt{2}}{4} \qquad \text{Use 4 as the denominator.}$$

$$= \frac{\sqrt{2}}{4} \qquad \text{Subtract.}$$

b. We first (1) make the denominators perfect cubes and find the cube root, (2) find the LCD of the resulting fractions, and then (3) subtract.

$$\sqrt[3]{\frac{3}{16}} - \sqrt[3]{\frac{3}{2}} = \sqrt[3]{\frac{3 \cdot 4}{16 \cdot 4}} - \sqrt[3]{\frac{3 \cdot 4}{2 \cdot 4}}$$

$$= \sqrt[3]{\frac{12}{64}} - \sqrt[3]{\frac{12}{8}} \qquad \text{(1) Make the denominators perfect cubes and find the cube root.}$$

$$= \frac{\sqrt[3]{12}}{4} - \frac{\sqrt[3]{12}}{2}$$

$$= \frac{\sqrt[3]{12}}{4} - \frac{\sqrt[3]{12} \cdot 2}{2 \cdot 2} \qquad \text{(2) Since the LCD is 4, multiply numerator and denominator of the second fraction by 2.}$$

$$= \frac{\sqrt[3]{12} - 2\sqrt[3]{12}}{4}$$

$$= \frac{-\sqrt[3]{12}}{4} \qquad \text{(3) Subtract.}$$

PROBLEM 2

Perform the indicated operations:

a. $5\sqrt{\dfrac{1}{2}} - 7\sqrt{\dfrac{1}{18}}$

b. $4\sqrt[3]{\dfrac{5}{4}} - \sqrt[3]{\dfrac{5}{32}}$

Calculate It

Verifying Radical Addition and Subtraction

Example 2(b) can be verified by evaluating the original problem, evaluating the answer, and comparing the results to be sure they yield the same value.

Original problem:

$$\sqrt[3]{(3/16)} - \sqrt[3]{(3/2)}$$

$$-0.5723571213$$

Answer:

$$-\sqrt[3]{(12)}/4$$

$$-0.5723571213$$

B   **Multiplying and Dividing Radical Expressions**

The distributive property, in conjunction with $\sqrt{a} \cdot \sqrt{b} = \sqrt{ab}$, can be used to simplify radical expressions that contain parentheses. For example,

$$\sqrt{2} \cdot (\sqrt{3} + \sqrt{5}) = \sqrt{2} \cdot \sqrt{3} + \sqrt{2} \cdot \sqrt{5} \qquad \text{Distributive property}$$

$$= \sqrt{6} + \sqrt{10}$$

Answers

2. a. $\dfrac{4\sqrt{2}}{3}$ **b.** $\dfrac{7\sqrt[3]{10}}{4}$

Web It

For a tutorial on operations with radical expressions, go to link 7-3-2 at mhhe.com/ bello.

Similarly, if $x \geq 0$, then

$$\sqrt{2x} \cdot (\sqrt{x} + \sqrt{3}) = \sqrt{2x} \cdot \sqrt{x} + \sqrt{2x} \cdot \sqrt{3}$$
$$= \sqrt{2x^2} + \sqrt{6x}$$
$$= \sqrt{2}\sqrt{x^2} + \sqrt{6x} \qquad \sqrt{ab} = \sqrt{a} \cdot \sqrt{b}, \sqrt{2x^2} = \sqrt{2} \cdot \sqrt{x^2}$$
$$= x\sqrt{2} + \sqrt{6x} \qquad \sqrt{x^2} = x \text{ if } x \geq 0.$$

Notice in the example that when $x \geq 0$, $(\sqrt{x})^2 = \sqrt{x} \cdot \sqrt{x} = \sqrt{x^2} = x$. This means that $(\sqrt{x})^2 = x$, and in general, $(\sqrt[n]{x})^n = x$ **when** $x \geq 0$.

EXAMPLE 3	**Multiplying radical expressions**

Perform the indicated operations:

a. $\sqrt{3}(\sqrt{5} + \sqrt{12})$

b. $\sqrt{3x}(\sqrt{x} - \sqrt{5})$, $x \geq 0$

c. $\sqrt[3]{3x}(\sqrt[3]{9x^2} - \sqrt[3]{18x})$

SOLUTION

a. $\sqrt{3}(\sqrt{5} + \sqrt{12}) = \sqrt{3} \cdot \sqrt{5} + \sqrt{3} \cdot \sqrt{12} = \sqrt{15} + \sqrt{36}$
$$= \sqrt{15} + 6$$

b. $\sqrt{3x}(\sqrt{x} - \sqrt{5}) = \sqrt{3x}\sqrt{x} - \sqrt{3x}\sqrt{5} = \sqrt{3x^2} - \sqrt{15x}$
$$= x\sqrt{3} - \sqrt{15x}$$

c. $\sqrt[3]{3x}(\sqrt[3]{9x^2} - \sqrt[3]{18x}) = \sqrt[3]{3x}\sqrt[3]{9x^2} - \sqrt[3]{3x}\sqrt[3]{18x} = \sqrt[3]{27x^3} - \sqrt[3]{54x^2}$
$$= \sqrt[3]{3^3 \cdot x^3} - \sqrt[3]{3^3 \cdot 2 \cdot x^2}$$
$$= 3x - 3\sqrt[3]{2x^2}$$

PROBLEM 3

Perform the indicated operations:

a. $\sqrt{2}(\sqrt{3} + \sqrt{10})$

b. $\sqrt{5x}(\sqrt{x} - \sqrt{3})$, $x > 0$

c. $\sqrt[3]{2x}(\sqrt[3]{4x^2} - \sqrt[3]{12x})$

Teaching Tip

Students need to see the contrast between adding and multiplying with radicals.

Add	Multiply
$2\sqrt{3} + 5\sqrt{3}$	$2\sqrt{3} \cdot 5\sqrt{3}$
$= 7\sqrt{3}$	$= 10\sqrt{9}$
	$= 10 \cdot 3$
	$= 30$

Add coefficients and keep radical. Multiply coefficients and multiply radicands. Simplify, if possible (radicals do not have to be similar or "like" radical terms).

Teaching Tip

Remind students that the word FOIL has four letters in it, each standing for a multiplication step. Therefore, you should have four terms as a result of that multiplication before you attempt to combine any like terms.

Answers

3. a. $\sqrt{6} + 2\sqrt{5}$
b. $x\sqrt{5} - \sqrt{15x}$, $x > 0$
c. $2x - 2\sqrt[3]{3x^2}$

If we wish to obtain the product of two binomials that contain radicals, we first simplify the radicals involved (if possible) and then use FOIL. For example, to find the product $(\sqrt{98} + \sqrt{27})(\sqrt{72} + \sqrt{75})$, we proceed as follows.

$$(\sqrt{98} + \sqrt{27})(\sqrt{72} + \sqrt{75})$$
$$= (\sqrt{49 \cdot 2} + \sqrt{9 \cdot 3})(\sqrt{36 \cdot 2} + \sqrt{25 \cdot 3}) \qquad \text{Factor under each radical.}$$
$$= (7\sqrt{2} + 3\sqrt{3})(6\sqrt{2} + 5\sqrt{3}) \qquad \text{Simplify.}$$

$$\overset{F}{= 7 \cdot 6 \cdot \sqrt{2} \cdot \sqrt{2}} + \overset{O}{7 \cdot 5 \cdot \sqrt{2} \cdot \sqrt{3}} + \overset{I}{3 \cdot 6 \cdot \sqrt{3} \cdot \sqrt{2}} + \overset{L}{3 \cdot 5 \cdot \sqrt{3} \cdot \sqrt{3}}$$

Use FOIL.

$$= 42\sqrt{2^2} + 35\sqrt{2} \cdot \sqrt{3} + 18\sqrt{3} \cdot \sqrt{2} + 15\sqrt{3^2} \qquad \text{Simplify.}$$
$$= 42 \cdot 2 + 35\sqrt{6} + 18\sqrt{6} + 15 \cdot 3 \qquad \sqrt{2^2} = 2, \sqrt{3}\sqrt{2} = \sqrt{6} \text{ and } \sqrt{3^2} = 3$$
$$= 84 + 53\sqrt{6} + 45 \qquad \text{Multiply.}$$
$$= 129 + 53\sqrt{6} \qquad \text{Combine like terms.}$$

EXAMPLE 4   **Using FOIL to multiply binomials containing radicals**

Find the product: $(\sqrt{63} + \sqrt{75})(\sqrt{28} - \sqrt{27})$

SOLUTION   We first simplify the radicals and then use FOIL.

$$(\sqrt{63} + \sqrt{75})(\sqrt{28} - \sqrt{27}) = (\sqrt{9 \cdot 7} + \sqrt{25 \cdot 3})(\sqrt{4 \cdot 7} - \sqrt{9 \cdot 3})$$

$$= (3\sqrt{7} + 5\sqrt{3})(2\sqrt{7} - 3\sqrt{3})$$

$$\overset{\text{F}}{=} 6\sqrt{7^2} \overset{\text{O}}{-} 9\sqrt{21} \overset{\text{I}}{+} 10\sqrt{21} \overset{\text{L}}{-} 15\sqrt{3^2}$$

$$= 6 \cdot 7 + \sqrt{21} - 15 \cdot 3$$

$$= 42 + \sqrt{21} - 45$$

$$= -3 + \sqrt{21}$$

PROBLEM 4

Find the product:

$$(\sqrt{27} - \sqrt{28})(\sqrt{75} + \sqrt{63})$$

Teaching Tip

Before doing Example 5, have students do some polynomial examples of squaring a binomial and finding the product of the sum and difference of the same two terms.

Examples:

$(x + 6)^2 = x^2 + 12x + 36$
$(x - 3)^2 = x^2 - 6x + 9$
$(x + 5)(x - 5) = x^2 - 25$

Calculate It   Verifying Radical Multiplication

Let's verify the results of Example 4. Enter $(\sqrt{(63)} + \sqrt{(75)})(\sqrt{(28)} - \sqrt{(27)})$ [ENTER]. It helps to know that $\sqrt{n} = n^{1/2}$, so you can enter $\sqrt{63}$ as $63^{1/2}$, (or [2nd] [x²] 63); $\sqrt{75}$ as $75^{1/2}$ (or [2nd] [x²] 75); and so on as you wish. Now enter the answer $-3 + \sqrt{21}$ [ENTER]. In both cases, the result is 1.582575695 as shown in the window.

```
(√(63)+√(75))(√(28)
-√(27))
            1.582575695
-3+√(21)
            1.582575695
```

EXAMPLE 5   **Multiplying radical expressions**

Multiply:

a. $(\sqrt{3} + 2)^2$ **b.** $(3 - \sqrt{2})^2$ **c.** $(\sqrt{3} + \sqrt{2})(\sqrt{3} - \sqrt{2})$

SOLUTION

a. Since $(x + a)^2 = x^2 + 2ax + a^2$,

$$(\sqrt{3} + 2)^2 = (\sqrt{3})^2 + 2 \cdot 2 \cdot \sqrt{3} + 2^2$$

$$= 3 + 4\sqrt{3} + 4$$

$$= 7 + 4\sqrt{3}$$

b. Since $(x - a)^2 = x^2 - 2ax + a^2$,

$$(3 - \sqrt{2})^2 = 3^2 - 2 \cdot \sqrt{2} \cdot 3 + (\sqrt{2})^2$$

$$= 9 - 6\sqrt{2} + 2$$

$$= 11 - 6\sqrt{2}$$

c. Since $(x + a)(x - a) = x^2 - a^2$,

$$(\sqrt{3} + \sqrt{2})(\sqrt{3} - \sqrt{2}) = (\sqrt{3})^2 - (\sqrt{2})^2$$

$$= 3 - 2$$

$$= 1$$

PROBLEM 5

Multiply:

a. $(\sqrt{2} + 5)^2$

b. $(2 - \sqrt{7})^2$

c. $(\sqrt{5} + \sqrt{3})(\sqrt{5} - \sqrt{3})$

Answers

4. $3 - \sqrt{21}$
5. a. $27 + 10\sqrt{2}$
b. $11 - 4\sqrt{7}$
c. 2

In Chapter 8 some of the answers will be of the form

$$\frac{6 + \sqrt{8}}{2}$$

We can simplify this expression in two ways.

Method 1.   Write

$$\frac{6 + \sqrt{8}}{2}$$

in lowest terms by ① writing $\sqrt{8}$ as $\sqrt{4 \cdot 2} = 2\sqrt{2}$, ② factoring, and then ③ reducing. The procedure looks like this:

$$\frac{6 + \sqrt{8}}{2} \overset{①}{=} \frac{6 + 2\sqrt{2}}{2} \overset{②}{=} \frac{2(3 + \sqrt{2})}{2} \overset{③}{=} 3 + \sqrt{2}$$

Web It

For a slide show summary on radicals and rational exponents, go to link 7-3-3 at mhhe.com/bello.

Method 2.   If we view

$$\frac{6 + \sqrt{8}}{2}$$

as a division of a binomial by a monomial, we can solve the problem by ① writing $\sqrt{8}$ as $2\sqrt{2}$; then we ② write each term in the numerator over the common denominator and ③ reduce. The procedure looks like this:

$$\frac{6 + \sqrt{8}}{2} \overset{①}{=} \frac{6 + 2\sqrt{2}}{2} \overset{②}{=} \frac{6}{2} + \frac{2\sqrt{2}}{2} \overset{③}{=} 3 + \sqrt{2} \quad \text{As before}$$

EXAMPLE 6 Simplifying radical expressions	**PROBLEM 6**
Simplify:	Simplify:

EXAMPLE 6   **Simplifying radical expressions**

Simplify:

$$\frac{6 + \sqrt{18}}{3}$$

PROBLEM 6

Simplify:

$$\frac{10 + \sqrt{75}}{5}$$

SOLUTION

Method 1.   Since $\sqrt{18} = \sqrt{9 \cdot 2} = 3\sqrt{2}$, we have

$$\frac{6 + \sqrt{18}}{3} \overset{①}{=} \frac{6 + 3\sqrt{2}}{3} \overset{②}{=} \frac{3(2 + \sqrt{2})}{3} \overset{③}{=} 2 + \sqrt{2}$$

Method 2.   We can also do this problem by dividing individual terms. Thus

$$\frac{6 + \sqrt{18}}{3} \overset{①}{=} \frac{6 + 3\sqrt{2}}{3} \overset{②}{=} \frac{6}{3} + \frac{3\sqrt{2}}{3} \overset{③}{=} 2 + \sqrt{2}$$

C   Rationalizing Denominators

We know how to rationalize the denominator in expressions of the form

$$\frac{a}{\sqrt{b}}$$

We now show you how to rationalize radical expressions that contain sums or differences involving square root radicals in the denominator. The procedure involves the concept of *conjugate expressions.*

Answer

6. $2 + \sqrt{3}$

Web It

For a 10 question quiz on sim-plifying radical expressions, go to link 7-3-4 at mhhe.com/bello.

Teaching Tip

Ask students to describe $a^2 - b^2$ in words. They should recognize that $a^2 - b^2$ is the difference of two perfect squares. Then have them explain how $(a + b)(a - b)$ will help rationalize a denominator for

$$\frac{3}{3 + \sqrt{3}}$$

Now challenge them to find out if this process will work to rationalize $\frac{5}{2 + \sqrt[3]{7}}$.

Teaching Tip

$$\frac{3 - \sqrt{3}}{2}$$

may be written as

$$\frac{3}{2} - \frac{\sqrt{3}}{2}$$

CONJUGATE

The expressions $a + b$ and $a - b$ are **conjugates** of each other.

Here are some numbers and their conjugates.

Number	Conjugate
$3 + \sqrt{2}$	$3 - \sqrt{2}$
$-4 + \sqrt{5}$	$-4 - \sqrt{5}$
$7 - \sqrt{3}$	$7 + \sqrt{3}$
$-8 - \sqrt{6}$	$-8 + \sqrt{6}$

Since $(a + b)(a - b) = a^2 - b^2$, the product of a number and its conjugate is $a^2 - b^2$. Now suppose we want to rationalize the denominator in

$$\frac{2}{5 + \sqrt{3}}$$

We use the fundamental property of fractions and multiply the numerator and denominator of

$$\frac{2}{5 + \sqrt{3}}$$

by the conjugate of $(5 + \sqrt{3})$, that is, by $(5 - \sqrt{3})$, to obtain

$$\frac{2}{5 + \sqrt{3}} = \frac{2 \cdot (5 - \sqrt{3})}{(5 + \sqrt{3})(5 - \sqrt{3})}$$

$$= \frac{2 \cdot (5 - \sqrt{3})}{5^2 - (\sqrt{3})^2}$$

$$= \frac{2 \cdot (5 - \sqrt{3})}{25 - 3}$$

$$= \frac{\overset{1}{2} \cdot (5 - \sqrt{3})}{\underset{11}{\cancel{22}}}$$

$$= \frac{5 - \sqrt{3}}{11}$$

$(a + b)(a - b) = a^2 - b^2,$
$(5 + \sqrt{3})(5 - \sqrt{3}) = 5^2 - (\sqrt{3})^2$

$(\sqrt{3})^2 = \sqrt{3} \cdot \sqrt{3}$
$\qquad = \sqrt{9}$
$\qquad = 3$

NOTE

To rationalize radical expressions that contain sums or differences and that involve square root radicals in the denominator, multiply numerator and denominator by the conjugate of the denominator.

EXAMPLE 7	**Rationalizing denominators**	**PROBLEM 7**

Rationalize the denominator:

a. $\dfrac{13}{4 + \sqrt{3}}$

b. $\dfrac{\sqrt{x}}{\sqrt{x} - 2}$

where x represents a positive number

Rationalize the denominator:

$$\frac{\sqrt{y}}{\sqrt{y} + 4}$$

Answer

7. $\dfrac{y - 4\sqrt{y}}{y - 16}$

SOLUTION

a. We first multiply numerator and denominator by $(4 - \sqrt{3})$, the conjugate of $(4 + \sqrt{3})$.

$$\frac{13}{4 + \sqrt{3}} = \frac{13}{(4 + \sqrt{3})} \cdot \frac{(4 - \sqrt{3})}{(4 - \sqrt{3})} \qquad \text{Fundamental property of fractions}$$

$$= \frac{13(4 - \sqrt{3})}{(4)^2 - (\sqrt{3})^2} \qquad (4 + \sqrt{3})(4 - \sqrt{3}) = (4)^2 - (\sqrt{3})^2$$

$$= \frac{13(4 - \sqrt{3})}{16 - 3} \qquad (4)^2 = 16 \text{ and } (\sqrt{3})^2 = 3$$

$$= \frac{\cancel{13}(4 - \sqrt{3})}{\cancel{13}} \qquad \text{Simplify.}$$

$$= 4 - \sqrt{3} \qquad \text{Reduce.}$$

b. We first multiply numerator and denominator by $(\sqrt{x} + 2)$, the conjugate of $(\sqrt{x} - 2)$.

$$\frac{\sqrt{x}}{\sqrt{x} - 2} = \frac{\sqrt{x}(\sqrt{x} + 2)}{(\sqrt{x} - 2)(\sqrt{x} + 2)} \qquad \text{Fundamental property of fractions}$$

$$= \frac{\sqrt{x}(\sqrt{x} + 2)}{(\sqrt{x})^2 - (2)^2} \qquad (\sqrt{x} - 2)(\sqrt{x} + 2) = (\sqrt{x})^2 - (2)^2$$

$$= \frac{\sqrt{x}(\sqrt{x} + 2)}{x - 4} \qquad (\sqrt{x})^2 = x \text{ and } (2)^2 = 4$$

$$= \frac{\sqrt{x^2} + 2\sqrt{x}}{x - 4} \qquad \text{Use the distributive law.}$$

$$= \frac{x + 2\sqrt{x}}{x - 4} \qquad \sqrt{x^2} = x \text{ for } x > 0$$

You can use a similar procedure to rationalize the *numerator* of a radical expression. We discuss how to do this in the *Using Your Knowledge* section of the Exercises.

Calculate It Exercises

1. Verify the result in Example 5(a).

2. Verify the result in Example 6.

3. Can you graphically verify the result in Example 7(b)? Explain.

4. The age A of a human fetus (in weeks) can be approximated by $A = 25W^{1/3}$, where W is the weight of the fetus in kilograms. (A kilogram, kg, is about 2.2 lb.) If you want to make a table for $A = 25W^{1/3}$ using Y1 for the age A and X for the weight W, press [Y=] and enter $25x^{1/3}$, then press [2nd] [WINDOW] on a TI-83 Plus. Now answer these questions.

a. What would you use for the minimum weight?

b. What kind of increment (ΔTbl) would you use? (*Hint:* Look at the table shown in the window.)

c. If a fetus weighs about 2.3 kg, how old is it?

d. Assuming a 36-wk pregnancy, what birth weight would you predict based on the table?

X	Y1
0	0
.1	11.604
.2	14.62
.3	16.736
.4	18.42
.5	19.843
.6	21.086
X=0	

Exercises 7.3

A In Problems 1–28, perform the indicated operations. (Where the index is even, assume all variables are positive.)

1. $12\sqrt{2} + 3\sqrt{2}$ $15\sqrt{2}$

2. $15\sqrt{3} + 2\sqrt{3}$ $17\sqrt{3}$

3. $\sqrt{80a} + \sqrt{125a}$ $9\sqrt{5a}$

4. $\sqrt{98a} + \sqrt{32a}$ $11\sqrt{2a}$

5. $\sqrt{50} - 4\sqrt{32}$ $-11\sqrt{2}$

6. $\sqrt{75} + 7\sqrt{12}$ $19\sqrt{3}$

7. $\sqrt{50a^2} - \sqrt{200a^2}$ $-5a\sqrt{2}$

8. $\sqrt{48a^2} - \sqrt{363a^2}$ $-7a\sqrt{3}$

9. $2\sqrt{300} - 9\sqrt{12} - 7\sqrt{48}$ $-26\sqrt{3}$

10. $\sqrt{175} + \sqrt{567} - \sqrt{63}$ $11\sqrt{7}$

11. $3x\sqrt{20x} - \sqrt{24x} + \sqrt{45x^3}$
$9x\sqrt{5x} - 2\sqrt{6x}$

12. $a\sqrt{18} + 4a\sqrt{8} - 5a\sqrt{3}$
$11a\sqrt{2} - 5a\sqrt{3}$

13. $\sqrt[3]{40} + 3\sqrt[3]{625}$
$17\sqrt[3]{5}$

14. $\sqrt[3]{54} + 2\sqrt[3]{16}$ $7\sqrt[3]{2}$

15. $\sqrt[3]{81} - 3\sqrt[3]{375}$ $-12\sqrt[3]{3}$

16. $\sqrt[3]{24} - \sqrt[3]{81}$ $-\sqrt[3]{3}$

17. $2\sqrt[3]{-24} - 4\sqrt[3]{-81} - \sqrt[3]{375}$
$3\sqrt[3]{3}$

18. $10\sqrt[3]{-40} - 2\sqrt[3]{-135} + 4\sqrt[3]{-320}$
$-30\sqrt[3]{5}$

19. $\sqrt[3]{3a} - \sqrt[3]{24a} + \sqrt[3]{375a}$
$4\sqrt[3]{3a}$

20. $\sqrt[3]{r^5} - \sqrt[3]{8r^5} - r\sqrt[3]{64r^2}$ $-5r\sqrt[3]{r^2}$

21. $\dfrac{3\sqrt{3}}{2} - \dfrac{\sqrt{3}}{3}$ $\dfrac{7\sqrt{3}}{6}$

22. $\dfrac{4}{5} - \dfrac{\sqrt{2}}{2}$ $\dfrac{8 - 5\sqrt{2}}{10}$

23. $\sqrt{\dfrac{1}{2}} + \sqrt{\dfrac{1}{3}} + \sqrt{\dfrac{1}{6}}$
$\dfrac{3\sqrt{2} + 2\sqrt{3} + \sqrt{6}}{6}$

24. $\sqrt{\dfrac{25}{3}} - 2\sqrt{\dfrac{16}{3}} + 2\sqrt{\dfrac{4}{3}}$ $\dfrac{\sqrt{3}}{3}$

25. $\sqrt{\dfrac{2}{3}} - \sqrt{\dfrac{1}{6}} + \sqrt{\dfrac{1}{2}}$
$\dfrac{\sqrt{6} + 3\sqrt{2}}{6}$

26. $\sqrt[3]{\dfrac{1}{5}} + \sqrt[3]{\dfrac{1}{40}}$ $\dfrac{3\sqrt[3]{25}}{10}$

27. $6\sqrt[3]{\dfrac{3}{5}} + 6\sqrt[3]{\dfrac{81}{40}}$ $3\sqrt[3]{75}$

28. $2a\sqrt[3]{\dfrac{a}{5}} + 6\sqrt[3]{\dfrac{a^4}{40}}$ $a\sqrt[3]{25a}$

B In Problems 29–62, perform the indicated operations. (Where the index is even, assume all variables are positive.)

29. $3(5 - \sqrt{2})$ $15 - 3\sqrt{2}$

30. $-2(\sqrt{2} - 3)$ $6 - 2\sqrt{2}$

31. $\sqrt[3]{2}(\sqrt[3]{4} + 3)$ $2 + 3\sqrt[3]{2}$

32. $\sqrt[3]{3}(\sqrt[3]{9} + 2)$
$3 + 2\sqrt[3]{3}$

33. $2\sqrt{3}(7\sqrt{5} + 5\sqrt{3})$
$14\sqrt{15} + 30$

34. $2\sqrt{5}(5\sqrt{2} + 3\sqrt{5})$
$10\sqrt{10} + 30$

35. $3\sqrt[3]{5}(2\sqrt[3]{3} - \sqrt[3]{25})$
$6\sqrt[3]{15} - 15$

36. $4\sqrt[3]{2}(3\sqrt[3]{4} - 3\sqrt[3]{2})$
$24 - 12\sqrt[3]{4}$

37. $-4\sqrt{7}(2\sqrt{3} - 5\sqrt{2})$
$-8\sqrt{21} + 20\sqrt{14}$

38. $-3\sqrt{2}(5\sqrt{7} - 2\sqrt{3})$
$-15\sqrt{14} + 6\sqrt{6}$

39. $(5\sqrt{3} + \sqrt{5})(3\sqrt{3} + 2\sqrt{5})$
$55 + 13\sqrt{15}$

40. $(2\sqrt{2} + 5\sqrt{3})(3\sqrt{2} + \sqrt{3})$
$27 + 17\sqrt{6}$

41. $(3\sqrt{6} - 2\sqrt{3})(4\sqrt{6} + 5\sqrt{3})$
$42 + 21\sqrt{2}$

42. $(3\sqrt{5} - 2\sqrt{3})(2\sqrt{5} + 3\sqrt{3})$
$12 + 5\sqrt{15}$

43. $(7\sqrt{5} - 11\sqrt{7})(5\sqrt{5} + 8\sqrt{7})$
$-441 + \sqrt{35}$

44. $(2\sqrt{3} - 5\sqrt{2})(3\sqrt{3} + 2\sqrt{2})$
$-2 - 11\sqrt{6}$

45. $(1 + \sqrt{2})(1 - \sqrt{2})$
-1

46. $(2 + \sqrt{3})(2 - \sqrt{3})$
1

47. $(2 + 3\sqrt{3})(2 - 3\sqrt{3})$ -23

48. $(5 + 5\sqrt{2})(5 - 5\sqrt{2})$ -25

49. $(\sqrt{3} + \sqrt{2})^2$ $5 + 2\sqrt{6}$

50. $(\sqrt{2} + \sqrt{5})^2$ $7 + 2\sqrt{10}$

51. $(a + \sqrt{b})^2$ $a^2 + 2a\sqrt{b} + b$

52. $(\sqrt{a} + b)^2$ $a + 2b\sqrt{a} + b^2$

53. $(\sqrt{3} - \sqrt{2})^2$ $5 - 2\sqrt{6}$ **54.** $(\sqrt{2} - \sqrt{5})^2$ $7 - 2\sqrt{10}$ **55.** $(a - \sqrt{b})^2$ $a^2 - 2a\sqrt{b} + b$

56. $(\sqrt{b} - a)^2$ $b - 2a\sqrt{b} + a^2$ **57.** $(\sqrt{a} - \sqrt{b})^2$ $a - 2\sqrt{ab} + b$ **58.** $(\sqrt{b} - \sqrt{a})^2$ $a - 2\sqrt{ab} + b$

59. $\dfrac{3 + \sqrt{18}}{3}$ $1 + \sqrt{2}$ **60.** $\dfrac{5 + \sqrt{50}}{5}$ $1 + \sqrt{2}$ **61.** $\dfrac{6 - \sqrt{27}}{12}$ $\dfrac{2 - \sqrt{3}}{4}$

62. $\dfrac{8 - \sqrt{32}}{4}$ $2 - \sqrt{2}$

C In Problems 63–73, rationalize the denominator. (Assume all variables represent positive numbers.)

63. $\dfrac{3 + \sqrt{3}}{\sqrt{2}}$ $\dfrac{3\sqrt{2} + \sqrt{6}}{2}$ **64.** $\dfrac{2 + \sqrt{5}}{\sqrt{3}}$ $\dfrac{2\sqrt{3} + \sqrt{15}}{3}$ **65.** $\dfrac{2}{3 - \sqrt{2}}$ $\dfrac{6 + 2\sqrt{2}}{7}$

66. $\dfrac{6}{2 - \sqrt{2}}$ $6 + 3\sqrt{2}$ **67.** $\dfrac{4a}{3 - \sqrt{5}}$ $3a + a\sqrt{5}$ **68.** $\dfrac{3a}{4 - \sqrt{3}}$ $\dfrac{12a + 3a\sqrt{3}}{13}$

69. $\dfrac{3a + 2b}{3 + \sqrt{2}}$ $\dfrac{9a - 3a\sqrt{2} + 6b - 2b\sqrt{2}}{7}$ **70.** $\dfrac{5a + b}{2 + \sqrt{3}}$ $10a - 5a\sqrt{3} + 2b - b\sqrt{3}$

71. $\dfrac{\sqrt{a} + b}{\sqrt{a} - b}$ $\dfrac{a + 2b\sqrt{a} + b^2}{a - b^2}$ **72.** $\dfrac{a + \sqrt{b}}{a - \sqrt{b}}$ $\dfrac{a^2 + 2a\sqrt{b} + b}{a^2 - b}$

73. $\dfrac{\sqrt{a} + \sqrt{2b}}{\sqrt{a} - \sqrt{2b}}$ $\dfrac{a + 2\sqrt{2ab} + 2b}{a - 2b}$

SKILL CHECKER

Try the "Skill Checker" exercises so you'll be ready for the next section.

In Problems 74–78, solve the equation.

74. $x + 5 = 9$ 4 **75.** $2x + 3 = 25$ 11 **76.** $x^2 - 15x + 50 = 0$ 5 or 10

77. $x^2 - 3x + 2 = 0$ 2 or 1 **78.** $x^2 + 6x + 5 = 0$ -1 or -5

USING YOUR KNOWLEDGE

Rationalizing Numerators in Calculus

In this section we learned how to rationalize the denominator of a fraction involving the sum or difference of radical expressions. In calculus, we sometimes have to rationalize the *numerator* of a fraction that involves the sum or difference of radical expressions. The idea is the same: Multiply numerator and denominator by the *conjugate* of the numerator. To rationalize the numerator in

$$\frac{\sqrt{3} + \sqrt{2}}{5}$$

we proceed as follows.

$$\frac{\sqrt{3} + \sqrt{2}}{5} = \frac{(\sqrt{3} + \sqrt{2})(\sqrt{3} - \sqrt{2})}{5(\sqrt{3} - \sqrt{2})}$$

Multiply numerator and denominator by the conjugate of the numerator.

$$= \frac{3 - 2}{5(\sqrt{3} - \sqrt{2})}$$

$(\sqrt{3} + \sqrt{2})(\sqrt{3} - \sqrt{2}) = (\sqrt{3})^2 - (\sqrt{2})^2$

$$= \frac{1}{5(\sqrt{3} - \sqrt{2})}$$

In Problems 79–88, rationalize the numerator.

79. $\dfrac{\sqrt{5} + \sqrt{2}}{3}$ $\dfrac{1}{\sqrt{5} - \sqrt{2}}$

80. $\dfrac{\sqrt{5} + \sqrt{3}}{4}$ $\dfrac{1}{2\sqrt{5} - 2\sqrt{3}}$

81. $\dfrac{\sqrt{x} - \sqrt{2}}{5}$ $\dfrac{x - 2}{5\sqrt{x} + 5\sqrt{2}}$

82. $\dfrac{\sqrt{5} - \sqrt{x}}{5}$ $\dfrac{5 - x}{5\sqrt{5} + 5\sqrt{x}}$

83. $\dfrac{\sqrt{x} + \sqrt{y}}{x}$ $\dfrac{x - y}{x\sqrt{x} - x\sqrt{y}}$

84. $\dfrac{\sqrt{x} - \sqrt{y}}{x}$ $\dfrac{x - y}{x\sqrt{x} + x\sqrt{y}}$

85. $\dfrac{\sqrt{x} + \sqrt{y}}{\sqrt{x}}$ $\dfrac{x - y}{x - \sqrt{xy}}$

86. $\dfrac{\sqrt{x} + \sqrt{y}}{\sqrt{y}}$ $\dfrac{x - y}{\sqrt{xy} - y}$

87. $\dfrac{\sqrt{x} - \sqrt{y}}{\sqrt{x}}$ $\dfrac{x - y}{x + \sqrt{xy}}$

88. $\dfrac{\sqrt{x} - \sqrt{y}}{\sqrt{y}}$ $\dfrac{x - y}{\sqrt{xy} + y}$

WRITE ON

89. Why is it impossible to combine $\sqrt{3x} + \sqrt[3]{3x}$ into a single term? Answers may vary.

90. Explain why $\sqrt{a + b} \neq \sqrt{a} + \sqrt{b}$ and give examples. Answers may vary.

91. Explain why $\sqrt[3]{x} \cdot \sqrt[3]{x} \neq x$. What factor do you need in the box, $\sqrt[3]{x} \cdot \sqrt[3]{x} \cdot \square = x$, to make the statement true? Answers may vary.

92. What does it mean when we say "rationalize the denominator"? Answers may vary.

93. State what conditions have to be met for a radical expression to be simplified. Answers may vary.

MASTERY TEST

If you know how to do these problems, you have learned your lesson!

Rationalize the denominator (assume variables are positive):

94. $\dfrac{\sqrt{y}}{\sqrt{y} + \sqrt{x}}$ $\dfrac{y - \sqrt{xy}}{y - x}$

95. $\dfrac{\sqrt{y}}{\sqrt{y} - \sqrt{x}}$ $\dfrac{y + \sqrt{xy}}{y - x}$

Reduce to lowest terms:

96. $\dfrac{10 + \sqrt{50}}{5}$ $2 + \sqrt{2}$

97. $\dfrac{20 + \sqrt{32}}{4}$ $5 + \sqrt{2}$

Perform the indicated operations (assume variables are positive):

98. $(\sqrt{27} + \sqrt{28})(\sqrt{75} - \sqrt{112})$ $-11 - 2\sqrt{21}$

99. $(\sqrt{28} - \sqrt{27})(\sqrt{112} + \sqrt{75})$ $11 - 2\sqrt{21}$

100. $\sqrt{2}(\sqrt{3} + \sqrt{10})$ $\sqrt{6} + 2\sqrt{5}$

101. $\sqrt{5x}(\sqrt{x} - \sqrt{3})$ $x\sqrt{5} - \sqrt{15x}$

102. $\sqrt[3]{2x}(\sqrt[3]{4x^2} - \sqrt[3]{16x})$ $2x - 2\sqrt[3]{4x^2}$

103. $5\sqrt{\dfrac{1}{2}} - 7\sqrt{\dfrac{1}{8}}$ $\dfrac{3\sqrt{2}}{4}$

104. $4\sqrt[3]{\dfrac{5x}{4x^2}} - \sqrt[3]{\dfrac{5}{32x}}$ $\dfrac{7\sqrt[3]{10x^2}}{4x}$

105. $2\sqrt[3]{250x} - 4\sqrt[3]{16x}$ $2\sqrt[3]{2x}$

106. $3\sqrt{20x} - 5\sqrt{45x}$ $-9\sqrt{5x}$

107. $\sqrt{98} - \sqrt{50}$ $2\sqrt{2}$

108. $\sqrt{44} + \sqrt{99}$ $5\sqrt{11}$

7.4 SOLVING EQUATIONS CONTAINING RADICALS

To Succeed, Review How To ...

1. Solve linear and quadratic equations (pp. 79, 399–403).

2. Square a radical expression (pp. 543–545).

Objectives

A Solve equations involving radicals.

B Solve applications requiring the solution of radical equations.

Radical Curves

If a traffic engineer wants the speed limit v on this curve to be 45 mi/hr, what radius should the curve have? The speed v (in mi/hr) a car can travel on a concrete highway curve without skidding is $v = \sqrt{9r}$, where r is the radius of the curve in feet. Since $v = 45$, to find the answer we must find r in the equation

$$45 = \sqrt{9r}$$

$$(45)^2 = (\sqrt{9r})^2 \qquad \text{Square both sides.}$$

$$2025 = 9r \qquad (\sqrt{9r})^2 = 9r$$

$$\frac{2025}{9} = r \qquad \text{Divide by 9.}$$

$$225 = r$$

The curve must have a radius r of 225 ft or more. We can check this by substituting 225 for r in the equation

$$45 = \sqrt{9r}$$

to obtain $\quad 45 = \sqrt{9 \cdot 225} = \sqrt{9} \cdot \sqrt{225} = 3 \cdot 15$

which is a true statement. Thus the curve must have at least a 225-ft radius.

This problem gives rise to the notion that sometimes it is necessary to know how to solve equations with radical terms, and that is what we will study in this section.

A   Solving Equations Containing Radicals

In algebra, the equation $45 = \sqrt{9r}$ is called a **radical equation** and can be solved by squaring both sides of the equation. Sometimes, however, squaring both sides introduces **extraneous solutions**—solutions that do not satisfy the original equation. For example, the equation $x = 2$ has one solution, 2. Squaring both sides gives $x^2 = 4$. The equation $x^2 = 4$ has two solutions, 2 and -2.

We introduced the extraneous solution -2 when we squared both sides. However, all solutions of the equation $x = 2$ are solutions of $x^2 = 4$.

Power Rule for Equations	All solutions of the equation $P = Q$ are solutions of the equation $P^n = Q^n$, where n is a natural number.

This rule tells us that when we raise both sides of an equation to a power, the solutions of the *original* equation are *always* solutions of the new equation. However, the new equation may have extraneous solutions that have to be discarded. Because of this, *the solutions of the new equations must be checked in the original equation and extraneous solutions discarded.* Here is the procedure we need to solve equations containing radicals.

Teaching Tip

These six steps could be shortened to the **IRS** method of solving radical equations:

Isolate one radical at a time.

Raise both sides to the power of the index and simplify.

Solve the resulting equation and check in the original equation.

PROCEDURE

To Solve Equations Containing Radicals

1. **Isolate** one radical that contains variables on one side of the equation.

2. **Raise** each side of the equation to a power that is the same as the index of the radical.

3. **Simplify.**

4. **Repeat** steps 1–3 *if* the equation still contains a radical term.

5. **Solve** the resulting linear or quadratic equation using the appropriate methods.

6. **Check** all proposed (trial) solutions in the original equation.

EXAMPLE 1   **Solving equations containing one radical**

Solve:

a. $\sqrt{x + 5} = 3$ 　　　**b.** $\sqrt{x + 1} = x - 1$ 　　　**c.** $\sqrt{x - 1} - x = -1$

SOLUTION

a. We use the six-step procedure. In this case, the radical containing the variable is already isolated.

1.	$\sqrt{x + 5} = 3$	Given.
2.	$(\sqrt{x + 5})^2 = 3^2$	Square each side.
3.	$x + 5 = 9$	$(\sqrt{x + 5})^2 = x + 5$
4.	There are no radical terms left.	
5.	$x = 4$	Linear equation—solve with CRAM. Subtract 5 on both sides.

6. Substituting the proposed solution 4 for x in the original equation gives

$$\sqrt{4 + 5} \overset{?}{=} 3$$
$$\sqrt{9} = 3$$

a true statement. The solution of $\sqrt{x + 5} = 3$ is 4.

b. Using the six-step procedure, we note that the radical is already isolated, so we square both sides to eliminate the radical.

1.	$\sqrt{x + 1} = x - 1$	Given.
2.	$(\sqrt{x + 1})^2 = (x - 1)^2$	Square each side.
3.	$x + 1 = x^2 - 2x + 1$	Expand $(x - 1)^2$, which equals $x^2 - 2x + 1$
4.	There are no radical terms left.	
5.	$0 = x^2 - 3x$	Quadratic equation—solve with OFF.

　　　　O　Subtract x and 1 to write the resulting quadratic equation in standard form.

$$0 = x(x - 3)$$ 　　F　Factor.

$x = 0$ 　or 　$x - 3 = 0$ 　　F　Factor = 0. Use the zero-factor property.

$x = 0$ 　or 　　　$x = 3$ 　　Solve $x - 3 = 0$.

Thus the proposed solutions are 0 and 3.

　　Note: In step 5 do not divide by x. It would eliminate one of the possible solutions to the equation, $x = 0$.

6. Substituting 0 for x in the original equation, we have

$$\sqrt{0 + 1} \overset{?}{=} 0 - 1$$
$$1 \overset{?}{=} -1$$ 　　A false statement

PROBLEM 1

Solve:

a. $\sqrt{x + 2} = 5$

b. $\sqrt{2x - 1} = x - 2$

c. $\sqrt{3 - x} + 1 = x$

Web It

For more details and examples on how to solve radical equations, go to link 7-4-1 at mhhe.com/bello.

Teaching Tip

In Example 1 students need to be reminded that after eliminating the radical, the resulting equation might be *linear* or *quadratic*. Solve a simple example of each type before doing Example 1.

Linear	Quadratic
$2x + 3 = 5$	$x^2 + x = 6$
$2x = 2$	$x^2 + x - 6 = 0$
$x = 1$	$(x - 2)(x + 3) = 0$
	$x - 2 = 0$ $x + 3 = 0$
	$x = 2$ $x = -3$

Answers

1. a. 23 　**b.** 5 　**c.** 2

Thus zero is not a solution. Substituting 3 for x in $\sqrt{x + 1} = x - 1$, we have

$$\sqrt{3 + 1} \stackrel{?}{=} 3 - 1$$
$$\sqrt{4} = 2 \qquad \text{A true statement}$$

The solution of $\sqrt{x + 1} = x - 1$ is 3. (Discard the extraneous solution, 0!)

c. We first have to isolate the radical on one side (step 1), so we start by adding x on both sides. Then we proceed as before.

$\sqrt{x - 1} - x = -1$ Given.

1. $\sqrt{x - 1} = x - 1$ Add x on both sides. (Now $\sqrt{x - 1}$ is isolated.)

2. $(\sqrt{x - 1})^2 = (x - 1)^2$ Square each side.

3. $x - 1 = x^2 - 2x + 1$ Expand on the right.

4. There are no radicals left.

5. $0 = x^2 - 3x + 2$ **0**   Quadratic equation—solve with OFF. Subtract x and add 1.

$0 = (x - 2)(x - 1)$ **F**   Factor.

$x - 2 = 0 \quad \text{or} \quad x - 1 = 0$ **F**   Factors = 0. Use the zero-factor property.

$x = 2 \quad \text{or} \qquad x = 1$ Solve.

6. The proposed solutions are 2 and 1. Substituting 2 for x, we have

$$\sqrt{x - 1} - x = -1$$
$$\sqrt{2 - 1} - 2 \stackrel{?}{=} -1$$
$$\sqrt{1} - 2 = -1 \qquad \text{A true statement}$$

2 is a solution. Now we check the proposed solution 1 by substitution.

$$\sqrt{x - 1} - x = -1$$
$$\sqrt{1 - 1} - 1 \stackrel{?}{=} -1$$
$$\sqrt{0} - 1 = -1 \qquad \text{A true statement}$$

1 is also a solution. The solutions are 1 and 2.

Web It

For information on how to solve radical equations with a TI-83 Plus, go to link 7-4-2 at mhhe.com/bello.

Web It

For more explanation on solving equations with radical terms, go to link 7-4-3 at mhhe.com/bello.

Web It

For a practice worksheet on solving radical equations, with answers, go to link 7-4-4 at mhhe.com/bello.

Sometimes we have radicals on both sides of the equation. In such cases, we must isolate one of the radicals and raise both sides of the equation to the appropriate power. Since we still have radicals on one side of the equation, we isolate them and square again (step 4 in the procedure). To solve $\sqrt{x - 11} = \sqrt{x} - 1$, we first square both sides of the equation. The right-hand side then contains an expression of the form $(x - a)^2 = x^2 - 2ax + a^2$. Using our six-step procedure, we have

1. $\sqrt{x - 11} = \sqrt{x} - 1$ $\sqrt{x - 11}$ is isolated.

2. $(\sqrt{x - 11})^2 = (\sqrt{x} - 1)^2$ Square each side.

3. $x - 11 = x - 2 \cdot 1 \cdot \sqrt{x} + 1$ Simplify.

$x - 11 = x - 2\sqrt{x} + 1$

4. $-11 = -2\sqrt{x} + 1$ Subtract x.

$-12 = -2\sqrt{x}$ Subtract 1.

$6 = \sqrt{x}$ Divide by -2.

Since we still have a radical term ($\sqrt{x}$), we isolate $\sqrt{x}$. This is step 4 in the procedure.

5. $36 = x$ Square each side.

6. The solution is 36. You can verify this by substituting in the original equation.

| EXAMPLE 2 | Solving equations containing two radicals | PROBLEM 2 |

Solve: $\sqrt{x-5} - \sqrt{x} = -1$

Solve:

$$\sqrt{x-3} - \sqrt{x} = -1$$

SOLUTION   In step 1, we add $\sqrt{x}$ to both sides of the equation so that $\sqrt{x-5}$ is isolated.

$$\sqrt{x-5} - \sqrt{x} = -1 \qquad \text{Given.}$$

1. $\sqrt{x-5} = \sqrt{x} - 1$ Add $\sqrt{x}$ to isolate $\sqrt{x-5}$.

2. $(\sqrt{x-5})^2 = (\sqrt{x} - 1)^2$ Square each side.

3. $x - 5 = x - 2\sqrt{x} + 1$ $(\sqrt{x} - 1)^2 = x - 2\sqrt{x} + 1$

4. $-5 = -2\sqrt{x} + 1$ Subtract x.

      $-6 = -2\sqrt{x}$ Subtract 1.   We have to isolate $\sqrt{x}$.

      $3 = \sqrt{x}$ Divide both sides by -2.

5. $9 = x$ Square each side.

6. The proposed solution is 9. Since $\sqrt{9-5} - \sqrt{9} = \sqrt{4} - \sqrt{9} = 2 - 3 = -1$, 9 is the correct solution.

The sides of an equation can be raised to powers greater than 2. For example, to solve the equation $\sqrt[4]{x} = 2$, we raise both sides of the equation to the fourth power. (Step 2 says that we have to *raise* both sides of the equation to a power that is the same as the index of the radical, 4.) Thus

$$(\sqrt[4]{x})^4 = 2^4$$
$$x = 16$$

Here is another example.

| EXAMPLE 3 | Solving equations containing a cube root | PROBLEM 3 |

Solve: $\sqrt[3]{x-2} = 3$

Solve:

$$\sqrt[3]{x-5} = -2$$

SOLUTION   The radical is isolated, so we go to step 2.

2. $(\sqrt[3]{x-2})^3 = 3^3$ Cube each side because 3 is the index.

3. $x - 2 = 27$ Simplify.

4. There are no radicals left.

5. $x = 29$ Linear equation—solve with CRAM. Add 2.

6. Substituting 29 for x in the original equation gives

$$\sqrt[3]{29 - 2} \stackrel{?}{=} 3$$

$$\sqrt[3]{27} = 3$$

Since this statement is true, 29 is the solution of $\sqrt[3]{x-2} = 3$.

| EXAMPLE 4 | Solving equations containing a fourth root | PROBLEM 4 |

Solve: $\sqrt[4]{x-1} + 3 = 0$

Solve:

$$\sqrt[4]{x+3} + 4 = 6$$

SOLUTION   We use our six-step procedure.

$$\sqrt[4]{x-1} + 3 = 0 \qquad \text{Given.}$$

1. $\sqrt[4]{x-1} = -3$ Subtract 3 to isolate $\sqrt[4]{x-1}$.

2. $(\sqrt[4]{x-1})^4 = (-3)^4$ Raise to the fourth power.

3. $x - 1 = 81$ Simplify.

Answers

2. 4   **3.** -3 **4.** 13

4. There are no radicals left.

5. $x = 82$ Linear equation—solve with CRAM.
                        Add 1 to solve $x - 1 = 81$.

6. Substitute 82 for x in $\sqrt[4]{x - 1} + 3 = 0$ to obtain

$$\sqrt[4]{82 - 1} + 3 = 0$$
$$\sqrt[4]{81} + 3 = 0$$
$$3 + 3 = 0 \quad \text{(False!)}$$

Thus 82 is an extraneous solution. There is no real-number solution for this equation. The solution set is $\varnothing$.

Now look at step 1. Can you tell that there is no solution to the equation? How can you tell?

B   Solving Applications Involving Radicals

A common use of radicals occurs in problems involving free-falling objects. (See Problems 43 and 44.) Here is an example.

EXAMPLE 5   Free-falling object

If an object is dropped from a height of h feet, the relationship between its velocity v (in ft/sec) when it hits the ground and the height h is $v^2 = 2gh$, where $g = 32$ ft/sec^2.

a. Solve for v.

b. If an object is dropped from a height of 81 ft, what is its velocity when it hits the ground?

PROBLEM 5

The distance a free-falling object has fallen from a position of rest is dependent upon the time of fall. The distance can be computed by this formula; the distance fallen after a time of t seconds is given by the formula

$$d = \frac{1}{2} g t^2$$

where g is the acceleration of gravity (approximately 10 m/sec^2 on Earth).

a. Solve for t.

b. How long does it take a free-falling object to fall 80 meters?

SOLUTION

a. We have to solve the equation $v^2 = 2gh$ for v. Taking the square root of both sides of $v^2 = 2gh$, we have $v = \sqrt{2gh}$.

b. Here we have to find v when $h = 81$ and $g = 32$. Substituting 81 for h and 32 for g in $v = \sqrt{2gh}$, we obtain

$$v = \sqrt{2gh} = \sqrt{2 \cdot 32 \cdot 81} = \sqrt{64 \cdot 81} = 8 \cdot 9 = 72$$

Thus the velocity v of the object when it hits the ground is 72 ft/sec. (Remember that v is in ft/sec.)

EXAMPLE 6   Call lengths in cell phones

Do you have a cell phone? How long are your calls? According to the Cellular Telecommunications Industry Association, the average length of a call for 1999, 2000, and 2001, was 2.38, 2.56, and 2.74 min, respectively. The average length can be approximated by $L(t) = \sqrt{t} + 5.5$, where $L(t)$ represents the length of the call, in minutes, t years after 1999.

a. Use the formula to approximate the average length of a call for 2000. (The actual length was 2.56 min.)

b. In what year (to the nearest year) would you expect the average length of a call to be 4 min?

PROBLEM 6

Use Example 6 to do the following:

a. Use the formula to approximate the average length of a call for 2007.

b. In what year (to the nearest year) would you expect the average length of a call to be 5 minutes?

Answers

5. a. $t = \sqrt{\frac{2d}{g}}$ or $t = \frac{\sqrt{2dg}}{g}$ **b.** 4 seconds

6. a. ≈ 3.7 min   **b.** 2019

SOLUTION

a. Since t is the number of years after 1999, in 2000 $t = 1$ and
$L(1) = \sqrt{1 + 5.5} = \sqrt{6.5} \approx 2.55$ min.

b. To predict when the call length will be 4 min, we have to find t when
$L(t) = 4$. Thus we have to solve the equation

$$\sqrt{t + 5.5} = 4$$

$t + 5.5 = 16$ Square both sides.

$t = 10.5$ Subtract 5 from both sides.

$t \approx 11$ To the nearest year

Approximately 11 years after 1999, in 2010, the average length
of a cellular call is expected to be 4 min.

Calculate It Exercises   Using the Solve Feature of Your Calculator

In this section we shall use the solve feature of your TI-83 Plus. This feature is found in your MATH menu, under number 0. But even with a "solver" you have to be careful.

Let's try Example 1(c). To solve an equation with the solve feature, you must write the equation in the form $A(x) = 0$. The given equation is $\sqrt{x - 1} - x = -1$, so add 1 to both sides to obtain $\sqrt{x - 1} - x + 1 = 0$.

From the home screen start by pressing , which brings you to a screen that says

EQUATION SOLVER:
eqn: 0 =

Then enter $\sqrt{x - 1} - x + 1$ ENTER , which brings you to a screen saying

$\sqrt{x - 1} - x + 1 = 0$
x =
bound = {−1E99, 1 . . .

Use the arrow keys to place the cursor at the end of x = and enter your guess of 3. Now press ALPHA ENTER and x should say 2.

Now we need a new guess to get the other possible answers. Use the arrow keys to place the cursor over the 2 and enter 0 ALPHA ENTER .

What did you get? An error message! Why? Let's look at the graph of $Y_1 = \sqrt{x - 1} - x + 1$ using a $[-1, 3]$ by $[-1, 2]$ window, as shown in Window 1. The graph is defined for values of x greater than or equal to 1. You can see this by looking at the term $\sqrt{x - 1}$, defined only when $x - 1$ is nonnegative—when x is greater than or equal to 1! This means that our guess of 0 is not an acceptable value. (When $x = 0$, $\sqrt{x - 1} = \sqrt{0 - 1} = \sqrt{-1}$, which is not a real number.) When solving an equation using the solve feature, it's a good idea to graph the equation and make sure that your guesses for x are based on the graph.

Now let's pick a number near 1, but less than 2, say, 1.1. You get the answer 2 again. Since the graph indicates that one of the

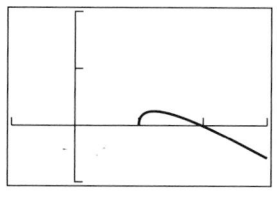

Window 1

answers may be 1, pick 1 for your guess. Finally, you get the second answer, 1. If this method is not to your liking, there's another possibility. First, isolate the radical in Example 1(c), to obtain $\sqrt{x - 1} = x - 1$. Now graph $Y_1 = \sqrt{x - 1}$ and $Y_2 = x - 1$. The value of x where the two curves intersect is a solution of the equation. Can you see from Window 2 that the two points are $x = 1$ and $x = 2$? (With a TI-83 Plus, you can find the point of intersection by pressing 2nd TRACE 5 with 3 ENTER 's. Try it!)

Our moral here is: It's probably easier to do these problems algebraically!

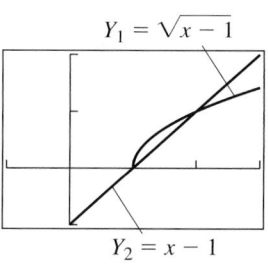

Window 2

1. To graphically solve the equation $\sqrt{x - 5} - \sqrt{x} = -1$ of Example 2, graph $Y_1 = \sqrt{x - 5} - \sqrt{x}$ and $Y_2 = -1$ and find the point of intersection. With a TI-83 Plus, use 2nd TRACE 5 to find the point of intersection. (Make sure that your guesses for the intersection are near the point at which the graphs intersect.) The x-value at the point of intersection is the solution for the equation. What is this x-value?

2. Use the graphical method of Exercise 1 to check the results of Example 3. (*Hint:* A standard window *does not* show the intersection of the two graphs.)

3. Have you ridden your bicycle lately? The greatest speed s at which a bicyclist can safely turn a corner of radius r feet is $s = 4\sqrt{r}$ (in miles per hour).
   a. Graph $s = 4\sqrt{r}$.
   b. Find the greatest speed at which a bicyclist can safely turn a corner with a 20-ft radius.
   c. Find the speed at which the radius r is the same as the speed s. (*Hint:* Graph $Y_2 = x$ and find the intersection.)

Exercises 7.4

A In Problems 1–20, solve the given equation.

1. $\sqrt{x} = 4$ 16

2. $\sqrt{3x} = 6$ 12

3. $\sqrt{x + 6} = 7$ 43

4. $\sqrt{x - 3} = 10$ 103

5. $\sqrt{\dfrac{x}{2}} = 3$ 18

6. $\sqrt{\dfrac{3x}{2}} = 3$ 6

7. $\sqrt[4]{x + 1} + 2 = 0$
No real-number solution

8. $\sqrt[4]{x + 3} + 5 = 0$
No real-number solution

9. $\sqrt[3]{3x - 1} = \sqrt[3]{5x - 7}$ 3

10. $\sqrt[3]{5x - 3} = \sqrt[3]{7x - 5}$ 1

11. $\sqrt{x + 4} = x + 2$ 0

12. $\sqrt{x + 3} = x + 1$ 1

13. $\sqrt{x + 3} = x - 3$ 6

14. $\sqrt{x + 9} = x - 3$ 7

15. $\sqrt[3]{y + 8} = -2$ −16

16. $\sqrt[3]{y + 4} = -1$ −5

17. $\sqrt{x + 5} - x = -7$ 11

18. $\sqrt{x + 5} - x = -1$ 4

19. $\sqrt{x - 5} - x = -7$ 9

20. $\sqrt{x - 1} - x = -3$ 5

In Problems 21–30, you are required to square twice to eliminate all radicals.

21. $\sqrt{y + 1} = \sqrt{y} + 1$ 0

22. $\sqrt{y - 4} = 2 + \sqrt{y}$
No real-number solution

23. $\sqrt{y + 8} - \sqrt{y} = 2$ 1

24. $\sqrt{y + 5} - \sqrt{y} = 1$ 4

25. $\sqrt{x + 3} = \sqrt{x} + \sqrt{3}$ 0

26. $\sqrt{x + 5} = \sqrt{x} + \sqrt{5}$ 0

27. $\sqrt{5x - 1} + \sqrt{x + 3} = 4$ 1

28. $\sqrt{2x - 1} + \sqrt{x + 3} = 3$ 1

29. $\sqrt{x - 3} + \sqrt{2x + 1} = 2\sqrt{x}$ 4

30. $\sqrt{x + 4} + \sqrt{3x + 9} = \sqrt{x + 25}$ 0

In Problems 31–40, solve for x or y.

31. $\sqrt{x - a} = b$ $x = a + b^2$

32. $\sqrt{x + a} = b$ $x = b^2 - a$

33. $\sqrt[3]{a - by} = c$ $y = \dfrac{a - c^3}{b}$

34. $\sqrt[3]{a^3 + y} = b$ $y = b^3 - a^3$

35. $\sqrt{\dfrac{x}{a}} = b$ $x = ab^2$

36. $\sqrt{\dfrac{x}{b}} = \dfrac{a}{b}$ $x = \dfrac{a^2}{b}$

37. $\sqrt{\dfrac{a}{b - x}} = \sqrt{b}$ $x = \dfrac{b^2 - a}{b}$

38. $\sqrt{\dfrac{b}{a - x}} = \sqrt{a}$ $x = \dfrac{a^2 - b}{a}$

39. $\sqrt[3]{3x - a} = \sqrt[3]{b - a}$ $x = \dfrac{b}{3}$

40. $\sqrt[3]{2x - b} = \sqrt[3]{b - 2a}$ $x = b - a$

APPLICATIONS

41. The radius r of a sphere is given by

$$r = \sqrt{\dfrac{S}{4\pi}}$$

where S is the surface area. If the surface area of a sphere is 942 ft^2, find its radius. (Use 3.14 for π.)
$5\sqrt{3}$ ft

42. The radius r of a cone is given by

$$r = \sqrt{\dfrac{3V}{\pi h}}$$

where V is the volume and h is the height. If a 10-cm-high cone contains 94.26 cm^3 of ice cream, what is its radius? (Use 3.142 for π.) 3 cm

43. The time t (in seconds) it takes a body to fall d feet is given by

$$t = \sqrt{\frac{2d}{g}}$$

where g is the gravitational acceleration.

a. Solve for d. $d = \frac{gt^2}{2}$

b. How far would a body fall in 3 sec? (Use 32.2 ft/sec^2 for g.) 144.9 ft

44. After traveling d feet, the velocity v (in ft/sec) of a falling body starting from rest is given by $v = \sqrt{2gd}$.

a. Solve for d. $d = \frac{v^2}{2g}$

b. If a body that started from rest is traveling at 44 ft/sec, how far has it fallen? (Use 32 ft/sec^2 for g.)   30.25 ft

45. A pendulum of length L (in feet) takes

$$t = 2\pi\sqrt{\frac{L}{g}}$$

seconds to go through a complete cycle.

a. Solve for L. $L = \frac{gt^2}{4\pi^2}$

b. If a pendulum takes 2 sec to go through one complete cycle, how long is the pendulum? (Use 32 ft/sec^2 for g and $\frac{22}{7}$ for π.) $L = \frac{392}{121}$ ft ≈ 3.2 ft

SKILL CHECKER

Try the "Skill Checker" exercises so you'll be ready for the next section.

Simplify by removing parentheses and collecting like terms:

46. $(5 + 4x) + (7 - 2x)$ $12 + 2x$

47. $(3 + 4x) + (8 + 2x)$ $11 + 6x$

48. $(9 + 2x) - (2 + 4x)$ $7 - 2x$

49. $(6 + 5x) - (7 - 3x)$ $-1 + 8x$

50. $(8 + 3x) - (5 - 4x)$ $3 + 7x$

Rationalize the denominator:

51. $\dfrac{2 + 3\sqrt{2}}{4 + \sqrt{2}}$ $\dfrac{1 + 5\sqrt{2}}{7}$

52. $\dfrac{4 + 4\sqrt{3}}{3 + 2\sqrt{3}}$ $\dfrac{12 - 4\sqrt{3}}{3}$

53. $\dfrac{2 - \sqrt{2}}{5 - 3\sqrt{2}}$ $\dfrac{4 + \sqrt{2}}{7}$

54. $\dfrac{3 - \sqrt{3}}{5 - \sqrt{3}}$ $\dfrac{6 - \sqrt{3}}{11}$

55. $\dfrac{\sqrt{x} - \sqrt{y}}{\sqrt{x} + \sqrt{y}}$ $\dfrac{x - 2\sqrt{xy} + y}{x - y}$

USING YOUR KNOWLEDGE

The Roads with Radicals

Suppose you are the engineer designing several roads. We mentioned at the beginning of this section that the velocity v (mi/hr) that a car can travel on a concrete highway curve without skidding is $v = \sqrt{9r}$, where r (in feet) is the radius of the curve.

Use your knowledge to determine the radius of the curve on a highway exit in which you want the speed to be as follows.

56. 25 mi/hr $\frac{625}{9} \approx 69.4$ ft

57. 30 mi/hr   100 ft

58. 35 mi/hr $\frac{1225}{9} \approx 136.1$ ft

59. 40 mi/hr $\frac{1600}{9} \approx 177.8$ ft

WRITE ON

60. Consider the equation $\sqrt{x + 1} + 2 = 0$.

a. What should be the first step in solving this equation?   Subtract 2 from both sides

b. List reasons that show that this equation has no real-number solutions.   Answers may vary.

61. Consider the equation $\sqrt{x + 3} = -\sqrt{2x - 3}$. List reasons that show that this equation has no real-number solutions.   Answers may vary.

62. What is your definition of a "proposed" or "trial" solution when you solve equations involving radicals?   Answers may vary.

63. Why is it necessary to check proposed solutions in the original equation when you solve equations involving radicals?   Answers may vary.

MASTERY TEST

If you know how to do these problems, you have learned your lesson!

Solve, if possible:

64. $\sqrt[3]{x-5}=2$
13

65. $\sqrt[4]{x+3}+16=0$
No real-number solution

66. $\sqrt{x-3}-\sqrt{x}=-1$
4

67. $\sqrt{x+1}-x=1$
-1 or 0

68. $\sqrt{x+1}=x-5$
8

69. $\sqrt{x+2}+3=0$
No real-number solution

70. $\sqrt{x-2}-2=0$
6

71. $\sqrt{x}+\sqrt{2x+1}=1$
0

72. The power used by an appliance is given by

$$I = \sqrt{\dfrac{P}{R}}$$

where I is the current (in amps), R is the resistance (in ohms), and P is the power (in watts).

a. Solve for R. $R = \dfrac{P}{I^2}$

b. Find the resistance R for an electric oven rated at 1500 watts and drawing $I = 10$ amps of current.   15 ohms

---

7.5   COMPLEX NUMBERS

To Succeed, Review How To ...

1. Remove parentheses and collect like terms in an expression (pp. 51–55).

2. Rationalize the denominator of an expression (pp. 534–536).

Objectives

A   Write the square root of a negative integer in terms of i.

B   Add and subtract complex numbers.

C   Multiply and divide complex numbers.

D   Find powers of i.

GETTING STARTED   Sums and Products

Can you find two numbers whose sum is 10 and whose product is 40? Girolamo Cardan (1501–1576), an Italian mathematician (at right), claimed the answer is

$$5 + \sqrt{-15} \text{ and } 5 - \sqrt{-15}$$

Obviously, the sum is 10, but what about the product? If we let $a = 5$ and $b = \sqrt{-15}$, we can use the product rule

$$(a + b)(a - b) = a^2 - b^2$$

and obtain $5^2 - (\sqrt{-15})^2 = 25 - (-15) = 40$

Too Complex?

What is the problem? Well, $\sqrt{-15}$ is not a real number! To solve this, Carl Friedrich Gauss (1777–1855) (at left) developed a new set of numbers containing elements that are square roots of negative numbers. One of these numbers is i, and it is defined as follows.

> i is a number such that $i^2 = -1$; that is, $i = \sqrt{-1}$

We will learn how to operate with complex numbers in this section.

 **Writing Square Roots of Negative Numbers in Terms of *i***

With the preceding definition of *i*, the square root of any negative real number can be written as the product of a real number and *i*. Thus

$$\sqrt{-4} = \sqrt{-1} \cdot \sqrt{4} = i2 \quad \text{or} \quad 2i$$
$$\sqrt{-3} = \sqrt{-1 \cdot 3} = \sqrt{-1} \cdot \sqrt{3} = i\sqrt{3} \quad \text{or} \quad \sqrt{3}i$$

Note: Radical sign is not over *i*.

It is easy to confuse $\sqrt{3}i$ and $\sqrt{3i}$. When possible, we write products involving radicals and *i* as factors with the *i* in front; that is, we may write $i\sqrt{5}$ instead of $\sqrt{5}i$. Both notations are acceptable.

EXAMPLE 1 **Writing expressions in terms of *i***	PROBLEM 1

Write the given expression in terms of *i*.

a. $\sqrt{-9}$ **b.** $\sqrt{-18}$

Write the given expression in terms of *i*:

a. $\sqrt{-25}$ **b.** $\sqrt{-28}$

SOLUTION

a. $\sqrt{-9} = \sqrt{-1 \cdot 9} = \sqrt{-1}\,\sqrt{9} = 3i$

b. $\sqrt{-18} = \sqrt{-1 \cdot 18} = \sqrt{-1}\,\sqrt{18} = i\sqrt{18} = i\sqrt{9 \cdot 2} = 3i\,\sqrt{2} \text{ or } 3\sqrt{2}i$

The numbers $3i$ and $3i\sqrt{3} = 3\sqrt{3}i$ are called **pure imaginary numbers.** We can form a new set of numbers by adding these imaginary numbers to real numbers as follows.

Teaching Tip

Real numbers and pure imaginary numbers may be written in $(a + bi)$ complex number form as follows:

$$\sqrt{2} = \sqrt{2} + 0i$$
$$-\frac{1}{5} = -\frac{1}{5} + 0i$$
$$\frac{4}{5}i = 0 + \frac{4}{5}i$$
$$-2i = 0 - 2i$$

COMPLEX NUMBER

If *a* and *b* are real numbers, then any number of the form

$$\underline{a + bi}$$
$$\quad\uparrow\quad\ \uparrow$$
Real part   Imaginary part

is called a **complex number.**

In the complex number $a + bi$, *a* is called the *real* part and *bi* the *imaginary* part. The number $-3 + 4i$ is a complex number whose real part is -3 and whose imaginary part is $4i$. Similarly, $2 - 3i$ is a complex number with 2 as its real part and $-3i$ as its imaginary part.

NOTE

Real numbers and pure imaginary numbers are also complex numbers. For example, the real numbers $\sqrt{2}$, 0, $-0.\overline{3}$, and $\frac{1}{5}$ are complex numbers. The pure imaginary numbers $\sqrt{2}i$, $-3i$, and $-\frac{4}{5}i$ are also complex numbers.

Web It

For another explanation and examples of operating with complex numbers, go to link 7-5-1 at mhhe.com/bello.

Answers

1. a. $5i$ **b.** $2\sqrt{7}i$ or $2i\sqrt{7}$

B Adding and Subtracting Complex Numbers

To add (or subtract) complex numbers, we add (or subtract) the real parts and the imaginary parts separately. The rules for these operations are as follows.

Rules for Adding and Subtracting Complex Numbers

For a, b, c, and d real numbers,

$$(a + bi) + (c + di) = (a + c) + (b + d)i$$

$$(a + bi) - (c + di) = (a - c) + (b - d)i$$

You will find that these operations are similar to combining like terms in a polynomial. For example, $(3 + 4i) + (8 + 2i) = (3 + 8) + (4 + 2)i = 11 + 6i$, and $(9 + 2i) - (2 + 4i) = (9 - 2) + (2 - 4)i = 7 - 2i$.

NOTE

The sum or difference of two complex numbers is always a complex number and should be written in the form $a + bi$.

EXAMPLE 2 **Adding and subtracting complex numbers**

Find:

a. $(5 + 4i) + (7 - 2i)$ **b.** $(6 + 5i) - (7 - 3i)$

SOLUTION

a. $(5 + 4i) + (7 - 2i) = (5 + 7) + [4 + (-2)]i$

$$= 12 + 2i$$

b. $(6 + 5i) - (7 - 3i) = (6 - 7) + [5 - (-3)]i$

$$= -1 + 8i$$

PROBLEM 2

Find:

a. $(7 + 3i) + (2 + 4i)$

b. $(2 + 3i) - (4 + 5i)$

C Multiplying and Dividing Complex Numbers

The commutative, associative, and distributive properties of real numbers also apply to complex numbers. These properties can be used to find the product and quotient of complex numbers. In practice, we multiply complex numbers using the rule to multiply binomials (FOIL) and replacing i^2 by -1 (defined in *Getting Started*). Thus

$$\overset{\text{F\quad O\quad I\quad L}}{(3 + 4i)(2 + 3i) = 6 + 9i + 8i + 12i^2}$$

$$= 6 + 9i + 8i - 12 \qquad i^2 = -1,\ 12i^2 = -12$$

$$= -6 + 17i$$

The answer is written in the form $a + bi$ because the product of two complex numbers is always a complex number.

Answers

2. a. $9 + 7i$ **b.** $-2 - 2i$

| **EXAMPLE 3** | **Multiplying complex numbers** | **PROBLEM 3** |

Find the product:

a. $(2 - 5i)(3 + 7i)$ **b.** $-3i(4 - 7i)$

PROBLEM 3

Find the product:

a. $(2 - 4i)(2 + 6i)$

b. $-4i(5 - 8i)$

SOLUTION

$$\begin{array}{c}\;\;\;\;\text{F}\quad\;\;\text{O}\quad\;\;\;\text{I}\quad\;\;\;\text{L}\end{array}$$

a. $(2 - 5i)(3 + 7i) = 6 + 14i - 15i - 35i^2$

$\qquad\qquad\qquad\qquad = 6 - i + 35 \qquad i^2 = -1, -35i^2 = 35$

$\qquad\qquad\qquad\qquad = 41 - i$

b. $-3i(4 - 7i) = -12i + 21i^2 \qquad$ Use the distributive property

$\qquad\qquad\qquad = -12i - 21 \qquad\quad i^2 = -1, 21i^2 = -21$

$\qquad\qquad\qquad = -21 - 12i \qquad\quad a + bi$ form

Expressions such as $\sqrt{-9}$ and $\sqrt{-4}$ should be written in the form bi before any other operations are carried out. Let's see why. For example, to multiply $\sqrt{-9} \cdot \sqrt{-4}$, we write in bi form first,

$$\sqrt{-9} \cdot \sqrt{-4} = 3i \cdot 2i = 6i^2 = -6$$

However, if we use the product rule for radicals before changing to bi form, we get $\sqrt{-9} \cdot \sqrt{-4} = \sqrt{36} = 6$, which is *not correct*.

It will be important to remember this next procedure.

> ### *bi* FORM
>
> If $a < 0$, then $\sqrt{a}$ must be written in bi form before any operations are performed.

| **EXAMPLE 4** | **Multiplying square roots of negative numbers** | **PROBLEM 4** |

Find:

a. $\sqrt{-16}(3 + \sqrt{-8})$ **b.** $\sqrt{-36}(\sqrt{-3} - \sqrt{-18})$

PROBLEM 4

Find:

a. $\sqrt{-25}(6 + \sqrt{-8})$

b. $\sqrt{-49}(\sqrt{-3} - \sqrt{-27})$

SOLUTION

a. We first write the square roots of negative numbers in terms of i and then proceed as usual. Since $\sqrt{-16} = \sqrt{-1 \cdot 16} = 4i$ and $\sqrt{-8} = \sqrt{-1 \cdot 4 \cdot 2} = 2i\sqrt{2}$, we write

$$\sqrt{-16}(3 + \sqrt{-8}) = 4i(3 + 2i\sqrt{2}) \qquad \sqrt{-16} = 4i \text{ and } \sqrt{-8} = 2i\sqrt{2}.$$

$$\qquad\qquad\qquad\qquad = 12i + 8i^2\sqrt{2} \qquad \text{Use the distributive property.}$$

$$\qquad\qquad\qquad\qquad = 12i - 8\sqrt{2} \qquad\quad i^2 = -1, 8i^2\sqrt{2} = -8\sqrt{2}$$

$$\qquad\qquad\qquad\qquad = -8\sqrt{2} + 12i \qquad \text{Write in the form } a + bi.$$

b. Since $\sqrt{-36} = \sqrt{-1 \cdot 36} = 6i$, $\sqrt{-3} = \sqrt{-1 \cdot 3} = i\sqrt{3}$, and $\sqrt{-18} = \sqrt{-1 \cdot 9 \cdot 2} = 3i\sqrt{2}$, we write

$$\sqrt{-36}(\sqrt{-3} - \sqrt{-18}) = 6i(i\sqrt{3} - 3i\sqrt{2})$$

$$\qquad\qquad\qquad\qquad\qquad = 6i^2\sqrt{3} - 18i^2\sqrt{2}$$

$$\qquad\qquad\qquad\qquad\qquad = -6\sqrt{3} + 18\sqrt{2}$$

Web It

For more information on why there is a need for the number i and the complex numbers in mathematics, go to link 7-5-2 at mhhe.com/bello.

To find the *quotient* of two complex numbers, we use the rationalizing process developed in Section 7.3 and the assumption that

$$\frac{a + bi}{c} = \frac{a}{c} + \frac{bi}{c}$$

For example, to find

$$\frac{2 + 3i}{4 - i}$$

Answers

3. a. $28 + 4i$ **b.** $-32 - 20i$

4. a. $-10\sqrt{2} + 30i$ **b.** $14\sqrt{3}$

Teaching Tip

Have students do the following radical examples before dividing complex numbers:

1. $\dfrac{5+\sqrt{3}}{\sqrt{2}} = \dfrac{(5+\sqrt{3})\cdot\sqrt{2}}{\sqrt{2}\cdot\sqrt{2}}$

$= \dfrac{5\sqrt{2}+\sqrt{6}}{2}$

2. $\dfrac{3}{4+\sqrt{5}}$

$= \dfrac{3\cdot(4-\sqrt{5})}{(4+\sqrt{5})\cdot(4-\sqrt{5})}$

$= \dfrac{12-3\sqrt{5}}{11}$ or $\dfrac{12}{11}-\dfrac{3}{11}\sqrt{5}$

See if students can remember why we multiply by the conjugate in the second example and not in the first.

we proceed as follows.

$\dfrac{2+3i}{4-i} = \dfrac{(2+3i)(4+i)}{(4-i)(4+i)}$ — Multiply the numerator and denominator by the conjugate of $(4-i)$, that is, $(4+i)$.

$= \dfrac{8+2i+12i+3i^2}{4^2-i^2}$ — $(4-i)(4+i)=4^2-i^2$

$= \dfrac{8+2i+12i-3}{16-(-1)}$ — $i^2=-1$

$= \dfrac{5+14i}{17}$ — Simplify.

$= \dfrac{5}{17}+\dfrac{14}{17}i$ — Write in $(a+bi)$ form.

PROCEDURE

Dividing One Complex Number by Another

To divide one complex number by another, multiply the numerator and denominator by the conjugate of the denominator. The conjugate of $a+bi$ is $a-bi$ and $(a+bi)(a-bi)=a^2-(bi)^2=a^2+b^2$.

EXAMPLE 5 **Dividing complex numbers**

Find:

a. $\dfrac{5+4i}{3+2i}$ **b.** $\dfrac{2-4i}{5-3i}$ **c.** $\dfrac{3-2i}{i}$

SOLUTION

a. $\dfrac{5+4i}{3+2i} = \dfrac{(5+4i)(3-2i)}{(3+2i)(3-2i)}$ — Multiply by the conjugate of $(3+2i)$.

$= \dfrac{15-10i+12i-8i^2}{3^2+2^2}$ — $(3+2i)(3-2i)=3^2+2^2$

$= \dfrac{15-10i+12i+8}{13}$ — $i^2=-1$

$= \dfrac{23+2i}{13}$ — Simplify.

$= \dfrac{23}{13}+\dfrac{2}{13}i$ — Write in $(a+bi)$ form.

b. $\dfrac{2-4i}{5-3i} = \dfrac{(2-4i)(5+3i)}{(5-3i)(5+3i)}$ — Multiply by the conjugate of $(5-3i)$.

$= \dfrac{10+6i-20i-12i^2}{5^2+3^2}$ — $(5-3i)(5+3i)=5^2+3^2$

$= \dfrac{10+6i-20i+12}{34}$ — $i^2=-1$

$= \dfrac{22-14i}{34}$ — Simplify.

$= \dfrac{22}{34}-\dfrac{14}{34}i$ — Write in $(a+bi)$ form.

$= \dfrac{11}{17}-\dfrac{7}{17}i$ — Reduce.

PROBLEM 5

Find:

a. $\dfrac{3+5i}{2+3i}$

b. $\dfrac{3-5i}{4-3i}$

c. $\dfrac{2-3i}{i}$

Web It

For another resource on complex numbers, go to link 7-5-3 at mhhe.com/bello.

Answers

5. a. $\dfrac{21}{13}+\dfrac{1}{13}i$ **b.** $\dfrac{27}{25}-\dfrac{11}{25}i$
c. $-3-2i$

c. The conjugate of $a + bi$ is $a - bi$, so the conjugate of $0 + 1i$ is $0 - 1i = -i$. Multiplying numerator and denominator of the fraction by $-i$, we have

$$\frac{3 - 2i}{i} = \frac{(3 - 2i)(-i)}{i \cdot (-i)} \qquad \text{Multiply by } (-i).$$

$$= \frac{-3i + 2i^2}{-i^2} \qquad \text{Distributive Law}$$

$$= \frac{-3i - 2}{1} \qquad i^2 = -1$$

$$= -2 - 3i \qquad \text{Write in } (a + bi) \text{ form.}$$

D Finding Powers of i

We already know that, by definition, $i^2 = -1$. If we assume that the laws of exponents hold, we can write any power of i as one of the numbers $1, -1, i,$ or $-i$. Thus

$$i^1 = i \qquad\qquad\qquad i^5 = i \cdot i^4 = i \cdot (1) = i$$

$$i^2 = -1 \qquad\qquad\qquad i^6 = i \cdot i^5 = i \cdot i = -1$$

$$i^3 = i \cdot i^2 = i(-1) = -i \qquad i^7 = i \cdot i^6 = i(-1) = -i$$

$$i^4 = i^2 \cdot i^2 = (-1)(-1) = 1 \qquad i^8 = i \cdot i^7 = i \cdot (-i) = 1$$

Web It

For some basic facts about complex numbers, including a table of the powers of i, go to link 7-5-4 at mhhe.com/bello.

Since $i^4 = 1$, the easiest way to simplify higher powers of i is to write them in terms of i^4. To find i^{20}, we write

$$i^{20} = (i^4)^5 = (1)^5 = 1$$

Similarly,

$$i^{21} = (i^4)^5 \cdot i = 1 \cdot i = i$$

$$i^{22} = (i^4)^5 \cdot i^2 = 1 \cdot (-1) = -1$$

$$i^{23} = (i^4)^5 \cdot i^3 = 1 \cdot i^3 = -i$$

Dividing the exponent, 20 in this case, by 4 will give you the answer.

If the remainder is 0, the answer is 1 (as in i^{20}).

If the remainder is 1, the answer is i (as in i^{21}).

If the remainder is 2, the answer is -1 (as in i^{22}).

If the remainder is 3, the answer is $-i$ (as in i^{23}).

After that, the answers repeat.

EXAMPLE 6 **Finding powers of i**	**PROBLEM 6**
Find:	Find:
a. i^{53} **b.** i^{47} **c.** i^{-3} **d.** i^{-1} **e.** i^{-34}	**a.** i^{42} **b.** i^{27} **c.** i^{-8} **d.** i^{-2}

SOLUTION

a. Dividing 53 by 4, we obtain 13 with a remainder of 1. Thus the answer is i. To show this, write

$$i^{53} = (i^4)^{13} \cdot i$$

$$= 1 \cdot i = i$$

Answers

6. a. -1 **b.** $-i$ **c.** $\frac{1}{1} = 1$

d. $\frac{1}{-1} = -1$

b. If we divide 47 by 4, the remainder is 3. Thus the answer is $-i$.

$$i^{47} = (i^4)^{11} \cdot i^3$$
$$= 1 \cdot i^3$$
$$= -i$$

c. By definition of negative exponents,

$$i^{-3} = \frac{1}{i^3}$$

$$i^{-3} = \frac{1 \cdot i}{i^3 \cdot i} \qquad \text{Multiply by } i.$$

$$= \frac{i}{i^4} \qquad \text{Write with denominator of } i^4 = 1.$$

$$= \frac{i}{1} = i \qquad \text{Simplify.}$$

d. $i^{-1} = \dfrac{1}{i} = \dfrac{1 \cdot i^3}{i \cdot i^3} = \dfrac{i^3}{i^4} = \dfrac{i^3}{1} = -i$ 　　Multiply by i^3 so denominator is $i^4 = 1$.

e. $i^{-34} = \dfrac{1}{i^{34}} = \dfrac{1 \cdot i^2}{i^{34} \cdot i^2} = \dfrac{i^2}{i^{36}} = \dfrac{i^2}{1} = -1$ 　　Multiply by i^2 so denominator is $i^{36} = 1$.

Web It

For more on the origin of complex numbers and the notation i, go to link 7-5-5 at mhhe.com/bello.

The introduction of the complex numbers completes the development of our number system. We started with the natural numbers (N) and whole numbers (W), studied the integers (I) and the rational numbers (Q), and then concentrated on the real numbers (R). Now we have developed the complex numbers (C), which include all the other numbers we have discussed. In set language, the relationship is

$$N \subseteq W \subseteq I \subseteq Q \subseteq R \subseteq C$$

as shown in Figure 1. A diagram for the complex numbers is shown in Figure 2.

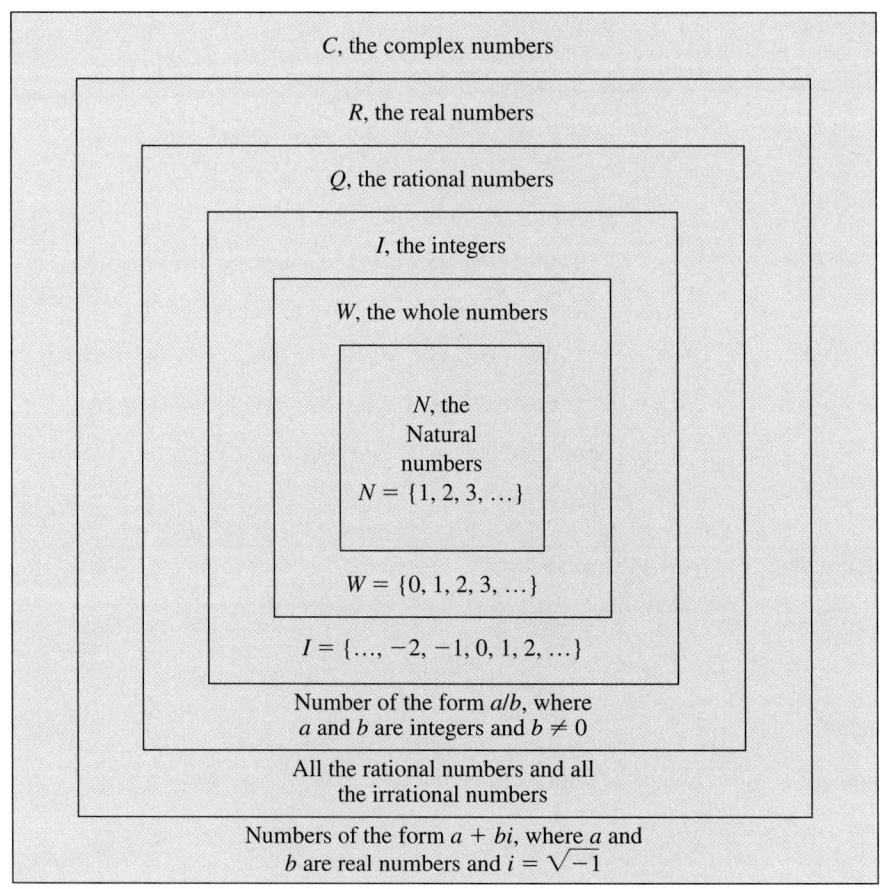

Figure 1
The Number System

Contents of diagram:

C, the complex numbers

R, the real numbers

Q, the rational numbers

I, the integers

W, the whole numbers

N, the Natural numbers
$N = \{1, 2, 3, \ldots\}$

$W = \{0, 1, 2, 3, \ldots\}$

$I = \{\ldots, -2, -1, 0, 1, 2, \ldots\}$

Number of the form a/b, where a and b are integers and $b \neq 0$

All the rational numbers and all the irrational numbers

Numbers of the form $a + bi$, where a and b are real numbers and $i = \sqrt{-1}$

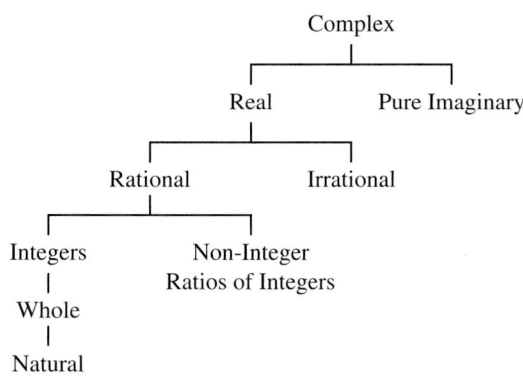

Figure 2

Exercises 7.5

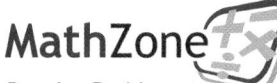
A In Problems 1–10, write the given expression in terms of i.

1. $\sqrt{-25}$ $5i$

2. $\sqrt{-81}$ $9i$

3. $\sqrt{-50}$ $5i\sqrt{2}$

4. $\sqrt{-98}$ $7i\sqrt{2}$

5. $4\sqrt{-72}$ $24i\sqrt{2}$

6. $3\sqrt{-200}$ $30i\sqrt{2}$

7. $-3\sqrt{-32}$ $-12i\sqrt{2}$

8. $-5\sqrt{-64}$ $-40i$

9. $4\sqrt{-28} + 3$ $3 + 8i\sqrt{7}$

10. $7\sqrt{-18} + 5$ $5 + 21i\sqrt{2}$

B In Problems 11–30, perform the indicated operations. (Write the answer in the form $a + bi$.)

11. $(4 + i) + (2 + 3i)$ $6 + 4i$

12. $(7 + 3i) + (2 + i)$ $9 + 4i$

13. $(3 - 2i) - (5 + 4i)$ $-2 - 6i$

14. $(4 - 5i) - (2 + 3i)$ $2 - 8i$

15. $(-3 - 5i) + (-2 - i)$ $-5 - 6i$

16. $(-7 - 3i) + (-2 - i)$ $-9 - 4i$

17. $(3 + \sqrt{-4}) - (5 - \sqrt{-9})$
$-2 + 5i$

18. $(-2 - \sqrt{-16}) - (3 - \sqrt{-25})$
$-5 + i$

19. $(-5 + \sqrt{-1}) + (-2 + 3\sqrt{-1})$
$-7 + 4i$

20. $(-3 + 2\sqrt{-1}) + (-4 + 5\sqrt{-1})$
$-7 + 7i$

21. $(3 - 4i) + (5 + 3i)$ $8 - i$

22. $(3 - 7i) + (3 + 4i)$ $6 - 3i$

23. $(4 + \sqrt{-9}) + (6 + \sqrt{-4})$
$10 + 5i$

24. $(-3 - \sqrt{-25}) + (5 - \sqrt{-16})$
$2 - 9i$

25. $(2 - \sqrt{-2}) - (5 + \sqrt{-2})$
$-3 - 2i\sqrt{2}$

26. $(3 + \sqrt{-50}) - (7 + \sqrt{-2})$
$-4 + 4i\sqrt{2}$

27. $(-5 - \sqrt{-2}) - (-4 - \sqrt{-18})$
$-1 + 2i\sqrt{2}$

28. $(-8 - \sqrt{-125}) - (-2 - \sqrt{-5})$
$-6 - 4i\sqrt{5}$

29. $(-4 + \sqrt{-20}) + (-3 + \sqrt{-5})$
$-7 + 3i\sqrt{5}$

30. $(-7 + \sqrt{-24}) + (-3 + \sqrt{-6})$
$-10 + 3i\sqrt{6}$

C In Problems 31–70, perform the indicated operations. (Write the answer in the form $a + bi$.)

31. $3(4 + 2i)$ $12 + 6i$

32. $5(4 + 3i)$ $20 + 15i$

33. $-4(3 - 5i)$ $-12 + 20i$

34. $-3(7 - 4i)$ $-21 + 12i$

35. $\sqrt{-4}(3 + 2i)$ $-4 + 6i$

36. $\sqrt{-9}(2 + 5i)$ $-15 + 6i$

37. $\sqrt{-3}(3 + \sqrt{-3})$ $-3 + 3i\sqrt{3}$

38. $\sqrt{-5}(2 - \sqrt{-5})$ $5 + 2i\sqrt{5}$

39. $3i(3 + 2i)$ $-6 + 9i$

40. $7i(4 + 3i)$ $-21 + 28i$

41. $4i(3 - 7i)$ $28 + 12i$

42. $-5i(2 - 3i)$ $-15 - 10i$

43. $-\sqrt{-16}(-5 - \sqrt{-25})$
$-20 + 20i$

44. $-\sqrt{-25}(-3 - \sqrt{-9})$
$-15 + 15i$

45. $(3 + i)(2 + 3i)$ $3 + 11i$

46. $(2 + 3i)(4 + 5i)$ $-7 + 22i$

47. $(3 - 2i)(3 + 2i)$ $13 + 0i$

48. $(4 - 3i)(4 + 3i)$ $25 + 0i$

49. $(3 + 2\sqrt{-4})(4 - \sqrt{-9})$
$24 + 7i$

50. $(-3 + 3\sqrt{-9})(-2 + 5\sqrt{-4})$
$-84 - 48i$

51. $(2 + 3\sqrt{-3})(2 - 3\sqrt{-3})$
$31 + 0i$

52. $(4 + 2\sqrt{-5})(4 - 2\sqrt{-5})$
$36 + 0i$

53. $\dfrac{3}{i}$ $0 - 3i$

54. $\dfrac{5}{i}$ $0 - 5i$

55. $\dfrac{6}{-i}$ $0 + 6i$

56. $\dfrac{3}{-2i}$ $0 + \dfrac{3}{2}i$

57. $\dfrac{i}{1 + 2i}$ $\dfrac{2}{5} + \dfrac{1}{5}i$

58. $\dfrac{2i}{1 + 3i}$ $\dfrac{3}{5} + \dfrac{1}{5}i$

59. $\dfrac{3i}{1 - 2i}$ $\dfrac{-6}{5} + \dfrac{3}{5}i$

60. $\dfrac{4i}{2 - 3i}$ $\dfrac{-12}{13} + \dfrac{8}{13}i$

61. $\dfrac{3 + 4i}{1 - 2i}$ $-1 + 2i$

62. $\dfrac{3 + 5i}{1 - 3i}$ $\dfrac{-6}{5} + \dfrac{7}{5}i$

63. $\dfrac{4 + 3i}{2 + 3i}$ $\dfrac{17}{13} - \dfrac{6}{13}i$

64. $\dfrac{5 + 4i}{3 + 2i}$ $\dfrac{23}{13} + \dfrac{2}{13}i$

65. $\dfrac{3}{\sqrt{-4}}$ $0 - \dfrac{3}{2}i$

66. $\dfrac{-4}{\sqrt{-9}}$ $0 + \dfrac{4}{3}i$

67. $\dfrac{3 + \sqrt{-5}}{4 + \sqrt{-2}}$ $\dfrac{12 + \sqrt{10}}{18} + \dfrac{4\sqrt{5} - 3\sqrt{2}}{18}i$

68. $\dfrac{2 + \sqrt{-2}}{1 + \sqrt{-3}}$ $\dfrac{2 + \sqrt{6}}{4} + \dfrac{\sqrt{2} - 2\sqrt{3}}{4}i$

69. $\dfrac{-1 - \sqrt{-2}}{-3 - \sqrt{-3}}$ $\dfrac{3 + \sqrt{6}}{12} + \dfrac{3\sqrt{2} - \sqrt{3}}{12}i$

70. $\dfrac{-1 - \sqrt{-3}}{-2 - \sqrt{-2}}$ $\dfrac{2 + \sqrt{6}}{6} + \dfrac{2\sqrt{3} - \sqrt{2}}{6}i$

D In Problems 71–82, write the answer as 1, -1, i, or $-i$.

71. i^{40} 1

72. i^{28} 1

73. i^{19} $-i$

74. i^{38} -1

75. i^{21} i

76. i^{-44} 1

77. i^{-32} 1

78. i^{53} i

79. i^{65} i

80. i^{16} 1

81. i^{-10} -1

82. i^{-27} i

APPLICATIONS

Impedance

83. The impedance in a circuit is the measure of how much a circuit impedes (hinders) the flow of current through it. If the impedance of a resistor is $Z_1 = 5 + 3i$ ohms and the impedance of another resistor is $Z_2 = 3 - 2i$ ohms, what is the total impedance (sum) of the two resistors when placed in series? $Z_1 + Z_2 = (8 + i)$ ohms

84. Repeat Problem 83 where the impedance of the first resistor is $3 + 7i$ and the impedance of the second is $4 - 5i$. $(7 + 2i)$ ohms

85. If two resistors Z_1 and Z_2 are connected in parallel, their total impedance is given by

$$Z_T = \frac{Z_1 \cdot Z_2}{Z_1 + Z_2}$$ $Z_T = \dfrac{167}{65} - \dfrac{29}{65}i$

Find the total impedance of the resistors of Problem 83.

86. Use the formula in Problem 85 and find the total impedance of the resistors of Problem 84. $Z_T = \dfrac{355}{53} - \dfrac{3}{53}i$

SKILL CHECKER

Try the "Skill Checker" exercises so you'll be ready for the next section.

Solve:

87. $x^2 - 4 = 0$
  $x = 2$ or -2

88. $x^2 - 16 = 0$
  $x = 4$ or -4

89. $x^2 = 25$
  $x = 5$ or -5

90. $x^2 = 36$
  $x = 6$ or -6

USING YOUR KNOWLEDGE

Absolute Values of Complex Numbers

If x is a real number, the absolute value of x is defined as follows.

$$|x| = \begin{cases} x, & \text{if } x > 0 \\ 0, & \text{if } x = 0 \\ -x, & \text{if } x < 0 \end{cases}$$

Thus

$$|5| = 5 \qquad\qquad 5 > 0$$

$$|0| = 0$$

$$|-8| = -(-8) = 8 \qquad -8 < 0$$

How can we define the absolute value of a complex number? The definition is

$$|a + bi| = \sqrt{a^2 + b^2}$$

In Problems 91–94, use this definition to find each value.

91. $|3 + 4i|$ 5

92. $|12 + 5i|$ 13

93. $|2 - 3i|$ $\sqrt{13}$

94. $|5 - 7i|$ $\sqrt{74}$

WRITE ON

95. Explain why it is incorrect to use the product rule for radicals to multiply $\sqrt{-4} \cdot \sqrt{-9}$. What answer do you get if you do? What should the answer be? Under what conditions is $\sqrt{a} \cdot \sqrt{b} = \sqrt{a \cdot b}$?   Answers may vary.

96. Write a short paragraph explaining the relationship among the real numbers, the imaginary numbers, and the complex numbers.   Answers may vary.

97. Write in your own words the procedure you use to
  a. find the conjugate of a complex number.
  b. add (or subtract) two complex numbers.
  c. find the quotient of two complex numbers.
  Answers may vary.

MASTERY TEST

If you know how to do these problems, you have learned your lesson!

Find:

98. i^{42} -1

99. i^{23} $-i$

100. i^{-8} 1

101. i^{-2} -1

102. $\dfrac{3 + 5i}{2 + 3i}$ $\dfrac{21}{13} + \dfrac{1}{13}i$

103. $\dfrac{3 - 5i}{4 - 3i}$ $\dfrac{27}{25} - \dfrac{11}{25}i$

104. $\dfrac{2 - 3i}{i}$ $-3 - 2i$

105. $\sqrt{-25}(6 + \sqrt{-8})$
  $-10\sqrt{2} + 30i$

106. $\sqrt{-49}(\sqrt{-3} - \sqrt{-27})$
  $14\sqrt{3} + 0i$

107. $(2 - 4i)(2 + 6i)$
  $28 + 4i$

108. $-4i(5 - 8i)$
  $-32 - 20i$

109. $(7 + 3i) - (4 + 5i)$ $3 - 2i$

110. $(9 - 2i) - (12 - 3i)$ $-3 + i$

111. $(2 + 3i) + (-3 + 4i)$ $-1 + 7i$

112. $(-3 - 4i) + (-5 - 7i)$ $-8 - 11i$

Write in terms of i:

113. $\sqrt{-50}$ $5i\sqrt{2}$

114. $\sqrt{-25}$ $5i$

COLLABORATIVE LEARNING 7A

1. Complete the table to find the missing lengths.

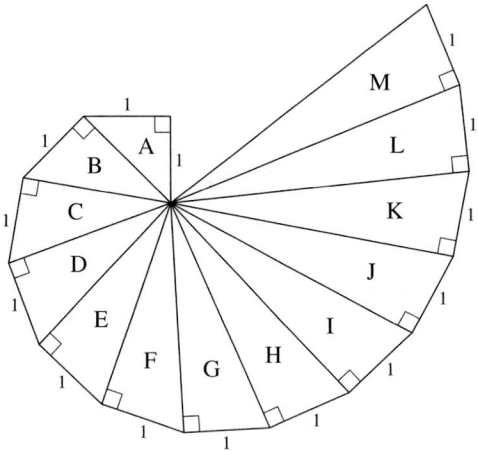

Source: Annenberg/CPB Learner.org.

Right Triangle	A	B	C	D	E	F	G	H	I	J	K	L	M
Leg lengths	1,1	1, $\sqrt{2}$	1, ?	1, ?	1, ?	1, ?	1, ?	1, ?	1, ?	1, ?	1, ?	1, ?	1, ?
Hypotenuse length	$\sqrt{2}$	$\sqrt{3}$	?	?	?	?	?	?	?	?	?	?	?

2. Find the pattern in the lengths for the hypotenuses.

3. Make a table to find the missing length for the diagonal of each of the following five squares.

 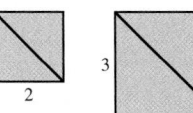 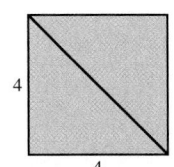 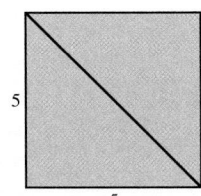

4. Find the pattern in the lengths of the diagonals of the squares.

5. Baseball "diamonds" are squares in which one of the diagonals of the square is the distance from home plate to second base. If the major league fields have 90 ft between bases, use the pattern found in number 4 to find how far the catcher has to throw from home plate to attempt to get a player out who is trying to steal second base.

6. If the adult fast pitch softball "diamond" has 60 ft between its bases, how far is it from home plate to second base?

7. Research and discuss other areas where patterns can help facilitate problem solving.

COLLABORATIVE LEARNING 7B

Divide into groups. You are going to investigate a car accident. We have the formula to find the rate of speed R of the vehicle based on the length L of the skid marks before the car stops. The formula is:

$$R = \sqrt{20L - 600}$$

1. Using the formula, complete the table to find the rate of speed R for the various skid mark lengths L.

2. Use the table of values from number 1 to make a two-dimensional graph where the horizontal axis represents the length of the skid marks L and the vertical axis represents the rate of speed R of the vehicle.

3. Your graph should indicate a linear model. Using your skills from Section 3.4, write the equation for the line of best fit for this graph.

Length of Skid Marks	Rate of Speed in mi/hr
60 ft	
75 ft	
90 ft	
110 ft	
130 ft	

4. Complete the following table by copying the rate results from column 2 of the previous table into column 2 in the following table. Use the equation from number 3 to complete column 3.

Length of Skid Marks	Rate Using $R = \sqrt{20L - 600}$	Rate Using Linear Equation
60 ft		
75 ft		
90 ft		
110 ft		
130 ft		

5. Comparing the results of columns 2 and 3 in the table in number 4, discuss which formula would be best to use to find the rate of speed R given the length of the skid marks L and why?

6. Research other formulas for finding the speed of a vehicle given the length of the skid marks including the one in the *Getting Started* in Section 7.2. How do they compare to your results and why?

Research Questions

1. Who invented the square root sign $\sqrt{}$ (called a radix)?

2. What mathematician coined the word *imaginary*, and in what book did the term first appear?

3. Who was the first mathematician to use the notation i for $\sqrt{-1}$, and in what book did this symbol first appear?

4. The author's *Algebra* (1770) remarks that "such numbers, which by their nature are impossible, are ordinarily called imaginary or fanciful numbers because they exist only in the imagination." Write a short paragraph about this author and his other contributions to mathematics.

5. Who was the first person to attempt (not successfully) to graphically represent the complex numbers, and what was the name of the book in which he tried?

6. The graphical representation of complex numbers occurred at nearly the same time to three mathematicians with no connection or knowledge of each other. Who were these three mathematicians?

7. Write a short paragraph about the mathematical contributions of each of the three people named in Question 6.

Summary

SECTION	ITEM	MEANING	EXAMPLE
7.1	nth root	If a and x are real numbers and n is a positive integer, x is an nth root of a if $x^n = a$.	The square root of 25 is 5, the cube root of -8 is -2, and the fourth root of -16 is not a real number.
7.1A	$\sqrt{}$ Radicand Index (order) $\sqrt[n]{a}$, (a radical expression)	A radical sign In $\sqrt[n]{a}$, a is the radicand. In $\sqrt[n]{a}$, the index is n. The nth root of a	In $\sqrt[3]{22}$, 22 is the radicand. In $\sqrt[3]{22}$, the index is 3. $\sqrt[3]{64} = 4$, $\sqrt[5]{-32} = -2$, and $\sqrt{-25}$ is not a real number.
7.1B	$a^{1/n}$ $a^{m/n}$ $a^{-m/n}$	$a^{1/n} = \sqrt[n]{a}$, if it exists. $a^{m/n} = (\sqrt[n]{a})^m = \sqrt[n]{a^m}$ $a^{-m/n} = \dfrac{1}{a^{m/n}}$	$8^{1/3} = \sqrt[3]{8} = 2$ and $32^{1/5} = \sqrt[5]{32} = 2$ $16^{3/2} = (\sqrt{16})^3 = 4^3 = 64$ $4^{-3/2} = \dfrac{1}{4^{3/2}} = \dfrac{1}{(\sqrt{4})^3} = \dfrac{1}{8}$
7.1C	$(a^m)^{1/n}$ (m and n positive, even integers)	$\lvert a \rvert^{m/n}$ If r, s, and t are rational numbers and a is a real number, the following properties apply: I. $a^r \cdot a^s = a^{r+s}$ II. $\dfrac{a^r}{a^s} = a^{r-s}$ $(a \neq 0)$ III. $(a^r)^s = a^{rs}$ IV. $(a^r b^s)^t = a^{rt}b^{st}$ V. $\left(\dfrac{a^r}{b^s}\right)^t = \dfrac{a^{rt}}{b^{st}}$ $(a \neq 0, b \neq 0)$	$[(-4)^2]^{1/2} = \lvert -4 \rvert = 4$ I. $x^4 \cdot x^{-1} = x^{4+(-1)} = x^3$ II. $\dfrac{x^7}{x^5} = x^{7-5} = x^2$ III. $(x^2)^3 = x^{2\cdot3} = x^6$ IV. $(x^3 y^4)^5 = x^{3\cdot5} \cdot y^{4\cdot5}$ $= x^{15}y^{20}$ V. $\left(\dfrac{x^3}{y^4}\right)^5 = \dfrac{x^{3\cdot5}}{y^{4\cdot5}} = \dfrac{x^{15}}{y^{20}}$
7.2A	Law I Law II Law III	$\sqrt[n]{a^n} = a$, for $a \geq 0$ $\sqrt[n]{ab} = \sqrt[n]{a} \cdot \sqrt[n]{b}$ $\sqrt[n]{\dfrac{a}{b}} = \dfrac{\sqrt[n]{a}}{\sqrt[n]{b}}$	$\sqrt[4]{2^4} = 2$ $\sqrt{18} = \sqrt{9} \cdot \sqrt{2} = 3\sqrt{2}$ $\sqrt[3]{\dfrac{8}{27}} = \dfrac{\sqrt[3]{8}}{\sqrt[3]{27}} = \dfrac{2}{3}$
7.2C	Simplified form of a radical	A radical is in simplified form if the radicand has no factors with exponents greater than or equal to the index, there are no fractions under the radical sign or radicals in the denominator, and the index is as low as possible.	$\dfrac{\sqrt[3]{2ac}}{2c^2}$ is in simplified form.
7.3A	Like radicals	Radicals with the same index and the same radicand	$\sqrt[3]{3x^2}$ and $-7\sqrt[3]{3x^2}$ are like radicals.

SECTION	ITEM	MEANING	EXAMPLE
7.3C	Conjugates	$a + b$ and $a - b$ are conjugates.	$3 + \sqrt{2}$ and $3 - \sqrt{2}$ are conjugates.
7.4A	Extraneous solution	A proposed solution that does not satisfy the original equation	0 is an extraneous solution of $\sqrt{x + 1} = x - 1$.
7.5A	i Complex number	$i = \sqrt{-1}; i^2 = -1$ A number that can be written in the form $a + bi$, where a and b are real numbers	$3 + 7i$ and $-4 - 8i$ are complex numbers.
7.5B	Addition and subtraction of complex numbers	$(a + bi) \pm (c + di) = (a \pm c) + (b \pm d)i$	$(3 + 2i) + (5 + 3i) = 8 + 5i$ $(4 - 2i) - (5 + 3i) = -1 - 5i$
7.5C	Multiplication of complex numbers	$(a + bi)(c + di)$ Use the rule to multiply binomials, FOIL, and replace i^2 by -1.	$(2 + 3i)(4 - 5i)$ $= 8 - 10i + 12i - 15i^2$ $= 8 + 2i + 15$ $= 23 + 2i$

Review Exercises

(If you need help with these exercises, look in the section indicated in brackets.)

In Problems 1–8, evaluate if possible.

1. [7.1A]
  a. $\sqrt{-8}$ Not a real number
  b. $\sqrt[3]{-64}$ -4

2. [7.1A]
  a. $\sqrt{-9}$ Not a real number
  b. $\sqrt[3]{-125}$ -5

3. [7.1B]
  a. $(-27)^{1/3}$ -3
  b. $(-64)^{1/3}$ -4

4. [7.1B]
  a. $\left(\dfrac{1}{16}\right)^{1/4}$ $\dfrac{1}{2}$
  b. $\left(\dfrac{1}{256}\right)^{1/4}$ $\dfrac{1}{4}$

5. [7.1B]
  a. $125^{2/3}$ 25
  b. $64^{2/3}$ 16

6. [7.1B]
  a. $(-25)^{3/2}$ Not a real number
  b. $(-36)^{3/2}$ Not a real number

7. [7.1B]
  a. $(-8)^{-2/3}$ $\dfrac{1}{4}$
  b. $(-64)^{-2/3}$ $\dfrac{1}{16}$

8. [7.1B]
  a. $27^{-2/3}$ $\dfrac{1}{9}$
  b. $64^{-2/3}$ $\dfrac{1}{16}$

In Problems 9–18, simplify (x and y are positive).

9. [7.1C]
  a. $x^{1/5} \cdot x^{1/3}$ $x^{8/15}$
  b. $x^{1/5} \cdot x^{1/4}$ $x^{9/20}$

10. [7.1C]
  a. $\dfrac{x^{-1/4}}{x^{1/5}}$ $\dfrac{1}{x^{9/20}}$
  b. $\dfrac{x^{-1/3}}{x^{1/5}}$ $\dfrac{1}{x^{8/15}}$

11. [7.1C]
  a. $(x^{1/3}y^{2/5})^{-15}$ $\dfrac{1}{x^5 y^6}$
  b. $(x^{1/6}y^{2/5})^{-30}$ $\dfrac{1}{x^5 y^{12}}$

12. [7.1C]
  a. $x^{3/5}(x^{-2/5} + y^{3/5})$ $x^{1/5} + x^{3/5}y^{3/5}$
  b. $x^{4/5}(x^{-1/5} + y^{3/5})$ $x^{3/5} + x^{4/5}y^{3/5}$

13. [7.2A]
  a. $\sqrt[4]{(-7)^4}$ 7
  b. $\sqrt[4]{(-6)^4}$ 6

14. [7.2A]
  a. $\sqrt[8]{(-x)^8}$ $|-x| = |x|$
  b. $\sqrt[4]{(-x)^4}$ $|-x| = |x|$

15. [7.2A]
  a. $\sqrt[3]{48}$ $2\sqrt[3]{6}$
  b. $\sqrt[3]{56}$ $2\sqrt[3]{7}$

16. [7.2A]
  a. $\sqrt[3]{16x^4 y^6}$ $2xy^2\sqrt[3]{2x}$
  b. $\sqrt[3]{16x^8 y^{15}}$ $2x^2 y^5\sqrt[3]{2x^2}$

17. [7.2A]
  a. $\sqrt{\dfrac{5}{243}}$ $\dfrac{\sqrt{15}}{27}$
  b. $\sqrt{\dfrac{5}{1024}}$ $\dfrac{\sqrt{5}}{32}$

18. [7.2A]
  a. $\sqrt[3]{\dfrac{1}{x^3}}$ $\dfrac{1}{x}$
  b. $\sqrt[3]{\dfrac{5}{x^3}}$ $\dfrac{\sqrt[3]{5}}{x}$

In Problems 19–22, rationalize the denominator.

19. [7.2B]

a. $\dfrac{\sqrt{5}}{\sqrt{11}}$ $\dfrac{\sqrt{55}}{11}$

b. $\dfrac{\sqrt{5}}{\sqrt{13}}$ $\dfrac{\sqrt{65}}{13}$

20. [7.2B]

a. $\dfrac{\sqrt{2}}{\sqrt{5x}}, x > 0$ $\dfrac{\sqrt{10x}}{5x}$

b. $\dfrac{\sqrt{3}}{\sqrt{5x}}, x > 0$ $\dfrac{\sqrt{15x}}{5x}$

21. [7.2B]

a. $\dfrac{1}{\sqrt[3]{5x}}$ $\dfrac{\sqrt[3]{25x^2}}{5x}$

b. $\dfrac{1}{\sqrt[3]{7x}}$ $\dfrac{\sqrt[3]{49x^2}}{7x}$

22. [7.2B]

a. $\dfrac{\sqrt[4]{1}}{\sqrt[4]{16x}}$ $\dfrac{\sqrt[4]{x^3}}{2x}$

b. $\dfrac{\sqrt[4]{5}}{\sqrt[4]{8x^3}}$ $\dfrac{\sqrt[4]{10x}}{2x}$

In Problems 23–24, reduce the index of the expression.

23. [7.2C]

a. $\sqrt[4]{\dfrac{256}{81}}$ $\dfrac{4}{3}$

b. $\sqrt[4]{\dfrac{625}{81}}$ $\dfrac{5}{3}$

24. [7.2C]

a. $\sqrt[6]{81c^4d^4}$ $\sqrt[3]{9c^2d^2}$

b. $\sqrt[6]{625c^4d^4}$ $\sqrt[3]{25c^2d^2}$

25. [7.2C] Simplify.

a. $\sqrt[6]{\dfrac{3a^2}{243c^{16}}}$ $\dfrac{\sqrt[3]{3ac}}{3c^3}$

b. $\sqrt[6]{\dfrac{9a^2}{81c^{16}}}$ $\dfrac{\sqrt[3]{9ac}}{3c^3}$

In Problems 26–35, perform the indicated operations.

26. [7.3A]

a. $\sqrt{8} + \sqrt{32}$ $6\sqrt{2}$

b. $\sqrt{18} + \sqrt{32}$ $7\sqrt{2}$

27. [7.3A]

a. $\sqrt{63} - \sqrt{28}$ $\sqrt{7}$

b. $\sqrt{112} - \sqrt{63}$ $\sqrt{7}$

28. [7.3A]

a. $3\sqrt{\dfrac{1}{2}} - 3\sqrt{\dfrac{1}{8}}$ $\dfrac{3\sqrt{2}}{4}$

b. $6\sqrt{\dfrac{1}{2}} - 3\sqrt{\dfrac{1}{8}}$ $\dfrac{9\sqrt{2}}{4}$

29. [7.3A]

a. $7\sqrt[3]{\dfrac{3}{4x}} - \sqrt[3]{\dfrac{3}{32x}}$ $\dfrac{13\sqrt[3]{6x^2}}{4x}$

b. $6\sqrt[3]{\dfrac{3}{4x}} - \sqrt[3]{\dfrac{3}{32x}}$ $\dfrac{11\sqrt[3]{6x^2}}{4x}$

30. [7.3B]

a. $\sqrt{2}(\sqrt{18} + \sqrt{3})$ $6 + \sqrt{6}$

b. $\sqrt{2}(\sqrt{32} + \sqrt{3})$ $8 + \sqrt{6}$

31. [7.3B]

a. $\sqrt[3]{2x}(\sqrt[3]{24x^2} - \sqrt[3]{81x})$ $2x\sqrt[3]{6} - 3\sqrt[3]{6x^2}$

b. $\sqrt[3]{3x}(\sqrt[3]{16x^2} - \sqrt[3]{54x})$ $2x\sqrt[3]{6} - 3\sqrt[3]{6x^2}$

32. [7.3B]

a. $(\sqrt{27} + \sqrt{18})(\sqrt{12} + \sqrt{8})$
$30 + 12\sqrt{6}$

b. $(\sqrt{12} + \sqrt{18})(\sqrt{12} + \sqrt{8})$
$24 + 10\sqrt{6}$

33. [7.3B]

a. $(\sqrt{3} + 4)(\sqrt{3} + 4)$
$19 + 8\sqrt{3}$

b. $(\sqrt{3} + 5)(\sqrt{3} + 5)$
$28 + 10\sqrt{3}$

34. [7.3B]

a. $(7 - \sqrt{3})^2$ $52 - 14\sqrt{3}$

b. $(4 - \sqrt{3})^2$ $19 - 8\sqrt{3}$

35. [7.3B]

a. $(\sqrt{8} + \sqrt{3})(\sqrt{8} - \sqrt{3})$ 5

b. $(\sqrt{7} + \sqrt{3})(\sqrt{7} - \sqrt{3})$ 4

36. [7.3B] Reduce.

a. $\dfrac{20 - \sqrt{50}}{5}$ $4 - \sqrt{2}$

b. $\dfrac{30 - \sqrt{50}}{5}$ $6 - \sqrt{2}$

37. [7.3C] Rationalize the denominator.

a. $\dfrac{5}{\sqrt{2} - 1}$ $5\sqrt{2} + 5$

b. $\dfrac{\sqrt{x}}{\sqrt{x} + 4}$ $\dfrac{x - 4\sqrt{x}}{x - 16}$

38. [7.4A] Solve.

a. $\sqrt{x - 2} = -2$
No real-number solution

b. $\sqrt{x - 2} = -3$
No real-number solution

39. [7.4A] Solve.

a. $\sqrt{x + 5} = x - 1$ 4

b. $\sqrt{x + 6} = x - 6$ 10

40. [7.4A] Solve.

a. $\sqrt{x - 7} - x = -7$ 7 or 8

b. $\sqrt{x - 4} - x = -4$ 4 or 5

41. [7.4A] Solve.

a. $\sqrt{x - 3} - \sqrt{x} = -1$ 4

b. $\sqrt{x - 5} - \sqrt{x} = -1$ 9

42. [7.4A] Solve.

a. $\sqrt[3]{x - 5} = 3$ 32

b. $\sqrt[3]{x - 3} = 4$ 67

43. [7.4A] Solve.

a. $\sqrt[4]{x + 1} + 1 = 0$
No real-number solution

b. $\sqrt[4]{x - 2} + 16 = 0$
No real-number solution

44. [7.4B] The distance d (in feet) from a light source of intensity I (in foot-candles) is

$$d = \sqrt{\frac{k}{I}}$$

where k is a constant.

  a. Solve for I. $I = \frac{k}{d^2}$

  b. Solve for k. $k = d^2 I$

45. [7.5A] Write in terms of i.

  a. $\sqrt{-100}$ $10i$

  b. $\sqrt{-121}$ $11i$

46. [7.5A] Write in terms of i.

  a. $\sqrt{-72}$ $6i\sqrt{2}$

  b. $\sqrt{-50}$ $5i\sqrt{2}$

47. [7.5B] Find.

  a. $(3 + 5i) + (7 - 2i)$ $10 + 3i$

  b. $(4 + 7i) + (2 - 4i)$ $6 + 3i$

48. [7.5B] Find.

  a. $(3 + 5i) - (7 - 2i)$ $-4 + 7i$

  b. $(4 + 7i) - (2 - 4i)$ $2 + 11i$

49. [7.5C] Multiply.

  a. $(3 + 2i)(5 - 3i)$ $21 + i$

  b. $(4 + 5i)(2 - 3i)$ $23 - 2i$

50. [7.5C] Multiply.

  a. $\sqrt{-16}(4 - \sqrt{-72})$

    $24\sqrt{2} + 16i$

  b. $\sqrt{-36}(4 - \sqrt{-72})$

    $36\sqrt{2} + 24i$

51. [7.5C] Divide.

  a. $\dfrac{2 + 3i}{4 + 3i}$ $\dfrac{17}{25} + \dfrac{6}{25}i$

  b. $\dfrac{3 - 5i}{4 - 3i}$ $\dfrac{27}{25} - \dfrac{11}{25}i$

52. [7.5D] Write the answer as 1, -1, i, or $-i$.

  a. i^{38} -1

  b. i^{75} $-i$

53. [7.5D] Write the answer as 1, -1, i, or $-i$.

  a. i^{-14} -1 **b.** i^{-27} i

Practice Test 7

(Answers on pages 576–577)

1. Find, if possible.

  a. $\sqrt[3]{-64}$ **b.** $\sqrt{-36}$

2. Find, if possible.

  a. $(-27)^{1/3}$ **b.** $\left(\dfrac{1}{16}\right)^{1/4}$

3. Evaluate, if possible.

  a. $8^{2/3}$ **b.** $(-25)^{3/2}$

4. Evaluate, if possible.

  a. $(-27)^{-2/3}$ **b.** $8^{-2/3}$

5. If x is positive, simplify.

  a. $x^{1/2} \cdot x^{2/3}$ **b.** $\dfrac{x^{-1/3}}{x^{2/5}}$

6. If x is positive, simplify.

  a. $(x^{1/5}y^{3/4})^{-20}$ **b.** $x^{3/5}(x^{-1/5} + y^{3/5})$

7. Simplify.

  a. $\sqrt[4]{(-5)^4}$ **b.** $\sqrt[4]{(-x)^4}$

8. Simplify.

  a. $\sqrt[3]{40}$ **b.** $\sqrt[3]{54a^4b^{12}}$

9. Simplify.

  a. $\sqrt{\dfrac{5}{48}}$ **b.** $\sqrt[3]{\dfrac{5}{x^6}}$

10. Rationalize the denominator.

  a. $\dfrac{\sqrt{5}}{\sqrt{6}}$ **b.** $\dfrac{\sqrt{2}}{\sqrt{5x}}, x > 0$

11. Rationalize the denominator.

  a. $\dfrac{1}{\sqrt[3]{3x}}$ **b.** $\dfrac{\sqrt[5]{7}}{\sqrt[5]{16x^3}}$

12. Reduce the index.

  a. $\sqrt[4]{\dfrac{16}{81}}$ **b.** $\sqrt[6]{81c^4d^4}$

13. Simplify. $\sqrt[6]{\dfrac{a^2}{16c^{16}}}$

14. Perform the indicated operations.

  a. $\sqrt{32} + \sqrt{98}$ **b.** $\sqrt{112} - \sqrt{28}$

15. Perform the indicated operations.

   a. $4\sqrt{\dfrac{1}{2}} - 3\sqrt{\dfrac{1}{8}}$ **b.** $5\sqrt[3]{\dfrac{3}{4x}} - \sqrt[3]{\dfrac{3}{32x}}$

16. Perform the indicated operations.

   a. $\sqrt{2}(\sqrt{8} + \sqrt{5})$

   b. $\sqrt[3]{3x}(\sqrt[3]{9x^2} - \sqrt[3]{16x})$

17. Find the product.

   a. $(\sqrt{27} + \sqrt{50})(\sqrt{12} + \sqrt{18})$

   b. $(\sqrt{3} + 3)(\sqrt{3} + 3)$

18. Find the product.

   a. $(3 - \sqrt{3})^2$

   b. $(\sqrt{6} + \sqrt{3})(\sqrt{6} - \sqrt{3})$

19. Reduce. $\dfrac{10 - \sqrt{75}}{5}$

20. Rationalize the denominator in $\dfrac{2}{\sqrt{x} - 5}$

21. a. Solve $\sqrt{x + 2} = -2$.

   b. Solve $\sqrt{x + 2} = x - 10$.

22. Solve $\sqrt{x - 3} - x = -3$.

23. Solve $\sqrt{x - 9} - \sqrt{x} = -1$.

24. Solve $\sqrt[3]{x - 3} - 3 = 0$.

25. The time t (in seconds) it takes a free-falling object to fall d ft is given by

$$t = \frac{\sqrt{d}}{4}$$

   a. Solve for d.

   b. An object dropped from the top of a building hits the ground after 4 sec. How tall is the building?

26. Write in terms of i.

   a. $\sqrt{-64}$

   b. $\sqrt{-98}$

27. Find.

   a. $(4 + 5i) + (6 - 14i)$

   b. $(5 + 2i) - (7 - 8i)$

28. Multiply.

   a. $(2 + 2i)(4 - 3i)$

   b. $\sqrt{-9}(3 - \sqrt{-8})$

29. Find.

   a. $\dfrac{4 + 3i}{2 + 3i}$ **b.** $\dfrac{4 - 3i}{3 - 5i}$

30. Write the answer as 1, -1, i, or $-i$.

   a. i^{56}

   b. i^{-9}

Answers to Practice Test

	ANSWER		IF YOU MISSED	REVIEW		
			QUESTION	SECTION	EXAMPLES	PAGE
1. a. -4	**b.** Not a real number		1	7.1A	1	522
2. a. -3	**b.** $\frac{1}{2}$		2	7.1B	2	523
3. a. 4	**b.** Not a real number		3	7.1B	3	523–524
4. a. $\frac{1}{9}$	**b.** $\frac{1}{4}$		4	7.1B	4	524
5. a. $x^{7/6}$	**b.** $\frac{1}{x^{11/15}}$		5	7.1C	5	525
6. a. $\frac{1}{x^4 y^{15}}$	**b.** $x^{2/5} + x^{3/5}y^{3/5}$		6	7.1C	6	525–526
7. a. 5	**b.** $|-x| = |x|$		7	7.2A	1	531
8. a. $2\sqrt[3]{5}$	**b.** $3ab^4\sqrt[3]{2a}$		8	7.2A	2	532–533

ANSWER		IF YOU MISSED		REVIEW		
		QUESTION	SECTION	EXAMPLES	PAGE	

ANSWER	IF YOU MISSED QUESTION	SECTION	EXAMPLES	PAGE
9. a. $\frac{\sqrt{15}}{12}$ **b.** $\frac{\sqrt[3]{5}}{x^2}$	9	7.2A	3	533
10. a. $\frac{\sqrt{30}}{6}$ **b.** $\frac{\sqrt{10x}}{5x}$	10	7.2B	4	535
11. a. $\frac{\sqrt[3]{9x^2}}{3x}$ **b.** $\frac{\sqrt[5]{14x^2}}{2x}$	11	7.2B	5	535–536
12. a. $\frac{2}{3}$ **b.** $\sqrt[3]{9c^2d^2}$	12	7.2C	6	536
13. $\frac{\sqrt[3]{2ac}}{2c^3}$	13	7.2C	7	537
14. a. $11\sqrt{2}$ **b.** $2\sqrt{7}$	14	7.3A	1	542
15. a. $\frac{5\sqrt{2}}{4}$ **b.** $\frac{9\sqrt[3]{6x^2}}{4x}$	15	7.3A	2	543
16. a. $4 + \sqrt{10}$ **b.** $3x - 2\sqrt[3]{6x^2}$	16	7.3B	3	544
17. a. $48 + 19\sqrt{6}$ **b.** $12 + 6\sqrt{3}$	17	7.3B	4, 5	545
18. a. $12 - 6\sqrt{3}$ **b.** 3	18	7.3B	5	545
19. $2 - \sqrt{3}$	19	7.3B	6	546
20. $\frac{2\sqrt{x} + 10}{x - 25}$	20	7.3C	7	547–548
21. a. No real-number solution **b.** 14	21	7.4A	1	553–554
22. 3 or 4	22	7.4A	1	553–554
23. 25	23	7.4A	2	555
24. 30	24	7.4A	3, 4	555–556
25. a. $d = 16t^2$ **b.** 256 ft	25	7.4B	5	556
26. a. $8i$ **b.** $7i\sqrt{2}$	26	7.5A	1	561
27. a. $10 - 9i$ **b.** $-2 + 10i$	27	7.5B	2	562
28. a. $14 + 2i$ **b.** $6\sqrt{2} + 9i$	28	7.5C	3, 4	563
29. a. $\frac{17}{13} - \frac{6}{13}i$ **b.** $\frac{27}{34} + \frac{11}{34}i$	29	7.5C	5	564–565
30. a. 1 **b.** $-i$	30	7.5D	6	565–566

Cumulative Review Chapters 1–7

1. The number $\sqrt{19}$ belongs to which of these sets?
natural numbers, whole numbers, integers, rational
numbers, irrational numbers, real numbers
Name all that apply. Irrational numbers, real numbers

2. Simplify: $(2x^2y^{-4})^{-4} \dfrac{y^{16}}{16x^8}$

3. Evaluate: $[-9(9 + 6)] + 7$ -128

4. Solve: $\dfrac{7}{6} - \dfrac{x}{18} = \dfrac{3(x + 9)}{54}$ 6

5. Graph: $-7 \le -7x - 14 < 7$

6. The perimeter of a rectangle is $P = 2L + 2W$, where
L is the length and W is the width. If the perimeter is
100 ft and the length is 30 ft more than the width, what
are the dimensions? 10 ft by 40 ft

7. Find the x- and y-intercepts of $y = 8x + 6$.
x-intercept: $\frac{-3}{4}$; y-intercept: 6

8. The line through $A(5, -2)$ and $B(-3, y)$ is perpendicular to a line with slope -4. Find y. -4

9. Find the standard form of the equation of the line that passes through the points $(-6, 4)$ and $(8, 7)$.
$3x - 14y = -74$

10. Graph: $y \leq x - 2$

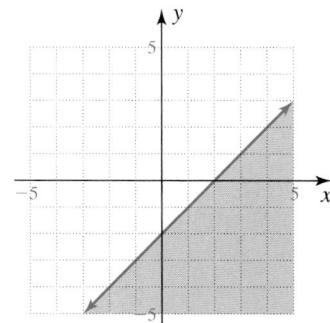

11. Graph: $|y| \leq 2$

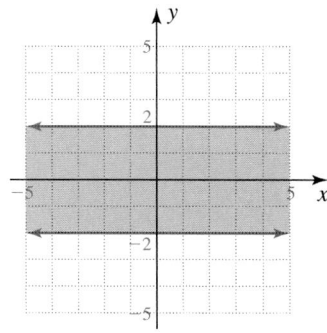

12. Solve the system by substitution:
$x - 2y = 3$
$x = -3 - y$ $(-1, -2)$

13. Solve the system:
$$\frac{x}{4} + \frac{y}{3} = -6$$
$$\frac{x}{12} + \frac{y}{9} = -2$$
Infinitely many solutions such that $y = \frac{-3}{4}x - 18$

14. Solve the system:
$2x + y + z = -5$
$-4x + y + z = -3$
$3y + 3z = -2$
No solution

15. Use Cramer's rule to solve for x:
$3x + 2y + 3z = 29$
$2x + y + z = 15$
$x + 2y - 2z = -1$ $x = 5$

16. Bart has three investments totaling $90,000. These investments earn interest at 7%, 9%, and 11%, respectively. Bart's total income from these investments is $8500. The income from the 11% investment exceeds the total income from the other two investments by $300. Find how much Bart has invested at each rate. $20,000 at 7%; $30,000 at 9%; $40,000 at 11%

17. Find the range of the relation $\{(-3, -5), (3, -3), (-5, 0)\}$. $R = \{-5, -3, 0\}$

18. Find $f(2)$ given $f(x) = 4x - 3$. 5

19. Multiply: $(x - 4)(x^2 + 4x + 2)$ $x^3 - 14x - 8$

20. Factor: $9x^6 - 12x^5 + 15x^4 + 12x^3$
$3x^3(3x^3 - 4x^2 + 5x + 4)$

21. Factor: $9x^3y - 15x^2y^2 - 6xy^3$ $3xy(3x + y)(x - 2y)$

22. Solve for x: $6x^2 - 7x = -2$ $\frac{1}{2}; \frac{2}{3}$

23. Reduce to lowest terms: $\frac{x^2 - 9}{27 - x^3}$ $\frac{-(x + 3)}{9 + 3x + x^2}$

24. Factor $3x^3 + 16x^2 + 17x + 4$ if $(x + 4)$ is one of its factors. $(x + 4)(x + 1)(3x + 1)$

25. Solve for x: $\frac{x}{x + 5} - \frac{x}{x - 5} = \frac{x^2 + 25}{x^2 - 25}$ No solution

26. Find two consecutive even integers such that the sum of their reciprocals is $\frac{11}{60}$. 10 and 12

27. If the temperature of a gas is held constant, the pressure P varies inversely as the volume V. A pressure of 1720 lb per in.^{2} is exerted by 4 ft^3 of air in a cylinder fitted with a piston. Find k. 6880

28. Evaluate: $9^{3/2}$ 27

29. Assume x and y are positive and simplify:
$x^{2/5}(x^{-1/5} + y^{4/5})$ $x^{1/5} + x^{2/5}y^{4/5}$

30. Simplify: $\sqrt{175} + \sqrt{112}$ $9\sqrt{7}$

31. Find the product: $(7 - \sqrt{7})^2$ $56 - 14\sqrt{7}$

32. Rationalize the denominator: $\frac{\sqrt{x}}{\sqrt{x} - \sqrt{5}}$ $\frac{x + \sqrt{5x}}{x - 5}$

33. Solve: $\sqrt{x + 34} = x + 14$ -9

34. Find: $\frac{7 + 6i}{1 + 9i}$ $\frac{61}{82} - \frac{57}{82}i$

35. Find: i^{49} i

Quadratic Equations and Inequalities

8

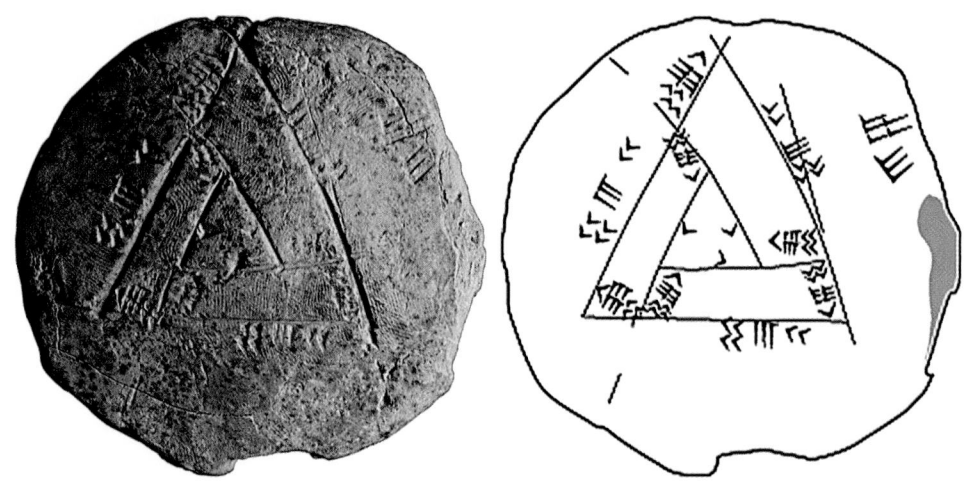

Problem with two concentric and parallel equilateral triangles. Babylonia, 19th c. B.C.

The Human Side of Algebra

Many civilizations, starting with the Babylonians in 2000 B.C., were familiar with quadratic equations, whose solutions were found by following verbal instructions given in a text. We believe that the Arabian mathematician al-Khowarizmi (ca. 820) was the first to classify quadratic equations into three types: $x^2 + ax = b$, $x^2 + b = ax$, and $x^2 = ax + b$, where a and b are positive. These three types of equations were solved by using a few general rules, and their correctness was demonstrated geometrically. In general, the quadratic equation $x^2 + px = q$ was solved by the method of "completion of squares."

Although the Arabs recognized the existence of two solutions for quadratic equations, they listed only the positive ones. In the twelfth century A.D., the Hindu mathematician Bhaskara affirmed the existence and validity of negative as well as positive solutions. What types of quadratic equations were being solved? A Babylonian clay tablet from 2000 B.C. states, "I have added the area and two-thirds of my square and it is 0;36 [$\frac{36}{60}$]. What is the side of my square?" Can you translate this into an equation? The problem was solved by using a quadratic formula equivalent to the one we discuss in this chapter.

Pretest for Chapter 8

(Answers on page 581)

1. Solve.

   a. $4x^2 - 25 = 0$

   b. $3x^2 + 18 = 0$

2. Solve.

   a. $(x - 2)^2 = 45$

   b. $2(x - 1)^2 = 49$

3. Solve $2x^2 + 8x + 8 = 5$.

4. Solve by completing the square: $x^2 + 6x - 5 = 0$.

5. Solve by completing the square: $4x^2 - 4x - 1 = 0$.

6. Solve by the quadratic formula: $5x^2 + 4x - 1 = 0$.

7. Solve by the quadratic formula: $2x^2 = 2x + 3$.

8. Solve by the quadratic formula: $8x = x^2$.

9. Solve by the quadratic formula: $\dfrac{x^2}{9} + \dfrac{5}{3}x = \dfrac{16}{9}$.

10. Solve by the quadratic formula: $2x^2 + 3x = -2$.

11. Solve $27x^3 - 64 = 0$.

12. If the demand (d) is given by $d = \dfrac{100}{p}$ and the supply is given by $s = 200p - 100$, find the equilibrium point (at which $s = d$).

13. Find.

   a. Given $2x^2 + x = 4$, find the value of the discriminant and classify the solutions.

   b. Given $9x^2 - kx = -4$, find the discriminant and then find k so that the equation has exactly one rational solution.

14. **a.** Determine whether $15x^2 - 11x - 14$ is factorable into factors with integer coefficients.

   b. Factor $12x^2 - 32x - 35$ into factors with integer coefficients, if possible.

15. Write a quadratic equation whose solutions are $\{-4, \frac{1}{2}\}$.

16. **a.** Find the sum and the product of the roots of the equation $6x^2 - 13x - 5 = 0$.

   b. Use the sum and product properties to check if the solutions are $-\frac{1}{3}$ and $\frac{5}{2}$.

17. Solve.

   a. $\dfrac{6}{x^2 - 9} + \dfrac{1}{x + 3} = 1$

   b. $\dfrac{8}{x^2 - 4} - \dfrac{6}{x - 2} = 1$

18. Solve $x^4 - 6x^2 + 5 = 0$.

19. Solve $(x^2 - 2x)^2 + (x^2 - 2x) - 12 = 0$.

20. Solve $x^{1/2} - 9x^{1/4} + 18 = 0$.

21. Solve $x - 3\sqrt{x} + 2 = 0$.

For Problems 22–25, write the solutions in interval notation.

22. Solve $(x - 1)(x + 2) < 0$.

23. Solve $x^2 + 2x \geq 8$.

24. Solve $(x - 1)(x - 2)(x + 3) \leq 0$.

25. **a.** Solve $\dfrac{x - 4}{x + 5} > 0$.

   b. Solve $\dfrac{x + 1}{x - 1} \geq 2$.

Answers to Pretest

ANSWER	IF YOU MISSED		REVIEW	
	QUESTION	SECTION	EXAMPLES	PAGE
1. a. $\dfrac{5}{2}, -\dfrac{5}{2}$ **b.** $i\sqrt{6}, -i\sqrt{6}$	1	8.1	1	584
2. a. $2 \pm 3\sqrt{5}$ **b.** $1 \pm \dfrac{7\sqrt{2}}{2}$ or $\dfrac{2 \pm 7\sqrt{2}}{2}$	2	8.1	2	585–586
3. $-2 \pm \dfrac{\sqrt{10}}{2}$ or $\dfrac{-4 \pm \sqrt{10}}{2}$	3	8.1	3	586–587
4. $-3 \pm \sqrt{14}$	4	8.1	4	588–589
5. $\dfrac{1 \pm \sqrt{2}}{2}$ or $\dfrac{1}{2} \pm \dfrac{\sqrt{2}}{2}$	5	8.1	5	589–590
6. $\dfrac{1}{5}, -1$	6	8.2	1	595–596
7. $\dfrac{1 \pm \sqrt{7}}{2}$	7	8.2	2	596–597
8. $0, 8$	8	8.2	3	597
9. $1, -16$	9	8.2	4	598
10. $-\dfrac{3}{4} \pm \dfrac{\sqrt{7}}{4}i$	10	8.2	5	598–599
11. $\dfrac{4}{3}, -\dfrac{2}{3} \pm \dfrac{2\sqrt{3}}{3}i$	11	8.2	6	600–601
12. When price is \$1	12	8.2	7	602
13. a. 33; two real, irrational solutions **b.** $k^2 - 144; k = \pm 12$	13	8.3	1	607–608
14. a. Yes. $D = 31^2$. **b.** $(2x - 7)(6x + 5)$	14	8.3	2, 3	608–609
15. $2x^2 + 7x - 4 = 0$	15	8.3	4	609–610
16. a. Sum: $\dfrac{13}{6}$, product: $-\dfrac{5}{6}$ **b.** Yes	16	8.3	5	610–611
17. a. 4 **b.** $0, -6$	17	8.4	1	615
18. $\pm 1, \pm\sqrt{5}$	18	8.4	2	616
19. $3, -1, 1 \pm i\sqrt{3}$	19	8.4	3	617
20. $81, 1296$	20	8.4	4	617
21. $1, 4$	21	8.4	5	618
22. $(-2, 1)$	22	8.5	1	623
23. $(-\infty, -4] \cup [2, \infty)$	23	8.5	2	624
24. $(-\infty, -3] \cup [1, 2]$	24	8.5	3	626
25. a. $(-\infty, -5) \cup (4, \infty)$ **b.** $(1, 3]$	25	8.5	4	627–628

8.1 SOLVING QUADRATICS BY COMPLETING THE SQUARE

To Succeed, Review How To ...

1. Take the square root of a number (pp. 521–522).

2. Rationalize the denominator of a fraction (pp. 534–536).

3. Add, subtract, multiply, and divide complex numbers (pp. 562–565).

4. Expand $(x \pm a)^2$ (pp. 562–563).

Objectives

A Solve quadratic equations of the form $ax^2 + c = 0$.

B Solve quadratic equations of the form $a(x + b)^2 = c$.

C Solve quadratic equations by completing the square.

GETTING STARTED

Baseball Diamonds and Quadratic Equations

Look at the picture of the baseball "diamond." If you pay close attention to the infield you will notice that it is a square with the four corners representing home plate, first

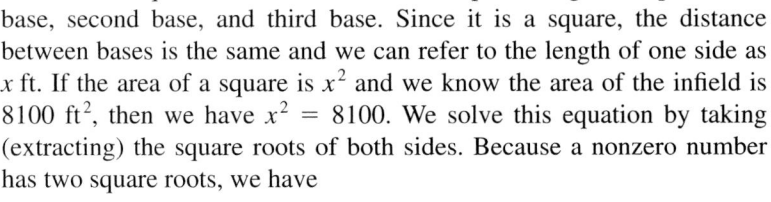

base, second base, and third base. Since it is a square, the distance between bases is the same and we can refer to the length of one side as x ft. If the area of a square is x^2 and we know the area of the infield is 8100 ft^2, then we have $x^2 = 8100$. We solve this equation by taking (extracting) the square roots of both sides. Because a nonzero number has two square roots, we have

$$x = \sqrt{8100} = 90 \quad \text{or} \quad x = -\sqrt{8100} = -90$$

The solutions (or roots) are 90 and -90. This procedure is usually shortened to

$$x^2 = 8100$$
$$x = \pm\sqrt{8100} = \pm 90$$

where the notation ± 90 (read "plus or minus 90") means that $x = 90$ or $x = -90$. If x represents the length of a side, the answer -90 must be discarded because the length of an object cannot be negative.

In Chapter 2, we studied methods of solving *linear equations*—equations in which the variable involved has an exponent of 1 (the first power). We are now ready to discuss equations containing the *second* (but no higher) power of the unknown.

Such equations are called *second-degree*, or *quadratic*, equations; these can be written in *standard form*. Here is the definition.

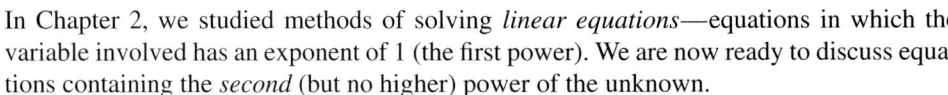

QUADRATIC EQUATION IN STANDARD FORM

$$ax^2 + bx + c = 0 \quad (a, b, \text{ and } c \text{ are real numbers}, a \neq 0)$$

is a **quadratic equation in standard form.**

The procedure used to solve these equations is similar to that employed in Chapter 2 and consists of applying certain transformations to obtain equivalent equations whose solution set is evident.

In Section 5.7 we learned how to solve quadratic equations with factoring (using the OFF method). For example,

$$x^2 - 25 = 0 \qquad \text{O \quad Given (equation is already set equal to 0).}$$

$$(x + 5)(x - 5) = 0 \qquad \text{F \quad Factor (difference of squares).}$$

$$x + 5 = 0 \quad \text{or} \quad x - 5 = 0 \qquad \text{F \quad Factors = 0 (Zero-factor property).}$$

$$x = -5 \quad \text{or} \qquad x = 5$$

This equation is a special kind of quadratic equation because it is missing the bx term. In this case we can use a different approach for solving. It will require that the x^2 term be isolated and then take the square roots of both sides.

$$x^2 - 25 = 0 \qquad \text{Given.}$$

$$x^2 = 25 \qquad \text{Isolate } x^2 \text{ by adding 25 to both sides.}$$

$$\sqrt{x^2} = \pm\sqrt{25} \qquad \text{Take the square roots of both sides.}$$

$$x = \pm 5 \qquad \text{Simplify.}$$

The equation has two solutions, $x = 5$ or $x = -5$, as before.

For the special case of the quadratic equation $x^2 = a$, we have

---

Solutions of $x^2 = k$

The solutions of $x^2 = k$, where k is a real number, are $\sqrt{k}$ and $-\sqrt{k}$, abbreviated as $\pm\sqrt{k}$.

If we compare the equation $x^2 = k$ to the standard form equation, $ax^2 + bx + c = 0$, we see that the bx term is missing.

A Solving Quadratic Equations of the Form $ax^2 + c = 0$

To solve the equation $9x^2 - 5 = 0$, we notice that the bx term is missing. We need to isolate the x^2 and take the square roots of both sides. Here's how we do it:

$$9x^2 - 5 = 0 \qquad \text{Given.}$$

$$9x^2 = 5 \qquad \text{Add 5 to both sides.}$$

$$x^2 = \frac{5}{9} \qquad \text{Divide both sides by 9 (now } x^2 \text{ is by itself).}$$

$$x = \pm\sqrt{\frac{5}{9}} = \pm\frac{\sqrt{5}}{3} \qquad \begin{array}{l}\text{Take square roots of both sides.} \\ \text{Simplify the radical } (\sqrt{\frac{5}{9}} = \frac{\sqrt{5}}{\sqrt{9}} = \frac{\sqrt{5}}{3}).\end{array}$$

The solutions are

$$\frac{\sqrt{5}}{3} \quad \text{and} \quad -\frac{\sqrt{5}}{3}$$

and the solution set is

$$\left\{ \frac{\sqrt{5}}{3}, -\frac{\sqrt{5}}{3} \right\}$$

Teaching Tip

Review the quotient rule.

$$\sqrt{\frac{5}{9}} = \frac{\sqrt{5}}{\sqrt{9}} = \frac{\sqrt{5}}{3}$$

Web It

For another explanation of extracting the root and completing the square to solve quadratic equations, go to link 8-1-1 at mhhe.com/bello.

EXAMPLE 1   Solving $ax^2 + c = 0$

Solve:

a. $4x^2 - 7 = 0$ **b.** $3x^2 + 24 = 0$ **c.** $5x^2 - 4 = 0$

SOLUTION

a. The idea is to isolate x^2 and then take the square roots of both sides.

$4x^2 - 7 = 0$	Given.
$4x^2 = 7$	Add 7 to both sides.
$x^2 = \dfrac{7}{4}$	Divide both sides by 4 (now x^2 is isolated).
$x = \pm\sqrt{\dfrac{7}{4}} = \pm\dfrac{\sqrt{7}}{2}$	Take square roots of both sides ($\sqrt{\frac{7}{4}} = \frac{\sqrt{7}}{\sqrt{4}} = \frac{\sqrt{7}}{2}$).

The solutions are

$$\frac{\sqrt{7}}{2} \quad \text{and} \quad -\frac{\sqrt{7}}{2}$$

and the solution set is

$$\left\{ \frac{\sqrt{7}}{2}, -\frac{\sqrt{7}}{2} \right\}$$

b.

$3x^2 + 24 = 0$	Given.
$3x^2 = -24$	Add -24 to both sides.
$x^2 = -8$	Divide both sides by 3.
$x = \pm\sqrt{-8}$	Take square roots of both sides.
$x = \pm 2i\sqrt{2}$	Simplify: $\sqrt{-8} = \sqrt{4}\sqrt{-2} = 2i\sqrt{2}$.

The solutions are $2i\sqrt{2}$ and $-2i\sqrt{2}$. The answers are non-real **complex numbers** given in the *simplified* form, $\pm 2i\sqrt{2}$ instead of $\pm i\sqrt{8}$.

c.

$5x^2 - 4 = 0$	Given.
$5x^2 = 4$	Add 4 to both sides.
$x^2 = \dfrac{4}{5}$	Divide both sides by 5.
$x = \pm\sqrt{\dfrac{4}{5}} = \pm\dfrac{2}{\sqrt{5}}$	Take square roots of both sides ($\sqrt{\frac{4}{5}} = \frac{\sqrt{4}}{\sqrt{5}} = \frac{2}{\sqrt{5}}$).
$= \pm\dfrac{2 \cdot \sqrt{5}}{\sqrt{5} \cdot \sqrt{5}}$	Rationalize the denominator.
$= \pm\dfrac{2\sqrt{5}}{5}$	

The solutions are

$$\frac{2\sqrt{5}}{5} \quad \text{and} \quad -\frac{2\sqrt{5}}{5}$$

and the solution set is

$$\left\{ \frac{2\sqrt{5}}{5}, -\frac{2\sqrt{5}}{5} \right\}$$

The answer should be written with a *rationalized* denominator.

PROBLEM 1

Solve:

a. $9x^2 - 4 = 0$

b. $3x^2 + 54 = 0$

c. $3x^2 - 16 = 0$

Teaching Tip

In Example 1(b) you may write the answer as $\pm 2\sqrt{2}i$.

Answers

1. a. $\frac{2}{3}$ and $-\frac{2}{3}$ **b.** $3i\sqrt{2}$ and $-3i\sqrt{2}$ **c.** $\frac{4\sqrt{3}}{3}$ and $-\frac{4\sqrt{3}}{3}$

B   Solving Quadratic Equations of the Form $a(x + b)^2 = c$

The method used to solve equations of the form $ax^2 + c = 0$ can also be used to solve equations of the form $(x + a)^2 = b$. Thus to solve the equation $(x + 3)^2 = 16$, we proceed as follows:

$(x + 3)^2 = 16$	Given.
$\sqrt{(x + 3)^2} = \pm\sqrt{16} = \pm 4$	Take square roots of both sides.
$x + 3 = 4 \quad$ or $\quad x + 3 = -4$	
$x = 1 \quad$ or $\quad\quad x = -7$	Solve $x + 3 = 4$ and $x + 3 = -4$.

The solutions are 1 and -7 and the solution set is $\{1, -7\}$.

To solve the equation $(x - 2)^2 - 9 = 0$, we add 9 to both sides and then take the square roots to obtain

$$(x - 2)^2 - 9 = 0$$
$$(x - 2)^2 = 9$$
$$\sqrt{(x - 2)^2} = \pm\sqrt{9} = \pm 3$$
$$x - 2 = 3 \quad \text{or} \quad x - 2 = -3$$
$$x = 5 \quad \text{or} \quad\quad x = -1$$

The solutions for $(x - 2)^2 - 9 = 0$ are 5 and -1, and the solution set is $\{5, -1\}$.

Teaching Tip

Before doing Example 2(a), review adding radical terms using the following examples:

$2a + 3a = 5a \quad \vert \quad 2\sqrt{7} + 3\sqrt{7} = 5\sqrt{7}$
$2 + 3a = 2 + 3a \quad \vert \quad 2 + 3\sqrt{7} = 2 + 3\sqrt{7}$

EXAMPLE 2   Solving $a(x + b)^2 = c$

Solve:

a. $(x - 1)^2 = 27$ **b.** $9(x + 5)^2 = 11$ **c.** $3(x - 2)^2 + 25 = 0$

SOLUTION

a.

$(x - 1)^2 = 27$	Given.
$\sqrt{(x - 1)^2} = \pm\sqrt{27}$	Take the square root of both sides.
$x - 1 = \pm 3\sqrt{3}$	Simplify the radical ($\sqrt{27} = 3\sqrt{3}$).
$x - 1 = 3\sqrt{3} \quad$ or $\quad x - 1 = -3\sqrt{3}$	
$x = 1 + 3\sqrt{3} \quad$ or $\quad x = 1 - 3\sqrt{3}$	Solve for x by adding 1 to both sides.

The solutions are $1 + 3\sqrt{3}$ and $1 - 3\sqrt{3}$, and the solution set is $\{1 + 3\sqrt{3}, 1 - 3\sqrt{3}\}$.

This can be verified by substituting the solutions into the original equation.

$(x - 1)^2 = 27$	Original equation	$(x - 1)^2 = 27$
$[(1 + 3\sqrt{3}) - 1]^2 = 27$	Substitute ($1 \pm 3\sqrt{3}$) for x.	$[(1 - 3\sqrt{3}) - 1]^2 = 27$
$[1 + 3\sqrt{3} - 1]^2 = 27$	Remove parentheses.	$[1 - 3\sqrt{3} - 1]^2 = 27$
$[3\sqrt{3}]^2 = 27$	Combine like terms.	$[-3\sqrt{3}]^2 = 27$
$9 \cdot 3 = 27$	Square $[3\sqrt{3}]$.	$9 \cdot 3 = 27$
$27 = 27$		$27 = 27$

Both solutions resulted in a true statement when placed in the original equation, so the solutions are verified.

PROBLEM 2

Solve:

a. $(x - 2)^2 = 24$

b. $4(x + 7)^2 = 3$

c. $5(x - 3)^2 + 36 = 0$

Answers

2. a. $2 \pm 2\sqrt{6}$ **b.** $\frac{-14 \pm \sqrt{3}}{2}$

c. $3 \pm \frac{6\sqrt{5}}{5}i$

b. $9(x + 5)^2 = 11$ Given.

$(x + 5)^2 = \dfrac{11}{9}$ Divide both sides by 9 [now the $(x + 5)^2$ is isolated].

$x + 5 = \pm\sqrt{\dfrac{11}{9}}$ Take the square root of both sides.

$x + 5 = \dfrac{\pm\sqrt{11}}{3}$ Simplify the radical.

$x = -5 \pm \dfrac{\sqrt{11}}{3}$ Subtract 5 from both sides.

$x = \dfrac{-15 \pm \sqrt{11}}{3}$ Write as one fraction with denominator 3 (multiply $-\frac{5}{1} \cdot \frac{3}{3}$).

c. $3(x - 2)^2 + 25 = 0$ Given.

$3(x - 2)^2 = -25$ Subtract 25 from both sides.

$(x - 2)^2 = \dfrac{-25}{3}$ Divide both sides by 3 [now the $(x - 2)^2$ is isolated].

$x - 2 = \pm\sqrt{\dfrac{-25}{3}}$ Take the square root of both sides.

$x - 2 = \pm\dfrac{5i}{\sqrt{3}}$ Simplify the radical.

$x - 2 = \pm\dfrac{5i \cdot \sqrt{3}}{\sqrt{3} \cdot \sqrt{3}}$ Rationalize the denominator.

$x = 2 \pm \dfrac{5\sqrt{3}}{3}i$ Solve for x by adding 2.

The solutions are

$$2 + \frac{5\sqrt{3}}{3}i \quad \text{and} \quad 2 - \frac{5\sqrt{3}}{3}i$$

and the solution set is

$$\left\{2 + \frac{5\sqrt{3}}{3}i, 2 - \frac{5\sqrt{3}}{3}i\right\}$$

We leave the verification to you.

EXAMPLE 3   **Writing as $a(x + b)^2 = c$ and solving**

Solve: $x^2 + 4x + 4 = 20$

SOLUTION   If we subtract 20 from both sides, we obtain

$$x^2 + 4x - 16 = 0$$

Answer

3. $-3 \pm 2\sqrt{6}$

Web It

For another resource on solving quadratic equations with examples, go to link 8-1-2 at mhhe.com/bello.

Teaching Tip

Before doing Example 3, review how to recognize a perfect square trinomial and then factor it.

Examples:

$$x^2 + 6x + 9 = (x + 3)^2$$
$$x^2 - 8x + 16 = (x - 4)^2$$

PROBLEM 3

Solve:

$$x^2 + 6x + 9 = 24$$

This is *not* factorable using integer coefficients, so we try to factor the left side first.

$$x^2 + 4x + 4 = 20$$ Given.

$$(x + 2)^2 = 20$$ Factor $x^2 + 4x + 4$ as $(x + 2)^2$ (perfect square trinomial).

$$x + 2 = \pm\sqrt{20}$$ Take the square root of both sides.

$$x + 2 = \pm 2\sqrt{5}$$ Simplify the radical ($\sqrt{20} = \sqrt{4 \cdot 5} = 2\sqrt{5}$).

$$x = -2 \pm 2\sqrt{5}$$ Solve for x by subtracting 2 from both sides.

The solutions are

$$x = -2 + 2\sqrt{5} \quad \text{and} \quad x = -2 - 2\sqrt{5}$$

and the solution set is

$$\{-2 + 2\sqrt{5}, -2 - 2\sqrt{5}\}$$

We leave the verification to you.

C   Solving Quadratic Equations by Completing the Square

Web It

For examples of solving quadratic equations by completing the square, go to link 8-1-3 at mhhe.com/bello.

The solutions of Examples 2 and 3 were obtained by writing the equation in the form $a(x + b)^2 = c$. Suppose we have an equation not of this form. We can make it of this form if we learn a technique called **completing the square.** As you recall,

$$(x + a)^2 = x^2 + 2ax + a^2 \quad \text{and} \quad (x - a)^2 = x^2 - 2ax + a^2$$

In both cases, the last term is the square of one-half the coefficient of x. How can we make $x^2 + 10x$ a perfect square trinomial? Since the coefficient of x is 10, and in a perfect square trinomial, the coefficient of x is $2a$, $10 = 2a$ and a must be 5. Thus we make $x^2 + 10x$ a perfect square trinomial by adding 5^2. We then have

$$x^2 + 10x + 5^2 = (x + 5)^2$$

Now consider the equation $x^2 + 8x = 12$. To make $x^2 + 8x$ a perfect square trinomial, we note that the coefficient of x is 8, so the number to be added is the square of one-half of 8, or 4^2. Adding 4^2 to both sides, we have

$$x^2 + 8x + 4^2 = 12 + 4^2$$

$$(x + 4)^2 = 12 + 4^2$$ Factor $x^2 + 8x + 4^2$ as $(x + 4)^2$.

$$(x + 4)^2 = 12 + 16$$

And, $(x + 4)^2 = 28$, the form we want. Now take the square roots of both sides:

$$\sqrt{(x + 4)^2} = \pm\sqrt{28}$$ Take the square root of both sides.

$$x + 4 = \pm 2\sqrt{7}$$ Simplify the radical.

$$x = -4 \pm 2\sqrt{7}$$ Subtract 4 from both sides.

The solutions are $-4 + 2\sqrt{7}$ and $-4 - 2\sqrt{7}$, and the solution set is $\{-4 + 2\sqrt{7}, -4 - 2\sqrt{7}\}$. Note that in the equation $x^2 + 8x = 12$, the variables x and x^2 are *isolated* on one side of the equation.

EXAMPLE 4 Solving by completing the square	**PROBLEM 4**

Solve by completing the square: $x^2 + 10x - 2 = 0$

Solve by completing the square:

$$x^2 + 12x - 8 = 0$$

SOLUTION

$x^2 + 10x - 2 = 0$	Given.
$x^2 + 10x = 2$	Add 2 to both sides so that only the variable terms are on one side.
$x^2 + 10x + 5^2 = 2 + 5^2$	Add 5^2, the square of one-half of the coefficient of x—add $[\frac{1}{2} \cdot 10]^2$—to both sides.
$(x + 5)^2 = 27$	Factor on the left; simplify on the right.
$x + 5 = \pm\sqrt{27}$	Take the square root of both sides.
$x + 5 = \pm 3\sqrt{3}$	Simplify the radical.
$x = -5 \pm 3\sqrt{3}$	Subtract 5.

The solutions are $-5 + 3\sqrt{3}$ and $-5 - 3\sqrt{3}$, and the solution set is $\{-5 + 3\sqrt{3}, -5 - 3\sqrt{3}\}$.

When the coefficient of x^2 is not 1 and the left-hand side of the equation is not factorable using integer coefficients, we first isolate the terms containing the variable on one side of the equation and then divide each term by the coefficient of x^2. For example, to solve $3x^2 + 12x - 6 = 0$, we add 6 to both sides (so all the terms containing the variable are isolated on one side), and then divide by 3, the coefficient of x^2. Here are the steps.

Teaching Tip

Remind students that when dividing both sides of the equation by the same number, each term must be divided by the number.

Example:

$$\frac{3x^2 + 2x}{3} \text{ is } \frac{3x^2}{3} + \frac{2x}{3} \text{ or } x^2 + \frac{2x}{3}$$

$3x^2 + 12x - 6 = 0$	Given.

Step 1. Add 6 to isolate the variables on the left-hand side.

$3x^2 + 12x = 6$	Add 6 to both sides.

Step 2. Divide each term by 3, the coefficient of x^2.

$x^2 + 4x = 2$	Divide each term by 3.

Step 3. Add the square of one-half the first-degree term's coefficient to both sides—add

$$\left[\tfrac{1}{2}(4)\right]^2 = (2)^2$$

to both sides

$x^2 + 4x + (2)^2 = 2 + (2)^2$	The first-degree term's coefficient is 4, so add $(\frac{1}{2} \cdot 4)^2 = (2)^2$ to both sides.

Step 4. Factor the left-hand side, and simplify the right-hand side.

$$(x + 2)^2 = 6$$

Step 5. Take the square root of both sides.

$$x + 2 = \pm\sqrt{6}$$

Answer

4. $-6 \pm 2\sqrt{11}$

Step 6. Solve the resulting equation by subtracting 2 from both sides and simplifying.

$x = -2 \pm \sqrt{6}$	Subtract 2.

The solution set is

$$\{-2 + \sqrt{6}, -2 - \sqrt{6}\}$$

Here is a summary of this procedure.

PROCEDURE

Solving a Quadratic Equation by Completing the Square ($ax^2 + bx + c = 0$)

1. Write the equation with the variables in descending order on the left and the numbers on the right. $(ax^2 + bx = -c)$

2. If the coefficient of the square term is not 1, divide each term by this coefficient. $(x^2 + \frac{b}{a}x = -\frac{c}{a})$

3. Add the square of one-half of the coefficient of the first-degree term to both sides. $(\frac{1}{2} \cdot \frac{b}{a})^2$

4. Rewrite the left-hand side as a perfect square in factored form and simplify the right-hand side. $(x - \frac{b}{2a})^2$

5. Take the square root of both sides and rationalize the denominator if necessary.

6. Solve the resulting equation.

Teaching Tip

Before doing Example 5, have students practice writing rational numbers with the same denominator. **Example:**

$$\frac{1}{2} + \frac{1}{6} = \frac{3}{6} + \frac{1}{6}$$

$$\frac{2}{3} + \frac{1}{4} = \frac{8}{12} + \frac{3}{12}$$

EXAMPLE 5   **Completing the square when the coefficient of x^2 is not 1**

Solve by completing the square: $3x^2 - 3x - 1 = 0$

SOLUTION   We use the six-step procedure.

$3x^2 - 3x - 1 = 0$ Given.

Step 1. $3x^2 - 3x = 1$ Add 1 so the variables are isolated.

Step 2. $x^2 - x = \dfrac{1}{3}$ Divide each term by 3.

Step 3. $x^2 - x + \left(\dfrac{1}{2}\right)^2 = \dfrac{1}{3} + \left(\dfrac{1}{2}\right)^2$ The first-degree term's coefficient is -1, so add $[\frac{1}{2}(-1)]^2 = (-\frac{1}{2})^2 = (\frac{1}{2})^2$ to each side.

Step 4. $\left(x - \dfrac{1}{2}\right)^2 = \dfrac{7}{12}$ Factor the left-hand side, simplify the right-hand side.

Step 5. $x - \dfrac{1}{2} = \pm\sqrt{\dfrac{7}{12}} = \pm\dfrac{\sqrt{7} \cdot \sqrt{3}}{\sqrt{12} \cdot \sqrt{3}}$ Take the square root of both sides and rationalize the denominator.

$x - \dfrac{1}{2} = \pm\dfrac{\sqrt{21}}{6}$ Simplify.

PROBLEM 5

Solve by completing the square:

$$5x^2 - 5x - 1 = 0$$

Answer

5. $\dfrac{5 \pm 3\sqrt{5}}{10}$ or $\dfrac{1}{2} \pm \dfrac{3\sqrt{5}}{10}$

Step 6. $x - \dfrac{1}{2} = \dfrac{\sqrt{21}}{6}$ or $x - \dfrac{1}{2} = -\dfrac{\sqrt{21}}{6}$

$x = \dfrac{1}{2} + \dfrac{\sqrt{21}}{6}$ or $x = \dfrac{1}{2} - \dfrac{\sqrt{21}}{6}$ Solve by adding $\frac{1}{2}$ to both sides.

The solutions are

$\dfrac{1}{2} + \dfrac{\sqrt{21}}{6}$ and $\dfrac{1}{2} - \dfrac{\sqrt{21}}{6}$,   or $\dfrac{3 + \sqrt{21}}{6}$ and $\dfrac{3 - \sqrt{21}}{6}$

The solution set is thus

$$\left\{ \dfrac{3 + \sqrt{21}}{6}, \dfrac{3 - \sqrt{21}}{6} \right\}$$

The verification is left to you.

Calculate It Exercises   Using Graphs to Verify Solutions to Quadratic Equations

To graphically solve Example 5, place all variables on the left (Y_1) and all numbers on the right (Y_2) of the equation. You have to do this algebraically. We then find the intersection of Y_1 and Y_2. The value of x thus obtained is the solution of the equation. Start by rewriting the equation as $3x^2 - 3x = 1$; then graph $Y_1 = 3x^2 - 3x$ and $Y_2 = 1$ using a $[-1, 3]$ by $[-2, 3]$ window. To find the intersection with a TI-83 Plus, press (2nd) (TRACE) (5) : Move the cursor to a point on Y_1 and near the left intersection point. The calculator prompt is "First curve?" Press (ENTER). Press (ENTER) again when the calculator asks "Second Curve?" and when it asks for a "Guess?" The intersection is given as X = −.2637626 and Y = 1. The solution is $x = -0.2637626$. Check to see that this decimal approximation corresponds to the solution

$$x = \dfrac{3 - \sqrt{21}}{6}$$

Now press (2nd) (TRACE) (5) again and move the cursor to a point near the right intersection. Press (ENTER) three times. The intersection, as shown in the window, occurs when X = 1.2637626, which corresponds to

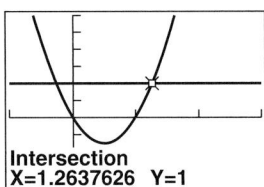

Intersection
X=1.2637626 Y=1

$$\dfrac{3 + \sqrt{21}}{6}$$

1. To solve Example 3 using the intersect feature of your calculator,

   a. What would you use for Y_1?

   b. What would you use for Y_2?

2. How can you tell if a quadratic equation has no real-number solutions by looking at its graph?

3. In some instances, we want to find the maximum or minimum of a quadratic function. For example, the number of

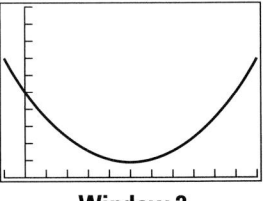

QuadReg
y=ax²+bx+c
a=1.763882863
b=-17.2356833
c=250.8031453

Window 1   **Window 2**

robberies per 100,000 inhabitants for 1980, 1985, 1990, and 1991 was 251, 208, 257, and 273, respectively. Enter this information in two tables by pressing (STAT) (1) and then using L_1 for the year (1980 = 0, 1985 = 5, etc.) and L_2 for the number of robberies per 100,000. Now to find our quadratic equation, press (STAT) (▷) (5) (ENTER) (see Window 1). You don't want to copy all this to graph it. So we tell the calculator to copy the equation produced from our statistical data by pressing (Y=) (VARS) (5) (▷) (▷) (1). Finally, since we are covering a period of 11 yr and the number of robberies varies from 208 to 273, use a $[-1, 11]$ by $[200, 300]$ window and press (GRAPH). The graph is shown in Window 2.

   a. In what time interval is the number of robberies decreasing?

   b. In what time interval is the number of robberies increasing?

   c. What was the minimum number of robberies per year? (Answer to the closest whole number.)

   d. In what year did they have the minimum number of robberies? (Answer to the closest whole number.)

4. If you have a TI-83 Plus, the minimum of a function can be found by pressing (2nd) (TRACE) (3) and answering the calculator queries for "Left bound?", "Right bound?", and "Guess?". Use this method to find X and Y in Exercise 3 and explain what they mean.

Exercises 8.1

A In Problems 1–20, solve the equation.

1. $x^2 = 64$ ± 8

2. $x^2 = 81$ ± 9

3. $x^2 = -121$ $\pm 11i$

4. $x^2 = -144$ $\pm 12i$

5. $x^2 - 169 = 0$ ± 13

6. $x^2 - 100 = 0$ ± 10

7. $x^2 + 4 = 0$ $\pm 2i$

8. $x^2 + 25 = 0$ $\pm 5i$

9. $36x^2 - 49 = 0$ $\pm\frac{7}{6}$

10. $36x^2 - 81 = 0$ $\pm\frac{3}{2}$

11. $4x^2 + 81 = 0$ $\pm\frac{9}{2}i$

12. $9x^2 + 64 = 0$ $\pm\frac{8}{3}i$

13. $3x^2 - 25 = 0$ $\pm\frac{5\sqrt{3}}{3}$

14. $5x^2 - 16 = 0$ $\pm\frac{4\sqrt{5}}{5}$

15. $5x^2 + 36 = 0$ $\pm\frac{6\sqrt{5}}{5}i$

16. $11x^2 + 49 = 0$ $\pm\frac{7\sqrt{11}}{11}i$

17. $3x^2 - 100 = 0$ $\pm\frac{10\sqrt{3}}{3}$

18. $4x^2 - 13 = 0$ $\pm\frac{\sqrt{13}}{2}$

19. $13x^2 + 81 = 0$ $\pm\frac{9\sqrt{13}}{13}i$

20. $11x^2 + 4 = 0$ $\pm\frac{2\sqrt{11}}{11}i$

B In Problems 21–40, solve the equation.

21. $(x + 5)^2 = 4$ $-3, -7$

22. $(x + 3)^2 = 9$ $0, -6$

23. $x^2 + 4x + 4 = -25$ $-2 \pm 5i$

24. $x^2 + 2x + 1 = -16$ $-1 \pm 4i$

25. $(x - 6)^2 = 18$ $6 \pm 3\sqrt{2}$

26. $(x - 2)^2 = 50$ $2 \pm 5\sqrt{2}$

27. $x^2 - 2x + 1 = -28$ $1 \pm 2i\sqrt{7}$

28. $x^2 - 6x + 9 = -32$ $3 \pm 4i\sqrt{2}$

29. $(x - 1)^2 - 50 = 0$ $1 \pm 5\sqrt{2}$

30. $(x - 2)^2 - 18 = 0$ $2 \pm 3\sqrt{2}$

31. $(x - 5)^2 - 32 = 0$ $5 \pm 4\sqrt{2}$

32. $(x - 2)^2 - 4 = 0$ $0, 4$

33. $(x - 9)^2 + 64 = 0$ $9 \pm 8i$

34. $(x - 2)^2 + 25 = 0$ $2 \pm 5i$

35. $3x^2 + 6x + 3 = 96$ $-1 \pm 4\sqrt{2}$

36. $3x^2 + 30x + 75 = 72$ $-5 \pm 2\sqrt{6}$

37. $7(x - 2)^2 - 350 = 0$ $2 \pm 5\sqrt{2}$

38. $3(x - 1)^2 - 54 = 0$ $1 \pm 3\sqrt{2}$

39. $7(x - 5)^2 + 189 = 0$ $5 \pm 3i\sqrt{3}$ **40.** $5(x - 3)^2 + 250 = 0$ $3 \pm 5i\sqrt{2}$

C In Problems 41–70, solve by completing the square.

41. $x^2 + 6x + 5 = 0$ $-1, -5$

42. $x^2 + 4x + 3 = 0$ $-3, -1$

43. $x^2 + 8x + 15 = 0$ $-5, -3$

44. $x^2 + 8x + 7 = 0$ $-7, -1$

45. $x^2 + 6x + 10 = 0$ $-3 \pm i$

46. $x^2 + 12x + 37 = 0$ $-6 \pm i$

47. $x^2 - 10x + 24 = 0$ $4, 6$

48. $x^2 + 12x - 28 = 0$ $-14, 2$

49. $x^2 - 10x + 21 = 0$ $3, 7$

50. $x^2 - 2x - 143 = 0$ $-11, 13$

51. $x^2 - 8x + 17 = 0$ $4 \pm i$

52. $x^2 - 14x + 58 = 0$ $7 \pm 3i$

53. $2x^2 + 4x + 3 = 0$ $-1 \pm \frac{\sqrt{2}}{2}i$

54. $2x^2 + 7x + 6 = 0$ $-\frac{3}{2}, -2$

55. $3x^2 + 6x + 78 = 0$ $-1 \pm 5i$

56. $9x^2 + 6x + 2 = 0$ $-\frac{1}{3} \pm \frac{1}{3}i$

57. $25y^2 - 25y + 6 = 0$ $\frac{3}{5}, \frac{2}{5}$

58. $4y^2 - 16y + 15 = 0$ $\frac{3}{2}, \frac{5}{2}$

59. $4y^2 - 4y + 5 = 0$ $\frac{1}{2} \pm i$

60. $9x^2 - 12x + 13 = 0$ $\frac{2}{3} \pm i$

61. $4x^2 - 7 = 4x$ $\frac{1 \pm 2\sqrt{2}}{2}$

62. $2x^2 - 18 = -9x$ $-6, \frac{3}{2}$

63. $2x^2 + 1 = 4x$ $\frac{2 \pm \sqrt{2}}{2}$

64. $2x^2 + 3 = 6x$ $\frac{3 \pm \sqrt{3}}{2}$

65. $(x + 3)(x - 2) = -4$ $1, -2$

66. $(x + 4)(x - 1) = -6$ $-2, -1$

67. $2x(x + 5) - 1 = 0$ $\frac{-5 \pm 3\sqrt{3}}{2}$

68. $2x(x - 4) = 2(9 - 8x) - x$ $-6, \frac{3}{2}$ **69.** $2x(x + 3) - 10 = 0$ $\frac{-3 \pm \sqrt{29}}{2}$

70. $4x(x + 1) - 5 = 0$ $\frac{-1 \pm \sqrt{6}}{2}$

APPLICATIONS

Distance and Dollars

71. The distance traveled in t seconds by an object dropped from a height h is given by $h = 16t^2$. How long would it take an object dropped from a height of 64 ft to hit the ground?   2 sec

72. Use the formula given in Problem 71 to find the time it takes an object dropped from a height of 32 ft to hit the ground. $\sqrt{2} \approx 1.4$ sec

73. The amount of money A received at the end of 2 yr when P dollars are invested at a compound rate r is $A = P(1 + r)^2$. Find the rate of interest r (written as a percent) if a person invested \$100 and received \$121 at the end of 2 yr.   10%

74. Use the formula of Problem 73 to find the rate of interest r (written as a percent) if a person invested \$100 and received \$144 at the end of 2 yr.   20%

SKILL CHECKER

Try the "Skill Checker" exercises so you'll be ready for the next section.

Evaluate

$$\frac{-b \pm \sqrt{b^2 - 4ac}}{2a}$$

for the given values of a, b, and c:

75. $a = 1, b = -9, c = 0$ $0, 9$

76. $a = 1, b = -6, c = 0$ $0, 6$

77. $a = 1, b = -2, c = -2$ $1 \pm \sqrt{3}$

78. $a = 1, b = -4, c = -4$ $2 \pm 2\sqrt{2}$

79. $a = 8, b = 7, c = -1$ $\frac{1}{8}, -1$

80. $a = 3, b = -2, c = -5$ $\frac{5}{3}, -1$

81. $a = 3, b = -8, c = 7$ $\frac{4}{3} \pm \frac{\sqrt{5}}{3}i$

82. $a = 4, b = -3, c = 5$ $\frac{3}{8} \pm \frac{\sqrt{71}}{8}i$

83. $a = 1, b = 2, c = 6$ $-1 \pm i\sqrt{5}$

84. $a = 1, b = 2, c = 5$ $-1 \pm 2i$

USING YOUR KNOWLEDGE

Average, Demand, and Bacteria

Many applications of mathematics require finding the maximum or the minimum of certain algebra expressions. A certain business may wish to find the price at which a product will bring *maximum* profits, while a team of engineers may be interested in *minimizing* the amount of carbon monoxide produced by automobiles. Now suppose you are the manufacturer of a product whose average manufacturing cost $\bar{C}$ (in dollars), based on producing x (thousand) units, is given by the expression

$$\bar{C} = x^2 - 8x + 18$$

How many units should be produced to minimize the cost per unit? If we consider the right-hand side of the equation, we can complete the square and leave the equation

unchanged by adding and subtracting the appropriate number. Thus

$$\begin{aligned}
\bar{C} &= x^2 - 8x + 18 \\
&= (x^2 - 8x +) + 18 \\
&= (x^2 - 8x + 4^2) + 18 - 4^2
\end{aligned}$$

Then

$$\bar{C} = (x - 4)^2 + 2$$

Now for $\bar{C}$ to be as small as possible (minimizing the cost), we make $(x - 4)^2$ zero by letting $x = 4$; then $\bar{C} = 2$. This tells us that when 4 (thousand) units are produced, the minimum cost is 2. That is, the minimum average cost per unit is \$2.

Use your knowledge about completing the square to solve the following problems.

85. A manufacturer's average cost $\bar{C}$ (in dollars), based on manufacturing x (thousand) items, is given by

$$\bar{C} = x^2 - 4x + 6$$

a. How many units should be produced to minimize the cost per unit? 2000

b. What is the minimum average cost per unit? $2

87. Have you seen people adding chlorine to their swimming pools? This is done to reduce the number of bacteria present in the water. Suppose that after t days, the number of bacteria per cubic centimeter is given by the expression

$$B = 20t^2 - 120t + 200$$

In how many days will the number of bacteria be at its lowest? 3 days

86. The demand D for a certain product depends on x (thousand) units produced and is given by

$$D = x^2 - 2x + 3$$

For what number of units is the demand at its lowest? 1000

WRITE ON

88. Explain why $\sqrt{49}$ has only one answer, 7, but $x^2 = 49$ has two solutions, 7 and -7. Answers may vary.

89. We have solved $x^2 - 49 = 0$ by adding 49 to both sides and then extracting roots. Describe another procedure you can use to solve $x^2 - 49 = 0$. Does your method work when solving $x^2 - 2 = 0$? Factoring; no

90. What is the first step in solving a quadratic equation by completing the square? Answers may vary.

MASTERY TEST

If you know how to do these problems, you have learned your lesson!

Solve by completing the square:

91. $5x^2 - 5x = 1$ $\frac{5 \pm 3\sqrt{5}}{10}$

92. $2x^2 - 2x - 1 = 0$ $\frac{1 \pm \sqrt{3}}{2}$

93. $x^2 + 12x = 8$ $-6 \pm 2\sqrt{11}$

94. $x^2 + 6x - 1 = 0$ $-3 \pm \sqrt{10}$

Solve:

95. $2x^2 + 12x + 18 = 27$ $\frac{-6 \pm 3\sqrt{6}}{2}$

96. $3x^2 + 6x + 3 = 20$ $\frac{-3 \pm 2\sqrt{15}}{3}$

97. $5(x - 3)^2 + 36 = 0$ $3 \pm \frac{6\sqrt{5}}{5}i$

98. $2(x - 1)^2 + 49 = 0$ $-1 \pm \frac{7\sqrt{2}}{2}i$

99. $(x - 2)^2 = 24$ $2 \pm 2\sqrt{6}$

100. $(x + 1)^2 = 18$ $-1 \pm 3\sqrt{2}$

101. $3x^2 - 16 = 0$ $\pm \frac{4\sqrt{3}}{3}$

102. $3x^2 + 54 = 0$ $\pm 3i\sqrt{2}$

103. $5x^2 + 60 = 0$ $\pm 2i\sqrt{3}$

104. $9x^2 - 4 = 0$ $\pm \frac{2}{3}$

8.2 THE QUADRATIC FORMULA: APPLICATIONS

To Succeed, Review How To . . .

1. Find the square root of a number (pp. 521–522).

2. Simplify square roots (pp. 530–534).

3. Write fractions in lowest terms (pp. 428–431).

4. Solve a quadratic equation by completing the square (pp. 587–590).

5. Factor cubes (pp. 387–389).

Objectives

A Solve equations using the quadratic formula.

B Solve factorable cubic equations.

C Solve applications involving quadratic equations.

GETTING STARTED

Rockets and Quadratics

The man standing by the first liquid-fueled rocket is Dr. Robert Goddard. His rocket went up 41 ft. Do you know how long it took it to go up that high? The height (in feet) of the rocket is given by

$$h = -16t^2 + v_0 t$$

where v_0 is the initial velocity (51.225 ft/sec). We can substitute 41 for h, the height, and 51.225 for v_0, the initial velocity, and solve for t in the equation

$$-16t^2 + 51.225t = 41$$

Unfortunately, this equation is not factorable, and to complete the square, we would have to divide by -16 and add

$$\left(-\frac{51.225}{32}\right)^2$$

to both sides. There must be a better way to solve this equation. In this section we derive a formula that will give us a more efficient method for solving this type of problem.

To derive a formula for solving quadratic equations we start with the general equation $ax^2 + bx + c = 0$ $(a \neq 0)$ and use the six-step procedure we learned for solving by completing the square in Section 8.1. To facilitate the understanding of each step we will parallel the solving of the specific equation $2x^2 + 5x + 1 = 0$ along side the solving of the general quadratic equation.

	Specific Equation	*General Equation*	
	$2x^2 + 5x + 1 = 0$	$ax^2 + bx + c = 0 \quad (a \neq 0)$	Given.
Add -1 to both sides.	$2x^2 + 5x = -1$	$ax^2 + bx = -c$	Add $-c$ to both sides.
Divide each term by 2.	$x^2 + \frac{5}{2}x = \frac{-1}{2}$	$x^2 + \frac{b}{a}x = -\frac{c}{a}$	Divide each term by a.
Add $\left(\frac{1}{2}\cdot\frac{5}{2}\right)^2$ to both sides.	$x^2 + \frac{5}{2}x + \left(\frac{5}{4}\right)^2 = \left(\frac{5}{4}\right)^2 - \frac{1}{2}$	$x^2 + \frac{b}{a}x + \left(\frac{b}{2a}\right)^2 = \left(\frac{b}{2a}\right)^2 - \frac{c}{a}$	Add the square of one-half the coefficient of x, that is, $\left(\frac{1}{2}\cdot\frac{b}{a}\right)^2$ to both sides.

Factor the left-hand side; simplify the right-hand side.	$\left(x + \dfrac{5}{4}\right)^2 = \dfrac{25}{16} - \dfrac{1}{2}$	$\left(x + \dfrac{b}{2a}\right)^2 = \dfrac{b^2}{4a^2} - \dfrac{c}{a}$	Factor the left-hand side, simplify the right-hand side. Since the LCD on the right is $4a^2$; write
Write $\dfrac{1}{2}$ as $\dfrac{8}{16}$.	$\left(x + \dfrac{5}{4}\right)^2 = \dfrac{25}{16} - \dfrac{8}{16}$	$\left(x + \dfrac{b}{2a}\right)^2 = \dfrac{b^2}{4a^2} - \dfrac{4ac}{4a^2}$	$-\dfrac{c}{a}$ as $-\dfrac{4ac}{4a^2}$
Write as one fraction.	$\left(x + \dfrac{5}{4}\right)^2 = \dfrac{25 - 8}{16}$	$\left(x + \dfrac{b}{2a}\right)^2 = \dfrac{b^2 - 4ac}{4a^2}$	Then combine $\dfrac{b^2}{4a^2}$ and $-\dfrac{4ac}{4a^2}$.
Take the square root of both sides.	$x + \dfrac{5}{4} = \dfrac{\pm\sqrt{25 - 8}}{4}$	$x + \dfrac{b}{2a} = \dfrac{\pm\sqrt{b^2 - 4ac}}{2a}$	Take the square root of both sides.
Add $-\dfrac{5}{4}$ to both sides.	$x = -\dfrac{5}{4} \pm \dfrac{\sqrt{17}}{4}$	$x = -\dfrac{b}{2a} \pm \dfrac{\sqrt{b^2 - 4ac}}{2a}$	Add $-\dfrac{b}{2a}$.
Write as one fraction.	$x = \dfrac{-5 \pm \sqrt{17}}{4}$	$x = \dfrac{-b \pm \sqrt{b^2 - 4ac}}{2a}$	Write as one fraction.

Remember, we only have to do this once. Now we have the **quadratic formula** to use!

A   Solving Equations Using the Quadratic Formula

Teaching Tip

Before using the quadratic formula, do several examples of writing the equations in standard form and identifying a, b, and c.

If an equation in one variable has 2 as the highest exponent for any variable term in the equation, it is considered a quadratic or second degree equation. The standard form for the quadratic equation is $ax^2 + bx + c = 0$. If it is in standard form, we can find the solutions by substituting the values for a, b, and c in the quadratic formula.

| **Solutions of a Quadratic Equation in Standard Form** | The solutions of $ax^2 + bx + c = 0$ are $$x = \frac{-b \pm \sqrt{b^2 - 4ac}}{2a} \qquad a \neq 0$$ |

Calculate It   The Two Points on the Graph of the Quadratic Equation That Are the Solutions

So far, we've used calculators to verify our work. They are even more efficient in predicting the *type* of solutions we will get when we solve quadratic equations. If you graph the equations in Examples 1–4, you will notice that the graphs cross the horizontal axis at two points and the equations have two solutions.

The calculator gives decimal approximations for these two points, and they should have the same value as the solutions.

EXAMPLE 1   **Using the quadratic formula to solve equations**

Solve $8x^2 + 7x + 1 = 0$ by using the quadratic formula.

SOLUTION   The equation is second degree and is written in standard form:

$$\overbrace{8x^2}^{a=8} + \overbrace{7x}^{b=7} + \overbrace{1}^{c=1} = 0$$

It is clear that $a = 8$, $b = 7$, and $c = 1$. Substituting the values of a, b, and c in the formula, we obtain

$$x = \frac{-7 \pm \sqrt{(7)^2 - 4(8)(1)}}{2(8) \leftarrow a}$$ Let $a = 8$, $b = 7$, and $c = 1$.

$$= \frac{-7 \pm \sqrt{49 - 32}}{16}$$ Since $(7)^2 = 49$ and $-4(8)(1) = -32$

PROBLEM 1

Solve $3x^2 + 2x - 5 = 0$ by using the quadratic formula.

Answer

1. $-\dfrac{5}{3}$, 1

$$= \frac{-7 \pm \sqrt{17}}{16} \qquad\qquad \sqrt{49 - 32} = \sqrt{17}$$

Thus

$$x = \frac{-7 + \sqrt{17}}{16} \quad \text{or} \quad x = \frac{-7 - \sqrt{17}}{16}$$

The solution set is

$$\left\{ \frac{-7 + \sqrt{17}}{16}, \frac{-7 - \sqrt{17}}{16} \right\}$$

Calculate It   Solving a Quadratic Equation by Finding the *x*-Intercepts

To solve Example 1, we graph $Y_1 = 8x^2 + 7x + 1$ using a $[-3, 3]$ by $[-3, 3]$ window.

To find where Y_1 intersects the horizontal axis, we can enter $Y_2 = 0$ (which is the horizontal axis) and use `2nd` `TRACE` `5`, the intersect feature of your calculator, and `ENTER` `ENTER` `ENTER`. You will get one solution -0.1798059. This will confirm that

$$-0.1798059 \approx \frac{-7 + \sqrt{17}}{6}$$

is indeed one of the solutions. The other solution is given as

$$-0.6951941 \approx \frac{-7 - \sqrt{17}}{16}$$

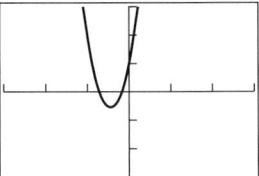

and can be found by pressing `TRACE` and using the left arrow key to move the cursor to the 2nd point of intersection. Once there, again use `2nd` `TRACE` `5` `ENTER` `ENTER` `ENTER`.

EXAMPLE 2   More practice using the quadratic formula

Solve $2x^2 = 2x + 1$ by using the quadratic formula.

SOLUTION   Since it is a second degree equation, we proceed by steps as before. We write the equation in standard form by subtracting $2x$ and 1 from both sides of $2x^2 = 2x + 1$ to obtain

$$\overbrace{2x^2}^{a=2} - \overbrace{2x}^{b=-2} - \overbrace{1}^{c=-1} = 0 \qquad \text{In standard form}$$

where $a = 2$, $b = -2$, and $c = -1$. Substituting these values in the quadratic formula, we have

$$x = \frac{-(-2) \pm \sqrt{(-2)^2 - 4(2)(-1)}}{2(2) \leftarrow a} \qquad \text{Let } a = 2, b = -2, \text{ and } c = -1.$$

$$= \frac{2 \pm \sqrt{4 + 8}}{4} \qquad \text{Since } (-2)^2 = 4 \text{ and } -4(2)(-1) = 8$$

$$= \frac{2 \pm \sqrt{12}}{4} \qquad \sqrt{4 + 8} = \sqrt{12}$$

$$= \frac{2 \pm \sqrt{4 \cdot 3}}{4} \qquad \sqrt{12} = \sqrt{4 \cdot 3}$$

PROBLEM 2

Solve $x^2 = 4x + 4$ by using the quadratic formula.

Teaching Tip

For Example 2, have the students use the quadratic formula without putting it in standard form and compare their results.

Example:

$$2x^2 = 2x + 1$$

using $a = 2$, $b = 2$, $c = 1$

Answer

2. $2 + 2\sqrt{2}, 2 - 2\sqrt{2}$

$$= \frac{2 \pm 2\sqrt{3}}{4}$$ $\sqrt{4 \cdot 3} = \sqrt{4} \cdot \sqrt{3} = 2\sqrt{3}$

$$= \frac{\overset{1}{2}(1 \pm \sqrt{3})}{\underset{2}{4}}$$ Reduce.

Thus

$$x = \frac{1 + \sqrt{3}}{2} \quad \text{or} \quad x = \frac{1 - \sqrt{3}}{2}$$

The solution set is

$$\left\{ \frac{1 + \sqrt{3}}{2}, \frac{1 - \sqrt{3}}{2} \right\}$$

EXAMPLE 3   **Using the quadratic formula when there is no constant term**

Solve $9x = x^2$ by using the quadratic formula.

SOLUTION   Subtracting $9x$ from both sides of $9x = x^2$, we have

$$0 = x^2 - 9x$$

or

$$\underbrace{x^2}_{a = 1} - \underbrace{9x}_{b = -9} + \underbrace{0}_{c = 0} = 0 \qquad \text{In standard form}$$

where $a = 1$, $b = -9$, and $c = 0$ (because the c term is missing). Substituting these values in the quadratic formula, we obtain

$$x = \frac{-(\overset{b}{-9}) \pm \sqrt{(\overset{b}{-9})^2 - 4(\overset{a}{1})(\overset{c}{0})}}{2(1) \leftarrow a}$$ Let $a = 1$, $b = -9$, and $c = 0$.

$$= \frac{9 \pm \sqrt{81 - 0}}{2}$$ Since $(-9)^2 = 81$ and $-4(1)(0) = 0$

$$= \frac{9 \pm \sqrt{81}}{2}$$ $\sqrt{81 - 0} = \sqrt{81}$

$$= \frac{9 \pm 9}{2}$$ $\pm\sqrt{81} = \pm 9$

Thus

$$x = \frac{9 + 9}{2} = \frac{18}{2} = 9 \quad \text{or} \quad x = \frac{9 - 9}{2} = \frac{0}{2} = 0$$

The solutions are 9 and 0, and the solution set is $\{9, 0\}$.

PROBLEM 3

Solve $x^2 = 15x$ by using the quadratic formula.

Teaching Tip

Solve $9x = x^2$ by dividing both sides by x, resulting in $x = 9$. Ask students to discuss why this does not lead to the complete solution. Have them make a generalization about what dividing by the variable will do to the solution set.

Web It

For more examples and practice problems on how to solve quadratic equations by using the quadratic formula go to link 8-2-1 at mhhe.com/bello.

Answer

3. 0, 15

NOTE

The expression $x^2 - 9x = x(x - 9) = 0$ could have been solved by factoring. Try factoring the equation *before* you use the quadratic formula.

EXAMPLE 4 **Clearing fractions before using the quadratic formula**

Solve $\dfrac{x^2}{4} + \dfrac{2}{3}x = -\dfrac{1}{3}$ by using the quadratic formula.

SOLUTION We have to write the equation in standard form, but first we clear fractions by multiplying each term by the LCM of 4 and 3, namely 12:

$$\overset{3}{\cancel{12}} \cdot \dfrac{x^2}{4} + \overset{4}{\cancel{12}} \cdot \dfrac{2}{3}x = -\dfrac{1}{\underset{1}{\cancel{3}}} \cdot \overset{4}{\cancel{12}}$$

$$3x^2 + 8x = -4$$

We then add 4 to obtain the equivalent equation

$$\underset{a=3}{\underbrace{3x^2}} + \underset{b=8}{\underbrace{8x}} + \underset{c=4}{\underbrace{4}} = 0 \qquad \text{In standard form}$$

where $a = 3$, $b = 8$, and $c = 4$. Substituting in the quadratic formula

$$x = \dfrac{-8 \pm \sqrt{64 - 4(3)(4)}}{2(3)} \qquad \text{Let } a = 3, b = 8, \text{ and } c = 4.$$

$$= \dfrac{-8 \pm \sqrt{64 - 48}}{6} \qquad \text{Since } -4(3)(4) = -48$$

$$= \dfrac{-8 \pm \sqrt{16}}{6} \qquad \sqrt{64 - 48} = \sqrt{16}$$

$$= \dfrac{-8 \pm 4}{6} \qquad \pm\sqrt{16} = \pm 4$$

Thus

$$x = \dfrac{-8 + 4}{6} = \dfrac{-4}{6} = -\dfrac{2}{3} \quad \text{or} \quad x = \dfrac{-8 - 4}{6} = \dfrac{-12}{6} = -2$$

The solutions are $-\dfrac{2}{3}$ and -2, and the solution set is $\{-\dfrac{2}{3}, -2\}$.

Some quadratic equations have non-real complex number solutions. Such solutions can be obtained by using the quadratic formula, as shown in Example 5.

EXAMPLE 5 **Solving quadratic equations with non-real complex number solutions**

Solve: $3x^2 + 3x = -2$

SOLUTION We add 2 to both sides of $3x^2 + 3x = -2$ so the equation is in standard form. We then have

$$\underset{a=3}{\underbrace{3x^2}} + \underset{b=3}{\underbrace{3x}} + \underset{c=2}{\underbrace{2}} = 0$$

where $a = 3$, $b = 3$, and $c = 2$. Now we have

$$x = \dfrac{-3 \pm \sqrt{(3)^2 - 4(3)(2)}}{2(3)}$$

$$= \dfrac{-3 \pm \sqrt{9 - 24}}{6}$$

PROBLEM 4

Solve $\dfrac{x^2}{4} - \dfrac{3}{8}x = \dfrac{1}{4}$ by using the quadratic formula.

Teaching Tip

$3x^2 + 8x + 4 = 0$ may also be solved by factoring:

$$3x^2 + 8x + 4 = 0$$
$$(3x + 2)(x + 2) = 0$$

and so on.

Teaching Tip

Before Example 5, do a few other examples like the following:

1. $x^2 = -16$
   $$x = \pm\sqrt{-16} = \pm 4i$$
2. $x^2 = -12$
   $$x = \pm\sqrt{-12} = \pm 2i\sqrt{3}$$
3. $(x + 2)^2 = -9$
   $$x + 2 = \pm 3i$$
   $$x = -2 \pm 3i$$

PROBLEM 5

Solve:

$$3x^2 + 2x = -1$$

Answers

4. $-\dfrac{1}{2}, 2$

5. $-\dfrac{1}{3} + \dfrac{\sqrt{2}}{3}i, -\dfrac{1}{3} - \dfrac{\sqrt{2}}{3}i$

$$= \frac{-3 \pm \sqrt{-15}}{6}$$

$$= \frac{-3 \pm i\sqrt{15}}{6} \qquad \sqrt{-15} = i\sqrt{15}$$

The solutions are thus

$$\frac{-3 + i\sqrt{15}}{6} \quad \text{and} \quad \frac{-3 - i\sqrt{15}}{6}$$

or

$$-\frac{1}{2} + \frac{\sqrt{15}}{6}i \quad \text{and} \quad -\frac{1}{2} - \frac{\sqrt{15}}{6}i$$

and the solution set is

$$\left\{ -\frac{1}{2} + \frac{\sqrt{15}}{6}i, \; -\frac{1}{2} - \frac{\sqrt{15}}{6}i \right\}$$

Calculate It   Recognizing Complex Number Solutions from the Graph of a Quadratic Equation

What happens if you try to solve $3x^2 + 3x = -2$ in Example 5? First, write the equation as $3x^2 + 3x + 2 = 0$ and graph $Y_1 = 3x^2 + 3x + 2$ using a $[-3, 3]$ by $[-3, 3]$ window. The curve does not cut or touch the horizontal axis, as shown in the window.

This means that there are no real-number solutions to this equation. The solutions are *complex* numbers, and you have to find them algebraically. (Remember, even with the best calculator available, you have to know your algebra!)

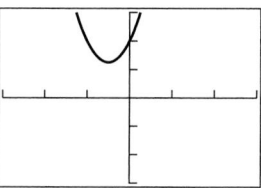

B   Solving Factorable Cubic Equations

Now let's consider $x^3 - 27 = 0$, which is *not* a quadratic equation. We will solve it by factoring (using the OFF method). First, recall the procedure for factoring sums and differences of cubes.

$$x^3 + a^3 = (x + a)(x^2 - ax + a^2)$$

$$x^3 - a^3 = (x - a)(x^2 + ax + a^2)$$

Here's how we do it.

$$x^3 - 27 = 0 \qquad \text{0   Given (equation is set equal to 0).}$$

$$(x - 3)(x^2 + 3x + 9) = 0 \qquad \text{F   Factor.}$$

$$x - 3 = 0 \quad \text{or} \quad x^2 + 3x + 9 = 0 \qquad \text{F   Factors = 0 (zero-factor property).}$$

$$x = 3 \quad \text{or} \quad x^2 + 3x + 9 = 0$$

Since the second equation does not factor, we use the quadratic formula to solve it. For this equation, $a = 1$, $b = 3$, and $c = 9$. Substituting in the quadratic formula, we have

$$x = \frac{-3 \pm \sqrt{3^2 - 4(1)(9)}}{2 \cdot 1}$$

$$= \frac{-3 \pm \sqrt{9 - 36}}{2}$$

$$= \frac{-3 \pm \sqrt{-27}}{2}$$

$$= \frac{-3 \pm \sqrt{9 \cdot 3 \cdot (-1)}}{2}$$

$$= \frac{-3 \pm 3i\sqrt{3}}{2}$$

The solutions of $x^3 - 27 = 0$ are thus

$$3, \quad \frac{-3 + 3i\sqrt{3}}{2}, \quad \text{and} \quad \frac{-3 - 3i\sqrt{3}}{2}$$

and the solution set is

$$\left\{ 3, \frac{-3}{2} + \frac{3\sqrt{3}}{2}i, \frac{-3}{2} - \frac{3\sqrt{3}}{2}i \right\}$$

$x^3 - 27 = 0$ *cannot* be completely solved by taking the cube roots—that is, by writing

$$x^3 - 27 = 0$$
$$x^3 = 27 \qquad \text{Add 27.}$$
$$x = \sqrt[3]{27} \qquad \text{Take cube roots of both sides.}$$
$$x = 3$$

As you can see, this method yields only one real-number solution when there are actually three solutions: one real-number and two non-real complex number solutions.

Web It

For another resource on solving quadratic equations using the quadratic formula and a visual display of what type and how many solutions you can obtain, go to link 8-2-2 at mhhe.com/bello.

EXAMPLE 6   **Solving a factorable cubic equation**

Solve: $8x^3 - 27 = 0$

SOLUTION   We factor the equation, use the zero-factor property, and then use the quadratic formula.

$$8x^3 - 27 = 0 \qquad \text{0   Given. Equation is set = 0.}$$

$$(2x - 3)(4x^2 + 6x + 9) = 0 \qquad \text{F   Factor.}$$

$$2x - 3 = 0 \quad \text{or} \quad 4x^2 + 6x + 9 = 0 \qquad \text{F   Factors = zero (zero-factor property).}$$

$$x = \frac{3}{2} \quad \text{or} \quad 4x^2 + 6x + 9 = 0$$

The second equation is a quadratic equation with $a = 4$, $b = 6$, and $c = 9$. We solve it with the quadratic formula:

$$x = \frac{-6 \pm \sqrt{6^2 - 4(4)(9)}}{2 \cdot 4}$$

$$= \frac{-6 \pm \sqrt{36 - 144}}{8}$$

PROBLEM 6

Solve:

$$64x^3 + 1 = 0$$

Answer

6. $\frac{1}{8} + \frac{\sqrt{3}}{8}i, \frac{1}{8} - \frac{\sqrt{3}}{8}i, -\frac{1}{4}$

$$= \frac{-6 \pm \sqrt{-108}}{8}$$

$$= \frac{-6 \pm \sqrt{36 \cdot (3) \cdot (-1)}}{8}$$

$$= \frac{-6 \pm 6i\sqrt{3}}{8}$$

$$= \frac{2(-3 \pm 3i\sqrt{3})}{2 \cdot 4}$$

$$= \frac{-3 \pm 3i\sqrt{3}}{4}$$

The solutions of $8x^3 - 27 = 0$ are

$$\frac{3}{2}, \quad \frac{-3 + 3i\sqrt{3}}{4} \quad \text{and} \quad \frac{-3 - 3i\sqrt{3}}{4}$$

and the solution set is

$$\left\{ \frac{3}{2}, -\frac{3}{4} + \frac{3\sqrt{3}}{4}i, -\frac{3}{4} - \frac{3\sqrt{3}}{4}i \right\}$$

> **Teaching Tip**
>
> After solving Example 6, have students discuss the number of solutions obtained in comparison to the number of solutions for a quadratic or linear equation. Have them make some conjectures about the number of solutions and the degree of the equation.

Calculate It   Finding the Real Solutions for a Cubic Equation

Try to find the solutions of $8x^3 - 27 = 0$. The graph is shown in the window and, as before, only one solution, $x = 1.5 = \frac{3}{2}$, is a real number.

    You have to use the quadratic equation to get the other two solutions! (By the way, if you want to see a more complete graph, try a $[-2, 2]$ by $[-50, 50]$ window.)

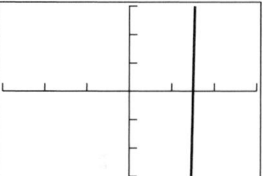

By the way, we left Dr. Goddard's rocket up in the air! How long *did* it fly to reach the height of 41 ft? The equation was

$$-16t^2 + 51.225t = 41$$

or, in standard form,

$$-16t^2 + 51.225t - 41 = 0$$

Here $a = -16$, $b = 51.225$, and $c = -41$, so the quadratic formula gives

$$t = \frac{-51.225 \pm \sqrt{51.225^2 - 4(-16)(-41)}}{2 \cdot (-16)}$$

With a calculator, we get $t \approx 1.6$. Thus it took the rocket about 1.6 sec to reach 41 ft.

Solving Applications Involving Quadratic Equations

Quadratic equations are used in many fields: rockets, economics, and marketing, to name a few. Here are some examples.

EXAMPLE 7   What's the price?

In business, when the price p (in dollars) of a product increases, the demand d decreases and is given by $d = 300/p$. On the other hand, when the price p increases, the supply s producers are willing to sell increases and is given by $s = 100p - 50$. In economic theory, the point at which the supply equals the demand, $s = d$, is called the **equilibrium point.** Find the price p at the equilibrium point.

SOLUTION   Since $s = d$ at equilibrium, we have

$$100p - 50 = \frac{300}{p}$$

$$p(100p - 50) = p \cdot \frac{300}{p} \qquad \text{Multiply by } p.$$

$$100p^2 - 50p = 300$$

$$2p^2 - p = 6 \qquad \text{Divide by 50.}$$

$$2p^2 - p - 6 = 0 \qquad \text{Subtract 6.}$$

Since $a = 2$, $b = -1$, and $c = 6$, we have

$$p = \frac{-(-1) \pm \sqrt{(-1)^2 - 4(2)(-6)}}{2 \cdot 2}$$

$$= \frac{1 \pm \sqrt{1 + 48}}{4}$$

$$= \frac{1 \pm \sqrt{49}}{4}$$

$$= \frac{1 \pm 7}{4}$$

Thus

$$p = \frac{1 + 7}{4} = 2 \quad \text{or} \quad p = \frac{1 - 7}{4} = -\frac{3}{2}$$

Since the price must be positive, we use $p = \$2$. In this case, we obtain two rational roots, which means that the original equation was factorable. Before you use the quadratic formula, always try to factor. You will often save time!

EXAMPLE 8   Where's the beef?

According to the U.S. Department of Agriculture, per capita consumption of beef declined from 1985 to 1992, while poultry consumption rose in the same period. If $B(t)$ and $P(t)$ represent per capita consumption of beef and poultry (in pounds), respectively, and t represents the number of years after 1985, beef and poultry consumption can be described by these equations.

Beef: $B(t) = 0.2t^2 - 3.23t + 75$

Poultry: $P(t) = 2.12t + 45$

a. Was the consumption of beef and poultry ever the same?

b. In what year would this happen?

PROBLEM 7

Find the price p at the equilibrium point if the demand and supply are given by

$$d = \frac{50}{p} \quad \text{and} \quad s = 100p - 50$$

Calculate It Exercises

1. Solve Example 7 by finding the point of intersection for

$$Y_1 = 100x - 50$$

and

$$Y_2 = \frac{300}{x}$$

2. Verify the results of Example 8 using a graphical method. [Use the intersect feature to find the intersection of $B(t)$ and $P(t)$.]

PROBLEM 8

If the supply equation for manufacturing a certain item is $s = p^2 - 8$ and the demand equation is $d = -p + 4$, find the equilibrium point.

Answer

7. $p = \$1$ **8.** $p = 3$

SOLUTION

a. To find out whether the consumption of beef and poultry would ever be the same, we have to determine whether $B(t) = P(t)$ has a real-number solution. So we assume $B(t) = P(t)$, or

$$0.2t^2 - 3.23t + 75 = 2.12t + 45$$

$$0.2t^2 - 5.35t + 30 = 0 \qquad \text{Subtract } 2.12t \text{ and } 45.$$

Here $a = 0.2$, $b = -5.35$, and $c = 30$. Substituting in the quadratic formula, we have

$$t = \frac{-(-5.35) \pm \sqrt{(-5.35)^2 - 4 \cdot (0.2)(30)}}{2 \cdot (0.2)}$$

$$= \frac{5.35 \pm \sqrt{28.6225 - 24}}{0.4}$$

$$= \frac{5.35 \pm \sqrt{4.6225}}{0.4}$$

$$= \frac{5.35 \pm 2.15}{0.4}$$

Thus

$$t = \frac{7.5}{0.4} = 18.75 \quad \text{or} \quad t = \frac{3.2}{0.4} = 8$$

which means that the consumption of beef and poultry would be the same for the specified values of t.

b. The consumption of beef and poultry would have been the same $t = 18.75$ years from 1985, that is, in 2003.75 (2004) or $t = 8$ yr from 1985, in 1993.

Exercises 8.2

A In Problems 1–30, use the quadratic formula to solve the equation.

1. $x^2 + x - 2 = 0$ $1, -2$

2. $x^2 + 4x - 1 = 0$ $-2 \pm \sqrt{5}$

3. $x^2 + 4x = -1$ $-2 \pm \sqrt{3}$

4. $x^2 + 6x = -5$ $-1, -5$

5. $x^2 - 3x = 2$ $\frac{3 \pm \sqrt{17}}{2}$

6. $x^2 - 4x = 12$ $-2, 6$

7. $7y^2 = 12y - 5$ $\frac{5}{7}, 1$

8. $7x^2 = 6x - 1$ $\frac{3 \pm \sqrt{2}}{7}$

9. $5y^2 + 8y = -5$ $-\frac{4}{5} \pm \frac{3}{5}i$

10. $5y^2 + 6y = -5$ $-\frac{3}{5} \pm \frac{4}{5}i$

11. $7y + 6 = -2y^2$ $-\frac{3}{2}, -2$

12. $7y + 3 = -2y^2$ $-\frac{1}{2}, -3$

13. $\frac{x^2}{5} - \frac{x}{2} = -\frac{3}{10}$ $\frac{3}{2}, 1$

14. $\frac{x^2}{4} - \frac{x}{2} = -\frac{1}{8}$ $\frac{2 \pm \sqrt{2}}{2}$

15. $\frac{x^2}{7} + \frac{x}{2} = -\frac{3}{14}$ $-\frac{1}{2}, -3$

16. $\frac{x^2}{8} + \frac{x}{2} = -\frac{1}{8}$ $-2 \pm \sqrt{3}$

17. $\frac{x^2}{2} - \frac{3x}{4} = -\frac{1}{8}$ $\frac{3 \pm \sqrt{5}}{4}$

18. $\frac{x^2}{10} - \frac{x}{5} = \frac{3}{2}$ $5, -3$

19. $\frac{x^2}{8} = -\frac{x}{4} - \frac{1}{8}$ -1

20. $\frac{x^2}{12} = -\frac{x}{4} - \frac{1}{3}$ $-\frac{3}{2} \pm \frac{\sqrt{7}}{2}i$

21. $6x = 4x^2 + 1$ $\frac{3 \pm \sqrt{5}}{4}$

22. $6x = 9x^2 - 4$ $\frac{1 \pm \sqrt{5}}{3}$

23. $3x = 1 - 3x^2$ $\frac{-3 \pm \sqrt{21}}{6}$

24. $3x = 2x^2 - 5$ $\frac{5}{2}, -1$

25. $x(x + 2) = 2x(x + 1) - 4$ ± 2

26. $x(4x - 7) - 10 = 6x^2 - 7x$ $\pm i\sqrt{5}$ **27.** $6x(x + 5) = (x + 15)^2$ $\pm 3\sqrt{5}$ **28.** $6x(x + 1) = (x + 3)^2$ $\pm\frac{3\sqrt{5}}{5}$

29. $(x - 2)^2 = 4x(x - 1)$ $\pm\frac{2\sqrt{3}}{3}$ **30.** $(x - 4)^2 = 4x(x - 2)$ $\pm\frac{4\sqrt{3}}{3}$

B In Problems 31–36, solve the equation.

31. $x^3 - 8 = 0$ $2, -1 \pm i\sqrt{3}$ **32.** $x^3 - 1 = 0$ $1, -\frac{1}{2} \pm \frac{\sqrt{3}}{2}i$ **33.** $8x^3 - 1 = 0$ $\frac{1}{2}, -\frac{1}{4} \pm \frac{\sqrt{3}}{4}i$

34. $27x^3 - 1 = 0$ $\frac{1}{3}, -\frac{1}{6} \pm \frac{\sqrt{3}}{6}i$ **35.** $x^3 + 27 = 0$ $-3, \frac{3}{2} \pm \frac{3\sqrt{3}}{2}i$ **36.** $x^3 + 64 = 0$ $-4, 2 \pm 2i\sqrt{3}$

APPLICATIONS

Milk Consumption, Price, Bending Moment

37. According to the U.S. Department of Agriculture, annual per capita consumption of whole milk declined from 1970 to 1989 while that of lowfat milk increased in the same period. If $W(t)$ and $L(t)$ represent the annual per capita consumption of whole milk and lowfat milk (in gallons), respectively, and t represents the number of years after 1970, milk consumption can be described by the following equations:

Whole: ↓ $W(t) = 0.013t^2 - 0.96t + 25.4$

Lowfat: ↑ $L(t) = 0.4t + 6.03$

a. Can the consumption of whole and lowfat milk ever be the same? Yes

b. In what year could this happen? 1987

38. Find the price p (dollars) at the equilibrium point if the supply is $s = 30p - 50$ and the demand is $d = \frac{20}{p}$. $2

39. Find the price p (dollars) at the equilibrium point if the supply is $s = 30p - 50$ and the demand is $d = \frac{10}{p}$. $1.85

40. Find the price p (dollars) at the equilibrium point if the supply is $s = 20p - 60$ and the demand is $d = \frac{30}{p}$. $3.44

41. The bending moment M of a simple beam is given by $M = 20x - x^2$. For what values of x is $M = 40$? $10 \pm 2\sqrt{15}$

42. Use the formula given in Problem 41 to find the values of x for which $M = 60$. $10 \pm 2\sqrt{10}$

The maximum safe length L for which a beam will support a load d is given by $aL^2 + bL + c = d$, where a, b, c, and d depend on the materials and structures used. In Problems 43 and 44, find:

43. L when $a = 400$, $b = 200$, $c = 200$, and $d = 800$ 1   **44.** L when $a = 5$, $b = 0$, $c = 100$, and $d = 180$ 4

SKILL CHECKER

Try the "Skill Checker" exercises so you'll be ready for the next section.

Simplify $\sqrt{b^2 - 4ac}$ using:

45. $a = 3$, $b = -2$, $c = -1$ 4   **46.** $a = 2$, $b = 3$, $c = -1$ $\sqrt{17}$

47. $a = 3$, $b = -5$, $c = 4$ $i\sqrt{23}$ **48.** $a = 3$, $b = -1$, $c = 1$ $i\sqrt{11}$

Find the product:

49. $(2x + 1)(3x - 4)$ $6x^2 - 5x - 4$ **50.** $(3x + 1)(2x - 5)$ $6x^2 - 13x - 5$

51. $(3x - 7)(4x + 3)$ $12x^2 - 19x - 21$ **52.** $(4x - 8)(3x + 5)$ $12x^2 - 4x - 40$

USING YOUR KNOWLEDGE

A Different Way of Finding the Quadratic Formula

In this section we derived the quadratic formula by completing the square. The procedure depends on making the x^2 coefficient 1. But there is another way to derive the quadratic formula. See if you can state the reasons for each of the steps, given consecutively in Problems 53–59. ($ax^2 + bx + c = 0$ is given.)

53. $4a^2x^2 + 4abx + 4ac = 0$ Multiply both sides by $4a$.   **54.** $4a^2x^2 + 4abx = -4ac$ Add $-4ac$ to both sides.

55. $4a^2x^2 + 4abx + b^2 = b^2 - 4ac$ Add b^2 to both sides.   **56.** $(2ax + b)^2 = b^2 - 4ac$ Factor the left-hand side.

57. $2ax + b = \pm\sqrt{b^2 - 4ac}$ Take square root of both sides.   **58.** $2ax = -b \pm\sqrt{b^2 - 4ac}$ Add $-b$ to both sides.

59. $x = \dfrac{-b \pm \sqrt{b^2 - 4ac}}{2a}$ Divide both sides by $2a$.

CALCULATOR CORNER

Using a Calculator with the Quadratic Formula

Any non-graphing calculator that has store (STO▸) and re-call (RCL) keys can be extremely helpful in finding the solutions of a quadratic equation with the quadratic formula. The solutions you obtain are being approximated by decimals. It's especially convenient to start with the radical part in the solution of the quadratic equation and then store this value so you can evaluate both solutions without having to back-track or write down any intermediate steps. Let's look at the equation of Example 1:

$$8x^2 + 7x + 1 = 0$$

Using the quadratic formula, one solution is obtained by entering

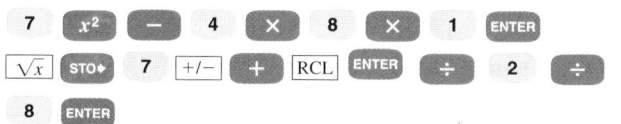

The display shows -0.1798059 (given as

$$\frac{-7 + \sqrt{17}}{16}$$

in the example). To obtain the other solution, enter

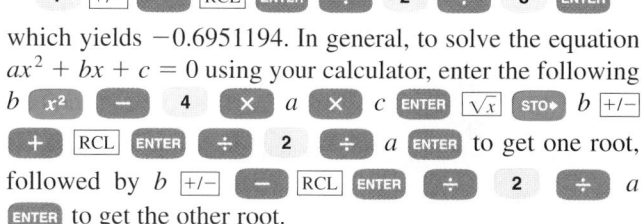

which yields -0.6951194. In general, to solve the equation $ax^2 + bx + c = 0$ using your calculator, enter the following

b [x²] [−] 4 [×] a [×] c [ENTER] [√x] [STO▸] b [+/−] [+] [RCL] [ENTER] [÷] 2 [÷] a [ENTER] to get one root, followed by b [+/−] [−] [RCL] [ENTER] [÷] 2 [÷] a [ENTER] to get the other root.

If $b^2 - 4ac < 0$, the calculator will give you an error message when you press [√x]. In such cases, you will have to change the sign before pressing [√x] and calculate the real and imaginary parts separately. (Try this in Example 6.)

WRITE ON

60. Why do we have the restriction $a \neq 0$ when solving the equation $ax^2 + bx + c = 0$? Answers may vary.

61. Explain the difference between linear, quadratic, and cubic equations. Answers may vary.

62. The term $\sqrt{a}$ is a real number only when a is nonnegative. Use this to write a procedure that enables you to determine whether a quadratic equation with real coefficients has real-number solutions or non-real complex number solutions. Answers may vary.

MASTERY TEST

If you know how to do these problems, you have learned your lesson!

Solve:

63. $27x^3 - 8 = 0$ $\frac{2}{3}, -\frac{1}{3} \pm \frac{\sqrt{3}}{3}i$ **64.** $64x^3 - 1 = 0$ $\frac{1}{4}, -\frac{1}{8} \pm \frac{\sqrt{3}}{8}i$ **65.** $3x^2 + 2x = -1$ $-\frac{1}{3} \pm \frac{\sqrt{2}}{3}i$

66. $2x^2 + 3x = -2$ $-\frac{3}{4} \pm \frac{\sqrt{7}}{4}i$ **67.** $\dfrac{x^2}{4} - \dfrac{3}{8}x = \dfrac{1}{4}$ $-\frac{1}{2}, 2$ **68.** $\dfrac{x^2}{2} - \dfrac{1}{4}x = \dfrac{3}{2}$ $-\frac{3}{2}, 2$

69. $6x = x^2$ $0, 6$ **70.** $7x - x^2 = 0$ $0, 7$ **71.** $x^2 = 4x + 4$ $2 \pm 2\sqrt{2}$

72. $x^2 = 2 + 2x$ $1 \pm \sqrt{3}$ **73.** $3x^2 + 2x - 5 = 0$ $-\frac{5}{3}, 1$ **74.** $2x^2 - 3x = 4$ $\frac{3 \pm \sqrt{41}}{4}$

75. The supply s of a certain product that producers are willing to sell is given by $s = 100p - 50$, where p is the price in dollars. If the demand d for this product is given by $d = \frac{50}{p}$, find the price p when $s = d$, the equilibrium point. $\$1$

THE DISCRIMINANT AND ITS APPLICATIONS

To Succeed, Review How To . . .

1. Evaluate and simplify expressions that contain radicals (pp. 530–534).

2. Multiply two binomials (pp. 359–362).

Objectives

A Use the discriminant to determine the number and type of solutions of a quadratic equation.

B Use the discriminant to determine whether a quadratic expression is factorable and then factor it.

C Find a quadratic equation with specified solutions.

D Verify the solutions of a quadratic equation.

GETTING STARTED

Breaking Even Through Discriminants

A merchant in this mall wants to know if she is going to "break even." First, her daily costs, C, are represented by

$$C = 0.001x^2 + 10x + 100$$

where x is the number of items sold, and the corresponding revenue is $R = 20x - 0.01x^2$. The break-even point occurs when the cost C equals the revenue R—that is, when

$$C = R$$
$$0.001x^2 + 10x + 100 = 20x - 0.01x^2$$
$$0.011x^2 - 10x + 100 = 0 \qquad \text{In standard form}$$

She can break even only *if* this equation has real-number solutions. How do we ascertain that? By using the $b^2 - 4ac$ under the radical in the quadratic formula. Keep reading to see how.

The quadratic formula

$$x = \frac{-b \pm \sqrt{b^2 - 4ac}}{2a}$$

gives the solutions to any quadratic equation in standard form, so we can find out what type of solutions the equation has by looking at the expression under the radical, $b^2 - 4ac$. To do this, we need the following definition.

DISCRIMINANT

The expression $b^2 - 4ac$ under the radical is called the **discriminant** D; that is, $D = b^2 - 4ac$.

To find the discriminant of $0.011x^2 - 10x + 100 = 0$, $a = 0.011$, $b = -10$, and $c = 100$. Thus

$$b^2 - 4ac = (-10)^2 - 4(0.011)(100)$$
$$= 100 - 4.4$$
$$= 95.6$$

This means that the solutions of the equation are

$$\frac{-(-10) + \sqrt{95.6}}{2(0.011)} \quad \text{and} \quad \frac{-(-10) - \sqrt{95.6}}{2(0.011)}$$

There are two real-number positive solutions to the equation, and our merchant can break even.

Using the Discriminant to Classify Solutions

Calculate It

Determining Solutions from Graphs

Can you use your calculator to determine what type of solutions a quadratic equation will have *without* calculating the discriminant? Almost! If the graph of a quadratic expression crosses the horizontal axis at *two* points, you know that the corresponding quadratic equation has *two* solutions, but in general, you will *not* know if they are rational or irrational. For example, $x^2 - 4$ cuts the horizontal axis twice: $x^2 - 4 = 0$ has two rational solutions, 2 and -2. On the other hand, $x^2 - 3 = 0$ also crosses the horizontal axis twice: $x^2 - 3 = 0$ has two solutions, but they happen to be irrational.

If we calculate the discriminant of a quadratic equation, we can predict whether the solutions will be rational, irrational, or non-real complex numbers, as well as the number of solutions we will have. This is important in applied problems where irrational or non-real number solutions are not always acceptable. The different possibilities are listed in Table 1.

Table 1 Solutions to $ax^2 + bx + c = 0$ Based on the Discriminant

The discriminant of $ax^2 + bx + c = 0$ is $b^2 - 4ac$. If a, b, and c are real numbers ($a \neq 0$), the type and number of solutions are as follows:

Discriminant, $b^2 - 4ac$	Solutions, $\dfrac{-b \pm \sqrt{b^2 - 4ac}}{2a}$
Positive, not the square of an integer	*Two* different real, *irrational* solutions
Positive, and the square of an integer	*Two* different real, *rational* solutions
Negative	*Two* different *non-real complex* solutions
Zero	*One* real, *rational* solution

Here are some examples.

Equation	$b^2 - 4ac$	D	Solutions (Roots)
$4x^2 - 3x - 5 = 0$	$(-3)^2 - 4(4)(-5) = 89$	Positive, not perfect square	Two real, irrational numbers
$x^2 - 2x - 3 = 0$	$(-2)^2 - 4(1)(-3) = 16$	Positive, perfect square	Two real, rational numbers
$4x^2 - 3x + 5 = 0$	$(-3)^2 - 4(4)(5) = -71$	Negative	Two non-real complex numbers
$4x^2 - 4x + 1 = 0$	$(-4)^2 - 4(4)(1) = 0$	Zero	One real, rational number

---

EXAMPLE 1   Using the discriminant to classify the solution

Find and classify:

a. Given $3x^2 - 5x = 2$, find the value of the discriminant and classify the solutions.

b. Given $4x^2 - kx = -1$, find the discriminant and then find k so that the equation has exactly one real, rational solution.

SOLUTION

a. First write the equation in standard form by adding -2 to both sides to obtain $3x^2 - 5x - 2 = 0$. Thus $a = 3$, $b = -5$, $c = -2$, and

$$b^2 - 4ac = (-5)^2 - 4(3)(-2)$$
$$= 25 + 24$$
$$= 49$$

Since 49 is a positive perfect square, the equation will have two real, rational solutions.

PROBLEM 1

Find and classify:

a. Given $2x^2 = 6x - 7$, find the discriminant and classify the solutions.

b. Given $x^2 + kx = -9$, find the discriminant and then find k so that the equation has exactly one real, rational solution.

Answers

1. a. -20; 2 non-real complex solutions   **b.** $k^2 - 36$; $-6, 6$

b. We first write the equation in standard form by adding 1 to both sides to obtain $4x^2 - kx + 1 = 0$. Thus $a = 4$, $b = -k$, $c = 1$, and

$$b^2 - 4ac = (-k)^2 - 4(4)(1)$$
$$= k^2 - 16$$

For this equation to have one rational solution, the discriminant $k^2 - 16$ must be zero:

$$0 = k^2 - 16$$
$$16 = k^2$$
$$\pm\sqrt{16} = k$$
$$\pm 4 = k$$

When k is 4 or -4, the discriminant is zero, and the equation has one real, rational number as its solution.

Teaching Tip

Have students try solving $4x^2 - kx + 1 = 0$ with $k = 4$ and see if they get one rational number as its solution.

Web It

For an interactive tutorial on the discriminant and the quadratic formula, go to link 8-3-1 at mhhe.com/bello.

B Determining Whether a Quadratic Expression Is Factorable

We have now learned that if the discriminant D is a positive, perfect square, $ax^2 + bx + c = 0$ has two real, rational solutions, say r and s. Thus $a(x - r)(x - s) = 0$, which means that $ax^2 + bx + c$ is factorable into factors with *integer* coefficients. Here is our result.

> **Factorable Quadratic Expressions**
>
> If $b^2 - 4ac$ is a positive perfect square, then $ax^2 + bx + c$ is factorable using integer coefficients.

To find out whether $12x^2 + 20x - 25$ is factorable, we must find $b^2 - 4ac = (20)^2 - 4(12)(-25) = 1600$. Since 1600 is a positive perfect square [$(40)^2 = 1600$], $12x^2 + 20x - 25$ is factorable. (Contrast this technique with the ac test we learned in Section 5.4 or with trial and error.)

EXAMPLE 2 **Using the discriminant to determine whether the expression is factorable**

Use the discriminant to determine whether $20x^2 + 10x - 32$ is factorable.

SOLUTION Here $a = 20$, $b = 10$, and $c = -32$.

$$b^2 - 4ac = (10)^2 - 4(20)(-32)$$
$$= 100 + 2560$$
$$= 2660$$

Since 2660 is not a perfect square, $20x^2 + 10x - 32$ is not factorable.

PROBLEM 2

Use the discriminant to determine whether the following quadratic expression is factorable.

$$20x^2 + 10x - 30$$

Now that we know how to use the discriminant to determine whether a quadratic is factorable, we shall learn how to factor it. To factor $12x^2 + 20x - 25$, we first solve the corresponding quadratic equation $12x^2 + 20x - 25 = 0$. Recall that $b^2 - 4ac = 1600$; thus

$$x = \frac{-20 \pm \sqrt{1600}}{2(12)} = \frac{-20 \pm 40}{24}$$

The solutions are

Answer

2. Yes; $D = 2500 = (50)^2$

$$\frac{-20 + 40}{24} = \frac{5}{6} \quad \text{and} \quad \frac{-20 - 40}{24} = -\frac{5}{2}$$

We now *reverse* the steps for solving a quadratic equation by factoring.

	Solution 1 $= \dfrac{5}{6}$	Solution 2 $= -\dfrac{5}{2}$	
	$x = \dfrac{5}{6}$	$x = -\dfrac{5}{2}$	
Multiply by 6.	$6x = 5$	$2x = -5$	Multiply by 2.
Subtract 5.	$6x - 5 = 0$	$2x + 5 = 0$	Add 5.

That is, $(6x - 5)(2x + 5) = 0$. (You can check this by multiplying.) Thus the expression, $12x^2 + 20x - 25 = (6x - 5)(2x + 5)$.

EXAMPLE 3 Factoring a quadratic expression	**PROBLEM 3**
Factor if possible: $12x^2 + x - 35$	Factor if possible:

SOLUTION

$$b^2 - 4ac = (1)^2 - 4(12)(-35) = 1 + 1680 = 1681$$

Factor if possible:
$$21x^2 + x - 10$$

Since $1681 = 41^2$, the expression is factorable. We use the quadratic formula to solve the related equation $12x^2 + x - 35 = 0$.

$$x = \frac{-1 + \sqrt{1681}}{2(12)} = \frac{-1 \pm 41}{24}$$

$$x = \frac{5}{3} \quad \text{and} \quad x = -\frac{7}{4}$$

Now we reverse the steps for solving a quadratic equation by factoring.

$$x = \frac{5}{3} \quad \text{or} \qquad x = -\frac{7}{4}$$

Multiply by 3.	$3x = 5$ or	$4x = -7$	Multiply by 4.
Subtract 5.	$3x - 5 = 0$ or	$4x + 7 = 0$	Add 7.

Multiplying, $(3x - 5)(4x + 7) = 0$. Thus the expression, $12x^2 + x - 35 = (3x - 5)(4x + 7)$. You can check that the factorization is correct by multiplying the factors to obtain $12x^2 + x - 35$.

C   Finding Quadratic Equations with Specified Solutions

The process just discussed can be used to find a quadratic equation with specified solutions as shown in Example 4.

EXAMPLE 4 Testing, testing . . .	**PROBLEM 4**
A professor wants to create a test question involving a quadratic equation whose solutions are $\{\frac{4}{5}, -\frac{2}{3}\}$. What should the quadratic equation be?	Write the quadratic equation whose solutions are $\{-2, \frac{3}{4}\}$.

SOLUTION   In essence, you have to work backward. Since the solutions are $\frac{4}{5}$ and $-\frac{2}{3}$, the professor wants:

$$x = \frac{4}{5} \quad \text{or} \qquad x = -\frac{2}{3}$$

Multiply by 5.	$5x = 4$ or	$3x = -2$	Multiply by 3.
Subtract 4.	$5x - 4 = 0$ or	$3x + 2 = 0$	Add 2.

Answers

3. $(3x - 2)(7x + 5)$
4. $4x^2 + 5x - 6 = 0$

$$(5x - 4)(3x + 2) = 0 \quad \text{By the zero-factor property.}$$
$$15x^2 + 10x - 12x - 8 = 0 \quad \text{Multiply.}$$
$$15x^2 - 2x - 8 = 0 \quad \text{Simplify.}$$

Thus $15x^2 - 2x - 8 = 0$ is a quadratic equation whose solution set is $\{\frac{4}{5}, -\frac{2}{3}\}$.

D Verifying the Solutions of a Quadratic Equation

In the process of solving Example 3, we found out that the solutions of the equation $12x^2 + x - 35 = 0$ are $\frac{5}{3}$ and $-\frac{7}{4}$. To verify this, we can substitute these values in the original equation. But there is another way. If the equation is

$$ax^2 + bx + c = 0$$

then dividing by a, we can rewrite it as

$$x^2 + \frac{b}{a}x + \frac{c}{a} = 0 \qquad \textbf{(1)}$$

Web It

For another explanation and examples of writing the quadratic equation given the solutions and using the sum and product formulas for verifying quadratic equation roots, go to link 8-3-2 at mhhe.com/bello.

If the solutions are r_1 and r_2, we can also write

$$(x - r_1)(x - r_2) = 0$$
$$x^2 - r_1 x - r_2 x + r_1 r_2 = 0 \qquad \text{Multiply.}$$
$$x^2 - (r_1 + r_2)x + r_1 r_2 = 0 \qquad \text{Factor } [-r_1 x - r_2 x = -(r_1 + r_2)x]. \qquad \textbf{(2)}$$

Comparing equations (1) and (2), we see that

$$\frac{b}{a} = -(r_1 + r_2) \quad \text{and} \quad \frac{c}{a} = r_1 r_2$$

This discussion can be summarized as follows.

Sum and Product of the Solutions of a Quadratic Equation	If r_1 and r_2 are the solutions of the equation $ax^2 + bx + c = 0$, then $$r_1 + r_2 = -\frac{b}{a} \quad \text{and} \quad r_1 r_2 = \frac{c}{a}$$ That is, the sum of the solutions of a quadratic equation is $-\frac{b}{a}$, and the product of the solutions is $\frac{c}{a}$.

We can now verify that $\frac{5}{3}$ and $-\frac{7}{4}$ are solutions of $12x^2 + x - 35 = 0$. The sum of the solutions is

$$\frac{5}{3} + \left(-\frac{7}{4}\right) = \frac{20}{12} + \left(-\frac{21}{12}\right) = -\frac{1}{12} = -\frac{b}{a}$$

The product is

$$\frac{5}{3} \cdot \left(-\frac{7}{4}\right) = -\frac{35}{12} = \frac{c}{a}$$

So our results are correct.

EXAMPLE 5 **Using the sum and product properties to verify solutions**

Use the sum and product properties to see whether the solutions of $3x^2 + 5x - 2 = 0$ are

a. $-\frac{1}{3}$ and 2

b. $\frac{1}{3}$ and -2

PROBLEM 5

Use the sum and product properties to see whether the solutions of $4x^2 - 12x + 5 = 0$ are:

a. $\frac{1}{2}$ and $-\frac{5}{2}$ **b.** $\frac{1}{2}$ and $\frac{5}{2}$

Answers

5. a. No **b.** Yes

SOLUTION

a. In the equation $3x^2 + 5x - 2 = 0$, $a = 3$, $b = 5$, and $c = -2$. Using $-\frac{b}{a}$ for the sum of the solutions we get

$$-\frac{b}{a} = -\frac{5}{3}$$

Finding the sum of the proposed solutions we get

$$-\frac{1}{3} + 2 = -\frac{1}{3} + \frac{6}{3} = \frac{5}{3}$$

Since $-\frac{b}{a} = -\frac{5}{3}$ and $r_1 + r_2 = \frac{5}{3}$, $-\frac{1}{3}$ and 2 cannot be the solutions.

b. The sum of the proposed solutions is $\frac{1}{3} + (-\frac{6}{3}) = -\frac{5}{3}$, which also equals $-\frac{b}{a}$. The product of the solutions of $3x^2 + 5x - 2 = 0$ must be

$$\frac{c}{a} = -\frac{2}{3}$$

The product of the proposed solutions is $\frac{1}{3} \cdot (-2) = -\frac{2}{3}$. Since $\frac{c}{a}$ is the same as the product, $-\frac{2}{3}$, $\frac{1}{3}$ and -2 are the correct solutions.

Calculate It Exercises   Using the Graph to Determine the Number of Solutions to a Quadratic Equation

Here's how you use a calculator to determine the number of solutions of a quadratic equation:

Case 1. If the graph of a quadratic cuts the horizontal axis in *two* places, the corresponding quadratic equation has *two* solutions, but you *still* have to use the discriminant to determine if the solutions are rational or irrational.

Case 2. If the graph cuts the horizontal axis in *one* place, there is *one* solution, but we don't know if it is rational or irrational. [Try $x^2 - 4x + 4 = 0$ whose solution is $x = 2$, which is rational, and $(x - \sqrt{2})^2 = 0$ whose solution is $x = \sqrt{2}$, which is irrational.]

Case 3. If the graph does not touch the horizontal axis, there are no real-number solutions. The *two* resulting solutions are non-real complex numbers. (Try $x^2 + 4 = 0$.)

In Example 8 of the previous section, we were asked to determine whether the consumption of beef and poultry could ever be the same. The answer was obtained by solving the equation $B(t) = P(t)$. We could answer the question more quickly by looking at the discriminant of $B(t) - P(t) = 0$. If the discriminant $b^2 - 4ac > 0$, the equation has two solutions. Use this idea to solve the following problem. According to the U.S. Department of Agriculture, the consumption of citrus and noncitrus fruits (in pounds) t years after 1979 can be approximated by

$$\text{Citrus:} \quad C(x) = -0.013x^2 - 0.24x + 29$$
$$\text{Noncitrus:} \quad N(x) = x + 50$$

1. Use the discriminant of $C(x) - N(x)$ to find out whether citrus and noncitrus consumption can ever be the same.

2. Verify your result by graphing $C(x)$ and $N(x)$.

Exercises 8.3

A In Problems 1–10, find the discriminant and determine the number and type of solutions.

1. $3x^2 + 5x - 2 = 0$ $D = 49$;
two rational numbers

2. $3x^2 - 2x + 5 = 0$ $D = -56$;
two non-real complex numbers

3. $4x^2 = 4x - 1$ $D = 0$;
one rational number

4. $x^2 - 10x = -25$ $D = 0$;
one rational number

5. $2x^2 = 2x + 5$ $D = 44$;
two irrational numbers

6. $x^2 - 5x = 5$ $D = 45$;
two irrational numbers

7. $4x^2 - 5x + 3 = 0$ $D = -23$;
two non-real complex numbers

8. $5x^2 - 7x + 8 = 0$ $D = -111$;
two non-real complex numbers

9. $x^2 - 2 = \frac{5}{2}x$ $D = \frac{57}{4}$;
two irrational numbers

10. $x^2 + \frac{1}{5} = \frac{2}{5}x$ $D = -\frac{16}{25}$;
two non-real complex numbers

In Problems 11–20, determine the value of k that will make the given equation have exactly one rational solution.

11. $x^2 - 4kx + 64 = 0$ ± 4　　　　**12.** $3x^2 + kx + 3 = 0$ ± 6　　　　**13.** $kx^2 - 10x = 5$ -5

14. $2kx^2 - 12x = -9$ 2　　　　**15.** $2x^2 = kx - 8$ ± 8　　　　**16.** $3x^2 = kx - 3$ ± 6

17. $25x^2 - kx = -4$ ± 20　　　　**18.** $4x^2 + 9kx = -1$ $\pm\frac{4}{9}$　　　　**19.** $x^2 + 8x = k$ -16

20. $2x^2 - 4x = k$ -2

B In Problems 21–30, use the discriminant to determine whether the given polynomial is factorable into factors with integer coefficients. If it is, use the technique of Example 3 to factor it.

21. $10x^2 - 7x + 8$ Not factorable　　**22.** $10x^2 - 7x + 1$ $(2x - 1)(5x - 1)$　　**23.** $12x^2 - 17x + 6$ $(4x - 3)(3x - 2)$

24. $27x^2 + 51x - 56$
$(9x - 7)(3x + 8)$

25. $12x^2 - 17x + 2$
Not factorable

26. $15x^2 + 52x - 83$
Not factorable

27. $15x^2 + 52x - 84$
$(5x - 6)(3x + 14)$

28. $27x^2 - 57x - 40$
$(3x - 8)(9x + 5)$

29. $12x^2 - 61x + 60$
$(4x - 15)(3x - 4)$

30. $30x^2 - 19x - 140$
$(2x - 5)(15x + 28)$

C In Problems 31–40, find a quadratic equation with integer coefficients and the given solution set.

31. $\{3, 4\}$ $x^2 - 7x + 12 = 0$　　**32.** $\{-1, 3\}$ $x^2 - 2x - 3 = 0$　　**33.** $\{-5, -7\}$ $x^2 + 12x + 35 = 0$

34. $\{-3, -4\}$ $x^2 + 7x + 12 = 0$　　**35.** $\left\{3, -\frac{2}{3}\right\}$ $3x^2 - 7x - 6 = 0$　　**36.** $\left\{-5, -\frac{2}{7}\right\}$ $7x^2 + 37x + 10 = 0$

37. $\left\{\frac{1}{2}, -\frac{1}{2}\right\}$ $4x^2 - 1 = 0$　　**38.** $\left\{\frac{1}{3}, -\frac{1}{3}\right\}$ $9x^2 - 1 = 0$　　**39.** $\left\{0, -\frac{1}{5}\right\}$ $5x^2 + x = 0$

40. $\left\{-\frac{3}{4}, 0\right\}$ $4x^2 + 3x = 0$

D In Problems 41–45, use the sum and product properties to (a) find the sum of the solutions, (b) find the product of the solutions, and (c) determine if the two given values are the solutions of the given equation.

41. $4x^2 - 6x + 5 = 0$; the proposed solutions are $\frac{1}{2}$ and $\frac{5}{2}$.
a. $\frac{6}{4} = \frac{3}{2}$ **b.** $\frac{5}{4}$ **c.** No

42. $2x^2 + 9x = 35$; the proposed solutions are $-\frac{7}{2}$ and -1.
a. $-\frac{9}{2}$ **b.** $-\frac{35}{2}$ **c.** No

43. $5x^2 + 13x = 6$; the proposed solutions are $\frac{2}{5}$ and -3.
a. $-\frac{13}{5}$ **b.** $-\frac{6}{5}$ **c.** Yes

44. $4 - 3x = 7x^2$; the proposed solutions are $\frac{7}{4}$ and 1.
a. $-\frac{3}{7}$ **b.** $-\frac{4}{7}$ **c.** No

45. $-2 - 5x = 2x^2$; the proposed solutions are $\frac{1}{2}$ and 2.
a. $-\frac{5}{2}$ **b.** 1 **c.** No

46. If d is a constant and 3 is one solution of the equation $2x^2 - dx + 5 = 0$, use the product property to find the other solution. $\frac{5}{6}$

47. If k is a constant and -5 is one solution of $3x^2 + kx = 40$, use the product property to find the other solution. $\frac{8}{3}$

48. If the sum of the solutions of the equation $2x^2 - kx = 4$ is 3, find the value of k. $k = 6$

49. If the sum of the solutions of the equation $10x^2 + (k - 2)x = 3$ is $-\frac{13}{10}$, find k. $k = 15$

50. If the sum of the solutions of $3x^2 + (2k - 5)x + 8 = 0$ is 4, find k. $k = -\frac{7}{2}$

SKILL CHECKER

Try the "Skill Checker" exercises so you'll be ready for the next section.

Solve:

51. $x + 2\sqrt{x} - 3 = 0$ 1

52. $x + 4\sqrt{x} - 12 = 0$ 4

53. $x^2 + 6x + 5 = 0$ $-1, -5$

54. $x^2 - 14x - 15 = 0$ $-1, 15$

USING YOUR KNOWLEDGE

Take a Dive

The highest regularly performed head-first dives are made at La Quebrada in Acapulco, Mexico. The height h (in meters) above the water of the diver after t seconds is given by $h = -t^2 + 2t + 27$. Use the discriminant to find out whether:

55. The diver will ever be 27.5 m above the water. How many times will this occur?   Yes; occurs twice

56. The diver will ever be 28 m above the water. How many times will this occur?   Yes; occurs once

57. The diver will ever be 29 m above the water.   Does not occur ($b^2 - 4ac$ is negative)

WRITE ON

58. Why do you think the expression $b^2 - 4ac$ is called the *discriminant?*   Answers may vary.

59. Can a quadratic equation with integer coefficients have exactly one complex solution? Explain why or why not.   No; answers may vary.

60. Can a quadratic equation with integer coefficients have exactly one irrational solution? Explain why or why not.   No; answers may vary.

61. Can a quadratic equation with integer coefficients have exactly one rational and one irrational solution? Explain why or why not.   No; answers may vary.

MASTERY TEST

If you know how to do these problems, you have learned your lesson!

Use the sum and product properties to see whether the solutions of $4x^2 - 12x + 5 = 0$ are

62. $\frac{1}{2}$ and $-\frac{5}{2}$ No

63. $\frac{1}{2}$ and $\frac{5}{2}$ Yes

Use the discriminant to determine whether the expression is factorable. If it is, factor it.

64. $12x^2 + 23x + 10$ $(4x + 5)(3x + 2)$

65. $12x^2 + x - 35$ $(3x - 5)(4x + 7)$

Find a quadratic equation with integer coefficients whose solution set is

66. $\{-2, 4\}$ $x^2 - 2x - 8 = 0$

67. $\left\{-1, \frac{2}{3}\right\}$ $3x^2 + x - 2 = 0$

68. Consider $2x^2 + 3x = 7$.

    a. Find the discriminant. $D = 65$

    b. Classify the solutions without solving.   Two real, irrational solutions

SOLVING EQUATIONS IN QUADRATIC FORM

To Succeed, Review How To ...

1. Find the LCD of two or more rational expressions (pp. 447–450).

2. Solve quadratic equations by factoring or by using the quadratic formula (pp. 397–403, 595–599).

Objectives

A Solve equations involving rational expressions by converting them to quadratic equations.

B Solve equations that are quadratic in form by substitution.

GETTING STARTED

Work Project

This man and his son can split a supply of firewood in two days. If each works alone, the son takes three days more than the father. How long would it take each of them working alone to finish the job? If we assume that the man can finish in d days, he will do $1/d$ of the work each day. The son will finish in $d + 3$ days and do

$$\frac{1}{d + 3}$$

of the work each day. The work done each day is

work done by father	+	work done by son	=	work done together
$\dfrac{1}{d}$	+	$\dfrac{1}{d + 3}$	=	$\dfrac{1}{2}$

To solve this equation, we multiply each term by the LCD, $2d(d + 3)$, to obtain

$$2d(d + 3) \cdot \frac{1}{d} + 2d(d + 3) \cdot \frac{1}{d + 3} = 2d(d + 3) \cdot \frac{1}{2}$$

$2(d + 3) + 2d = d(d + 3)$	Simplify.
$2d + 6 + 2d = d^2 + 3d$	Remove parentheses.
$0 = d^2 - d - 6$	0 Set equation equal to 0 (standard form).
$0 = (d - 3)(d + 2)$	F Factor.
$d - 3 = 0 \quad \text{or} \quad d + 2 = 0$	F Factors = 0 by the zero-factor property.
$d = 3 \quad \text{or} \quad d = -2$	Solve.

Since d is the number of days, $d = -2$ has to be discarded. Thus it takes the father working alone three days to finish and the son working alone three days more— six days—to finish.

A

Solving Equations That Contain Rational Expressions

As we have seen, equations involving rational expressions can lead to quadratic equations that can be solved by factoring or by using the quadratic formula. When solving such equations, make sure that the proposed solutions are checked in the original equations to avoid zero denominators. If a zero denominator occurs, discard the proposed solution as an *extraneous* solution.

EXAMPLE 1   **Solving equations that contain rational expressions**

Solve:

a. $\dfrac{4}{x^2 - 4} - \dfrac{1}{x - 2} = 1$ **b.** $\dfrac{-12}{x^2 - 9} + \dfrac{1}{x - 3} = 1$

SOLUTION

a. Since $x^2 - 4 = (x + 2)(x - 2)$, the LCD is $(x + 2)(x - 2)$. Multiplying each term by the LCD $(x + 2)(x - 2)$ of the fractions, we have

$$\cancel{(x + 2)}\cancel{(x - 2)} \cdot \dfrac{4}{\cancel{(x^2 - 4)}} - (x + 2)\cancel{(x - 2)} \cdot \dfrac{1}{\cancel{x - 2}} = (x + 2)(x - 2) \cdot 1$$

	Simplify.	$4 - (x + 2) = x^2 - 4$
	Remove parentheses.	$4 - x - 2 = x^2 - 4$
	Combine like terms.	$2 - x = x^2 - 4$
	Subtract 2 and add x.	$0 = x^2 + x - 6$
0	Set equation = 0 (standard form).	$x^2 + x - 6 = 0$
F	Factor.	$(x + 3)(x - 2) = 0$
F	Factors = 0 by zero-factor property.	$x + 3 = 0 \quad$ or $\quad x - 2 = 0$
	Solve.	$x = -3 \quad$ or $\quad x = 2$

The proposed solutions are -3 and 2. However, $x = 2$ is not a solution, since

$$\dfrac{1}{x - 2}$$

is not defined for $x = 2$. Discard $x = 2$ as an extraneous solution. The only solution is -3, which you can check in the original equation.

b. Since $x^2 - 9 = (x + 3)(x - 3)$, the LCD is $(x + 3)(x - 3)$. Multiplying each term by the LCD, we have

$$\cancel{(x + 3)}\cancel{(x - 3)} \cdot \dfrac{-12}{\cancel{x^2 - 9}} + (x + 3)\cancel{(x - 3)} \cdot \dfrac{1}{\cancel{x - 3}} = (x + 3)(x - 3) \cdot 1$$

	Simplify.	$-12 + x + 3 = x^2 - 9$
0	Set equation = 0 (standard form).	$x^2 - x = 0$
F	Factor.	$x(x - 1) = 0$
F	Factors = 0.	$x = 0 \quad$ or $\quad x - 1 = 0$
	Solve.	$x = 0 \quad$ or $\quad x = 1$

This time both solutions satisfy the original equation. Thus the solutions are 0 and 1. Check this.

PROBLEM 1

Solve:

a. $\dfrac{10}{x^2 - 25} - \dfrac{1}{x - 5} = 1$

b. $\dfrac{10}{x^2 - 16} + \dfrac{1}{x - 4} = 1$

Teaching Tip

Suggest that students find the values of x for each fraction that will lead to undefined terms *before* solving the equation. For Example 1(a):

$x^2 - 4 \neq 0$	$x - 2 \neq 0$
$x^2 \neq 4$	$x \neq 2$
$x \neq \pm 2$	

Thus, exclude 2 and -2 as possible solutions.

Web It

For some more notes on solving equations that are quadratic, go to link 8-4-1 at mhhe.com/bello.

B   Solving Equations by Substitution

Some equations that are not quadratic equations can be written in quadratic form and solved by appropriate substitutions. The way to recognize if an equation can be written in quadratic form is to see if the exponents on the variable are in a 2 to 1 ratio like the

Answers

1. a. -6 **b.** $-5, 6$

exponents in a quadratic equation. That is, one exponent is twice the other. Examples of variables with such exponents would be,

$$x^4 \quad \text{and} \quad x^2$$

$$x^{1/2} \quad \text{and} \quad x^{1/4}$$

$$x \quad \text{and} \quad \sqrt{x}, \text{ which mean } x^1 \quad \text{and} \quad x^{1/2}$$

$$x^{-4} \quad \text{and} \quad x^{-2}$$

In each of these cases if we substitute u^2 and u for the two variable terms, the resulting quadratic equation in u can be solved. Then, using substitution again, the original equation can be solved. The procedure follows.

PROCEDURE

Solving a Quadratic-Type Equation by Substitution

1. Substitute u^2 and u for the appropriate variables.

2. Solve the resulting quadratic equation for u, either by the factoring (OFF) method or the quadratic formula.

3. Substitute the variable that represented u into the equations with the two solutions for u.

4. Solve the resulting equations.

5. Verify the solutions in the original equation.

We illustrate this procedure of using substitution in the next examples.

EXAMPLE 2 **Solving equations by substitution**

Solve: $x^4 - 10x^2 + 9 = 0$

SOLUTION Recognizing that the exponents on the variables are in a 2 to 1 ratio, we can solve this as a quadratic-type equation by substituting u^2 and u for x^4 and x^2, respectively.

$x^4 - 10x^2 + 9 = 0$		Given.
$(x^2)^2 - 10(x^2) + 9 = 0$		Substitute $u = x^2$.
$u^2 - 10u + 9 = 0$	O	Set equation $= 0$ (standard form).
$(u - 9)(u - 1) = 0$	F	Factor.
$u - 9 = 0$ or $u - 1 = 0$	F	Factors $= 0$.
$u = 9$ or $u = 1$		Solve.
$x^2 = 9$ or $x^2 = 1$		Substitute $x^2 = u$.
$x = \pm 3$ or $x = \pm 1$		Take the square root of both sides.

The solutions of $x^4 - 10x^2 + 9 = 0$ are 3, 1, -3, and -1. The verification is left to you.

PROBLEM 2

Solve:

$$x^4 - 17x^2 + 16 = 0$$

Answer

2. $\pm 1, \pm 4$

| **EXAMPLE 3**   Solving equations that are quadratic in form | **PROBLEM 3** |

Solve: $(x^2 - x)^2 - (x^2 - x) - 30 = 0$

Solve:

$$(x^2 + x)^2 - 11(x^2 + x) - 12 = 0$$

SOLUTION   This equation is quadratic in form. If we let $u = x^2 - x$, we can write

$$u^2 - u - 30 = 0$$

$$(u - 6)(u + 5) = 0 \qquad \text{Factor.}$$

$$u = 6 \quad \text{or} \qquad u = -5 \qquad \text{Solve } u - 6 = 0 \text{ and } u + 5 = 0.$$

$$x^2 - x = 6 \quad \text{or} \qquad x^2 - x = -5 \qquad \text{Substitute } u = x^2 - x.$$

$$x^2 - x - 6 = 0 \quad \text{or} \quad x^2 - x + 5 = 0 \qquad \text{In standard form}$$

The first equation can be solved by factoring:

$$x^2 - x - 6 = (x - 3)(x + 2) = 0$$

Thus $x = 3$ or $x = -2$. We use the quadratic formula for $x^2 - x + 5 = 0$. Here $a = 1$, $b = -1$, and $c = 5$. Thus

$$x = \frac{-(-1) \pm \sqrt{(-1)^2 - 4(1)(5)}}{2(1)} = \frac{1 \pm \sqrt{-19}}{2}$$

$$= \frac{1 \pm i\sqrt{19}}{2}$$

The solutions of $(x^2 - x)^2 - (x^2 - x) - 30 = 0$ are

$$3, \quad -2, \quad \frac{1}{2} + \frac{\sqrt{19}}{2}i, \quad \text{and} \quad \frac{1}{2} - \frac{\sqrt{19}}{2}i$$

| **EXAMPLE 4**   Solving equations containing rational exponents | **PROBLEM 4** |

Solve: $x^{1/2} - 8x^{1/4} + 15 = 0$

Solve:

$$x^{1/2} - 7x^{1/4} + 10 = 0$$

SOLUTION   Recognizing that the exponents on the variables are in a 2 to 1 ratio, we can solve this as a quadratic-type equation by substituting u^2 and u for $x^{1/2}$ and $x^{1/4}$, respectively.

$$x^{1/2} - 8x^{1/4} + 15 = 0 \qquad \text{Given.}$$

$$(x^{1/4})^2 - 8(x^{1/4}) + 15 = 0 \qquad \text{Substitute } u = x^{1/4}.$$

$$u^2 - 8u + 15 = 0 \qquad \text{Set equation} = 0 \text{ (standard form).}$$

$$(u - 3)(u - 5) = 0 \qquad \text{Factor.}$$

$$u - 3 = 0 \quad \text{or} \quad u - 5 = 0 \qquad \text{Factors} = 0.$$

$$u = 3 \quad \text{or} \qquad u = 5 \qquad \text{Solve.}$$

$$x^{1/4} = 3 \quad \text{or} \qquad x^{1/4} = 5 \qquad \text{Substitute } x^{1/4} = u.$$

$$(x^{1/4})^4 = (3)^4 \quad \text{or} \quad (x^{1/4})^4 = (5)^4 \qquad \text{Raise each side to the fourth power.}$$

$$x = 81 \quad \text{or} \qquad x = 625 \qquad \text{Simplify.}$$

The solutions are 81 and 625. The verification is left to you.

Answers

3. $-4, 3, -\frac{1}{2} \pm \frac{\sqrt{3}}{2}i$

4. $16, 625$

Teaching Tip

Remind students that

$$\sqrt{x} = x^{1/2}$$

and

$$(\sqrt{x})^2 = (x^{1/2})^2 = x$$

Finally, we solve some equations involving radicals. For example, the equation $x + 2\sqrt{x} - 3 = 0$ can be written as a quadratic equation if we let $u = \sqrt{x}$. This makes $u^2 = x$ so that we can write

$$x + 2\sqrt{x} - 3 = 0$$
$$u^2 + 2u - 3 = 0 \qquad \text{Substitute } u \text{ for } \sqrt{x} \text{ and } u^2 \text{ for } x.$$
$$(u + 3)(u - 1) = 0 \qquad \text{Factor.}$$
$$u = -3 \quad \text{or} \quad u = 1 \qquad \text{Solve } u + 3 = 0 \text{ and } u - 1 = 0.$$
$$\sqrt{x} = -3 \quad \text{or} \quad \sqrt{x} = 1 \qquad \text{Substitute } u = \sqrt{x}.$$

But the square root of x is never negative, so the equation $\sqrt{x} = -3$ has no solution. The solution of $\sqrt{x} = 1$ is 1. Thus the equation $x + 2\sqrt{x} - 3 = 0$ has only one solution, 1. You can verify that this solution is correct by direct substitution into the equation.

| **EXAMPLE 5** | **Using substitution to solve equations containing radicals** |

Solve: $x - 4\sqrt{x} + 3 = 0$

SOLUTION We let $u = \sqrt{x}$, which makes $u^2 = x$.

$$x - 4\sqrt{x} + 3 = 0$$
$$u^2 - 4u + 3 = 0 \qquad \text{Substitute.}$$
$$(u - 3)(u - 1) = 0 \qquad \text{Factor.}$$
$$u = 3 \quad \text{or} \quad u = 1 \qquad \text{Solve } u - 3 = 0 \text{ and } u - 1 = 0.$$
$$\sqrt{x} = 3 \quad \text{or} \quad \sqrt{x} = 1 \qquad \text{Substitute } u = \sqrt{x}.$$
$$x = 9 \quad \text{or} \quad x = 1 \qquad \text{Square both sides.}$$

We leave it to you to verify that both solutions satisfy the original equation.

PROBLEM 5

Solve:

$$x - 3\sqrt{x} - 10 = 0$$

| **EXAMPLE 6** | **Solving equations containing negative exponents** |

Solve: $5x^{-4} - 4x^{-2} - 1 = 0$

SOLUTION We can solve this equation by making the substitution $u = x^{-2}$, but this time we first convert the equation to one with positive exponents. Since

$$x^{-4} = \frac{1}{x^4} \quad \text{and} \quad x^{-2} = \frac{1}{x^2}$$

we have

$$5x^{-4} - 4x^{-2} - 1 = \frac{5}{x^4} - \frac{4}{x^2} - 1 = 0$$

Multiplying each term by the LCD, x^4, we obtain

$$x^4 \cdot \frac{5}{x^4} - x^4 \cdot \frac{4}{x^2} - x^4 \cdot 1 = 0$$
$$5 - 4x^2 - x^4 = 0 \qquad \text{Simplify.}$$
$$-x^4 - 4x^2 + 5 = 0 \qquad \text{Write in descending order.}$$
$$x^4 + 4x^2 - 5 = 0 \qquad \text{Multiply by } -1.$$
$$u^2 + 4u - 5 = 0 \qquad \text{Let } u = x^2.$$
$$(u + 5)(u - 1) = 0 \qquad \text{Factor.}$$
$$u = -5 \quad \text{or} \quad u = 1 \qquad \text{Solve } u + 5 = 0 \text{ and } u - 1 = 0.$$
$$x^2 = -5 \quad \text{or} \quad x^2 = 1 \qquad \text{Substitute } x^2 \text{ for } u.$$
$$x^2 = \pm\sqrt{5}i \quad \text{or} \quad x = \pm 1 \qquad \text{Take the square root of both sides.}$$

We leave it to you to verify that the four solutions satisfy the original equation.

PROBLEM 6

Solve:

$$2x^{-2} - 5x^{-1} - 3 = 0$$

Answers

5. 25 **6.** $\frac{1}{3}$, -2

Calculate It Exercises   Solve Equations by Graphing

Your calculator can be used in several ways to solve the equations we have studied. One way is to use the solve feature. When doing so, it's usually easy to obtain one solution, but in many cases, the second solution is hard to obtain. Try Example 1(b) if you don't believe this. Moreover, there

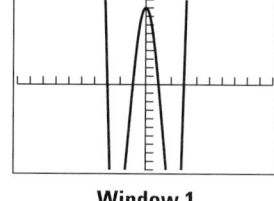

Window 1

may be cases in which the graph does not cross the horizontal axis. This means that there are no real-number solutions.

Now let's try to solve Example 2 by graphing. It will be easier to graph $Y_1 = x^4 - 10x^2 + 9$ and $Y_2 = 0$ using a standard window and then pressing (2nd) (TRACE) (5) (ENTER) (ENTER) (ENTER) to find one intersection. To find the other intersections press (TRACE) and use the arrow keys to move the cursor near the intersection point. Then press (2nd) (TRACE) (5) and (ENTER) three times. The solutions are $-3, -1, 1,$ and 3, as shown in Window 1.

What about Example 6? Since some of the solutions are real and some imaginary, what does this graph look like? Enter $Y_1 = 5x^{-4} - 4x^{-2} - 1$ and graph. You can see only the real solutions, -1 and 1, as shown in Window 2

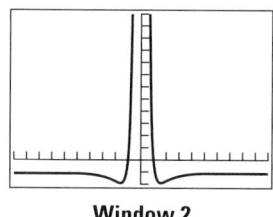

Window 2

using a window of $[-10, 10]$ by $[-2, 12]$. How do you know there are more solutions? You need more information! The original equation is equivalent to $x^4 + 4x^2 - 5 = 0$, which has degree 4, so the equation must have four solutions (some may repeat). Since we found only two solutions, we suspect that the other two solutions may be imaginary. (An nth-degree equation has n solutions but some of them may repeat. At this time, the only way to check for repeated solutions is to solve the problem algebraically.)

1. Verify the results of Example 4 using your calculator. If you have a root feature, use it.

2. Verify the results of Example 5 using your calculator.

Exercises 8.4

A In Problems 1–10, solve the equation.

1. $\dfrac{x}{x+4} + \dfrac{x}{x+1} = 0$ $0, -\dfrac{5}{2}$

2. $\dfrac{x}{x+2} + \dfrac{x}{x+3} = 0$ $0, -\dfrac{5}{2}$

3. $\dfrac{x-1}{x+11} - \dfrac{2}{x-1} = 0$ $-3, 7$

4. $\dfrac{x+1}{x-2} - \dfrac{8}{x-1} = 0$ $3, 5$

5. $\dfrac{x}{x-1} - \dfrac{x}{x+1} = 0$ 0

6. $\dfrac{x}{x+4} + \dfrac{x}{x-2} = -\dfrac{1}{2}$ $-2, \dfrac{4}{5}$

7. $\dfrac{x}{x+2} - \dfrac{x}{x+1} = -\dfrac{1}{6}$ $2, 1$

8. $\dfrac{x}{x+1} - \dfrac{x}{x-1} = -\dfrac{3}{4}$ $-\dfrac{1}{3}, 3$

9. $\dfrac{x}{x+4} + \dfrac{x}{x+2} = -\dfrac{4}{3}$ $-\dfrac{16}{5}, -1$

10. $\dfrac{2x}{x-2} + \dfrac{x}{x-1} = \dfrac{7}{6}$ $-1, \dfrac{14}{11}$

B In Problems 11–39, solve by substitution.

11. $x^4 - 13x^2 + 36 = 0$ $\pm 3, \pm 2$

12. $x^4 - 5x^2 + 4 = 0$ $\pm 2, \pm 1$

13. $4x^4 + 35x^2 = 9$ $\pm \dfrac{1}{2}, \pm 3i$

14. $3x^4 + 2x^2 = 8$ $\pm \dfrac{2\sqrt{3}}{3}, \pm i\sqrt{2}$

15. $3y^4 = 5y^2 + 2$ $\pm\sqrt{2}, \pm\dfrac{\sqrt{3}}{3}i$

16. $6y^4 = 7y^2 - 2$ $\pm\dfrac{\sqrt{6}}{3}, \pm\dfrac{\sqrt{2}}{2}$

17. $x^6 + 7x^3 - 8 = 0$ $1, -2, 1 \pm i\sqrt{3}, -\dfrac{1}{2} \pm \dfrac{\sqrt{3}}{2}i$

18. $x^6 - 26x^3 - 27 = 0$ $-1, 3, \dfrac{1}{2} \pm \dfrac{\sqrt{3}}{2}i, -\dfrac{3}{2} \pm \dfrac{3\sqrt{3}}{2}i$

19. $(x + 1)^2 - 3(x + 1) = 40$ $7, -6$ **20.** $(x + 2)^2 - 2(x + 2) = 8$ $2, -4$

21. $(y^2 - y)^2 - 8(y^2 - y) = 9$ $\frac{1 \pm \sqrt{37}}{2}, \frac{1}{2} \pm \frac{\sqrt{3}}{2}i$ **22.** $(y^2 - y)^2 - 4(y^2 - y) = 12$ $3, -2, \frac{1}{2} \pm \frac{\sqrt{7}}{2}i$

23. $x^{1/2} + 3x^{1/4} - 10 = 0$ 16 **24.** $x^{1/2} + 4x^{1/4} - 12 = 0$ 16

25. $y^{2/3} - 5y^{1/3} = -6$ $8, 27$ **26.** $y^{2/3} + 5y^{1/3} = -6$ $-8, -27$

27. $x + \sqrt{x} - 6 = 0$ 4 **28.** $x - \sqrt{x} - 30 = 0$ 36

29. $(x^2 - 4x) - 8\sqrt{x^2 - 4x} + 15 = 0$ $2 \pm \sqrt{29}, 2 \pm \sqrt{13}$ **30.** $(x^2 + 3x) + 5\sqrt{x^2 + 3x} - 14 = 0$ $-4, 1$

31. $z + 3 - \sqrt{z + 3} - 6 = 0$ 6 **32.** $z + 4 - \sqrt{z + 4} - 12 = 0$ 12

33. $3\sqrt{x} - 5\sqrt[4]{x} + 2 = 0$ $1, \frac{16}{81}$ **34.** $x^{-2} + 2x^{-1} - 3 = 0$ $-\frac{1}{3}, 1$

35. $x^{-2} + 2x^{-1} - 8 = 0$ $\frac{1}{2}, -\frac{1}{4}$ **36.** $8x^{-4} - 9x^{-2} + 1 = 0$ $\pm 1, \pm 2\sqrt{2}$

37. $3x^{-4} - 5x^{-2} - 2 = 0$ $\pm i\sqrt{3}, \pm \frac{\sqrt{2}}{2}$ **38.** $6x^{-4} + x^{-2} - 1 = 0$ $\pm\sqrt{3}, \pm i\sqrt{2}$

39. $6x^{-4} + 5x^{-2} - 4 = 0$ $\pm\sqrt{2}, \pm\frac{\sqrt{3}}{2}i$

APPLICATIONS

Working Together and Alone

40. Working together, two workers can complete a job in 6 hr. If they work alone, one of them takes 9 hr more than the other to finish. How long does it take for each of them working alone to finish the job? 9 hr, 18 hr

41. Working together, Jack and Jill can shovel the snow in the driveway in 6 hr. It takes Jack 5 hr more than Jill to do the job by himself. How long does it take each of them working alone to finish the job? 10 hr, 15 hr

42. Given the equation $x^4 - 6x^2 + 9 = 0$ of the graph, find the exact value of the x-intercepts.

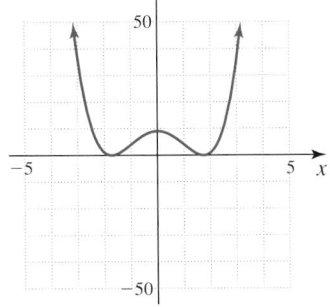

x-intercepts: $(\sqrt{3}, 0)$; $(-\sqrt{3}, 0)$

43. Given the equation $x^4 - 10x^2 + 25 = 0$ of the graph, find the exact value of the x-intercepts.

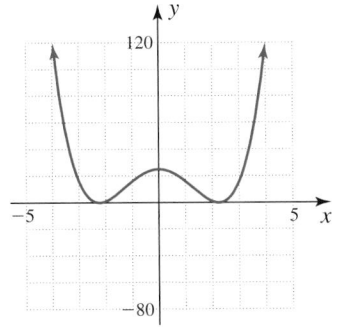

x-intercepts: $(\sqrt{5}, 0)$; $(-\sqrt{5}, 0)$

SKILL CHECKER

Try the "Skill Checker" exercises so you'll be ready for the next section.

Solve:

44. $\frac{x}{5} - \frac{x}{3} \leq \frac{x - 5}{5}$ $x \geq 3$

45. $\frac{7x + 2}{6} \geq \frac{3x - 2}{4}$ $x \geq -2$

46. $\frac{8x - 23}{6} + \frac{1}{3} \geq \frac{5x}{2}$ $x \leq -3$

USING YOUR KNOWLEDGE

A Rental Problem

47. A group of students rented a cabin for $1600. When two of the group failed to pay their shares, the cost to each of the remaining students was $40 more. Use your knowledge of the RSTUV procedure to find how many students were in the group.

n = number of students in the group

$n - 2$ = number of students who paid

$\dfrac{1600}{n}$ = amount each student should have paid in

dollars 10 students

48. A group of students rented a bus for $720. If there had been six more students, the price per student would have been $6 less. How many students were in the group? 24 students

WRITE ON

49. Find the number of solutions for $x - 1 = 0$, $x^2 - 1 = 0$, and $x^3 - 1 = 0$. Make a conjecture regarding the number of solutions for $x^4 - 1 = 0$. (If you want to solve the equation $x^4 - 1 = 0$ first, do it by factoring.) 1; 2; 3; 4

50. Which method would you use to solve $x + 2\sqrt{x} - 3 = 0$, the substitution method or isolating the radical and squaring both sides? Explain. Which method has fewer steps? Answers may vary.

51. In Example 6, we solved the problem by transforming the given equation into an equivalent one with positive exponents. Now solve Example 6 using the substitution $u = x^{-2}$. Which method do you prefer? Explain why. Answers may vary.

52. The equation $x^4 - 1 = 0$ has four solutions, two real and two non-real complex. (Solve $x^4 - 1 = 0$ by factoring if you don't believe this.) How many solutions do you think the equation $x^6 - 1 = 0$ has? How many do you think are real? How many do you think are non-real complex? Make a conjecture about the number and nature of the solutions (real or non-real complex) of $x^{2n} - 1 = 0$. 6; Answers may vary.

MASTERY TEST

If you know how to do these problems, you have learned your lesson!

Solve:

53. $x - 5\sqrt{x} + 4 = 0$ 1, 16

54. $2x - 3\sqrt{x} = -1$ $\frac{1}{4}$, 1

55. $x^{1/2} - 5x^{1/4} + 6 = 0$ 16, 81

56. $x^{1/4} - 5x^{1/2} = -4$ 1

57. $x^{-4} - 9x^{-2} + 14 = 0$ $\pm\frac{\sqrt{2}}{2}, \pm\frac{\sqrt{7}}{7}$

58. $x^{-4} - 7x^{-2} = -12$ $\pm\frac{1}{2}, \pm\frac{\sqrt{3}}{3}$

59. $(x^2 - x)^2 - (x^2 - x) - 2 = 0$ $2, -1, \frac{1}{2} \pm \frac{\sqrt{3}}{2}i$

60. $4(x^2 + 1)^2 - 7(x^2 + 1) = 2$ $\pm\frac{\sqrt{5}}{2}i, \pm 1$

61. $\dfrac{12}{x^2 - 36} - \dfrac{1}{x - 6} = 1$ -7

62. $\dfrac{6}{x^2 - 4} + \dfrac{1}{x - 2} = 1$ $4, -3$

63. $x^4 - 5x^2 + 4 = 0$ $\pm 1, \pm 2$

64. $x^4 - 6x^2 + 5 = 0$ $\pm 1, \pm\sqrt{5}$

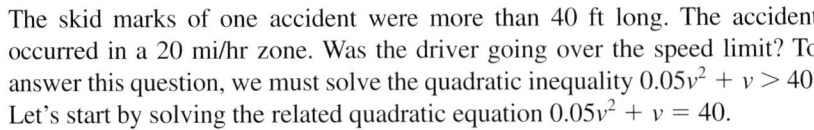

8.5 NONLINEAR INEQUALITIES

To Succeed, Review How To . . .

1. Solve a quadratic equation
   (pp. 399–403, 595–599).

2. Solve and graph linear inequalities
   (pp. 121–127).

Objectives

A Solve quadratic inequalities.

B Solve inequalities of degree 3 or higher.

C Solve rational inequalities.

D Solve an application involving inequalities.

GETTING STARTED

Skidding to Quadratics

Have you seen an officer measuring skid marks at the scene of an accident? The distance d (in ft) in which a car traveling v mi/hr can be stopped is given by

$$d = 0.05v^2 + v$$

The skid marks of one accident were more than 40 ft long. The accident occurred in a 20 mi/hr zone. Was the driver going over the speed limit? To answer this question, we must solve the quadratic inequality $0.05v^2 + v > 40$. Let's start by solving the related quadratic equation $0.05v^2 + v = 40$.

$0.05v^2 + v = 40$	Given.
$5v^2 + 100v = 4000$	Multiply by 100.
$v^2 + 20v = 800$	Divide by 5.
$v^2 + 20v - 800 = 0$	Standard form
$(v + 40)(v - 20) = 0$	Factor.
$v = -40$ or $v = 20$	Solve $v + 40 = 0$ and $v - 20 = 0$.

Now divide a number line into three regions, A, B, and C, using the solutions (*critical values*) -40 and 20 as boundaries (see the following table). To find the regions where $0.05v^2 + v > 40$, choose a test point in each of the regions, and test whether $0.05v^2 + v > 40$ for each of the points. Let's select the points -50, 0, and 30 from intervals A, B, and C, respectively.

Test

In original:

For $v = -50$	For $v = 0$	For $v = 30$
$0.05v^2 + v > 40$	$0.05v^2 + v > 40$	$0.05v^2 + v > 40$
$0.05(-50)^2 + (-50) > 40$	$0.05(0)^2 + 0 > 40$	$0.05(30)^2 + 30 > 40$
$125 - 50 > 40$	$0 > 40$ (FALSE)	$45 + 30 > 40$
$75 > 40$ (TRUE)		$75 > 40$ (TRUE)
Thus $0.05v^2 + v > 40$.	Thus $0.05v^2 + v < 40$.	Thus $0.05v^2 + v > 40$.
Use this interval: $(-\infty, -40)$.	Discard this interval.	Use this interval: $(20, \infty)$.

The solution set is the union of the two intervals where the inequality $0.05v^2 + v > 40$ is true, that is, $(-\infty, -40) \cup (20, \infty)$. The end points -40 and 20 are *not* included in the solution set. Also, since the velocity must be positive, we discard the interval $(-\infty, -40)$ and conclude that the car was going more than 20 mi/hr.

A Solving Quadratic Inequalities

Web It

For a tutorial on solving quadratic inequalities, go to link 8-5-1 at mhhe.com/bello.

The procedure used in *Getting Started* is based on the fact that, if you select a test point in one of the chosen intervals and the inequality is satisfied by the test point, then *every* point in the interval satisfies the inequality. The procedure we just used to solve the speed-limit problem can be generalized to solve a quadratic inequality. Here are the steps.

Teaching Tip

Remind students that standard form means set equal to 0. Also, in step 4, if replacing a test point into factors, the sign is all that is necessary to conclude true or false.

$(-5)(-1) < 0$	$(-)(-) < 0$
$5 < 0$	$+ < 0$
False	False because all positive numbers are greater than 0.

PROCEDURE

Procedure to Solve Quadratic Inequalities

1. Write the quadratic inequality in standard form—the quadratic expression is on the left side and zero is on the right side.

2. Find the *critical values* by solving the related equation.

3. Separate the number line into intervals using the critical values found in step 2 as boundaries.

4. Choose a test point in each interval and determine whether the point satisfies the **original quadratic inequality.**

5. State the solution set, which consists of all the intervals in which the test point satisfies the original quadratic inequality.

EXAMPLE 1 Using the five-step procedure to solve a quadratic inequality

Solve and write the solution in interval notation: $(x - 1)(x + 3) < 0$

SOLUTION We use the five-step procedure.

1. The inequality $(x - 1)(x + 3) < 0$ is already in standard form.
2. The related equation is $(x - 1)(x + 3) = 0$, and the critical values are $x = 1$ and $x = -3$.
3. Separate the number line into three regions using -3 and 1 as boundaries.
4. Select the test points: -4 from A, 0 from B, and 2 from C.

PROBLEM 1

Solve and write the solution in interval notation:

$$(x - 4)(x + 2) \leq 0$$

Test	For $x = -4$	For $x = 0$	For $x = 2$
In original:	$(x - 1)(x + 3) < 0$	$(x - 1)(x + 3) < 0$	$(x - 1)(x + 3) < 0$
	$(-4 - 1)(-4 + 3) < 0$	$(0 - 1)(0 + 3) < 0$	$(2 - 1)(2 + 3) < 0$
	$(-5)(-1) < 0$	$(-1)(3) < 0$	$(1)(5) < 0$
	$5 < 0$ (FALSE)	$-3 < 0$ (TRUE)	$5 < 0$ (FALSE)
	Thus $(x - 1)(x + 3) > 0$.	Thus $(x - 1)(x + 3) < 0$.	Thus $(x - 1)(x + 3) > 0$.
	Discard this interval.	Use this interval: $(-3, 1)$.	Discard this interval.

5. The solution set is the interval $(-3, 1)$.

Answer

1. $[-2, 4]$

EXAMPLE 2	**More practice solving quadratic inequalities**

Solve, graph, and write the solution in interval notation: $x^2 - 6x \geq -7$

SOLUTION Again, we use the five-step procedure.

1. We first add 7 to both sides so that the inequality is in the standard form, $x^2 - 6x + 7 \geq 0$.

2. The related equation is $x^2 - 6x + 7 = 0$. Since $x^2 - 6x + 7$ is *not* factorable, we use the quadratic formula with $a = 1$, $b = -6$, and $c = 7$, and we obtain

$$x = \frac{-(-6) \pm \sqrt{(-6)^2 - 4(1)(7)}}{2(1)} = \frac{6 \pm \sqrt{8}}{2} = \frac{6 \pm 2\sqrt{2}}{2} = 3 \pm \sqrt{2}$$

The critical values are $x = 3 + \sqrt{2}$ and $x = 3 - \sqrt{2}$.

3. Separate the number line into three regions using $3 + \sqrt{2}$ and $3 - \sqrt{2}$ as boundaries. Approximating $\sqrt{2}$, as 1.4, the boundaries are $3 + 1.4 = 4.4$ and $3 - 1.4 = 1.6$.

4. Select the test points: 0 from A, 3 from B, and 6 from C.

PROBLEM 2

Solve, graph, and write the solution in interval notation:

$$x^2 - 5 > -2$$

Teaching Tip

Review an example of graphing the following two linear inequalities to know when to include the end points.
Example:
Graph:

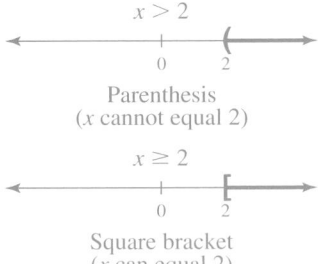

Parenthesis
(x cannot equal 2)

Square bracket
(x can equal 2)

Test	**For x = 0, we want to know if $x^2 - 6x \geq -7$.**	**For x = 3, we want to know if $x^2 - 6x \geq -7$.**	**For x = 6, we want to know if $x^2 - 6x \geq -7$.**
In original:	$x^2 - 6x \geq -7$	$x^2 - 6x \geq -7$	$x^2 - 6x \geq -7$
	$(0)^2 - 6(0) \geq -7$	$(3)^2 - 6(3) \geq -7$	$(6)^2 - 6(6) \geq -7$
	$0 - 0 \geq -7$	$9 - 18 \geq -7$	$36 - 36 \geq -7$
	$0 \geq -7$ (TRUE)	$-9 \geq -7$ (FALSE)	$0 \geq -7$ (TRUE)
	Thus $x^2 - 6x = 0 \geq -7$.	Thus $x^2 - 6x = -9 < -7$.	Thus $x^2 - 6x = 0 \geq -7$.
	Use this interval: $(-\infty, 3 - \sqrt{2}]$.	Discard this interval.	Use this interval: $[3 + \sqrt{2}, \infty)$.

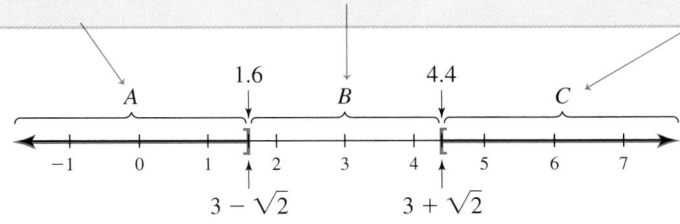

5. The solution set is $(-\infty, 3 - \sqrt{2}] \cup [3 + \sqrt{2}, \infty)$. The critical values $3 - \sqrt{2}$ and $3 + \sqrt{2}$ are part of the solution set because they satisfy the inequality $x^2 - 6x \geq -7$. (Check this, but first read the caution note for a good rule of thumb.)

CAUTION

The end points of an interval are included in the solution set when the original inequality sign is $\leq$ or $\geq$; they are omitted when the original inequality sign is $>$ or $<$.

Answer

2. $(-\infty, -\sqrt{3}) \cup (\sqrt{3}, \infty)$

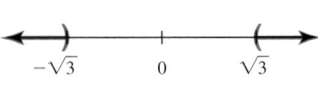

Web It

For another resource on solving quadratic inequalities, go to link 8-5-2 at mhhe.com/bello.

 **Solving Polynomial Inequalities**

Suppose we wish to solve the inequality $(x + 3)(x - 2)(x - 4) \geq 0$, which is not a quadratic inequality. This would be called a *polynomial inequality* and the steps to solving it are similar to the steps for solving the quadratic inequality. The procedure follows.

PROCEDURE

Procedure to Solve Polynomial Inequalities

1. Write the polynomial inequality in standard form—the polynomial expression is on the left side and zero is on the right side.

2. Find the *critical values* by solving the related equation.

3. Separate the number line into intervals using the critical values found in step 2 as boundaries.

4. Choose a test point in each interval and determine whether the point satisfies the **original polynomial inequality.**

5. State the solution set, which consists of all the intervals in which the test point satisfies the original polynomial inequality.

The original inequality, $(x + 3)(x - 2)(x - 4) \geq 0$, is in standard form and the polynomial on the left is in factored form. This makes it easy to set each factor equal to zero and solve for the three critical numbers, $x = -3$, $x = 2$, and $x = 4$. We separate the number line into *four* regions using -3, 2, and 4 as boundaries. Now select the points -4 from A, 0 from B, 3 from C, and 5 from D and determine which of these points satisfies the original polynomial inequality.

$$(x + 3)(x - 2)(x - 4) \geq 0$$

Test	**For x = −4**	**For x = 0**	**For x = 3**	**For x = 5**
In original:	$(x + 3)(x - 2)(x - 4) \geq 0$	$(x + 3)(x - 2)(x - 4) \geq 0$	$(x + 3)(x - 2)(x - 4) \geq 0$	$(x + 3)(x - 2)(x - 4) \geq 0$
	$(-4 + 3)(-4 - 2)(-4 - 4) \geq 0$	$(0 + 3)(0 - 2)(0 - 4) \geq 0$	$(3 + 3)(3 - 2)(3 - 4) \geq 0$	$(5 + 3)(5 - 2)(5 - 4) \geq 0$
	$(-1)(-6)(-8) \geq 0$	$(3)(-2)(-4) \geq 0$	$(6)(1)(-1) \geq 0$	$(8)(3)(1) \geq 0$
	$-48 \geq 0$ (FALSE)	$24 \geq 0$ (TRUE)	$-6 \geq 0$ (FALSE)	$24 \geq 0$ (TRUE)
	Thus $(x + 3)(x - 2)(x - 4)$	Thus $(x + 3)(x - 2)(x - 4)$	Thus $(x + 3)(x - 2)(x - 4)$	Thus $(x + 3)(x - 2)(x - 4)$
	$= -48 < 0$.	$= 24 \geq 0$.	$= -6 < 0$.	$= 24 \geq 0$.
	Discard this interval.	Use this interval: $[-3, 2]$.	Discard this interval.	Use this interval: $[4, \infty)$.

The solution set is the union of the two intervals selected, $[-3, 2] \cup [4, \infty)$. -3, 2, and 4 are part of the solution set because they satisfy the inequality $(x + 3)(x - 2)(x - 4) \geq 0$.

EXAMPLE 3	Using the five-step procedure to solve polynomial inequalities

Solve, graph, and write the solution in interval notation:

$$(x + 1)(x - 3)(x + 4) \leq 0$$

SOLUTION We follow our five-step procedure.

1. The inequality is in standard form.
2. The related equation is $(x + 1)(x - 3)(x + 4) = 0$, and the critical values are -1, 3, and -4.
3. Separate the number line into four regions using -1, 3, and -4 as boundaries.
4. Select the test points: -5 from A, -3 from B, 0 from C, and 4 from D.

Test	For x = −5	For x = −3	For x = 0	For x = 4
In original:	$(x + 1)(x - 3)(x + 4) \leq 0$	$(x + 1)(x - 3)(x + 4) \leq 0$	$(x + 1)(x - 3)(x + 4) \leq 0$	$(x + 1)(x - 3)(x + 4) \leq 0$
	$(-5 + 1)(-5 - 3)(-5 + 4) \leq 0$	$(-3 + 1)(-3 - 3)(-3 + 4) \leq 0$	$(0 + 1)(0 - 3)(0 + 4) \leq 0$	$(4 + 3)(4 - 3)(4 + 4) \leq 0$
	$(-4)(-8)(-1) \leq 0$	$(-2)(-6)(1) \leq 0$	$(1)(-3)(4) \leq 0$	$(7)(1)(8) \leq 0$
	$-32 \leq 0$ (TRUE)	$12 \leq 0$ (FALSE)	$-12 \leq 0$ (TRUE)	$56 \leq 0$ (FALSE)
	Thus $(x + 1)(x - 3)(x + 4)$	Thus $(x + 1)(x - 3)(x + 4)$	Thus $(x + 1)(x - 3)(x + 4)$	Thus $(x + 1)(x - 3)(x + 4)$
	$= -32 \leq 0.$	$= 12 > 0.$	$= -12 \leq 0.$	$= 56 > 0.$
	Use this interval: $(-\infty, -4]$.	Discard this interval.	Use this interval: $[-1, 3]$.	Discard this interval.

5. The solution set is the union of the two intervals selected, $(-\infty, -4] \cup [-1, 3]$.

PROBLEM 3

Solve, graph, and write the solution in interval notation:

$$(x + 3)(x - 2)(x + 1) \leq 0$$

C Solving Rational Inequalities

In Chapter 6 we studied quotients of polynomials called *rational expressions*. We are now ready to consider *rational inequalities* such as

$$\frac{x - 3}{x + 2} \leq 0$$

This time, the **critical values** are the numbers that make the *numerator* $(x - 3)$ and *denominator* $(x + 2)$ zero, that is, 3 and -2. The critical values that make the denominator zero are excluded from the solution set. As before, we separate the number line into three regions, A, B, and C using -2 and 3 as boundaries, and then select test points from each of these regions to determine whether they satisfy our inequality. Let's select -3 from A, 0 from B, and 4 from C.

Teaching Tip

The interval $(-2, 3]$ must be open at -2 because the value -2 will make the fraction meaningless. That takes precedence over the inequality symbol of $\leq$.

Answer

3. $(-\infty, -3] \cup [-1, 2]$

Web It

For an analytical approach to solving rational inequalities, go to link 8-5-3 at mhhe.com/bello.

Test

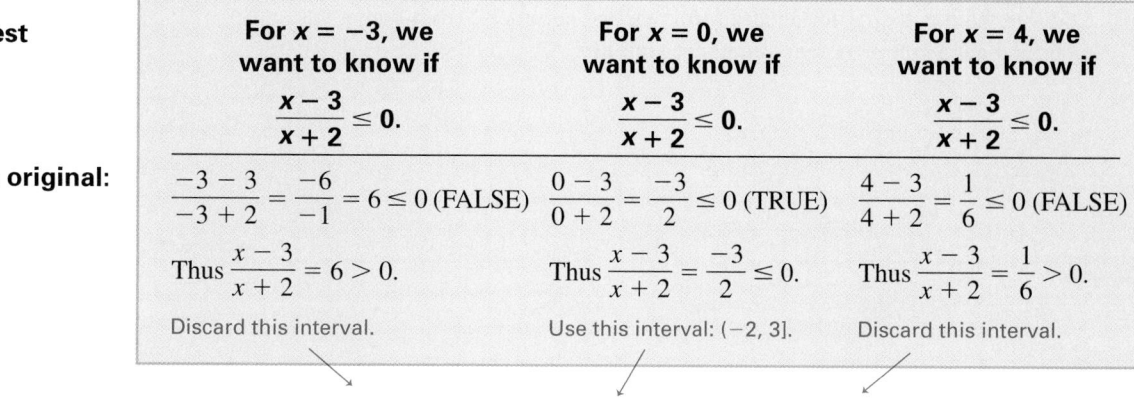

For $x = -3$, we want to know if	For $x = 0$, we want to know if	For $x = 4$, we want to know if
$\dfrac{x-3}{x+2} \le 0.$	$\dfrac{x-3}{x+2} \le 0.$	$\dfrac{x-3}{x+2} \le 0.$

In original:
$$\frac{-3-3}{-3+2} = \frac{-6}{-1} = 6 \le 0 \text{ (FALSE)} \qquad \frac{0-3}{0+2} = \frac{-3}{2} \le 0 \text{ (TRUE)} \qquad \frac{4-3}{4+2} = \frac{1}{6} \le 0 \text{ (FALSE)}$$

Thus $\dfrac{x-3}{x+2} = 6 > 0.$ Thus $\dfrac{x-3}{x+2} = \dfrac{-3}{2} \le 0.$ Thus $\dfrac{x-3}{x+2} = \dfrac{1}{6} > 0.$

Discard this interval.   Use this interval: $(-2, 3]$.   Discard this interval.

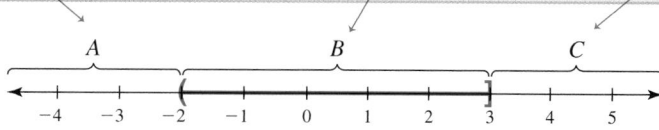

The solution set is the interval $(-2, 3]$.

Notice we enclosed -2 with a parenthesis and not a bracket like we used to enclose 3. This means we will not be able to include -2 as part of the solution set. The fraction would be undefined if we allowed x to equal -2 because it would give the denominator the value of 0.

EXAMPLE 4 **Using the five-step procedure to solve rational inequalities**

Solve, graph, and write the solution in interval notation:

a. $\dfrac{3x}{x+4} > 0$ **b.** $\dfrac{x}{x-2} \ge 2$

SOLUTION   We still follow our five-step procedure.

a. 1. The inequality is in standard form.

  2. The critical points of 0 and -4 are obtained by setting the numerator and denominator equal to 0; then solve each.

  3. Separate the number line into three regions using 0 and -4 as the boundaries.

  4. Select the test points: -5 from A, -1 from B, and 2 from C.

PROBLEM 4

Solve, graph, and write the solution in interval notation:

a. $\dfrac{x+3}{2x-1} \le 0$ **b.** $\dfrac{x}{x-1} > 1$

Web It

For a tutorial with examples on solving rational inequalities, go to link 8-5-4 at mhhe.com/bello.

Test

In original:

For $x = -5$	For $x = -1$	For $x = 2$
$\dfrac{3x}{x+4} > 0$	$\dfrac{3x}{x+4} > 0$	$\dfrac{3x}{x+4} > 0$
$\dfrac{3(-5)}{-5+4} = \dfrac{-15}{-1} = 15 > 0$	$\dfrac{3(-1)}{-1+4} = \dfrac{-3}{3} = -1 > 0$	$\dfrac{3(2)}{2+4} = \dfrac{6}{6} = 1 > 0$
(TRUE)	(FALSE)	(TRUE)
Use this interval: $(-\infty, -4)$.	Discard this interval.	Use this interval: $(0, \infty)$.

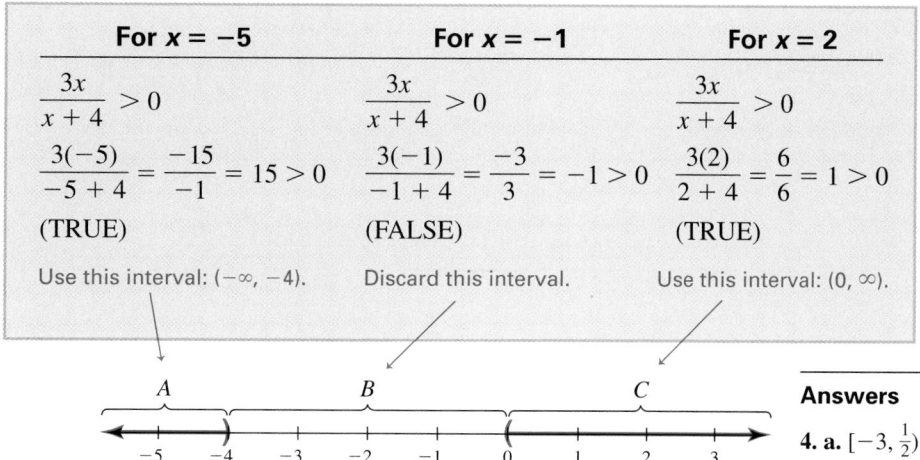

5. The solution set is $(-\infty, -4) \cup (0, \infty)$.

Answers

4. a. $\left[-3, \frac{1}{2}\right)$

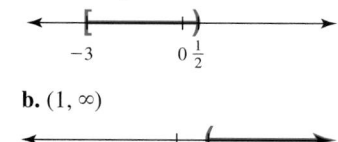

b. $(1, \infty)$

b. 1. To write the inequality in standard form, we first subtract 2 from both sides and get a common denominator to simplify.

$$\frac{x}{x-2} \geq 2 \qquad \text{Given.}$$

$$\frac{x}{x-2} - 2 \geq 0 \qquad \text{Subtract 2.}$$

$$\frac{x}{x-2} - \frac{2(x-2)}{x-2} \geq 0 \qquad 2 = \frac{2(x-2)}{x-2}$$

$$\frac{x - 2x + 4}{x - 2} \geq 0 \qquad \text{Remove parentheses in the numerator and write as one fraction.}$$

$$\frac{4 - x}{x - 2} \geq 0 \qquad \text{Simplify.}$$

2. The critical points are 4 and 2. The 2 will cause the denominator to be zero, so it is not included in the solution set. The 4 is included because it will make the rational expression zero, thus yielding a true statement.

3. Separate the number line into three regions using 2 and 4 as boundaries.

4. Select the test points: 0 from *A*, 3 from *B*, and 5 from *C*.

Test	**For x = 0, we want to know if**	**For x = 3, we want to know if**	**For x = 5, we want to know if**
	$\dfrac{x}{x-2} \geq 2.$	$\dfrac{x}{x-2} \geq 2.$	$\dfrac{x}{x-2} \geq 2.$
In original:	$\dfrac{0}{0-2} = 0 \geq 2$ (FALSE)	$\dfrac{3}{3-2} = 3 \geq 2$ (TRUE)	$\dfrac{5}{5-2} = \dfrac{5}{3} \geq 2$ (FALSE)
	Thus $\dfrac{x}{x-2} = 0 < 2.$	Thus $\dfrac{x}{x-2} = 3 \geq 2.$	Thus $\dfrac{x}{x-2} = \dfrac{5}{3} < 2.$
	Discard this interval.	Use this interval: (2, 4].	Discard this interval.

5. The solution set is (2, 4], which does not include 2 but does include 4, because 4 satisfies the inequality, $\frac{x}{x-2} \geq 2$, and 2 does not.

You have probably noticed that all the solution sets we've obtained are either intervals or unions of intervals. There are other possibilities, and they are based on the fact that for any real number a, $a^2 \geq 0$. Table 2 gives you some of these unusual solution sets, where a and x represent real numbers.

Table 2 Unusual Solution Sets

Explanation	Example	Solution Set
$a^2 \geq 0$ for every real number x.	$(x-1)^2 \geq 0$ for every real number x.	The solution set of $(x-1)^2 \geq 0$ is the set of all real numbers.
$a^2 \leq 0$ only when $a = 0$.	$(x-1)^2 \leq 0$ only when $x - 1 = 0$, that is, when $x = 1$.	The solution set of $(x-1)^2 \leq 0$ is $\{1\}$.
a^2 is never negative, so $a < 0$ is false.	$(x-1)^2$ is never negative, so $(x-1)^2 < 0$ is false.	The solution set of $(x-1)^2 < 0$ is the empty set $\varnothing$.

You can check all these examples by using the same five-step procedure we've been using.

D **Solving an Application Involving Inequalities**

Example 5 shows how to apply the material we've been studying to a projectile motion problem. Remember our RSTUV method? We will use it again here.

EXAMPLE 5 **Problem for a rocket scientist**

The height h (in ft) of a rocket is given by $h = 64t - 16t^2$, where t is the time in seconds. During what time interval will the rocket be more than 48 ft above the ground?

SOLUTION We'll use the RSTUV method.

1. Read the problem.
The problem asks for a certain time interval in which $h > 48$.

2. Select the unknown.
The unknown is the time t.

3. Think of a plan.
We have to find values of t for which $h = 64t - 16t^2 > 48$.

4. Use the five-step procedure to solve the inequality.
   1. Subtract 48 from both sides and rewrite in descending order. $\quad -16t^2 + 64t - 48 > 0$

   2. The related equation is $-16t^2 + 64t - 48 = 0$

$$t^2 - 4t + 3 = 0 \quad \text{Divide by } -16.$$
$$(t-1)(t-3) = 0 \quad \text{Factor.}$$

The critical values are 1 and 3.

48 ft 48 ft

PROBLEM 5

The bending moment of a beam is $M = 30 - x^2$, where x is the distance, in ft, from one end of the beam. At what distances will the bending moment be more than 5?

Answer

5. Between 0 and 5 ft

3. Separate the number line into three regions, A, B, and C with 1 and 3 as boundaries.
4. Select the test points: 0 from A, 2 from B, and 4 from C. Test in original inequality, $64t - 16t^2 > 48$

For $t = 0$, $64(0) - 16(0)^2 = 0 > 48$. (FALSE)   Discard interval A.

For $t = 2$, $64(2) - 16(2)^2 = 128 - 64 = 64 > 48$. (TRUE)   Use interval B: $(1, 3)$.

For $t = 4$, $64(4) - 16(4)^2 = 256 - 256 = 0 > 48$. (FALSE)   Discard interval C.

$$
\begin{array}{c}
\quad A \qquad\qquad B \qquad\qquad C \\
\overbrace{}\quad\overbrace{}\quad\overbrace{} \\
\longleftarrow\!\!+\quad+\quad(\quad+\quad)\quad+\quad+\!\!\longrightarrow \\
-1\quad 0\quad 1\quad 2\quad 3\quad 4\quad 5
\end{array}
$$

5. The solution set is $(1, 3)$, which means that the rocket is more than 48 ft high when the time t is between 1 and 3 sec.

5. Verify the solution.
When $t = 1$, the height of the rocket is $h = 64(1) - 16(1)^2 = 48$, exactly 48 ft. If t is more than 1 and less than 3, the height is more than 48. Now on the way down, when $t = 3$, the height is $h = 64(3) - 16(3)^2 = 48$ ft again.

Calculate It **Solve a Polynomial Inequality**

Your calculator simplifies the work in this section more than in any other work we've encountered. To solve a polynomial inequality with a calculator,

1. Write the inequality in standard form.
2. Graph the related equation.
3. The solution set consists of the intervals on the **horizontal** axis, the **x-axis,** for which the corresponding portions of the graph satisfy the inequality (above the x-axis if the inequality is $>$ or $\geq$ below the x-axis if the inequality is $<$ or $\leq$).

Thus to do Example 1, graph the related equation $Y_1 = (x - 1)(x + 3)$ in a standard window. The portion of the graph *below* the horizontal axis (less than zero) has corresponding x-values between -3 and 1. Hence the solution set is as before, the interval $(-3, 1)$ (see Window 1).

In Example 2, enter $Y_1 = x^2 - 6x + 7$. Enter $Y_2 = 0$, then press `2nd` `TRACE` `5` with `ENTER` `ENTER` `ENTER` to get the points at which the graph cuts the x-axis, which can be approximated by 1.59 and 4.41. Since we want the points for which the graph is ≥ 0, the solution set consists of the points in the interval $(-\infty, 1.59) \cup (4.41, \infty)$ (see Window 2). Now you try Example 3.

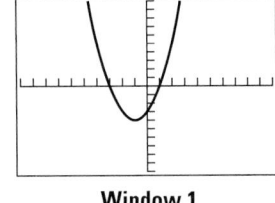

Window 1

$Y_1 = (x - 1)(x + 3)$

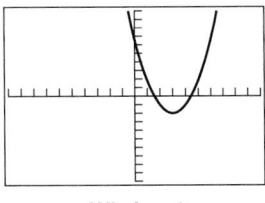

Window 2

$Y_1 = x^2 - 6x + 7$

For Example 4(b), graph

$$Y_1 = \frac{(4 - x)}{(x - 2)}$$

using a $[-5, 5]$ by $[-5, 5]$ window or, if you want to avoid some of the algebra, graph

$$Y_1 = \left(\frac{x}{(x - 2)}\right) - 2$$

an equivalent form of

$$\frac{(4 - x)}{(x - 2)}$$

In either case, be very careful with the placement of parentheses! Clearly, the solution set consists of the x-values whose corresponding y-values (vertical) are above the x-axis—the values between 2 and 4. The 2 is excluded because it will yield a zero denominator, so the solution set is as before, $(2, 4]$ (see Window 3).

You can also graph

$$Y_1 = \frac{x}{(x - 2)}$$

and $Y_2 = 2$ and find the interval where Y_1 is above Y_2 as shown in Window 4.

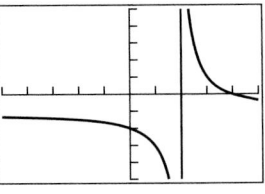

Window 3

$Y_1 = \dfrac{4 - x}{x - 2}$

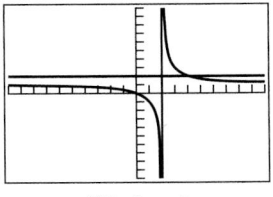

Window 4

Exercises 8.5

A In Problems 1–16, solve and graph the given inequality.

1. $(x + 1)(x - 3) > 0$

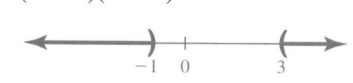

$x < -1$ or $x > 3$; $(-\infty, -1) \cup (3, \infty)$

2. $(x - 1)(x + 2) < 0$

$-2 < x < 1$; $(-2, 1)$

3. $x(x + 4) \leq 0$

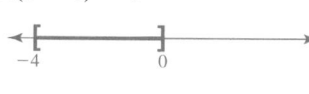

$-4 \leq x \leq 0$; $[-4, 0]$

4. $(x - 1)x \geq 0$

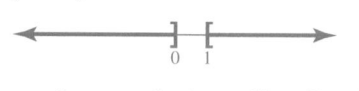

$x \leq 0$ or $x \geq 1$; $(-\infty, 0] \cup [1, \infty)$

5. $x^2 - x - 2 \leq 0$

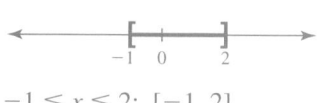

$-1 \leq x \leq 2$; $[-1, 2]$

6. $x^2 - x - 6 \leq 0$

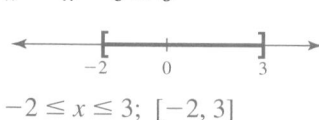

$-2 \leq x \leq 3$; $[-2, 3]$

7. $x^2 - 3x \geq 0$

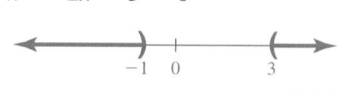

$x \leq 0$ or $x \geq 3$; $(-\infty, 0] \cup [3, \infty)$

8. $x^2 + 2x \leq 0$

$-2 \leq x \leq 0$; $[-2, 0]$

9. $x^2 - 3x + 2 < 0$

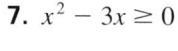

$1 < x < 2$; $(1, 2)$

10. $x^2 - 2x - 3 > 0$

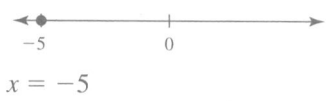

$x < -1$ or $x > 3$; $(-\infty, -1) \cup (3, \infty)$

11. $x^2 + 2x - 3 < 0$

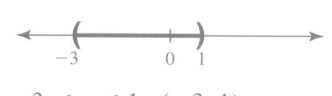

$-3 < x < 1$; $(-3, 1)$

12. $x^2 + x - 2 < 0$

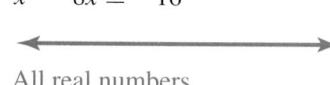

$-2 < x < 1$; $(-2, 1)$

13. $x^2 + 10x \leq -25$

$x = -5$

14. $x^2 + 8x \leq -16$

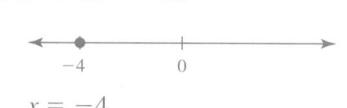

$x = -4$

15. $x^2 - 8x \geq -16$

All real numbers

16. $x^2 - 6x \geq -9$

All real numbers

In Problems 17–20, use the quadratic formula to approximate the critical values.

17. $x^2 - x \geq 1$ (use 2.2 for $\sqrt{5}$) 1.6, −0.6

18. $x^2 - x \geq 3$ (use 3.6 for $\sqrt{13}$) 2.3, −1.3

19. $x^2 - x \leq 4$ (use 4.1 for $\sqrt{17}$) 2.55, −1.55

20. $x^2 - x \leq 5$ (use 4.6 for $\sqrt{21}$) 2.8, −1.8

B In Problems 21–24, solve and graph the inequality.

21. $(x + 1)(x - 2)(x + 3) \geq 0$

$-3 \leq x \leq -1$ or $x \geq 2$; $[-3, -1] \cup [2, \infty)$

22. $(x - 1)(x + 2)(x - 3) \geq 0$

$-2 \leq x \leq 1$ or $x \geq 3$; $[-2, 1] \cup [3, \infty)$

23. $(x - 1)(x - 2)(x - 3) \leq 0$

$x \leq 1$ or $2 \leq x \leq 3$; $(-\infty, 1] \cup [2, 3]$

24. $(x - 2)(x - 3)(x - 4) \leq 0$

$x \leq 2$ or $3 \leq x \leq 4$; $(-\infty, 2] \cup [3, 4]$

C In Problems 25–34, solve and graph the inequality.

25. $\dfrac{2}{x-2} \geq 0$ $x > 2$; $(2, \infty)$

26. $\dfrac{3}{x-1} \leq 0$ $x < 1$; $(-\infty, 1)$

27. $\dfrac{x+5}{x-1} > 0$ $x < -5$ or $x > 1$; $(-\infty, -5) \cup (1, \infty)$

28. $\dfrac{2x-3}{x+3} < 0$ $-3 < x < \dfrac{3}{2}$; $\left(-3, \dfrac{3}{2}\right)$

29. $\dfrac{3x-4}{2x-1} < 0$ $\dfrac{1}{2} < x < \dfrac{4}{3}$; $\left(\dfrac{1}{2}, \dfrac{4}{3}\right)$

30. $\dfrac{x-1}{x+5} < 1$ $x > -5$; $(-5, \infty)$

31. $\dfrac{x+5}{x-1} > 2$ $1 < x < 7$; $(1, 7)$

32. $\dfrac{3x-1}{x+4} > 1$ $x < -4$ or $x > \dfrac{5}{2}$; $(-\infty, -4) \cup \left(\dfrac{5}{2}, \infty\right)$

33. $\dfrac{1}{x-1} < \dfrac{1}{x-2}$ $x < 1$ or $x > 2$; $(-\infty, 1) \cup (2, \infty)$

34. $\dfrac{3}{x} + 1 < \dfrac{1}{x} - 2$ $-\dfrac{2}{3} < x < 0$; $\left(-\dfrac{2}{3}, 0\right)$

In Problems 35–38, find all values of x for which the given expression is a real number. (*Hint:* $\sqrt{a}$ is a real number if $a \geq 0$.)

35. $\sqrt{x^2 - 9}$ $x \leq -3$ or $x \geq 3$; $(-\infty, -3] \cup [3, \infty)$

36. $\sqrt{x^2 - 4x + 4}$ All real numbers

37. $\sqrt{x^2 - 6x + 5}$ $x \leq 1$ or $x \geq 5$; $(-\infty, 1] \cup (5, \infty)$

38. $\sqrt{3x - 8}$ $x \geq \dfrac{8}{3}$; $\left[\dfrac{8}{3}, \infty\right)$

APPLICATIONS

More, Time, and Speed

39. The equivalent resistance of two electric circuits is given by $R^2 - 3R + 1$. When is this resistance more than 5 ohms? $R > 4$

40. The bending moment of a beam is $M = 20 - x^2$, where x is the distance, in ft, from one end of the beam. At what distances will the bending moment be more than 4? $0 < x < 4$

41. The number N of water mites in a water sample depends on the temperature T in degrees Fahrenheit and is given by $N = 110T - T^2$. At what temperatures will the number of mites exceed 1000? $10° < T < 100°$

42. The profit P in a business varies in an 8-hr day according to the formula $P = 15t - 5t^2$, where t is the time in hours. During what hours is there a profit; that is, during what hours is $P > 0$? $0 < t < 3$

43. The height h (in ft) of a projectile is $h = 48t - 16t^2$, where t is the time in seconds. During what time interval will the projectile be more than 32 ft above the ground? $1 < t < 2$

44. The distance d (in ft) in which a car traveling v mi/hr can be stopped is given by

$$d = 0.05v^2 + v$$

At what speed will it take more than 120 ft to stop the car? $v > 40$ mi/hr

SKILL CHECKER

Try the "Skill Checker" exercises so you'll be ready for the next section.

Find y for the given value of x:

45. $y = 3x + 6$, $x = 2$ 12

46. $y = 3x + 6$, $x = -2$ 0

47. $y = -\dfrac{2}{3}x + 4$, $x = 3$ 2

48. $y = -\dfrac{2}{3}x + 4$, $x = -3$ 6

Find x for the given value of y:

49. $y = 3x + 6$, $y = 0$ -2

50. $y = 2x + 8$, $y = 0$ -4

USING YOUR KNOWLEDGE

Stopping Jaguars

In Problems 51–53, give your answers to the nearest tenth of a unit. We already know that the distance d (in feet) in which a car traveling v mi/hr can be stopped is given by

$$d = 0.05v^2 + v$$

51. For what values of v is $50 \le d \le 60$? (*Hint:* $\sqrt{11} \approx 3.32$ and $\sqrt{13} \approx 3.61$.) $23.2 \le v \le 26.1$

52. For what values of v is $80 \le d \le 90$? (*Hint:* $\sqrt{17} \approx 4.12$ and $\sqrt{19} \approx 4.36$.) $31.2 \le v \le 33.6$

53. According to the *Guinness Book of World Records,* the longest skid mark on a public road was made by a Jaguar automobile involved in an accident in England. The skid mark was 950 ft long. If you assume that the car actually stopped after the brakes were applied and that it traveled for 950 ft, how fast was the car traveling? (Use a calculator to solve this problem.) $v = 128.2$ mi/hr

WRITE ON

Explain in your own words:

54. Why the solution set of $(x - 1)^2 \ge 0$ is the set of all real numbers. Answers may vary.

55. Why the solution set of $(x - 1)^2 \le 0$ is the single point $\{1\}$. Answers may vary.

56. Why the solution set of $(x - 1)^2 < 0$ is the empty set $\varnothing$. Answers may vary.

57. Why the solution set of the rational inequality

$$\frac{ax + b}{cx + d} \le 0$$

cannot include more than one end point (where a, b, and c are integers). Answers may vary.

MASTERY TEST

If you know how to do these problems, you have learned your lesson!

Solve and graph the solution set:

58. $\dfrac{3}{x - 2} \le 0$ $x < 2$; $(-\infty, 2)$

59. $\dfrac{x + 2}{x} > 0$ $x < -2$ or $x > 0$; $(-\infty, -2) \cup (0, \infty)$

60. $(x + 3)(x - 2)(x + 1) \ge 0$
  $-3 \le x \le -1$ or $x \ge 2$; $[-3, -1] \cup [2, \infty)$

61. $(x - 4)(x + 3)(x - 1) \le 0$
  $x \le -3$ or $1 \le x \le 4$; $(-\infty, -3] \cup [1, 4]$

62. $x^2 + x \geq 2$ $x \leq -2$ or $x \geq 1$; $(-\infty, -2] \cup [1, \infty)$

63. $x^2 + x \leq 6$ $-3 \leq x \leq 2$; $[-3, 2]$

64. $(x + 3)(x - 2) < 0$ $-3 < x < 2$; $(-3, 2)$

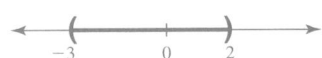

65. $(x + 1)(x - 3) \geq 0$ $x \leq -1$ or $x \geq 3$; $(-\infty, -1] \cup [3, \infty)$

66. The profit P for a restaurant varies in an 8-hr day according to the formula $P = 3t - t^2$, where t is the time in hours. During what hours is there a profit; that is, during what hours is $P > 0$? During the first 3 hr

COLLABORATIVE LEARNING 8A

Is the data in Graph 1 and Graph 2 related to a quadratic or linear function? Use Graph 1 and Graph 2 to answer the following questions.

1. Which graph (1 or 2) would best be modeled as a quadratic function?

2. Write the quadratic equation that would approximate this data.

3. Which graph (1 or 2) would best be modeled as a linear function?

4. Write the linear equation of the line of best fit.

5. Discuss how both graphs can indicate rainfall measures and yet have different shaped graphs and equation models.

Divide into groups and research the monthly rainfall in your area over the period of a year and then over the period of one month.

6. Make a table of your data, draw a line graph, and discuss your findings.

7. Discuss how your findings compare to Graph 1 and Graph 2.

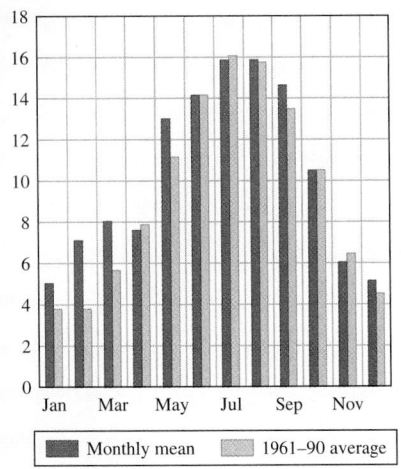

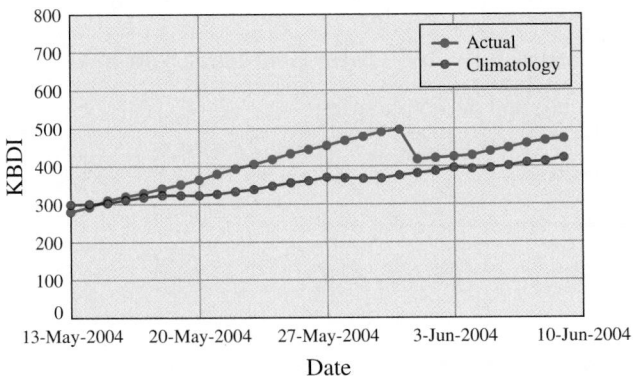

Keetch-Byram Drought Index (KBDI) at Americus From May 13, 2004 to Jun 9, 2004

COLLABORATIVE LEARNING 8B

Exercise and Pulse Rate

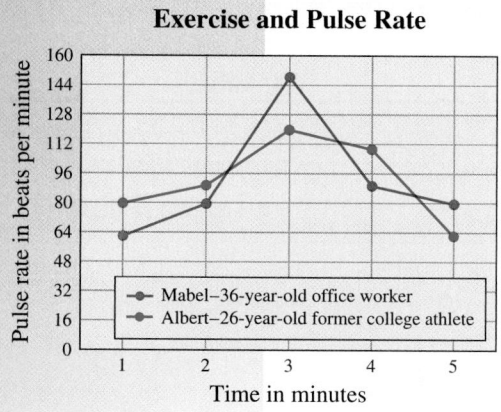

Time in minutes

How does a few minutes of exercise affect your pulse rate? That depends on whether you are Mabel, a 36-year-old office worker, or Albert, a 26-year-old former college athlete. The graph indicates the pulse rate in beats per minute of the two, Mabel and Albert, over a 3-minute exercise period.

Divide into groups, answer the questions, and then conduct your own research regarding pulse rate and exercise with respect to age.

1. Comparing pulse rates for Mabel and Albert,

   a. Whose pulse rate was the highest in beats per minute? What was the rate?

   b. Whose pulse rate was the lowest? What was the rate?

2. The graphs resemble those of a quadratic function. Write the equations for the two graphs based on the data in the following table. Using your calculator you can enter the list and calculate a quadratic regression. If you do not have a calculator, you can solve the system of three equations created by using $y = ax^2 + bx + c$ and entering each ordered pair as the respective x and y values each time.

Albert		Mabel	
Time in Minutes	Pulse Rate in Beats per Minute	Time in Minutes	Pulse Rate in Beats per Minute
1	80	1	62
3	120	3	150
5	62	5	80

3. Using the equations found in Question 2 for Albert and Mabel's pulse rates, what is the maximum of each parabola? How does that compare with the actual maximum point for each in the graph?

4. Measure the pulse rate in beats per minute of two people with significant age differences before, during, and after an exercise workout. Put the data in a table, graph it, and then discuss how it compares to Albert and Mabel's pulse rates during their workout. How do you account for any discrepancies?

Research Questions

1. Write a paper explaining how Euclid used theorems on areas to solve the quadratic equation $x^2 + b^2 = ax$.

2. Write a paper detailing Al-Khowarizmi's geometric demonstration of his rules to solve the quadratic equation $x^2 + 10x = 39$.

3. Why did they call Al-Khowarizmi's method of solving quadratics "completing the square"?

4. We have mentioned that Al-Khowarizmi was probably the foremost Arabic algebraist. The other one was dubbed "The Reckoner from Egypt." Who was this person, what did he write, and on what subjects?

Summary

SECTION	ITEM	MEANING	EXAMPLE
8.1	Quadratic equation in standard form $\pm\sqrt{a}$	An equation of the form $$ax^2 + bx + c = 0 \; (a \neq 0)$$ The solutions of $x^2 = a$	$2x^2 - 3x + 7 = 0$ is a quadratic equation in standard form. $\pm\sqrt{3}$ are the solutions of $x^2 = 3$.
8.1C	Completing the square	1. Write the equation with the variables in descending order on the left and numbers on the right. 2. Divide each term by the coefficient of x^2 ($\neq 1$). 3. Add the square of one-half the coefficient of x to both sides. 4. Factor the perfect square trinomial on the left; simplify on the right. 5. Solve the resulting equation.	$5x^2 - 5x - 1 = 0$ Given. 1. $5x^2 - 5x = 1$ 2. $x^2 - x = \dfrac{1}{5}$ 3. $x^2 - x + \left(-\dfrac{1}{2}\right)^2$ $= \dfrac{1}{5} + \left(-\dfrac{1}{2}\right)^2$ 4. $\left(x - \dfrac{1}{2}\right)^2 = \dfrac{9}{20}$ 5. $x - \dfrac{1}{2} = \pm\sqrt{\dfrac{9}{20}}$ $x - \dfrac{1}{2} = \pm\dfrac{3\sqrt{5}}{10}$ $x = \dfrac{1}{2} \pm \dfrac{3\sqrt{5}}{10} = \dfrac{5 \pm 3\sqrt{5}}{10}$
8.2A	Quadratic formula	The solutions of $ax^2 + bx + c = 0$ are $$x = \frac{-b \pm \sqrt{b^2 - 4ac}}{2a}$$	The solutions of $3x^2 + 2x - 5 = 0$ are $\dfrac{-2 \pm \sqrt{4 + 60}}{6}$ which equal 1 and $-\dfrac{5}{3}$.
8.3	Discriminant	The discriminant of $ax^2 + bx + c = 0$ is $D = b^2 - 4ac$.	The discriminant of $3x^2 + 2x - 5 = 0$ is $b^2 - 4ac = 64$.
8.3A	Types of solutions for $ax^2 + bx + c = 0$ where a, b, and c are rational numbers	1. If $D > 0$ and D is not a perfect square, two different real, irrational solutions 2. If $D > 0$ and $D = N^2$ (a perfect square), two different real, rational solutions 3. If $D < 0$, two different non-real complex solutions 4. If $D = 0$, one rational solution	1. $4x^2 - 3x - 5 = 0 \; (D = 89)$ has two different irrational solutions. 2. $x^2 - 2x - 3 = 0 \; (D = 16)$ has two different rational solutions. 3. $4x^2 - 3x + 5 = 0 \; (D = -71)$ has two different non-real solutions. 4. $4x^2 - 4x + 1 = 0 \; (D = 0)$ has one rational solution.

SECTION	ITEM	MEANING	EXAMPLE
8.3B	Factorable quadratic expressions	If D is a perfect square, $ax^2 + bx + c$ is factorable.	$20x^2 + 10x - 30$ is factorable ($D = 2500$). $20x^2 + 10x - 30 = 10(2x + 3)(x - 1)$
8.3D	Sum and product of the solutions	The sum and product of the solutions of $ax^2 + bx + c = 0$ are $-\frac{b}{a}$ and $\frac{c}{a}$, respectively.	The sum and product of the solutions of $4x^2 - 12x + 5 = 0$ are 3 and $\frac{5}{4}$, respectively.
8.4A	Equations that are quadratic in form	Equations that can be written as quadratics by use of appropriate substitutions	$x^4 - 5x^2 + 4 = 0$, $\Rightarrow u^2 - 5u + 4 = 0$ $(x^2 - x)^2 - (x^2 - x) - 2 = 0$, $\Rightarrow u^2 - u - 2 = 0$
8.5B	Polynomial inequality	An inequality that can be written with the polynomial on the left side and zero on the right side. The symbol can be $<, >, \le,$ or $\ge$.	$x^2 - x - 2 < 0$ $(x + 1)(x - 2) < 0$ Critical numbers: $-1, 2$ Solution: $(-1, 2)$

Review Exercises

(If you need help with these exercises, look in the section indicated in brackets.)

1. [8.1A] Solve.
   a. $16x^2 - 49 = 0$ $\pm\frac{7}{4}$
   b. $25x^2 - 16 = 0$ $\pm\frac{4}{5}$

2. [8.1A] Solve.
   a. $5x^2 + 30 = 0$ $\pm i\sqrt{6}$
   b. $6x^2 + 42 = 0$ $\pm i\sqrt{7}$

3. [8.1B] Solve.
   a. $(x - 3)^2 = 32$ $3 \pm 4\sqrt{2}$
   b. $(x - 5)^2 = 50$ $5 \pm 5\sqrt{2}$

4. [8.1B] Solve.
   a. $2(x - 2)^2 + 25 = 0$ $2 \pm \frac{5\sqrt{2}}{2}i$
   b. $3(x - 3)^2 + 64 = 0$ $3 \pm \frac{8\sqrt{3}}{3}i$

5. [8.1B] Solve.
   a. $5x^2 - 10x + 5 = 12$ $\frac{5 \pm 2\sqrt{15}}{5}$
   b. $12x^2 + 12x + 3 = 16$ $\frac{-3 \pm 4\sqrt{3}}{6}$

6. [8.1C] Solve by completing the square.
   a. $x^2 - 8x - 9 = 0$ $9, -1$
   b. $x^2 + 12x + 32 = 0$ $-4, -8$

7. [8.1C] Solve by completing the square.
   a. $4x^2 + 4x - 3 = 0$ $\frac{1}{2}, -\frac{3}{2}$
   b. $16x^2 - 24x + 7 = 0$ $\frac{3 \pm \sqrt{2}}{4}$

8. [8.2A] Solve by the quadratic formula.
   a. $3x^2 + 5x - 2 = 0$ $-2, \frac{1}{3}$
   b. $5x^2 - 9x - 2 = 0$ $-\frac{1}{5}, 2$

9. [8.2A] Solve by the quadratic formula.
   a. $3x^2 = 2x + 4$ $\frac{1 \pm \sqrt{13}}{3}$
   b. $4x^2 = 6x + 3$ $\frac{3 \pm \sqrt{21}}{4}$

10. [8.2A] Solve by the quadratic formula.
   a. $16x = x^2$ $0, 16$
   b. $12x = x^2$ $0, 12$

11. [8.2A] Solve by the quadratic formula.
   a. $x^2 + \frac{x}{15} = \frac{1}{3}$ $\frac{-1 \pm \sqrt{301}}{30}$
   b. $\frac{x^2}{2} + \frac{9x}{10} = \frac{1}{5}$ $\frac{1}{5}, -2$

12. [8.2A] Solve by the quadratic formula.
   a. $3x^2 - 2x = -1$ $\frac{1}{3} \pm \frac{\sqrt{2}}{3}i$
   b. $5x^2 - 2x = -4$ $\frac{1}{5} \pm \frac{\sqrt{19}}{5}i$

13. [8.2B] Solve.

a. $8x^3 - 125 = 0$ $\frac{5}{2}, -\frac{5}{4} \pm \frac{5\sqrt{3}}{4}i$

b. $125x^3 - 8 = 0$ $\frac{2}{5}, -\frac{1}{5} \pm \frac{\sqrt{3}}{5}i$

14. [8.2C] The demand d is given by

$$d = \frac{450}{p}$$

and the supply s is given by $s = 100p - 150$.

a. Find the equilibrium point (at which $d = s$). $p = 3$

b. Find the equilibrium point when

$$d = \frac{50}{p}$$

and $s = 150p - 100$. $p = 1$

15. [8.3A]

a. Given $16x^2 - 3x = -1$, find the value of the discriminant and classify the solutions. $D = -55$; two non-real complex solutions

b. Given $8x^2 - kx + 2 = 0$, find the discriminant and then find k so that the equation has exactly one rational solution. $D = k^2 - 64$; $k = \pm 8$

16. [8.3B] Use the discriminant to determine whether the given quadratic is factorable into factors with integer coefficients. If so, factor it.

a. $3x^2 - 11x - 6$ Not factorable

b. $18x^2 + 13x + 2$ $(2x + 1)(9x + 2)$

17. [8.3B] Factor into factors with integer coefficients if possible.

a. $18x^2 - 9x - 5$ $(6x - 5)(3x + 1)$

b. $18x^2 + 13x + 1$ Not factorable

18. [8.3C] Find a quadratic equation with integer coefficients whose solution set is

a. $\{-2, 3\}$ $x^2 - x - 6 = 0$

b. $\left\{\frac{1}{4}, -\frac{2}{3}\right\}$ $12x^2 + 5x - 2 = 0$

19. [8.3D] Without solving the equation, find the sum, using $-\frac{b}{a}$, and the product, using $\frac{c}{a}$, of the solutions of

a. $15x^2 + 4x - 3 = 0$
Sum $= -\frac{4}{15}$; product $= -\frac{1}{5}$

b. $9x^2 - 12x - 5 = 0$
Sum $= \frac{4}{3}$; product $= -\frac{5}{9}$

20. [8.3D] Use the sum and product properties to check whether

a. $\frac{1}{3}$ and $-\frac{3}{5}$ are the solutions of the equation in Problem 19, part **a.** Yes

b. $\frac{1}{3}$ and $-\frac{5}{3}$ are the solutions of the equation in Problem 19, part **b.** No

21. [8.4A] Solve.

a. $\frac{8}{x^2 - 16} - \frac{1}{x - 4} = 1$ -5

b. $\frac{-24}{x^2 - 36} + \frac{2}{x + 6} = 1$ $0, 2$

22. [8.4B] Solve.

a. $(x^2 + x)^2 + 2(x^2 + x) - 8 = 0$
$1, -2, -\frac{1}{2} \pm \frac{\sqrt{15}}{2}i$

b. $(x^2 - 3x)^2 - 4(x^2 - 3x) - 12 = 0$
$1, 2, \frac{3 \pm \sqrt{33}}{2}$

23. [8.4B] Solve.

a. $x^{1/2} - 4x^{1/4} = -3$ $1, 81$

b. $x^{1/2} + x^{1/4} = 6$ 16

24. [8.4B] Solve.

a. $x^{2/3} + x^{1/3} = 12$ $27, -64$

b. $x^{2/3} - 5x^{1/3} = 6$ $216, -1$

25. [8.4B] Solve.

a. $x - 2\sqrt{x} = 3$ 9

b. $x - 4\sqrt{x} = 5$ 25

26. [8.4B] Solve.

a. $3x^{-4} - 4x^{-2} + 1 = 0$ $\pm 1, \pm\sqrt{3}$

b. $3x^{-4} - 2x^{-2} - 1 = 0$ $\pm 1, \pm i\sqrt{3}$

For Exercises 27–32, write the solutions in interval notation.

27. [8.5A] Solve and graph.

a. $(x - 2)(x + 3) < 0$ $(-3, 2)$

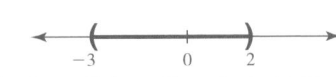

b. $(x + 2)(x - 3) < 0$ $(-2, 3)$

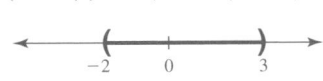

28. [8.5A] Solve and graph.

a. $x^2 + 4x \geq 0$ $(-\infty, -4] \cup [0, \infty)$

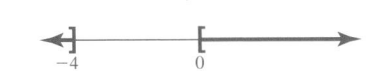

b. $x^2 - 3x \geq 0$ $(-\infty, 0] \cup [3, \infty)$

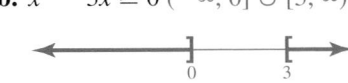

29. [8.5A] Solve and graph.

a. $x^2 + 4x < 45$ $(-9, 5)$

b. $x^2 - 2x < 2$
$(1 - \sqrt{3}, 1 + \sqrt{3})$

30. [8.5B] Solve and graph.

    a. $(x - 1)(x - 2)(x - 3) \leq 0$

      $(-\infty, 1] \cup [2, 3]$

    b. $(x + 1)(x + 2)(x - 3) \leq 0$

      $(-\infty, -2] \cup [-1, 3]$

31. [8.5B] Solve and graph.

    a. $(x + 1)(x + 2)(x - 3) > 0$

      $(-2, -1) \cup (3, \infty)$

    b. $(x - 1)(x + 2)(x + 3) > 0$

      $(-3, -2) \cup (1, \infty)$

32. [8.5C] Solve and graph.

    a. $\dfrac{x + 2}{x - 2} \leq 0 \ \ [-2, 2)$

    b. $\dfrac{-12}{x - 6} < 3 \ \ (-\infty, 2) \cup (6, \infty)$

Practice Test 8

(Answers on page 640)

1. Solve.

    a. $25x^2 - 4 = 0$

    b. $18x^2 + 3 = 0$

3. Solve $3x^2 + 6x + 3 = 5$.

5. Solve by completing the square: $9x^2 - 6x - 1 = 0$.

7. Solve by the quadratic formula: $3x^2 = 3x + 2$.

9. Solve by the quadratic formula:

$$\frac{x^2}{16} + \frac{5x}{4} = \frac{11}{4}$$

11. Solve $64x^3 - 27 = 0$.

13. a. Given $4x^2 - 5x = -6$, find the value of the discriminant and classify the solutions.

    b. Given $4x^2 - kx + 9 = 0$, find k so that the equation has exactly one rational solution.

15. Factor $35x^2 - 32x - 12$ into factors with integer coefficients if possible.

17. a. Find the sum and the product of the solutions of the equation $6x^2 + 7x - 5 = 0$.

    b. Use the sum and product properties to check whether the solutions are $-\frac{5}{3}$ and $\frac{1}{2}$.

19. Solve.

    a. $x^4 - 7x^2 + 6 = 0$

    b. $(x^2 - 2x)^2 - (x^2 - 2x) - 6 = 0$

21. Solve $3x^{-4} - 2x^{-2} - 1 = 0$.

For Exercises 22–25, write the solutions in interval notation.

22. Solve and graph $(x + 1)(x - 2) < 0$.

24. Solve and graph $(x + 1)(x - 2)(x - 3) \leq 0$.

2. Solve.

    a. $(x - 1)^2 = 45$

    b. $2(x - 2)^2 = 49$

4. Solve by completing the square: $x^2 - 6x + 5 = 0$.

6. Solve by the quadratic formula: $7x^2 + 5x - 2 = 0$.

8. Solve by the quadratic formula: $32x = x^2$.

10. Solve by the quadratic formula: $4x^2 + 3x = -2$.

12. If the demand d is given by

$$d = \frac{500}{p}$$

and the supply s is given by $s = 200p - 150$, find the equilibrium point (at which $s = d$).

14. Use the discriminant to determine whether $12x^2 - 4x - 21$ is factorable into factors with integer coefficients.

16. Find a quadratic equation with integer coefficients whose solution set is $\left\{\frac{2}{3}, -\frac{1}{4}\right\}$.

18. Solve.

    a. $\dfrac{6}{x^2 - 9} - \dfrac{1}{x - 3} = 1$

    b. $\dfrac{2}{x^2 - 1} - \dfrac{3}{x - 1} = 1$

20. Solve.

    a. $x^{1/2} - 3x^{1/4} + 2 = 0$

    b. $x - 5\sqrt{x} + 6 = 0$

23. Solve and graph $x^2 + 2x \geq 15$.

25. Solve and graph: $\dfrac{x + 4}{x - 7} \leq 0$

Answers to Practice Test

ANSWER	IF YOU MISSED	REVIEW		
	QUESTION	SECTION	EXAMPLES	PAGE
1. a. $\frac{2}{5}, -\frac{2}{5}$ **b.** $\frac{\sqrt{6}}{6}i, -\frac{\sqrt{6}}{6}i$	1	8.1A	1	584
2. a. $1 \pm 3\sqrt{5}$ **b.** $2 \pm \frac{7\sqrt{2}}{2}$ or $\frac{4 \pm 7\sqrt{2}}{2}$	2	8.1B	2	585–586
3. $-1 \pm \frac{\sqrt{15}}{3}$ or $\frac{-3 \pm \sqrt{15}}{3}$	3	8.1B	3	586–587
4. 1, 5	4	8.1C	4	588–589
5. $\frac{1 \pm \sqrt{2}}{3}$	5	8.1C	5	589–590
6. $\frac{2}{7}, -1$	6	8.2A	1	595–596
7. $\frac{3 \pm \sqrt{33}}{6}$	7	8.2A	2	596–597
8. 0, 32	8	8.2A	3	597
9. 2, −22	9	8.2A	4	598
10. $-\frac{3}{8} \pm \frac{\sqrt{23}}{8}i$	10	8.2A	5	598–599
11. $\frac{3}{4}, -\frac{3}{8} \pm \frac{3\sqrt{3}}{8}i$	11	8.2B	6	600–601
12. When price = \$2	12	8.2C	7	602
13. a. −71; two non-real complex solutions **b.** $k = \pm 12$	13	8.3A	1	607–608
14. Yes, $D = 32^2$.	14	8.3B	2	608
15. $(5x - 6)(7x + 2)$	15	8.3B	3	609
16. $12x^2 - 5x - 2 = 0$	16	8.3C	4	609–610
17. a. Sum: $-\frac{7}{6}$; product: $-\frac{5}{6}$ **b.** Yes	17	8.3D	5	610–611
18. a. −4 **b.** 0, −3	18	8.4A	1	615
19. a. $\pm 1, \pm \sqrt{6}$ **b.** 3, −1, 1 ± i	19	8.4B	2, 3	616–617
20. a. 1, 16 **b.** 4, 9	20	8.4B	4, 5	617–618
21. $\pm 1, \pm \sqrt{3}i$	21	8.4B	6	618
22. (−1, 2)	22	8.5A	1	623
23. (−∞, −5] ∪ [3, ∞)	23	8.5A	2	624
24. (−∞, −1] ∪ [2, 3]	24	8.5B	3	626
25. [−4, 7)	25	8.5C	4	627–628

Cumulative Review Chapters 1–8

1. Use braces to list the elements of the set of even natural numbers less than 14. $\{2, 4, 6, 8, 10, 12\}$

2. Write $\frac{11}{15}$ as a decimal. $0.7\overline{3}$

3. Find: $|-11|$ 11

4. Evaluate: $[-3(7 + 6)] + 7$ -32

5. Solve: $x + 4 = 2(5x - 3)$ $\frac{10}{9}$

6. Graph $\{x \mid x > -1 \text{ and } x < 1\}$

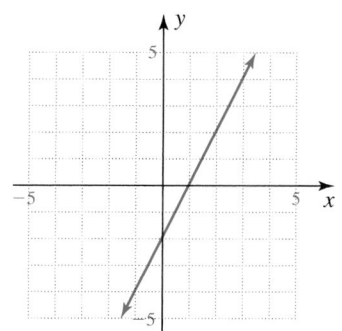

7. Graph: $|3x + 1| > 2$

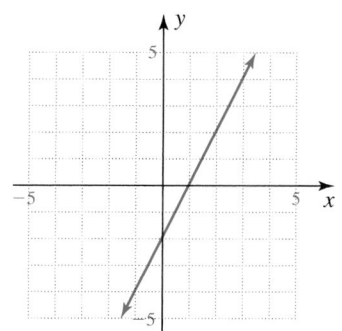

8. Martin purchased some municipal bonds yielding 7% annually and some certificates of deposit yielding 9% annually. If Martin's total investment amounts to $18,000 and the annual income is $1420, how much money is invested in bonds and how much is invested in certificates of deposit?
$10,000 in bonds; $8000 in CDs

9. Graph: $2x - y = 2$

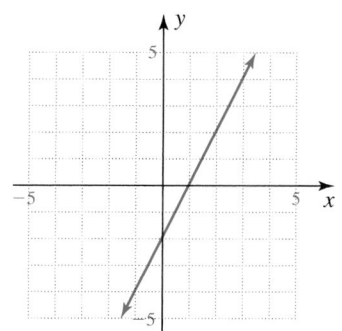

10. A line L_1 has slope $\frac{3}{11}$. Find whether the line through $(-7, 5)$ and $(-4, -6)$ is parallel or perpendicular to line L_1.   Perpendicular

11. Find an equation of the line with slope -1 and passing through the point $(-3, 2)$. $x + y = -1$

12. Find an equation of the line that passes through the point $(-6, 4)$ and is parallel to the line $8x - y = 3$.
$8x - y = -52$

13. Find the domain and range of $\{(x, y) \mid y = 5 + x\}$.
$D = \{\text{all real numbers}\}; R = \{\text{all real numbers}\}$

14. Use the graphical method to solve the system:
$x + y \leq -1$
$y \leq 3x - 5$

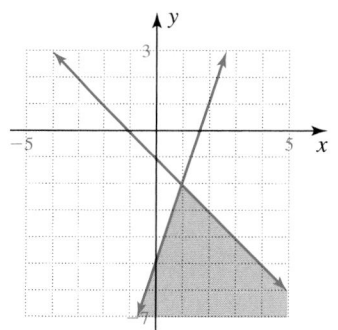

15. Solve by the elimination method:
$3x + 4y = -1$
$x + \ \ y = -1$ $(-3, 2)$

16. Use Cramer's rule to solve for x:
$x + 2y = 6$
$4x + 2y = 7$ $x = \frac{1}{3}$

17. Jose has $3.20 in nickels and dimes. He has 10 more nickels than dimes. How many of each coin does he have?   28 nickels; 18 dimes

18. Subtract $6x^3 - 6x^2 + 4x - 9$ from $4x^3 + 3x^2 + 9$.
$-2x^3 + 9x^2 - 4x + 18$

19. Factor: $36x^2 + 60xy + 25y^2$ $(6x + 5y)^2$

20. Factor: $2x^4 + 8x^2$ $2x^2(x^2 + 4)$

21. Solve for x: $x^3 + x^2 - 9x - 9 = 0$ $x = -1$ or -3 or 3

22. Divide: $\dfrac{-12x^7 + 12x^3 + 6x^2}{6x^5}$ $-2x^2 + \dfrac{2}{x^2} + \dfrac{1}{x^3}$

23. Perform the indicated operations: $\dfrac{\dfrac{x}{x^2 - 81} - \dfrac{9}{x^2 - 81}}{\dfrac{1}{x + 9}}$

24. Simplify: $\dfrac{x - \dfrac{1}{x^2}}{x - \dfrac{1}{x^3}}$ $\dfrac{x(x^2 + x + 1)}{(x + 1)(x^2 + 1)}$

25. Solve for x: $1 - \dfrac{1}{x + 1} = \dfrac{2}{x^2 - 1}$ $x = 2$

26. Find two consecutive even integers such that the sum of their reciprocals is $\dfrac{9}{40}$. 8 and 10

27. Find, if possible in the real numbers: $(-16)^{1/2}$
Not a real number

28. Simplify: $\sqrt[3]{192a^7b^9}$ $4a^2b^3\sqrt[3]{3a}$

29. Rationalize the denominator: $\sqrt{\dfrac{5}{6k}}$ $\dfrac{\sqrt{30k}}{6k}$

30. Perform the indicated operations: $\sqrt[3]{3x}\left(\sqrt[3]{9x^2} + \sqrt[3]{16x}\right)$ $3x + 2\sqrt[3]{6x^2}$

31. Solve: $\sqrt{x + 8} - x = 8$ $x = -8$ or -7

32. Find: $(1 - 4i) - (4 + i)$ $-3 - 5i$

33. Multiply: $(-1 - 8i)(-4 - 3i)$ $-20 + 35i$

34. Solve for x: $16x^2 - 9 = 0$ $x = \dfrac{3}{4}$ or $-\dfrac{3}{4}$

35. Solve by completing the square: $x^2 + 2x - 3 = 0$ $x = -3$ or 1

36. Solve by the quadratic formula: $x^2 = -4x + 2$ $x = -2 \pm \sqrt{6}$

37. If the demand (d) is given by $d = \dfrac{480}{p}$ and the supply is given by $s = 200p - 680$, find the equilibrium point (at which $s = d$). $p = 4$

38. Determine whether $12x^2 + 7x - 10$ is factorable into factors with integer coefficients. Yes

39. Solve for x: $x^4 - 8x^2 + 7 = 0$ $x = \pm 1$ or $x = \pm\sqrt{7}$

40. Solve for x: $(x + 6)(x - 4) < 0$ $-6 < x < 4$

Quadratic Functions and the Conic Sections

9

9.1 Quadratic Functions and Their Graphs

9.2 Circles and Ellipses

9.3 Hyperbolas and Identification of Conics

9.4 Nonlinear Systems of Equations

9.5 Nonlinear Systems of Inequalities

The Human Side of Algebra

Why do we study conic sections? The answer is a tangled tale based in antiquity. The first 300 years of Greek mathematics yielded three famous problems: the squaring of the circle, the trisection of an angle, and the duplication of the cube. The third problem was first mentioned by a Greek poet describing mythical king Minos as dissatisfied with the size of the tomb erected for his son Glaucus. "You have embraced too little space; quickly double it without spoiling its beautiful cubical form." When the construction was done incorrectly, legend says, local geometers were hastily summoned to solve the problem. A different version has the Athenians appealing to the oracle at Delos to get rid of pestilence. "Apollo's cubical altar must be doubled in size" was the reply. When workmen did the work, again incorrectly, "the indignant god made the pestilence even worse than before."

Hippocrates of Chios (ca. 440–380 B.C.) finally showed that the duplication of the cube could be done by using curves with certain properties, but it was the Greek mathematician Menaechmus (350 B.C.), the tutor of Alexander the Great, reputed to have discovered the curves later known as the ellipse, the parabola, and the hyperbola (the conics) to solve the problem.

Pretest for Chapter 9

(Answers on pages 645–646)

1. Find the vertex and graph the parabola $y = x^2 + 4x - 1$.

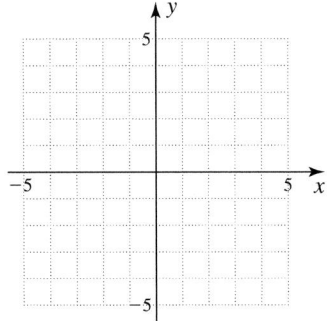

2. Find the vertex and graph the parabola $x = 4(y - 1)^2 - 1$.

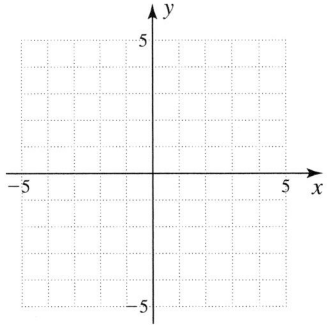

3. If the revenue is given by $R = 40x - 0.02x^2$, find the value of x that yields the maximum revenue.

4. Find an equation of the circle of radius 3 and with center at $(-1, 2)$.

5. Find an equation of the circle of radius 2 and with center at the origin.

6. Find the center and the radius and sketch the graph of $(x - 1)^2 + (y + 2)^2 = 7$.

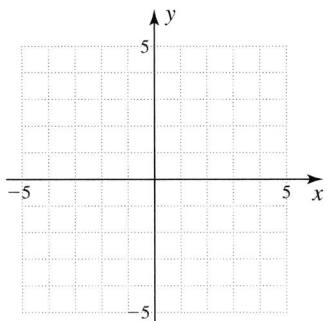

7. Find the coordinates of the center, the vertices, and graph $\frac{(x - 2)^2}{9} + \frac{(y + 1)^2}{4} = 1$.

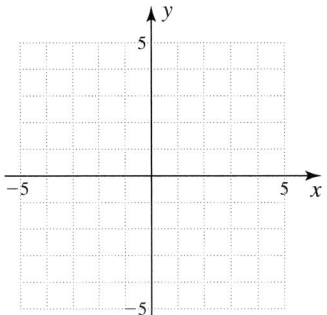

8. Find the coordinates of the center, the vertices, and graph $\frac{y^2}{9} - \frac{x^2}{16} = 1$.

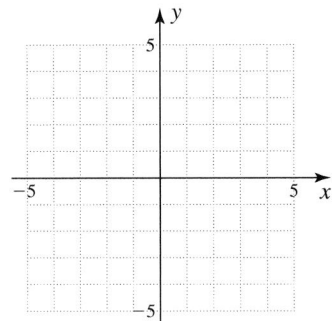

9. Identify each of the following curves.

  a. $x^2 = 4 - y^2$

  b. $y = x^2 - 9$

  c. $9x^2 = 144 - 16y^2$

  d. $16x^2 = 144 + 9y^2$

10. Use the substitution method to solve the system.

$$x^2 + y^2 = 9$$
$$x + y = 3$$

11. Use the substitution method to solve the system.

$$x^2 + y^2 = 4$$
$$x + y = 4$$

12. Solve the system.

$$x^2 - y^2 = 2$$
$$x^2 + 2y^2 = 8$$

13. The cost C of manufacturing and selling x units of a product is $C = 20x + 75$, and the corresponding revenue R is $R = x^2 - 50$. Find the break-even value of x.

14. Graph the inequality $y \le -x^2 + 4$.

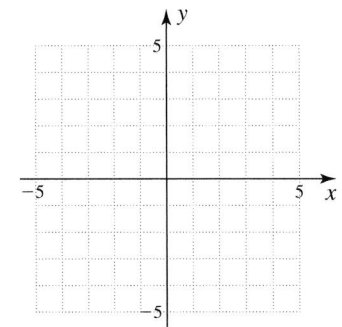

15. Graph the system $x^2 + y^2 \le 9$ and $y \le x^2$.

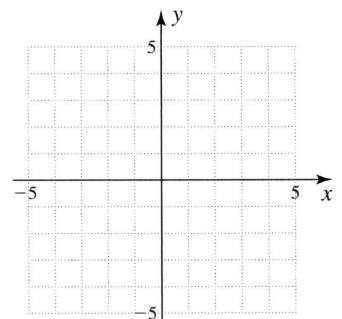

Answers to Pretest

ANSWER	IF YOU MISSED	REVIEW		
	QUESTION	**SECTION**	**EXAMPLES**	**PAGE**

1. Vertex $(-2, -5)$

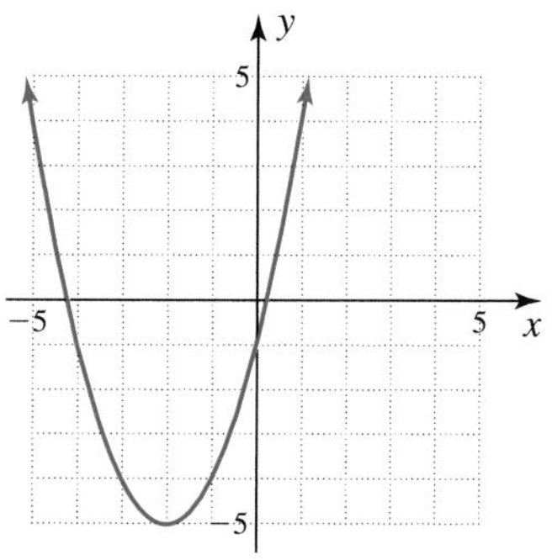

| | 1 | 9.1 | 5, 6 | 654–657 |

2. Vertex $(-1, 1)$

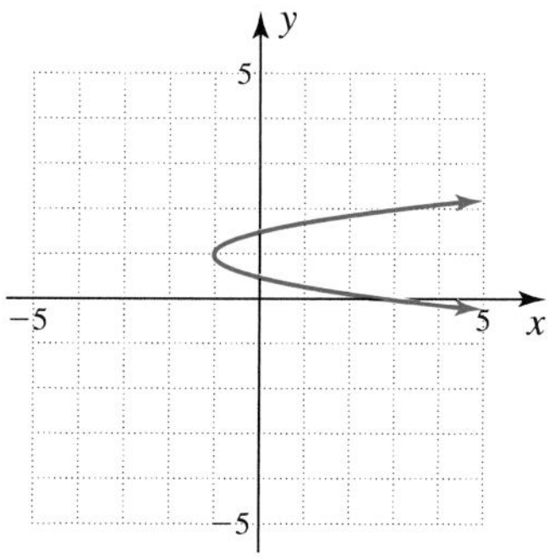

| | 2 | 9.1 | 7 | 658 |

3. $x = 1000$

| | 3 | 9.1 | 9 | 659 |

4. $(x + 1)^2 + (y - 2)^2 = 9$

| | 4 | 9.2 | 2 | 671 |

5. $x^2 + y^2 = 4$

| | 5 | 9.2 | 3 | 671 |

6. Center, $(1, -2)$; $r = \sqrt{7} \approx 2.6$

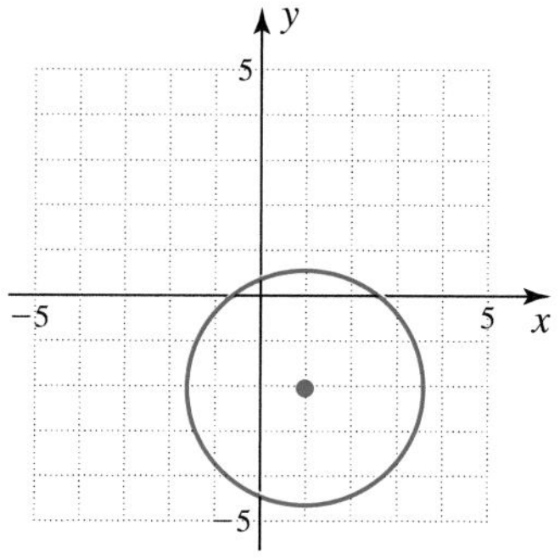

| | 6 | 9.2 | 4 | 672 |

7. Center at $(2, -1)$; $V_1 = (5, -1)$; $V_2 = (-1, -1)$

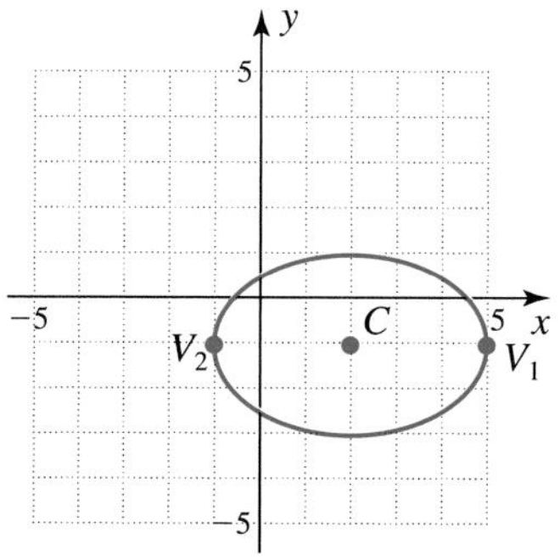

| | 7 | 9.2 | 8 | 675–676 |

ANSWER	IF YOU MISSED	REVIEW		
	QUESTION	SECTION	EXAMPLES	PAGE
8. Center at $(0, 0)$; $V_1 = (0, 3)$; $V_2 = (0, -3)$ 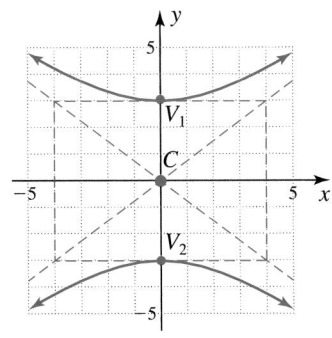	8	9.3	1a	688
9. a. Circle **b.** Parabola **c.** Ellipse **d.** Hyperbola	9	9.3	3	692
10. $(3, 0)$ and $(0, 3)$	10	9.4	1	700–701
11. $(2 + i\sqrt{2}, 2 - i\sqrt{2})$, $(2 - i\sqrt{2}, 2 + i\sqrt{2})$	11	9.4	2	701–702
12. $(2, \sqrt{2})$, $(2, -\sqrt{2})$, $(-2, \sqrt{2})$, $(-2, -\sqrt{2})$	12	9.4	3	702
13. $x = 25$ units to break even	13	9.4	4	703
14. 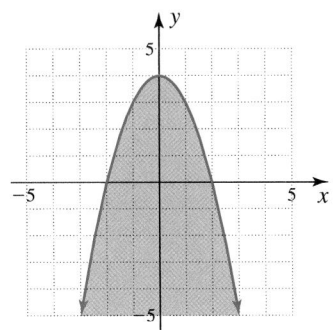	14	9.5	1	710–711
15. 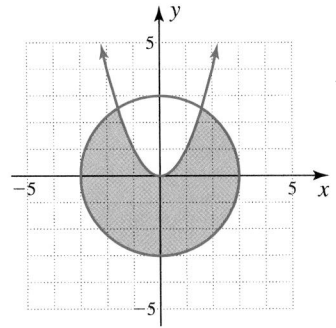	15	9.5	4	712

9.1   QUADRATIC FUNCTIONS AND THEIR GRAPHS

To Succeed, Review How To . . .

1. Graph points in the Cartesian coordinate system (pp. 163–165).

2. Find x- and y-intercepts (pp. 167–168).

3. Use the quadratic formula (pp. 594–599).

4. Complete the square in a quadratic equation (pp. 587–590).

5. Find the discriminant of a quadratic equation (pp. 606–608).

Objectives

A   Graph a parabola of the form $y = ax^2 + k$.

B   Graph a parabola of the form $y = a(x - h)^2 + k$.

C   Graph a parabola of the form $y = ax^2 + bx + c$.

D   Graph parabolas that are not functions with form $x = a(y - k)^2 + h$ and $x = ay^2 + by + c$.

E   Solve applications involving parabolas.

GETTING STARTED   The Fountain of Parabolas

The streams of water are in the shapes of a parabola.

Have you seen any parabolas lately? They are as near as your fountain. The streams of water follow the path of a quadratic function called a parabola. Parabolas, ellipses, circles, and hyperbolas are called **conic sections** because they can be obtained by intersecting (slicing) a cone with a plane, as shown in Figure 1. As you will see later, these conic sections occur in many practical applications. For example, your satellite dish, your flashlight lens, and your telescope lens have a parabolic shape, while comets travel in orbits that are either elliptical (those may be seen from Earth more than once) or hyperbolic (once in a lifetime viewing!). We shall begin our study of conic sections by examining the parabola.

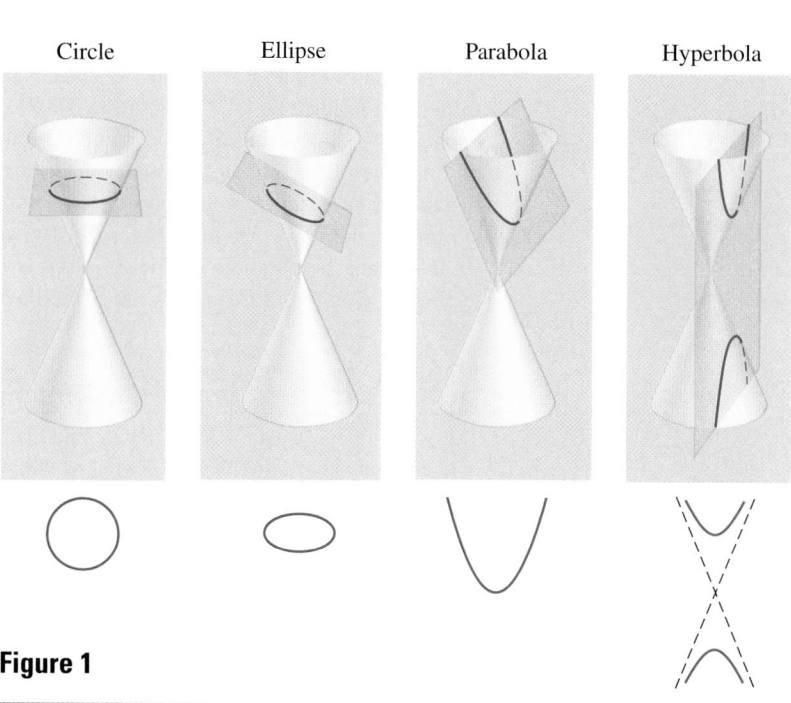

Circle   Ellipse   Parabola   Hyperbola

Figure 1

Graphing the Parabola $y = f(x) = ax^2 + k$

In Chapter 3, we studied *linear functions* like $f(x) = 2x + 5$ and $g(x) = -3x - 4$ whose graphs were straight lines. In this chapter we shall study equations (functions) defined by a quadratic (second-degree) polynomial of the form

$$f(x) = ax^2 + bx + c$$

These functions are **quadratic functions** and their graphs are **parabolas.**

Just as the "simplest" *line* to graph is $y = f(x) = x$, the "simplest" *parabola* to graph is $y = f(x) = x^2$. To draw this graph, we select values for x and find the corresponding values of y:

x-Value	$f(x) = y$-Value
$x = -2$	$f(-2) = (-2)^2 = 4$
$x = -1$	$f(-1) = (-1)^2 = 1$
$x = 0$	$f(0) = (0)^2 = 0$
$x = 1$	$f(1) = (1)^2 = 1$
$x = 2$	$f(2) = (2)^2 = 4$

Then we make a table of ordered pairs, plot the ordered pairs on a coordinate system, and draw a smooth curve through the plotted points as in Figure 2.

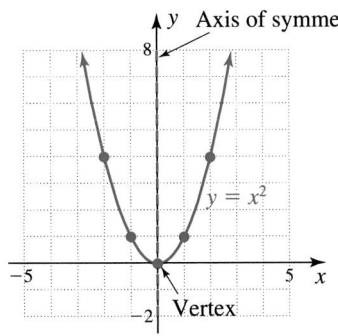

Figure 2

x	y	(x, y)
-2	4	$(-2, 4)$
-1	1	$(-1, 1)$
0	0	$(0, 0)$
1	1	$(1, 1)$
2	4	$(2, 4)$

Web It

An important feature of this parabola is its **symmetry** with respect to the y-axis. If you folded the graph of $y = x^2$ along the y-axis, the two halves of the graph would coincide because the same value of y is obtained for any value of x and its opposite $-x$. The points on one side of the axis of symmetry are mirror images of the points on the other side. For instance, $x = 2$ and $x = -2$ both give $y = 4$. (See the preceding tables.) Because of this symmetry, the y-axis is called the **axis of symmetry** or the **axis** of the parabola. The point $(0, 0)$, where the parabola crosses its axis, is called the **vertex** of the curve. The arrows on the curve in Figure 2 mean that the parabola goes on without end.

Calculate It   Selecting Window for $y = ax^2 + k$

The graph of $f(x) = ax^2 + k$ can be obtained with a calculator. The only difficulty here is in selecting a window that will show the vertex $(0, k)$. Thus to graph $f(x) = 2x^2 + 6$, whose vertex is at $(0, 6)$, make sure that $(0, 6)$ is part of the graph by selecting a standard window. If you were graphing $f(x) = 2x^2 + 10$, the vertex would be at $(0, 10)$, so you would have to adjust the window to see more of the graph. Try a $[-15, 15]$ by $[-15, 15]$ window. The graph of $y = 2x^2 + 6$ in a standard window is shown.

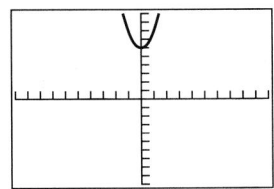

Use this idea and a standard window to check the graphs in Examples 1, 2, 3, and 4.

EXAMPLE 1   Graphing a parabola that opens downward

Graph and name the coordinates of the vertex: $y = -x^2$

SOLUTION   We could make a table of x- and y-values as before. However, for any x-value, the y-value will be the *negative* of the y-value on the parabola $y = x^2$. (If you don't believe this, go ahead and make the table and check it, but it's easier to copy the table for $y = x^2$ with the negatives of the y-values entered as shown.) Thus, the parabola $y = -x^2$ has the same shape as $y = x^2$, but it is turned in the *opposite* direction (opens *downward*). The graph of $y = -x^2$ is shown in Figure 3.

x	y	(x, y)
-2	-4	$(-2, -4)$
-1	-1	$(-1, -1)$
0	0	$(0, 0)$
1	-1	$(1, -1)$
2	-4	$(2, -4)$

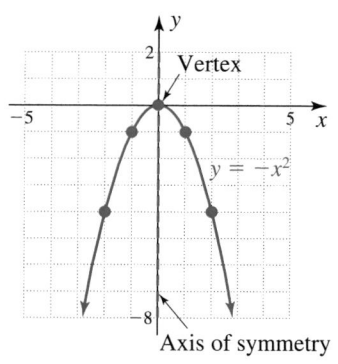

Figure 3

PROBLEM 1

Graph and name the coordinates of the vertex: $y = -2x^2$

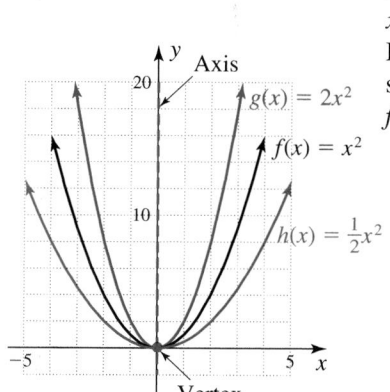

Figure 4

As you can see from the preceding examples, when the coefficient of x^2 is positive (as in $y = x^2 = 1x^2$), the parabola opens upward (is **concave up**), but when the coefficient of x^2 is negative (as in $y = -x^2 = -1x^2$), the parabola opens downward (is **concave down**). In either case, the vertex is at $(0, 0)$. To see the effect of a in $f(x) = ax^2$ in general, let's plot some points and see how the graphs of $g(x) = 2x^2$ and $h(x) = \frac{1}{2}x^2$ compare to the graph of $f(x) = x^2$. All three graphs are shown in Figure 4.

x	$g(x) = 2x^2$	x	$h(x) = \frac{1}{2}x^2$
-2	$2(-2)^2 = 8$	-2	$\frac{1}{2}(-2)^2 = 2$
-1	$2(-1)^2 = 2$	-1	$\frac{1}{2}(-1)^2 = \frac{1}{2}$
0	$2(0)^2 = 0$	0	$\frac{1}{2}(0)^2 = 0$
1	$2(1)^2 = 2$	1	$\frac{1}{2}(1)^2 = \frac{1}{2}$
2	$2(2)^2 = 8$	2	$\frac{1}{2}(-2)^2 = 2$

The graph of $g(x) = 2x^2$ is narrower than that of $f(x) = x^2$, while the graph of $h(x) = \frac{1}{2}x^2$ is wider. The vertex and line of symmetry is the same for the three curves. In general, we have the following.

Answer

1.

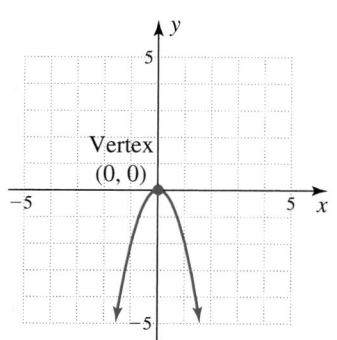

Properties of the Parabola $g(x) = ax^2$

The graph of $g(x) = ax^2$ is a parabola with the vertex at the origin and the y-axis as its line of symmetry.

If a is *positive*, the parabola opens *upward*, if a is *negative*, the parabola opens *downward*.

If $|a|$ is greater than 1 ($|a| > 1$), the parabola is narrower than the parabola $f(x) = x^2$.

If $|a|$ is between 0 and 1 ($0 < |a| < 1$), the parabola is wider than the parabola $f(x) = x^2$.

Using this information, you can draw the graph of any parabola of the form $g(x) = ax^2$, as we illustrate in Example 2.

EXAMPLE 2	**Graphing a parabola of the form** $y = ax^2$

Graph and name the coordinates of the vertex:

a. $f(x) = 3x^2$ **b.** $g(x) = -3x^2$ **c.** $h(x) = \dfrac{1}{3}x^2$

SOLUTION

a. By looking at the properties of the parabola $y = ax^2$, we know that the vertex of the parabola $f(x) = 3x^2$ is at the origin and that the y-axis is its line of symmetry. Since $3 > 0$, we also know that the parabola $f(x) = 3x^2$ opens upward and because $3 > 1$ the graph is narrower than the parabola $y = x^2$. We pick three easy points to complete our graph, shown in Figure 5.

x	$y = 3x^2$	(x, y)
-1	$y = 3(-1)^2 = 3$	$(-1, 3)$
0	$y = 3(0)^2 \;\;\;= 0$	$(0, 0)$ ⟵ Vertex
1	$y = 3(1)^2 \;\;\;= 3$	$(1, 3)$

We picked three points, but we could have picked the first two points and obtained the third by using the principles of symmetry. There must be a point on the other side of $x = 0$ that is 1 unit away and on the same horizontal line with $(-1, 3)$. Thus, the mirror-image point of $(-1, 3)$ is **(1, 3)**.

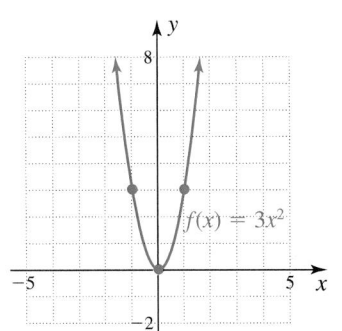

b. The parabola $g(x) = -3x^2$ opens downward but is again narrower than the parabola $y = x^2$. The parabola $g(x) = -3x^2$ is the reflection of the parabola $f(x) = 3x^2$ across the x-axis. Again, we pick three points to complete the graph, shown in Figure 6.

Figure 5

x	$y = -3x^2$	(x, y)
-1	$y = -3(-1)^2 = -3$	$(-1, -3)$
0	$y = -3(0)^2 \;\;\;= -0$	$(0, 0)$ ⟵ Vertex
1	$y = -3(1)^2 \;\;\;= -3$	$(1, -3)$

We picked three points, but we could have picked the first two points and obtained the third by using the principles of symmetry. There must be a point on the other side of $x = 0$ that is 1 unit away and on the same horizontal line with $(-1, -3)$. Thus, the mirror-image point of $(-1, -3)$ is **(1, -3)**.

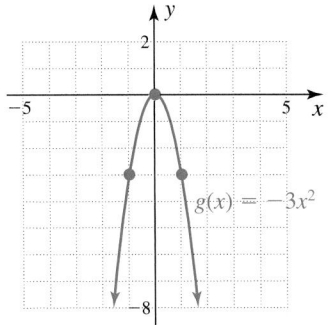

Figure 6

PROBLEM 2

Graph and name the coordinates of the vertices:

a. $f(x) = -4x^2$

b. $g(x) = 4x^2$

c. $h(x) = \dfrac{1}{4}x^2$

Answers

2.

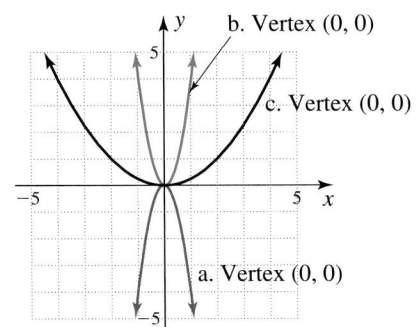

b. Vertex $(0, 0)$

c. Vertex $(0, 0)$

a. Vertex $(0, 0)$

c. The parabola $h(x) = \frac{1}{3}x^2$ opens upward, since $\frac{1}{3} > 0$, but is wider than the parabola $y = x^2$. This time, instead of selecting $x = -1, 0$, and 1, we select $x = -3, 0$, and 3 for ease of computation; the completed graph is shown in Figure 7.

x	$y = \frac{1}{3}x^2$	(x, y)
-3	$y = \frac{1}{3}(-3)^2 = 3$	$(-3, 3)$
0	$y = \frac{1}{3}(0)^2 = 0$	$(0, 0)$ ← Vertex
3	$y = \frac{1}{3}(3)^2 = 3$	$(3, 3)$

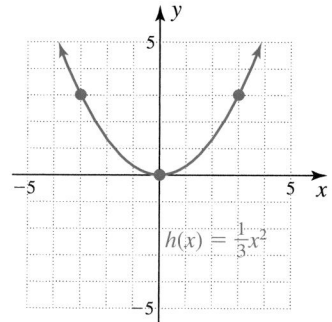

Figure 7

We picked three points, but we could have picked the first two points and obtained the third by using the principles of symmetry. There must be a point on the other side of $x = 0$ that is 3 units away and on the same horizontal line with $(-3, 3)$. Thus, the mirror-image point of $(-3, 3)$ is **(3, 3)**.

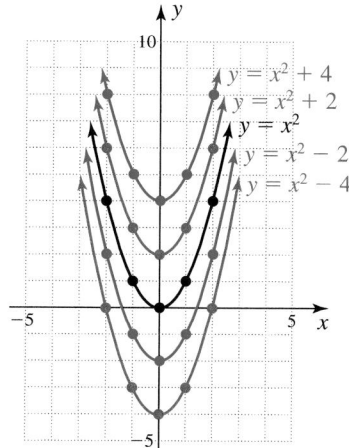

Figure 8

What do you think will happen if we graph the parabola $y = x^2 + 2$? Two things: first, the parabola opens upward because the coefficient of x^2 is understood to be $+1$. Second, all of the points will be 2 units higher than those for the same value of x on the parabola $y = x^2$. Thus we can make the graph of $y = x^2 + 2$ by following the pattern of $y = x^2$. The graphs of $y = x^2 + 2$, $y = x^2 + 4$, $y = x^2 - 2$, and $y = x^2 - 4$ are shown in Figure 8. The points used to make the graphs are listed in the following table.

For $y = x^2 + 2$		For $y = x^2 + 4$		For $y = x^2 - 2$		For $y = x^2 - 4$	
x	y	x	y	x	y	x	y
Vertex → 0	2	0	4	0	-2	0	-4
± 1	3	± 1	5	± 1	-1	± 1	-3
± 2	6	± 2	8	± 2	2	± 2	0

Adding or subtracting a positive number k on the right-hand side of the equation $y = x^2$ raises or lowers the graph (and the vertex) by k units.

EXAMPLE 3   **Graphing a parabola opening downward**

Graph and name the coordinates of the vertex: $y = -x^2 - 2$

SOLUTION Since the coefficient of x^2 (understood to be -1) is negative, the parabola opens downward. It is also 2 units lower than the graph of $y = -x^2$. Thus the graph of $y = -x^2 - 2$ is a parabola opening downward with its vertex at $(0, -2)$. Letting $x = 1$, we get $y = -3$ and for $x = 2$, $y = -6$. Graph the two points $(1, -3)$ and $(2, -6)$ and, by symmetry, the points $(-1, -3)$ and $(-2, -6)$. The parabola passing through all these points is shown in Figure 9.

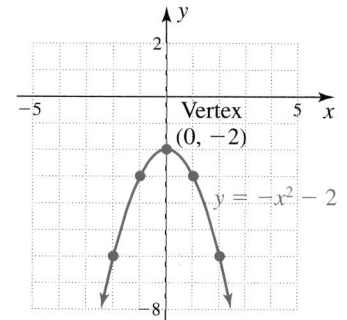

Figure 9

PROBLEM 3

Graph and name the coordinates of the vertex: $y = -x^2 - 1$

Answer

3.

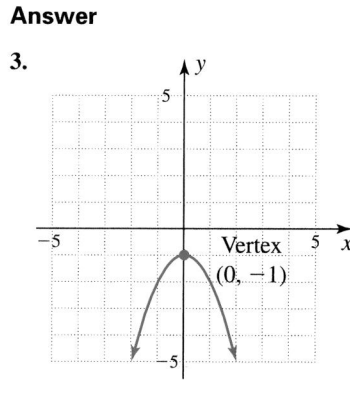

B ## Graphing a Parabola of Form $y = f(x) = a(x - h)^2 + k$

So far, we have graphed only parabolas of the form $y = ax^2 + k$. What do you think the graph of $y = (x - 1)^2$ looks like? As before, we make a table of values.

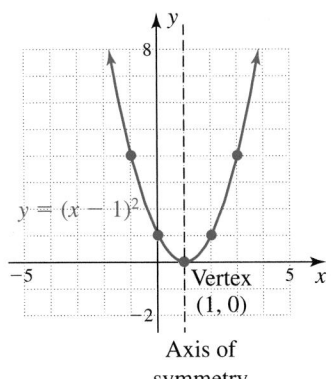

x	$y = (x - 1)^2$	
$x = -1,$	$y = (-1 - 1)^2 = (-2)^2 = 4$	
$x = 0,$	$y = (0 - 1)^2 = (-1)^2 = 1$	← y-intercept
$x = 1,$	$y = (1 - 1)^2 = (0)^2 = 0$	← Vertex
$x = 2,$	$y = (2 - 1)^2 = 1^2 = 1$	
$x = 3,$	$y = (3 - 1)^2 = 2^2 = 4$	

or

x	y
-1	4
0	1
1	0
2	1
3	4

Figure 10

The graph appears in Figure 10.

The shape of the graph is identical to that of $y = x^2$ but it is shifted 1 unit to the *right*. Thus the vertex is at $(1, 0)$ and the axis of symmetry is as shown in Figure 10.

Similarly, the graph of $y = -(x + 1)^2$ is identical to that of $y = -x^2$ but shifted 1 unit to the *left*. Thus the vertex is at $(-1, 0)$ and the axis of symmetry is as shown in Figure 11. Some easy points to plot are $x = 0$, $y = -(1)^2 = -1$ and $x = 1$, $y = -(1 + 1)^2 = -2^2 = -4$. When we plot the points $(0, -1)$ and $(1, -4)$, then by symmetry the points $(-2, -1)$ and $(-3, -4)$ are also on the graph.

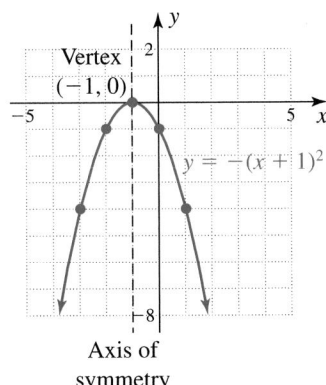

Figure 11

EXAMPLE 4 ### Graphing a parabola of the form $y = a(x - h)^2 + k$

Graph and name the coordinates of the vertex: $y = (x - 1)^2 - 2$

SOLUTION The graph of this equation is identical to the graph of $y = x^2$ except for its position. The new parabola is shifted 1 unit to the right (because of the -1) and 2 units down (because of the -2). Thus the vertex is $(1, -2)$. Note the axis of symmetry. Figure 12 indicates these two facts and shows the finished graph of $y = (x - 1)^2 - 2$.

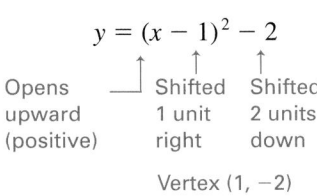

Figure 12

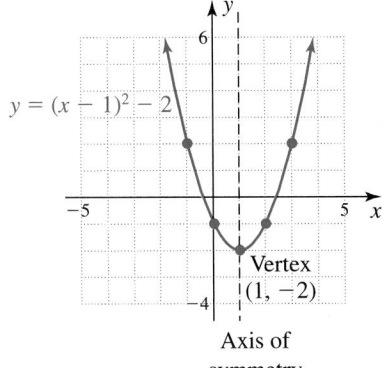

PROBLEM 4

Graph and name the coordinates of the vertex: $y = (x - 2)^2 - 1$

Answer

4.

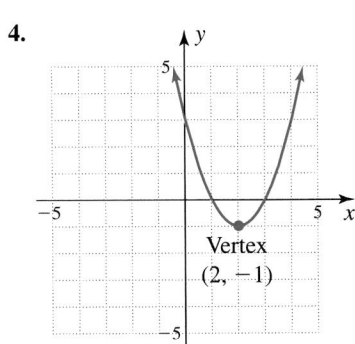

Teaching Tip

Write several examples on the board or overhead projector and have students name the shift of the graph of $y = x^2$ without graphing.

$y = x^2 + 2$	up 2
$y = (x - 3)^2$	right 3
$y = -4x^2$	no shift, but down and thin
$y = (x + 1)^2 - 1$	left 1 and down 1
etc.	

From these examples we can see that

1. The graph of $y = -x^2 - 2$ (Example 3) is exactly the same as the graph of $y = -x^2$ (Example 1) but moved 2 units *down*. In general, the graph of $y = ax^2 + k$ is the same as the graph of $y = ax^2$ but moved vertically k units. The vertex is at $(0, k)$.

2. The graph of $y = (x - 1)^2$ is the same as that of $y = x^2$ but moved 1 unit *right*. The vertex is at $(1, 0)$.

3. The graph of $y = (x - 1)^2 - 2$ (Example 4) is exactly the same as the graph of $y = (x - 1)^2$ but moved 2 units *down*. The vertex is at $(1, -2)$.

Here is the summary of this discussion.

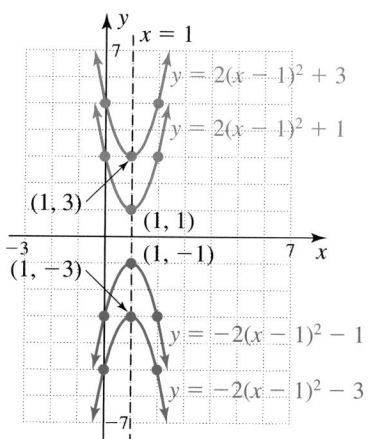

Figure 13

> **Properties of the Parabola $y = a(x - h)^2 + k$**
>
> The graph of the parabola $y = a(x - h)^2 + k$ is the same as that of $y = ax^2$ but moved h units horizontally and k units vertically. The *vertex* is at the point (h, k), and the axis of symmetry is $x = h$.

In conclusion, follow the given directions to graph an equation of the form

$$y = a(x - h)^2 + k \qquad \text{Vertex } (h, k)$$

Opens upward for $a > 0$, downward for $a < 0$

Shifts the graph right or left

Moves the graph up or down

The graphs of $y = 2(x - 1)^2 + 1$, $y = 2(x - 1)^2 + 3$, $y = -2(x - 1)^2 - 1$, and $y = -2(x - 1)^2 - 3$ are shown in Figure 13. The axis of symmetry is $x = 1$.

C Graphing the Parabola $y = f(x) = ax^2 + bx + c$

How can we graph $f(x) = ax^2 + bx + c$? If we learn to write $f(x) = ax^2 + bx + c$ as $f(x) = a(x - h)^2 + k$, we can do it by using the techniques we just learned.

We do this by completing the square. This was previously done in Section 8.2, pages 594 and 595, paralleling a numerical example beside the standard form of the quadratic equation, $ax^2 + bx + c = 0$. The difference here is the equation we start with is $f(x) = ax^2 + bx + c$. Here's how:

Web It

For an interactive application that will graph a parabola by inputting a, b, and c, go to link 9-1-2 at mhhe.com/bello.

Teaching Tip

Do a numerical example while developing this theory.

Example:
$$\begin{aligned} f(x) &= 2x^2 - 12x + 23 \\ &= 2(x^2 - 6x +) + 23 \\ &= 2(x^2 - 6x + 9 - 9) + 23 \\ &= 2[(x - 3)^2 - 9] + 23 \\ &= 2(x - 3)^2 - 2 \cdot 9 + 23 \\ &= 2(x - 3)^2 + 5 \end{aligned}$$

$$f(x) = ax^2 + bx + c \qquad \text{Given.}$$

$$= (ax^2 + bx) + c \qquad \text{Group.}$$

$$= a\left[x^2 + \frac{b}{a}x + \right] + c \qquad \text{Factor } a.$$

$$= a\left[x^2 + \frac{b}{a}x + \left(\frac{b}{2a}\right)^2 - \left(\frac{b}{2a}\right)^2\right] + c \qquad \text{To complete the square, add and subtract } \left(\frac{1}{2} \cdot \frac{b}{a}\right)^2 = \left(\frac{b}{2a}\right)^2 \text{ inside the brackets.}$$

$$= a\left[\left(x + \frac{b}{2a}\right)^2 - \left(\frac{b}{2a}\right)^2\right] + c \qquad \text{Factor inside the brackets.}$$

$$= a\left(x + \frac{b}{2a}\right)^2 - a\left(\frac{b}{2a}\right)^2 + c \qquad \text{Use the distributive property.}$$

$$= a\left(x + \frac{b}{2a}\right)^2 - a \cdot \frac{b^2}{4a^2} + c \qquad \text{Square } \frac{b}{2a}.$$

Web It

For an interactive application that will graph a parabola by inputting a, h, and k, go to link 9-1-3 at mhhe.com/bello.

$$= a\left(x + \frac{b}{2a}\right)^2 - \frac{b^2}{4a} + c$$

$$= a\left(x + \frac{b}{2a}\right)^2 + \frac{4ac - b^2}{4a}$$

Multiply $-a \cdot \frac{b^2}{4a^2} = -\frac{b^2}{4a}$.

Find the LCD of $\frac{-b^2}{4a}$ and c and write as one fraction.

Thus to write

$$f(x) = a\left(x + \frac{b}{2a}\right)^2 + \frac{4ac - b^2}{4a}$$

as

$$f(x) = \underbrace{a(x - h)^2}_{} + \underbrace{k}_{}$$

we must have

$$h = -\frac{b}{2a} \quad \text{and} \quad k = \frac{4ac - b^2}{4a}$$

the coordinates of the vertex. You *do not* have to memorize the y-coordinate of the vertex. After you find the x-coordinate, substitute in the equation and find y.

Here is a summary of our discussion.

Teaching Tip

To keep the two forms of the quadratic equation straight, have students think of $y = a(x - h)^2 + k$ as the graphing form and $y = ax^2 + bx + c$ as standard form. Recognize the form first and then proceed with the appropriate method.

PROCEDURE

Graphing the Parabola $y = f(x) = ax^2 + bx + c$

1. To find the vertex use one of the following methods:

    Method 1. Let $x = -\frac{b}{2a}$ in the equation and solve for y,

    or

    Method 2. Complete the square and compare with $y = a(x - h)^2 + k$.

2. Let $x = 0$. The result, c, is the y-intercept.

3. Since the parabola is symmetric with respect to its axis, use this symmetry to find additional points.

4. Let $y = 0$. Find x by solving $ax^2 + bx + c = 0$. If the solutions are real numbers, they are the x-intercepts. If not, the parabola does not intersect the x-axis.

5. Draw a smooth curve through the points found in steps 1–4. Remember that if $a > 0$, the parabola opens *upward;* if $a < 0$, the parabola opens *downward.*

We demonstrate this procedure in Example 5.

EXAMPLE 5   **Graphing a parabola of the form $y = ax^2 + bx + c$**

Graph and name the coordinates of the vertex: $y = x^2 + 3x + 2$

SOLUTION

1. We first find the vertex using either of the two methods.

    Method 1
    Use the vertex formula for the x-coordinate. Since $a = 1$, $b = 3$, and $c = 2$,

    $$x = -\frac{b}{2a} = -\frac{3}{2}$$

    Method 2
    Complete the square.

    $$y = [x^2 + 3x + \quad] + 2$$

    $$= \left[x^2 + 3x + \left(\frac{3}{2}\right)^2\right] + 2 - \left(\frac{3}{2}\right)^2$$

PROBLEM 5

Graph and name the coordinates of the vertex: $y = x^2 + 2x - 3$

Answer

5.

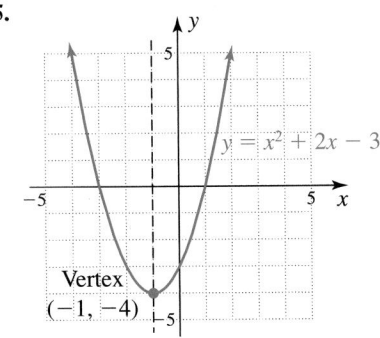

$y = x^2 + 2x - 3$

Vertex $(-1, -4)$

Substituting for x in the equation gives

$y = x^2 + 3x + 2$

$= \left(-\dfrac{3}{2}\right)^2 + 3\left(-\dfrac{3}{2}\right) + 2$

$= \dfrac{9}{4} - \dfrac{9}{2} + 2$

$= \dfrac{9}{4} - \dfrac{18}{4} + \dfrac{8}{4} = -\dfrac{1}{4}$

The vertex is at $(-\frac{3}{2}, -\frac{1}{4})$.

$= \left(x + \dfrac{3}{2}\right)^2 + 2 - \dfrac{9}{4}$

$= \left(x + \dfrac{3}{2}\right)^2 - \dfrac{1}{4}$

The vertex is at $(-\frac{3}{2}, -\frac{1}{4})$.

2. Let $x = 0$; then $y = x^2 + 3x + 2 = (0)^2 + 3(0) + 2 = 2$. The y-intercept is at $(0, 2)$.

3. By symmetry, the point $(-3, 2)$ is also on the graph.

4. Let $y = 0$; $y = x^2 + 3x + 2$ becomes

$$0 = x^2 + 3x + 2$$
$$0 = (x + 2)(x + 1)$$

Thus $x = -2$ or $x = -1$. The graph intersects the x-axis at $(-2, 0)$ and $(-1, 0)$.

5. Since the coefficient of x^2 is 1, $a > 0$ and the parabola opens upward.

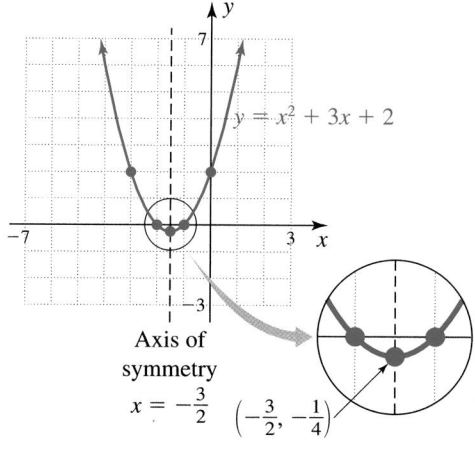

Axis of
symmetry
$x = -\dfrac{3}{2}$ $\left(-\dfrac{3}{2}, -\dfrac{1}{4}\right)$

Figure 14

We draw a smooth curve through these points to obtain the graph of the parabola as shown in Figure 14.

In Example 5, $x^2 + 3x + 2$ can be factored, and we can find the points at which the parabola crosses the x-axis. If the equation of the parabola cannot be factored, look at the discriminant $D = b^2 - 4ac$ and determine what kind of solutions the equation has.

PROCEDURE

Using the Discriminant, D, to Graph Quadratics, $y = f(x) = ax^2 + bx + c$
Compute $D = b^2 - 4ac$

1. If $D < 0$, there are no real solutions and the graph will *not* cross the x-axis.

2. If $D = 0$, there is one real solution and the graph will have one x-intercept, the vertex.

3. If $D \geq 0$, either factor or use the quadratic formula to find x and approximate the answers so you can graph them as the x-intercepts.

The possibilities are shown in Figures 15–17.

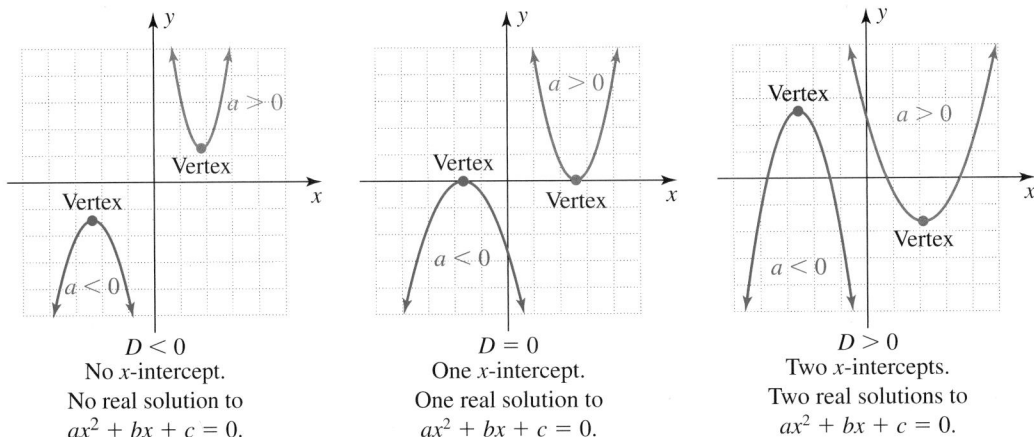

$D < 0$	$D = 0$	$D > 0$
No x-intercept.	One x-intercept.	Two x-intercepts.
No real solution to	One real solution to	Two real solutions to
$ax^2 + bx + c = 0$.	$ax^2 + bx + c = 0$.	$ax^2 + bx + c = 0$.
Figure 15	**Figure 16**	**Figure 17**

EXAMPLE 6 **Graphing a parabola whose equation cannot be factored**

Graph and name the coordinates of the vertex: $y = -2x^2 + 4x - 3$

SOLUTION

1. To find the vertex, we can use either of these methods:

Method 1

Use the vertex formula.

Here $a = -2$ and $b = 4$, so

$$x = -\frac{b}{2a}$$

$$= \frac{-4}{2(-2)}$$

$$= 1$$

If we substitute $x = 1$ in $y = -2x^2 + 4x - 3$,

$$y = -2(1)^2 + 4(1) - 3$$

$$= -2 + 4 - 3$$

$$= -1$$

The vertex is at $(1, -1)$.

Method 2

Complete the square.

$$y = -2x^2 + 4x - 3$$

$$= -2(x^2 - 2x + \quad) - 3$$

$$= -2(x^2 - 2x + 1 - 1) - 3$$

$$= -2[(x - 1)^2 - 1] - 3$$

$$= -2(x - 1)^2 + 2 - 3$$

$$= -2(x - 1)^2 - 1$$

The vertex is at $(1, -1)$.

2. If $x = 0$, $y = -2x^2 + 4x - 3 = -2(0)^2 + 4(0) - 3 = -3$. The y-intercept is at $(0, -3)$.

3. We graph the vertex $(1, -1)$ and the y-intercept -3. To make a more accurate graph, we need some more points.

   The parabola is symmetric, so we know there must be a point on the other side of the axis of symmetry from the y-intercept that is the same distance away, 1 unit. The axis of symmetry is at $x = 1$ and one more unit from that would be $x = 2$. The mirror image point of the y-intercept, $(0, -3)$, is $(2, -3)$.

PROBLEM 6

Graph and name the coordinates of the vertex: $y = -x^2 - 4x - 3$

Teaching Tip

Use $y = x^2 - 4x + 4$ where $D = 0$ and have students discuss the x-intercepts for this case.

Answer

6.

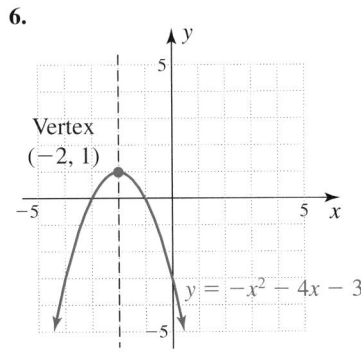

4. For $y = 0, 0 = -2x^2 + 4x - 3$.
However, the right-hand side is not factorable. The discriminant of the equation is $4^2 - 4(-2)(-3) = 16 - 24 = -8$. This means that this equation has no real number solutions and there are no x-intercepts. The graph does not cross the x-axis.

5. Since $a = -2 < 0$, the parabola opens downward. The completed graph is shown in Figure 18.

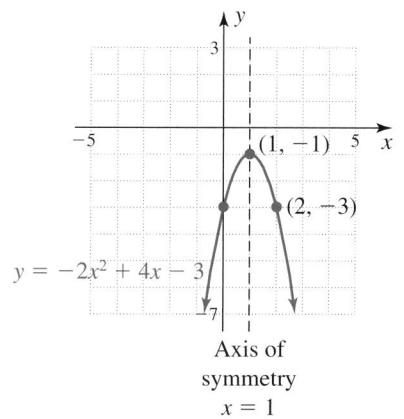

Figure 18

D   Graphing Parabolas that are not Functions: $x = a(y - k)^2 + h$ or $x = ay^2 + by + c$

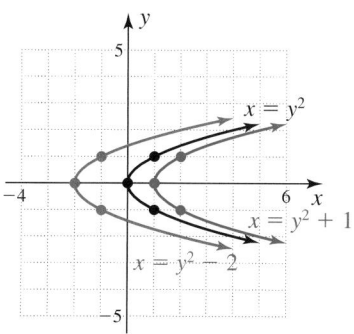

Figure 19

Do you remember how the graph of $y = x^2$ looks? The graph of $x = y^2$ has a similar shape but opens *horizontally to the right*. Similarly, the graph of $x = y^2 + 1$ looks like that of $x = y^2$ but is shifted *right* 1 unit, as shown in Figure 19. When $y = 0$, $x = 1$ so the vertex is at $(1, 0)$. The graph of $x = y^2 - 2$ is similar to that of $x = y^2$ but is shifted *left* 2 units. Here, when $y = 0$, $x = -2$ so the vertex is at $(-2, 0)$.

Similarly, the graphs of $x = -y^2$, $x = -y^2 + 1$, and $x = -y^2 - 2$ look like the graph of $x = -y^2$, opening *horizontally to the left* and then shifted right or left the correct number of units. Thus for $x = -y^2 + 1$, when $y = 0$, $x = 1$ and the vertex is at $(1, 0)$. For $x = -y^2 - 2$, when $y = 0$, $x = -2$ so the vertex is at $(-2, 0)$. The graphs are shown in Figure 20. None of these are graphs of functions. (Remember the vertical line test? You can find several vertical lines that will intersect all of these graphs at more than one point.)

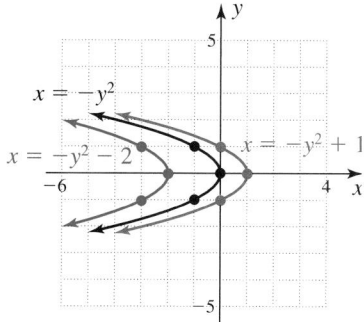

Figure 20

> **Properties of the Parabola $x = a(y - k)^2 + h$**
>
> The graph of the parabola $x = a(y - k)^2 + h$ is the same as that of $x = ay^2$ but moved h units horizontally and k units vertically. The *vertex* is at the point (h, k), and the axis of symmetry is $y = k$.

In conclusion, follow the given directions to graph an equation of the form

$$x = a(y - k)^2 + h \qquad \text{Vertex } (h, k)$$

Opens right for $a > 0$, left for $a < 0$

Shifts the graph up or down

Moves the graph right or left

Calculate It   Graphing Parabolas That Are Not Functions

To graph $x = 2(y - 1)^2 - 3$ in Example 7, first solve for y to obtain

$$y = 1 \pm \sqrt{\frac{x + 3}{2}}$$

Since the graph consists of two branches, graph

$$Y_1 = 1 + \sqrt{\frac{x + 3}{2}}$$

and

$$Y_2 = 1 - \sqrt{\frac{x + 3}{2}}$$

The two graphs together form the graph of $x = 2(y - 1)^2 - 3$ as shown in the window.

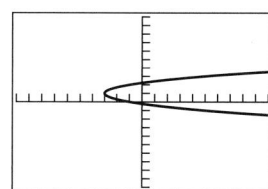

EXAMPLE 7   **Graphing a parabola of the form**
$x = a(y - k)^2 + h$

Graph and name the coordinates of the vertex: $x = 2(y - 1)^2 - 3$

SOLUTION   In this problem, the roles of x and y are reversed, so the graph will look like that of $y = 2(x - 1)^2 - 3$ but opening horizontally.

The vertex of $x = 2(y - 1)^2 - 3$ is at $(-3, 1)$. The curve opens to the right, the positive x-direction. The graph is shown in Figure 21. You can verify that the graph is correct by letting $x = -1$, which gives $y = 0$ or $y = 2$.

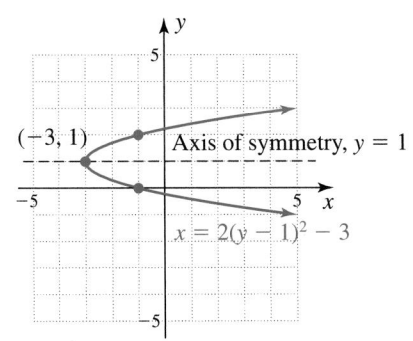

Figure 21

PROBLEM 7

Graph and name the coordinates of the vertex: $x = (y + 1)^2 - 2$

EXAMPLE 8   **Graphing a parabola of the form**
$x = ay^2 + by + c$

Graph and name the coordinates of the vertex: $x = y^2 + 3y + 2$

SOLUTION   The graph is similar to that of $y = x^2 + 3x + 2$, but it opens horizontally (see Example 5). The vertex occurs where

$$y = -\frac{b}{2a} = -\frac{3}{2}$$

Substituting for y in the equation gives $x = (-\frac{3}{2})^2 + 3(-\frac{3}{2}) + 2 = -\frac{1}{4}$. The vertex is at $(-\frac{1}{4}, -\frac{3}{2})$. The x-intercept is 2 and the y-intercepts are where $x = 0$. Thus,

$$0 = y^2 + 3y + 2$$
$$= (y + 2)(y + 1)$$

That is, $y = -2$ or $y = -1$. The parabola opens to the right because the coefficient of y^2 is $1 > 0$, and the completed graph is shown in Figure 22.

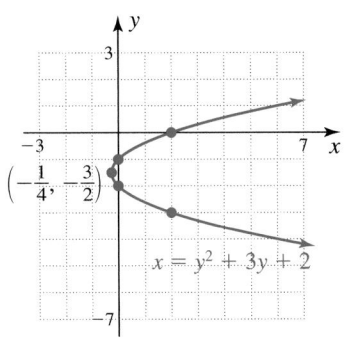

Figure 22

PROBLEM 8

Graph and name the coordinates of the vertex: $x = y^2 + 2y - 3$

Web It

To read about some occurrences of parabolas in real life, go to link 9-1-4 at mhhe.com/bello.

Answers

7.

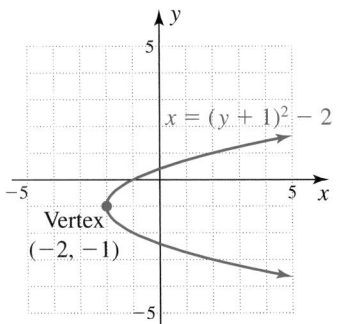

8.

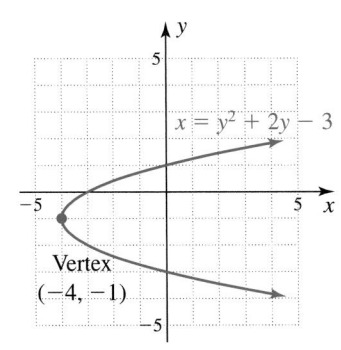

Calculate It   How to Solve for *y* when Parabolas are not Functions

How would you graph $x = y^2 + 3y + 2$ in Example 8? You have to complete the square and then solve for *y* or use the quadratic formula. Try it and see if you agree that it's easier to do it algebraically.

E   Solving an Application Involving Parabolas

Every parabola of the form $y = ax^2 + bx + c$ that we have graphed has its vertex at either its maximum (highest) or minimum (lowest) point on the graph. If the graph opens upward, the vertex is the minimum (Figure 23) and if the graph opens downward, the vertex is the maximum (Figure 24).

If we are dealing with a quadratic function, we can find its maximum or minimum by finding the vertex of the corresponding parabola. This idea can be used to solve many real-world applications. For example, suppose that a CD company manufactures and sells *x* CDs per week. If the revenue is given by $R = 10x - 0.01x^2$, we can use the techniques we've just studied to maximize the revenue. We do that next.

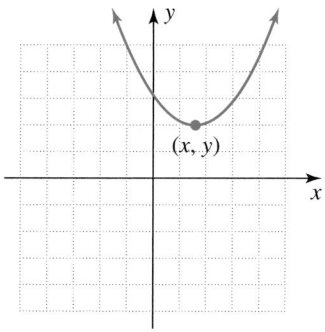

Figure 23
$f(x)$ has a minimum at the vertex
(x, y).

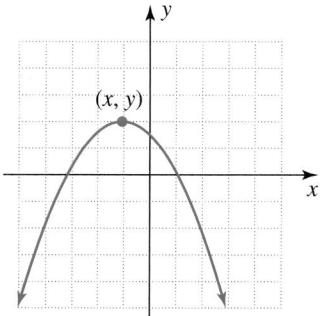

Figure 24
$f(x)$ has a maximum at the vertex
(x, y).

EXAMPLE 9   **Recording a maximum revenue**

If $R = 10x - 0.01x^2$, how many CDs does the company have to sell to obtain maximum revenue? (*R* is revenue and *x* is the number of CDs made and sold per week.)

SOLUTION   We first write the equation as $R = -0.01x^2 + 10x$. Since the coefficient of x^2 is negative, the parabola opens downward (is concave down), and the vertex is its highest point. Letting $a = -0.01$ and $b = 10$, we have

$$x = -\frac{b}{2a} = -\frac{10}{-0.02} = 500$$

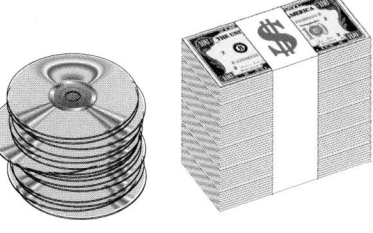

$R = 10(500) - 0.01(500)^2 = 5000 - 2500 = 2500$. When the company sells $x = 500$ CDs a week, the revenue is a maximum: \$2500.

PROBLEM 9

A farmer has 200 feet of fencing he wants to use to enclose a rectangular plot of land for his wife's vegetable garden. What dimensions will give her the maximum area?

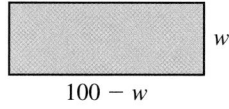

$100 - w$

Answer

9. 50 ft × 50 ft

We end this section by presenting in Table 1 a summary of the material we have studied.

Table 1 Summary of Graphing Parabolas

Description	Graph when $a > 0$	Graph when $a < 0$
$f(x) = ax^2 + k$ A parabola with vertex at $(0, k)$. When $|a| > 1$, the graph is narrower than the graph of $y = x^2$. When $0 < |a| < 1$, it is wider.	Vertex at $(0, k)$ For $a > 0$, the parabola opens upward.	Vertex at $(0, k)$ For $a < 0$, the parabola opens downward.
$f(x) = a(x - h)^2 + k$ A parabola with vertex at (h, k). When $|a| > 1$, the graph is narrower than the graph of $y = x^2$. When $0 < |a| < 1$, it is wider.	Vertex at (h, k) For $a > 0$, the parabola opens upward.	Vertex at (h, k) For $a < 0$, the parabola opens downward.
$f(x) = ax^2 + bx + c$ A parabola with vertex at $\left(\dfrac{-b}{2a}, f\left(\dfrac{-b}{2a}\right)\right)$. When $|a| > 1$, the graph is narrower than the graph of $y = x^2$. When $0 < |a| < 1$, it is wider.	Vertex at $\left(\dfrac{-b}{2a}, f\left(\dfrac{-b}{2a}\right)\right)$ For $a > 0$, the parabola opens upward.	Vertex at $\left(\dfrac{-b}{2a}, f\left(\dfrac{-b}{2a}\right)\right)$ For $a < 0$, the parabola opens downward.
$x = a(y - k)^2 + h$ A parabola with vertex at (h, k) and is not the graph of a function. When $|a| > 1$, the graph is narrower than the graph of $x = y^2$. When $0 < |a| < 1$, it is wider.	Vertex at (h, k) For $a > 0$, the parabola opens to the right.	Vertex at (h, k) For $a < 0$, the parabola opens to the left.

Calculate It Exercises

1. Graph the horizontal parabola of Example 8, $x = y^2 + 3y + 2$.

2. Graph the parabolas $g(x) = ax^2$ for $a = \frac{1}{4}, \frac{1}{2}, 2$, and 4. *Note:* Some calculators can graph "families" of curves. If your calculator is one of these, enter $Y_1 = \{\frac{1}{4}, \frac{1}{2}, 2, 4\}x^2$

   a. What happens to the graph as a increases?

   b. What are the domain and range of $g(x) = ax^2$ for $a > 0$?

3. Graph the parabolas $g(x) = ax^2$ for $a = -\frac{1}{4}, -\frac{1}{2}, -2$, and -4.

   a. What happens to the graph as a decreases?

   b. What are the domain and range of $g(x) = ax^2$ for $a < 0$?

   c. How does the sign of a in $g(x) = ax^2$ affect the graph?

4. Graph the parabolas $u(x) = x^2 + k$ for $k = -2, -1, 1$, and 2.

   a. What effect does k have on the graph?

   b. What are the domain and range of $u(x)$?

5. Graph the parabolas $v(x) = (x - h)^2$ for $h = -2, -1, 1$, and 2.

   a. What effect does h have on the graph?

   b. What are the domain and range of $v(x)$?

Exercises 9.1

A In Problems 1–8, graph the given equations on the same coordinate axes and name the coordinates of the vertices.

1. **a.** $y = 2x^2$ $(0, 0)$
   b. $y = 2x^2 + 2$ $(0, 2)$
   c. $y = 2x^2 - 2$ $(0, -2)$

2. **a.** $y = 3x^2 + 1$ $(0, 1)$
   b. $y = 3x^2 + 3$ $(0, 3)$
   c. $y = 3x^2 - 2$ $(0, -2)$

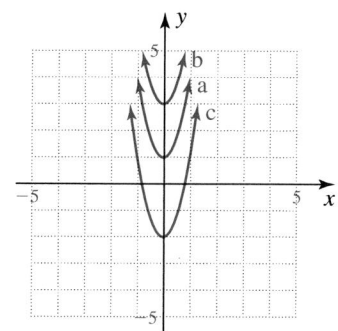

3. **a.** $y = -2x^2$ $(0, 0)$
   b. $y = -2x^2 + 1$ $(0, 1)$
   c. $y = -2x^2 - 1$ $(0, -1)$

4. **a.** $y = -4x^2$ $(0, 0)$
   b. $y = -4x^2 + 1$ $(0, 1)$
   c. $y = -4x^2 - 1$ $(0, -1)$

5. **a.** $y = \frac{1}{4}x^2$ $(0, 0)$

   b. $y = -\frac{1}{4}x^2$ $(0, 0)$

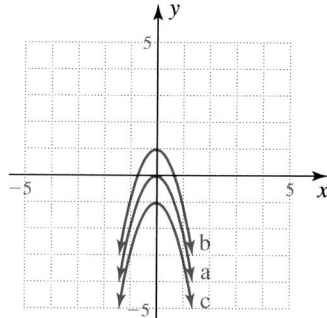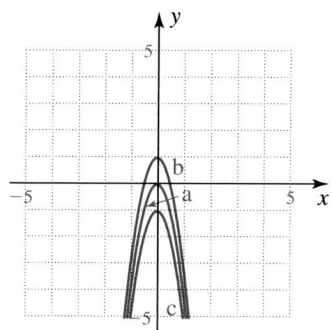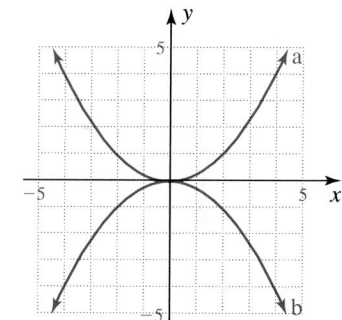

6. **a.** $y = \frac{1}{5}x^2$ $(0, 0)$

   b. $y = -\frac{1}{5}x^2$ $(0, 0)$

7. **a.** $y = \frac{1}{3}x^2 + 1$ $(0, 1)$

   b. $y = -\frac{1}{3}x^2 + 1$ $(0, 1)$

8. **a.** $y = \frac{1}{4}x^2 + 1$ $(0, 1)$

   b. $y = -\frac{1}{4}x^2 + 1$ $(0, 1)$

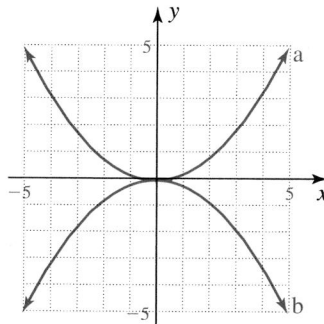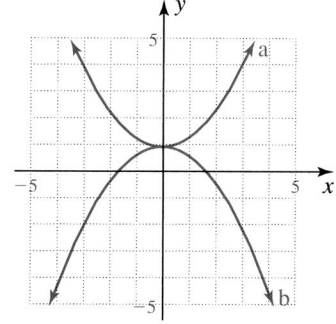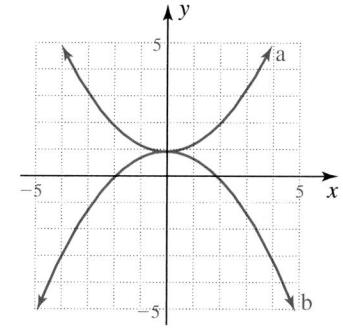

B In Problems 9–16, graph the given equations on the same coordinate axes and name the coordinates of the vertices.

9. a. $y = (x + 2)^2 + 3$ $(-2, 3)$
   b. $y = (x + 2)^2$ $(-2, 0)$
   c. $y = (x + 2)^2 - 2$ $(-2, -2)$

10. a. $y = (x - 2)^2 + 2$ $(2, 2)$
   b. $y = (x - 2)^2$ $(2, 0)$
   c. $y = (x - 2)^2 - 2$ $(2, -2)$

11. a. $y = -(x + 2)^2 - 2$ $(-2, -2)$
   b. $y = -(x + 2)^2$ $(-2, 0)$
   c. $y = -(x + 2)^2 - 4$ $(-2, -4)$

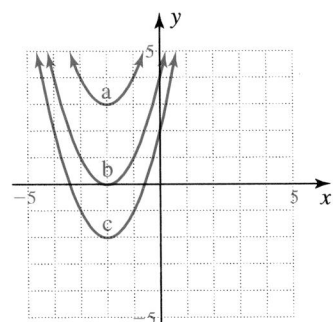
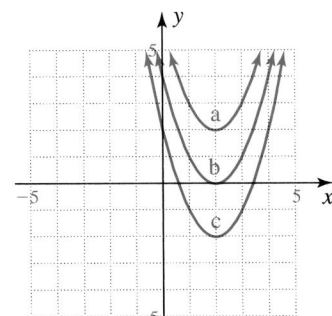
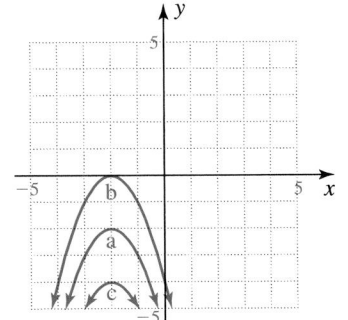

12. a. $y = -(x - 1)^2 + 1$ $(1, 1)$
   b. $y = -(x - 1)^2$ $(1, 0)$
   c. $y = -(x - 1)^2 + 2$ $(1, 2)$

13. a. $y = -2(x + 2)^2 - 2$ $(-2, -2)$
   b. $y = -2(x + 2)^2$ $(-2, 0)$
   c. $y = -2(x + 2)^2 - 4$ $(-2, -4)$

14. a. $y = -2(x - 1)^2 + 1$ $(1, 1)$
   b. $y = -2(x - 1)^2$ $(1, 0)$
   c. $y = -2(x - 1)^2 + 2$ $(1, 2)$

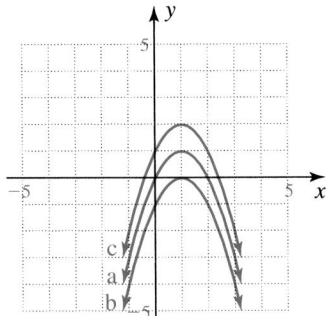
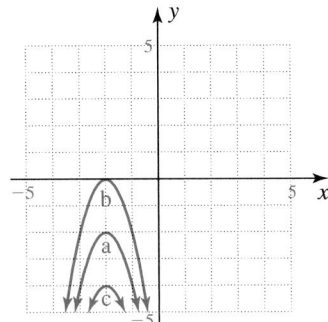
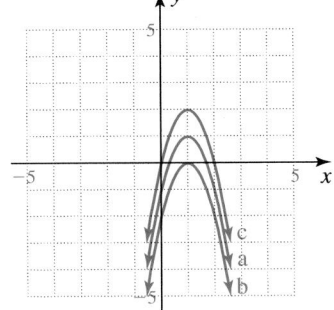

15. a. $y = 2(x + 1)^2 + \dfrac{1}{2}$ $\left(-1, \dfrac{1}{2}\right)$
   b. $y = 2(x + 1)^2$ $(-1, 0)$

16. a. $y = 2(x + 1)^2 - \dfrac{1}{2}$ $\left(-1, -\dfrac{1}{2}\right)$
   b. $y = 2(x + 1)^2$ $(-1, 0)$

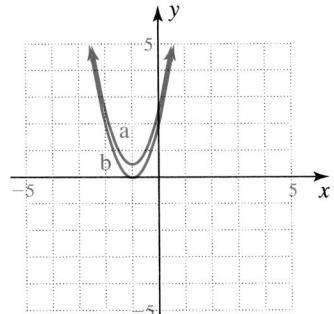
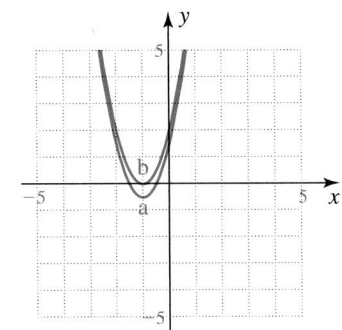

C In Problems 17–28, use the five-step procedure in the text to sketch the graph. Label the vertex and the intercepts. For irrational intercepts, approximate the values to one decimal place.

17. $y = x^2 + 2x + 1$

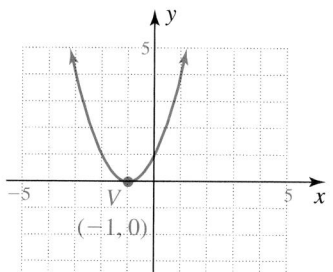

$V: (-1, 0)$; Int: $(-1, 0), (0, 1)$

18. $y = x^2 + 4x + 4$

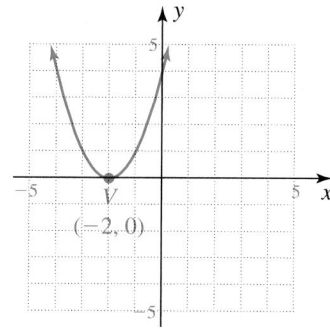

$V: (-2, 0)$; Int: $(-2, 0), (0, 4)$

19. $y = -x^2 + 2x + 1$

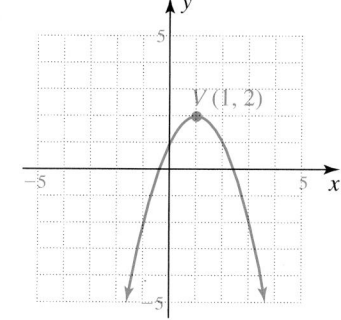

$V: (1, 2)$; Int: $(2.4, 2), (-0.4, 0), (0, 1)$

20. $y = -x^2 + 4x - 2$

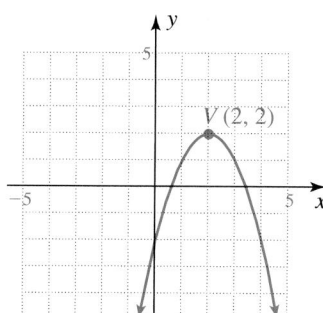

$V: (2, 2)$; Int: $(0.6, 0), (3.4, 0), (0, -2)$

21. $y = -x^2 + 4x - 5$

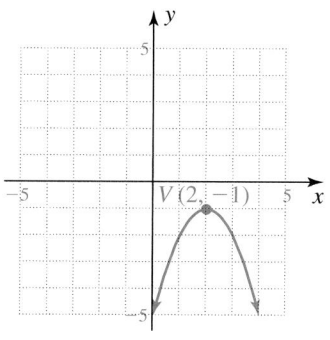

$V: (2, -1)$; Int: $(0, -5)$

22. $y = -x^2 + 4x - 3$

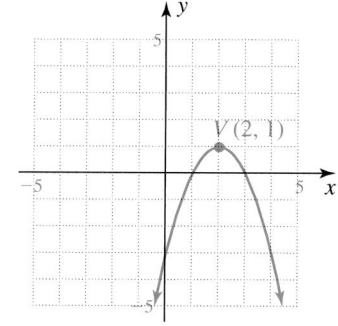

$V: (2, 1)$; Int: $(1, 0), (3, 0), (0, -3)$

23. $y = 3 - 5x + 2x^2$

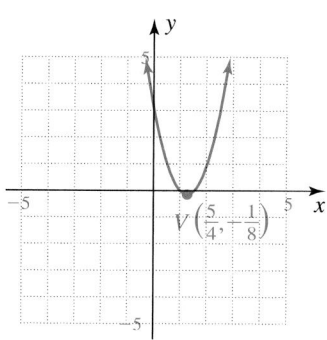

$V: (\frac{5}{4}, -\frac{1}{8})$;
Int: $(\frac{3}{2}, 0), (1, 0), (0, 3)$

24. $y = 3 + 5x + 2x^2$

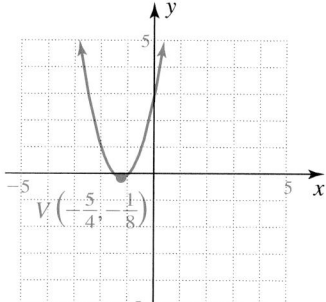

$V: (-\frac{5}{4}, -\frac{1}{8})$;
Int: $(-1, 0), (-\frac{3}{2}, 0), (0, 3)$

25. $y = 5 - 4x - 2x^2$
(*Hint:* $\sqrt{56} \approx 7.5$)

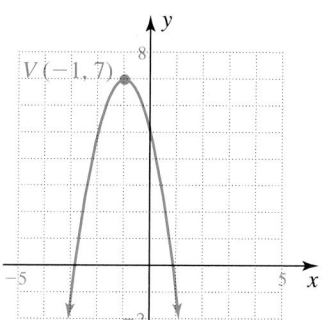

$V: (-1, 7)$;
Int: $(-2.9, 0), (0.9, 0), (0, 5)$

26. $y = 3 - 4x - 2x^2$
(*Hint:* $\sqrt{40} \approx 6.3$)

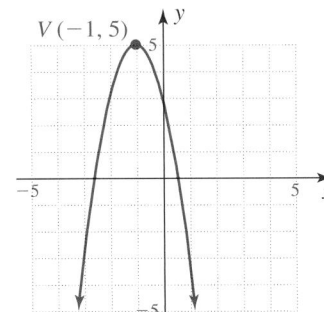

27. $y = -3x^2 + 3x + 2$
(*Hint:* $\sqrt{33} \approx 5.7$)

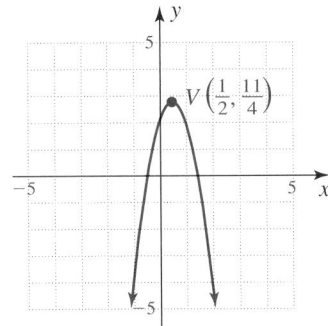

28. $y = -3x^2 + 3x + 1$
(*Hint:* $\sqrt{21} \approx 4.6$)

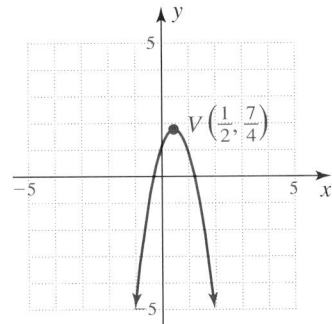

V: $(-1, 5)$; Int: $(0.6, 0)$, $(-2.6, 0)$, $(0, 3)$ V: $(\frac{1}{2}, \frac{11}{4})$; Int: $(1.5, 0)$, $(-0.5, 0)$, $(0, 2)$ V: $(\frac{1}{2}, \frac{7}{4})$; Int: $(1.3, 0)$, $(-0.3, 0)$, $(0, 1)$

D In Problems 29–34, graph on the same coordinate axes and name the coordinates of the vertices.

29. a. $x = (y + 2)^2 + 3$ $(3, -2)$
  b. $x = (y + 2)^2$ $(0, -2)$

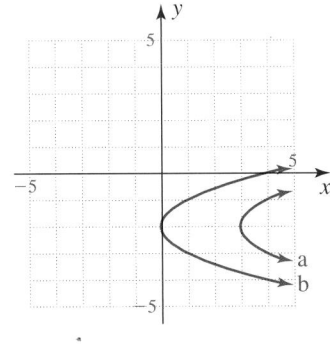

30. a. $x = (y - 2)^2 + 2$ $(2, 2)$
  b. $x = (y - 2)^2$ $(0, 2)$

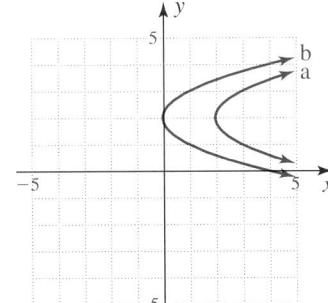

31. a. $x = -(y + 2)^2 - 2$ $(-2, -2)$
  b. $x = -(y + 2)^2$ $(0, -2)$

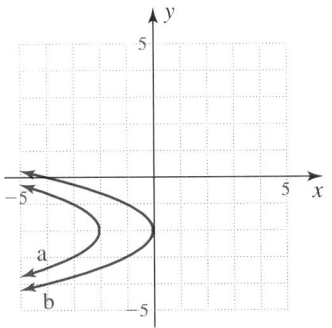

32. a. $x = -(y - 1)^2 + 1$ $(1, 1)$
  b. $x = -(y - 1)^2$ $(0, 1)$

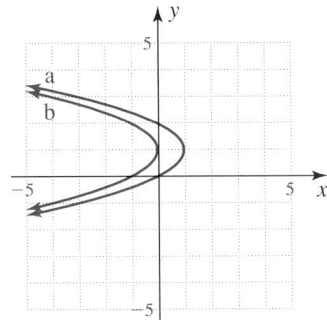

33. a. $x = -y^2 + 2y + 1$ $(2, 1)$
  b. $x = -y^2 + 2y + 4$ $(5, 1)$

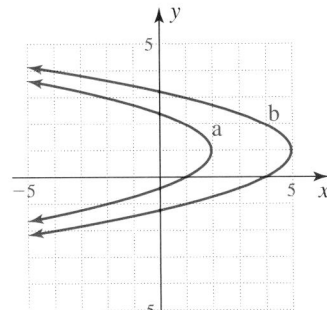

34. a. $x = -y^2 + 4y - 5$ $(-1, 2)$
  b. $x = -y^2 + 4y - 3$ $(1, 2)$

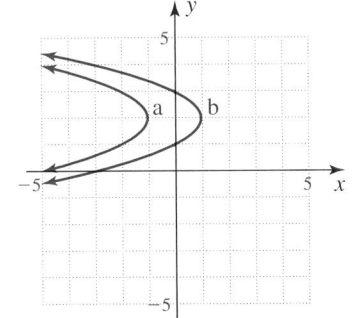

APPLICATIONS

Finding the Maximum

35. The profit P (in dollars) for a company is $P = -5000 + 8x - 0.001x^2$, where x is the number of items produced each month. How many items does the company have to produce to obtain maximum profit? What is this profit? $x = 4000$; $P = \$11,000$

36. The revenue R for Shady Glasses is given by $R = 1500p - 75p^2$, where p is the price of each pair of sunglasses (R and p in dollars). What should the price be to maximize revenue? $p = \$10$

37. After spending x thousand dollars on an advertising campaign, the number of units N sold is given by $N = 50x - x^2$. How much should be spent in the campaign to obtain maximum sales?
$25 thousand ($25,000)

38. The number N of units of a product sold after a television commercial blitz is $N = 40x - x^2$, where x is the amount spent in thousands of dollars. How much should be spent on television commercials to obtain maximum sales? $20 thousand ($20,000)

39. If a ball is batted upward at 160 ft/sec, its height h feet after t seconds is given by $h = -16t^2 + 160t$. Find the maximum height reached by the ball. 400 ft

40. If a ball is thrown upward at 20 ft/sec, its height h feet after t seconds is given by $h = -16t^2 + 20t$. How many seconds does it take for the ball to reach its maximum height, and what is this height? $\frac{5}{8}$ sec; $6\frac{1}{4}$ ft

41. If a farmer digs potatoes today, she will have 600 bushels worth $1 per bushel. Every week she waits, the crop increases by 100 bushels, but the price decreases 10¢ a bushel. Show that she should dig and sell her potatoes at the end of 2 weeks.
$P = (600 + 100W)(1 - 0.10W)$; $P =$ price, $W =$ weeks elapsed. Max P occurs when $W = 2$ (at end of 2 weeks).

42. A man has a large piece of property along Washington Street. He wants to fence the sides and back of a rectangular plot. If he has 400 ft of fencing, what dimensions will give him the maximum area?
200 ft by 100 ft ($x = 100$)

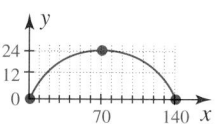

43. Have you read the story "The Jumping Frog of Calaveras County"? According to the *Guinness Book of World Records,* the second greatest distance covered by a frog in a triple jump is 21 ft, $5\frac{3}{4}$ in. at the annual Calaveras Jumping Jubilee; this occurred on May 18, 1986.

  a. If Rosie the Ribiter's (the winner) path in her first jump is approximated by $R = -\frac{1}{98}x^2 + \frac{6}{7}x$ (where x is the horizontal distance covered in inches), what are the coordinates of the vertex of Rosie's path? (42, 18)

  b. Find the maximum height attained by Rosie in her first jump. 18 in.

  c. Use symmetry to find the horizontal length of Rosie's first jump. 84 in.

  d. Make a sketch for R showing the initial position (0, 0), the vertex, and Rosie's ending position after her first jump.

44. Amazingly, Rosie's is not the best triple jump on record. (See Problem 43.) That distinction belongs to Santjie, a South African frog who jumped 33 ft, $5\frac{1}{2}$ in. on May 21, 1977.

  a. If Santjie's path in his first jump is approximated by $S = -\frac{1}{200}x^2 + \frac{7}{10}x$ (where x is the horizontal distance covered in inches), what are the coordinates of the vertex of Santjie's path? $(70, 24\frac{1}{2})$

  b. Find Santjie's maximum height in his first jump. $24\frac{1}{2}$ in.

  c. Use symmetry to find the horizontal length of Santjie's first jump. 140 in.

  d. Make a sketch for S showing the initial position (0, 0), the vertex, and Santjie's ending position after his first jump.

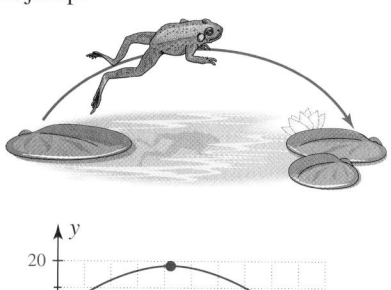

45. A baseball hit at an angle of 35° has a velocity of 130 mi/hr. Its trajectory can be approximated by the equation $d = -\frac{1}{400}x^2 + x$, where x is the distance the ball travels in feet.

     a. What are the coordinates of the vertex of the trajectory? (200, 100)

     b. Find the maximum height attained by the ball. 100 ft

     c. Use symmetry to find how far the ball travels horizontally. 400 ft

     d. Make a sketch for d showing the initial position (0, 0), the vertex, and the ending position of the baseball.

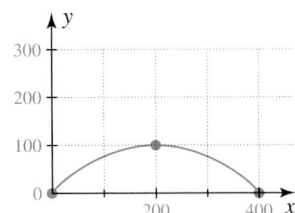

46. From 1983 to 1990, the graph of the average men's SAT verbal scores is nearly a parabola.

     a. What is the maximum average verbal score for men in this period? about 438

     b. What is the minimum average verbal score for men in this period? about 425

     c. If the function approximating the average men's verbal score is

$$f(x) = ax^2 + bx + c$$

     what can you say about a? $a < 0$

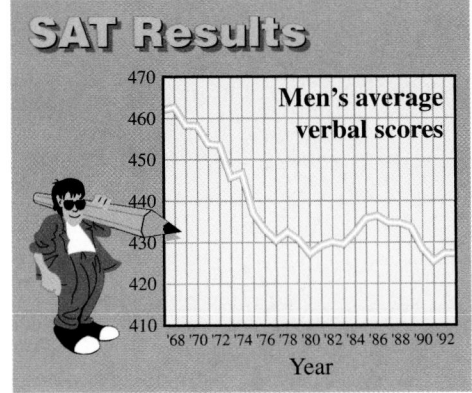

Source: Data from USA TODAY, 1993.

SKILL CHECKER

Try the "Skill Checker" exercises so you'll be ready for the next section.

Find the distance between each pair of points:

47. $A(3, 4)$ and $B(6, 8)$ 5

48. $A(1, 4)$ and $B(-2, 3)$ $\sqrt{10} \approx 3.2$

49. $A(2, -3)$ and $B(4, 2)$ $\sqrt{29} \approx 5.4$

50. $A(3, 2)$ and $B(3, -4)$ 6

USING YOUR KNOWLEDGE

Parabolas Revisited

Here is another way of defining a parabola.

PARABOLA

A **parabola** is the set of all points equidistant from a fixed point $F(0, p)$ (called the **focus**) and a fixed line $y = -p$ (called the **directrix**).

     If $P(x, y)$ is a point on the parabola, this definition says that $FP = DP$; that is, the distance from F to P is the same as the distance from D to P:

51. Find FP. $FP = \sqrt{x^2 + (y - p)^2}$

52. Find DP. $DP = y + p$

53. Set $FP = DP$ and solve for x^2. $x^2 = 4py$

54. For the parabola $x^2 = 4y$,

     a. Locate the focus.   Focus at (0, 1)

     b. Write the equation of the directrix.
         Directrix: $y = -1$

Many applications of the parabola depend on an important focal property of the curve. If the parabola is a mirror, a ray of light parallel to the axis reflects to the focus, and a ray originating at the focus reflects parallel to the axis. (This can be proved by methods of calculus.)

If the parabola is revolved about its axis, a surface called a *paraboloid of revolution* is formed. This is the shape used for automobile headlights and searchlights that throw a parallel beam of light when the light source is placed at the focus; it's also the shape of a radar dish or a reflecting telescope mirror that collects parallel rays of energy (light) and reflects them to the focus.

We can find the equation of the parabola needed to generate a paraboloid of revolution by using the equation $x^2 = 4py$ as follows: Suppose a parabolic mirror has a diameter of 6 ft and a depth of 1 ft. Then, we find the value of p that makes the parabola pass through the point $(3, 1)$. This means that we substitute into the equation and solve for p. Thus we have

$$3^2 = 4p(1)$$

so that $4p = 9$ and $p = 2.25$. The equation of the parabola is $x^2 = 9y$ and the focus is at $(0, 2.25)$.

55. A radar dish has a diameter of 10 ft and a depth of 2 ft. The dish is in the shape of a paraboloid of revolution. Find an equation for a parabola that would generate this dish and locate the focus. $x^2 = 12.5y$; Focus: $(0, 3.125)$

56. The cables of a suspension bridge hang very nearly in the shape of a parabola. A cable on such a bridge spans a distance of 1000 ft and sags 50 ft in the middle. Find an equation for this parabola. $x^2 = 5000y$

WRITE ON

Explain:

57. How you determine whether the graph of a quadratic function opens up or down. $f(x) = ax^2 + bx + c$ opens up if $a > 0$, down if $a < 0$.

58. What causes the graph of the function $f(x) = ax^2$ to be wider or narrower than the graph of $f(x) = x^2$. Narrower if $|a| > 1$; wider if $0 < |a| < 1$

59. The effect of the constant k on the graph of the function $f(x) = ax^2 + k$. Moves the graph up or down

60. How a parabola that has two x-intercepts and vertex at $(1, 1)$ opens—that is, does it open up or down? Why? Down; answers may vary.

61. Why the graph of a function never has two y-intercepts. Answers may vary.

62. How you can tell if the vertex of a parabola is the maximum or the minimum point on the graph of a parabola with a vertical axis. Maximum if the parabola opens down ($a < 0$); minimum if it opens up ($a > 0$)

MASTERY TEST

If you know how to do these problems, you have learned your lesson!

Graph and name the coordinates of the vertex.

63. $x = y^2 + 2y - 3$ $V(-4, -1)$

64. $x = -y^2 + 2y + 3$ $V(4, 1)$

65. $x = 2(y - 1)^2 + 3$ $V(3, 1)$

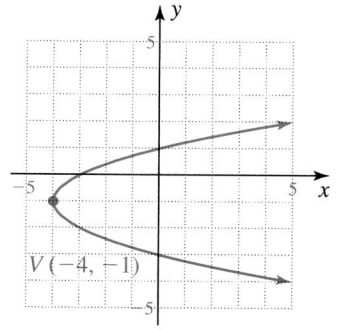

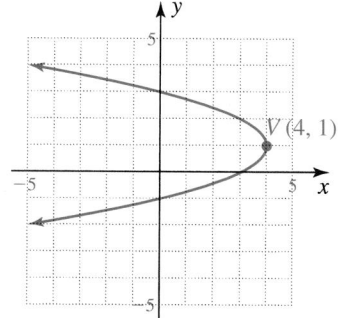

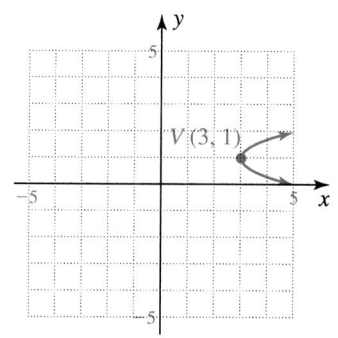

66. $x = -2(y - 1)^2 + 3$ $V(3, 1)$

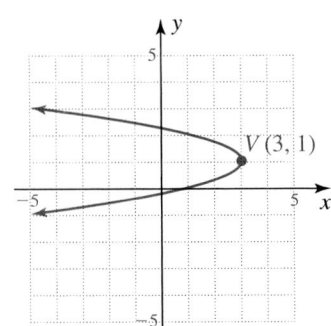

67. $y = -2x^2 - 4x - 3$ $V(-1, -1)$

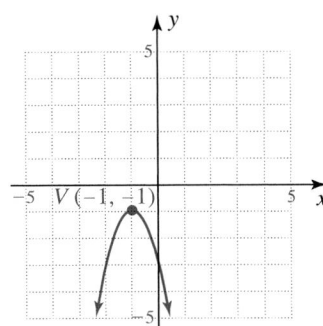

68. $y = x^2 + 2x - 3$ $V(-1, -4)$

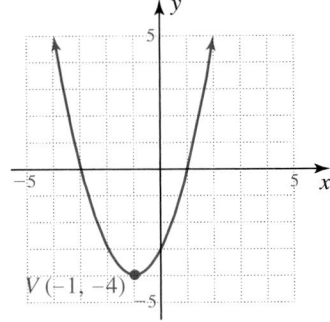

69. $y = (x - 2)^2 + 3$ $V(2, 3)$

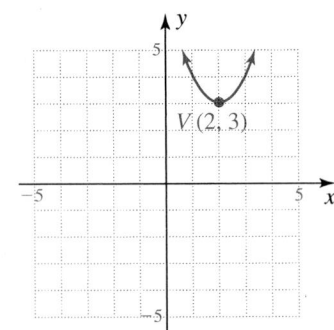

70. $y = -(x - 3)^2 + 2$ $V(3, 2)$

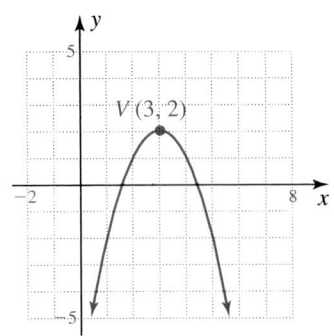

71. $f(x) = -x^2 + 4$ $V(0, 4)$

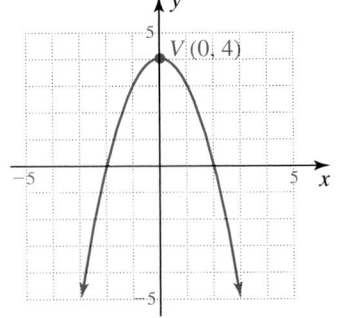

72. $f(x) = x^2 - 4$ $V(0, -4)$

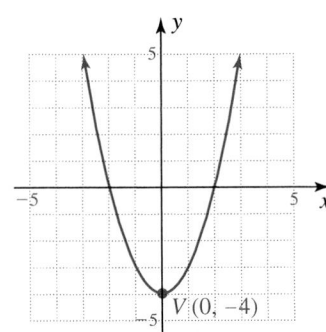

73. $f(x) = 2x^2$ $V(0, 0)$

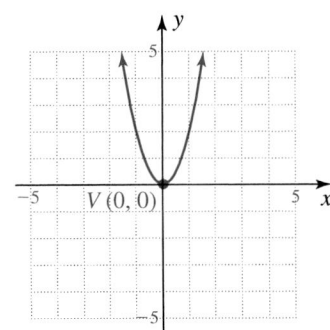

74. $g(x) = \dfrac{1}{2}x^2$ $V(0, 0)$

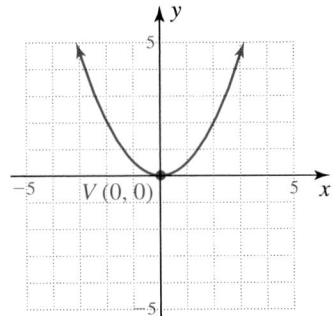

75. $h(x) = -2x^2$ $V(0, 0)$

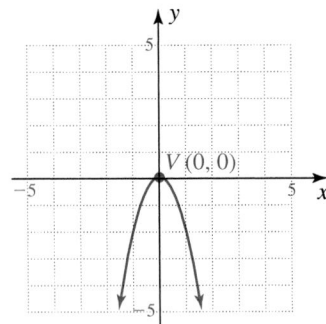

76. $f(x) = -\dfrac{1}{2}x^2$ $V(0, 0)$

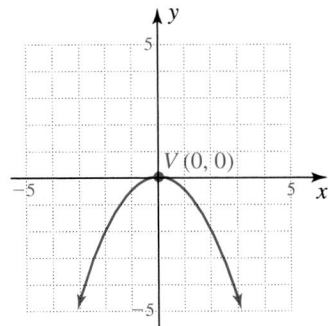

77. The revenue R for a company is $R = 300p - 15p^2$, where p is the price of each unit (R and p are in dollars). What should the price p be so that the revenue R is maximized? $\$10$

9.2   CIRCLES AND ELLIPSES

To Succeed, Review How To . . .

1. Simplify radicals (pp. 530–536).

2. Find the hypotenuse of a right triangle using the Pythagorean theorem (pp. 404–406).

3. Complete the square for a quadratic equation (pp. 587–590).

Objectives

A Find the distance between two points.

B Find the equation of a circle with a given center and radius.

C Find the center and radius and sketch the graph of a circle when its equation is given.

D Graph an ellipse when its equation is given.

E Solve applications involving circles and ellipses.

GETTING STARTED

Comets and Conics

This diagram shows the orbit of the comet Kohoutek with respect to the orbit of the earth. The comet's orbit is an *ellipse,* whereas the earth's orbit is nearly a perfect *circle.* In this section we continue our study of the conic sections by discussing circles and ellipses.

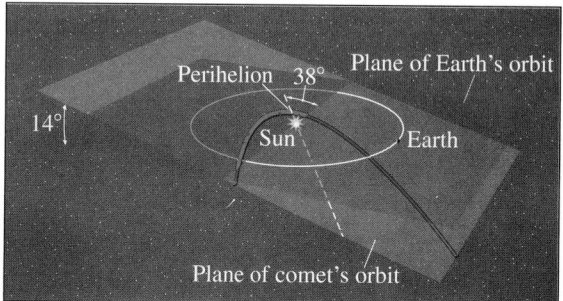

Architecture and Conics

Usually we think of the conic sections as the results of intersecting a cone with a plane. However any cylinder sliced on an angle will reveal an ellipse in the cross-section as seen in the Tycho Brahe Planetarium in Copenhagen.

A   **The Distance Formula**

In the triangle of Figure 25, the distance a between $(-2, -1)$ and $(6, -1)$ is $6 - (-2) = 8$, and the distance b between $(6, 5)$ and $(6, -1)$ is $5 - (-1) = 6$. To find the distance c, we need to use the Pythagorean theorem. According to that theorem, if the legs of a right triangle are a and b and the hypotenuse is c, then

$$c^2 = a^2 + b^2$$

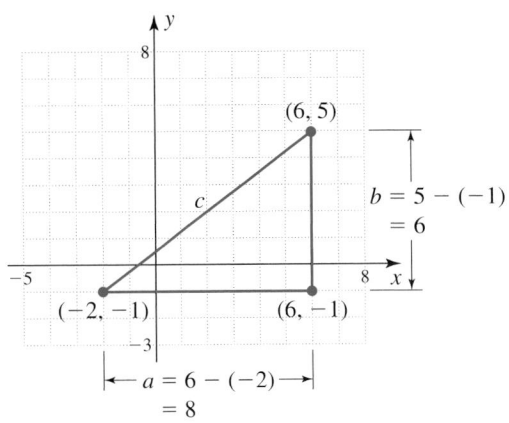

Figure 25

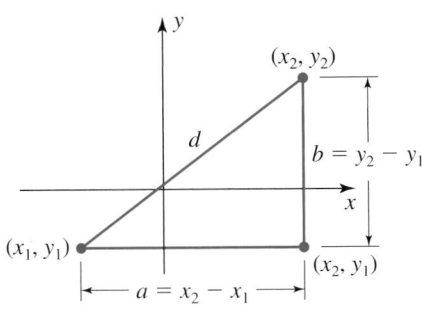

Figure 26

Thus

$$c^2 = 8^2 + 6^2 = 64 + 36$$
$$c^2 = 100$$
$$c = \pm\sqrt{100} = \pm10$$

Since c represents a distance, we discard the negative answer and conclude that $c = 10$. We can repeat a similar argument to find the distance d between any two points (x_1, y_1) and (x_2, y_2), as shown in Figure 26. As before, the distance a is $|x_2 - x_1|$ and the distance b is $|y_2 - y_1|$. By the Pythagorean theorem, we have

$$d^2 = |x_2 - x_1|^2 + |y_2 - y_1|^2$$
$$d = \sqrt{(x_2 - x_1)^2 + (y_2 - y_1)^2}$$

The square of a number is the same as the square of its opposite, so we don't need the absolute-value signs when we square a quantity. Also, the order of x_1 and x_2 does not matter. That is $(x_2 - x_1)^2 = (x_1 - x_2)^2$. Here is a summary of what we have done.

The Distance Formula	The distance between the points (x_1, y_1) and (x_2, y_2) is

$$d = \sqrt{(x_2 - x_1)^2 + (y_2 - y_1)^2}$$

EXAMPLE 1   **Using the distance formula**

Find the distance between the given points:

a. $A(1, 1)$ and $B(5, 4)$

b. $C(2, -2)$ and $D(-2, 5)$

c. $E(-2, 1)$ and $F(-2, 3)$

SOLUTION

a. If we let $x_1 = 1$, $y_1 = 1$, $x_2 = 5$, and $y_2 = 4$, then

$$d = \sqrt{(5 - 1)^2 + (4 - 1)^2} = \sqrt{(4)^2 + (3)^2} = \sqrt{25} = 5$$

b. Here $x_1 = 2$, $y_1 = -2$, $x_2 = -2$, and $y_2 = 5$. Thus

$$d = \sqrt{(-2 - 2)^2 + (5 + 2)^2} = \sqrt{(-4)^2 + (7)^2} = \sqrt{65}$$

c. Now $x_1 = -2$, $y_1 = 1$, $x_2 = -2$, and $y_2 = 3$. Hence

$$d = \sqrt{[-2 - (-2)]^2 + (3 - 1)^2} = \sqrt{(0)^2 + (2)^2} = \sqrt{4} = 2$$

EF is a vertical line, so that $d = |3 - 1| = 2$.

PROBLEM 1

Find the distance between the given points:

a. $A(2, 2)$ and $B(8, 10)$

b. $C(-3, 2)$ and $D(5, 4)$

c. $E(-4, 3)$ and $F(-4, 7)$

Answers

1. a. 10   **b.** $\sqrt{68} = 2\sqrt{17}$
c. 4

B   Finding the Equation of a Circle

Can you define a circle? A **circle** is defined as a set of points in a plane equidistant from a fixed point. The fixed point is called the c*enter* and the given distance is the *radius*. To find the equation of a circle of radius r, suppose the center is at a point $C(h, k)$; see Figure 27. The distance from C to any point $P(x, y)$ on the circle is found by the distance formula. Since r is the radius, this distance must be r. Thus

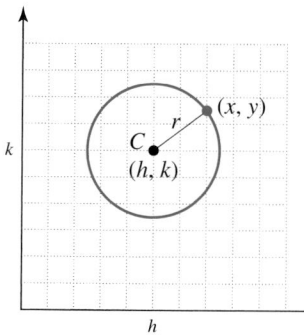

Figure 27
The distance from (h, k) to (x, y) is r.

$$\sqrt{(x_1 - x_2)^2 + (y_1 - y_2)^2} = d \qquad \text{Distance formula.}$$

$$\sqrt{(x - h)^2 + (y - k)^2} = r \qquad \text{Substitute } (h, k) \text{ for } (x_2, y_2).$$

$$(x - h)^2 + (y - k)^2 = r^2 \qquad \text{Square both sides.}$$

We then have the following.

> **Graphing Form of the Equation of a Circle Centered at (h, k)**
>
> The equation of a circle with **radius** r and with **center** at $C(h, k)$ is
>
> $$(x - h)^2 + (y - k)^2 = r^2$$

EXAMPLE 2   Finding the equation of a circle

Find the equation of the circle with center at $(3, -5)$ and radius 2.

SOLUTION   Here, the center $(h, k) = (3, -5)$ and $r = 2$. This means $h = 3$, $k = -5$, and $r = 2$. Using the formula, we have

$$(x - h)^2 + (y - k)^2 = r^2$$

$$(x - 3)^2 + [(y - (-5)]^2 = 2^2 \qquad \text{Substitute } h = 3, k = -5, r = 2.$$

$$(x - 3)^2 + (y + 5)^2 = 4 \qquad \text{Simplify.}$$

PROBLEM 2

Find the equation of the circle with center at $(-1, 2)$ and radius 3.

EXAMPLE 3   Finding the equation of a circle centered at the origin, (0, 0)

Find the equation of a circle of radius 3 and with center at the origin, $(0, 0)$.

SOLUTION   The center is at $(h, k) = (0, 0)$. Thus $h = 0$, $k = 0$, and $r = 3$. Substituting $h = 0$, $k = 0$, $r = 3$ in $(x - h)^2 + (y - k)^2 = r^2$ gives

$$(x - 0)^2 + (y - 0)^2 = 3^2$$

$$x^2 + y^2 = 9$$

PROBLEM 3

Find the equation of a circle of radius $\sqrt{3}$ and with center at the origin.

In general, we have the following.

> **Graphing Form of a Circle Centered at the Origin, (0, 0)**
>
> The equation of a circle of radius r with center at the origin, $(0, 0)$, is
>
> $$x^2 + y^2 = r^2$$

Answers

2. $(x + 1)^2 + (y - 2)^2 = 9$
3. $x^2 + y^2 = 3$

C Finding the Center and Radius of a Circle

If we have the equation of a circle, we can write it in graphing form $(x - h)^2 + (y - k)^2 = r^2$ and find the center and radius. For example, if a circle has equation $(x - 3)^2 + (y - 4)^2 = 5^2$, then $h = 3$, $k = 4$, and $r = 5$. Thus the equation $(x - 3)^2 + (y - 4)^2 = 5^2$ is the equation of a circle of radius 5 with center at $(3, 4)$.

EXAMPLE 4 Finding the center and radius of a circle

Find the center and radius and sketch the graph of the circle whose equation is

$$(x + 2)^2 + (y - 1)^2 = 9$$

SOLUTION We first write the equation in graphing form

$$(x - h)^2 + (y - k)^2 = r^2$$

Thus

$$[x - (-2)]^2 + (y - 1)^2 = 3^2$$

We have $h = -2$, $k = 1$, and $r = 3$. The center is at $(h, k) = (-2, 1)$ and the radius is 3. The sketch is shown in Figure 28.

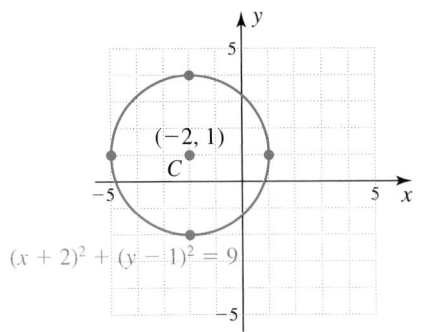

Figure 28

PROBLEM 4

Find the center and radius and sketch the graph of the circle whose equation is

$$(x - 3)^2 + (y + 5)^2 = 1$$

EXAMPLE 5 Sketching the graph of a circle centered at the origin

Find the center and radius and sketch the graph of

$$x^2 + y^2 = 25$$

SOLUTION This equation can be written as $x^2 + y^2 = 5^2$, the equation of a circle of radius 5 centered at the origin, $(0, 0)$. The graph is shown in Figure 29.

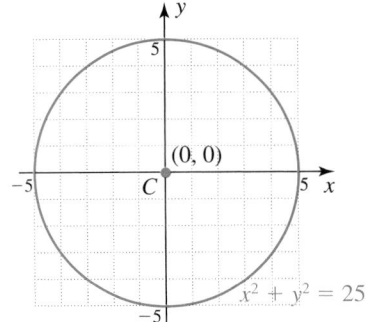

Figure 29

PROBLEM 5

Find the center and radius and sketch the graph of

$$x^2 + y^2 = 7$$

Answers

4. Center $(3, -5)$; $r = 1$ **5.** Center $(0, 0)$; $r = \sqrt{7} \approx 2.6$

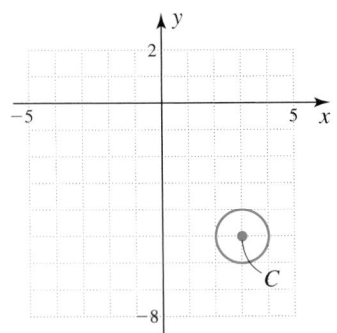

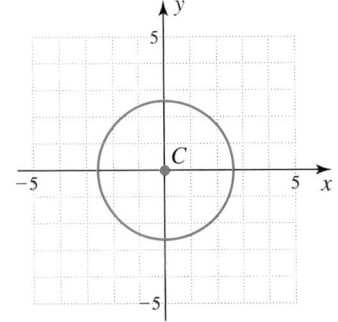

Calculate It   Graphing Circles

Most calculators graph only certain types of algebraic expressions. Do you know what types? Functions! Would your calculator graph $x^2 + y^2 = 25$ from Example 5? The vertical line test should convince you that $x^2 + y^2 = 25$ doesn't represent a function. How can we graph it? We do it by first solving for y to obtain $y = \pm\sqrt{25 - x^2}$. Then we graph each of the two halves of the circle, $Y_1 = \sqrt{25 - x^2}$ (the top half) and $Y_2 = -\sqrt{25 - x^2}$ (the bottom half). Unfortunately, the result looks like an ellipse (Window 1).

Do you know why? Because the units on the x-axis are longer than the units on the y-axis; the graph is wider than it is high. To fix it, use a "square window" (ZOOM 5) on a TI-83 Plus) to obtain the graph shown in Window 2.

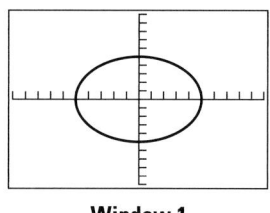
Window 1

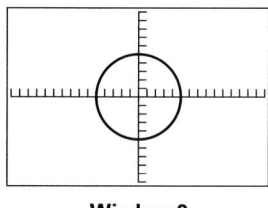
Window 2

(You can save time when you graph $y = \pm\sqrt{25 - x^2}$ if you enter $Y_1 = \sqrt{25 - x^2}$ and then $Y_2 = -Y_1$. With a TI-83 Plus, place the cursor by $Y_2 =$ and press (−) VARS ▶ 1 1 .)

EXAMPLE 6

Sketching the graph of a circle not centered at the origin

Find the center and radius and sketch the graph of

$$x^2 - 4x + y^2 + 6y + 8 = 0$$

SOLUTION   We must find the center and the radius by writing the equation in graphing form $(x - h)^2 + (y - k)^2 = r^2$. We can do this by subtracting 8 from both sides and then completing the squares on x and y.

$x^2 - 4x + y^2 + 6y + 8 = 0$ Given.

$x^2 - 4x + \underline{} + y^2 + 6y + \underline{} = -8$ To complete the squares add $(\frac{1}{2} \cdot 4)^2$ and $(\frac{1}{2} \cdot 6)^2$.

$x^2 - 4x + 4 + y^2 + 6y + 9 = -8 + 4 + 9$

$(x - 2)^2 + (y + 3)^2 = 5$

$(x - 2)^2 + (y + 3)^2 = (\sqrt{5})^2$ Remember that $(\sqrt{5})^2 = 5$.

Thus the center is at $(2, -3)$, and the radius is $\sqrt{5} \approx 2.2$. See Figure 30.

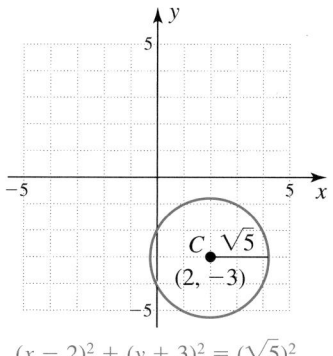

$(x - 2)^2 + (y + 3)^2 = (\sqrt{5})^2$

Figure 30

PROBLEM 6

Find the center and radius and sketch the graph of

$$x^2 - 4x + y^2 - 6y + 9 = 0$$

Web It

For more examples and practice problems on writing the equation of a circle, go to link 9-2-1 at mhhe.com/bello.

Answer

6. Center $(2, 3)$; $r = 2$

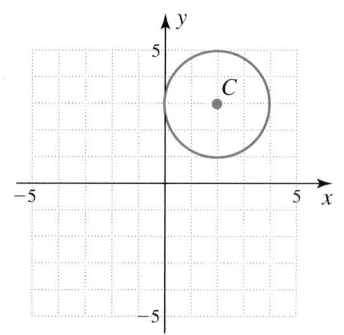

D Graphing Ellipses

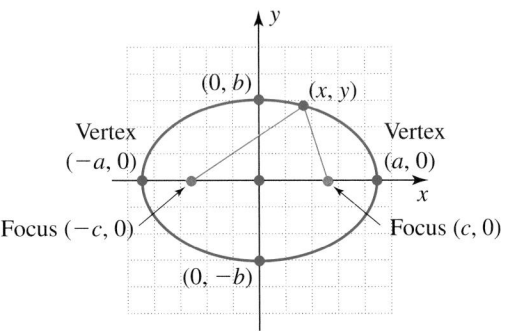

Figure 31

If you look at the diagram in the *Getting Started* at the beginning of this section, you will see the drawing of part of an ellipse. An *ellipse* is the set of points in a plane such that the sum of the distances of each point from two fixed points (called the **foci;** singular, *focus*) is a constant. If the coordinates of the foci are $(c, 0)$ and $(-c, 0)$, then the center of the ellipse is at the origin, and the x- and y-intercepts are given by $x = \pm a$ and $y = \pm b$. The points $(a, 0)$ and $(-a, 0)$ are called the vertices (singular, vertex). The graph is shown in Figure 31.

We can use the distance formula to find the equation of an ellipse (see *Using Your Knowledge* in the Exercises) but for the time being, we assume the following.

Graphing Form of the Equation of an Ellipse with Center at (0, 0)	The equation of the ellipse with center at $(0, 0)$ whose x-intercepts are $(a, 0)$ and $(-a, 0)$ and whose y-intercepts are $(0, b)$ and $(0, -b)$ is $$\frac{x^2}{a^2} + \frac{y^2}{b^2} = 1, \quad \text{where } a^2 > b^2, a \neq 0, b \neq 0$$ The vertices are at $(a, 0)$ and $(-a, 0)$. The equation of the ellipse with center at $(0, 0)$ whose x-intercepts are $(b, 0)$ and $(-b, 0)$ and whose y-intercepts are $(0, a)$ and $(0, -a)$ is $$\frac{x^2}{b^2} + \frac{y^2}{a^2} = 1, \quad \text{where } a^2 > b^2, a \neq 0, b \neq 0$$ The vertices are at $(0, a)$ and $(0, -a)$. If a and b are equal, the ellipse is a circle.

EXAMPLE 7   **Graphing an ellipse with center at (0, 0)**

Find the coordinates of the center, the x- and y-intercepts, and the vertices, and sketch the graph of

$$4x^2 + 25y^2 = 100$$

SOLUTION   To make sure we have an ellipse with center at $(0, 0)$, we write the equation in graphing form

$$\frac{x^2}{a^2} + \frac{y^2}{b^2} = 1$$

If we divide each term by 100 (to make the right side 1), we have

$$\frac{x^2}{25} + \frac{y^2}{4} = 1$$

where $a^2 = 25$ with $a = 5$ and $b^2 = 4$ with $b = 2$.

The x-intercepts are $(5, 0)$ and $(-5, 0)$, the y-intercepts are $(0, 2)$ and $(0, -2)$, and the center is at $(0, 0)$. The vertices are at $V_1 = (5, 0)$ and $V_2 = (-5, 0)$. We then pass the ellipse through the four intercepts, as shown in Figure 32.

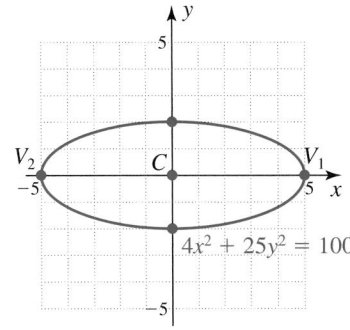

Figure 32

PROBLEM 7

Find the center, x- and y-intercepts, and the vertices, and sketch the graph

$$9x^2 + 4y^2 = 36$$

Answer

7. Center $(0, 0)$; x-int $(\pm 2, 0)$, y-int $(0, \pm 3)$; $V_1 = (0, 3)$, $V_2 = (0, -3)$

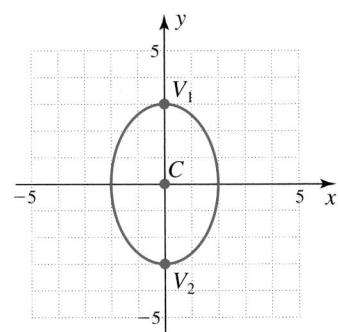

We can graph the original equation $4x^2 + 25y^2 = 100$ by letting $x = 0$ to obtain

$$25y^2 = 100$$
$$y^2 = 4$$
$$y = \pm\sqrt{4} = \pm 2$$

Letting $y = 0$ will yield $x = \pm 5$, as before.

As in the case of circles, ellipses can be centered away from the origin, as stated here and detailed in Example 8.

Graphing Form of the Equation of an Ellipse with Center at (h, k)	The equation of the ellipse with center at (h, k) is $$\frac{(x - h)^2}{a^2} + \frac{(y - k)^2}{b^2} = 1, \quad \text{where } a^2 > b^2, a \neq 0, b \neq 0$$ The vertices are horizontally $\pm a$ units from (h, k). The equation of the ellipse with center at (h, k) is $$\frac{(x - h)^2}{b^2} + \frac{(y - k)^2}{a^2} = 1, \quad \text{where } a^2 > b^2, a \neq 0, b \neq 0$$ The vertices are vertically $\pm a$ units from (h, k).

EXAMPLE 8   Graphing an ellipse not centered at the origin

Find the coordinates of the center and the vertices and graph

$$\frac{(x - 3)^2}{4} + \frac{(y + 1)^2}{9} = 1$$

SOLUTION   The center of this ellipse is at $(3, -1)$. Use the values of a and b to determine the dimensions of the ellipse, as shown in Figure 33. The vertices are vertically ± 3 units from $(3, -1)$. The coordinates of the vertices are $V_1 = (3, 2)$ and $V_2 = (3, -4)$.

The graph of the ellipse is shown in Figure 34.

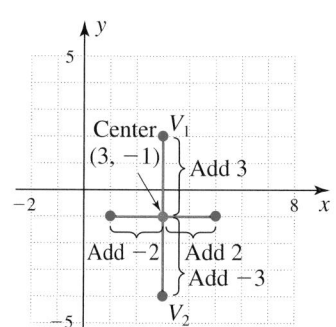

Figure 33

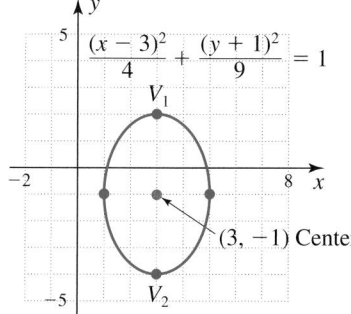

Figure 34

PROBLEM 8

Name the coordinates of the center and the vertices and graph

$$\frac{(x + 2)^2}{4} + \frac{(y - 1)^2}{9} = 1$$

Answer

8. Center $(-2, 1)$; $V_1 = (-2, 4)$, $V_2 = (-2, -2)$

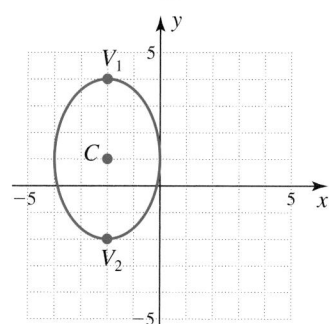

Calculate It   Graphing Ellipses

The ellipses in Examples 7 and 8 can be graphed using the same technique we used to graph the circle of Example 5. Thus if we solve for y in 7, we have

$$y = \pm\sqrt{\frac{100 - 4x^2}{25}}$$

Graphing the top and bottom halves of the ellipse in a square window produces the graph of $4x^2 + 25y^2 = 100$ as shown in the window. Now try Example 8.

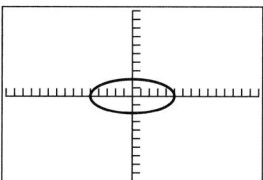

We can find the equation of an ellipse with foci on one of the axes and with center at the origin if we know its x- and y-intercepts. The ellipse of Example 7 passes through $(\pm 5, 0)$ and $(0, \pm 2)$ and has equation

$$\frac{x^2}{5^2} + \frac{y^2}{2^2} = 1$$

Similarly, if the ellipse passes through $(\pm 3, 0)$ and $(0, \pm 4)$, its equation would be

$$\frac{x^2}{3^2} + \frac{y^2}{4^2} = 1$$

We will use this idea in Problems 62–65. Finally, make sure you know how to tell the difference between the graph of a circle and that of an ellipse. The equation

$$Ax^2 + By^2 = C \qquad (A, B, \text{ and } C \text{ positive})$$

has a **circle** as its graph when $A = B$ and an **ellipse** when $A \neq B$.

Solve an Application

| EXAMPLE 9 | Solving an application with a conic equation |

To protect the baseball mound in rainy weather it is covered with a tarp. The tarp in Figure 35 has a diameter of 20 ft. Assuming the center of the tarp is placed at the center of an x-y coordinate system, write the equation that describes the circular edge of the mound.

Figure 35

SOLUTION The outer edge of the mound is a circle. The assumption that its center coincides with $(0, 0)$ allows us to use the graphing form of the equation of a circle, $x^2 + y^2 = r^2$. To complete the equation of the circle we need to know the radius. The radius can be obtained by taking half of the diameter or $r = \frac{1}{2}$ of $20 = 10$. Thus, the equation of the outer edge of the circular tarp is $x^2 + y^2 = 100$.

| PROBLEM 9 |

The shapes of leaf types vary. The one below is roughly the shape of an ellipse. Assuming the center of the leaf is placed at the center of an x-y coordinate system with the longest measurement coinciding with the x-axis, write the equation of the ellipse. The longer measurement or axis of the leaf is 4 in. and the shorter axis is 2 in.

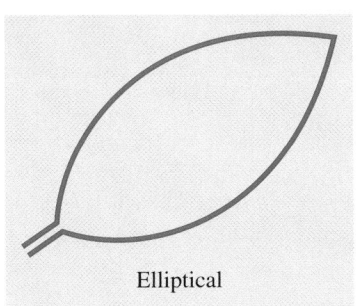

Elliptical

Answer

9. $\frac{x^2}{4} + \frac{y^2}{1} = 1$

Exercises 9.2

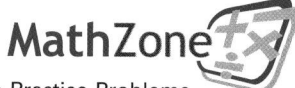

A In Problems 1–10, find the distance between the points.

1. $A(2, 4)$, $B(-1, 0)$ 5

2. $A(3, -2)$, $B(8, 10)$ 13

3. $C(-4, -5)$, $D(-1, 3)$ $\sqrt{73}$

4. $C(5, 7)$, $D(-2, 3)$ $\sqrt{65}$

5. $E(4, 8)$, $G(1, -1)$
$\sqrt{90} = 3\sqrt{10}$

6. $H(-2, -2)$, $I(6, -4)$
$\sqrt{68} = 2\sqrt{17}$

7. $A(3, -1)$, $B(-2, -1)$ 5

8. $C(-2, 3)$, $D(4, 3)$ 6

9. $E(-1, 2)$, $F(-1, -4)$ 6

10. $G(-3, 2)$, $H(-3, 5)$ 3

B In Problems 11–20, find the equation of a circle with the given center and radius.

11. Center $(3, 8)$, radius 2
$(x - 3)^2 + (y - 8)^2 = 4$

12. Center $(2, 5)$, radius 3
$(x - 2)^2 + (y - 5)^2 = 9$

13. Center $(-3, 4)$, radius 5
$(x + 3)^2 + (y - 4)^2 = 25$

14. Center $(-5, 2)$, radius 5
$(x + 5)^2 + (y - 2)^2 = 25$

15. Center $(-3, -2)$, radius 4
$(x + 3)^2 + (y + 2)^2 = 16$

16. Center $(-1, -7)$, radius 9
$(x + 1)^2 + (y + 7)^2 = 81$

17. Center $(2, -4)$, radius $\sqrt{5}$
$(x - 2)^2 + (y + 4)^2 = 5$

18. Center $(3, -5)$, radius $\sqrt{7}$
$(x - 3)^2 + (y + 5)^2 = 7$

19. Radius 3, center at the origin
$x^2 + y^2 = 9$

20. Radius 2, center at the origin
$x^2 + y^2 = 4$

C In Problems 21–42, find the center and radius of the circle and sketch the graph.

21. $(x - 1)^2 + (y - 2)^2 = 9$

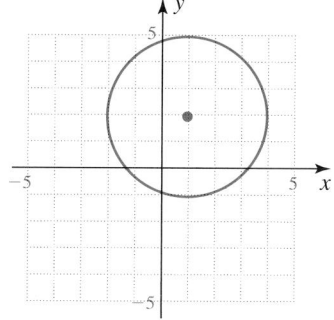

$C(1, 2)$; $r = 3$

22. $(x - 2)^2 + (y - 1)^2 = 4$

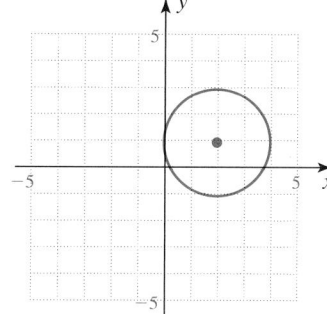

$C(2, 1)$; $r = 2$

23. $(x + 1)^2 + (y - 2)^2 = 4$

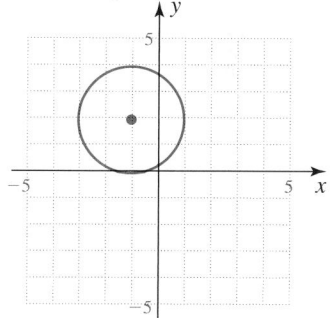

$C(-1, 2)$; $r = 2$

24. $(x + 2)^2 + (y - 1)^2 = 9$

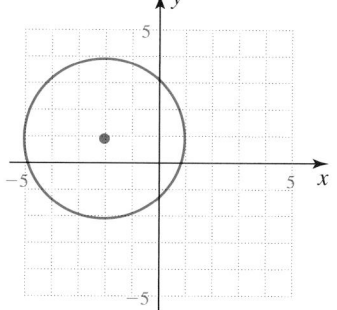

$C(-2, 1)$; $r = 3$

25. $(x - 1)^2 + (y + 2)^2 = 1$

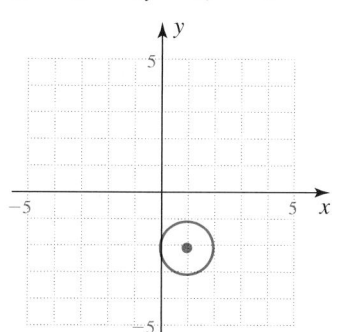

$C(1, -2)$; $r = 1$

26. $(x - 2)^2 + (y + 1)^2 = 4$

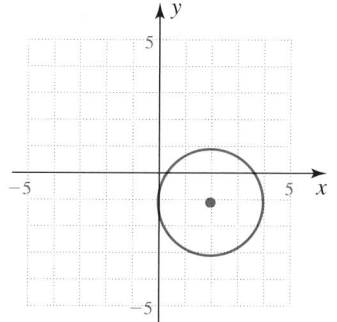

$C(2, -1)$; $r = 2$

27. $(x + 2)^2 + (y + 1)^2 = 9$

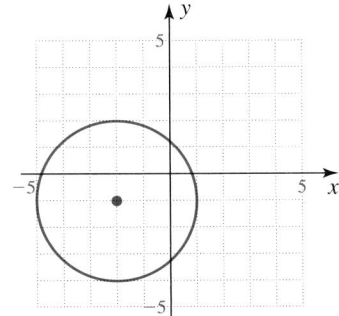

$C(-2, -1); r = 3$

28. $(x + 3)^2 + (y + 1)^2 = 4$

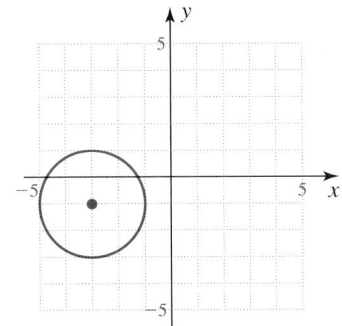

$C(-3, -1); r = 2$

29. $(x - 1)^2 + (y - 1)^2 = 7$

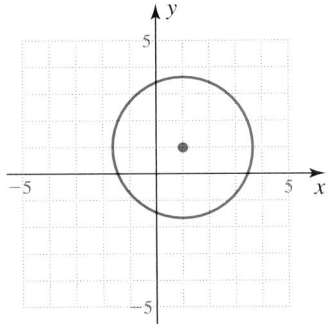

$C(1, 1); r = \sqrt{7} \approx 2.6$

30. $(x - 1)^2 + (y - 1)^2 = 3$

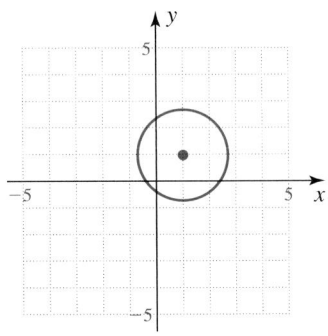

$C(1, 1); r = \sqrt{3} \approx 1.7$

31. $x^2 - 6x + y^2 - 4y + 9 = 0$

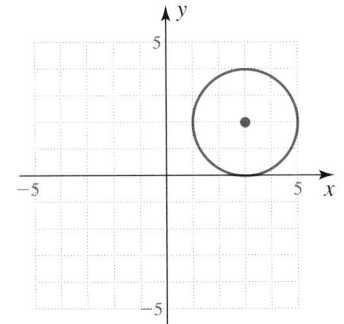

$C(3, 2); r = 2$

32. $x^2 - 6x + y^2 - 2y + 9 = 0$

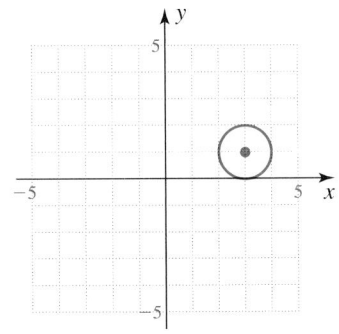

$C(3, 1); r = 1$

33. $x^2 + y^2 - 4x + 2y - 4 = 0$

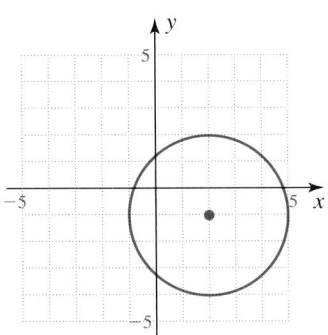

$C(2, -1); r = 3$

34. $x^2 + y^2 + 2x - 4y - 4 = 0$

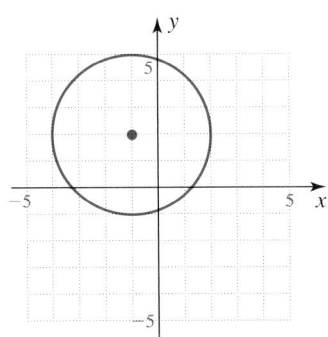

$C(-1, 2); r = 3$

35. $x^2 + y^2 - 25 = 0$

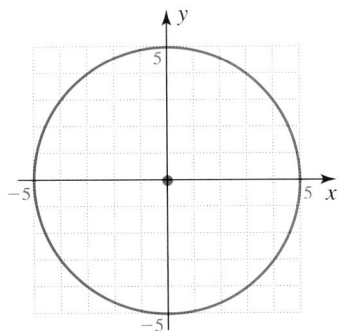

$C(0, 0); r = 5$

36. $x^2 + y^2 - 9 = 0$

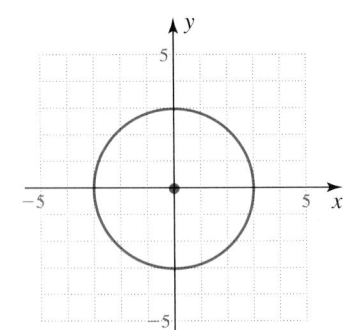

$C(0, 0); r = 3$

37. $x^2 + y^2 - 7 = 0$

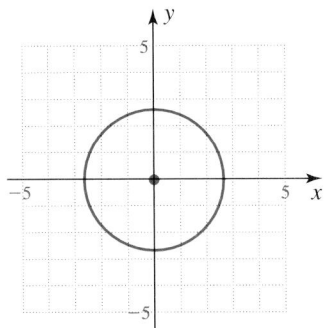

$C(0, 0); r = \sqrt{7} \approx 2.6$

38. $x^2 + y^2 - 3 = 0$

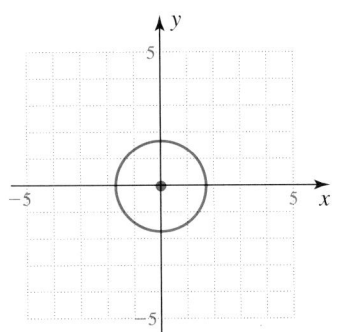

$C(0, 0); r = \sqrt{3} \approx 1.7$

39. $x^2 + y^2 + 6x - 2y = -6$

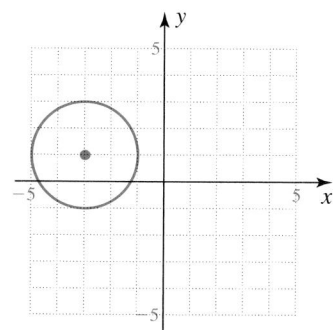

$C(-3, 1); r = 2$

40. $x^2 + y^2 + 4x - 2y = -4$

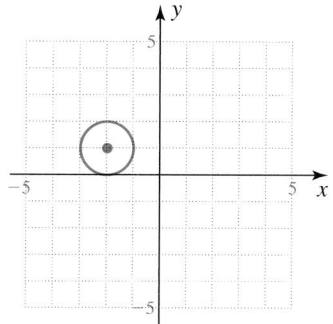

$C(-2, 1); r = 1$

41. $x^2 + y^2 - 6x - 2y + 6 = 0$

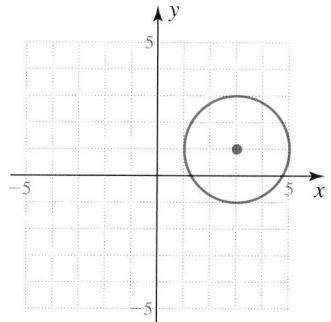

$C(3, 1); r = 2$

42. $x^2 + y^2 - 4x - 6y + 12 = 0$

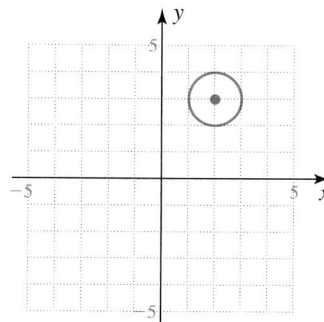

$C(2, 3); r = 1$

D In Problems 43–56, graph the ellipse. Give the coordinates of the center and the values of a and b.

43. $25x^2 + 4y^2 = 100$

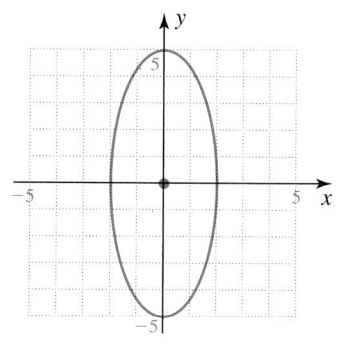

$C(0, 0); a = 5, b = 2$

44. $9x^2 + 4y^2 = 36$

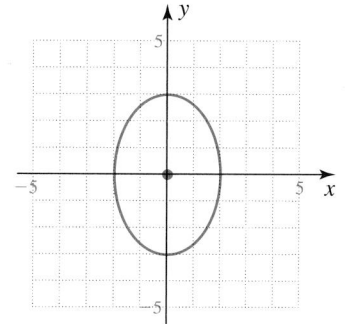

$C(0, 0); a = 3, b = 2$

45. $x^2 + 4y^2 = 4$

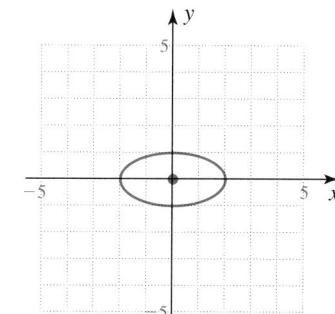

$C(0, 0); a = 2, b = 1$

46. $x^2 + 9y^2 = 9$

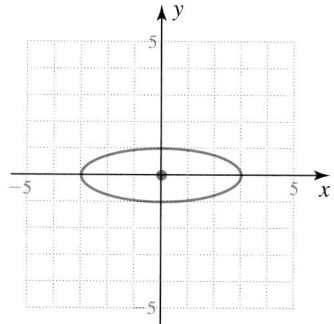

$C(0, 0); a = 3, b = 1$

47. $x^2 + 4y^2 = 16$

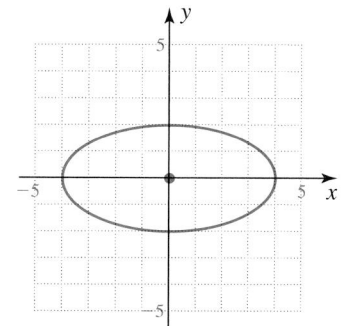

$C(0, 0); a = 4, b = 2$

48. $x^2 + 9y^2 = 25$

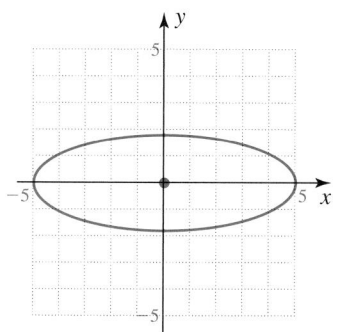

$C(0, 0); a = 5, b = \frac{5}{3}$

49. $\dfrac{x^2}{9} + \dfrac{y^2}{16} = 1$

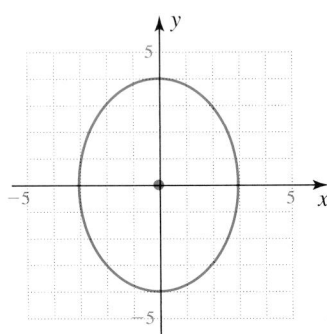

$C(0, 0); a = 4, b = 3$

50. $\dfrac{x^2}{4} + \dfrac{y^2}{1} = 1$

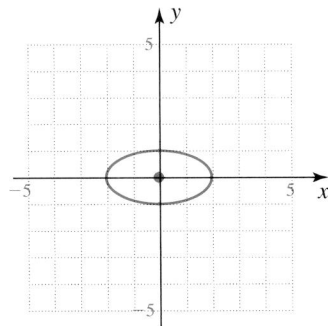

$C(0, 0); a = 2, b = 1$

51. $\dfrac{(x - 1)^2}{4} + \dfrac{(y - 2)^2}{9} = 1$

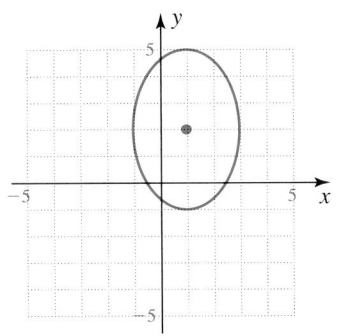

$C(1, 2); a = 3, b = 2$

52. $\dfrac{(x - 2)^2}{9} + \dfrac{(y - 1)^2}{4} = 1$

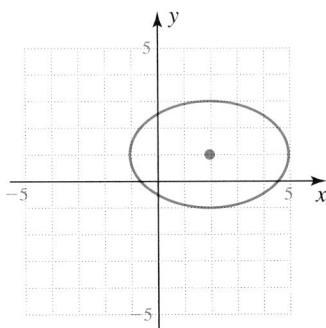

$C(2, 1); a = 3, b = 2$

53. $\dfrac{(x - 2)^2}{9} + \dfrac{(y + 3)^2}{4} = 1$

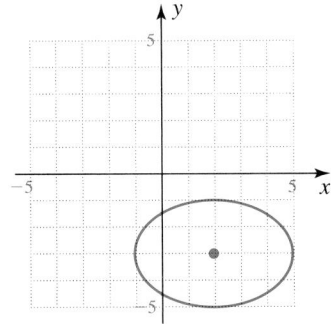

$C(2, -3); a = 3, b = 2$

54. $\dfrac{(x - 1)^2}{4} + \dfrac{(y + 2)^2}{9} = 1$

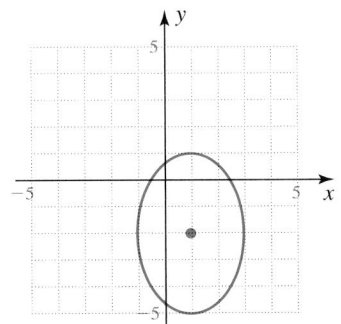

$C(1, -2); a = 3, b = 2$

55. $\dfrac{(x - 1)^2}{16} + \dfrac{(y - 1)^2}{9} = 1$

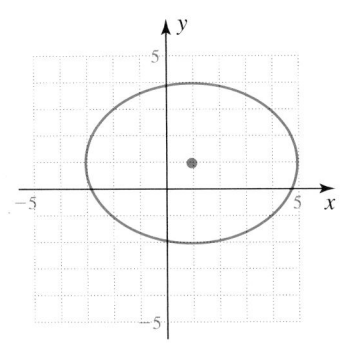

$C(1, 1); a = 4, b = 3$

56. $\dfrac{(x - 2)^2}{9} + \dfrac{(y - 1)^2}{16} = 1$

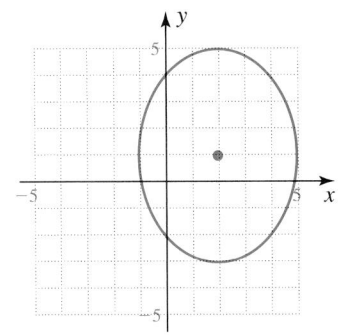

$C(2, 1); a = 4, b = 3$

In Problems 57–61, find an equation of the circle with center at the origin and:

57. Passing through the point $(4, 3)$ $x^2 + y^2 = 25$

58. Passing through the point $(3, 4)$ $x^2 + y^2 = 25$

59. Passing through the point $(-5, -12)$ $x^2 + y^2 = 169$

60. x-intercepts ± 5 $x^2 + y^2 = 25$

61. y-intercepts ± 3 $x^2 + y^2 = 9$

In Problems 62–65, find an equation of the ellipse centered at the origin and passing through:

62. Points $(\pm 7, 0)$ and $(0, \pm 2)$ $\dfrac{x^2}{49} + \dfrac{y^2}{4} = 1$

63. Points $(\pm 2, 0)$ and $(0, \pm 6)$ $\dfrac{x^2}{4} + \dfrac{y^2}{36} = 1$

64. Points $(\pm 3, 0)$ and $(0, \pm 7)$ $\dfrac{x^2}{9} + \dfrac{y^2}{49} = 1$

65. Points $(\pm 6, 0)$ and $(0, \pm 4)$ $\dfrac{x^2}{36} + \dfrac{y^2}{16} = 1$

APPLICATIONS

Rounded Objects

66. A circular arch for a bridge has a 100-ft span. If the height of the arch above the water is 25 ft, find an equation of the circle containing the arch if the center is at the origin as shown. [*Hint:* If the radius is r, $(50, r - 25)$ must satisfy the equation $x^2 + y^2 = r^2$.]

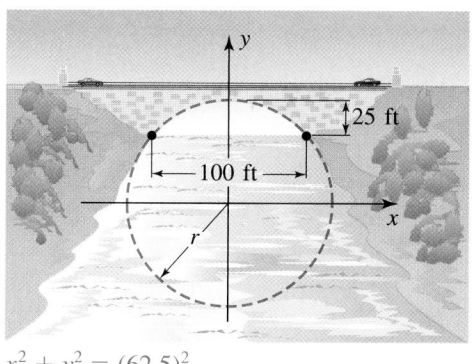

$x^2 + y^2 = (62.5)^2$

67. A cylindrical drum is cut to make a barbecue grill. The end of the resulting grill is 5 in. high and 20 in. wide at the top. What is the radius r of the original drum?

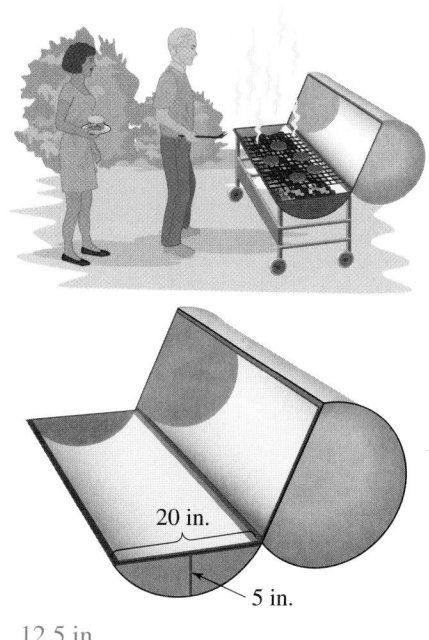

12.5 in.

68. The larger semicircle has a radius of 15 ft.

a. If the x-axis is placed at ground level, what is the equation of the circle? $x^2 + y^2 = 225$

b. If the 20 vertical bars inside the smaller semicircle are 1 ft apart, what is the length of the longest vertical bar?   10 ft

c. What is the length of the bar 1 ft to the right of the longest bar? $3\sqrt{11} \approx 9.9$ ft

69. A portion of a circle with a 15-ft radius (blue) is shown. If the vertical bar (red) is 5 ft, how long is the horizontal bar (green)? (*Hint:* Look at the diagram.)
$10\sqrt{5} \approx 22.4$ ft

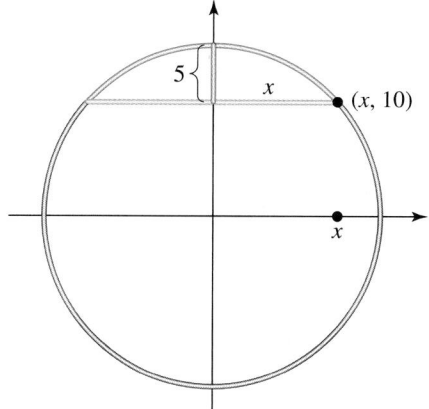

70. The elliptical drain pipe is 150 cm wide and 100 cm high. If the origin is placed at the center of the pipe, what is the equation in standard form of the elliptical opening?

$$\frac{x^2}{75^2} + \frac{y^2}{50^2} = 1 \text{ or } \frac{x^2}{5625} + \frac{y^2}{2500} = 1$$

71. The elliptical portion of the sign is 8 ft wide and 5 ft high. What is the standard form of the equation of this ellipse if the origin is at the center of the sign?

$$\frac{x^2}{4^2} + \frac{y^2}{2.5^2} = 1 \text{ or } \frac{x^2}{16} + \frac{y^2}{6.25} = 1$$

72. The top half of a sign is half of an ellipse 7 ft high and 13 ft wide. If the origin is at the center of the sign, what is the equation of the complete ellipse in standard form?

$$\frac{x^2}{6.5^2} + \frac{y^2}{3.5^2} = 1 \text{ or } \frac{x^2}{42.25} + \frac{y^2}{12.25} = 1$$

73. The elliptical portion of the tanker truck is 8 ft wide and 6 ft high. What is the equation of the ellipse whose origin is at the center of the elliptical portion of the truck?

$$\frac{x^2}{16} + \frac{y^2}{9} = 1$$

74. An elliptical running track is 100 yd long and 50 yd wide.

a. If the origin is at its center and the longer side is along the x-axis, what is the equation of the ellipse?
$$\frac{x^2}{50^2} + \frac{y^2}{25^2} = 1 \text{ or } \frac{x^2}{2500} + \frac{y^2}{625} = 1$$

b. A running strip for pole vaulting is parallel to the y-axis 20 yd from the right-hand end of the track. Both ends of this strip are 5 yd from the running track. How long is the strip? 30 yd

75. The orbit of the earth around the sun (one of the foci) is an ellipse as shown. The equation of the ellipse is written as

$$\frac{x^2}{a^2} + \frac{y^2}{b^2} = 1$$

where x and y are in millions of miles.

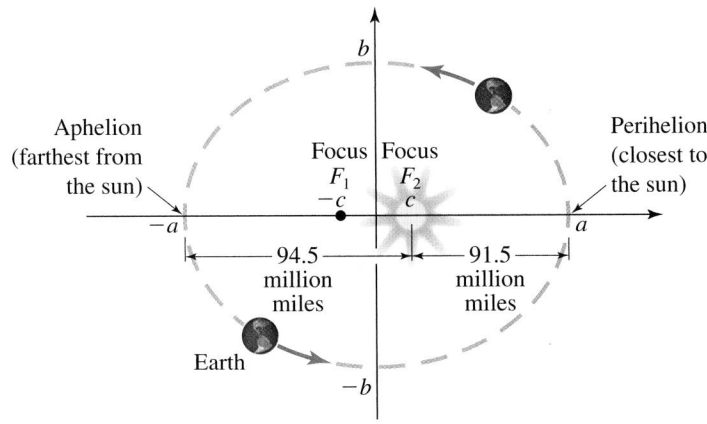

a. Find a. 93 million mi   **b.** Find c. 1.5 million mi

c. If $b^2 = a^2 - c^2$, find b. Round the answer to two decimal places. 92.99 million mi

76. A semielliptic arch is supporting a bridge spanning a river 50 ft wide. The center of the arch is 20 ft above the center of the river. Write an equation of the ellipse in which the *x*-axis coincides with the water level and the *y*-axis passes through the center of the arch.

$$\frac{x^2}{25^2} + \frac{y^2}{20^2} = 1 \text{ or } \frac{x^2}{625} + \frac{y^2}{400} = 1$$

77. Have you eaten from plastic food plates lately? These plates are made using an elliptical mold 12 in. long and 9 in. wide.

   a. Write in standard form the equation of the outside ellipse of the mold. $\frac{x^2}{6^2} + \frac{y^2}{4.5^2} = 1$ or $\frac{x^2}{36} + \frac{y^2}{20.25} = 1$

   b. Find the width of the dish at a distance of 4 in. from the center. (This is the width in the direction perpendicular to the *x*-axis.) $\frac{6\sqrt{5}}{20} \approx 6.71$ in.

78. A semielliptic arch spanning a river has the dimensions shown. What is the height *h* of the arch at a distance of 10 ft from the center? $\frac{20\sqrt{5}}{3} \approx 14.9$ ft

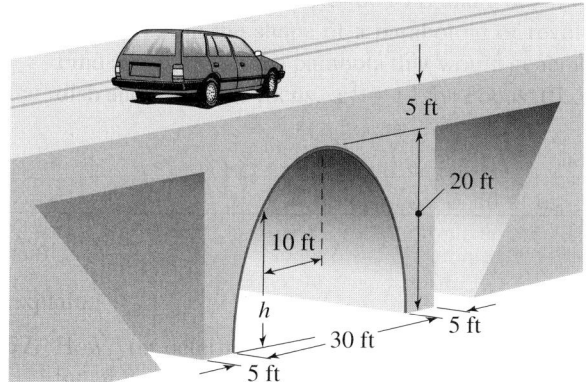

79. An 8-ft-wide boat with a mast whose top is 15 ft above the water is about to go under the bridge of Problem 76. How close can it get to the bank on the right side of the river and still fit under the bridge?

$21 - \frac{25\sqrt{7}}{4} \approx 4.5$ ft between side of boat and riverbank

---

SKILL CHECKER

Try the "Skill Checker" exercises so you'll be ready for the next section.

80. If you solve the equation $x^2 + y^2 = 25$ for *x*, you obtain two answers. What are these answers?

$x = \pm\sqrt{25 - y^2}$

81. One of the answers in Problem 80 is always nonnegative. Look at the graph of $x^2 + y^2 = 25$ in Example 5. To what part of the graph does the positive answer correspond? Right half of circle

82. One of the answers in Problem 80 is always nonpositive. Look at the graph of $x^2 + y^2 = 25$ in Example 5. To what part of the graph does the negative answer correspond? Left half of circle

---

USING YOUR KNOWLEDGE

Ellipses Revisited

The definition of an ellipse is as follows.

ELLIPSE

An **ellipse** is the set of all points, the sum of whose distances from two fixed points $(c, 0)$ and $(-c, 0)$ is a constant $2a$ $(a > c)$. Each fixed point is called a **focus**.

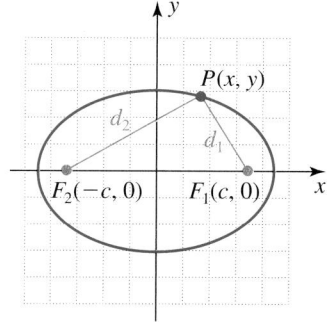

In Problems 83–88, we will prove that this definition leads to the equation of the ellipse that we gave in the text.

83. Suppose $P(x, y)$ is a point on the ellipse. Find the distance d_1 from *P* to F_1. $PF_1 = \sqrt{(x - c)^2 + y^2}$

84. Find the distance d_2 from *P* to F_2.

$PF_2 = \sqrt{(x + c)^2 + y^2}$

85. The sum of the distances found in Problems 83 and 84 must be 2a. Thus

$$\sqrt{(x - c)^2 + y^2} + \sqrt{(x + c)^2 + y^2} = 2a$$

Rewrite this equation with one radical on each side. Then square both sides and simplify. What is your answer?

$a^2 + cx = a\sqrt{(x + c)^2 + y^2}$ or
$a^2 - cx = a\sqrt{(x - c)^2 + y^2}$

86. Rewrite your answer for Problem 85 with the radical on one side and then square both sides again and simplify. What is your answer?

$x^2(a^2 - c^2) + a^2y^2 = a^2(a^2 - c^2)$

87. In your answer for Problem 86, since $a > c$, let $a^2 - c^2 = b^2$. Isolate all the variables on the left side. What is your answer? $b^2x^2 + a^2y^2 = a^2b^2$

88. Divide all terms of the answer you obtained in Problem 87 by a^2b^2. What is your answer? If everything went well, you should have

$$\frac{x^2}{a^2} + \frac{y^2}{b^2} = 1$$

the equation of an ellipse. $\frac{x^2}{a^2} + \frac{y^2}{b^2} = 1$

WRITE ON

If we remove the restriction that $a^2 > b^2$, then the general equation for an ellipse centered at the origin and with foci on one of the axes is

$$\frac{x^2}{a^2} + \frac{y^2}{b^2} = 1$$

Discuss the resulting graph:

89. When $a > b$. Answers may vary.

90. When $a < b$. Answers may vary.

91. When $a = b$. Answers may vary.

92. Can you explain why a circle is a special case of an ellipse? Answers may vary.

MASTERY TEST

If you know how to do these problems, you have learned your lesson!

Graph:

93. $\dfrac{(x + 3)^2}{4} + \dfrac{(y + 1)^2}{9} = 1$

94. $\dfrac{(x - 3)^2}{4} + \dfrac{(y - 1)^2}{9} = 1$

95. $4x^2 + 9y^2 = 36$

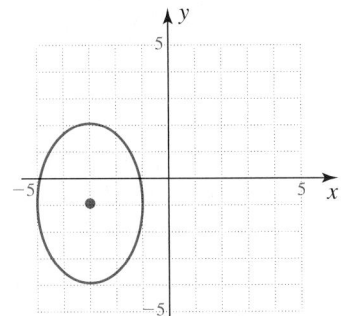
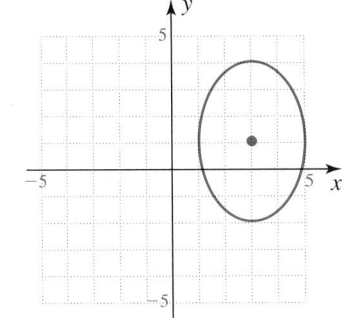
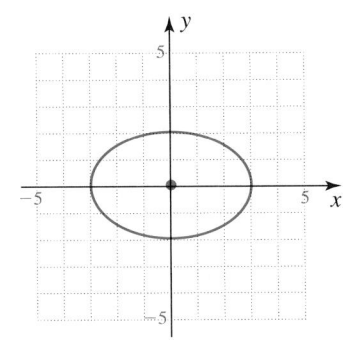

96. $9x^2 + 4y^2 = 36$

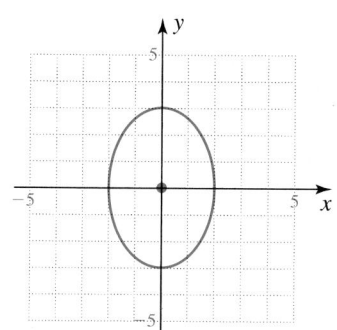

97. $x^2 - 6x + y^2 - 4y + 9 = 0$

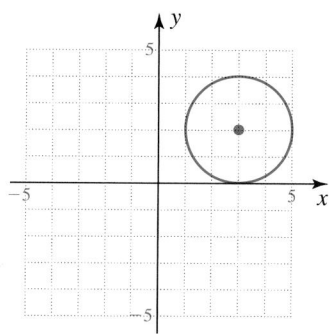

98. $x^2 + 6x + y^2 + 2y + 7 = 0$

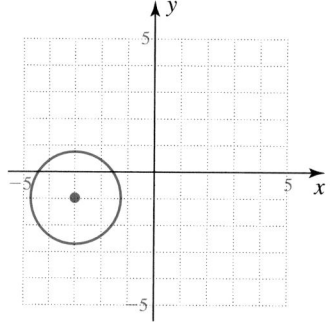

99. $x^2 + y^2 = 4$

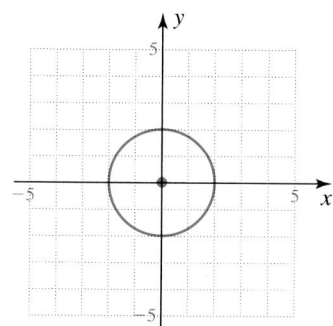

100. $x^2 + y^2 - 6 = 0$

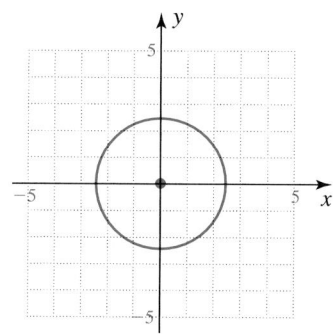

Find the center and radius and sketch the graph of:

101. The circle whose equation is $(x - 3)^2 + (y - 1)^2 = 4$.

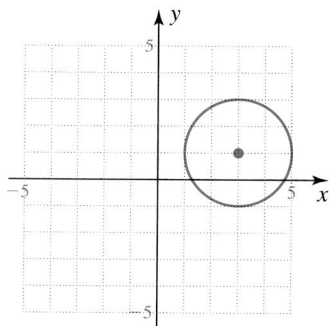

$C(3, 1); r = 2$

102. The circle whose equation is $(x + 3)^2 + (y + 1)^2 - 5 = 0$.

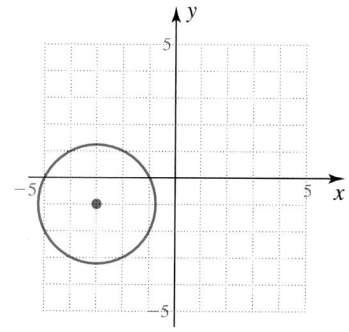

$C(-3, -1); r = \sqrt{5} \approx 2.2$

Find the equation of:

103. A circle of radius 5 and with center at the origin.
$x^2 + y^2 = 25$

104. A circle of radius 4 and with center at the origin.
$x^2 + y^2 = 16$

105. A circle with center at $(-3, 6)$ and radius 3.
$(x + 3)^2 + (y - 6)^2 = 9$

106. A circle with center at $(-2, -3)$ and radius 2.
$(x + 2)^2 + (y + 3)^2 = 4$

9.3 HYPERBOLAS AND IDENTIFICATION OF CONICS

To Succeed, Review How To . . .

1. Graph points on the Cartesian plane (pp. 163–165).

2. Find the *x*- and *y*-intercepts of a curve (pp. 167–168).

Objectives

A Graph hyperbolas.

B Identify conic sections by examining their equations.

GETTING STARTED

Hyperbolas in the Night

Have you seen any hyperbolas lately? Next time you are outside at night, look at building lights. Many of the beams of light you see are hyperbolas. If you've studied chemistry or physics, you might also know that alpha particles (one of three types of radiation resulting from natural radioactivity) have trajectories (paths) that are hyper-

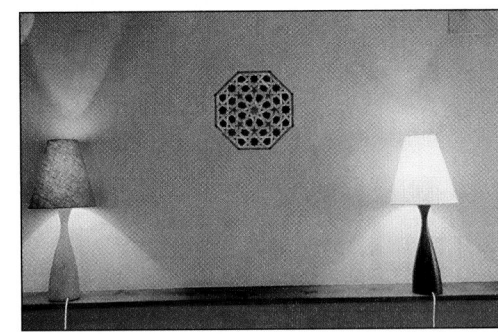

bolas. In this section we shall study hyperbolas by examining their equations and then graphing them.

Hyperbolas in the Day

When traveling during the day, look for buildings or towers that might be hyperbolic. At the left is a hyperboloid tower in Kobe, Japan. On the right is St. Louis Science Center's James S. McDonnell Planetarium. The roof is a hyperboloid.

A Graphing Hyperbolas

A **hyperbola** is the set of points in a plane such that the difference of the distances of each point from two fixed points (called the **foci**) is a constant (See *Using Your Knowledge* in the Exercises.) Consider the equation

$$\frac{x^2}{4} - \frac{y^2}{9} = 1$$

Web It

For the description, history, and properties of a hyperbola, go to link 9-3-1 at mhhe.com/bello.

When $x = 0$, $y^2 = -9$, so there are no *y*-intercepts because $y^2 = -9$ has no real-number solution. When $y = 0$, $x^2 = 4$, and $x = \pm 2$ are the *x*-intercepts, called the vertices of the hyperbola. The graph is shown in Figure 36.

Similarly, the graph of

$$\frac{y^2}{9} - \frac{x^2}{4} = 1$$

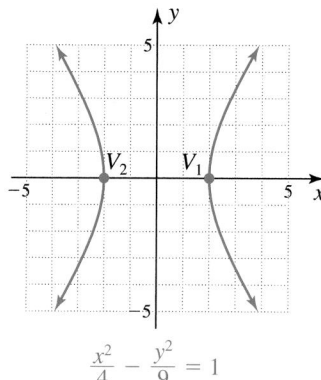

$$\frac{x^2}{4} - \frac{y^2}{9} = 1$$

Figure 36

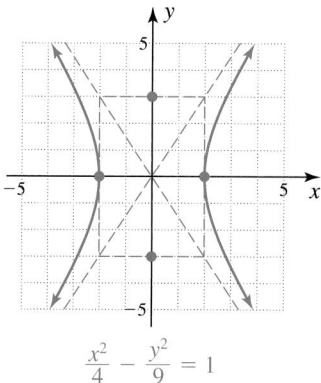

$$\frac{x^2}{4} - \frac{y^2}{9} = 1$$

Figure 38

has no *x*-intercept because $y = 0$ yields $x^2 = -4$, which has no real-number solution. The *y*-intercepts are ± 3, the vertices. The graph is shown in Figure 37.

The hyperbola

$$\frac{x^2}{4} - \frac{y^2}{9} = 1$$

has $a^2 = 4$ and $a = 2$. The vertices are $(\pm 2, 0)$.

We can use the denominator of y^2 to help us with the graph. If we draw an auxiliary rectangle with sides parallel to the *x*- and *y*-axes and passing through the *x*-intercepts and the points on the *y*-axis corresponding to the square root of the denominator of

$$\frac{y^2}{9}$$

in this case ± 3, and then connect opposite corners of the rectangle with lines, the graph of the hyperbola will approach these lines, called *asymptotes*. The asymptotes are *not* part of the hyperbola, but are used to help graph it. The hyperbola never touches the asymptotes but gets closer and closer to them as *x* and *y* get larger and larger in absolute value. The graphs of the hyperbolas

$$\frac{x^2}{4} - \frac{y^2}{9} = 1 \quad \text{and} \quad \frac{y^2}{9} - \frac{x^2}{4} = 1$$

are shown in Figures 38 and 39.

Here is a summary of this discussion.

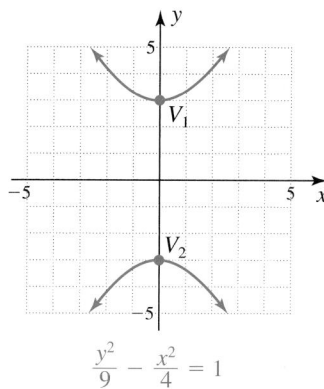

$$\frac{y^2}{9} - \frac{x^2}{4} = 1$$

Figure 37

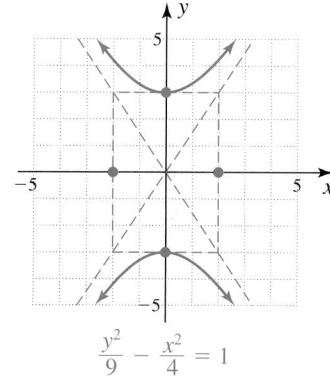

$$\frac{y^2}{9} - \frac{x^2}{4} = 1$$

Figure 39

Graphing Forms of Equations of Hyperbolas with Center at (0, 0)

The graph of the equation

$$\frac{x^2}{a^2} - \frac{y^2}{b^2} = 1 \qquad \textbf{(1)}$$

is a **hyperbola** centered at $(0, 0)$ with vertices $x = (\pm a, 0)$

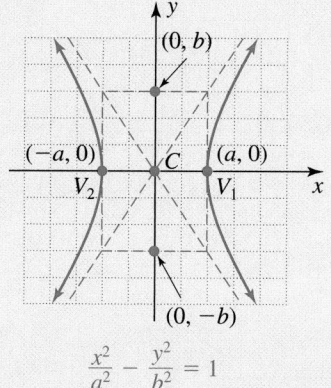

$$\frac{x^2}{a^2} - \frac{y^2}{b^2} = 1$$

The graph of the equation

$$\frac{y^2}{a^2} - \frac{x^2}{b^2} = 1 \qquad \textbf{(2)}$$

is a **hyperbola** centered at $(0, 0)$ with vertices $y = (0, \pm a)$

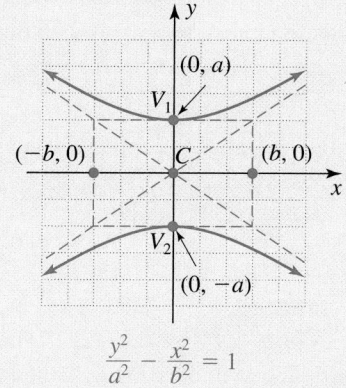

$$\frac{y^2}{a^2} - \frac{x^2}{b^2} = 1$$

Note: a^2 is always the denominator of the positive variable, and either $a^2 = b^2$, $a^2 > b^2$, or $a^2 < b^2$.

The **asymptotes** of either hyperbola are the lines through opposite corners of the auxiliary rectangle whose sides pass through $(\pm a, 0)$ and $(0, \pm b)$ for hyperbola (1), and $(0, \pm a)$ and $(\pm b, 0)$ for hyperbola (2).

Teaching Tip

Remind students that in the equation for an ellipse, a^2 is always greater than b^2, while in the equation for a hyperbola, a^2 is always the denominator of the positive variable.

PROCEDURE

Graphing a Hyperbola with Center at (0, 0)

1. Find and graph the points

$$(\pm a, 0) \text{ and } (0, \pm b) \text{ for } \frac{x^2}{a^2} - \frac{y^2}{b^2} = 1$$

or $$(0, \pm a) \text{ and } (\pm b, 0) \text{ for } \frac{y^2}{a^2} - \frac{x^2}{b^2} = 1$$

   (*Note: a^2 is always the denominator of the positive variable.*)

2. Connect the opposite corners of the auxiliary rectangle passing through $\pm a$ and $\pm b$ with lines called asymptotes.

3. Start the graph from the vertices $(\pm a, 0)$ or $(0, \pm a)$ and draw the hyperbola so that it approaches (but does not touch) the asymptotes.

EXAMPLE 1   **Graphing hyperbolas with center at (0, 0)**

Find the coordinates of the center and the vertices and graph:

a. $\dfrac{y^2}{4} - \dfrac{x^2}{25} = 1$ **b.** $25x^2 - 4y^2 = 100$

SOLUTION

a. The hyperbola is centered at $(0, 0)$, the origin. Since the y^2 term is positive, $a^2 = 4$ and $a = 2$. It has vertices at $V_1 = (0, 2)$ and $V_2 = (0, -2)$. (There are no x-intercepts.) Our auxiliary rectangle will pass through $y = \pm 2$ and through $x = \pm 5$, the square root of the denominator of x^2. We then connect opposite corners to complete our asymptotes, as shown in Figure 40. Since our hyperbola has vertices $(0, \pm 2)$, we start our graph from the vertex $y = 2$ and approach the asymptotes, obtaining the top half of the hyperbola. The bottom half is obtained similarly by starting at the vertex $y = -2$. (See Figure 41.)

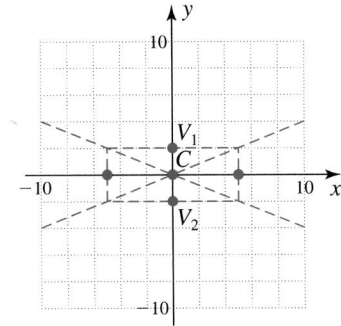

Figure 40

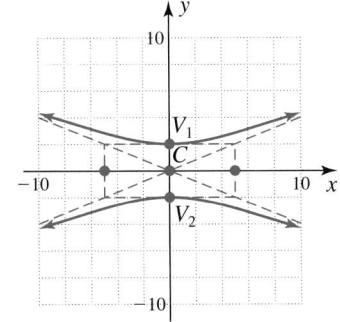

Figure 41

b. Divide each term by 100 to obtain a 1 on the right-hand side. We then have

$$\frac{x^2}{4} - \frac{y^2}{25} = 1,$$

which indicates the center at $(0, 0)$, $a^2 = 4$ with $a = 2$ and $b^2 = 25$ with $b = 5$.

PROBLEM 1

Graph and name the coordinates of the vertices:

a. $\dfrac{y^2}{16} - \dfrac{x^2}{9} = 1$

b. $\dfrac{x^2}{16} - \dfrac{y^2}{9} = 1$

Answers

1. a. $a = \pm 4, b = \pm 3$; vertices $(0, \pm 4)$

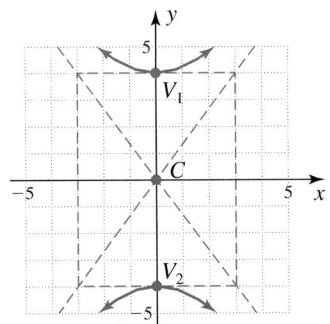

b. $a = \pm 4, b = \pm 3$; vertices $(\pm 4, 0)$

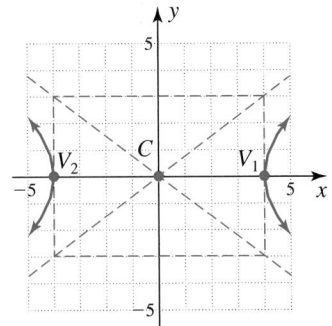

This time, we will show our auxiliary rectangle and the hyperbola on the same graph. Since the x^2 term is positive, the hyperbola has vertices at $V_1 = (2, 0)$ and $V_2 = (-2, 0)$. Our auxiliary rectangle will pass through $x = \pm 2$ and through $y = \pm 5$. We then complete the auxiliary rectangle, the asymptotes, and the graph of the hyperbola with vertices $(\pm 2, 0)$, as shown in Figure 42.

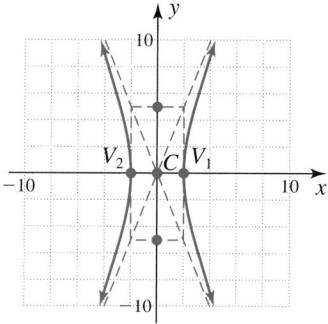

Figure 42

Calculate It   Graphing a Hyperbola

As we've mentioned, calculators only graph functions. Is a hyperbola the graph of a function? If you apply the vertical line test to the hyperbolas shown in Example 1, you will see that the graphs are *not* graphs of functions. To graph

$$\frac{y^2}{4} - \frac{x^2}{25} = 1$$

we first solve for y by adding

$$\frac{x^2}{25}$$

to both sides of the equation, multiplying both sides of the equation by 4, and taking the square root of both sides of the equation to obtain

$$y = \pm \sqrt{4 + \frac{4x^2}{25}}$$

We then use a square window and graph

$$Y_1 = \sqrt{4 + \frac{4x^2}{25}}$$

and

$$Y_2 = -\sqrt{4 + \frac{4x^2}{25}}$$

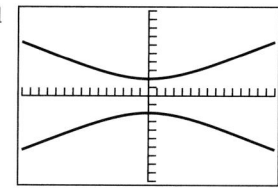

The graphs for Y_1 (top part) and Y_2 (bottom part) are shown in the window. Remember, you can save time when you enter the expressions to be graphed by entering

$$Y_1 = \sqrt{4 + \frac{4x^2}{25}} \quad \text{and} \quad Y_2 = -Y_1$$

Now try to graph Example 1(b).

| **Graphing Forms of Equations of Hyperbolas with Center at (h, k)** | The equation of the hyperbola with center at (h, k) is $$\frac{(x - h)^2}{a^2} - \frac{(y - k)^2}{b^2} = 1$$ The vertices are horizontally $\pm a$ units from (h, k). The equation of the hyperbola with center at (h, k) is $$\frac{(y - k)^2}{a^2} - \frac{(x - h)^2}{b^2} = 1$$ The vertices are vertically $\pm a$ units from (h, k). *Note:* a^2 is always the denominator of the positive variable and either $a^2 = b^2$, $a^2 > b^2$, or $a^2 < b^2$. |

EXAMPLE 2   **Graphing a hyperbola not centered at the origin**

Name the coordinates of the center and the vertices and graph:

$$\frac{(x-2)^2}{16} - \frac{(y+1)^2}{4} = 1$$

SOLUTION   The center of this hyperbola is at $(2, -1)$. Since the x^2 term is positive, $a^2 = 16$ with $a = 4$, and $b^2 = 4$ with $b = 2$. Now construct the auxiliary rectangle with sides parallel to $x = 2$ that are ± 4 units away from $x = 2$, and sides parallel to $y = -1$ that are ± 2 units away from $y = -1$, as shown in Figure 43. The vertices of the hyperbola are horizontally ± 4 units from $(2, -1)$, thus the hyperbola has vertices at $V_1 = (6, -1)$ and $V_2 = (-2, -1)$. Use the auxiliary rectangle to draw the asymptotes. Use the vertices to sketch the graph of the hyperbola, as shown in Figure 44.

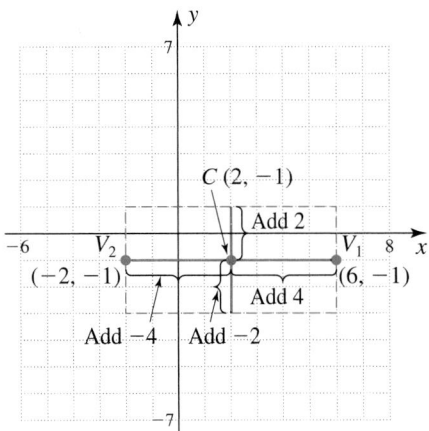

Figure 43

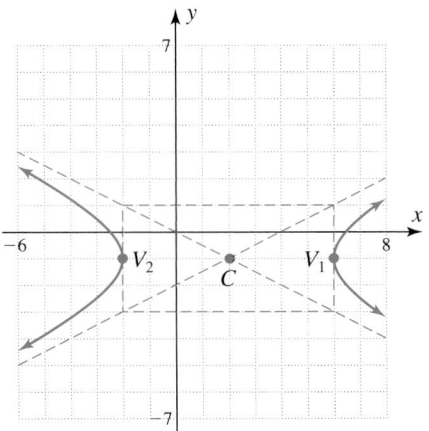

Figure 44

PROBLEM 2

Name the coordinates of the center and the vertices and graph:

$$\frac{(y+3)^2}{9} - \frac{(x+1)^2}{9} = 1$$

Answer

2. Center at $(-1, -3)$;
$V_1 = (2, -3)$ and $V_2 = (-4, -3)$

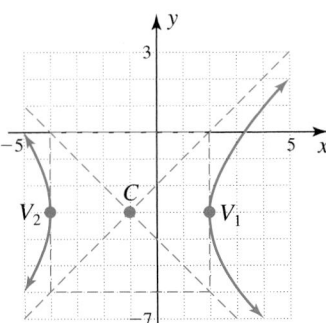

You have probably noticed that the equations whose graphs are hyperbolas are similar to the equations whose graphs are ellipses (when written in standard form, hyperbolas are written as differences). We examine this idea in more detail next.

B   **Identifying Conic Sections by Their Equations**

How would you know the shape of the graph of a conic by studying its equation? Table 2 will help you with this.

Table 2 Identifying Conic Sections

Equation	Graph	Description	Identification
$y = a(x - h)^2 + k$		Parabola with vertex at (h, k). Opens upward for $a > 0$, downward for $a < 0$.	y is not squared.
$y = ax^2 + bx + c$		Parabola with vertex at $x = -\frac{b}{2a}$. Opens upward for $a > 0$, downward for $a < 0$.	y is not squared.
$x = a(y - k)^2 + h$		Parabola with vertex at (h, k). Opens right if $a > 0$, left if $a < 0$.	x is not squared.
$x = ay^2 + by + c$		Parabola with vertex at $y = -\frac{b}{2a}$. Opens right if $a > 0$, left if $a < 0$.	x is not squared.
$(x - h)^2 + (y - k)^2 = r^2$		Circle of radius r, centered at (h, k).	The coefficients of $(x - h)^2$ and $(y - k)^2$ are positive and equal when the variables are on the same side of the equation.
$\dfrac{x^2}{a^2} + \dfrac{y^2}{b^2} = 1$		Ellipse with x-intercepts $\pm a$, y-intercepts $\pm b$.	The coefficients of x^2 and y^2 are positive and not equal when variables are on the same side.
$\dfrac{x^2}{b^2} + \dfrac{y^2}{a^2} = 1$		Ellipse with x-intercepts $\pm b$, y-intercepts $\pm a$.	The coefficients of x^2 and y^2 are positive and not equal when variables are on the same side.
$\dfrac{x^2}{a^2} - \dfrac{y^2}{b^2} = 1$		Hyperbola with vertices $(\pm a, 0)$. Auxiliary rectangle passing through $(\pm a, 0)$ and $(0, \pm b)$. Asymptotes drawn through the corners of the auxiliary rectangle.	x^2 has positive coefficient, y^2 has negative coefficient when variables are on the same side.
$\dfrac{y^2}{a^2} - \dfrac{x^2}{b^2} = 1$		Hyperbola with vertices $(0, \pm a)$. Auxiliary rectangle passing through $(0, \pm a)$ and $(\pm b, 0)$. Asymptotes drawn through the corners of the auxiliary rectangle.	y^2 has positive coefficient, x^2 has negative coefficient when variables are on the same side.

EXAMPLE 3   **Identifying conic sections**

Identify:

a. $x^2 = 9 - y^2$

b. $y = x^2 - 4$

c. $4x^2 = 36 - 9y^2$

d. $9x^2 = 36 + 4y^2$

SOLUTION   If both variables appear to the second power, we write all variables on the left to make the identification easier.

a. In $x^2 = 9 - y^2$, both variables appear to the second power. Thus $x^2 = 9 - y^2$ is written as $x^2 + y^2 = 9$. The square terms have the same coefficient (1) and are both positive. The equation $x^2 = 9 - y^2$ represents a circle centered at the origin and with radius 3.

b. In this case, only one variable is squared, the x. Thus the conic is a parabola with the vertex at $(0, -4)$ and opening upward.

c. Both variables are squared. We then write $4x^2 = 36 - 9y^2$ as $4x^2 + 9y^2 = 36$. Here the square terms have different coefficients and are both positive. The equation corresponds to an ellipse centered at the origin. The x-intercepts are found by letting $y = 0$ to obtain $x = \pm 3$. Similarly, the y-intercepts are $y = \pm 2$. To confirm that $4x^2 = 36 - 9y^2$ is an ellipse, we write the equation in standard form by dividing each term in the equation by 36 so we obtain a 1 on the right-hand side of the equation $4x^2 + 9y^2 = 36$. We then have

$$\frac{4x^2}{36} + \frac{9y^2}{36} = \frac{36}{36}$$

$$\frac{x^2}{9} + \frac{y^2}{4} = 1$$

$$\frac{x^2}{3^2} + \frac{y^2}{2^2} = 1$$

confirming that the conic is an ellipse with x-intercepts at $(\pm 3, 0)$ and y-intercepts at $(0, \pm 2)$.

d. Again, both variables are squared. Thus we write $9x^2 = 36 + 4y^2$ as $9x^2 - 4y^2 = 36$. The minus sign indicates that the conic is a hyperbola with vertices at $(\pm 2, 0)$. To confirm that $9x^2 - 4y^2 = 36$ corresponds to a hyperbola, divide each term by 36 to obtain

$$\frac{9x^2}{36} - \frac{4y^2}{36} = \frac{36}{36}$$

$$\frac{x^2}{2^2} - \frac{y^2}{3^2} = 1$$

which confirms that the conic is a hyperbola.

PROBLEM 3

Identify:

a. $4x^2 = 36 + 9y^2$

b. $y = x^2 + 3$

c. $y^2 = 9 - x^2$

d. $9x^2 = 36 - 4y^2$

Answers

3. a. Hyperbola; center $(0, 0)$, vertices $(3, 0)$ and $(-3, 0)$
b. Parabola; vertex $(0, 3)$, opens upward   **c.** Circle; center $(0, 0)$, radius 3   **d.** Ellipse; center $(0, 0)$, x-intercepts $(\pm 2, 0)$ and y-intercepts $(0, \pm 3)$

Calculate It   **Graphing Conics**

You can identify the graphs corresponding to the equations in Example 3 if you solve each of the equations for y and graph the results. To make sure you don't mistake a circle for an ellipse, use a square window (press **ZOOM** 5 on a TI-83 Plus) and check the intercepts. For Example 3(a), we obtain $y = \pm\sqrt{9 - x^2}$. Graphing

$$Y_1 = \sqrt{9 - x^2} \quad \text{and} \quad Y_2 = -\sqrt{9 - x^2}$$

we obtain the graph of $x^2 = 9 - y^2$, which can be easily identified as a circle, as shown in the window.

Now, solve for y in parts **b**, **c**, and **d**, graph the results, and identify each of the graphs.

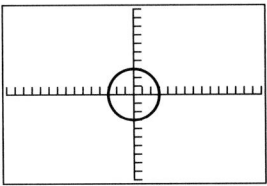

Exercises 9.3

A In Problems 1–12, draw the auxiliary rectangle, name the vertices, and graph.

1. $\dfrac{x^2}{25} - \dfrac{y^2}{9} = 1$ $V: (\pm 5, 0)$

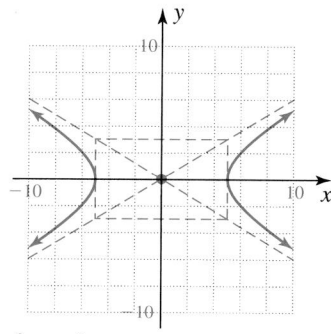

2. $\dfrac{y^2}{9} - \dfrac{x^2}{25} = 1$ $V: (0, \pm 3)$

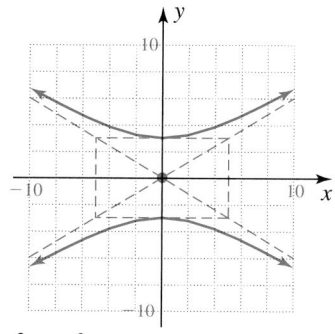

3. $\dfrac{y^2}{9} - \dfrac{x^2}{9} = 1$ $V: (0, \pm 3)$

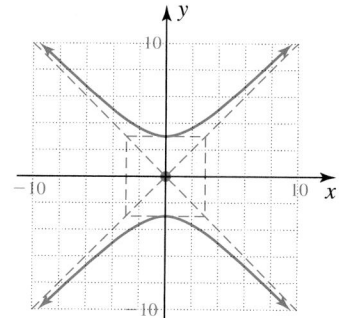

4. $\dfrac{x^2}{9} - \dfrac{y^2}{9} = 1$ $V: (\pm 3, 0)$

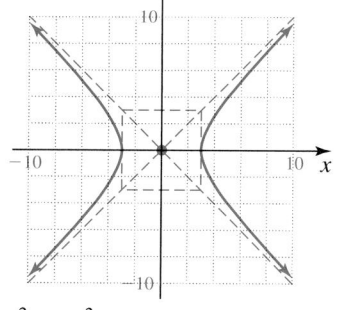

5. $\dfrac{x^2}{9} - \dfrac{y^2}{1} = 1$ $V: (\pm 3, 0)$

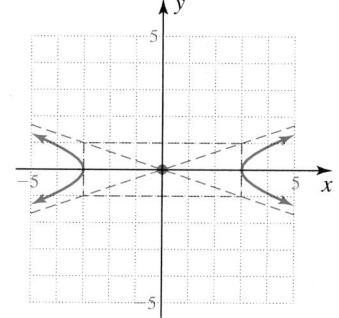

6. $\dfrac{y^2}{16} - \dfrac{x^2}{1} = 1$ $V: (0, \pm 4)$

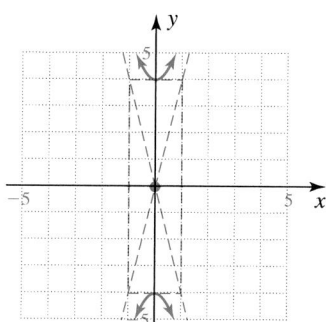

7. $\dfrac{x^2}{64} - \dfrac{y^2}{49} = 1$ $V: (\pm 8, 0)$

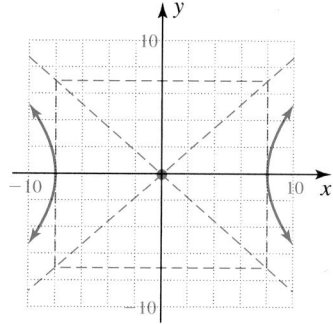

8. $\dfrac{y^2}{49} - \dfrac{x^2}{64} = 1$ $V: (0, \pm 7)$

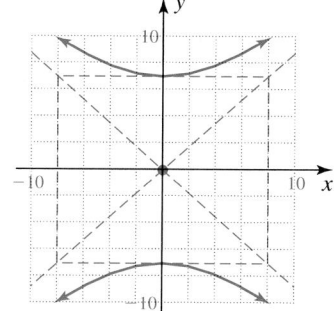

9. $\dfrac{y^2}{\frac{16}{9}} - \dfrac{x^2}{\frac{9}{16}} = 1$ $V: \left(0, \pm \dfrac{4}{3}\right)$

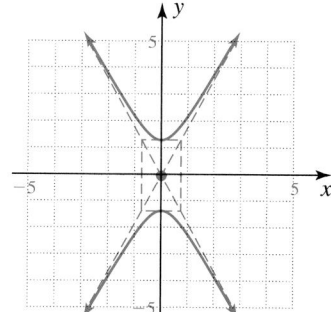

10. $\dfrac{x^2}{\frac{9}{4}} - \dfrac{y^2}{\frac{4}{9}} = 1$ $V: \left(\pm \dfrac{3}{2}, 0\right)$

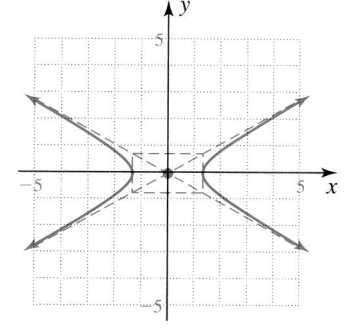

11. $y^2 - 9x^2 = 9$ $V: (0, \pm 3)$

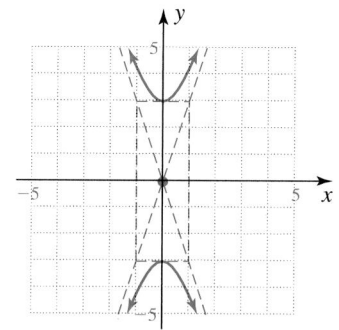

12. $x^2 - 16y^2 = 16$ $V: (\pm 4, 0)$

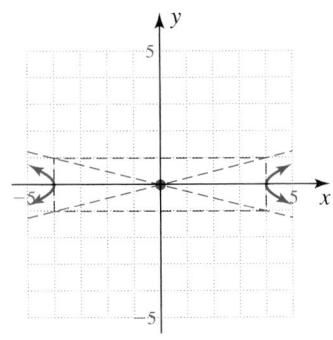

In Problems 13–16, sketch the hyperbola. Give the coordinates of the center and the values of a and b. (*Hint:* They are not centered at the origin.)

13. $\dfrac{(x-1)^2}{4} - \dfrac{(y+1)^2}{9} = 1$

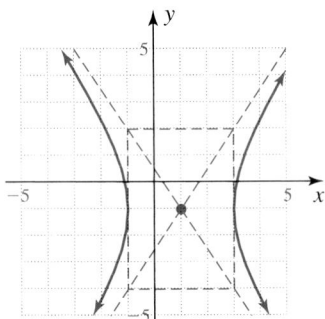

$C(1, -1);\ a = 2,\ b = 3$

14. $\dfrac{(x-2)^2}{9} - \dfrac{(y+1)^2}{4} = 1$

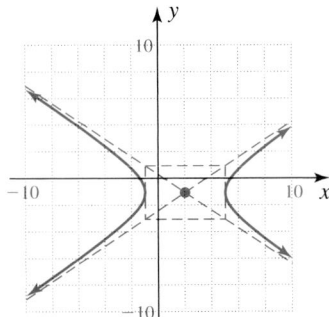

$C(2, -1);\ a = 3,\ b = 2$

15. $\dfrac{(y-1)^2}{9} - \dfrac{(x-2)^2}{4} = 1$

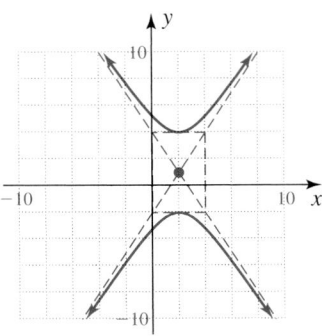

$C(2, 1);\ a = 3,\ b = 2$

16. $\dfrac{(y-2)^2}{4} - \dfrac{(x-1)^2}{9} = 1$

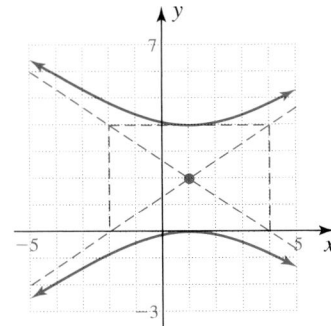

$C(1, 2);\ a = 2,\ b = 3$

B In Problems 17–30, identify the conic and name the intercepts. If the conic is a parabola, name only the vertex.

17. $x^2 + y^2 = 25$
Circle; $(5, 0), (0, 5), (-5, 0), (0, -5)$

18. $x^2 - y^2 = 25$
Hyperbola; $(5, 0), (-5, 0)$

19. $x^2 - y^2 = 36$
Hyperbola; $(6, 0), (-6, 0)$

20. $x^2 + y^2 = 36$
Circle; $(6, 0), (0, 6), (-6, 0), (0, -6)$

21. $x^2 - y = 9$
Parabola; $(0, -9)$

22. $x^2 + y = 9$
Parabola; $(0, 9)$

23. $y^2 - x = 4$
Parabola; $(-4, 0)$

24. $y^2 + x = 4$
Parabola; $(4, 0)$

25. $9x^2 = 36 - 9y^2$
Circle; $(2, 0), (0, 2), (-2, 0), (0, -2)$

26. $4x^2 = 16 - 4y^2$
Circle; $(2, 0), (0, 2), (-2, 0), (0, -2)$

27. $9x^2 = 36 + 9y^2$
Hyperbola; $(2, 0), (-2, 0)$

28. $4x^2 = 36 - 9y^2$
Ellipse; $(3, 0), (0, 2), (-3, 0), (0, -2)$

29. $x^2 = 9 - 9y^2$
Ellipse; $(3, 0), (0, 1), (-3, 0), (0, -1)$

30. $y^2 = 4 - 4x^2$
Ellipse; $(1, 0), (0, 2), (-1, 0), (0, -2)$

APPLICATIONS

Graphing Equations

31. A semicircular plate of diameter D with a circular opening of diameter d is to be constructed. If the area of the plate is π square inches, the relationship between D and d is given by

$$\frac{D^2}{8} - \frac{d^2}{4} = 1$$

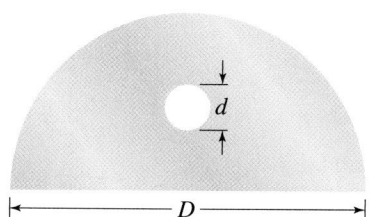

  a. What type of conic corresponds to this equation?
    Hyperbola

  b. Sketch the graph of

$$\frac{D^2}{8} - \frac{d^2}{4} = 1$$

  (Use 2.8 for $\sqrt{8}$.)

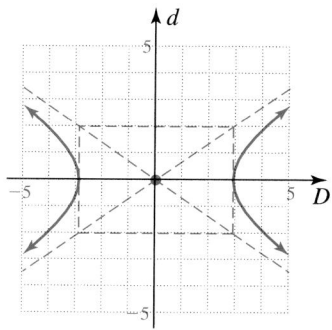

32. If three holes of diameter d were drilled in a semicircular plate of diameter D and the remaining area was π square inches, the relationship between D and d would be

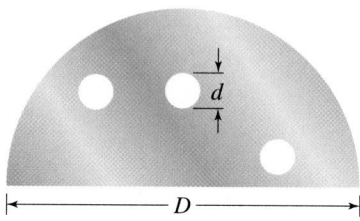

$$\frac{D^2}{8} - \frac{3d^2}{4} = 1$$

  a. What type of conic corresponds to this equation?
    Hyperbola

  b. Show that you can write the equation of the conic as

$$\frac{D^2}{8} - \frac{d^2}{\frac{4}{3}} = 1 \quad \frac{3d^2}{4} = \frac{3}{4}d^2 = d^2 \div \frac{4}{3} = \frac{d^2}{\frac{4}{3}},$$

$$\text{so } \frac{D}{8} - \frac{d^2}{\frac{4}{3}} = 1$$

  c. Sketch the graph of

$$\frac{D^2}{8} - \frac{3d^2}{4} = 1$$

  (Use 2.8 for $\sqrt{8}$ and 1.15 for $\sqrt{\frac{4}{3}}$.)

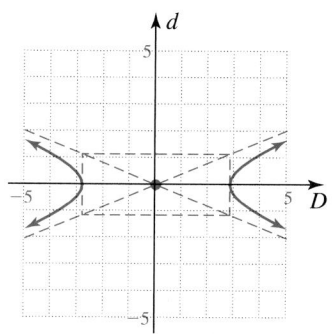

33. The total kinetic energy of a spinning body moving through the air is 144 foot-pound (ft-lb). The velocity v through the air and the spinning velocity ω are related by the equation $4v^2 + 9\omega^2 = 144$. Graph this equation.

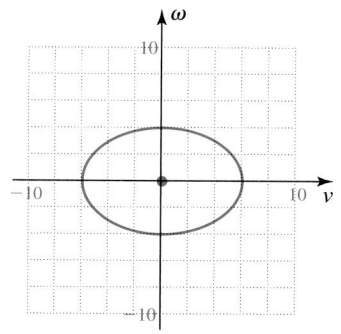

34. If the equation in Problem 33 is $16v^2 + 4\omega^2 = 256$, graph the equation.

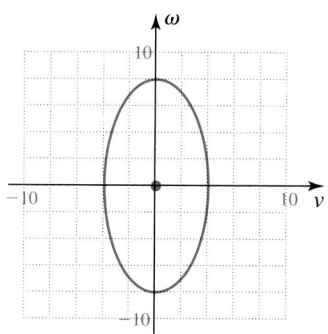

SKILL CHECKER

Try the "Skill Checker" exercises so you'll be ready for the next section.

Graph:

35. $y = x - 4$

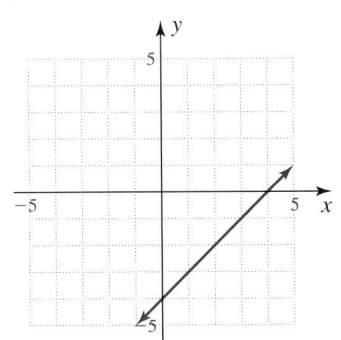

36. $x^2 + y^2 = 4$

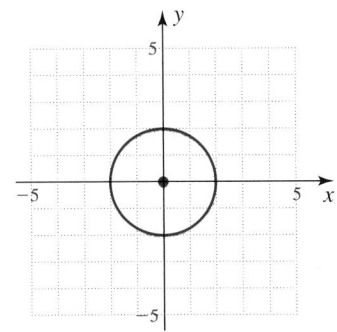

37. $y = x^2 + 1$

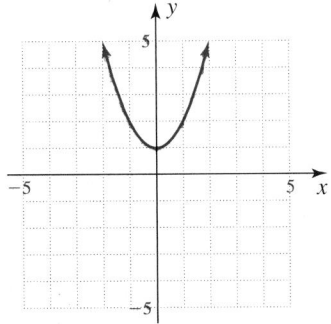

38. $y = x^2 - 1$

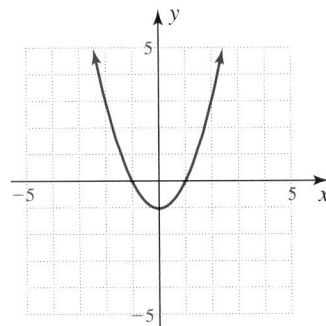

39. $4x^2 + 9y^2 = 36$

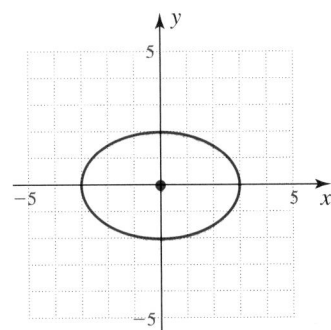

40. $9x^2 - 4y^2 = 36$

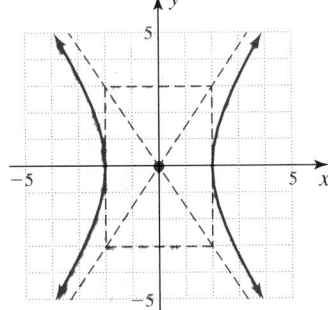

USING YOUR KNOWLEDGE

Hyperbolas Revisited

The definition for a hyperbola is as follows.

HYPERBOLA

A **hyperbola** is the set of points in a plane such that the difference of the distances of each point from two fixed points (called the **foci**) is a constant.

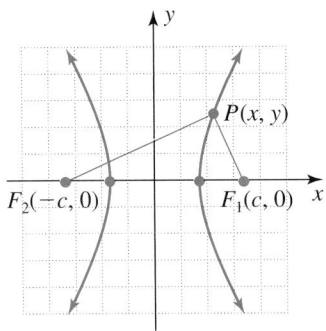

In Problems 41–46, we will prove that this definition leads to the equation of the hyperbola we gave in the text.

41. Suppose (x, y) is a point on the hyperbola. Find the distance from P to F_1. $PF_1 = \sqrt{(x - c)^2 + y^2}$

42. Find the distance from P to F_2. $PF_2 = \sqrt{(x + c)^2 + y^2}$

43. Let the difference between the distances found in Problems 41 and 42 be $2a$. Thus

$$\sqrt{(x - c)^2 + y^2} - \sqrt{(x + c)^2 + y^2} = 2a$$

Rewrite this equation with one radical on each side. Then square both sides and simplify. What is your answer?
$a^2 + cx = a\sqrt{(x + c)^2 + y^2}$

44. Rewrite your answer for Problem 43 with the radical on one side. Then square both sides again and simplify. What is your answer?
$x^2(c^2 - a^2) - a^2y^2 = a^2(c^2 - a^2)$

45. In your answer for Problem 44, let $c^2 - a^2 = b^2$, where $a < c$. Isolate all the variables on the left side. What is your answer? $b^2x^2 - a^2y^2 = a^2b^2$

46. Divide all terms of the answer you obtained in Problem 45 by a^2b^2. What is your answer? If everything went well, you should have

$$\frac{x^2}{a^2} - \frac{y^2}{b^2} = 1$$

the equation of a hyperbola. $\frac{x^2}{a^2} - \frac{y^2}{b^2} = 1$

We have shown you how to graph the asymptotes of a hyperbola. We are now ready to use this knowledge to find the equation of these asymptotes. Consider the hyperbola

$$\frac{x^2}{a^2} - \frac{y^2}{b^2} = 1$$

47. The expression on the left is the difference of two squares. Factor it. $\left(\frac{x}{a} + \frac{y}{b}\right)\left(\frac{x}{a} - \frac{y}{b}\right) = 1$

48. Isolate $\frac{x}{a} - \frac{y}{b}$ on the left. The expression on the right is a complex fraction with 1 as the numerator. What is it?

$$\frac{x}{a} - \frac{y}{b} = \frac{1}{\frac{x}{a} + \frac{y}{b}}$$

49. Look at the denominator of the complex fraction in Problem 48. If x and y are positive and very large, what happens to the denominator? What happens to the complex fraction? The denominator becomes very large. The fraction approaches zero.

50. If you answered that the complex fraction is very small, you are correct. In mathematics we say that $\frac{x}{a} - \frac{y}{b} \to 0$ (the expression approaches zero). Thus for very large positive x and y, $\frac{x}{a} - \frac{y}{b} \approx 0$. This means that $\frac{x}{a} - \frac{y}{b} = 0$ is an asymptote. Solve for y and find its equation.
$y = \frac{b}{a}x$

51. We can show in the same way that $\frac{x}{a} + \frac{y}{b} = 0$ is an asymptote. Solve for y and find its equation. $y = -\frac{b}{a}x$

52. a. In summary, what are the equations of the asymptotes for the hyperbola

$$\frac{x^2}{a^2} - \frac{y^2}{b^2} = 1 \quad y = \pm\frac{b}{a}x$$

b. What about the asymptotes for the hyperbola

$$\frac{y^2}{a^2} - \frac{x^2}{b^2} = 1 \quad y = \pm\frac{a}{b}x$$

WRITE ON

53. Write an explanation of the procedure you would use to determine whether the graph of an equation is an ellipse or a hyperbola. Answers may vary.

54. Write your own definition of the asymptote of a hyperbola. Answers may vary.

55. Consider the equation $Ax^2 + By^2 = C$, where A, B, and C are real numbers. Under what conditions will the graph of this equation be

a. A circle?   **b.** An ellipse?   **c.** A hyperbola?
Answers may vary.

MASTERY TEST

If you know how to do these problems, you have learned your lesson!

Identify:

56. $9x^2 = 36 - 4y^2$ Ellipse

57. $y^2 = 9 - x^2$ Circle

58. $y^2 = 9 - 4x^2$ Ellipse

59. $y = x^2 + 3$ Parabola

60. $4x^2 = 36 + 9y^2$ Hyperbola

Graph:

61. $\dfrac{y^2}{16} - \dfrac{x^2}{9} = 1$

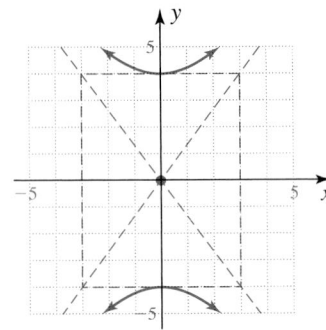

62. $\dfrac{x^2}{16} - \dfrac{y^2}{9} = 1$

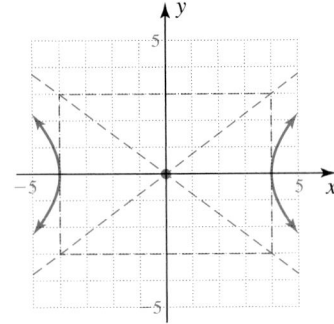

63. $9x^2 - 25y^2 = 225$

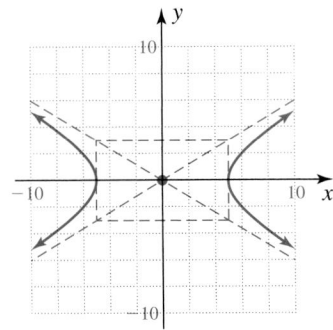

64. $25y^2 - 9x^2 = 225$

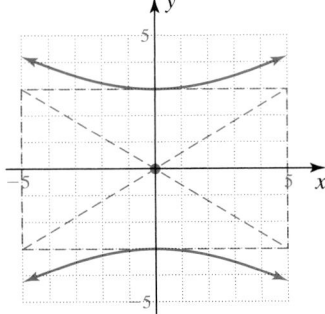

9.4 # NONLINEAR SYSTEMS OF EQUATIONS

To Succeed, Review How To ...

1. Graph lines and conics
   (pp. 165–168, 648–657, 671–676,
   686–692).

2. Use the graphical and substitution
   methods to solve systems of
   equations (pp. 253–259).

Objectives

A Solve a nonlinear system by
substitution.

B Solve a system with two second-
degree equations by elimination.

C Solve applications involving
nonlinear systems.

GETTING STARTED

Consumer's Demand and Supply

How do supply and demand determine
the market price and quantity of wheat
available for sale? As the price of wheat
decreases, the quantity demanded by consumers *increases.* If the price *increases,*
the demand *decreases.* On the other hand,
as the price *increases,* the amount the suppliers are willing to sell also *increases.*
The point *C* of intersection of the two
curves is called the **equilibrium point.**
At this point, the
price of a bushel
of wheat is $3,
and the amount
demanded by the
consumers   (12
million   bushels
per   month)   ex-
actly equals the amount supplied by producers. The equations
that describe this system are an example of a nonlinear system
of equations. Graphical results may be difficult to confirm so
we use the substitution method instead. We show you how to do
this next.

**How Supply and Demand Determine
Market Price and Quantity**

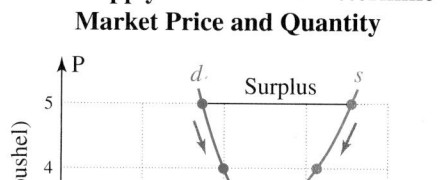
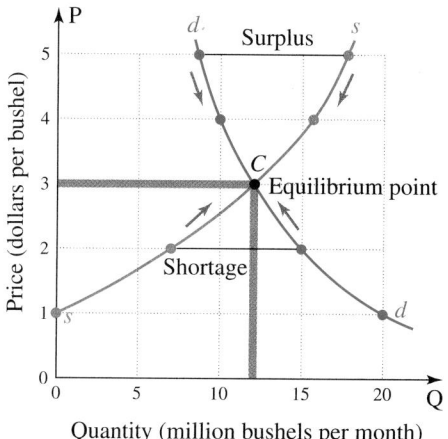

Quantity (million bushels per month)

A   ## Solving Nonlinear Systems by Substitution

The systems of equations we studied in Chapter 4 consisted of linear equations. An equation in which some terms have more than one variable or have a variable of degree 2 or
higher is a **nonlinear equation.** A system of equations including at least one nonlinear
equation is called a **nonlinear system of equations.** Such systems may have **one, more
than one,** or **no** real solutions. Let's start with a system consisting of two parabolas, solve
them graphically, and confirm the results by substitution.

Suppose the demand curve for a product is given by $y = (x - 5)^2$ (a parabola), where
x is the number of units produced and y is the price, and the supply curve is given by
$y = x^2 + 2x + 13$ (another parabola), where y is the price and x the number of units
available. To find the equilibrium point, we sketch both curves to find the intersection, as
shown in Figure 45.

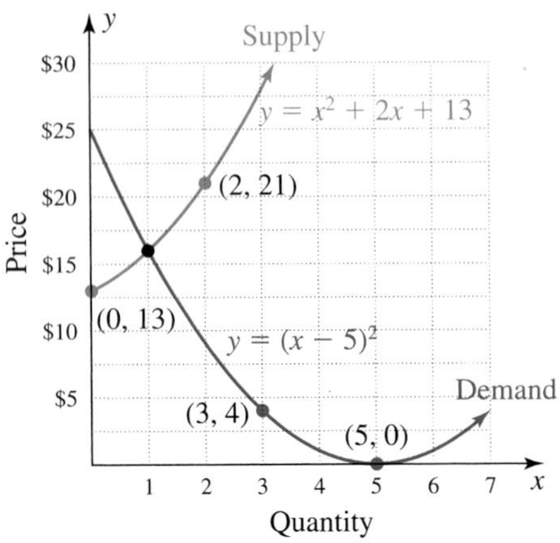

Figure 45

The equilibrium point seems to be the point $(1, 16)$. How can we be sure? We use the substitution method to solve the system and confirm our result.

$$y = (x - 5)^2 \qquad \text{A parabola} \qquad \textbf{(1)}$$
$$y = x^2 + 2x + 13 \qquad \text{A parabola} \qquad \textbf{(2)}$$

We wish to find an ordered pair (x, y) that is a solution of *both* equations (1) and (2). Using the substitution method, we substitute $(x - 5)^2$ for y on the left-hand side of equation (2) to obtain

$$(x - 5)^2 = x^2 + 2x + 13$$
$$x^2 - 10x + 25 = x^2 + 2x + 13 \qquad \text{Multiply.}$$
$$-10x + 25 = 2x + 13 \qquad \text{Subtract } x^2.$$
$$-12x = -12 \qquad \text{Subtract } 2x \text{ and } 25.$$
$$x = 1 \qquad \text{Divide by } -12.$$

If $x = 1$ in equation (2), then

$$y = (1)^2 + 2(1) + 13 = 16$$

Hence the solution for the given system is $(1, 16)$, as you can verify by substituting $x = 1$ and $y = 16$ in the two equations.

Web It

For some more examples on solving nonlinear systems, go to link 9-4-1 at mhhe.com/bello.

EXAMPLE 1 **Solving a nonlinear system by substitution**

Find the solution of the given system by the substitution method. Check the solution by sketching the graphs of the equations.

$$x^2 + y^2 = 25 \qquad \text{A circle} \qquad \textbf{(1)}$$
$$x + y = 5 \qquad \text{A line} \qquad \textbf{(2)}$$

SOLUTION We first rewrite equation (2) in the equivalent form $y = 5 - x$ to obtain

$$x^2 + y^2 = 25 \qquad \textbf{(1)}$$
$$y = 5 - x \qquad \textbf{(3)}$$

Replacing y in equation (1) by $(5 - x)$, we get

$$x^2 + (5 - x)^2 = 25 \qquad \text{Substitute } (5 - x) \text{ in equation (1).} \quad \textbf{(4)}$$
$$x^2 + 25 - 10x + x^2 = 25 \qquad \text{Multiply.}$$
$$2x^2 - 10x = 0 \qquad \text{Simplify; subtract 25.}$$
$$x^2 - 5x = 0 \qquad \text{Divide by 2.}$$
$$x(x - 5) = 0 \qquad \text{Factor.}$$
$$x = 0 \quad \text{or} \quad x - 5 = 0$$
$$x = 0 \quad \text{or} \qquad x = 5 \qquad \text{Solve for } x.$$

We now let $x = 0$ and $x = 5$ in equation (3) to obtain the corresponding y-values:

$$y = 5 - 0 = 5 \quad \text{and} \quad y = 5 - 5 = 0$$

Thus when $x = 0$, $y = 5$, and when $x = 5$, $y = 0$. Therefore, the solutions of the system are $(0, 5)$ and $(5, 0)$.

If we had substituted $x = 0$ and $x = 5$ in equation (1) rather than in equation (3), we would have obtained

$$0^2 + y^2 = 25 \quad \text{and} \quad 5^2 + y^2 = 25$$

PROBLEM 1

Find the solutions:

$$x + y = 4$$
$$x^2 + y^2 = 16$$

Teaching Tip

Review the three possible solution sets to a linear system—one, none, many. Have students sketch the graph for an example of each case. Then have them consider and sketch graphs for all possible cases of solutions for the following nonlinear systems.

$$\begin{cases} \text{Parabola} \\ \text{Parabola} \end{cases} \begin{cases} \text{Line} \\ \text{Parabola} \end{cases} \begin{cases} \text{Circle} \\ \text{Parabola} \end{cases} \begin{cases} \text{Circle} \\ \text{Line} \end{cases}$$

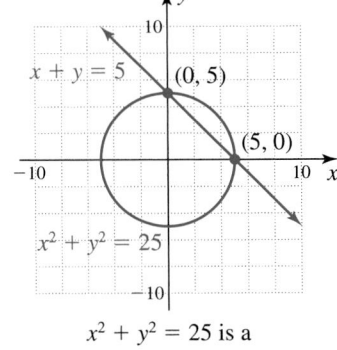

$x^2 + y^2 = 25$ is a circle of radius 5.

Figure 46

Answer

1. $(0, 4)$ and $(4, 0)$

That is, $y = \pm5$ and $y = 0$. In this case, the solutions obtained would have been $(0, 5)$, $(0, -5)$, and $(5, 0)$. However, $(0, -5)$ is *not* a solution of equation (3), since $-5 \neq 5 - 0$. Therefore, the only solutions are $(0, 5)$ and $(5, 0)$, as before.

> **CAUTION**
>
> If the degrees of the equations are different, one component of a solution should be substituted in the *lower-degree* equation to find the ordered pairs satisfying *both* equations.

For this reason, we double-check our work by graphing the given system (see Figure 46) and verify that our solutions are correct.

Calculate It   Graphing to Check Solutions of a Nonlinear System

You can use your calculator to check the solutions of systems of nonlinear equations. In Example 1, start by graphing $x^2 + y^2 = 25$. First, solve for y to obtain $y = \pm\sqrt{25 - x^2}$ and then graph $y = 5 - x$ using a square window. As you can see, the graphs intersect at two points, so there are two solutions.

To find them with a TI-83 Plus, press [2nd] [TRACE] [5], and three [ENTER]'s. (To make things easier, "turn off" $Y_2 = -Y_1 = -\sqrt{25 - x^2}$.) The calculator will show the points of intersection $(0, 5)$ and $(5, 0)$. Now try Example 2.

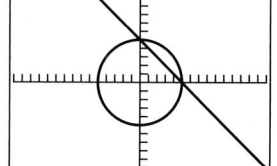

| **EXAMPLE 2** | **Solving nonlinear systems by substitution and checking with a graph** |

Find the solution of the given system by the substitution method. Check the solution by sketching the graphs of the equations.

$$x^2 + y^2 = 9 \tag{1}$$
$$x + y = 5 \tag{2}$$

SOLUTION Rewriting equation (2) in the form $y = 5 - x$, we obtain the equivalent system

$$x^2 + y^2 = 9 \tag{1}$$
$$y = 5 - x \tag{3}$$

Substituting $y = 5 - x$ in equation (1), we get

$$x^2 + (5 - x)^2 = 9$$
$$x^2 + 25 - 10x + x^2 = 9 \qquad \text{Multiply.}$$
$$2x^2 - 10x + 25 = 9 \qquad \text{Simplify.}$$
$$2x^2 - 10x + 16 = 0 \qquad \text{Subtract 9.}$$
$$x^2 - 5x + 8 = 0 \qquad \text{Divide by 2.}$$

Using the quadratic formula with $a = 1$, $b = -5$, $c = 8$, we get

$$x = \frac{5 \pm \sqrt{25 - 4 \cdot 8}}{2} = \frac{5 \pm \sqrt{-7}}{2} = \frac{5 \pm i\sqrt{7}}{2}$$

Substituting these values in equation (3), we obtain

$$y = 5 - \frac{5 + i\sqrt{7}}{2} \quad \text{and} \quad y = 5 - \frac{5 - i\sqrt{7}}{2}$$

PROBLEM 2

Find the solutions:

$$x + y = 3$$
$$x^2 + y^2 = 3$$

Web It

For a video on solving non-linear systems, go to link 9-4-2 at mhhe.com/bello.

Answer

2. $\left(\frac{3}{2} + \frac{\sqrt{3}}{2}i, \frac{3}{2} - \frac{\sqrt{3}}{2}i\right)$ and $\left(\frac{3}{2} - \frac{\sqrt{3}}{2}i, \frac{3}{2} + \frac{\sqrt{3}}{2}i\right)$

That is,

$$y = \frac{5}{2} - \frac{\sqrt{7}}{2}i \quad \text{and} \quad y = \frac{5}{2} + \frac{\sqrt{7}}{2}i$$

Hence the solutions of the system are

$$\left(\frac{5}{2} + \frac{\sqrt{7}}{2}i, \frac{5}{2} - \frac{\sqrt{7}}{2}i\right) \quad \text{and} \quad \left(\frac{5}{2} - \frac{\sqrt{7}}{2}i, \frac{5}{2} + \frac{\sqrt{7}}{2}i\right)$$

as can be checked in the original equations. The graphs of the two equations are shown in Figure 47. As you can see, the graphs *do not intersect.* When the solutions of a system of equations are non-real complex numbers, there are no points of intersection for the graphs. This is because the coordinates of points in the real plane are *real* numbers.

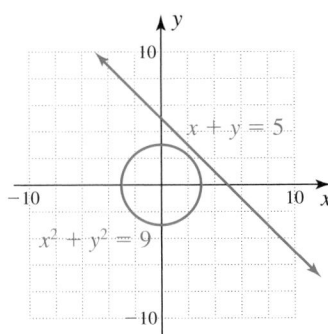

Figure 47

Teaching Tip

Have students identify the graph of each equation before solving Example 3. Then have them discuss all possible solutions for the graph of this system.

B | Solving Systems with Two Second-Degree Equations

When both equations in a system are of degree 2, it's easier to use the elimination method, as in Example 3.

EXAMPLE 3 | **Solving a system with two second-degree equations**

Find the solutions:

$$x^2 - 2y^2 = 1 \tag{1}$$
$$x^2 + 4y^2 = 25 \tag{2}$$

Verify the solution by graphing.

SOLUTION To eliminate y^2, we multiply the first equation by 2 and add the result to the second equation:

$$
\begin{array}{r}
2x^2 - 4y^2 = 2 \\
x^2 + 4y^2 = 25 \\
\hline
3x^2 \quad\quad = 27 \\
x^2 = 9 \\
x = \pm 3
\end{array}
$$

The x-coordinates of the point of intersection are 3 and -3. Substituting in the second equation,

$$(\pm 3)^2 + 4y^2 = 25$$
$$4y^2 = 16$$
$$y^2 = 4$$
$$y = \pm 2$$

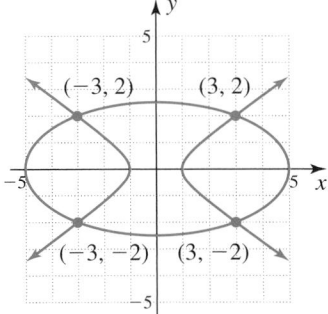

Figure 48

Thus the four points of intersection are $(3, 2)$, $(3, -2)$, $(-3, 2)$, and $(-3, -2)$, as you can check in the original equations. The graph for the two equations is shown in Figure 48.

PROBLEM 3

Find the solutions:

$$x^2 + y^2 = 9$$
$$x^2 - 9y^2 = 9$$

Calculate It

Solving a Nonlinear System

To do Example 3, graph

$$y = \pm\sqrt{\frac{x^2-1}{2}} \quad \text{and} \quad y = \pm\sqrt{\frac{25-x^2}{4}}$$

Press [2nd] [TRACE] [5]. Then you will need to use the [◆] key to select the part of each graph that the calculator is looking at. For instance, the top half of the ellipse (Y_1) intersects the top half of the hyperbola (Y_3) twice. Press [ENTER] to select the part of each graph, then once more to find a point of intersection. The graphs and one of the points of intersection, point $(-3, 2)$, are shown in the window.

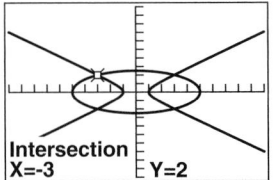

NOTE

An independent system of equations with at least one being quadratic can have four, three, two, one, or no real-number solutions.

Answer

3. $(3, 0)$ and $(-3, 0)$

 **Solving Applications of Nonlinear Systems**

The break-even point is the point at which enough units have been sold so that the cost C and the revenue R are equal. Example 4 shows how to use the methods we've studied to find the break-even point.

EXAMPLE 4   Breaking even

The total cost C for manufacturing and selling x units of a product each week is given by $C = 30x + 100$, whereas the revenue R is given by $R = 81x - 0.5x^2$. How many items must be manufactured and sold for the company to break even—that is, for C to equal R?

SOLUTION   We are asked to find the value of x for which $C = R$, that is,

$$\underbrace{C}_{30x + 100} = \underbrace{R}_{81x - 0.5x^2}$$

or, in standard form,

$$0.5x^2 - 51x + 100 = 0$$

where $a = 0.5$, $b = -51$, and $c = 100$. Using the quadratic formula, we get

$$x = \frac{51 \pm \sqrt{(-51)^2 - 4(0.5)(100)}}{2(0.5)}$$

$$= \frac{51 \pm \sqrt{2601 - 200}}{1}$$

$$= \frac{51 \pm \sqrt{2401}}{1}$$

$$= 51 \pm 49$$

Thus x is $51 - 49 = 2$ or $x = 51 + 49 = 100$; the company will break even when 2 or 100 items are sold (see Figure 49). The company makes a profit when they sell between 2 and 100 items.

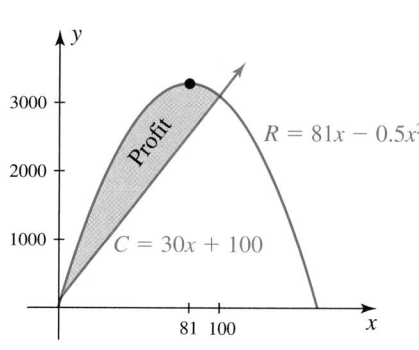

Figure 49

PROBLEM 4

If the total cost for manufacturing and selling x units of a product each year is represented by the equation $C = 40x - 500$ and the revenue is represented by the equation $R = 135x - x^2$, find how many units must be manufactured and sold for the company to break even.

Answer

4. 100 units

Calculate It   Using a Graph to Find Break-Even Points

In Example 4, we want to find the points at which $C = 30x + 100$ is the same as $R = 81x - 0.5x^2$. To do this, let

$$C = Y_1 = 30x + 100,$$
$$R = Y_2 = 81x - 0.5x^2$$

and find the points at which $C = R$, the points at which the graphs intersect.

To obtain the complete graph, we must be careful with the window we use. Start with a $[-10, 100]$ by $[-10, 3500]$ window. When $x = 100$.

$$R = 81(100) - 0.5 \cdot 100^2 = 3100$$

Since we didn't get a complete graph (see Window 1), change the domain for x to $[-10, 200]$.

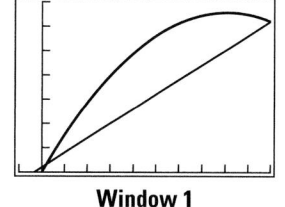

Window 1

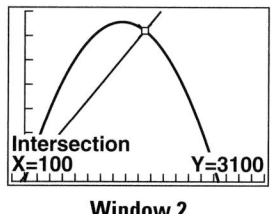

Window 2

We now have a complete graph and are able to find the points of intersection. With a TI-83 Plus, use the intersect feature to find these points. The graph with one of the points of intersection, point (100, 3100), is shown in Window 2.

You should find the second point to convince yourself that the answers are as before.

EXAMPLE 5   A new dimension in desserts

Can you guess what the dimensions of one of the largest shortcakes must have been? It covered 360 square feet and had a 106 ft perimeter. Instead of guessing, we will use the RSTUV method to find the dimensions of this rectangular shortcake.

SOLUTION

1. Read the problem.
We are asked to find the dimensions, and we know that the area is 360 ft^2 and the perimeter is 106 ft.

2. Select the unknowns.
Let L be the length and W the width.

3. Think of a plan.
Translate the problem.

Area A of the rectangle: $A = LW$

Perimeter P of the rectangle: $P = 2L + 2W$

In our case $A = 360$ and $P = 106$, thus $360 = LW$

$$106 = 2L + 2W$$

4. Use the techniques we have studied to solve this system.

First, solve $360 = LW$ for W to obtain $W = \dfrac{360}{L}$

Substitute $\frac{360}{L}$ for W in $106 = 2L + 2W$

We then have $106 = 2L + 2\left(\dfrac{360}{L}\right)$

$106L = 2L^2 + 720$ Multiply each term by L.

$L^2 - 53L + 360 = 0$ Divide by 2 and write in standard form.

$(L - 45)(L - 8) = 0$ Factor.

$L = 45$ or $L = 8$

If $L = 45$, $W = \frac{360}{L} = \frac{360}{45} = 8$. If $L = 8$, $W = \frac{360}{8} = 45$.

Since the length is usually longer than the width, we select the first case in which $L = 45$ and $W = 8$. Thus the shortcake is 45 ft long and 8 ft wide.

5. Verify the answer.
Since $L = 45$ and $W = 8$, the area is $A = 45(8) = 360$ ft^2, and the perimeter is $P = 2(45) + 2(8) = 106$ ft, as required.

PROBLEM 5

If the area of a picture frame is 88 in.^{2} and its perimeter is 38 in., find the dimensions of this picture frame.

Answer

5. 8 in. $\times$ 11 in.

Exercises 9.4

A In Problems 1–16, solve the system and check by graphing.

1. $x^2 + y^2 = 16$
　$x + y = 4$ $(0, 4), (4, 0)$

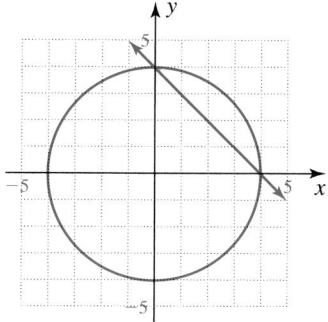

2. $x^2 + y^2 = 9$
　$x + y = 3$ $(0, 3), (3, 0)$

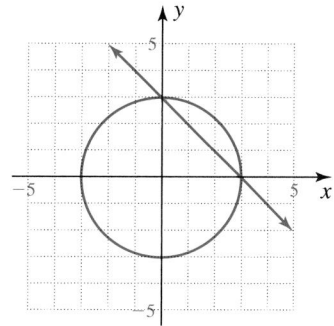

3. $x^2 + y^2 = 25$
　$y - x = 5$ $(-5, 0), (0, 5)$

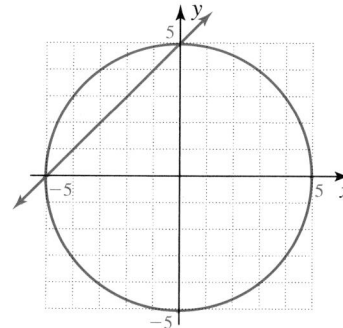

4. $x^2 + y^2 = 9$
　$y - x = 3$ $(-3, 0), (0, 3)$

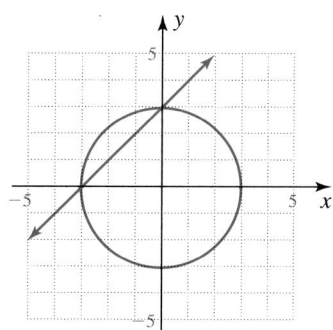

5. $x^2 + y^2 = 25$
　$y - x = 1$ $(-4, -3), (3, 4)$

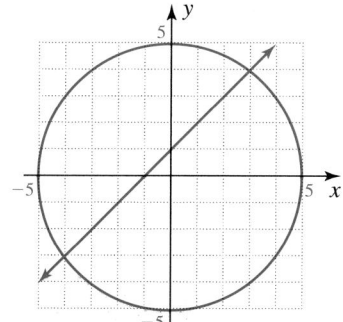

6. $x^2 + y^2 = 5$
　$y - x = 1$ $(-2, -1), (1, 2)$

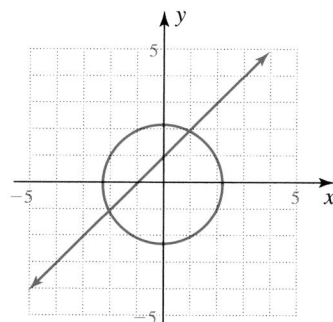

7. 　$y = x^2 - 5x + 4$
　$x - y = 1$ $(1, 0), (5, 4)$

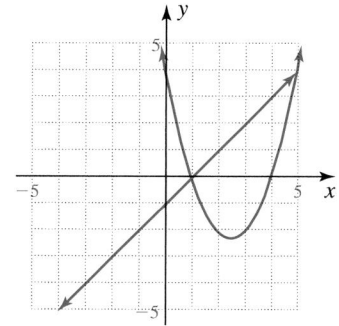

8. 　$y = x^2 - 2x + 1$
　$x - y = 1$ $(2, 1), (1, 0)$

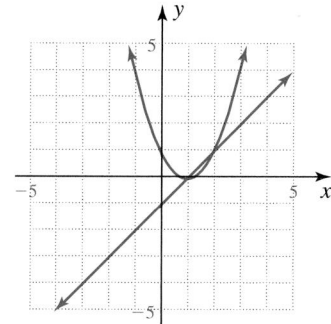

9. $y = (x - 1)^2$
$y - x = 1$ $(0, 1), (3, 4)$

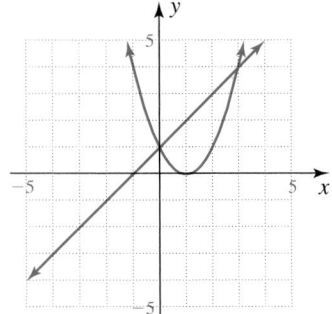

10. $y = (x + 3)^2$
$x + y = -1$ $(-5, 4), (-2, 1)$

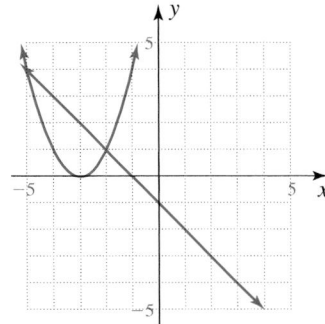

11. $4x^2 + 9y^2 = 36$
$3y - 2x = 6$ $(-3, 0), (0, 2)$

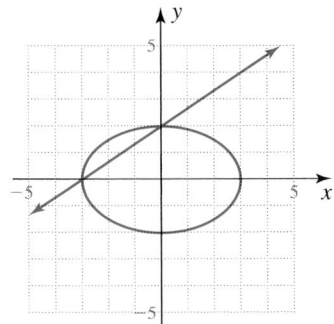

12. $4x^2 + 9y^2 = 36$
$3y + 2x = 6$ $(0, 2), (3, 0)$

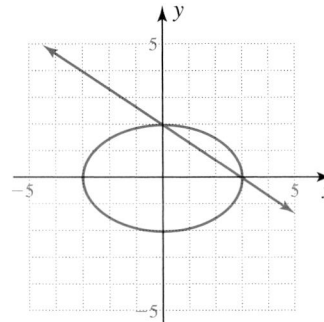

13. $x^2 - y^2 = 16$
$x + 4y = 4$ $(4, 0), (-\frac{68}{15}, \frac{32}{15})$

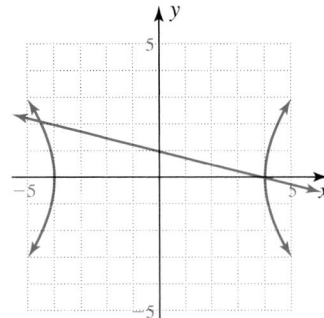

14. $x^2 - y^2 = 9$
$x + 3y = 3$ $(3, 0), (-\frac{15}{4}, \frac{9}{4})$

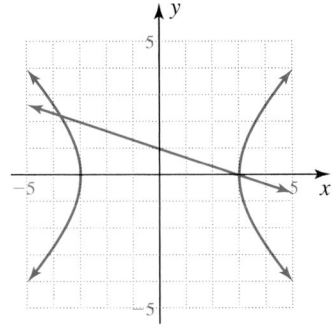

15. $x^2 + y^2 = 4$ $(-\frac{5}{2} + \frac{\sqrt{17}}{2}i, \frac{5}{2} + \frac{\sqrt{17}}{2}i)$,

$y - x = 5$ $(-\frac{5}{2} - \frac{\sqrt{17}}{2}i, \frac{5}{2} - \frac{\sqrt{17}}{2}i)$

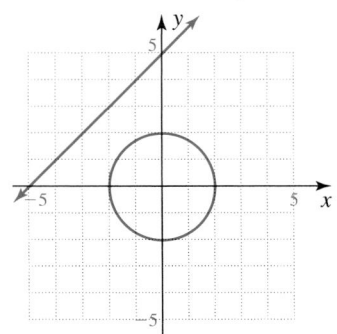

16. $x^2 + y^2 = 4$

$y - x = 3$ $(-\frac{3}{2} + \frac{1}{2}i, \frac{3}{2} + \frac{1}{2}i), (-\frac{3}{2} - \frac{1}{2}i, \frac{3}{2} - \frac{1}{2}i)$

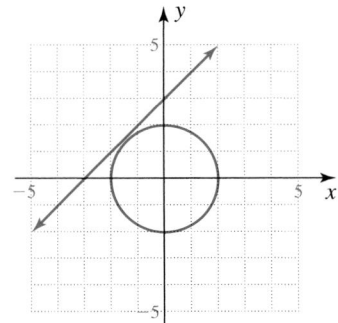

B In Problems 17–30, solve the system and check by graphing.

17. $y = 4 - x^2$
$y = x^2 - 4$ $(-2, 0), (2, 0)$

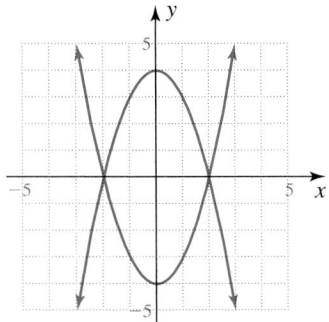

18. $y = 2 - x^2$
$y = x^2 - 2$ $(\sqrt{2}, 0), (-\sqrt{2}, 0)$

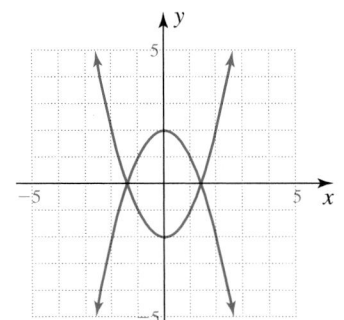

19. $x^2 + y^2 = 25$ $(4, 3), (4, -3)$,
$x^2 - y^2 = 7$ $(-4, 3), (-4, -3)$

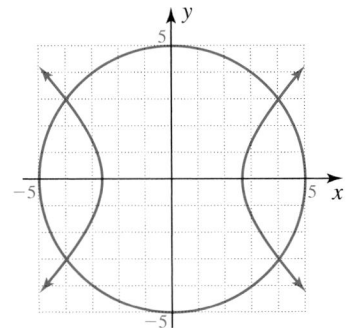

20. $x^2 + y^2 = 20$ $(\sqrt{11}, 3), (-\sqrt{11}, 3),$
$x^2 - y^2 = 2$ $(\sqrt{11}, -3), (-\sqrt{11}, -3)$

21. $x^2 + y^2 = 16$
$x^2 + 16y^2 = 16$ $(-4, 0), (4, 0)$

22. $x^2 + y^2 = 9$
$x^2 + 9y^2 = 9$ $(3, 0), (-3, 0)$

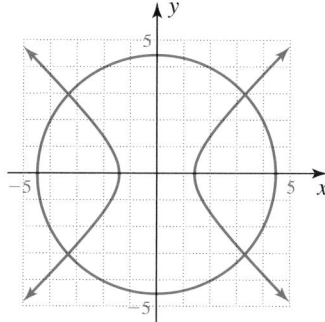

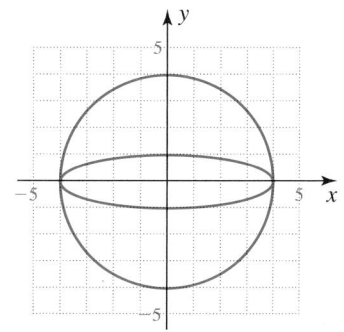

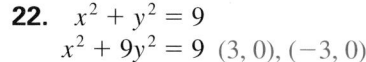

23. $3x^2 - y^2 = 2$ $(1, -1), (-1, -1),$
$x^2 + 2y^2 = 3$ $(1, 1), (-1, 1)$

24. $x^2 - 4y^2 = 4$
$9x^2 + 4y^2 = 36$ $(2, 0), (-2, 0)$

25. $x^2 + 2y^2 = 11$ $(-3, 1), (-3, -1),$
$2x^2 + y^2 = 19$ $(3, 1), (3, -1)$

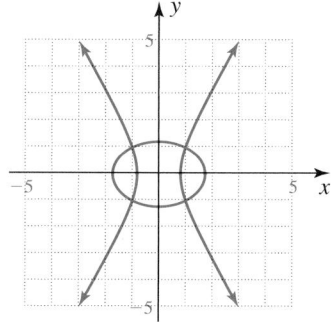

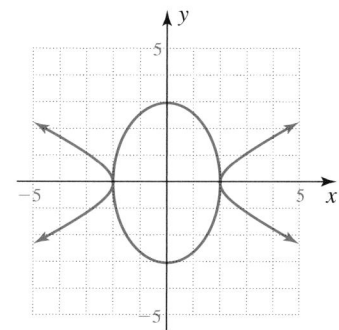

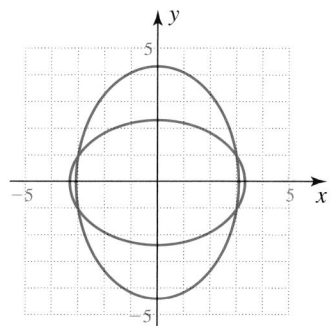

26. $4x^2 + 9y^2 = 52$ $(2, -2), (-2, -2),$
$9x^2 + 4y^2 = 52$ $(2, 2), (-2, 2)$

27. $x^2 + y^2 = 4$ $(\frac{\sqrt{26}}{2}, \frac{\sqrt{10}}{2}i),$
$x^2 - y^2 = 9$ $(\frac{\sqrt{26}}{2}, -\frac{\sqrt{10}}{2}i),$

28. $x^2 + y^2 = 9$ $(\frac{5\sqrt{2}}{2}, \frac{\sqrt{14}}{2}i),$
$x^2 - y^2 = 16$ $(\frac{5\sqrt{2}}{2}, -\frac{\sqrt{14}}{2}i),$

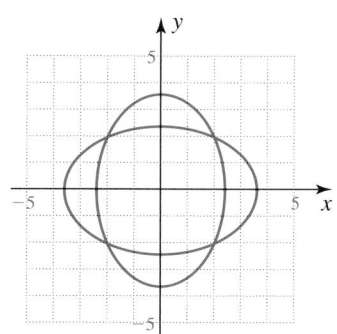

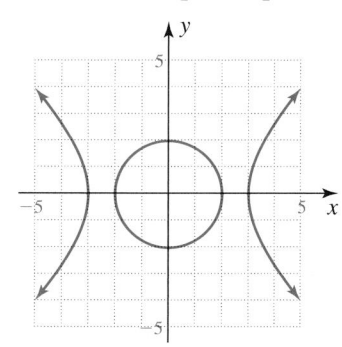

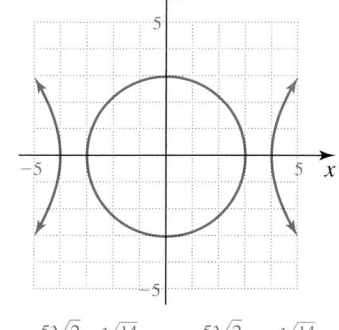

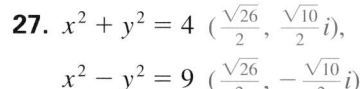

$(-\frac{\sqrt{26}}{2}, \frac{\sqrt{10}}{2}i), (-\frac{\sqrt{26}}{2}, -\frac{\sqrt{10}}{2}i)$

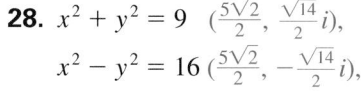

$(-\frac{5\sqrt{2}}{2}, \frac{\sqrt{14}}{2}i), (-\frac{5\sqrt{2}}{2}, -\frac{\sqrt{14}}{2}i)$

29. $x^2 + y^2 = 1$ $(\frac{3\sqrt{15}}{5}i, \frac{4\sqrt{10}}{5}), (\frac{3\sqrt{15}}{5}i, -\frac{4\sqrt{10}}{5}),$
$4x^2 + 9y^2 = 36$ $(-\frac{3\sqrt{15}}{5}i, \frac{4\sqrt{10}}{5}), (-\frac{3\sqrt{15}}{5}i, -\frac{4\sqrt{10}}{5})$

30. $x^2 + y^2 = 25$ $(\frac{3\sqrt{105}}{5}, \frac{8\sqrt{5}}{5}i), (\frac{3\sqrt{105}}{5}, -\frac{8\sqrt{5}}{5}i),$
$4x^2 + 9y^2 = 36$ $(-\frac{3\sqrt{105}}{5}, \frac{8\sqrt{5}}{5}i), (-\frac{3\sqrt{105}}{5}, -\frac{8\sqrt{5}}{5}i)$

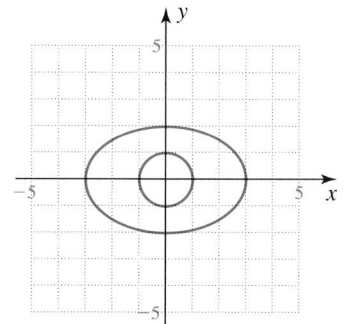

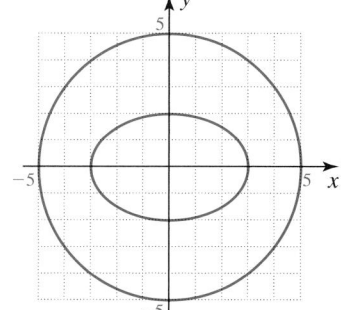

APPLICATIONS

Dollars and Dimensions

31. The total cost C (in thousands of dollars) for manufacturing and selling x (thousand) items of a product each month is given by $C = x + 4$, and the revenue is $R = 6x - x^2$. How many items must be manufactured and sold for the company to break even? 4000 or 1000

32. The total cost C (in thousands of dollars) for manufacturing and selling x (thousand) items of a product each month is given by $C = 2x + 6$, and the revenue is $R = 150 + 20x - x^2$. How many items must be manufactured and sold for the company to break even? 24,000

33. There are two very interesting numbers. Their sum is 15 and the difference of their squares is also 15. Find the numbers. 7 and 8

34. Two squares differ in area by 108 ft^2 and their sides differ by 6 ft. Find the lengths of the sides of these squares. (*Hint:* Make a picture!) 12 ft and 6 ft

35. Find two numbers such that their product is 176 and the sum of their squares is 377. 11 and 16 or -11 and -16

36. Find the dimensions of a rectangle whose area is 96 in.^{2} and whose perimeter is 44 in. 16 in. by 6 in.

37. Have you written any big checks lately? The world's *physically* largest check had an area of 2170 ft^2 and a perimeter of 202 ft. If the check was rectangular, what were its dimensions? 70 ft by 31 ft

38. The largest ancient carpet (A.D. 743) was a gold-enriched silk carpet that covered 54,000 ft^2. If the carpet was rectangular and its perimeter was 960 ft, what were the dimensions of the carpet? 300 ft by 180 ft

39. A person receives \$340 interest from an amount loaned at simple interest for 1 yr. If the interest rate had been 1% higher, she would have received \$476. What was the amount loaned and what was the interest rate? (*Hint:* $I = Pr$, where I is the interest, P is the principal, and r is the rate.) $P = \$13,600; r = 2.5\%$

40. There are two interesting positive numbers such that their product, their difference, and the difference of their squares are all equal. Find the numbers. (*Hint:* They are irrational.) $\frac{-1 + \sqrt{5}}{2}$ and $\frac{3 - \sqrt{5}}{2}$

SKILL CHECKER

Try the "Skill Checker" exercises so you'll be ready for the next section.

Graph:

41. $x - y < 4$

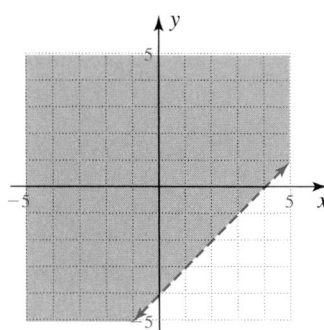

42. $y - x < 4$

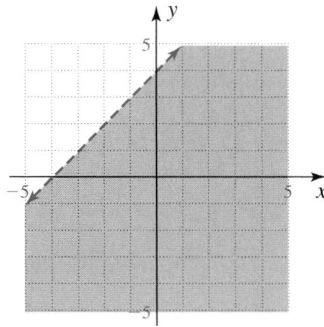

43. $2x - 3y \geq 6$

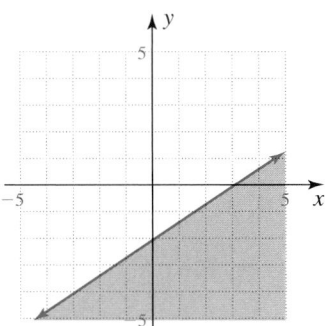

44. $3x - 2y \geq 6$

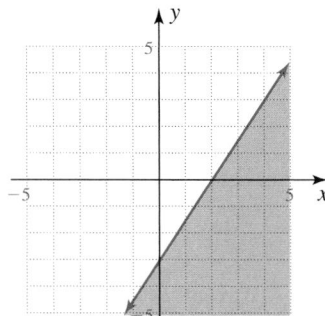

45. $y \geq 2x + 4$

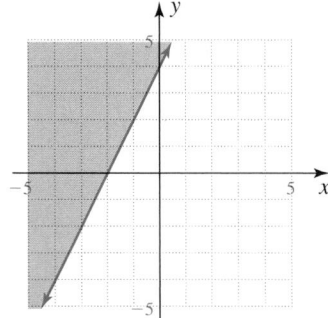

46. $y \geq 3x + 6$

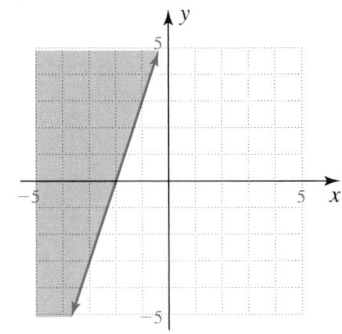

USING YOUR KNOWLEDGE

100-year-old Problems!

Do you think the algebra we have been studying is relatively new? What does it have to do with the price of eggs? Here are a couple of problems from *Elementary Algebra,* published by Macmillan in 1894. See if you have the knowledge to solve these problems.

47. "The number of eggs which can be bought for 25 cents is equal to twice the number of cents which 8 eggs cost. How many eggs can be bought for 25 cents?" (*Hint:* If you let p be the price of eggs, then $\frac{25}{p}$ is the number of eggs you can buy for 25¢.) 20 eggs

48. "One half of the number of cents which a dozen apples cost is greater by 2 than twice the number of apples which can be bought for 30 cents. How many can be bought for $2.50?" (*Hint:* If you let d be the cost of a dozen apples, then $\frac{d}{12}$ is the cost of one apple and

$$\frac{30}{\frac{d}{12}}$$

is the number of apples that can be bought for 30¢.)
75 apples

WRITE ON

49. Consider the system

$$y = ax + b$$
$$y = cx + d$$

a. If $b \neq d$, what is the maximum number of solutions for this system? Write an explanation to support your answer. How many solutions are possible if b can be equal to d? Answers may vary.

b. Write a set of conditions under which this system will have no solution. What will the relationship between a and c be then? Answers may vary.

50. Consider the system

$$y = ax + b$$
$$x^2 + y^2 = r^2$$

a. What is the maximum number of real solutions for this system? Write an explanation to support your answer. Answers may vary.

b. Make sketches showing a system similar to the given one and that has no real solution, one solution, and two solutions. What is the geometric interpretation for the number of solutions in a system? Answers may vary.

Explain:

51. What is the maximum number of solutions you can have for a system of equations consisting of

a. Two hyperbolas?   **b.** Two circles?   **c.** Two ellipses?   **d.** A hyperbola and a circle?
    Answers may vary.   Answers may vary.   Answers may vary.   Answers may vary.

MASTERY TEST

If you know how to do these problems, you have learned your lesson!

52. The total cost C for manufacturing and selling x units of a product is given by $C = 30x + 100$, whereas the revenue is given by $R = 57x - 0.5x^2$. How many items must be manufactured and sold for the company to break even? 4 or 50

53. The perimeter of a rectangle is 44 cm and its area is 120 cm^2. What are the dimensions of this rectangle?
12 cm by 10 cm

Solve each system:

54. $x^2 - 9y^2 = 9$
    $x^2 + y^2 = 9$ $(3, 0), (-3, 0)$

55. $4y^2 - x^2 = 4$
    $x^2 + y^2 = 1$ $(0, 1), (0, -1)$

56. $x^2 + y^2 = 3$ $(\frac{3}{2} + \frac{\sqrt{3}}{2}i, \frac{3}{2} - \frac{\sqrt{3}}{2}i)$,
    $x + y = 3$ $(\frac{3}{2} - \frac{\sqrt{3}}{2}i, \frac{3}{2} + \frac{\sqrt{3}}{2}i)$

57. $x^2 + y^2 = 4$ $(2 + i\sqrt{2}, 2 - i\sqrt{2})$,
    $x + y = 4$ $(2 - i\sqrt{2}, 2 + i\sqrt{2})$

58. $x^2 + 4y^2 = 16$
    $x + 2y = 4$ $(4, 0), (0, 2)$

59. $x^2 + y^2 = 1$
    $x + y = 1$ $(1, 0), (0, 1)$

9.5 NONLINEAR SYSTEMS OF INEQUALITIES

To Succeed, Review How To . . .

1. Graph linear inequalities in two variables (pp. 165–168).

2. Graph conic sections (pp. 648–657, 671–676, 686–692).

Objectives

A Graph second-degree inequalities.

B Graph the solution set of a system of nonlinear inequalities.

GETTING STARTED Profits and Losses

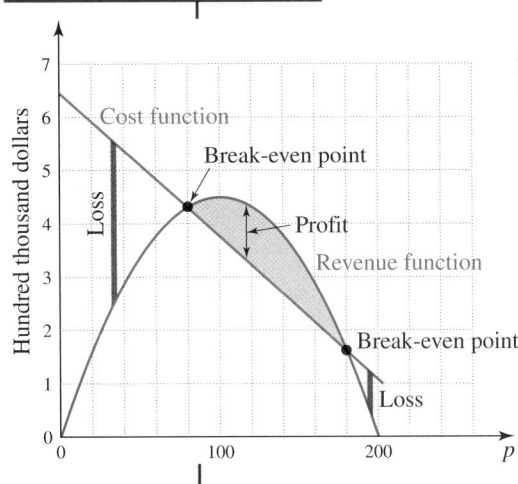

This graph offers a wealth of information if you know how to read it. The line represents the cost C of a product, whereas the parabola represents the revenue R obtained when the product is sold. Here are some facts about the graph:

1. Where the graph representing the cost C lies above the graph of the revenue R, there is a loss.

2. The revenue equals the cost at a point (or points) called a break-even point.

3. Where the graph representing the revenue R lies above the graph of the cost C, there is a profit.

This is an example of a nonlinear system of inequalities, a concept we shall investigate in detail in this section.

A Graphing Second-Degree Inequalities

Teaching Tip

Before starting this section, graph examples of solid and broken boundary lines for a linear inequality.

Example:

$y \leq x + 2$ and $y < x + 2$

In Section 3.5, we graphed linear inequalities by graphing the boundary, selecting a test point not on the boundary, and determining the regions corresponding to the solution set. The region under the parabola in the graph above is obtained by graphing a **second-degree inequality** using a similar procedure. Let's do an example.

EXAMPLE 1 Solving a second-degree inequality involving a parabola

Graph: $y \leq -x^2 + 3$

SOLUTION The boundary of the required region is the parabola $y = -x^2 + 3$ with its vertex at $(0, 3)$, as shown in Figure 50. Since the inequality sign is $\leq$, the boundary *is included* in the solution set. To determine which region represents the solution set, select a test point not on the boundary and test the original inequality. A convenient point is $(0, 0)$. When $x = 0$ and $y = 0$, $y \leq -x^2 + 3$ becomes $0 \leq 0 + 3$, a true statement. Thus all points on the

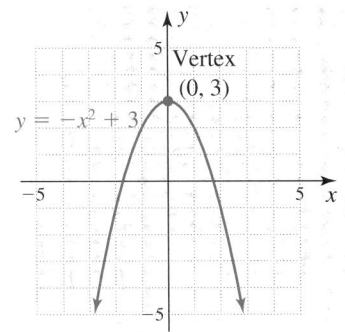

Figure 50

PROBLEM 1

Graph:

$$y \geq x^2 - 2$$

Answer

1.

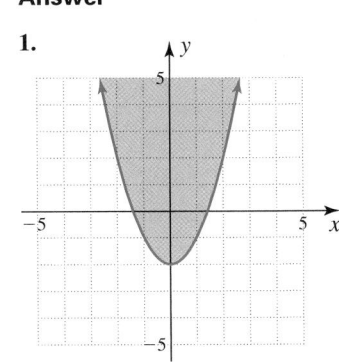

same side as $(0, 0)$ will be in the solution set. The graph of the solution set is shown shaded in Figure 51.

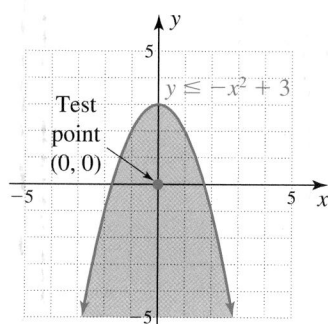

Figure 51

EXAMPLE 2   **Solving a second-degree inequality involving a circle**

Graph: $x^2 + y^2 > 4$

SOLUTION   This time, the boundary is the circle $x^2 + y^2 = 4$ with radius 2 and centered at the origin. The inequality sign is $>$, so that the boundary is *not* included in the solution set. It is shown dashed in Figure 52. For a test point, select the point $(0, 0)$, $x = 0$ and $y = 0$, and $x^2 + y^2 > 4$ becomes $0 + 0 > 4$, a false statement. This means that the region containing $(0, 0)$ is *not* in the solution set. We then shade the region *outside* the circle to represent the solution set, as shown in Figure 52.

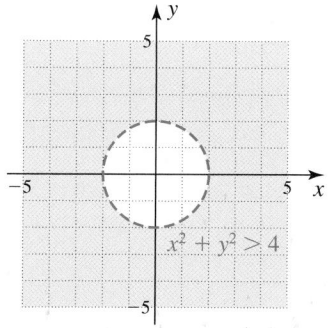

Figure 52

PROBLEM 2

Graph:

$$x^2 + y^2 < 9$$

EXAMPLE 3   **Solving a second-degree inequality involving a hyperbola**

Graph: $9y^2 - 4x^2 \leq 36$

SOLUTION   Since all the variables are on the left and there is a minus sign between the terms, the boundary is a hyperbola, with equation $\frac{y^2}{4} - \frac{x^2}{9} = 1$. The center is at $(0, 0)$ and the vertices are $(0, \pm 2)$. Since the inequality is $\leq$, the boundary *is* part of the solution set. If we use $(0, 0)$ for our test point,

PROBLEM 3

Graph:

$$4x^2 - 4y^2 \geq 16$$

Answers

2.   3.

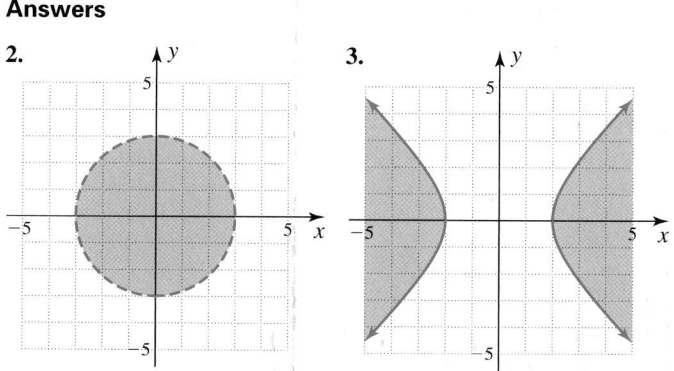

$9y^2 - 4x^2 \le 36$ becomes $0 - 0 \le 36$, which is true. Thus the point $(0, 0)$ is part of the complete solution set, shown shaded in Figure 53.

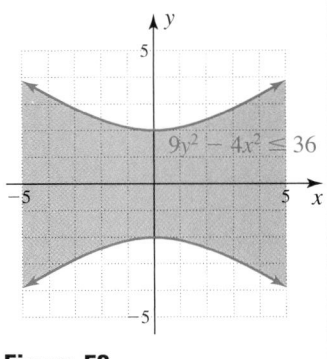

Figure 53

B   Graphing Systems of Inequalities

We have already solved a linear system of inequalities in two variables using the graphical method. To solve a *nonlinear system of inequalities* graphically, we use the following procedure.

> **PROCEDURE**
>
> **Graphing a Nonlinear System of Inequalities**
>
> 1. Graph each of the inequalities on the same set of axes.
>
> 2. Find the region common to both graphs, the intersection of both graphs. The result is the solution set of the system.

We illustrate this procedure next.

EXAMPLE 4   **Graphing a nonlinear system of inequalities**

Graph: The system $x^2 + y^2 \le 4$ and $y \ge x^2 - 3$

SOLUTION   The boundary for the first inequality is a circle of radius 2 centered at the origin. The solution set of this inequality includes the boundary and the points inside the circle and is shown in red in Figure 54. The boundary for the second inequality is a parabola with vertex at $(0, -3)$, and the solution set includes the boundary and all the points above the parabola, shown in blue.

The solution set for the system is the intersection of the two solution sets, the region inside *both* the circle and the parabola and the points on the circle *above* or *on* the parabola where the red and blue regions overlap.

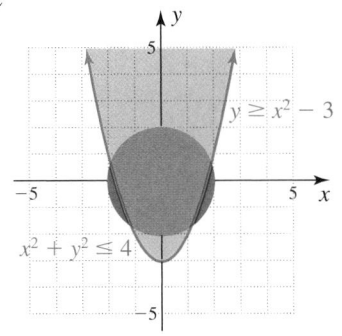

Figure 54

PROBLEM 4

Graph the system:

$$x^2 + y^2 \le 9$$
$$y \ge x^2$$

Answer

4.

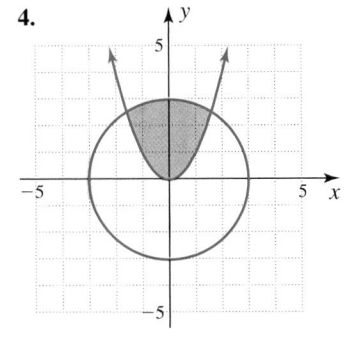

Calculate It   Graphing Nonlinear System of Inequalities

To do Example 4 with your calculator, solve for y in $x^2 + y^2 = 4$ to obtain $y = \pm\sqrt{4 - x^2}$. The corresponding inequality is $y \le \pm\sqrt{4 - x^2}$—that is, the points *on* the boundary and *inside* the circle $x^2 + y^2 = 4$. Remember to use a square window (**ZOOM** 5 on a TI-83 Plus) to make the circle appear round. The solution set of $y \ge x^2 - 3$ consists of the points *on* the boundary and *above* the parabola $y = x^2 - 3$. The intersection of these two sets consists of the points *inside* the circle $x^2 + y^2 = 4$ and *above* the parabola $y = x^2 - 3$, as before, and the points *on* the circle *above* or *on* the parabola, as shown in the window.

Exercises 9.5

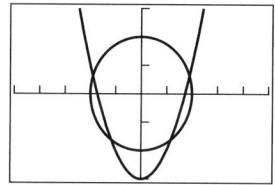
A In Problems 1–12, graph the solution set of the inequality.

1. $x^2 + y^2 > 16$

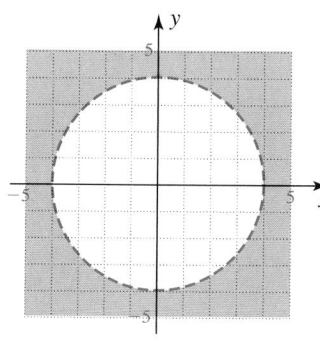

2. $x^2 + y^2 < 16$

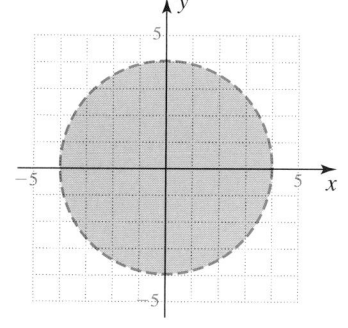

3. $x^2 + y^2 \le 1$

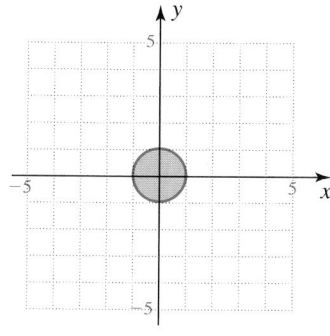

4. $x^2 + y^2 \ge 1$

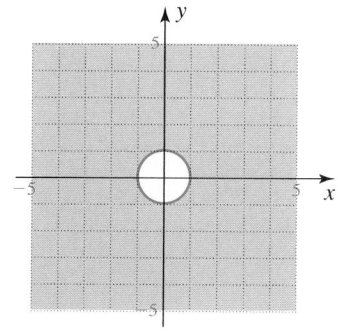

5. $y < x^2 - 2$

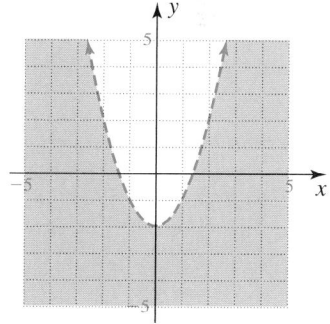

6. $y > x^2 - 2$

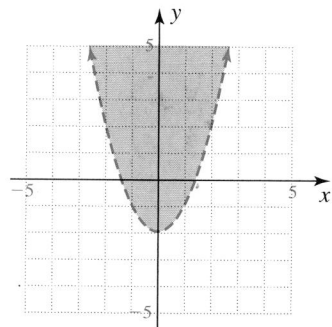

7. $y \le -x^2 + 3$

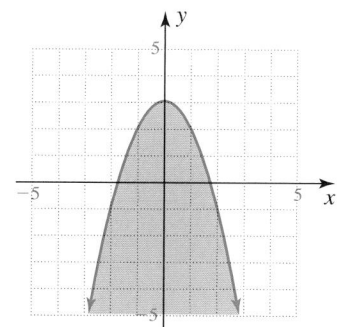

8. $y \ge -x^2 + 3$

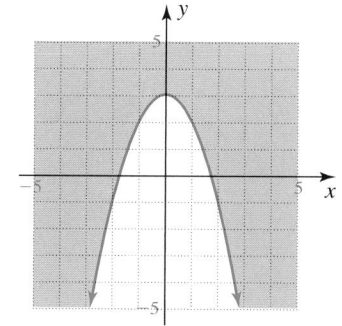

9. $4x^2 - 9y^2 > 36$

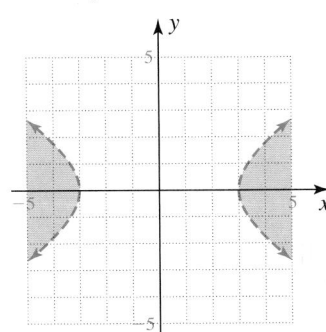

10. $4x^2 - 9y^2 < 36$

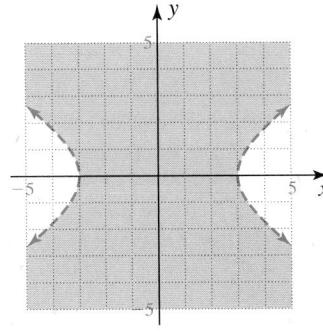

11. $x^2 - y^2 \geq 1$

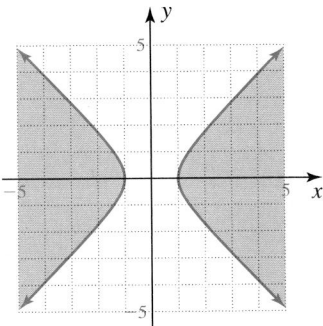

12. $x^2 - y^2 \leq 1$

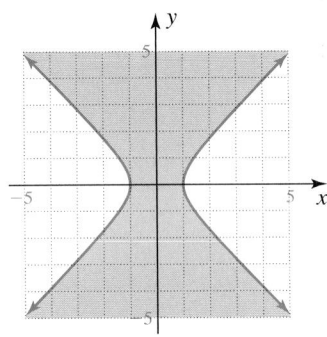

B In Problems 13–24, graph the solution set of the system.

13. $x^2 + y^2 \leq 25$
$\quad\ y \geq x^2$

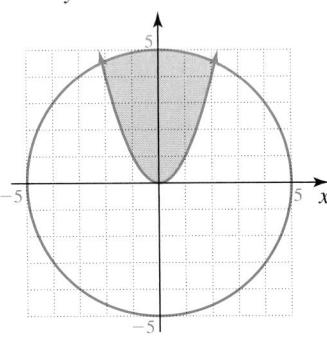

14. $x^2 + y^2 \leq 25$
$\quad\ y \leq x^2$

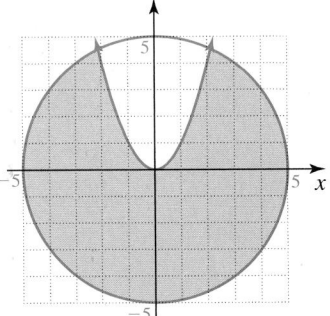

15. $x^2 + y^2 \geq 25$
$\quad\ y \leq x^2$

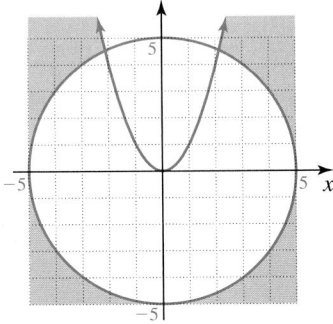

16. $x^2 + y^2 \geq 25$
$\quad\ y \geq x^2$

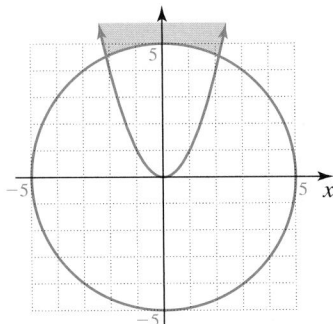

17. $y < x^2 + 2$
$\quad y > x^2 - 2$

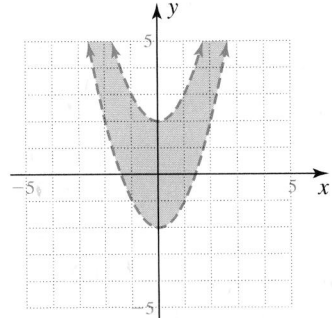

18. $y < x^2 + 2$
$\quad y < x^2 - 2$

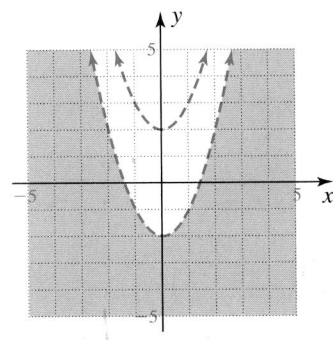

19. $y \geq x^2 + 2$
$y \geq x^2 - 2$

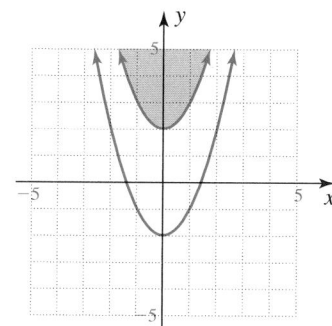

20. $y \geq x^2 + 2$
$y \leq x^2 - 2$ No solution

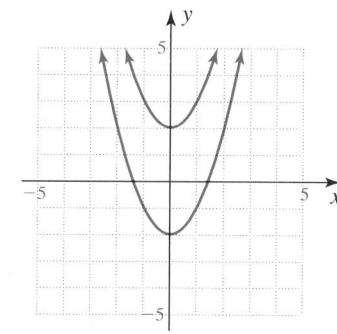

21. $\dfrac{x^2}{4} - \dfrac{y^2}{4} \geq 1$

$\dfrac{x^2}{25} + \dfrac{y^2}{4} \leq 1$

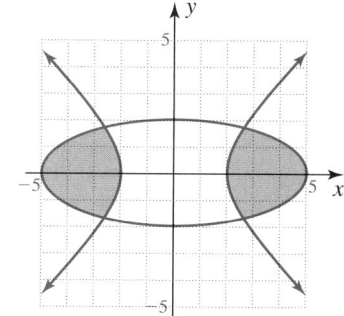

22. $\dfrac{x^2}{4} - \dfrac{y^2}{4} \geq 1$

$\dfrac{x^2}{25} + \dfrac{y^2}{4} \geq 1$

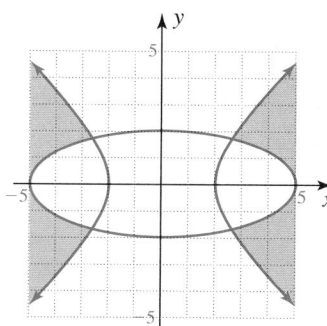

23. $\dfrac{x^2}{36} + \dfrac{y^2}{16} < 1$

$\dfrac{x^2}{16} + \dfrac{y^2}{36} < 1$

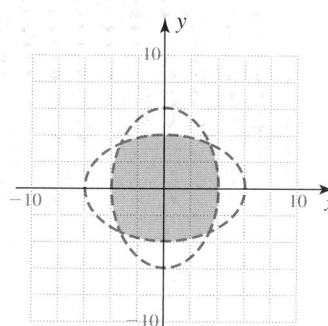

24. $\dfrac{x^2}{36} + \dfrac{y^2}{16} > 1$

$\dfrac{x^2}{16} + \dfrac{y^2}{36} < 1$

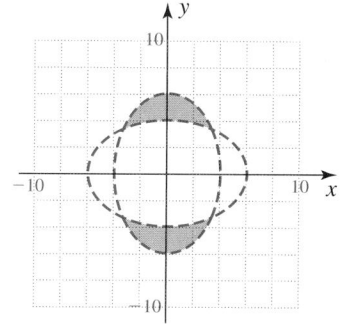

25. Fill in each blank with an inequality symbol so that the system

$\dfrac{x^2}{36} + \dfrac{y^2}{16}$ ——— 1   (ellipse)

$\dfrac{x^2}{16} + \dfrac{y^2}{16}$ ——— 1   (circle)

will have

a. No solution. $\frac{x^2}{36} + \frac{y^2}{16} > 1; \frac{x^2}{16} + \frac{y^2}{16} < 1$
b. Two solutions. $\frac{x^2}{36} + \frac{y^2}{16} \geq 1; \frac{x^2}{16} + \frac{y^2}{16} \leq 1$

26. Can you fill in each blank with an inequality symbol so that the system

$\dfrac{x^2}{25} - \dfrac{y^2}{4} \underline{\geq} 1$ (hyperbola)

$\dfrac{x^2}{25} + \dfrac{y^2}{4} \underline{\leq} 1$ (ellipse)

will have no solution?

SKILL CHECKER

Try the "Skill Checker" exercises so you'll be ready for the next section.

Find the value of $y = 3x + 5$ when:

27. $x = -2$ -1

28. $x = -5$ -10

If $P(x) = x^2 - 9$ and $Q(x) = x + 3$, find:

29. The sum and the difference of $P(x)$ and $Q(x)$.
$P(x) + Q(x) = x^2 + x - 6$; $P(x) - Q(x) = x^2 - x - 12$

30. The product and the quotient of $P(x)$ and $Q(x)$.
$P(x) \cdot Q(x) = x^3 + 3x^2 - 9x - 27$; $\dfrac{P(x)}{Q(x)} = x - 3$

USING YOUR KNOWLEDGE

Give Me a Break-Even Point!

You may have wondered about the equations representing the cost C and the revenue R in the graph in *Getting Started* at the beginning of this section. Both the cost C and the revenue R are given in terms of p, the price per unit. Can we find the break-even point—the point at which the revenue R equals the cost C? Let's try.

31. The demand equation—the number x of units retailers are likely to buy at p dollars per unit—is given by $x = 6000 - 30p$, and the cost C is given by $C = 72{,}000 + 60x$. Substitute $x = 6000 - 30p$ in $C = 72{,}000 + 60x$ and find C in terms of the price p.
$C = 432{,}000 - 1800p$

32. The revenue is $R = xp$—the revenue is the product of the number of units retailers are likely to buy and the price p per unit. To find R in terms of p, substitute $x = 6000 - 30p$ into $R = xp$. What is R in terms of p? What shape does the graph of R have?
$R = 6000p - 30p^2$; The graph is a parabola.

33. The graph in *Getting Started* on page 710 shows two points at which the revenue R equals the cost—$R = C$. Substitute the expressions for R and C from Problems 31 and 32 in the equation $R = C$ and solve for p.
$p = 180$ or $p = 80$

WRITE ON

34. Write your own definition of a second-degree inequality. Answers may vary.

35. How do you decide whether the boundary of a region should be included in the solution set? Answers may vary.

36. Write the procedure you use to solve a system of non-linear inequalities graphically. Answers may vary.

37. The system of inequalities we solved in Example 4 has a solution set with infinitely many points. Can you find a system of nonlinear inequalities with

a. A solution set consisting of only one point? Make a sketch to show the system. Answers may vary.

b. A solution set consisting of exactly two points? Make a sketch to show the system. Answers may vary.

MASTERY TEST

If you know how to do these problems, you have learned your lesson!

Graph:

38. $x^2 + y^2 \le 9$ and $y \ge x^2 + 2$

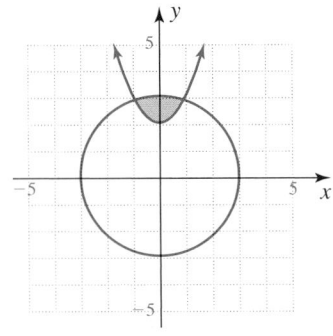

39. $x^2 + 4y^2 \ge 4$ and $y \ge x^2 + 1$

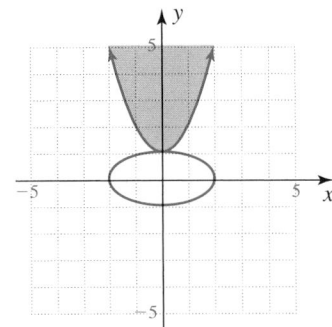

40. $4y^2 - 9x^2 \le 36$

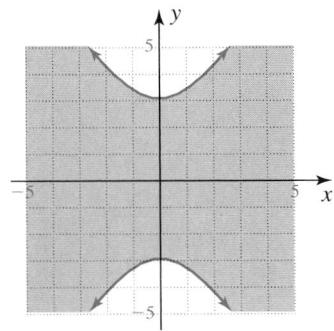

41. $x^2 - 9y^2 \ge 9$

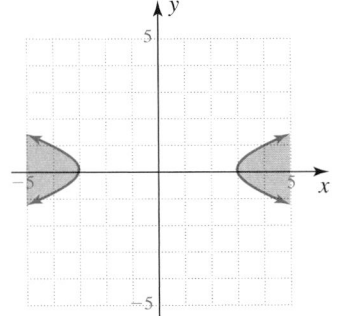

42. $x^2 + y^2 > 9$

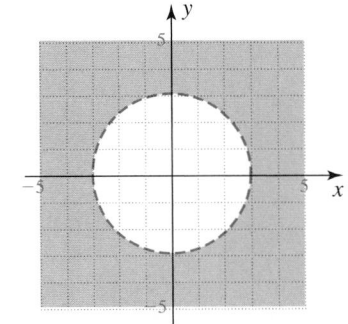

43. $x^2 > 4 - y^2$

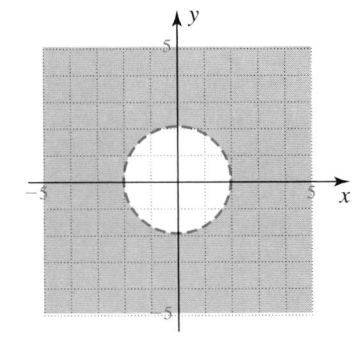

44. $y \leq -x^2 - 3$

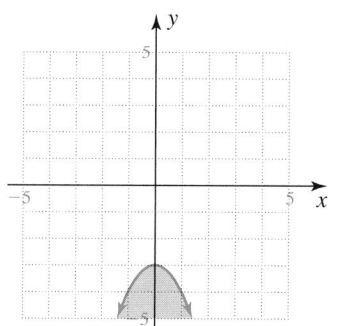

45. $y \geq -x^2 + 2$

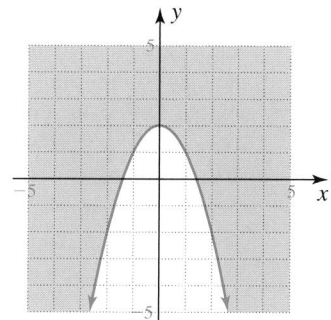

COLLABORATIVE LEARNING 9A

A resort rents 40 bicycles each day (on the average) and charges an $8 rental fee for each. They are considering raising their rental fee. After investigating studies done on bicycle rentals, there is an indication that for every $1 increase in the rental fee, they will lose two rentals. Divide into groups and answer the questions.

Assuming the data from these studies is accurate, how much should the resort charge to maximize their revenue (total income)?

1. Complete the table.

Price Increase	Rental Fee Per Bicycle	Number of Rentals	Revenue
Present Fee	$8	40	$8(40) = $320
1 rental increase	$8 + 1($1)	40 − 1(2)	$9(38) = $342
2 rental increases	$8 + 2($1)	40 − 2(2)	$10(36) = $360
3 rental increases			
4 rental increases			
x rental increases			(?)(?) = R

2. Use the information from the table to write the equation that will find the revenue, R, given x rental increases.

3. Draw the graph of the equation and find the vertex of the curve.

4. What is the maximum point on the curve?

5. How much should the resort charge for each bicycle rental to maximize the revenue?

6. Each group should find an application of maximizing or minimizing and present the results to the class for discussion. Include in your discussion why the equation of a parabola is appropriate for this type of problem.

COLLABORATIVE LEARNING 9B

The National Statuary Hall in the Capitol building in Washington, D.C. is a semicircular room with a ceiling in the shape of an ellipse. It was the meeting place for the House of Representatives from 1819 to 1857. Now it is an exhibition hall commemorating outstanding United States citizens chosen by each state.

When John Quincy Adams was a member of the House of Representatives he took advantage of the Hall's acoustics to eavesdrop on other members conversing on the opposite side of the room. He placed his desk at a focal point of the elliptical ceiling and was able to eavesdrop on the private conversations of other House members sitting near the other focal point. Can you calculate where he placed his desk? Divide into groups and answer the questions.

The equation of an ellipse with center at the origin is $\frac{x^2}{a^2} + \frac{y^2}{b^2} = 1$, where $a > b$. One-half the length of the major axis is a, one-half the length of the minor axis is b, and one-half the distance between the two foci is c. To find c, use the formula, $c = \sqrt{a^2 - b^2}$.

1. Find the distance a, b, and c for the given ellipses.

   a. $\frac{x^2}{9} + \frac{y^2}{16} = 1$ **b.** $\frac{x^2}{4} + \frac{y^2}{1} = 1$

   c. $\frac{x^2}{16} + \frac{y^2}{4} = 1$ **d.** $\frac{x^2}{1} + \frac{y^2}{9} = 1$

2. Sketch the graphs for the ellipses in Question 1 and name the coordinates for c.

3. If the estimated dimensions for Statuary Hall indicate a major axis of approximately 240 ft and a minor axis of approximately 100 ft, find c, one-half the length between the foci.

4. Using the origin of an x-y coordinate system as the center of the ellipse that approximates Statuary Hall, sketch its graph. Be sure to label and name the coordinates of the foci.

5. How far from the center of Statuary Hall did John Quincy Adams place his desk so he could hear the Congressmen talking at the other focal point?

6. Each group should research other acoustical phenomena to see if they have any connection to the ellipse or other conic sections. Be sure to include information about the acoustics at the Mayan ruins found at http://www.tomzap.com/sounds.html.

7. Present your research findings to the class and have a discussion about which may be related to the conic sections. If they are not, then is there a scientific explanation for the acoustical phenomena?

Source: Architect of the Capitol; National Statuary Hall (gives most of the information about John Quincy Adams).

Research Questions

1. Write a paper detailing the origin and meaning of the words *ellipse, hyperbola,* and *parabola.*

2. Who was the first person that used the terminology "ellipse, hyperbola, and parabola" and from what book or paper were these terms adapted?

3. Who was "The Great Geometer"?

4. Write a short paper about Hippocrates of Chios, his discoveries, his works, and his connection to the conics.

5. Write a short paper about Menaechmus, his discoveries, his works, and his connection to the conics.

6. Write a short paper about Apollonius of Perga, his discoveries, his works, and his connection to the conics.

7. One of the three famous problems in the first 300 years of Greek mathematics was "the duplication of the cube." What were the other two?

8. How are conics used in mathematics, science, art, and other areas?

9. Write a paper detailing Girard Desargues' contribution to the study of the conic sections.

10. A famous French mathematician wrote *A Complete Work on Conics.* Who was this mathematician and what did the book contain?

Summary

SECTION	ITEM	MEANING	EXAMPLE
9.1	Conic sections	Curves obtained by slicing a cone with a plane	Parabolas, circles, ellipses, and hyperbolas
9.1A	Vertex of a parabola	The highest or lowest point of a parabola opening vertically	The vertex of $y = x^2 + 2$ is at $(0, 2)$.
9.1B, C	Parabola (vertical axis)	A curve with equation $y = a(x - h)^2 + k$ or $y = ax^2 + bx + c$	$y = x^2$, $y = 3(x - 1)^2 + 2$, and $y = -2x^2 + 3x - 7$
9.1D	Parabola (horizontal axis)	A curve with equation $x = a(y - k)^2 + h$ or $x = ay^2 + by + c$	$x = y^2$, $x = 3(y - 1)^2 + 2$, and $x = -4(y + 1)^2 - 3$
9.2A	Distance formula	The distance between (x_1, y_1) and (x_2, y_2) is $d = \sqrt{(x_2 - x_1)^2 + (y_2 - y_1)^2}$	For $(x_1, y_1) = (3, 4)$ and $(x_2, y_2) = (-3, 12)$ $d = \sqrt{(-3 - 3)^2 + (12 - 4)^2}$ $= \sqrt{36 + 64}$ $= \sqrt{100}$ $= 10$
9.2B	Circle centered at (h, k) with radius r	A curve with the equation $(x - h)^2 + (y - k)^2 = r^2$	$(x - 3)^2 + (y + 4)^2 = 9$ is a circle with center at $(3, -4)$ and radius 3.

(Continued)

SECTION	ITEM	MEANING	EXAMPLE
9.2D	Ellipse centered at the origin	A curve with equation $\frac{x^2}{a^2} + \frac{y^2}{b^2} = 1$ or $\frac{x^2}{b^2} + \frac{y^2}{a^2} = 1$, where $a^2 > b^2$	$\frac{x^2}{16} + \frac{y^2}{9} = 1$ is an ellipse.
9.3A	Hyperbola centered at the origin	A curve with equation $\frac{x^2}{a^2} - \frac{y^2}{b^2} = 1$ or $\frac{y^2}{a^2} - \frac{x^2}{b^2} = 1$, where a^2 is always the denominator of the positive variable.	$\frac{x^2}{9} - \frac{y^2}{16} = 1$ is a hyperbola.
	Asymptotes	Lines associated with a hyperbola, going through the opposite corners of the rectangle whose sides pass through $\pm a$ and $\pm b$	The asymptotes for $\frac{x^2}{a^2} - \frac{y^2}{b^2} = 1$ are shown
9.4A	Nonlinear systems	A system of equations containing at least one second-degree equation	$x^2 + y^2 = 9$ $x - y = 3$ is a nonlinear system.
	Substitution method	A method of solving nonlinear systems in which substitution is made from one of the equations into the other	To solve $x^2 + y^2 = 9$ $x = y + 3$ by substitution, replace x by $y + 3$ in $x^2 + y^2 = 9$.
9.5	Second-degree inequality	An inequality containing at least one second-degree term	$y \le x^2 + 2$ and $x^2 + y^2 > 9$ are second-degree inequalities.

Review Exercises

(If you need help with these exercises, look in the section indicated in brackets.)

1. [9.1A] Graph.

a. $y = 9x^2$

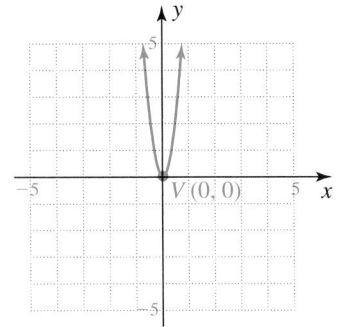

b. $y = -9x^2$

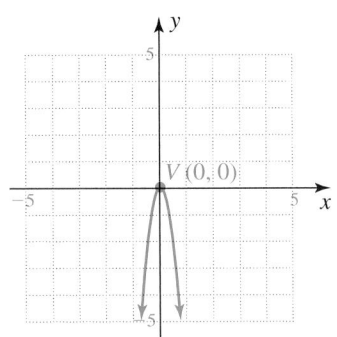

2. [9.1B] Find the vertex and graph.

a. $y = (x - 1)^2 - 2$ $V(1, -2)$

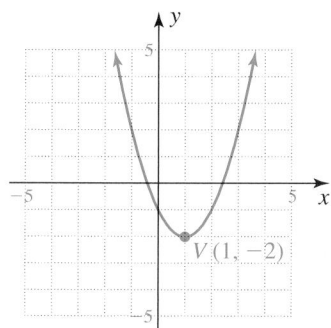

b. $y = -(x - 1)^2 + 2$ $V(1, 2)$

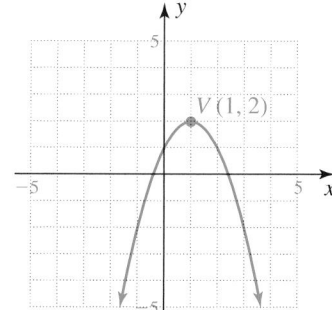

3. [9.1C] Find the vertex and graph.

a. $y = x^2 - 4x + 2$ $V(2, -2)$

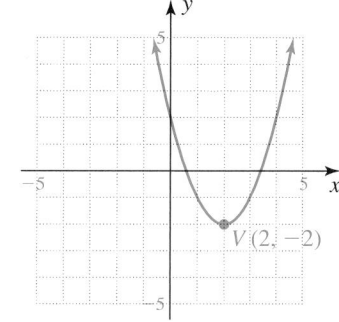

b. $y = -x^2 + 6x - 5$ $V(3, 4)$

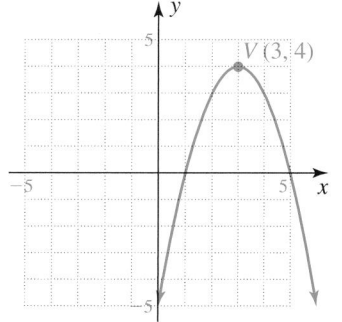

4. [9.1C] Find the vertex and graph.

a. $y = 2x^2 - 4x + 3$ $V(1, 1)$

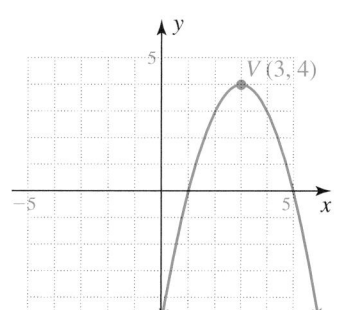

b. $y = -2x^2 + 4x - 5$ $V(1, -3)$

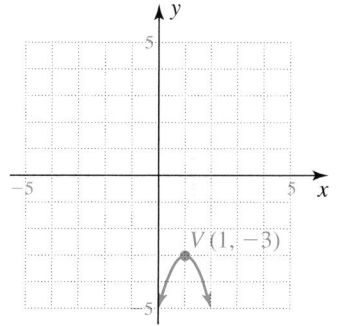

5. [9.1D] Find the vertex and graph.

a. $x = 2(y - 2)^2 - 2$ $V(-2, 2)$

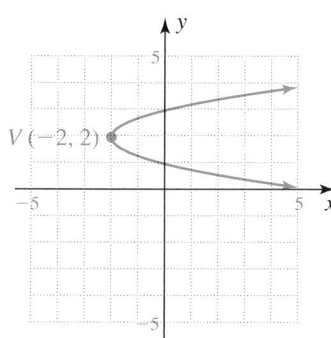

b. $x = -2(y - 3)^2 + 1$ $V(1, 3)$

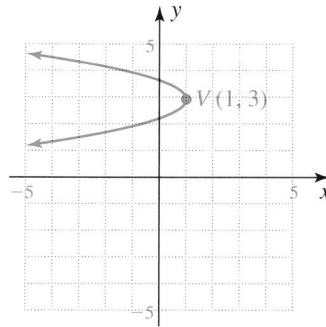

6. [9.1D] Find the vertex and graph.

a. $x = y^2 - 4y + 1$ $V(-3, 2)$

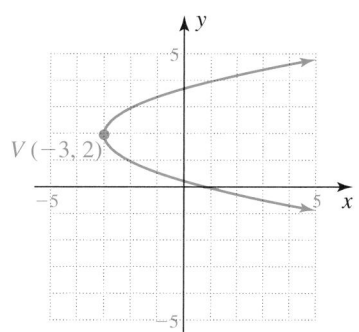

b. $x = y^2 - 2y + 3$ $V(2, 1)$

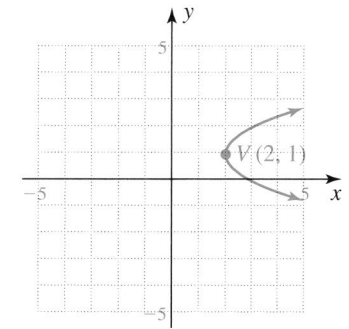

7. [9.1E] Find the value of x that gives the maximum revenue R if

a. $R = 20x - 0.01x^2$ $x = 1000$

b. $R = 10x - 0.02x^2$ $x = 250$

8. [9.1E] If a ball is thrown upward at 24 ft/sec, its height h, in feet, after t seconds is given by $h = -16t^2 + 24t$. How many seconds does it take for the ball to reach its maximum height, and what is this height? $\frac{3}{4}$ sec; 9 ft

9. [9.2A] Find the distance between each pair of points.

a. (5, -3) and (2, 8) $\sqrt{130}$

b. (10, 10) and (2, -4) $2\sqrt{65}$

c. (-3, 3) and (-3, 8) 5

10. [9.2B] Find an equation of the circle of radius 3 and with center at

a. (-2, 2)

$(x + 2)^2 + (y - 2)^2 = 9$

b. (3, -2)

$(x - 3)^2 + (y + 2)^2 = 9$

11. [9.2C] Find the center and the radius and sketch the graph.

a. $(x + 2)^2 + (y - 1)^2 = 4$

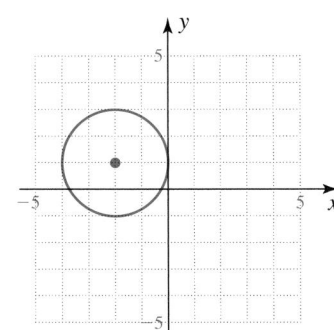

$C(-2, 1); r = 2$

b. $(x - 1)^2 + (y + 2)^2 = 9$

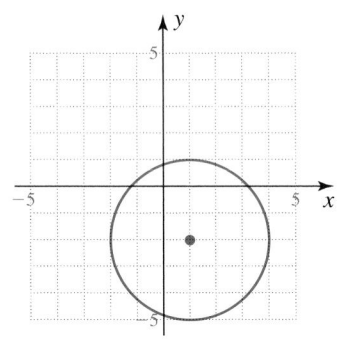

$C(1, -2); r = 3$

12. [9.2C] Sketch the graph.

a. $x^2 + y^2 = 4$

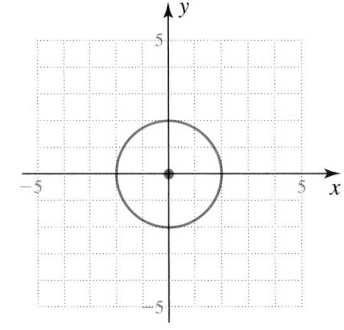

b. $x^2 + y^2 = 25$

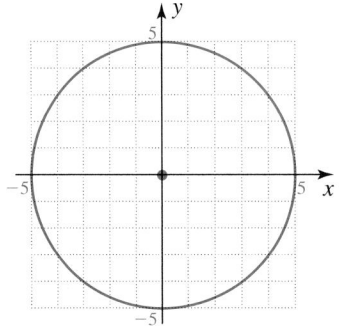

13. [9.2C] Find the center and the radius and sketch the graph.

a. $x^2 + y^2 + 2x + 2y - 2 = 0$

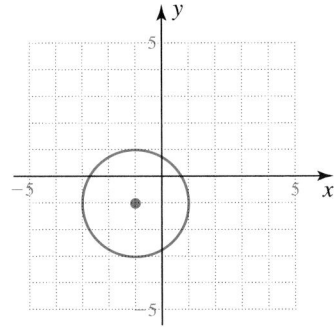

$C(-1, -1); r = 2$

b. $x^2 + y^2 - 4x + 6y + 9 = 0$

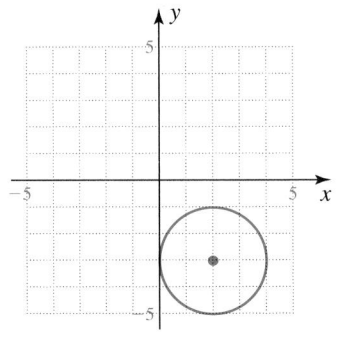

$C(2, -3); r = 2$

14. [9.2D] Graph.

a. $4x^2 + 9y^2 = 36$

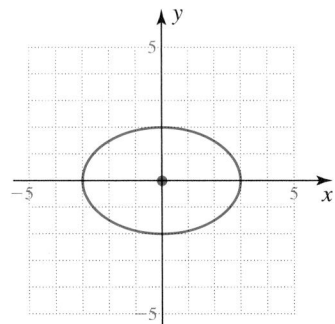

b. $9x^2 + y^2 = 9$

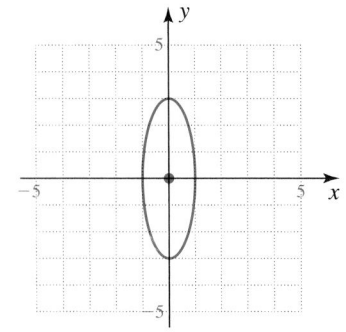

15. [9.2D] Graph.

a. $\dfrac{(x - 1)^2}{4} + \dfrac{(y - 2)^2}{9} = 1$

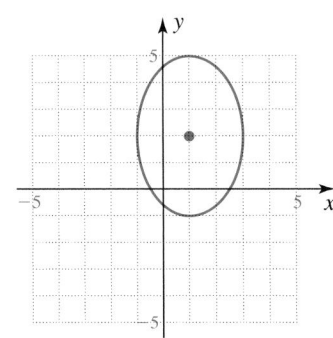

b. $\dfrac{(x + 2)^2}{9} + \dfrac{(y - 2)^2}{4} = 1$

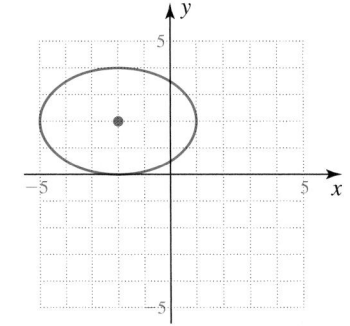

16. [9.3A] Graph.

a. $\dfrac{x^2}{9} - \dfrac{y^2}{16} = 1$

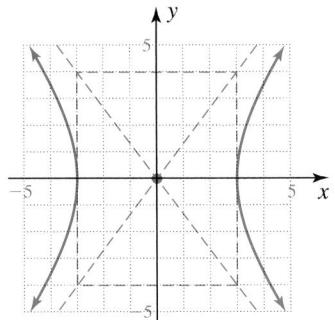

b. $\dfrac{x^2}{16} - \dfrac{y^2}{9} = 1$

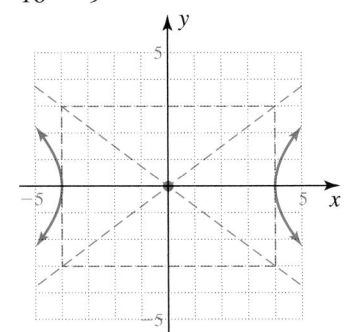

17. [9.3A] Graph.

a. $\dfrac{y^2}{9} - \dfrac{x^2}{16} = 1$

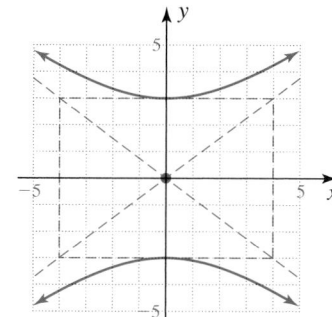

b. $\dfrac{y^2}{16} - \dfrac{x^2}{9} = 1$

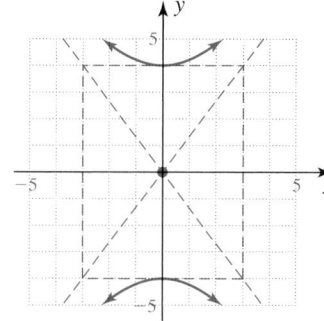

18. [9.3B] Identify each of the curves.

a. $x = 1 - y^2$ Parabola

b. $x^2 = 4y^2 - 4$ Hyperbola

c. $y^2 = 9 - x^2$ Circle

d. $4y^2 = 36 - 9x^2$ Ellipse

19. [9.4A] Solve the system by the substitution method.

a. $x^2 + y^2 = 1$

$x + y = 1$ $(0, 1), (1, 0)$

b. $x^2 + y^2 = 10$

$x + y = 4$ $(1, 3), (3, 1)$

20. [9.4A] Solve the system by the substitution method.

a. $x^2 - y^2 = 16$

$2x = y$ $\left(\dfrac{4\sqrt{3}}{3}i, \dfrac{8\sqrt{3}}{3}i\right), \left(-\dfrac{4\sqrt{3}}{3}i, -\dfrac{8\sqrt{3}}{3}i\right)$

b. $x^2 + y^2 = 4$

$x + y = 3$ $\left(\dfrac{3}{2} + \dfrac{1}{2}i, \dfrac{3}{2} - \dfrac{1}{2}i\right), \left(\dfrac{3}{2} - \dfrac{1}{2}i, \dfrac{3}{2} + \dfrac{1}{2}i\right)$

21. [9.4B] Solve the system.

a. $x^2 - y^2 = 5$

$x^2 + 2y^2 = 17$ $(3, 2), (3, -2), (-3, 2), (-3, -2)$

b. $x^2 - y^2 = 3$

$2x^2 + y^2 = 9$ $(2, 1), (2, -1), (-2, 1), (-2, -1)$

22. [9.4C]

a. The cost C of manufacturing and selling x units of a product is $C = 10x + 400$, and the corresponding revenue R is $R = x^2 - 200$. Find the break-even point.

$x = 30$

b. Repeat part **a** if $C = 6x + 80$ and $R = x^2 - 200$.

$x = 20$

23. [9.5A] Graph the inequality.

a. $y \le 1 - x^2$

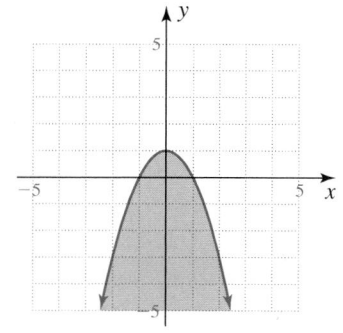

b. $x \le 4 - y^2$

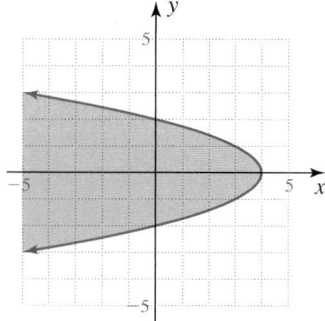

24. [9.5A] Graph the inequality.

a. $x^2 + y^2 \le 4$

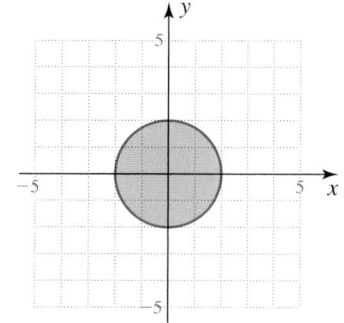

b. $x^2 + y^2 > 9$

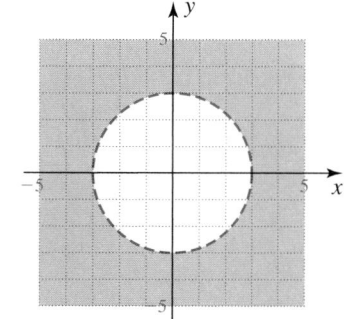

25. [9.5A] Graph the inequality.

a. $4x^2 - y^2 \leq 4$

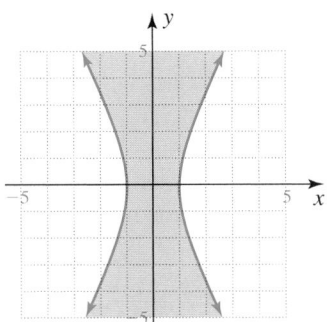

b. $x^2 - 4y^2 \leq 4$

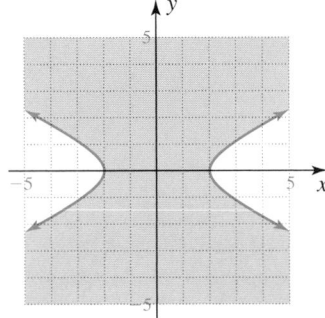

26. [9.5B] Graph the system.

a. $x^2 + y^2 \leq 4$ and $y \leq 2 - x^2$

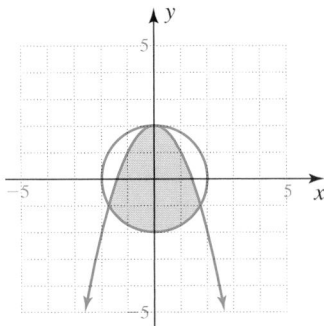

b. $x^2 + y^2 \leq 4$ and $y \geq 4x^2$

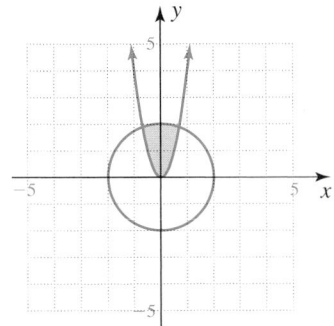

Practice Test 9

(Answers on pages 725–729)

1. Graph the parabola $y = -x^2 - 4$.

2. Find the vertex and graph the parabola $y = (x - 2)^2 + 2$.

3. Find the vertex and graph the parabola $y = -x^2 - 4x - 1$.

4. Find the vertex and graph the parabola $y = -2x^2 + 4x + 1$.

5. Find the vertex and graph the parabola $x = 2(y + 1)^2 - 1$.

6. Find the vertex and graph the parabola $x = y^2 + 2y - 1$.

7. If the revenue is given by $R = 60x - 0.03x^2$, find the value of x that yields the maximum revenue.

8. Find an equation of the circle of radius 2 with its center at $(1, -2)$.

9. Find an equation of the circle of radius $\sqrt{3}$ with its center at the origin, $(0, 0)$.

10. Find the center and the radius and sketch the graph of $(x + 1)^2 + (y - 2)^2 = 9$.

11. Sketch the graph of $x^2 + y^2 = 16$.

12. Find the center and the radius and sketch the graph of $x^2 + y^2 + 4x - 2y - 4 = 0$.

13. Graph $x^2 + 9y^2 = 9$.

14. Graph $\dfrac{(x - 2)^2}{4} + \dfrac{(y + 1)^2}{9} = 1$.

15. Graph $\dfrac{y^2}{9} - \dfrac{x^2}{25} = 1$.

16. Graph $\dfrac{x^2}{9} - \dfrac{y^2}{25} = 1$.

17. Identify each of the following curves.
   a. $x = y^2 - 4$ **b.** $16y^2 = 144 - 9x^2$
   c. $9y^2 = 144 + 16x^2$ **d.** $x^2 = 16 - y^2$

18. Use the substitution method to solve the system.
$$x^2 + y^2 = 4$$
$$x + y = 2$$

19. Use the substitution method to solve the system.
$$x^2 + y^2 = 1$$
$$x + y = 2$$

20. Solve the system.
$$x^2 + y^2 = 20$$
$$x^2 - 2y^2 = 8$$

21. The cost C of manufacturing and selling x units of a product is $C = 20x + 50$, and the corresponding revenue R is $R = x^2 - 75$. Find the break-even value of x.

22. Graph the inequality $y \le -x^2 - 1$.

23. Graph the inequality $x^2 + y^2 < 9$.

24. Graph the inequality $y^2 - 4x^2 \le 4$.

25. Graph the system $x^2 + y^2 \le 4$ and $y \le 2 - x^2$.

Answers to Practice Test

ANSWER	IF YOU MISSED	REVIEW		
	QUESTION	SECTION	EXAMPLES	PAGE
1.	1	9.1A	1, 2, 3	649–651
2. Vertex (2, 2)	2	9.1B	4	652

ANSWER	IF YOU MISSED	REVIEW		
	QUESTION	SECTION	EXAMPLES	PAGE
3. Vertex $(-2, 3)$	3	9.1C	5, 6	654–657

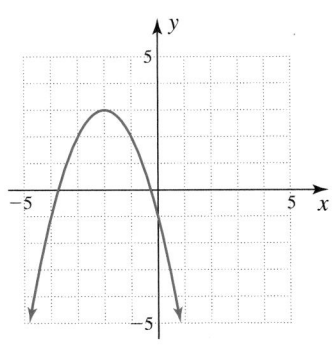

4. Vertex $(1, 3)$	4	9.1C	5, 6	654–657

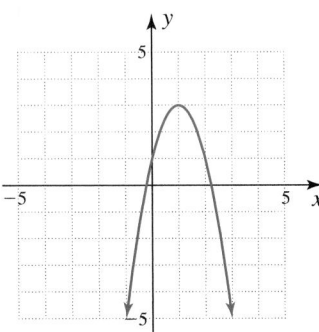

5. Vertex $(-1, -1)$	5	9.1D	7	658

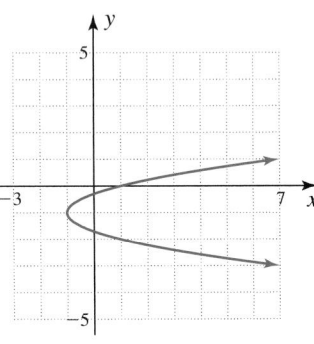

6. Vertex $(-2, -1)$	6	9.1D	8	658

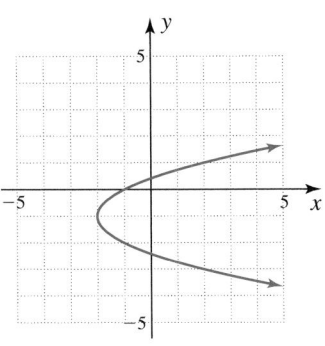

7. $x = 1000$	7	9.1E	9	659

ANSWER	IF YOU MISSED	REVIEW		
	QUESTION	SECTION	EXAMPLES	PAGE
8. $(x - 1)^2 + (y + 2)^2 = 4$	8	9.2B	2	671
9. $x^2 + y^2 = 3$	9	9.2B	3	671
10. Center $(-1, 2)$; $r = 3$	10	9.2C	4	672

11.	11	9.2C	5	672

12. Center $(-2, 1)$; $r = 3$	12	9.2C	6	673

13.	13	9.2D	7	674–675

ANSWER	IF YOU MISSED	REVIEW		
	QUESTION	SECTION	EXAMPLES	PAGE
14. 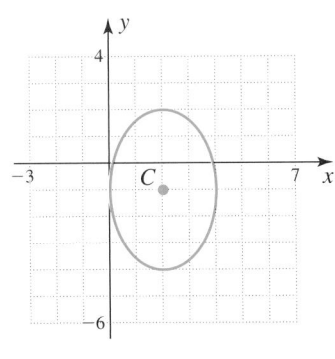	14	9.2D	8	675–676
15. 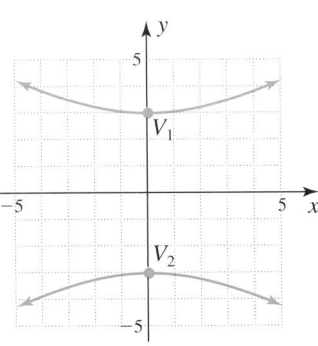	15	9.3A	1a	688
16. 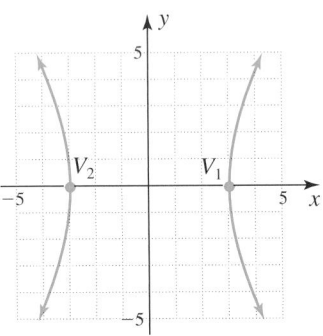	16	9.3A	1b	688–689
17. a. Parabola **b.** Ellipse **c.** Hyperbola **d.** Circle	17	9.3B	3	692
18. $(2, 0)$ and $(0, 2)$	18	9.4A	1	700–701
19. $\left(1 + \dfrac{\sqrt{2}}{2}i, 1 - \dfrac{\sqrt{2}}{2}i\right),$ $\left(1 - \dfrac{\sqrt{2}}{2}i, 1 + \dfrac{\sqrt{2}}{2}i\right)$	19	9.4A	2	701–702
20. $(4, 2), (4, -2), (-4, 2), (-4, -2)$	20	9.4B	3	702
21. $x = 25$	21	9.4C	4	703

ANSWER	IF YOU MISSED	REVIEW		
	QUESTION	SECTION	EXAMPLES	PAGE
22.	22	9.5A	1	710–711
23.	23	9.5A	2	711
24.	24	9.5A	3	711–712
25.	25	9.5B	4	712

Cumulative Review Chapters 1–9

1. Graph the additive inverse of $\frac{4}{3}$ on a number line.

2. Perform the indicated operation and simplify: $\dfrac{45x^8}{15x^{-7}} \quad 3x^{15}$

3. Evaluate: $-4^3 + \dfrac{12 - 8}{2} + 15 \div 3 \quad -57$

4. Solve: $\left|\frac{3}{2}x + 7\right| + 3 = 8 \quad x = -\frac{4}{3} \text{ or } -8$

5. Graph: $x + 1 \le 4$ and $-2x < 8$

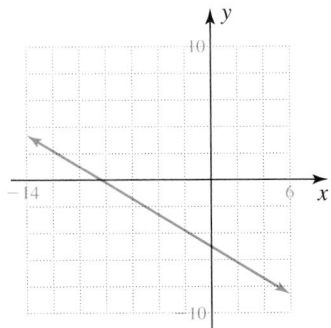

6. Solve for A in $B = \frac{7}{8}(A - 10)$. $\quad A = \dfrac{8B + 70}{7}$

7. A woman's salary was increased by 10% to $31,900. What was her salary before the increase? $29,000

8. Find the x- and y-intercepts of $y = -2x - 5$.
x-intercept: $\left(-\frac{5}{2}, 0\right)$; y-intercept: $(0, -5)$

9. A line goes through the point $(-5, -2)$ and has slope $-\frac{1}{2}$. Graph this line.

10. Graph: $2x - 5y \le -10$

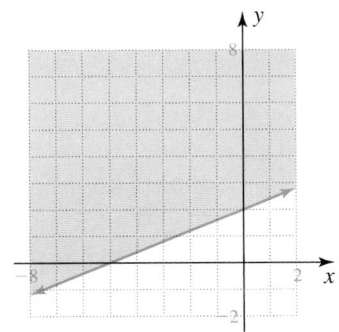

11. Find the slope of the line passing through the points $A(3, 6)$ and $B(7, 3)$. $\quad -\frac{3}{4}$

12. Find the domain of the relation $\{(-2, -3), (5, -2), (-3, 1)\}$. $\quad D = \{-2, 5, -3\}$

13. Find $f(3)$ given $f(x) = -2x + 2$. $\quad -4$

14. Solve by the elimination method:
$$5x + 2y = 14$$
$$10x + 4y = 4 \quad \text{No solution}$$

15. Solve the system:
$$x + y + z = 4$$
$$-x + 2y + z = 3$$
$$2x + y + z = 6 \quad (2, 3, -1)$$

16. Evaluate: $\begin{vmatrix} 5 & 2 & -4 \\ 5 & 5 & -1 \\ 4 & 4 & 5 \end{vmatrix} \quad 87$

17. A motorboat can go 12 mi downstream on a river in 20 min. It takes 30 min for this boat to go back upstream the same 12 mi. Find the speed of the boat.
30 mi/hr

18. Multiply: $(3j + 1)(3j - 1) \quad 9j^2 - 1$

19. Factor: $12x^6 - 3x^4 + 8x^3 - 2x$
$x(3x^3 + 2)(2x + 1)(2x - 1)$

20. Factor: $32x^2y^2 - 48xy^3 + 18y^4 \quad 2y^2(4x - 3y)^2$

21. Solve for x: $2x^2 + 9x = -9 \quad x = -\frac{3}{2}, -3$

22. Multiply: $\dfrac{x + 1}{3x + y} \cdot \dfrac{9x^2 - y^2}{2x^2 - x - 3} \quad \dfrac{3x - y}{2x - 3}$

23. Solve for x: $\dfrac{x}{x+7} - \dfrac{x}{x-7} = \dfrac{x^2+49}{x^2-49}$ No solution

24. Janet can paint a kitchen in 5 hr and Charles can paint the same kitchen in 4 hr. How long would it take for both working together to paint the kitchen? $2\frac{2}{9}$ hr

25. An enclosed gas exerts a pressure P on the walls of the container. This pressure is directly proportional to the temperature T of the gas. If the pressure is 6 lb/in.^{2} when the temperature is 360°F, find k. $\;k = \frac{1}{60}$

26. Assume x is positive and simplify: $x^{2/3} \cdot x^{5/2}$ $\;x^{19/6}$

27. Reduce the order (index) of: $\sqrt[6]{16c^4d^4}$ $\;\sqrt[3]{4c^2d^2}$

28. Find the product: $(\sqrt{125} + \sqrt{18})(\sqrt{20} + \sqrt{50})$
$80 + 31\sqrt{10}$

29. Solve over the real numbers: $\sqrt{x+3} = -2$
No real-number solution

30. Find: $\dfrac{1+2i}{9+4i}$ $\;\dfrac{17}{97} + \dfrac{14}{97}i$

31. Solve for x: $7x^2 + 42x + 63 = 11$ $\;x = \dfrac{-21 \pm \sqrt{77}}{7}$

32. Solve for x: $27x^3 - 64 = 0$ $\;x = \frac{4}{3}, -\frac{2}{3} \pm \frac{2\sqrt{3}}{3}i$

33. Find the sum and the product of the roots of the equation $4x^2 - 5x - 4 = 0$. Sum: $\frac{5}{4}$; product: -1

34. Solve for x: $x^2 + x \geq 56$ $\;x \leq -8$ or $x \geq 7$

35. Graph the parabola and name the vertex:
$y = -3x^2 - 4x + 3$

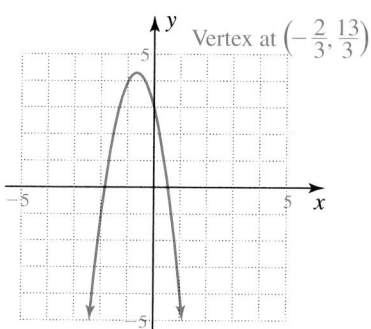

Vertex at $\left(-\frac{2}{3}, \frac{13}{3}\right)$

36. If the revenue is given by $R = 20x - 0.02x^2$, find the value of x that yields the maximum revenue. $\;x = 500$

37. Find an equation of the circle of radius 9 and with its center at the origin. $\;x^2 + y^2 = 81$

38. Graph and name the center: $\dfrac{(x-2)^2}{25} + \dfrac{(y+4)^2}{16} = 1$

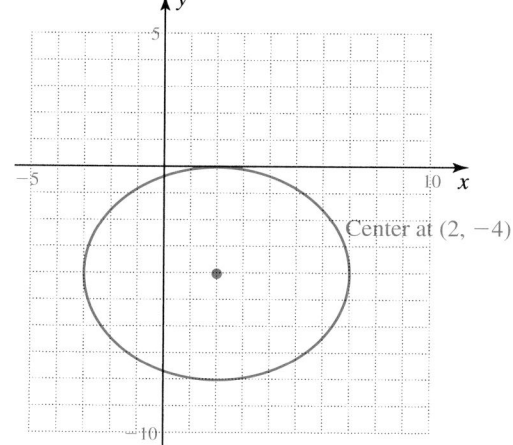

Center at $(2, -4)$

39. Use the substitution method to solve the system:
$$x^2 + y^2 = 25$$
$$x + y = 1 \quad (-3, 4), (4, -3)$$

40. Graph the inequality: $x^2 + y^2 < 4$

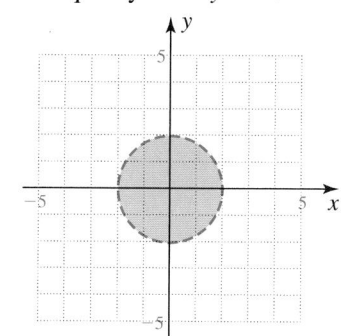

Functions—Inverse, Exponential, and Logarithmic

10

John Napier

The Human Side of Algebra

Who invented logarithms? It's sometimes implied that the invention of logarithms was the work of one man: John Napier. Interestingly enough, Napier was not a professional mathematician but a Scottish laird (landowner) and Baron of Murchiston who dabbled in many controversial topics. (In a commentary on the Book of Revelations, he argued that the pope at Rome was the Antichrist!) Napier claimed that he worked on the invention of logarithms for 20 years before he published his *Mirifici logarithmorum canonis descriptio* (*A description of the Marvelous Rule of Logarithms*) in 1614.

In 1617, the year of Napier's death, Henry Briggs constructed the first table of Briggsian or common logarithms. But there is one more participant in the invention of logarithms. Jobst Bürgi of Switzerland independently developed the idea of logarithms as early as 1588. Unfortunately, his results were not published until 1620. In his scheme, Bürgi used "red" and "black" numbers, with the "red" numbers appearing on the side of the page and "black" numbers in the body of the table, thus creating what could now be described as an antilogarithm table.

Pretest for Chapter 10

(Answers on pages 735–736)

1. Let $f(x) = 1 - x^2$ and $g(x) = 1 - x$. Find the following.

   a. $(f + g)(x)$ **b.** $(f - g)(x)$

   c. $(fg)(x)$ **d.** $(\frac{f}{g})(x)$

2. $f(x) = 5x + 1$, find

$$\frac{f(x) - f(a)}{x - a}, \text{ where } x \neq a$$

3. If $f(x) = x^3$ and $g(x) = 1 + x$, find:

   a. $(g \circ f)\,(2)$ **b.** $(f \circ g)\,(x)$ **c.** $(g \circ f)(x)$

4. If $f(x) = \frac{-8}{x}$ and $g(x) = \frac{x+3}{x-2}$, find the domain of $f + g$, $f - g$, and fg.

5. Find the domain of $\frac{f}{g}$ if $f(x) = \frac{3}{x-2}$ and $g(x) = \frac{4}{x+2}$.

6. Let $S = \{(4, 2), (6, 4), (9, 7)\}$ and find

   a. The domain and range of S **b.** S^{-1}

   c. The domain and range of S^{-1}

   d. The graph of S and S^{-1}

7. Let $f(x) = y = 3x + 6$

   a. Find $f^{-1}(x)$.   **b.** Graph f and its inverse.

8. Find the inverse of $f(x) = y = 5x^2$. Is the inverse a function?

9. Graph $y = 5^x$.

   a. Is the inverse a function?

   b. Is $y = 5^x$ increasing or decreasing?

10. A radioactive substance decays so that the number of grams present after t years is

$$G = 500e^{-1.4t}$$

Find, to the nearest gram, the amount of the substance present

   a. At the start.   **b.** In 2 yr.

11. Graph on the same coordinate axes.

   a. $f(x) = 10^x$ **b.** $g(x) = \log_{10} x$

12. Write the equation

   a. $64 = 4^x$ in logarithmic form.

   b. $\log_9 81 = x$ in exponential form.

13. Solve.

   a. $\log_4 x = -2$ **b.** $\log_x 64 = 2$

14. Use the properties of logarithms to show that $\log(\frac{x^2}{15}) = 2 \log x - \log 15$.

15. Use the properties of logarithms to show that $\log \sqrt{ab} = \frac{1}{2} \log a + \frac{1}{2} \log b$.

16. Find.

   a. log 325 (round to 4 decimal digits)

   b. inv log 3.5502 (round to nearest integer)

17. Find.

   a. ln 325 (round to 4 decimal digits)

   b. inv ln 1.1618 (round to 4 decimal digits)

18. a. Use the change-of-base formula to fill in the blank:

     $\log_5 20 = $ _____

   b. Use the result of part **a** to find a numerical approximation for $\log_5 20$ (round to 4 decimal digits).

19. Graph.

   a. $f(x) = e^{2x}$ **b.** $g(x) = -e^{2x}$

20. Graph.

   a. $f(x) = \ln(x + 2)$ **b.** $g(x) = \ln x + 2$

21. Solve.

   a. $3^{2x+1} = 27$ **b.** $32^{x-2} = 2^{3x+2}$

22. Solve.

   a. $4^x = 3$ (round to 4 decimal digits)

   b. $40 = e^{0.30k}$ (round to 1 decimal digit)

23. Solve.

   a. $\log(x - 4) + \log(x + 5) = 1$

   b. $\log_2(x + 3) - \log_2(x + 1) = 1$

24. The compound amount with continuous compounding is given by $A = Pe^{rt}$, where P is the principal, r is the interest rate, and t is the time in years. If the rate is 6%, find how long it takes for the money to double—for A to equal $2P$ (use 0.69315 for ln 2).

25. A radioactive substance decays so that the amount A present at time t (years) is $A = A_0 e^{-0.4t}$. Find the half-life (time for half to decay) of this substance (use -0.69315 for ln 0.5).

Answers to Pretest

ANSWER	IF YOU MISSED	REVIEW		
	QUESTION	SECTION	EXAMPLES	PAGE
1. a. $-x^2 - x + 2$ **b.** $-x^2 + x$ **c.** $x^3 - x^2 - x + 1$ **d.** $1 + x, x \neq 1$	1	10.1	1	738
2. $5, x \neq a$	2	10.1	2	739
3. a. 9 **b.** $(1 + x)^3$ **c.** $1 + x^3$	3	10.1	3	740
4. $\{x \mid x$ is a real number and $x \neq 0$ and $x \neq 2\}$	4	10.1	4	741
5. $\{x \mid x$ is a real number and $x \neq -2$ and $x \neq 2\}$	5	10.1	5, 6	741–742
6. a. $D = \{4, 6, 9\}; R = \{2, 4, 7\}$ **b.** $\{(2, 4), (4, 6), (7, 9)\}$ **c.** $D = \{2, 4, 7\}; R = \{4, 6, 9\}$ **d.**	6	10.2	1	751
7. a. $f^{-1}(x) = \frac{x - 6}{3} = \frac{1}{3}x - 2$ **b.**	7	10.2	2	753
8. $y = \pm\sqrt{\frac{x}{5}}$ or $\pm\frac{\sqrt{5x}}{5}$; no	8	10.2	3	755
9. **a.** Yes **b.** Increasing	9	10.3	1, 2	765–767

ANSWER	IF YOU MISSED	REVIEW		
	QUESTION	SECTION	EXAMPLES	PAGE
10. a. 500 g **b.** 30 g	10	10.3	4	769
11. (graph: $f(x) = 10^x$, $g(x) = \log_{10} x$)	11	10.4	1	776
12. a. $\log_4 64 = x$ **b.** $9^x = 81$	12	10.4	2, 3	777
13. a. $\frac{1}{16}$ **b.** 8	13	10.4	4	778
14. $\log\left(\frac{x^2}{15}\right) = 2\log x - \log 15$ $= \log x^2 - \log 15$ $= \log\left(\frac{x^2}{15}\right)$	14	10.4	6	780
15. $\log\sqrt{ab} = \frac{1}{2}\log a + \frac{1}{2}\log b$ $= \log a^{1/2} + \log b^{1/2}$ $= \log\sqrt{a} + \log\sqrt{b}$ $= \log\sqrt{ab}$	15	10.4	7	780
16. a. 2.5119 **b.** 3550	16	10.5	1, 2	788–789
17. a. 5.7838 **b.** 3.1957	17	10.5	3, 4	789
18. a. $\frac{\log 20}{\log 5}$ or $\frac{\ln 20}{\ln 5}$ **b.** 1.8614	18	10.5	5	790
19. (graph: $f(x) = e^{(2x)}$, $g(x) = -e^{(2x)}$)	19	10.5	7	792–793
20. (graph: $f(x) = \ln(x + 2)$, $g(x) = \ln x + 2$)	20	10.5	8	793–794
21. a. 1 **b.** 6	21	10.6	1	803
22. a. $\frac{\log 3}{\log 4} \approx 0.7925$ **b.** 12.3	22	10.6	2, 3	804–805
23. a. 5 **b.** 1	23	10.6	6	807
24. About 11.6 yr	24	10.6	7	808
25. About 1.7 yr	25	10.6	10	810

10.1 THE ALGEBRA OF FUNCTIONS

To Succeed, Review How To . . .

1. Evaluate expressions (pp. 46–49).

2. Add, subtract, multiply, and divide polynomials (pp. 348–350, 356–362).

Objectives

A Find the sum, difference, product, and quotient of two functions.

B Find the composite of two functions.

C Find the domain of $(f + g)(x)$, $(f - g)(x)$, $(fg)(x)$, and $(\frac{f}{g})(x)$.

D Solve an application.

GETTING STARTED

A Lot of Garbage!

How many pounds of garbage does each person generate each day? The average is more than 4 lb and growing! Some of this waste is recovered and recycled into other products. The amount of solid waste recovered (in millions of tons) is a **function** of time and can be approximated by

$$R(t) = 0.04t^2 - 0.59t + 7.42$$

where t is the number of years after 1960.

On the other hand, the amount of solids *not* recovered can be approximated by

$$N(t) = 2.93t + 82.58$$

Can we find the total amount of solid waste generated? This amount is the **sum** of $R(t)$ and $N(t)$,

$$R(t) + N(t) = (0.04t^2 - 0.59t + 7.42) + (2.93t + 82.58)$$
$$= 0.04t^2 + 2.34t + 90$$

The items recovered most often are paper and paperboard. The amount of paper and paperboard is also a function of time and can be approximated by $P(t) = 0.02t^2 - 0.25t + 6$. You may be surprised to learn what fraction of the total recovered waste $R(t)$ is actually paper and paperboard. That fraction is the **quotient:**

$$\frac{P(t)}{R(t)}$$

Now suppose you want to know what this fraction was in 1990. Since 1990 is 30 $(1990 - 1960)$ years after 1960, this fraction for 1990 is

$$\frac{P(t)}{R(t)} = \frac{0.02(30)^2 - 0.25(30) + 6}{0.04(30)^2 - 0.59(30) + 7.42} = \frac{16.5}{25.72} = 0.64$$

Thus 64% of the total amount of recovered waste in 1990 was paper and paperboard. We can perform the fundamental operations of addition, subtraction, multiplication, and division using functions. We shall study such operations in this section.

 **Using Operations with Functions**

Here are the definitions we need to add, subtract, multiply, and divide functions.

OPERATIONS WITH FUNCTIONS

If f and g are functions and x is in the domain of both functions, then

$$(f + g)(x) = f(x) + g(x)$$ The sum of f and g

$$(f - g)(x) = f(x) - g(x)$$ The difference of f and g

$$(fg)(x) = f(x) \cdot g(x)$$ The product of f and g

$$\left(\frac{f}{g}\right)(x) = \frac{f(x)}{g(x)}, \quad g(x) \neq 0$$ The quotient of f and g

---

EXAMPLE 1   **Performing operations with functions**

If $f(x) = x^2 + 4$ and $g(x) = x + 2$, find:

a. $(f + g)(x)$ **b.** $(f - g)(x)$ **c.** $(fg)(x)$ **d.** $\left(\frac{f}{g}\right)(x)$

SOLUTION

a. $(f + g)(x) = \quad f(x) \quad + \quad g(x)$
$$\downarrow \qquad \downarrow$$
$$= (x^2 + 4) + (x + 2)$$
$$= x^2 + x + 6 \qquad \text{Simplify.}$$

b. $(f - g)(x) = \quad f(x) \quad - \quad g(x)$
$$\downarrow \qquad \downarrow$$
$$= (x^2 + 4) - (x + 2) \qquad \text{Recall that } -(a + b) = -a - b.$$
$$= x^2 + 4 - x - 2 \qquad \text{Simplify.}$$
$$= x^2 - x + 2$$

c. $(fg)(x) = \quad f(x) \quad \cdot \quad g(x)$
$$\downarrow \qquad \downarrow$$
$$= (x^2 + 4)(x + 2) \qquad \text{Multiply (use FOIL method).}$$
$$= x^3 + 2x^2 + 4x + 8$$

d. $\left(\frac{f}{g}\right)(x) = \frac{f(x)}{g(x)}$
$$= \frac{x^2 + 4}{x + 2}$$

The denominator can't be zero, so $x + 2 \neq 0$, that is, $x \neq -2$.

We shall talk about domains of quotients later in this section.

PROBLEM 1

If $f(x) = x^2 - 9$ and $g(x) = x - 3$, find:

a. $(f + g)(x)$

b. $(f - g)(x)$

c. $(fg)(x)$

d. $\left(\frac{f}{g}\right)(x)$

Teaching Tip

In Example 1(a), have students find $f(3)$, $g(3)$, and $f(3) + g(3)$. Then have them use $(f + g)(x)$ to find $(f + g)(3)$. Are the results the same? In general, is it easier to find $f(x) + g(x)$ or $(f + g)(x)$?

Answers

1. **a.** $x^2 + x - 12$ **b.** $x^2 - x - 6$
c. $x^3 - 3x^2 - 9x + 27$
d. $x + 3, x \neq 3$

Web It

For a tutorial practicing the operations with functions, go to link 10-1-1 at mhhe.com/bello.

EXAMPLE 2 **More practice in performing operations with functions**

If $f(x) = 3x + 1$, find:

$$\frac{f(x) - f(a)}{x - a}, \quad x \neq a$$

SOLUTION Since $f(x) = 3x + 1$, and $f(a) = 3a + 1$, we have

$$\frac{f(x) - f(a)}{x - a} = \frac{(3x + 1) - (3a + 1)}{x - a}$$

$$= \frac{3x + 1 - 3a - 1}{x - a} \qquad \text{Simplify.}$$

$$= \frac{3x - 3a}{x - a}$$

$$= \frac{3(x - a)}{x - a} \qquad \text{Factor.}$$

$$= 3, \quad x \neq a$$

PROBLEM 2

If $f(x) = 2x - 3$, find:

$$\frac{f(x) - f(a)}{x - a}, \quad x \neq a$$

B Finding Composite Functions

There are many instances in which some quantity depends on a variable that, in turn, depends on another variable. For instance, the tax you pay on your house depends on the assessed value and the millage rate. (A 1-mill rate means that you pay $1 for each $1000 of assessed value.) Many states offer a Homestead Exemption, exempting a certain amount of the house value from taxes. In Florida, this exemption is $25,000. Thus the function g giving the correspondence between the assessed value x of a home in Florida and its taxable value is $g(x) = x - 25,000$.

A house assessed at $150,000 will have a taxable value given by $g(150,000) = 150,000 - 25,000 = \$125,000$. When the tax rate is 5 mills, the function f that computes the tax on the house is

$$f(V) = \frac{5}{1000} \cdot V = 0.005V$$

where V is the taxable value of the house. Thus a house valued at $150,000 with a $25,000 Homestead Exemption will pay $0.005 \cdot (150,000 - 25,000) = 0.005 \cdot 125,000 = \$625$ in taxes. If you look at the table, you will see that a house assessed at $150,000 should indeed pay $625 in taxes. Can you find a function h that will find the tax? If you guessed $h(x) = 0.005(x - 25,000)$, you guessed correctly.

$$g(x) = x - 25,000 \qquad f(x) = 0.005g(x)$$

Assessed Value (A)	Taxable Value (V)	Tax T	
$50,000	$25,000	0.005(25,000)	= $125
$100,000	$75,000	0.005(75,000)	= $375
$150,000	$125,000	0.005(125,000)	= $625

$h(x)$

Answer

2. $2, \ x \neq a$

Teaching Tip

See if students can recognize that composition is not commutative. Have them show examples to demonstrate that this is not the case.

The taxable value of a house with a $25,000 exemption is $V = g(x) - 25,000$. To find the actual tax, we find

$$f(V) = f(g(x)) = f(x - 25,000) = 0.005(x - 25,000)$$

This gives a formula for h: $h(x) = 0.005(x - 25,000)$. The function h is called the composite of f and g and is denoted by $f \circ g$ (read "f of g"). Here is the definition.

COMPOSITE FUNCTION

If f and g are functions, then

$$(f \circ g)(x) = f(g(x))$$

is the **composite of f with g.**

NOTE

The domain of $f \circ g$ is the set of all x in the domain of g such that $g(x)$ is in the domain of f.

Thus if $f(x) = x^2$ and $g(x) = x + 2$,

$$(f \circ g)(x) = f(g(x)) = f(x + 2)$$

$$= (x + 2)^2$$

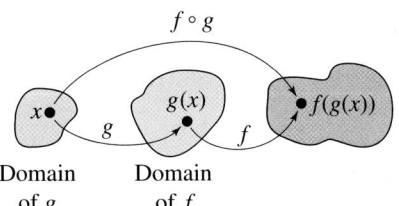

Domain of g Domain of f

EXAMPLE 3   Finding composite functions

If $f(x) = x^3$ and $g(x) = x - 1$, find:

a. $(f \circ g)(3)$ **b.** $(f \circ g)(x)$ **c.** $(g \circ f)(x)$

SOLUTION

a. Substitute 3 for x into "g." The result is substituted for x into "f."

$$(f \circ g)(3) = f(g(3)) = f(3 - 1) = f(2) = 2^3 = 8$$

b. Substituting $x - 1$ for $g(x)$ in $f(g(x))$, we have

$$(f \circ g)(x) = f(g(x)) = f(x - 1)$$

$$= (x - 1)^3$$

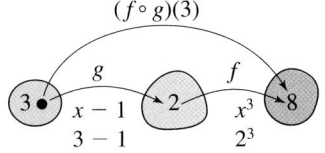

c. Substituting x^3 for $f(x)$ in $g(f(x))$, we have

$$(g \circ f)(x) = g(f(x)) = g(x^3) = x^3 - 1$$

NOTE

$(f \circ g)(x) = (x - 1)^3$ and $(g \circ f)(x) = x^3 - 1$. Thus, $(f \circ g)(x) \neq (g \circ f)(x)$.

PROBLEM 3

If $f(x) = x^2 - 1$ and $g(x) = 3x + 1$, find:

a. $(f \circ g)(-1)$ **b.** $(f \circ g)(x)$

c. $(g \circ f)(x)$

Answers

3. a. 3   **b.** $9x^2 + 6x$ **c.** $3x^2 - 2$

C   Finding the Domains of Combinations of Functions

To find

$$(f + g)(a), \quad (f - g)(a), \quad (fg)(a), \quad \text{and} \quad \left(\frac{f}{g}\right)(a)$$

we first have to find $f(a)$ and $g(a)$. To do this, we must determine whether a is in the domain of f and g.

| EXAMPLE 4   **Determining whether a is in the domain of f and g** | PROBLEM 4 |

Let $f(x) = \frac{3}{x}$ and $g(x) = \frac{x+2}{x-1}$.   Find the domain of $f + g, f - g$, and fg.

SOLUTION   Since division by zero is not defined,

$$f(x) = \frac{3}{x}$$

is not defined when $x = 0$, and

$$g(x) = \frac{x+2}{x-1}$$

is not defined when $x - 1 = 0$ (when $x = 1$). Thus,

Domain of $f = \{x \mid x \text{ is a real number and } x \neq 0\}$

and   Domain of $g = \{x \mid x \text{ is a real number and } x \neq 1\}$

To find $f(a) + g(a), f(a) - g(a)$, and $fg(a)$, we have to be sure that a is in *both* the domain of f and the domain of g. Hence

Domain of $f + g$ = domain of $f - g$ = domain of fg

$$= \{x \mid x \text{ is a real number and } x \neq 0 \text{ and } x \neq 1\}$$

PROBLEM 4

Let $f(x) = \frac{1}{2x}$ and $g(x) = \frac{3x-1}{x+4}$.
Find the domain of $f + g, f - g$, and fg.

Now let's talk about quotients of functions. Suppose that in Example 1(d), we wanted to find

$$\left(\frac{f}{g}\right)(-2)$$

We can easily find $f(-2)$ and $g(-2)$. Since $f(x) = x^2 + 4$, $f(-2) = (-2)^2 + 4 = 8$, and $g(x) = x + 2$, so $g(-2) = -2 + 2 = 0$. But then

$$\left(\frac{f}{g}\right)(-2) = \frac{f(-2)}{g(-2)} = \frac{8}{0}$$

which is not defined. Even though -2 is in the domain of *both* f and g, -2 is *not* in the domain of $\frac{f}{g}$.

Answer

4. Domain of $f + g$ = domain of $f - g$ = domain of fg = $\{x \mid x$ is a real number and $x \neq 0$ and $x \neq -4\}$

| EXAMPLE 5   **Finding the domain of $\frac{f}{g}$** | PROBLEM 5 |

Find the domain of $\frac{f}{g}$, if

$$f(x) = \frac{2}{x - 3} \quad \text{and} \quad g(x) = \frac{4}{x + 1}$$

SOLUTION   Since the domain of $f = \{x \mid x \text{ is a real number and } x \neq 3\}$
and the domain of $g = \{x \mid x \text{ is a real number and } x \neq -1\}$, we conclude that

PROBLEM 5

Find the domain of $\frac{f}{g}$, if

$$f(x) = \frac{5}{x + 6} \quad \text{and} \quad g(x) = \frac{10}{x - 1}$$

the domain of $\frac{f}{g}$ is the set of all real numbers *except* 3, −1, and any other values of x for which $g(x) = 0$. Since

$$g(x) = \frac{4}{x + 1}$$

is never zero, there are no other values of x such that $g(x) = 0$. Hence

$$\text{Domain of } \frac{f}{g} = \{x \,|\, x \text{ is a real number and } x \neq 3 \text{ and } x \neq -1\}$$

NOTE

Don't try to compute $\frac{f}{g}$ and then exclude the values that make the denominator zero. If we do this in Example 5, we would have

$$\left(\frac{f}{g}\right)(x) = \frac{\frac{2}{x - 3}}{\frac{4}{x + 1}} = \frac{2}{x - 3} \div \frac{4}{x + 1}$$

$$= \frac{2}{x - 3} \cdot \frac{x + 1}{4}$$

$$= \frac{x + 1}{2(x - 3)}$$

and only $x = 3$ would be excluded from the domain of $\frac{f}{g}$.

EXAMPLE 6   **More practice finding the domain of $\frac{f}{g}$**

Find: The domain of $\frac{f}{g}$, if

$$f(x) = \frac{2}{x - 3} \quad \text{and} \quad g(x) = \frac{4x}{x + 1}$$

SOLUTION   As before, the domain of $f = \{x \,|\, x \text{ is a real number and } x \neq 3\}$ and the domain of $g = \{x \,|\, x \text{ is a real number and } x \neq -1\}$. Thus the domain of $\frac{f}{g}$ is the set of all real numbers *except* 3, −1, and any other values of x for which $g(x) = 0$. Since

$$g(x) = \frac{4x}{x + 1} = 0$$

when $x = 0$, we conclude that

$$\text{Domain of } \frac{f}{g} = \{x \,|\, x \text{ is a real number and } x \neq 3, x \neq -1, \text{ and } x \neq 0\}$$

PROBLEM 6

Find the domain of $\frac{f}{g}$, if

$$f(x) = \frac{1}{x + 7} \quad \text{and} \quad g(x) = \frac{x + 4}{x - 1}$$

In general, we have the following procedure.

PROCEDURE

Finding the domain of a sum, difference, product, or quotient of two functions

1. Find the domain of each function.

2. The domain of the **sum, difference,** or **product** is the set of all values common to *both* domains.

3. The domain of the **quotient** is the set of all values common to both domains excluding any value that would result in division by zero.

Answers

5. $\{x \,|\, x \text{ is a real number and } x \neq -6 \text{ and } x \neq 1\}$

6. $\{x \,|\, x \text{ is a real number and } x \neq -7, x \neq 1, \text{ and } x \neq -4\}$

 **Solving an Application**

Functions and their operations are used frequently in business. For example, suppose the cost C of making x items and the resulting revenue R are given as functions of x.

When does one make a profit? Since the profit (or loss) is the difference between the revenue and the cost, the profit P is

$$P(x) = R(x) - C(x)$$

EXAMPLE 7 **The profit function**	**PROBLEM 7**

The revenue (in dollars) obtained from selling x units of a product is given by

$$R(x) = 200x - \frac{x^2}{30}$$

and the cost is given by $C(x) = 72,000 + 60x$.

a. Find the profit function, $P(x)$.

b. How many units must be made and sold to yield the maximum profit? Find this profit.

SOLUTION

a. $P(x) = R(x) - C(x)$

$$= \left(200x - \frac{x^2}{30}\right) - (72,000 + 60x)$$

$$= -\frac{x^2}{30} + 140x - 72,000$$

b. To find the value of x that gives the maximum profit, note that the graph of the profit function is a parabola opening downward. Thus the maximum value of P is at the vertex. The vertex is at the point where

$$x = -\frac{b}{2a} = -\frac{140}{-\frac{2}{30}} = (15)(140) = 2100$$

2100 units must be made and sold to give the maximum profit. This profit is P dollars, where P is given by

$$P(2100) = -\frac{(2100)^2}{30} + 140(2100) - 72,000$$

$$= 75,000$$

So the maximum profit is $75,000.

Do Example 7 if $C(x) = 50,000 + 50x$.

Teaching Tip

Before doing Example 7, review the equation of a parabola to recognize if the graph will have a maximum or minimum. Have students discuss the two techniques used to find the vertex of the parabola and then practice a few examples.

Example:

Find the vertex of

$$f(x) = -x^2 + 4x + 1$$

Answers

7. a. $P(x) = -\frac{x^2}{30} + 150x - 50,000$
b. 2250 units must be made and sold to give a maximum profit of $118,750.

Calculate It   **Finding the Maximum or Minimum Value of a Function**

You can use the TI-83 Plus to find the maximum or minimum values of the functions discussed in this section. In Example 7, the profit function can be graphed using an appropriate window (try $[-1000, 4000]$ by $[-1000, 75,000]$ with Xscl = 1000, Yscl = 1000). Using the Y= , enter

$$Y_1 = -\frac{x^2}{30} + 140x - 72,000$$

To find the maximum, press 2nd TRACE 4 . Use the arrow key to move the blinking cursor to a point on the curve on the left side of the maximum and press ENTER . Now use the arrow

key to move the blinking cursor to a point on the curve that is on the right side of the maximum and press ENTER . When prompted with "GUESS," press ENTER . The calculator gives the coordinates of the maximum for the function, X = 2100.0001 and Y = 75000, which you can approximate to 2100 and 75,000, as before.

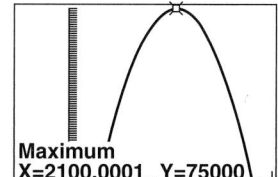

Exercises 10.1

A In Problems 1–6, use $f(x) = x + 4$, $g(x) = x^2 - 5x + 4$, and $h(x) = x^2 + 16$, to find the following.

1. $(f + g)(x)$ $x^2 - 4x + 8$

2. $(f - g)(x)$ $-x^2 + 6x$

3. $(hf)(x)$ $x^3 + 4x^2 + 16x + 64$

4. $\left(\dfrac{h}{f}\right)(x)$ $\dfrac{x^2 + 16}{x + 4}$

5. $\left(\dfrac{f}{h}\right)(x)$ $\dfrac{x + 4}{x^2 + 16}$

6. $(f + g - h)(x)$ $-4x - 8$

In Problems 7–12, find

$$\frac{f(x) - f(a)}{x - a}, \quad x \neq a$$

7. $f(x) = 3x - 2$ 3

8. $f(x) = 5x - 1$ 5

9. $f(x) = x^2$ $x + a$

10. $f(x) = x^3$ $x^2 + ax + a^2$

11. $f(x) = x^2 + 3x$ $x + a + 3$

12. $f(x) = x^2 - 2x$ $x + a - 2$

B In Problems 13–20, find the following.

a. $(f \circ g)(1)$ **b.** $(f \circ g)(x)$ **c.** $(g \circ f)(x)$

13. $f(x) = x^2$, $g(x) = \sqrt{x}, x > 0$
   a. 1   **b.** x **c.** $|x|$

14. $f(x) = x - 1$, $g(x) = x^2$
   a. 0   **b.** $x^2 - 1$ **c.** $(x - 1)^2$

15. $f(x) = 3x - 2$, $g(x) = x + 1$
   a. 4   **b.** $3x + 1$ **c.** $3x - 1$

16. $f(x) = x^2$, $g(x) = x - 1$
   a. 0   **b.** $(x - 1)^2$ **c.** $x^2 - 1$

17. $f(x) = \sqrt{x + 1}, x > -1$, $g(x) = x^2 - 1$
   a. 1   **b.** $|x|$ **c.** x

18. $f(x) = \sqrt{x^2 + 1}$, $g(x) = 2x + 1$
   a. $\sqrt{10}$ **b.** $\sqrt{4x^2 + 4x + 2}$ **c.** $2\sqrt{x^2 + 1} + 1$

19. $f(x) = 3$, $g(x) = -1$ **a.** 3   **b.** 3   **c.** -1

20. $f(x) = ax$, $g(x) = bx$ **a.** ab **b.** abx **c.** abx

C In Problems 21–30, evaluate the indicated combination of f and g for the given values of the independent variable. (If this is not possible, state the reason.)

21. $f(x) = \sqrt{x}, x > 0$, $g(x) = x^2 - 1$

   a. $(f + g)(4)$ 17

   b. $(f - g)(4)$ -13

22. $f(x) = \sqrt{x - 2}, x > 2$, $g(x) = x^2 + 1$

   a. $(f + g)(1)$ $2 + \sqrt{-1}$

   b. $(f - g)(1)$ $-2 + \sqrt{-1}$

23. $f(x) = |x|$, $g(x) = 3$

   a. $\left(\dfrac{f}{g}\right)(3)$ 1   **b.** $\left(\dfrac{f}{g}\right)(0)$ 0

24. $f(x) = x^2 - 4$, $g(x) = x + 2$

   a. $\left(\dfrac{f}{g}\right)(2)$ 0   **b.** $\left(\dfrac{f}{g}\right)(-2)$ Undefined; $x \neq -2$

25. $f(x) = x - 3$, $g(x) = (x + 3)(x - 3)$

   a. $\left(\dfrac{f}{g}\right)(3)$ Undefined; $x \neq 3$

   b. $\left(\dfrac{g}{f}\right)(3)$ Undefined; $x \neq 3$

26. $f(x) = x + 5$, $g(x) = (x + 5)(x - 5)$

   a. $\left(\dfrac{f}{g}\right)(5)$ Undefined; $x \neq 5$

   b. $\left(\dfrac{f}{g}\right)(-5)$ Undefined; $x \neq -5$

27. $f(x) = \sqrt{x}, x > 0,$ $g(x) = x^2 + 1$

    a. $(f \circ g)(1)$ $\sqrt{2}$

    b. $(g \circ f)(-1)$ 0

    c. $(f \circ g)(x)$ $\sqrt{x^2 + 1}$

    d. $(g \circ f)(x)$ $x + 1$

28. $f(x) = \sqrt{x + 1}, x > -1,$ $g(x) = x^2$

    a. $(f \circ g)(-1)$ $\sqrt{2}$

    b. $(g \circ f)(-1)$ 0

    c. $(f \circ g)(x)$ $\sqrt{x^2 + 1}$

    d. $(g \circ f)(x)$ $x + 1$

29. $f(x) = \dfrac{1}{x^2 - 2},$ $g(x) = \sqrt{x}, x > 0$

    a. $(f \circ g)(2)$ Undefined; $x \neq 2$

    b. $(g \circ f)(2)$ $\sqrt{\dfrac{1}{2}} = \dfrac{\sqrt{2}}{2}$

    c. $(f \circ g)(x)$ $\dfrac{1}{x - 2}, x > 0$

    d. $(g \circ f)(x)$ $\sqrt{\dfrac{1}{x^2 - 2}}$ or $\dfrac{\sqrt{x^2 - 2}}{x^2 - 2}$

30. $f(x) = \dfrac{1}{x^2},$ $g(x) = \sqrt{x}, x > 0$

    a. $(f \circ g)(-1)$ -1

    b. $(g \circ f)(-1)$ 1

    c. $(f \circ g)(x)$ $\dfrac{1}{x}, x > 0$

    d. $(g \circ f)(x)$ $\dfrac{1}{x}, x \neq 0$

C In Problems 31–40, determine the domain of the sum, difference, product, and quotient of f and g.

31. $f(x) = x^2$ The domain of the sum, difference, and product is: $\{x \mid x$ is a real number$\}$

    $g(x) = x - 1$ The domain of the quotient is: $\{x \mid x$ is a real number and $x \neq 1\}$

32. $f(x) = -2x^2$ The domain of the sum, difference, and product is: $\{x \mid x$ is a real number$\}$

    $g(x) = x - 2$ The domain of the quotient is: $\{x \mid x$ is a real number and $x \neq 2\}$

33. $f(x) = 2x + 1$ The domain of the sum, difference, and product is: $\{x \mid x$ is a real number$\}$

    $g(x) = -2x + 2$ The domain of the quotient is: $\{x \mid x$ is a real number and $x \neq 1\}$

34. $f(x) = -3x + 1$ The domain of the sum, difference, and product is: $\{x \mid x$ is a real number$\}$

    $g(x) = 3x - 2$ The domain of the quotient is: $\{x \mid x$ is a real number and $x \neq \frac{2}{3}\}$

35. $f(x) = \dfrac{1}{x - 1}$ The domain of the sum, difference, and product is: $\{x \mid x$ is a real number and $x \neq 1$ and $x \neq -2\}$

    $g(x) = \dfrac{-3}{x + 2}$ The domain of the quotient is: $\{x \mid x$ is a real number and $x \neq 1$ and $x \neq -2\}$

36. $f(x) = \dfrac{-3}{x - 2}$ The domain of the sum, difference, and product is: $\{x \mid x$ is a real number and $x \neq 2$ and $x \neq -5\}$

    $g(x) = \dfrac{-4}{x + 5}$ The domain of the quotient is: $\{x \mid x$ is a real number and $x \neq 2$ and $x \neq -5\}$

37. $f(x) = \dfrac{2x}{x + 4}$ The domain of the sum, difference, and product is: $\{x \mid x$ is a real number and $x \neq 1$ and $x \neq -4\}$

    $g(x) = \dfrac{x}{x - 1}$ The domain of the quotient is: $\{x \mid x$ is a real number and $x \neq 1$ and $x \neq -4$ and $x \neq 0\}$

38. $f(x) = \dfrac{5x}{x + 2}$ The domain of the sum, difference, and product is: $\{x \mid x$ is a real number and $x \neq -2$ and $x \neq -3\}$

    $g(x) = \dfrac{x}{x + 3}$ The domain of the quotient is: $\{x \mid x$ is a real number and $x \neq -2$ and $x \neq -3$ and $x \neq 0\}$

39. $f(x) = \dfrac{x - 2}{x + 3}$ The domain of the sum, difference, and product is: $\{x \mid x$ is a real number and $x \neq -3$ and $x \neq 2\}$

    $g(x) = \dfrac{x - 1}{x - 2}$ The domain of the quotient is: $\{x \mid x$ is a real number and $x \neq -3$ and $x \neq 2$ and $x \neq 1\}$

40. $f(x) = \dfrac{x + 3}{x - 4}$ The domain of the sum, difference, and product is: $\{x \mid x$ is a real number and $x \neq 4$ and $x \neq -2\}$

    $g(x) = \dfrac{x + 3}{x + 2}$ The domain of the quotient is: $\{x \mid x$ is a real number and $x \neq 4$ and $x \neq -2$ and $x \neq -3\}$

APPLICATIONS

Functions

41. The dollar revenue obtained from selling x copies of a textbook is $R(x) = 40x - 0.0005x^2$. The production cost C is $C(x) = 120{,}000 + 6x$. Find the profit function $P(x)$.
$P(x) = -0.0005x^2 + 34x - 120{,}000$

42. Repeat Problem 41 if the production cost increases by $20,000. $P(x) = -0.0005x^2 + 34x - 140{,}000$

43. The clam and crab catch (in millions of pounds) in New England between 1980 and 1990 can be approximated by $L(x) = -0.28x^2 + 2.8x + 15$ and $R(x) = 0.2x + 7$, respectively, where x is the number of years after 1980.

   a. What is the total catch of clams and crabs? $L(x) + R(x) = -0.28x^2 + 3x + 22$

   b. How many pounds of clams and crabs were caught in 1980? $L(0) + R(0) = 22$ million

   c. How many pounds of clams and crabs were caught in 1990? $L(10) + R(10) = 24$ million

   d. How many more pounds of clams than of crabs were caught in 1990? $L(10) - R(10) = 6$ million

44. The total fish catch (in millions of pounds) in Hawaii from 1980 to 1990 can be approximated by $C(x) = 1.5x + 10.5$ where x is the number of years after 1980. The tuna catch during the same period was $T(x) = 0.7x + 7$.

   a. How many pounds of fish other than tuna were caught? $C(x) - T(x) = 0.8x + 3.5$

   b. How many pounds of tuna did they catch in Hawaii in 1990? $T(10) = 14$ million

   c. How many pounds of fish other than tuna were caught in Hawaii in 1990? $C(10) - T(10) = 11.5$ million

   d. What fraction of the total catch in Hawaii in 1990 was tuna? $T(10) \div C(10) = \frac{28}{51}$

   e. What percent of the total catch in Hawaii in 1990 was tuna? Round the answer to one decimal digit. 54.9%

45. The total Medicare costs (in billions of dollars) between 1990 and 1995 can be approximated by the function $C(t) = 2.5t^2 + 8.5t + 111$, where t is the number of years after 1990. The total U.S. population age 65 and older in the same period is

$$P(t) = -0.46t^2 + 1.14t + 31.08 \text{ (in millions)}$$

   a. Find a function that represents the average cost of Medicare for persons 65 and older during the years 1990–1995. $\dfrac{C(t)}{P(t)} = \dfrac{2.5t^2 + 8.5t + 111}{-0.46t^2 + 1.14t + 31.08}$ (in thousands)

   b. Find the average cost of Medicare for persons 65 and older for 1990. Round to the nearest dollar. $3571

   c. Repeat part **b** for 1995. $8544

46. The reaction distance $R(v) = 0.75v$ (in feet) is the distance a car moving at v miles per hour travels while a driver with a reaction time of 0.5 sec is reacting to apply the brakes. If the braking distance for the car is $B(v) = 0.06v^2$, find a function that gives the total distance (in feet) a car moving at v miles per hour travels during a panic stop. What is this distance if the car is moving at 30 mi/hr?

$B(v) + R(v) = 0.06v^2 + 0.75v$; $B(30) + R(30) = 76.5$ ft

47. The function $C(F) = \frac{5}{9}(F - 32)$ converts the temperature F in degrees Fahrenheit to C in degrees Celsius. The function $K(C) = C + 273$ converts degrees Celsius to kelvins.

   a. Find a composite function that converts degrees Fahrenheit to kelvins.

   $(K \circ C)(F) = \dfrac{5}{9}(F - 32) + 273$

   b. If the temperature is 41°F, what is the Celsius temperature? $C(41) = 5°$

   c. If water boils at 212°F, at what kelvin temperature does water boil? $(K \circ C)(212) = 373$ K

48. A parent realizes that her son's demand for birthday toys (in dollars) depends on the average number of hours of TV that he watches each week. If h is the average number of hours of TV that he watches each week, the cost of the birthday toys is $D(h) = 5h + 8$.

   a. Find $D(7)$. $43

   b. If the number of hours of TV he is allowed to watch each week is $h(A) = A + 2$, where A is his age, find a composite function that gives the son's demand for birthday toys as a function of his age. $(D \circ h)(A) = 5A + 18$

   c. What is the cost of the toys that the son demands when he is 11? $73

49. The function F giving the correspondence between dress sizes in the United States and France is $F(x) = x + 32$, where x is the U.S. size and $F(x)$ the French size. The function $E(F) = F - 30$ gives the correspondence between dress sizes in France and those in England.

a. What French size corresponds to a U.S. size 6?   38

b. Find a function that will give a correspondence between U.S. and English dress sizes. $(E \circ F)(x) = x + 2$

c. What English size corresponds to a U.S. size 8?   10

50. The correspondence between dress sizes in the United States and Italy is $I(x) = 2x + 22$ where x is the U.S. dress size and the correspondence between Italian and English dress sizes is $f(I) = \frac{1}{2}I - 9$.

a. What Italian size corresponds to a U.S. size 8?   38

b. What English size corresponds to a U.S. size 10?   12

c. Find a composite function that will give the correspondence between U.S. and English dress sizes. $(f \circ I)(x) = x + 2$

51. There are many interesting functions that can be defined using the ideas of this section. For example, we have mentioned that the frequency with which a cricket chirps is a function of the temperature. The table shows the number of chirps per minute and the temperature in degrees Fahrenheit. If f is the function that relates the number of chirps per minute, c, and the temperature x, find

a. $f(40)$ 0   **b.** $f(42)$ 8   **c.** $f(44)$ 16

Temperature (°F)	40	41	42	43	44
Chirps per minute	0	4	8	12	16

52. The function relating the number of chirps per minute of the cricket and the temperature x (in degrees Fahrenheit) is given by $f(x) = 4(x - 40)$. If the temperature is 80°F, how many chirps per minute will you hear from your friendly house cricket?   160

53. The function $F(x) = \frac{9}{5}x + 32$ converts the temperature from degrees Celsius to degrees Fahrenheit:

a. Find a composite function that would relate the number of chirps per minute a cricket makes (Problem 52) to the temperature in degrees Celsius. $(f \circ F)(x) = 4(\frac{9}{5}x - 8)$

b. How many chirps per minute would a cricket make when the temperature is 10°C?   40

54. The distance (in centimeters per second) traveled by a certain type of ant when the temperature is C (in degrees Celsius) is $d(C) = \frac{1}{6}(C - 4)$. The function $C(F) = \frac{5}{9}(F - 32)$ converts the temperature from degrees Fahrenheit to Celsius.

a. Find a composite function that would relate the distance the ant travels to the temperature in degrees Fahrenheit. $(d \circ C)(F) = \frac{1}{6}\left(\frac{5F - 160}{9}\right)$

b. How fast is the ant traveling when the temperature is 50°F?   1 cm/sec

SKILL CHECKER

Try the "Skill Checker" exercises so you'll be ready for the next section.

Graph:

55. $x + y = 3$

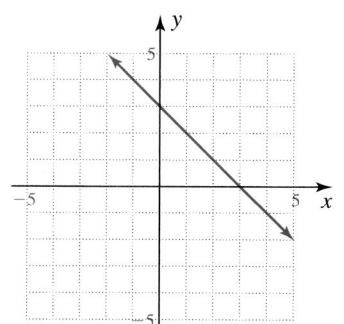

56. $2x - y = 2$

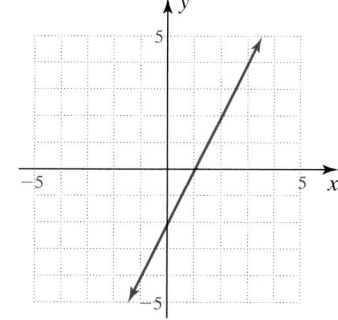

57. $2x + \frac{1}{2}y = 2$

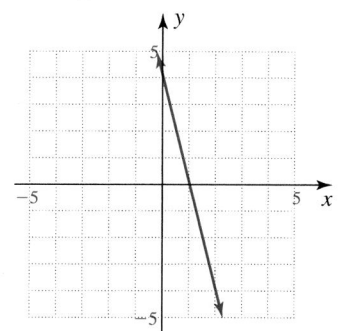

58. $y = -x - 3$

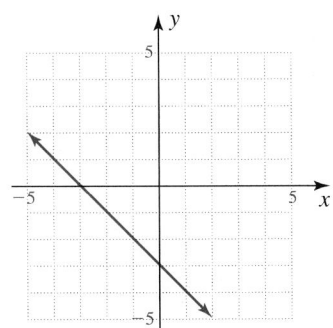

59. $y = -2x + 4$

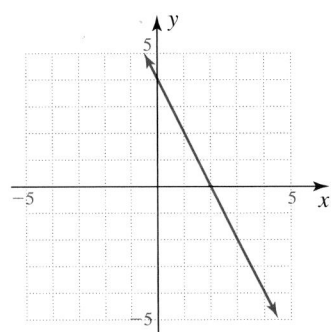

60. $y = -3x + 6$

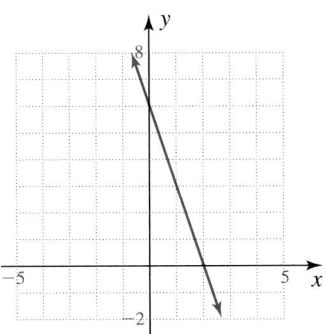

USING YOUR KNOWLEDGE

Projecting Domains

Look at the graphs of the functions f and g shown in Figure 1. As you can see,

$$\text{Domain of } f = \{x \mid -1 \le x \le 4\}$$
$$\text{Domain of } g = \{x \mid 1 \le x \le 5\}$$

These domains can be regarded as the "projections" of f and g on the x-axis. Thus the domains of $f + g, f - g,$ and fg are $\{x \mid 1 \le x \le 4\}$—the values common to the domains of f and g. To find the domain of $\frac{f}{g}$, $g(2) = 0$, thus

$$\text{Domain of } \frac{f}{g} = \{x \mid 1 \le x \le 4 \text{ and } x \ne 2\}$$

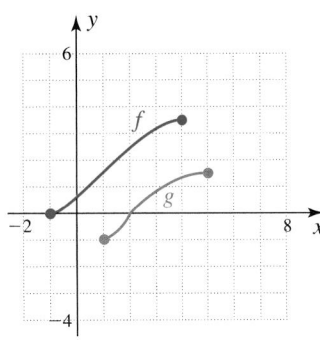

Figure 1

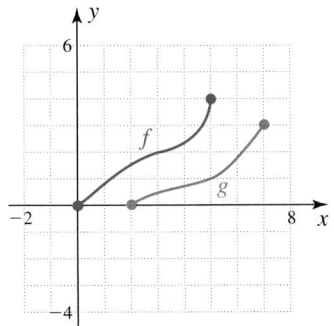

Figure 2

Now refer to Figure 2 and find:

61. The domain of f $\{x \mid 0 \le x \le 5\}$

62. The domain of g $\{x \mid 2 \le x \le 7\}$

63. The domain of $f + g, f - g,$ and fg $\{x \mid 2 \le x \le 5\}$

64. The domain of $\frac{f}{g}$ $\{x \mid 2 < x \le 5\}$

WRITE ON

65. Explain why $(f + g)(x) = (g + f)(x)$ for every value of x but $(f - g)(x) \ne (g - f)(x)$. Answers may vary.

66. Explain why

$$\left(\frac{f}{g}\right)(x) \ne \left(\frac{g}{f}\right)(x)$$ Answers may vary.

67. Is $(f \circ g)(x) = (g \circ f)(x)$ for every value of x? Explain why or why not. Answers may vary.

68. If $f(x) = \frac{1}{2}x$ and $g(x) = 2x$,

   a. Find $(f \circ g)(x)$ and $(g \circ f)(x)$. Are they the same? Yes.

   b. Can you find other functions f and g so that $(f \circ g)(x) = (g \circ f)(x)$? Answers may vary.

MASTERY TEST

If you know how to do these problems, you have learned your lesson!

69. The revenue (in dollars) obtained by selling x units of a product is given by

$$R(x) = 300x - \frac{x^2}{30}$$

and the cost $C(x)$ of making x units is given by $C(x) = 72{,}000 + 60x$. Find the profit function $P(x)$.

$P(x) = -\dfrac{x^2}{30} + 240x - 72{,}000$

If $f(x) = x^3$ and $g(x) = x + 1$, find:

70. $(f \circ g)(x)$ $(x + 1)^3$

71. $(g \circ f)(x)$ $x^3 + 1$

72. $(f \circ g)(3)$ 64

73. $(g \circ f)(-3)$ -26

74. Find $\dfrac{f(x) - f(a)}{x - a}$ if $f(x) = 2x + 1$. 2

75. Find $\dfrac{f(x) - f(a)}{x - a}$ if $f(x) = x^2 + 1$. $x + a, x \neq a$

If $f(x) = x^2 + 4$ and $g(x) = x - 2$, find (if possible):

76. $(f + g)(x)$ $x^2 + x + 2$

77. $(f - g)(x)$ $x^2 - x + 6$

78. $(g - f)(x)$ $-x^2 + x - 6$

79. $\left(\dfrac{f}{g}\right)(-2)$ -2

80. $\left(\dfrac{g}{f}\right)(x)$ $\dfrac{x - 2}{x^2 + 4}$

81. $\left(\dfrac{g}{f}\right)(-2)$ $-\dfrac{1}{2}$

82. $\left(\dfrac{f}{g}\right)(2)$ Undefined; $x \neq 2$

Find the domains of:

83. $f + g, f - g, fg,$ and $\dfrac{f}{g}$ if $f(x) = \dfrac{1}{x}$ and $g(x) = \dfrac{3}{x + 1}$

The domain of the sum, difference, product, and quotient is: $\{x \mid x$ is a real number and $x \neq 0$ and $x \neq -1\}$

84. $\dfrac{f}{g}$ if $f(x) = \dfrac{3}{x}$ and $g(x) = \dfrac{x - 1}{x + 2}$

$\{x \mid x$ is a real number and $x \neq 0$ and $x \neq -2$ and $x \neq 1\}$

10.2 INVERSE FUNCTIONS

To Succeed, Review How To...

1. Find the range and domain of a relation (pp. 176–178).

2. Solve an equation for a specified variable (pp. 88–92).

3. Graph linear and quadratic equations (pp. 165–168, 648–657).

Objectives

A Find the inverse of a function when the function is given as a set of ordered pairs.

B Find the equation of the inverse of a function.

C Graph a function and its inverse and determine whether the inverse is a function.

D Solve applications involving inverse functions.

GETTING STARTED **Money, Money, Money**

	Currency Unit	USD per Unit	Units per USD
DZD	Algeria Dinars	0.0144009	69.4400
USD	United States Dollars	1.00000	1.00000
ARS	Argentina Pesos	0.334784	2.98700
AUD	Australia Dollars	0.723700	1.38179
ATS	Austria Schillings**	0.0871783	11.4707
BSD	Bahama Dollars	1.00000	1.00000
BBD	Barbados Dollars	0.502513	1.99000
BEF	Belgium Francs**	0.0297373	33.6278
BMD	Bermuda Dollars	1.00000	1.00000
BRL	Brazil Reais	0.339386	2.94650
GBP	United Kingdom Pounds	1.72150	0.580889
BGN	Bulgaria Leva	0.613572	1.62980
CAD	Canada Dollars	0.770060	1.29660
CIP	Chile Pesos	0.00160231	624.100
CNY	China Yuan Renminbi	0.120817	8.27700
CYP	Cyprus Pounds	2.06612	0.484000
CZK	Czech Republic Koruny	0.0374616	26.6940
DKK	Denmark Kroner	0.161160	6.20500

**Note:* Currencies marked with a double asterisk are Euro-zone currencies that may not have been legal tender on the date in question.
Source: The Interactive Currency Table.

If you're planning on traveling somewhere, it's a good idea to know how much your dollars are worth in the country you plan to visit. The fourth column in the table shows the value of $1 (U.S.) in different currencies in a recent year. If you look at the third column, the value of a Canadian dollar in U.S. dollars is above $0.770. This means that the number of dollars D you get for C Canadian dollars is given by

$$D = 0.770C$$

To find how many Canadian dollars you get for one U.S. dollar, solve for C in $D = 0.770C$ to obtain

$$C = \frac{1D}{0.770} \approx 1.299\,D$$

You get 1.299 Canadian dollars for every U.S. dollar. You can check this in the table. The table says 1.29660. In this section we learn how to find the inverse of a function, which uses a similar procedure.

Finding the Inverse of a Function

Let's look again at how we found the exchange rate for Canadian dollars in the *Getting Started*. If we think of *D* as the definition for the function *f*(*C*), we can make a table and write the function *f* as

C Canadian Dollars	D U.S. Dollars
1	0.77
10	7.70
100	77.00
1000	770.00

$$f = \{(1, 0.77), (10, 7.70), (100, 77.00), (1000, 770.00)\}$$

On the other hand, if you exchange U.S. currency for Canadian, the number of Canadian dollars you get for $0.770, $7.70, $77.00, and $770 (U.S.), respectively, corresponds to the set of ordered pairs (*D*, *C*) and is given by

$$g = \{(0.770, 1), (7.70, 10), (77.00, 100), (770, 1000)\}$$

The relation *g* obtained by reversing the order of the coordinates in each ordered pair in *f* is called the *inverse of f*. As you can see, the domain of *f* is the range of *g* and the range of *f* is the domain of *g*. Here is the definition we need.

INVERSE OF A FUNCTION

If *f* is a relation, then the **inverse of *f*,** denoted by f^{-1} (read "*f* inverse," or "the inverse of *f*") is the relation obtained by reversing the order of *x* and *y* in each ordered pair in *f*.

> **NOTE**
>
> The −1 in f^{-1} is *not* an exponent. Here it denotes the inverse of *f*.

EXAMPLE 1 **Finding the inverse of a function**

Let $S = \{(1, 2), (3, 4), (5, 4)\}$ and find:

a. The domain and range of *S*

b. S^{-1}

c. The domain and range of S^{-1}

d. The graphs of *S* and S^{-1} on the same coordinate axes

SOLUTION

a. The domain of *S* is $\{1, 3, 5\}$. The range is $\{2, 4\}$.

b. $S^{-1} = \{(2, 1), (4, 3), (4, 5)\}$

c. The domain of S^{-1} is $\{2, 4\}$; the range is $\{1, 3, 5\}$.

d. The graphs of *S* (in blue) and S^{-1} (in red) are shown in Figure 3. As you can see, the two graphs are symmetric with respect to the line $y = x$ (shown dashed in Figure 3).

Figure 3

PROBLEM 1

Let $S = \{(4, 3), (3, 2), (2, 1)\}$ and find:

a. The domain and range of *S*

b. S^{-1}

c. The domain and range of S^{-1}

d. The graph of *S* and S^{-1}

Answers

1. a. $D = \{2, 3, 4\}; R = \{1, 2, 3\}$

b. $\{(3, 4), (2, 3), (1, 2)\}$

c. $D = \{1, 2, 3\}; R = \{2, 3, 4\}$

d.

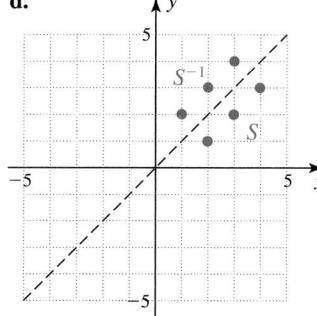

Calculate It Graphing Ordered Pairs

With some calculators (the TI-83 Plus, for example), you can graph ordered pairs by entering the domain and range using the list feature. To do Example 1(a), clear any lists in the calculator, then press [STAT] [1] 1, 3, and 5 (the domain) under L1 and 2, 4, and 4 (the range) under L2. The result is shown in Window 1.

To tell the calculator to plot these points, first press [ZOOM] 6 for a standard window, then press [2nd] [Y=] [1] and select [ON], [|.:] (the first type of graph), L1, L2, and [□]. Finally, press [GRAPH]. The result using a standard window is shown in Window 2.

Follow this procedure to graph S^{-1} using lists L3 (2, 4, 4) and L4 (1, 3, 5) for the domain and range of S^{-1}. Press [2nd] [Y=] [2] and select [ON], [|.:], L3, L4, and [+]. Now press [GRAPH] to obtain the graph shown in Window 3. We used a square window so that S and S^{-1} appear as reflections of each other across the line $y = x$.

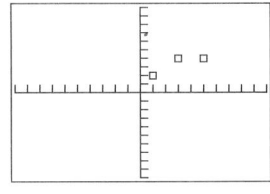

Window 1   **Window 2**

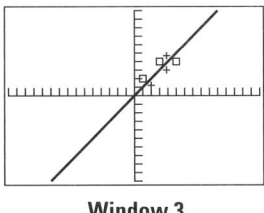

Window 3

B Finding the Equation of the Inverse Function

Previously, we found the inverse of

$$f = \{(1, 0.77), (10, 7.70), (100, 77.00), (1000, 770.00)\}$$

by reversing the *order* of the coordinates in each ordered pair in f. To find the inverse of the function $y = f(x) = 0.770x$, we interchange the *variables* in $y = 0.770x$ to obtain

$$x = 0.770y$$

$$y = \frac{1}{0.770} x \qquad \text{Solve for } y.$$

$$y = 1.299x \qquad \text{Divide 1 by 0.770.}$$

Thus $f^{-1}(x) = 1.299x$. The procedure is summarized below.

Web It

For an excellent lesson on how to find the inverse of a function given as ordered pairs, go to link 10-2-1 at mhhe.com/bello.

Teaching Tip

Have students discuss why we expect symmetry with respect to $y = x$ when graphing inverse functions.

> **PROCEDURE**
>
> **Finding the equation of an inverse function**
>
> **1.** Interchange the roles of x and y in the equation for f.
>
> **2.** Solve for y.

For example, consider the relation $f(x) = 4x - 4$ or

$$y = 4x - 4 \qquad \qquad \textbf{(1)}$$

The inverse of this relation, f^{-1}, is obtained by first interchanging the x- and y-coordinates—by writing $x = 4y - 4$ and then solving for y:

$$x + 4 = 4y \qquad \text{Add 4.}$$

$$\frac{x + 4}{4} = y \qquad \text{Divide by 4.}$$

Web It

For an introduction dealing with inverse functions written in functional notation, go to link 10-2-2 at mhhe.com/bello.

We have

$$f^{-1}(x) = y = \frac{x+4}{4}, \text{ which can be written } y = \frac{1}{4}x + 1 \qquad (2)$$

The graphs of equation (1) (in blue) and its inverse, equation (2) (in red), are shown in Figure 4. Clearly, the graphs are symmetric to each other with respect to the line $y = x$, shown dashed. This is to be expected, because one relation was obtained from the other by interchanging x and y.

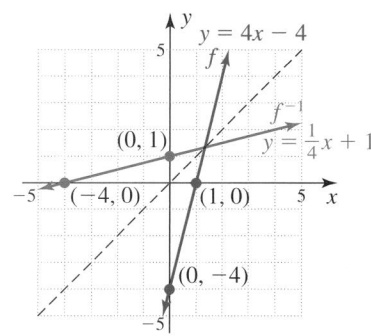

Figure 4

EXAMPLE 2   Finding and graphing an inverse function

Let $f(x) = y = 4x - 2$:

a. Find $f^{-1}(x)$.   **b.** Graph f and its inverse.

SOLUTION

a. Since $y = 4x - 2$, we interchange the variables x and y and solve for y to obtain

$$x = 4y - 2$$
$$x + 2 = 4y \qquad \text{Add 2.}$$
$$y = \frac{x+2}{4} \qquad \text{Divide by 4.}$$

Thus the inverse of f is

$$f^{-1}(x) = \frac{x+2}{4} = \frac{1}{4}x + \frac{1}{2}$$

b. We can graph $y = 4x - 2$ by putting the y-intercept at $(0, -2)$ and using the slope of $\frac{4}{1}$; going up 4 from $(0, -2)$ and to the right one to the point $(1, 2)$. Connect the points. The graph is shown in blue in Figure 5. We then graph

$$y = \frac{1}{4}x + \frac{1}{2}$$

in a similar manner. This graph is shown in red and is symmetric to the graph of $y = 4x - 2$ with respect to the line $y = x$, shown dashed. We could have obtained the graph of the inverse function by reflecting the graph of $y = 4x - 2$ about the line $y = x$.

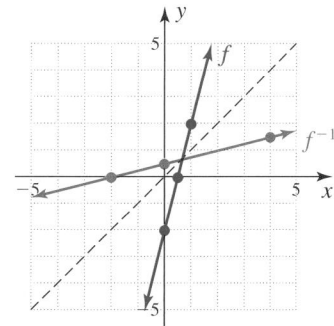

Figure 5

PROBLEM 2

Let $f(x) = y = 2x - 3$

a. Find $f^{-1}(x)$

b. Graph f and its inverse

Answers

2. a. $f^{-1}(x) = \frac{x+3}{2} = \frac{1}{2}x + \frac{3}{2}$

b.

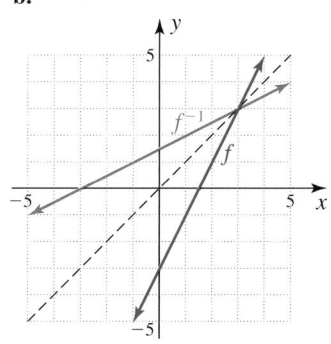

C Graphing Functions and Their Inverses

Web It

How do you know if a graph has an inverse? Go to link 10-2-3 at mhhe.com/bello to find out.

Every function is a set of ordered pairs (a relation), so every function has an inverse. Is this inverse always a function? We can see that it is not. For example, if S is the function defined by $S = \{(1, 2), (3, 2)\}$, the inverse of S is $S^{-1} = \{(2, 1), (2, 3)\}$, which is *not* a function, because two distinct ordered pairs have the same first component, 2. On the other hand, if G is the function defined by $G = \{(3, 4), (5, 6)\}$, the inverse is $G^{-1} = \{(4, 3), (6, 5)\}$, which *is* a function. The reason that the inverse of S is not a function is that S has two ordered pairs with the same *second* component. A function in which no two distinct ordered pairs have the same second component is called a **one-to-one function.** The inverse of such a function is always a function. We summarize this discussion as follows.

One-to-One Function	If the function $y = f(x)$ is one-to-one, then the inverse of f is also a function and is denoted by $y = f^{-1}(x)$.

To determine whether the inverse of a function is a function, we must ascertain whether the original function is one-to-one. To do this, we must return to the definition of a one-to-one function—a one-to-one function *cannot* have two ordered pairs with the same second component. Since any two points with the same second coordinate will be on a *horizontal* line parallel to the x-axis, if any horizontal line intersects the graph of a function more than once, the function will not be one-to-one. Its inverse will not be a function. Thus we have the following **horizontal line test.**

Horizontal Line Test to Determine if a Function is One-to-One	If a horizontal line intersects the graph of a function more than once, then it is not a one-to-one function and the inverse of the function is not a function.

Passes Horizontal Line Test

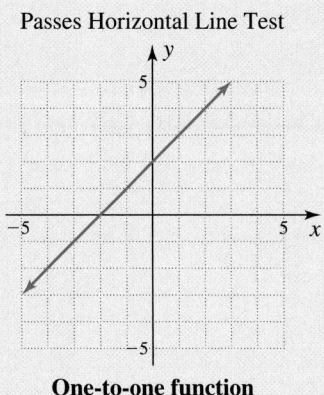

One-to-one function

Does Not Pass Horizontal Line Test

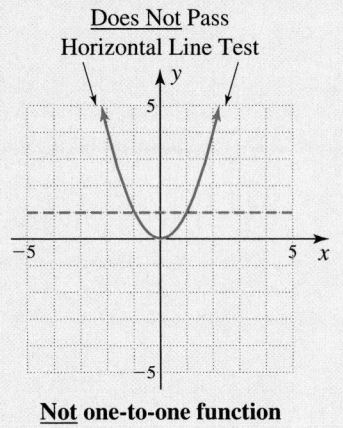

<u>**Not**</u> **one-to-one function**

Web It

For more on inverse functions, go to link 10-2-4 at mhhe. com/bello.

Non-constant linear functions have inverses that are functions (horizontal lines will intersect the graph only once), but quadratic functions do not have inverses that are functions. This is because quadratic functions have graphs that can be intersected by a horizontal line more than once.

Here are the steps we need to find the inverse of a function.

PROCEDURE

Finding the inverse of a function $y = f(x)$

1. Replace $f(x)$ by y, if necessary.
2. Interchange the roles of x and y.
3. Solve the resulting equation for y, if possible.
4. Replace y by $f^{-1}(x)$. (If the original function was one-to-one, f^{-1} is a function.)

| **EXAMPLE 3** | **Finding inverses and using the horizontal line test** |

Find the inverse of $f(x) = x^2$. Is the inverse a function?

SOLUTION   We use the four-step procedure.

1. Replace $f(x)$ by y. $y = x^2$

2. Interchange the roles of x and y. $x = y^2$

3. Solve for y. $y = \pm\sqrt{x}$

4. Replace y by $f^{-1}(x)$. $f^{-1}(x) = \pm\sqrt{x}$

The inverse of $f(x)$ is *not* a function, because we can draw a horizontal line that intersects the graph of $y = x^2$ at more than one point.

 The function $f(x) = x^2$ (in blue) and its inverse (in red) are shown in the graph in Figure 6.

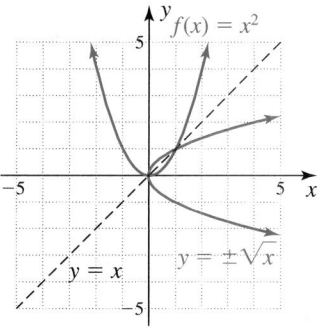

Figure 6

| **PROBLEM 3** |

Find the inverse of $f(x) = y = x^3$. Is the inverse a function?

| **EXAMPLE 4** | **More practice using the horizontal line test** |

Graph the function $f(x) = y = 3^x$. Is the inverse a function?

SOLUTION   Following is a table of values to graph $y = 3^x$. If we examine the graph of $y = 3^x$ (in blue in Figure 7), we can see that any horizontal line will intersect the graph only once. Thus the inverse is a function. To try to find the inverse, we interchange the x and y in $y = 3^x$ to obtain $x = 3^y$.

 Unfortunately, we are not able to solve for y at this time (we will have to wait until later in this chapter to do this). However, we can still graph the inverse $x = 3^y$ by giving values to y and finding the corresponding x-values, as shown in the following table on the right. The graph of the inverse relation $x = 3^y$ is in red.

Figure 7

| **PROBLEM 4** |

Graph $f(x) = y = 2^x$. Is the inverse a function?

x	$y = 3^x$		$x = 3^y$	y
-1	$3^{-1} = \frac{1}{3}$		$3^{-1} = \frac{1}{3}$	-1
0	$3^0 = 1$		$3^0 = 1$	0
1	$3^1 = 3$		$3^1 = 3$	1

Answers

3. $f^{-1}(x) = \sqrt[3]{x}$; yes

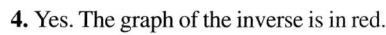
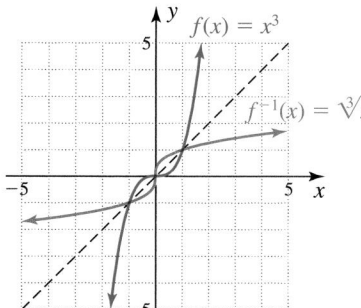

4. Yes. The graph of the inverse is in red.

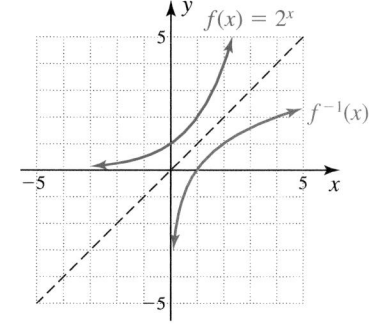

Calculate It Graphing Inverses

To do Example 4, you need a calculator with a draw feature. Enter $Y_1 = 3^x$ and graph it using a square window. To graph the inverse using the draw feature on a TI-83 Plus, press 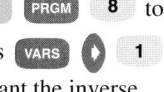 to tell the calculator you want an inverse. Then press VARS ▶ 1 1 ENTER to tell the calculator you specifically want the inverse of Y_1. The result is shown in the window. The calculator still doesn't tell you how to find the inverse or what the equation for this inverse is. To find out, you have to learn the techniques in the next section.

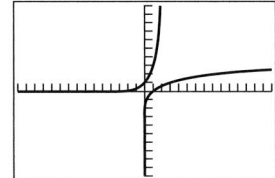

Web It

To see a step-by-step Power-Point presentation of how to find the inverse, go to link 10-2-5 at mhhe.com/bello.

D Solving Applications Involving Functions

Can you tell yet what the temperature is by listening to crickets? Let's return to our cricket problem one more time and solve it yet another way.

| EXAMPLE 5 "Functional" cricket chirping | PROBLEM 5 |

The relationship between the temperature F (in degrees Fahrenheit) and the number of chirps c a cricket makes in 1 minute is given by the function

$$c = n(F) = 4(F - 40)$$

a. Find the inverse of n.

b. If a cricket is chirping 120 times a minute, what is the temperature in degrees Fahrenheit?

70°F

The speed S (in cm/sec) of a certain type of ant is $S = n(C) = \frac{1}{6}(C - 4)$, where C is the temperature in degrees Celsius.

a. Find the inverse of n.

b. If an ant is moving at 2 cm/sec, what is the temperature in degrees Celsius?

SOLUTION

a. We again use our four-step procedure, noting that the inverse of n will give the temperature F as a function of the number of chirps c. To find $n^{-1}(F)$,

1. Replace $n(F)$ by c. $c = 4(F - 40)$

2. Interchange the roles of F and c. $F = 4(c - 40)$

3. Solve for c. $c = \dfrac{F}{4} + 40$ Divide by 4, add 40.

4. Replace c by $n^{-1}(c)$. $n^{-1}(c) = \dfrac{c}{4} + 40$

b. If the cricket is chirping 120 times a minute, the temperature is

$$n^{-1}(120) = \frac{120}{4} + 40 = 70°F$$

CHECK Does the cricket make 120 chirps per minute when the temperature is 70°F? Using the equation $n(F) = 4(F - 40)$, $n(70) = 4(70 - 40) = 120$ as expected.

Answers

5. a. $n^{-1}(S) = 6S + 4$ **b.** 16°C

Calculate It Exercises

1. Window 1 shows the coordinates of six different points. Let f be the set of ordered pairs whose coordinates are shown using the symbol □ and g be the set of ordered pairs whose coordinates are shown using the symbol ◇.

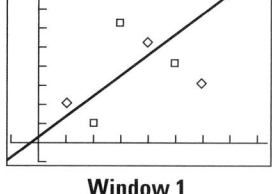
Window 1

a. Write f as a set of ordered pairs.

b. Write g as a set of ordered pairs.

c. Is g the inverse of f?

2. Window 2 shows the graphs of two linear functions and the graph of the function $f(x) = y = x$.

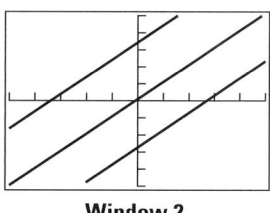
Window 2

a. Are the functions symmetric with respect to the line $y = x$?

b. Are the functions inverses of each other?

c. Can you find the equation of each of the functions?

3. You can check your work in this section's exercise set by graphing. In Problem 36, for example, $S = f(L) = 3L - 22$, and you are asked to find $f^{-1}(S)$. To check this problem graphically, use a $[-1, 30]$ by $[-1, 30]$ window as shown in Window 3.

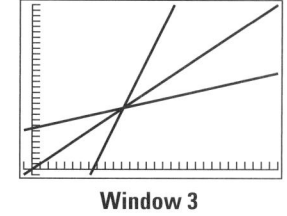
Window 3

a. What should Y_1 be?

b. If Y_2 is the inverse of Y_1, what should Y_2 be?

c. You have graphed your own Y_1 and Y_2 and the line $y = x$. Does the graph confirm that the two functions shown are inverses of each other? Explain.

Exercises 10.2

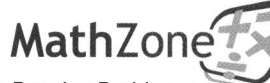
A **B** In Problems 1–4, find f^{-1}, draw the graphs of f and f^{-1} on the same axes, and determine whether f^{-1} is a function.

1. $f = \{(1, 3), (2, 4), (3, 5)\}$

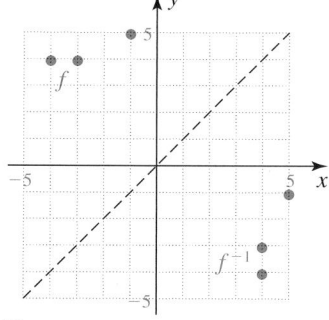

Yes

2. $f = \{(2, 3), (3, 4), (4, 5)\}$

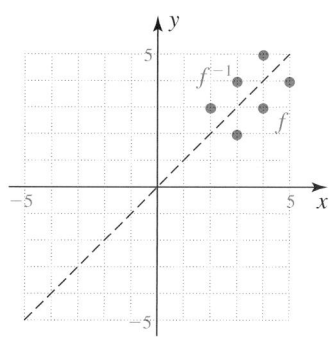

Yes

3. $f = \{(-1, 5), (-3, 4), (-4, 4)\}$

No

4. $f = \{(-2, 4), (-3, 3), (-5, 3)\}$

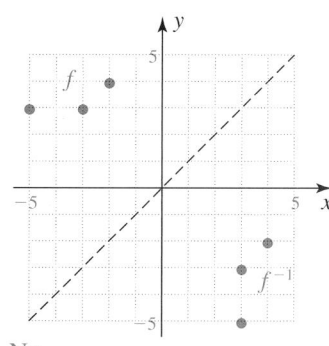

No

B C In Problems 5–14, find the equation of the inverse, graph it, and state whether the inverse is a function.

5. $\{(x,y) \mid y = 3x + 3\}$

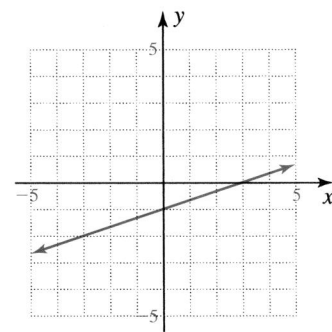

$\{(x,y) \mid y = \frac{1}{3}x - 1\}$; yes

6. $\{(x,y) \mid y = 2x + 4\}$

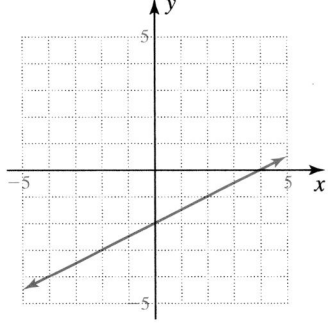

$\{(x,y) \mid y = \frac{1}{2}x - 2\}$; yes

7. $\{(x,y) \mid y = 2x - 4\}$

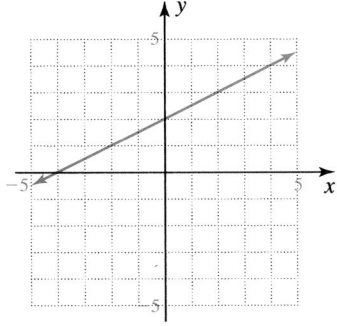

$\{(x,y) \mid y = \frac{1}{2}x + 2\}$; yes

8. $\{(x,y) \mid y = 3x - 3\}$

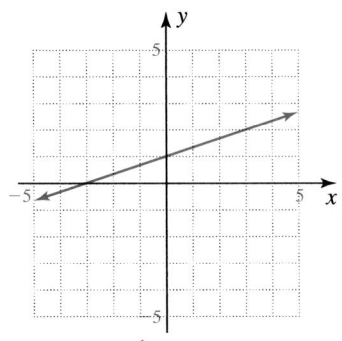

$\{(x,y) \mid y = \frac{1}{3}x + 1\}$; yes

9. $\{(x,y) \mid y = 2x^2\}$

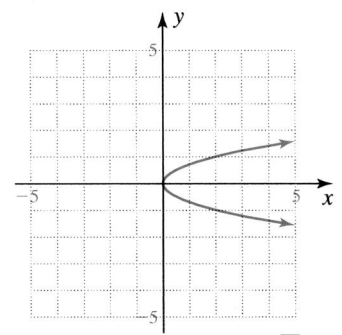

$\{(x,y) \mid y^2 = \frac{1}{2}x;\ y = \pm\frac{\sqrt{2x}}{2}\}$; no

10. $\{(x,y) \mid y = x^2 + 1\}$

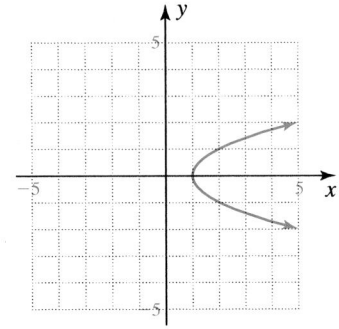

$\{(x,y) \mid y^2 = x - 1;\ y = \pm\sqrt{x-1}\}$; no

11. $\{(x,y) \mid y = x^2 - 1\}$

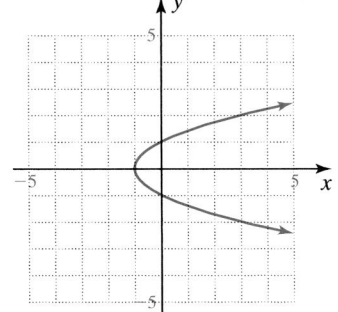

$\{(x,y) \mid y^2 = x + 1;$
$y = \pm\sqrt{x+1}\}$; no

12. $\{(x,y) \mid y = x^3 - 1\}$

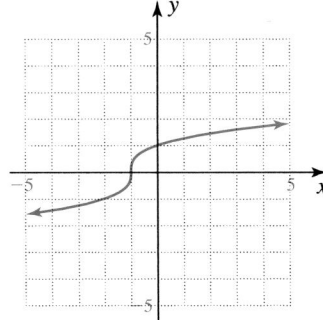

$\{(x,y) \mid y^3 = x + 1;$
$y = \sqrt[3]{x+1}\}$; yes

13. $\{(x,y) \mid y = -x^3\}$

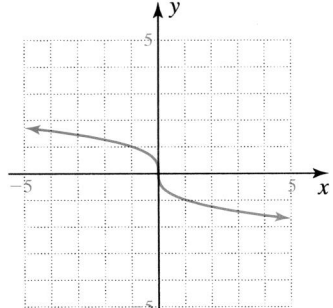

$\{(x,y) \mid y^3 = -x;$
$y = -\sqrt[3]{x}\}$; yes

14. $\{(x,y) \mid y = -2x^3\}$

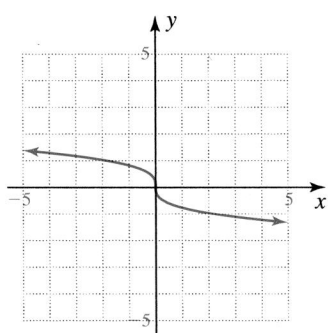

$\{(x,y) \mid y^3 = -\frac{1}{2}x;\ y = -\frac{1}{2}\sqrt[3]{4x}\}$; yes

C In Problems 15–20, use a table of values (see Example 4) to graph the function and its inverse and determine whether the inverse is a function.

15. $y = f(x) = 2^{x+1}$

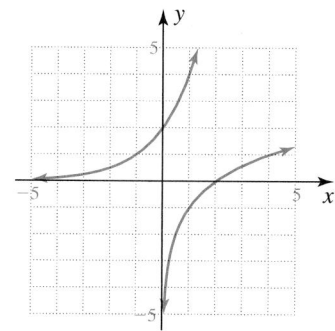

$y = 2^{x+1}$; $x = 2^{y+1}$; yes

16. $y = f(x) = 3^{x+1}$

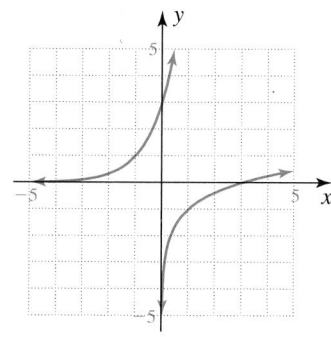

$y = 3^{x+1}$; $x = 3^{y+1}$; yes

17. $y = f(x) = \left(\dfrac{1}{3}\right)^x$

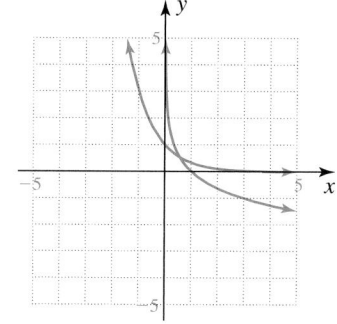

$y = (\frac{1}{3})^x$; $x = (\frac{1}{3})^y$; yes

18. $y = f(x) = \left(\dfrac{1}{2}\right)^x$

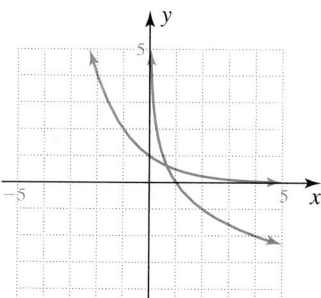

$y = (\frac{1}{2})^x$; $x = (\frac{1}{2})^y$; yes

19. $y = f(x) = 2^{-x}$

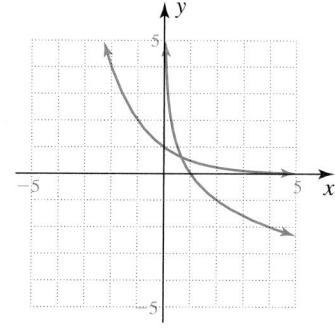

$y = 2^{-x}$; $x = 2^{-y}$; yes

20. $y = f(x) = 3^{-x}$

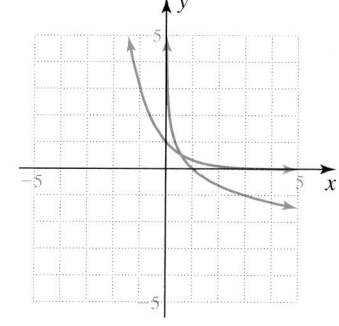

$y = 3^{-x}$; $x = 3^{-y}$; yes

21. If $f(x) = 4x + 4$, $f^{-1}(x) = \dfrac{x-4}{4}$. Find:

    a. $f(f^{-1}(3))$ 3   **b.** $f^{-1}(f(-1))$ -1

22. If $f(x) = 2x - 2$, $f^{-1}(x) = \dfrac{x+2}{2}$. Find:

    a. $f(f^{-1}(-1))$ -1 **b.** $f^{-1}(f(x))$ x

23. If $y = f(x) = \dfrac{1}{x}$, find $f^{-1}(x)$. $\dfrac{1}{x}$

24. If $y = f(x) = \dfrac{2}{x}$, find $f^{-1}(x)$. $\dfrac{2}{x}$

In Problems 25–35, graph the given function and determine the equation of its inverse. Graph the inverse by reflecting the given function along the line $y = x$ (shown dashed).

25. $y = 2x$

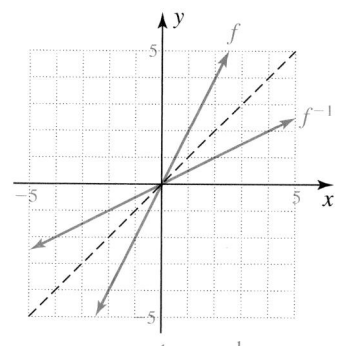

$f(x) = 2x$; $f^{-1}(x) = \frac{1}{2}x$

26. $y = -3x$

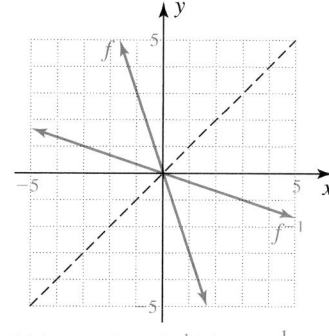

$f(x) = -3x$; $f^{-1}(x) = -\frac{1}{3}x$

27. $y = -x^2 + 2$

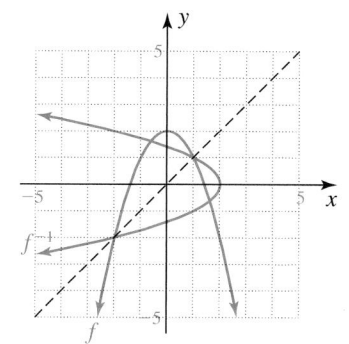

$f(x) = -x^2 + 2$; $f^{-1}(x) = \pm\sqrt{2 - x}$

or $x - 2 = -y^2$

28. $y = x^2 + 1$

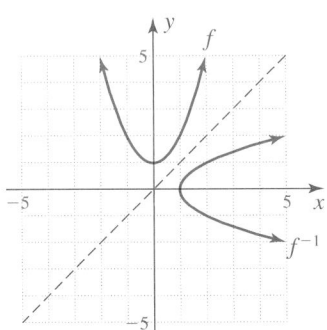

$f(x) = x^2 + 1; \; f^{-1}(x) = \pm\sqrt{x-1}$

29. $y = \sqrt{4 - x^2}$

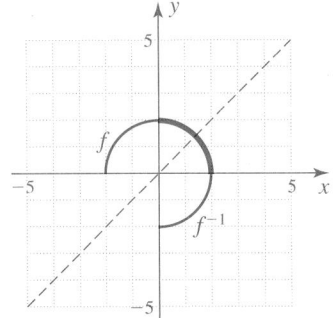

$f(x) = \sqrt{4 - x^2}; \; f^{-1}: x = \sqrt{4 - y^2}$

30. $y = -\sqrt{4 - x^2}$

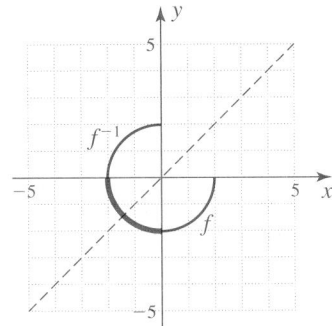

$f(x) = -\sqrt{4 - x^2};$
$f^{-1}: x = -\sqrt{4 - y^2}$

31. $y = \sqrt{9 - x^2}$

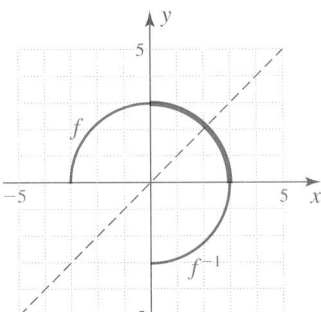

$f(x) = \sqrt{9 - x^2}; \; f^{-1}: x = \sqrt{9 - y^2}$

32. $y = -\sqrt{9 - x^2}$

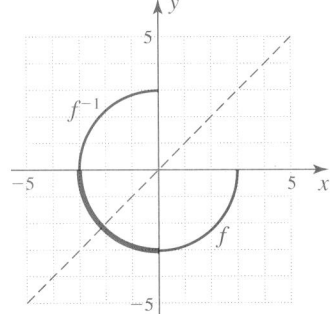

$f(x) = -\sqrt{9 - x^2};$
$f^{-1}: x = -\sqrt{9 - y^2}$

33. $y = x^3$

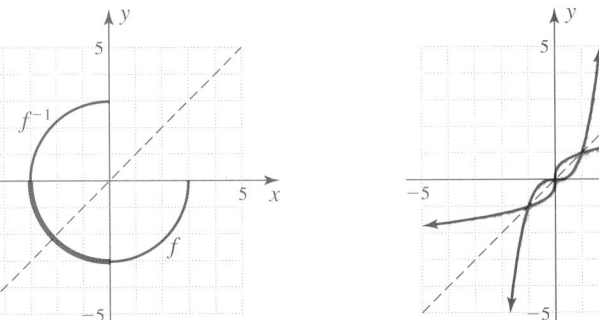

$f(x) = x^3; \; f^{-1}(x) = \sqrt[3]{x}$

34. $y = -x^3$

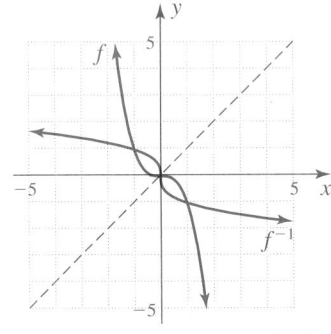

$f(x) = -x^3; \; f^{-1}(x) = \sqrt[3]{-x}$

35. $y = |x|$

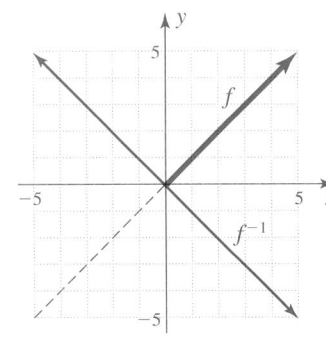

$f(x) = |x|; \; f^{-1}: x = |y|$

APPLICATIONS

Sizes and Sports

36. Your shoe size S is a function of the length of your foot L in inches. For men, the function giving this correspondence is

$$S = f(L) = 3L - 22$$

a. Find $f^{-1}(S)$. $f^{-1}(S) = \dfrac{S + 22}{3}$

b. If a man's shoe size is 7, what is the length of his foot? $9\frac{2}{3}$ in.

37. For women, the function giving the correspondence between S and L is

$$S = f(L) = 3L - 21$$

a. Find $f^{-1}(S)$. $f^{-1}(S) = \dfrac{S + 21}{3}$

b. If a woman's shoe size is 7, what is the length of her foot? $9\frac{1}{3}$ in.

38. There is a correspondence between dress sizes d in the United States and France. If the U.S. dress size is x, the corresponding size in France is

$$d = f(x) = x + 32$$

a. Find $f^{-1}(d)$. $f^{-1}(d) = d - 32$

b. What U.S. dress size corresponds to a French size 40? 8

40. There is also a relationship between bust size b (in inches) and dress size d. The correspondence is defined by the function

$$d = f(b) = b - 24$$

a. If $b = 38$, what is d? $d = 14$

b. Find $f^{-1}(d)$. $f^{-1}(d) = d + 24$

c. If $d = 16$, what is b? $b = 40$

39. In the United States, dress sizes d are a function of waist sizes w (in inches) and are given by

$$d = f(w) = w - 16$$

a. What dress size corresponds to a 32-in. waist? 16

b. Find $f^{-1}(d)$. $f^{-1}(d) = d + 16$

c. If a woman wears a size 12, what is her waist size? 28 in.

41. Can you predict Olympic outcomes? You can come close if you start with the right function. The winning time in the Women's Olympic 400-m track relay w is a function of the year x in which the event was run and is approximately given by

$$w = f(x) = -0.12x + 280 \quad \text{(seconds)}$$

a. Predict the winning time for the 1988 Olympics (the actual time was 41.98 sec). $f(1988) = 41.44$ sec

b. Find $f^{-1}(w)$. $f^{-1}(w) = \frac{280 - w}{0.12}$

c. Use $f^{-1}(w)$ to predict in what year the winning time was 40 sec. In the year 2000

42. The winning time in the Women's Olympic 200-m dash w is a function of the year x in which the event was run, starting with 1948, and is given by

$$w = f(x) = -0.0661x + 152.8 \quad \text{(seconds)}$$

a. Predict the winning time for the 1988 Olympics. (The actual time was 21.34 sec.) $f(1988) = 21.3932$ sec

b. Find $f^{-1}(w)$. $f^{-1}(w) = \frac{152.8 - w}{0.0661}$

c. Use $f^{-1}(w)$ to predict in approximately what year the winning time will be 20 sec. In the year 2009

SKILL CHECKER

Try the "Skill Checker" exercises so you'll be ready for the next section.

Evaluate:

43. 3^2 9

44. 3^{-2} $\frac{1}{9}$

45. $\left(\frac{1}{3}\right)^{-3}$ 27

46. $\left(\frac{1}{2}\right)^{x/3}$ when $x = 6$ $\frac{1}{4}$

47. $\left(\frac{1}{2}\right)^{x/3}$ when $x = -6$ 4

48. $\left(\frac{1}{3}\right)^{x/2}$ when $x = -4$ 9

USING YOUR KNOWLEDGE

The "Undoer"

What does the inverse of a function do? You can think of the inverse as "undoing" the operations performed on the variable by the function. We will demonstrate this with the function from Example 2, $f(x) = 4x - 2$. The following table shows how.

To construct the inverse, we start with x at the bottom of column (4) and work our way up. Columns (2) and (3) respectively show the operations performed on x and their "undoing."

	Operations		
Function (1)	**Function (2)**	**Inverse Operation (3)**	**Inverse Function (4)**
			Finish here.
$f(x) = 4x - 2$	Multiply by 4.	$\xrightarrow{\text{undo}}$ Divide by 4.	$f^{-1}(x) = \dfrac{x + 2}{4}$
	Subtract 2.	$\xrightarrow{\text{undo}}$ Add 2.	$x + 2$
			x
			Start here.

Now let's examine the function from Example 5, $c = n(F) = 4(F - 40)$, which as you will remember became $F = 4(c - 40)$. So we shall "undo" the operations on c. We proceed in a similar manner.

	Operations		
Function (1)	**Function (2)**	**Inverse Operation (3)**	**Inverse Function (4)**
			Finish here.
$c = n(F) = 4(F - 40)$	Subtract 40.	$\xrightarrow{\text{undo}}$ Add 40.	$n^{-1}(c) = \dfrac{c}{4} + 40$
	Multiply by 4.	$\xrightarrow{\text{undo}}$ Divide by 4.	$\dfrac{c}{4}$
			c
			Start here.

Use this method to construct the inverse for the given functions.

49. $f(x) = 3x - 2$ $f^{-1}(x) = \dfrac{x + 2}{3}$ **50.** $f(x) = 5x + 2$ $f^{-1}(x) = \dfrac{x - 2}{5}$ **51.** $f(x) = \dfrac{x + 1}{2}$ $f^{-1}(x) = 2x - 1$

52. $f(x) = \dfrac{x - 1}{2}$ $f^{-1}(x) = 2x + 1$ **53.** $f(x) = x^3 + 1$ $f^{-1}(x) = \sqrt[3]{x - 1}$ **54.** $f(x) = x^3 - 1$ $f^{-1}(x) = \sqrt[3]{x + 1}$

55. $f(x) = \sqrt{x}$ $f^{-1}(x) = x^2, x \geq 0$ **56.** $f(x) = \sqrt[3]{x}$ $f^{-1}(x) = x^3$

WRITE ON

57. Explain why the function $f(x) = ax + b$ $(a \neq 0)$ *always* has an inverse that is a function. Answers may vary.

58. Explain why the function $f(x) = ax^2 + bx + c$ $(a \neq 0)$ *never* has an inverse that is a function. Answers may vary.

59. If f and f^{-1} are functions that are inverses of each other, explain the result of the composition of f and f^{-1}; that is, what happens when you take $(f \circ f^{-1})(x)$? Answers may vary.

60. Under the same conditions as in Problem 59, what happens when you take $(f^{-1} \circ f)(x)$? Answers may vary.

61. In view of your answers for Problems 59 and 60, how could you verify that f and f^{-1} are inverses of each other? Answers may vary.

MASTERY TEST

If you know how to do these problems, you have learned your lesson!

62. Graph $y = f(x) = 2^x$. Is the inverse a function?

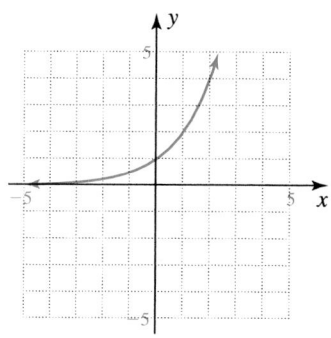

Yes

63. Find the inverse of $f(x) = x^3$. Is the inverse a function? Graph it.

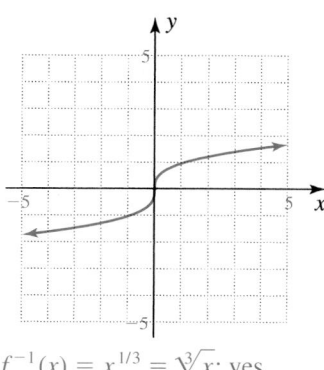

$f^{-1}(x) = x^{1/3} = \sqrt[3]{x}$; yes

64. Let $y = f(x) = 2x - 4$.

a. Find $f^{-1}(x)$. $f^{-1}(x) = \frac{1}{2}x + 2$

b. Graph f and its inverse.

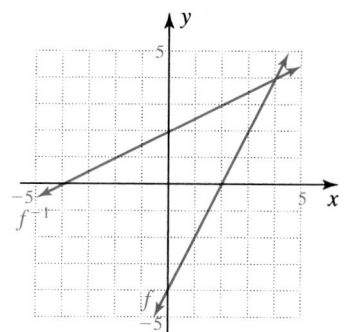

Let $S = \{(4, 3), (3, 2), (2, 1)\}$:

65. Find the domain and range of S.
$D = \{2, 3, 4\}; R = \{1, 2, 3\}$

66. Find S^{-1}.
$S^{-1} = \{(3, 4), (2, 3), (1, 2)\}$

67. Find the domain and range of S^{-1}.
$D = \{1, 2, 3\}; R = \{2, 3, 4\}$

68. Graph S and S^{-1} on the same coordinate axes.

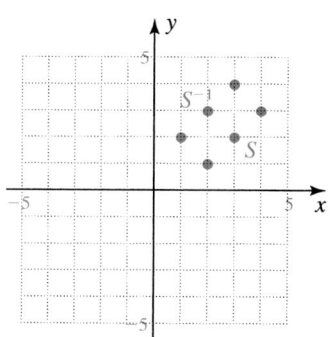

10.3 EXPONENTIAL FUNCTIONS

To Succeed, Review How To . . .

1. Understand the concept of base and exponent (pp. 31–33).

2. Interpret and evaluate expressions containing rational exponents (pp. 522–528).

Objectives

A Graph exponential functions of the form a^x or a^{-x} ($a > 0$ and $a \neq 1$).

B Determine whether an exponential function is increasing or decreasing.

C Solve applications involving exponential functions.

GETTING STARTED   What Do Cells Know about Exponents

Are you taking biology? Have you studied cell reproduction? The photographs show a cell reproducing by a process called *mitosis*. In mitosis, a single cell or bacterium divides and forms two identical daughter cells. Each daughter cell then doubles in size and

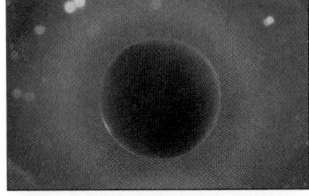
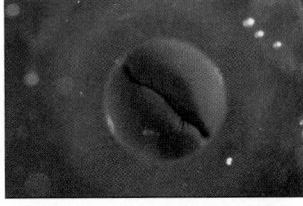
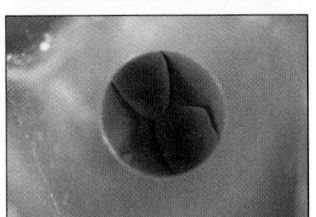

divides. As you can see, the number of bacteria present is a function of time. If we start with one cell and assume that each cell divides after 10 min, then the number of bacteria present at the end of the first 10-min period ($t = 10$) is

$$2 = 2^1 = 2^{10/10}$$

At the end of the second 10-min period ($t = 20$), the two cells divide, and the number of bacteria present is

$$4 = 2^2 = 2^{20/10}$$

Similarly, at the end of the third 10-min period ($t = 30$), the number is

$$8 = 2^3 = 2^{30/10}$$

Thus we can see that the number of bacteria present at the end of t minutes is given by the function

$$f(t) = 2^{t/10}$$

This also gives the correct result for $t = 0$, because $2^0 = 1$.

The function $f(t) = 2^{t/10}$ is called an *exponential function* because the variable t is in the exponent. In this section we shall learn more about graphing exponential functions, and we shall see how such functions can be used to solve real-world problems.

A   Graphing Exponential Functions

Exponential functions take many forms. For example, the following functions are also exponential functions:

$$f(x) = 3^x, \quad F(y) = \left(\frac{1}{2}\right)^y, \quad H(z) = (1.02)^{z/2}$$

In general, we have the following definition.

Web It

For a great lesson about exponential functions, go to link 10-3-1 at mhhe.com/bello.

EXPONENTIAL FUNCTION

An **exponential function** is a function defined for all real values of x by

$$f(x) = b^x \qquad (b > 0, b \neq 1)$$

Web It

For an introduction to exponential functions with some applications, go to link 10-3-2 at mhhe.com/bello.

NOTE

The variable b *must not* equal 1 because $f(x) = 1^x = 1$ is a constant function, *not* an exponential function.

In this definition, b is a constant called the **base,** and the **exponent** x is the variable. It is proved in more advanced courses that for $b > 0$, b^x has a unique real value for each real value of x. We assume this in all the following work.

An exponential function is frequently not in the form given in our definition, but it can be put in that form. For example, $f(t) = 2^{t/10}$ can be written as $(2^{1/10})^t$ so that the base is $2^{1/10}$ and the exponent is t.

The exponential function defined by $f(t) = 2^{t/10}$ can be graphed and used to predict the number of bacteria present after time t. To make this graph, we first construct a table giving the value of the function for certain convenient times:

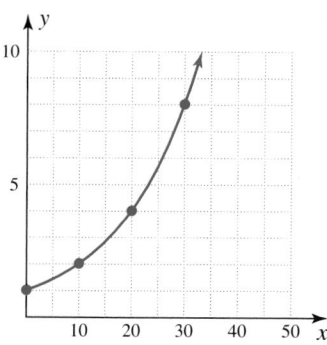

Figure 8

t	0	10	20	30
$f(t) = 2^{t/10}$	1	2^1	2^2	2^3

The corresponding points can then be graphed and joined with a smooth curve, as shown in Figure 8. In general, we graph an exponential function by plotting several points calculated from the function and then drawing a smooth curve through these points.

EXAMPLE 1	**Graphing exponential functions**

Graph on the same coordinate system:

a. $f(x) = 2^x$ **b.** $g(x) = \left(\dfrac{1}{2}\right)^x$

SOLUTION

a. We first make a table with convenient values for x and find the corresponding values for $f(x)$. We then graph the points and connect them with a smooth curve, as shown in blue in Figure 9.

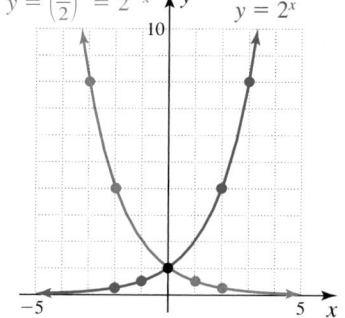

Figure 9

x	-2	-1	0	1	2
$f(x) = 2^x$	$2^{-2} = \frac{1}{4}$	$2^{-1} = \frac{1}{2}$	$2^0 = 1$	$2^1 = 2$	$2^2 = 4$

b. If we let $x = -2$,

$$g(-2) = \left(\frac{1}{2}\right)^{-2} = \frac{1}{\left(\frac{1}{2}\right)^2} = 4$$

Similarly, for $x = -1$,

$$g(-1) = \left(\frac{1}{2}\right)^{-1} = \frac{1}{\left(\frac{1}{2}\right)^1} = 2$$

For $x = 0$, 1, and 2, the function values are $\left(\frac{1}{2}\right)^0 = 1$, $\left(\frac{1}{2}\right)^1 = \frac{1}{2}$, and $\left(\frac{1}{2}\right)^2 = \frac{1}{4}$, as shown in the table.

x	-2	-1	0	1	2
$g(x) = \left(\frac{1}{2}\right)^x$	4	2	1	$\frac{1}{2}$	$\frac{1}{4}$

PROBLEM 1

Graph $f(x) = 3^x$ and $g(x) = (\frac{1}{3})^x$ on the same coordinate system.

Answer

1.

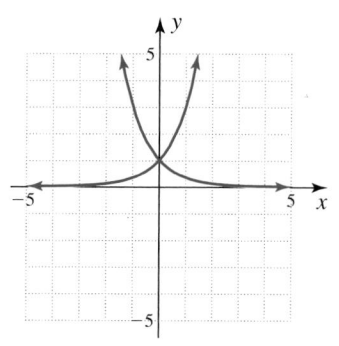

We can save time if we realize that $(\frac{1}{2})^x = (2^{-1})^x$, whose values are shown in the table for $f(x)$ in part **a.** The graph of $g(x) = (\frac{1}{2})^x = 2^{-x}$ is shown in red in Figure 9.

The two graphs in Figure 9 are symmetric to each other with respect to the y-axis. In general, we have the following fact.

> **Symmetric Graphs**
>
> The graphs of $y = b^x$ and $y = b^{-x}$ are **symmetric** to each other with respect to the y-axis.

> **CAUTION**
>
> The two functions graphed in Figure 9, $f(x) = 2^x$ and $g(x) = (\frac{1}{2})^x$ are NOT inverse functions.

Calculate It   Graphing Exponential Functions

To do Example 1, enter $Y_1 = 2^x$. (Recall that to enter 2^x with a TI-83 Plus, you press 2 [∧] [X,T,θ,n] .) Enter $Y_2 = (\frac{1}{2})^x$ by pressing [( ] 1 [÷] 2 [ )] [∧] [X,T,θ,n] . Now press [GRAPH]. The result is shown in the window.

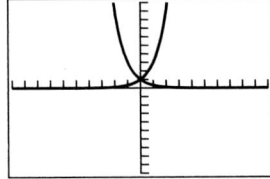

B   Determining Whether Functions Are Increasing or Decreasing

As you have seen, the graphs of some functions increase and some decrease. We'll make the idea more precise next.

INCREASING AND DECREASING FUNCTIONS

If the graph of a function rises from left to right, the function is an **increasing function.** If the graph falls from left to right, the function is a **decreasing function.**

Thus we see that $f(x) = 2^x$ is an increasing function and $g(x) = (\frac{1}{2})^x$ is a decreasing function. (See Figure 9.)

In our definition of the function $y = b^x$, it was required only that $b > 0$ and $b \neq 1$. For many practical applications, however, there is a particularly important base, the irrational number e. The value of e is approximately 2.7182818. When e is used as the base in an exponential function, the function is referred to as the natural exponential function.

> **The Natural Exponential Function, Base e**
>
> The natural exponential function is defined by
>
> $$f(x) = e^x$$
>
> The irrational real number e has the approximate value 2.7182818.

Web It

To use a calculator to determine if a function is increasing or decreasing, go to link 10-3-3 at mhhe.com/bello.

The reasons for using this base are made clear in more advanced mathematics courses, but for our purposes we need only note that e is defined as the value that the quantity $(1 + \frac{1}{n})^n$ approaches as n increases indefinitely. In symbols,

$$\left(1 + \frac{1}{n}\right)^n \rightarrow e \approx 2.7182818 \quad \text{as} \quad n \rightarrow \infty$$

To show this, we use increasing values of n (1000, 10,000, 100,000, 1,000,000) and evaluate the expression $(1 + \frac{1}{n})^n$ using a calculator with a $\boxed{x^y}$ key. (See the *Calculate It* on page 770 to find out how to do this with your calculator.)

For $n = 1000$, $\left(1 + \dfrac{1}{n}\right)^n = (1.001)^{1000} \approx 2.7169239$ Enter 1.001 $\boxed{x^y}$ 1000 $\boxed{=}$.

For $n = 10{,}000$, $\left(1 + \dfrac{1}{n}\right)^n = (1.0001)^{10{,}000} \approx 2.7181459$

For $n = 100{,}000$, $\left(1 + \dfrac{1}{n}\right)^n = (1.00001)^{100{,}000} \approx 2.7182682$

For $n = 1{,}000{,}000$, $\left(1 + \dfrac{1}{n}\right)^n = (1.000001)^{1{,}000{,}000} \approx 2.7182805$

As you can see, the value of $(1 + \frac{1}{n})^n$ is indeed getting closer to $e \approx 2.7182818$.

To graph the functions $f(x) = e^x$ and $g(x) = e^{-x}$, we make a table giving x different values (say $-2, -1, 0, 1, 2$) and finding the corresponding $y = e^x$ and $y = e^{-x}$ values. This can be done with a calculator with an $\boxed{e^x}$ key. On such calculators, you usually have to enter $\boxed{\text{INV}}$ or $\boxed{\text{2nd}}$ to find the value of e^x. [Enter 1 $\boxed{\text{2nd}}$ (or $\boxed{\text{INV}}$) $\boxed{e^x}$, and the calculator will give the value 2.7182818.] We use these ideas next.

Teaching Tip

Point out in Example 2 that because e is approximately 3, the graph of $y = e^x$ and $y = e^{-x}$ should approximate those of $y = 3^x$ and $y = 3^{-x}$.

Web It

For an interesting application relating increasing and decreasing functions and birth weight, go to link 10-3-4 at mhhe.com/bello.

EXAMPLE 2   **Graphing increasing and decreasing functions**

Use the values in the table to graph: $f(x) = e^x$ and $g(x) = e^{-x}$ and determine which of these is increasing and which is decreasing. (Use the same coordinate system.)

x	-2	-1	0	1	2
e^x	0.1353	0.3679	1	2.7183	7.3891

x	-2	-1	0	1	2
e^{-x}	7.3891	2.7183	1	0.3679	0.1353

SOLUTION   Plotting the given values, we obtain the graphs of $f(x) = e^x$ and $g(x) = e^{-x}$ shown in Figure 10. e^x and e^{-x} are symmetric to each other with respect to the y-axis. Also, e^x is increasing and e^{-x} is decreasing.

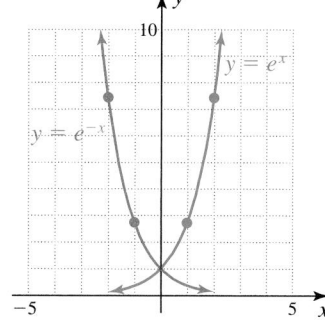

Figure 10

PROBLEM 2

Use the table to graph $f(x) = -e^x$ and $g(x) = -e^{-x}$. Which is increasing and which is decreasing?

Answer

2.

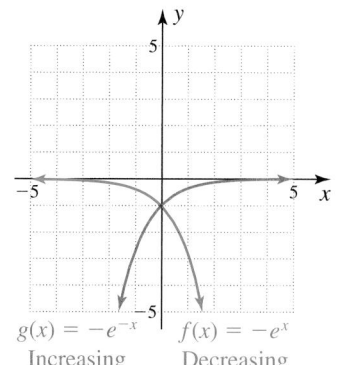

$g(x) = -e^{-x}$ $f(x) = -e^x$
Increasing   Decreasing

C Solving Applications Involving Exponential Functions

Do you have some money invested? Is it earning interest compounded annually, quarterly, monthly, or daily? Does the frequency of compounding make a difference? Some banks have instituted *continuous interest compounding*. We can compare continuous compounding and *n* compoundings per year by examining their formulas:

Continuous compounding: $A = Pe^{rt}$

n compoundings per year: $A = P\left(1 + \dfrac{r}{n}\right)^{nt}$

where A = compound amount

P = principal

r = interest rate

t = time in years

and n = periods per year (second formula only)

Web It

To learn more about compound interest, go to link, 10-3-5 at mhhe.com/bello.

As the number of times the interest is compounded increases, n increases, and it can be shown that

$$\left(1 + \frac{r}{n}\right)^{nt}$$

gets closer to e^{rt}. Let's see what the two formulas earn.

| EXAMPLE 3 Maximizing your investment | PROBLEM 3 |

EXAMPLE 3 **Maximizing your investment**

Find the compound amount:

a. For $100 compounded continuously for 18 months at 6%.

b. For $100 compounded quarterly for 18 months at 6%.

SOLUTION

a. Use the formula, $A = Pe^{rt}$, for compounding continuously. Substitute $P = 100$, $r = 0.06$, and $t = 1.5$ (18 months), so

$$A = Pe^{rt}$$
$$= 100e^{(0.06)(1.5)}$$
$$= 100e^{0.09}$$

A calculator gives the value

$$e^{0.09} \approx 1.0942$$

Thus

$$A \approx (100)(1.0942) = 109.42$$

and the compound amount is $109.42.

b. Use the formula, $A = P(1 + \frac{r}{n})^{nt}$, for compounding quarterly. As before, $P = 100$, $r = 0.06$, $t = 1.5$, and $n = 4$, so

$$A = P\left(1 + \frac{r}{n}\right)^{nt} = 100\left(1 + \frac{0.06}{4}\right)^{4(1.5)}$$
$$= 100(1 + 0.015)^6$$
$$= 100(1.015)^6$$
$$\approx 109.34$$

Thus the compound amount for $100 at the same rate, compounded quarterly, is given by $A = 100(1.015)^6 \approx 109.34$. At 18 months, the difference between continuous and quarterly compounding is only 8¢. For more comparisons, see the *Using Your Knowledge* in the exercises.

PROBLEM 3

Find the compound amount for $100 compounded

a. continuously for 30 months at 6%.

b. quarterly for 30 months at 6%.

Answers

3. a. $116.18 **b.** $116.05

EXAMPLE 4   Calculating radioactive decay

A radioactive substance decays so that G, the number of grams present, is given by

$$G = 1000e^{-1.2t}$$

where t is the time in years. Find, to the nearest gram, the amount of the substance present:

a. At the start   **b.** In 2 yr

SOLUTION

a. Here, $t = 0$, so $G = 1000e^{0} = 1000(1) = 1000$, 1000 g of the substance are present at the start.

b. Since $t = 2$, $G = 1000e^{-1.2(2)} = 1000e^{-2.4}$. To evaluate G, we use a calculator to obtain

$$e^{-2.4} \approx 0.090718$$

so that

$$G \approx (1000)(0.090718)$$
$$= 90.718$$

So, after starting with 1000 g, there are about 91 g present in 2 yr.

PROBLEM 4

Repeat Example 4(b) if the time is 18 months.

Calculate It Radioactive Decay

In Example 4, let $Y_1 = 1000e^{-1.2x}$ using a $[-1, 10]$ by $[-100, 1000]$ window and Yscl = 100. Do you now see that the substance is *decaying?* Now let's calculate the value of the function when $x = 2$ as required in the example. Press [2nd] [TRACE] [1]. Answer the calculator question by entering $X = 2$ and pressing [ENTER]. The result is $Y = 90.717953$ as shown in the window. Rounding the answer to the nearest gram, the answer is 91, as before.

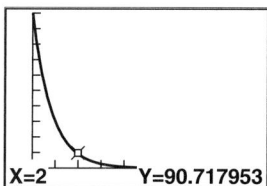

EXAMPLE 5   Predicting U.S. population

Thomas Robert Malthus invented a model for predicting population based on the idea that, when the birth rate (B) and the death rate (D) are constant and no other factors are considered, the population P is given by

$$P = P_0 e^{kt}$$

where $P =$ population at any time t

                    $P_0 =$ initial population

                    $k =$ annual growth rate ($B - D$)

                    $t =$ time in years after a given base year

According to the *Statistical Abstract of the United States,* the population in 1980 was 226,546,000, the birth rate was 0.016, and the death rate 0.0086. Use this information to predict the number of people in the United States in the year

a. 2000

b. 2010

PROBLEM 5

Using the population formula for Example 5, what would the prediction be for 2020?

Answers

4. About 165 g

5. About 304,584,000

SOLUTION

a. To predict the population in the year 2000, we identify the values to use in the formula. The population for 1980 is given, so 1980 is the base year. The initial population is $P_0 = 226{,}546{,}000$, $k = 0.016 - 0.0086 = 0.0074$, and the number of years from 1980 to 2000 is $t = 20$.

$$P = P_0 \cdot e^{kt}$$
$$= 226{,}546{,}000 e^{0.0074(20)}$$
$$= 226{,}546{,}000 e^{0.148}$$
$$\approx 262{,}683{,}009$$

The predicted population for 2000 is 262,683,009. The actual population in 2000 was about 281,000,000. Why do you think there is a discrepancy?

b. The number of years from 1980 to 2010 is $t = 30$.

$$P = P_0 \cdot e^{kt}$$
$$= 226{,}546{,}000 e^{0.0074(30)}$$
$$= 226{,}546{,}000 e^{0.222}$$
$$\approx 282{,}858{,}851$$

Thus, the predicted population is near 282,858,000. The Census Bureau predicts about 300 million. Why the discrepancy?

Web It

You can make more accurate predictions after you go to link 10-3-6 at mhhe.com/bello.

Calculate It Exercises   Approximating e

The numerical work preceding Example 2 can be done with a calculator (we used a TI-83 Plus). To find $(1.001)^{1000}$, press `2nd` `MODE` `CLEAR` (to clear your home screen) and then 1 `.` `0` `0` `1` `^` `1` `0` `0` `0` `ENTER`. For a more dramatic approximation for $e = 2.718281828$, let's tell the calculator to graph the sequence of numbers $(1 + \frac{1}{n})^n$ as n increases. Press `MODE`, move the cursor three lines down to the line starting with "FUNC"; select "SEQ." Go down to the next line and select "DOT" and then press `ENTER`. (We have told the calculator we are about to graph a sequence.) Now press `Y=`. The symbols $u(n)$ and $v(n)$ are on your screen. Let's define the sequence $u(n)$ by entering $(1 + \frac{1}{n})^n$. (The n is entered by pressing `X,T,θ,n`.) To adjust the window, press `WINDOW`, scroll down and select Xmin $= -1$, Xmax $= 10$, Xscl $= 1$, Ymin $= -1$, and Ymax $= 3$. Now press `GRAPH`. The result is shown in Window 1.

To show you that the sequence of dots are getting closer to e, use the draw feature to graph $y = e$. [Since we have entered a sequence, we can't enter the function $f(x) = e$.] Press `2nd` `PRGM` 6 and then enter `2nd` `LN` 1 `)` `ENTER`. The result is shown in Window 2. Do you now see how the sequence of dots representing $(1 + \frac{1}{n})^n$ approaches e?

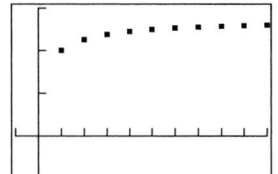

Window 1

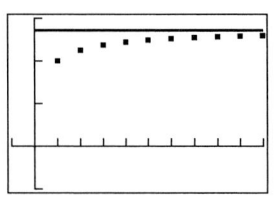

Window 2

1. Graph $Y_1 = 2^{ax}$, for $a = 1$, 2, and 3 and $Y_2 = \left(\frac{1}{2}\right)^{ax}$ for $a = 1$, 2, and 3. With some calculators you can graph the family of curves:
$$Y_1 = 2^{(\{1,\,2,\,3\}x)} \quad \text{and} \quad Y_2 = \left(\frac{1}{2}\right)^{(\{1,\,2,\,3\}x)}$$

a. What is the line of symmetry between Y_1 and Y_2?

b. Based on these graphs, make a conjecture regarding the graphs of
$$f(x) = n^x \quad \text{and} \quad g(x) = \left(\frac{1}{n}\right)^x$$

2. Graph $f(x) = b^x$ when $b = 2$, 3, and 4. (With some calculators you can graph $Y_1 = \{2, 3, 4\}^x$.)

a. Why do you think we didn't use $b = 1$?

b. Based on these graphs, what is your conjecture about the function $f(x) = b^x$ as b increases?

3. If $b > 1$, what happens to the graph of
$$f(x) = \left(\frac{1}{b}\right)^x$$
as b increases?

4. Graph $f(x) = 2^x$, $g(x) = 2^x + 3$ and $h(x) = 2^x - 3$. What happens to the graph of $f(x) = 2^x$ when a constant k is added to $f(x)$?

5. Graph $f(x) = 2^x$, $g(x) = 2^{x+3}$, and $h(x) = 2^{x-3}$. What happens to the graph of $f(x) = 2^x$ when a constant k is added to the exponent x?

Exercises 10.3

A In Problems 1–6, find the value of the given exponential for the indicated values of the variable.

1. 5^x

  a. $x = -1$ $\frac{1}{5}$

  b. $x = 0$ 1

  c. $x = 1$ 5

2. 5^{-x}

  a. $x = -1$ 5

  b. $x = 0$ 1

  c. $x = 1$ $\frac{1}{5}$

3. 3^t

  a. $t = -2$ $\frac{1}{9}$

  b. $t = 0$ 1

  c. $t = 2$ 9

4. 3^{-t}

  a. $t = -2$ 9

  b. $t = 0$ 1

  c. $t = 2$ $\frac{1}{9}$

5. $10^{t/2}$

  a. $t = -2$ $\frac{1}{10}$

  b. $t = 0$ 1

  c. $t = 2$ 10

6. $10^{-t/2}$

  a. $t = -2$ 10

  b. $t = 0$ 1

  c. $t = 2$ $\frac{1}{10}$

A, **B** In Problems 7–16, graph the functions given in parts **a** and **b** on the same coordinate system. State whether the function is increasing or decreasing.

7. **a.** $f(x) = 5^x$ Increasing

   b. $g(x) = 5^{-x}$ Decreasing

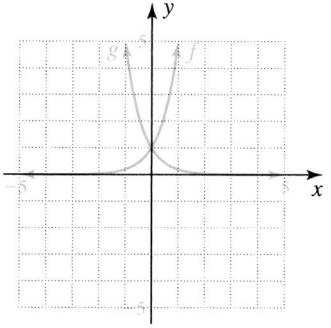

8. **a.** $f(t) = 3^t$ Increasing

   b. $g(t) = 3^{-t}$ Decreasing

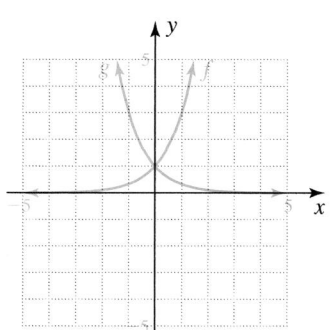

9. **a.** $f(x) = 10^x$ Increasing

   b. $g(x) = 10^{-x}$ Decreasing

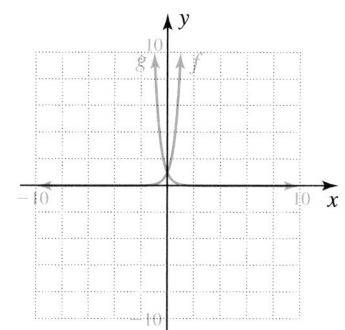

10. **a.** $f(t) = 10^{t/2}$ Increasing

    b. $g(t) = 10^{-t/2}$ Decreasing

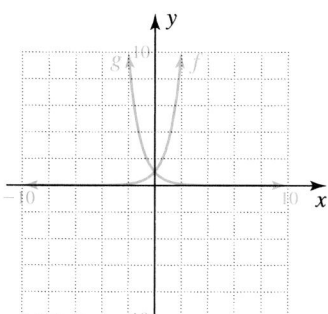

11. **a.** $f(x) = e^{2x}$ Increasing

    b. $g(x) = e^{-2x}$ Decreasing

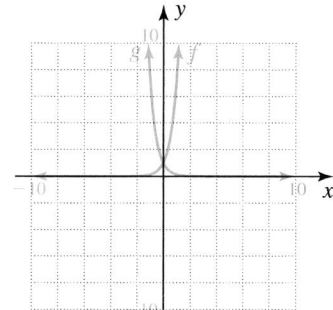

12. **a.** $f(t) = e^{t/4}$ Increasing

    b. $g(t) = e^{-t/4}$ Decreasing

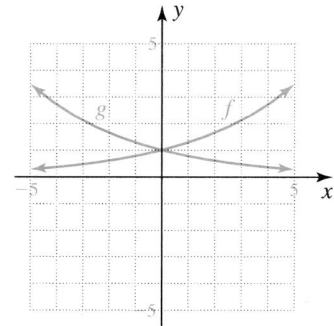

13. a. $f(x) = 3^x + 1$ Increasing

   b. $g(x) = 3^{-x} + 1$ Decreasing

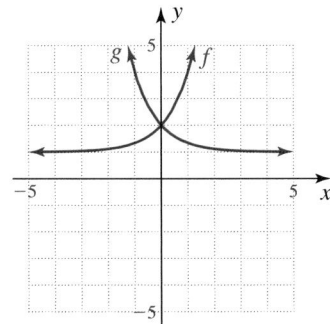

14. a. $f(x) = \left(\dfrac{1}{3}\right)^x + 1$ Decreasing

   b. $g(x) = \left(\dfrac{1}{3}\right)^{-x} + 1$ Increasing

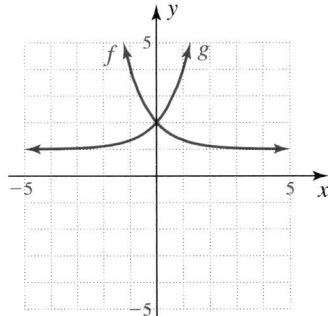

15. a. $f(x) = 2^{x+1}$ Increasing

   b. $g(x) = 2^{-x+1}$ Decreasing

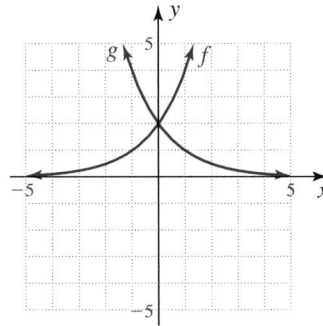

16. a. $f(x) = \left(\dfrac{1}{2}\right)^{x+1}$ Decreasing

   b. $g(x) = \left(\dfrac{1}{2}\right)^{-x+1}$ Increasing

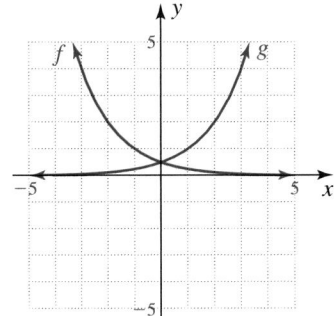

APPLICATIONS

Find and Predict

In Problems 17–20, find the compound amount if the compounding is (a) continuous or (b) quarterly.

17. $1000 at 9% for 10 yr
   a. $2459.60 b. $2435.19

18. $1000 at 9% for 20 yr
   a. $6049.65 b. $5930.15

19. $1000 at 6% for 10 yr
   a. $1822.12 b. $1814.02

20. $1000 at 6% for 20 yr
   a. $3320.12 b. $3290.66

21. The population of a town is given by the equation $P = 2000(2^{0.2t})$, where t is the time in years from 1985. Find the population in

   a. 1985 2000   **b.** 1990 4000   **c.** 1995 8000

22. A colony of bacteria grows so that their number, B, is given by the equation $B = 1200(2^t)$, where t is in days. Find the number of bacteria

   a. At the start ($t = 0$). **b.** In 5 days. **c.** In 10 days.
      1200   38,400   1,228,800

23. A radioactive substance decays so that G, the number of grams present, is given by $G = 2000e^{-1.05t}$, where t is the time in years. To the nearest tenth of a gram, find the amount of the substance present

   a. At the start. **b.** In 1 yr. **c.** In 2 yr.
      2000 g   699.9 g   244.9 g

24. Solve Problem 23 where the equation is $G = 2000e^{-1.1t}$.

   a. 2000 g   **b.** 665.7 g   **c.** 221.6 g

25. In 1980, the number of persons of Hispanic origin living in the United States was 14,609,000. If their birth rate is 0.0232 and their death rate is 0.004, predict the number of persons of Hispanic origin living in the United States for the year 2000. Round to the nearest thousand. (*Hint:* See Example 5.) 21,448,000

26. In 1980, the number of African-Americans living in the United States was 26,683,000. If their birth rate is 0.0221 and their death rate is 0.0088, predict the number of African-Americans living in the United States in the year 2000. Round to the nearest thousand. (*Hint:* See Example 5.) 34,814,000

27. The number of compact discs (CDs) sold in the United States (in millions) since 1985 can be approximated by the exponential function $S(t) = 32(10)^{0.19t}$, where t is the number of years after 1985.

  a. Predict the number of CDs sold in 1990.
  285.2 million

  b. Predict the number that will be sold in the year 2010. 1,799,492 million

28. The number of cellular phones sold in the United States (in thousands) since 1989 can be approximated by the exponential function $S(t) = 900(10)^{0.27t}$, where t is the number of years after 1989.

  a. Predict the number of cellular phones sold in 1990.
  1675.8 thousands ≈ 1.68 million

  b. Predict the number that will be sold in the year 2010.
  421,000 million thousands = 421 billion

29. According to the *Statistical Abstract of the United States,* about $\frac{2}{3}$ of all aluminum cans distributed are recycled. If a company distributes 500,000 cans, the number still in use after t years is given by the exponential function $N(t) = 500,000(\frac{2}{3})^t$. How many cans are still in use after

  a. 1 yr? 333,333 **b.** 2 yr? 222,222 **c.** 10 yr? 8671

30. If the value of an item each year is about 60% of its value the year before, after t years the salvage value of an item originally costing C dollars is given by $S(t) = C(0.6)^t$. Find the salvage value of a computer costing \$10,000

  a. 1 yr after it was bought. \$6000

  b. 10 yr after it was bought. \$60.47

31. The atmospheric pressure A (in pounds per square inch) can be approximated by the exponential function $A(a) = 14.7(10)^{-0.000018a}$, where a is the altitude in feet.

  a. The highest mountain in the world is Mount Everest, about 29,000 ft high. Find the atmospheric pressure at the top of Mount Everest. 4.42 lb/in.^{2}

  b. In the United States, the highest mountain is Mount McKinley in Alaska, whose highest point is about 20,000 ft. Find the atmospheric pressure at the top of Mount McKinley. 6.42 lb/in.^{2}

32. The atmospheric pressure A (in pounds per square inch) can also be approximated by $A(a) = 14.7e^{-0.21a}$, where a is the altitude in miles. (See Problem 31.)

  a. If we assume that the altitude of Mount Everest is about 6 mi, what is the atmospheric pressure at the top of Mount Everest? 4.17 lb/in.^{2}

  b. If we assume that the altitude of Mount McKinley is about 4 mi, what is the atmospheric pressure at the top of Mount McKinley? 6.35 lb/in.^{2}

SKILL CHECKER

Try the "Skill Checker" exercises so you'll be ready for the next section.

State the property of exponents being applied:

33. $(10^x)(10^y) = 10^{x+y}$
   Product property

34. $\dfrac{10^x}{10^y} = 10^{x-y}$
   Quotient property

35. $(10^x)^3 = 10^{3x}$
   Power property

36. $[(2)(10^x)]^y = (2^y)(10^{xy})$
   Power of a product property

USING YOUR KNOWLEDGE

Compounding Your Money

In this section you learned that for continuous compounding, the compound amount is given by $A = Pe^{rt}$. For ordinary compound interest, the compound amount is given by $A = P\left(1 + \frac{r}{n}\right)^{nt}$, where r is the annual interest rate and n is the number of periods per year. Suppose you have \$1000 to put into an account where the interest rate is 6%. How much more would you have at the end of 2 yr for continuous compounding than for monthly compounding?

For continuous compounding, the amount is given by

$$A = 1000e^{(0.06)(2)} = 1000e^{0.12}$$
$$\approx (1000)(1.1275) \quad \text{Use a calculator.}$$
$$= 1127.50$$

The amount is \$1127.50. For monthly compounding,

$$\frac{r}{n} = \frac{0.06}{12} = 0.005 \quad \text{and} \quad nt = 24$$

The amount is given by $A = 1000(1 + 0.005)^{24} = 1000(1.005)^{24}$. Compound interest tables or your calculator will give the value

$$(1.005)^{24} \approx 1.1271598$$

so that $A = 1127.16$ (to the nearest hundredth). The amount is $1127.16. Continuous compounding earns you only 34¢ more. But, see what happens when the period t is extended in Problems 37 and 38.

37. Make the same comparison where the time is 10 yr.
  Continuous = $1822.12; Monthly = $1819.40;
  Continuous is $2.72 more.

38. Make the same comparison where the time is 20 yr.
  Continuous = $3320.12; Monthly = $3310.20;
  Continuous is $9.92 more.

WRITE ON

39. The definition of the exponential function $f(x) = b^x$ does not allow $b = 1$.

   a. What type of graph will $f(x)$ have when $b = 1$?
     A horizontal line

   b. Is $f(x) = b^x$ a function when $b = 1$? Explain.
     Yes; answers may vary.

   c. Does $f(x) = b^x$ have an inverse when $b = 1$? Explain.
     No; answers may vary.

40. List some reasons to justify the condition $b > 0$ in the definition of the exponential function $f(x) = b^x$.
  Answers may vary.

41. Discuss the relationship between the graphs of $f(x) = b^x$ and $g(x) = b^{-x}$. Answers may vary.

42. In Example 5, we predicted the U.S. population for the year 2000 to be 262,683,000. The actual population in 2000 was near 281,000,000. Can you give some reasons for this discrepancy? Answers may vary.

MASTERY TEST

If you know how to do these problems, you have learned your lesson!

43. A radioactive substance decays so that the number of grams present, G, is given by $G = 1000e^{-1.2t}$, where t is the time in years. Find, to the nearest gram, the amount of substance present in 18 months. 165 g

44. The compound amount A for a principal P at rate r compounded continuously for t years is $A = Pe^{rt}$. Find the compound amount for $100 compounded continuously for 36 months where the rate is 6%. $119.72

Graph:

45. $f(x) = e^{x/2}$

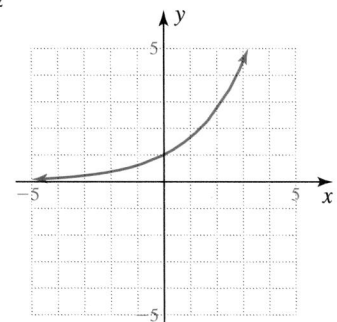

46. $f(x) = e^{-x/2}$

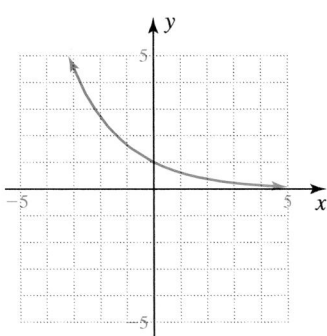

47. $f(x) = 6^x$

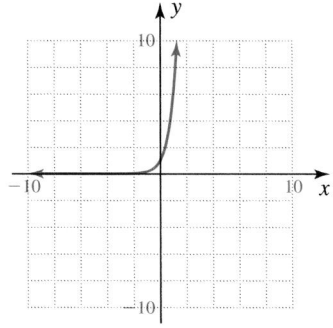

48. $f(x) = 6^{-x}$

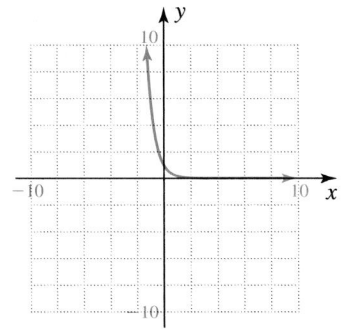

LOGARITHMIC FUNCTIONS AND THEIR PROPERTIES

To Succeed, Review How To ...

1. Use the rules of exponents in multiplication, division, and raising variables to a power (pp. 31–40).

2. Simplify algebraic expressions (pp. 51–55).

Objectives

A Graph logarithmic functions.

B Write an exponential equation in logarithmic form and a logarithmic equation in exponential form.

C Solve logarithmic equations.

D Use the properties of logarithms to simplify logarithms of products, quotients, and powers.

E Solve applications involving logarithmic functions.

GETTING STARTED

Shake, Rattle, and Roll

The damage done by the 1989 earthquake in San Francisco was tremendous, as can be seen in the photo. If I_0 denotes the minimum intensity of an earthquake (used for comparison purposes), the intensity of this quake was $10^{7.1} \cdot I_0$ or $10^{7.1}$ times as intense as the minimum intensity.

The magnitude of an earthquake is usually measured on the Richter scale. The Richter scale is logarithmic. In this scale the magnitude R of an earthquake of intensity $I = 10^{7.1} \cdot I_0$ is reported as

$$R = \log_{10} \frac{I}{I_0} = \log_{10} \frac{10^{7.1} \cdot I_0}{I_0} = \log_{10} 10^{7.1} = 7.1$$

The magnitude was reported as 7.1 on the Richter scale. 7.1 is the *exponent* of 10. The exponent 7.1 is called the logarithm, base 10, of $10^{7.1}$ and is written as

$$\log_{10} 10^{7.1} = 7.1$$

In this section we shall study logarithmic functions and their properties.

A

Graphing Logarithmic Functions

Teaching Tip

To change $y = \log_b x$ to exponential form, begin with the base (subscript) and move clockwise:

$$y = \log_b x$$

To change $b^y = x$ to logarithmic form, remember that the base is always the subscript and the exponent (y) is always on the opposite side of the equals sign from the word *log*.

In Section 10.2, we graphed the function $f(x) = 3^x$ and its inverse $x = 3^y$ and promised to show how to find $f^{-1}(x)$. Here are the steps we shall use:

$$f(x) = 3^x \qquad \text{Given.}$$

1. Replace $f(x)$ by y. $y = 3^x$

2. Interchange x and y. $x = 3^y$

3. Solve for y. $y =$ the exponent to which we raise 3 to get x

4. Replace y by $f^{-1}(x)$. $f^{-1}(x) =$ the exponent to which we raise 3 to get x

Since a logarithm is an exponent, we define "the exponent to which we raise 3 to get x" as follows.

Web It

For a presentation similar to the one given here, go to link 10-4-1 at mhhe.com/bello.

DEFINITION OF LOG₃x

$\log_3 x$ ("the logarithm base 3 of x") means "the exponent to which we raise 3 to get x."

If we use this definition in step 4, $f^{-1}(x) = \log_3 x$. $f^{-1}(9) = \log_3 9 = 2$ because 2 is the exponent to which we raise the base 3 to get 9; $\log_3 9 = 2$ means $3^2 = 9$.

In general, for any exponential function $f(x) = b^x$, the inverse is $f^{-1}(x) = \log_b x$. The graph of $f^{-1}(x) = \log_b x$ can be drawn by reflecting the graph of $f(x) = b^x$ across the line $y = x$ (see Figure 11). (Recall that the definition of exponential function requires $b > 0$ and $b \neq 1$.) Using these ideas, we have the following definition.

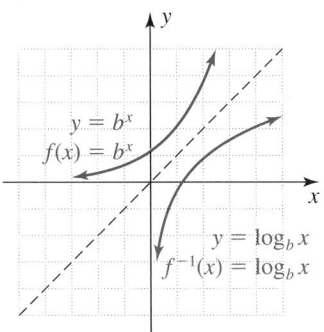

Figure 11

DEFINITION OF LOGARITHMIC FUNCTION

$f(x) = y = \log_b x$ is equivalent to $b^y = x$ $(b > 0, b \neq 1,$ and $x > 0)$

This definition means that the logarithm of a number is an exponent. For example,

$4 = \log_2 16$	is equivalent to	$2^4 = 16$	The logarithm is the exponent.
$2 = \log_5 25$	is equivalent to	$5^2 = 25$	The logarithm is the exponent.
$-3 = \log_{10} 0.001$	is equivalent to	$10^{-3} = 0.001$	The logarithm is the exponent.

EXAMPLE 1   Graphing a logarithmic function

Graph: $y = f(x) = \log_4 x$

SOLUTION   By the definition of logarithm, $y = \log_4 x$ is equivalent to $4^y = x$. We can graph $4^y = x$ by first assigning values to y and calculating the corresponding x-values.

y	$x = 4^y$	x	Ordered Pair (x, y)
0	$x = 4^0 =$	1	$(1, 0)$
1	$x = 4^1 =$	4	$(4, 1)$
2	$x = 4^2 =$	16	$(16, 2)$
-1	$x = 4^{-1} =$	$\frac{1}{4}$	$(\frac{1}{4}, -1)$
-2	$x = 4^{-2} =$	$\frac{1}{16}$	$(\frac{1}{16}, -2)$

We then graph the ordered pairs and connect them with a smooth curve, the graph of $y = \log_4 x$. To confirm that our graph is correct, graph $y = 4^x$ on the same coordinate axes (see Figure 12). As expected, the graphs are reflections of each other across the line $y = x$.

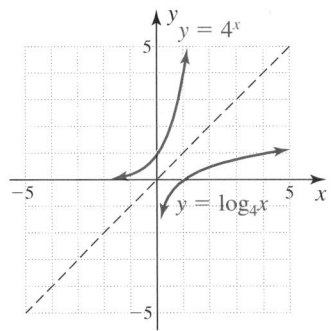

Figure 12

PROBLEM 1

Graph $y = f(x) = \log_3 x$

Teaching Tip

In Example 1, be sure students note that although we assigned y-values first and computed x next, we still plot the resulting point as the ordered pair (x, y).

Answer

1.

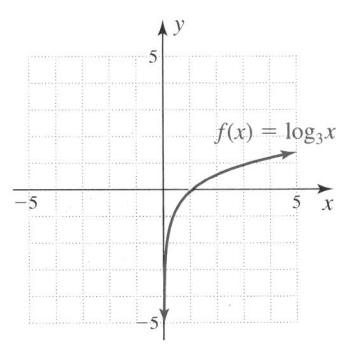

Calculate It   Graphing Log Functions

If your calculator has a **LOG** key, you can graph functions of the form $f(x) = \log x$, where the base is understood to be 10. If this is the case, how would you graph Example 1, $f(x) = \log_4 x$? To do so, you need the following property, which we cover in the next section:

$$\log_b x = \frac{\log x}{\log b}$$

Letting $b = 4$,

$$Y_1 = \log_4 x = \frac{\log x}{\log 4}$$

The graph, using a $[-5, 5]$ by $[-5, 5]$ window, is shown in the window.

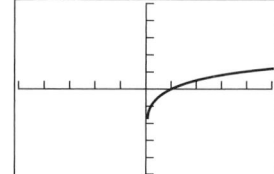

B   Converting Exponential to Logarithmic Equations and Vice Versa

The definition of logarithm states "$y = \log_b x$ is equivalent to $b^y = x$." We can use this definition to write logarithmic equations as exponential equations, and vice versa. *The logarithm is an exponent for a given base.* Thus

Web It

For a good lesson on converting exponential to logarithmic equations, go to link 10-4-2 at mhhe.com/bello.

$8 = 2^3$	is equivalent to	$\log_2 8 = 3$	The logarithm 3 is the exponent for the base 2.
$25 = 5^2$	is equivalent to	$\log_5 25 = 2$	The logarithm 2 is the exponent for the base 5.
$9 = b^a$	is equivalent to	$\log_b 9 = a$	The logarithm a is the exponent for the base b.

Similarly,

$y = \log_5 3$	is equivalent to	$5^y = 3$	The logarithm y is the exponent for the base 5.
$-2 = \log_b 4$	is equivalent to	$b^{-2} = 4$	The logarithm -2 is the exponent for the base b.
$a = \log_b c$	is equivalent to	$b^a = c$	The logarithm a is the exponent for the base b.

EXAMPLE 2   Converting an exponential equation to logarithmic form

Write in logarithmic form: $125 = 5^x$

SOLUTION   By the definition of logarithm,

$$\overset{\text{exponent}}{x = b^{\overset{\downarrow}{y}}} \quad \text{equivalent to} \quad \log_{\underset{\text{base}}{b}} x = y$$

$125 = 5^x$ is equivalent to $\log_5 125 = x$.

PROBLEM 2

Write $2^x = 1024$ in logarithmic form.

EXAMPLE 3   Converting a logarithmic equation to exponential form

Write in exponential form and check its accuracy: $\log_{32} 64 = \frac{6}{5}$

SOLUTION   By the definition of logarithm, $\log_b x = y$ means $x = b^y$. Hence $\log_{32} 64 = \frac{6}{5}$ means $32^{6/5} = 64$. Because $32^{6/5} = (\sqrt[5]{32})^6 = 2^6 = 64$, the equation $\log_{32} 64 = \frac{6}{5}$ is correct.

PROBLEM 3

Write in exponential form and check: $\log_4 32 = \frac{5}{2}$.

Answers

2. $\log_2 1024 = x$
3. $4^{5/2} = 32$ or $(\sqrt{4})^5 = 32$ (true)

C Solving Logarithmic Equations

In Example 2, we pointed out that $125 = 5^x$ is equivalent to $\log_5 125 = x$. Can we find x? Yes, if we write the equation $125 = 5^x$ as *exponentials with the same base,* set the exponents equal, and then solve the resulting equation. This can be done because the exponential function $f(x) = b^x$ is *one-to-one,* so when $b^x = b^y$, $x = y$. Thus if we rewrite the equation $125 = 5^x$ as $5^3 = 5^x$, it's easy to see that $3 = x$. In doing this, we have used the following property.

Teaching Tip

Before doing Example 4, remind students that in the definition of a logarithmic function (p. 776), the base must be greater than 0 and not equal to 1.

Equivalence Property

For any $b > 0$, $b \neq 1$, $b^x = b^y$ is equivalent to $x = y$.

| **EXAMPLE 4** Solving logarithmic equations | **PROBLEM 4** |

EXAMPLE 4 Solving logarithmic equations

Solve:

a. $\log_3 x = -2$ 　　　　　　**b.** $\log_x 9 = 2$

SOLUTION

a. The equation $\log_3 x = -2$ is equivalent to

$$3^{-2} = x$$

$$\frac{1}{9} = x$$

CHECK Substitute $x = \frac{1}{9}$ in the original equation to obtain $\log_3 \frac{1}{9} = -2$, equivalent to $3^{-2} = \frac{1}{9}$, a true statement.

b. The equation $\log_x 9 = 2$ is equivalent to

$$x^2 = 9$$

$$x = \pm 3 \quad \text{Take the square roots.}$$

But x represents the base, which must be positive, so $x = 3$ is the only answer. (We discard $x = -3$, which is negative and cannot be the base.)

CHECK Substitute $x = 3$ in the original equation to obtain $\log_3 9 = 2$, equivalent to $3^2 = 9$, a true statement.

PROBLEM 4

Solve:

a. $\log_3 x = -5$

b. $\log_x 81 = 2$

Calculate It Solving Logarithmic Equations

You can solve logarithmic equations with the same techniques used to solve other equations; graph Y_1 (the left-hand side of the equation), and Y_2 (the right-hand side of the equation) and find the point of intersection by using the ⦅ZOOM⦆ and ⦅TRACE⦆ features or the intersection feature of your calculator. The point of intersection is the solution of the equation. In Example 4(a), graph

$$Y_1 = \frac{\log x}{\log 3} = \log_3 x$$

and $Y_2 = -2$. If you have the intersection feature, press ⦅2nd⦆ ⦅TRACE⦆ ⦅5⦆, move the cursor so that it shows a positive value for x, and press ⦅ENTER⦆ three times. The answer is $x = 0.11111111$, as shown in the window. (A $[-5, 5]$ by $[-5, 5]$ window is shown.)

How do you know this is the same answer we obtained before— $\frac{1}{9}$? Divide 1 by 9, and you get $0.11111111\ldots$ the same answer.

Answers

4. a. $x = \frac{1}{243}$ 　**b.** $x = 9$

EXAMPLE 5	**Finding logarithms**

Find:

a. $\log_2 32$

b. $\log_2\left(\dfrac{1}{4}\right)$

c. $\log_{10} 1000$

d. $\log_{10} 0.01$

SOLUTION

a. We know that $\log_2 32$ means the exponent to which we must raise the base 2 to get 32. Since $2^5 = 32$, the exponent is 5 and $\log_2 32 = 5$. Alternatively, since we are looking for $\log_2 32$, we can write

$$\log_2 32 = x$$

which is equivalent to $2^x = 32$

or $2^x = 2^5$

Since the exponents must be equal, $x = 5$, as before.

b. Since $\log_2\left(\frac{1}{4}\right)$ is the exponent to which we must raise the base 2 to get $\frac{1}{4}$ and since $2^{-2} = \frac{1}{4}$, the exponent is -2 and $\log_2\left(\frac{1}{4}\right) = -2$. Alternatively, since we are looking for $\log_2\left(\frac{1}{4}\right)$, we can write

$$\log_2\left(\dfrac{1}{4}\right) = x$$

which is equivalent to $2^x = \dfrac{1}{4}$

or $2^x = 2^{-2}$

Since the exponents must be equal, $x = -2$, as before.

c. Since $10^3 = 1000$, $\log_{10} 1000 = 3$. Alternatively, let

$$\log_{10} 1000 = x$$

which is equivalent to $10^x = 1000 = 10^3$ Since the exponents

and $x = 3$ must be equal

d. Since $10^{-2} = \frac{1}{100} = 0.01$, $\log_{10} 0.01 = -2$. Alternatively, let

$$\log_{10} 0.01 = x$$

which is equivalent to $10^x = 0.01 = 10^{-2}$ Since the exponents

and $x = -2$ must be equal

PROBLEM 5

Find:

a. $\log_2 64$ **b.** $\log_2\left(\dfrac{1}{8}\right)$

c. $\log_{10} 100$ **d.** $\log_{10} 0.1$

Teaching Tip

For Example 5(b), remind students about expressing values with prime number bases like $4 = 2^2$. Also, practice using the definition of negative exponents like $2^{-2} = \frac{1}{2^2}$.

Web It

For a lesson covering logarithms, go to link 10-4-3 at mhhe.com/bello.

If you want a lesson with practice, answers, and an instructional video go to link 10-4-4 at mhhe.com/bello.

D	**Using Properties of Logarithms**

Web It

To see a lesson explaining the properties of logarithms go to links 10-4-5 and 10-4-6 at mhhe.com/bello.

Logarithms have three important properties that are the counterparts of the corresponding properties of exponents. By the definition of logarithm, if $x = \log_b M$ and $y = \log_b N$, then $M = b^x$ and $N = b^y$. From the properties of exponents, it follows that

$$MN = b^x b^y = b^{x+y}$$

so that

$$\log_b MN = x + y = \log_b M + \log_b N$$

This means that the *logarithm of a product is the sum of the logarithms of its factors.* Similarly,

$$\frac{M}{N} = \frac{b^x}{b^y} = b^{x-y}$$

Answers

5. a. 6   **b.** -3 **c.** 2   **d.** -1

which shows that

$$\log_b \frac{M}{N} = x - y = \log_b M - \log_b N$$

Thus the *logarithm of M divided by N is the logarithm of M minus the logarithm of N*. If we have a power of a number such as M^r, then that

$$M^r = (b^x)^r = b^{rx}$$

means that $\log_b M^r = rx = r \log_b M$. In words, the *logarithm of M^r is r times the logarithm of M*. We summarize these results as follows.

Properties of Logarithms		
	$\log_b MN = \log_b M + \log_b N$	Product property
	$\log_b \dfrac{M}{N} = \log_b M - \log_b N$	Quotient property
	$\log_b M^r = r \log_b M$	Power property

EXAMPLE 6   **Using the power and quotient properties**

Use the properties of logarithms to show that

$$\log_b\left(\frac{x^4}{16}\right) = 4 \log_b x - \log_b 16$$

SOLUTION   Use the properties to rewrite the right side

$$\log_b\left(\frac{x^4}{16}\right) = 4 \log_b x - \log_b 16$$

$$= \log_b x^4 - \log_b 16 \qquad \text{Power property}$$

$$= \log_b\left(\frac{x^4}{16}\right) \qquad \text{Quotient property}$$

PROBLEM 6

Use the properties of logarithms to show that

$$\log_b\left(\frac{a^2}{c^3}\right) = 2 \log_b a - 3 \log_b c$$

EXAMPLE 7   **Using the product and quotient properties**

Assume the logarithms are all to base 10 and use the properties of logarithms to show that

$$\log\left(\frac{\sqrt[3]{pq}}{r}\right) = \frac{1}{3} \log p + \frac{1}{3} \log q - \log r$$

where base 10 is understood throughout.

SOLUTION   Since

$$\log\left(\frac{\sqrt[3]{pq}}{r}\right) = \frac{1}{3} \log p + \frac{1}{3} \log q - \log r$$

$$= \log p^{1/3} + \log q^{1/3} - \log r \qquad \text{Power property}$$

$$= \log(pq)^{1/3} - \log r \qquad \text{Product property}$$

$$= \log\left(\frac{\sqrt[3]{pq}}{r}\right) \qquad \text{Quotient property}$$

PROBLEM 7

Assume the logarithms are all to base 10 and use the properties of logarithms to show that

$$\log\left(\frac{\sqrt{xy}}{z^3}\right) = \frac{1}{2} \log x + \frac{1}{2} \log y - 3 \log z$$

Teaching Tip

For Example 7, remind students that radicals can be expressed with rational exponents. Thus, $\sqrt[3]{pq} = (pq)^{1/3}$.

Answers

6. $\log_b\left(\frac{a^2}{c^3}\right) = 2 \log_b a - 3 \log_b c$

$= \log_b a^2 - \log_b c^3$

$= \log_b\left(\frac{a^2}{c^3}\right)$

7. $\log\left(\frac{\sqrt{xy}}{z^3}\right) = \frac{1}{2} \log x + \frac{1}{2} \log y - 3 \log z$

$= \log x^{1/2} + \log y^{1/2} - \log z^3$

$= \log(xy)^{1/2} - \log z^3$

$= \log\left(\frac{\sqrt{xy}}{z^3}\right)$

There are three more properties of logarithms worth mentioning. We have stated why these properties are true by writing the given logarithmic equation as an equivalent exponential equation.

Other Properties of Logarithms

$$\log_b 1 = 0 \qquad \text{Because } b^0 = 1$$

$$\log_b b = 1 \qquad \text{Because } b^1 = b$$

$$\log_b b^x = x \qquad \text{Because } b^x = b^x$$

 **Solving Applications Involving Logarithmic Functions**

We can use logarithmic functions to solve a variety of applications, from measuring earthquake intensity to compounding interest.

EXAMPLE 8   **An earthquake of logarithmic proportion**

In the *Getting Started* we mentioned the 1989 San Francisco earthquake, but earthquakes are commonplace in California. One California earthquake was $10^{6.4}$ times as intense as that of a quake of minimum intensity, I_0. What was the magnitude of that quake on the Richter scale?

SOLUTION   We are given that $I = 10^{6.4} I_0$, so the magnitude on the Richter scale is

$$R = \log_{10} \frac{I}{I_0} = \log_{10} 10^{6.4} = 6.4$$

PROBLEM 8

Do Example 8 with $10^{6.4}$ replaced by $10^{7.2}$.

Teaching Tip

For Example 8, given that $I = 10^{6.4} I_0$, dividing both sides by I_0 yields

$$\frac{I}{I_0} = 10^{6.4}$$

Thus

$$\log \frac{I}{I_0} = \log 10^{6.4}$$

EXAMPLE 9   **Compounding the purchase price of Manhattan**

In 1626, Peter Minuit bought the island of Manhattan from the Native Americans for the equivalent of about $24. If this money had been invested at 5% compounded annually, in 2001 the money would have been worth $24(1.05)^{375}$. If it's known that $\log_{10} 24 \approx 1.3802$ and $\log_{10} 1.05 \approx 0.0212$, find $\log_{10} 24(1.05)^{375}$.

SOLUTION

$$\log_{10} 24(1.05)^{375} = \log_{10} 24 + \log_{10} (1.05)^{375}$$

$$= \log_{10} 24 + 375(\log_{10} 1.05)$$

$$\approx 1.3802 + 375(0.0212)$$

$$= 1.3802 + 7.9500$$

$$= 9.3302 \qquad \text{Log of the accumulated amount}$$

Now suppose the accumulated amount is A. We have

$$\log_{10} A \approx 9.3302$$

or

$$10^{9.3302} = A \approx 2,139,000,000 \qquad \text{Rounded from 2,138,946,884}$$

Over two billion dollars. Do you think you could buy Manhattan for two billion dollars today?

PROBLEM 9

Do Example 9 when the rate is changed to 6% if it is given that $\log_{10} 1.06 \approx 0.0253$.

Web It

For more applications of logarithms, go to link 10-4-7 at mhhe.com/bello.

Answers

8. 7.2   **9.** 10.8677

Calculate It Exercises

1. Assuming the logarithms are all to base 10 in Example 6, confirm graphically by making sure that both sides of the equation

$$\log\left(\frac{x^4}{16}\right) = 4 \log x - \log 16$$

produce the same graph. To do this, you need to make some decisions.

a. What should Y_1 be?
b. What should Y_2 be?

2. Graph $f(x) = \log x$ using a $[-5, 5]$ by $[-5, 5]$ window and, based on the graph, answer the following questions regarding the logarithmic function $f(x) = \log x$.

a. What is the domain of $f(x)$?
b. What is the range of $f(x)$?
c. What is the x-intercept of $f(x)$?
d. Is there a y-intercept?
e. What happens to $f(x)$ as x increases?

Exercises 10.4

A In Problems 1–8, graph the equation.

1. $y = \log_2 x$

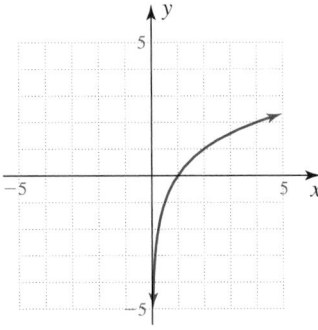

2. $y = \log_3 x$

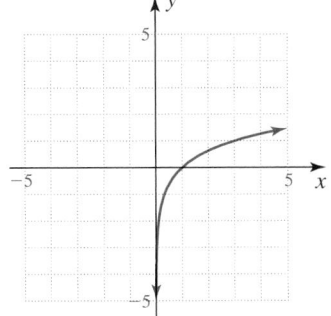

3. $y = \log_5 x$

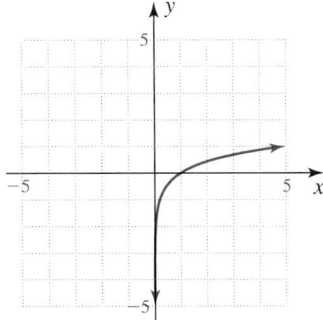

4. $f(x) = \log_6 x$

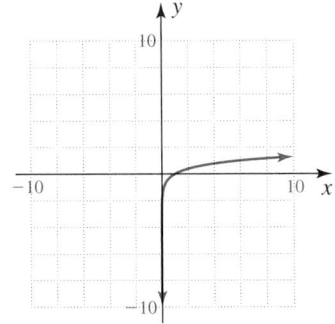

5. $f(x) = \log_{1/2} x$

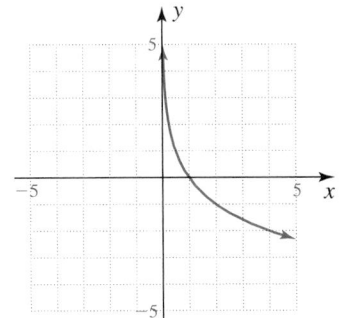

6. $f(x) = \log_{1/3} x$

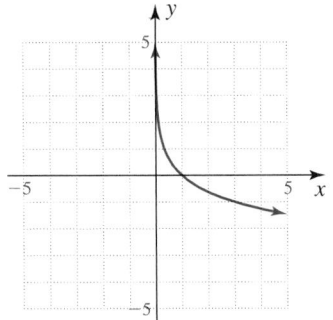

7. $y = \log_{1.5} x$

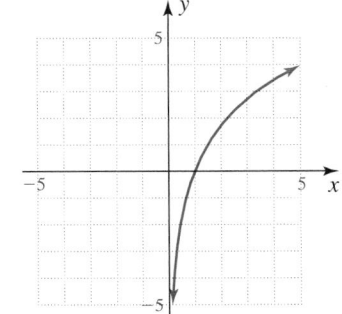

8. $y = \log_{2.5} x$

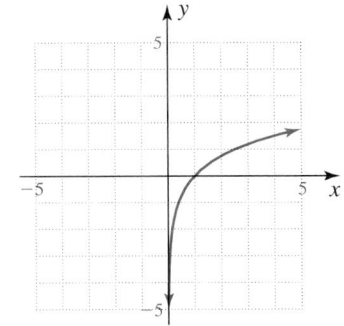

B In Problems 9–18, write the equation in logarithmic form.

9. $2^x = 128$ $\log_2 128 = x$ **10.** $3^x = 81$ $\log_3 81 = x$ **11.** $10^t = 1000$ $\log_{10} 1000 = t$

12. $10^{-t} = 0.001$ $\log_{10} 0.001 = -t$ **13.** $81^{1/2} = 9$ $\log_{81} 9 = \frac{1}{2}$ **14.** $16^{1/2} = 4$ $\log_{16} 4 = \frac{1}{2}$

15. $216^{1/3} = 6$ $\log_{216} 6 = \frac{1}{3}$ **16.** $64^{1/6} = 2$ $\log_{64} 2 = \frac{1}{6}$ **17.** $e^3 = t$ $\log_e t = 3$

18. $e^2 = 7.389056$ $\log_e 7.389056 = 2$

In Problems 19–32, write the equation in exponential form and check its accuracy (if possible).

19. $\log_9 729 = 3$ $9^3 = 729$; true   **20.** $\log_7 343 = 3$ $7^3 = 343$; true   **21.** $\log_2 \frac{1}{256} = -8$ $2^{-8} = \frac{1}{256}$; true

22. $\log_5 \frac{1}{125} = -3$ $5^{-3} = \frac{1}{125}$; true **23.** $\log_{81} 27 = \frac{3}{4}$ $81^{3/4} = 27$; true **24.** $\log_{64} 32 = \frac{5}{6}$ $64^{5/6} = 32$; true

25. $x = \log_4 16$ $4^x = 16$ **26.** $t = \log_5 10$ $5^t = 10$ **27.** $-2 = \log_{10} 0.01$ $10^{-2} = 0.01$; true

28. $-3 = \log_{10} 0.001$ **29.** $\log_e 30 = 3.4012$ **30.** $\log_e 40 = 3.6889$
   $10^{-3} = 0.001$; true $e^{3.4012} = 30$; true $e^{3.6889} = 40$; true

31. $\log_e 0.3166 = -1.15$ **32.** $\log_e 0.2592 = -1.35$
   $e^{-1.15} = 0.3166$; true $e^{-1.35} = 0.2592$; true

C In Problems 33–50, solve the equation.

33. $\log_3 x = 2$ $x = 9$ **34.** $\log_2 x = 3$ $x = 8$ **35.** $\log_3 x = -3$ $x = \frac{1}{27}$

36. $\log_2 x = -3$ $x = \frac{1}{8}$ **37.** $\log_4 x = \frac{1}{2}$ $x = 2$ **38.** $\log_9 x = \frac{1}{2}$ $x = 3$

39. $\log_8 x = \frac{1}{3}$ $x = 2$ **40.** $\log_x 4 = 2$ $x = 2$ **41.** $\log_x 16 = 4$ $x = 2$

42. $\log_x 8 = 3$ $x = 2$ **43.** $\log_x 27 = 3$ $x = 3$ **44.** $\log_8 \frac{1}{8} = x$ $x = -1$

45. $\log_2 \frac{1}{4} = x$ $x = -2$ **46.** $\log_3 1 = x$ $x = 0$ **47.** $\log_{16} \frac{1}{2} = x$ $x = -\frac{1}{4}$

48. $\log_{32} \frac{1}{2} = x$ $x = -\frac{1}{5}$ **49.** $\log_3 \frac{1}{9} = x$ $x = -2$ **50.** $\log_2 \frac{1}{8} = x$ $x = -3$

In Problems 51–74, find the value of the logarithm.

51. $\log_2 256$ 8   **52.** $\log_2 128$ 7   **53.** $\log_3 81$ 4   **54.** $\log_3 243$ 5

55. $\log_2 \frac{1}{8}$ -3 **56.** $\log_3 \frac{1}{27}$ -3 **57.** $\log_{10} 1{,}000{,}000$ 6   **58.** $\log_{10} 0.001$ -3

59. $\log_3 1$ 0   **60.** $\log_{10} 1$ 0   **61.** $\log_{10} 10$ 1   **62.** $\log_2 2$ 1

63. $\log_e e$ 1   **64.** $\log_e \frac{1}{e}$ -1 **65.** $\log_5 \frac{1}{5}$ -1 **66.** $\log_{27} 3$ $\frac{1}{3}$

67. $\log_8 2$ $\frac{1}{3}$ **68.** $\log_e e^2$ 2   **69.** $\log_e e^{-3}$ -3 **70.** $\log_e e^{-5}$ -5

71. $\log_{10} 10^t$ t **72.** $\log_e e^x$ x **73.** $\log_4 4^t$ t **74.** $\log_5 5^t$ t

D In Problems 75–84, use the properties of logarithms to transform the left-hand side into the right-hand side of the stated equation. Assume the logarithms are all to base 10.

75. $\log \frac{26}{7} - \log \frac{15}{63} + \log \frac{5}{26} = \log 3$

$\log \frac{26}{7} - \log \frac{15}{63} + \log \frac{5}{26} = \log \left(\frac{26}{7} \div \frac{15}{63} \cdot \frac{5}{26} \right)$

$= \log \left(\frac{\cancel{26}^1}{\cancel{7}_1} \cdot \frac{\cancel{63}^9}{\cancel{15}_3} \cdot \frac{\cancel{5}^1}{\cancel{26}_1} \right) = \log \left(\frac{9}{3} \right) = \log 3$

76. $\log 9 - \log 8 - \log\sqrt{75} + \log \sqrt{\frac{25}{27}} = -3 \log 2$

$\log 9 - \log 8 - \log\sqrt{75} + \log \sqrt{\frac{25}{27}}$

$= \log \left(9 \div 8 \div \sqrt{75} \cdot \sqrt{\frac{25}{27}} \right) = \log \left(9 \cdot \frac{1}{8} \cdot \frac{1}{\sqrt{75}} \cdot \sqrt{\frac{25}{27}} \right)$

$= \log \left(\cancel{9} \cdot \frac{1}{8} \cdot \frac{1}{5\cancel{\sqrt{3}}} \cdot \frac{5\cancel{\sqrt{3}}}{\cancel{9}} \right) = \log \frac{1}{8} = \log (2^{-3})$

$= -3 \log 2$

77. $\log b^3 + \log 2 - \log \sqrt{b} + \log \dfrac{\sqrt{b^3}}{2} = 4 \log b$

$\log b^3 + \log 2 - \log \sqrt{b} + \log \dfrac{\sqrt{b^3}}{2}$

$= \log\left(b^3 \cdot 2 \div \sqrt{b} \cdot \dfrac{\sqrt{b^3}}{2}\right) = \log\left(b^3 \cdot \cancel{2} \cdot \dfrac{1}{\sqrt{b}} \cdot \dfrac{b\sqrt{b}}{\cancel{2}}\right)$

$= \log b^4 = 4 \log b$

78. $\log k^2 - \log k^{-2} - \log \sqrt{k} - \log k^{-1} = \dfrac{9}{2} \log k$

$\log k^2 - \log k^{-2} - \log \sqrt{k} - \log k^{-1}$

$= \log(k^2 \div k^{-2} \div \sqrt{k} \div k^{-1})$

$= \log\left[k^{(2+2-\frac{1}{2}+1)}\right] = \log(k^{9/2})$

$= \dfrac{9}{2} \log k$

79. $\log k^{3/2} + \log r - \log k - \log r^{3/4} = \dfrac{1}{4}(\log k^2 r)$

$\log k^{3/2} + \log r - \log k - \log r^{3/4}$

$= \log(k^{3/2} \cdot r \div k \div r^{3/4}) = \log(k^{3/2-1} \cdot r^{1-3/4})$

$= \log(k^{1/2} r^{1/4}) = \log(k^{2/4} r^{1/4}) = \log(k^2 r)^{1/4}$

$= \dfrac{1}{4} \log k^2 r$

80. $\log a - \dfrac{1}{6} \log b - \dfrac{1}{2} \log a + \dfrac{1}{3} \log b = \dfrac{1}{6} \log a^3 b$

$\log a - \dfrac{1}{6} \log b - \dfrac{1}{2} \log a + \dfrac{1}{3} \log b$

$= \log(a \div b^{1/6} \div a^{1/2} \cdot b^{1/3})$

$= \log(a^{1-1/2} \cdot b^{-1/6+1/3}) = \log(a^{1/2} b^{1/6})$

$= \log(a^{3/6} b^{1/6}) = \log(a^3 b)^{1/6}$

$= \dfrac{1}{6} \log(a^3 b)$

81. $2 \log b + 6 \log a - 3 \log b = \log\left(\dfrac{a^6}{b}\right)$

$2 \log b + 6 \log a - 3 \log b$

$= \log b^2 + \log a^6 - \log b^3$

$= \log(b^2 \cdot a^6 \div b^3)$

$= \log\left(\dfrac{b^2}{1} \cdot \dfrac{a^6}{1} \cdot \dfrac{1}{b^3}\right)$

$= \log\left(\dfrac{a^6}{b}\right)$

82. $2 \log b + 2 \log a - 4 \log b = \log\left(\dfrac{a^2}{b^2}\right)$

$2 \log b + 2 \log a - 4 \log b$

$= \log b^2 + \log a^2 - \log b^4$

$= \log(b^2 \cdot a^2 \div b^4)$

$= \log\left(\dfrac{b^2}{1} \cdot \dfrac{a^2}{1} \cdot \dfrac{1}{b^4}\right)$

$= \log\left(\dfrac{a^2}{b^2}\right)$

83. $\dfrac{1}{3} \log x + \dfrac{1}{3} \log y^2 - \log z = \log\left(\dfrac{\sqrt[3]{xy^2}}{z}\right)$

$\dfrac{1}{3} \log x + \dfrac{1}{3} \log y^2 - \log z$

$= \log x^{1/3} + \log y^{2/3} - \log z$

$= \log(x^{1/3} \cdot y^{2/3}) - \log z$

$= \log(\sqrt[3]{xy^2} \div z)$

$= \log\left(\dfrac{\sqrt[3]{xy^2}}{z}\right)$

84. $\dfrac{1}{4} \log x^3 + \dfrac{1}{4} \log y - 2 \log z = \log\left(\dfrac{\sqrt[4]{x^3 y}}{z}\right)$

$\dfrac{1}{4} \log x^3 + \dfrac{1}{4} \log y - 2 \log z$

$= \log x^{3/4} + \log y^{1/4} - \log z^2$

$= \log(x^{3/4} \cdot y^{1/4}) - \log z^2$

$= \log(\sqrt[4]{x^3 y} \div z^2)$

$= \log\left(\dfrac{\sqrt[4]{x^3 y}}{z^2}\right)$

APPLICATIONS

Earthquakes, Money, and Height

85. The worst earthquake ever recorded occurred in the Pacific Ocean near Colombia in 1906. The intensity of this earthquake was $10^{8.9}$ times as great as that of an earthquake of minimum intensity I_0. What was the magnitude of this earthquake on the Richter scale? $R = 8.9$

86. The San Francisco earthquake of 1906 was $10^{8.3}$ times as intense as an earthquake of minimum intensity I_0. What was the magnitude of the 1906 San Francisco earthquake on the Richter scale? $R = 8.3$

87. When Johnny was born, his father deposited $5000 in an account that paid 6% compounded monthly. This was to help pay Johnny's college expenses starting on his eighteenth birthday. The compound amount in this account was $A = \$5000(1.005)^{216}$. If $\log_{10} 5000 \approx 3.69897$ and $\log_{10} 1.005 \approx 0.00217$, find $\log_{10} A$ to five decimal places using the properties of logarithms.
$\log A = 4.16769$

88. Suppose that the account in Problem 87 paid 10% compounded quarterly. The compound amount would then be given by $A = \$5000(1.025)^{72}$. If $\log_{10} 1.025 \approx 0.01072$, find $\log_{10} A$ to five decimal places using the properties of logarithms.
$\log A = 4.47081$

89. By the age of 2, most children have reached about 50% of their mature height. Thus if you measure the height of a 2-yr-old child and double this height, the result should be close to the child's mature height. The percent P of adult height attained by a boy can be approximated by the function

$$P(A) = 29 + 50 \log_{10}(A + 1)$$

where A is the age in years ($0 < A < 17$).

Use a calculator to answer the following questions. (Approximate the answers to the nearest percent.)

a. What percent of his mature height is a 2-yr-old boy? 53%

b. What percent of his mature height is an 8-yr-old boy? 77%

c. What percent of his mature height is a 16-yr-old boy? 91%

d. Graph $P(A)$.

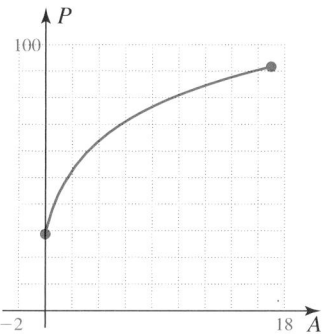

e. If a 12-yr-old boy is 60 in. tall, how tall, to the nearest inch, would you expect him to be when he is an adult? 71 in.

90. The height H of a girl (in inches) can be approximated by the function $H(A) = 11 + 19.44 \log_e A$, where A is the age in years and $6 < A < 15$.

a. What should the height of a 7-yr-old girl be? 49 in.

b. What should the height of an 11-yr-old girl be? 58 in.

c. What should the height of a 13-yr-old girl be? 61 in.

d. Graph $H(A)$. Remember that the domain is $\{A \mid 6 < A < 15\}$.

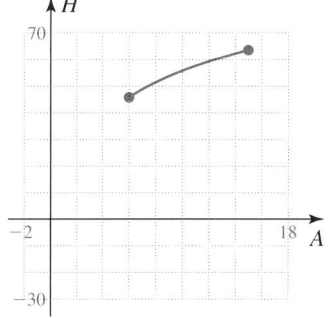

SKILL CHECKER

Try the "Skill Checker" exercises so you'll be ready for the next section.

Write in scientific notation:

91. 32.68 3.268×10^1

92. 326.8 3.268×10^2

93. 0.002387 2.387×10^{-3}

94. 0.0004392 4.392×10^{-4}

95. 0.0000569 5.69×10^{-5}

96. 0.000006731 6.731×10^{-6}

USING YOUR KNOWLEDGE

Which Is Better?

One way of comparing the compound amounts in Problems 87 and 88 is to find their ratio. If we denote this ratio by R, then

$$R = \frac{5000(1.005)^{216}}{5000(1.025)^{72}} = \frac{(1.005)^{216}}{(1.025)^{72}}$$

so that

$$\log_{10} R = \log_{10}(1.005)^{216} - \log_{10}(1.025)^{72}$$
$$= 216 \log_{10} 1.005 - 72 \log_{10} 1.025$$

Using the values of the logarithms given in Problems 87 and 88, we get

$$\log R \approx (216)(0.00217) - (72)(0.01072) = -0.30312$$

The negative value for $\log_{10} R$ means that R is less than 1. Can you see why? This means that the compound amount in Problem 87 is less than that in Problem 88. (A calculator gives the value of R as about 0.49631 so that the first amount is less than half the second.)

97. Compare an investment of $1000 at 8% compounded quarterly with the same amount invested at 8.5% compounded annually for a period of 10 yr. You will need the values $\log_{10} 1.02 \approx 0.00860$ and $\log_{10} 1.085 \approx 0.03543$.
$R = \frac{(1.02)^{40}}{(1.085)^{10}}$; $\log_{10} R = -0.01030$
8.5% annually is greater.

98. Compare an investment of $1000 at 6% compounded quarterly with the same amount invested at 6.25% compounded annually for a period of 10 yr. You will need the values $\log_{10} 1.015 \approx 0.00647$ and $\log_{10} 1.0625 \approx 0.02633$.
$R = \frac{(1.015)^{40}}{(1.0625)^{10}}$; $\log_{10} R = -0.0045$
6.25% annual interest is greater.

WRITE ON

Explain:

99. The relationship between the exponential function $f(x) = b^x$ and the logarithmic function $g(x) = \log_b x$. Answers may vary.

100. Why the logarithm of a negative number is not defined. Answers may vary.

101. Why the number 1 is not allowed as a base for the logarithmic function $f(x) = \log_b x$. Answers may vary.

102. Why $\log_b 1 = 0$. Answers may vary.

MASTERY TEST

If you know how to do these problems, you have learned your lesson!

103. A recent California earthquake was $10^{6.4}$ times as intense as that of an earthquake of minimum intensity I_0. What was the magnitude of this earthquake on the Richter scale? 6.4

In Problems 104–105, use the properties of logarithms to show:

104. $\log 30 = \log 6 - \log \sqrt{2} + \log \sqrt{50}$
$\log 30 = \log 6 - \log \sqrt{2} + \log \sqrt{50}$
$= \log(6 \div \sqrt{2} \cdot \sqrt{50})$
$= \log(\frac{6}{1} \cdot \frac{1}{\sqrt{2}} \cdot \frac{5\sqrt{2}}{1})$
$= \log 30$

105. $\log\left(\frac{xy^2}{z^3}\right)^2 = 2 \log x + 4 \log y - 6 \log z$
$\log(\frac{xy^2}{z^3})^2 = 2[\log(\frac{xy^2}{z^3})]$
$= 2[\log x + \log y^2 - \log z^3]$
$= 2[\log x + 2 \log y - 3 \log z]$
$= 2 \log x + 4 \log y - 6 \log z$

Write the equation:

106. $\log_4 32 = \frac{5}{2}$ in exponential form $4^{5/2} = 32$

107. $2^{10} = 1024$ in logarithmic form $\log_2 1024 = 10$

Find:

108. $\log_2 64$ 6

109. $\log_2 \frac{1}{8}$ -3

110. $\log_{10} 100$ 2

111. $\log_{10} 0.1$ -1

Solve:

112. $\log_3 x = 2$
$x = 9$

113. $\log_3 x = -3$
$x = \frac{1}{27}$

114. $\log_x 16 = 2$
$x = 4$

115. $\log_3 \frac{1}{27} = x$
$x = -3$

116. $\log_{16} x = \frac{1}{4}$
$x = 2$

Graph:

117. $y = \log_7 x$

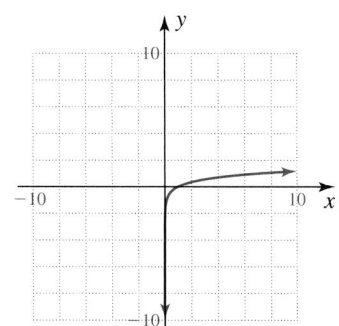

118. $f(x) = \log_8 x$

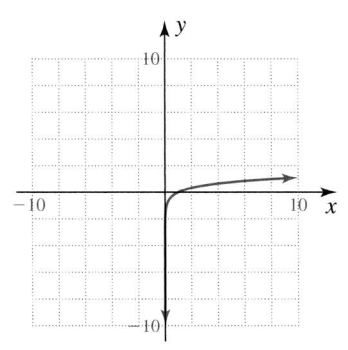

10.5 COMMON AND NATURAL LOGARITHMS

To Succeed, Review How To . . .

1. Write numbers in scientific notation (pp. 40–42).

2. Write a radical expression using the exponential form (pp. 522–524).

Objectives

A Find logarithms and antilogarithms base 10.

B Find logarithms and antilogarithms base e.

C Change the base of a logarithm.

D Graph exponential and logarithmic functions base e.

E Solve applications involving common and natural logarithms.

GETTING STARTED

How Many Decibels Was That?

The scale for measuring the loudness of sound is called the *decibel* scale. When the loudness L of a sound of intensity I is measured in decibels (dB), it is expressed as

$$L = 10 \log_{10} \frac{I}{I_0}$$

where I_0 is the minimum intensity detectable by the human ear. For example, the sound of a riveting machine 30 ft away is 10^{10} times as intense as the minimum intensity I_0, and hence its loudness in decibels is expressed as

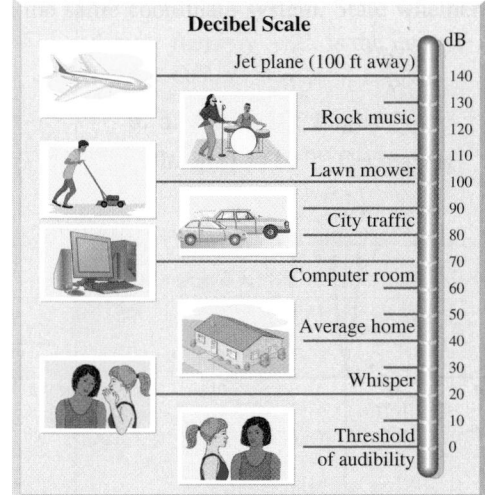

Decibel Scale

	dB
Jet plane (100 ft away)	140
	130
Rock music	120
	110
Lawn mower	100
	90
City traffic	80
	70
Computer room	60
	50
Average home	40
	30
Whisper	20
	10
Threshold of audibility	0

$$L = 10 \log_{10} \frac{10^{10} I_0}{I_0} = 10 \log_{10} 10^{10} = (10 \cdot 10) \log_{10} 10$$

$$= 100 \text{ dB}$$

The decibel scale uses logarithms to the base 10; these are called *common* logarithms and we shall study them and another type of logarithm called *natural* logarithms in this section.

A Finding Common Logarithms

The decibel scale is just one example of the many applications of logarithms that use the base 10. Logarithms to base 10 are called **common logarithms** because they use the same base as our "common" decimal system; it's customary to omit the base when working with these logarithms. Be sure to keep in mind that when the base is omitted,

$$\log M = \log_{10} M$$

↑ no base (10 is understood)

If you have a scientific calculator, you can find the logarithm of a number by using the **LOG** key. To find log 396 using a scientific calculator, enter 396 and press **LOG**. The result is 2.5976952 or, to four decimal places, 2.5977. Before calculators were widely used, the logarithm of a number was found using a table.

We should be able to estimate the common log for a number like 396 even without a calculator or table if we compare it to the following list of powers of 10 and their related logarithmic fact.

Web It

For a concise well written lesson regarding logarithms, go to link 10-5-1 at mhhe. com/bello.

$$10^0 = 1 \qquad\qquad \log 1 = 0$$
$$10^1 = 10 \qquad\qquad \log 10 = 1$$
$$10^2 = 100 \qquad\qquad \log 100 = 2$$
$$10^3 = 1000 \qquad\qquad \log 1000 = 3$$

Since 396 is a number between 100 and 1000 we anticipate its common logarithm to be a number between 2 and 3.

$$10^{\textcircled{2}} = 100 \qquad\qquad \log 100 = 2$$
$$10^? = \textcircled{396} \longrightarrow \log 396 \approx ? \qquad 2.\underline{\ \ }$$
$$10^{\textcircled{3}} = 1000 \qquad\qquad \log 1000 = 3$$

This is a good way to see if your answer is reasonable when using a calculator to find common logarithms.

EXAMPLE 1   Finding common logarithms

Find the value, rounded to 4 decimal places:

a. log 42,500   **b.** log 0.000425

SOLUTION

a. Enter 42,500 and press **LOG** to obtain

$$\log 42{,}500 \approx 4.6284 \qquad \text{Rounded to four decimal places}$$

$10^4 = 10{,}000$ and $10^5 = 100{,}000$. Since 42,500 is between 10,000 and 100,000, its logarithm should be 4.__

b. Enter 0.000425 and press **LOG** to obtain

$$\log 0.000425 \approx -3.3716$$

PROBLEM 1

Find the value, rounded to 4 decimal places:

a. log 734,000

b. log 0.0000734

Calculate It

Find Common Logarithms

To do Example 1(a), go to your home screen. Now to find log 42,500, press **LOG**; then enter 42,500 and press **ENTER**. You will obtain 4.62838893, which can be rounded to the desired number of decimal places.

Answers

1. a. 5.8657   **b.** −4.1343

As you recall, the inverse of a logarithmic function is an exponential function. The inverse of a logarithm is an **antilogarithm** or **inverse logarithm.** To find the inverse logarithm, we find the power of the base. Thus

$$\text{if} \quad f(x) = \log x, \quad f^{-1}(x) = \text{inverse log } x = 10^x$$

For example, if log $x = 2.5105$, then inverse log $2.5105 = 10^x = 10^{2.5105}$. Most calculators don't have an antilog key, so you have to know that to find the inverse, you must use the $\boxed{10^x}$ key, if your calculator has one. If it doesn't you can use the $\boxed{x^y}$ key. Sometimes, the **LOG** key is used as the $\boxed{10^x}$ key after using the **2nd** or $\boxed{\text{INV}}$ key. In either case, to find inverse log 2.5105, enter 2.5105 and press **2nd** (or $\boxed{\text{INV}}$) $\boxed{10^x}$. The result is 324.0 (rounded to the nearest whole number).

EXAMPLE 2   **Finding inverse logarithms**

Find (round to 4 decimal places):

a. inv log 0.8176   **b.** inv log(−2.1824)

SOLUTION

a. Using a scientific calculator, enter 0.8176 and press [2nd] (or [INV]) [10ˣ] to obtain

    inverse log 0.8176 ≈ 6.5705   Rounded to four decimal places

b. Using a scientific calculator, enter −2.1824 [2nd] (or [INV]) [10ˣ] to obtain

    inverse log(−2.1824) ≈ 0.0066   Rounded to four decimal places

To enter −2.1824, you have to enter 2.1824 and press the [+/−] key (if available) to change the sign.

PROBLEM 2

Find (round to 4 decimal places):

a. inv log 0.6243

b. inv log(−3.1963)

Calculate It

Find Inverse Logarithm

To find the inv $\log_{10}$ on a graphing calculator, we use the 10^x feature. Thus to find inv $\log_{10}$ 0.8176, press [2nd] [LOG] [.] [8] [1] [7] [6] and [ENTER] to obtain 6.57052391.

B   Finding Natural Logarithms

The irrational number $e \approx 2.718281828$ was discussed in Section 10.3. This number is used as the base of an important system of logarithms called **natural logarithms** or **Napierian logarithms** in honor of John Napier (1550–1617), the discoverer of logarithms (see *The Human Side of Algebra* at the beginning of this chapter). To distinguish natural from common logarithms, the abbreviation ln (pronounced *el-en*) is used instead of $\log_e$.

> **Natural Logarithmic Function**
>
> The natural logarithmic function is defined by $f(x) = \ln x$, where ln x means $\log_e x$ and $x > 0$.

The calculator key [LN] or [ln x] is used for natural logarithms.

EXAMPLE 3   **Finding a natural logarithm**

Find (round to 4 decimal places): ln 3252

SOLUTION   Enter 3252 and press [LN] to obtain

    ln 3252 ≈ 8.0870   Rounded to four decimal places

PROBLEM 3

Find ln 4563 (round to 4 decimal places).

Inverse logs base e can be found using the [eˣ] key, if your calculator has one. If it doesn't, use the [xʸ] key and approximate e. Sometimes, the [LN] key is used as the [eˣ] key after using the [2nd] or [INV] key.

EXAMPLE 4   **Finding the inverse log of a natural logarithm**

Find (round to 4 decimal places): inv ln(−3.4865)

SOLUTION   Enter −3.4865 (you may have to enter 3.4865 and press [+/−]). Now, to find the inv ln, press [2nd] (or [INV]) [LN] to obtain

    inv ln(−3.4865) ≈ 0.0306   Rounded to four decimal places

PROBLEM 4

Find inv ln(−4.3874) (round to 4 decimal places).

Answers

2. a. 4.2102   **b.** 0.0006
3. 8.4257   **4.** 0.0124

Calculate It   **Find Natural Logarithms**

You can do Examples 3 and 4 by using the [eˣ] and [LN] keys if your calculator has them.

 **Changing Logarithmic Bases**

Calculators are used to find log x and ln x, but can we find $\log_4 x$ or $\log_5 x$? To do this, we need the following conversion formula.

Change-of-Base Formula

For any logarithms with base a and b and any number $M > 0$,

$$\log_b M = \frac{\log_a M}{\log_a b}$$

This can be proved as follows: Let $x = \log_b M$. By the definition of logarithm,

$$b^x = M$$

$$\log_a b^x = \log_a M \qquad \text{Take the logarithm base } a \text{ of both sides.}$$

$$x \log_a b = \log_a M \qquad \text{Use the power rule on the left side.}$$

$$x = \frac{\log_a M}{\log_a b} \qquad \text{Solve for } x.$$

$$\log_b M = \frac{\log_a M}{\log_a b} \qquad \text{Since we let } x = \log_b M$$

Web It

For a lesson dealing with changing bases with examples and answers, go to link 10-5-2 at mhhe.com/bello.

EXAMPLE 5   **Using the change-of-base formula with common logarithms**

Using common logarithms, find $\log_4 20$ (round to 4 decimal places at each step).

SOLUTION   Using the change-of-base formula with $b = 4$, $a = 10$,

$$\log_4 20 = \frac{\log_{10} 20}{\log_{10} 4}$$

We then use a calculator to find $\log_{10} 20 \approx 1.3010$ and $\log_{10} 4 \approx 0.6021$ and substitute the values in the equation. Thus

$$\log_4 20 = \frac{\log_{10} 20}{\log_{10} 4}$$

$$\approx \frac{1.3010}{0.6021}$$

$$\approx 2.1608$$

PROBLEM 5

Use common logarithms to find $\log_5 20$ (round to 4 decimal places at each step).

EXAMPLE 6   **Using the change-of-base formula with natural logarithms**

Using natural logarithms, find $\log_4 20$ (round to 4 decimal places at each step).

SOLUTION   Using the change-of-base formula with $b = 4$, $a = e$,

$$\log_4 = \frac{\ln 20}{\ln 4}$$

PROBLEM 6

Use natural logarithms to find $\log_5 20$ (round to 4 decimal places at each step).

Answers

5. 1.8612   **6.** 1.8614

We then use a calculator to find $\ln 20 \approx 2.9957$ and $\ln 4 \approx 1.3863$ and substitute the values in the equation. Thus

$$
\log_4 20 = \frac{\ln 20}{\ln 4}
$$

$$
\approx \frac{2.9957}{1.3863}
$$

$$
\approx 2.1609
$$

Because of rounding, the answer differs from the previous one in the last decimal digit.

To avoid rounding errors, **do not** round the intermediate values $\log_{10} 20$ and $\log_{10} 4$ in Example 5, or $\ln 20$ and $\ln 4$ in Example 6; in Example 5 enter

> **2**   **0**   **LOG**   **÷**   **4**   **LOG**   **ENTER**

and *then* round the answer 2.160964047 to 2.1610. In Example 6, enter

> **2**   **0**   **LN**   **÷**   **4**   **LN**   **ENTER**

and you will obtain the same answer as before.

Teaching Tip

After doing Examples 5 and 6, have students decide if it matters whether base 10 or base e is used.

NOTE

It's best to wait until the final step to round off the answer.

D Graphing Exponential and Logarithmic Functions Base e

The logarithmic function $f(x) = \ln x$ is the **inverse** of the exponential function $f(x) = e^x$. Because of that, they are reflections across the line $y = x$ as shown in Figure 13. What will the graphs of $f(x) = e^{ax}$, $f(x) = -e^{ax}$, and $f(x) = e^x + b$ look like? The answers are given in Table 1.

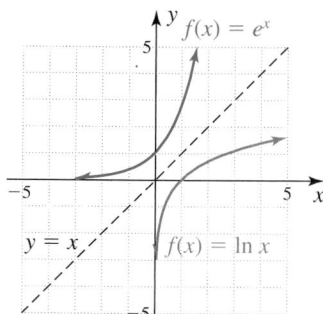

Figure 13

Web It

To see a lesson on how to graph exponential and logarithmic functions, including many labor saving graphing techniques, go to link 10-5-3 at mhhe.com/bello.

For lessons on graphing exponential and logarithmic functions, go to links 10-5-4 and 10-5-5 at mhhe.com/bello.

Teaching Tip

Before Examples 7 and 8, review how to recognize shifts in the graph by noticing certain parts of the equation.

Example:

$y = x^2$ is a parabola

$y = x^2 + ③$ is a shift (up) by 3

$y = (x - ④)^2$ is a shift (right) by 4

$y = ⊝x^2$ is a (reflection down)

Calculate It   Graphing Exponential Functions

To do Example 7(a) with a TI-83 Plus, press **WINDOW** and set $\text{Xmin} = -5$, $\text{Xmax} = 5$, $\text{Xscl} = 1$, $\text{Ymin} = -5$, $\text{Ymax} = -5$, and $\text{Yscl} = 1$. To select the function to be entered, press **Y=** and enter **2nd** **LN** **(** **1** **÷** **3** **)** **X,T,θ,n** **)** . Now press **GRAPH**. The result is shown in the window.

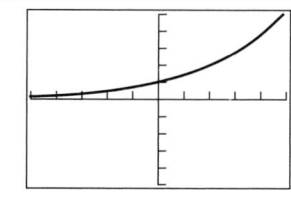

Table 1 Graphs of $f(x) = e^{ax}$, $f(x) = -e^{ax}$, and $f(x) = e^x + b$

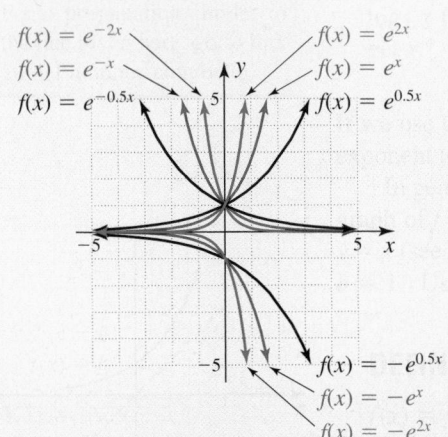

1. The graph of $f(x) = e^{ax}$, where a is positive, looks like that of $f(x) = e^x$. The larger the value of a, the "steeper" the graph. The graph of $f(x) = e^{2x}$ is steeper than the graph of $f(x) = e^{0.5x}$.

2. The graph of $f(x) = -e^{ax}$, where a is positive, is the reflection across the x-axis of the graph of $f(x) = e^{ax}$. The graph of $f(x) = -e^{2x}$ is the reflection across the x-axis of the graph of $f(x) = e^{2x}$.

3. The graph of $f(x) = e^{-ax}$, where a is positive, is the reflection across the y-axis of the graph of $f(x) = e^{ax}$. The graph of $f(x) = e^{-2x}$ is the reflection across the y-axis of the graph of $f(x) = e^{2x}$.

The graph of $f(x) = e^x + b$ is identical to that of $f(x) = e^x$ shifted b units (upward when b is positive, downward when b is negative).

EXAMPLE 7   **Graphing exponential functions**

Graph:

a. $f(x) = e^{(1/3)x}$　　　　**b.** $f(x) = -e^{(1/3)x}$　　　　**c.** $f(x) = e^x + 3$

SOLUTION

a. We use convenient values for x (say $x = -3, 0$, and 3) and find the $f(x) = y$-values using a calculator; then we plot the resulting points and draw the graph (Figure 14). For example, when $x = 3, f(3) = e^1 \approx 2.72$.

x	$e^{(1/3)x}$
-3	0.37
0	1
3	2.72

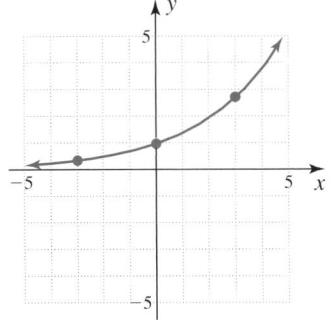

Figure 14

b. Before you draw the graph, recall that $f(x) = e^{(1/3)x}$ and $f(x) = -e^{(1/3)x}$ are reflections of each other across the x-axis. As before, use convenient points such as $x = -3, 0$, and 3 but note that the y-values are the *negatives* of the

---

PROBLEM 7

Graph:

a. $f(x) = e^{(1/4)x}$

b. $f(x) = -e^{(1/4)x}$

c. $f(x) = e^{(1/4)x} + 2$

Answer

7.

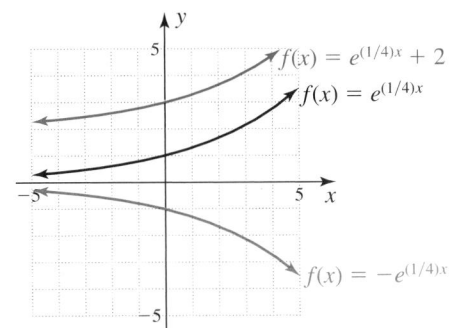

y-values obtained in part **a.** Plot the resulting points and draw the graph (Figure 15).

x	$-e^{(1/3)x}$
-3	-0.37
0	-1
3	-2.72

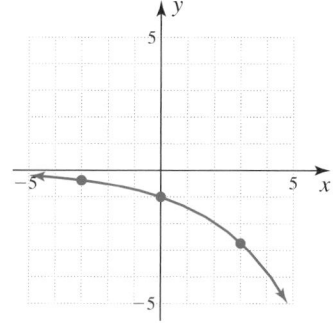

Figure 15

c. The graph of $f(x) = e^x + 3$ is identical to that of $f(x) = e^x$ but shifted 3 units up. Using $x = -1, 0$, and 1 and a calculator, we find the points shown in the table. We plot these points to obtain the graph shown in Figure 16.

x	$e^x + 3$
-1	3.37
0	4
1	5.72

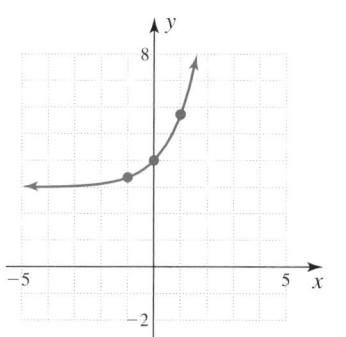

Figure 16

Calculate It

Graphing Exponential Functions

You can graph Example 7(b), $f(x) = -e^{(1/3)x}$ by entering $Y_2 =$ (−) VARS ▶ 1 1 GRAPH, that is, by making $Y_2 = -Y_1 = -e^{(1/3)x}$. $f(x) = e^{(1/3)x}$ and $f(x) = -e^{(1/3)x}$ are reflections of each other across the x-axis, as shown in the window.

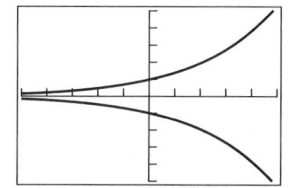

The functions $f(x) = \ln(x + 1)$ and $f(x) = \ln x + 1$ have graphs similar to that of $f(x) = \ln x$. The graph of $f(x) = \ln(x + 1)$ is the graph of $f(x) = \ln x$ shifted 1 unit to the *left,* while the graph of $f(x) = \ln x + 1$ is the graph of $f(x) = \ln x$ shifted 1 unit *up.*

CAUTION

Remember that $\ln(x + 1)$ and $\ln x + 1$ are **different.** The 1 added inside the parentheses in $\ln(x + 1)$ shifts the curve $\ln x$ 1 unit *left.* The 1 added to $\ln x$ in $\ln x + 1$ shifts the curve 1 unit *up.* In general, the a in $\ln(x + a)$ shifts the curve a units *right* or *left* (right if a is *negative,* left if a is *positive*). The a in $\ln x + a$ shifts the curve *up* or *down* (up if a is *positive,* down if a is *negative*).

We use these ideas next.

EXAMPLE 8   **Graphing logarithmic functions**

Graph:

a. $f(x) = \ln(x + 2)$ **b.** $f(x) = \ln x + 2$

SOLUTION

a. The graph of $f(x) = \ln(x + 2)$ is the same as that of $f(x) = \ln x$ shifted 2 units left. Since $\ln x$ is only defined for positive values of x, we must select x's that make $x + 2$ positive so that $\ln(x + 2)$ is defined. This means that $x + 2 > 0$, so $x > -2$. Using convenient values for x, say $x = 0, 1$, and 2 and a calculator to find the corresponding y-values, we construct a table

PROBLEM 8

Graph:

a. $f(x) = \ln(x + 1)$

b. $f(x) = \ln x + 1$

Answer

Answer on page 794.

and then plot the resulting points to obtain the graph shown in Figure 17.

x	$\ln(x + 2)$
0	0.69
1	1.10
2	1.39

b. The graph of $f(x) = \ln x + 2$ is the same as that of $f(x) = \ln x$ shifted 2 units up. Using the values $x = 1, 2,$ and 3 and a calculator, we obtain the numbers in the table, plot the resulting points, and draw the graph (Figure 18).

x	$\ln x + 2$
1	2
2	2.69
3	3.10

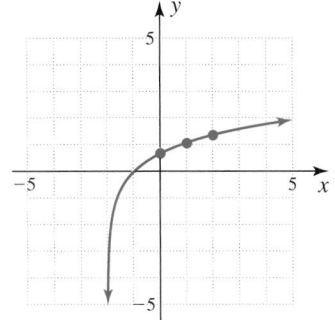

Figure 17

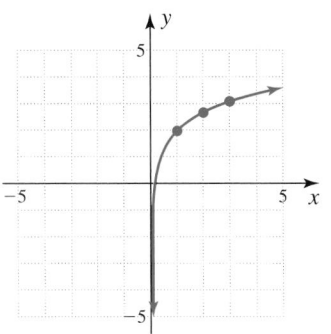

Figure 18

Calculate It

Graphing Logarithmic Functions

To do Example 8(a), enter $Y_1 = \ln(x + 2)$, making sure you use parentheses. The result using a $[-5, 5]$ by $[-5, 5]$ window is shown in the window.

The domain consists of all real numbers x greater than -2. The line $x = -2$ is an *asymptote* for the function.

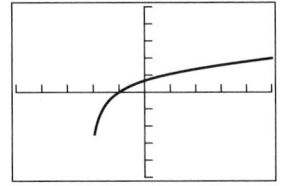

E   Solving Applications Involving Common and Natural Logarithms

Here are some examples of the many problems that involve logarithms for their solution.

EXAMPLE 9   **Finding the pH of a solution**

In chemistry, the pH (a measure of acidity) of a solution is defined by the formula $pH = -\log[H^+]$, where $[H^+]$ is the hydrogen ion concentration of the solution in moles per liter. Find the pH of a solution for which $[H^+] = 8 \times 10^{-6}$ (round to 4 decimal places).

SOLUTION   The pH is defined to be $-\log[H^+]$, so for this solution

$$pH = -\log(8 \times 10^{-6})$$
$$= -(\log 8 + \log 10^{-6})$$
$$= -(\log 8 - 6)$$
$$= 6 - \log 8$$
$$= 6 - 0.9031$$
$$\approx 5.0969$$

The pH of the solution is 5.0969.

PROBLEM 9

Find the pH of a solution for which $[H^+] = 4.2 \times 10^{-7}$ (round to 4 decimal places).

Answers

8.

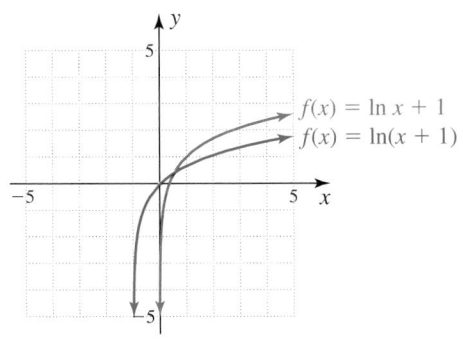

9. 6.3768

EXAMPLE 10	Diastolic pressure

Have you taken your blood pressure lately? Normal blood pressures are stated as 120/80, where 120 is the *systolic* pressure (when the heart is contracting) and 80 is the *diastolic* pressure (when the heart is relaxed). Over short periods, the diastolic pressure in the aorta of a normal adult is a function of time and can be approximated by the equation

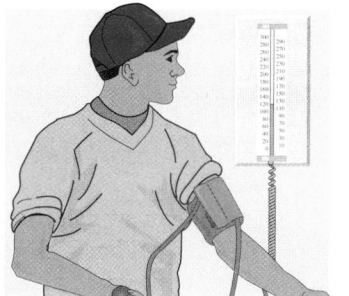

$$P = 90e^{-0.5t}$$

a. What is the diastolic pressure when the aortic valve is closed ($t = 0$)?

b. What is the diastolic pressure when $t = 0.3$ sec (round to 2 decimal places)?

SOLUTION

a. At $t = 0$, $P = 90e^{-0.5(0)} = 90e^{0} = 90(1) = 90$

b. When $t = 0.3$,

$$P = 90e^{-0.5(0.3)}$$
$$= 90e^{-0.15}$$
$$= 90(0.8607) \quad \text{Use a calculator.}$$
$$\approx 77.46$$

PROBLEM 10

What is the diastolic pressure when:

a. $t = 0.35$?

b. $t = 0.4$?

Web It

To study four types of word problems involving logarithms, go to link 10-5-6 at mhhe.com/bello.

Answers

10. **a.** 75.55   **b.** 73.69

Calculate It Exercises

1. Use your calculator to confirm the results of Example 9.

2. Graph the function $f(x) = 90e^{-0.5x}$ of Example 10.
   a. If you have a TI-83 Plus, find the value of $f(x)$ at $x = 0.3$ by pressing 2nd TRACE 1 and entering 0.3 for the x-value.
   b. What happens to $f(x)$ as x increases?
   c. What does that mean for your diastolic blood pressure?
   d. What is the y-intercept for $f(x)$, and what does that mean for your diastolic pressure?

3. Graph the functions $f(x) = e^x$ and $f(x) = e^x + k$ for $k = 1, 2,$ and 3. What happens to the graph of $f(x)$ when a constant k is added to it? (*Hint:* Consider two cases, k positive and k negative.)

4. What happens to the graph of $f(x) = e^x$ when a positive constant k is multiplied by $f(x)$ to obtain $g(x) = ke^x$?

Exercises 10.5

A In Problems 1–12, find the common logarithm of the given number. (Round to 4 decimal places.)

1. 74.5 1.8722

2. 952 2.9786

3. 1840 3.2648

4. 3.05 0.4843

5. 0.0437 -1.3595

6. 0.0673 -1.1720

7. 50.18 1.7005

8. 94.44 1.9752

9. 0.01238 -1.9073

10. 0.01004 -1.9983

11. 0.008606 -2.0652

12. 0.0004632 -3.3342

Boost *your* GRADE at mathzone.com!

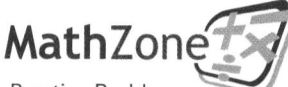

MathZone

• Practice Problems
• Self-Tests
• Videos
• NetTutor
• e-Professors

In Problems 13–20, find the inverse logarithm base 10 (inv log). (Round to 2 decimal places.)

13. inv log 1.2672 18.50

14. inv log 2.4409 275.99

15. inv log(−2.2328) 0.01

16. inv log(−1.4045) 0.04

17. inv log 1.4630 29.04

18. inv log 2.9408 872.57

19. inv log(−0.134) 0.73

20. inv log(−1.2275) 0.06

B In Problems 21–30, find the natural logarithm. (Round to 4 decimal places.)

21. ln 3 1.0986

22. ln 4 1.3863

23. ln 52 3.9512

24. ln 62 4.1271

25. ln 2356 7.7647

26. ln 208.3 5.3390

27. ln 0.054 −2.9188

28. ln 0.049 −3.0159

29. ln 0.00062 −7.3858

30. ln 0.00132 −6.6301

In Problems 31–40, find the inverse logarithm (inv ln). (Round to 2 decimal places.)

31. inv ln 1.2528 3.50

32. inv ln 2.2925 9.90

33. inv ln 4.1744 65.00

34. inv ln 4.6052 100.00

35. inv ln 0.0392 1.04

36. inv ln 0.0198 1.02

37. inv ln(−2.3025) 0.10

38. inv ln(−0.3566) 0.70

39. inv ln(−4.6051) 0.01

40. inv ln(−4.8281) 0.01

C In Problems 41–50, use the change-of-base formula to find the logarithm using (a) common logarithms and (b) natural logarithms. (Round to 4 decimal places.)

41. $\log_3 20$ 2.7268

42. $\log_3 40$ 3.3578

43. $\log_{100} 40$ 0.8010

44. $\log_{200} 50$ 0.7384

45. $\log_{0.2} 3$ −0.6826

46. $\log_{0.1} 4$ −0.6021

47. $\log_4 0.4$ −0.6610

48. $\log_2 0.16$ −2.6439

49. $\log_{\sqrt{2}} 0.8$ −0.6439

50. $\log_{\sqrt{2}} 0.16$ −5.2877

D In Problems 51–70, graph the function.

51. $f(x) = e^{3x}$

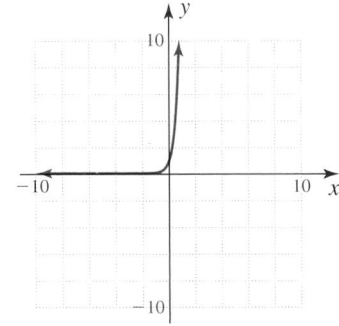

52. $f(x) = e^{4x}$

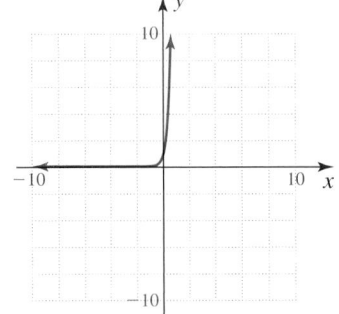

53. $f(x) = -e^{3x}$

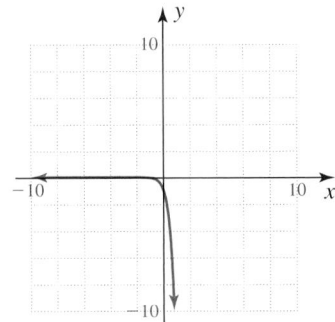

54. $f(x) = -e^{4x}$

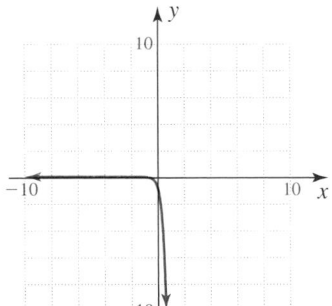

55. $f(x) = e^{-3x}$

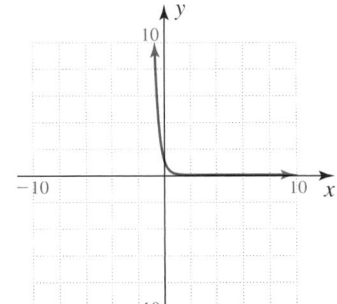

56. $f(x) = e^{-4x}$

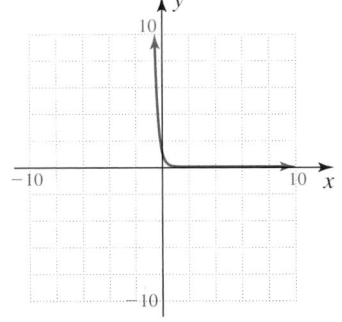

57. $f(x) = e^{(1/2)x}$

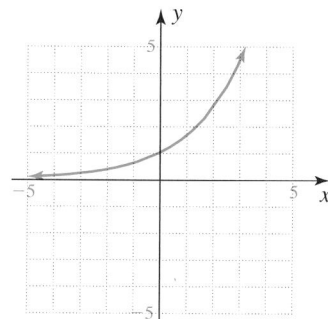

58. $f(x) = -e^{(1/2)x}$

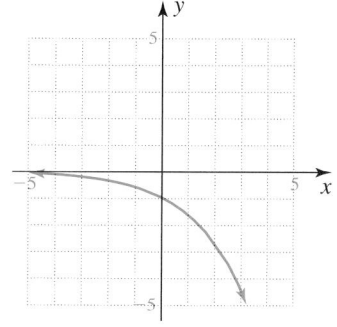

59. $f(x) = e^x + 1$

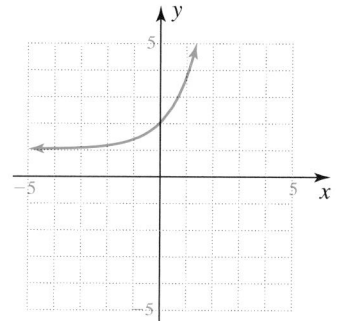

60. $f(x) = e^x - 1$

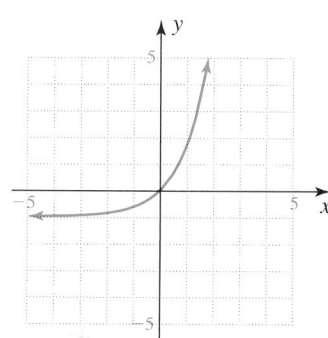

61. $f(x) = e^{0.5x} + 1$

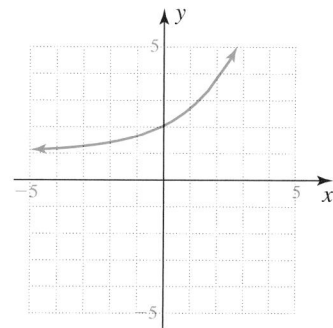

62. $f(x) = e^{0.5x} - 1$

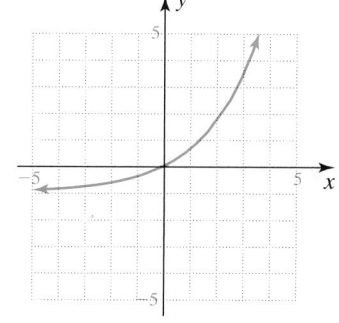

63. $f(x) = 2e^x$

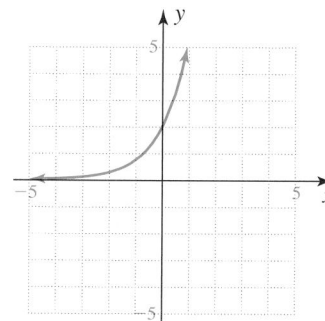

64. $f(x) = -2e^x$

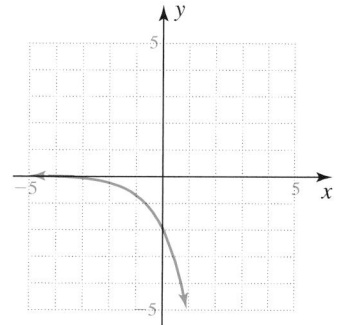

65. $f(x) = \ln(x + 1)$

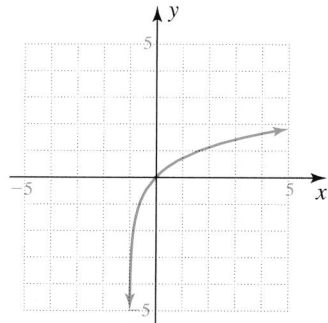

66. $f(x) = \ln x + 1$

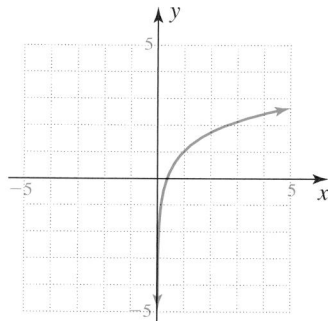

67. $f(x) = \ln x + 4$

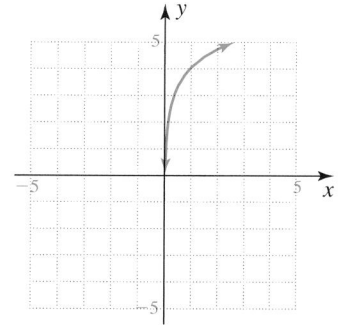

68. $f(x) = \ln(x + 4)$

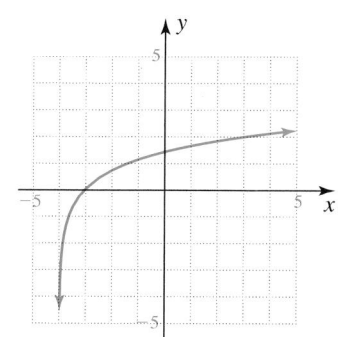

69. $f(x) = \ln(x - 1)$

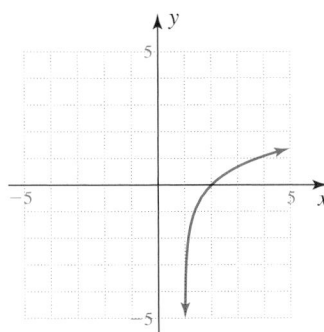

70. $f(x) = \ln x - 1$

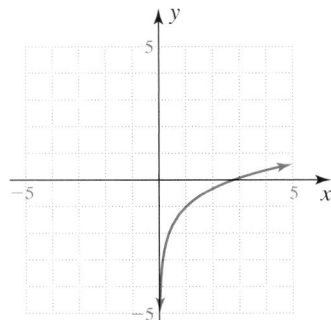

APPLICATIONS

pH, Approximations, and Height

In Problems 71–76, find the pH of a solution with the given [H$^+$]. Round to 1 decimal place. (*Hint:* pH $= -\log[$H$^+]$.)

71. [H$^+$] $= 7 \times 10^{-7}$ 6.2

72. [H$^+$] $= 1.5 \times 10^{-9}$ 8.8

73. Eggs whose [H$^+$] is 1.6×10^{-8} 7.8

74. Tomatoes whose [H$^+$] is 6.3×10^{-5} 4.2

75. Milk whose [H$^+$] is 4×10^{-7} 6.4

76. [H$^+$] $= 5 \times 10^{-8}$ 7.3

77. The number of Hispanic persons in the United States can be approximated by

$$H = 33{,}000{,}000 e^{0.03t}$$

where t is the number of years after 2000

  a. What was the number of Hispanic persons in 2000?
33,000,000

  b. Using this approximation, what will be the number of Hispanic persons in the United States in the year 2010? 44,545,341

78. The number of African-Americans in the United States can be approximated by

$$B = 36{,}000{,}000 e^{0.01t}$$

where t is the number of years after 2000

  a. What was the number of African-Americans in 2000?
36,000,000

  b. Using this approximation, what will be the number of African-Americans in the United States in the year 2010? 39,786,153

79. The number of bacteria present in a certain culture can be approximated by

$$B = 50{,}000 e^{0.2t}$$

where t is measured in hours and $t = 0$ corresponds to 12 noon. Find the number of bacteria present at

  a. noon. 50,000

  b. 2 P.M. 74,591

  c. 6 P.M. 166,006

80. If a bactericide (a bacteria killer) is introduced into a bacteria culture, the number of bacteria can be approximated by

$$B = 50{,}000 e^{-0.1t}$$

where t is measured in hours. Find the number of bacteria present

  a. when $t = 0$. 50,000

  b. when $t = 1$. 45,242

  c. when $t = 10$. 18,394

81. Sales begin to decline d days after the end of an advertising campaign and can be approximated by

$$S = 1000e^{-0.1d}$$

a. How many sales will be made on the last day of the campaign—when $d = 0$? 1000

b. How many sales will be made 10 days after the end of the campaign? 368

82. The demand function for a certain commodity is approximated by

$$p = 100e^{-q/2}$$

where q is the number of units demanded at a price of p dollars per unit.

a. If there is a 100-unit demand for the product, what will be its price? $p = 100e^{-50} \approx 0$

b. If there is no demand for the product, what will its price be? $100

83. The concentration C of a drug in the bloodstream at time t (in hours) can be approximated by

$$C = 100(1 - e^{-0.5t})$$

a. What will the concentration be when $t = 0$? 0

b. What will the concentration be after 1 hr? (Round to 1 decimal place.) 39.3

84. The number of people $N(t)$ reached by a particular rumor at time t is approximated by

$$N(t) = \frac{5050}{1 + 100e^{-0.06t}}$$

a. Find $N(0)$. 50

b. Find $N(10)$. 90

85. The stellar magnitude M of a star is defined by

$$M = -2.5 \log\left(\frac{B}{B_0}\right)$$

where B is the brightness of the star and B_0 the minimum of brightness.

a. Find the stellar magnitude of the North Star, 2.1 times as bright as B_0. (Round to 4 decimal places.) -0.8055

b. Find the stellar magnitude of Venus, 36.2 times as bright as B_0. (Round to 4 decimal places.) -3.8968

86. The percent P of adult height a male has reached at age A $(13 \leq A \leq 18)$ is

$$P = 16.7 \log(A - 12) + 87$$

a. What percent of adult height has a 13-yr-old male reached? (Round to the nearest percent.) 87%

b. What percent of adult height has an 18-yr-old male reached? (Round to the nearest thousandth of a percent.) 99.995%

SKILL CHECKER

Try the "Skill Checker" exercises so you'll be ready for the next section.

Solve for k. (Round to 4 decimal places.)

87. $\log 3 = k \log 2$ $k = 1.5850$

88. $\log 25 = 0.15k$ $k = 9.3196$

89. $25 = 10^k$ $k = 1.3979$

90. $100 = 5^k$ $k = 2.8614$

USING YOUR KNOWLEDGE

Exponential and Logarithmic Functions

Is there a relationship between the graph of an exponential function and the graph of the logarithm of the function? Let's use what we learned in this section to find out.

91. a. Graph $f(x) = 2^x$.

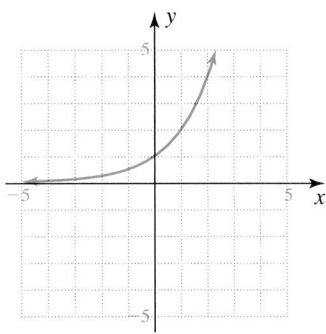

b. Graph $g(x) = \log 2^x$.

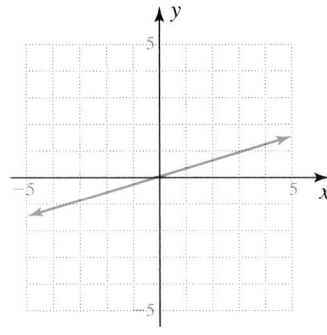

c. What is the slope of $g(x) = \log 2^x$? (*Hint:* Use the properties of logarithms to simplify $\log 2^x$.)
Slope is $\log 2 \approx 0.3$.

92. a. Graph $f(x) = 3^x$.

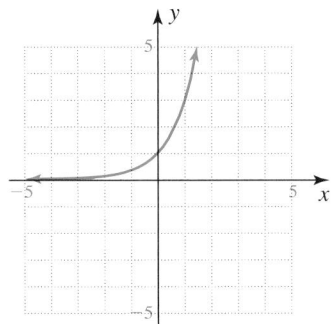

b. Graph $g(x) = \log 3^x$.

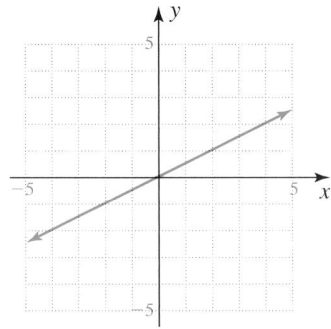

c. What is the slope of $g(x) = \log 3^x$?
Slope is $\log 3 \approx 0.5$.

93. Compare the base used for the function f in Problems 91 and 92 with the slope of the function g. What is the relationship between the two? The slope of $g(x) = \log$ of base of $f(x)$.

WRITE ON

Explain:

94. The difference between **common** logarithms and **natural** logarithms. Answers may vary.

95. What the antilog$_{10}$ of a number is. Answers may vary.

96. The usefulness of the change-of-base formula.
Answers may vary.

97. The relationship between the graphs of $f(x) = 10^x$ and $g(x) = \log x$. Answers may vary.

98. The relationship between the graphs of $f(x) = e^x$ and $g(x) = \ln x$. Answers may vary.

99. The meaning of "the graph of $f(x) = e^{2x}$ is steeper than the graph of $g(x) = e^{0.5x}$." Answers may vary.

MASTERY TEST

If you know how to do these problems, you have learned your lesson!

Find each value to 4 decimal places:

100. $\ln 3120$ 8.0456

101. inv ln(-1.5960) 0.2027

102. $\log_4 40$ 2.6610

103. $\log_6 20$ 1.6720

104. log 41,500   4.6180

105. log 0.000415 -3.3820

106. inv log 0.8432   6.9695

107. inv log(-2.4683) 0.0034

Graph:

108. $f(x) = e^{(1/4)x}$

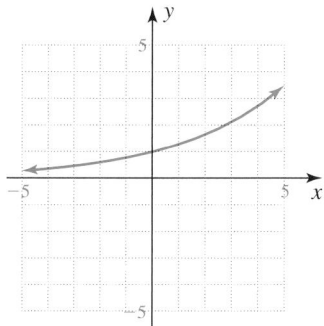

109. $f(x) = -e^{(1/4)x}$

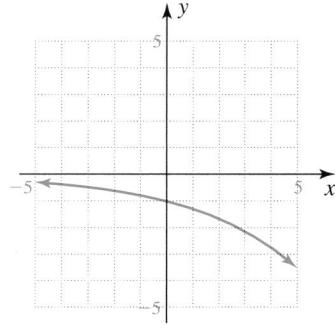

110. $f(x) = e^{(-1/4)x}$

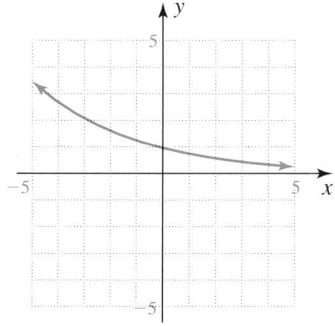

111. $f(x) = e^{(1/4)x} + 1$

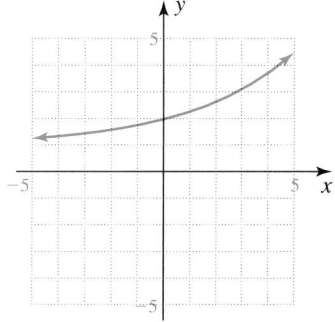

112. $f(x) = \ln(x + 3)$

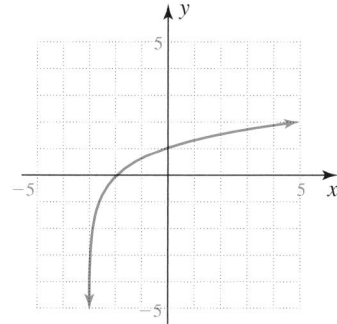

113. $f(x) = \ln x + 3$

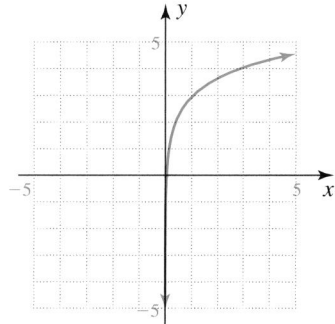

114. Over short periods, the systolic pressure of a normal adult can be approximated by

$$P = 130e^{-0.5t}$$

   a. What is the systolic pressure when $t = 0$?   130

   b. What is the systolic pressure when $t = 0.5$ sec?   101

10.6 EXPONENTIAL AND LOGARITHMIC EQUATIONS AND APPLICATIONS

To Succeed, Review How To . . .

1. Use the properties of logarithms (pp. 779–781).

2. Use the laws of exponents (pp. 31–40).

3. Solve linear equations (pp. 74–79).

4. Evaluate logarithms (pp. 788–790).

Objectives

A Solve exponential equations.

B Solve logarithmic equations.

C Solve applications involving exponential or logarithmic equations.

GETTING STARTED

Don't Drink and Drive!

Is there a relationship between blood alcohol level (BAC) and the probability of having an accident? Absolutely! As the chart shows, the probability $P(b)$ of having an accident, written as a percent, is a function of your blood alcohol level b. The formula is given by

$$P(b) = e^{kb} \tag{1}$$

As you can see from the chart, this probability is 25% when the BAC is 0.15%. Can we find k using this information? To do this, let $P(b) = 25$ and $b = 0.15$ in equation (1). We then have

$$25 = e^{0.15k} \tag{2}$$

Equation (2) is an example of an *exponential* equation, because the variable k occurs in the exponent. We shall solve this equation in Examples 3 and 4. We will even be able to predict what BAC theoretically leads to certain disaster—what alcohol level b corresponds to a 100% probability of an accident.

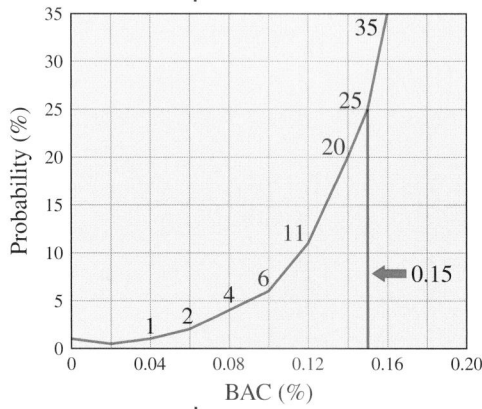

A   Solving Exponential Equations

In all the equations we have solved, we have seldom used variables as exponents. When this happens, we have an exponential equation.

EXPONENTIAL EQUATION

An **exponential equation** is an equation in which the variable occurs in an exponent.

$6^x = 14$, $3^{5x} = 20$, and $2^{6x} = 32$ are exponential equations. The equation $2^{6x} = 32$ can be rewritten using powers of the base 2, as

$$2^{6x} = 2^5$$

Because we are using the same base, the exponents must be equal;

$$6x = 5 \quad \text{and} \quad x = \frac{5}{6}$$

Remember the equivalence property from Section 10.4? We restate it here for your convenience.

Equivalence Property

For $b > 0$, $b \neq 1$, $b^x = b^y$ is equivalent to $x = y$.

We can use this equivalence property to help us solve exponential equations in which the exponential terms on each side can be expressed in the same base. The procedure follows.

PROCEDURE

To solve exponential equations when it is possible to write each side as a power of the same base.

1. Be sure the exponential terms are on opposite sides.   ($b^x = c$ or $b^x = c^t$)

2. Rewrite each side as a power of the same base.   ($b^x = b^y$)

3. Equate the exponents. (Equivalence property)   ($x = y$)

4. Solve the equation from step 3.

5. Verify the solution.

EXAMPLE 1   **Solving exponential equations**

Solve:

a. $3^{2x-1} = 81$ **b.** $2^{x+1} = 8^{x-1}$

SOLUTION

a. Since $3^4 = 81$, we write each side of $3^{2x-1} = 81$ as a power of 3.

$$3^{2x-1} = 3^4$$

Since the base is the same (3), the exponents must be equal. Thus

$$2x - 1 = 4 \qquad \text{Equate the exponents.}$$
$$2x = 5 \qquad \text{Add 1.}$$
$$x = \frac{5}{2} \qquad \text{Divide by 2.}$$

CHECK   Letting $x = \frac{5}{2}$ in $3^{2x-1} = 81$, we obtain

$$3^{2 \cdot (5/2) - 1} = 81$$

or $$3^4 = 81 \qquad \text{A true statement}$$

The solution is $\frac{5}{2}$.

b. The idea is to write both sides of the equation using the same base. Since $8 = 2^3$, the equation can be rewritten as

$$2^{x+1} = (2^3)^{x-1}$$
$$2^{x+1} = 2^{3x-3} \qquad \text{Simplify.}$$

Since the base is the same, 2, the exponents must be equal. Thus

$$x + 1 = 3x - 3 \qquad \text{Equate the exponents.}$$
$$-2x + 1 = -3 \qquad \text{Subtract } 3x.$$
$$-2x = -4 \qquad \text{Subtract 1.}$$
$$x = 2 \qquad \text{Divide by } -2.$$

PROBLEM 1

Solve:

a. $3^{3x-4} = 9$ **b.** $2^{x+3} = 8^{x-1}$

Teaching Tip

For Example 1(b), some students may need to see the "$x - 1$" enclosed in parentheses to remember to distribute the 3. Thus, $(2^3)^{(x-1)}$ becomes 2^{3x-3} when the 3 is distributed.

Answers

1. a. $x = 2$ **b.** $x = 3$

CHECK If we let $x = 2$ in the original equation, we obtain

$$2^{2+1} = 8^{2-1}$$

$$2^3 = 8^1 \qquad \text{A true statement}$$

The solution is 2.

Web It

For a lesson on solving exponential equations, go to link 10-6-1 at mhhe.com/bello.

For another lesson dealing with exponential equations, go to link 10-6-2 at mhhe.com/bello.

To learn how to solve exponential and logarithmic equations step-by-step, giving the reasons for each step, go to link 10-6-3 at mhhe.com/bello.

How can we solve $6^x = 14$? Since it isn't possible to write each side of $6^x = 14$ as a power of the same base, we make use of a fundamental property of logarithms.

Equivalence Property for Logarithms

If M, N, and b are all positive numbers and $b \neq 1$, then

$$\log_b M = \log_b N \quad \text{is equivalent to} \quad M = N$$

This means that

$$\text{if } \log_b M = \log_b N, \quad \text{then} \quad M = N$$

and conversely

$$\text{if } M = N, \quad \text{then} \quad \log_b M = \log_b N$$

To solve exponential equations when it is not possible to write each side as a power of the same base, we use the following procedure.

PROCEDURE

To solve exponential equations when it is not possible to write each side as a power of the same base.

1. Be sure the base and power are isolated on one side and the constant is on the other side. $(b^x = c)$

2. Take the log of both sides. (Equivalence property for logarithms)

3. Rewrite $\log_b M^r$ as $r \log_b M$. (Power property of logarithms)

4. Solve.

5. Verify the solution.

EXAMPLE 2 **More practice solving exponential equations**

Solve: $6^x = 14$

SOLUTION

$$6^x = 14 \qquad \text{Given.}$$

$$\log 6^x = \log 14 \qquad \text{Take the log of both sides.}$$

$$x \log 6 = \log 14 \qquad \text{Since } \log 6^x = x \log 6 \text{ (Power property of logarithms)}$$

$$x = \frac{\log 14}{\log 6} \qquad \text{Divide by log 6.}$$

This is the *exact* answer. If we wish to approximate it, use a calculator to obtain

$$x = \frac{\log 14}{\log 6} \approx 1.4729 \qquad \text{(Rounded to 4 decimal places)}$$

Remember to wait until the final step to round your answer.

PROBLEM 2

Solve (round to 4 decimal places):
$5^x = 12$

Teaching Tip

Help students recognize that 6^x and 14 cannot be expressed with the same base, so we need a new method to solve the equation. Also, have them note that this can be done using either common or natural logs.

Answer

2. $x = \frac{\log 12}{\log 5} \approx 1.5440$

Now let's return to the *Getting Started* and solve equation (2), $25 = e^{0.15k}$. Keep in mind that the property "$\log_b M = \log_b N$ is equivalent to $M = N$" works when the base $b = e$, that is

$$M = N \quad \text{is equivalent to} \quad \ln M = \ln N$$

EXAMPLE 3 Using an equivalency property to solve an exponential equation	**PROBLEM 3**

Solve (round to the nearest tenth): $25 = e^{0.15k}$

Solve (round to the nearest hundredth): $20 = e^{0.15k}$

SOLUTION

$25 = e^{0.15k}$	Given.
$\ln 25 = \ln e^{0.15k}$	Take the natural logarithm of both sides.
$\ln 25 = 0.15k$	Since $\log_b b^x = x$, $\ln e^{0.15k} = 0.15k$ (Power property of logarithms).
$\dfrac{\ln 25}{0.15} = k$	Divide both sides by 0.15.

Thus

$$k = \frac{\ln 25}{0.15} \approx \frac{3.2189}{0.15} \approx 21.5 \quad \text{(to the nearest tenth)}$$

EXAMPLE 4 Accident rate and blood alcohol level	**PROBLEM 4**

If we substitute $k = 21.5$ into equation (2) in *Getting Started*, the formula for $P(b)$ becomes $P(b) = e^{21.5b}$. At what blood alcohol level b will the probability of having an accident be 100% (round to the nearest hundredth)?

At what alcohol level b will the probability of having an accident be 50% (round to the nearest hundredth)?

SOLUTION   We use the RSTUV method.

1. Read the problem.

2. Select the unknown.
We are asked to find b so that $P(b) = 100$. To do this we must solve the equation

$$100 = e^{21.5b}$$

3. Think of a plan.
First, we take natural logarithms of both sides of the equation, then solve for b.

4. Use algebra to solve the problem.

$\ln 100 = \ln e^{21.5b}$	Take the natural logarithm of both sides.
$\ln 100 = 21.5b$	Power property of logarithms ($\log_b b^x = x$)
$\dfrac{\ln 100}{21.5} = b$	Divide by 21.5.

or $b = \dfrac{\ln 100}{21.5}$

This is the exact value. Using a calculator to approximate b to the nearest hundredth, $b \approx 0.21\%$.

5. Verify the answer.
The verification that $100 = e^{21.5(0.214)}$ is left to you (use your calculator). When the blood alcohol level is about 0.21%, the probability of an accident is 100%. If your blood alcohol level is 0.21%, you are probably too drunk to even get in your car.

Answers

3. $k = \frac{\ln 20}{0.15} \approx 19.97$
4. About 0.18%

B   Solving Logarithmic Equations

Teaching Tip

Remind students that when solving logarithmic equations they must check their answers for extraneous solutions. Thus, solving $y = \log_b x$ requires x to be positive. Hence, in Example 5, $(2x - 3)$ must be positive.

We have already solved certain types of **logarithmic equations**—equations containing *logarithmic expressions*—in Section 10.4.

Let's recall the definition of a logarithmic function.

$$y = \log_b x \quad \text{is equivalent to} \quad b^y = x,$$

where $b > 0$, $b \neq 1$, and $x > 0$

When we are solving logarithmic equations we must discard any values of the variable that do not satisfy the part of the definition that says $x > 0$. We will use this definition of a logarithm and convert the given equation into an exponential equation. This technique is used in Example 5.

EXAMPLE 5 Solving logarithmic equations	**PROBLEM 5**
Solve: $\log_5(2x - 3) = 2$	Solve: $\log_5(2x - 1) = 2$

SOLUTION

$$\log_5(2x - 3) = 2 \qquad \text{Given.}$$
$$2x - 3 = 5^2 \qquad \text{Since } \log_b x = y \text{ is equivalent to } b^y = x$$
$$2x - 3 = 25 \qquad \text{Simplify.}$$
$$2x = 28 \qquad \text{Add 3.}$$
$$x = 14 \qquad \text{Divide by 2.}$$

The solution is 14. You can check this by letting $x = 14$ in $\log_5(2x - 3) = 2$ to obtain

$$\log_5(2 \cdot 14 - 3) = 2$$
$$\log_5(25) = 2$$
$$5^2 = 25 \qquad \text{A true statement}$$

Web It

For two lessons on solving logarithmic equations, go to links 10-6-4 and 10-6-5 at mhhe.com/bello.

In general, we use the following procedure to solve logarithmic equations.

PROCEDURE

Solving Logarithmic Equations

1. Write the equation as an equivalent one with a single logarithmic expression on one side; write the equation in the form $\log_b M = N$.

2. Write the equivalent exponential equation $b^N = M$ and solve.

3. Always check your answer. Discard any values of the variable for which $M \leq 0$.

Web It

For more practice and examples solving exponential and logarithmic equations, go to link 10-6-6 at mhhe.com/bello.

Answer

5. $x = 13$

| **EXAMPLE 6** | **More practice solving logarithmic equations** | **PROBLEM 6** |

Solve:

a. $\log(x + 3) + \log x = 1$ **b.** $\log_3(x - 1) - \log_3(x - 3) = 1$

SOLUTION

a.

$$\log(x + 3) + \log x = 1 \qquad \text{Given.}$$

$$\log(x + 3)x = 1 \qquad \text{Use the product property of logarithms.}$$

$$(x + 3)x = 10^1 \qquad \text{Write as an equivalent exponential equation.}$$

$$x^2 + 3x = 10 \qquad \text{Simplify.}$$

$$x^2 + 3x - 10 = 0 \qquad \text{0 Subtract 10 to set equation = 0.}$$

$$(x + 5)(x - 2) = 0 \qquad \text{F Factor.}$$

$$x + 5 = 0 \quad \text{or} \quad x - 2 = 0 \qquad \text{F Factors = 0. By the zero-factor property}$$

$$x = -5 \quad \text{or} \qquad x = 2$$

CHECK   For $x = -5$, the expression $\log x$ becomes $\log(-5)$, which isn't defined because we cannot find the logarithm of a negative number. We discard that value and check $x = 2$.

$$\log(x + 3) + \log x = 1$$

becomes $\log(2 + 3) + \log 2 = 1$

$$\log 5 + \log 2 = 1$$

or $\log(5 \cdot 2) = 1 \qquad \text{Use the product property of logarithms.}$

$$10^1 = 5 \cdot 2 \qquad \text{Use the definition of logarithm.}$$

Since the result is a true statement, the solution is 2.

b. $\log_3(x - 1) - \log_3(x - 3) = 1 \qquad \text{Given.}$

$$\log_3\left(\frac{x - 1}{x - 3}\right) = 1 \qquad \text{Use the quotient property of logarithms.}$$

$$\frac{x - 1}{x - 3} = 3^1 \qquad \text{Write as an equivalent exponential equation.}$$

$$x - 1 = 3(x - 3) \qquad \text{Multiply both sides by } (x - 3).$$

$$x - 1 = 3x - 9 \qquad \text{By the distributive property.}$$

$$-1 = 2x - 9 \qquad \text{Subtract } x \text{ from both sides.}$$

$$8 = 2x \qquad \text{Add 9 to both sides.}$$

$$4 = x \qquad \text{Divide both sides by 2.}$$

CHECK   For $x = 4$,

$$\log_3(x - 1) - \log_3(x - 3) = 1$$

becomes $\log_3(4 - 1) - \log_3(4 - 3) = 1$

or $\log_3 3 - \log_3 1 = 1$

$$1 - 0 = 1 \qquad \text{Since } \log_3 3 = 1 \text{ and } \log_3 1 = 0$$

The resulting statement $1 - 0 = 1$ is true, so the solution is 4.

PROBLEM 6

Solve:

a. $\log(x - 3) + \log x = 1$

b. $\log_3(x + 1) - \log_3(x - 3) = 1$

Answers

6. a. $x = 5$ **b.** $x = 5$

Solving Applications Involving Exponential or Logarithmic Equations

Exponential and logarithmic equations have many applications in such areas as business, engineering, social science, psychology, and science. The following examples will give you an idea of the variety and range of their use.

EXAMPLE 7 **Doubling your money at 6% interest**	PROBLEM 7
With continuous compounding, a principal of P dollars accumulates to an amount A given by the equation	In Example 7, how long would it take if the rate is 8%?

$$A = Pe^{rt}$$

where r is the interest rate and t is the time in years. If the interest rate is 6%, how long would it take for the money in your bank account to double?

SOLUTION   We use the RSTUV method.

1. **Read the problem.**

2. **Select the unknown.**

3. **Think of a plan.**
With $A = 2P$ and $r = 0.06$, the equation becomes

$$2P = Pe^{0.06t}$$

or

$$2 = e^{0.06t} \qquad \text{Divide by } P.$$

4. **Use algebra to solve the problem.**
We want to solve this equation for t, so we take natural logarithms of both sides:

$$\ln 2 = \ln e^{0.06t} \qquad \text{Take natural logarithm of both sides.}$$

$$\ln 2 = 0.06t \qquad \text{Power property of logarithms, } \log_b b^x = x$$

$$\frac{\ln 2}{0.06} = t \qquad \text{Divide by 0.06.}$$

$$t = \frac{\ln 2}{0.06}$$

Using a calculator,

$$t \approx 11.6$$

This means that it would take about 11.6 yr for your money to double.

5. **Verify the solution.**
The verification is left to you.

Answer

7. About 8.7 yr

| **EXAMPLE 8**   World population in the year 2000 | **PROBLEM 8** |

In 2000, the population of the world exceeded 6 billion for the first time reaching about 6.1 billion with a yearly growth rate of 1.2%. The equation giving the population P in terms of the time t is

$$P = 6.1e^{0.012t}$$

Estimate the world population P in the year 2010.

SOLUTION   We use the RSTUV method.

1. Read the problem.

2. Select the unknown.
We are looking for the population P in the year 2010.

3. Think of a plan.
Since $P = 6.1$ for $t = 0$, the equation shows that t is measured from the year 2000. To estimate the population in 2010, we use $t = 10$ in the equation:

$$P = 6.1e^{(0.012)(10)} = 6.1e^{0.12}$$

4. Use arithmetic to solve the problem.
Using a calculator and approximating to the nearest tenth,

$$P = (6.1)(e^{0.12}) \approx 6.9$$

Our estimate for the population in 2010 is about 6.9 billion.

5. Verify the answer.
The verification is left to you.

Estimate the world population in the year 2020.

Web It

For the predictions from the United Nations Population Division, go to link 10-6-7 at mhhe.com/bello.

| **EXAMPLE 9**   Population explosion in a bacteria culture | **PROBLEM 9** |

If B is the number of bacteria present in a laboratory culture after t minutes, then, under ideal conditions,

$$B = Ke^{0.05t}$$

where K is a constant. If the initial number of bacteria is 1000, how long would it take for there to be 50,000 bacteria present?

SOLUTION   We use the RSTUV method.

1. Read the problem.

2. Select the unknown.
Since $B = 1000$ for $t = 0$, we have

$$1000 = Ke^0 = K$$

The equation for B is then

$$B = 1000e^{0.05t}, \text{ where } t \text{ is the unknown number of minutes.}$$

3. Think of a plan.
Now let $B = 50,000$:

$$50,000 = 1000e^{0.05t}$$

$$50 = e^{0.05t} \qquad \text{Divide by 1000.}$$

In Example 9, how long would it take to have 100,000 bacteria?

Answers

8. About 7.8 billion
9. About 92.1 min

4. Use algebra to solve the problem.

To solve for t, take natural logarithms of both sides:

$\ln 50 = \ln e^{0.05t}$ Take natural logarithm of both sides.

$\ln 50 = 0.05t$ Power property of logarithms, $\log_b b^x = x$

$\dfrac{\ln 50}{0.05} = t$ Divide by 0.05.

Using a calculator,

$$t = \frac{\ln 50}{0.05} \approx 78.2 \text{ min}$$

It will take approximately 78.2 min for there to be 50,000 bacteria present.

5. Verify the solution.

The verification is left to you.

EXAMPLE 10   **Half-life of cesium-137**

The element cesium-137 decays at the rate of 2.3% per year. Find the half-life of this element. (Round to 1 decimal place.)

SOLUTION   We use the RSTUV method.

1. Read the problem.

2. Select the unknown.

The half-life of a substance is found by using the equation

$$A(t) = A_0 e^{-kt}$$

where $A(t)$ is the amount present at time t (years), k is the decay rate, and A_0 is the initial amount of the substance present. In this problem,

$$k = 2.3\% = 0.023, \quad A(t) = \frac{1}{2} A_0$$

and we want to find t.

3. Think of a plan.

With this information, the basic equation becomes

$\dfrac{1}{2} A_0 = A_0 e^{-0.023t}$

$\dfrac{1}{2} = e^{-0.023t}$ Divide by A_0.

$0.5 = e^{-0.023t}$ Rewrite $\frac{1}{2}$ as 0.5.

4. Use algebra to solve the problem.

Now take natural logarithms of both sides:

$\ln 0.5 = \ln e^{-0.023t}$ Take natural logarithm of both sides.

$\ln 0.5 = -0.023t$ Power property of logarithms, $\log_b b^x = x$

$\dfrac{\ln 0.5}{-0.023} = t$ Divide by -0.023.

Using a calculator,

$$t = \frac{\ln 2}{0.023} \approx 30.1 \text{ yr}$$

The half-life of cesium-137 is about 30.1 yr.

5. Verify the solution.

The verification is left to you.

PROBLEM 10

Find the half-life of an element that decays at the rate of 4% per year.

Answer

10. About 17.3 yr

Calculate It   Solving Equations

Do you remember how to solve equations using your calculator? The idea is to graph both sides of the equation, call them Y_1 and Y_2 and use ⟨TRACE⟩ and ⟨ZOOM⟩ (or the intersect feature) to find the point of intersection (the solution). In Example 1(a), graph $Y_1 = 3^{(2x-1)}$ (note the parentheses around the $2x - 1$) and $Y_2 = 81$ using a $[-3, 3]$ by $[-10, 100]$ window with the Yscl $= 10$. On the TI-83 Plus, press ⟨2nd⟩ ⟨TRACE⟩ 5 and ⟨ENTER⟩ three times to answer the calculator questions "First Curve?" "Second Curve?" "Guess?". The intersection is $X = 2.5$, as shown in Window 1. Now try Example 1(b).

For Example 2, let $Y_1 = 6^x$ and $Y_2 = 14$ and follow the procedure of Example 1(a) using a $[-1, 10]$ by $[-1, 20]$ window. The intersection is $X = 1.4728859$, as shown in Window 2. To do Example 3, let $Y_1 = 25$ and $Y_2 = e^{0.15x}$ with a $[-5, 30]$ by $[-5, 30]$ window and Xscl $=$ YScl $= 5$. As usual, press ⟨2nd⟩ ⟨TRACE⟩ 5 and ⟨ENTER⟩ three times to find the intersection $X = 21.459172$, as shown in Window 3.

What type of window do you need for Example 4? Note that $Y_1 = 100$. After you find the appropriate window, you can check the results of Example 4.

In Example 5, we have to graph $Y_1 = \log_5(2x - 3)$, but calculators don't have a $\boxed{\log_5}$ key. Here, *you* have to remember that the change-of-base formula states that

$$Y_1 = \log_5(2x - 3) = \frac{\log(2x - 3)}{\log 5}$$

Then let $Y_2 = 2$. If you use a standard window, the point of intersection **does not** show (try it!), so we suggest a $[0, 20]$ by $[-1, 3]$ window. Press ⟨2nd⟩ ⟨TRACE⟩ 5 and ⟨ENTER⟩ three times to obtain the intersection $X = 14$, as shown in Window 4. To do Example 6(a), use the ⟨LOG⟩ function in your calculator. Let

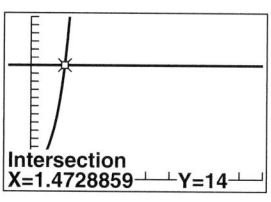

Intersection
X=2.5   Y=81
Window 1

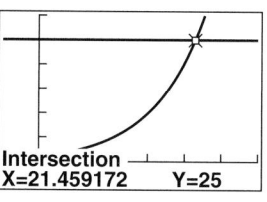

Intersection
X=1.4728859   Y=14
Window 2

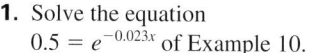

Intersection
X=21.459172   Y=25
Window 3

$Y_1 = \log(x + 3) + \log x$ (there is no need to simplify Y_1) and $Y_2 = 1$ and use a standard window. Press ⟨2nd⟩ ⟨TRACE⟩ 5 and move the cursor so that it shows a positive value for x. Then press ⟨ENTER⟩ three times. The intersection $X = 2$ is shown in Window 5.

The equations of Examples 6(b), 7, 8, and 9 can be solved with your calculator using the same techniques as those employed to solve Examples 3 and 4. You must know how to set up the equations first!

1. Solve the equation $0.5 = e^{-0.023x}$ of Example 10.

2. Has a virus ever attacked your computer? The cost (in billions of dollars) of computer viruses in the United States for 4 successive years can be approximated by $f(t) = 0.12e^t$, where t is the number of years after year 0.

   a. Use the function $f(t)$ to find the cost of virus infections in year 0. How close is your answer to the one given by the graph?

   b. In what year will the cost of virus infections reach 16 billion dollars? (Answer to the nearest year.)

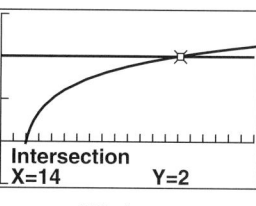

Intersection
X=14   Y=2
Window 4

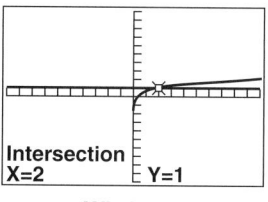

Intersection
X=2   Y=1
Window 5

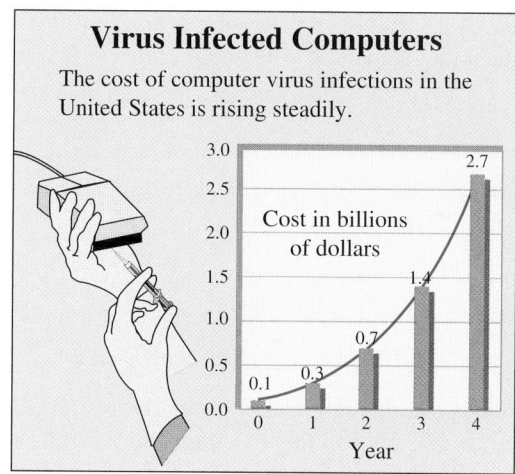

Virus Infected Computers
The cost of computer virus infections in the United States is rising steadily.

Exercises 10.6

A In Problems 1–26, solve the equation. Round the final answer to 4 decimal places where necessary (do not round intermediate values).

1. $5^x = 25$ $x = 2$

2. $3^x = 81$ $x = 4$

3. $2^x = 32$ $x = 5$

4. $3^{2x} = 81$ $x = 2$

5. $5^{3x} = 625$ $x = \frac{4}{3}$

6. $6^{-2x} = 216$ $x = -\frac{3}{2}$

7. $7^{-3x} = 343$ $x = -1$

8. $5^x = 4$ $x = 0.8614$

9. $7^x = 512$ $x = 3.2059$

10. $3^{x+1} = 729$ $x = 5$

11. $5^{x-2} = 625$ $x = 6$

12. $2^{3x+1} = 128$ $x = 2$

13. $3^x = 2$ $x = 0.6309$

14. $3^x = 20$ $x = 2.7268$

15. $2^{3x-2} = 32$ $x = \frac{7}{3}$

16. $3^{4x-3} = 27$ $x = \frac{3}{2}$

17. $5^{3x} \cdot 5^{x^2} = 25$ $x = \frac{-3 \pm \sqrt{17}}{2}$

18. $3^{4x} \cdot 3^{x^2} = 243$ $x = 1$ or $x = -5$

19. $e^x = 10$ $x = 2.3026$

20. $e^x = 100$ $x = 4.6052$

21. $e^{-x} = 0.1$ $x = 2.3026$

22. $e^{-x} = 0.01$ $x = 4.6052$

23. $30 = e^{2k}$ $k = 1.7006$

24. $40 = e^{3k}$ $k = 1.2296$

25. $10 = e^{-2k}$ $k = -1.1513$

26. $20 = e^{-3k}$ $k = -0.9986$

B In Problems 27–50, solve the equation. Round the final answer to 4 decimal places where necessary (do not round intermediate values).

27. $\log_2 x = 3$ $x = 8$

28. $\log_3 x = 2$ $x = 9$

29. $\log_2 x = -3$ $x = \frac{1}{8}$

30. $\log_3 x = -2$ $x = \frac{1}{9}$

31. $\ln x = 1$ $x = e \approx 2.7183$

32. $\ln x = -1$ $x = e^{-1} \approx 0.3679$

33. $\ln x = 3$ $x = e^3 \approx 20.0855$

34. $\ln x = -3$ $x = e^{-3} \approx 0.0498$

35. $\log_2(3x - 5) = 1$ $x = \frac{7}{3}$

36. $\log_3(2x - 1) = 4$ $x = 41$

37. $\log_4(3x - 1) = 2$ $x = \frac{17}{3}$

38. $\log_6(2x + 1) = 2$ $x = \frac{35}{2}$

39. $\log_5(3x + 1) = 2$ $x = 8$

40. $\log_2(4x - 1) = 4$ $x = \frac{17}{4}$

41. $\log x + \log(x - 3) = 1$ $x = 5$

42. $\log x + \log(x - 11) = 1$
  $x = \frac{11 + \sqrt{161}}{2}$

43. $\log_2(x + 1) + \log_2(x + 3) = 3$
  $x = 1$

44. $\log_3(x + 4) + \log_3(x - 2) = 3$
  $x = 5$

45. $\log(x + 1) - \log x = 1$
  $x = \frac{1}{9}$

46. $\log(x - 1) - \log x = 1$
  No solution

47. $\log_2(3 + x) - \log_2(7 - x) = 2$
  $x = 5$

48. $\log_3(2 + x) - \log_3(8 - x) = 2$
  $x = 7$

49. $\log_2(x^2 + 4x + 7) = 2$
  $x = -3$ or $x = -1$

50. $\log_2(x^2 + 4x + 3) = 3$
  $x = -5$ or $x = 1$

APPLICATIONS

In Problems 51–54, assume continuous compounding and follow the procedure in Example 7 to find how long it takes a given amount to double at the given interest rate. Round to 2 decimal places.

51. $r = 5\%$ 13.86 yr

52. $r = 7\%$ 9.90 yr

53. $r = 6.5\%$ 10.66 yr

54. $r = 7.5\%$ 9.24 yr

55. Suppose that the population of the world grows at the rate of 1.5% and that the population in 1984 was about 4.8 billion. Follow the procedure of Example 8 to estimate the population in the year 2000.
About 6.1 billion

56. Repeat Problem 55 for growth rate of 1.75%.
About 6.4 billion

In Problems 57–60, assume that the number of bacteria present in a culture after t minutes is given by $B = 1000e^{0.04t}$. Find the time it takes (to the nearest tenth of a minute) for the number of bacteria present to be

57. 2000.
17.3 min

58. 5000.
40.2 min

59. 25,000.
80.5 min

60. 50,000.
97.8 min

61. When a bacteria-killing solution is introduced into a certain culture, the number of live bacteria is given by the equation $B = 100,000e^{-0.2t}$, where t is the time in hours. Find the number of live bacteria present at the following times.
  a. $t = 0$ **b.** $t = 2$ **c.** $t = 10$ **d.** $t = 20$
  100,000   67,032   13,534   1832

62. The number of honey bees in a hive is growing according to the equation $N = N_0 e^{0.015t}$, where t is the time in days. If the bees swarm when their number is tripled, find how many days until this hive swarms. 73 days

In Problems 63–66, follow the procedure of Example 10 to find the half-life of the substance.

63. Plutonium, whose decay rate is 0.003% per year
23,104.9 yr

64. Krypton, whose decay rate is 6.3% per year 11 yr

65. A radioactive substance whose decay rate is 5.2% per year 13.3 yr

66. A radioactive substance whose decay rate is 0.2% per year 346.6 yr

67. The atmospheric pressure P in pounds per square inch at an altitude of h feet above the earth is given by the equation $P = 14.7e^{-0.00005h}$. Find the pressure at an altitude of

  a. 0 ft.   **b.** 5000 ft.   **c.** 10,000 ft.
     14.7 lb/in.^{2} 11.4 lb/in.^{2} 8.9 lb/in.^{2}

68. If the atmospheric pressure in Problem 67 is measured in inches of mercury, then $P = 30e^{-0.207h}$, where h is the altitude in miles. Find the pressure

  a. At sea level.   **b.** At 5 mi above sea level.
     30 in.   10.66 in.

69. According to the National Football League Players Association (NFLPA), average NFL salaries (in thousands of dollars) are as shown in the graph and can be approximated by

$$S = 540,000(1.09)^t$$

where t is the number of years after 1992.

  a. In how many years will average salaries be $800,000 (answer to the nearest year)?
     5 yr after 1992 (1997)

  b. Based on this approximation formula, in how many years will salaries reach the 1 million dollar mark (answer to the nearest year)?
     7 yr after 1992 (1999)

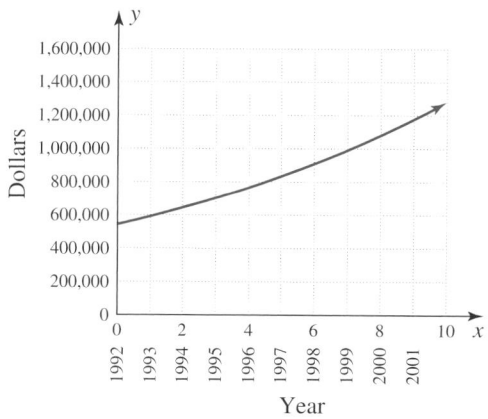

70. According to the *Statistical Abstract of the United States,* about $\frac{2}{3}$ of all aluminum cans distributed are recycled. If a company distributes 500,000 cans, the number in use after t years is

$$N(t) = 500,000\left(\frac{2}{3}\right)^t$$

How many years will it take for the number of cans to reach 100,000 (answer to the nearest year)? 4 yr

71. Do you have a fear of flying? The U.S. Department of Transportation has good news for nervous fliers: The number of general aviation accidents A has gone down significantly in the last 30 years. It can be approximated by

$$A = 5000e^{-0.04t}$$

where t is the number of years after 1970.

  a. How many accidents were there in 1970? 5000

  b. How many accidents were there in 1990? 2247

  c. In what year do you predict the number of accidents to be 1000 (answer to the nearest year)? In 2010

72. After exercise the diastolic blood pressure of normal adults is a function of time and can be approximated by

$$P = 90e^{-0.5t}$$

where t is the time in minutes. How long would it be before the diastolic pressure comes down to 80 (answer to two decimal places)? 0.24 min

73. How much do you spend on your credit cards? According to the Federal Reserve Board and the Bankcard Holders of America, credit card spending (in millions of dollars) is on the increase and can be approximated by

$$S = 54e^{0.15t}$$

where t is the number of years after 1980.

  a. How many millions of dollars were spent in 1980? 54 million

  b. In what year did the amount spent on credit cards reach 500 million dollars (answer to the nearest year)? In 1995

74. The percentage of adult height P attained by a boy can be approximated by

$$P = h(A) = 29 + 50 \log(A + 1)$$

where P is the percentage of adult height and A is the age in years ($0 < A < 17$). At what age will a boy reach 89% of his adult height? 14.8 yr

75. The height H of a girl (in inches) can be approximated by the function

$$H = h(A) = 11 + 19.44 \ln A$$

where A is the age in years ($6 < A < 15$). To the nearest year, at what age would you expect a girl to be

a. 60 in. tall? 12 yr old

b. 50 in. tall? 7 yr old

76. According to the National Football League, paid attendance (in millions) is as shown in the graph and can be approximated by

$$A = 17.4 + 0.32 \ln t$$

where t is the number of years after 1988.

a. Based on the equation, in what year did attendance reach 18 million (answer to the nearest year)? In 1995

b. Based on the equation, in what year will attendance reach 18.5 million (answer to the nearest year)? In 2019

c. What will the estimated attendance be in 2010? About 18.4 million

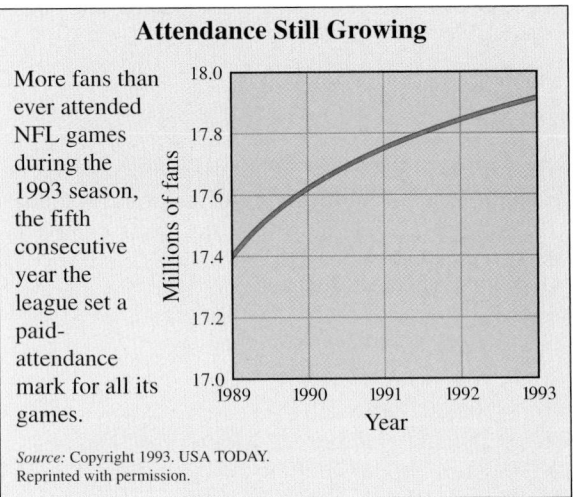

Attendance Still Growing

More fans than ever attended NFL games during the 1993 season, the fifth consecutive year the league set a paid-attendance mark for all its games.

Source: Copyright 1993. USA TODAY. Reprinted with permission.

77. If $1000 is invested in an account earning 1% interest each month, the number of months t it takes the account to grow to an amount A can be approximated by

$$t = -694.2 + 231.4 \log A$$

where $A \geq 1000$. How long would it take for this account to grow to 1 million dollars? About 694 months

SKILL CHECKER

Try the "Skill Checker" exercises so you'll be ready for the next section.

78. Evaluate the expression $\frac{1}{2}n(n - 1)$ when:

a. $n = 1$ 0   **b.** $n = 3$ 3   **c.** $n = 10$ 45

79. Find the value of the function $f(n) = 2n + 1$ when:

a. $n = 1$ 3   **b.** $n = 5$ 11   **c.** $n = 10$ 21

USING YOUR KNOWLEDGE

Just the Fax, Ma'am

Have you ever wondered how the approximations in this exercise set are done? To construct a growth equation based on time t, we start with the basic equation

$$A = A_0 e^{kt}$$

When $t = 0$, $A = A_0 e^0 = A_0$. Now, if we know the value A for a certain time t, we can solve the resulting equation and find k. Use this idea and the fact that the sale of fax machines soared from 580,000 in 1987 to 11 million in 1992 to solve Problems 80–85.

80. If the number N of fax machines sold (in thousands) is given by $N = N_0 e^{kt}$, where t is the number of years after 1987, what is N_0? $N_0 = 580$ thousand

81. Using the N_0 obtained in Problem 80, we can write $N = 580 e^{kt}$. We know that $N = 11,000$ in 1992, when $t = 5$ (1992 − 1987 = 5). Thus $11,000 = 580 e^{5k}$. Solve for k (to two decimal places) in this equation. $k = 0.59$

82. Use the values obtained in Problems 80 and 81 to find an exponential equation $N = N_0 e^{kt}$ for the number of fax machines sold from 1987 to 1992. $N = 580 e^{0.59t}$

83. According to the equation obtained in Problem 82, how many fax machines were sold in 1990? 3,405,095

84. a. According to the equation obtained in Problem 82, how many fax machines were sold in 1992? 11,081,453

b. What was the actual number of fax machines sold in 1992? 11,000,000

c. What was the percent error between your approximation and actual number of fax machines sold? 0.74%

85. How many fax machines would you predict were sold in the year 2000? 1,242,987,238

WRITE ON

In Problems 80–85, you constructed your own mathematical model, $A = A_0 e^{kt}$, based on a given set of data. How do you know what type of model to use for a particular set of data? Two of the possibilities are logarithmic or exponential:

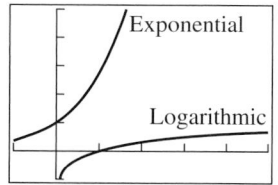

86. The number of cellular phones sold from 1985 to 1994 is shown in the graph. What type of mathematical model would you use to approximate the number of phones sold based on the information provided in the graph? Explain why. Exponential; answers may vary.

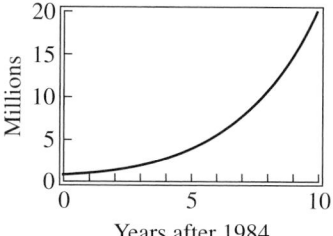

Years after 1984

87. If you use the exponential model $S = S_0 e^{kt}$ to approximate the information in the graph for Problem 86, explain the procedure you would use to find S_0. Answers may vary.

88. Explain how you would find k in $S = S_0 e^{kt}$. Answers may vary.

89. The number of cigarettes produced per capita from 1969 to 1990 is shown in the graph. What type of model would you use to approximate this information? Explain. Logarithmic; answers may vary.

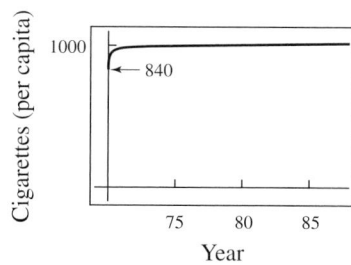

Year

90. If you assume that the equation modeling the graph in Problem 89 is of the form $N = a \ln x + b$, what information do you need to find b? Explain. Answers may vary.

91. For the equation in Problem 90, what other information do you need to find a? Answers may vary.

92. Explain the procedure you would use to find a in the equation in Problem 90. Answers may vary.

MASTERY TEST

If you know how to do these problems, you have learned your lesson!

93. The number of bacteria B present in a laboratory culture after t minutes is given by $B = Ke^{0.05t}$. If the initial number of bacteria is 1000, how long (to the nearest tenth of a minute) would it take for there to be 20,000 bacteria present? 59.9 min

94. After a bactericide is introduced, the number of bacteria present in a laboratory culture after t minutes is given by $B = Ke^{-0.02t}$. If the initial number of bacteria is 50,000, how long (to the nearest tenth of a minute) would it be before this number is reduced to 10,000? 80.5 min

Solve. Round the final answer to 4 decimal places where necessary (do not round intermediate values).

95. $\log(x + 9) + \log x = 1$ $x = 1$
96. $\log_2(x + 3) - \log_2 x = 1$ $x = 3$
97. $\log_6(4x - 4) = 2$ $x = 10$

98. $50 = e^{0.15k}$ $k = 26$
99. $5^x = 10$ $x = 1.4307$
100. $2^{3x-1} = 4$ $x = 1$

101. $2^{2x+1} = 8^{x-1}$ $x = 4$

COLLABORATIVE LEARNING 10A

What are the trends in the population growth for the United States? The growth rate of the population takes into consideration the births, deaths, and immigration. It seems the U.S. birth rate is on the decline since 1970. However, immigration has been rapidly increasing since 1930 as shown in the bar graph.

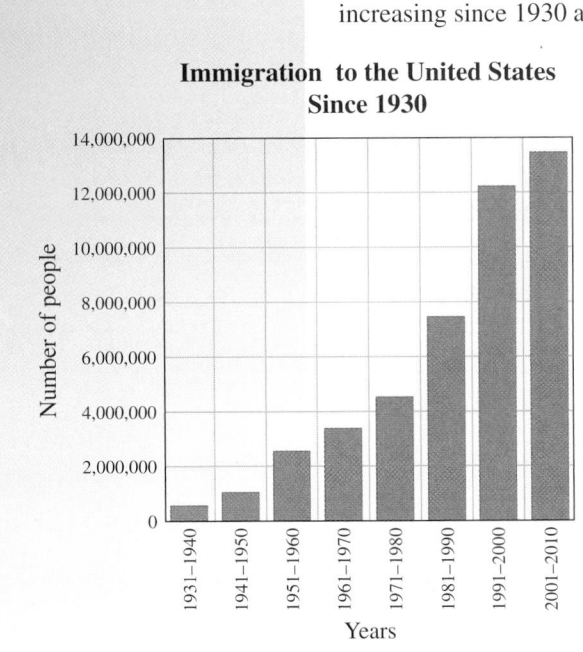

Immigration to the United States Since 1930

The population growth will have an impact on the environment and other issues in our American society. For the government to be able to plan for, keep up with, and possibly limit immigration, it would be important to be able to approximate the immigration population for a certain period.

Divide into groups, answer the questions, and then conduct your own research.

1. Make a table that contains as its first column the numbers 10–80 representing the 8 decades of immigration data. The second column should be the approximation for the amount of immigration population for each of the bars.

2. Use the data in the table to make a curved line graph.

3. The graph resembles an exponential function. If the exponential equation, $A = 800{,}000\, e^{0.35t}$, is used to approximate the U.S. immigration population since 1930 where t is the time in 10-year increments and A is the immigration population, find the immigration population for each of the 8 decades. Make a new column on your table for your results.

4. How do the answers, found using the exponential equation for the immigration population in the new column, compare to the values you approximated for the table from the bar graph? How do you account for any discrepancies?

5. Using the exponential equation, approximate what the immigration population will be in the decade 2011–2020.

6. Research the immigration population growth for other major countries in the world and see how it compares to the United States.

7. Each group should present their findings to the class and then discuss how this growth may impact the environment and other issues in society.

COLLABORATIVE LEARNING 10B

Fresh Egg

Note the small whitish spot on the germ from which the life of the chick begins.

Do you know how quickly a baby chick embryo can grow? The way a baby chick embryo grows is by repeated cell division beginning with one cell. Divide into groups, answer the questions, do the research, and then have a class discussion.

1. Make a two-column table that will describe the number of cells after successive divisions, the entries for the first line being 1 division and 2 cells. Continue the table for 4 more lines.

2. Graph the ordered pairs on an x-y coordinate plane.

3. Write the equation that would describe this cell division.

4. Soon after fertilization of the egg occurs, cell division starts. As long as the egg remains warmer than 67°, cell division will continue. If cell division were to occur every 20 minutes for one day, how many cells would be produced during that day?

1st Division $\qquad 2 = 2^1$

2nd Division $\qquad 4 = 2^2$

3rd Division $\qquad 8 = 2^3$

And so on ...

The growth of cell populations can be modeled using the assumption that each cell divides into two. However, this could be misleading. The population growth rate may be reduced if the length of the cell cycle increases, and it could be reduced if only a fraction of the cells are dividing.

5. Do some research and find exponential equations that describe other cell divisions. How do they compare to the equation that you found in Question 3. Discuss the differences.

Research Questions

1. Who was the first person to publish a book describing the rules of logarithms, what was the name of the book he published, and what does "logarithm" mean?

2. Who wrote the book from which our words "mantissa" and "characteristic" are derived, and what is the meaning of each of these words?

3. Write a few paragraphs about Jobst Bürgi and describe how he used "red" and "black" numbers in his logarithmic table.

4. Name the author and the title of his book published in 1748 that "gave his approval for e to represent the base of natural logarithms."

5. The symbol e was first used in 1731. Name the circumstances under which the symbol e was used.

Summary

SECTION	ITEM	MEANING	EXAMPLE
10.1A	$(f + g)(x)$ $(f - g)(x)$ $(fg)(x)$ $\left(\frac{f}{g}\right)(x)$	$f(x) + g(x)$ $f(x) - g(x)$ $f(x) \cdot g(x)$ $\frac{f(x)}{g(x)}, g(x) \neq 0$	If $f(x) = x^2$ and $g(x) = x + 1$, then: $(f + g)(x) = x^2 + x + 1$ $(f - g)(x) = x^2 - x - 1$ $(fg)(x) = x^3 + x^2$ $\left(\frac{f}{g}\right)(x) = \frac{x^2}{x+1}, x \neq -1$
10.1B	$(f \circ g)(x)$	$f(g(x))$, the composite of f with g	If $f(x) = x^2$ and $g(x) = x + 1$, then $(f \circ g)(x) = f(g(x)) = (x + 1)^2$
10.2A	f^{-1}	The inverse of a relation, obtained by reversing the order of the coordinates in each ordered pair in f	If $f = \{(1, 2), (4, 6), (6, 9)\}$, then $f^{-1} = \{(2, 1), (6, 4), (9, 6)\}$
10.2C	Horizontal line test	If any horizontal line intersects the graph of $f(x)$ more than once, then $f(x)$ is not a one-to-one function and f^{-1} is not a function.	$f(x) = x^2$. The graph is a parabola that opens upward. Thus any horizontal line $y = b > 0$ cuts the graph in two points. $f(x)$ is not a one-to-one function and the inverse, $f^{-1}(x) = \pm\sqrt{x}$, is not a function.
10.3A	Exponential function	A function defined for all real values of x by $f(x) = b^x$, $b > 0$, $b \neq 1$	$f(x) = 2^x$
10.3B	Increasing function Decreasing function Natural exponential function	A function whose graph rises from left to right A function whose graph falls from left to right $f(x) = e^x$, $e \approx 2.7182818$	$f(x) = 2^x$ $f(x) = 2^{-x}$ or $f(x) = \left(\frac{1}{2}\right)^x$ $f(x) = e^x$ is an increasing function.
10.4A	Logarithm	$\log_b x = y$ means $x = b^y$, where $b > 0$, $b \neq 1$, and $x > 0$	$\log_2 8 = 3$ because $2^3 = 8$.
10.4D	Logarithm of a product Logarithm of a quotient Logarithm of a power	$\log_b MN = \log_b M + \log_b N$ $\log_b \frac{M}{N} = \log_b M - \log_b N$ $\log_b M^r = r \log_b M$	$\log_2(4 \times 8) = \log_2 4 + \log_2 8$ $\log_2 \frac{4}{8} = \log_2 4 - \log_2 8$ $\log_2 4^3 = 3 \log_2 4$
10.5A	Common logarithm Inverse logarithm	The logarithm to base 10 The number that corresponds to a given logarithm	$\log 10 = 1$, $\log 100 = 2$ inv log $2.3010 \approx 200$
10.5B	Natural logarithms	Logarithms to base e	$\ln 2 \approx 0.69315$
10.5C	Change-of-base formula	$\log_b M = \frac{\log_a M}{\log_a b}$	$\log_5 20 = \frac{\log 20}{\log 5} \approx 1.86135$
10.6A	Exponential equation	An equation in which the variable occurs in an exponent	$10^{2x} = 5$
10.6B	Logarithmic equation	Equations containing logarithmic expressions	$\log(2x - 1) = 3$ and $\log_2(3x - 1) = 2$ are logarithmic equations.

Review Exercises

(If you need help with these exercises, look in the section indicated in brackets.)

1. [10.1A] Let $f(x) = 2 - x^2$ and $g(x) = 2 + x$. Find the following.

   a. $(f + g)(x)$ $4 + x - x^2$

   b. $(f - g)(x)$ $-x - x^2$

   c. $(fg)(x)$ $4 + 2x - 2x^2 - x^3$

   d. $\left(\dfrac{f}{g}\right)(x)$ $\dfrac{2 - x^2}{2 + x}$

2. [10.1A] Let $f(x) = 3 - x^2$ and $g(x) = 3 + x$. Find the following.

   a. $(f + g)(x)$ $6 + x - x^2$

   b. $(f - g)(x)$ $-x - x^2$

   c. $(fg)(x)$ $9 + 3x - 3x^2 - x^3$

   d. $\left(\dfrac{f}{g}\right)(x)$ $\dfrac{3 - x^2}{3 + x}$

3. [10.1A] Find $\dfrac{f(x) - f(a)}{x - a}$, $(x \neq a)$ if:

   a. $f(x) = 6x + 1$ 6

   b. $f(x) = 7x + 1$ 7

4. [10.1B] If $f(x) = x^3$ and $g(x) = 2 - x$, find:

   a. $(g \circ f)(2)$ -6

   b. $(f \circ g)(x)$ $(2 - x)^3$

   c. $(g \circ f)(x)$ $2 - x^3$

5. [10.1B] If $f(x) = x^3$ and $g(x) = 3 - x$, find

   a. $(f \circ g)(x)$ $(3 - x)^3$

   b. $(g \circ f)(x)$ $3 - x^3$

   c. $(g \circ f)(2)$ -5

6. [10.1C] Let $f(x) = \dfrac{9}{x - 1}$ and $g(x) = \dfrac{x + 3}{x - 4}$. Find the domain of $f + g$, $f - g$, and fg. $\{x \mid x$ is a real number and $x \neq 1$ and $x \neq 4\}$

7. [10.1C] Let $f(x) = \dfrac{x + 1}{x - 3}$ and $g(x) = \dfrac{x - 1}{x + 4}$. Find the domain of $\dfrac{f}{g}$. $\{x \mid x$ is a real number and $x \neq 1$ and $x \neq 3$ and $x \neq -4\}$

8. [10.1D]

   a. The revenue (in dollars) obtained from selling x units of a product is $R(x) = 100x - 0.02x^2$, and the cost per unit is $C(x) = 30,000 + 30x$. Find the profit function $P(x) = R(x) - C(x)$.
   $P(x) = -0.02x^2 + 70x - 30,000$

   b. If the revenue $R(x) = 100x - 0.02x^2$ and the cost per unit is $C(x) = 40,000 + 40x$, find the profit function $P(x)$. $P(x) = -0.02x^2 + 60x - 40,000$

9. [10.2A] Let $S = \{(4, 4), (6, 6), (8, 8)\}$ and find

   a. The domain and range of S
   $D = \{4, 6, 8\}$; $R = \{4, 6, 8\}$

   b. S^{-1} $S^{-1} = \{(4, 4), (6, 6), (8, 8)\}$

   c. The domain and range of S^{-1}
   $D = \{4, 6, 8\}$; $R = \{4, 6, 8\}$

   d. The graphs of S and S^{-1}

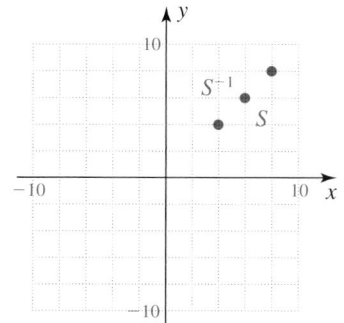

10. [10.2A] Let $S = \{(4, 5), (6, 7), (8, 9)\}$ and find

   a. The domain and range of S
   $D = \{4, 6, 8\}$; $R = \{5, 7, 9\}$

   b. S^{-1} $S^{-1} = \{(5, 4), (7, 6), (9, 8)\}$

   c. The domain and range of S^{-1}
   $D = \{5, 7, 9\}$; $R = \{4, 6, 8\}$

   d. The graphs of S and S^{-1}

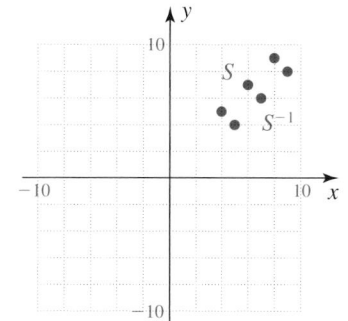

11. [10.2B] Let $f(x) = y = 3x - 3$.

    a. Find $f^{-1}(x)$. $f^{-1}(x) = \frac{x + 3}{3}$

    b. Graph f and its inverse.

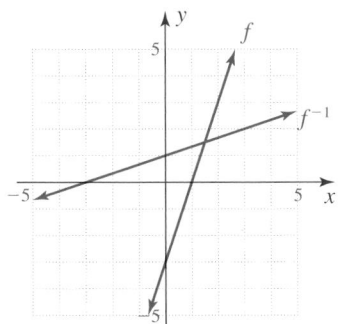

12. [10.2B] Let $f(x) = y = 4x - 4$.

    a. Find $f^{-1}(x)$. $f^{-1}(x) = \frac{x + 4}{4}$

    b. Graph f and its inverse.

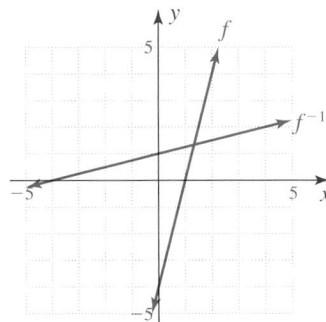

13. [10.2C] Find the inverse of $f(x) = y = 4x^2$. Is the inverse a function?

    $y^2 = \frac{1}{4}x$; $f^{-1}(x) = \pm\frac{\sqrt{x}}{2}$; no

14. [10.2C] Find the inverse of $f(x) = y = 5x^2$. Is the inverse a function?

    $y^2 = \frac{1}{5}x$; $f^{-1}(x) = \pm\frac{\sqrt{5x}}{5}$; no

15. [10.2C] Graph $y = 3^x$. Is the inverse a function? Yes

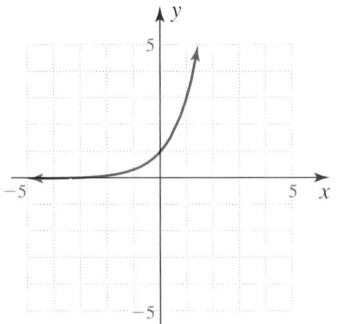

16. [10.2C] Graph $y = 4^x$. Is the inverse a function? Yes

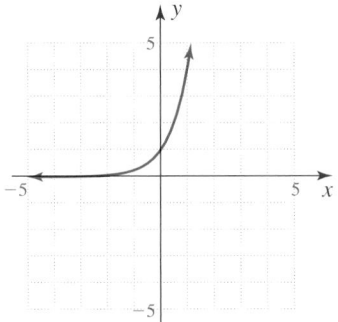

17. [10.2D] The relationship between dress size d and waist size w (in inches) is given by $d = f(w) = w - 16$.

    a. What dress size corresponds to a 28-in. waist? 12

    b. Find $f^{-1}(d)$. $f^{-1}(d) = d + 16$

    c. If a woman wears a size 10 dress, what is her waist size? 26 in.

18. [10.3A] Graph the function:

    a. $f(x) = 2^{x/2}$

    b. $f(x) = 2^{-x/2}$

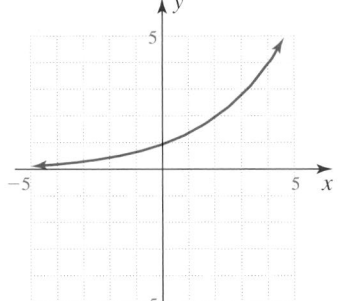

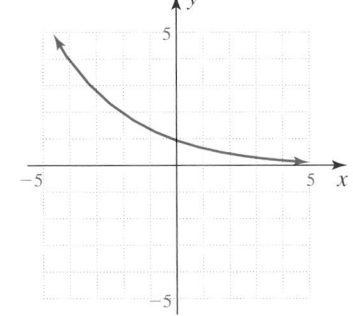

19. [10.3A] Graph the function:

a. $g(x) = \left(\dfrac{1}{2}\right)^{x/2}$

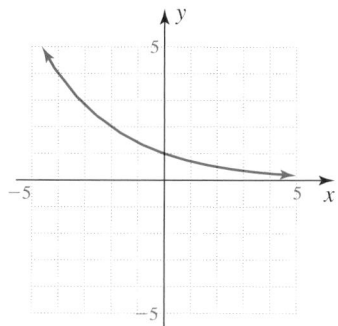

b. $g(x) = \left(\dfrac{1}{2}\right)^{-x/2}$

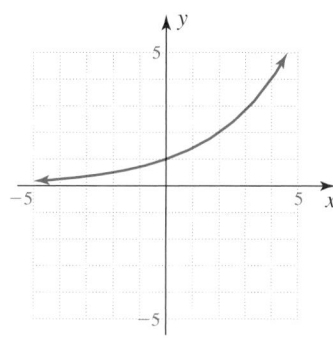

20. [10.3C] A radioactive substance decays so that G, the number of grams present, is given by

$$G = 1000e^{-1.4t}$$

where t is the time in years. Find, to the nearest gram, the amount of the substance present

a. At the start. 1000 g

b. In 2 yr. 61 g

21. [10.4A] Graph $f(x) = \log_5 x$.

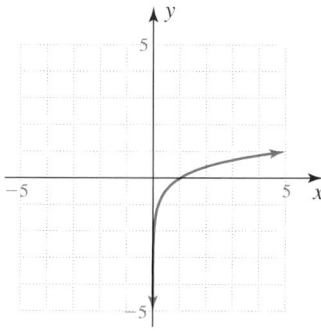

22. [10.4B] Write in logarithmic form.

a. $243 = 3^5$ $\log_3 243 = 5$

b. $\dfrac{1}{8} = 2^{-3}$ $\log_2\left(\dfrac{1}{8}\right) = -3$

23. [10.4B] Write in exponential form.

a. $\log_2 32 = 5$ $2^5 = 32$

b. $\log_3 \dfrac{1}{81} = -4$ $3^{-4} = \dfrac{1}{81}$

24. [10.4C] Solve.

a. $\log_4 x = -2$ $x = \dfrac{1}{16}$

b. $\log_x 16 = 2$ $x = 4$

25. [10.4C] Find.

a. $\log_2 16$ 4

b. $\log_3 \dfrac{1}{27}$ -3

26. [10.4D] Fill in the blank with the correct expression.

a. $\log_b MN = \underline{\log_b M + \log_b N}$

b. $\log_b M - \log_b N = \underline{\quad \log_b \frac{M}{N} \quad}$

c. $\log_b M^r = \underline{\quad r \log_b M \quad}$

In Problems 27–31, use a calculator and round to 4 decimal places where necessary.

27. [10.5A]

a. log 975 2.9890

b. log 837 2.9227

28. [10.5 A]

a. log 0.00759 -2.1198

b. log 0.000648 -3.1884

29. [10.5A]

a. inv log 2.8215 662.9793

b. inv log -3.3904 0.0004

30. [10.5B]

a. ln 2850 7.9551

b. ln 0.345 -1.0642

31. [10.5B]

a. inv ln 2.0855 8.0486

b. inv ln 2.7183 15.1545

32. [10.5C] Use the change of base formula to find each value. Round the final answer to 4 decimal places (do not round intermediate values).

  a. $\log_3 10$ 2.0959

  b. $\log_3 100$ 4.1918

33. [10.5D] Graph.

  a. $f(x) = e^{(1/2)x}$

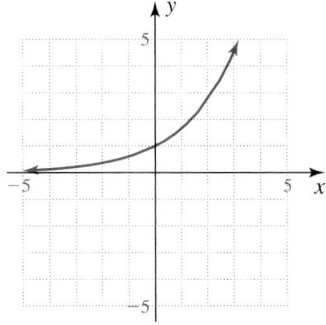

  b. $g(x) = -e^{(1/2)x}$

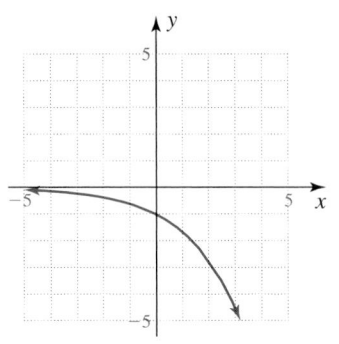

34. [10.5D] Graph.

  a. $f(x) = \ln(x + 1)$

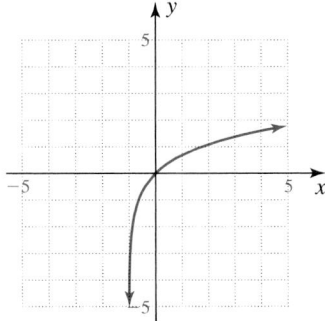

  b. $g(x) = \ln x + 1$

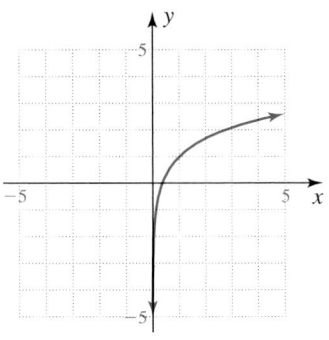

35. [10.5E] The pH of a solution is defined by pH $= -\log[H^+]$, where $[H^+]$ is the hydrogen ion concentration of the solution in moles per liter. Find the pH, to 3 decimal places, of a solution for which $[H^+] = 4 \times 10^{-6}$. 5.398

36. [10.6A] Solve.

  a. $2^{2x-1} = 32$ $x = 3$

  b. $3^{x+1} = 9^{x-1}$ $x = 3$

37. [10.6A] Solve. Round the final answer to 4 decimal places (do not round intermediate values).

  a. $2^x = 3$ $x = 1.5850$

  b. $5^{2x} = 2.5$ $x = 0.2847$

38. [10.6A] Solve. Round the final answer to 4 decimal places (do not round intermediate values).

  a. $e^{5.6x} = 2$ $x = 0.1238$

  b. $e^{-0.33x} = 2$ $x = -2.1004$

39. [10.6B] Solve.

  a. $\log x + \log(x - 10) = 1$ $x = 5 + \sqrt{35} \approx 10.9161$

  b. $\log_3(x + 1) - \log_3(x - 1) = 1$ $x = 2$

40. [10.6C] The compound amount with continuous compounding is given by $A = Pe^{rt}$, where P is the principal, r the rate, and t the time in years. Using 0.69315 for ln 2, find how long it takes for the money to double—for A to equal $2P$—if the rate is

  a. 5% 13.863 yr

  b. 8.5% About 8.2 yr

41. [10.6C] The number of bacteria in a culture after t minutes is given by $N = 1000e^{kt}$. If there are 1804 bacteria after the given time, find k to 4 decimal places.

  a. 2 min $k = 0.2950$

  b. 4 min $k = 0.1475$

42. [10.6C] A radioactive substance decays so that the amount present in t years is given by $A = A_0e^{-kt}$. Use -0.69315 for ln 0.5 and find the half-life if

  a. $k = 0.5$ 1.3863 yr

  b. $k = 0.02$ 34.6575 yr

Practice Test 10

(Answers on pages 824–826)

1. Let $f(x) = x^2 + 16$ and $g(x) = 4 - x$. Find the following.

    a. $(f + g)(x)$ **b.** $(f - g)(x)$ **c.** $(fg)(x)$ **d.** $\left(\dfrac{f}{g}\right)(x)$

2. $f(x) = 7x - 2$, find $\dfrac{f(x) - f(a)}{x - a}$, where $x \neq a$

3. If $f(x) = x^2 + 2$ and $g(x) = x + 3$, find:

    a. $(g \circ f)(-2)$ **b.** $(f \circ g)(x)$ **c.** $(g \circ f)(x)$

4. If $f(x) = \dfrac{-1}{5x}$ and $g(x) = \dfrac{3x + 1}{2x - 2}$, find the domain of $f + g$, $f - g$, and fg.

5. Find the domain of $\dfrac{f}{g}$ if $f(x) = \dfrac{3}{x + 3}$ and $g(x) = \dfrac{4x}{x - 6}$

6. Let $S = \{(3, 5), (5, 7), (7, 9)\}$ and find

    a. The domain and range of S **b.** S^{-1}

    c. The domain and range of S^{-1}

    d. The graph of S and S^{-1}

7. Let $f(x) = y = 4x - 4$.

    a. Find $f^{-1}(x)$.   **b.** Graph f and its inverse.

8. Find the inverse of $f(x) = y = 3x^2$. Is the inverse a function?

9. Graph $y = 3^x$.

    a. Is the inverse a function?

    b. Is $y = 3^x$ increasing or decreasing?

10. A radioactive substance decays so that the number of grams present after t years is

$$G = 1000e^{-1.4t}$$

Find, to the nearest gram, the amount of the substance present

    a. At the start.   **b.** In 2 yr.

11. Graph on the same coordinate axes.

    a. $f(x) = 2^x$

    b. $f(x) = \log_2 x$

12. Write the equation

    a. $27 = 3^x$ in logarithmic form.

    b. $\log_5 25 = x$ in exponential form.

13. Solve.

    a. $\log_4 x = -1$ **b.** $\log_x 16 = 2$

14. Use the properties of logarithms to show that $\log\left(\dfrac{x^3}{12}\right) = 3 \log x - \log 12$.

15. Use the properties of logarithms to show that $\log \sqrt[4]{rt} = \dfrac{1}{4} \log r + \dfrac{1}{4} \log t$.

16. Find.

    a. $\log 325$ **b.** inv log 3.5502

17. Find.

    a. $\ln 325$ **b.** inv ln 1.1618

18. **a.** Use the change-of-base formula to fill in the blank: $\log_3 10 = $ _____

    b. Use the result of part **a** to find a numerical approximation for $\log_3 10$.

19. Graph.

    a. $f(x) = e^{(1/2)x}$ **b.** $g(x) = -e^{(1/2)x}$

20. Graph.

    a. $f(x) = \ln(x + 1)$ **b.** $g(x) = \ln x + 1$

21. Solve.

    a. $5^{2x+1} = 25$

    b. $3^{x+1} = 9^{2x-1}$

22. Solve.

    a. $3^x = 2$

    b. $50 = e^{0.20k}$

23. Solve.

    a. $\log(x + 2) + \log(x - 7) = 1$

    b. $\log_3(x + 5) - \log_3(x - 1) = 1$

24. The compound amount with continuous compounding is given by $A = Pe^{rt}$, where P is the principal, r is the interest rate, and t is the time in years. If the rate is 8%, find how long it takes for the money to double—for A to equal $2P$ (use 0.69315 for $\ln 2$).

25. A radioactive substance decays so that the amount A present at time t (years) is $A = A_0 e^{-0.5t}$. Find the half-life (time for half to decay) of this substance (use 0.69315 for $\ln 2$).

Answers to Practice Test

ANSWER	IF YOU MISSED	REVIEW		
	QUESTION	SECTION	EXAMPLES	PAGE
1. a. $x^2 - x + 20$ **b.** $x^2 + x + 12$ **c.** $-x^3 + 4x^2 - 16x + 64$ **d.** $\frac{x^2 + 16}{4 - x}$, $x \neq 4$	1	10.1A	1	738
2. 7, $x \neq a$	2	10.1A	2	739
3. a. 9 **b.** $x^2 + 6x + 11$ **c.** $x^2 + 5$	3	10.1B	3	740
4. $\{x \mid x \text{ is a real number and } x \neq 0 \text{ and } x \neq 1\}$	4	10.1C	4	741
5. $\{x \mid x \text{ is a real number and } x \neq -3, x \neq 6, \text{ and } x \neq 0\}$	5	10.1C	5, 6	741–742
6. a. $D = \{3, 5, 7\}$; $R = \{5, 7, 9\}$ **b.** $\{(5, 3), (7, 5), (9, 7)\}$ **c.** $D = \{5, 7, 9\}$; $R = \{3, 5, 7\}$ **d.**	6	10.2A	1	751
7. a. $f^{-1}(x) = \frac{x + 4}{4}$ **b.**	7	10.2B	2	753
8. $y = \pm\sqrt{\frac{x}{3}}$ or $\pm\frac{\sqrt{3x}}{3}$; No	8	10.2C	3	755

ANSWER	IF YOU MISSED	REVIEW		
	QUESTION	SECTION	EXAMPLES	PAGE
9. a. Yes	9	10.3A, B	1, 2	765–767

b. Increasing

ANSWER	IF YOU MISSED	REVIEW		
10. a. 1000 g **b.** 61 g	10	10.3C	4	769
11.	11	10.4A	1	776

ANSWER	IF YOU MISSED	REVIEW		
12. a. $\log_3 27 = x$ **b.** $5^x = 25$	12	10.4B	2, 3	777
13. a. $\dfrac{1}{4}$ **b.** 4	13	10.4C	4	778

14. $\log\left(\dfrac{x^3}{12}\right) = 3 \log x - \log 12$
$= \log x^3 - \log 12$
$= \log\left(\dfrac{x^3}{12}\right)$

	IF YOU MISSED	REVIEW		
14		10.4D	6	780

15. $\log \sqrt[4]{rt} = \frac{1}{4}\log r + \frac{1}{4}\log t$
$= \log r^{1/4} + \log t^{1/4}$
$= \log \sqrt[4]{r} + \log \sqrt[4]{t}$
$= \log \sqrt[4]{rt}$

	IF YOU MISSED	REVIEW		
15		10.4D	7	780

ANSWER	IF YOU MISSED	REVIEW		
16. a. 2.5119 **b.** 3550	16	10.5A	1, 2	788–789
17. a. 5.7838 **b.** 3.1957	17	10.5B	3, 4	789
18. a. $\dfrac{\log 10}{\log 3}$ or $\dfrac{\ln 10}{\ln 3}$ **b.** 2.0959	18	10.5C	5	790

ANSWER	IF YOU MISSED	REVIEW		
	QUESTION	SECTION	EXAMPLES	PAGE
19.	19	10.5D	7	792–793

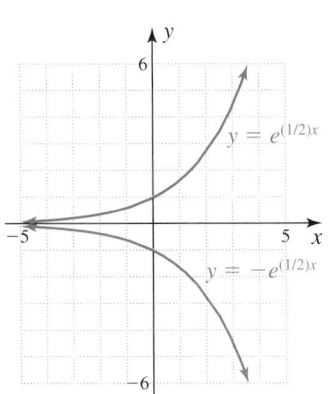

	IF YOU MISSED	REVIEW		
20.	20	10.5D	8	793–794

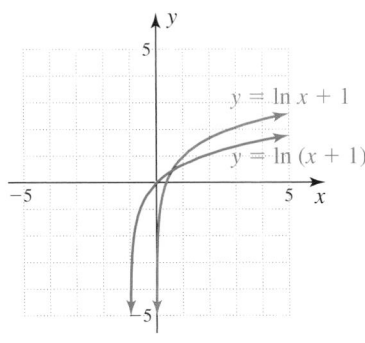

	IF YOU MISSED	REVIEW		
21. a. $\frac{1}{2}$ **b.** 1	21	10.6A	1	803
22. a. $\frac{\log 2}{\log 3} \approx 0.6309$ **b.** $\frac{\ln 50}{0.2} \approx 19.56$	22	10.6A	2, 3	804–805
23. a. 8 **b.** 4	23	10.6B	6	807
24. About 8.66 yr	24	10.6C	7	808
25. About 1.386 yr	25	10.6C	10	810

Cumulative Review Chapters 1–10

1. Simplify:
$[(3x^2 - 2) + (8x + 3)] - [(x - 2) + (2x^2 - 6)]$
$x^2 + 7x + 9$

2. Simplify: $(4x^4y^{-2})^2 \; \frac{16x^8}{y^4}$

3. Solve: $0.02P + 0.04(1700 - P) = 65 \;\; P = 150$

4. Solve: $|x - 4| = |x - 8| \;\; x = 6$

5. Graph on a number line: $\{\, x \mid x < -4 \text{ or } x \geq 4\}$

6. Graph on a number line: $|6x - 9| \leq 3$

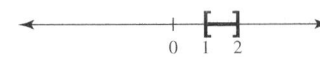

7. If $H = 2.85h + 72.69$, find h when $H = 135.39$.
$h = 22$

8. The perimeter of a rectangle is $P = 2L + 2W$, where L is the length and W is the width. If the perimeter is 160 ft and the length is 20 ft more than the width, what are the dimensions? 30 ft $\times$ 50 ft

9. Find the distance between the points $E(5, 2)$ and $F(-2, 1)$. $5\sqrt{2}$

10. The line through $A(3, -4)$ and $B(-1, y)$ is perpendicular to a line with slope $\frac{4}{9}$. Find y. $y = 5$

11. Find an equation of the line that passes through the point $(-6, 6)$ and is parallel to the line $8x + 2y = 2$.
$4x + y = -18$

12. Graph on an x-y coordinate system: $|x + 3| > 2$

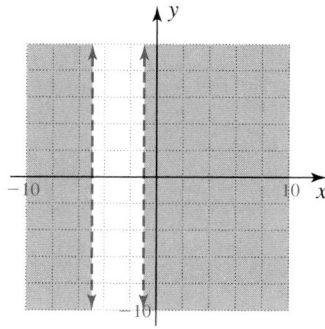

13. If the temperature of a gas is held constant, the pressure P varies inversely as the volume V. A pressure of 1850 lb/in.^{2} is exerted by 5 ft^3 of air in a cylinder fitted with a piston. Find k. $k = 9250$

14. Use the substitution method to solve the system:
$$x - 4y = -19$$
$$-3x = -12y + 61 \quad \text{No solution}$$

15. Solve the system:
$$2x = 5y - 28$$
$$2y = 5x + 28 \quad (-4, 4)$$

16. Solve the system:
$$3x + y + z = -17$$
$$x + 2y - z = -15 \quad (-5, -4, 2)$$
$$3x + y - z = -21$$

17. Evaluate: $\begin{vmatrix} 5 & -4 \\ -5 & 1 \end{vmatrix}$ -15

18. The total height of a building and the flagpole on the roof is 252 ft. The building is 8 times as tall as the flagpole. How tall is the building? 224 ft

19. If $P(x) = x^2 + 5x + 1$, find $P(-4)$. -3

20. Multiply: $(3h + 5)(6h - 7)$ $18h^2 + 9h - 35$

21. Factor completely: $48x^4y + 20x^3y^2 - 12x^2y^3$
$4x^2y(3x - y)(4x + 3y)$

22. Factor completely: $27n^3 + 8$ $(3n + 2)(9n^2 - 6n + 4)$

23. Solve for x: $x^3 + 4x^2 - x - 4 = 0$ $-4, -1, 1$

24. Factor $3x^3 + 22x^2 + 37x + 10$ if $(x + 5)$ is one of its factors. $(x + 5)(x + 2)(3x + 1)$

25. Divide: $\dfrac{x + 7}{x - 7} \div (x^2 + 14x + 49)$ $\dfrac{1}{x^2 - 49}$

26. Perform the indicated operations:

$$\dfrac{x - 3}{x^2 - 5x + 6} - \dfrac{x - 2}{x^2 - 4} \quad \dfrac{4}{(x + 2)(x - 2)}$$

27. Find two consecutive even integers such that the sum of their reciprocals is $\frac{7}{24}$. 6 and 8

28. Evaluate: $(27)^{-4/3}$ $\dfrac{1}{81}$

29. Rationalize the denominator: $\dfrac{\sqrt[5]{7}}{\sqrt[5]{16d^3}}$ $\dfrac{\sqrt[5]{14d^2}}{2d}$

30. Simplify: $\sqrt{18} + \sqrt{50}$ $8\sqrt{2}$

31. Reduce: $\dfrac{4 + \sqrt{8}}{2}$ $2 + \sqrt{2}$

32. Solve: $\sqrt{x + 7} = x + 5$ $x = -3$

33. Multiply: $(-5 + 9i)(-3 - 5i)$ $60 - 2i$

34. Solve by the quadratic formula: $\frac{x^2}{72} + \frac{x}{9} = \frac{1}{8}$ $-9, 1$

35. Solve for x: $x^{1/2} - 3x^{1/4} + 2 = 0$ $16, 1$

36. Solve for x: $(x + 5)(x - 2)(x - 7) \le 0$
$x \le -5$ or $2 \le x \le 7$; $(-\infty, -5] \cup [2, 7]$

37. Solve for x: $\dfrac{-2x + 10}{x - 1} \le 0$
$x < 1$ or $x \ge 5$; $(-\infty, 1) \cup [5, \infty)$

38. Graph the parabola:
$y = -(x - 3)^2 + 2$

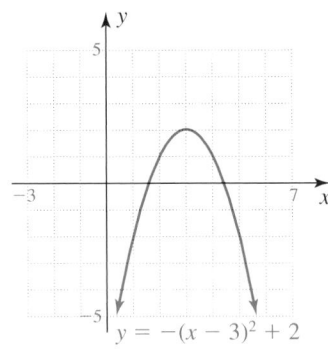

39. Find the center and the radius of $x^2 + y^2 - 6x - 10y + 30 = 0$.
Center $(3, 5)$; $r = 2$

40. Graph: $\dfrac{y^2}{16} - \dfrac{x^2}{4} = 1$

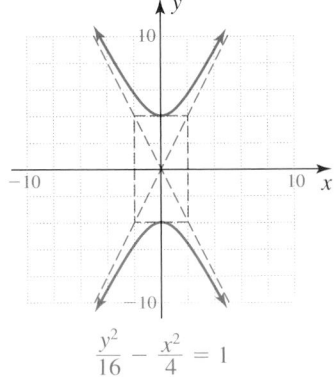

41. Identify the following curve:
$x = y^2 - 4$ Parabola

42. The cost C of manufacturing and selling x units of a product is $C = 23x + 85$, and the corresponding revenue R is $R = x^2 - 55$. Find the break-even value of x.   28

43. Find the domain of $y = \sqrt{x + 9}$.
$\{x \mid x \ge -9\}$

44. If $f(x) = x^4$ and $g(x) = 4 - x$, find $(f \circ g)(x)$. $(4 - x)^4$

45. Let $f(x) = y = 4x - 6$. Graph f and its inverse.

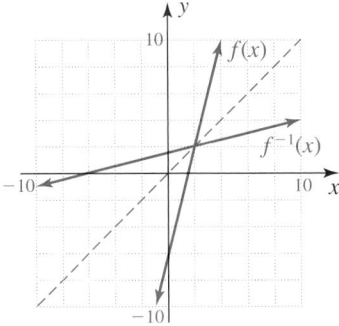

46. Graph $f(x) = 2^x$. Is the inverse a function?

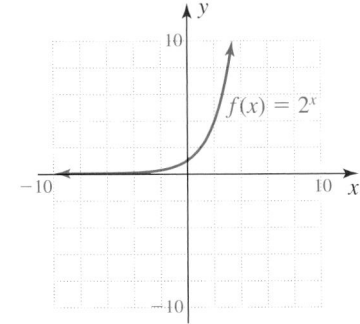

Yes, the inverse is a function.

47. Show that $\log_b \sqrt{\dfrac{21}{83}} = \dfrac{1}{2} \log_b 21 - \dfrac{1}{2} \log_b 83$

$\log_b \sqrt{\dfrac{21}{83}} = \log_b \left(\dfrac{21}{83}\right)^{1/2} = \dfrac{1}{2}[\log 21 - \log 83]$

$= \dfrac{1}{2}\log 21 - \dfrac{1}{2} \log 83$

48. Find x to 4 decimal places if $e^{9.6x} = 7$, and you are given $\ln 7 \approx 1.94591$. $x \approx 0.2027$

49. Use the change of base formula to find $\log_3 5$ to 4 decimal places.
1.4650

50. The number of bacteria present in a culture after t minutes is given as $B = 1000e^{kt}$. If there are 8392 bacteria present after 7 min, find k to 4 decimal places.
$k = 0.3040$

Appendix A:
Sequences and Series

A.1 # SEQUENCES AND SERIES

To Succeed, Review How To . . .

1. Simplify algebraic expressions (pp. 51–55).

2. Evaluate a formula (pp. 88–92).

Objectives

A Find specific and general terms in a sequence.

B Find specified terms in a sequence when the general term is given.

C Use summation notation to find partial sums of a series.

D Solve an application involving sequences.

GETTING STARTED

Rabbity Numbers

Leonardo Fibonacci was one of the greatest mathematicians of the Middle Ages. Here is a problem that greatly interested him.

> Let's suppose you have a 1-month-old pair of rabbits. Assume that in the second month, and every month thereafter, they produce a new pair. If the new pair does the same and none of the rabbits die, how many pairs of rabbits will there be at the beginning of each month?

You can check the solution just by counting:

Month	m_1	m_2	m_3	m_4	m_5	m_6	m_7	. . .
Number of Pairs	1	1	2	3	5	8	13	. . .

The number of pairs 1, 1, 2, 3, 5, 8, 13, . . . form an *infinite sequence* called the **Fibonacci sequence.** The three dots (called an ellipsis) indicate that the sequence continues without stopping. If we stop after a certain number of months, say 6, we obtain the *finite sequence* 1, 1, 2, 3, 5, 8.

In this section we shall study sequences like this one and learn how to find terms in them.

A   ## Finding Specific and General Terms in a Sequence

We can think of the Fibonacci sequence as a function that pairs 1 with 1, 2 with 1, 3 with 2, 4 with 3, and so on. In general, we have this definition.

SEQUENCE

An infinite **sequence** is a function whose domain is the set of natural numbers.

Web It

For practice on how to find the next term in a sequence, go to link A-1-1 at mhhe.com/bello.

For a more advanced lesson, with interactive examples using more sophisticated notation, go to link A-1-2 at mhhe.com/bello.

Calculate It

Finding the Term of a Sequence

If you know the general term of a sequence, most calculators will give you any term you wish. In Example 1, the general term is $a_n = 2^n$. With a TI-83 Plus, start by putting the calculator in sequence mode. To do this, press MODE, go down to the row that begins with "Func," scroll across to "Seq" and press ENTER. Return to and clear the home screen (press 2nd MODE CLEAR). Tell the calculator that the general term is 2^n (press ALPHA + 2 ^ X,T,θ,n ALPHA +). Next define your sequence U_n by storing 2^n into $u(n)$. This is done by pressing STO▸ 2nd 7. Finally, tell the calculator that you want the second term by pressing ALPHA . 2nd 7 (2). Now press ENTER, and the answer, 4, will be on the next line, as shown in the window.

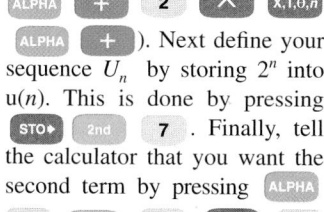

```
"2^n"→u:u(2)
                        4
```

Fortunately, you don't have to reenter all the information to find the fifth term. Press 2nd ENTER and move the cursor left until it is on top of the 2. Press 5 ENTER and the answer 32 will appear.

Intuitively, you can think of a sequence as a set of numbers arranged according to some pattern. The numbers in a sequence are called the *first term, second term, third term,* and so on. These **terms** are usually denoted by subscripts. Thus, a_1 (read "a sub-one"), a_2, and a_3 are the first three terms in a sequence. The expression a_n, which defines a sequence, is called the **general term.** The notation a_n means the same as $a(n)$. Here are some examples of sequences.

$$\text{Positive multiples of 2:} \quad 2, 4, 6, \ldots$$
$$\text{Powers of 10:} \quad 10^1, 10^2, 10^3, \ldots$$
$$-1 \text{ and } 1 \text{ alternating:} \quad -1, 1, -1, \ldots$$

To find the first three terms and the general term in the sequence consisting of the positive multiples of 2, write the terms a_1, a_2, a_3, and so on in one line and the sequence $2, 4, 6, \ldots$ on the next line and find the pattern associating (linking) them:

$$
\begin{array}{ccccc}
a_1 & a_2 & a_3 & \cdots & a_n & \cdots \\
\downarrow & \downarrow & \downarrow & & \downarrow & \\
2 & 4 & 6 & \cdots & ? & \cdots \\
a_1 = 2 \cdot 1 & a_2 = 2 \cdot 2 & a_3 = 2 \cdot 3 & \cdots & a_n = 2 \cdot n & \cdots
\end{array}
$$

The first three terms are $a_1 = 2$, $a_2 = 4$, $a_3 = 6$, and the general term is $a_n = 2n$.

Similarly, the first three terms of the sequence 10^1, 10^2, 10^3, ... are $a_1 = 10$, $a_2 = 10^2 = 100$, and $a_3 = 10^3 = 1000$. The general term is $a_n = 10^n$.

The first three terms in the sequence $-1, 1, -1, \ldots$ are $a_1 = -1$, $a_2 = 1$, and $a_3 = -1$.

$$
\begin{array}{ccccc}
a_1 & a_2 & a_3 & \cdots & a_n & \cdots \\
\downarrow & \downarrow & \downarrow & & \downarrow & \\
-1 & 1 & -1 & \cdots & ? & \cdots \\
a_1 = (-1)^1 = -1 & a_2 = (-1)^2 = 1 & a_3 = (-1)^3 = -1 & \cdots & a_n = (-1)^n & \cdots
\end{array}
$$

From the pattern, you can see that $a_n = (-1)^n$.

EXAMPLE 1   **Finding specific terms in a sequence**

For the sequence $2, 4, 8, \ldots$, where each term after a_1 is double the preceding term, find: a_2, a_5, and a_n

SOLUTION   As before, we write

$$
\begin{array}{ccccc}
a_1 & a_2 & a_3 & \cdots & a_n & \cdots \\
\downarrow & \downarrow & \downarrow & & \downarrow & \\
2 & 4 & 8 & \cdots & ? & \cdots \\
a_1 = 2^1 & a_2 = 2^2 & a_3 = 2^3 & \cdots & a_n = 2^n & \cdots
\end{array}
$$

Following the pattern, the second term is $a_2 = 2^2 = 4$, the fifth term is $a_5 = 2^5 = 32$, and the general term is $a_n = 2^n$.

PROBLEM 1

For the sequence $4, 8, 12, 16, \ldots$ where each term is a multiple of four, find a_2, a_4, a_{10}, and a_n.

Answer

1. $a_2 = 8$, $a_4 = 16$, $a_{10} = 40$, and $a_n = 4n$

EXAMPLE 2 More practice finding terms	PROBLEM 2

For the sequence of even integers with alternating signs $2, -4, 6, -8, \ldots$, find: a_{10} and a_n

For the sequence of multiples of four with alternating signs $-4, 8, -12, 16, \ldots$ find a_{10} and a_n.

SOLUTION   The only difference between this problem and the sequence $2, 4, 6, 8, \ldots$ is the sign of the terms of the sequence. To get the correct sign, we use the factor $(-1)^{n-1}$, which gives alternately $+1$ and -1. Then we write

$$
\begin{array}{ccc}
a_1 & a_2 & a_3 \quad \cdots \\
\downarrow & \downarrow & \downarrow \\
2 & -4 & 6 \quad \cdots
\end{array}
$$

$a_1 = (-1)^{1-1}2 \cdot 1 = 2 \quad a_2 = (-1)^{2-1}2 \cdot 2 = -4 \quad a_3 = (-1)^{3-1}2 \cdot 3 = 8 \ldots$

$$
\begin{array}{c}
a_n \quad \cdots \\
\downarrow \\
? \quad \cdots
\end{array}
$$

$a_n = (-1)^{n-1}2n \ldots$

Thus $a_{10} = (-1)^{10-1}2 \cdot 10 = (-1)^9 2 \cdot 10 = (-1)20 = -20$, and $a_n = (-1)^{n-1} 2n$.

EXAMPLE 3 Sequential folios	PROBLEM 3

In the publishing industry large printed pages are folded to make the pages of a book. If sheets are folded once, this makes 2 pages or a *folio*. If the sheets are folded twice, this makes 4 pages or a *quarto*. If sheets are folded three times, this makes 8 pages, or an *octavo*. Further folding produces units called 16mo, 32mo, and so on.

a. Write the sequence that gives the number of pages after each fold.

b. If a sheet is folded 6 times, how many pages are there?

c. How many pages result after n folds?

Suppose each card in a pack of six index cards is cut in half and all the halves are put in a single pile. Then, the procedure is repeated again and again.

a. Write the sequence that gives the number of cards in the pack after each cut.

b. How many cards are in the pack after the fourth cut?

c. How many cards are in the pack after the nth cut?

SOLUTION   Since 1 fold produces 2 pages, $a_1 = 2$. Then 2 folds produce 4 pages, so $a_2 = 4$, and 3 folds produce 8 pages, so $a_3 = 8$. We then write

$$
\begin{array}{ccccc}
a_1 & a_2 & a_3 & \cdots & a_n \quad \cdots \\
\downarrow & \downarrow & \downarrow & & \downarrow \\
2 & 4 & 8 & \cdots & ? \quad \cdots
\end{array}
$$

$a_1 = 2^1 \quad a_2 = 2^2 \quad a_3 = 2^3 \quad \cdots \quad a_n = 2^n \quad \cdots$

a. The sequence is $2, 4, 8, \ldots, 2^n, \ldots$.

b. If a sheet is folded 6 times, the number of pages is $2^6 = 64$.

c. After n folds, there are 2^n pages.

Answers

2. $a_{10} = 40$, $a_n = (-1)^n(4n)$
3. a. $12, 24, 48, \ldots$ **b.** 96
c. $6 \cdot 2^n$

Finding Specified Terms When the General Term Is Given

Web It

For a lesson on how to find specified terms when general terms are given, specifically, when the general term for a special sequence called the Fibonacci sequence is given, go to link A-1-3 at mhhe.com/bello.

In Examples 1 and 2, we found the general term of a given sequence. However, if only a finite number of successive terms are given *without* a rule that defines the general term, a *unique* general term cannot be obtained. Let's see why.

Consider two sequences with general terms:

$$a_n = 2n \qquad \text{and} \quad a_n = 2n + \tfrac{1}{2}(n-1)(n-2)(n-3)(n-4)$$

$$a_1 = 2 \cdot 1 = 2 \quad \text{and} \quad a_1 = 2 \cdot 1 + \tfrac{1}{2}(1-1)(1-2)(1-3)(1-4) = 2$$

$$a_2 = 2 \cdot 2 = 4 \quad \text{and} \quad a_2 = 2 \cdot 2 + \tfrac{1}{2}(2-1)(2-2)(2-3)(2-4) = 4$$

$$a_3 = 2 \cdot 3 = 6 \quad \text{and} \quad a_3 = 2 \cdot 3 + \tfrac{1}{2}(3-1)(3-2)(3-3)(3-4) = 6$$

$$a_4 = 2 \cdot 4 = 8 \quad \text{and} \quad a_4 = 2 \cdot 4 + \tfrac{1}{2}(4-1)(4-2)(4-3)(4-4) = 8$$

$$a_5 = 2 \cdot 5 = 10 \quad \text{but} \quad a_5 = 2 \cdot 5 + \tfrac{1}{2}(5-1)(5-2)(5-3)(5-4)$$

$$= 10 + \tfrac{1}{2}(24) = 10 + 12 = 22$$

Thus examining the first four terms in the sequence 2, 4, 6, 8, . . . may lead you to use $a_n = 2n$ or $a_n = 2n + \tfrac{1}{2}(n-1)(n-2)(n-3)(n-4)$ as the general term. They are both correct. However, the fifth terms are not equal. You cannot find a unique general term from a finite number of terms.

> **NOTE**
>
> There may be more than one general term that produces the same first three or four terms in a sequence and, consequently, there are no rules for finding the general term of a sequence from the first few terms.

EXAMPLE 4   **Finding specified terms when the general term is given**

Find the first three terms and the ninth term of the sequence whose general term is

$$a_n = \frac{1}{2}n(n-1)$$

SOLUTION   We find the required terms by substituting the corresponding values of n into the given formula. Thus

$$a_1 = \frac{1}{2}(1)(1-1) = 0$$

$$a_2 = \frac{1}{2}(2)(2-1) = 1$$

$$a_3 = \frac{1}{2}(3)(3-1) = 3$$

$$a_9 = \frac{1}{2}(9)(9-1) = 36$$

PROBLEM 4

Find the first three terms and the tenth term of the sequence whose general term is $a_n = \frac{1}{3}(n^2 - 1)$.

Answer

4. $a_1 = 0$, $a_2 = 1$, $a_3 = \frac{8}{3}$; $a_{10} = 33$

Sometimes function notation rather than subscript notation is used to define the terms of a sequence. For instance, in Example 4, we could use function notation and write:

$$a(n) = \frac{1}{2}n(n-1) \qquad \text{Instead of } a_n = \tfrac{1}{2}n(n-1)$$

so that

$$a(1) = \frac{1}{2}(1)(1-1) = 0$$

$$a(2) = \frac{1}{2}(2)(2-1) = 1, \quad \text{and so on}$$

EXAMPLE 5   Finding the sequence given a function

Consider the function

$$a(n) = 2n + 1, \quad n = 1, 2, 3, \ldots$$

Find: The sequence corresponding to this function

SOLUTION

for $n = 1$, $a(1) = 2(1) + 1 = 3$

for $n = 2$, $a(2) = 2(2) + 1 = 5$

for $n = 3$, $a(3) = 2(3) + 1 = 7$

for $n = 4$, $a(4) = 2(4) + 1 = 9$

The sequence is $3, 5, 7, 9, \ldots$.

PROBLEM 5

Consider the function

$$a(n) = n(n-1), \quad n = 1, 2, 3, \ldots$$

Find the sequence corresponding to this function.

C   Using Summation Notation

There is an Old English rhyme that reads as follows:

As I was going to St. Ives
I met a man with seven wives;
Every wife had seven sacks;
Every sack had seven cats;
Every cat had seven kits,
Kits, cats, sacks and wives,
How many were going to St. Ives?

The answer is really 1. If you don't believe this, then read the rhyme again. A less misleading question would be, How many were *leaving* St. Ives? The sequence of numbers involved in this second question is

1	$1 \cdot 7 = 7$	$7 \cdot 7 = 49$	$7 \cdot 7 \cdot 7 = 343$	$7 \cdot 7 \cdot 7 \cdot 7 = 2401$
↓	↓	↓	↓	↓
Man	Wives	Sacks	Cats	Kits

Thus the number leaving St. Ives is $1 + 7 + 49 + 343 + 2401 = 2801$.

The sequence 1, 7, 49, 343, 2401 is a *finite sequence,* and the indicated sum for the sequence $1 + 7 + 49 + 343 + 2401$ is called a *series*.

Web It

To explore the relationship between infinite series and summation notation (coming up shortly), go to link A-1-4 at mhhe.com/bello.

INFINITE AND FINITE SERIES

Given the infinite sequence

$$a_1, a_2, a_3, \ldots, a_n, \ldots$$

the sum of the terms is called an **infinite series:**

$$a_1 + a_2 + a_3 + \cdots + a_n + \cdots$$

The **partial sum**

$$a_1 + a_2 + a_3 + \cdots + a_n$$

is called a **finite series** and is denoted by S_n.

Answer

5. $0, 2, 6, 12, 20, \ldots$

Thus the sequence $7, 7^2, 7^3, 7^4, \ldots, 7^n \ldots$ has the following partial sums:

$S_1 = 7$	The **first** term of the sequence
$S_2 = 7 + 49 = 56$	The sum of the first **two** terms
$S_3 = 7 + 49 + 343 = 399$	The sum of the first **three** terms
$S_4 = 7 + 49 + 343 + 2401 = 2800$	The sum of the first **four** terms

---

EXAMPLE 6   **Finding partial sums**

For the sequence $-1, 3, -5, 7, -9, 11, -13$, find the sums:

a. S_4 **b.** S_7

SOLUTION

a. $S_4 = -1 + 3 + (-5) + 7 = 4$

b. $S_7 = -1 + 3 + (-5) + 7 + (-9) + 11 + (-13) = -7$

PROBLEM 6

For the sequence $-2, 2, -6, 6, -10, 10, -14$, find:

a. S_4 **b.** S_7

Calculate It

Summation

To do sums (as in Example 6) with a TI-83 Plus, press `2nd` `STAT`, move the cursor to the "MATH" (third) column and press 5. Now enter the numbers to be added (separate them by commas and enclose them in braces). To find S_7 for the sequence in Example 6, we enter $\{-1, 3, -5, 7, -9, 11, -13\}$; pressing `ENTER` gives the answer shown in the window.

```
sum ({-1,3,-5,7,-
9,11, -13})
                   -7
```

If we know the general term of a sequence, we can represent a sum of terms using *summation (sigma) notation*. In this notation, the Greek letter Σ (capital sigma), which corresponds to the English letter S, indicates that we are to add the given terms. Thus

$$\sum_{i=1}^{n} a_i \qquad \text{Read "the sum of } a_i \text{ from } i = 1 \text{ to } n."$$

is defined by

$$\sum_{i=1}^{n} a_i = a_1 + a_2 + a_3 + \cdots + a_n$$

The sum need not start at $i = 1$ and end at $i = n$. Furthermore, any letter may be used in place of the *index i*. The following examples illustrate some of the possibilities.

$$\sum_{i=3}^{6} a_i = a_3 + a_4 + a_5 + a_6$$

$$\sum_{j=1}^{5} j = 1 + 2 + 3 + 4 + 5$$

$$\sum_{k=1}^{6} k^2 = 1^2 + 2^2 + 3^2 + 4^2 + 5^2 + 6^2$$

$$\sum_{n=1}^{5} (-1)^n n a_n = -a_1 + 2a_2 - 3a_3 + 4a_4 - 5a_5$$

EXAMPLE 7   **Evaluating sums given in summation notation**

Find and evaluate the sum:

a. $\displaystyle\sum_{n=1}^{4} n^2$ **b.** $\displaystyle\sum_{k=0}^{3} (2k + 1)$

SOLUTION

a. $\displaystyle\sum_{n=1}^{4} n^2 = 1^2 + 2^2 + 3^2 + 4^2 = 1 + 4 + 9 + 16 = 30$

Evaluate n^2 for $n = 1, 2, 3,$ and 4 and then add.

b. $\displaystyle\sum_{k=0}^{3} (2k + 1) = (2 \cdot 0 + 1) + (2 \cdot 1 + 1) + (2 \cdot 2 + 1) + (2 \cdot 3 + 1)$

$= 1 + 3 + 5 + 7 = 16$

PROBLEM 7

Find and evaluate:

a. $\displaystyle\sum_{n=1}^{3} n^3$

b. $\displaystyle\sum_{k=0}^{3} (3k + 1)$

Answers

6. a. 0 **b.** -14

7. a. 36 **b.** 22

| EXAMPLE 8 | More practice using summation notation | PROBLEM 8 |

EXAMPLE 8 — **More practice using summation notation**

Write using summation notation:

a. $2 + 4 + 6 + \cdots + 20$

b. $\dfrac{1}{2} + \dfrac{1}{3} + \dfrac{1}{4} + \cdots + \dfrac{1}{30}$

SOLUTION

a. The finite series $2 + 4 + 6 + \cdots + 20$ is a sum of even numbers with general term $2n$ starting with $n = 1$ and ending with $n = 10$. The summation notation is

$$\sum_{n=1}^{10} 2n$$

b. The finite series

$$\frac{1}{2} + \frac{1}{3} + \frac{1}{4} + \cdots + \frac{1}{30}$$

is a sum of fractions whose denominators are consecutive numbers starting with 2 and ending with 30. The summation notation is

$$\sum_{n=2}^{30} \frac{1}{n}$$

PROBLEM 8

Write using summation notation:

a. $3 + 6 + 9 + 12 + 15$

b. $\dfrac{1}{3} + \dfrac{1}{4} + \dfrac{1}{5} + \cdots + \dfrac{1}{20}$

Calculate It   Summing Sequences

Many calculators use the summation (sigma) language to find the sum of a sequence. To do Example 7(a), you must tell the calculator you want the sum of a sequence. With a TI-83 Plus, press 2nd STAT , move the cursor to the "MATH" column, and press 5. This tells the calculator you want a sum. Then press 2nd STAT move to the "OPS" column and press 5 to make it clear that you want the sum of a sequence. Enter the sequence as N^2 by pressing ALPHA LOG ^ 2 , ; the variable you are using (press ALPHA LOG ,); the starting point (press 1 ,); the ending point (press 4 ,); and the increments you want for N (press 1). Now close the parentheses and press ENTER . The answer is 30, as shown in the window.

```
sum (seq(N^2,N,1,
4,1))
                  30
```

D   Solving an Application Involving Sequences

Did you know that sequences were used to find some of the planets of our solar system? In 1772, a German astronomer named Johann Bode discovered a pattern in the distances of the planets from the Sun. We examine his sequence in Example 9.

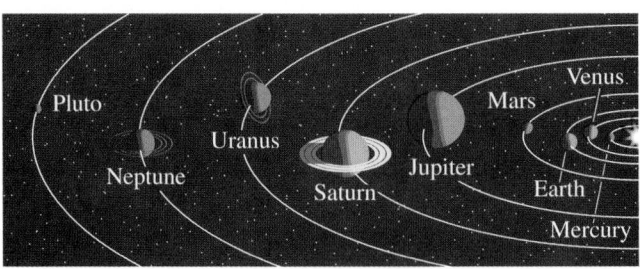

Answers

8. a. $\displaystyle\sum_{n=1}^{5} 3n$ **b.** $\displaystyle\sum_{n=3}^{20} \frac{1}{n}$

EXAMPLE 9 **Finding missing planets**

Bode's sequence is as follows:

p_1	p_2	p_3	p_4	p_5	p_6	p_7	p_8
Mercury	Venus	Earth	Mars	?	Jupiter	Saturn	?
↓	↓	↓	↓	↓	↓	↓	
$0 + 4 = 4$	$3 + 4 = 7$	$6 + 4 = 10$	$12 + 4 = 16$	?	$48 + 4 = 52$	$96 + 4 = 100$	?

a. What number corresponds to the missing fifth planet?

b. What number corresponds to the eighth planet?

SOLUTION

a. Starting with p_3, the number in front of 4 is twice the number in front of 4 in the preceding term. Thus the number for the unknown planet is $2 \cdot 12 + 4 = 28$. (It turns out that this planet is really Ceres, a planetoid or asteroid.)

b. The number for the planet after Saturn is $2 \cdot 96 + 4 = 196$, which corresponds to Uranus, discovered by William Herschel in 1781.

PROBLEM 9

What number would correspond to

a. The 9th planet?

b. The 10th planet?

Calculate It Exercises

Use your calculator to do:

1. Example 2

2. Example 3   **4.** Example 5

3. Example 4   **5.** Example 7(b)

Answers

9. a. 388   **b.** 772

Exercises A.1

A In Problems 1–22, find a tenth term and an nth term to fit the given sequence.

1. $1, 4, 7, 10, \ldots$
$a_{10} = 28$; $a_n = 3n - 2$

2. $5, 7, 9, 11, \ldots$
$a_{10} = 23$; $a_n = 2n + 3$

3. $5, 8, 11, 14, \ldots$
$a_{10} = 32$; $a_n = 3n + 2$

4. $3, 8, 13, 18, \ldots$
$a_{10} = 48$; $a_n = 5n - 2$

5. $20, 25, 30, 35, \ldots$
$a_{10} = 65$; $a_n = 5n + 15$

6. $15, 18, 21, 24, \ldots$
$a_{10} = 42$; $a_n = 3n + 12$

7. $50, 45, 40, 35, \ldots$
$a_{10} = 5$; $a_n = 55 - 5n$

8. $30, 28, 26, 24, \ldots$
$a_{10} = 12$; $a_n = 32 - 2n$

9. $\dfrac{1}{2}, \dfrac{1}{3}, \dfrac{1}{4}, \dfrac{1}{5}, \cdots$
$a_{10} = \dfrac{1}{11}$; $a_n = \dfrac{1}{n+1}$

10. $\dfrac{1}{2}, \dfrac{2}{3}, \dfrac{3}{4}, \dfrac{4}{5}, \cdots$
$a_{10} = \dfrac{10}{11}$; $a_n = \dfrac{n}{n+1}$

11. $1, -1, 1, -1, \ldots$
$a_{10} = -1$;
$a_n = (-1)^{n-1}$

12. $-1, 2, -4, 8, \ldots$
$a_{10} = 512$;
$a_n = (-1)^n 2^{n-1}$

13. $x, x^2, x^3, x^4, \ldots$
$a_{10} = x^{10}$; $a_n = x^n$

14. $x^2, x^4, x^6, x^8, \ldots$
$a_{10} = x^{20}$; $a_n = x^{2n}$

15. $x, -x^3, x^5, -x^7, \ldots$
$a_{10} = -x^{19}$;
$a_n = (-1)^{n-1} x^{2n-1}$

16. $-x, x^2, -x^4, x^8, \ldots$
$a_{10} = x^{512}$;
$a_n = (-1)^n x^{2^{n-1}}$

17. $x, -x, x, -x, \ldots$
$a_{10} = -x$;
$a_n = (-1)^{n-1} x$

18. $-x, x, -x, x, \ldots$
$a_{10} = x$;
$a_n = (-1)^n x$

19. $x, \dfrac{x^2}{2}, \dfrac{x^3}{3}, \dfrac{x^4}{4}, \ldots$
$a_{10} = \dfrac{x^{10}}{10}$; $a_n = \dfrac{x^n}{n}$

20. $\dfrac{x}{5}, \dfrac{x^2}{10}, \dfrac{x^3}{15}, \dfrac{x^4}{20}, \cdots$
$a_{10} = \dfrac{x^{10}}{50}$; $a_n = \dfrac{x^n}{5n}$

21. $\dfrac{x}{2}, -\dfrac{x^2}{4}, \dfrac{x^3}{8}, -\dfrac{x^4}{16}, \ldots$
$a_{10} = -\dfrac{x^{10}}{1024}$;
$a_n = (-1)^{n-1}\left(\dfrac{x}{2}\right)^n$

22. $\dfrac{x}{2}, -\dfrac{x^3}{4}, \dfrac{x^5}{8}, -\dfrac{x^7}{16}, \cdots$
$a_{10} = -\dfrac{x^{19}}{1024}$;
$a_n = (-1)^{n-1} \dfrac{x^{2n-1}}{2^n}$

B In Problems 23–36, find the first three terms of the sequence with the given general term.

23. $a_n = 2n - 3$
$-1, 1, 3$

24. $a_n = 2n + 3$
$5, 7, 9$

25. $a_n = \dfrac{n(n-2)}{2}$
$-\dfrac{1}{2}, 0, \dfrac{3}{2}$

26. $a_n = \dfrac{n(n+2)}{2}$
$\dfrac{3}{2}, 4, \dfrac{15}{2}$

27. $a(n) = 1 - \dfrac{1}{n}$
$0, \dfrac{1}{2}, \dfrac{2}{3}$

28. $a(n) = 1 + \dfrac{2}{n}$
$3, 2, \dfrac{5}{3}$

29. $a_n = n^2$
$1, 4, 9$

30. $a_n = -n^3$
$-1, -8, -27$

31. $a(n) = \dfrac{n}{2n+1}$

$\dfrac{1}{3}, \dfrac{2}{5}, \dfrac{3}{7}$

32. $a(n) = \dfrac{n}{3n-1}$

$\dfrac{1}{2}, \dfrac{2}{5}, \dfrac{3}{8}$

33. $a_n = (-1)^n$

$-1, 1, -1$

34. $a_n = (-2)^{n-1}$

$1, -2, 4$

35. $a(n) = (-1)^n 2^{-n}$

$-\dfrac{1}{2}, \dfrac{1}{4}, -\dfrac{1}{8}$

36. $a(n) = (-1)^{n-1} 3^n$

$3, -9, 27$

C In Problems 37–48, compute the indicated sums.

37. $\displaystyle\sum_{k=1}^{6} k^2$ 91

38. $\displaystyle\sum_{i=1}^{8} i$ 36

39. $\displaystyle\sum_{k=1}^{4} k^3$ 100

40. $\displaystyle\sum_{n=1}^{5} 2n$ 30

41. $\displaystyle\sum_{i=1}^{7} 3$ 21

42. $\displaystyle\sum_{k=3}^{8} 5$ 30

43. $\displaystyle\sum_{j=1}^{4} \dfrac{1}{2j}$

$\dfrac{25}{24}$

44. $\displaystyle\sum_{j=0}^{6} \dfrac{1}{j+1}$

$\dfrac{363}{140}$

45. $\displaystyle\sum_{k=1}^{7} \dfrac{k+1}{k}$

$\dfrac{1343}{140}$

46. $\displaystyle\sum_{n=1}^{6} (-1)^n$

0

47. $\displaystyle\sum_{k=1}^{5} (-1)^{k+1}$

1

48. $\displaystyle\sum_{k=1}^{6} (-1)^k 3^{k+1}$

1638

In Problems 49–56, write each expression using the sigma notation.

49. $1 + 2 + 3 + \cdots + 200$ $\displaystyle\sum_{n=1}^{200} n$

50. $1 + 4 + 9 + 16 + \cdots + 49$ $\displaystyle\sum_{n=1}^{7} n^2$

51. $1 + \dfrac{1}{2} + \dfrac{1}{3} + \dfrac{1}{4} + \cdots + \dfrac{1}{50}$ $\displaystyle\sum_{n=1}^{50} \dfrac{1}{n}$

52. $x_1^2 + x_2^2 + x_3^2 + \cdots + x_{100}^2$ $\displaystyle\sum_{n=1}^{100} x_n^2$

53. $1 - 2 + 3 - 4 + 5 - 6 + \cdots - 50$ $\displaystyle\sum_{n=1}^{50} (-1)^{n-1} n$

54. $2 - 4 + 8 - 16 + 32$ $\displaystyle\sum_{n=1}^{5} (-1)^{n-1} 2^n$

55. $1 + 6 + 11 + 16 + 21$ $\displaystyle\sum_{n=1}^{5} (5n - 4)$

56. $\dfrac{1}{2} + 1 + \dfrac{3}{2} + 2 + \dfrac{5}{2} + 3$ $\displaystyle\sum_{n=1}^{6} \dfrac{n}{2}$

APPLICATIONS

Money and Measurements

57. A property valued at $30,000 will depreciate $1380 the first year, $1340 the second year, $1300 the third year, and so on. What will be the depreciation during

   a. The eighth year? $a_8 = \$1100$

   b. The tenth year? $a_{10} = \$1020$

58. Strikers at a plant were ordered to return to work and were told they would be fined $50 the first day they failed to do so, $75 the second day, $100 the third day, and so on. If the strikers stayed out for 6 days, what was the fine for the sixth day? $175

59. When dropped on a hard surface, a Super Ball takes a sequence of bounces, each one about $\frac{9}{10}$ as high as the preceding one. If a Super Ball is dropped from a height of 10 ft, find how high it will bounce on the

   a. First bounce 9 ft

   b. Third bounce $\frac{729}{100}$ ft

   c. nth bounce $\frac{9^n}{10^{n-1}}$

60. An ancient legend says that the Shah of Persia offered the inventor of chess anything he wished as a reward for his invention. The man asked for 1 grain of wheat to be placed on the first square of the chessboard, 2 grains on the second, 4 grains on the third, and so on. How many grains would there be on

   a. The fifth square? 2^4 or 16

   b. The ninth square? 2^8 or 256

   c. The nth square? 2^{n-1}

61. A colony of bacteria starts with 100 members and doubles every hour. How many bacteria are there at the end of

   a. 2 hr? 400   **b.** 4 hr? 1600   **c.** n hr? $100(2)^n$

62. A free-falling body falls about 16 ft the first second, 48 ft the next second, 80 ft the third second, 112 ft the fourth second, and so on. How far does it fall during

   a. The eighth second? 240 ft

   b. The nth second? $16(2n - 1)$ ft

63. A salesman sold $100 worth of goods on Monday and doubled his sales each day thereafter for a week. What was the amount of sales on Saturday? $3200

64. A sprinter runs 6 meters in the first second of a certain race and increases her speed by 25 cm/sec in each succeeding second. (This means that she goes 6 m 25 cm the second second, 6 m 50 cm the third second, and so on.) How far does she go during

   a. The eighth second? 6 m 175 cm or 775 cm

   b. The nth second?
   6 m + $25(n - 1)$ cm or $(575 + 25n)$ cm

SKILL CHECKER

Try the "Skill Checker" exercises so you'll be ready for the next section.

Simplify:

65. $7 + (n - 1)(3)$ $4 + 3n$

66. $16 + (n - 1)(2)$ $14 + 2n$

67. $\frac{1}{2}n(16 + 32n - 16)$ $16n^2$

68. $\frac{n}{2}(3 + 5n - 2)$ $\frac{5n^2 + n}{2}$

Factor:

69. $5n^2 + n - 328$
   $(5n + 41)(n - 8)$

70. $7n^2 - n - 336$
   $(7n + 48)(n - 7)$

USING YOUR KNOWLEDGE

The Fibonacci Rabbits

Let's return to the rabbit problem introduced in the *Getting Started.* Here is what happens in the first 5 months:

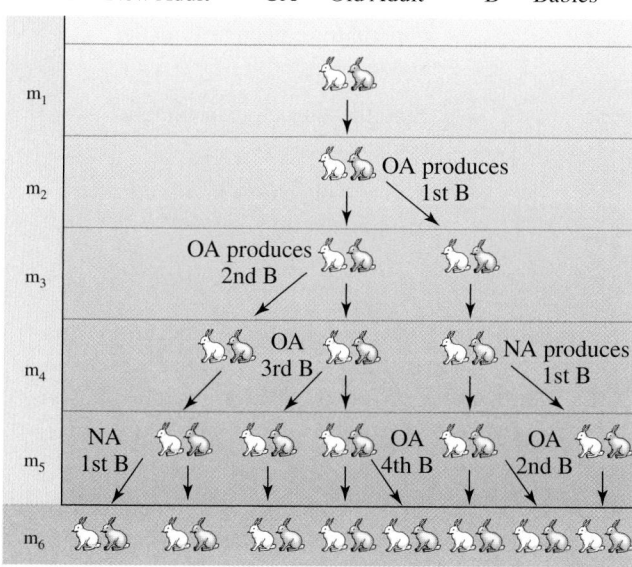

NA = New Adult   OA = Old Adult   B = Babies

As we can see, the number of rabbit pairs at the beginning of each month are the terms of the Fibonacci sequence 1, 1, 2, 3, 5, 8, 13, ...

Starting with the third term, each term is the sum of the two preceding terms: $2 = 1 + 1$, $3 = 1 + 2$, $5 = 2 + 3$, and so on. This leads us to the general formula

$$a_n = a_{n-2} + a_{n-1}$$

Use these ideas to write the following terms of the Fibonacci sequence.

71. The eighth term $a_8 = 21$

72. The ninth term $a_9 = 34$

73. The tenth term $a_{10} = 55$

74. The eleventh term $a_{11} = 89$

75. The twelfth term $a_{12} = 144$

WRITE ON

76. Write your own definition of a sequence.
   Answers may vary.

77. What is the difference between a finite sequence and an infinite sequence?   Answers may vary.

78. Given the sequence 1, 3, 5, . . . can you find a *unique* general term a_n? Explain why or why not.
   Answers may vary.

79. If the general term a_n for a sequence is known, is the resulting sequence unique? Explain why or why not.
   Answers may vary.

MASTERY TEST

If you know how to do these problems, you have learned your lesson!

80. A famous painting doubles in value every 50 yr. Find the value of the painting in the year 2000 if it was worth $1000 in the year 1500. $1,024,000

81. Find the first three terms and the tenth term of the sequence whose general term is $a_n = \frac{1}{2}(n^2 - 1)$.
   $0, \frac{3}{2}, 4; a_{10} = \frac{99}{2}$

82. Suppose each card in a pack of five index cards is cut in half and all the halves are put in a single pile and cut again. This procedure is repeated again and again.

     a. Write the sequence that gives the number of cards in the pack after each cut. $10, 20, 40, \ldots$

     b. How many cards are in the pack after the fourth cut? 80

     c. How many cards are in the pack after the nth cut? $5 \cdot 2^n$

85. Write using summation (sigma) notation:

     a. $6 + 12 + 18 + 24$ $\displaystyle\sum_{n=1}^{4} 6n$

     b. $2 - 4 + 8 - 16 + 32 - 64$ $\displaystyle\sum_{n=1}^{6} (-1)^{n-1} 2^n$

83. For the sequence of multiples of 3 with alternating signs, $-3, 6, -9, 12, \ldots$ find a_{10} and a_n.
$a_{10} = 30; \; a_n = (-1)^n \cdot 3n$

84. For the sequence of positive multiples of 3—3, 6, 9, 12, . . . —find:

     a. a_2 and a_4 $a_2 = 6; \; a_4 = 12$ **b.** a_{10} $a_{10} = 30$

     c. a_n $a_n = 3n$

86. Find and evaluate the sum:

     a. $\displaystyle\sum_{k=1}^{6} (2^k + 1)$ 132 **b.** $\displaystyle\sum_{k=1}^{5} \frac{1}{k^2}$ $\dfrac{5269}{3600}$

A.2   ARITHMETIC SEQUENCES AND SERIES

To Succeed, Review How To . . .

1. Recognize the terms of a sequence (pp. A-2–A-4).

2. Find the nth term of a sequence when the general term is given (pp. A-5–A-6).

Objectives

A Find the common difference and general term in an arithmetic sequence.

B Find the sum of an arithmetic sequence.

C Solve applications involving arithmetic sequences.

GETTING STARTED   Falling Sequences

A skydiver plunges toward the ground. Do you know how far he will fall in the first 5 sec? A free-falling body travels about 16 ft in the first second, 48 ft in the next second, 80 ft in the third second, and so on. The number of feet traveled in each successive second is

$$16, 48, 80, 112, 144, \ldots$$

Can you find the next number in the sequence? The second term (48) is obtained by adding 32 to the first term (16). Similarly, the third term (80) is obtained by adding 32 to the second term (48), and so on. Thus the term after 144 is found by adding 32 to 144 to obtain 176. Sequences in which successive terms are found by adding a constant to the preceding term are called *arithmetic sequences* or *arithmetic progressions,* and we shall discuss them in this section.

A   Arithmetic Sequences

Getting Started gives an application of arithmetic sequences. You will find other applications in the problem set.

Web It

For an excellent lesson on arithmetic sequences, go to link A-2-1 at mhhe.com/bello.

ARITHMETIC SEQUENCE

An **arithmetic sequence** or **arithmetic progression** is a sequence in which each term after the first is obtained by adding a quantity d, called the *common difference,* to the preceding term.

The sequence $16, 48, 80, 112, 144, \ldots$ mentioned in *Getting Started* is an arithmetic sequence in which each term is obtained by adding the common difference 32 to the preceding term. This means that the common difference for an arithmetic sequence is just the difference between any two consecutive terms.

COMMON DIFFERENCE

The **common difference d** is defined by

$$d = a_{n+1} - a_n$$

EXAMPLE 1 **Finding the common difference**	**PROBLEM 1**
Find the common difference in each sequence:	Find the common difference for the sequence:
a. $7, 37, 67, 97, \ldots$ **b.** $10, 5, 0, -5, \ldots$	**a.** $5, 9, 13, 17, \ldots$
SOLUTION	**b.** $12, 6, 0, -6, \ldots$
a. The common difference is $37 - 7 = 30$ (or $67 - 37$, or $97 - 67$).	
b. The common difference is $5 - 10 = -5$ (or $0 - 5$, or $-5 - 0$).	

Web It

To learn how to find the common difference and indicated terms of a sequence, go to link A-2-2 at mhhe.com/bello.

It's customary to denote the first term of an arithmetic sequence by a_1 (read, "a sub 1"), the common difference by d, and the nth term by a_n. Thus in the sequence $16, 48, 80, 112, 144, \ldots$, we have $a_1 = 16$ and $d = 32$. The second term of the sequence, a_2, is

$$a_2 = a_1 + 32 = 16 + 32 = 48$$

Since each term is obtained from the preceding one by adding 32,

$$a_3 = a_2 + 32 = (a_1 + 32) + 32 \quad = a_1 + 2 \cdot 32 = 80$$
$$a_4 = a_3 + 32 = (a_1 + 2 \cdot 32) + 32 = a_1 + 3 \cdot 32 = 112$$
$$a_5 = a_4 + 32 = (a_1 + 3 \cdot 32) + 32 = a_1 + 4 \cdot 32 = 144$$

By following this pattern, we make the following definition.

General Term of an Arithmetic Sequence

$$a_n = a_1 + (n - 1) \cdot d$$

EXAMPLE 2 **Working with an arithmetic sequence**	**PROBLEM 2**
Consider the sequence $7, 10, 13, 16, \ldots$ and find:	Consider the sequence $3, 8, 13, 18, \ldots$ Find:
a. a_1, the first term **b.** d, the common difference	**a.** a_1 **b.** d **c.** a_8 **d.** a_n
c. a_{11}, the eleventh term **d.** a_n, the nth term	
SOLUTION	
a. The first term a_1 is 7.	
b. The common difference d is $10 - 7 = 3$.	**Answers**
c. The eleventh term is $a_{11} = 7 + (11 - 1) \cdot 3 = 7 + 10 \cdot 3 = 37$.	**1. a.** 4 **b.** -6
d. $a_n = a_1 + (n - 1) \cdot d = 7 + (n - 1) \cdot 3 = 7 + 3n - 3 = 4 + 3n$	**2. a.** 3 **b.** 5 **c.** 38 **d.** $5n - 2$

B   Finding the Sum of an Arithmetic Sequence

Web It

For a lesson reviewing common differences, general terms, and deriving the formula for the partial sum of a sequence, go to link A-2-3 at mhhe.com/bello.

Let's return to the problem of the skydiver. How far does he fall in 5 sec? The first five terms of the sequence are 16, 48, 80, 112, and 144; thus we need to find the sum

$$16 + 48 + 80 + 112 + 144$$

Since successive terms of an arithmetic sequence are obtained by adding the common difference d, the sum S_n of the first n terms is

$$S_n = a_1 + (a_1 + d) + (a_1 + 2d) + (a_1 + 3d) + \cdots + a_n \tag{1}$$

We can also start with a_n and obtain successive terms by subtracting the common difference d. Thus with the terms written in reverse order,

$$S_n = a_n + (a_n - d) + (a_n - 2d) + \cdots + a_1 \tag{2}$$

Adding equations (1) and (2), we find that the d's drop out, and we obtain

$$2S_n = (a_1 + a_n) + (a_1 + a_n) + \cdots + (a_1 + a_n)$$
$$= n(a_1 + a_n)$$

We find the following formula.

Sum of an Arithmetic Sequence

The sum S_n of the first n terms starting with a_1 and ending with a_n is given by

$$S_n = \frac{n}{2}(a_1 + a_n)$$

We are now able to determine the sum S_5, the distance the skydiver dropped in 5 sec:

$$S_5 = \frac{5}{2}(16 + 144) = 400 \text{ ft}$$

Calculate It   Finding Terms and Sums of a Sequence

As we mentioned in Section A.1, if you know the general term of a sequence, most calculators can find any term you wish. Moreover, they can also find the sum of the terms of an arithmetic sequence (a series). In Example 3(b), the general term is $a_n = 32n - 16$. To find a_{10}, tell the calculator what the general term is. With a TI-83 Plus, press [ALPHA] [+] [3] [2] [X,T,θ,n] [−] [1] [6] [ALPHA] [+]. Define the sequence U_n by storing $32n - 16$ into $u(n)$ by pressing [STO▸] [2nd] [7] and then tell the calculator that you want the tenth term by pressing [ALPHA] [.] [2nd] [7] [(] [1] [0] [)]. Now press [ENTER], and the answer, 304, will be on the next line, as shown in the window.

What about the sum S_{10}? Since you know that

$$S_{10} = \frac{n}{2}(a_1 + a_n) \quad \text{and} \quad a_1 = 16$$

and we found $a_{10} = 304$, you can use the calculator feature of your calculator to find the answer.

What about the rest of the problems? There's not much a calculator can do for you in those. Remember, you still have to know your algebra!

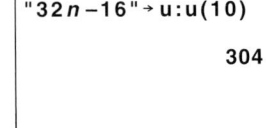

```
"32n−16"→u:u(10)
                304
```

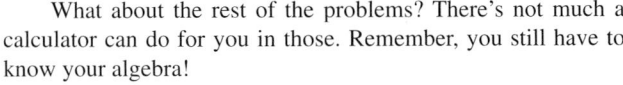

| **EXAMPLE 3** | **Finding the sum of an arithmetic sequence** | **PROBLEM 3** |

Find the distance the skydiver falls in:

a. 10 sec   **b.** n sec

Find the distance the skydiver falls in

a. 20 sec   **b.** x sec

Answers

3. a. 6400 ft   **b.** $16x^2$

SOLUTION   To do this problem, we first find the distance fallen in the nth second. Since $a_1 = 16$ and $d = 32$, we get

$$a_n = a_1 + (n - 1)d$$
$$= 16 + (n - 1)(32)$$
$$= 16 + 32n - 32$$

Thus

$$a_n = 32n - 16$$

a. For $n = 10$, $a_{10} = (32)(10) - 16 = 320 - 16 = 304$ so that

$$S_{10} = \frac{10}{2}(16 + 304) = 5(320) = 1600$$

The skydiver falls 1600 ft in 10 sec.

b. Here, we use the general formulas for S_n and a_n.

$$S_n = \frac{n}{2}(a_1 + a_n) \qquad \text{and} \qquad a_n = 32n - 16$$

Now we substitute $a_1 = 16$ and $a_n = 32n - 16$ into the formula for S_n to get

$$S_n = \frac{n}{2}(16 + 32n - 16) = \frac{n}{2}(32n)$$

or

$$S_n = 16n^2$$

You can check that this surprisingly simple formula gives the same answers that we found in part **a**.

EXAMPLE 4   **Using the sum to find a term and the common difference**

The sum of the first 10 terms of an arithmetic sequence is 205 and the tenth term is 34. Find:

a. a_1, the first term

b. d, the common difference

SOLUTION

a. We use the formula for the sum with $n = 10$:

$$S_{10} = \frac{10}{2}(a_1 + a_{10}) = 5(a_1 + a_{10})$$

We then substitute $S_{10} = 205$ and $a_{10} = 34$ to get

$$205 = 5(a_1 + 34)$$
$$41 = a_1 + 34 \qquad \text{Divide by 5.}$$
$$7 = a_1 \qquad \text{Subtract 34.}$$

The first term a_1 is 7.

b. Now we use the formula for the nth term,

$$a_n = a_1 + (n - 1)d$$

with $n = 10$, $a_1 = 7$, and $a_{10} = 34$, to get

$$34 = 7 + 9d$$
$$27 = 9d \qquad \text{Subtract 7.}$$
$$3 = d \qquad \text{Divide by 9.}$$

The common difference d is 3.

PROBLEM 4

The sum of the first eight terms of an arithmetic sequence is 104 and the eighth term is 20. Find:

a. a_1, the first term

b. d, the common difference

Web It

For a site with many applications of arithmetic sequences, go to link A-2-4 at mhhe.com/bello.

Answers

4. a. 6   **b.** 2

C   Solving Applications of Arithmetic Sequences

EXAMPLE 5   Depreciating a truck

A heavy-duty truck valued at $50,000 depreciates $5000 the first year, $4800 the second year, $4600 the third year, and so on. What will be the value of the truck at the end of 8 yr?

SOLUTION   The yearly depreciations form an arithmetic sequence

$$5000, 4800, 4600, \ldots$$

so that $a_1 = 5000$, $d = -200$, and $n = 8$. Since

$$a_n = a_1 + (n-1)d$$
$$a_8 = 5000 + (7)(-200)$$
$$= 3600$$

The total depreciation will be the sum of the first eight terms of the sequence:

$$S_8 = \frac{8}{2}(a_1 + a_8)$$
$$= 4(5000 + 3600)$$
$$= 4(8600)$$
$$= 34{,}400$$

So the total depreciation is $34,400 and the remaining value is

$$\$50{,}000 - \$34{,}400 = \$15{,}600$$

PROBLEM 5

Repeat Example 5 if the truck depreciates $4000 the first year, $3800 the second year, $3600 the third year, and so on.

Web It

To see more applications with worked out solutions go to link A-2-5 at mhhe.com/bello.

EXAMPLE 6   A long time saving

Alice started a savings campaign. She put aside 3¢ the first day, 8¢ the second day, 13¢ the third day, and so on in arithmetic sequence. After a few days, Alice found that she had saved $1.64. How many days had she been saving?

SOLUTION

1. Read the problem.
Here, we know that $a_1 = 3$, $d = 5$, and $S_n = 164$, and we want to find n. So we use the formula for the sum

$$S_n = \frac{n}{2}(a_1 + a_n)$$

and the formula for the nth term

$$a_n = a_1 + (n-1)d$$

2. Select the unknown.
Substituting the known values into the preceding two formulas, we get

$$164 = \frac{n}{2}(3 + a_n)$$
$$a_n = 3 + (n-1)(5)$$
$$= 5n - 2$$

PROBLEM 6

If in Example 6 Alice increased her daily savings amount by 10 cents, so that the sequence is 3, 13, 23, . . . , how many days would it take her to get a total of $304?

Answers

5. $23,600   **6.** 8 days

3. Think of a plan.

We next substitute for a_n in the first equation to obtain

$$164 = \frac{n}{2}(3 + 5n - 2)$$

$$328 = n(5n + 1) \qquad \text{Multiply by 2 and simplify.}$$

4. Use algebra to solve the problem.

We rewrite this equation in standard quadratic form:

$$5n^2 + n - 328 = 0$$

$$(5n + 41)(n - 8) = 0 \qquad \text{Factor.}$$

5. Verify the solution.

Since n must be positive, the solution is $n = 8$, so it took 8 days for Alice to save up $1.64.

Exercises A.2

A In Problems 1–10, an arithmetic sequence is given. Find:

    a. a_1, the first term   **b.** d, the common difference

    c. a_n, the nth term

1. 5, 8, 11, 14, . . .
    a. $a_1 = 5$; **b.** $d = 3$;
    c. $a_n = 3n + 2$

2. 5, 10, 15, 20, . . .
    a. $a_1 = 5$; **b.** $d = 5$;
    c. $a_n = 5n$

3. 11, 6, 1, −4, . . .
    a. $a_1 = 11$; **b.** $d = -5$;
    c. $a_n = 16 - 5n$

4. 43, 32, 21, 10, . . .
    a. $a_1 = 43$; **b.** $d = -11$;
    c. $a_n = 54 - 11n$

5. 3, −1, −5, −9, . . .
    a. $a_1 = 3$; **b.** $d = -4$;
    c. $a_n = 7 - 4n$

6. 0.6, 0.2, −0.2, −0.6, . . .

    a. $a_1 = 0.6$; **b.** $d = -0.4$;

    c. $a_n = 1.0 - 0.4n$

7. $\frac{1}{2}, \frac{1}{4}, 0, -\frac{1}{4}, \ldots$

    a. $a_1 = \frac{1}{2}$; **b.** $d = -\frac{1}{4}$;

    c. $a_n = \frac{3}{4} - \frac{1}{4}n$ or $\frac{3 - n}{4}$

8. $\frac{2}{3}, \frac{5}{6}, 1, \frac{7}{6}, \ldots$

    a. $a_1 = \frac{2}{3}$; **b.** $d = \frac{1}{6}$;

    c. $a_n = \frac{1}{2} + \frac{1}{6}n$ or $\frac{3 + n}{6}$

9. $-\frac{5}{6}, -\frac{1}{3}, \frac{1}{6}, \frac{2}{3}, \ldots$

    a. $a_1 = -\frac{5}{6}$; **b.** $d = \frac{1}{2}$;

    c. $a_n = \frac{1}{2}n - \frac{4}{3}$ or $\frac{3n - 8}{6}$

10. $-\frac{1}{4}, \frac{1}{4}, \frac{3}{4}, \frac{5}{4}, \ldots$

    a. $a_1 = -\frac{1}{4}$; **b.** $d = \frac{1}{2}$;

    c. $a_n = \frac{1}{2}n - \frac{3}{4}$ or $\frac{2n - 3}{4}$

B In Problems 11–20, some values for an arithmetic sequence are given. Find the other indicated values.

11. Given $a_1 = 7$, $n = 15$, $d = 6$; find a_{15}, S_{15}
    $a_{15} = 91$; $S_{15} = 735$

12. Given $a_1 = -2$, $d = -5$, $a_n = -72$; find n, S_n
    $n = 15$; $S_{15} = -555$

13. Find a_1, d, S_8 for the sequence 4, 10, 16, 22, . . .
    $a_1 = 4$; $d = 6$; $S_8 = 200$

14. Find a_1, d, S_n for the sequence 3, −1, −5, −9, . . .
    $a_1 = 3$; $d = -4$; $S_n = 5n - 2n^2$

15. Given $a_1 = 3$, $a_6 = 8$; find d, S_6
    $d = 1$; $S_6 = 33$

16. Given $a_1 = -1$, $a_{10} = -4$; find d, S_{10}
    $d = -\frac{1}{3}$; $S_{10} = -25$

17. Given $a_1 = 6$, $S_{14} = -280$; find d, a_{14}
    $d = -4$; $a_{14} = -46$

18. Given $a_1 = 15$, $a_n = -25$, $S_n = -85$; find d, n
    $d = -\frac{5}{2}$; $n = 17$

19. Given $d = 40$, $S_{40} = 40$; find a_1, a_{40}
    $a_1 = -779$; $a_{40} = 781$

20. Given $a_1 = 4$, $d = 2$, $a_n = 30$; find n, S_n
    $n = 14$; $S_{14} = 238$

APPLICATIONS

Sequences and Series

21. A certain property valued at $30,000 will depreciate $1380 the first year, $1340 the second year, $1300 the third year, and so on, the annual depreciation decreasing $40 per year. What will the property be worth at the end of 20 yr? $10,000

22. Strikers at a certain plant were ordered to return to work and told that their union would be fined $50 the first day they refused to do so, $60 the second day, $70 the third day, and so on. If their union paid a $680 fine, after how many days did they go back to work? 8 days

23. The diagram shows a sequence of crosses. Starting with a single square, a cross is constructed by adding a square to each side. Then the crosses are extended by adding a square to both ends of the vertical and the horizontal pieces. Find the total number of squares in the first 10 elements of this sequence. 190

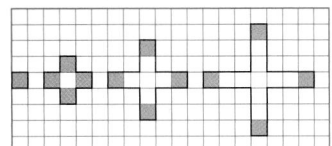

24. The diagram shows a sequence of staircases. Starting with a single square, more squares are added at each stage, as shown by the shaded squares. Find the number of squares in the tenth staircase. (*Hint:* the number of squares is the sum of a sequence.) 55

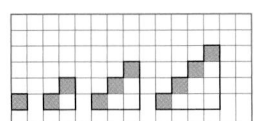

25. Show that the sum of the first n natural numbers is

$$\frac{n(n+1)}{2}$$

The natural numbers are $1, 2, 3, \ldots, n, \ldots$;

$S_n = \frac{n}{2}(1+n) = \frac{n(n+1)}{2}$

26. Show that the sum of the first n odd natural numbers is n^2.

The odd numbers are $1, 3, 5, \ldots, 2n-1, \ldots$;

$S_n = \frac{n}{2}(1+[2n-1]) = \frac{n}{2}(2n) = n^2$

27. Show that the sum of the first n even natural numbers is $n^2 + n$.

The even numbers are $2, 4, 6, \ldots, 2n, \ldots$;

$S_n = \frac{n}{2}(2+2n) = \frac{n}{2} \cdot 2(n+1) = n(n+1) = n^2 + n$

28. Find the sum of the natural numbers between 50 and 100 that are divisible by 3.

$51, 54, 57, \ldots, 99; S_{17} = \frac{17}{2}(51+99) = 1275$

SKILL CHECKER

Try the "Skill Checker" exercises so you'll be ready for the next section.

Evaluate:

29. r^{n-1} for $r = 2, n = 6$ 32

30. r^{n-1} for $r = 3, n = 4$ 27

31. $\dfrac{1-r^n}{1-r}$ for $r = 2, n = 6$ 63

32. $\dfrac{1-r^n}{1-r}$ for $r = 3, n = 4$ 40

33. $\dfrac{10}{1-r}$ for $r = \dfrac{1}{2}$ 20

34. $\dfrac{10}{1-r}$ for $r = \dfrac{1}{4}$ $\dfrac{40}{3}$

USING YOUR KNOWLEDGE

Depressing Depreciation

The ideas in this section can be applied to many areas. For example, the IRS code lists many types of depreciation. We will do some depreciation problems here.

35. You have a piece of property valued at $35,000, which for tax purposes is to be depreciated to a value of $5000 in 5 yr. Suppose that the first year's depreciation is $10,000, and the depreciation for the successive years decreases by a fixed amount each year. Find the depreciation for each of the remaining 4 yr.
$8000, $6000, $4000, $2000

36. In Problem 35, suppose the first year's depreciation is $5000 and the depreciation for the successive years increases by a fixed amount. Find the depreciation for each of the remaining 4 yr.
$5500, $6000, $6500, $7000

37. The book value after t years for an asset depreciated using the **straight-line method,** decreasing in value by a fixed amount each year, is an arithmetic sequence given by

$$b_t = C - t\left(\frac{C - S}{N}\right)$$

where C is the cost of the asset, S is the salvage value (trade-in) of the asset, and N is the years of expected life for the asset. A $12,000 machine has an expected life of 5 yr and a salvage value of $6000 at the end of 5 yr.

   a. Find the formula for b_t. $b_t = 12{,}000 - 1200t$

   b. Find the salvage value of the machine after 0, 1, 2, 3, 4, and 5 yr. $b_0 = \$12{,}000$; $b_1 = \$10{,}800$; $b_2 = \$9600$; $b_3 = \$8400$; $b_4 = \$7200$; $b_5 = \$6000$

WRITE ON

Write your own definition of:

38. An arithmetic sequence Answers may vary.

39. An arithmetic series Answers may vary.

40. What is the difference between an arithmetic sequence and an arithmetic series? Answers may vary.

Describe a situation you have encountered that can be modeled by:

41. An arithmetic sequence Answers may vary.

42. An arithmetic series Answers may vary.

MASTERY TEST

If you know how to do these problems, you have learned your lesson!

43. Consider the sequence 5, 9, 13, 17, . . . and find:

   a. a_1, the first term $a_1 = 5$

   b. d, the common difference $d = 4$

   c. a_{10}, the tenth term $a_{10} = 41$

   d. a_n, the nth term $a_n = 4n + 1$

44. Do you remember the song "The Twelve Days of Christmas"? On the first day, you get 1 gift. On the second day, you get $1 + 2$ gifts, on the third day you get $1 + 2 + 3$ gifts, and so on. (*Hint:* the number of gifts is the sum of a sequence.)

   a. How many gifts would you get on the tenth day? $a_{10} = 55$

   b. How many gifts would you get on the nth day? (*Hint:* You are adding the terms in the sequence $1, 2, 3, \ldots, n$.) $\frac{n^2 + n}{2}$

45. The sum of the first eight terms in an arithmetic sequence is 136 and the eighth term is 24. Find:

   a. The first term $a_1 = 10$

   b. The common difference $d = 2$

46. A heavy-duty truck valued at $50,000 depreciates $6000 the first year, $5800 the second year, $5600 the third year, and so on. What will be the value of the truck at the end of 8 yr? $7600

47. Pedro started a bank account with $30. He then saved $40 the next month, $50 the next, and so on in an arithmetic sequence. How many months did it take for Pedro to accumulate $750? 10 months

A.3 ## GEOMETRIC SEQUENCES AND SERIES

To Succeed, Review How To . . .

1. Define a sequence (pp. A-2–A-3).

2. Recognize arithmetic sequences (pp. A-12–A-13).

Objectives

A Find the common ratio and the general term in a geometric sequence.

B Find the sum of a geometric sequence.

C Find the sum of an infinite geometric series if it exists.

D Solve applications involving geometric sequences and series.

GETTING STARTED

Games, Games, Games

The game of chess is said to have originated in Persia. Legend has it that the Shah of Persia offered the inventor of the game anything he wanted as a reward. The inventor asked that 1 grain of wheat be placed on the first square of the chessboard, 2 grains on the second, 4 on the third, and so on. The sequence enumerating the number of grains on each square is given by

$$1, 2, 4, 8, 16, \ldots$$

and the inventor was to receive the sum of the first 64 terms in this sequence (there are 64 squares on a chessboard). As you can see, each *term* after the first in this sequence is obtained by *doubling* (multiplying by 2) the preceding term.

Clearly, the sequence $1, 2, 4, 8, 16, \ldots$ is **not** an arithmetic sequence. It's a different type of sequence called a *geometric sequence* or *geometric progression*. How many grains of wheat should the man have received? More than is produced in the entire world in 1 year! But for a better answer, you have to learn how to sum geometric sequences, so we'll wait until Example 8 to find the exact answer.

A ### Geometric Sequences

Many everyday applications involve geometric sequences. We will discuss some of them in the exercises.

> **GEOMETRIC SEQUENCE**
>
> A **geometric sequence** or **geometric progression** is a sequence in which each term after the first is obtained by *multiplying* the preceding term by a constant r called the **common ratio.**

Web It

For a well-developed lesson that includes the identification of geometric sequences, finding the nth term, and the sum of a series, go to link A-3-1 at mhhe.com/bello.

Web It

For a lesson on how to find the common ratio for a geometric sequence, with examples and solutions, visit A-3-2 at mhhe.com/bello.

Multiplying each term by the number r produces a fixed ratio between any two consecutive terms. We can find r by finding the ratio of two successive terms. For example, in the sequence 1, 2, 4, 8, 16, ... The ratio r is given by

$$r = \frac{2}{1} = \frac{4}{2} = \frac{8}{4} = \frac{16}{8} = 2$$

COMMON RATIO

The **common ratio** r is defined by

$$r = \frac{a_{n+1}}{a_n}$$

Since in a geometric sequence each term after the first is obtained by multiplying the preceding term by r, the first n terms in such a sequence are

$$a_1, a_1r, a_1r^2, a_1r^3, \ldots, a_1r^{n-1}$$
$$\uparrow \quad \uparrow \quad \uparrow \quad \uparrow \qquad \qquad \uparrow$$
$$a_1 \quad a_2 \quad a_3 \quad a_4 \qquad \qquad a_n$$

General Term of a Geometric Sequence	The nth term of the geometric sequence is expressed as $$a_n = a_1 r^{n-1}$$

EXAMPLE 1   **Finding terms in a geometric sequence**

For the geometric sequence 1, 3, 9, 27, . . . , find:

a. a_1 **b.** r **c.** a_6 **d.** a_n

SOLUTION

a. By inspection, we see that the first term is $a_1 = 1$.

b. The common ratio r can be found by taking the ratio of any term to the preceding term. Using the ratio of the second term to the first, we find

$$r = \frac{3}{1} = 3$$

c. The formula $a_n = a_1 r^{n-1}$ gives, for $n = 6$,

$$a_6 = (1)(3^{6-1}) = 3^5 = 243$$

d. The general term is obtained with $r = 3$.

$$a_n = (1)(3^{n-1}) = 3^{n-1}$$

PROBLEM 1

For the geometric sequence $3, 1, \frac{1}{3}, \frac{1}{9}, \ldots$, find:

a. a_1 **b.** r **c.** a_6 **d.** a_n

Answers

1. a. 3   **b.** $\frac{1}{3}$ **c.** $\frac{1}{81}$

d. $\frac{1}{3^{n-2}}$ or 3^{2-n}

Calculate It   Find a Term in a Geometric Sequence

In Example 1, you can find a_n for any n if you do the algebra involved in parts **a**, **b**, and **c**. After that, graph $Y_1 = 3^{(x-1)}$ using an integer window. If you want to find a_6, use the TRACE key until you get to $X = 6$.

The calculator shows the corresponding Y-value even though the point *is not shown* in the window!

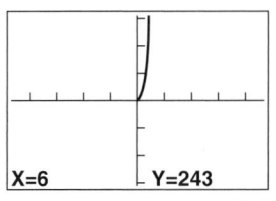

X=6   Y=243

B   Finding the Sum of a Geometric Sequence

Can we find the sum S_n of the first n terms in a geometric sequence?

$$S_n = a_1 + a_1 r + a_1 r^2 + a_1 r^3 + \cdots + a_1 r^{n-1} \qquad \text{By definition}$$

$$rS_n = a_1 r + a_1 r^2 + a_1 r^3 + \cdots + a_1 r^{n-1} + a_1 r^n \qquad \text{Multiply by } r.$$

$$S_n - rS_n = a_1 - a_1 r^n = a_1(1 - r^n) \qquad \text{Subtract.}$$

$$S_n(1 - r) = a_1(1 - r^n) \qquad \text{By the distributive property}$$

$$S_n = \frac{a_1(1 - r^n)}{1 - r} \qquad \text{Divide by } 1 - r.$$

We have the following result.

> **Sum of a Geometric Sequence**
>
> The sum S_n of the first n terms starting with a_1 and ending with a_n is given by
>
> $$S_n = \frac{a_1(1 - r^n)}{1 - r}$$

Web It

For a colorful lesson on how to find the sum of a geometric sequence, go to link A-3-3 at mhhe.com/bello.

EXAMPLE 2   Finding the sum of a geometric sequence

For the geometric sequence $4, -8, 16, -32, \ldots$, find:

a. r **b.** a_{10} **c.** S_{10}

SOLUTION

a. Since r is the common ratio,

$$r = \frac{-8}{4} = -2$$

b. We have $a_1 = 4$ and $r = -2$, so

$$a_{10} = a_1 r^{n-1}$$

$$= (4)(-2)^9 = (4)(-512)$$

$$= -2048$$

c. Using the formula for S_n with $n = 10$, we get

$$S_{10} = \frac{4[1 - (-2)^{10}]}{1 - (-2)} = \frac{4(1 - 2^{10})}{1 + 2}$$

$$= \frac{4(1 - 1024)}{3} = -1364$$

PROBLEM 2

For the sequence in Example 2, find:

a. a_n **b.** S_n

Answers

2. a. $4(-2)^{n-1} = (-1)^{n-1}2^{n+1}$
b. $\frac{4}{3}[1 - (-2)^n]$

EXAMPLE 3 **Finding a specific term and the common ratio of a geometric sequence**

Given a geometric sequence with $a_1 = 2$ and $S_3 = 26$, find a_3 and r.

SOLUTION Using the formulas for the nth term and the sum of n terms, we first find r and use it to find a_3. So

$$a_3 = a_1 r^2 = 2r^2 \quad \text{and} \quad S_3 = \frac{a_1(1 - r^3)}{1 - r} = \frac{2(1 - r^3)}{1 - r}$$

Since $1 - r^3 = (1 - r)(1 + r + r^2)$, the formula for S_3 can be simplified to

$$S_3 = \frac{2(1 - r)(1 + r + r^2)}{1 - r}$$
$$= 2(1 + r + r^2)$$

We have

$$26 = 2(1 + r + r^2)$$
$$13 = 1 + r + r^2$$

which gives a quadratic equation for r:

$$r^2 + r - 12 = 0$$

By factoring, we get

$$(r + 4)(r - 3) = 0$$

so that $r = -4$ or $r = 3$. If $r = -4$, then $a_3 = 2(-4)^2 = 32$, and if $r = 3$, then $a_3 = 2(3^2) = 18$.

We can check these results by writing the three terms of the sequence.

$$\text{For } r = -4: \quad 2, -8, 32, \text{ which add to } 26$$

$$\text{For } r = 3: \quad 2, 6, 18, \text{ which add to } 26$$

Hence the correct answers are $a_3 = 32$, $r = -4$ or $a_3 = 18$, $r = 3$.

PROBLEM 3

Do Example 3 if $S_3 = 62$.

Calculate It   Sums of Sequences

If you understand the terminology of sequences, you can find the sum of a geometric sequence using a calculator. First, recall that

$$S_n = a_1 + a_1 r + a_1 r^2 + \cdots + a_1 r^{n-1}$$
$$= \frac{a_1(1 - r^n)}{1 - r}$$

In Example 2, we want to find S_n. Let's graph

$$Y_1 = \frac{a_1(1 - r^n)}{1 - r}$$

by replacing a_1 by 4, n by x, r by -2, and using the dot mode with an integer window. We graph

$$Y_1 = \frac{4(1 - (-2)^x)}{1 - (-2)} = \frac{4(1 - (-2)^x)}{3}$$

Now, use the TRACE key to find S_{10}, as shown in the window.

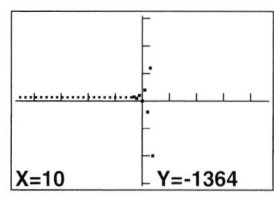

X=10   Y=-1364

Answer

3. $a_3 = 50$, $r = 5$ or $a_3 = 72$, $r = -6$

C Finding the Sum of an Infinite Geometric Series

In the preceding examples, we found the sum of the first n terms of a geometric sequence. We now consider what happens if n is allowed to increase without bound. We indicate that the number of terms is *infinite*—unlimited—by writing

$$\sum_{n=1}^{\infty} a_1 r^{n-1} = a_1 + a_1 r + a_1 r^2 + \cdots + a_1 r^{n-1} + \cdots$$

This expression is called an *infinite geometric series*. We can find the sum of the first n terms of this series by using the formula

$$S_n = \frac{a_1(1 - r^n)}{1 - r} = \frac{a_1}{1 - r}(1 - r^n)$$

If r is less than 1 in absolute value, $|r| < 1$, or, equivalently, $-1 < r < 1$, then r^n becomes smaller and smaller as n increases. For example, if $r = 0.6$, then a calculation with logarithms or with a calculator gives the following approximate results:

$$r^{10} = (0.6)^{10} \approx 6.05 \times 10^{-3}$$

$$r^{100} = (0.6)^{100} \approx 6.53 \times 10^{-23}$$

$$r^{1000} = (0.6)^{1000} \approx 1.4 \times 10^{-222}$$

(The first of these has 2 zeros before the first significant digit, and the second and third have 22 and 221 zeros, respectively, before the first significant digit.) So we can make r^n as small as we like by taking n large enough.

You can see that if $|r| < 1$, the factor $(1 - r^n)$ in the sum formula can be made as close to 1 as we wish by taking n large enough. Hence as n becomes greater and greater, the sum

$$S_n = \frac{a_1}{1 - r}(1 - r^n)$$

is more and more closely approximated by the expression

$$\frac{a_1}{1 - r}$$

We summarize this discussion by making the following definition.

Web It

For a short, simple lesson on how to find the sum of an infinite geometric series, go to link A-3-4 at mhhe.com/bello.

Sum of an Infinite Geometric Series

If $|r| < 1$, then the sum S of the geometric series with first term a_1 and common ratio r is defined to be

$$S = a_1 + a_1 r + a_1 r^2 + \cdots + a_1 r^{n-1} + \cdots$$

$$= \frac{a_1}{1 - r}$$

If $|r| \geq 1$, the sum of n terms does not get closer and closer to any number as n becomes larger and larger without bound. In this case, we say that S *does not exist.*

NOTE

If $|r| \geq 1$, the sum of an infinite geometric series **does not exist.**

For $r = \frac{1}{2}$ and $a_1 = 1$, we can give a graphical interpretation of the behavior of S_n as n increases without bound. We consider the series

$$\sum_{n=1}^{\infty}\left(\frac{1}{2}\right)^{n-1} = 1 + \frac{1}{2} + \frac{1}{4} + \frac{1}{8} + \cdots + \left(\frac{1}{2}\right)^{n-1} + \cdots$$

For this series,

$$S_1 = 1$$

$$S_2 = 1 + \frac{1}{2} = 1\frac{1}{2} = \frac{3}{2}$$

$$S_3 = 1 + \frac{1}{2} + \frac{1}{4} = 1\frac{3}{4} = \frac{7}{4}$$

$$S_4 = 1 + \frac{1}{2} + \frac{1}{4} + \frac{1}{8} = 1\frac{7}{8} = \frac{15}{8}$$

The sums are as shown here:

As you can see, each step cuts in half the remaining distance to the point marked 2. By making n sufficiently large, we can make the value of S_n as close as we like to 2. For the series $1 + \frac{1}{2} + \frac{1}{4} + \frac{1}{8} + \cdots + \left(\frac{1}{2}\right)^{n-1} + \cdots$

$$S = \frac{a_1}{1 - r} = \frac{1}{1 - \frac{1}{2}} = 2$$

EXAMPLE 4   **Finding the sum of an infinite geometric series**

Find the sum of the geometric series $4 - 2 + 1 - \frac{1}{2} + \cdots$.

SOLUTION   In this series, $a_1 = 4$ and $r = -\frac{1}{2}$, so

$$S = \frac{a_1}{1 - r} = \frac{4}{1 - \left(-\frac{1}{2}\right)} = \frac{8}{2 + 1} = \frac{8}{3}$$

PROBLEM 4

Find the sum of the geometric series $2 - 1 + \frac{1}{2} - \frac{1}{4} + \cdots$.

EXAMPLE 5   **More practice finding the sum of an infinite geometric series**

If the following geometric series has a sum, find it.

$$\sum_{n=1}^{\infty}(1.01)^n = (1.01) + (1.01)^2 + (1.01)^3 + \cdots$$

SOLUTION   For this series the ratio r is

$$\frac{(1.01)^2}{(1.01)} = 1.01$$

which is greater than 1. So the sum of this series does not exist.

PROBLEM 5

Find the sum of

$$\sum_{n=1}^{\infty}(0.99)^n = (0.99)^1 + (0.99)^2 + \cdots$$

if it exists.

Answers

4. $\frac{4}{3}$ **5.** $r = 0.99 < 1$; the sum is 99

Web It

For a vast collection of links dealing with applications to geometric series, go to link A-3-5 at mhhe.com/bello.

D   Solving Applications of Geometric Sequences and Series

Geometric series can be used to express nonterminating repeating decimals as fractions. For example, the decimal

$$0.333\ldots = 0.\overline{3}$$

can be written as

$$\frac{3}{10} + \frac{3}{100} + \frac{3}{1000} + \cdots$$

which is an infinite geometric series with

$$a_1 = \frac{3}{10} \quad \text{and} \quad r = \frac{1}{10}$$

Thus

$$S = \frac{\frac{3}{10}}{1 - \frac{1}{10}} = \frac{3}{9} = \frac{1}{3}$$

EXAMPLE 6   Finding fraction equivalents

Find the fraction equivalent to the repeating decimal $0.414141\ldots$.

SOLUTION   We can write this decimal as

$$\frac{41}{100} + \frac{41}{(100)^2} + \frac{41}{(100)^3} + \cdots$$

which is an infinite geometric series with

$$a_1 = \frac{41}{100} \quad \text{and} \quad r = \frac{1}{100}$$

Thus the sum of this series is

$$S = \frac{a_1}{1 - r} = \frac{\frac{41}{100}}{1 - \frac{1}{100}}$$

$$= \frac{\frac{41}{100}}{\frac{99}{100}} = \frac{41}{99}$$

PROBLEM 6

Find the fraction equivalent to the repeating decimal $0.373737\ldots$

EXAMPLE 7   Job offers and series

Suppose you have two job offers for a 2-week (14-day) trial period. Job A starts at $50 per day with a $50 raise each day. Job B starts at 50¢ per day and your salary is doubled every day. Find the total amount paid by each of the jobs at the end of the 14 days.

SOLUTION

1. Read the problem.
We must find the total amount paid by each of the jobs at the end of the 14 days.

2. Select the unknown.
We are asked to find the total amount paid, S_{14}, for each job.

PROBLEM 7

Find the total amount paid by each of the jobs at the end of one week (7 days).

Answers

6. $\frac{37}{99}$ **7.** A pays $1400; B pays $63.50

3. Think of a plan.

Let's find the amount job A pays at the end of 14 days. Then we'll find the amount job B pays at the end of 14 days.

The pay for job A starts at \$50 ($a_1 = 50$) and increases by \$50 each day ($d = 50$). The salary for the fourteenth day is $a_{14} = 50 + 13 \cdot 50 = 700$. The pay for job B starts at \$0.50 ($a_1 = 0.50$) and doubles every day ($r = 2$).

4. Use the formula for the sum of an arithmetic and a geometric sequence to solve the problem.

For job A, the sum of the arithmetic sequence for 14 days is

$$S_{14} = \frac{n(a_1 + a_n)}{2} = \frac{14(50 + 700)}{2} = \frac{14 \cdot 750}{2} = \$5250$$

For job B, the sum of the geometric sequence for 14 days is

$$S_{14} = \frac{a_1(1 - r^n)}{1 - r} = \frac{0.50(1 - 2^{14})}{1 - 2}$$

$$= \frac{0.50(1 - 2^{14})}{-1}$$

$$= 0.50(2^{14} - 1)$$

$$= 0.50\,(16{,}383)$$

$$= \$8191.50$$

Job B pays much more.

5. Verify the solution.

We leave the verification to you.

EXAMPLE 8   **Take it with a grain of wheat**

As you recall from *Getting Started,* the Shah of Persia so liked the game of chess that he offered its inventor anything he wanted. The inventor asked that 1 grain of wheat be placed on the first square of the chessboard, 2 grains on the second, 4 on the third, and so on. If there are 64 squares on a chessboard:

a. How many grains were to be placed on the 64th square?

b. What is the total number of grains the inventor should have received?

SOLUTION

a. The number of grains in each square is

Square 1	Square 2	Square 3	Square 4	. . .	Square 64
1	2	$4 = 2^{3-1}$	$8 = 2^{4-1}$	. . .	$2^{63} = 2^{64-1}$

b. The sum of the geometric sequence $1, 2, 4, \ldots, 2^{63}$ where $a_1 = 1$ and $r = 2$ (since the number of grains is doubled on each succeeding square) is

$$S_{64} = \frac{1(1 - 2^{64})}{1 - 2} = \frac{1 - 2^{64}}{-1}$$

$$= 2^{64} - 1$$

Since 2^{10} is about 1000, $2^{60} = (2^{10})^6$ is about $(1000)^6$ or $1{,}000{,}000{,}000{,}000{,}000{,}000$ (one quintillion)! If you do this with a calculator, you will find that the answer is closer to 18 quintillion.

Web It

If you want to save the world and do geometric sequences at the same time, go to link A-3-6 at mhhe.com/bello.

The site will give you the first four terms of a geometric sequence. To save the world, you have to find the next four terms and write them on the next line. Go now, and save us!

To download a summary dealing with geometric sequences go to link A-3-7 at mhhe.com/bello. The information will be automatically delivered to your computer.

PROBLEM 8

a. How many grains were to be placed on the 60th square?

b. What is the total number of grains in the 60 squares?

Calculate It Exercises

1. Find a_{10} and a_{11} in Example 1.

2. In Example 2, find S_{12} and S_{13}. As n gets larger, does S_n get close to any specific value? In Example 4, use the technique from Example 2 to find S_{12} and S_{13}. As n gets larger, does S_n get close to any specific value?

3. Use the procedure of Example 2 to find the sum S_{14} in Example 7, which represents the 14-day salary for job B, which started at 50¢ per day and doubled every day.

Answers

8. a. 2^{60-1} **b.** $2^{60} - 1$

Exercises A.3

A **B** In Problems 1–10, a geometric sequence is given. Find the indicated values.

a. a_1 **b.** r **c.** a_n **d.** S_n

1. $3, 6, 12, 24, \ldots$ **a.** $a_1 = 3$; **b.** $r = 2$; **c.** $a_n = 3(2^{n-1})$; **d.** $S_n = 3(2^n - 1)$

2. $\dfrac{1}{3}, 1, 3, 9, \ldots$ **a.** $a_1 = \dfrac{1}{3}$; **b.** $r = 3$; **c.** $a_n = 3^{n-2}$; **d.** $S_n = \dfrac{1}{6}(3^n - 1)$

3. $8, 24, 72, 216, \ldots$ **a.** $a_1 = 8$; **b.** $r = 3$; **c.** $a_n = 8(3^{n-1})$; **d.** $S_n = 4(3^n - 1)$

4. $\dfrac{1}{5}, \dfrac{1}{10}, \dfrac{1}{20}, \dfrac{1}{40}, \ldots$ **a.** $a_1 = \dfrac{1}{5}$; **b.** $r = \dfrac{1}{2}$; **c.** $a_n = \dfrac{1}{5}\left(\dfrac{1}{2}\right)^{n-1}$; **d.** $S_n = \dfrac{2}{5}\left(1 - \left[\dfrac{1}{2}\right]^n\right)$

5. $16, -4, 1, -\dfrac{1}{4}, \ldots$ **a.** $a_1 = 16$; **b.** $r = -\dfrac{1}{4}$; **c.** $a_n = (-4)^{3-n}$; **d.** $S_n = \dfrac{64}{5}\left(1 - \left[-\dfrac{1}{4}\right]^n\right)$

6. $3, -1, \dfrac{1}{3}, -\dfrac{1}{9}, \ldots$ **a.** $a_1 = 3$; **b.** $r = -\dfrac{1}{3}$; **c.** $a_n = (-1)^{n-1}3^{2-n}$; **d.** $S_n = \dfrac{9}{4}\left(1 - \left[-\dfrac{1}{3}\right]^n\right)$

7. $-\dfrac{3}{5}, \dfrac{3}{2}, -\dfrac{15}{4}, \dfrac{75}{8}, \ldots$ **a.** $a_1 = -\dfrac{3}{5}$; **b.** $r = -\dfrac{5}{2}$; **c.** $a_n = \left(-\dfrac{3}{5}\right)\left(-\dfrac{5}{2}\right)^{n-1}$; **d.** $S_n = -\dfrac{6}{35}\left(1 - \left[-\dfrac{5}{2}\right]^n\right)$

8. $60, -6, \dfrac{6}{10}, -\dfrac{6}{100}, \ldots$ **a.** $a_1 = 60$; **b.** $r = -\dfrac{1}{10}$; **c.** $a_n = 60\left(-\dfrac{1}{10}\right)^{n-1}$; **d.** $S_n = \dfrac{600}{11}\left(1 - \left[-\dfrac{1}{10}\right]^n\right)$

9. $-\dfrac{3}{4}, -\dfrac{1}{4}, -\dfrac{1}{12}, -\dfrac{1}{36}, \ldots$ **a.** $a_1 = -\dfrac{3}{4}$; **b.** $r = \dfrac{1}{3}$; **c.** $a_n = \left(-\dfrac{1}{4}\right)(3^{2-n})$; **d.** $S_n = -\dfrac{9}{8}\left(1 - \left[\dfrac{1}{3}\right]^n\right)$

10. $-\dfrac{5}{6}, -\dfrac{1}{3}, -\dfrac{2}{15}, -\dfrac{4}{75}, \ldots$ **a.** $a_1 = -\dfrac{5}{6}$; **b.** $r = \dfrac{2}{5}$; **c.** $a_n = \left(-\dfrac{5}{6}\right)\left(\dfrac{2}{5}\right)^{n-1}$; **d.** $S_n = -\dfrac{25}{18}\left(1 - \left[\dfrac{2}{5}\right]^n\right)$

In Problems 11–20, some values for a geometric sequence are given. Find the remaining indicated values.

11. Given $a_1 = 1$, $S_3 = \dfrac{7}{4}$; find a_3, r
$a_3 = \frac{1}{4}, r = \frac{1}{2}$ or $a_3 = \frac{9}{4}, r = -\frac{3}{2}$

12. Given $a_1 = 4$, $S_3 = 7$; find a_3, r
$a_3 = 1, r = \frac{1}{2}$ or $a_3 = 9, r = -\frac{3}{2}$

13. Given $a_1 = 3$, $S_3 = 21$; find a_3, r
$a_3 = 27, r = -3$ or $a_3 = 12, r = 2$

14. Given $a_1 = \dfrac{1}{2}$, $S_3 = \dfrac{39}{50}$; find a_3, r
$a_3 = \frac{49}{50}, r = -\frac{7}{5}$ or $a_3 = \frac{2}{25}, r = \frac{2}{5}$

15. Given $r = 2$, $S_8 = 1785$; find a_1, a_8
$a_1 = 7, a_8 = 896$

16. Given $a_6 = -\dfrac{16}{27}$, $r = -\dfrac{1}{3}$; find a_1, S_6
$a_1 = 144, S_6 = \frac{2912}{27}$

17. Given $a_1 = -4$, $a_n = 108$, $S_n = 80$; find r, n
$r = -3, n = 4$

18. Given $a_1 = \dfrac{3}{4}$, $a_n = 192$, $S_n = 255\dfrac{3}{4}$; find r, n
$r = 4, n = 5$

19. Given $a_1 = \dfrac{16}{125}$, $r = \dfrac{5}{2}$, $a_n = \dfrac{25}{2}$; find n, S_n
$n = 6, S_6 = \frac{5187}{250}$

20. Given $a_1 = 7$, $r = 2$, $a_n = 896$; find n, S_n
$n = 8, S_8 = 1785$

C In Problems 21–30, an infinite geometric series is given. Find the sum if it exists.

21. $6 + 3 + 1\dfrac{1}{2} + \cdots$ $S = 12$

22. $12 + 4 + 1\dfrac{1}{3} + \cdots$ $S = 18$

23. $(-6) + (-3) + \left(-\dfrac{3}{2}\right) + \cdots$ $S = -12$

24. $(-8) + (-4) + (-2) + \cdots$ $S = -16$

25. $\displaystyle\sum_{n=1}^{\infty}(-1)^{n-1}2^{2-n} = 2 - 1 + \frac{1}{2} - \frac{1}{4} + \cdots$ $S = \frac{4}{3}$

26. $\displaystyle\sum_{n=1}^{\infty}(-1)^{n-1}3^{3-n} = 9 - 3 + 1 - \frac{1}{3} + \cdots$ $S = \frac{27}{4}$

27. $4 - 8 + 16 - 32 + \cdots$
Sum does not exist; $|r| = 2 > 1$

28. $(-5) + (-10) + (-20) + \cdots$
Sum does not exist; $|r| = 2 > 1$

29. $\dfrac{1}{10} + \dfrac{1}{5} + \dfrac{2}{5} + \cdots$
Sum does not exist; $|r| = 2 > 1$

30. $0.0001 - 0.001 + 0.01 - \cdots$
Sum does not exist; $|r| = 10 > 1$

D In Problems 31–40, find a fraction equivalent to the given repeating decimal.

31. $0.555\ldots$ $\dfrac{5}{9}$

32. $0.666\ldots$ $\dfrac{2}{3}$

33. $0.181818\ldots$ $\dfrac{2}{11}$

34. $0.242424\ldots$ $\dfrac{8}{33}$

35. $4.050505\ldots$ $\dfrac{401}{99}$

36. $2.313131\ldots$ $\dfrac{229}{99}$

37. $2.3161616\ldots$ $\dfrac{2293}{990}$

38. $4.1272727\ldots$ $\dfrac{227}{55}$

39. $0.140140140\ldots$ $\dfrac{140}{990}$

40. $1.123123123\ldots$ $\dfrac{374}{333}$

APPLICATIONS

41. The population of a certain town increases at the rate of 4% per year.
   a. Write the sequence associated with this population.
   $P_0,\ 1.04P_0,\ (1.04)^2 P_0, \ldots$
   b. If the present population is 20,000, what will the population be at the end of 5 yr?   24,333

42. The number of bacteria in a culture increased from 320,000 at the beginning of the first day to 2,430,000 at the beginning of the sixth day. Find the daily rate of increase if this rate is assumed to be constant—if the starting number of bacteria on successive days form the geometric sequence
$$a_1, a_1 r, a_1 r^2, a_1 r^3, a_1 r^4, a_1 r^5 \quad 150\%$$

43. The distance traveled in any swing by a point on a compound pendulum is 20% less than in the preceding swing. If the length of the first swing is 62.5 cm, find the total distance the point has traveled at the end of the fourth swing. (*Hint:* Write the sequence associated with the four swings.)   184.5 cm

44. A small business makes a net profit of $10,000 in its first year. If the net profit increases by 25% each year for the next 4 yr, what is the total net profit for these 5 yr? $82,070.31

45. If the rate of increase in Problem 44 is 50%, what is the total net profit for the 5 yr? $131,875

46. A polluted tank holds 100 gal of a poisonous chemical that mixes readily with water. After 25 gal of the chemical are drawn off, the tank is refilled with water. Then 25 gal of the mixture are drawn off, and the tank is again refilled with water. If this operation is performed until five batches have been drawn from the tank, how much of the original chemical remains?   About 23.7 gal

47. If the first three terms in an arithmetic sequence are increased by 1, 3, and 13, respectively, the resulting numbers are in geometric sequence. Find the original terms if their sum is 9.   1, 3, 5 or 17, 3, −11

48. If the first three terms in an arithmetic sequence are increased by 9, 7, and 9, respectively, the resulting numbers are in geometric sequence. Find the original terms if their sum is 3.   7, 1, −5 or −5, 1, 7

49. Roberto and Jimmy are golf pals. Yesterday, Roberto persuaded Jimmy to bet 1¢ on the first hole, 2¢ on the second hole, 4¢ on the third hole, and so on (doubling the bet on each successive hole). Roberto did not have very good luck. He won the first hole, lost the second hole, and continued winning a hole and losing a hole in that order for the remainder of the game. How much did Roberto lose overall? (*Hint:* You will need to use the number $2^{18} = 262,144$.) $873.81

50. Refer to Problem 49. Suppose Roberto and Jimmy play again, making the same bets as before. This time Roberto wins the first two holes, loses the next two holes, and continues winning two holes and losing two holes in that order for the remainder of the game. How did Roberto come out this time? (*Hint:* Use one sequence for the odd-numbered holes and one for the even-numbered ones.) He won $1572.87

51. A rubber ball is dropped from a height of 8 ft. It makes a sequence of bounces, each three-fourths the height of the preceding bounce. Find the total vertical distance the ball travels before coming to rest. (*Hint:* Use one sequence for the distance traveled downward and one for the distance traveled upward.) 56 ft

52. A pendulum on each separate swing describes an arc whose length is 98% of the length of the preceding swing. If the first arc is 12 in. long, about how far does the pendulum travel before it comes to rest? 600 in.

SKILL CHECKER

Try the "Skill Checker" exercises so you'll be ready for the next section.

Expand:

53. $(a + b)^2$ $a^2 + 2ab + b^2$

54. $(x - y)^2$ $x^2 - 2xy + y^2$

55. $(x + y)^3$ $x^3 + 3x^2y + 3xy^2 + y^3$

56. $(a - b)^3$ $a^3 - 3a^2b + 3ab^2 - b^3$

57. $(y - 2z)^3$ $y^3 - 6y^2z + 12yz^2 - 8z^3$

58. $(2a + b)^3$ $8a^3 + 12a^2b + 6ab^2 + b^3$

59. $\left(\dfrac{1}{x} - \dfrac{1}{y}\right)^3$ $\dfrac{1}{x^3} - \dfrac{3}{x^2y} + \dfrac{3}{xy^2} - \dfrac{1}{y^3}$

60. $\left(\dfrac{2}{y} - \dfrac{1}{2}\right)^3$ $\dfrac{8}{y^3} - \dfrac{6}{y^2} + \dfrac{3}{2y} - \dfrac{1}{8}$

USING YOUR KNOWLEDGE

What Is the Sequence?

In the preceding sections you have learned what we mean by a sequence, an arithmetic sequence, and a geometric sequence. You can use this knowledge in the following problems.

In Problems 61–72, the first four terms of a series are given. Determine whether these terms form an *arithmetic* sequence, a *geometric* sequence, or *neither*. If the terms form a geometric sequence and the series is extended to form an infinite geometric series, find the sum if it has one.

61. $1 + \dfrac{2}{5} + \dfrac{4}{25} + \dfrac{8}{125} + \cdots$ Geometric; $r = \dfrac{2}{5}$; $S = \dfrac{5}{3}$

62. $1 - \dfrac{1}{2} + \dfrac{1}{3} - \dfrac{1}{4} + \cdots$ Neither

63. $1 + \dfrac{2}{5} - \dfrac{1}{5} - \dfrac{4}{5} - \cdots$ Arithmetic; $d = -\dfrac{3}{5}$

64. $1 + 3 + 4 + 7 + \cdots$ Neither

65. $2 + \dfrac{7}{4} + \dfrac{49}{32} + \dfrac{343}{256} + \cdots$ Geometric; $r = \dfrac{7}{8}$; $S = 16$

66. $2 + 1\dfrac{3}{4} + 1\dfrac{1}{2} + 1\dfrac{1}{4} + \cdots$ Arithmetic; $d = -\dfrac{1}{4}$

67. $5 + 4 + 2 - 1 - \cdots$ Neither

68. $4 + 2 + 0 - 2 - \cdots$ Arithmetic; $d = -2$

69. $\dfrac{1}{2} + \dfrac{2}{3} + \dfrac{3}{4} + \dfrac{4}{5} + \cdots$ Neither

70. $-12 + 4 - \dfrac{4}{3} + \dfrac{4}{9} - \cdots$ Geometric; $r = -\dfrac{1}{3}$; $S = -9$

71. $4 - 2 + 1 - \dfrac{1}{2} + \cdots$ Geometric; $r = -\dfrac{1}{2}$; $S = \dfrac{8}{3}$

72. $-6 - 3 + 0 + 3 + \cdots$ Arithmetic; $d = 3$

WRITE ON

73. Explain the difference between an arithmetic sequence and a geometric sequence. Answers may vary.

74. The Fibonacci sequence is $1, 1, 2, 3, 5, 8, \ldots$. Is this sequence an arithmetic sequence, a geometric sequence, or neither? Explain your answer. Answers may vary.

75. Explain the difference between an infinite geometric series and an infinite geometric sequence.
Answers may vary.

76. What conditions do you need for
$$S_n = \frac{a_1}{1 - r}(1 - r^n)$$
to be very close to
$$\frac{a_1}{1 - r}$$
Answers may vary.

MASTERY TEST

If you know how to do these problems, you have learned your lesson!

77. For the geometric sequence $2, 1, \frac{1}{2}, \frac{1}{4}, \ldots$, find:

    a. a_1 **b.** r **c.** a_6 **d.** a_n

    a. $a_1 = 2$; **b.** $r = \frac{1}{2}$; **c.** $a_6 = \frac{1}{16}$; **d.** $a_n = \frac{1}{2^{n-2}}$

78. For the geometric sequence $3, -6, 12, -24, \ldots$, find:

    a. a_n $a_n = 3(-2)^{n-1}$ **b.** S_n $S_n = 1 - (-2)^n$

79. Given a geometric sequence with $a_1 = 2$, $S_3 = 42$, find a_3 and r. $a_3 = 50$, $r = -5$ or $a_3 = 32$, $r = 4$

80. Find the sum of the geometric series
$9 - 3 + 1 - \frac{1}{3} + \cdots$. $S = \frac{27}{4}$

81. If the geometric series defined by
$$\sum_{n=1}^{\infty}(-1)^{n+1}(1.01)^n = (1.01) - (1.01)^2 + (1.01)^3 - \cdots$$
has a sum, find it. Sum does not exist; $|r| = 1.01 > 1$.

82. Find the fraction equivalent to the repeating decimal $0.432432432 \ldots$. $\frac{48}{111}$

83. An investment firm claims that you can double your money every year if you invest with them. Suppose you invest $100.

    a. Write the first five terms of the sequence associated with this claim, starting with $a_1 = 100$. 100, 200, 400, 800, 1600

    b. How much money would you have at the end of 6 years? $6400

    c. How much money would you have at the end of the nth year? $100 \cdot 2^n$

    d. How many years would it take for you to have over $100,000 (answer to the nearest year)? 10 yr

A.4   THE BINOMIAL EXPANSION

To Succeed, Review How To . . .

1. Simplify an algebraic expression (pp. 51–55).

2. Expand binomials of the form $(x + y)^n$, where $n < 6$ (pp. 357–360).

Objectives

A Use the binomial expansion to expand and simplify a power of a binomial.

B Find the coefficient of a term in a binomial expansion.

C Solve applications involving binomial expansions.

GETTING STARTED

It's Chinese to Me

This drawing made in A.D. 1303 by Chu Shi-kie turns out to be a nice depiction of Pascal's triangle (Pascal having lived in the 1600s). As you can see, every row starts and ends with the symbol ⊖. If we make ⊖ = 1, we can say the triangle starts with the number 1. The next row has the symbols ⊖ and ⊖, or in our translation, the numbers 1 and 1. The next row has the symbols ⊖, ⊜, and ⊖— that is, the numbers 1, 2, and 1. What are the "numbers" in the next row? Can you see a pattern emerging? It looks like this:

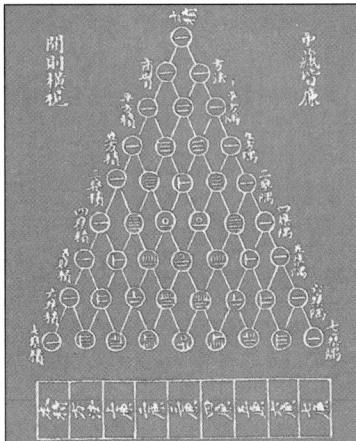

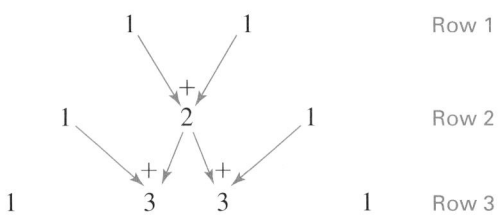

Row 1

Row 2

Row 3

The 2 in the second row is the sum of the two elements in the first row. The first 3 in the third row is the sum of the first two elements in the second row, and so on. If we continue to construct rows in this way, we get the following triangular array

$$
\begin{array}{ccccccccccc}
&&&&& 1 && 1 &&&& \\
&&&& 1 && 2 && 1 &&& \\
&&& 1 && 3 && 3 && 1 && \\
&& 1 && 4 && 6 && 4 && 1 & \\
& 1 && 5 && 10 && 10 && 5 && 1 \\
&&&&& \vdots &&&&&
\end{array}
$$

This array is known as Pascal's triangle, in honor of the French mathematician Blaise Pascal. We shall examine Pascal's triangle as our introduction to binomial expansions.

Pascal's Triangle and Binomial Expansions

The seventeenth-century mathematician Blaise Pascal wrote a book about the triangle presented in *Getting Started*. In the book he stated that if we have 5 coins, the relationship between the number of heads that can come up and the number of ways in which these heads can occur is as follows:

Number of Heads	5	4	3	2	1	0
Number of Occurrences	1	5	10	10	5	1

Web It

For an introduction to the binomial theorem, go to link A-4-1 at mhhe.com/bello.

To study the relationship between the binomial theorem and Pascal's triangle, go to link A-4-2 at mhhe.com/bello.

For example, if 5 coins are tossed, 5 heads can only occur one way (HHHHH), but 4 heads can occur in 5 ways (THHHH, HTHHH, HHTHH, HHHTH, HHHHT). What do these results have to do with algebra? As it turns out, there is a somewhat different way to arrive at the numbers in Pascal's triangle. From our previous work, you can verify that

$$(x + y)^1 = x + y$$
$$(x + y)^2 = x^2 + 2xy + y^2$$
$$(x + y)^3 = x^3 + 3x^2y + 3xy^2 + y^3$$
$$(x + y)^4 = x^4 + 4x^3y + 6x^2y^2 + 4xy^3 + y^4$$
$$(x + y)^5 = x^5 + 5x^4y + 10x^3y^2 + 10x^2y^3 + 5xy^4 + y^5$$

Look at the coefficients (the numbers that multiply x and y) of the terms in the expansion of $(x + y)^5$. They are 1, 5, 10, 10, 5, and 1. Now compare this with the fifth row in Pascal's triangle. The numbers are identical.

Can we predict the expansion of $(x + y)^6$ by studying these expansions? The pattern for the powers of x and y is easy. In going from left to right, the powers of x decrease from n to 0 and the powers of y increase from 0 to n. Ignoring the coefficients, the expansion of $(x + y)^n$ has the form

$$(x + y)^n = x^n y^0 + \Box x^{n-1}y^1 + \Box x^{n-2}y^2 + \cdots + \Box x^{n-k}y^k + \cdots + y^n x^0$$

where $\Box$ represents the missing coefficients. (The sum of the powers of x and y is always n.) To write the expansion of $(x + y)^n$, we need only know how to find these missing coefficients. In the case of $(x + y)^6$, the coefficients should be 1, 6, 15, 20, 15, 6, and 1, and they are obtained by constructing the sixth row of Pascal's triangle, starting and ending with a 1 and finding each coefficient by the addition pattern we have just discussed. In going from left to right, the powers of x decrease from $n = 6$ to 0, the powers of y increase from 0 to 6, and the sum of the exponents of x and y is always 6. Thus we have

$$(x + y)^6 = 1x^6y^0 + 6x^5y^1 + 15x^4y^2 + 20x^3y^3 + 15x^2y^4 + 6x^1y^5 + 1x^0y^6$$
$$= x^6 + 6x^5y + 15x^4y^2 + 20x^3y^3 + 15x^2y^4 + 6xy^5 + y^6$$

Here is the generalization for the expression of $(x + y)^n$.

PROCEDURE

The Expansion of $(x + y)^n$

1. The first term is x^n and the last term is y^n.

2. The powers of x *decrease* by 1 from term to term, the powers of y *increase* by 1 from term to term, and the sum of the powers of x and y in any term is always n.

3. The coefficients for the expansion correspond to those in the nth row of Pascal's triangle and can be written as $\binom{n}{r}$ where r is the power of x.

Web It

To learn the general form of the binomial theorem, go to link A-4-3 at mhhe.com/bello.

Using this notation,

$$(x + y)^6 = \binom{6}{6}x^6 + \binom{6}{5}x^5y + \binom{6}{4}x^4y^2 + \binom{6}{3}x^3y^3 + \binom{6}{2}x^2y^4 + \binom{6}{1}xy^5 + \binom{6}{0}y^6$$

$$= x^6 + 6x^5y + 15x^4y^2 + 20x^3y^3 + 15x^2y^4 + 6xy^5 + y^6$$

where

$$\binom{6}{6} = 1, \binom{6}{5} = 6, \binom{6}{4} = 15, \binom{6}{3} = 20, \binom{6}{2} = 15, \binom{6}{1} = 6, \text{ and } \binom{6}{0} = 1$$

To facilitate defining $\binom{n}{r}$, we first introduce a special symbol, $n!$ (read "n factorial"), for which we have the following definition.

Factorial

For any natural number n,

$$n! = n(n - 1)(n - 2) \cdots (1)$$

Thus

$$5! = 5 \cdot 4 \cdot 3 \cdot 2 \cdot 1 = 120$$

$$7! = 7 \cdot 6 \cdot 5 \cdot 4 \cdot 3 \cdot 2 \cdot 1 = 5040$$

This notation can also be used to denote the product of consecutive integers beginning with integers different from 1. For example,

$$\frac{7!}{4!} = 7 \cdot 6 \cdot 5$$

since

$$\frac{7!}{4!} = \frac{7 \cdot 6 \cdot 5 \cdot 4 \cdot 3 \cdot 2 \cdot 1}{4 \cdot 3 \cdot 2 \cdot 1}$$

Similarly,

$$\frac{9!}{5!} = 9 \cdot 8 \cdot 7 \cdot 6 \quad \text{and} \quad \frac{10!}{6!} = 10 \cdot 9 \cdot 8 \cdot 7$$

Also,

$$n! = n[(n - 1)(n - 2)(n - 3) \cdots (1)]$$
$$= n(n - 1)!$$

Web It

To study and work practice problems with factorials, go to link A-4-4 at mhhe.com/bello.

Thus we can write

$$n! = n(n - 1)!$$

With this notation,

$$9! = 9 \cdot 8!$$

$$13! = 13 \cdot 12!$$

If we let $n = 1$ in $n! = n(n - 1)!$, we obtain

$$1! = 1 \cdot 0!$$

Hence we define 0! as follows.

Zero Factorial

$$0! = 1$$

EXAMPLE 1 **Calculations with factorials**	**PROBLEM 1**

Find:

a. 4!   **b.** $\dfrac{8!}{5!}$

a. 8!   **b.** $\dfrac{9!}{7!}$

SOLUTION

a. Since $n! = n(n-1)(n-2)\cdots(1)$

$$4! = 4 \cdot 3 \cdot 2 \cdot 1 = 24$$

b.
$$8! = 8 \cdot 7 \cdot 6 \cdot 5 \cdot 4 \cdot 3 \cdot 2 \cdot 1$$
$$\text{and } 5! = 5 \cdot 4 \cdot 3 \cdot 2 \cdot 1$$

Thus

$$\frac{8!}{5!} = \frac{8 \cdot 7 \cdot 6 \cdot \cancel{5} \cdot \cancel{4} \cdot \cancel{3} \cdot \cancel{2} \cdot \cancel{1}}{\cancel{5} \cdot \cancel{4} \cdot \cancel{3} \cdot \cancel{2} \cdot \cancel{1}} = 336$$

Calculate It   Factorials

You can use the numerical capabilities of your calculator in this section. Suppose you want to do Example 1. With a TI-83 Plus, go to the home screen by pressing [2nd] [MODE] and enter the number 4. To get 4!, press [MATH], move the cursor three places to the right so you are in the PRB column and then press 4. The home screen now shows 4!. Press [ENTER] and the answer 24 appears as shown in Window 1.

Now suppose you want to find $\binom{8}{4}$. Most calculators use the notation nCr instead of $\binom{n}{r}$. Thus to find $\binom{8}{4}$, start at the home screen, enter 8, then press [MATH], move to the PRB column, and press [3] [4] [ENTER]. The answer 70 is obtained, as shown in Window 2.

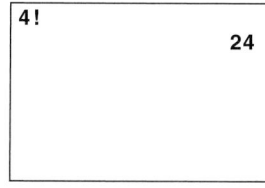

Window 1

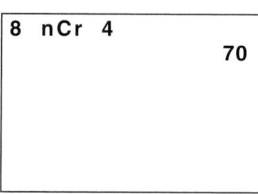

Window 2

We are now ready to use factorial notation to define the binomial coefficient.

Binomial Coefficient

The **binomial coefficient,** denoted $\binom{n}{r}$, is defined as

$$\binom{n}{r} = \frac{n!}{r!(n-r)!}$$

Thus

$$\binom{5}{2} = \frac{5!}{2!(5-2)!} = \frac{5!}{2!3!} = \frac{5 \cdot 4 \cdot 3!}{2!3!} = 10$$

$$\binom{5}{5} = \frac{5!}{5!(5-5)!} = \frac{5!}{5!0!} = 1$$

Web It

For some interesting properties of the binomial coefficient, go to link A-4-5 at mhhe.com/bello.

Answers

1. a. 40,320   **b.** 72

EXAMPLE 2   **Evaluating a binomial coefficient**

Evaluate:

$$\binom{6}{4}$$

Evaluate: $\binom{7}{5}$

SOLUTION   By definition,

$$\binom{n}{r} = \frac{n!}{r!(n-r)!}$$

$$\binom{6}{4} = \frac{6!}{4!2!} = \frac{6 \cdot 5 \cdot 4!}{4!2!} = \frac{6 \cdot 5}{2} = 15$$

Now that we know the meaning of $\binom{n}{r}$, we can use it to write the coefficients of a binomial expansion.

EXAMPLE 3   **Writing the coefficients of a binomial expansion**

Given that $(a + b)^4 = a^4 + 4a^3b + 6a^2b^2 + 4ab^3 + b^4$, write each coefficient in the expansion using the $\binom{n}{r}$ notation.

Given that $(a + b)^5 = a^5 + 5a^4b + 10a^3b^2 + 10a^2b^3 + 5ab^4 + b^5$ write each coefficient using $\binom{n}{r}$ notation.

SOLUTION

$$(a + b)^4 = a^4 + 4a^3b + 6a^2b^2 + 4ab^3 + b^4$$

$$= \binom{4}{4}a^4 + \binom{4}{3}a^3b + \binom{4}{2}a^2b^2 + \binom{4}{1}ab^3 + \binom{4}{0}b^4$$

You can easily verify that

$$\binom{4}{4} = 1, \binom{4}{3} = 3, \binom{4}{2} = 6, \binom{4}{1} = 4, \text{ and } \binom{4}{0} = 1$$

Following the pattern in Example 3, the general formula for $(x + y)^n$ can be written using the $\binom{n}{r}$ notation, where n represents the exponent of the binomial being expanded and r is the power of x. The powers of x *decrease* by 1 from term to term, while those of y *increase* by 1 from term to term. Thus we have

The General Binomial Expansion	$(x + y)^n = \binom{n}{n}x^n + \binom{n}{n-1}x^{n-1}y^1 + \binom{n}{n-2}x^{n-2}y^2 + \cdots + \binom{n}{n-k}x^{n-k}y^k + \cdots + \binom{n}{0}y^n$

Using this formula, we have

$$(x + y)^5 = \binom{5}{5}x^5 + \binom{5}{4}x^4y^1 + \binom{5}{3}x^3y^2 + \binom{5}{2}x^2y^3 + \binom{5}{1}x^1y^4 + \binom{5}{0}y^5$$

$$= x^5 + 5x^4y + 10x^3y^2 + 10x^2y^3 + 5xy^4 + y^5$$

To avoid confusion, we first write the expansion completely, and *then* substitute the coefficients

Answers

2. 21

3. $\binom{5}{5} = 1; \binom{5}{4} = 5; \binom{5}{3} = 10;$

$\binom{5}{2} = 10; \binom{5}{1} = 5; \binom{5}{0} = 1$

$$\binom{5}{5} = 1, \binom{5}{4} = 5, \binom{5}{3} = 10, \binom{5}{2} = 10, \binom{5}{1} = 5, \text{ and } \binom{5}{0} = 1$$

in the expansion. Note that

$$\binom{5}{5} = \binom{5}{0}, \binom{5}{4} = \binom{5}{1}, \text{ and } \binom{5}{3} = \binom{5}{2}$$

Web It

To read some more about the general binomial theorem and see examples, go to link A-4-6 at mhhe.com/bello.

In general,

$$\binom{n}{k} = \binom{n}{n-k}$$

This tells you that the coefficients of the first and last terms, second and next to last, and so on, are equal, thus saving you time and computation when you expand binomials.

EXAMPLE 4   **Using the binomial expansion**

Expand: $(a - 2b)^5$

SOLUTION   We use the binomial expansion with $x = a$, $y = -2b$, and $n = 5$. This gives

$$(a - 2b)^5 = \binom{5}{5}a^5 + \binom{5}{4}a^4(-2b)^1 + \binom{5}{3}a^3(-2b)^2 + \binom{5}{2}a^2(-2b)^3$$
$$+ \binom{5}{1}a^1(-2b)^4 + \binom{5}{0}(-2b)^5$$

We know that

$$\binom{5}{5} = \binom{5}{0} = 1, \binom{5}{4} = \binom{5}{1} = \frac{5!}{1!4!} = 5 \text{ and } \binom{5}{3} = \binom{5}{2} = \frac{5!}{2!3!} = 10$$

Thus

$(a - 2b)^5$
$= a^5 + 5a^4(-2b)^1 + 10a^3(-2b)^2 + 10a^2(-2b)^3 + 5a^1(-2b)^4 + (-2b)^5$
$= a^5 - 10a^4b \quad + 40a^3b^2 \quad - 80a^2b^3 \quad + 80ab^4 \quad - 32b^5$

PROBLEM 4

Expand: $(2x - y)^4$

B   Finding the Coefficient of a Term in a Binomial Expansion

As you recall, when we expand $(x + y)^n$, in the coefficient $\binom{n}{r}$, n represents the exponent of the binomial being expanded and r is the power of x. We shall use this to find the coefficient of a particular term in an expansion without writing out the whole expansion.

EXAMPLE 5   **Finding the coefficient of a term in a binomial expansion**

Find the coefficient of a^4b^3 in the expansion of $(a - 3b)^7$.

SOLUTION   In our case, $n = 7$, $x = a$, $y = -3b$, and the power of a is $r = 4$. Thus the coefficient of a^4b^3 can be obtained by simplifying

$$\binom{7}{4}a^4(-3b)^3 = \frac{7!}{4!3!}a^4(-3b)^3 = \frac{7 \cdot 6 \cdot 5}{3 \cdot 2 \cdot 1}a^4(-27b^3) = -945a^4b^3$$

Hence the coefficient of a^4b^3 is -945.

PROBLEM 5

Find the coefficient of x^3 in the expansion of $(x - 2)^7$.

Answers

4. $16x^4 - 32x^3y + 24x^2y^2 - 8xy^3 + y^4$ **5.** 560

C   Solving Applications Involving Binomial Expansions

The binomial coefficients for $(x + y)^n$ furnish us with a row of Pascal's triangle. As you recall, the numbers in such a row also tell us how many ways a stated number of heads can come up if n coins are tossed. For example, the first coefficient $\binom{n}{n} = 1$ tells us that the number of ways that n heads can occur when n coins are tossed is 1. The second coefficient,

$$\binom{n}{n-1} = n$$

tells us that the number of ways that $n - 1$ heads can occur when n coins are tossed is n, the third coefficient

$$\binom{n}{n-2} = \frac{n(n-1)}{1 \cdot 2}$$

tells us that the number of ways that $n - 2$ heads can occur when n coins are tossed is

$$\frac{n(n-1)}{1 \cdot 2}$$

and so on. We use this idea in Example 6.

EXAMPLE 6   Using the binomial expansion

Six coins are tossed. In how many ways can exactly 2 heads come up?

SOLUTION   In this case, $n = 6$ and $r = 2$. Thus the number of ways in which exactly 2 heads can come up when 6 coins are tossed is

$$\binom{6}{2} = \frac{6!}{2!4!} = \frac{6 \cdot 5}{2 \cdot 1} = 15$$

PROBLEM 6

Seven coins are tossed. In how many ways can exactly 3 heads come up?

It is shown in probability theory that if p is the probability of an event occurring favorably in *one* trial, then the probability of its occurring favorably r times in n trials is

$$\binom{n}{r} p^r (1 - p)^{n-r}$$

For example, if a fair coin is tossed, the probability of its coming up heads is $\frac{1}{2}$. So, if the coin is tossed 6 times, the probability of getting exactly 4 heads is

$$\binom{6}{4}\left(\frac{1}{2}\right)^4\left(1 - \frac{1}{2}\right)^2 = \left(\frac{6 \cdot 5}{2 \cdot 1}\right)\left(\frac{1}{16}\right)\left(\frac{1}{4}\right) = \frac{30}{128} = \frac{15}{64}$$

EXAMPLE 7   More practice using binomial expansion

A fair coin is tossed 8 times. Find the probability of getting exactly 4 heads.

SOLUTION   Here $n = 8$, $r = 4$, and $p = \frac{1}{2}$. Substituting in the probability formula, we get

$$\binom{8}{4}\left(\frac{1}{2}\right)^4\left(1 - \frac{1}{2}\right)^4 = 70\left(\frac{1}{2}\right)^8 = \frac{70}{256} = \frac{35}{128}$$

PROBLEM 7

A fair coin is tossed 10 times. Find the probability of getting exactly 5 heads.

Answers

6. 35   **7.** $\frac{252}{1024} = \frac{63}{256}$

Calculate It Exercises

Use your calculator:

   1. To find 10! and 12!.

   2. To find $\binom{10}{6}$ and $\binom{14}{8}$.

   3. To verify the results of Example 2.

Exercises A.4

A In Problems 1–14, evaluate the given expression.

1. 3! 6

2. 9! 362,880

3. 10! 3,628,800

4. 2! 2

5. $\frac{6!}{2!}$ 30

6. $\frac{11!}{10!}$ 11

7. $\frac{9!}{6!}$ 504

8. $\frac{3!}{0!}$ 6

9. $\binom{6}{2}$ 15

10. $\binom{6}{5}$ 6

11. $\binom{11}{1}$ 11

12. $\binom{11}{0}$ 1

13. $\binom{4}{0}$ 1

14. $\binom{7}{4}$ 35

In Problems 15–26, use the binomial expansion to expand these expressions.

15. $(a+3b)^4$ $a^4+12a^3b+54a^2b^2+108ab^3+81b^4$

16. $(a-3b)^4$ $a^4-12a^3b+54a^2b^2-108ab^3+81b^4$

17. $(x+4)^4$ $x^4+16x^3+96x^2+256x+256$

18. $(4x-1)^4$ $256x^4-256x^3+96x^2-16x+1$

19. $(2x-y)^5$
$32x^5-80x^4y+80x^3y^2-40x^2y^3+10xy^4-y^5$

20. $(x+2y)^5$
$x^5+10x^4y+40x^3y^2+80x^2y^3+80xy^4+32y^5$

21. $(2x+3y)^5$
$32x^5+240x^4y+720x^3y^2+1080x^2y^3+810xy^4+243y^5$

22. $(3x-2y)^5$
$243x^5-810x^4y+1080x^3y^2-720x^2y^3+240xy^4-32y^5$

23. $\left(\frac{1}{x}-\frac{y}{2}\right)^4$ $\frac{1}{x^4}-\frac{2y}{x^3}+\frac{3y^2}{2x^2}-\frac{y^3}{2x}+\frac{y^4}{16}$

24. $\left(\frac{x}{2}+\frac{3}{y}\right)^4$ $\frac{x^4}{16}+\frac{3x^3}{2y}+\frac{27x^2}{2y^2}+\frac{54x}{y^3}+\frac{81}{y^4}$

25. $(x+1)^6$ $x^6+6x^5+15x^4+20x^3+15x^2+6x+1$

26. $(y-1)^6$ $y^6-6y^5+15y^4-20y^3+15y^2-6y+1$

B In Problems 27–34, find the coefficient of the indicated term in the expansion.

27. Given $(x-3)^6$; find the coefficient of x^3 -540

28. Given $(x+2)^6$; find the coefficient of x^3 160

29. Given $(x+2y)^7$; find the coefficient of x^2y^5 672

30. Given $(y-2z)^7$; find the coefficient of y^2z^5 -672

31. Given $(2x-1)^8$; find the coefficient of x^4 1120

32. Given $(y+2z)^8$; find the coefficient of y^4z^4 1120

33. Given $\left(\frac{a}{2}-1\right)^5$; find the coefficient of a^2 $-\frac{5}{2}$

34. Given $\left(y+\frac{z}{2}\right)^5$; find the coefficient of y^3z^2 $\frac{5}{2}$

APPLICATIONS

Tossing Coins and Rolling Dice

35. Six coins are tossed. In how many ways can exactly 3 heads come up? 20

36. Six coins are tossed. In how many ways can exactly 2 heads come up? 15

37. Nine coins are tossed. In how many ways can exactly 3 heads come up? 84

38. Nine coins are tossed. In how many ways can exactly 4 heads come up? 126

39. A fair coin is tossed 6 times. Find the probability of getting exactly 2 heads. $\frac{15}{64}$

40. A fair coin is tossed 6 times. Find the probability of getting exactly 3 heads. $\frac{5}{16}$

41. A fair coin is tossed 9 times. Find the probability of getting exactly 3 heads. $\frac{21}{128}$

42. A fair coin is tossed 9 times. Find the probability of getting exactly 4 heads. $\frac{63}{256}$

43. A single die (plural, dice) is rolled 4 times. Find the probability that a 4 comes up exactly twice. $\left(Note:\text{ the probability that a specified number, 1, 2, 3, 4, 5, or 6, comes up in a single throw is }\frac{1}{6}\cdot\right)$ $\frac{25}{216}$

USING YOUR KNOWLEDGE

Follow the Pattern

How do you relate the pattern used to construct Pascal's triangle with the $\binom{n}{r}$ notation? As you recall, the numbers in Pascal's triangle are obtained by adding the numbers above and to the left and right of the number in question. Using $\binom{n}{r}$ notation, this fact is written as

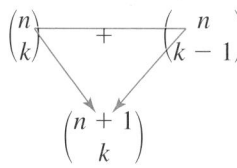

$$\binom{n}{k} + \binom{n}{k-1} \rightarrow \binom{n+1}{k}$$

44. Prove that

$$\binom{n}{k} + \binom{n}{k-1} = \binom{n+1}{k}$$

$$\binom{n}{k} + \binom{n}{k+1} = \frac{n!}{k!(n-k)!} + \frac{n!}{(k-1)!(n-k+1)!}$$

$$= \frac{n!(n-k+1) + k \cdot n!}{k!(n-k+1)!}$$

$$= \frac{n!(n-k+1+k)}{k!(n-k+1)!} = \frac{n!(n+1)}{k!(n-k+1)!}$$

$$= \frac{(n+1)!}{k!(n-k+1)!} = \binom{n+1}{k}$$

45. In Example 4, we used $\binom{5}{5} = \binom{5}{0}$, $\binom{5}{4} = \binom{5}{1}$ and so on. Prove that

$$\binom{n}{k} = \binom{n}{n-k}$$

where $k \leq n$.

$$\binom{n}{k} = \frac{n!}{k!(n-k)!} = \frac{n!}{(n-k)!k!} = \binom{n}{n-k}$$

WRITE ON

46. Write an explanation of how Pascal's triangle is constructed. Answers may vary.

47. Give your own definition of $n!$. Answers may vary.

48. In the binomial expansion, which part is the "binomial"? Answers may vary.

MASTERY TEST

If you know how to do these problems, you have learned your lesson!

Evaluate:

49. $6!$ 720

50. $\dfrac{7!}{3!}$ 840

51. $\dbinom{6}{2}$ 15

52. $\dbinom{7}{3}$ 35

Expand:

53. $(2a - b)^4$ $16a^4 - 32a^3b + 24a^2b^2 - 8ab^3 + b^4$

54. $(3a - 2b)^5$
$243a^5 - 810a^4b + 1080a^3b^2 - 720a^2b^3 + 240ab^4 - 32b^5$

Find the coefficient of:

55. a^2b^4 in the expansion of $(2a - 3b)^6$ 4860

56. ab^3 in the expansion of $(3a - b)^4$ -12

57. Five coins are tossed. In how many ways can exactly 4 heads come up? 5

58. A fair coin is tossed 7 times. Find the probability of getting exactly 5 tails. $\dfrac{21}{128}$

Research Questions

1. Write a short essay about Gauss's childhood.

2. Find out and write a report about Gauss's inventions.

3. Aside from being a superb mathematician, Gauss did some work in the field of astronomy. Report on some of his discoveries in this field.

4. In 1807, a famous French mathematician made Gauss pay an involuntary 2000-franc contribution to the French government. Find out who this famous mathematician was and the circumstances of the payment.

5. Write a report on the correspondence between Blaise Pascal and Pierre Fermat and its influence on the development of the theory of probability.

6. The binomial coefficients studied in this chapter date back to the year A.D. 1050. Write a report about the binomial coefficients and their connection to the Chinese mathematicians Chu Shih-Chien and Chia Hsien.

7. Write a report about the first triangular arrangement of binomial coefficients printed in Europe. (In what book did they appear and who was the author?)

Summary

SECTION	ITEM	MEANING	EXAMPLE
A.1A	Sequence	A set of numbers arranged according to some given law	2, 4, 6, . . .
	Terms $a_1, a_2, a_3, . . .$	The numbers of a sequence	The terms of 2, 4, 6, . . . are $a_1 = 2$, $a_2 = 4$, $a_3 = 6$, and so on.
	General term	The formula that generates the terms of the sequence	In the sequence 2, 4, 6, . . . , the general term is $a_n = 2n$.
A.1C	Σ	Summation notation	$\displaystyle\sum_{i=1}^{4} 2^i = 2 + 2^2 + 2^3 + 2^4$
A.2A	Arithmetic sequence	A sequence in which each term after the first is obtained by adding a quantity d to the preceding term	2, 4, 6, . . . is an arithmetic sequence.
	Common difference d	The difference between two successive terms of an arithmetic sequence	In the arithmetic sequence 3, 6, 9, the common difference is $d = 3$.
	General term of an arithmetic sequence	$a_n = a_1 + (n - 1)d$	In the sequence 8, 12, 16, . . . , the general term is $a_n = 8 + (n - 1)(4)$.
A.2B	Sum S_n of an arithmetic sequence	$S_n = \dfrac{n}{2}(a_1 + a_n)$	In the sequence 2, 4, 6, . . . , $S_6 = \dfrac{6(2 + 12)}{2} = 42$.
A.3A	Geometric sequence	A sequence in which each term after the first is formed by multiplying the preceding term by a constant r	3, 6, 12, . . . is a geometric sequence.
	Common ratio r	The ratio of two successive terms of a geometric sequence	In the geometric sequence 3, 6, 12, . . . , $r = \dfrac{6}{3} = 2$.
	General term of a geometric sequence	$a_n = a_1 r^{n-1}$	In the geometric sequence 3, 6, 12, . . . , $a_n = 3 \cdot 2^{n-1}$.
A.3B	Sum S_n of a geometric sequence	$S_n = \dfrac{a_1(1 - r^n)}{1 - r}$	For the geometric sequence 3, 6, 12, . . . , $S_6 = \dfrac{3(1 - 2^6)}{1 - 2}$ $= \dfrac{3 \cdot (-63)}{-1}$ $= 189$
A.3C	Infinite geometric series	A series of the form $a_1 + a_1 r + a_1 r^2 + \cdots + a_1 r^{n-1} + \cdots$	$2 + 1 + \frac{1}{2} + \cdots$, where $a_1 = 2$ and $r = \frac{1}{2}$
	Sum S of a geometric series with $|r| < 1$	$S = \dfrac{a_1}{1 - r}$	For $2 + 1 + \frac{1}{2} + \cdots$, $S = \dfrac{2}{1 - \frac{1}{2}} = 4$

SECTION	ITEM	MEANING	EXAMPLE
A.4A	Pascal's triangle	An arrangement of numbers of the form $\quad\quad 1 \quad\quad 1$ $\quad 1 \quad 2 \quad 1$ $1 \quad 3 \quad 3 \quad 1$ $\quad\quad \vdots$ $\quad\quad \vdots$	
	Binomial coefficient	$\dbinom{n}{r} = \dfrac{n!}{r!(n-r)!}$	$\dbinom{5}{2} = \dfrac{5!}{2!3!}$ $= \dfrac{5 \cdot 4 \cdot 3 \cdot 2 \cdot 1}{2 \cdot 1 \cdot 3 \cdot 2 \cdot 1} = 10$
	Binomial expansion of $(x + y)^n$	An expansion in which the general term is $\dbinom{n}{r}x^r y^{n-r}$	$(x + y)^3 = \dbinom{3}{3}x^3 + \dbinom{3}{2}x^2 y$ $\quad + \dbinom{3}{1}xy^2 + \dbinom{3}{0}y^3$ $= x^3 + 3x^2 y + 3xy^2 + y^3$

Review Exercises

(If you need help with these exercises, look in the section indicated in brackets.)

1. [A.1A] For the given sequence, find a_6 and a_n.

   a. The sequence of odd counting numbers:
1, 3, 5, 7, . . . $a_6 = 11$; $a_n = 2n - 1$

   b. The sequence of multiples of 3: 3, 6, 9, 12, . . .
$a_6 = 18$; $a_n = 3n$

2. [A.1A] Each card in a pack of four index cards is cut in half, and the halves are put in a single pile. This step is repeated again and again.

   a. How many cards are in the pack after the fourth cut? $a_4 = 64$

   b. How many cards are in the pack after the nth cut?
$a_n = 2^{n+2}$

3. [A.1B] Find the first three terms, the tenth term, and S_3 for the sequence whose general term is

   a. $\dfrac{1}{3}(n^2 + 1)$ $\dfrac{2}{3}, \dfrac{5}{3}, \dfrac{10}{3}$; $a_{10} = \dfrac{101}{3}$; $S_3 = \dfrac{17}{3}$

   b. $\dfrac{1}{4}(n^2 + n)$ $\dfrac{1}{2}, \dfrac{3}{2}, 3$; $a_{10} = \dfrac{55}{2}$; $S_3 = 5$

4. [A.1B, C] Find the sequence that corresponds to the function.

   a. $a(n) = n^2 + 2, n = 1, 2, 3, \ldots$; then

      find $S_4 = \displaystyle\sum_{n=1}^{4}(n^2 + 2)$ 3, 6, 11, . . . ; $S_4 = 38$

   b. $a(n) = 2n^2 - 1, n = 1, 2, 3, \ldots$, then

      find $S_3 = \displaystyle\sum_{n=1}^{3}(2n^2 - 1)$ 1, 7, 17, . . . ; $S_3 = 25$

5. [A.1D] A painting doubles in value every 100 yr. Find the value of this painting in the year 2000 if it was worth $1000 in the year

   a. 1400 $64,000

   b. 1600 $16,000

6. [A.2A] For the given arithmetic sequence, find d and a_{10}.

   a. 3, 6, 9, 12, . . . $d = 3$; $a_{10} = 30$

   b. 4, 8, 12, 16, . . . $d = 4$; $a_{10} = 40$

7. [A.2A] For the given arithmetic sequence, find a_n.

   a. 3, 6, 9, 12, . . . $a_n = 3n$

   b. 4, 8, 12, 16, . . . $a_n = 4n$

8. [A.2A] The distance (in feet) that a free-falling body falls in each second, starting with the first second, is given by the arithmetic sequence 16, 48, 80, 112, ... Find what distance the body falls in the

   a. Sixth second 176 ft

   b. Eighth second 240 ft

9. [A.2B] The sixth term of an arithmetic sequence is 24. Find the first term and the common difference if the sum of the first six terms is

   a. 114 $a_1 = 14; d = 2$

   b. 99 $a_1 = 9; d = 3$

10. [A.2B] The sum of the first six terms of an arithmetic sequence is 54. Find the first term and the common difference if the sixth term is

    a. 14 $a_1 = 4; d = 2$

    b. 16.5 $a_1 = 1.5; d = 3$

11. [A.2C] A machine valued at $80,000 depreciates $8000 the first year, $7800 the second year, $7600 the third year, and so on. Find the value of the machine at the end of

    a. 8 yr $21,600

    b. 10 yr $9000

12. [A.2C] Juan is saving his pennies. He saved 5¢ the first week, 9¢ the second week, 13¢ the third week, and so on, in arithmetic sequence. How many weeks will it take Juan to save

    a. $2.30? 10 weeks

    b. $3.24? 12 weeks

13. [A.3A] For the given geometric sequence, find r and a_6.

    a. 3, 6, 12, 24, . . . $r = 2; a_6 = 96$

    b. 4, 2, 1, $\frac{1}{2}$, . . . $r = \frac{1}{2}; a_6 = \frac{1}{8}$

14. [A.3A] For the given geometric sequence, find r and a_n.

    a. 1, 4, 16, 64, . . . $r = 4; a_n = 4^{n-1}$

    b. 4, -2, 1, $-\frac{1}{2}$, . . .

    $r = -\frac{1}{2}; a_n = \frac{(-1)^{n-1}}{2^{n-3}}$

15. [A.3B] For the given geometric sequence, find a_n and S_n.

    a. 2, -4, 8, -16, . . .

    $a_n = (-1)^{n-1}\, 2^n; S_n = \frac{2}{3}[1 - (-2)^n]$

    b. 1, $-\dfrac{1}{2}$, $\dfrac{1}{4}$, $-\dfrac{1}{8}$, . . .

    $a_n = (-\frac{1}{2})^{n-1}; S_n = \frac{2}{3}[1 - (-\frac{1}{2})^n]$

16. [A.3C] Find the sum (if it exists) of the geometric series.

    a. $32 - 16 + 8 - 4 + \cdots$ $\frac{64}{3}$

    b. $18 - 6 + 2 - \frac{2}{3} + \cdots$ $\frac{27}{2}$

17. [A.3C] Find the sum (if it exists) of the geometric series.

    a. $1 + 1.005 + (1.005)^2 + (1.005)^3 + \cdots$
       Sum does not exist; $|r| = 1.005 > 1.$

    b. $1 - 1.001 + (1.001)^2 - (1.001)^3 + \cdots$
       Sum does not exist; $|r| = 1.001 > 1.$

18. [A.3D] Find the fraction equivalent to the repeating decimal.

    a. $0.313131\ldots$ $\frac{31}{99}$

    b. $0.324324324\ldots$ $\frac{12}{37}$

19. [A.3D] A rubber ball is dropped and takes a sequence of bounces, each one 0.5 as high as the preceding one. If the ball continues bouncing indefinitely, find the total distance it travels, given that it was dropped from a height of

    a. 8 ft 24 ft   **b.** 6 ft 18 ft

20. [A.4A] Find:

    a. 8! 40,320

    b. $\dfrac{8!}{4!}$ 1680

21. [A.4A] Find:

    a. $\dbinom{9}{3}$ 84   **b.** $\dbinom{8}{8}$ 1

22. [A.4A] Expand:

    a. $(a - 3b)^4$
    $a^4 - 12a^3 b + 54a^2 b^2 - 108ab^3 + 81b^4$
    b. $(3a + 2b)^4$
    $81a^4 + 216a^3 b + 216a^2 b^2 + 96ab^3 + 16b^4$

23. [A.4B] Find the coefficient of x^4 in the expansion of

    a. $(x + 2)^8$ 1120

    b. $(3x - 1)^7$ -2835

24. [A.4C] Seven coins are tossed. In how many ways can exactly

    a. 4 heads come up? 35

    b. 3 heads come up? 35

25. [A.4C] A fair coin is tossed 7 times. Find the probability of getting exactly

    a. 4 heads $\frac{35}{128}$ **b.** 3 heads $\frac{35}{128}$

Practice Test A

(Answers on pages A-46–A-47)

1. For the sequence of odd natural numbers $1, 3, 5, 7, \ldots$, find

    a. a_{10}, the tenth term

    b. a_n, the general term

2. Each card in a pack of three index cards is cut in half, and the halves are put in a single pile. This step is repeated again and again.

    a. Write the sequence that gives the number of cards in the deck after each cut.

    b. How many cards are in the pack after the fourth cut?

    c. How many cards are in the pack after the nth cut?

3. Find the first three terms and the eighth term of the sequence whose general term is

$$a_n = \frac{n(n + 1)}{2}$$

4. Find the sequence that corresponds to the function $a(n) = 3n - 1, n = 1, 2, 3, \ldots$; then find

$$\sum_{n=1}^{5} (3n - 1)$$

5. A sculpture by a famous artist doubles in value every 50 yr. Find the value of the sculpture in the year 2000 if it was worth $500 in the year 1500.

6. For the arithmetic sequence $5, 8, 11, 14, \ldots$, find

    a. d **b.** a_{10}

7. For the arithmetic sequence $5, 8, 11, 14, \ldots$, find a_n.

8. Sally draws a sequence of circles. She starts with a row of three circles, then adds two circles to get the second row, and adds two more circles to get the third row shown. Find how many circles she would have in the

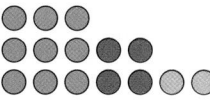

    a. Fifth row   **b.** Tenth row

    c. nth row

9. The distance (in feet) that a free-falling body falls in each second, starting with the first second, is given by the arithmetic sequence $16, 48, 80, 112, \ldots$. Find the distance that the body falls in 7 sec.

10. The sum of the first eight terms of an arithmetic sequence is 172 and the eighth term is 32. Find:

    a. a_1, the first term

    b. d, the common difference

11. A piece of machinery valued at $50,000 depreciates $8000 the first year, $7500 the second year, $7000 the third year, and so on. Find the value of this piece of machinery at the end of 6 yr.

12. Natasha is trying to save her pennies. She saved 5¢ the first week, 8¢ the second week, 11¢ the third week, and so on, in arithmetic sequence. After a few weeks, Natasha has saved a total of $1.85. For how many weeks has she been saving?

13. For the geometric sequence $3, 1, \frac{1}{3}, \frac{1}{9}, \ldots$, find:

    a. r **b.** a_6

14. For the sequence in Problem 13, find a_n.

15. For the geometric sequence $2, -4, 8, -16, \ldots$, find:

    a. r **b.** a_8 **c.** S_8

16. Find the sum of the geometric series
$8 - 4 + 2 - 1 + \cdots$.

17. Find the sum of the geometric series
$\sum_{n=1}^{\infty} (1.001)^n = 1.001 + (1.001)^2 + (1.001)^3 + \cdots$,
if the sum exists.

18. Find the fraction equivalent to the repeating decimal
$0.312312312\ldots$.

19. A rubber ball, dropped on a hard surface, takes a sequence of bounces, each one-half as high as the preceding one. The ball is dropped from a height of 12 ft, and it is assumed to continue bouncing indefinitely. Find the total distance it would travel.

20. Find:

    a. $9!$ **b.** $\dfrac{9!}{7!}$

21. Find:

    a. $\dbinom{9}{8}$ **b.** $\dbinom{9}{9}$

22. Expand $(2a - b)^4$.

23. Find the coefficient of a^4b^3 in the expansion of $(a - 2b)^7$.

24. Seven coins are tossed. In how many ways can exactly 4 heads turn up?

25. A fair coin is tossed 12 times. Find the probability of getting exactly 6 heads.

Answers to Practice Test

ANSWER		IF YOU MISSED	REVIEW		
		QUESTION	SECTION	EXAMPLES	PAGE
1. a. $a_{10} = 19$	**b.** $a_n = 2n - 1$	1	A.1A	1	A-3
2. a. $6, 12, 24, \ldots$	**b.** 48 **c.** $3 \cdot 2^n$	2	A.1A	2, 3	A-4
3. $a_1 = 1, a_2 = 3, a_3 = 6; a_8 = 36$		3	A.1B	4	A-5
4. $2, 5, 8, 11, 14, \ldots ; 40$		4	A.1B	5	A-6
5. $\$512{,}000$		5	A.1C	6	A-7
6. a. $d = 3$	**b.** $a_{10} = 32$	6	A.2A	1, 2	A-13
7. $a_n = 3n + 2$		7	A.2A	2	A-13
8. a. 11	**b.** 21 **c.** $2n + 1$	8	A.2A	2	A-13
9. 784 ft		9	A.2B	3	A-14–A-15
10. a. $a_1 = 11$	**b.** $d = 3$	10	A.2B	4	A-15
11. $\$9500$		11	A.2C	5	A-16
12. 10		12	A.2C	6	A-16
13. a. $r = \frac{1}{3}$	**b.** $a_6 = \frac{1}{81}$	13	A.3A	1	A-21
14. $a_n = \left(\frac{1}{3}\right)^{n-2}$		14	A.3A	1	A-21
15. a. $r = -2$	**b.** $a_8 = -256$	15	A.3B	2	A-22
c. $S_8 = -170$					
16. $\frac{16}{3}$, or $5\frac{1}{3}$		16	A.3C	4	A-25

ANSWER	IF YOU MISSED		REVIEW		
	QUESTION	SECTION	EXAMPLES		PAGE
17. Sum does not exist.	17	A.3C	5		A-25
18. $\frac{104}{333}$	18	A.3D	6		A-26
19. 36 ft	19	A.3D	7, 8		A-26–A-27
20. a. 362,880 **b.** 72	20	A.4A	1		A-35
21. a. 9 **b.** 1	21	A.4A	2		A-36
22. $16a^4 - 32a^3b + 24a^2b^2 - 8ab^3 + b^4$	22	A.4A	4		A-37
23. -280	23	A.4B	5		A-37
24. 35	24	A.4C	6		A-38
25. $\frac{231}{1024}$	25	A.4C	7		A-38

Selected Answers

The brackets preceding answers for the Chapter Review Exercises indicate the Chapter, Section, and Objective for you to review for further study. For example, [3.4C] appearing before answers means those exercises correspond to Chapter 3, Section 4, Objective C.

Chapter 1

Exercises 1.1

1. $\{1, 2\}$ **3.** $\{5, 6, 7\}$ **5.** $\{-1, -2, -3\}$ **7.** $\{0, 1, 2, 3\}$
9. $\{1, 2, 3, \ldots\}$ **11.** $\{x \mid x$ is an integer between 0 and 4$\}$
13. $\{x \mid x$ is an integer between -3 and 3$\}$
15. $\{x \mid x$ is an even number between 19 and 78$\}$ **17.** False **19.** True
21. True **23.** True **25.** True **27.** $0.\overline{6}$ **29.** 0.875 **31.** 2.5 **33.** $1.1\overline{6}$

Set	**35.** $\dfrac{-3}{8}$	**37.** $\sqrt{8}$	**39.** $0.\overline{3}$	**41.** 0.9	**43.** 3.1416
Natural numbers					
Whole numbers					
Integers					
Rational numbers	✓		✓	✓	✓
Irrational numbers		✓			
Real numbers	✓	✓	✓	✓	✓

45. True **47.** False **49.** -8 **51.** 7 **53.** $-\dfrac{3}{4}$ **55.** $\dfrac{1}{5}$ **57.** -0.5
59. $-0.\overline{2}$ **61.** $1.\overline{36}$ **63.** $-\pi$ **65.** 10 **67.** 17 **69.** $\dfrac{3}{5}$ **71.** $0.\overline{5}$
73. $3.\overline{61}$ **75.** $-\sqrt{2}$ **77.** π **79.** $<$ **81.** $>$ **83.** $>$ **85.** $<$
87. $<$ **89.** 0.11; Burger King **91.** The advanced degree, since
$0.25 > 0.0757...$ **93.** (b) Monjane $8'\ 0.75''$ **101.** 0.125 **103.** -8
105. 2 **107.** $\{x \mid x$ is 3 or an integer obtained when 3, 4, 5 . . . is added in
succession$\}$ **109.** $>$

Exercises 1.2

1. $\dfrac{2}{5}$ **3.** -0.1 **5.** 2 **7.** -0.8 **9.** $\dfrac{1}{5}$ **11.** -7 **13.** -0.3
15. $-\dfrac{13}{56}$ **17.** -14 **19.** -0.6 **21.** $-\dfrac{41}{63}$ **23.** -1 **25.** -4 **27.** -0.1
29. $\dfrac{22}{21}$ **31.** -40 **33.** -12 **35.** 50 **37.** 60 **39.** 40 **41.** 30 **43.** -40
45. -7.26 **47.** 2.86 **49.** $-\dfrac{25}{42}$ **51.** $\dfrac{1}{4}$ **53.** $-\dfrac{15}{4}$ **55.** -2 **57.** -4
59. 2 **61.** 0 **63.** Undefined **65.** -2 **67.** 9 **69.** 3 **71.** 1 **73.** -4
75. -7 **77.** $-\dfrac{21}{20}$ **79.** $\dfrac{4}{7}$ **81.** $-\dfrac{5}{7}$ **83.** $-\dfrac{1}{2}$ **85.** $\dfrac{1}{6}$
87. Commutative property of addition **89.** Commutative property of
multiplication **91.** Commutative property of multiplication
93. Multiplicative identity **95.** $A = a(b + c) = ab + ac$
97. $3500°C$ **99.** $\$46$ **101.** $14°C$

	Number	Additive Inverse	Reciprocal
103.	7	-7	$\dfrac{1}{7}$
105.	0	0	Undefined

107. -1.35 **109.** -3.045 **111.** 3.875 **115.** $\dfrac{16}{25}$ **117.** 2
119. $-\dfrac{59}{56}$ **121.** -12.8 **123.** 16 **125.** Inverse property of addition
127. Associative property of multiplication
129. Commutative property of multiplication

Exercises 1.3

1. -16 **3.** 25 **5.** -125 **7.** 1296 **9.** -32 **11.** $\dfrac{1}{16}$ **13.** $\dfrac{1}{125}$
15. $\dfrac{1}{81}$ **17.** $\dfrac{1}{x^6}$ **19.** $\dfrac{1}{a^8}$ **21.** $\dfrac{1}{64}$ **23.** $12x^2$ **25.** $-15y^2$ **27.** $\dfrac{20}{a^5}$
29. $-\dfrac{30y^3}{x^2}$ **31.** $-\dfrac{24y^6}{x^4}$ **33.** $-\dfrac{40}{a^2b^3}$ **35.** -30 **37.** $2x^4$ **39.** $\dfrac{a^2}{2}$
41. $-2x^3y^2$ **43.** $-\dfrac{x}{2}$ **45.** $\dfrac{2}{3a^3}$ **47.** $\dfrac{3}{4}$ **49.** $\dfrac{3b^3}{2a^6}$ **51.** $\dfrac{8x^9}{y^6}$ **53.** $\dfrac{4y^6}{x^4}$
55. $-\dfrac{1}{27x^9y^6}$ **57.** $\dfrac{1}{x^{12}y^6}$ **59.** $x^{12}y^{12}$ **61.** $\dfrac{a^2}{b^6}$ **63.** $-\dfrac{8b^6}{27a^3}$
65. a^8b^4 **67.** $\dfrac{1}{x^{15}y^6}$ **69.** $x^{27}y^6$ **71.** 2.68×10^8 **73.** 2.4×10^{-4}
75. $8{,}000{,}000$ **77.** 0.23 **79.** 2×10^3 **81.** 3×10^8 **83.** 31 yr
85. (a) -9.856; (b) -2.772 **87.** 3.34×10^5 **89.** (a) $\boxed{7.3\quad 10}$;
(b) $\boxed{1.23\quad -07}$ **97.** $\dfrac{8}{x^4y^2}$ **99.** $-\dfrac{1}{5x^{14}}$ **101.** -64 **103.** $\dfrac{x^8}{9y^{10}}$
105. $-\dfrac{1}{x^5}$ **107.** $\dfrac{49x^{10}}{25y^8}$ **109.** 2.4×10^{-2}

Exercises 1.4

1. (a) -26; (b) -70 **3.** (a) 27; (b) 3 **5.** -47 **7.** 20 **9.** -15
11. -13 **13.** 57 **15.** 3 **17.** -6 **19.** 1 **21.** -36 **23.** -10 **25.** -4
27. 0 **29.** 1 **31.** -20 **33.** 8 **35.** -24 **37.** -33 **39.** 11,800
41. 37 **43.** $\$1050$ **45.** 13 **47.** 32 **49.** 60 **51.** $4x - 4y$
53. $-9a + 9b$ **55.** $1.2x - 0.6$ **57.** $-\dfrac{3a}{2} + \dfrac{6}{7}$ or $\dfrac{-21a + 12}{14}$
59. $-2x + 6y$ **61.** $-2.1 - 3y$ **63.** $-4a - 20$ **65.** $-6x - xy$
67. $-8x + 8y$ **69.** $-6a + 21b$ **71.** $0.5x + 0.5y - 1$
73. $-\dfrac{6}{5}a + \dfrac{6}{5}b - 6$ **75.** $-2x + 2y - 6z - 10$
77. $-0.3x - 0.3y + 0.6z + 1.8$ **79.** $-\dfrac{5}{2}a + 5b - \dfrac{5}{2}c - 5d + 5$
81. $9x - 6$ **83.** $-9x - 8$ **85.** $11L - 4W$ **87.** $-2x$ **89.** $\dfrac{x}{9} + 2$
91. $6a + 2b$ **93.** $3x - 4y$ **95.** $x - 5y + 36$
97. $-2x^3 + 9x^2 - 3x + 12$ **99.** $\dfrac{2}{7}x^2 + \dfrac{4}{5}x - \dfrac{3}{4}$ **101.** $4a - 11$
103. $-7a + 10b - 3$ **105.** $-4.8x + 3.4y + 5$ **107.** 16 **109.** -2
111. $v_a = \dfrac{1}{2}v_1 + \dfrac{1}{2}v_2$ **113.** $KE = \dfrac{1}{2}mv_1^2 + \dfrac{1}{2}mv_2^2$ **117.** -3
119. $-2x + 6y - 4z + 8$ **121.** $\dfrac{1}{4}x + 5$ **123.** $-2a^3 + 2a^2 - a - 17$
125. 6

Review Exercises

1. [1.1A] **(a)** {4, 5, 6, 7, 8}; **(b)** {5, 6, 7}   **2.** [1.1B] **(a)** 0.2; **(b)** 0.4
3. [1.1B] **(a)** $0.\overline{1}$; **(b)** $0.\overline{2}$ **4.** [1.1C]

5. [1.1D] **(a)** 3.5; **(b)** $-\dfrac{3}{4}$ **6.** [1.1E] **(a)** 9; **(b)** 4.2   **7.** [1.1E] **(a)** $\dfrac{1}{8}$;

(b) $0.\overline{4}$ **8.** [1.1F] **(a)** <; **(b)** <   **9.** [1.1F] **(a)** >; **(b)** =
10. [1.1F] **(a)** <; **(b)** =   **11.** [1.2A] **(a)** -11; **(b)** -3

12. [1.2A] **(a)** $-\dfrac{2}{7}$; **(b)** -0.6 **13.** [1.2A] **(a)** 12; **(b)** 4

14. [1.2A] **(a)** $\dfrac{19}{20}$; **(b)** $\dfrac{13}{12}$ **15.** [1.2A] **(a)** -36; **(b)** -14.4

16. [1.2A] **(a)** $-\dfrac{21}{32}$; **(b)** $\dfrac{5}{21}$ **17.** [1.2A] **(a)** 0; **(b)** Undefined

18. [1.2A] **(a)** $-\dfrac{5}{3}$; **(b)** $\dfrac{1}{0.3}$ or $\dfrac{10}{3}$ **19.** [1.2A] **(a)** $-\dfrac{9}{4}$; **(b)** -3

Set	0.3	0	$\dfrac{-3}{4}$	-5	$\sqrt{3}$
Natural numbers					
Whole numbers		✓			
Integers		✓		✓	
Rational numbers	✓	✓	✓	✓	
Irrational numbers					✓
Real numbers	✓	✓	✓	✓	✓

20. [1.2B] **(a)** Commutative property of addition;

(b) Associative property of addition   **21.** [1.3A] **(a)** 81;

(b) -81 **22.** [1.3B] **(a–b)** 1   **23.** [1.3B] **(a)** $-\dfrac{1}{512}$; **(b)** $\dfrac{1}{x^{10}}$

24. [1.3C] **(a)** $-\dfrac{15y^{10}}{x^4}$; **(b)** $-\dfrac{24}{x^{11}y^8}$ **25.** [1.3C] **(a)** $\dfrac{3}{x^2}$; **(b)** $-4x^{11}$

26. [1.3C] **(a)** $-\dfrac{x}{3}$; **(b)** $-\dfrac{2}{x^{11}}$ **27.** [1.3D] **(a)** $-\dfrac{8x^{21}}{y^{18}}$; **(b)** $\dfrac{16}{x^{24}y^{24}}$

28. [1.3E] **(a)** $\dfrac{1}{x^{24}y^{12}}$; **(b)** $x^{25}y^{15}$ **29.** [1.3F] **(a)** 3.4×10^5;

(b) 4.7×10^{-5} **30.** [1.3F] **(a)** 37,000; **(b)** 0.0078
31. [1.4A] **(a)** -75; **(b)** 50   **32.** [1.4B] **(a)** 9; **(b)** 20
33. [1.4B] **(a)** -7; **(b)** -73 **34.** [1.4C] **(a)** 352 sq in.; **(b)** -33
35. [1.4D] **(a)** $-3x + 21$; **(b)** $2x + 17$
36. [1.4E, F] **(a)** $-x + 10y - 16$; **(b)** $3x^2 + 3x + 8$

Chapter 2

Exercises 2.1

1. Yes   **3.** Yes   **5.** Yes   **7.** No   **9.** No   **11.** 4   **13.** 1   **15.** 2   **17.** 2

19. 7   **21.** -1 **23.** -8 **25.** -4 **27.** 0   **29.** 0   **31.** $\dfrac{1}{13}$ **33.** 8

35. $-\dfrac{11}{80}$ **37.** 24   **39.** 1   **41.** 15   **43.** 10   **45.** 4   **47.** 10   **49.** -5

51. 0   **53.** -1 **55.** $-\dfrac{10}{11}$ **57.** $\dfrac{5}{2}$ **59.** An identity   **61.** 4   **63.** -6

65. 57   **67.** 0   **69.** 16   **71.** 21.5   **73.** 1500   **75.** 6   **77.** $-3x - 7$
79. 2   **81.** 50   **83.** 5   **85.** 11 in.   **87.** 8   **89.** 12 in.
91. (a) 47; **(b)** 48   **95.** 10   **97.** 16   **99.** -1 **101.** No

Exercises 2.2

1. $h = \dfrac{V}{\pi r^2}$ **3.** $W = \dfrac{V}{LH}$ **5.** $b = P - s_1 - s_2$ **7.** $s = \dfrac{A - \pi r^2}{\pi r}$ or

$s = \dfrac{A}{\pi r} - r$ **9.** $V_2 = \dfrac{P_1 V_1}{P_2}$ **11.** $y = 4 - \dfrac{2}{3}x$ or $y = \dfrac{12 - 2x}{3}$

13. (a) $T = \dfrac{D}{R}$; **(b)** 4 hr   **15. (a)** $A = 34 - 2H$; **(b)** 18 yr
17. (a) $C = (O)(N) - E$; **(b)** $29,500   **19. (a)** $B = 180 - A - C$;

(b) $14°$ **21. (a)** $L = \dfrac{H - 32}{1.88}$; **(b)** No; **(c)** Yes; **(d)** $L = \dfrac{H - 29}{1.95}$;

(e) Yes   **23. (a)** $C = 2000 + 309m$; **(b)** $14,051
25. (a) $C = 0.36 + 0.08(t - 1)$, $t \geq 1$; **(b)** 47 min
27. (a) $F = 20 + 10m$; **(b)** 12   **29.** $x = -5$; angles $= 35°$ each
31. $x = 20$; angles $= 140°$ each   **33.** $x = 3$; angles $= 50°$ each
35. $x = -5.5$; $(42 - 6x)° = 75°$; $(50 - 10x)° = 105°$ **37.** $3n + 6$
39. $p = 25,000$ **41.** $m = 70$ **43.** $x = 18$ **45.** 2025   **47.** 3.04 ft
51. $x = -20$; angles $= 170°$ each   **53. (a)** $C = 0.37 + 0.23(w - 1)$,

$w > 1$; **(b)** $w = \dfrac{C - 0.14}{0.23}$; **(c)** $2.44   **55. (a)** $C = 6S + 4$; **(b)** $16°C$

Exercises 2.3

1. $4m = m + 18$ **3.** $\dfrac{1}{2}x - 12 = \dfrac{2x}{3}$ **5.** $4x + 5 = 29$; $x = 6$
7. $3x + 8 = 35$; $x = 9$ **9.** $3x - 2 = 16$; $x = 6$ **11.** $5x = 12 + 2x$;
$x = 4$ **13.** $\dfrac{1}{3}x - 2 = 10$; $x = 36$ **15.** 75%   **17.** 8 yr

19. 130 mi   **21.** 44; 46; 48   **23.** -11; -9; -7 **25.** 87; 92
27. (a) 72 mi/hr; **(b)** 316 mi/hr   **29.** 213 lb   **31.** 47 awards
33. 12 ft by 5000 ft   **35.** 518 ft by 716 ft   **37.** $72°$ **39.** $135°$
41. $x = 19$; angles: $67°$, $23°$ **43.** $x = 39$; angles: $127°$, $53°$
45. $30°$; $60°$; $90°$ **47.** $n = 20$ **49.** $W = 20$ **51.** $x = 1200$ **53.** $T = 8$
55. 5   **63.** 31; 33; 35   **65.** $x - 12 = 7x$ **67.** $x = 27$; angles: $85°$, $95°$
69. 12.60 million

Exercises 2.4

1. (a) 64.5; **(b)** 9.75   **3.** $10.81   **5.** 98.8 million   **7.** $23.13
9. Markup: $24; 44.4% of selling price   **11.** 40%   **13.** 50
15. 20%   **17.** 30 mi/gal   **19.** 650   **21.** $1100
23. $5000 at 6%; $15,000 at 8%   **25.** $8000   **27. (a)** 2 hr;
(b) 288 km   **29.** 360 km   **31.** 10 mi   **33.** 25.2   **35.** 10   **37. (a)** <;
(b) >; **(c)** <   **39.** $-290.4 billion   **41.** $381.675 billion   **47.** $3000 at
5%; $5000 at 10%   **49.** 24 million   **51.** 20%

Exercises 2.5

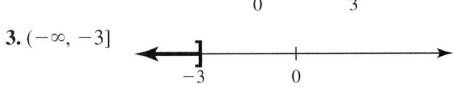

1. $(3, \infty)$

3. $(-\infty, -3]$

5. $[3, \infty)$; $x \geq 3$

7. $[-1, \infty)$; $x \geq -1$

9. $(-\infty, 2)$; $x < 2$

11. $(-\infty, 1]$; $x \leq 1$

13. $(-\infty, 3]$; $y \leq 3$

15. $(5, \infty)$; $x > 5$

17. $(-\infty, 2]$; $a \leq 2$

19. $(-\infty, -4)$; $z < -4$

21. $(-\infty, -1]$; $x \le -1$

23. $(-\infty, 0]$; $x \le 0$

25. $\left(-\infty, -\dfrac{11}{80}\right)$; $x < -\dfrac{11}{80}$

27. $[-20, \infty)$; $x \ge -20$

29. $[-2, \infty)$; $x \ge -2$

31. $x < -4$ or $x > 3$; $(-\infty, -4) \cup (3, \infty)$

33. $x < 1$ or $x > 2$; $(-\infty, 1) \cup (2, \infty)$

35. $x \le 4$ or $x > 5$; $(-\infty, 4] \cup (5, \infty)$

37. $x < 2$ or $x > 2$; $(-\infty, 2) \cup (2, \infty)$

39. $x \le 2$ or $x > 3$; $(-\infty, 2] \cup (3, \infty)$

41. $-2 \le x \le 4$; $[-2, 4]$

43. $2 < x \le 6$; $(2, 6]$

45. $1 < x < 3$; $(1, 3)$

47. No solution; $\varnothing$

49. $1 \le x \le 4$; $[1, 4]$

51. $2 < x < 3$; $(2, 3)$

53. $1 < x < 3$; $(1, 3)$

55. $-2 < x < 5$; $(-2, 5)$

57. $-5 < x < 1$; $(-5, 1)$

59. $3 < x < 4$; $(3, 4)$

61. $-2 < x < 4$; $(-2, 4)$

63. $-6 < y < 1$; $(-6, 1)$

65. $4 \le y \le 6$; $[4, 6]$

67. $-2 < x < 4$; $(-2, 4)$

69. $-3 < a < 3$; $(-3, 3)$

71. $h \le 29{,}028$ **73.** $e \ge 2$ **75.** $n \ge 4 \times 10^{25}$ **77.** 20,001

79. when $h < 13$ hr   **81.** $<$ **83.** $<$ **85.** $\dfrac{1}{5}$ **87.** $\sqrt{2}$

89. $x > 46$; 2006   **97.** $p \ge 45$

99.

101. $-5 \le x < 5$; $[-5, 5)$

103. $x > 2$; $(2, \infty)$

105. $x < -2$ or $x > -1$; $(-\infty, -2) \cup (-1, \infty)$

107. $2 < x < 5$; $(2, 5)$

Exercises 2.6

1. -13; 13   **3.** -2.3; 2.3   **5.** 0   **7.** No solution   **9.** -9; -5 **11.** -2; 6

13. $-\dfrac{6}{5}$; 2   **15.** -20; 4   **17.** -9; 18   **19.** -3 **21.** -5; $-\dfrac{1}{3}$ **23.** -7; 5

25. No solution   **27.** All real numbers   **29.** All real numbers

31. $-4 < x < 4$; $(-4, 4)$

33. $-2 \le z \le 2$; $[-2, 2]$

35. $-4 \le a \le 4$; $[-4, 4]$

37. $-1 < x < 3$; $(-1, 3)$

39. No solution

41. $-2 \le x \le -1$; $[-2, -1]$

43. $-2 < x < 1$; $(-2, 1)$

45. $x < -2$ or $x > 2$; $(-\infty, -2) \cup (2, \infty)$

47. $z \le -1$ or $z \ge 1$; $(-\infty, -1] \cup [1, \infty)$

49. $a \le -3$ or $a \ge 3$; $(-\infty, -3] \cup [3, \infty)$

51. $x < 0$ or $x > 2$;
   $(-\infty, 0) \cup (2, \infty)$

53. All real numbers; $(-\infty, \infty)$

55. $x \le -2$ or $x \ge -1$;
   $(-\infty, -2] \cup [-1, \infty)$

57. $x < -1$ or $x > \frac{5}{2}$;
   $(-\infty, -1] \cup (\frac{5}{2}, \infty)$

59. $x \le -2$ or $x \ge 1$;
   $(-\infty, -2] \cup [1, \infty)$

61. $a < 2$ or $a > 6$;
   $(-\infty, 2) \cup (6, \infty)$

63. All real numbers; $(-\infty, \infty)$

65. No solution; $\varnothing$

67. $\frac{1}{3} \le x \le 1$; $[\frac{1}{3}, 1]$

69. $\$450 \le a \le \$550$ **71.** Yes   **73.** -59 **75.** $x^2 + 11x + 7$

77. $-10x^2 - 2x + 3$ **83.** $x = \frac{9}{2}$ **85.** $x = \frac{4}{3}$; $x = -2$

87. $x = -\frac{21}{2}$; $x = \frac{15}{2}$

89. $x < -2$ or $x > 3$;
   $(-\infty, -2) \cup (3, \infty)$

91. $x \le -1$ or $x \ge \frac{7}{3}$;
   $(-\infty, -1] \cup [\frac{7}{3}, \infty)$

Review Exercises

1. [2.1A] **(a)** No; **(b)** No; **(c)** Yes   **2.** [2.1B] **(a)** 12; **(b)** 15; **(c)** 18
3. [2.1B] **(a)** 2; **(b)** 3; **(c)** 4   **4.** [2.1B] **(a)** 14; **(b)** 27; **(c)** 28
5. [2.1C] **(a)** 3; **(b)** 4; **(c)** 3   **6.** [2.1D] **(a)** $P = 500$; **(b)** $P = 2000$;
(c) $P = 4000$ **7.** [2.2A] **(a)** $h = \dfrac{H - 72.48}{2.5}$; $h = 4$;
(b) $h = \dfrac{H - 77.48}{2.5}$; $h = 2$; **(c)** $h = \dfrac{H - 84.98}{2.5}$; $h = -1$
8. [2.2A] **(a)** $A = \dfrac{7B + 14}{2}$; **(b)** $A = \dfrac{7B + 15}{3}$; **(c)** $A = \dfrac{5B + 28}{4}$
9. [2.2A] **(a)** $L = \dfrac{P - 2W}{2}$; 40 ft by 50 ft;
(b) $L = \dfrac{P - 2W}{2}$; 50 ft by 60 ft; **(c)** $L = \dfrac{P - 2W}{2}$; 60 ft by 70 ft
10. [2.2C] **(a)** $x = 10$; angles: 50° each; **(b)** $x = 15$;
angles: 57° each; **(c)** $x = 20$; angles: 58° each   **11.** [2.3B] **(a)** 100 mi;
(b) 150 mi; **(c)** 200 mi   **12.** [2.3C] **(a)** 49; 51; 53; **(b)** 51; 53; 55;
(c) 67; 69; 71   **13.** [2.4A]   **(a)** \$20,000; **(b)** \$30,000;
(c) \$15,000   **14.** [2.4B] **(a)** \$5000 in bonds; \$15,000 in CDs;
(b) \$17,000 in bonds; \$3000 in CDs; **(c)** \$10,000 in bonds;
\$10,000 in CDs   **15.** [2.4C] **(a)** 200 mi; **(b)** 300 mi; **(c)** 120 mi
16. [2.4D] **(a)** 100 L; **(b)** 25 L; **(c)** 0 L
17. [2.5A] **(a)**
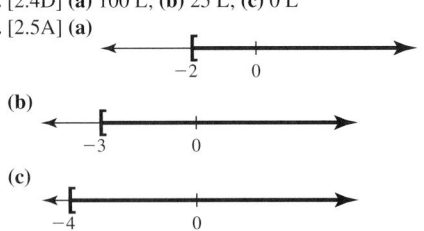
(b)

(c)

18. [2.5B] **(a)** $x \ge -3$; $[-3, \infty)$

(b) $x \ge -2$; $[-2, \infty)$

(c) $x \ge -2$; $[-2, \infty)$

19. [2.5B] **(a)** $x > 3$; $(3, \infty)$

(b) $x > 3$; $(3, \infty)$

(c) $x > 3$; $(3, \infty)$

20. [2.5C] **(a)** $-1 < x < 2$;
   $(-1, 2)$

(b) $-2 < x < 3$; $(-2, 3)$

(c) $-3 < x < 4$; $(-3, 4)$

21. [2.5C] **(a)** $(-\infty, -2) \cup [3, \infty)$

(b) $(-\infty, -3) \cup [2, \infty)$

(c) $(-\infty, -4) \cup [1, \infty)$

22. [2.5C] **(a)** $-2 < x \le 2$; $(-2, 2]$

(b) $-3 < x \le 3$; $(-3, 3]$

(c) $-2 < x \le 4$; $(-2, 4]$

23. [2.5C] **(a)** $-5 < x \le -1$; $(-5, -1]$

(b) $-4 < x \le -1$; $(-4, -1]$

(c) $-3 < x \le 1$; $(-3, -1]$

24. [2.6A] **(a)** $x = 7$; $x = -21$; **(b)** $x = 14$; $x = -28$; **(c)** $x = 21$;
$x = -35$ **25.** [2.6A] **(a)** $x = 2$; **(b)** $x = 4$; **(c)** $x = 6$

26. [2.6B] **(a)** $-\frac{1}{3} \le x \le 1$;
   $\left[-\frac{1}{3}, 1\right]$

(b) $-\frac{1}{2} \le x \le 1$; $\left[-\frac{1}{2}, 1\right]$

(c) $-\frac{3}{5} \le x \le 1$; $\left[-\frac{3}{5}, 1\right]$

27. [2.6B] **(a)** $x \leq -\frac{1}{3}$ or $x \geq 1$; $(-\infty, -\frac{1}{3}] \cup [1, \infty)$

(b) $x \leq -\frac{1}{2}$ or $x \geq 1$; $(-\infty, -\frac{1}{2}] \cup [1, \infty)$

(c) $x \leq -\frac{3}{5}$ or $x \geq 1$; $(-\infty, -\frac{3}{5}] \cup [1, \infty)$

Cumulative Review Chapters 1–2

1. $\{2, 4, 6\}$ **2.** $\frac{19}{100}$ **3.** -11 **4.** -20 **5.** $-5x^2 + 6x + 5$
6. $5x^{14}$ **7.** -9 **8.** No **9.** $\frac{8}{5}$ **10.** 58 **11.** 1850 **12.** $-\frac{14}{3}$; $-\frac{98}{3}$

13.

14.

15.

16.

17. $A = \dfrac{7B + 66}{6}$ **18.** 20 ft by 30 ft **19.** \$23,000 **20.** 174

Chapter 3

Exercises 3.1

1, 3, 5, 7.

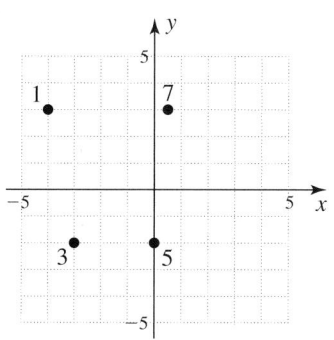

9. QI; QII **11.** QIII; QIV **13.** $(3, 5)$ **15.** $(-2, 3)$ **17.** $(-4, 0)$
19. $(0, -3)$
21. $(-2, 1)$; $(-1, 2)$; $(0, 3)$;
$(1, 4)$; $(2, 5)$

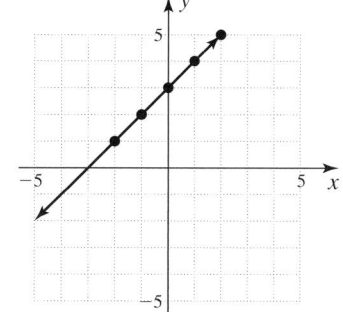

23. $(-1, -5)$; $(0, -4)$; $(1, -3)$

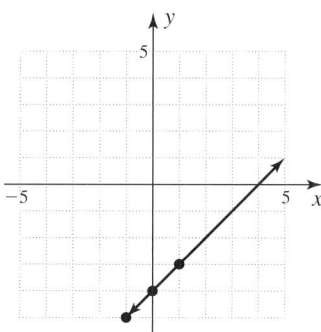

25. $(-1, -5)$; $(0, -3)$; $(1, -1)$

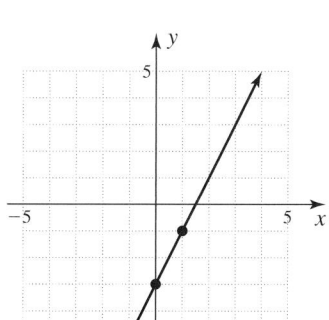

27. $(5, 0)$; $(0, -5)$

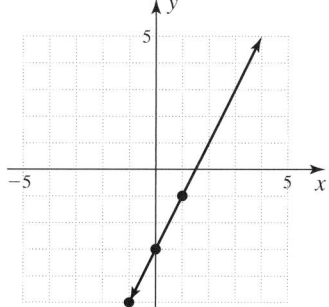

29. $(3, 0)$; $(0, 2)$

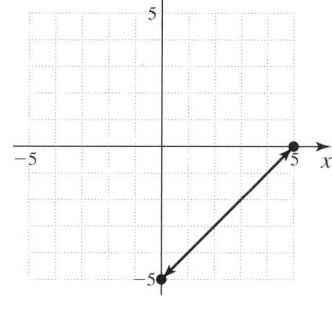

31. $(2, 0)$; $(0, -4)$

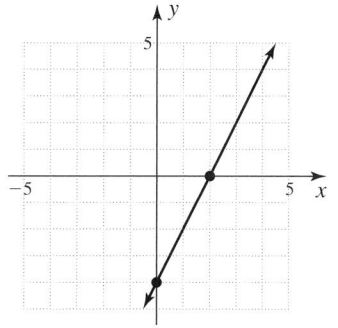

33. (2, 0); (0, 4)

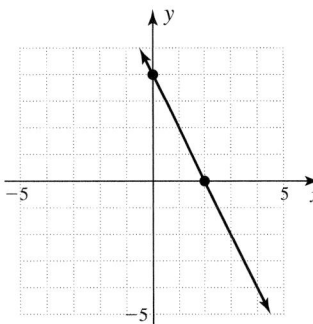

35. (0, 0)

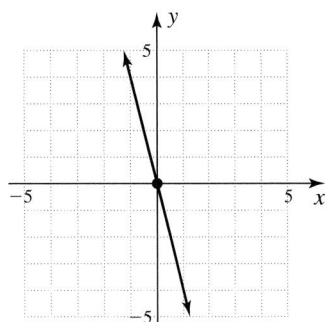

37. (−5, 0); (0, 2)

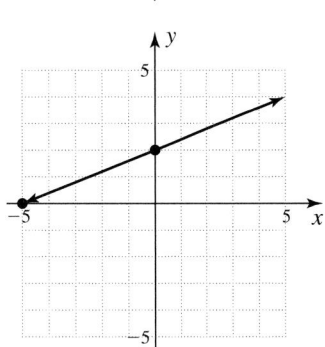

39. (5, 0); (0, −5)

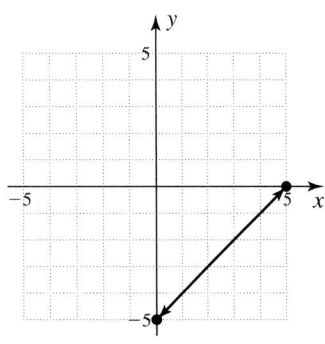

41. Vertical; $x = -4$

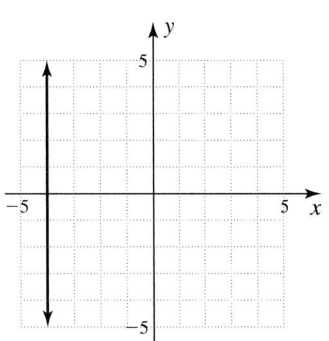

43. Vertical; $x = 4$

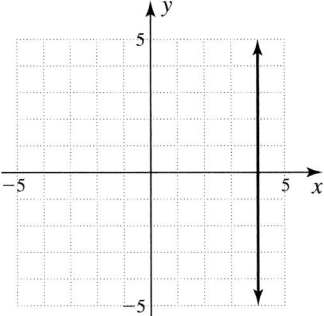

45. Vertical; $x = -4$

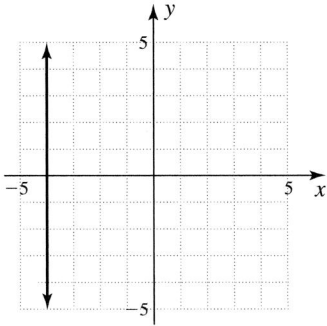

47. Horizontal; $y = 1$

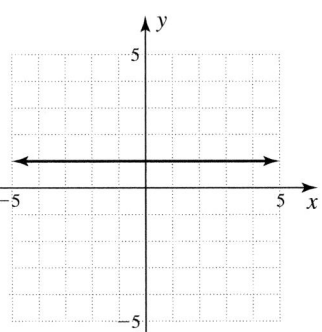

49. Vertical; $x = 2$

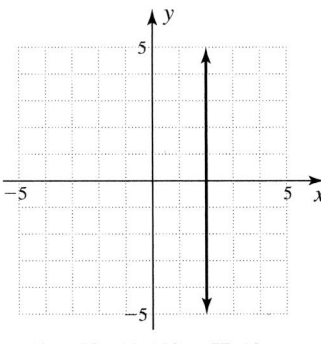

51. (a) Yes; **(b)** 80; **(c)** 160　**53.** (20, 140)　**55.** (45, 148)　**57.** 12
59. 2　**61.** −2　**63.** 86°F　**65.** Less than 10%　**67.** 11°F
77. $2y = -6$

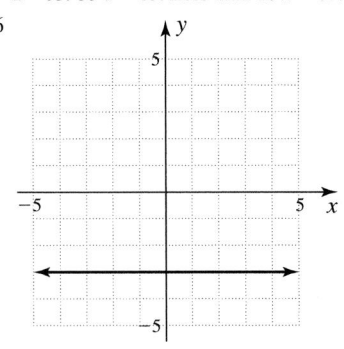

79. $-3x - y = -6$

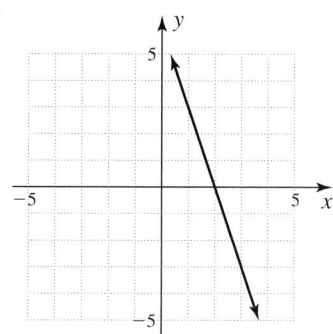

81. $(-2, 0); (0, 3)$

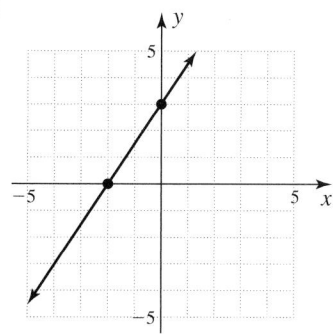

83. $A(-2, 3); B(-3, 0); C(4, -2)$

Exercises 3.2

1. $D = \{-3, -2, -1\}; R = \{0, 1, 2\}$; a function   **3.** $D = \{3, 4, 5\};$
$R = \{0\}$; a function   **5.** $D = \{1, 2\}; R = \{2, 3\}$; not a function
7. $D = \{1, 3, 5, 7\}; R = \{-1\}$; a function   **9.** $D = \{2\};$
$R = \{1, 0, -1, -2\}$; not a function   **11.** $D = \{x | -5 \leq x \leq 5\};$
$R = \{y | -5 \leq y \leq 5\}$; not a function   **13.** $D = \{x | -5 \leq x \leq 5\};$
$R = \{y | 0 \leq y \leq 5\}$; a function   **15.** $D = \{x | 0 \leq x \leq 5\};$
$R = \{y | -5 \leq y \leq 5\}$; not a function   **17.** $D = \{x | x \text{ is a real number}\};$
$R = \{y | y \geq 0\}$; a function   **19.** $D = \{x | x \text{ is a real number}\};$
$R = \{y | y \leq 0\}$; a function   **21.** $D = \{x | x \geq 0\}; R = \{y | y \text{ is a real}$
number$\}$; not a function   **23.** $D = \{x | x \text{ is a real number}\};$
$R = \{y | y \text{ is a real number}\}$; a function   **25.** $D = \{x | x \text{ is a real number}\};$
$R = \{y | y \text{ is a real number}\}$; a function   **27.** $D = \{x | -3 \leq x \leq 3\};$
$R = \{y | -2 \leq y \leq 2\}$; not a function   **29.** $D = \{x | x \geq 2 \text{ or } x \leq -2\};$
$R = \{y | y \text{ is a real number}\}$; not a function   **31.** A function
33. A function   **35.** A function   **37.** $D = \{x | x \geq 5\}$ **39.** $D = \{x | x \leq 2\}$
41. $D = \{x | x \text{ is a real number}\}$ **43.** $D = \{x | x \text{ is a real number and}$
$x \neq 5\}$ **45.** $D = \{x | x \text{ is a real number and } x \neq -5\}$
47. $D = \{x | x \text{ is a real number and } x \neq -2 \text{ or } x \neq -1\}$
49. $D = \{x | x \text{ is a real number and } x \neq \pm 4\}$ **51. (a)** 1; **(b)** 7; **(c)** -5
53. (a) 0; **(b)** 2; **(c)** 5   **55. (a)** $\frac{1}{4}$; **(b)** $\frac{3}{28}$; **(c)** $\frac{1}{28}$ **57. (a)** 5; **(b)** 19; **(c)** 24
59. (a) -10; **(b)** 7; **(c)** -70 **61. (a)** 3; **(b)** 4; **(c)** 7   **63. (a)** 7; **(b)** 8; **(c)** 56
65. (a) $P(x) = -0.0005x^2 + 24x - 100,000$; **(b)** \$90,000
67. (a) $U(50) = 140$; **(b)** $U(60) = 130$ **69. (a)** 160 lb; **(b)** 78 in.
71. (a) 639 lb/ft^2; **(b)** 6390 lb/ft^2 **73. (a)** L; **(b)** S; **(c)** Size 11; **(d)** Size 12
75. (a) 619.2; **(b)** 1352.8   **77. (a)** Yes; **(b)** $D = \{x | 1 \leq x \leq 6\};$
$R = \{y | 0 \leq y \leq 0.11\}$; **(c)** $D = \{x | 1 \leq x \leq 4\}; R = \{y | 0 \leq y \leq 0.09\}$
85. Yes   **87.** Yes   **89.** Yes   **91. (a)** 3; **(b)** -1; **(c)** 4
93. $D = \{x | x \text{ is a real number and } x \neq 2\}$ **95.** Not a function
97. $D = \{x | -3 \leq x \leq 3\}; R = \{y | -3 \leq y \leq 3\}$
99. $D = \{x | -3 \leq x \leq 3\}; R = \{y | -3 \leq y \leq 0\}$
101. $D = \{7, 8, 9\}; R = \{8, 9, 10\}$

Exercises 3.3

1. $m = \dfrac{4}{3}$

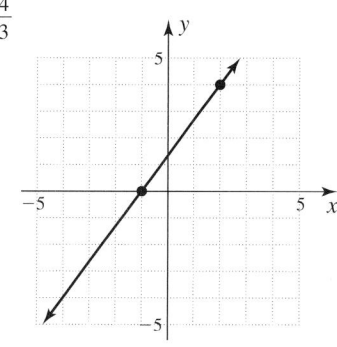

3. $m = \dfrac{8}{3}$

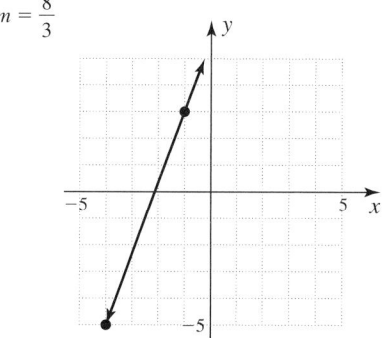

5. $m = 3$

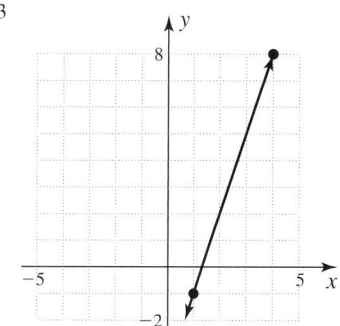

7. $m = 0$

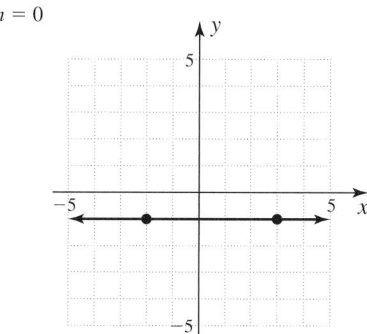

9. (a) $m = \dfrac{2000}{3}$; **(b)** The salary increased approximately \$667 per year.
11. Parallel   **13.** Perpendicular   **15.** Neither   **17.** Perpendicular
19. Parallel   **21.** $x = 2$ **23.** $x = 4$ **25.** $x = \dfrac{-16}{3}$ **27.** $y = -8$
29. $x = 3$

31.

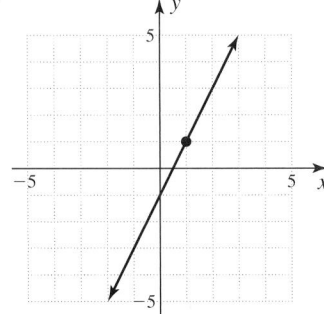

33.

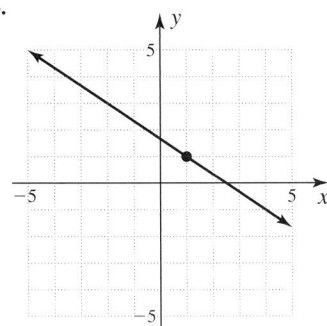

35.

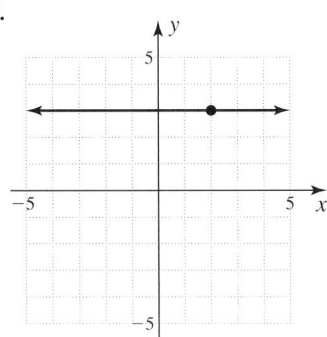

37.

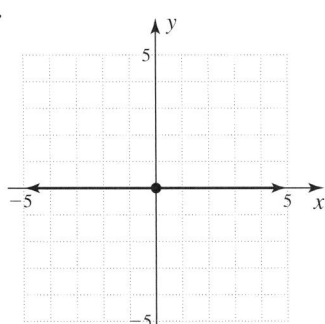

39.

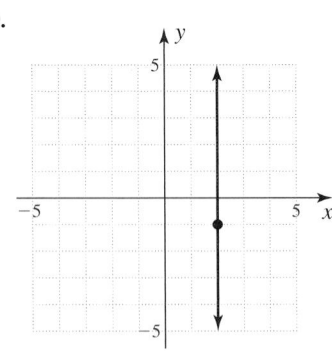

41. $(5, 3)$ **43.** $(-4, -4)$ **45.** $\left(\dfrac{-11}{2}, 0\right)$ **47.** $m = 1$; y-int: -4

49. $m = \dfrac{-1}{3}$; y-int: -2 **51.** $m = -2$; y-int: 4 **53.** $(83, 35)$ **55.** No

57. Yes   **59.** Yes   **61. (a)** 246 million; **(b)** 268 million; **(c)** 290 million;
(d) year 2020; **(e)**

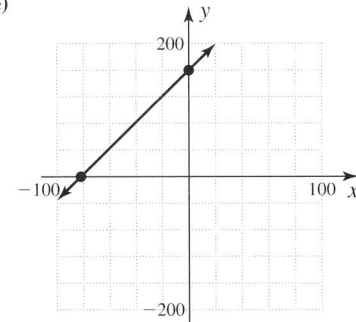

63. $y = -2x + 4$ **65.** $y = \dfrac{2}{3}x + 4$ **67.** $\dfrac{3}{2}$ **69.** -2.3 **71.** -3 **73.** 1

81.

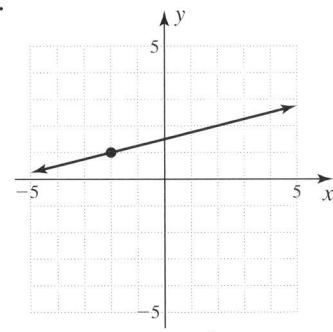

83. $x = 1$ **85. (a)** $m = \dfrac{-5}{2}$; **(b)** m undefined; **(c)** $m = 2$; **(d)** $m = 0$

Exercises 3.4

1. $3x - y = 4$ **3.** $x + y = 5$ **5.** $2x + y = 4$ **7.** $2x - y = -11$

9. $3x + y = -5$

11. $y = 5x + 2$

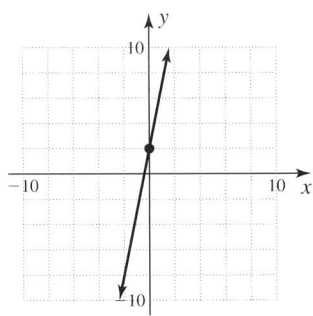

13. $y = \dfrac{-1}{5}x - \dfrac{1}{3}$

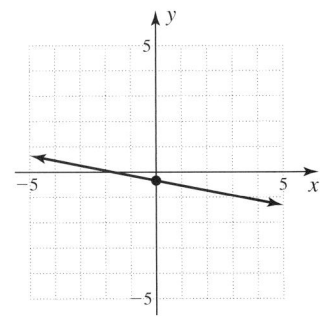

15. $y = 2x - 4$ **17.** $y = -3x - 12$ **19.** $y = -2x + 3$
21. $y = -x - 6$ **23.** $y = 4$ **25.** $x = -2$ **27.** $x = -2$ **29.** $y = 2$
31. (a)

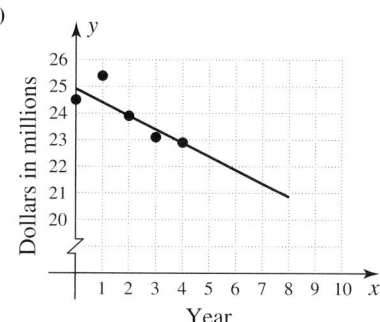

(b) $(2, 23.9)$; $(4, 22.9)$; **(c)** $y = \dfrac{-1}{2}x + 24.9$; **(d)** $20.9 million;

(e) Each year it decreases by $0.5 million.   **33.** $y = -4$
35. $2x - 3y = -6$ **37. (a)** $N = 0.25x + 5.5$; **(b)** Each year there is a
0.25 million increase in cases of diabetes.; **(c)** $x = 25$; **(d)** 11.75 million
39. (a) $C = 85x$; **(b)** $85 per sq foot; **(c)** $125,375; **(d)** 1624 sq feet
41. (a) $d + 2p = 14$; **(b)** 10 pairs   **43. (a)** $s - 2p = -20$; **(b)** $10; **(c)** 60
45. $4   **47.** x: 4; y: -2 **49.** x: -3; y: 6   **51.** $x = -2$ **53.** $y = 2x + 50$
55. (a) $2; **(b)** $75   **61.** $m = -2$; $b = 4$
63. $y = 3x + 2$

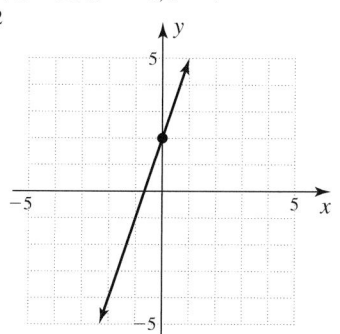

65. $y = -3x + 5$

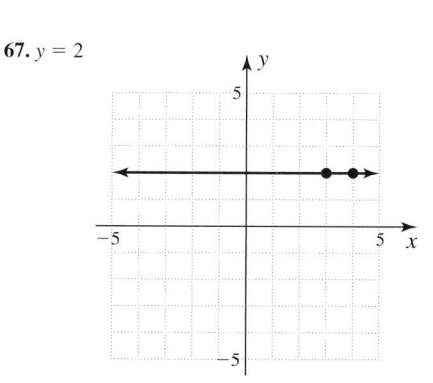

67. $y = 2$

Exercises 3.5
1. $x + 2y > 4$

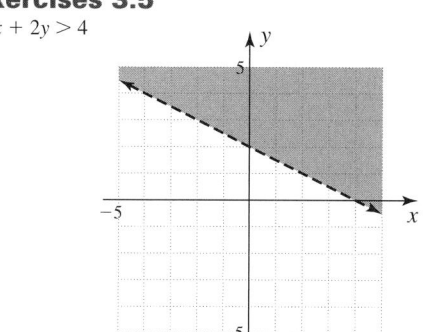

3. $-2x - 5y \le -10$

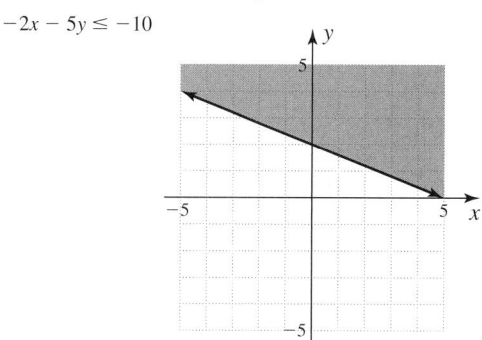

5. $y \ge 2x - 2$

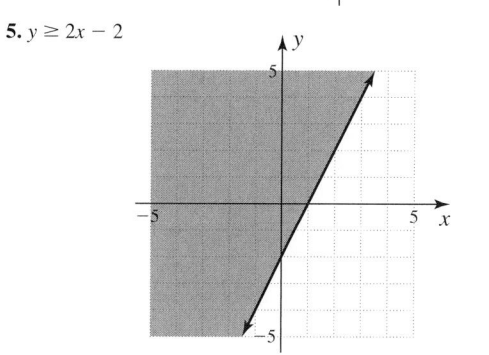

7. $6 < 3x - 2y$

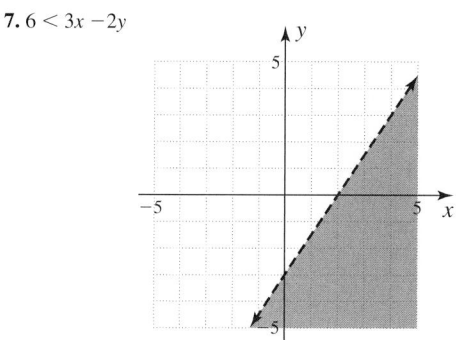

9. $4x + 3y \ge 12$

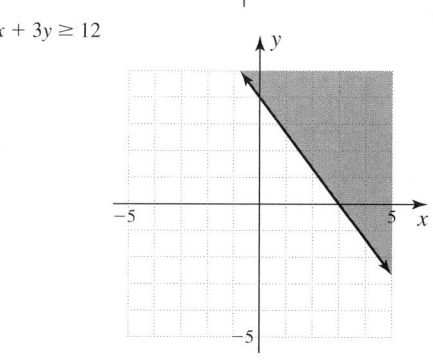

11. $10 < -5x + 2y$

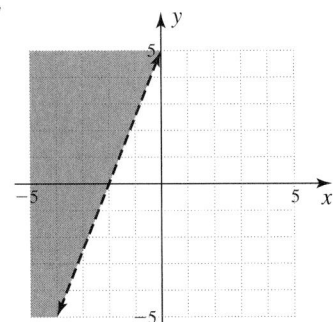

13. $2x \geq 2y - 4$

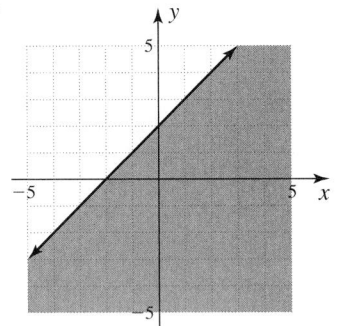

15. $2y < -4x + 8$

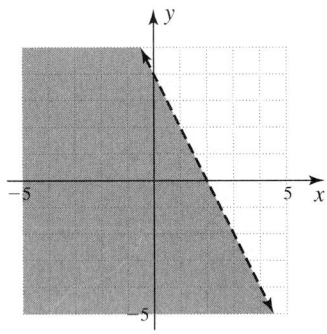

17. $x \geq -3$

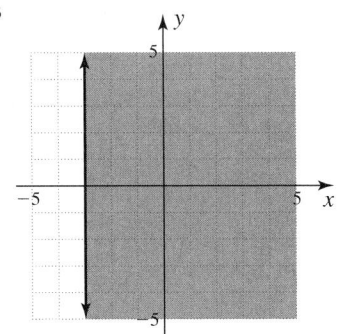

19. $y < 3$

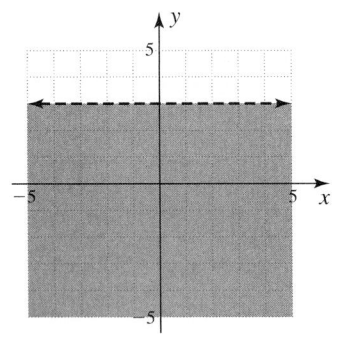

21. $|x| < 1$

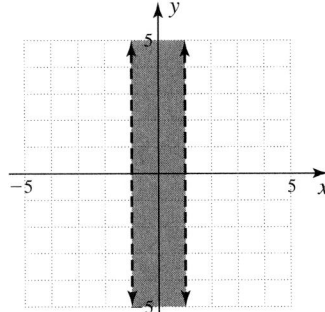

23. $|y| < 4$

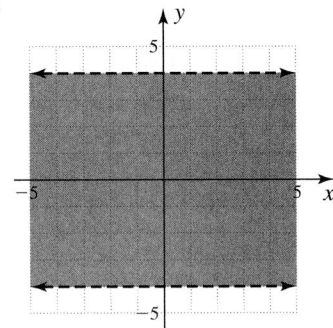

25. $|x| \geq 1$

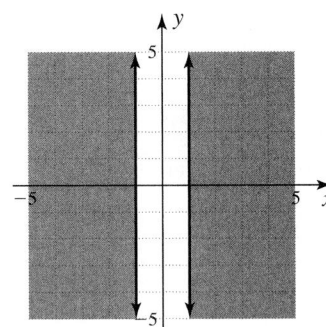

27. $|y| \geq 2$

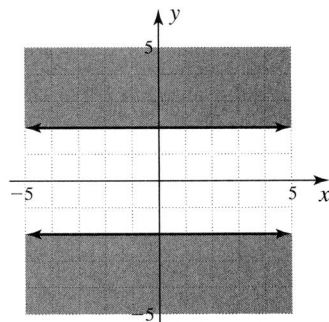

29. $|x + 2| < 1$

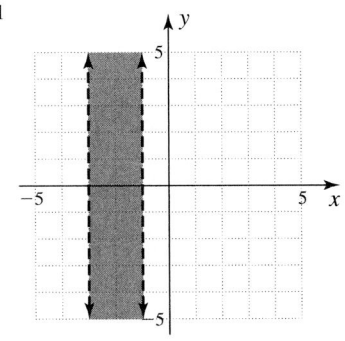

31. $|y + 2| < 1$

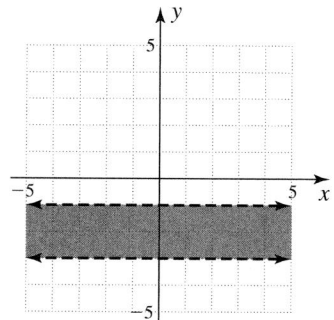

33. $|x + 1| \geq 3$

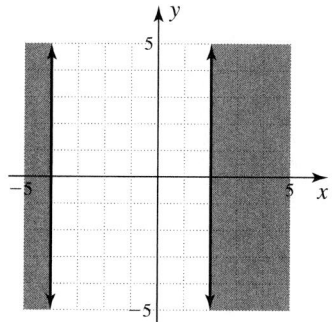

35. $|x - 1| \leq 2$

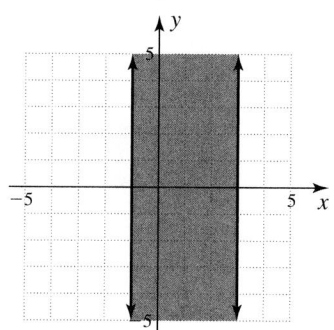

37. $|y - 2| < 1$

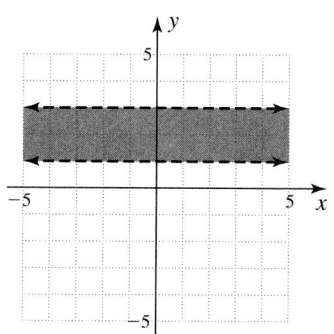

39. 4   **41.** 180   **43. (a)** 33 mi; **(b)** Rental A
45. If you plan to drive more than 33 mi, Rental A is cheaper.
51. $|x - 1| < 4$

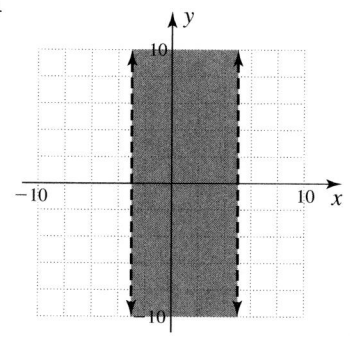

53. $|y| > 3$

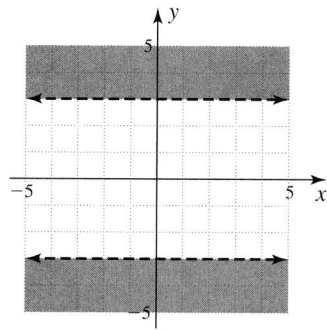

55. $x < 3$

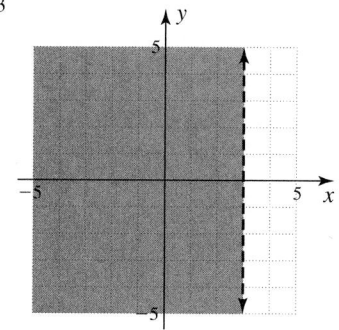

57. $-2x - 3y \geq -6$

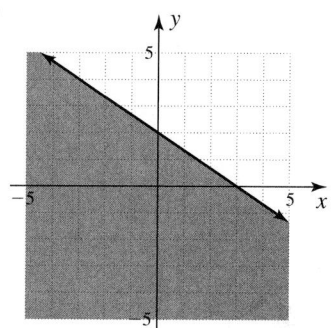

Review Exercises
1. [3.1A] **(a)**

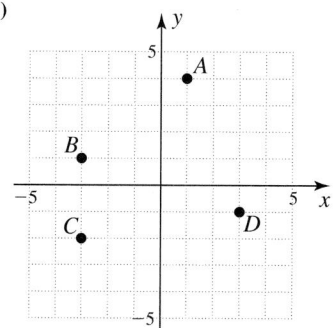

(b)

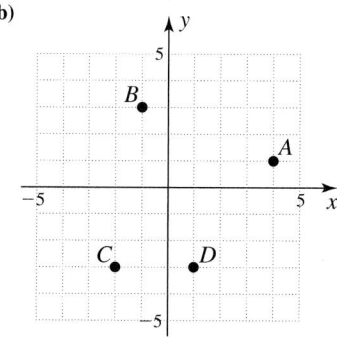

2. [3.1A] $A(2, 0)$; $B(-1, 1)$; $C(-3, -3)$; $D(4, -4)$
3. [3.1A] $A(2, 1)$; $B(0, 3)$; $C(-3, -1)$; $D(3, -3)$
4. [3.1B] **(a)** $x + 2y = 4$

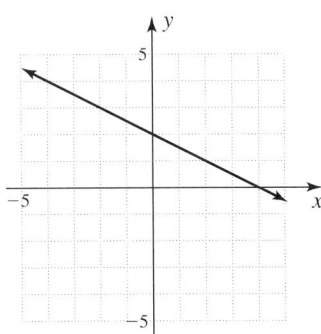

(b) $2x - y = 2$

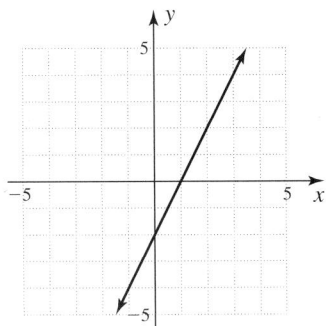

5. [3.1C] **(a)** $(-1, 0)$; $(0, 3)$

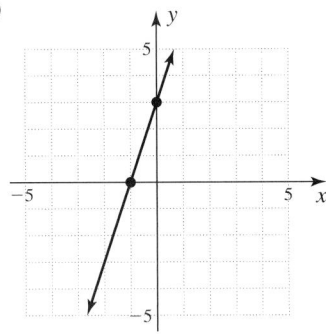

(b) $(2, 0)$; $(0, -4)$

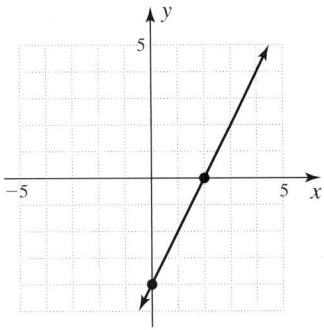

6. [3.1D] **(a–b)**

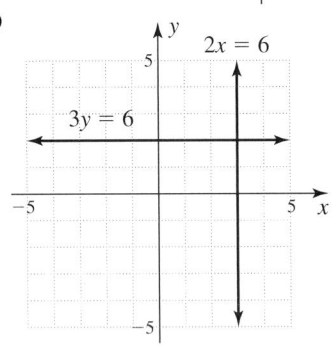

7. [3.1D] **(a–b)**

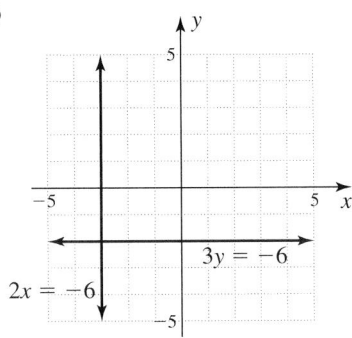

8. [3.2A] **(a)** $D = \{0, 2, 3, 5\}$; $R = \{5, 8, 9, 10\}$; **(b)** $D = \{0, 2, 3, 5\}$; $R = \{6, 9, 10, 11\}$
9. [3.2A] **(a)** $D = \{x \mid x \text{ is a real number}\}$;
    $R = \{y \mid y \text{ is a real number}\}$

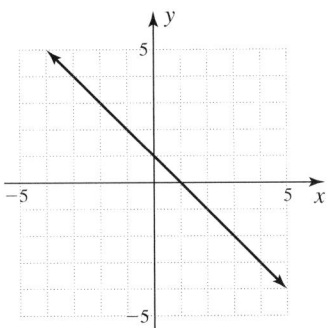

(b) $D = \{x \mid x \text{ is a real number}\}$;
    $R = \{y \mid y \text{ is a real number}\}$

10. [3.2A] **(a)** $D = \{x \mid -2 \le x \le 2\}$; $R = \{y \mid -2 \le y \le 2\}$;
(b) $D = \{x \mid -3 \le x \le 3\}$; $R = \{y \mid -3 \le y \le 3\}$
11. [3.2A] **(a)** $D = \{x \mid -2 \le x \le 2\}$; $R = \{y \mid 0 \le y \le 2\}$;
(b) $D = \{x \mid -3 \le x \le 3\}$; $R = \{y \mid 0 \le y \le 3\}$
12. [3.2A] **(a)** $D = \{x \mid -2 \le x \le 2\}$; $R = \{y \mid -2 \le y \le 0\}$;
(b) $D = \{x \mid -3 \le x \le 3\}$; $R = \{y \mid -3 \le y \le 0\}$
13. [3.2B] **(a)** Yes; **(b)** No   **14.** [3.2C] **(a)** $D = \{x \mid x \text{ is a real number}$
and $x \ne 1\}$; **(b)** $D = \{x \mid x \text{ is a real number and } x \ne 2\}$
15. [3.2C] **(a)** $D = \{x \mid x \ge 3\}$; **(b)** $D = \{x \mid x \ge 4\}$ **16.** [3.2D] **(a)** -2;
(b) -3; **(c)** 1   **17.** [3.2D] **(a)** 1; **(b)** -2; **(c)** 3   **18.** [3.2D] **(a)** 0; **(b)** -1;
(c) 1   **19.** [3.2D] **(a)** 1; **(b)** 0; **(c)** 1   **20.** [3.3A] **(a)** $m = \dfrac{-1}{2}$;
(b) Undefined   **21.** [3.3A] **(a)** $m = -1$; **(b)** $m = \dfrac{-3}{2}$
22. [3.3B] **(a)** Perpendicular; **(b)** Parallel   **23.** [3.3B] **(a)** Parallel;
(b) Neither   **24.** [3.3B] **(a)** $y = 2$; **(b)** $y = 5\dfrac{1}{2}$ or $\dfrac{11}{2}$

25. [3.3C] **(a)**

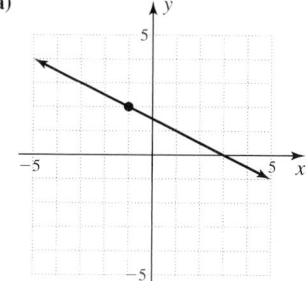

(b)

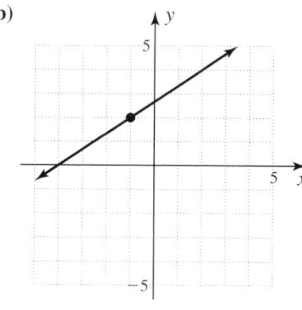

26. [3.3D] **(a)** $m = 6$; y-intercept $= -2$;

(b) $m = \dfrac{1}{2}$; y-intercept $= -2$ **27.** [3.4A] **(a)** $x - y = -3$;

(b) $5x + 2y = -14$ **28.** [3.4B] **(a)** $2x - y = 1$; **(b)** $2x - y = 10$

29. [3.4C] **(a)** $y = 3x + 2$; **(b)** $y = -3x + 4$

30. [3.4D] **(a)** $y = -2x + 5$; **(b)** $y = 3x - 5$

31. [3.4D] **(a)** $y = \dfrac{3}{2}x - 2$; **(b)** $y = \dfrac{-2}{3}x + \dfrac{7}{3}$

32. [3.4E] **(a)** $y = 7$; **(b)** $x = -3$

33. [3.4F] $y = \dfrac{3}{5}x + \dfrac{6}{5}$

(Answers may vary.)

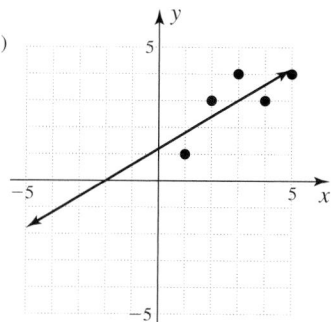

34. [3.4G] $y = \dfrac{-1}{3}x + 1$

35. [3.5A] **(a)** $2x - y < 1$

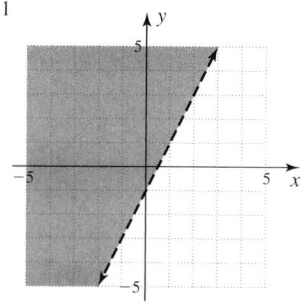

(b) $y \geq x - 3$

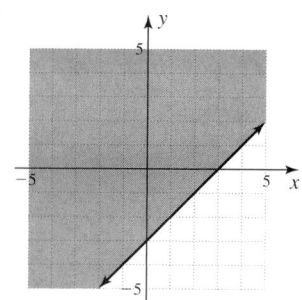

(c) $x \geq 4$

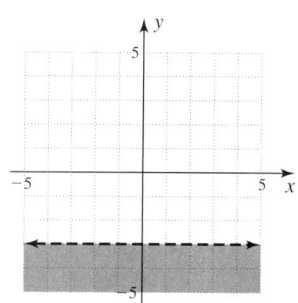

(d) $2y < -6$

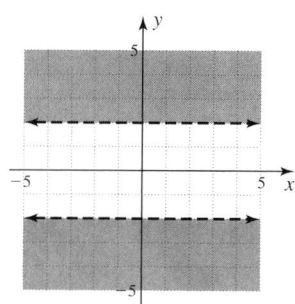

36. [3.5B] **(a)** $|y| > 2$

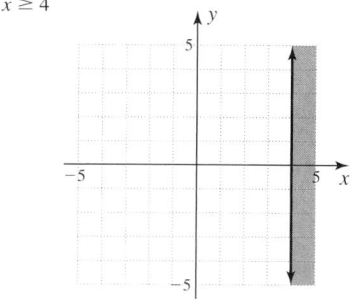

(b) $|x + 2| \leq 3$

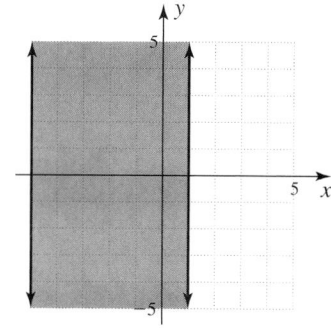

Cumulative Review Chapters 1–3

1. Irrational numbers; Real numbers

2.

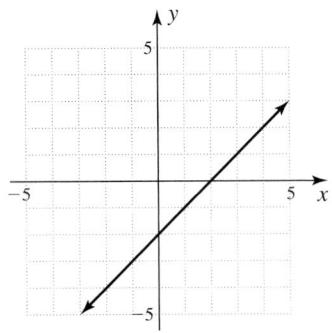

3. $\dfrac{2}{7}$ **4.** $\dfrac{8x^{12}}{y^{12}}$ **5.** -336 **6.** 14   **7.** 7

8.

9.

10. 26   **11.** 23; 25; 27   **12.** 240 mi   **13.** $\{-3, -2, -4\}$
14. $D = \{\text{all real numbers}\}; R = \{\text{all real numbers}\}$ **15.** $\{x \mid x \geq -25\}$
16. 1   **17.** 134

18. $x - y = 2$

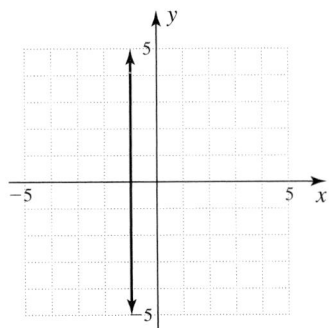

19. x-intercept: $\dfrac{2}{5}$; y-intercept: 2

20. $2x = -2$

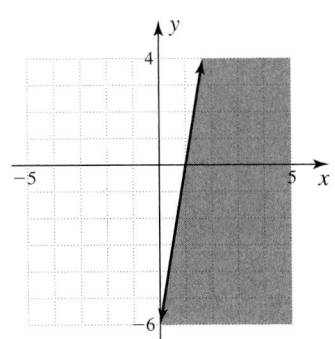

21. Perpendicular   **22.** $2x + y = -3$ **23.** $m = 2$; y-intercept: -12
24. $4x - 3y = -26$

25. $y \leq 6x - 6$

Chapter 4

Exercises 4.1

1. Solution: $(-2, -4)$; consistent
(1) $x - 2y = 6$; (2) $y = 2x$

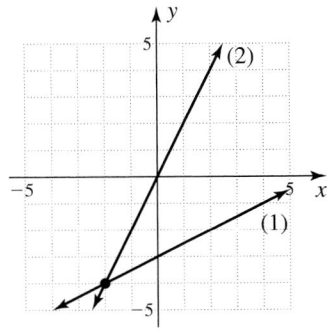

3. Solution: $(1, -2)$; consistent
(1) $y = x - 3$; (2) $y = 2x - 4$

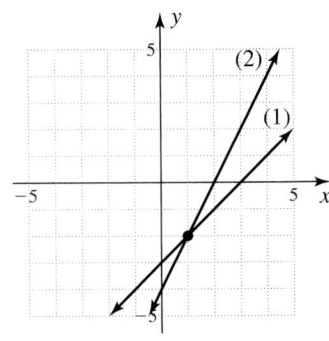

5. Solution: $(1\tfrac{1}{3}, 1\tfrac{1}{3})$; consistent
(1) $2y = -x + 4$; (2) $y = -2x + 4$

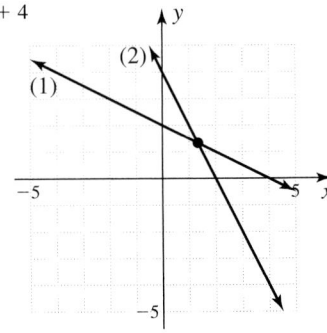

7. No solution; inconsistent
(1) $2x - y = -2$; (2) $y = 2x + 4$

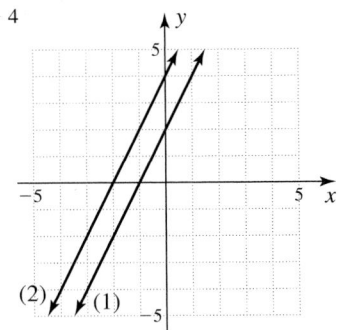

9. Infinitely many solutions; dependent
(1) $3x + 4y = 12$; (2) $8y = 24 - 6x$

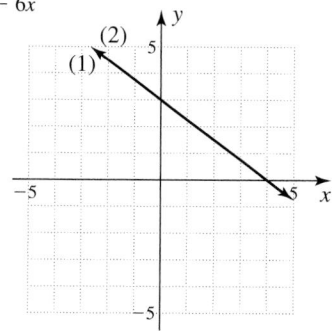

11. $(2, 0)$; consistent **13.** $(1, \frac{3}{2})$; consistent **15.** No solution; inconsistent **17.** No solution; inconsistent **19.** Infinitely many solutions; dependent **21.** $(-1, -1)$; consistent **23.** No solution; inconsistent **25.** $(4, 1)$; consistent **27.** $(5, 3)$; consistent **29.** $(0, \frac{1}{2})$; consistent **31.** $(0, 1)$; consistent **33.** $(2, 3)$; consistent **35.** $(\frac{5}{2}, -\frac{1}{2})$; consistent **37.** No solution; inconsistent **39.** $(3, 2)$; consistent **41.** Infinitely many solutions; dependent **43.** $(\frac{1}{3}, 2)$; consistent **45.** $(5, -2)$; consistent **47.** $(\frac{-1}{2}, \frac{-2}{3})$; consistent **49.** $(8, -12)$; consistent **51.** $(6, 8)$; consistent **53.** $(4, -3)$; consistent **55.** $(4, 2)$; consistent **57.** $p = 14$ **59.** $p = 200$ **61.** $p = 40$ **63.** $p = 28$; $D(28) = 512$ **65.** $p = 15$; $D(15) = S(15) = 435$ **67. (a)** $x + y = 90$; $y = x + 15$; **(b)** $x = 37.5$; $y = 52.5$ **69. (a)** $x + y = 180$; $y = 4x$; **(b)** $x = 36$; $y = 144$ **71. (a)** $x + y = 465$; $x = y + 15$; **(b)** 240 lb; 225 lb **73.** Japan: 2537; United States: 2100 **75.** 8.5 shekels **77.** 36 and 34 **79.** antenna: 222 ft; building: 1250 ft **81.** x: 3; y: 6 **83.** x: -4.5; y: 3 **85.** x: $\frac{-7}{2}$; y: $\frac{-7}{3}$ **91.** $(2, -1)$ **93.** Infinitely many solutions **95.** $(2, 1)$ **97.** Infinitely many solutions **99.** $(\frac{24}{5}, \frac{8}{5})$

101. No solution; inconsistent
(1) $2x + y = 4$; (2) $2y + 4x = 6$

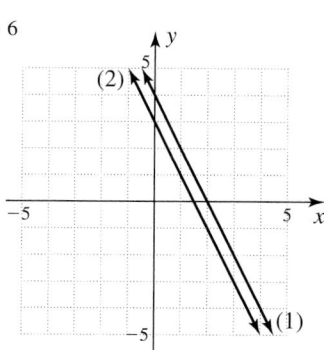

103. Infinitely many solutions; dependent
(1) $x + \frac{1}{2}y = -2$; (2) $y = -2x - 4$

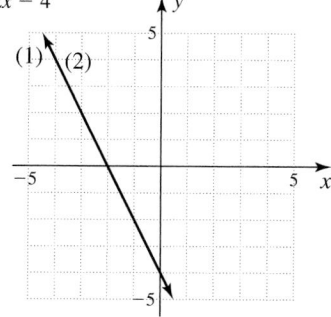

105. $(-1, 0)$; consistent
(1) $x - y = -1$; (2) $y = -x - 1$

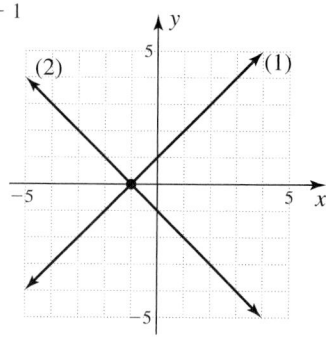

Exercises 4.2

1. $(5, 3, 4)$; consistent **3.** $(-1, 1, 4)$; consistent **5.** $(3, 4, 1)$; consistent **7.** No solution; inconsistent **9.** $(\frac{1}{2}, \frac{1}{4}, \frac{1}{3})$; consistent **11.** No solution; inconsistent **13.** No solution; inconsistent **15.** $(\frac{9}{2}, \frac{5}{2}, \frac{5}{2})$; consistent **17.** $(6, 3, -1)$; consistent **19.** $(-2, -3, -4)$; consistent **21.** $a = 1$; $b = 2$; $c = -3$ **23.** $\{(-1, 0, 2)\}$ **25.** $k = -2$; $(-1, 0, 2)$ **27.** 8; 16; 25 **29.** $30°$; $50°$; $100°$ **31.** Corns: 70; heel pain: 41; ingrown toenails: 39 **33.** Pizza Hut: 52%; Domino's: 12%; Papa John's: 8% **35.** Gates: \$6.3 billion; Kluge: \$5.5 billion; Walton family: \$5.1 billion **37.** \$12,000 at 5%; \$8000 at 7% **39.** 32 gal **41.** 250 mi/hr **43.** $d = \frac{5n + 16}{216}$ **51.** $(2, -1, 3)$; consistent **53.** No solution; inconsistent **55.** Infinitely many solutions; dependent

Exercises 4.3

1. 50 dimes; 25 nickels **3.** 10 nickels; 20 quarters **5.** 5 pennies; 5 nickels **7.** 20 tens; 5 twenties **9.** 13 nickels; 9 dimes; 22 quarters **11.** 43 and 59 **13.** 21 and 105 **15.** 4 and 20 **17.** Longs Peak: 14,255 ft; Pikes Peak: 14,110 ft **19.** Butterscotch: $2033\frac{1}{3}$ lb; caramel: $2033\frac{1}{3}$ lb; chocolate: $2633\frac{1}{3}$ lb **21.** Plane: 210 mi/hr; wind: 30 mi/hr **23.** Current: 3 mi/hr; boat: 12 mi/hr **25.** Wind: 50 mi/hr; plane: 450 mi/hr **27.** \$6000 at 8%; \$4000 at 6% **29.** \$10,000 at 6%; \$5000 at 8%; \$10,000 at 10% **31.** $L = 505$ ft; $W = 255$ ft **33.** $L = 134$ ft; $W = 85$ ft **35.** $(1, -2, -1)$ **37.** $15a + 5b + 20c = 170$ **39.** $10a + 15b + 10c = 110$ **43.** $L = 50$ cm; $W = 35$ cm **45.** Wind: 50 mi/hr; plane: 350 mi/hr **47.** 723 lb and 743 lb

Exercises 4.4

1. $(1, 2, 0)$ **3.** $(-1, -1, 3)$ **5.** $(-2, -1, 3)$ **7.** No solution **9.** $(1 - k, 2, k)$, k is any real number **11.** 30 dimes; 60 quarters; 40 one-dollar coins **13.** Type I: 8; type II: 10; type III: 12 **15.** 50% type I; 25% type II; 25% type III **17. (a)** Singular; **(b)** Nonsingular; **(c)** Nonsingular **19.** -150 **21.** -473 **23.** $\frac{1}{8}$ **29.** $(6, -3, 4)$ **31.** No solution **33.** $(1, k, k)$, k is any real number **35.** Dimes: 22; nickels: 8; quarters: 46

Exercises 4.5

1. 2 **3.** 7 **5.** 6 **7.** $\frac{1}{2}$ **9.** $\frac{-7}{40}$ **11.** $(2, 3)$ **13.** $(4, 5)$ **15.** $(3, -1)$ **17.** $(4, 5)$ **19.** $(x, \frac{-2x - 13}{3})$; dependent **21.** $(-2, -3)$ **23.** No solution; inconsistent **25.** $(10, 1)$ **27.** $(5, 2)$ **29.** $(-1, -1)$ **31.** -7 **33.** 0 **35.** -1 **37.** -4 **39.** -9 **41.** $(1, 2, 3)$ **43.** $(3, -1, -2)$ **45.** $(3, 0, 4)$ **47.** $(-5, 1, 5)$ **49.** $(-6, 2, 5)$

51. $\begin{vmatrix} a & b & 0 \\ c & d & 0 \\ e & f & 0 \end{vmatrix} = a\begin{vmatrix} d & 0 \\ f & 0 \end{vmatrix} - b\begin{vmatrix} c & 0 \\ e & 0 \end{vmatrix} + 0\begin{vmatrix} c & d \\ e & f \end{vmatrix} = a(0) - b(0) + 0 = 0$

53. $\begin{vmatrix} a & b & c \\ 1 & 2 & 3 \\ a & b & c \end{vmatrix} = -1\begin{vmatrix} b & c \\ b & c \end{vmatrix} + 2\begin{vmatrix} a & c \\ a & c \end{vmatrix} - 3\begin{vmatrix} a & b \\ a & b \end{vmatrix}$

$= -1(bc - bc) + 2(ac - ac) - 3(ab - ab)$

$= -1(0) + 2(0) - 3(0) = -0 + 0 - 0 = 0$

55. $\begin{vmatrix} 1 & 2 & 3 \\ 3 & 1 & 2 \\ 3k & 2k & k \end{vmatrix} = 3k\begin{vmatrix} 2 & 3 \\ 1 & 2 \end{vmatrix} - 2k\begin{vmatrix} 1 & 3 \\ 3 & 2 \end{vmatrix} + k\begin{vmatrix} 1 & 2 \\ 3 & 1 \end{vmatrix}$

$= 3k(1) - 2k(-7) + k(-5) = 3k + 14k - 5k$

$k\begin{vmatrix} 1 & 2 & 3 \\ 3 & 1 & 2 \\ 3 & 2 & 1 \end{vmatrix} = k\left[3\begin{vmatrix} 2 & 3 \\ 1 & 2 \end{vmatrix} - 2\begin{vmatrix} 1 & 3 \\ 3 & 2 \end{vmatrix} + 1\begin{vmatrix} 1 & 2 \\ 3 & 1 \end{vmatrix} \right]$

$= k[3(1) - 2(-7) + 1(-5)]$
$= k(3 + 14 - 5)$
$= 3k + 14k - 5k$

$\therefore \begin{vmatrix} 1 & 2 & 3 \\ 3 & 1 & 2 \\ 3k & 2k & k \end{vmatrix} = k\begin{vmatrix} 1 & 2 & 3 \\ 3 & 1 & 2 \\ 3 & 2 & 1 \end{vmatrix}$

57. $\begin{vmatrix} kb_1 & b_1 & 1 \\ kb_2 & b_2 & 2 \\ kb_3 & b_3 & 3 \end{vmatrix} = kb_1\begin{vmatrix} b_2 & 2 \\ b_3 & 3 \end{vmatrix} - kb_2\begin{vmatrix} b_1 & 1 \\ b_3 & 3 \end{vmatrix} + kb_3\begin{vmatrix} b_1 & 1 \\ b_2 & 2 \end{vmatrix}$

$= kb_1(3b_2 - 2b_3) - kb_2(3b_1 - b_3) + kb_3(2b_1 - b_2)$
$= 3kb_1b_2 - 2kb_1b_3 - 3kb_1b_2 + kb_2b_3 + 2kb_1b_3 - kb_2b_3$
$= 3kb_1b_2 - 3kb_1b_2 - 2kb_1b_3 + 2kb_1b_3 + kb_2b_3 - kb_2b_3$
$= 0$

59. $\begin{vmatrix} 1 & 1 & 1 \\ 2 & a & a \\ 3 & b & b \end{vmatrix} = 1\begin{vmatrix} a & a \\ b & b \end{vmatrix} - 2\begin{vmatrix} 1 & 1 \\ b & b \end{vmatrix} + 3\begin{vmatrix} 1 & 1 \\ a & a \end{vmatrix}$

$= 1(ab - ab) - 2(b - b) + 3(a - a)$
$= 1(0) - 2(0) + 3(0) = 0$

61. No   **63.** Yes   **65.** No   **67.** $2x - y + 3 = 0$

69. $2x + 9y - 34 = 0$ **71.** $bx + ay - ab = 0$

79. $2\begin{vmatrix} 1 & 0 \\ -1 & 1 \end{vmatrix} - (-1)\begin{vmatrix} 1 & 0 \\ 0 & 1 \end{vmatrix} + (-3)\begin{vmatrix} 1 & 1 \\ 0 & -1 \end{vmatrix} = 2(1) + 1(1) - 3(-1)$
$= 2 + 1 + 3$
$= 6$

81. No solution   **83.** -19 **85.** 8

Exercises 4.6

1. $x - y \geq 2; x + y \leq 6$

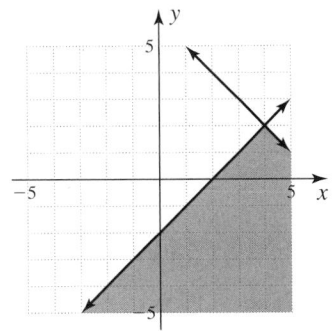

3. $2x - 3y \leq 6; 4x - 3y \geq 12$

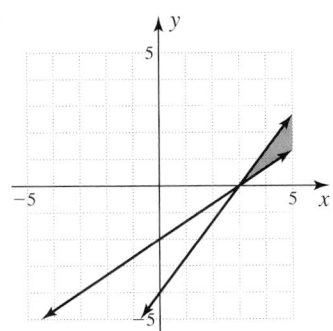

5. $2x - 3y \leq 5; x \geq y$

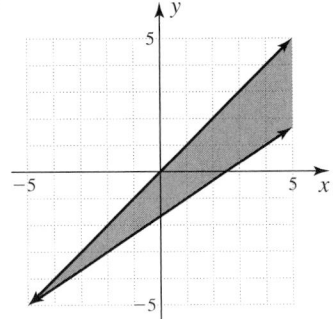

7. $x + 3y \leq 6; x \geq y$

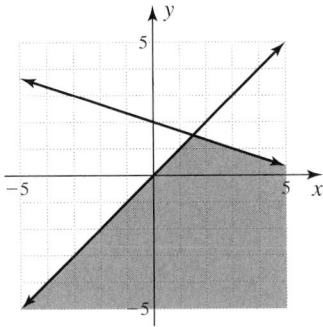

9. $x - y \leq 1; 3x - y < 3$

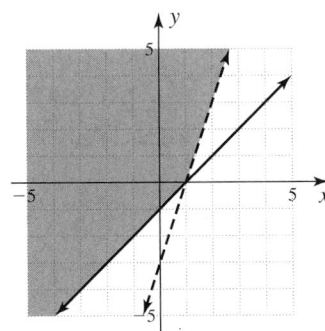

11. $x \geq 1; x \leq 4; y \leq 4; x - 3y \leq -2$

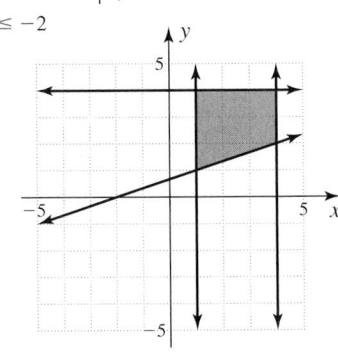

13. $x + y \geq 1$; $2y - x \leq 1$; $x \leq 1$

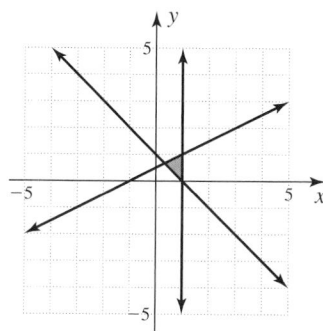

15. $x \geq 1$; $y \geq 2$; $4 \leq 2x + y$; $2x + y \leq 6$

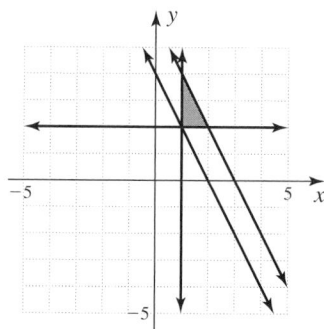

17. $y = \frac{1}{2}x - 3$ **19.** $y = x - 5$ **21.** 80 cars; 20 trucks

25. $y > 2x + 1$; $x \leq -1$

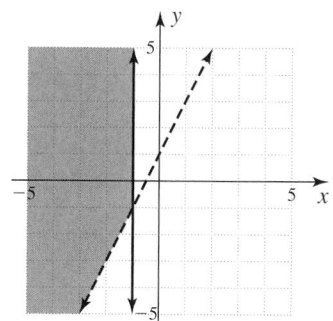

27. $x + y \leq 5$; $2x - y \leq 4$; $x \geq 0$; $y \geq 0$

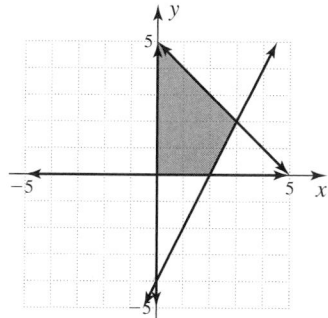

Review Exercises

1. [4.1A]
(a) (1) $2x - y = 2$; (2) $y = 3x - 4$

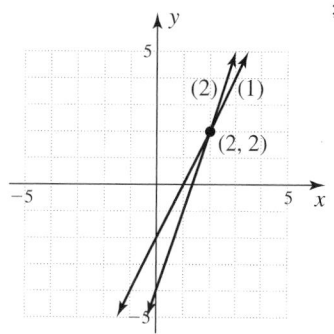

(b) (1) $x - 2y = 0$; (2) $y = x - 2$

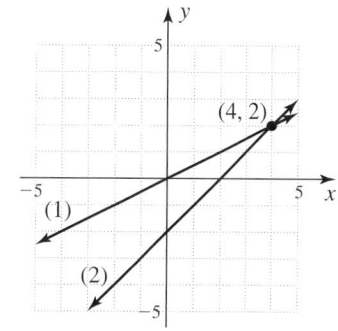

2. [4.1A]
(a) (1) $2y - x = 3$; (2) $4y = 2x + 8$

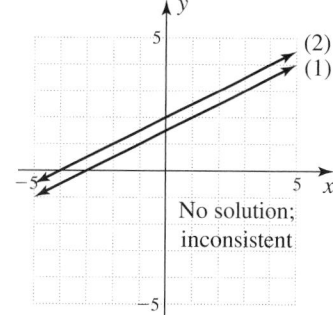

No solution; inconsistent

(b) (1) $3y + x = 5$; (2) $2x = 8 - 6y$

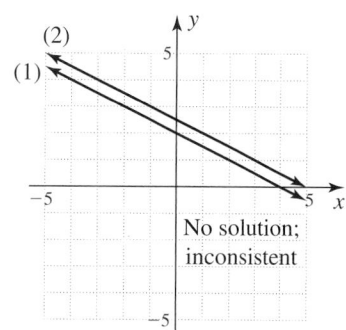

No solution; inconsistent

3. [4.1A]
(a) (1) $3x + 2y = 6$; (2) $y = 3 - \frac{3}{2}x$

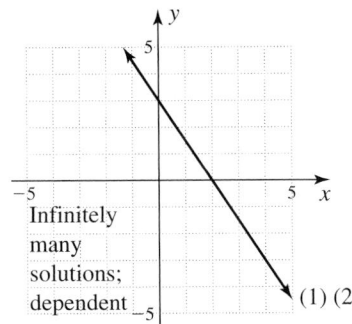

Infinitely many solutions; dependent

(b) (1) $x + 2y = 4$; (2) $2x = 8 - 4y$

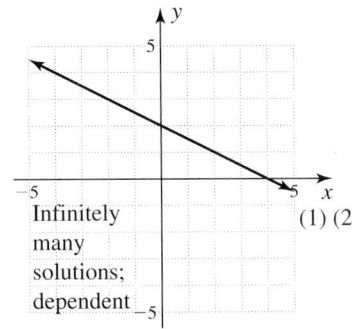

Infinitely many solutions; dependent

4. [4.1B] **(a)** $(3, 2)$; **(b)** $(\frac{7}{5}, \frac{12}{5})$ **5.** [4.1B] **(a)** No solution;
(b) No solution   **6.** [4.1B] **(a)** $(x, \frac{1}{2}x + \frac{5}{2})$; infinitely many solutions;
(b) $(x, 1 - \frac{1}{5}x)$; infinitely many solutions   **7.** [4.1C] **(a)** $(4, -1)$;
(b) $(-1, 2)$ **8.** [4.1C] **(a)** No solution; **(b)** No solution
9. [4.1C] **(a)** $(x, \frac{2 - 2x}{5})$; infinitely many solutions; **(b)** $(x, \frac{4}{3}x - 8)$;
infinitely many solutions   **10.** [4.1C] **(a)** $(2, -3)$; **(b)** $(1, 1)$
11. [4.2A] **(a)** $(4, 2, -1)$; **(b)** $(7, 2, -12)$ **12.** [4.2B] **(a)** No solution;
(b) No solution   **13.** [4.2B] **(a)** Infinitely many solutions;
$(2 - k, 2 - k, k)$, k any real number; **(b)** Infinitely many solutions;
$(4k - 8, 6 - 2k, k)$, k any real number   **14.** [4.3A] **(a)** 30 nickels;
25 dimes; **(b)** 10 nickels; 15 dimes   **15.** [4.3B] **(a)** 180 ft; **(b)** 160 ft
16. [4.3C] **(a)** 6 mi/hr; **(b)** 3 mi/hr   **17.** [4.3D] **(a)** \$10,000 at 4%;
\$10,000 at 6%; \$20,000 at 8%; **(b)** \$10,000 at 4%; \$15,000 at 6%;
\$20,000 at 8%   **18.** [4.3E] **(a)** 10 in. by 40 in.; **(b)** 10 in. by 30 in.
19. [4.4B] **(a)** $(3, -2, 5)$; **(b)** $(-4, 0, 1)$ **20.** [4.4B] **(a)** No solution;
(b) $(1, 2, -1)$ **21.** [4.5A] **(a)** -22; **(b)** 14   **22.** [4.5B] **(a)** $(1, -2)$;
$D = -23$; $D_x = -23$; $D_y = 46$; **(b)** $(\frac{1}{2}, -\frac{1}{2})$; $D = -28$; $D_x = -14$;
$D_y = 14$ **23.** [4.5C] **(a)** 15; **(b)** -18

24. [4.5C] **(a)** $1 \begin{vmatrix} 3 & 1 \\ 3 & 2 \end{vmatrix} + 1 \begin{vmatrix} 2 & 1 \\ 1 & 2 \end{vmatrix} + 1 \begin{vmatrix} 2 & 3 \\ 1 & 3 \end{vmatrix} = 1(3) + 1(3) + 1(3)$
$= 3 + 3 + 3 = 9$;

(b) $4 \begin{vmatrix} 5 & -2 \\ -2 & 2 \end{vmatrix} + 2 \begin{vmatrix} 2 & -2 \\ 1 & 2 \end{vmatrix} - 1 \begin{vmatrix} 2 & 5 \\ 1 & -2 \end{vmatrix} = 4(6) + 2(6) - 1(-9)$
$= 24 + 12 + 9 = 45$

25. [4.5C] **(a)** $0 \begin{vmatrix} 3 & 1 \\ 5 & -1 \end{vmatrix} + 2 \begin{vmatrix} 1 & 5 \\ 5 & -1 \end{vmatrix} - 3 \begin{vmatrix} 1 & 5 \\ 3 & 1 \end{vmatrix}$
$= 0 + 2(-26) - 3(-14)$
$= 0 - 52 + 42$
$= -10$;

(b) $-2 \begin{vmatrix} 0 & -2 \\ 3 & -2 \end{vmatrix} + 4 \begin{vmatrix} 1 & 1 \\ 3 & -2 \end{vmatrix} - 6 \begin{vmatrix} 1 & 1 \\ 0 & -2 \end{vmatrix} = -2(6) + 4(-5) - 6(-2)$
$= -12 - 20 + 12 = -20$

26. [4.5C] **(a)** $0 \begin{vmatrix} 0 & 1 \\ 2 & 4 \end{vmatrix} + 2 \begin{vmatrix} 1 & 3 \\ 2 & 4 \end{vmatrix} + 3 \begin{vmatrix} 1 & 3 \\ 0 & 1 \end{vmatrix} = 0 + 2(-2) + 3(1)$
$= 0 - 4 + 3 = -1$

(b) $5 \begin{vmatrix} 1 & 0 \\ 6 & 1 \end{vmatrix} + 2 \begin{vmatrix} 3 & 1 \\ 6 & 1 \end{vmatrix} + 3 \begin{vmatrix} 3 & 1 \\ 1 & 0 \end{vmatrix} = 5(1) + 2(-3) + 3(-1)$
$= 5 - 6 - 3 = -4$

27. [4.5D] **(a)** $(2, 3, -2)$; $D = -10$; $D_x = -20$; $D_y = -30$; $D_z = 20$;
(b) $(1, -2, -1)$; $D = 6$; $D_x = 6$; $D_y = -12$; $D_z = -6$
28. [4.6A, B]
(a) $y \geq x$; $x > -2$

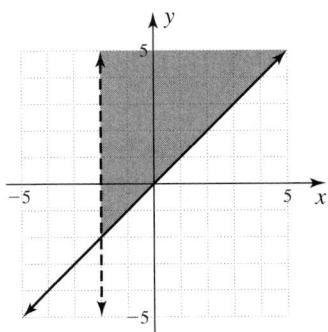

(b) $3x - 4y \geq -12$; $x < 1$; $y \geq 0$

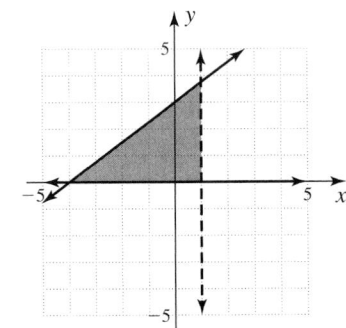

Cumulative Review Chapters 1–4

1. Irrational numbers; real numbers   **2.** $-2x^2 + 5x + 7$ **3.** $4x^{13}$
4. $\frac{9x^6}{y^8}$ **5.** -50 **6.** $x = \frac{6}{7}$ **7.** $P = 580$ **8.** $x = 7$
9.

10.

11.

12. 23   **13.** \$27,000   **14.** 42 gallons
15.

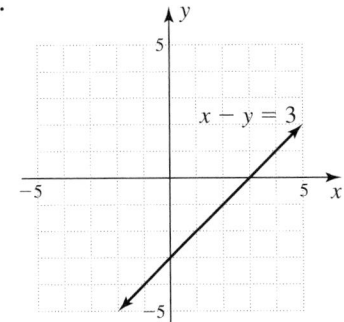

16. $\frac{15}{7}$

17.

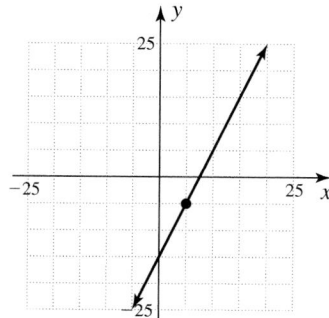

18. $m = -3$; y-intercept $= -18$ **19.** $4x + 3y = 30$

20. $y \le x + 1$

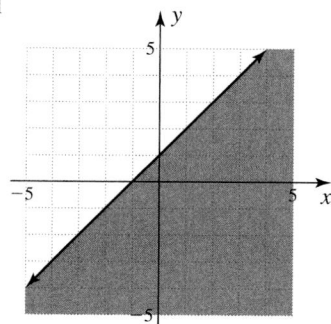

21. $|y| \le 4$

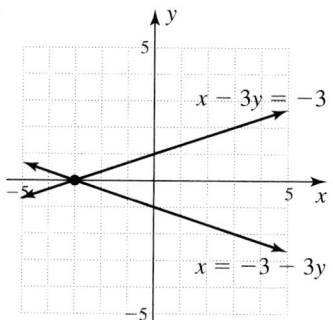

22. $D = \{\text{all real numbers}\}$; $R = \{\text{all real numbers}\}$ **23.** 4

24.   Solution: $(-3, 0)$

$x - 3y = -3$

$x = -3 - 3y$

25. Infinitely many solutions: $\{(x, y) \mid y = \frac{1}{2}x - 1\frac{1}{2}\}$ **26.** $(6, 15)$

27. No solution; inconsistent   **28.** 7   **29.** 3   **30.** 30 mi/hr

Chapter 5

Exercises 5.1

1. Monomial; 4   **3.** Binomial; 3   **5.** Trinomial; 3   **7.** Trinomial; 5
9. Zero polynomial; no degree   **11.** $-x^4 + 3x^2 + 5x + 7$
13. $-8x^2 + 9x + 20$ **15.** $-5x^2 + x + 2$ **17.** -8 **19.** 4   **21.** -1
23. 16   **25.** -136 **27.** (a) 0; (b) -9; (c) -9 **29.** (a) 8; (b) 2; (c) 6
31. $6x^2 - 5$ **33.** $x^2 - 2x - 4$ **35.** $3x^2 + 4x - 9$ **37.** $-11y^2 - y - 3$
39. $4x^3 - 12x^2 + 9x - 6$ **41.** $7y^2 + 12y - 11$ **43.** $3v^3 - 2v^2 + v - 7$
45. $-5u^3 - 5u^2 - 3u + 10$ **47.** $4x^3 + 2xy - 1$ **49.** $x^3 - x^2 - 4$
51. $-2a^2 + 5a$ **53.** $4y$ **55.** $3x^2 + 7y$ **57.** 2   **59.** 0   **61.** Commutative
property of addition   **63.** Distributive property   **65.** Associative
property of addition   **67.** Commutative property of addition
69. Distributive property   **71.** (a) 48 ft; (b) 64 ft   **73.** (a) \$42,500;
(b) \$70,000   **75.** (a) \$50,000; (b) \$0 (no value)   **77.** \$20   **79.** \$1000
81. \$30,000   **83.** $18x^3y^2$ **85.** $14x^4y^2$ **87.** $2x^2 + 6x$ **89.** $3x^2 + 4x$
91. $27x^2 + 10x$ **95.** (a) $P = -0.2x^2 + 47x - 100$; (b) \$2600
97. 4   **99.** (a) Binomial; (b) Polynomial; (c) Monomial; (d) Trinomial
101. -3 **103.** $-5x + 7$

Exercises 5.2

1. $12x^2 - 6x$ **3.** $-3x^3 + 9x^2$ **5.** $-24x^3 + 16x^2 - 8x$
7. $-18x^3y^2 - 9xy^4 + 21xy^2$ **9.** $6x^3y^6 - 10x^2y^5 + 2x^2y^4$
11. $x^3 + 4x^2 + 8x + 15$ **13.** $x^3 + 3x^2 - x + 12$
15. $x^3 + 2x^2 - 5x - 6$ **17.** $x^3 - 8$ **19.** $x^4 - x^3 + x^2 + x - 2$
21. $9x^2 + 9x + 2$ **23.** $5x^2 + 11x - 12$ **25.** $3a^2 + 14a - 5$
27. $2y^2 + 7y - 15$ **29.** $x^2 - 8x + 15$ **31.** $6x^2 - 7x + 2$
33. $4x^2 + 4ax - 15a^2$ **35.** $x^2 + 15x + 56$ **37.** $4a^2 + 10ab + 4b^2$
39. $16u^2 + 8uv + v^2$ **41.** $4y^2 + 4yz + z^2$ **43.** $9a^2 - 6ab + b^2$
45. $a^2 - b^2$ **47.** $25x^2 - 4y^2$ **49.** $b^2 - 9a^2$ **51.** $3x^3 + 9x^2 + 6x$
53. $-3x^3 + 12x^2 - 9x$ **55.** $x^3 + 6x^2 + 9x$ **57.** $-2x^3 + 4x^2 - 2x$
59. $4x^2y^2 - y^4$ **61.** $x^2 + \frac{3}{2}x + \frac{9}{16}$ **63.** $4y^2 - \frac{4}{5}y + \frac{1}{25}$
65. $\frac{9}{16}p^2 + \frac{3}{10}pq + \frac{1}{25}q^2$ **67.** $9x^2 + 24xy + 16y^2 + 6x + 8y + 1$
69. $9x^2 - 24xy + 16y^2 - 6x + 8y + 1$
71. $9x^2 + 12xy + 4y^2 - 6x - 4y + 1$
73. $16p^2 - 24pq + 9q^2 + 8p - 6q + 1$
75. (a) $R = 1000p - 30p^2$; (b) \$8000   **77.** $T_1^4 - T_2^4$
79. $Kt_n^2 - 2Kt_nt_a + Kt_a^2$ **81.** $5(x + y)$ **83.** $3a(b + c)$ **85.** (a) 9; (b) 5;
(c) No   **87.** (a) x^2; (b) xy; (c) y^2; (d) xy **89.** $(x + y)^2 = x^2 + 2xy + y^2$
95. $12x^2 + 17xy + 6y^2$ **97.** $25x^2 - 30xy + 9y^2$ **99.** $9x^2 - y^2$
101. $x^3 - 6x^2 + 5x + 6$ **103.** $20x^6 + 12x^5 - 8x^4 - 20x^3$
105. $9x^2 - 24xy + 16y^2 + 6x - 8y + 1$

Exercises 5.3

1. $8(x + 2)$ **3.** $9(y - 2)$ **5.** $-5(y - 5)$ **7.** $-8(x + 3)$ **9.** $4x(x + 9)$
11. $6x(1 - 7x^2)$ **13.** $-5x^2(1 + 7x^2)$ **15.** $3x(x^2 + 2x + 13)$
17. $9y(7y^2 - 2y + 3)$ **19.** $6x^2(6x^4 + 2x^3 - 3x^2 + 5)$
21. $8y^3(6y^5 + 2y^2 - 3y + 2)$ **23.** $\frac{1}{7}(4x^3 + 3x^2 - 9x + 3)$
25. $\frac{1}{8}y^2(7y^7 + 3y^4 - 5y^2 + 5)$ **27.** $(x + 2)(x^2 + 1)$
29. $(y - 3)(y^2 + 1)$ **31.** $(2x + 3)(2x^2 + 1)$ **33.** $(3x - 1)(2x^2 + 1)$
35. $(y + 2)(4y^2 + 1)$ **37.** $(2a^2 + 3)(a^4 + 1)$ **39.** $(x^2 + 4)(3x^3 + 1)$
41. $(2y^2 + 3)(3y^3 + 1)$ **43.** $y^2(y^2 + 3)(4y^3 + 1)$
45. $a^2(a^2 - 2)(3a^3 - 2)$ **47.** $a^2(2a - 3)(4a^2 - 5)$
49. $x^3(x - 2)(x^2 + 2)$ **51.** $(x - 4)(2x + 5)$ **53.** $\alpha L(t_2 - t_1)$
55. $(R - 1)(R - 1)$ or $(R - 1)^2$ **57.** $x^2 + 7x + 12$ **59.** $x^2 + 3x - 10$
61. $25x^2 + 20xy + 4y^2$ **63.** $25x^2 - 20xy + 4y^2$ **65.** $u^2 - 36$
67. $-w(l - z)$ **69.** $a(a + 2s)$ **71.** $-16(t^2 - 5t - 15)$
77. $x(2x^2 - 3)(3x^3 + 1)$ **79.** $(2x - 3)(3x^2 - 1)$
81. $3x^2(4x^5 + x^4 - 2x^3 + 9)$ **83.** $6(x - 8)$ **85.** $-\frac{1}{5}x^2(x^3 + 4x^2 - 2)$ or
$\frac{1}{5}x^2(2 - 4x^2 - x^3)$

Exercises 5.4

1. $(x + 2)(x + 3)$ **3.** $(a + 2)(a + 5)$ **5.** $(x + 4)(x - 3)$
7. $(x + 2)(x - 1)$ **9.** $(x + 1)(x - 2)$ **11.** $(x + 2)(x - 5)$
13. $(a - 7)(a - 9)$ **15.** $(y - 2)(y - 11)$ **17.** $(9x + 1)(x + 4)$
19. $(3a + 1)(a - 2)$ **21.** $(2y + 5)(y - 4)$ **23.** $(4x - 3)(x - 2)$
25. $(2x + 3)(3x - 4)$ **27.** $(3a + 2)(7a - 1)$ **29.** $(2x + 3y)(3x - y)$
31. $x^2(7x - 3y)(x - y)$ **33.** $y^3(3x + y)(5x - 2y)$
35. $xy(15x^2 - 2xy - 2y^2)$ **37.** $-(2b - 5)(b - 4)$
39. $-(4y - 3)(3y + 4)$ **41.** $(2y + 7)(y + 1)$ **43.** $(2x - 3)(x - 3)$
45. $-(a + 1)^4$ **47.** $(2g + 7)(g - 3)$ **49.** $(2R - 1)(R - 1)$
51. $4a^2 + 4ab + b^2$ **53.** $a^2 - 4ab + 4b^2$ **55.** $a^2 - b^2$ **57.** $4x^2 - 9y^2$
59. $(L - 3)(2L - 3)$ **61.** $(5t - 7)(t - 1)$ **65.** $(3x + 2)(2x + 3)$
67. $(x - 5)(x + 2)$ **69.** Not factorable   **71.** $2x^2y(2x - y)(3x + 2y)$
73. Not factorable   **75.** $-(4x - 7)(2x + 3)$ **77.** $(2y - 5)(y - 1)$

Exercises 5.5

1. $(x + 1)^2$ **3.** $(y + 11)^2$ **5.** $(1 + 2x)^2$ **7.** $(3x + 5y)^2$ **9.** $4(3a + 2)^2$
11. $(y - 1)^2$ **13.** $(7 - x)^2$ **15.** $(7a - 2x)^2$ **17.** $(4x - 3y)^2$
19. $(3x^2 + 2)^2$ **21.** $(4x^2 - 3)^2$ **23.** $(1 + x^2)^2$ **25.** $(y + 8)(y - 8)$
27. $(a + \frac{1}{3})(a - \frac{1}{3})$ **29.** $(8 + b)(8 - b)$ **31.** $(6a + 7b)(6a - 7b)$
33. $(\frac{x}{3} + \frac{y}{4})(\frac{x}{3} - \frac{y}{4})$ **35.** $(a + 2b + c)(a + 2b - c)$
37. $(2x - y + 1)(2x - y - 1)$ **39.** $(3y - 2x + 5)(3y - 2x - 5)$
41. $(4a + x + 3y)(4a - x - 3y)$ **43.** $(y + a - b)(y - a + b)$
45. $(x + 5)(x^2 - 5x + 25)$ **47.** $(1 + a)(1 - a + a^2)$
49. $(2x + y)(4x^2 - 2xy + y^2)$ **51.** $(x - 1)(x^2 + x + 1)$
53. $(5a - 2b)(25a^2 + 10ab + 4b^2)$
55. $(x + 2)(x - 2)(x^2 - 2x + 4)(x^2 + 2x + 4)$
57. $(x + \frac{1}{2})(x - \frac{1}{2})(x^2 - \frac{1}{2}x + \frac{1}{4})(x^2 + \frac{1}{2}x + \frac{1}{4})$
59. $(\frac{x}{2} + 1)(\frac{x}{2} - 1)(\frac{x^2}{4} - \frac{x}{2} + 1)(\frac{x^2}{4} + \frac{x}{2} + 1)$
61. $(x - y + 1)(x^2 - 2xy + y^2 - x + y + 1)$
63. $(1 + x + 2y)(1 - x - 2y + x^2 + 4xy + 4y^2)$
65. $(y - 2x - 1)(y^2 - 4xy + 4x^2 + y - 2x + 1)$
67. $(3 - x - 2y)(9 + 3x + 6y + x^2 + 4xy + 4y^2)$
69. $(4 + x^2 - y^2)(16 - 4x^2 + 4y^2 + x^4 - 2x^2y^2 + y^4)$
71. $x^2 - 2x - 15$ **73.** $x^2 - 6x - 16$ **75.** $4x^2 - 12xy + 9y^2$
77. $(10 + x)(10 - x)$ **79.** $(2x + 1)(4x^2 - 2x + 1)$
87. $(2x + 3y)(4x^2 - 6xy + 9y^2)$ **89.** $(x^2 + 1)(x^4 - x^2 + 1)$
91. $(4x^2 + 9y^2)(2x + 3y)(2x - 3y)$ **93.** $(x + 5y + 4)(x + 5y - 4)$
95. $(2x - 3y)^2$ **97.** $(x + 9)^2$ **99.** $4xy$
101. $(5 + x + y)(25 - 5x - 5y + x^2 + 2xy + y^2)$

Exercises 5.6

1. $3x^2(x + 2)(x - 3)$ **3.** $5x^2(x + 4y)(x - 2y)$ **5.** $-3x^4(x^2 + 2x + 7)$
7. $2x^4y(x^2 - 2xy - 5y^2)$ **9.** $-2x^4(2x^2 + 6xy + 9y^2)$
11. $2y^2(x + 2)(3x^2 + 1)$ **13.** $-3xy(x + 1)(3x^2 + 2)$
15. $-2x(x + y)(2x^2 - y)$ **17.** $3y^2(x + 4y)^2$ **19.** $-2k(3x + 2y)^2$
21. $4xy^2(2x - 3y)^2$ **23.** Not factorable   **25.** $3x^3(x + 2y)^2$
27. $2x^4(3x + y)^2$ **29.** $3x^2y^2(2x - 3y)^2$ **31.** $6(x + 2)(x + 1)(x - 1)$
33. $7(x^2 + y^2)(x + y)(x - y)$ **35.** $2x^2(x^2 + 4y^2)(x + 2y)(x - 2y)$
37. $-2(x + 3)^2$ **39.** $-3(x + 2)^2$ **41.** $-x^2(2x + y)^2$
43. $-y^2(3x + 2y)^2$ **45.** $-2y^2(2x - 3y)^2$ **47.** $-2x(3x + 2y)^2$
49. $-2x(3x + 5y)^2$ **51.** $-x(x + y)(x - y)$ **53.** $-x^2(x + 2y)(x - 2y)$
55. $-x^2(2x + 3y)(2x - 3y)$ **57.** $-2x(2x + 3y)(2x - 3y)$
59. $-2x^2(3x + 2y)(3x - 2y)$ **61.** $x^2(3 - x)(9 + 3x + x^2)$
63. $x^4(x - 2)(x^2 + 2x + 4)$ **65.** $x^4(3 + 2x)(9 - 6x + 4x^2)$
67. $x^4(3x + 4y)(9x^2 - 12xy + 16y^2)$ **69.** $(x + y + 2)(x - y + 2)$
71. Not factorable   **73.** $(x + y + 2)(x - y + 2)$ **75.** $-(3x - 5y)^2$
77. $2x(3x - 5y)^2$ **79.** $(2x + 1)(3x - 2)$ **81.** $(3x - 1)(4x + 1)$
83. $\frac{2\pi A}{360}$ $(R + Kt)$ **85.** $\frac{3S}{2bd^3}$ $(d + 2z)(d - 2z)$
91. $(x - 2)(3x + 1)(3x - 1)$ **93.** $x^2(2x^2 + xy + 2y^2)$
95. $(x + y - 5)(x - y - 5)$ **97.** $-(4y - 7)(2y + 3)$ **99.** $4x^3(3x + y)^2$
101. $8x^3(x^2 + 9)$

Exercises 5.7

1. $-1, -2$ **3.** $1, -4, -3$ **5.** $\frac{1}{2}, \frac{1}{3}$ **7.** $0, 3$ **9.** $8, -8$ **11.** $9, -9$
13. $0, -6$ **15.** $0, 3$ **17.** $3, 9$ **19.** $-1, -5$ **21.** $5, -3$
23. $-\frac{2}{3}, -1$ **25.** $1, \frac{1}{2}$ **27.** $1, -\frac{1}{2}$ **29.** $2, -6$ **31.** $1, \frac{1}{2}$ **33.** $-2, -4$
35. $4, 1$ **37.** $-2, -\frac{5}{2}$ **39.** 1 **41.** $2, -2, -4$ **43.** $5, 3, -3$
45. $2, -2, -1$ **47.** $6, 8, 10$ **49.** 10 in., 24 in., 26 in.   **51.** 1 sec
53. 1 sec   **55.** 40 **57.** 10 **59.** **(a)** $5.50 - x$; **(b)** $550 + 450x - 100x^2$;
(c) \$0.50 or \$4; **(d)** \$0.50   **61.** $\frac{1}{5x^2}$ **63.** $2x^{10}$ **65.** $\frac{2}{x^{12}}$
67. 4 sec $(3 + 1)$ **69.** 400 ft   **75.** $-\frac{1}{4}, \frac{1}{2}$ **77.** $1, 5$ **79.** $-4, 4$
81. 5 units, 12 units, 13 units

Review Exercises

1. [5.1A, B] **(a)** Binomial; degree 6; **(b)** Monomial; degree 8;
(c) Trinomial; degree 5   **2.** [5.1B] **(a)** $3x^4 - x^2 - 5x + 2$;
(b) $4x^3 - x^2 + 3x$; **(c)** $6x^2 + x - 2$ **3.** [5.1C, E] **(a)** \$20,000; **(b)** \$0
4. [5.1C] **(a)** 9; **(b)** 3; **(c)** 23   **5.** [5.1D] **(a)** $x^3 + 8x^2 - 10x + 7$;
(b) $5x^3 - 3x^2 - 3x + 8$; **(c)** $3x^3 + 5x^2 - 7x + 6$
6. [5.2D] **(a)** $-3x^3 + 7x^2 - 3x + 3$; **(b)** $10x^2 + x - 6$;
(c) $-7x^3 + 13x^2 - 12x + 4$ **7.** [5.2A] **(a)** $-2x^4y - 6x^3y^2 + 4x^2y^4$;
(b) $-3x^4y^2 - 9x^3y^3 + 6x^2y^5$; **(c)** $-4x^3y^2 - 12x^2y^3 + 8xy^5$
8. [5.2B] **(a)** $x^3 - 4x^2 + x + 2$; **(b)** $x^3 - 5x^2 + 4x + 4$;
(c) $x^3 - 11x - 6$ **9.** [5.2C] **(a)** $8x^2 + 2xy - 15y^2$;
(b) $6x^2 + 13xy + 6y^2$; **(c)** $10x^2 + 9xy - 9y^2$
10. [5.2D] **(a)** $4x^2 + 20xy + 25y^2$; **(b)** $9x^2 + 42xy + 49y^2$;
(c) $16x^2 + 72xy + 81y^2$ **11.** [5.2D] **(a)** $9x^2 - 12xy + 4y^2$;
(b) $16x^2 - 56xy + 49y^2$; **(c)** $25x^2 - 60xy + 36y^2$
12. [5.2E] **(a)** $9x^2 - 4y^2$; **(b)** $16x^2 - 9y^2$; **(c)** $25x^2 - 9y^2$
13. [5.3A] **(a)** $5x^2(3x^3 - 4x^2 + 2x + 5)$; **(b)** $3x^2(3x^3 - 4x^2 + 2x + 5)$;
(c) $2x^2(3x^3 - 4x^2 + 2x + 5)$ **14.** [5.3B] **(a)** $x(3x^2 - 1)(2x^3 + 5)$;
(b) $x(3x^2 - 4)(2x^3 + 5)$; **(c)** $x(3x^2 - 2)(2x^3 + 3)$
15. [5.4A] **(a)** $(x - 6y)(x + 3y)$; **(b)** $(x - 6y)(x + 2y)$; **(c)** $(x - 6y)(x + y)$
16. [5.4B] **(a)** $(2x + 5y)(x - 6y)$; **(b)** $(2x + 5y)(x - 4y)$;
(c) $(2x + 5y)(x - 5y)$ **17.** [5.4C] **(a)** $-3x^2y(3x + 2y)(2x - y)$;
(b) $-5x^2y(3x + 2y)(2x + y)$; **(c)** $6x^2y(3x + 2y)(2x - 3y)$
18. [5.5A] **(a)** $(2x - 7y)^2$; **(b)** $(3x - 7y)^2$; **(c)** $(4x - 7y)^2$
19. [5.5A] **(a)** $(3x + 4y)^2$; **(b)** $(3x + 5y)^2$; **(c)** $9(x + 2y)^2$
20. [5.5B] **(a)** $(9x^2 + y^2)(3x + y)(3x - y)$;
(b) $(x^2 + 4y^2)(x + 2y)(x - 2y)$; **(c)** $(9x^2 + 4y^2)(3x + 2y)(3x - 2y)$
21. [5.5B] **(a)** $(x + y - 2)(x - y - 2)$; **(b)** $(x + y - 3)(x - y - 3)$;
(c) $(x + y + 4)(x - y + 4)$ **22.** [5.5C] **(a)** $(3x + 2y)(9x^2 - 6xy + 4y^2)$;
(b) $(3x + 4y)(9x^2 - 12xy + 16y^2)$; **(c)** $(4x + 3y)(16x^2 - 12xy + 9y^2)$
23. [5.5C] **(a)** $(3x - 2y)(9x^2 + 6xy + 4y^2)$;
(b) $(3x - 4y)(9x^2 + 12xy + 16y^2)$; **(c)** $(4x - 3y)(16x^2 + 12xy + 9y^2)$
24. [5.6A] **(a)** $x^3(3x - 2y)(9x^2 + 6xy + 4y^2)$;
(b) $x^4(3x - 4y)(9x^2 + 12xy + 16y^2)$;
(c) $x^5(4x - 3y)(16x^2 + 12xy + 9y^2)$
25. [5.6A] **(a)** $3x^4(9x^2 + 1)$; **(b)** $4x^4(x^2 + 16)$; **(c)** $2x^4(x + 3)(x - 3)$
26. [5.6A] **(a)** $3x^2(3x + 2y)^2$; **(b)** $4x^2(3x - 2y)^2$; **(c)** $5x^2(3x + y)^2$
27. [5.6A] **(a)** $3x^2(3x - y)^2$; **(b)** $4x^2(3x - 2y)^2$; **(c)** $5x^2(3x + 2y)^2$
28. [5.6A] **(a)** $4xy(3x + y)(x - 4y)$; **(b)** $5xy(3x + y)(x + 4y)$;
(c) $6xy(3x + 2y)(x - 4y)$ **29.** [5.6A] **(a)** $(2x - 1)(x + 1)(x - 1)$;
(b) $(2x - 1)(3x + 1)(3x - 1)$; **(c)** $(2x - 1)(4x + 1)(4x - 1)$
30. [5.7A] **(a)** $3, -4$; **(b)** $4, -5$; **(c)** $4, -6$ **31.** [5.7A] **(a)** $\frac{1}{3}, -\frac{1}{2}$;
(b) $\frac{1}{4}, -\frac{1}{2}$; **(c)** $\frac{1}{5}, -\frac{1}{2}$ **32.** [5.7A] **(a)** $1, -1, -2$; **(b)** $1, -1, -4$;
(c) $3, -2, -3$ **33.** [5.7B] **(a)** 5 units, 12 units, 13 units;
(b) 9 units, 12 units, 15 units; **(c)** 12 units, 16 units, 20 units

Cumulative Review Chapters 1–5

1. $\{2, 4, 6\}$ **2.** $\frac{19}{100}$ **3.** 15 **4.** 6×10^{-3} **5.** -66 **6.** 14
7. $\frac{3}{4}$; $-\frac{39}{4}$

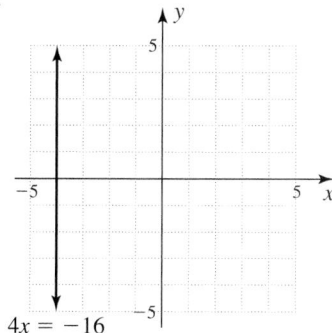

8. (number line)

9. (number line)

10. (number line)

11. $A = \dfrac{5B + 22}{2}$ **12.** 495 mi   **13.** x-intercept $= -\dfrac{5}{3}$; y-intercept $= -5$

14. (graph) $4x = -16$

15. Parallel   **16.** $5x - 4y = 19$ **17.** $y = -3x - 1$
18. $x \geq 3$

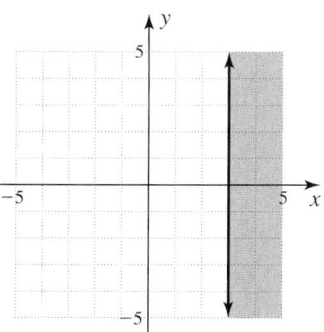

19. $(-3, -1)$ **20.** $(2, 1, 1)$ **21.** 60   **22.** $z = 0$ **23.** 33 nickels;
21 dimes   **24.** Binomial   **25.** 31   **26.** $-x^3 + 7x^2 - 7x + 14$
27. $4x^2 - 28xy + 49y^2$ **28.** $(25x^2 + y^2)(5x + y)(5x - y)$
29. $(2c + 5)(4c^2 - 10c + 25)$ **30.** $-\dfrac{3}{2}; -3$

Chapter 6

Exercises 6.1

1. $x = -3$ **3.** $m = -2$ **5.** $m = -1; 2$ **7.** $p = -3; 3$
9. $a = \dfrac{-1}{2}; 6$ **11.** None, defined for all values of v **13.** $\dfrac{4xy^2}{6y^3}$

15. $\dfrac{x(x - y)}{x^2 - y^2}$ or $\dfrac{x^2 - xy}{x^2 - y^2}$ **17.** $\dfrac{-x(y + x)}{y^2 - x^2}$ or $\dfrac{-xy - x^2}{y^2 - x^2}$

19. $\dfrac{-x(2x + 3y)}{4x^2 - 9y^2}$ or $\dfrac{-2x^2 - 3xy}{4x^2 - 9y^2}$ **21.** $\dfrac{4x(x - 2)}{x^2 - x - 2}$ or $\dfrac{4x^2 - 8x}{x^2 - x - 2}$

23. $\dfrac{-5x(x - 2)}{x^2 + x - 6}$ or $\dfrac{-5x^2 + 10x}{x^2 + x - 6}$ **25.** $\dfrac{3(x^2 - xy + y^2)}{x^3 + y^3}$ or

$\dfrac{3x^2 - 3xy + 3y^2}{x^3 + y^3}$ **27.** $\dfrac{x(x^2 + xy + y^2)}{x^3 - y^3}$ or $\dfrac{x^3 + x^2y + xy^2}{x^3 - y^3}$

29. $\dfrac{x(x + y)}{x^3 + y^3}$ or $\dfrac{x^2 + xy}{x^3 + y^3}$ **31.** $\dfrac{y}{2}$ **33.** $\dfrac{x}{5 - x}$ **35.** $\dfrac{-2x}{5y}$ **37.** $\dfrac{x + y}{x - y}$

39. $\dfrac{1}{x - 2}$ **41.** $\dfrac{x^3}{y^3}$ **43.** 3   **45.** $\dfrac{1}{3x + 2y}$ **47.** $\dfrac{(x - y)^2}{x + y}$ **49.** $y - 1$

51. $\dfrac{x + y}{x - y}$ **53.** $\dfrac{y - 5}{y + 6}$ **55.** -1 **57.** $-(x + 3)$ **59.** $-(y^2 + 2y + 4)$

61. -1 **63.** $-(x + 5)$ **65.** $2 - x$ **67.** $\dfrac{-1}{x + 6}$ **69.** $\dfrac{1}{x - 2}$ **71.** $2x^2$
73. $-2x^2$ **75.** $(x + 4)(x - 3)$ **77.** $(3x + 2y)(3x - 2y)$
79. $(x - 1)(x^2 + x + 1)$ **81.** $(2x + 1)(4x^2 - 2x + 1)$ **83.** $\dfrac{2}{3}$ **85.** $\dfrac{3}{2}$
87. $\dfrac{-3}{2}$ **89. (a)** \$525 million; **(b)** \$1400 million or \$1.4 billion;
(c) \$3150 million or \$3.15 billion; **(d)** No; Denominator $= 0$ for $p = 100$.
As p increases to 100, price increases without bound.
91. (a) $5(x - 1)$; **(b)** \$45; **(c)** \$495 (when $x = 100$)   **99.** x^4y^4
101. $\dfrac{-(x + y)}{x^2 + xy + y^2}$ **103.** $4 - x$ **105.** $x = -1; 1$ **107.** $\dfrac{4}{x}$ **109.** $\dfrac{14}{16}$
111. $\dfrac{4x^2 - 11x - 3}{x^2 - x - 6}$

Exercises 6.2

1. $\dfrac{3}{10}$ **3.** $\dfrac{2x}{3}$ **5.** $\dfrac{2y}{21x^2z^4}$ **7.** $\dfrac{4}{x + 1}$ **9.** 1   **11.** $\dfrac{-5}{x(x + y)}$ **13.** $\dfrac{3y - 2}{y - 1}$
15. 1   **17.** $1 - x$ **19.** $\dfrac{a + b}{a - b}$ **21.** $\dfrac{27}{50}$ **23.** $\dfrac{5x}{3}$ **25.** $\dfrac{9ad}{c}$ **27.** $\dfrac{3x}{x + 1}$
29. $\dfrac{4(y + 5)}{3(y + 2)}$ or $\dfrac{4y + 20}{3y + 6}$ **31.** $\dfrac{a + b}{a - b}$ **33.** $\dfrac{-(4a^2 + 2a + 1)}{2u^2w^2}$
35. $\dfrac{-(1 + x)}{5x}$ or $\dfrac{-1 - x}{5x}$ **37.** $\dfrac{2y + 3}{3y - 5}$ **39.** $\dfrac{2x - 3}{5 - 2x}$ **41.** $\dfrac{1}{x + 1}$
43. $\dfrac{x - 1}{x - 4}$ **45.** -1 **47.** $\dfrac{x}{x + y}$ **49.** $\dfrac{(x + 2y)(y + 2)}{(x - 3y)(y - 2)}$ or

$\dfrac{xy + 2x + 2y^2 + 4y}{xy - 2x - 3y^2 + 6y}$ **51.** $\dfrac{(x + 1)(x + 3)}{x^2 - x + 1}$ or $\dfrac{x^2 + 4x + 3}{x^2 - x + 1}$

53. $\dfrac{1}{x(x^2 + 2x + 4)}$ or $\dfrac{1}{x^3 + 2x^2 + 4x}$ **55.** $\dfrac{450}{x}$ **57.** $\dfrac{4t(t + 3)}{t^2 + 9}$ or

$\dfrac{4t^2 + 12t}{t^2 + 9}$ **59.** $\dfrac{x^2 + x - 2}{x^2 - x - 6}$ **61.** $\dfrac{x - 2}{x^3 - 8}$ **63.** $\dfrac{RR_T}{R - R_T}$

65. $R = \dfrac{60,000 + 9000x}{x}$ **67.** $\dfrac{2w^2 - Lw}{6}$ **71.** $\dfrac{-(x^2 - 3x + 9)}{x^2 + 2x + 4}$

73. $\dfrac{3(x - 1)}{x + 2}$ or $\dfrac{3x - 3}{x + 2}$ **75.** -1 **77.** $\dfrac{x - 4}{x^2 - x + 1}$

79. $\dfrac{x(x - 3)}{3(3x + 2y)}$ or $\dfrac{x^2 - 3x}{9x + 6y}$

Exercises 6.3

1. $\dfrac{3x}{5}$ **3.** $\dfrac{5x}{3}$ **5.** $\dfrac{3 + 2x}{5(x + 2)}$ **7.** $\dfrac{x + 2}{2(x + 1)}$ **9.** $\dfrac{2x + 5}{3(x - 1)}$
11. $\dfrac{2x^2 - 5x}{(x - 1)(x + 4)(x - 4)}$ **13.** $\dfrac{5x^2 + 19x}{(x + 5)(x - 2)(x + 3)}$
15. $\dfrac{6x - 4y}{(x + y)^2(x - y)}$ **17.** $\dfrac{10 - x}{(x - 5)(x + 5)}$ **19.** $\dfrac{-6x - 10}{(x + 1)(x + 2)(x + 3)}$
21. $\dfrac{2x - 5}{(x - 2)(x - 3)}$ **23.** $\dfrac{2}{x + 3}$ **25.** $\dfrac{2a + 6}{(a + 2)(a + 4)}$ **27.** $\dfrac{2}{a + 4}$
29. $\dfrac{15 - a^2}{(a + 5)(a + 3)}$ **31.** $\dfrac{y^2 + 2y}{(y + 1)(y - 1)}$ **33.** $\dfrac{-2y^2 - 4y + 5}{(y + 4)(y - 4)}$
35. $\dfrac{50x^2 + 8y^2}{(5x - 2y)(5x + 2y)}$ **37.** $\dfrac{5x^2}{(2x - y)(3x + y)}$ **39.** $\dfrac{2x^2 + x + 5}{(x - 2)(x + 1)^2}$
41. $\dfrac{4 - 3x}{(x - 2)(x - 4)(x - 1)}$ **43.** $\dfrac{-45}{(x + 5)(x - 5)}$ **45.** $\dfrac{x}{(x - y)(2 - x)}$
47. $\dfrac{18b^2 + 12ab + 12a - 18b}{(2a + 3b)(2a - 3b)}$ **49.** $\dfrac{1}{x^2 - 5x + 25}$
51. $\dfrac{-w_0x^3 + 3w_0L^2x - 2w_0L^3}{6L}$ **53.** $\dfrac{p^2 - 2gm^2rM}{2mr^2}$ **55.** 20
57. $24x + 18y$ **59.** $x^2 - 1$ **61.** $P(x + h) = x^2 + 2xh + h^2$
63. $\dfrac{P(x + h) - P(x)}{h} = 2x + h$ **69.** $\dfrac{-(5x + 9)}{(x + 1)(x - 2)(x + 2)}$
71. $\dfrac{1}{x + 3}$ **73.** $\dfrac{2x}{x - 2}$ **75.** $\dfrac{1}{x + 3}$

Exercises 6.4

1. $\dfrac{178}{33}$ **3.** $\dfrac{a}{c}$ **5.** $\dfrac{z}{xy}$ **7.** $\dfrac{2z}{5y}$ **9.** $\dfrac{1}{3}$ **11.** $\dfrac{ab-a}{b+a}$ **13.** $\dfrac{1+2x}{2x-1}$ **15.** $\dfrac{4+6x}{6x-3}$

17. $\dfrac{x}{xy-2}$ **19.** $\dfrac{x-y}{x^2y^2}$ **21.** $\dfrac{9}{5}$ **23.** $\dfrac{2a^2-a}{2a+1}$ **25.** $\dfrac{x^2}{2x-1}$ **27.** $\dfrac{-(x^2+1)}{2x}$

29. $\dfrac{x}{y}$ **31.** -1 **33.** $\dfrac{1-x}{1+x}$ **35.** $\dfrac{y+7}{y-2}$ **37.** $\dfrac{x^2-1}{5x^2-4x-2}$ **39.** $\dfrac{c+d}{c-d}$

41. $\dfrac{ab}{a^2+b^2}$ **43.** $R=\dfrac{R_1R_2}{R_2+R_1}$ **45.** $f=f_{\text{static}}\sqrt{\dfrac{c+v}{c-v}}$ **47.** $4x$ **49.** $-5x^2$

51. $\dfrac{-3}{x^2}$ **53.** $5x^3+5x^2$ **55.** $9x^5-15x^4$ **57.** $(3x-1)(2x+3)$

59. x^2+2x-8 **61.** $\text{APR}=\dfrac{288(NM-P)}{N(12P+NM)}$ **63.** $\dfrac{6}{25}$ yr **65.** $11\dfrac{43}{50}$ yr

71. $\dfrac{a^2+6a+12}{2(a+3)}$ **73.** $\dfrac{x(x-2)}{x-3}$ **75.** $\dfrac{4(2a-3b)}{2a+3b}$ **77.** $\dfrac{(x+1)(x^2+1)}{x(x^2+x+1)}$

79. -1

Exercises 6.5

1. x^2+3x-2 **3.** $-2x^2+x-3$ **5.** $-2y^2+8y-3$

7. $5x^2+4x-8+\dfrac{3}{x}$ **9.** $3xy-2+\dfrac{3}{xy}$ **11.** $x+3$ **13.** $y+5$

15. x^2-x-1 **17.** x^2+5x+6 **19.** x^2-2x-8

21. (x^2+4x+3) R 1 **23.** y^2-y-1 **25.** $(4x^2+3x+7)$ R 12

27. x^2-2x+4 **29.** $4y^2+8y+16$ **31.** a^2-2a-1

33. x^3+2x^2-x **35.** $(2x^3-3x^2+x-4)$ R -1

37. $(x+1)(x-3)(x-2)$ **39.** $(x-2)(x-2)(x+1)(x-1)$

41. $(x^2-3x+7)(x+5)(x+4)$ **43.** v^2+3v+1

45. (x^2+6x+5) R 15 **47.** z^2-6z+4

49. $(3y^3+12y^2+7y+15)$ R 52 **51.** $2y^3-y^2+5$

53. Rem. $=0$; 4 is a solution **55.** Rem. $=0$; -4 is a solution

57. Rem. $=0$; 5 is a solution **59.** Rem. $=0$; -1 is a solution

61. $\dfrac{500}{x}+4$ **63.** $x=4$ **65.** $x=3$ **67.** $x+1, x+2, x+3, x+6$

73. x^2 R 3 **75.** Rem. $=0$; 2 is a solution

77. $(x+3)(x-3)(2x-1)$ **79.** $(6x^2+12x+27)$ R 45

81. $4x^3-3x^2+2x$

Exercises 6.6

1. 6 **3.** -5 **5.** $\dfrac{1}{3}$ **7.** -4 **9.** $\dfrac{26}{9}$ **11.** $\dfrac{-1}{12}$ **13.** No solution

15. No solution **17.** -11 **19.** 7 **21.** 2 **23.** -3

25. $\dfrac{-3}{2}$ **27.** No solution **29.** 0 **31.** $\dfrac{4}{5}$ **33.** 2 **35.** -4 **37.** -4

39. $\dfrac{26}{9}$ **41.** $\dfrac{-1}{12}$ **43.** No solution **45.** No solution

47. $4\dfrac{4}{9}$ hr **49. (a)** Approximately $16\dfrac{1}{2}$ hr; **(b)** Approximately $2\dfrac{1}{2}$ days

51. Approximately 3 consecutive hits **53.** 21, 23, 25 **55.** 60 gallons

57. $h=\dfrac{2A}{b_1+b_2}$ **59.** $Q_1=\dfrac{PQ_2}{1+P}$ **61.** $f=\dfrac{ab}{a+b}$

67. $x=-7$ **69.** $x=3$ **71.** $x=-4$ and $x=1$ **73.** $x=5$

Exercises 6.7

1. 8 **3.** 5 and 10 **5.** 6 and 8 **7.** $\dfrac{2}{7}$ **9.** \$20,000 **11.** $1\dfrac{7}{8}$ hr

13. $4\dfrac{14}{19}$ hr **15.** 6 hr; 3 hr **17.** $15\dfrac{3}{4}$ hr **19.** $5\dfrac{1}{4}$ hr **21.** 36 sec

23. 18 hr **25.** 150 mi/hr **27.** 30 mi/hr **29.** Auto 25 mi/hr;

plane 125 mi/hr **31.** $\dfrac{1}{x^8}$ **33.** x^{20} **35.** $-8x^3y^6$ **37.** $\dfrac{1}{x^2}$ **39.** $\dfrac{1}{a^8b^6}$

41. $F=\dfrac{f_1f_2}{f_1+f_2}$ **43.** $R=\dfrac{2E-2ri}{i}$ **45.** 12 tablets a day

47. $\cos(2u)=2\cos^2 u-1$ **51.** 36 mi/hr **53.** 4 and 6

Exercises 6.8

1. $T=ks$ **3.** $W=kh^3$ **5.** $W=kB$ **7.** $R=\dfrac{k}{D^2}$ **9.** $I=kPr$

11. $A=ksv^2$ **13.** $V=kdw$ **15.** $I=\dfrac{ki}{d^2}$ **17.** $R=\dfrac{kL}{A}$

19. $W=\dfrac{k}{d^2}$ **21. (a)** $I=km$; **(b)** $k=0.055$ or 5.5%; **(c)** \$41.25

23. (a) $d=ks^2$; **(b)** $k=0.06$; **(c)** 216 ft **25. (a)** $S=\dfrac{k}{y}$; **(b)** 30 new songs

27. (a) $W=\dfrac{k}{d^2}$; **(b)** $k=121(3960)^2$; **(c)** 81 lb **29. (a)** $d=ks$;

(b) $k=17.63$; **(c)** The number of hours needed to travel d distance at s speed. **31. (a)** $C=4(F-37)$; **(b)** 212 chirps **33. (a)** $I=kn$;

(b) $k=1.365$; **(c)** 367.675 ppm **35.** \$208.33 **37.** $C=102.9$

39. $2x-y=2$

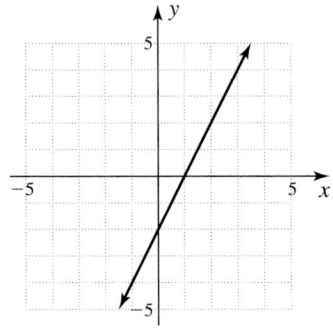

41. $y=-x-3$

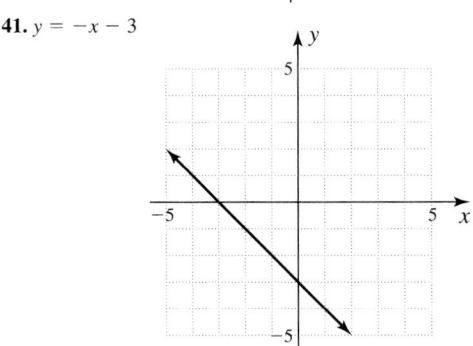

43.

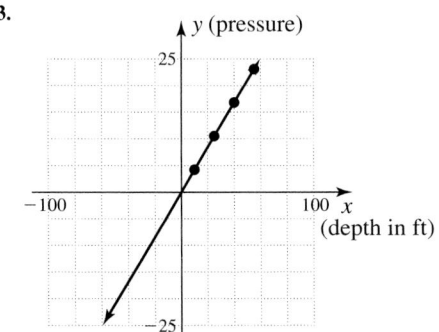

45. $p=kd$ **47.** They are equal. **53. (a)** $F=kAV^2$; **(b)** $k=0.0045$;

(c) 32.4 lb **55. (a)** $P=\dfrac{k}{r}$; **(b)** $k=10$

Review Exercises

1. [6.1A] **(a)** $x=4$; **(b)** $x=-1$; -9; **(c)** None, defined for all values

2. [6.1B] **(a)** $\dfrac{8x^2y^3}{36y^7}$; **(b)** $\dfrac{10x^2y^4}{45y^8}$; **(c)** $\dfrac{2x^2+11x+5}{x^2+6x+5}$; **(d)** $\dfrac{2x^2+13x+6}{x^2+7x+6}$

3. [6.1C] **(a)** $\dfrac{6}{y}$; **(b)** $\dfrac{7}{y}$; **(c)** $\dfrac{-8}{y}$ **4.** [6.1C] **(a)** $\dfrac{y-x}{6}$; **(b)** $\dfrac{7}{y-x}$; **(c)** $\dfrac{y-x}{8}$

5. [6.1D] **(a)** x^3y^5; **(b)** x^3y^6; **(c)** x^3y^7 **6.** [6.1D] **(a)** $\dfrac{y^2}{x-y}$; **(b)** $\dfrac{y^3}{x-y}$;

(c) $\dfrac{y^4}{x-y}$ **7.** [6.1D] **(a)** $\dfrac{2y-x}{x^2-2xy+4y^2}$; **(b)** $\dfrac{-(x+2y)}{x^2+2xy+4y^2}$;

(c) $\dfrac{-(x+3y)}{x^2+3xy+9y^2}$ **8.** [6.2A] **(a)** $\dfrac{3x+2y}{3x+1}$; **(b)** $\dfrac{3x+2y}{4x-3}$; **(c)** $\dfrac{3x+2y}{5x+2}$

9. [6.2B] **(a)** $\dfrac{1}{(x-2)(x+4)}$; **(b)** $\dfrac{1}{(x-2)(x+5)}$; **(c)** $\dfrac{1}{(x-2)(x+6)}$

10. [6.2C] **(a)** $\dfrac{-(x^2-3x+9)}{x^2+2x+4}$; **(b)** $\dfrac{-(x^2-3x+9)}{x^2+4x+16}$; **(c)** $\dfrac{-(x^2-5x+25)}{x^2+2x+4}$

11. [6.3A] **(a)** $\dfrac{1}{x-2}$; **(b)** $\dfrac{1}{x-3}$; **(c)** $\dfrac{1}{x-4}$ **12.** [6.3A] **(a)** $\dfrac{1}{x+3}$;

(b) $\dfrac{1}{x+4}$; **(c)** $\dfrac{1}{x+5}$ **13.** [6.3B] **(a)** $\dfrac{2x^2+9x+11}{(x-1)(x+2)(x+1)}$;

(b) $\dfrac{2x^2+10x+13}{(x-1)(x+2)(x+1)}$; **(c)** $\dfrac{2x^2+11x+15}{(x-1)(x+2)(x+1)}$

14. [6.3B] **(a)** $\dfrac{-4x-14}{(x-3)(x+2)(x+3)}$; **(b)** $\dfrac{-3x-11}{(x-3)(x+2)(x+3)}$;

(c) $\dfrac{-x-5}{(x-3)(x+2)(x+3)}$ **15.** [6.4A] **(a)** $\dfrac{x(x^2-x+1)}{(x^2+1)(x-1)}$;

(b) $\dfrac{x(x^2-x+1)}{(x^2+1)(x-1)}$; **(c)** $\dfrac{x(x^2-x+1)}{(x^2+1)(x-1)}$ **16.** [6.4A] **(a)** $\dfrac{a^2+20a+80}{4a+20}$;

(b) $\dfrac{a^2+30a+150}{5a+30}$; **(c)** $\dfrac{a^2+42a+252}{6a+42}$ **17.** [6.5A] **(a)** $3x^3-2x+1$;

(b) $3x^2-2+\dfrac{1}{x}$; **(c)** $3x-\dfrac{2}{x}+\dfrac{1}{x^2}$ **18.** [6.5B] **(a)** (x^2-x-1) R -6;

(b) (x^2-x-1) R -7; **(c)** (x^2-x-1) R -8

19. [6.5C] **(a)** $(x-1)(x-2)(x-3)$; **(b)** $(x-2)(x-1)(x-3)$;

(c) $(x-3)(x-1)(x-2)$ **20.** [6.5D] **(a)** $(x^3+9x^2+26x+24)$ R 4;

(b) $(x^3+8x^2+19x+12)$ R 4; **(c)** $(x^3+7x^2+14x+8)$ R 4

21. [6.5E] **(a)** Rem. $=0$, so -1 is a solution; **(b)** Rem. $=0$, so -2 is a solution; **(c)** Rem. $=0$, so -3 is a solution   **22.** [6.6A] **(a)** No solution; **(b)** $x=-9$; **(c)** $x=-11$ **23.** [6.6B] **(a)** 32 ft by 48 ft; **(b)** Approximately 7 or 8   **24.** [6.7A] **(a)** 10 and 12; **(b)** 12 and 14;

(c) 14 and 16   **25.** [6.7B] **(a)** $2\dfrac{2}{9}$ hr; **(b)** $2\dfrac{2}{5}$ hr; **(c)** $2\dfrac{6}{11}$ hr

26. [6.7C] **(a)** 225 mi/hr; **(b)** 275 mi/hr; **(c)** 350 mi/hr

27. [6.7D] **(a)** $a=2A-2b-3c$; **(b)** $b=\dfrac{2A-a-3c}{2}$ or

$b=A-\dfrac{a}{2}-\dfrac{3c}{2}$; **(c)** $c=\dfrac{2A-a-2b}{3}$ or $c=\dfrac{2A}{3}-\dfrac{a}{3}-\dfrac{2b}{3}$

28. [6.8A] **(a)** $P=kT$; **(b)** $k=\dfrac{1}{120}$ **29.** [6.8B] **(a)** $P=\dfrac{k}{V}$;

(b) $k=3200$ **30.** [6.8C] $g=kxt^2$

Cumulative Review Chapters 1–6

1. -14 **2.** Associative law of multiplication   **3.** $-3x^2+15$ **4.** -1

5. $x\geq-4$

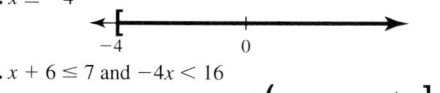

6. $x+6\leq7$ and $-4x<16$

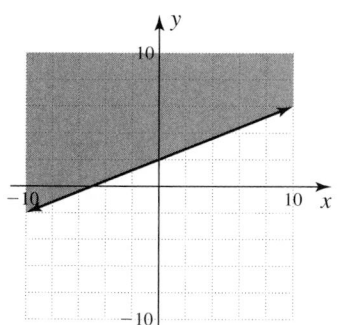

7. 23   **8.** 29, 31, 33   **9.** $\dfrac{-1}{3}$ **10.** $5x+y=6$ **11.** $m=2$; $y=-12$
12. $2x-5y\leq-10$

13. $|x+4|>2$

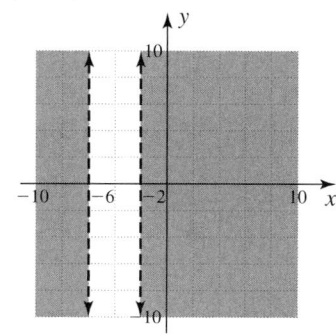

14. $y+x=-1$
$2y=-2x-4$ No solution

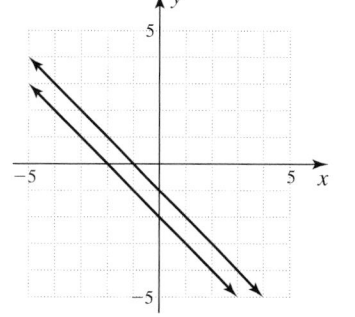

15. No solution   **16.** $(2,3)$ **17.** 24   **18.** No solution   **19.** 62   **20.** 21 ft
21. $D=\{\text{all real numbers}\}$; $R=\{\text{all real numbers}\}$ **22.** $-5x^3-x^2-9x$
23. $6n^2-13n+5$ **24.** $(3t+2)(4t-5)$

25. -3; -4; 4   **26.** $\dfrac{-(x^2-3x+9)}{x^2+x+1}$ **27.** $\dfrac{2x^2+x-11}{(x-3)(x-2)(x+2)}$

28. -1 **29.** $1\dfrac{7}{8}$ hr   **30.** $\dfrac{1}{60}$

Chapter 7

Exercises 7.1

1. 2   **3.** 2   **5.** -2 **7.** $\dfrac{-1}{4}$ **9.** 2   **11.** 2   **13.** 3   **15.** Not a real number

17. 3   **19.** 3   **21.** $\dfrac{-1}{2}$ **23.** Not a real number   **25.** 9   **27.** 25   **29.** $\dfrac{1}{4}$

31. 16   **33.** 16   **35.** -16 **37.** $\dfrac{1}{16}$ **39.** 7   **41.** $x^{3/7}$ **43.** $\dfrac{1}{x^{5/9}}$ **45.** $x^{2/5}$

47. z **49.** x^2 **51.** $\dfrac{1}{z^2}$ **53.** $\dfrac{1}{b^{4/5}}$ **55.** $\dfrac{1}{a^8 b^9}$ **57.** $\dfrac{b^9}{a^{10}}$ **59.** $x^{16}y^{30}$

61. $x+x^{1/3}y^{1/2}$ **63.** $x^{1/2}y^{3/4}-y^{5/4}$ **65.** $\dfrac{1}{x}$ **67.** $\dfrac{x^2}{y}$ **69.** x^3y^{11}

71. $v=15$ m/sec   **73.** $v=28$ ft/sec   **75.** 3   **77.** 13   **79.** $\dfrac{2}{3}$ **81.** $\dfrac{16x^3y}{8x^3y^3}$

83. $\dfrac{48x^3y^2}{16x^4y^4}$ **85.** $\dfrac{20x}{32x^5}$ **87. (a)** 2; **(b)** 8   **89.** $\dfrac{-1}{2}$ **91.** $\dfrac{1}{2}$ **97.** $\dfrac{1}{x^5y^{12}}$

99. $\dfrac{y^{1/3}}{x^{1/2}}$ **101.** $x^{1/5}+x^{2/5}y^{1/4}$ **103.** $x^{8/15}$ **105.** $\dfrac{1}{x^{11/12}}$ **107.** $\dfrac{1}{y^{3/20}}$

109. $\dfrac{1}{27}$ **111.** $\dfrac{1}{36}$ **113.** $\dfrac{1}{16}$ **115.** 7   **117.** $\dfrac{1}{3}$ **119.** -5

Exercises 7.2

1. 5   **3.** -4 **5.** $|-x|=|x|$ **7.** $|x+6|$ **9.** $|3x-2|$ **11.** $4|xy|\sqrt{xy}$

13. $2x\sqrt[3]{5xy}$ **15.** $|xy|\sqrt[4]{xy^3}$ **17.** $-3a^2b^3\sqrt[5]{b^2}$ **19.** $\dfrac{\sqrt{13}}{7}$

21. $\dfrac{\sqrt{17}}{2|x|}$ **23.** $\dfrac{\sqrt[3]{3}}{4x}$ **25.** $\dfrac{\sqrt{6}}{3}$ **27.** $\dfrac{-\sqrt{14}}{7}$ **29.** $\dfrac{\sqrt{10a}}{2a}$ **31.** $\dfrac{\sqrt{10ab}}{8ab}$

33. $\dfrac{-\sqrt{6ab}}{2a^2b^2}$ **35.** xy **37.** $\dfrac{-\sqrt[3]{21}}{3}$ **39.** $\dfrac{\sqrt[3]{12x}}{4x}$ **41.** $\dfrac{\sqrt[4]{x}}{2x}$ **43.** $\dfrac{\sqrt[4]{24}}{2}$

45. $\sqrt[3]{3}$ **47.** $\sqrt{2a}$ **49.** $x\sqrt{5xy}$ **51.** $x^2y\sqrt{7xy}$ **53.** $\sqrt{2ab}$

55. $\dfrac{\sqrt[3]{a^2b^2}}{b^2}$ **57.** $\dfrac{2\sqrt{6ab}}{3b^2}$ **59.** $\dfrac{\sqrt{2abx}}{2x}$ **61. (a)** $\dfrac{\sqrt[3]{6\pi^2 V}}{2\pi}$; **(b)** 3 ft

63. $m = \dfrac{m_0 c\sqrt{c^2 - v^2}}{c^2 - v^2}$ **65.** $x - y$ **67.** $2 + y$ **69.** $2xy^2 + 3y^2$ **71.** 11

73. 7   **75.** $\bar{v} = \dfrac{(3kTm)^{1/2}}{m}$ **77.** $P = \dfrac{k}{\sqrt[5]{V^7}}$; $P = \dfrac{k\sqrt[5]{V^3}}{V^2}$ **85.** $\dfrac{\sqrt[3]{6}}{x}$

87. $4\sqrt{2}$ **89.** $3a^2b\sqrt[3]{3b}$ **91.** $\dfrac{\sqrt{2ax}}{2a}$ **93.** x **95.** $x + 5$ **97.** $\sqrt[3]{2cd}$

99. $\dfrac{\sqrt[3]{10}}{2}$ **101.** $\dfrac{\sqrt{22x}}{8x^2}$

Exercises 7.3

1. $15\sqrt{2}$ **3.** $9\sqrt{5a}$ **5.** $-11\sqrt{2}$ **7.** $-5a\sqrt{2}$ **9.** $-26\sqrt{3}$
11. $9x\sqrt{5x} - 2\sqrt{6x}$ **13.** $17\sqrt[5]{5}$ **15.** $-12\sqrt[3]{3}$ **17.** $3\sqrt[3]{3}$ **19.** $4\sqrt[3]{3a}$
21. $\dfrac{7\sqrt[3]{3}}{6}$ **23.** $\dfrac{3\sqrt{2} + 2\sqrt{3} + \sqrt{6}}{6}$ **25.** $\dfrac{\sqrt{6} + 3\sqrt{2}}{6}$ **27.** $3\sqrt[3]{75}$
29. $15 - 3\sqrt{2}$ **31.** $2 + 3\sqrt[3]{2}$ **33.** $14\sqrt{15} + 30$ **35.** $6\sqrt[3]{15} - 15$
37. $-8\sqrt{21} + 20\sqrt{14}$ **39.** $55 + 13\sqrt{15}$ **41.** $42 + 21\sqrt{2}$
43. $-441 + \sqrt{35}$ **45.** -1 **47.** -23 **49.** $5 + 2\sqrt{6}$
51. $a^2 + 2a\sqrt{b} + b$ **53.** $5 - 2\sqrt{6}$ **55.** $a^2 - 2a\sqrt{b} + b$
57. $a - 2\sqrt{ab} + b$ **59.** $1 + \sqrt{2}$ **61.** $\dfrac{2 - \sqrt{3}}{4}$ **63.** $\dfrac{3\sqrt{2} + \sqrt{6}}{2}$
65. $\dfrac{6 + 2\sqrt{2}}{7}$ **67.** $3a + a\sqrt{5}$ **69.** $\dfrac{9a - 3a\sqrt{2} + 6b - 2b\sqrt{2}}{7}$
71. $\dfrac{a + 2b\sqrt{a} + b^2}{a - b^2}$ **73.** $\dfrac{a + 2\sqrt{2ab} + 2b}{a - 2b}$ **75.** 11   **77.** 2 or 1
79. $\dfrac{1}{\sqrt{5} - \sqrt{2}}$ **81.** $\dfrac{x - 2}{5\sqrt{x} + 5\sqrt{2}}$ **83.** $\dfrac{x - y}{x\sqrt{x} - x\sqrt{y}}$
85. $\dfrac{x - y}{x - \sqrt{xy}}$ **87.** $\dfrac{x - y}{x + \sqrt{xy}}$ **95.** $\dfrac{y + \sqrt{xy}}{y - x}$ **97.** $5 + \sqrt{2}$
99. $11 - 2\sqrt{21}$ **101.** $x\sqrt{5} - \sqrt{15x}$ **103.** $\dfrac{3\sqrt{2}}{4}$ **105.** $2\sqrt[3]{2x}$
107. $2\sqrt{2}$

Exercises 7.4

1. 16   **3.** 43   **5.** 18   **7.** No real-number solution   **9.** 3   **11.** 0   **13.** 6
15. -16 **17.** 11   **19.** 9   **21.** 0   **23.** 1   **25.** 0   **27.** 1   **29.** 4
31. $x = a + b^2$ **33.** $y = \dfrac{a - c^3}{b}$ **35.** $x = ab^2$ **37.** $x = \dfrac{b^2 - a}{b}$
39. $x = \dfrac{b}{3}$ **41.** $5\sqrt{3}$ ft   **43. (a)** $d = \dfrac{gt^2}{2}$; **(b)** 144.9 ft
45. (a) $L = \dfrac{gt^2}{4\pi^2}$; **(b)** $L = \dfrac{392}{121}$ ft ≈ 3.2 ft   **47.** $11 + 6x$ **49.** $-1 + 8x$
51. $\dfrac{1 + 5\sqrt{2}}{7}$ **53.** $\dfrac{4 + \sqrt{2}}{7}$ **55.** $\dfrac{x - 2\sqrt{xy} + y}{x - y}$ **57.** 100 ft
59. $\dfrac{1600}{9} \approx 177.8$ ft   **65.** No real-number solution   **67.** -1 or 0
69. No real-number solution   **71.** 0

Exercises 7.5

1. $5i$ **3.** $5i\sqrt{2}$ **5.** $24i\sqrt{2}$ **7.** $-12i\sqrt{2}$ **9.** $3 + 8i\sqrt{7}$
11. $6 + 4i$ **13.** $-2 - 6i$ **15.** $-5 - 6i$ **17.** $-2 + 5i$ **19.** $-7 + 4i$
21. $8 - i$ **23.** $10 + 5i$ **25.** $-3 - 2i\sqrt{2}$ **27.** $-1 + 2i\sqrt{2}$
29. $-7 + 3i\sqrt{5}$ **31.** $12 + 6i$ **33.** $-12 + 20i$ **35.** $-4 + 6i$
37. $-3 + 3i\sqrt{3}$ **39.** $-6 + 9i$ **41.** $28 + 12i$ **43.** $-20 + 20i$
45. $3 + 11i$ **47.** $13 + 0i$ **49.** $24 + 7i$ **51.** $31 + 0i$ **53.** $0 - 3i$
55. $0 + 6i$ **57.** $\dfrac{2}{5} + \dfrac{1}{5}i$ **59.** $\dfrac{-6}{5} + \dfrac{3}{5}i$ **61.** $-1 + 2i$ **63.** $\dfrac{17}{13} - \dfrac{6}{13}i$

65. $0 - \dfrac{3}{2}i$ **67.** $\dfrac{12 + \sqrt{10}}{18} + \dfrac{4\sqrt{5} - 3\sqrt{2}}{18}i$
69. $\dfrac{3 + \sqrt{6}}{12} + \dfrac{3\sqrt{2} - \sqrt{3}}{12}i$ **71.** 1   **73.** $-i$ **75.** i **77.** 1   **79.** i

81. -1 **83.** $Z_1 + Z_2 = (8 + i)$ ohms   **85.** $Z_T = \dfrac{167}{65} - \dfrac{29}{65}i$
87. $x = 2$ or -2 **89.** $x = 5$ or -5 **91.** 5   **93.** $\sqrt{13}$ **99.** $-i$ **101.** -1
103. $\dfrac{27}{25} - \dfrac{11}{25}i$ **105.** $-10\sqrt{2} + 30i$ **107.** $28 + 4i$ **109.** $3 - 2i$
111. $-1 + 7i$ **113.** $5i\sqrt{2}$

Review Exercises

1. [7.1A] **(a)** Not a real number; **(b)** -4 **2.** [7.1A] **(a)** Not a real
number; **(b)** -5 **3.** [7.1B] **(a)** -3; **(b)** -4 **4.** [7.1B] **(a)** $\dfrac{1}{2}$; **(b)** $\dfrac{1}{4}$
5. [7.1B] **(a)** 25; **(b)** 16   **6.** [7.1B] **(a)** Not a real number;
(b) Not a real number   **7.** [7.1B] **(a)** $\dfrac{1}{4}$; **(b)** $\dfrac{1}{16}$ **8.** [7.1B] **(a)** $\dfrac{1}{9}$; **(b)** $\dfrac{1}{16}$
9. [7.1C] **(a)** $x^{8/15}$; **(b)** $x^{9/20}$ **10.** [7.1C] **(a)** $\dfrac{1}{x^{9/20}}$; **(b)** $\dfrac{1}{x^{8/15}}$
11. [7.1C] **(a)** $\dfrac{1}{x^5y^6}$; **(b)** $\dfrac{1}{x^5y^{12}}$ **12.** [7.1C] **(a)** $x^{1/5} + x^{3/5}y^{3/5}$;
(b) $x^{3/5} + x^{4/5}y^{3/5}$ **13.** [7.2A] **(a)** 7; **(b)** 6   **14.** [7.2A] **(a)** $|-x| = |x|$;
(b) $|-x| = |x|$ **15.** [7.2A] **(a)** $2\sqrt[3]{6}$; **(b)** $2\sqrt[3]{7}$ **16.** [7.2A] **(a)** $2xy^2\sqrt[3]{2x}$;
(b) $2x^2y^5\sqrt[3]{2x^2}$ **17.** [7.2A] **(a)** $\dfrac{\sqrt{15}}{27}$; **(b)** $\dfrac{\sqrt{5}}{32}$ **18.** [7.2A] **(a)** $\dfrac{1}{x}$;
(b) $\dfrac{\sqrt[3]{5}}{x}$ **19.** [7.2B] **(a)** $\dfrac{\sqrt{55}}{11}$; **(b)** $\dfrac{\sqrt{65}}{13}$ **20.** [7.2B] **(a)** $\dfrac{\sqrt{10x}}{5x}$;
(b) $\dfrac{\sqrt{15x}}{5x}$ **21.** [7.2B] **(a)** $\dfrac{\sqrt[3]{25x^2}}{5x}$; **(b)** $\dfrac{\sqrt[3]{49x^2}}{7x}$
22. [7.2B] **(a)** $\dfrac{\sqrt[4]{x^3}}{2x}$; **(b)** $\dfrac{\sqrt[4]{10x}}{2x}$ **23.** [7.2C] **(a)** $\dfrac{4}{3}$; **(b)** $\dfrac{5}{3}$
24. [7.2C] **(a)** $\sqrt[3]{9c^2d^2}$; **(b)** $\sqrt[3]{25c^2d^2}$ **25.** [7.2C] **(a)** $\dfrac{\sqrt[3]{3ac}}{3c^3}$; **(b)** $\dfrac{\sqrt[3]{9ac}}{3c^3}$
26. [7.3A] **(a)** $6\sqrt{2}$; **(b)** $7\sqrt{2}$ **27.** [7.3A] **(a)** $\sqrt{7}$; **(b)** $\sqrt{7}$
28. [7.3A] **(a)** $\dfrac{3\sqrt{2}}{4}$; **(b)** $\dfrac{9\sqrt{2}}{4}$ **29.** [7.3A] **(a)** $\dfrac{13\sqrt[3]{6x^2}}{4x}$; **(b)** $\dfrac{11\sqrt[3]{6x^2}}{4x}$
30. [7.3B] **(a)** $6 + \sqrt{6}$; **(b)** $8 + \sqrt{6}$ **31.** [7.3B] **(a)** $2x\sqrt[3]{6} - 3\sqrt[3]{6x^2}$;
(b) $2x\sqrt[3]{6} - 3\sqrt[3]{6x^2}$ **32.** [7.3B] **(a)** $30 + 12\sqrt{6}$; **(b)** $24 + 10\sqrt{6}$
33. [7.3B] **(a)** $19 + 8\sqrt{3}$; **(b)** $28 + 10\sqrt{3}$ **34.** [7.3B] **(a)** $52 - 14\sqrt{3}$;
(b) $19 - 8\sqrt{3}$ **35.** [7.3B] **(a)** 5; **(b)** 4   **36.** [7.3B] **(a)** $4 - \sqrt{2}$;
(b) $6 - \sqrt{2}$ **37.** [7.3C] **(a)** $5\sqrt{2} + 5$; **(b)** $\dfrac{x - 4\sqrt{x}}{x - 16}$
38. [7.4A] **(a)** No real-number solution; **(b)** No real-number solution
39. [7.4A] **(a)** 4; **(b)** 10   **40.** [7.4A] **(a)** 7 or 8; **(b)** 4 or 5
41. [7.4A] **(a)** 4; **(b)** 9   **42.** [7.4A] **(a)** 32; **(b)** 67
43. [7.4A] **(a)** No real-number solution; **(b)** No real-number solution
44. [7.4B] **(a)** $I = \dfrac{k}{d^2}$; **(b)** $k = d^2I$ **45.** [7.5A] **(a)** $10i$; **(b)** $11i$
46. [7.5A] **(a)** $6i\sqrt{2}$; **(b)** $5i\sqrt{2}$ **47.** [7.5B] **(a)** $10 + 3i$; **(b)** $6 + 3i$
48. [7.5B] **(a)** $-4 + 7i$; **(b)** $2 + 11i$ **49.** [7.5C] **(a)** $21 + i$; **(b)** $23 - 2i$
50. [7.5C] **(a)** $24\sqrt{2} + 16i$; **(b)** $36\sqrt{2} + 24i$ **51.** [7.5C] **(a)** $\dfrac{17}{25} + \dfrac{6}{25}i$;
(b) $\dfrac{27}{25} - \dfrac{11}{25}i$ **52.** [7.5D] **(a)** -1; **(b)** $-i$ **53.** [7.5D] **(a)** -1; **(b)** i

Cumulative Review Chapters 1–7

1. Irrational numbers, real numbers   **2.** $\dfrac{y^{16}}{16x^8}$ **3.** -128 **4.** 6
5. $-7 \le -7x - 14 < 7$

6. 10 ft by 40 ft   **7.** x-intercept: $\dfrac{-3}{4}$; y-intercept: 6   **8.** -4
9. $3x - 14y = -74$

10. $y \leq x - 2$

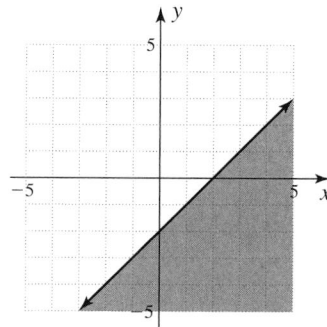

11. $|y| \leq 2$

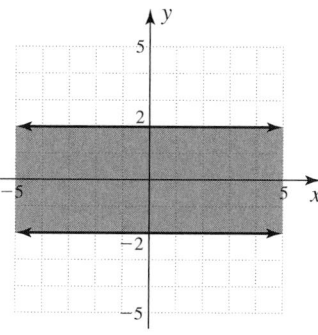

12. $(-1, -2)$ **13.** Infinitely many solutions such that $y = \dfrac{-3}{4}x - 18$

14. No solution   **15.** $x = 5$ **16.** \$20,000 at 7%; \$30,000 at 9%; \$40,000 at 11%   **17.** $R = \{-5, -3, 0\}$ **18.** 5   **19.** $x^3 - 14x - 8$

20. $3x^3(3x^3 - 4x^2 + 5x + 4)$ **21.** $3xy(3x + y)(x - 2y)$ **22.** $\dfrac{1}{2}; \dfrac{2}{3}$

23. $\dfrac{-(x + 3)}{9 + 3x + x^2}$ **24.** $(x + 4)(x + 1)(3x + 1)$ **25.** No solution

26. 10 and 12   **27.** 6880   **28.** 27   **29.** $x^{1/5} + x^{2/5}y^{4/5}$ **30.** $9\sqrt{7}$

31. $56 - 14\sqrt{7}$ **32.** $\dfrac{x + \sqrt{5x}}{x - 5}$ **33.** -9 **34.** $\dfrac{61}{82} - \dfrac{57}{82}i$ **35.** i

Chapter 8

Exercises 8.1

1. ± 8 **3.** $\pm 11i$ **5.** ± 13 **7.** $\pm 2i$ **9.** $\pm\dfrac{7}{6}$ **11.** $\pm\dfrac{9}{2}i$ **13.** $\pm\dfrac{5\sqrt{3}}{3}$

15. $\pm\dfrac{6\sqrt{5}}{5}i$ **17.** $\pm\dfrac{10\sqrt{3}}{3}$ **19.** $\pm\dfrac{9\sqrt{13}}{13}i$ **21.** $-3, -7$

23. $-2 \pm 5i$ **25.** $6 \pm 3\sqrt{2}$ **27.** $1 \pm 2i\sqrt{7}$ **29.** $1 \pm 5\sqrt{2}$

31. $5 \pm 4\sqrt{2}$ **33.** $9 \pm 8i$ **35.** $-1 \pm 4\sqrt{2}$ **37.** $2 \pm 5\sqrt{2}$

39. $5 \pm 3i\sqrt{3}$ **41.** $-1, -5$ **43.** $-5, -3$ **45.** $-3 \pm i$ **47.** $4, 6$

49. $3, 7$ **51.** $4 \pm i$ **53.** $-1 \pm \dfrac{\sqrt{2}}{2}i$ **55.** $-1 \pm 5i$ **57.** $\dfrac{3}{5}, \dfrac{2}{5}$

59. $\dfrac{1}{2} \pm i$ **61.** $\dfrac{1 \pm 2\sqrt{2}}{2}$ **63.** $\dfrac{2 \pm \sqrt{2}}{2}$ **65.** $1, -2$ **67.** $\dfrac{-5 \pm 3\sqrt{3}}{2}$

69. $\dfrac{-3 \pm \sqrt{29}}{2}$ **71.** 2 sec   **73.** 10%   **75.** $0, 9$ **77.** $1 \pm \sqrt{3}$

79. $\dfrac{1}{8}, -1$ **81.** $\dfrac{4}{3} \pm \dfrac{\sqrt{5}}{3}i$ **83.** $-1 \pm i\sqrt{5}$ **85. (a)** 2000; **(b)** \$2

87. 3 days   **91.** $\dfrac{5 \pm 3\sqrt{5}}{10}$ **93.** $-6 \pm 2\sqrt{11}$ **95.** $\dfrac{-6 \pm 3\sqrt{6}}{2}$

97. $3 \pm \dfrac{6\sqrt{5}}{5}i$ **99.** $2 \pm 2\sqrt{6}$ **101.** $\pm\dfrac{4\sqrt{3}}{3}$ **103.** $\pm 2i\sqrt{3}$

Exercises 8.2

1. $1, -2$ **3.** $-2 \pm \sqrt{3}$ **5.** $\dfrac{3 \pm \sqrt{17}}{2}$ **7.** $\dfrac{5}{7}, 1$ **9.** $-\dfrac{4}{5} \pm \dfrac{3}{5}i$

11. $-\dfrac{3}{2}, -2$ **13.** $\dfrac{3}{2}, 1$ **15.** $-\dfrac{1}{2}, -3$ **17.** $\dfrac{3 \pm \sqrt{5}}{4}$ **19.** -1

21. $\dfrac{3 \pm \sqrt{5}}{4}$ **23.** $\dfrac{-3 \pm \sqrt{21}}{6}$ **25.** ± 2 **27.** $\pm 3\sqrt{5}$ **29.** $\pm\dfrac{2\sqrt{3}}{3}$

31. $2, -1 \pm i\sqrt{3}$ **33.** $\dfrac{1}{2}, -\dfrac{1}{4} \pm \dfrac{\sqrt{3}}{4}i$ **35.** $-3, \dfrac{3}{2} \pm \dfrac{3\sqrt{3}}{2}i$

37. (a) Yes; **(b)** 1987   **39.** \$1.85   **41.** $10 \pm 2\sqrt{15}$ **43.** 1   **45.** 4

47. $i\sqrt{23}$ **49.** $6x^2 - 5x - 4$ **51.** $12x^2 - 19x - 21$

53. Multiply both sides by $4a$.   **55.** Add b^2 to both sides.

57. Take square root of both sides.   **59.** Divide both sides by $2a$.

63. $\dfrac{2}{3}, -\dfrac{1}{3} \pm \dfrac{\sqrt{3}}{3}i$ **65.** $-\dfrac{1}{3} \pm \dfrac{\sqrt{2}}{3}i$ **67.** $-\dfrac{1}{2}, 2$ **69.** $0, 6$

71. $2 \pm 2\sqrt{2}$ **73.** $-\dfrac{5}{3}, 1$ **75.** \$1

Exercises 8.3

1. $D = 49$; two rational numbers   **3.** $D = 0$; one rational number

5. $D = 44$; two irrational numbers   **7.** $D = -23$; two non-real complex

numbers   **9.** $D = \dfrac{57}{4}$; two irrational numbers   **11.** ± 4 **13.** -5

15. ± 8 **17.** ± 20 **19.** -16 **21.** Not factorable   **23.** $(4x - 3)(3x - 2)$

25. Not factorable   **27.** $(5x - 6)(3x + 14)$ **29.** $(4x - 15)(3x - 4)$

31. $x^2 - 7x + 12 = 0$ **33.** $x^2 + 12x + 35 = 0$ **35.** $3x^2 - 7x - 6 = 0$

37. $4x^2 - 1 = 0$ **39.** $5x^2 + x = 0$ **41. (a)** $\dfrac{6}{4} = \dfrac{3}{2}$; **(b)** $\dfrac{5}{4}$; **(c)** No

43. (a) $-\dfrac{13}{5}$; **(b)** $-\dfrac{6}{5}$; **(c)** Yes   **45. (a)** $-\dfrac{5}{2}$; **(b)** 1; **(c)** No   **47.** $\dfrac{8}{3}$

49. $k = 15$ **51.** 1   **53.** $-1, -5$ **55.** Yes; occurs twice

57. Does not occur ($b^2 - 4ac$ is negative)   **63.** Yes

65. $(3x - 5)(4x + 7)$ **67.** $3x^2 + x - 2 = 0$

Exercises 8.4

1. $0, -\dfrac{5}{2}$ **3.** $-3, 7$ **5.** 0 **7.** $2, 1$ **9.** $-\dfrac{16}{5}, -1$ **11.** $\pm 3, \pm 2$

13. $\pm\dfrac{1}{2}, \pm 3i$ **15.** $\pm\sqrt{2}, \pm\dfrac{\sqrt{3}}{3}i$ **17.** $1, -2, 1 \pm i\sqrt{3}, -\dfrac{1}{2} \pm \dfrac{\sqrt{3}}{2}i$

19. $7, -6$ **21.** $\dfrac{1 \pm \sqrt{37}}{2}, \dfrac{1}{2} \pm \dfrac{\sqrt{3}}{2}i$ **23.** 16   **25.** 8, 27   **27.** 4

29. $2 \pm \sqrt{29}, 2 \pm \sqrt{13}$ **31.** 6   **33.** $1, \dfrac{16}{81}$ **35.** $\dfrac{1}{2}, -\dfrac{1}{4}$

37. $\pm i\sqrt{3}, \pm\dfrac{\sqrt{2}}{2}$ **39.** $\pm\sqrt{2}, \pm\dfrac{\sqrt{3}}{2}i$ **41.** 10 hr, 15 hr

43. x-intercepts: $(\sqrt{5}, 0)$; $(-\sqrt{5}, 0)$ **45.** $x \geq -2$ **47.** 10 students

53. 1, 16   **55.** 16, 81   **57.** $\pm\dfrac{\sqrt{2}}{2}, \pm\dfrac{\sqrt{7}}{7}$ **59.** $2, -1, \dfrac{1}{2} \pm \dfrac{\sqrt{3}}{2}i$

61. -7 **63.** $\pm 1, \pm 2$

Exercises 8.5

1. $x < -1$ or $x > 3$;
$(-\infty, -1) \cup (3, \infty)$

3. $-4 \leq x \leq 0$; $[-4, 0]$

5. $-1 \leq x \leq 2$; $[-1, 2]$

7. $x \leq 0$ or $x \geq 3$;
$(-\infty, 0] \cup [3, \infty)$

9. $1 < x < 2$; $(1, 2)$

11. $-3 < x < 1$; $(-3, 1)$

13. $x = -5$

15. All real numbers ⟵————————⟶

17. 1.6, −0.6   **19.** 2.55, −1.55

21. −3 ≤ x ≤ −1 or x ≥ 2; [−3, −1] ∪ [2, ∞)

23. x ≤ 1 or 2 ≤ x ≤ 3; (−∞, 1] ∪ [2, 3]

25. x > 2; (2, ∞)

27. x < −5 or x > 1; (−∞, −5) ∪ (1, ∞)

29. $\frac{1}{2} < x < \frac{4}{3}; \left(\frac{1}{2}, \frac{4}{3}\right)$

31. 1 < x < 7; (1, 7)

33. x < 1 or x > 2; (−∞, 1) ∪ (2, ∞)

35. x ≤ −3 or x ≥ 3; (−∞, −3] ∪ [3, ∞)
37. x ≤ 1 or x ≥ 5; (−∞, 1] ∪ (5, ∞)   **39.** R > 4   **41.** 10° < T < 100°
43. 1 < t < 2   **45.** 12   **47.** 2   **49.** −2   **51.** 23.2 ≤ v ≤ 26.1
53. v = 128.2 mi/hr
59. x < −2 or x > 0; (−∞, −2) ∪ (0, ∞)

61. x ≤ −3 or 1 ≤ x ≤ 4; (−∞, −3] ∪ [1, 4]

63. −3 ≤ x ≤ 2; [−3, 2]

65. x ≤ −1 or x ≥ 3; (−∞, −1] ∪ [3, ∞)

23. [8.4B] **(a)** 1, 81; **(b)** 16   **24.** [8.4B] **(a)** 27, −64; **(b)** 216, −1
25. [8.4B] **(a)** 9; **(b)** 25   **26.** [8.4B] **(a)** ±1, ±√3; **(b)** ±1, ±i√3

27. [8.5A] **(a)** (−3, 2)

(b) (−2, 3)

28. [8.5A] **(a)** (−∞, −4] ∪ [0, ∞)

(b) (−∞, 0] ∪ [3, ∞)

29. [8.5A] **(a)** (−9, 5)

(b) (1 − √3, 1 + √3)

30. [8.5B] **(a)** (−∞, 1] ∪ [2, 3]

(b) (−∞, −2] ∪ [−1, 3]

31. [8.5B] **(a)** (−2, −1) ∪ (3, ∞)

(b) (−3, −2) ∪ (1, ∞)

32. [8.5C] **(a)** [−2, 2)

(b) (−∞, 2) ∪ (6, ∞)

Review Exercises

1. [8.1A] **(a)** $\pm\frac{7}{4}$; **(b)** $\pm\frac{4}{5}$ **2.** [8.1A] **(a)** $\pm i\sqrt{6}$; **(b)** $\pm i\sqrt{7}$

3. [8.1B] **(a)** $3 + 4\sqrt{2}$; **(b)** $5 \pm 5\sqrt{2}$ **4.** [8.1B] **(a)** $2 \pm \frac{5\sqrt{2}}{2}i$;

(b) $3 \pm \frac{8\sqrt{3}}{3}i$ **5.** [8.1B] **(a)** $\frac{5 \pm 2\sqrt{15}}{5}$; **(b)** $\frac{-3 \pm 4\sqrt{3}}{6}$

6. [8.1C] **(a)** 9, −1; **(b)** −4, −8   **7.** [8.1C] **(a)** $\frac{1}{2}, -\frac{3}{2}$; **(b)** $\frac{3 \pm \sqrt{2}}{4}$

8. [8.2A] **(a)** $-2, \frac{1}{3}$; **(b)** $-\frac{1}{5}, 2$ **9.** [8.2A] **(a)** $\frac{1 \pm \sqrt{13}}{3}$; **(b)** $\frac{3 \pm \sqrt{21}}{4}$

10. [8.2A] **(a)** 0, 16; **(b)** 0, 12   **11.** [8.2A] **(a)** $\frac{-1 \pm \sqrt{301}}{30}$; **(b)** $\frac{1}{5}, -2$

12. [8.2A] **(a)** $\frac{1}{3} \pm \frac{\sqrt{2}}{3}i$; **(b)** $\frac{1}{5} \pm \frac{\sqrt{19}}{5}i$

13. [8.2B] **(a)** $\frac{5}{2}, -\frac{5}{4} \pm \frac{5\sqrt{3}}{4}i$; **(b)** $\frac{2}{5}, -\frac{1}{5} \pm \frac{\sqrt{3}}{5}i$

14. [8.2C] **(a)** p = 3; **(b)** p = 1   **15.** [8.3A] **(a)** D = −55; two non-real complex solutions; **(b)** D = k² − 64; k = ±8
16. [8.3B] **(a)** Not factorable; **(b)** (2x + 1)(9x + 2)
17. [8.3B] **(a)** (6x − 5)(3x + 1); **(b)** Not factorable
18. [8.3C] **(a)** x² − x − 6 = 0; **(b)** 12x² + 5x − 2 = 0
19. [8.3D] **(a)** Sum = $-\frac{4}{15}$; product = $-\frac{1}{5}$; **(b)** Sum = $\frac{4}{3}$; product = $-\frac{5}{9}$
20. [8.3D] **(a)** Yes; **(b)** No   **21.** [8.4A] **(a)** −5; **(b)** 0, 2
22. [8.4B] **(a)** 1, −2, $-\frac{1}{2} \pm \frac{\sqrt{15}}{2}i$; **(b)** 1, 2, $\frac{3 \pm \sqrt{33}}{2}$

Cumulative Review Chapters 1–8

1. {2, 4, 6, 8, 10, 12}   **2.** $0.\overline{73}$ **3.** 11   **4.** −32   **5.** $\frac{10}{9}$
6. {x | x > −1 and x < 1}

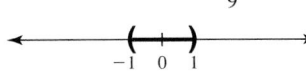

7. |3x + 1| > 2

8. $10,000 in bonds; $8000 in CDs

9. 2x − y = 2

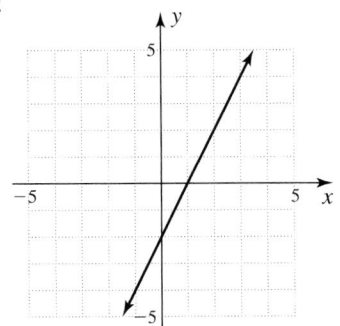

10. Perpendicular   **11.** x + y = −1   **12.** 8x − y = −52
13. D = {all real numbers}; R = {all real numbers}

14. $x + y \leq -1$; $y \leq 3x - 5$

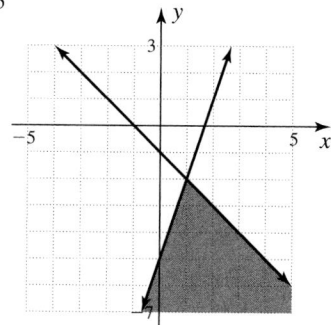

15. $(-3, 2)$ **16.** $x = \dfrac{1}{3}$ **17.** 28 nickels; 18 dimes

18. $-2x^3 + 9x^2 - 4x + 18$ **19.** $(6x + 5y)^2$ **20.** $2x^2(x^2 + 4)$

21. $x = -1$ or -3 or 3 **22.** $-2x^2 + \dfrac{2}{x^2} + \dfrac{1}{x^3}$ **23.** $\dfrac{1}{x + 9}$

24. $\dfrac{x(x^2 + x + 1)}{(x + 1)(x^2 + 1)}$ **25.** $x = 2$ **26.** 8 and 10 **27.** Not a real number

28. $4a^2b^3\sqrt[3]{3a}$ **29.** $\dfrac{\sqrt{30k}}{6k}$ **30.** $3x + 2\sqrt[3]{6x^2}$ **31.** $x = -8$ or -7

32. $-3 - 5i$ **33.** $-20 + 35i$ **34.** $x = \dfrac{3}{4}$ or $-\dfrac{3}{4}$ **35.** $x = -3$ or 1

36. $x = -2 \pm \sqrt{6}$ **37.** $p = 4$ **38.** Yes **39.** $x = \pm 1$ or $x = \pm\sqrt{7}$

40. $-6 < x < 4$

Chapter 9

Exercises 9.1

1. (a) $y = 2x^2$; $(0, 0)$; (b) $y = 2x^2 + 2$; $(0, 2)$;
 (c) $y = 2x^2 - 2$; $(0, -2)$

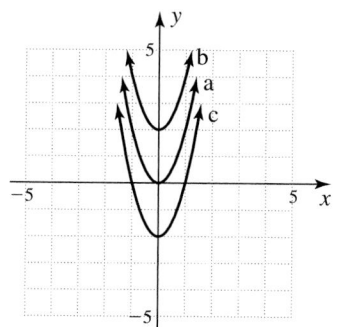

3. (a) $y = -2x^2$; $(0, 0)$; (b) $y = -2x^2 + 1$; $(0, 1)$;
 (c) $y = -2x^2 - 1$; $(0, -1)$

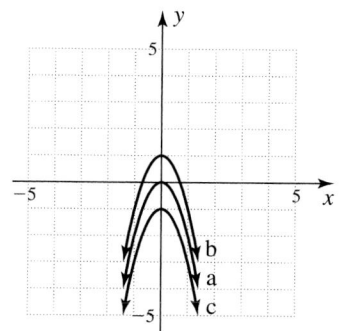

5. (a) $y = \dfrac{1}{4}x^2$; $(0, 0)$; (b) $y = -\dfrac{1}{4}x^2$; $(0, 0)$

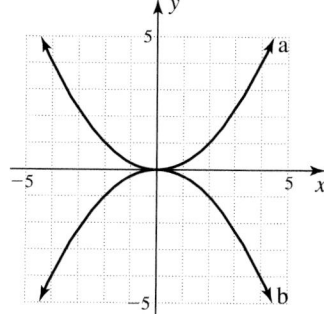

7. (a) $y = \dfrac{1}{3}x^2 + 1$; $(0, 1)$; (b) $y = -\dfrac{1}{3}x^2 + 1$; $(0, 1)$

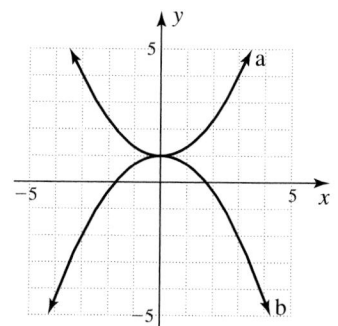

9. (a) $y = (x + 2)^2 + 3$; $(-2, 3)$;
 (b) $y = (x + 2)^2$; $(-2, 0)$;
 (c) $y = (x + 2)^2 - 2$; $(-2, -2)$

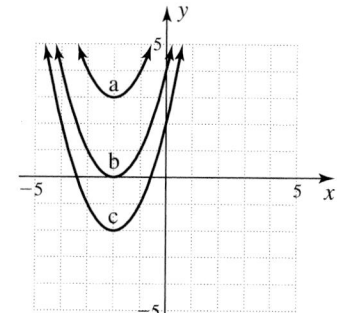

11. (a) $y = -(x + 2)^2 - 2$; $(-2, -2)$;
 (b) $y = -(x + 2)^2$; $(-2, 0)$;
 (c) $y = -(x + 2)^2 - 4$; $(-2, -4)$

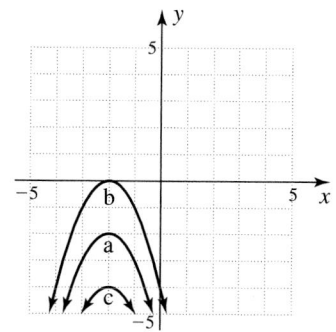

13. (a) $y = -2(x + 2)^2 - 2$; $(-2, -2)$;
  (b) $y = -2(x + 2)^2$; $(-2, 0)$;
  (c) $y = -2(x + 2)^2 - 4$; $(-2, -4)$

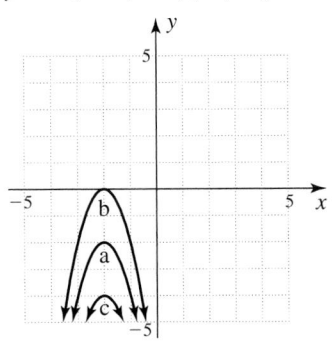

15. (a) $y = 2(x + 1)^2 + \dfrac{1}{2}$; $\left(-1, \dfrac{1}{2}\right)$;
  (b) $y = 2(x + 1)^2$; $(-1, 0)$

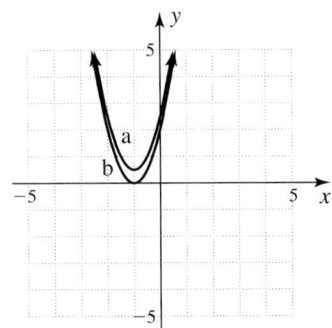

17. $y = x^2 + 2x + 1$; V: $(-1, 0)$; Int: $(-1, 0)$, $(0, 1)$

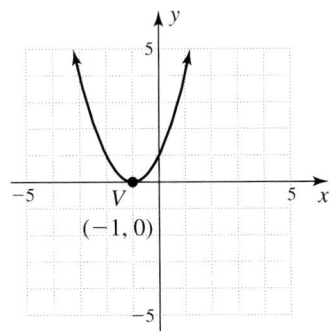

19. $y = -x^2 + 2x + 1$; V: $(1, 2)$; Int: $(2.4, 2)$, $(-0.4, 0)$, $(0, 1)$

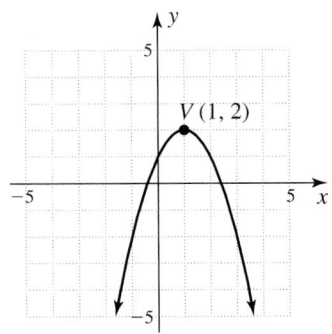

21. $y = -x^2 + 4x - 5$; V: $(2, -1)$; Int: $(0, -5)$

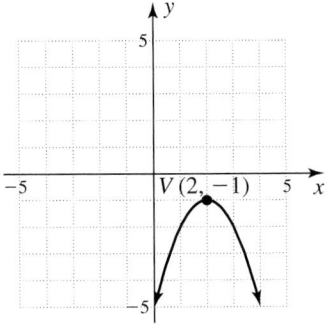

23. $y = 3 - 5x + 2x^2$; V: $\left(\dfrac{5}{4}, -\dfrac{1}{8}\right)$; Int: $\left(\dfrac{3}{2}, 0\right)$, $(1, 0)$, $(0, 3)$

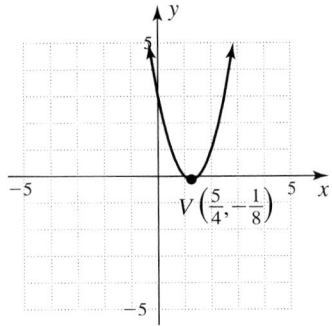

25. $y = 5 - 4x - 2x^2$; V: $(-1, 7)$; Int: $(-2.9, 0)$, $(0.9, 0)$, $(0, 5)$

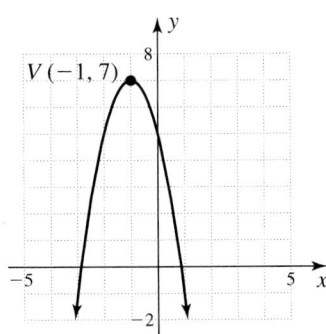

27. $y = -3x^2 + 3x + 2$; V: $\left(\dfrac{1}{2}, \dfrac{11}{4}\right)$; Int: $(1.5, 0)$, $(-0.5, 0)$, $(0, 2)$

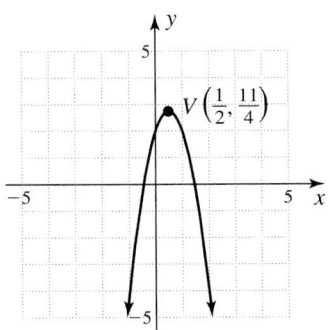

29. (a) $x = (y + 2)^2 + 3$; $(3, -2)$; **(b)** $x = (y + 2)^2$; $(0, -2)$

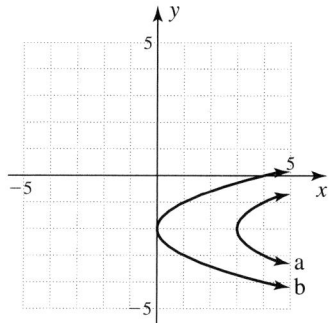

31. (a) $x = -(y + 2)^2 - 2$; $(-2, -2)$; **(b)** $x = -(y + 2)^2$; $(0, -2)$

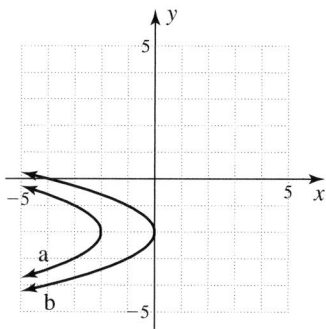

33. (a) $x = -y^2 + 2y + 1$; $(2, 1)$; **(b)** $x = -y^2 + 2y + 4$; $(5, 1)$

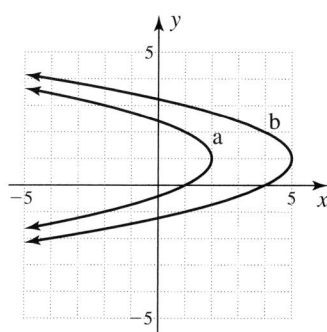

35. $x = 4000$; $P = \$11,000$ **37.** \$25 thousand (\$25,000)   **39.** 400 ft
41. $P = (600 + 100W)(1 - 0.10W)$; $P = $ price, $W = $ weeks elapsed.
Max P occurs when $W = 2$ (at end of 2 weeks).
43. (a) $(42, 18)$; **(b)** 18 in.; **(c)** 84 in.;
(d)

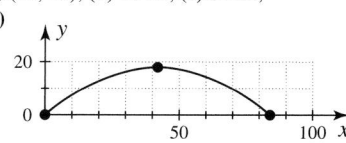

45. (a) $(200, 100)$; **(b)** 100 ft; **(c)** 400 ft;
(d)

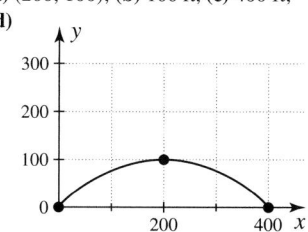

47. 5   **49.** $\sqrt{29} \approx 5.4$ **51.** $FP = \sqrt{x^2 + (y - p)^2}$ **53.** $x^2 = 4py$
55. $x^2 = 12.5y$; Focus: $(0, 3.125)$

63. $x = y^2 + 2y - 3$; $V(-4, -1)$

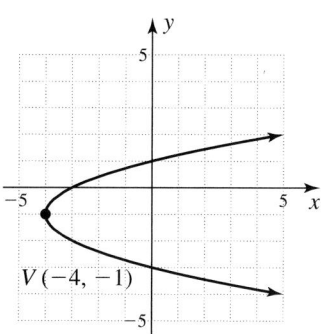

65. $x = 2(y - 1)^2 + 3$; $V(3, 1)$

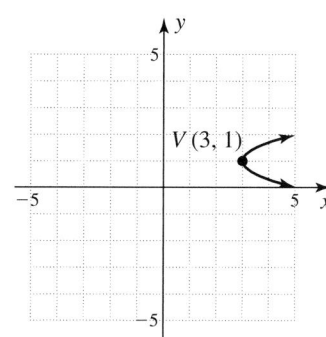

67. $y = -2x^2 - 4x - 3$; $V(-1, -1)$

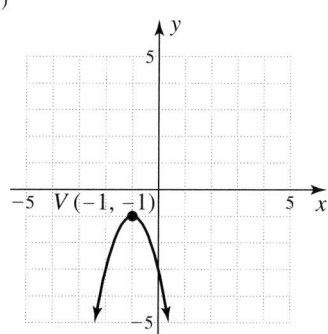

69. $y = (x - 2)^2 + 3$; $V(2, 3)$

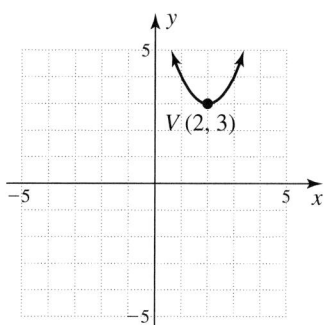

71. $f(x) = -x^2 + 4$; $V(0, 4)$

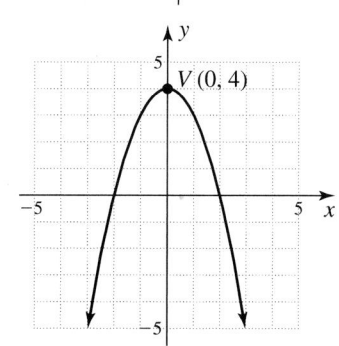

73. $f(x) = 2x^2$; $V(0, 0)$

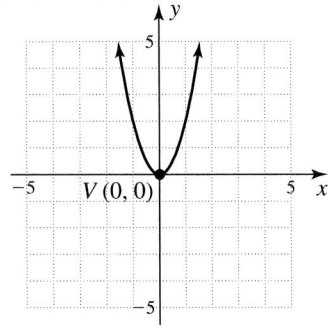

75. $h(x) = -2x^2$; $V(0, 0)$

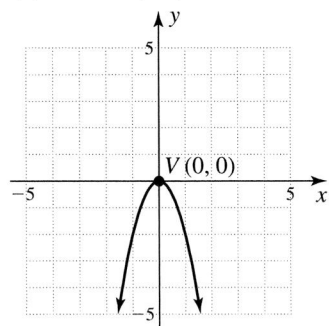

77. $10

Exercises 9.2

1. 5   **3.** $\sqrt{73}$ **5.** $\sqrt{90} = 3\sqrt{10}$ **7.** 5   **9.** 6
11. $(x - 3)^2 + (y - 8)^2 = 4$ **13.** $(x + 3)^2 + (y - 4)^2 = 25$
15. $(x + 3)^2 + (y + 2)^2 = 16$ **17.** $(x - 2)^2 + (y + 4)^2 = 5$
19. $x^2 + y^2 = 9$
21. $(x - 1)^2 + (y - 2)^2 = 9$; $C(1, 2)$; $r = 3$

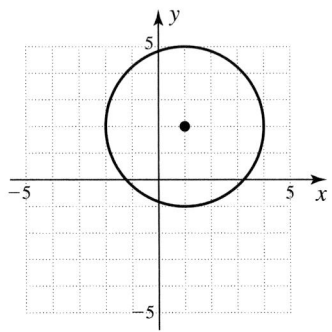

23. $(x + 1)^2 + (y - 2)^2 = 4$; $C(-1, 2)$; $r = 2$

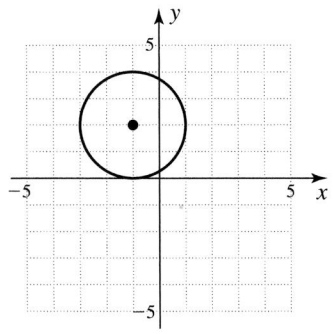

25. $(x - 1)^2 + (y + 2)^2 = 1$; $C(1, -2)$; $r = 1$

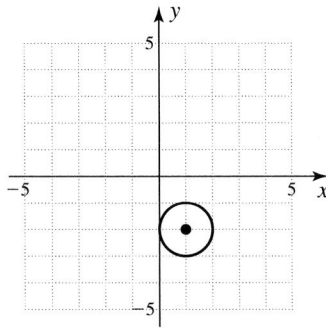

27. $(x + 2)^2 + (y + 1)^2 = 9$; $C(-2, -1)$; $r = 3$

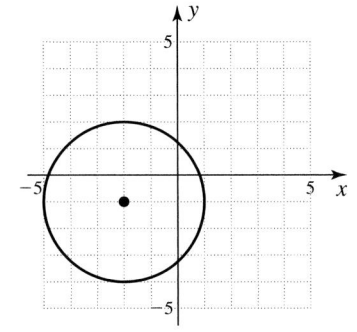

29. $(x - 1)^2 + (y - 1)^2 = 7$; $C(1, 1)$; $r = \sqrt{7} \approx 2.6$

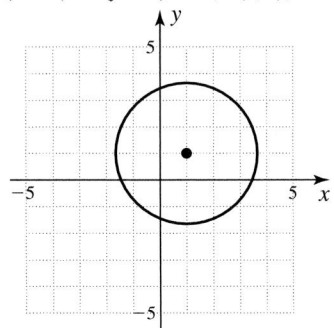

31. $x^2 - 6x + y^2 - 4y + 9 = 0$; $C(3, 2)$; $r = 2$

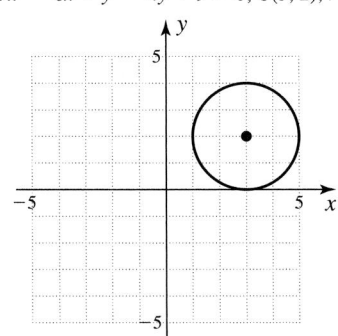

33. $x^2 + y^2 - 4x + 2y - 4 = 0$; $C(2, -1)$; $r = 3$

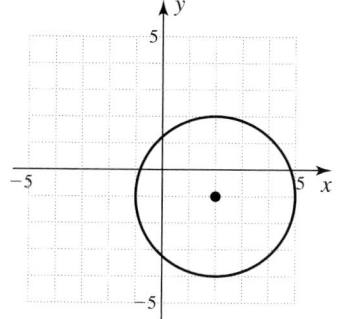

41. $x^2 + y^2 - 6x - 2y + 6 = 0$; $C(3, 1)$; $r = 2$

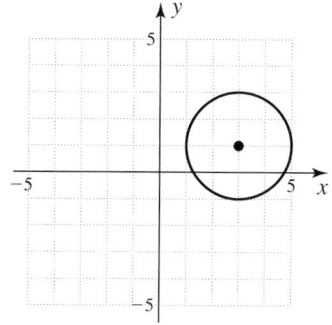

35. $x^2 + y^2 - 25 = 0$; $C(0, 0)$; $r = 5$

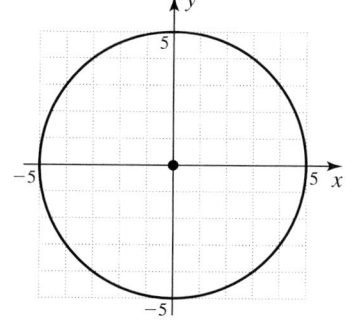

43. $25x^2 + 4y^2 = 100$; $C(0, 0)$; $a = 5$, $b = 2$

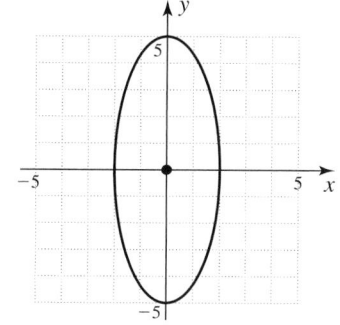

37. $x^2 + y^2 - 7 = 0$; $C(0, 0)$; $r = \sqrt{7} \approx 2.6$

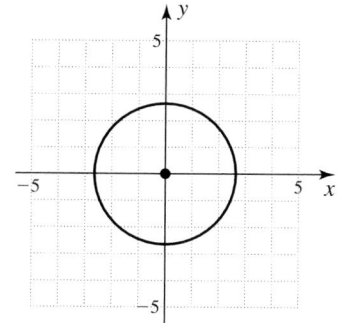

45. $x^2 + 4y^2 = 4$; $C(0, 0)$; $a = 2$, $b = 1$

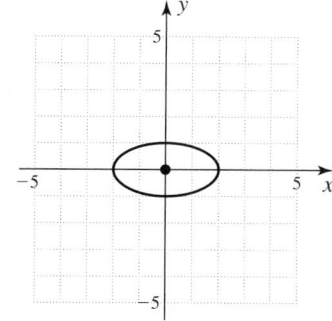

39. $x^2 + y^2 + 6x - 2y = -6$; $C(-3, 1)$; $r = 2$

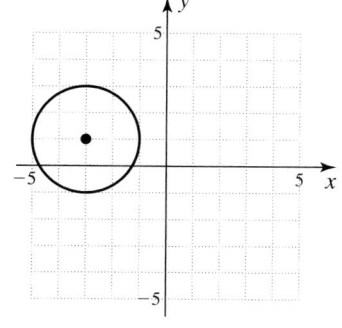

47. $x^2 + 4y^2 = 16$; $C(0, 0)$; $a = 4$, $b = 2$

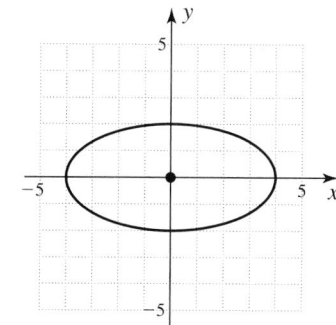

49. $\dfrac{x^2}{9} + \dfrac{y^2}{16} = 1$; $C(0, 0)$; $a = 4$, $b = 3$

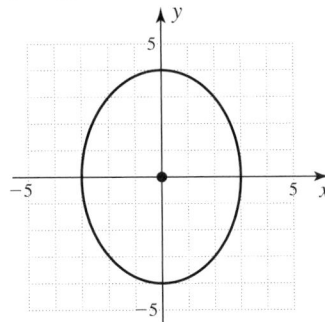

51. $\dfrac{(x-1)^2}{4} + \dfrac{(y-2)^2}{9} = 1$; $C(1, 2)$; $a = 3$, $b = 2$

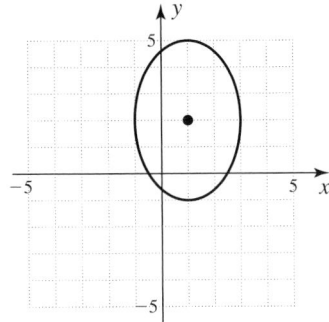

53. $\dfrac{(x-2)^2}{9} + \dfrac{(y+3)^2}{4} = 1$; $C(2, -3)$; $a = 3$, $b = 2$

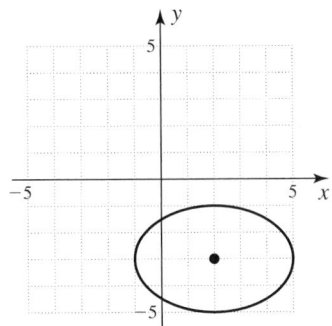

55. $\dfrac{(x-1)^2}{16} + \dfrac{(y-1)^2}{9} = 1$; $C(1, 1)$; $a = 4$, $b = 3$

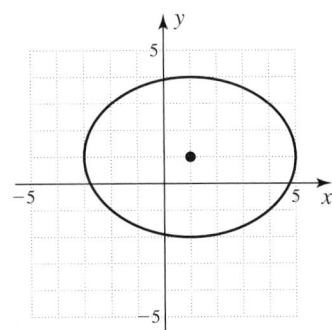

57. $x^2 + y^2 = 25$ **59.** $x^2 + y^2 = 169$ **61.** $x^2 + y^2 = 9$

63. $\dfrac{x^2}{4} + \dfrac{y^2}{36} = 1$ **65.** $\dfrac{x^2}{36} + \dfrac{y^2}{16} = 1$ **67.** 12.5 in.   **69.** $10\sqrt{5} \approx 22.4$ ft

71. $\dfrac{x^2}{4^2} + \dfrac{y^2}{2.5^2} = 1$ or $\dfrac{x^2}{16} + \dfrac{y^2}{6.25} = 1$ **73.** $\dfrac{x^2}{16} + \dfrac{y^2}{9} = 1$

75. (a) 93 million mi; **(b)** 1.5 million mi; **(c)** 92.99 million mi

77. (a) $\dfrac{x^2}{6^2} + \dfrac{y^2}{4.5^2} = 1$ or $\dfrac{x^2}{36} + \dfrac{y^2}{20.25} = 1$; **(b)** $\dfrac{6\sqrt{5}}{20} \approx 6.71$ in.

79. $21 - \dfrac{25\sqrt{7}}{4} \approx 4.5$ ft between side of boat and riverbank

81. Right half of circle   **83.** $PF_1 = \sqrt{(x-c)^2 + y^2}$

85. $a^2 + cx = a\sqrt{(x+c)^2 + y^2}$ or $a^2 - cx = a\sqrt{(x-c)^2 + y^2}$

87. $b^2x^2 + a^2y^2 = a^2b^2$

93. $\dfrac{(x+3)^2}{4} + \dfrac{(y+1)^2}{9} = 1$

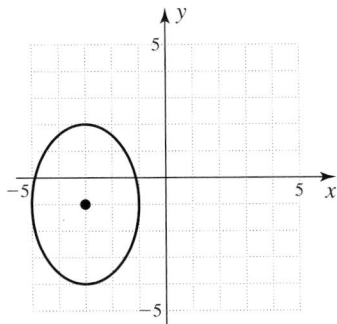

95. $4x^2 + 9y^2 = 36$

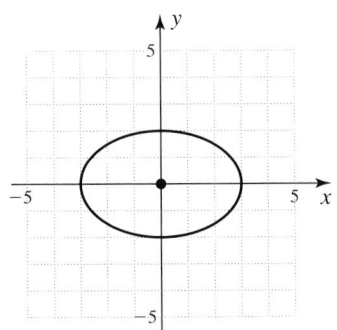

97. $x^2 - 6x + y^2 - 4y + 9 = 0$

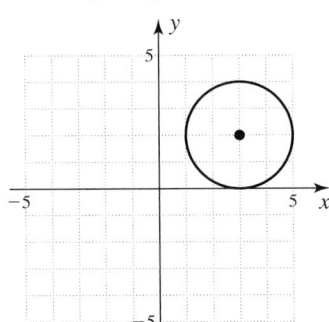

99. $x^2 + y^2 = 4$

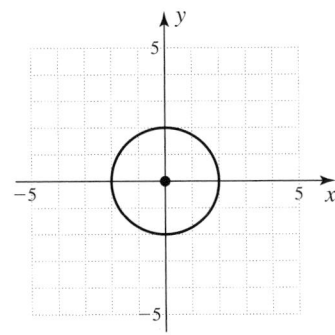

101. $(x - 3)^2 + (y - 1)^2 = 4$; $C(3, 1)$; $r = 2$

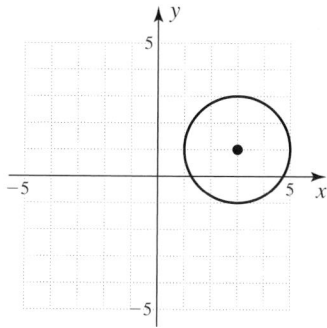

103. $x^2 + y^2 = 25$ **105.** $(x + 3)^2 + (y - 6)^2 = 9$

Exercises 9.3

1. $\dfrac{x^2}{25} - \dfrac{y^2}{9} = 1$; V: $(\pm 5, 0)$

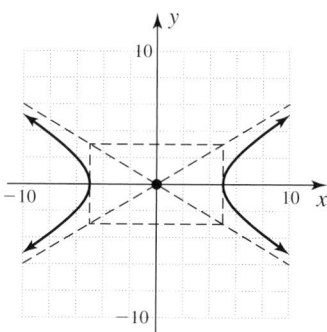

3. $\dfrac{y^2}{9} - \dfrac{x^2}{9} = 1$; V: $(0, \pm 3)$

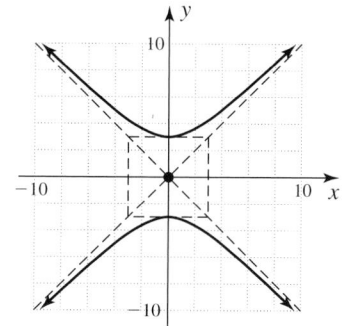

5. $\dfrac{x^2}{9} - \dfrac{y^2}{1} = 1$; V: $(\pm 3, 0)$

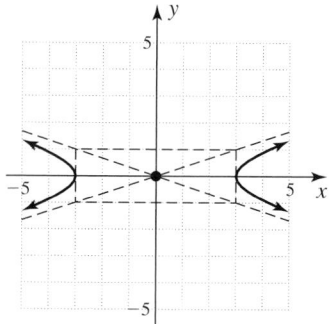

7. $\dfrac{x^2}{64} - \dfrac{y^2}{49} = 1$; V: $(\pm 8, 0)$

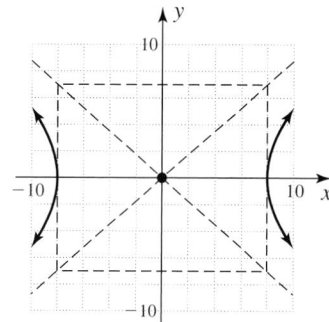

9. $\dfrac{y^2}{\frac{16}{9}} - \dfrac{x^2}{\frac{9}{16}} = 1$; V: $\left(0, \pm\dfrac{4}{3}\right)$

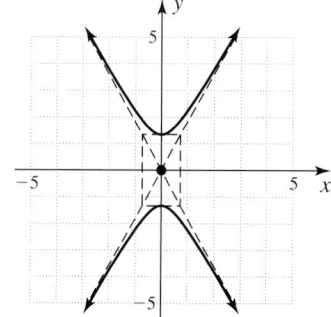

11. $y^2 - 9x^2 = 9$; V: $(0, \pm 3)$

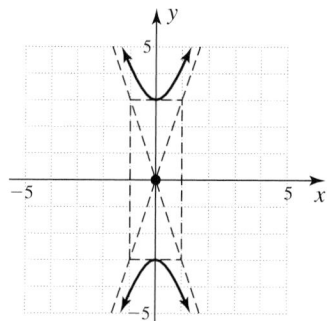

13. $\dfrac{(x - 1)^2}{4} - \dfrac{(y + 1)^2}{9} = 1$; $C(1, -1)$; $a = 2$, $b = 3$

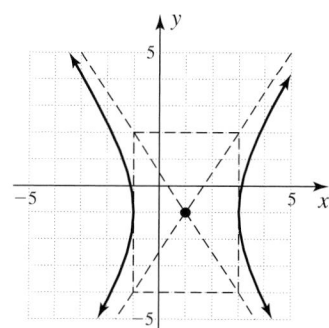

15. $\dfrac{(y-1)^2}{9} - \dfrac{(x-2)^2}{4} = 1$; $C(2, 1)$; $a = 3, b = 2$

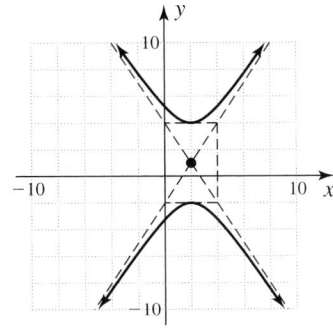

17. Circle; $(5, 0), (0, 5), (-5, 0), (0, -5)$
19. Hyperbola; $(6, 0), (-6, 0)$
21. Parabola; $(0, -9)$ **23.** Parabola; $(-4, 0)$
25. Circle; $(2, 0), (0, 2), (-2, 0), (0, -2)$
27. Hyperbola; $(2, 0), (-2, 0)$ **29.** Ellipse; $(3, 0), (0, 1), (-3, 0), (0, -1)$
31. (a) Hyperbola;
   (b) $\dfrac{D^2}{8} - \dfrac{d^2}{4} = 1$

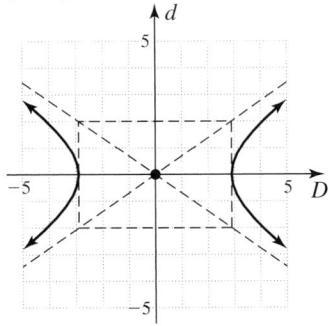

33. $4v^2 + 9\omega^2 = 144$

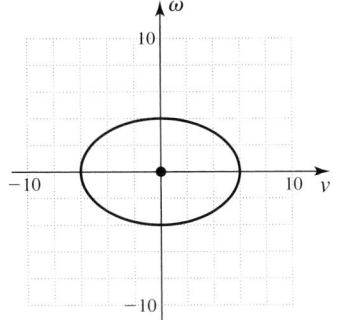

35. $y = x - 4$

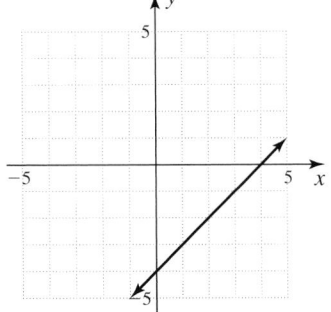

37. $y = x^2 + 1$

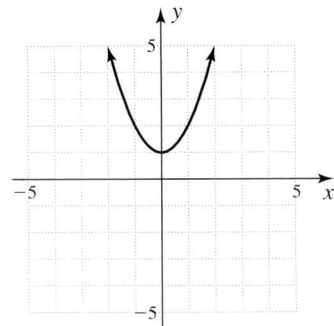

39. $4x^2 + 9y^2 = 36$

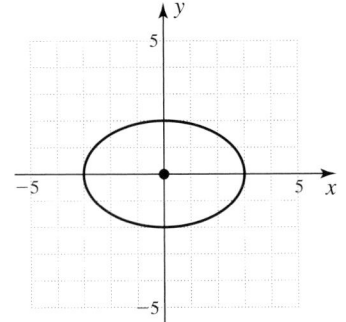

41. $PF_1 = \sqrt{(x - c)^2 + y^2}$ **43.** $a^2 + cx = a\sqrt{(x + c)^2 + y^2}$
45. $b^2x^2 - a^2y^2 = a^2b^2$ **47.** $\left(\dfrac{x}{a} + \dfrac{y}{b}\right)\left(\dfrac{x}{a} - \dfrac{y}{b}\right) = 1$
49. The denominator becomes very large. The fraction approaches zero.
51. $y = -\dfrac{b}{a}x$ **57.** Circle   **59.** Parabola
61. $\dfrac{y^2}{16} - \dfrac{x^2}{9} = 1$

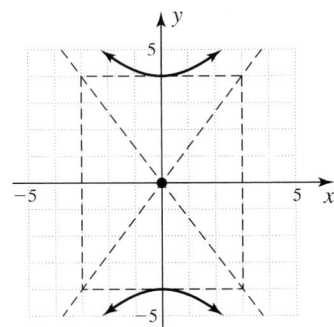

63. $9x^2 - 25y^2 = 225$

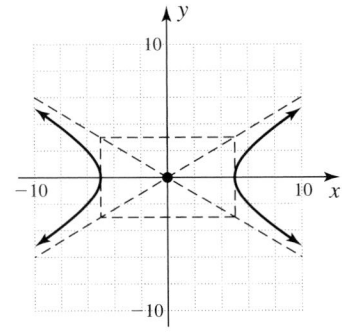

Exercises 9.4

1. $x^2 + y^2 = 16$; $x + y = 4$; $(0, 4)$, $(4, 0)$

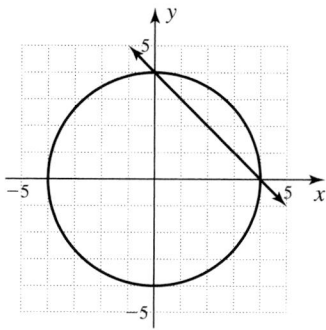

3. $x^2 + y^2 = 25$; $y - x = 5$; $(-5, 0)$, $(0, 5)$

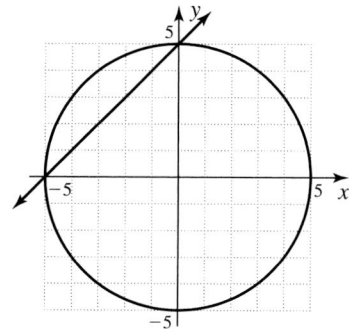

5. $x^2 + y^2 = 25$; $y - x = 1$; $(-4, -3)$, $(3, 4)$

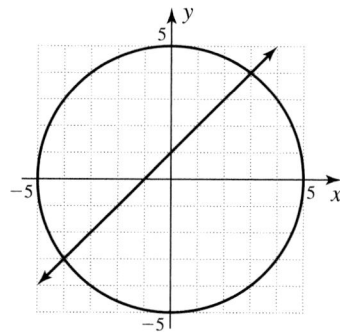

7. $y = x^2 - 5x + 4$; $x - y = 1$; $(1, 0)$, $(5, 4)$

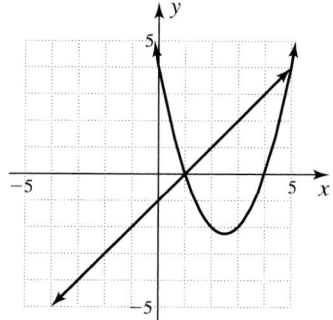

9. $y = (x - 1)^2$; $y - x = 1$; $(0, 1)$, $(3, 4)$

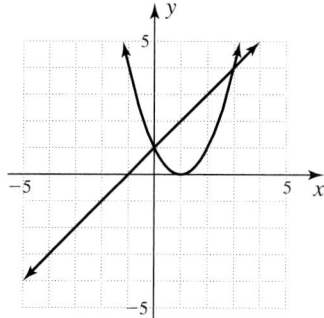

11. $4x^2 + 9y^2 = 36$; $3y - 2x = 6$; $(-3, 0)$, $(0, 2)$

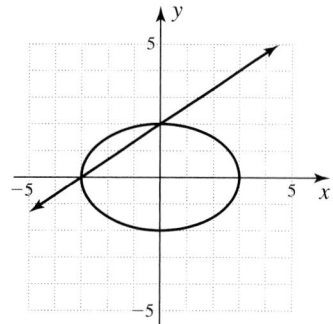

13. $x^2 - y^2 = 16$; $x + 4y = 4$; $(4, 0)$, $\left(-\dfrac{68}{15}, \dfrac{32}{15}\right)$

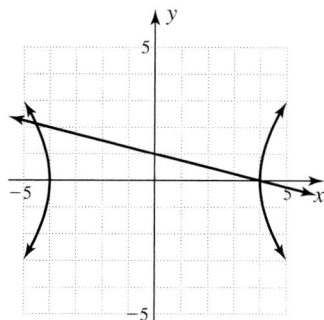

15. $x^2 + y^2 = 4$; $y - x = 5$; $\left(-\dfrac{5}{2} + \dfrac{\sqrt{17}}{2}i, \dfrac{5}{2} + \dfrac{\sqrt{17}}{2}i\right)$,

$\left(-\dfrac{5}{2} - \dfrac{\sqrt{17}}{2}i, \dfrac{5}{2} - \dfrac{\sqrt{17}}{2}i\right)$

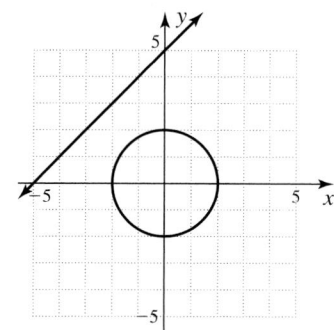

17. $y = 4 - x^2$; $y = x^2 - 4$; $(-2, 0)$, $(2, 0)$

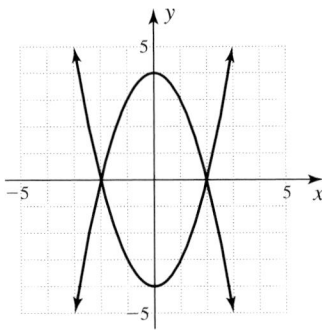

19. $x^2 + y^2 = 25$; $x^2 - y^2 = 7$; $(4, 3)$, $(4, -3)$, $(-4, 3)$, $(-4, -3)$

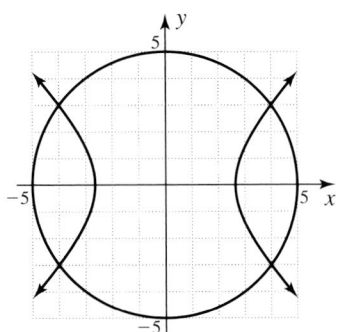

21. $x^2 + y^2 = 16$; $x^2 + 16y^2 = 16$; $(-4, 0)$, $(4, 0)$

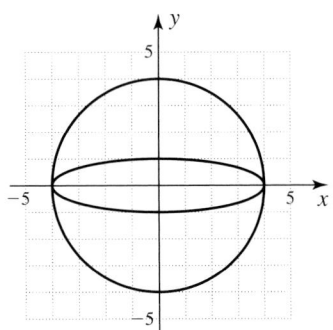

23. $3x^2 - y^2 = 2$; $x^2 + 2y^2 = 3$; $(1, -1)$, $(-1, -1)$, $(1, 1)$, $(-1, 1)$

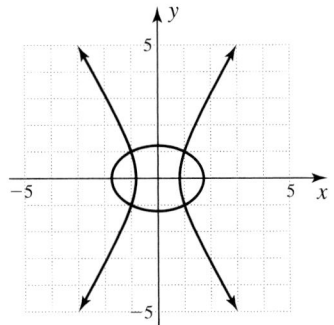

25. $x^2 + 2y^2 = 11$; $2x^2 + y^2 = 19$; $(-3, 1)$, $(-3, -1)$, $(3, 1)$, $(3, -1)$

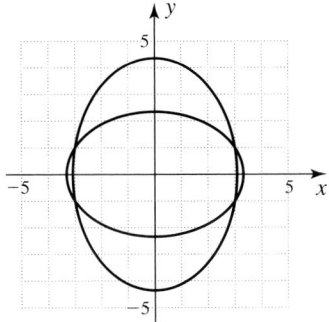

27. $x^2 + y^2 = 4$; $x^2 - y^2 = 9$; $\left(\dfrac{\sqrt{26}}{2}, \dfrac{\sqrt{10}}{2}i \right)$, $\left(\dfrac{\sqrt{26}}{2}, -\dfrac{\sqrt{10}}{2}i \right)$, $\left(-\dfrac{\sqrt{26}}{2}, \dfrac{\sqrt{10}}{2}i \right)$, $\left(-\dfrac{\sqrt{26}}{2}, -\dfrac{\sqrt{10}}{2}i \right)$

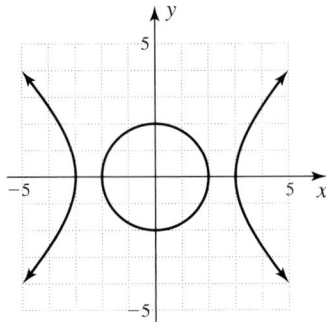

29. $x^2 + y^2 = 1$; $4x^2 + 9y^2 = 36$; $\left(\dfrac{3\sqrt{15}}{5}i, \dfrac{4\sqrt{10}}{5} \right)$, $\left(\dfrac{3\sqrt{15}}{5}i, -\dfrac{4\sqrt{10}}{5} \right)$, $\left(-\dfrac{3\sqrt{15}}{5}i, \dfrac{4\sqrt{10}}{5} \right)$, $\left(-\dfrac{3\sqrt{15}}{5}i, -\dfrac{4\sqrt{10}}{5} \right)$

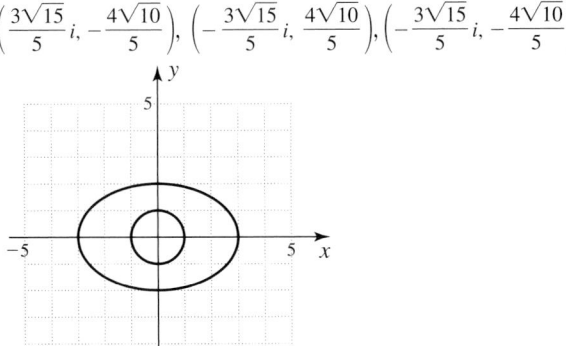

31. 4000 or 1000   **33.** 7 and 8   **35.** 11 and 16 or -11 and -16
37. 70 ft by 31 ft   **39.** $P = \$13{,}600$; $r = 2.5\%$
41. $x - y < 4$

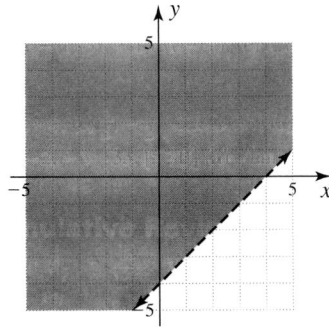

43. $2x - 3y \geq 6$

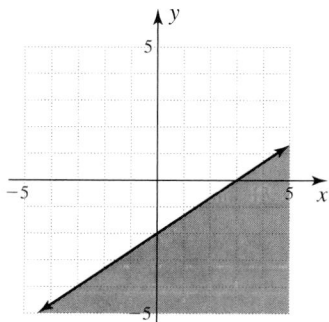

45. $y \geq 2x + 4$

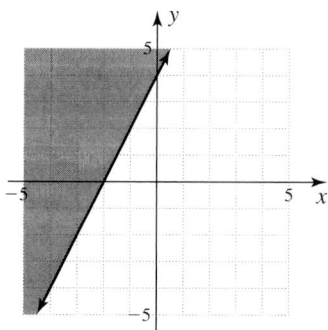

47. 20 eggs   **53.** 12 cm by 10 cm   **55.** (0, 1), (0, −1)
57. $(2 + i\sqrt{2}, 2 - i\sqrt{2}), (2 - i\sqrt{2}, 2 + i\sqrt{2})$ **59.** (1, 0), (0, 1)

Exercises 9.5
1. $x^2 + y^2 > 16$

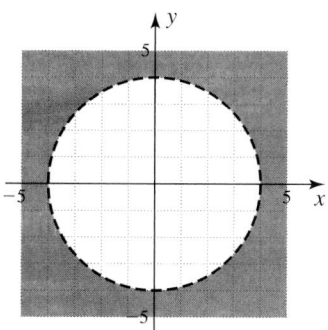

3. $x^2 + y^2 \leq 1$

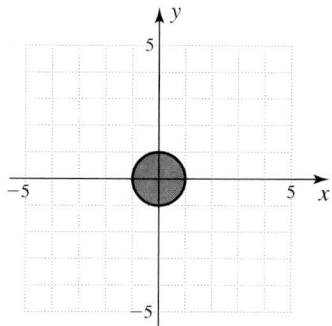

5. $y < x^2 - 2$

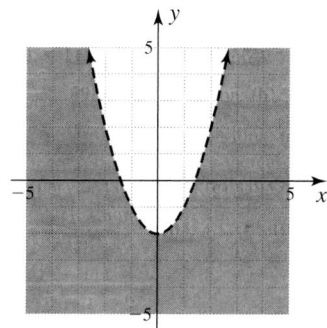

7. $y \leq -x^2 + 3$

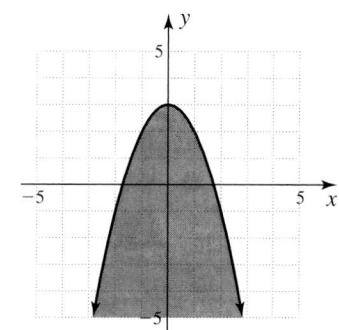

9. $4x^2 - 9y^2 > 36$

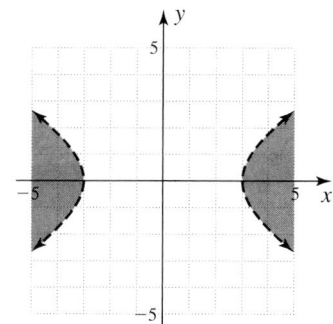

11. $x^2 - y^2 \geq 1$

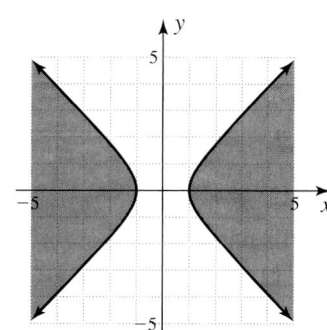

13. $x^2 + y^2 \le 25$; $y \ge x^2$

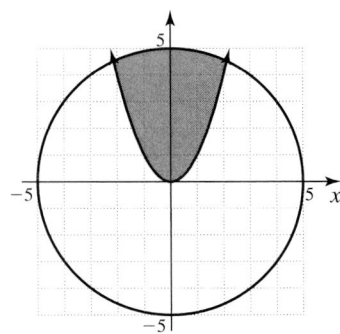

15. $x^2 + y^2 \ge 25$; $y \le x^2$

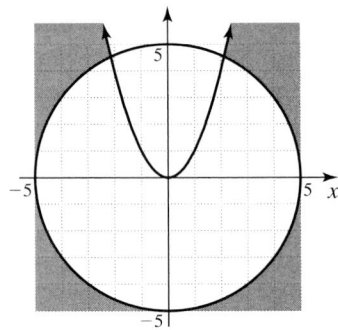

17. $y < x^2 + 2$; $y > x^2 - 2$

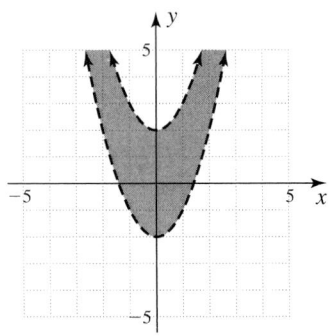

19. $y \ge x^2 + 2$; $y \ge x^2 - 2$

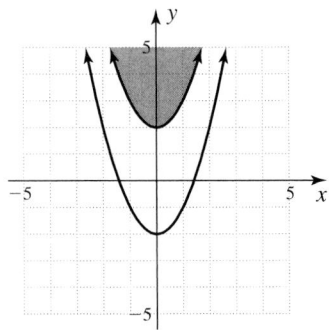

21. $\dfrac{x^2}{4} - \dfrac{y^2}{4} \ge 1$; $\dfrac{x^2}{25} + \dfrac{y^2}{4} \le 1$

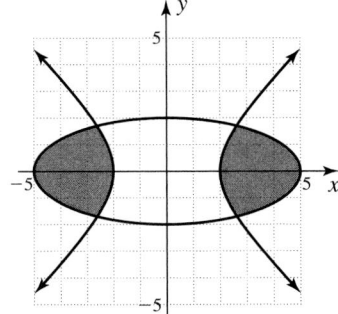

23. $\dfrac{x^2}{36} + \dfrac{y^2}{16} < 1$; $\dfrac{x^2}{16} + \dfrac{y^2}{36} < 1$

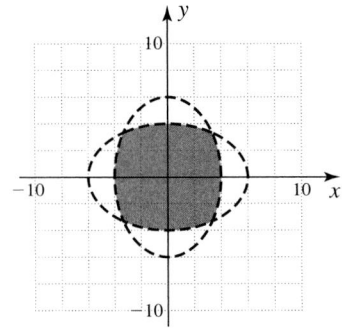

25. (a) $\dfrac{x^2}{36} + \dfrac{y^2}{16} > 1$; $\dfrac{x^2}{16} + \dfrac{y^2}{16} < 1$; **(b)** $\dfrac{x^2}{36} + \dfrac{y^2}{16} \ge 1$;
$\dfrac{x^2}{16} + \dfrac{y^2}{16} \le 1$ **27.** -1 **29.** $P(x) + Q(x) = x^2 + x - 6$;
$P(x) - Q(x) = x^2 - x - 12$ **31.** $C = 432{,}000 - 1800p$
33. $p = 180$ or $p = 80$
39. $x^2 + 4y^2 \ge 4$; $y \ge x^2 + 1$

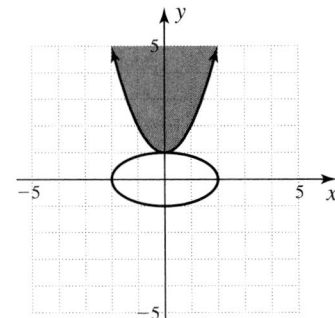

41. $x^2 - 9y^2 \ge 9$

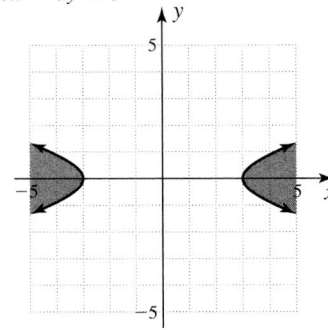

43. $x^2 > 4 - y^2$

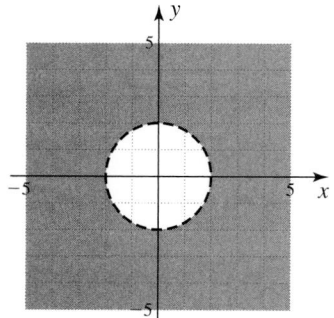

45. $y \geq -x^2 + 2$

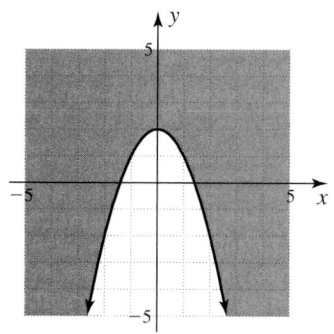

Review Exercises

1. [9.1A] **(a)** $y = 9x^2$

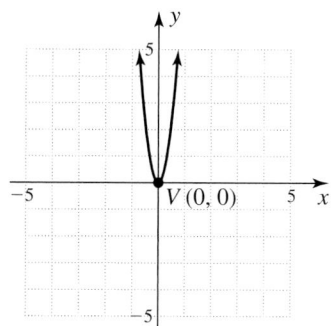

(b) $y = -9x^2$

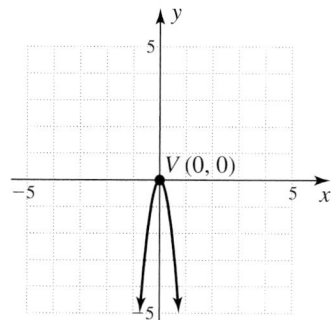

2. [9.1B] **(a)** $y = (x - 1)^2 - 2$; $V(1, -2)$

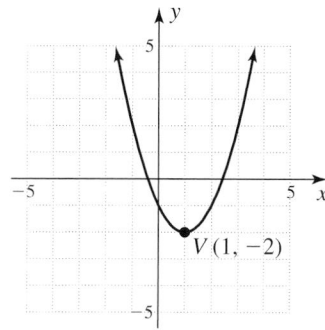

(b) $y = -(x - 1)^2 + 2$; $V(1, 2)$

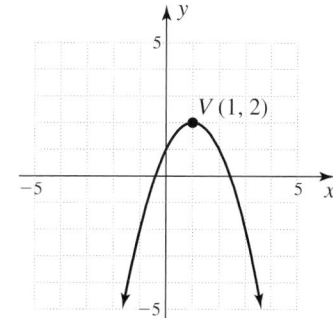

3. [9.1C] **(a)** $y = x^2 - 4x + 2$; $V(2, -2)$

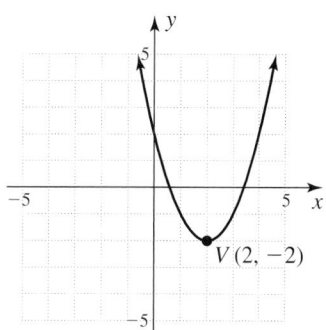

(b) $y = -x^2 + 6x - 5$; $V(3, 4)$

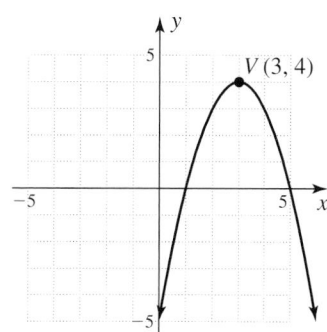

4. [9.1C] **(a)** $y = 2x^2 - 4x + 3$; $V(1, 1)$

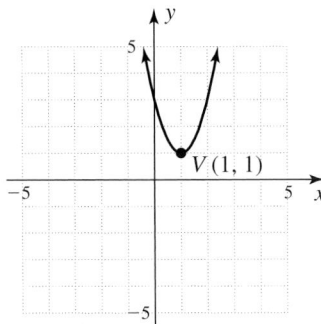

(b) $y = -2x^2 + 4x - 5$; $V(1, -3)$

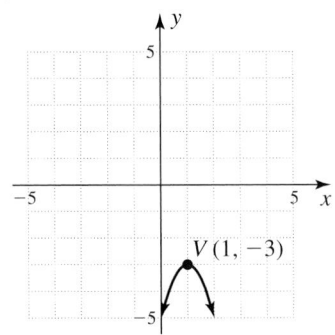

5. [9.1D] **(a)** $x = 2(y - 2)^2 - 2$; $V(-2, 2)$

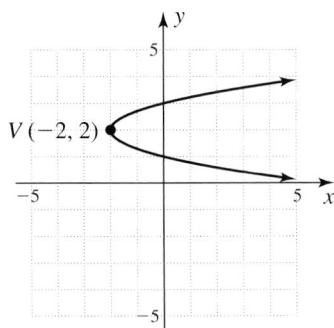

(b) $x = -2(y - 3)^2 + 1$; $V(1, 3)$

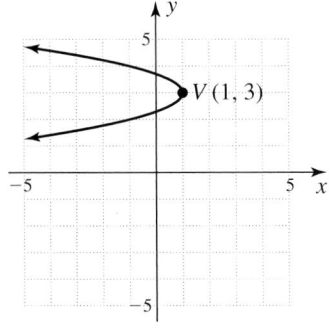

6. [9.1D] **(a)** $x = y^2 - 4y + 1$; $V(-3, 2)$

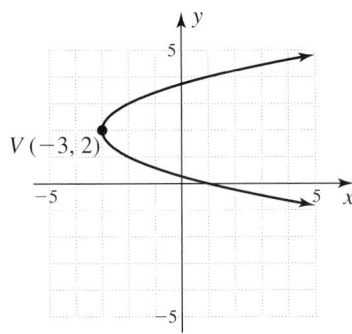

(b) $x = y^2 - 2y + 3$; $V(2, 1)$

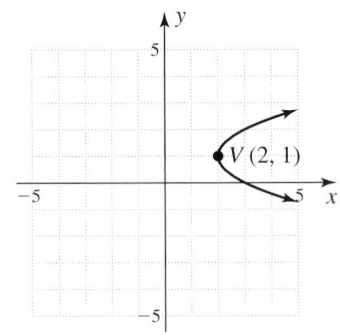

7. [9.1E] **(a)** $x = 1000$; **(b)** $x = 250$ **8.** [9.1E] $\dfrac{3}{4}$ sec; 9 ft
9. [9.2A] **(a)** $\sqrt{130}$; **(b)** $2\sqrt{65}$; **(c)** 5
10. [9.2B] **(a)** $(x + 2)^2 + (y - 2)^2 = 9$; **(b)** $(x - 3)^2 + (y + 2)^2 = 9$
11. [9.2C] **(a)** $(x + 2)^2 + (y - 1)^2 = 4$; $C(-2, 1)$; $r = 2$

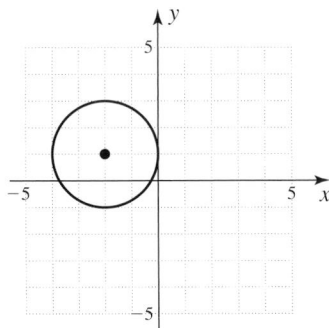

(b) $(x - 1)^2 + (y + 2)^2 = 9$; $C(1, -2)$; $r = 3$

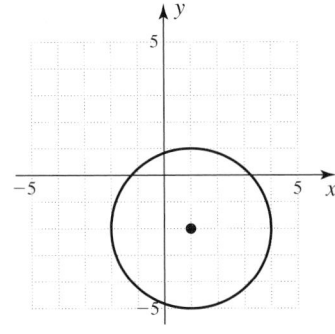

12. [9.2C] **(a)** $x^2 + y^2 = 4$

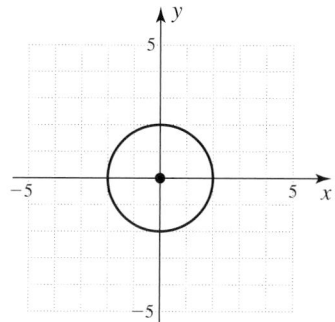

(b) $x^2 + y^2 = 25$

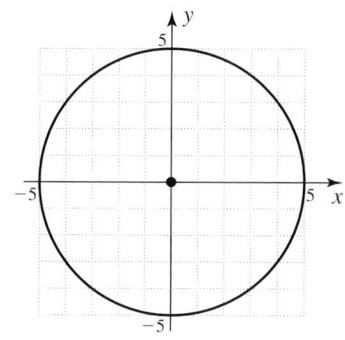

13. [9.2C] **(a)** $x^2 + y^2 + 2x + 2y - 2 = 0$; $C(-1, -1)$; $r = 2$

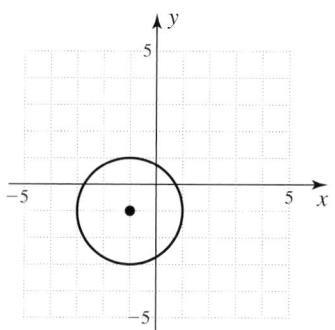

(b) $x^2 + y^2 - 4x + 6y + 9 = 0$; $C(2, -3)$; $r = 2$

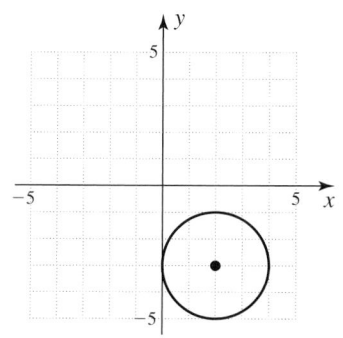

14. [9.2D] **(a)** $4x^2 + 9y^2 = 36$

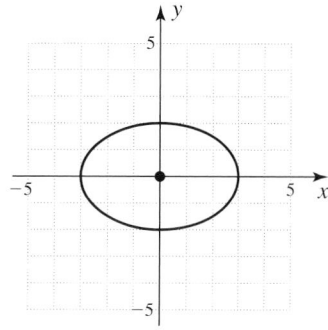

(b) $9x^2 + y^2 = 9$

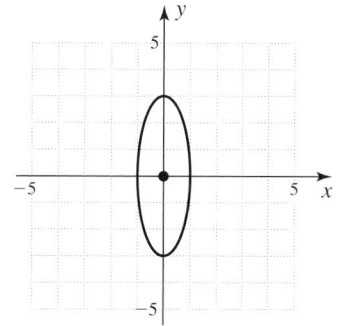

15. [9.2D] **(a)** $\dfrac{(x - 1)^2}{4} + \dfrac{(y - 2)^2}{9} = 1$

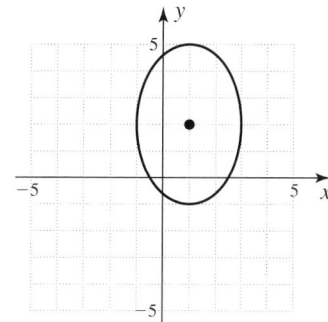

(b) $\dfrac{(x + 2)^2}{9} + \dfrac{(y - 2)^2}{4} = 1$

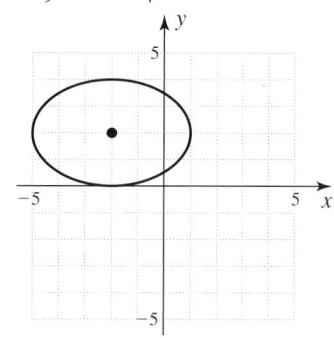

16. [9.3A] **(a)** $\dfrac{x^2}{9} - \dfrac{y^2}{16} = 1$

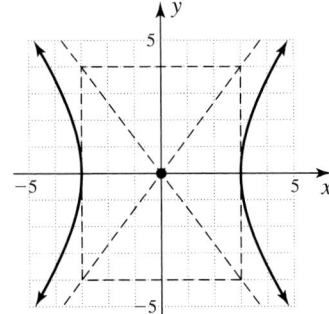

(b) $\dfrac{x^2}{16} - \dfrac{y^2}{9} = 1$

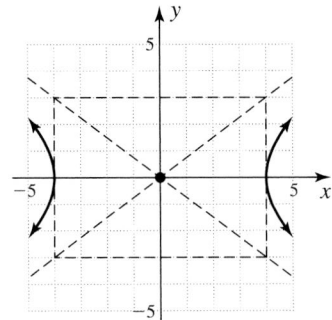

17. [9.3A] **(a)** $\dfrac{y^2}{9} - \dfrac{x^2}{16} = 1$

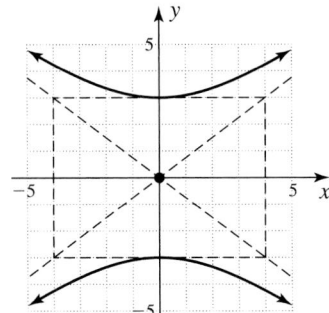

(b) $\dfrac{y^2}{16} - \dfrac{x^2}{9} = 1$

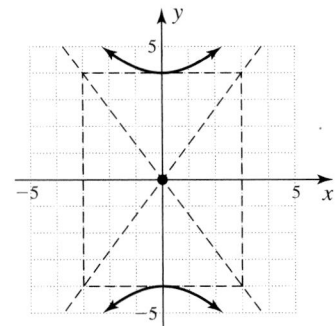

18. [9.3B] **(a)** Parabola; **(b)** Hyperbola; **(c)** Circle; **(d)** Ellipse
19. [9.4A] **(a)** (0, 1), (1, 0); **(b)** (1, 3), (3, 1)

20. [9.4A] **(a)** $\left(\dfrac{4\sqrt{3}}{3}i, \dfrac{8\sqrt{3}}{3}i\right), \left(-\dfrac{4\sqrt{3}}{3}i, -\dfrac{8\sqrt{3}}{3}i\right)$;

(b) $\left(\dfrac{3}{2} + \dfrac{1}{2}i, \dfrac{3}{2} - \dfrac{1}{2}i\right), \left(\dfrac{3}{2} - \dfrac{1}{2}i, \dfrac{3}{2} + \dfrac{1}{2}i\right)$
21. [9.4B] **(a)** (3, 2), (3, −2), (−3, 2), (−3, −2);
(b) (2, 1), (2, −1), (−2, 1), (−2, −1)   **22.** [9.4C] **(a)** $x = 30$; **(b)** $x = 20$
23. [9.5A] **(a)** $y \le 1 - x^2$

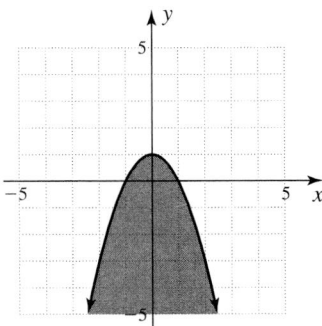

(b) $x \le 4 - y^2$

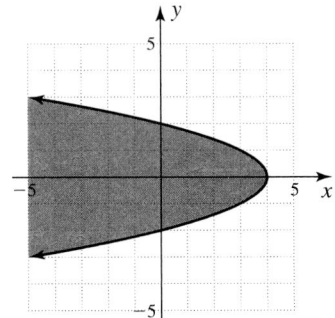

24. [9.5A] **(a)** $x^2 + y^2 \le 4$

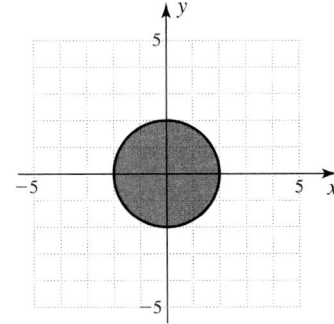

(b) $x^2 + y^2 > 9$

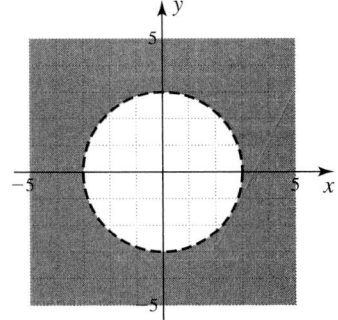

25. [9.5A] **(a)** $4x^2 - y^2 \le 4$

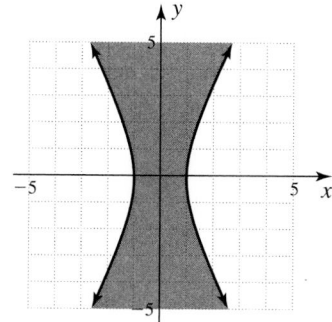

(b) $x^2 - 4y^2 \le 4$

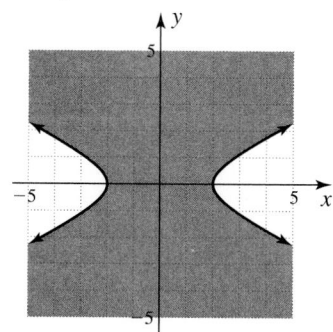

26. [9.5B] **(a)** $x^2 + y^2 \le 4$; $y \le 2 - x^2$

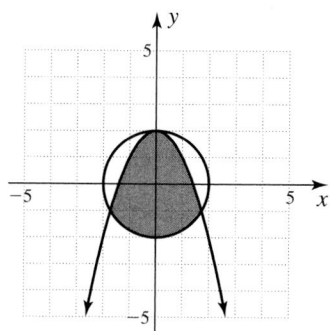

(b) $x^2 + y^2 \le 4$; $y \ge 4x^2$

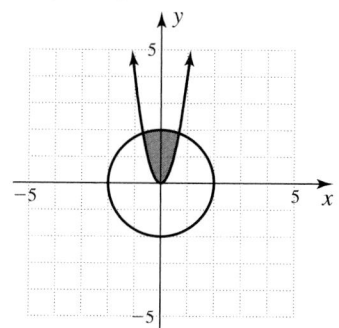

Cumulative Review Chapters 1–9

1.

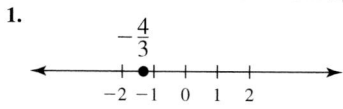

2. $3x^{15}$ **3.** -57 **4.** $x = -\dfrac{4}{3}$ or -8

5. $x + 1 \le 4$; $-2x < 8$

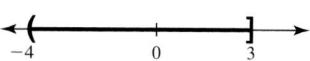

6. $A = \dfrac{8B + 70}{7}$ **7.** \$29,000   **8.** x-intercept: $\left(-\dfrac{5}{2}, 0\right)$; y-intercept: $(0, -5)$

9.

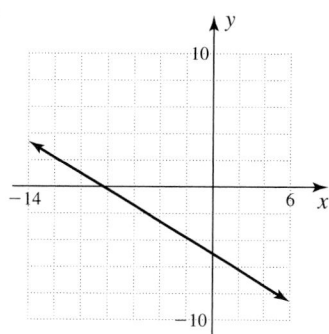

10. $2x - 5y \le -10$

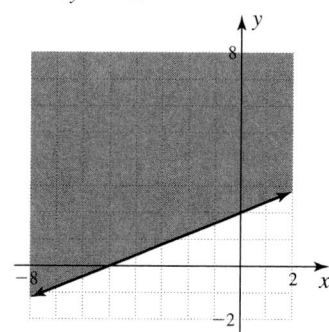

11. $-\dfrac{3}{4}$ **12.** $D = \{-2, 5, -3\}$ **13.** -4 **14.** No solution

15. $(2, 3, -1)$ **16.** 87   **17.** 30 mi/hr   **18.** $9j^2 - 1$

19. $x(3x^3 + 2)(2x + 1)(2x - 1)$ **20.** $2y^2(4x - 3y)^2$ **21.** $x = -\dfrac{3}{2}, -3$

22. $\dfrac{3x - y}{2x - 3}$ **23.** No solution   **24.** $2\dfrac{2}{9}$ hr   **25.** $k = \dfrac{1}{60}$ **26.** $x^{19/6}$

27. $\sqrt[3]{4c^2 d^2}$ **28.** $80 + 31\sqrt{10}$ **29.** No real-number solution

30. $\dfrac{17}{97} + \dfrac{14}{97}i$ **31.** $x = \dfrac{-21 \pm \sqrt{77}}{7}$ **32.** $x = \dfrac{4}{3}, -\dfrac{2}{3} \pm \dfrac{2\sqrt{3}}{3}i$

33. Sum: $\dfrac{5}{4}$; product: -1 **34.** $x \le -8$ or $x \ge 7$

35. $y = -3x^2 - 4x + 3$

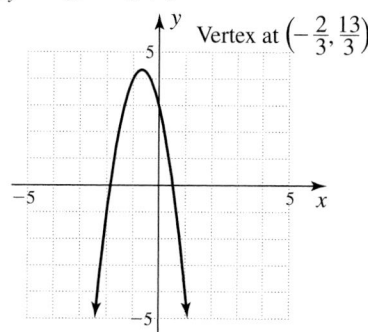

Vertex at $\left(-\dfrac{2}{3}, \dfrac{13}{3}\right)$

36. $x = 500$ **37.** $x^2 + y^2 = 81$

38. $\dfrac{(x-2)^2}{25} + \dfrac{(y+4)^2}{16} = 1$

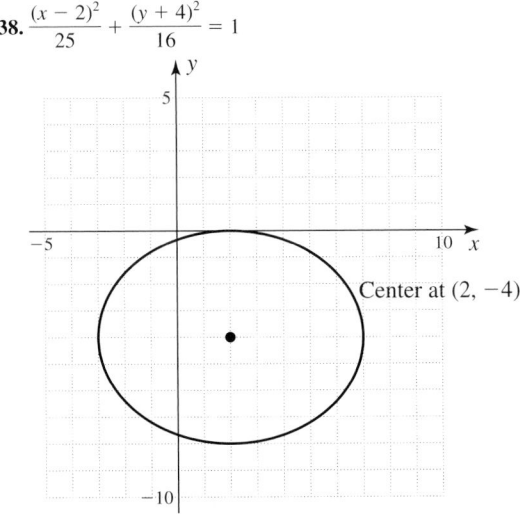

Center at $(2, -4)$

39. $(-3, 4), (4, -3)$

40. $x^2 + y^2 < 4$

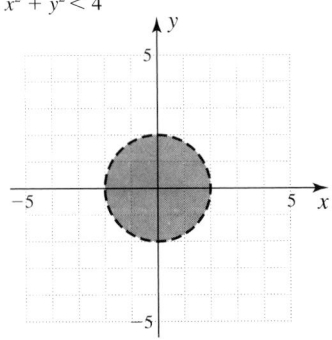

Chapter 10

Exercises 10.1

1. $x^2 - 4x + 8$ **3.** $x^3 + 4x^2 + 16x + 64$ **5.** $\dfrac{x+4}{x^2+16}$ **7.** 3   **9.** $x + a$
11. $x + a + 3$ **13. (a)** 1; **(b)** x; **(c)** $|x|$ **15. (a)** 4; **(b)** $3x + 1$; **(c)** $3x - 1$
17. (a) 1; **(b)** $|x|$; **(c)** x **19. (a)** 3; **(b)** 3; **(c)** -1 **21. (a)** 17; **(b)** -13
23. (a) 1; **(b)** 0   **25. (a)** Undefined; $x \neq 3$; **(b)** Undefined; $x \neq 3$
27. (a) $\sqrt{2}$; **(b)** 0; **(c)** $\sqrt{x^2+1}$; **(d)** $x + 1$ **29. (a)** Undefined; $x \neq 2$;
(b) $\sqrt{\dfrac{1}{2}} = \dfrac{\sqrt{2}}{2}$; **(c)** $\dfrac{1}{x-2}, x > 0$; **(d)** $\sqrt{\dfrac{1}{x^2-2}}$ or $\dfrac{\sqrt{x^2-2}}{x^2-2}$
31. The domain of the sum, difference, and product is: $\{x \mid x$ is a real number$\}$; The domain of the quotient is: $\{x \mid x$ is a real number and $x \neq 1\}$
33. The domain of the sum, difference, and product is: $\{x \mid x$ is a real number$\}$; The domain of the quotient is: $\{x \mid x$ is a real number and $x \neq 1\}$
35. The domain of the sum, difference, and product is: $\{x \mid x$ is a real number and $x \neq 1$ and $x \neq -2\}$; The domain of the quotient is: $\{x \mid x$ is a real number and $x \neq 1$ and $x \neq -2\}$ **37.** The domain of the sum, difference, and product is: $\{x \mid x$ is a real number and $x \neq 1$ and $x \neq -4\}$; The domain of the quotient is: $\{x \mid x$ is a real number and $x \neq 1$ and $x \neq -4$ and $x \neq 0\}$ **39.** The domain of the sum, difference, and product is: $\{x \mid x$ is a real number and $x \neq -3$ and $x \neq 2\}$; The domain of the quotient is: $\{x \mid x$ is a real number and $x \neq -3$ and $x \neq 2$ and $x \neq 1\}$
41. $P(x) = -0.0005x^2 + 34x - 120,000$
43. (a) $L(x) + R(x) = -0.28x^2 + 3x + 22$; **(b)** $L(0) + R(0) = 22$ million;
(c) $L(10) + R(10) = 24$ million; **(d)** $L(10) - R(10) = 6$ million
45. (a) $\dfrac{C(t)}{P(t)} = \dfrac{2.5t^2 + 8.5t + 111}{-0.46t^2 + 1.14t + 31.08}$ (in thousands);

(b) \$3571; **(c)** \$8544   **47. (a)** $(K \circ C)(F) = \dfrac{5}{9}(F - 32) + 273$;
(b) $C(41) = 5°$; **(c)** $(K \circ C)(212) = 373$ K   **49. (a)** 38;
(b) $(E \circ F)(x) = x + 2$; **(c)** 10   **51. (a)** 0; **(b)** 8; **(c)** 16
53. (a) $(f \circ F)(x) = 4\left(\dfrac{9}{5}x - 8\right)$; **(b)** 40   **55.** $x + y = 3$

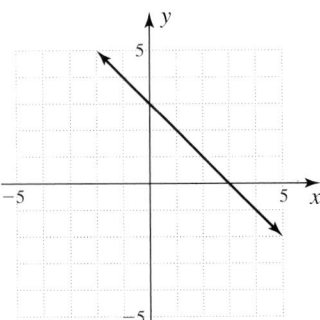

57. $2x + \dfrac{1}{2}y = 2$

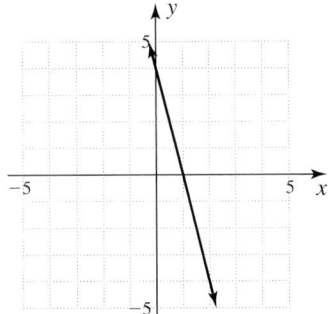

59. $y = -2x + 4$

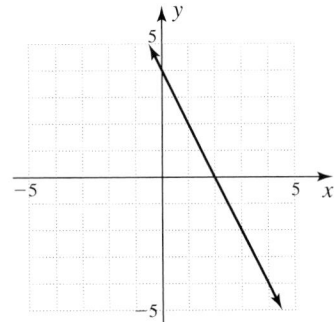

61. $\{x \mid 0 \leq x \leq 5\}$ **63.** $\{x \mid 2 \leq x \leq 5\}$
69. $P(x) = -\dfrac{x^2}{30} + 240x - 72,000$ **71.** $x^3 + 1$ **73.** -26
75. $x + a, x \neq a$ **77.** $x^2 - x + 6$ **79.** -2 **81.** $-\dfrac{1}{2}$
83. The domain of the sum, difference, product, and quotient is: $\{x \mid x$ is a real number and $x \neq 0$ and $x \neq -1\}$

Exercises 10.2
1. Yes

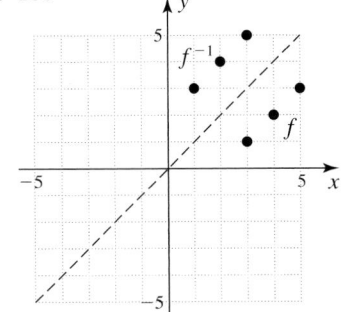

3. No

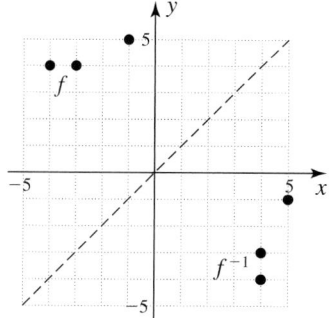

5. $\{(x, y) \mid y = \frac{1}{3}x - 1\}$; yes

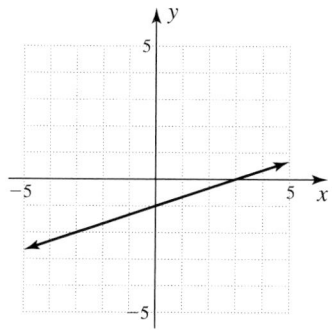

7. $\{(x, y) \mid y = \frac{1}{2}x + 2\}$; yes

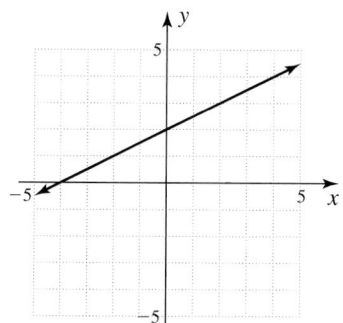

9. $\{(x, y) \mid y^2 = \frac{1}{2}x; y = \pm\frac{\sqrt{2x}}{2}\}$; no

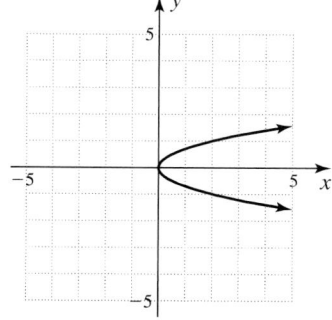

11. $\{(x, y) \mid y^2 = x + 1; y = \pm\sqrt{x + 1}\}$; no

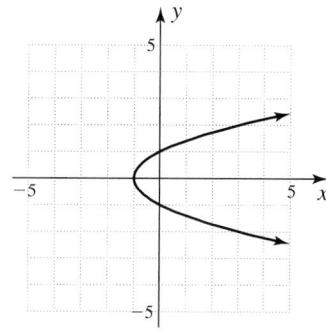

13. $\{(x, y) \mid y^3 = -x; y = -\sqrt[3]{x}\}$; yes

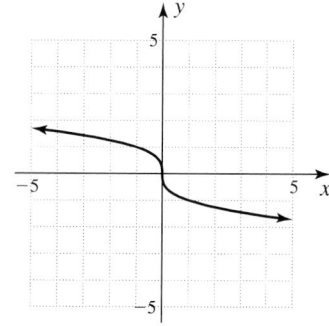

15. $y = 2^{x+1}; x = 2^{y+1}$; yes

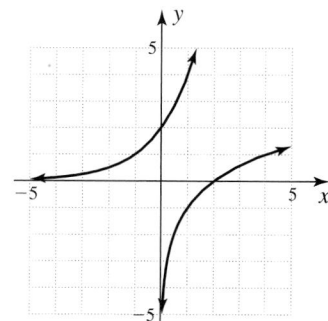

17. $y = \left(\frac{1}{3}\right)^x; x = \left(\frac{1}{3}\right)^y$; yes

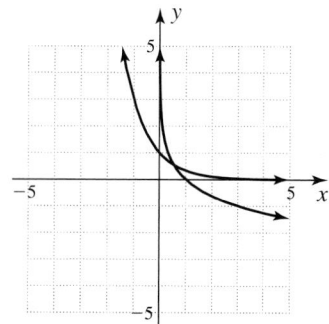

19. $y = 2^{-x}$; $x = 2^{-y}$; yes

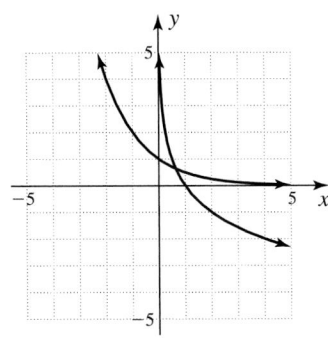

21. (a) 3; **(b)** -1 **23.** $\dfrac{1}{x}$

25. $f(x) = 2x$; $f^{-1}(x) = \dfrac{1}{2}x$

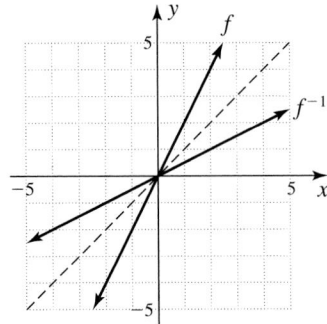

27. $f(x) = -x^2 + 2$; $f^{-1}(x) = \pm\sqrt{2-x}$
or $x - 2 = -y^2$

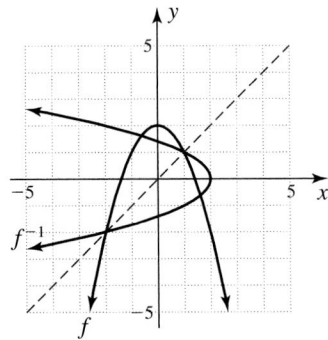

29. $f(x) = \sqrt{4 - x^2}$; f^{-1}: $x = \sqrt{4 - y^2}$

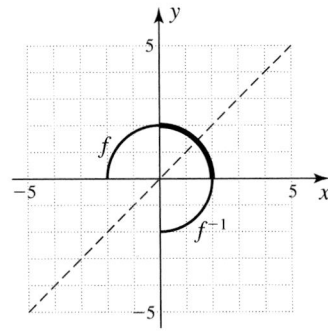

31. $f(x) = \sqrt{9 - x^2}$; f^{-1}: $x = \sqrt{9 - y^2}$

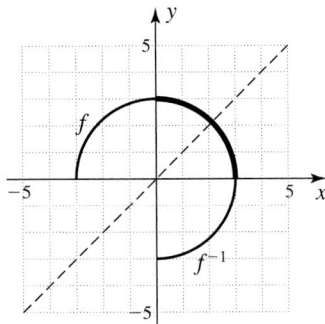

33. $f(x) = x^3$; $f^{-1}(x) = \sqrt[3]{x}$

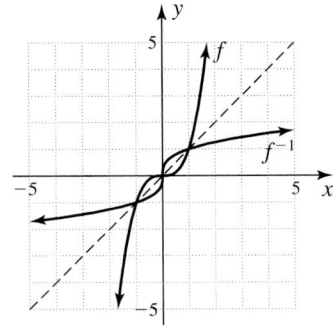

35. $f(x) = |x|$; f^{-1}: $x = |y|$

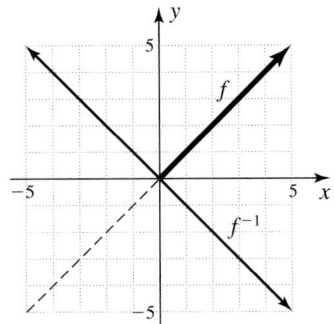

37. (a) $f^{-1}(S) = \dfrac{S + 21}{3}$; **(b)** $9\dfrac{1}{3}$ in.   **39. (a)** 16; **(b)** $f^{-1}(d) = d + 16$;
(c) 28 in.   **41. (a)** $f(1988) = 41.44$ sec; **(b)** $f^{-1}(w) = \dfrac{280 - w}{0.12}$;
(c) In the year 2000   **43.** 9   **45.** 27   **47.** 4   **49.** $f^{-1}(x) = \dfrac{x + 2}{3}$
51. $f^{-1}(x) = 2x - 1$ **53.** $f^{-1}(x) = \sqrt[3]{x - 1}$ **55.** $f^{-1}(x) = x^2, x \geq 0$
63. $f^{-1}(x) = x^{1/3} = \sqrt[3]{x}$; yes

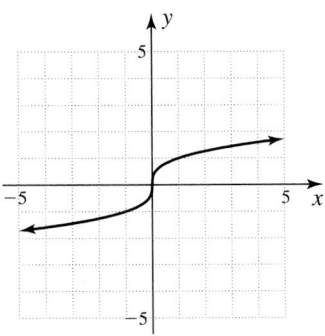

65. $D = \{2, 3, 4\}$; $R = \{1, 2, 3\}$ **67.** $D = \{1, 2, 3\}$; $R = \{2, 3, 4\}$

Exercises 10.3

1. (a) $\frac{1}{5}$; (b) 1; (c) 5   **3.** (a) $\frac{1}{9}$; (b) 1; (c) 9   **5.** (a) $\frac{1}{10}$; (b) 1; (c) 10

7. (a) Increasing; (b) Decreasing

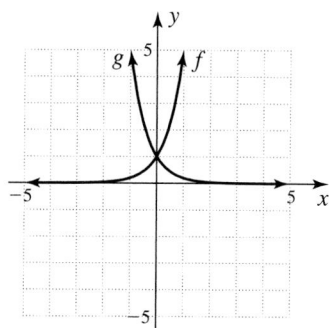

9. (a) Increasing; (b) Decreasing

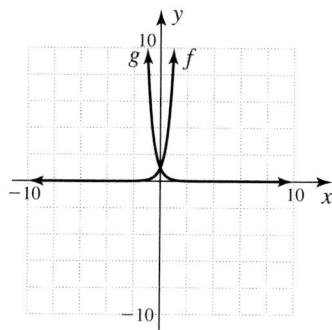

11. (a) Increasing; (b) Decreasing

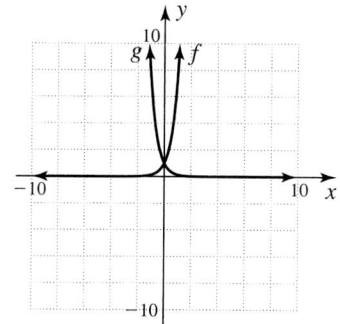

13. (a) Increasing; (b) Decreasing

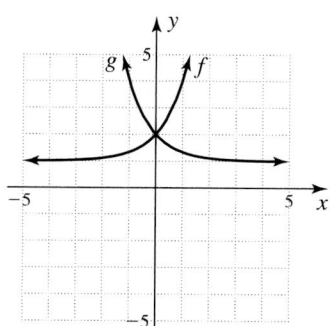

15. (a) Increasing; (b) Decreasing

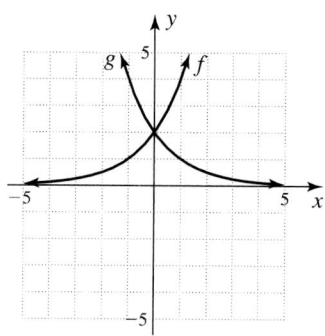

17. (a) $2459.60; (b) $2435.19   **19.** (a) $1822.12; (b) $1814.02
21. (a) 2000; (b) 4000; (c) 8000   **23.** (a) 2000 g; (b) 699.9 g; (c) 244.9 g
25. 21,448,000   **27.** (a) 285.2 million; (b) 1,799,492 million
29. (a) 333,333; (b) 222,222; (c) 8671   **31.** (a) 4.42 lb/in.^{2};
(b) 6.42 lb/in.^{2} **33.** Product property   **35.** Power property
37. Continuous = $1822.12; Monthly = $1819.40; Continuous is $2.72
more.   **43.** 165 g   **45.** $f(x) = e^{x/2}$

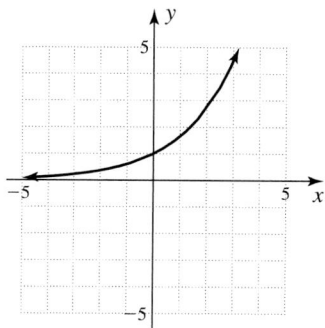

47. $f(x) = 6^x$

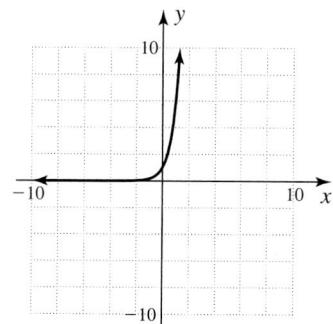

Exercises 10.4

1. $y = \log_2 x$

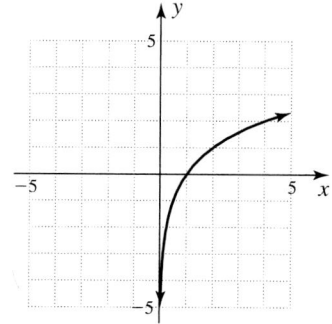

3. $y = \log_5 x$

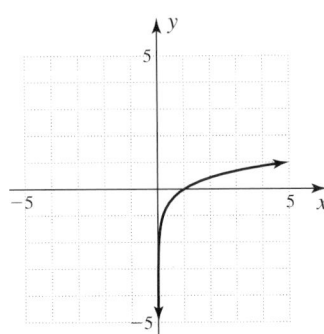

5. $f(x) = \log_{1/2} x$

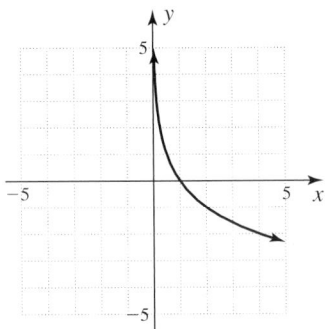

7. $y = \log_{1.5} x$

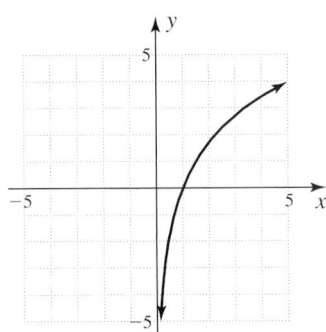

9. $\log_2 128 = x$ **11.** $\log_{10} 1000 = t$ **13.** $\log_{81} 9 = \dfrac{1}{2}$ **15.** $\log_{216} 6 = \dfrac{1}{3}$

17. $\log_e t = 3$ **19.** $9^3 = 729$; true   **21.** $2^{-8} = \dfrac{1}{256}$; true

23. $81^{3/4} = 27$; true   **25.** $4^x = 16$ **27.** $10^{-2} = 0.01$; true
29. $e^{3.4012} = 30$; true   **31.** $e^{-1.15} = 0.3166$; true   **33.** $x = 9$

35. $x = \dfrac{1}{27}$ **37.** $x = 2$ **39.** $x = 2$ **41.** $x = 2$ **43.** $x = 3$ **45.** $x = -2$

47. $x = -\dfrac{1}{4}$ **49.** $x = -2$ **51.** 8   **53.** 4   **55.** -3 **57.** 6   **59.** 0   **61.** 1

63. 1   **65.** -1 **67.** $\dfrac{1}{3}$ **69.** -3 **71.** t **73.** t

75. $\log \dfrac{26}{7} - \log \dfrac{15}{63} + \log \dfrac{5}{26} = \log\left(\dfrac{26}{7} \div \dfrac{15}{63} \cdot \dfrac{5}{26}\right)$

$= \log\left(\dfrac{\overset{1}{\cancel{26}}}{7} \cdot \dfrac{\overset{9}{\cancel{63}}}{\underset{3}{\cancel{15}}} \cdot \dfrac{5}{\underset{1}{\cancel{26}}}\right) = \log\left(\dfrac{9}{3}\right) = \log 3$

77. $\log b^3 + \log 2 - \log \sqrt{b} + \log \dfrac{\sqrt{b^3}}{2}$

$= \log\left(b^3 \cdot 2 \div \sqrt{b} \cdot \dfrac{\sqrt{b^3}}{2}\right) = \log\left(b^3 \cdot 2 \cdot \dfrac{1}{\cancel{\sqrt{b}}} \cdot \dfrac{b\cancel{\sqrt{b}}}{2}\right)$

$= \log b^4 = 4 \log b$

79. $\log k^{3/2} + \log r - \log k - \log r^{3/4}$

$= \log(k^{3/2}\, r \div k \div r^{3/4}) = \log(k^{3/2-1} \cdot r^{1-3/4})$

$= \log(k^{1/2}\, r^{1/4}) = \log(k^{2/4}\, r^{1/4}) = \log(k^2\, r)^{1/4}$

$= \dfrac{1}{4} \log k^2 r$

81. $2 \log b + 6 \log a - 3 \log b$

$= \log b^2 + \log a^6 - \log b^3$

$= \log(b^2 \cdot a^6 \div b^3)$

$= \log\left(\dfrac{b^2}{1} \cdot \dfrac{a^6}{1} \cdot \dfrac{1}{b^3}\right)$

$= \log\left(\dfrac{a^6}{b}\right)$

83. $\dfrac{1}{3} \log x + \dfrac{1}{3} \log y^2 - \log z$

$= \log x^{1/3} + \log y^{2/3} - \log z$

$= \log(x^{1/3} \cdot y^{2/3}) - \log z$

$= \log(\sqrt[3]{xy^2} \div z)$

$= \log\left(\dfrac{\sqrt[3]{xy^2}}{z}\right)$

85. $R = 8.9$ **87.** $\log A = 4.16769$
89. (a) 53%; **(b)** 77%; **(c)** 91%;

(d)

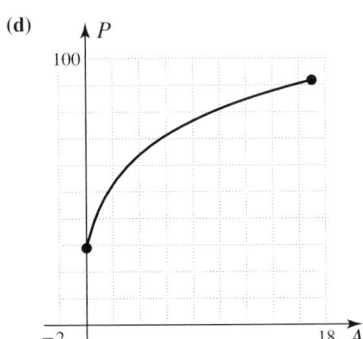

(e) 71 in.   **91.** 3.268×10^1 **93.** 2.387×10^{-3} **95.** 5.69×10^{-5}

97. $R = \dfrac{(1.02)^{40}}{(1.085)^{10}}$; $\log_{10} R = -0.01030$; 8.5% annually is greater.

103. 6.4
105. $\log\left(\dfrac{xy^2}{z^3}\right)^2 = 2\left[\log\left(\dfrac{xy^2}{z^3}\right)\right]$

$= 2\left[\log x + \log y^2 - \log z^3\right]$

$= 2\left[\log x + 2 \log y - 3 \log z\right]$

$= 2 \log x + 4 \log y - 6 \log z$

107. $\log_2 1024 = 10$ **109.** -3 **111.** -1 **113.** $x = \dfrac{1}{27}$ **115.** $x = -3$
117. $y = \log_7 x$

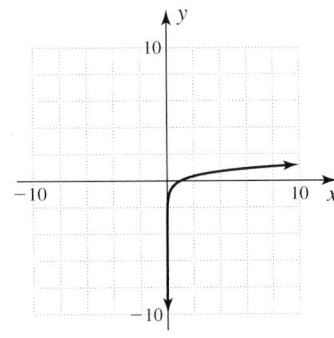

Exercises 10.5
1. 1.8722   **3.** 3.2648   **5.** -1.3595 **7.** 1.7005   **9.** -1.9073
11. -2.0652 **13.** 18.50   **15.** 0.01   **17.** 29.04   **19.** 0.73
21. 1.0986   **23.** 3.9512   **25.** 7.7647   **27.** -2.9188 **29.** -7.3858

31. 3.50   **33.** 65.00   **35.** 1.04   **37.** 0.10   **39.** 0.01   **41.** 2.7268
43. 0.8010   **45.** −0.6826   **47.** −0.6610   **49.** −0.6439
51. $f(x) = e^{3x}$

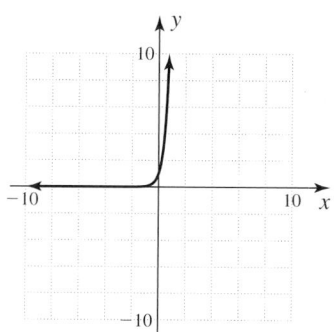

53. $f(x) = -e^{3x}$

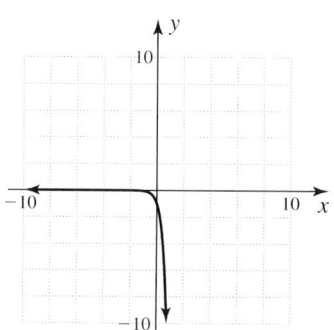

55. $f(x) = e^{-3x}$

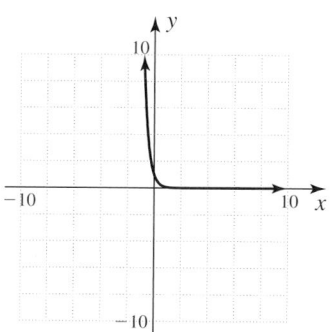

57. $f(x) = e^{(1/2)x}$

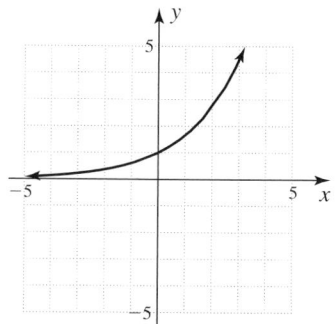

59. $f(x) = e^x + 1$

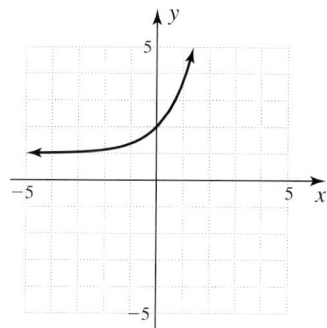

61. $f(x) = e^{0.5x} + 1$

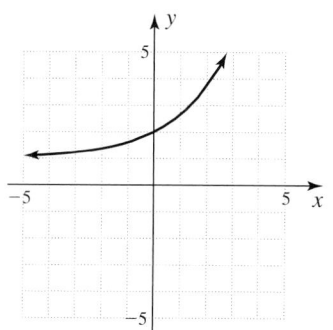

63. $f(x) = 2e^x$

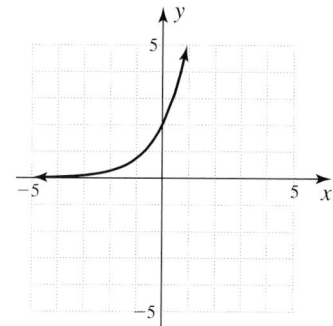

65. $f(x) = \ln(x + 1)$

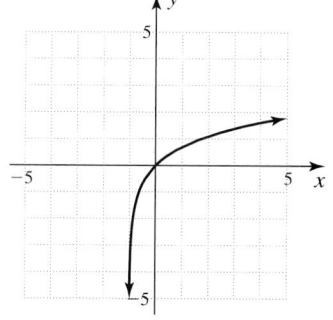

67. $f(x) = \ln x + 4$

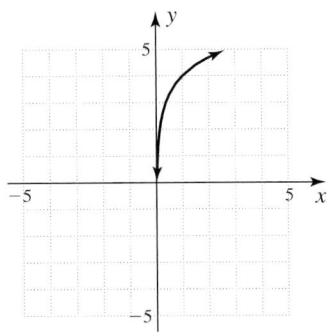

69. $f(x) = \ln(x - 1)$

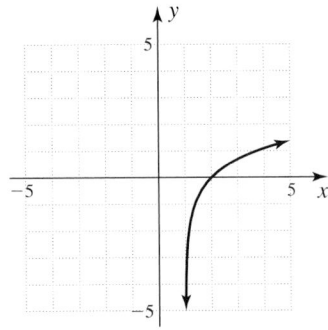

71. 6.2 **73.** 7.8 **75.** 6.4 **77. (a)** 33,000,000; **(b)** 44,545,341
79. (a) 50,000; **(b)** 74,591; **(c)** 166,006 **81. (a)** 1000; **(b)** 368
83. (a) 0; **(b)** 39.3 **85. (a)** −0.8055; **(b)** −3.8968 **87.** $k = 1.5850$
89. $k = 1.3979$
91. (a) Graph $f(x) = 2^x$.

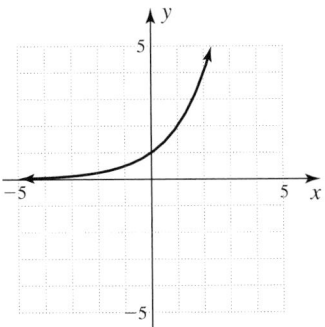

(b) Graph $g(x) = \log 2^x$.

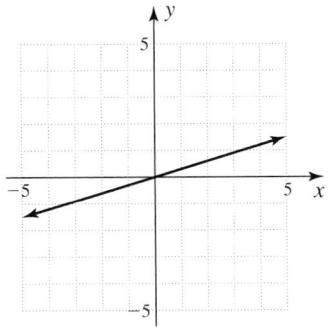

(c) Slope is $\log 2 \approx 0.3$. **93.** The slope of $g(x) = \log$ of base of $f(x)$.
101. 0.2027 **103.** 1.6720 **105.** −3.3820 **107.** 0.0034
109. $f(x) = -e^{(1/4)x}$

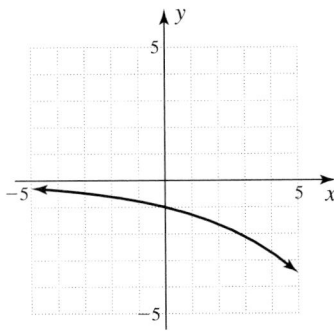

111. $f(x) = e^{(1/4)x} + 1$

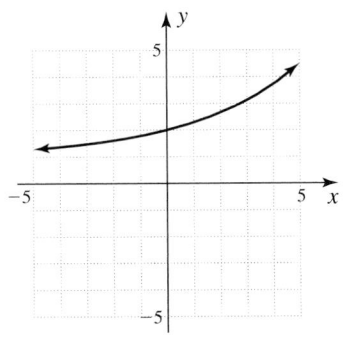

113. $f(x) = \ln x + 3$

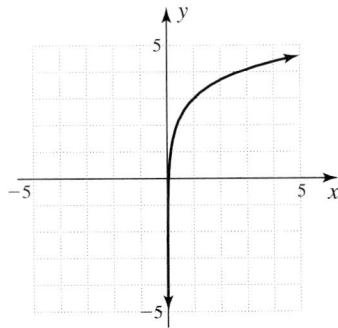

Exercises 10.6

1. $x = 2$ **3.** $x = 5$ **5.** $x = \dfrac{4}{3}$ **7.** $x = -1$ **9.** $x = 3.2059$ **11.** $x = 6$

13. $x = 0.6309$ **15.** $x = \dfrac{7}{3}$ **17.** $x = \dfrac{-3 \pm \sqrt{17}}{2}$ **19.** $x = 2.3026$

21. $x = 2.3026$ **23.** $k = 1.7006$ **25.** $k = -1.1513$ **27.** $x = 8$

29. $x = \dfrac{1}{8}$ **31.** $x = e \approx 2.7183$ **33.** $x = e^3 \approx 20.0855$ **35.** $x = \dfrac{7}{3}$

37. $x = \dfrac{17}{3}$ **39.** $x = 8$ **41.** $x = 5$ **43.** $x = 1$ **45.** $x = \dfrac{1}{9}$ **47.** $x = 5$

49. $x = -3$ or $x = -1$ **51.** 13.86 yr **53.** 10.66 yr **55.** About 6.1 billion
57. 17.3 min **59.** 80.5 min **61. (a)** 100,000; **(b)** 67,032; **(c)** 13,534;
(d) 1832 **63.** 23,104.9 yr **65.** 13.3 yr **67. (a)** 14.7 lb/in.²;

(b) 11.4 lb/in.²; **(c)** 8.9 lb/in.² **69. (a)** 5 yr after 1992 (1997);
(b) 7 yr after 1992 (1999) **71. (a)** 5000; **(b)** 2247; **(c)** In 2010
73. (a) 54 million; **(b)** In 1995 **75. (a)** 12 yr old;
(b) 7 yr old **77.** About 694 months **79. (a)** 3; **(b)** 11; **(c)** 21
81. $k = 0.59$ **83.** 3,405,095 **85.** 1,242,987,238 **93.** 59.9 min
95. $x = 1$ **97.** $x = 10$ **99.** $x = 1.4307$ **101.** $x = 4$

Review Exercises

1. [10.1A] **(a)** $4 + x - x^2$; **(b)** $-x - x^2$; **(c)** $4 + 2x - 2x^2 - x^3$; **(d)** $\dfrac{2 - x^2}{2 + x}$

2. [10.1A] **(a)** $6 + x - x^2$; **(b)** $-x - x^2$; **(c)** $9 + 3x - 3x^2 - x^3$; **(d)** $\dfrac{3 - x^2}{3 + x}$

3. [10.1A] **(a)** 6; **(b)** 7 **4.** [10.1B] **(a)** -6; **(b)** $(2 - x)^3$; **(c)** $2 - x^3$
5. [10.1B] **(a)** $(3 - x)^3$; **(b)** $3 - x^3$; **(c)** -5
6. [10.1C] $\{x \mid x$ is a real number and $x \neq 1$ and $x \neq 4\}$
7. [10.1C] $\{x \mid x$ is a real number and $x \neq 1$ and $x \neq 3$ and $x \neq -4\}$
8. [10.1D] **(a)** $P(x) = -0.02x^2 + 70x - 30{,}000$;
(b) $P(x) = -0.02x^2 + 60x - 40{,}000$ **9.** [10.2A]
(a) $D = \{4, 6, 8\}$; $R = \{4, 6, 8\}$; **(b)** $S^{-1} = \{(4, 4), (6, 6),$
$(8, 8)\}$; **(c)** $D = \{4, 6, 8\}$; $R = \{4, 6, 8\}$

(d)

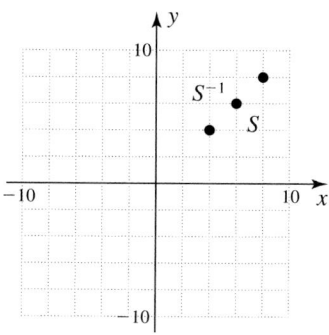

10. [10.2A] **(a)** $D = \{4, 6, 8\}$; $R = \{5, 7, 9\}$; **(b)** $S^{-1} = \{(5, 4), (7, 6),$
$(9, 8)\}$; **(c)** $D = \{5, 7, 9\}$; $R = \{4, 6, 8\}$;

(d)

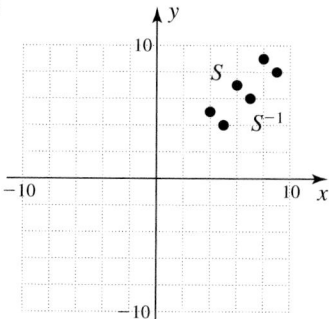

11. [10.2B] **(a)** $f^{-1}(x) = \dfrac{x + 3}{3}$;

(b)

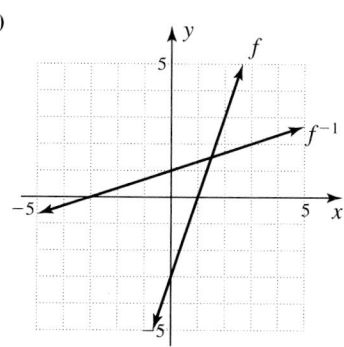

12. [10.2B] **(a)** $f^{-1}(x) = \dfrac{x + 4}{4}$;

(b)

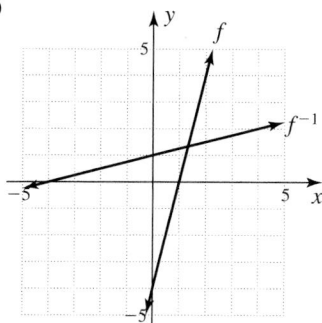

13. [10.2C] $y^2 = \dfrac{1}{4}x$; $f^{-1}(x) = \pm\dfrac{\sqrt{x}}{2}$; no

14. [10.2C] $y^2 = \dfrac{1}{5}x$; $f^{-1}(x) = \pm\dfrac{\sqrt{5x}}{5}$; no

15. [10.2C] Yes

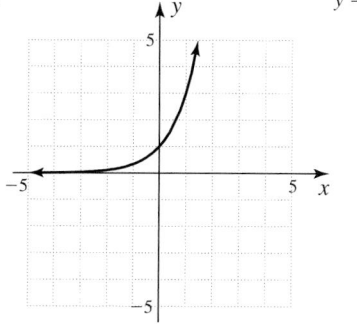

16. [10.2C] Yes

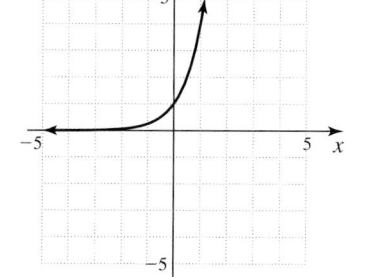

17. [10.2D] **(a)** 12; **(b)** $f^{-1}(d) = d + 16$; **(c)** 26 in.
18. [10.3A] **(a)** $f(x) = 2^{x/2}$

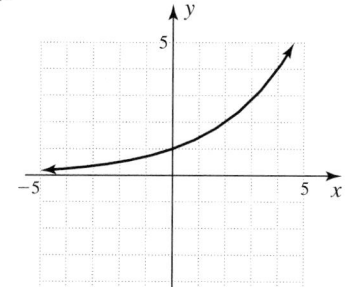

(b) $f(x) = 2^{-x/2}$

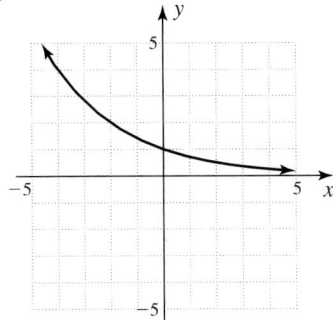

19. [10.3A] **(a)** $g(x) = \left(\dfrac{1}{2}\right)^{x/2}$

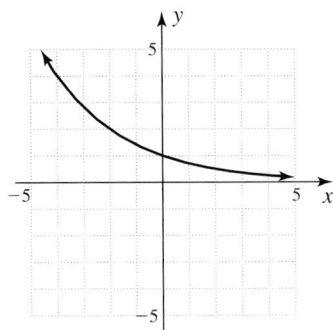

(b) $g(x) = \left(\dfrac{1}{2}\right)^{-x/2}$

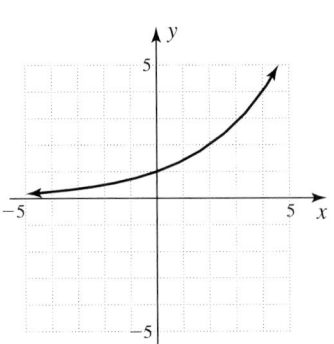

20. [10.3C] **(a)** 1000 g; **(b)** 61 g
21. [10.4A] $f(x) = \log_5 x$

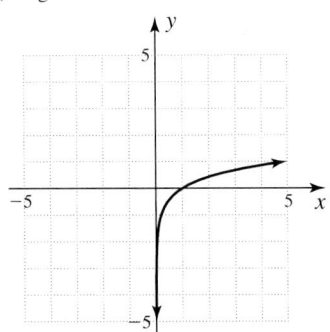

22. [10.4B] **(a)** $\log_3 243 = 5$; **(b)** $\log_2\left(\dfrac{1}{8}\right) = -3$

23. [10.4B] **(a)** $2^5 = 32$;   **(b)** $3^{-4} = \dfrac{1}{81}$ **24.** [10.4C] **(a)** $x = \dfrac{1}{16}$;

(b) $x = 4$ **25.** [10.4C] **(a)** 4; **(b)** -3 **26.** [10.4D] **(a)** $\log_b M + \log_b N$;

(b) $\log_b \dfrac{M}{N}$; **(c)** $r \log_b M$ **27.** [10.5A] **(a)** 2.9890; **(b)** 2.9227

28. [10.5A] **(a)** -2.1198; **(b)** -3.1884
29. [10.5A] **(a)** 662.9793; **(b)** 0.0004
30. [10.5B] **(a)** 7.9551; **(b)** -1.0642
31. [10.5B] **(a)** 8.0486; **(b)** 15.1545
32. [10.5C] **(a)** 2.0959; **(b)** 4.1918
33. [10.5D] **(a)** $f(x) = e^{(1/2)x}$

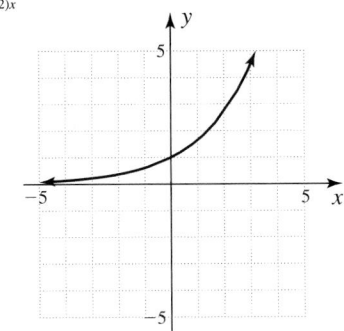

(b) $g(x) = -e^{(1/2)x}$

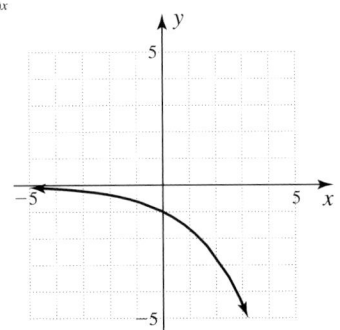

34. [10.5D] **(a)** $f(x) = \ln(x + 1)$

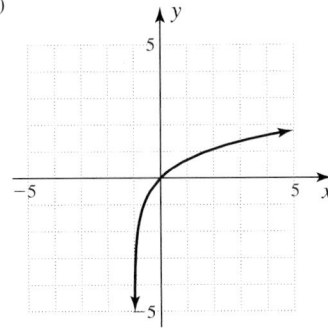

(b) $g(x) = \ln x + 1$

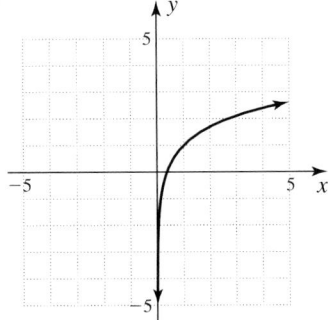

35. [10.5E] 5.398   **36.** [10.6A] **(a)** $x = 3$; **(b)** $x = 3$
37. [10.6A] **(a)** $x = 1.5850$; **(b)** $x = 0.2847$;   **38.** [10.6A] **(a)** $x = 0.1238$;
(b) $x = -2.1004$ **39.** [10.6B] **(a)** $x = 5 + \sqrt{35} \approx 10.9161$; **(b)** $x = 2$
40. [10.6C] **(a)** 13.863 yr; **(b)** About 8.2 yr
41. [10.6C] **(a)** $k = 0.2950$; **(b)** $k = 0.1475$
42. [10.6C] **(a)** 1.3863 yr; **(b)** 34.6575 yr

Cumulative Review Chapters 1–10

1. $x^2 + 7x + 9$ **2.** $\dfrac{16x^8}{y^4}$ **3.** $P = 150$ **4.** $x = 6$

5. $\{\,x \mid x < -4 \text{ or } x \geq 4\,\}$

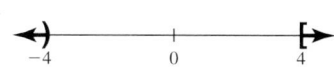

6. $|6x - 9| \leq 3$

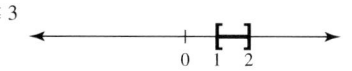

7. $h = 22$ **8.** $30 \text{ ft} \times 50 \text{ ft}$ **9.** $5\sqrt{2}$ **10.** $y = 5$ **11.** $4x + y = -18$
12. $|x + 3| > 2$

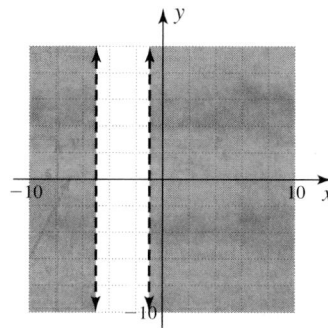

13. $k = 9250$ **14.** No solution **15.** $(-4, 4)$ **16.** $(-5, -4, 2)$
17. -15 **18.** 224 ft **19.** -3 **20.** $18h^2 + 9h - 35$
21. $4x^2y(3x - y)(4x + 3y)$ **22.** $(3n + 2)(9n^2 - 6n + 4)$

23. $-4, -1, 1$ **24.** $(x + 5)(x + 2)(3x + 1)$ **25.** $\dfrac{1}{x^2 - 49}$

26. $\dfrac{4}{(x + 2)(x - 2)}$ **27.** 6 and 8 **28.** $\dfrac{1}{81}$ **29.** $\dfrac{\sqrt[5]{14d^2}}{2d}$

30. $8\sqrt{2}$ **31.** $2 + \sqrt{2}$ **32.** $x = -3$ **33.** $60 - 2i$ **34.** $-9, 1$
35. $16, 1$ **36.** $x \leq -5$ or $2 \leq x \leq 7$; $(-\infty, -5] \cup [2, 7]$
37. $x < 1$ or $x \geq 5$; $(-\infty, 1) \cup [5, \infty)$
38.

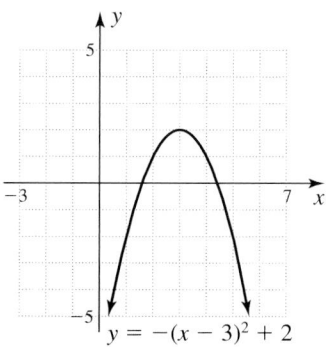

$y = -(x - 3)^2 + 2$

39. Center $(3, 5)$; $r = 2$
40.

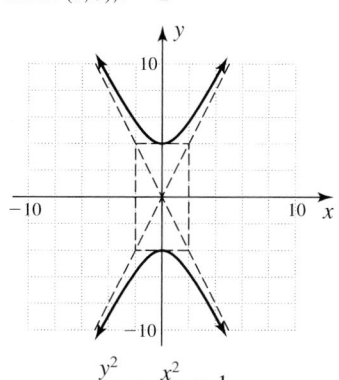

$\dfrac{y^2}{16} - \dfrac{x^2}{4} = 1$

41. Parabola **42.** 28 **43.** $\{\,x \mid x \geq -9\,\}$ **44.** $(4 - x)^4$
45.

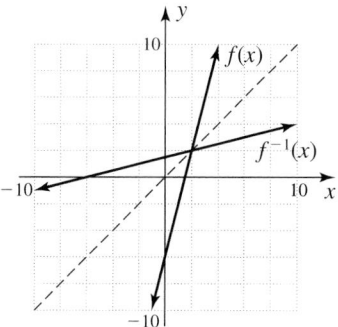

46.

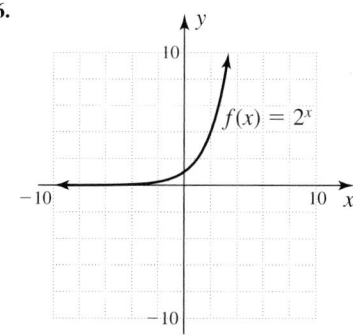

Yes, the inverse is a function.

47. $\log_b \sqrt{\dfrac{21}{83}} = \log_b \left(\dfrac{21}{83}\right)^{1/2}$

$\qquad\qquad = \dfrac{1}{2} \left[\log 21 - \log 83\right]$

$\qquad\qquad = \dfrac{1}{2} \log 21 - \dfrac{1}{2} \log 83$

48. $x \approx 0.2027$ **49.** 1.4650 **50.** $k = 0.3040$

Appendix A

Exercises A.1

1. $a_{10} = 28$; $a_n = 3n - 2$ **3.** $a_{10} = 32$; $a_n = 3n + 2$
5. $a_{10} = 65$; $a_n = 5n + 15$ **7.** $a_{10} = 5$; $a_n = 55 - 5n$

9. $a_{10} = \dfrac{1}{11}$; $a_n = \dfrac{1}{n + 1}$ **11.** $a_{10} = -1$; $a_n = (-1)^{n-1}$

13. $a_{10} = x^{10}$; $a_n = x^n$ **15.** $a_{10} = -x^{19}$; $a_n = (-1)^{n-1} x^{2n-1}$

17. $a_{10} = -x$; $a_n = (-1)^{n-1} x$ **19.** $a_{10} = \dfrac{x^{10}}{10}$; $a_n = \dfrac{x^n}{n}$

21. $a_{10} = -\dfrac{x^{10}}{1024}$; $a_n = (-1)^{n-1}\left(\dfrac{x}{2}\right)^n$ **23.** $-1, 1, 3$ **25.** $-\dfrac{1}{2}, 0, \dfrac{3}{2}$

27. $0, \dfrac{1}{2}, \dfrac{2}{3}$ **29.** $1, 4, 9$ **31.** $\dfrac{1}{3}, \dfrac{2}{5}, \dfrac{3}{7}$ **33.** $-1, 1, -1$

35. $-\dfrac{1}{2}, \dfrac{1}{4}, -\dfrac{1}{8}$ **37.** 91 **39.** 100 **41.** 21 **43.** $\dfrac{25}{24}$ **45.** $\dfrac{1343}{140}$ **47.** 1

49. $\displaystyle\sum_{n=1}^{200} n$ **51.** $\displaystyle\sum_{n=1}^{50} \dfrac{1}{n}$ **53.** $\displaystyle\sum_{n=1}^{50} (-1)^{n-1} n$ **55.** $\displaystyle\sum_{n=1}^{5} (5n - 4)$

57. (a) $a_8 = \$1100$; (b) $a_{10} = \$1020$ **59.** (a) 9 ft; (b) $\dfrac{729}{100}$ ft; (c) $\dfrac{9n}{10^{n-1}}$

61. (a) 400; (b) 1600; (c) $100(2)^n$ **63.** \$3200 **65.** $4 + 3n$ **67.** $16n^2$
69. $(5n + 41)(n - 8)$ **71.** $a_8 = 21$ **73.** $a_{10} = 55$ **75.** $a_{12} = 144$

81. $0, \dfrac{3}{2}, 4$; $a_{10} = \dfrac{99}{2}$ **83.** $a_{10} = 30$; $a_n = (-1)^n \cdot 3n$

85. (a) $\displaystyle\sum_{n=1}^{4} 6n$; (b) $\displaystyle\sum_{n=1}^{6} (-1)^{n-1} 2^n$

Exercises A.2

1. (a) $a_1 = 5$; **(b)** $d = 3$; **(c)** $a_n = 3n + 2$ **3. (a)** $a_1 = 11$; **(b)** $d = -5$;
(c) $a_n = 16 - 5n$ **5. (a)** $a_1 = 3$; **(b)** $d = -4$; **(c)** $a_n = 7 - 4n$

7. (a) $a_1 = \frac{1}{2}$; **(b)** $d = -\frac{1}{4}$; **(c)** $a_n = \frac{3}{4} - \frac{1}{4}n$ or $\frac{3-n}{4}$

9. (a) $a_1 = -\frac{5}{6}$; **(b)** $d = \frac{1}{2}$; **(c)** $a_n = \frac{1}{2}n - \frac{4}{3}$ or $\frac{3n-8}{6}$ **11.** $a_{15} = 91$;
$S_{15} = 735$ **13.** $a_1 = 4$; $d = 6$; $S_8 = 200$ **15.** $d = 1$; $S_6 = 33$ **17.** $d = -4$;
$a_{14} = -46$ **19.** $a_1 = -779$; $a_{40} = 781$ **21.** \$10,000   **23.** 190
25. The natural numbers are $1, 2, 3, \ldots, n, \ldots$;
$S_n = \frac{n}{2}(1 + n) = \frac{n(n + 1)}{2}$
27. The even numbers are $2, 4, 6, \ldots, 2n, \ldots$;
$S_n = \frac{n}{2}(2 + 2n) = \frac{n}{2} \cdot 2(n + 1) = n(n + 1) = n^2 + n$

29. 32   **31.** 63   **33.** 20   **35.** \$8000, \$6000, \$4000, \$2000
37. (a) $b_t = 12{,}000 - 1200t$; **(b)** $b_0 = \$12{,}000$; $b_1 = \$10{,}800$;
$b_2 = \$9600$; $b_3 = \$8400$; $b_4 = \$7200$; $b_5 = \$6000$ **43. (a)** $a_1 = 5$;
(b) $d = 4$; **(c)** $a_{10} = 41$; **(d)** $a_n = 4n + 1$ **45. (a)** $a_1 = 10$; **(b)** $d = 2$
47. 10 months

Exercises A.3

1. (a) $a_1 = 3$; **(b)** $r = 2$; **(c)** $a_n = 3(2^{n-1})$; **(d)** $S_n = 3(2^n - 1)$
3. (a) $a_1 = 8$; **(b)** $r = 3$; **(c)** $a_n = 8(3^{n-1})$; **(d)** $S_n = 4(3^n - 1)$

5. (a) $a_1 = 16$; **(b)** $r = -\frac{1}{4}$; **(c)** $a_n = (-4)^{3-n}$; **(d)** $S_n = \frac{64}{5}\left(1 - \left[-\frac{1}{4}\right]^n\right)$

7. (a) $a_1 = -\frac{3}{5}$; **(b)** $r = -\frac{5}{2}$; **(c)** $a_n = \left(-\frac{3}{5}\right)\left(-\frac{5}{2}\right)^{n-1}$;

(d) $S_n = -\frac{6}{35}\left(1 - \left[-\frac{5}{2}\right]^n\right)$ **9. (a)** $a_1 = -\frac{3}{4}$; **(b)** $r = \frac{1}{3}$;

(c) $a_n = \left(-\frac{1}{4}\right)(3^{2-n})$; **(d)** $S_n = -\frac{9}{8}\left(1 - \left[\frac{1}{3}\right]^n\right)$ **11.** $a_3 = \frac{1}{4}$, $r = \frac{1}{2}$

or $a_3 = \frac{9}{4}$, $r = -\frac{3}{2}$ **13.** $a_3 = 27$, $r = -3$ or $a_3 = 12$, $r = 2$

15. $a_1 = 7$, $a_8 = 896$ **17.** $r = -3$, $n = 4$ **19.** $n = 6$, $S_6 = \frac{5187}{250}$

21. $S = 12$ **23.** $S = -12$ **25.** $S = \frac{4}{3}$ **27.** Sum does not exist;

$|r| = 2 > 1$ **29.** Sum does not exist; $|r| = 2 > 1$ **31.** $\frac{5}{9}$

33. $\frac{2}{11}$ **35.** $\frac{401}{99}$ **37.** $\frac{2293}{990}$ **39.** $\frac{140}{990}$ **41. (a)** P_0, $1.04\,P_0$, $(1.04)^2\,P_0, \ldots$;

(b) 24,333   **43.** 184.5 cm   **45.** \$131,875   **47.** 1, 3, 5 or 17, 3, -11
49. \$873.81   **51.** 56 ft   **53.** $a^2 + 2ab + b^2$
55. $x^3 + 3x^2y + 3xy^2 + y^3$ **57.** $y^3 - 6y^2z + 12yz^2 - 8z^3$

59. $\frac{1}{x^3} - \frac{3}{x^2y} + \frac{3}{xy^2} - \frac{1}{y^3}$ **61.** Geometric; $r = \frac{2}{5}$; $S = \frac{5}{3}$

63. Arithmetic; $d = -\frac{3}{5}$ **65.** Geometric; $r = \frac{7}{8}$; $S = 16$ **67.** Neither

69. Neither   **71.** Geometric; $r = -\frac{1}{2}$; $S = \frac{8}{3}$ **77. (a)** $a_1 = 2$;

(b) $r = \frac{1}{2}$; **(c)** $a_6 = \frac{1}{16}$; **(d)** $a_n = \frac{1}{2^{n-2}}$ **79.** $a_3 = 50$, $r = -5$

or $a_3 = 32$, $r = 4$ **81.** Sum does not exist; $|r| = 1.01 > 1$.
83. (a) 100, 200, 400, 800, 1600; **(b)** \$6400; **(c)** $100 \cdot 2^n$; **(d)** 10 yr

Exercises A.4

1. 6   **3.** 3,628,800   **5.** 30   **7.** 504   **9.** 15   **11.** 11   **13.** 1
15. $a^4 + 12a^3b + 54a^2b^2 + 108ab^3 + 81b^4$
17. $x^4 + 16x^3 + 96x^2 + 256x + 256$
19. $32x^5 - 80x^4y + 80x^3y^2 - 40x^2y^3 + 10xy^4 - y^5$
21. $32x^5 + 240x^4y + 720x^3y^2 + 1080x^2y^3 + 810xy^4 + 243y^5$

23. $\frac{1}{x^4} - \frac{2y}{x^3} + \frac{3y^2}{2x^2} - \frac{y^3}{2x} + \frac{y^4}{16}$

25. $x^6 + 6x^5 + 15x^4 + 20x^3 + 15x^2 + 6x + 1$ **27.** -540 **29.** 672

31. 1120   **33.** $-\frac{5}{2}$ **35.** 20   **37.** 84   **39.** $\frac{15}{64}$ **41.** $\frac{21}{128}$ **43.** $\frac{25}{216}$

45. $\dbinom{n}{k} = \frac{n!}{k!(n - k)!} = \frac{n!}{(n - k)!k!} = \dbinom{n}{n - k}$ **49.** 720   **51.** 15

53. $16a^4 - 32a^3b + 24a^2b^2 - 8ab^3 + b^4$ **55.** 4860   **57.** 5

Review Exercises

1. [A.1A] **(a)** $a_6 = 11$; $a_n = 2n - 1$; **(b)** $a_6 = 18$; $a_n = 3n$

2. [A.1A] **(a)** $a_4 = 64$; **(b)** $a_n = 2^{n+2}$ **3.** [A.1B] **(a)** $\frac{2}{3}, \frac{5}{3}, \frac{10}{3}$;

$a_{10} = \frac{101}{3}$; $S_3 = \frac{17}{3}$; **(b)** $\frac{1}{2}, \frac{3}{2}, 3$; $a_{10} = \frac{55}{2}$; $S_3 = 5$

4. [A.1B, C] **(a)** $3, 6, 11, \ldots$; $S_4 = 38$; **(b)** $1, 7, 17, \ldots$; $S_3 = 25$
5. [A.1D] **(a)** \$64,000; **(b)** \$16,000   **6.** [A.2A] **(a)** $d = 3$;
$a_{10} = 30$; **(b)** $d = 4$; $a_{10} = 40$ **7.** [A.2A] **(a)** $a_n = 3n$; **(b)** $a_n = 4n$
8. [A.2A] **(a)** 176 ft; **(b)** 240 ft   **9.** [A.2B] **(a)** $a_1 = 14$; $d = 2$;
(b) $a_1 = 9$; $d = 3$ **10.** [A.2B] **(a)** $a_1 = 4$; $d = 2$;
(b) $a_1 = 1.5$; $d = 3$ **11.** [A.2C] **(a)** \$21,600; **(b)** \$9000
12. [A.2C] **(a)** 10 weeks; **(b)** 12 weeks   **13.** [A.3A] **(a)** $r = 2$;

$a_6 = 96$; **(b)** $r = \frac{1}{2}$; $a_6 = \frac{1}{8}$ **14.** [A.3A] **(a)** $r = 4$; $a_n = 4^{n-1}$;

(b) $r = -\frac{1}{2}$; $a_n = \frac{(-1)^{n-1}}{2^{n-3}}$ **15.** [A.3B] **(a)** $a_n = (-1)^{n-1}\,2^n$;

$S_n = \frac{2}{3}[1 - (-2)^n]$; **(b)** $a_n = \left(-\frac{1}{2}\right)^{n-1}$; $S_n = \frac{2}{3}\left[1 - \left(-\frac{1}{2}\right)^n\right]$

16. [A.3C] **(a)** $\frac{64}{3}$; **(b)** $\frac{27}{2}$ **17.** [A.3C] **(a)** Sum does not exist;

$|r| = 1.005 > 1$.; **(b)** Sum does not exist; $|r| = 1.001 > 1$.

18. [A.3D] **(a)** $\frac{31}{99}$; **(b)** $\frac{12}{37}$ **19.** [A.3D] **(a)** 24 ft; **(b)** 18 ft

20. [A.4A] **(a)** 40,320; **(b)** 1680   **21.** [A.4A] **(a)** 84; **(b)** 1
22. [A.4A] **(a)** $a^4 - 12a^3b + 54a^2b^2 - 108ab^3 + 81b^4$;
(b) $81a^4 + 216a^3b + 216a^2b^2 + 96ab^3 + 16b^4$
23. [A.4B] **(a)** 1120; **(b)** -2835 **24.** [A.4C] **(a)** 35; **(b)** 35

25. [A.4C] **(a)** $\frac{35}{128}$; **(b)** $\frac{35}{128}$

Photo Credits

Chapter 1

Opener(left): © Scala/Art Resource, NY; **Opener(right):** © Erich Lessing/Art Resource, NY; **p. 31:** © Benjamin Stein/All Too Flat; **p. 46:** © Royalty-Free/Corbis; **p. 50:** © Vol. 49/PhotoDisc.

Chapter 2

Opener(top): © Art Resource, NY; **Opener(bottom):** Photri-Microstock; **p. 72:** © Dennis Johnson; Papilio/CORBIS; **p. 87:** © Ignacio Bello; **p. 99:** ©LWA-Stephen Welstead/CORBIS; **p. 102(top):** Photri-Microstock; **p. 102(bottom):** © A. Ramey/Picture Quest; **p. 111(left):** © Vol. OS43/PhotoDisc; **p. 111(middle):** © Tony Freeman/PhotoEdit; **p. 111(right):** © Royalty-Free/CORBIS; **p. 120:** © Lawrence Manning/Corbis; **p. 137:** © Vol. 135/PhotoDisc.

Chapter 3

Opener: © Bettmann/CORBIS; **p. 163:** © John Bigelow Taylor/Art Resource, NY; **p. 176:** © Tim Laman/NG/Getty; **p. 181:** © Tom Pantages Stock Photos.

Chapter 4

Opener(left): The Art Archive/Academie des Sciences Paris/Dagli Orti; **Opener(right):** © Michael Nicholson/CORBIS; **p. 273:** © Vol. 126/PhotoDisc; **p. 278:** © PhotoLibrary; **p. 283:** © The McGraw-Hill Companies/David Tietz, photographer; **p. 285:** © Royalty Free/Tom Grill/Corbis; **p. 286:** © Vol. 60/PhotoDisc; **p. 289:** © Vol. 98/CORBIS; **p. 291:** © PIXTAL/age fotostock; **p. 292:** © Tom Pantages Stock Photos; **p. 307:** © Stefano Bianchetti/CORBIS.

Chapter 5

Opener: © 2004 age fotostock; **p. 344:** © Dallas and John Heaton/Picture Quest; **p. 356:** © Vol. 88/PhotoDisc; **p. 374:** © Philip Rostron/Masterfile; **p. 384:** © Fran Hopf; **p. 392:** © Images.com/Corbis; **p. 399:** © Gilles Mingasson/Liaison/Getty; **p. 407:** © Master File; **p. 407(top):** © Royalty-Free/Corbis; **p. 407(middle):** © Royalty Free/Image Source/Corbis; **p. 407(bottom):** © Royalty Free/Burke/Triolo/Brand X Pictures/Getty; **p. 411:** © Vol. 281/PhotoDisc.

Chapter 6

Opener: © The Granger Collection; **p. 435:** © The McGraw-Hill Companies/David Tietz, photographer; **p. 445:** © Vol. 144/PhotoDisc; **p. 454:** © Steve Finn/Getty Images; **p. 465:** © Royalty-Free/CORBIS; **p. 476:** © Fran Hopf; **p. 488:** © Scala/Art Resource, NY; **p. 492:** © Peter McBride/Aurora & Quanta Productions, Inc.; **p. 498(top):** © Tony Freeman/PhotoEdit; **p. 498(bottom):** © Irvine World News.

Chapter 7

Opener: © Bettmann/CORBIS; **p. 520:** © Mike Dobel/Masterfile; **p. 530:** © Ignacio Bello; **p. 541:** © NASA; **p. 552:** © PhotoDisk; **p. 560(top):** © Mary Evans Picture Library; **p. 560(bottom):** © Bettmann/CORBIS; **p. 571:** © Dennis MacDonald/PhotoEdit.

Chapter 8

Opener: © Martin Schøyen/The Schøyen Collection MS 2192; **p. 582:** © Chase Swift/CORBIS; **p. 594:** © Bettmann/CORBIS; **p. 606:** © PhotoDisc; **p. 614:** © Rob Crandal/The Image Works; **p. 622:** © David Young-Wolff/PhotoEdit.

Chapter 9

Opener(left): © Bob Krist/CORBIS; **Opener(right):** © Joseph Sohm; ChromoSohm Inc./CORBIS; **p. 647:** © Ignacio Bello; **p. 669:** © Bob Krist/CORBIS; **p. 676:** Patent Pending, Wind Weighted™ Baseball Tarps by Aer-Flo, Inc.; **p. 681, 682(all photos):** © Ignacio Bello; **p. 686(top):** © 2004 age fotostock; **p. 686(bottom left):** © Robert Essel NYC/CORBIS; **p. 686(bottom right):** © Joseph Sohm; ChromoSohm Inc./CORBIS; **p. 699:** © PhotoLink/Getty Images; **p. 717:** © Kelly-Mooney Photography/CORBIS; **p. 718:** © Royalty-Free/CORBIS.

Chapter 10

Opener: © Hulton Archive/Getty; **p. 737:** © Vol. 25/PhotoDisc; **p. 764(all photos):** © Hans Pfletschinger/Peter Arnold, Inc.; **p. 775:** © John Swart/AP Photo.

Appendix A

P. A-2: © Vol. OS25/PhotoDisc; **p. A-12:** © Altrendo images/Getty; **p. A-20:** © C Squared Studios/Getty Images.

Index

Index of Calculate It Topics

Intermediate Algebra